LINUS PAULING

Selected Scientific Papers

VOLUME II

Biomolecular Sciences

WORLD SCIENTIFIC SERIES IN 20TH CENTURY CHEMISTRY

Published:

Vol. 1: Molecular Structure and Statistical Thermodynamics
— Selected Papers of Kenneth S. Pitzer
by Kenneth S. Pitzer

Vol. 2: Modern Alchemy
— Selected Papers of Glenn T. Seaborg
by Glenn T. Seaborg

Vol. 3: Femtochemistry: Ultrafast Dynamics of the Chemical Bond
by Ahmed H. Zewail

Vol. 4: Solid State Chemistry
— Selected Papers of C. N. R. Rao
edited by S. K. Joshi and R. A. Mashelkar

Vol. 5: NMR in Structural Biology
— A Collection of Papers by Kurt Wüthrich
by Kurt Wüthrich

Vol. 6: Reason and Imagination: Reflections on Research in Organic Chemistry
— Selected Papers of Derek H. R. Barton
by Derek H. R. Barton

Vol. 7: Frontier Orbitals and Reaction Paths
— Selected Papers of Kenichi Fukui
edited by K. Fukui and H. Fujimoto

Vol. 8: Quantum Chemistry
— Classic Scientific Papers
translated and edited by Hinne Hettema

Vol. 9: The Oxidation of Oxygen and Related Chemistry
— Selected Papers of Neil Bartlett
edited by Neil Bartlett

World Scientific Series in 20th Century Chemistry – Vol. 10

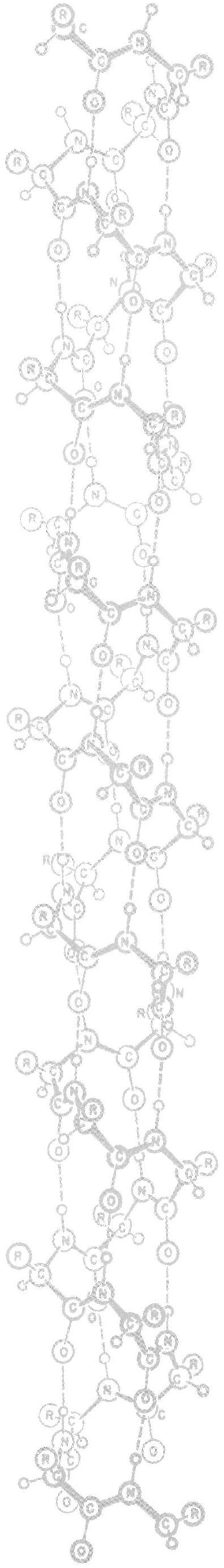

LINUS PAULING

Selected Scientific Papers

VOLUME II

Biomolecular Sciences

Editors

Barclay Kamb *(California Institute of Technology, USA)*

Linda Pauling Kamb *(LCProgeny Inc., USA)*

Peter Jeffress Pauling *(University College London, UK)*

Alexander Kamb *(Arcaris Inc., USA)*

Linus Pauling, Jr. *(Linus Pauling Institute of Science & Medicine, USA)*

World Scientific
New Jersey • London • Singapore • Hong Kong

Published by

World Scientific Publishing Co. Pte. Ltd.
P O Box 128, Farrer Road, Singapore 912805
USA office: Suite 1B, 1060 Main Street, River Edge, NJ 07661
UK office: 57 Shelton Street, Covent Garden, London WC2H 9HE

Library of Congress Cataloging-in-Publication Data
Pauling, Linus, 1901–1994
Linus Pauling : selected scientific papers / edited by Barclay Kamb ... [et al.].
p. cm. -- (World Scientific series in 20th century chemistry ; vol. 10)
Includes bibliographical references and index.
Contents: vol. I. Physical sciences -- vol. II. Biomolecular sciences.
ISBN 9810227841 (set) -- ISBN 9810229399 (vol. I) -- ISBN 9810229402 (vol. II)
1. Chemistry. 2. Biomolecules. 3. Medicine. I. Kamb, Barclay. II. Title. III. Series.

QD3 .P38 2001 v. 2
540--dc21 2001017782

British Library Cataloguing-in-Publication Data
A catalogue record for this book is available from the British Library.

Printed in Singapore by Mainland Press

BRIEF CONTENTS

VOLUME I — PHYSICAL SCIENCES

VOLUME II — BIOMOLECULAR SCIENCES

Introduction To Volume II

This book is Volume II of a two-volume anthology of the scientific writing of the famous American chemist Linus Pauling (1901–1994). Of the some 850 scientific papers and books that he wrote in his lifetime, we have selected 140 papers and 4 passages in books to convey the spirit and style of his writing and to highlight the publications that had such great impact and brought him such wide recognition, leading to the Nobel Prize in 1954 and to many other scientific prizes, medals, and awards.

The selected papers (and book passages) are aggregated into four broad groups, called Parts I, II, III, and IV, in accordance with general subject areas. Parts I and II, which are in Volume I, contain papers in the physical sciences, in particular the nature of the chemical bond in molecules and crystals, and the application of quantum mechanics and electron and x-ray diffraction to chemical bonding and analysis of physical properties. Parts III and IV are in Volume II. Part III covers the biochemical and biological areas of Pauling's scientific interests and writings—areas that began to develop in the late 1930's and reached culmination in the 1950's and 1960's. Part IV contains papers on biomedical science, which became the main area of Pauling's scientific studies and publications from the 1970's through the rest of his life.

To sharpen the focus on individual subject areas, the papers in Parts III and IV are grouped into seven chapters, designated Chapters 11 to 17, which is a continuation of the sequence of Chapters 1 to 10 in Parts I and II. These groupings are explained in introductions to the Parts. The introductions contain brief general comments on the contents of the chapters, but, with a few exceptions, do not endeavor to discuss the papers individually.

The sequence of the individual selected papers is governed primarily by the subject-matter sequence of the chapters, which in a general way follows the development of Pauling's scientific interests through time, but with some considerable departures from a chronological sequence. Within each chapter, the arrangement of selected papers is generally chronological but also somewhat arbitrary.

The individual papers are identified by a Selected Paper number (SP no.), numbered consecutively from SP 1 at the beginning of Part I (Volume I) to SP 144 at the end of Part IV. The selected papers are listed in a table of contents at the beginning of each chapter.

Included as Part V (Chapter 18) is a short, science-oriented biography of Pauling by Prof. Jack Dunitz. At the end of Part III, and again at the end of Part IV, is a group of photographs of Linus Pauling with colleagues/co-authors of the period.

Appendix I gives a conversion table between the SP numbers and the paper citation numbers in Chapter 18. Appendix III is a comprehensive list of Pauling's scientific publications, based on the full list of his publications assembled by Zelek Herman and Dorothy S. Munro (1996).

More detailed introductory information about the anthology is contained in the Foreword, Preface, and General Introduction at the beginning of Volume I.

Part III

BIOLOGICAL MACROMOLECULES

Introduction To Part III

Biological Macromolecules

In the realm of biology, into which he began to venture in the late 1930's, Linus Pauling came firmly to the conviction that life's complex processes could be understood as straight-forward consequences of chemical forces and structures. To achieve such understanding, he set out to find the structures and properties of biological macromolecules. A selection of his publications on these subjects is presented in Part III.

In the earliest stage of Pauling's attempts to reveal the structures of biological molecules, in which he concentrated on hemoglobin (Chapter 11) and antibodies (Chapter 12), he lacked sufficient information to guide creation of detailed structural models that could specify the actual arrangement of the individual atoms in these huge molecules. In order to proceed with these molecules at all, he chose the course of formulating speculative theories of their structures and interactions. Examples of such theories are in papers SP 88 (Chapter 12), SP 95 (Chapter 13), and SP 113 (Chapter 14).

By far the most important and influential of these theories is the role of *molecular complementariness* in biological processes. It is well expressed in the first four papers of Chapter 14, and it pervades Pauling's discussions of the molecular basis of biological phenomena. An example is his theory of *enzyme action* (SP 113). Based on principles of chemical catalysis, he theorized that enzymes are structurally complementary to the activated complexes ("transition states") of the reactions that they catalyze. This pioneering vision is, as we now know, largely correct, and explains the specificity of enzyme action in biochemical reactions. SP 113, which was originally published in 1946, is here presented in the form of its 1990 republication by the NIH with the addition of side bars describing interviews with Pauling and Alexander Rich on the significance of this paper for the origins of molecular biology.

Alongside the theoretical reasoning, many of the papers report experimental work carried out in the formulation and testing of theory—for example, SP 83 and following papers in Chapter 11, and SP 89 and following papers in Chapter 12. Particularly noteworthy are Pauling's trail-blazing *magnetic measurements on hemoglobin* (Chapter 11), which, in combination with the hemoglobin-oxygen binding curve, provided the basis for his formulation of a model for the cooperative interaction between hemoglobin and molecular oxygen (SP 82, SP 86). This work harks back to his early use of magnetic measurements in establishing the nature of the chemical bond (Chapter 1, SP 5). The significance of the work on hemoglobin is well summarized in SP 87, which shows how this work underlay Pauling's later discovery of molecular disease in hemoglobin (Part IV, Chapter 15).

Encouraged by the great immunologist (serologist) Karl Landsteiner, Pauling embarked in the early 1940's on an endeavor to understand the basis of antigen-antibody interaction and the extraordinary specificity of immunological reactions (Chapter 12). He started with a theoretical concept of the role of antigen-antibody complementariness, which was described in SP 88 and SP 93 and which had a great impact on immunologists. To test the theory, Pauling and his colleagues, led by Prof. Dan Campbell (Photo 17), undertook an extensive series of immunological experiments with rabbits. This resulted in a long sequence of detailed serological papers, here represented by SP 89, SP 90, SP 91, SP 92, and SP 94. In the end, the main contribution of these papers was to provide evidence for the "framework theory" of serological precipitation and for the "bivalent" character of the antibody-antigen interaction. The experiments were compatible with Pauling's theory that the interaction was of a *lock-and-key* type, made possible by *antigen-antibody molecular complementariness.* They also seemed to be compatible with his hypothesis that antibodies are manufactured in the body so as to conform in a complementary way to the shape of specific antigens — the so-called *"instructional theory"* of antibody formation and diversity (SP 88, SP 91, and SP 93). A vast amount of subsequent immunological work has ultimately shown that for antibodies the lock-and-key theory is correct but the instructional theory is not.

However, in a certain sense, the instructional theory describes an approximation to the mechanism by which another class of important immunological proteins, the major histocompatibility antigens, bind to peptides. Each MHC protein, as it folds, assumes a configuration that permits it to "clamp onto" any one of a large number of different peptides. The clamping mechanism permits a limited repertoire of MHC proteins to bind to diverse types of peptide molecules. This feature — the ability of a small number of protein molecules to bind to any one of a diversity of peptides — is something common to the MHC binding mechanism and Pauling's antigen-directed protein-folding mechanism, and in this particular respect, the two concepts are similar. They contrast in this regard with the now-known antibody generation mechanism (clonal selection), which is based on the body's ability to manufacture a vast number of different antibody molecules in order to provide binding to potential antigens of many kinds.

As x-ray techniques advanced further, and as more became known about the constituents of proteins, Pauling turned his attention (Chapter 13) to detailed atomic models of their secondary structure, i.e. the way the polypeptide chain — the primary structure—is rolled, looped, and coiled to create a stable molecular conformation that is all-important in the biological functioning of a protein molecule. It is this configuration, along with the tertiary structure—the arrangement of secondary structural elements in space—in combination with the placement and chemical character of the side chains, that determines the extent of complementariness of two protein molecules—that is, the extent to which the two will fit together geometrically and the extent to which bonds will be formed between them.

The experimental basis for Pauling's work on protein structure was established in a series of x-ray structure determinations, led by Prof. Robert B. Corey, of crystallized amino acids and other small molecules that are components of proteins. The results were summarized in SP 105 and SP 106 and in the extensive review papers [39-5] and [48-16] referenced in Appendix III, Group 6c. Pauling also had a hand later in interpreting x-ray diffraction diagrams of fibrous protein material (SP 110).

These efforts culminated in Pauling's celebrated proposal of the *α-helix* as a fundamental type of protein secondary structure (SP 97, SP 98). It was the first such structural feature to be later verified by x-ray diffraction studies of crystallized proteins, and it has proven to be a major structural component of most protein molecules. The α-helix demonstrates the central role of hydrogen bonding in establishing and stabilizing the secondary structure of protein molecules, with vast consequences for the biological functioning of proteins. Pauling's understanding of the hydrogen bond (Part I, Chapter 4) and the relation between covalent bond character and molecular geometry (i.e. directed valence—Chapter 1) played a decisive role in the α-helix discovery (see SP 105 and SP 106).

Almost simultaneously with his proposal of the α-helix, and based on the same structural principles, Pauling proposed a number of other patterns for protein secondary structure, including the γ-helix (SP 98), the *pleated sheet* (SP 100), the coiled coil (SP 108), and several others (other papers in Chapter 13). Of these various possible structures, several have been observed to occur in real proteins, while others remain hypothetical, or have proven incorrect (e.g. the collagen structure in SP 103, which, curiously, is a triple helix, foreshadowing the triple-helix DNA structure proposed two years later—see below). In selecting papers for Chapter 13, we have included most of the papers on these protein structures because of their actual or potential importance in biology and medicine. The α-helix and the pleated sheet, especially, were triumphs of the inspired-guess/model-building technique, which is closely related to the stochastic method of crystal structure determination that Pauling had invented earlier (see Part II, Chapter 5, SP 47). The story of the α-helix discovery is recounted in SP 111, a paper that was found by Dorothy Munro in Pauling's files after his death.

Grouped in Chapter 14 are papers of the broadest scope on the role of molecular structure in biology—essentially a distillation, synthesis, and generalization of the work in Chapters 11, 12, and 13. It includes the seminal papers on molecular evolution and molecular clocks (SP 119, SP 120 and SP 121), which built in a striking way upon the discovery of molecular disease (Part IV, Chapter 15). The papers in Chapter 14 had a great impact on biologists, and lead the way toward today's molecular biology.

Chapter 14 also contains the infamous SP 116, in which Pauling proposed a structure for the nucleic acids (i.e. DNA) that was erroneous. It was a triple helix, with the invariant phosphodiester backbone on the inside and the variable components (the purine and pyrimidine bases) on the outside. Thus it was an antithesis of the Watson-and-Crick DNA structure, published shortly thereafter, which is a double helix with the invariant backbone on the outside and the variable components on the inside. It seems that Pauling was predisposed to structural features of the type that he was familiar with in the α-helix: invariant polypeptide backbone on the inside, variable side chains sticking outward. Although in SP 112 he had postulated that the genetic material must have structural features with two complementary sides, so that it could reproduce itself, his three-strand helix did not incorporate such complementariness. The proposal of three strands came from calculation based on the published density of DNA, but the published density turned out later to be wrong. The feature for which Pauling was most criticized by chemists was putting the negatively charged phosphate groups close together on the inside of the helix, where they would have a large, destabilizing electrostatic repulsion. This may be explained by Pauling's belief that the acid treatment applied to the DNA fibers in preparation for x-ray diffraction protonated the phosphate groups, reducing their negative charge.

Beyond all this, at the time there were serious problems for Pauling in his life outside of science. The early 1950's was a time of considerable turmoil for intellectuals in the United States and for Pauling in particular. He was the target of various government efforts to investigate his involvement with alleged "unamerican" activities and organizations (Chapter 18, section entitled "Political Activism"). This harassment certainly interfered with Pauling's scientific work, and the denial of a passport by the State Department blocked an opportunity for him to go to a meeting in England where he would have seen the x-ray photographs of DNA fibers that were of crucial importance to Watson and Crick in developing their double-helix structure.

When the Watson-and-Crick structure was published, Pauling immediately saw the beauty and validity of it, and he undertook a small-molecule crystal-structure investigation (SP 117) that revealed the detailed configuration of the hydrogen bonding between pyrimidines and purines—the bonding feature that is the basis for complementariness between the two DNA strands in the double-helix structure.

In the mid to latter stages of Pauling's career, he increasingly turned to medical problems as the focus of his considerable energies. These problems are the subject of Part IV, but some of the work is as much a contribution to biology as it is to medicine. This is particularly true of one of his greatest achievements, the discovery of the molecular basis for a genetic disease, sickle cell anemia, and his generalization of this discovery to the broad concept of *molecular disease* (SP 122, in Chapter 15). From Pauling's vantage point in structural chemistry, he was able to hit upon the idea that the hemoglobin molecule itself was abnormal in sickle cell patients. This seminal insight, which is a concept both of molecular biology and of medicine, was ultimately proven correct in a collaboration with Harvey Itano, Jon Singer and I. C. Wells (SP 122). It led Pauling in turn to the concept of *molecular evolution* (e.g. evolution of the human hemoglobin molecule through time). Pauling realized that molecular evolution and molecular disease are very closely akin—they are two faces of the same coin (SP 119). Molecular evolution provides the basis for a molecular *evolutionary clock,* according to which the evolutionary history of different but related species can be traced backward in time to their common ancestor (SP 119, pages 200–206; also SP 120). Thus humans branched away from their ancestor in common with the great apes about 11 million years ago, based on the extent of similarity of their hemoglobin molecules. The concepts of molecular evolution are clear examples of Pauling's imaginative powers and the breadth of his thinking. Today they play a central role in interpreting DNA sequences revealed by genome decoding.

The scope of Pauling's knowledge, inventiveness, and boldness is nowhere more apparent than in the papers of Parts III and IV. As he is considered to be one of the greatest chemists, he should also be regarded as the first true molecular biologist. In this area, his genius lay in his ability to recognize the molecular basis for complex biological phenomena.

Chapter 11

HEMOGLOBIN: OXYGEN BONDING AND MAGNETIC PROPERTIES

Contents

SP 82 **The Oxygen Equilibrium of Hemoglobin and Its Structural Interpretation** **850**
Proc. Natl. Acad. Sci. USA **21**, 186–191 (1935)

83 **The Magnetic Properties and Structure of Hemoglobin, Oxyhemoglobin and Carbonmonoxyhemoglobin** **856**
(by Linus Pauling and Charles D. Coryell)
Proc. Natl. Acad. Sci. USA **22**, 210–216 (1936)

84 **The Magnetic Properties and Structure of Ferrihemoglobin (Methemoglobin) and Some of Its Compounds** **863**
(by C. D. Coryell, F. Stitt, and Linus Pauling)
J. Am. Chem Soc. **59**, 633–642 (1937)

85 **The Combining Power of Hemoglobin for Alkyl Isocyanides, and the Nature of the Heme-Heme Interactions in Hemoglobin** **873**
(by Robert C. C. St. George and Linus Pauling)
Science **114**, 629–634 (1951)

86 **The Nature of the Iron-Oxygen Bond in Oxyhaemoglobin** **879**
Nature **203**, 182–183 (1964)

87 **The Normal Hemoglobins and Hemoglobinopathies: Background** **882**
Texas Reports on Biology and Medicine **40**, 1–7 (1980–1981)

Reprinted from the Proceedings of the NATIONAL ACADEMY OF SCIENCES,
Vol. 21, No. 4, pp. 186–191. April, 1935.

THE OXYGEN EQUILIBRIUM OF HEMOGLOBIN AND ITS STRUCTURAL INTERPRETATION

BY LINUS PAULING

GATES CHEMICAL LABORATORY, CALIFORNIA INSTITUTE OF TECHNOLOGY

Communicated March 16, 1935

The work of the early investigators of blood (Bohr, Hasselbalch, Krogh, Barcroft, Haldane, Henderson) showed that the oxygen equilibrium of hemoglobin is affected by the hydrogen ion concentration of the solution, and, moreover, that it does not correspond to the simple equilibrium $Hb + O_2 \rightleftarrows HbO_2$ (in which the symbol Hb represents the portion of the hemoglobin molecule containing one heme and capable of combining with one oxygen molecule). After completing his osmotic pressure experiments showing that the hemoglobin molecule contains four hemes, Adair[1] suggested that the Hb_4 molecule adds four oxygen molecules successively, with different equilibrium constants. He and Ferry and Green[2] found that their data on the oxygen equilibrium of hemoglobin solutions at constant pH can be represented by a four-constant equation corresponding to this conception, the values of the constants changing in a regular way with changing pH.

On the basis of certain postulates regarding the structure of the hemoglobin molecule, I have derived an equation involving only two constants which satisfactorily represents the data on the oxygen equilibrium at constant pH, and also an equation involving two constants which represents the change in the oxygen equilibrium with change in pH. The validity of these equations provides considerable support for a particular structure of the molecule.

The most accurate and extensive data on the oxygen equilibrium are those obtained by Ferry and Green[2] for horse hemoglobin in phosphate and borate buffers. Ferry and Green remarked that their curves of y (the ratio

O_2 absorbed/O_2 content at saturation) versus p (the oxygen pressure) at different values of pH can be superimposed by making a change of scale for p. On doing this, the experimental points all fall near a smooth curve, as shown in figure 1. This curve can be reproduced roughly by the equation

$$y = \frac{Kp^n}{1 + Kp^n} \tag{1}$$

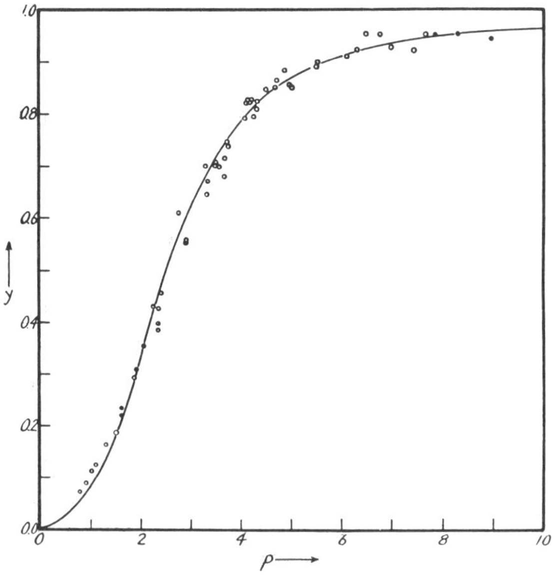

FIGURE 1

The experimental values of the fractional absorption of oxygen, y, plotted against the oxygen pressure p (the points obtained at various pH values being reduced to pH 8.30 by a change of scale in p), compared with the theoretical curve of Equation 3, corresponding to the square structure of the hemoglobin molecule.

which corresponds to the equilibrium $Hb_n + n\ O_2 \rightleftarrows Hb_n(O_2)_n$, the value of n being about 2.6. The curve can be accurately represented by Adair's four-constant equation

$$y = \frac{K_1p + 2K_2p^2 + 3K_3p^3 + 4K_4p^4}{4(1 + K_1p + K_2p^2 + K_3p^3 + K_4p^4)}, \tag{2}$$

and also by the two-constant Equation 3, which we shall now derive.

Let us make the following postulates regarding the structure of the hemoglobin molecule.

1. The hemoglobin molecule contains four hemes, each of which is

connected with k others, the connection being such as to permit the interaction of the hemes in pairs.

2. The four hemes are equivalent.
3. The interactions for connected pairs are equivalent.

We distinguish four cases, corresponding to $k = 0$, 1, 2 and 3.

No interaction of hemes, $k = 0$. This leads to Equation 1 with $n = 1$, and is hence not compatible with experiment.

Interaction of the hemes in pairs $\begin{pmatrix} \text{Hb} & \text{Hb} \\ | & | \\ \text{Hb} & \text{Hb} \end{pmatrix}$, $k = 1$. This leads to an equation which can be approximated by Equation 1 with $1 \leqslant n \leqslant 2$, and is hence not compatible with experiment.

Hemes at the corners of a square, interactions along the sides of the square $\begin{pmatrix} \text{Hb—Hb} \\ | \quad\;\; | \\ \text{Hb—Hb} \end{pmatrix}$, $k = 2$. This leads to Equation 3, in agreement with experiment.

Hemes at the corners of a tetrahedron, interactions along the edges of the tetrahedron, $k = 3$. This leads to Equation 4, in agreement with experiment; however, for other reasons the structure is less satisfactory than the square structure.

To derive Equation 3 on the basis of the square arrangement, we note that at the oxygen pressure p the relative amounts of the six molecular species present are the following:

Hb—Hb	Hb—HbO_2	Hb—HbO_2	Hb—HbO_2	Hb—HbO_2	HbO_2—HbO_2
| |	| |	| |	| |	| |	| |
Hb—Hb	Hb—Hb	Hb—HbO_2	HbO_2—Hb	HbO_2—HbO_2	HbO_2—HbO_2
1	$4K'p$	$4\alpha K'^2p^2$	$2K'^2p^2$	$4\alpha^2K'^3p^3$	$\alpha^4K'^4p^4$

The constants K' and α have the following significance, $RT \ln K'$ is the free energy change accompanying the addition of oxygen to heme, and $RT \ln \alpha$ is the additional free energy stabilizing two interacting HbO_2 groups; that is, the free energy is decreased by $RT \ln \alpha$ for each interaction shown by a heavy bar. The factors 2 and 4 result from the symmetry numbers.

These relative amounts $1:4K'p:(4\alpha + 2)\ K'^2p^2:4\alpha^2K'^3p^3:\alpha^4K'^4p^4$ of the substances Hb_4, Hb_4O_2, $Hb_4(O_2)_2$, $Hb_4(O_2)_3$ and $Hb_4(O_2)_4$ lead directly to the following equation for y:

$$y = \frac{K'p + (2\alpha + 1)K'^2p^2 + 3\alpha^2K'^3p^3 + \alpha^4K'^4p^4}{1 + 4K'p + (4\alpha + 2)K'^2p^2 + 4\alpha^2K'^3p^3 + \alpha^4K'^4p^4}. \qquad (3)$$

The curve shown in figure 1 is calculated by means of this equation, with $\alpha = 12$ and $K' = 0.033$. It is seen that the agreement with the experimental points is satisfactory. (The uniform deviation shown by the first five points may be due to experimental error, or may be real, indicating that the theory is only approximate.)

The similar discussion of the tetrahedral configuration leads to the equation

$$y = \frac{K'p + 3\alpha K'^2p^2 + 3\alpha^3K'^3p^3 + \alpha^6K'^4p^4}{1 + 4K'p + 6\alpha K'^2p^2 + 4\alpha^3K'^3p^3 + \alpha^6K'^4p^4}, \qquad (4)$$

in which K' and α have the same significance as before. This equation

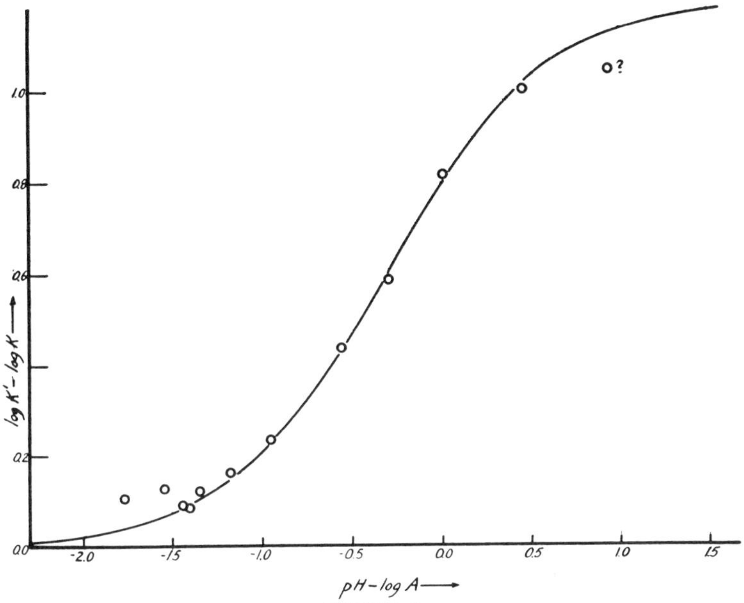

FIGURE 2

The dependence of the equilibrium constant K' of Equation 3 on pH. The circles are experimental values, and the curve is drawn according to Equation 6.

provides a curve almost indistinguishable from the one shown in the figure, the value of α being about $12^{2/3} = 5.3$.

Although the tetrahedral configuration is thus compatible with the oxygen equilibrium data, we reject it in favor of the square configuration for the following reasons. The hemoglobin molecule is roughly spherical in shape, with a radius of about 29 Å, and it is very probable that the hemes are located on the surface of the molecule; if arranged tetrahedrally, they would be about 47 Å apart. It is very difficult to imagine a connection between two hemes this far apart which would lead to an interaction energy as large as $RT \ln \alpha = 1000$ cal./mole. Moreover, the tetrahedral configuration would lead one to expect a three-fold axis of symmetry for the hemes rather than the four-fold axis possessed by the porphyrin nucleus.

To account for the dependence of the oxygen equilibrium on the hydrogen ion concentration, we give quantitative expression to the familiar qualitative explanation that there is an interaction between the oxygen molecules of oxyhemoglobin and some acid groups. The fact that an equation of the form of 3 can be made to account for the data for various pH values by changing K' alone indicates strongly that each heme contains its own acid groups, the interaction of one heme with the acid groups of another being negligible.[3] Let us first assume that there is one acid group for each heme, with interaction constant β such that the free energy change on dissociation of $HHbO_2$ to H^+ and HbO_2^- is $RT \ln \beta$ less than that of HHb to H^+ and Hb^-, oxyhemoglobin being a stronger acid than hemoglobin. The relative amounts of the four molecular species (considering only one heme) HHb, Hb^-, $HHbO_2$ and HbO_2^- are then $1:A/[H^+]:Kp:Kp\beta A/[H^+]$, and the oxygen equilibrium formula becomes

$$y = \frac{K'p}{1 + K'p}$$

with

$$K' = K\frac{(1 + \beta A/[H^+])}{(1 + A/[H^+])}, \tag{5}$$

in which A is an acid-strength constant. It is found on similarly considering the 55 molecular species H_4Hb_4, $H_4Hb_4O_2$, etc., corresponding to the square arrangement that the oxygen equilibrium is expressed by Equation 3, with K' dependent on the hydrogen ion concentration in the way given by Equation 5.

The dependence of K' on pH as found experimentally from the factors used in reducing the points of figure 1 to a single curve is shown[4] in figure 2. It is found that Equation 5 does not represent this dependence satisfactorily. If, however, we assume that there are two equivalent acid groups interacting with each heme, we derive the equation

$$K' = K\frac{(1 + \beta A/[H^+])^2}{(1 + A/[H^+])^2}, \tag{6}$$

which does agree satisfactorily with experiment,[5] as shown by the curve in the figure, calculated for $\beta = 4$, $-\log A = 7.94$ and $K = 0.0035$.

We conclude that the data on the oxygen equilibrium of hemoglobin indicate that the four hemes of the molecule are arranged at the corners of a square; each heme is connected with two others in such a way as to give rise to an interaction energy of $RT \ln \alpha = 1500$ cal./mole for each pair of adjacent oxyhemes, and each heme contains two acid groups, the acid group-oxyheme interaction energy being $RT \ln \beta = 820$ cal./mole. The total oxyheme-oxyheme interaction energy is about 6000 cal./mole, and the total acid group-oxyheme interaction energy is about 6600 cal./mole.

The none too exact experiments on the carbon monoxide equilibrium and the carbon monoxide-oxygen balance of hemoglobin suggest that the interactions α and β are effective for carboxyhemoglobin as well as for oxyhemoglobin. On the other hand, the results of Anson and Mirsky[6] on the carbon monoxide equilibrium of hemochromogen, which agree with Equation 1 with $n = 1$, show that there is no heme-heme interaction in the hemochromogen studied. It seems not improbable that the hemochromogens differ from hemoglobin mainly in that in the hemochromogens the hemes are independent and in hemoglobin four hemes form a conjugated system.

[1] G. S. Adair, *Proc. Roy. Soc.*, **A109**, 292 (1925); *Jour. Biol. Chem.*, **63**, 529 (1925).

[2] R. M. Ferry and A. A. Green, *Jour. Biol. Chem.*, **81**, 175 (1929).

[3] It is possible that the results of more accurate experiments would require this conclusion to be revised.

[4] The points for pH 5.8, 4.9 and 4.5 are not shown. These are obtained from only one or three measurements; they lie somewhat above the origin (0.1–0.2), which may indicate that some other interaction becomes effective in very acid solutions.

[5] We have neglected the mutual interaction of the two acid groups, the data not being sufficiently accurate and extensive to make a discussion of this interaction profitable.

[6] M. L. Anson and A. E. Mirsky, *Jour. Physiol.*, **60**, 50 (1925).

Reprinted from the Proceedings of the NATIONAL ACADEMY OF SCIENCES,
Vol. 22, No. 4, pp. 210–216. April, 1936.

THE MAGNETIC PROPERTIES AND STRUCTURE OF HEMOGLOBIN, OXYHEMOGLOBIN AND CARBONMONOXYHEMOGLOBIN

BY LINUS PAULING AND CHARLES D. CORYELL

GATES CHEMICAL LABORATORY, CALIFORNIA INSTITUTE OF TECHNOLOGY

Communicated March 19, 1936

Over ninety years ago, on November 8, 1845, Michael Faraday investigated the magnetic properties of dried blood and made a note "Must try recent fluid blood." If he had determined the magnetic susceptibilities of arterial and venous blood, he would have found them to differ by a large amount (as much as twenty per cent for completely oxygenated and completely deoxygenated blood); this discovery without doubt would have excited much interest and would have influenced appreciably the course of research on blood and hemoglobin.[1]

Continuing our investigations of the magnetic properties and structure of hemoglobin and related substances,[2] we have found oxyhemoglobin and carbonmonoxyhemoglobin to contain no unpaired electrons, and ferrohemoglobin (hemoglobin itself) to contain four unpaired electrons per heme. The description of our experiments and the interpretation and discussion of the results are given below.

Note on Nomenclature.—The current nomenclature of hemoglobin and related substances was formulated at a time when precise information about the chemical composition and structure of the substances was not available. Now that some progress has been made in gathering this information, especially in regard to chemical composition, it is possible to revise the nomenclature in such a way as to make the names of substances more descriptive than the older names, without introducing any radical changes. In formulating the following set of names we have profited by the continued advice of Dr. Alfred E. Mirsky.

The names whose use we advocate are given below, followed in some cases by acceptable synonyms. The expressions in parentheses are those whose use we consider to be undesirable.

Heme: an iron-porphyrin complex (generic term, used for either ferroheme or ferriheme).

Ferroheme (reduced heme): a complex of ferrous iron and a porphyrin.

Ferriheme (oxidized heme): a complex of ferric iron and a porphyrin.

Ferriheme chloride, hemin: a compound of ferriheme and chloride ion.

Ferriheme hydroxide, hematin: a compound of ferriheme and hydroxyl ion.

Ferrohemochromogen, hemochromogen: a complex of ferroheme and another substance, or two other substances, having the characteristic hemochromogen spectrum and involving covalent bonds from the iron atom to the porphyrin nitrogen atoms and the attached groups.[2] Individual hemochromogens may be designated by specifying the attached groups, as globin hemochromogen (ferroheme and denatured globin), dicyanide hemochromogen, dipyridine hemochromogen, carbonmonoxyhemochromogen, pyridine carbonmonoxyhemochromogen, etc.

Ferrihemochromogen (parahematin): a compound of ferriheme and another substance or two other substances, involving covalent bonds from the iron atom to the porphyrin nitrogen atoms and the attached groups.[3]

Hemoglobin: a conjugated protein containing heme and native globin (generic term, used for both ferrohemoglobin and ferrihemoglobin and also for closely related substances); specifically, ferrohemoglobin.

Ferrohemoglobin, hemoglobin (reduced hemoglobin): a conjugated protein formed by combination of ferroheme and native protein.

Oxyhemoglobin: a compound of ferrohemoglobin and oxygen.

Carbonmonoxyhemoglobin, carbon monoxide hemoglobin (carboxyhemoglobin): a compound of ferrohemoglobin and carbon monoxide.

Ferrihemoglobin (methemoglobin): a conjugated protein formed by combination of ferriheme and native globin.[3]

Carbonmonoxyhemoglobin.—The magnetic measurements are described in the experimental part below. The carbonmonoxyhemoglobin molecule is found to have zero magnetic moment, and hence to contain no unpaired electrons. This is to be interpreted as showing that at least two $3d$ orbitals of each ferrous iron atom are involved in covalent bond formation, the atom presumably forming six octahedral d^2sp^3 bonds, four to the porphyrin nitrogen atoms, one to an atom (probably nitrogen) of the globin, and one to the carbon monoxide molecule:

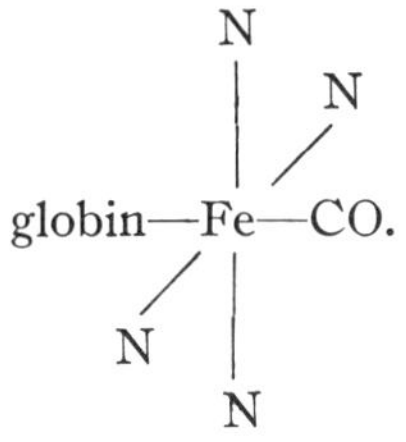

In view of the discovery of Brockway and Cross[4] that the nickel-carbon bond in nickel carbonyl has a large amount of double bond character, we may well expect this to be the case for the iron-carbon bond in carbonmonoxyhemoglobin also, the double bond being formed with the use of a pair of electrons conventionally assigned to the iron atom as $3d$ electrons. To carbonmonoxyhemoglobin there would then be ascribed the resonating structure:

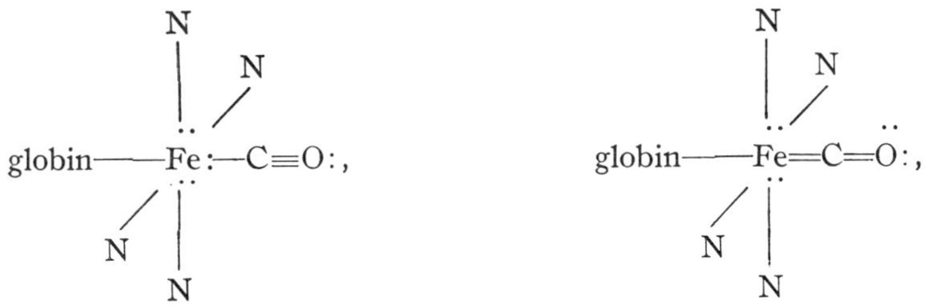

in which the dashes represent shared electron pairs and the dots unshared electrons.

Oxyhemoglobin.—The molecule of oxyhemoglobin, like that of carbonmonoxyhemoglobin, is found to have zero magnetic moment and to contain no unpaired electrons. Each iron atom is accordingly attached to the four porphyrin nitrogen atoms, the globin molecule, and the oxygen molecule by covalent bonds.

The free oxygen molecule in its normal state ($^3\Sigma$) contains two unpaired electrons. It might well have been expected, in view of the ease with which oxygen is attached to and detached from hemoglobin, that the oxygen molecule in oxyhemoglobin would retain these unpaired electrons, a pair of σ electrons of one oxygen atom, unshared in the free molecule, being used for the formation of the bond to hemoglobin:

$$\text{Hb} + :\overset{\cdots}{\text{O}}:\underset{\cdots}{\text{O}}: \rightleftarrows \text{Hb}:\text{O}:\overset{\cdots}{\text{O}}:.$$

However, this is shown not to be so by the magnetic data, there being no unpaired electrons in oxyhemoglobin. *The oxygen molecule undergoes a profound change in electronic structure on combination with hemoglobin.*

Of the structures of oxyhemoglobin compatible with the magnetic data, the most probable is the resonating structure analogous to that of carbonmonoxyhemoglobin:

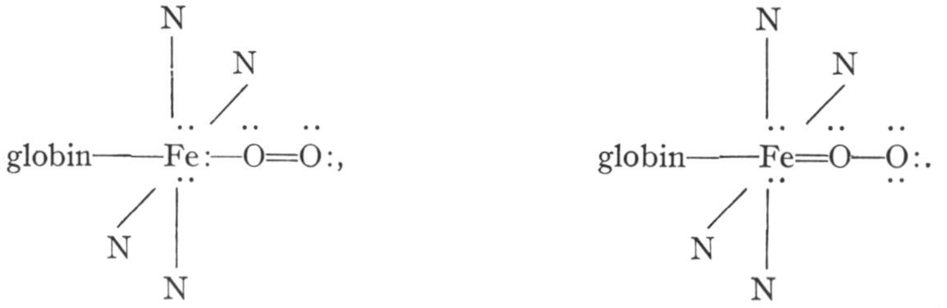

The great similarity in properties of oxyhemoglobin and carbonmonoxyhemoglobin provides strong support for this structure. The structure in which each of the two oxygen atoms (connected with one another by a single bond) is attached to the iron atom by a single bond is rendered improbable by the strain involved in the three-membered ring.

Ferrohemoglobin.—In contrast to oxyhemoglobin and carbonmonoxyhemoglobin, hemoglobin itself contains unpaired electrons, its magnetic susceptibility showing the presence of a pronounced paramagnetic contribution. The interpretation of the magnetic data can be made only in conjunction with a discussion of the nature and magnitude of the mutual interactions of the four hemes in the molecule.[5] One possibility is that the heme-heme interaction is sufficiently strong to couple the moments of all electrons in the molecule into a resultant moment, with the same value for all molecules. The magnetic data interpreted in this way lead to the value μ = 10.92 Bohr magnetons for the moment of the molecule. We reject this possibility on the following grounds. (1) The heme-heme interaction energy, as evaluated from the oxygen equilibrium data,[5] is hardly large enough to overcome the entropy advantage of independent heme moments. (2) The value 10.92 for the moment is not far from that (8.94) for eight unpaired electrons, two per heme, with parallel spins; however, it is about 22% larger, and this difference could be accounted for only as a surprisingly large contribution of orbital moment. (3) On this basis the magnetic susceptibility of partially oxygenated hemoglobin solutions would show large deviations from a linear dependence on the amount of uncombined heme; we have found large deviations not to occur. (These experiments will be described in a later paper.)

The other simple possibility, which we believe to be approximated in reality, is that the magnetic moments of the four hemes orient themselves in the applied magnetic field independently of one another. With the calculations made on this assumption, the experimental data lead to the value μ = 5.46 Bohr magnetons for the effective moment per heme. This shows that there are present in each heme four unpaired electrons, and that consequently the iron atom is not attached to the four porphyrin nitrogen atoms and the globin molecule by covalent bonds, but is present as a ferrous ion, the bonds to the neighboring atoms being essentially ionic bonds.

The resultant spin moment for four unpaired electrons is 4.90 magnetons. In compounds containing ferrous ion values of 4.9 to 5.4 are observed, the increase over the spin moment arising from a small orbital contribution. Complexes of ferrous iron with substances containing nitrogen (hydrazine, etc.) give values in the lower part of this range, the quenching of orbital moment being nearly complete.[6] It does not seem probable that the high value for ferrohemoglobin is to be accounted for as due to orbital moment, since the porphyrin nitrogen atoms should have a strong quenching effect on the orbital moment. We interpret this high value instead as due to a heme-heme interaction which tends to stabilize states with parallel heme moments relative to those with opposed heme moments, the oxygen-equilibrium value of the heme-heme interaction energy being of the order of magnitude required for this interpretation.

It is interesting and surprising that the hemoglobin molecule undergoes such an extreme structural change on the addition of oxygen or carbon monoxide; in the ferrohemoglobin molecule there are sixteen unpaired electrons and the bonds to iron are ionic, while in oxyhemoglobin and carbonmonoxyhemoglobin there are no unpaired electrons and the bonds are covalent. The change from ionic bonds to covalent bonds also occurs on formation of hemochromogen from ferroheme. Such a difference in bond type in very closely related substances has been observed so far only in hemoglobin derivatives.

It is not yet possible to discuss the significance of these structural differences in detail, but they are without doubt closely related to and in a sense responsible for the characteristic properties of hemoglobin. For example, the change in multiplicity of the system oxygen molecule–heme in hemoglobin on formation of oxyhemoglobin need be only as great as two (from the triplet corresponding to the opposed oxygen molecule triplet and ferroheme quintet to the singlet of oxyheme), whereas the change in multiplicity on formation of carbonmonoxyhemoglobin is four; in view of the infrequency of transitions involving a change in multiplicity, we might accordingly anticipate that the reactions of hemoglobin with carbon monoxide would be slower than those with oxygen, in agreement with observation. The change in multiplicity may be related also to the photochemical reactivity of carbonmonoxyhemoglobin. The difference in bond type in hemoglobin and its compounds is probably connected with the preferential affinity of hemoglobin for oxygen and carbon monoxide in contrast to other substances. Further experimental information is needed before these questions can be discussed in detail.

Experiments.—Solutions: Defibrinated bovine blood (provided through the courteous coöperation of Cornelius Bros., Ltd.) was used as the source of material. Preparations *A* and *B* consisted of whole blood, collected and separately oxygenated by rotating 20 minutes in air in a large open vessel, and then packed in ice and used as soon as possible. For preparations *C* and *D* oxygenated blood was centrifuged, and the corpuscles washed three times with equal volumes of physiological sodium chloride. Ether was used to hemolyze the collected corpuscles, the stromata-emulsions were separated by centrifuging, and the dissolved ether removed from the oxyhemoglobin solutions by a current of air. The solutions were kept on ice until used.

Analyses were made for oxygen content in a Van Slyke-Neill constant-volume blood gas apparatus. The transfer pipet was calibrated for content and retention on the walls of whole blood or concentrated oxyhemoglobin solution corresponding to conditions of use; the gas pipet was also calibrated for volume. Correction was made for dissolved oxygen on the assumption that the quantity dissolved is proportional to the water present in the solution.

Corrected results of analyses: Blood *A:* 100 ml. combine with 20.20 ml. O_2 S.T.P.; formality of heme-iron, 0.00902. Blood *B:* 100 ml. combine with 20.59 ml. O_2 S.T.P.; formality of heme-iron, 0.00919. Solution *C:* 100 ml. combine with 37.15 ml. O_2 S.T.P.; formality of heme-iron, 0.01658. Solution *D:* 100 ml. combine with 41.26 ml. O_2 S.T.P.; formality of heme-iron, 0.01841.

Apparatus: The apparatus for magnetic susceptibility determinations has already been described.[2] All hemoglobin solutions were measured against water in a tube of about 18 mm. internal diameter. Fields of 7640 and 8830 gauss were used, the forces being reported as average Δw (in milligrams) for the former. A small correction to the observed Δw has been applied for blank on the tube, so that reported forces are for solution against pure water. Solutions were measured at approximately 20°C.

Calibration of field and tube with water against air: $\Delta w = -49.59$. (For hemochromogen and $6N$ NaOH the tube with $\Delta w = -45.40$ for water against air was used.)

Carbonmonoxyhemoglobin.—Samples of blood *A* equilibrated with CO by rotation of 50 ml. in a liter tonometer filled with pure carbon monoxide: $\Delta w = -0.56$, -0.60, -0.76, -0.61, average -0.63. Samples of blood *B* equilibrated with CO: $\Delta w = -0.84$, -1.03, -0.80, -0.86, average -0.88. Samples of solution *C* equilibrated with CO: $\Delta w = -0.28$, -0.68, average -0.48. Samples of solution *D* equilibrated with CO: $\Delta w = -0.36$, -0.45, average, -0.41. (Completeness of saturation with carbon monoxide was generally tested by adding $Na_2S_2O_4.2H_2O$ to the magnetic tube and measuring the increase in susceptibility due to formation of hemoglobin.)

We have established in the previous paper[2] the presence of no unpaired electrons in globin hemochromogen and dicyanide hemochromogen. For globin hemochromogen made by denaturing 32 ml. of whole blood with 10 ml. of 6*N* NaOH after reduction of the heme: average $\Delta w = -1.71$; for dicyanide hemochromogen prepared in a similar manner: average $\Delta w = -1.53$; average for the two, $\Delta w = -1.62$. Measurement of the 6*N* NaOH against water in the same tube gives $\Delta w = -4.88$, -4.95. Assuming the additivity of atomic diamagnetism (Wiedemann's rule), whole blood without paramagnetic constituent should give $\Delta w = -0.58$ in the tube used for the hemoglobin series. This value is in satisfactory agreement with the Δw values given above. The calculated value for blood with two unpaired electrons per heme, and independent hemes, is $\Delta w = +1.52$, for four, $\Delta w = +5.69$; the calculated values for the hemoglobin solutions are about twice as great.

Conclusion: carbonmonoxyhemoglobin contains no unpaired electrons.

Oxyhemoglobin.—Samples of blood *A:* $\Delta w = -0.65$, -0.40, -0.44, average, -0.50. Blood *B:* $\Delta w = -0.58$, -0.62, -0.62, average, -0.61. Solution *C:* $\Delta w = -0.44$, -0.55, -0.50, -0.50, average, -0.50. Solution *D:* $\Delta w = -0.38$, -0.36, average, -0.37. Oxyhemoglobin relative to carbonmonoxyhemoglobin: *A*, $+0.13$; *B*, $+0.27$; *C*, $+0.02$; *D*, $+0.04$; calculated for two unpaired electrons on oxyhemoglobin: *A*, $+2.03$; *B*, $+2.07$; *C*, $+3.74$; *D*, $+4.11$.

Conclusion: oxyhemoglobin contains no unpaired electrons.

Hemoglobin.—35 ml. of blood *A* reduced in differential tube by addition of from 0.4 to 1.0 g. $Na_2S_2O_4.2H_2O$: $\Delta w = +7.32$, 6.85, 6.98, 6.97, average, $+7.03$. Taking the mean of the oxy- and carbonmonoxyhemoglobin values (-0.57) for Δw of diamagnetism of hemoglobin, the change on removing coördinating group (O_2, CO) is $+7.60$ gm., corresponding to paramagnetism and a magnetic moment of 5.48 Bohr magnetons per heme, assuming independent hemes. (The change in diamagnetism involved in loss of of CO or O_2 is negligible.)

Blood *B*, reduced: $\Delta w = +7.21$, 7.16, 7.51, 7.22, 7.50, 7.20, 7.51, 7.06, average, $+7.30$; diamagnetic value, -0.74; change, $+8.04$; $\mu = 5.58$.

Solution *C*, reduced: $\Delta w = +13.09$, 12.83, 12.89, 13.28, average, 13.02; diamagnetic value, -0.49; change, $+13.52$; $\mu = 5.38$.

Solution *D*, reduced: $\Delta w = +14.95$, 14.33, average, $+14.64$; diamagnetic value, -0.39; change, $+15.03$; $\mu = 5.40$.

Summary of results for hemoglobin: blood *A*, $\mu = 5.48$; blood *B*, 5.58; solution *C*, 5.38; solution *D*, 5.40; average of the four, $\mu = 5.46$.

Spin moment for four unpaired electrons, 4.9; for two, 2.83; for none, 0.00. Moment observed for ferrous ion in solution, about 5.3; moment observed for solid $Fe(N_2H_4)_2Cl_2$, 4.86. Conclusion: ferrohemoglobin has a susceptibility corresponding to four unpaired electrons per heme, with evidence for some magnetic interaction between the hemes.

Summary.—It is shown by magnetic measurements that oxyhemoglobin and carbonmonoxyhemoglobin contain no unpaired electrons; the oxygen molecule, with two unpaired electrons in the free state, accordingly undergoes a profound change in electronic structure on attachment to hemoglobin. The magnetic susceptibility of hemoglobin itself (ferrohemoglobin) corresponds to an effective magnetic moment of 5.46 Bohr magnetons per heme, calculated for independent hemes. This shows the presence of four unpaired electrons per heme, and indicates that the heme-heme interaction tends to stabilize to some extent the parallel configuration of the moments of the four hemes in the molecule. The bonds from iron to surrounding atoms are ionic in hemoglobin, and covalent in oxyhemoglobin and carbonmonoxyhemoglobin.

We have been helped a great deal by the advice and encouragement of Dr. Alfred E. Mirsky of the Hospital of the Rockefeller Institute. This investigation is part of a program of research on the structure of hemoglobin being carried on with the aid of a grant from the Rockefeller Foundation.

[1] A. Gamgee, *Proc. Roy. Soc. London*, **68**, 503–512 (1901), and one or two more recent investigators have reported blood to be about as diamagnetic as water, without discovering the difference between arterial and venous blood.

[2] L. Pauling and C. D. Coryell, These PROCEEDINGS, **22**, 159 (1936).

[3] We shall discuss the structure of ferrihemoglobin and the ferrihemochromogens in a later paper.

[4] L. O. Brockway and P. C. Cross, *Jour. Chem. Phys.*, **3**, 828 (1935).

[5] L. Pauling, These PROCEEDINGS, **21**, 186 (1935).

[6] L. Pauling, *Jour. Am. Chem. Soc.*, **53**, 1367 (1931).

[Reprint from the Journal of the American Chemical Society, 59, 633 (1937).]

[CONTRIBUTION FROM THE GATES AND CRELLIN LABORATORIES OF CHEMISTRY, CALIFORNIA INSTITUTE OF TECHNOLOGY, No. 580]

The Magnetic Properties and Structure of Ferrihemoglobin (Methemoglobin) and Some of its Compounds

BY CHARLES D. CORYELL, FRED STITT AND LINUS PAULING

Our studies of the magnetic properties of hemoglobin derivatives containing ferrous iron,[1,2] including ferroheme, several hemochromogens, hemoglobin, and carbonmonoxyhemoglobin, led to the discovery that in two of these substances (ferroheme, hemoglobin) there are four unpaired electrons per heme, indicating that the bonds attaching the iron atoms to the rest of the molecule are essentially ionic in character, whereas the others contain no unpaired electrons, each iron atom being attached to six adjacent atoms by essentially covalent bonds. We have now investigated ferrihemoglobin[3] (acid methemoglobin), ferrihemoglobin hydroxide (alkaline methemoglobin), ferrihemoglobin fluoride, ferrihemoglobin cyanide, and ferrihemoglobin hydrosulfide, and have found a variety in magnetic properties greater than that for the ferrohemoglobin derivatives; the magnetic susceptibilities

(1) L. Pauling and C. D. Coryell, *Proc. Nat. Acad. Sci.*, **22**, 159 (1936).

(2) L. Pauling and C. D. Coryell, *ibid.*, **22**, 210 (1936).

(3) The nomenclature used in this paper is described in ref. 2.

of ferrihemoglobin and its fluoride appear to correspond to five unpaired electrons per heme, those of the cyanide and hydrosulfide to one, and that of the hydroxide to three. The structural significance of these results is discussed in the last section of this paper.

Technique of the Magnetic Measurements.—The Gouy method was used to determine the magnetic susceptibilities of solutions of hemoglobin derivatives, the difference in susceptibility of solution and water being measured with use of a glass tube divided by a glass partition into two compartments, one containing solution and the other water.[4] The tubes used were about 18 mm. in internal diameter and 30 cm. long, and were provided with ground glass caps for the ends and with suitable supports for suspension from the balance arm. Fields of about 7640 and 8810 gausses were used; forces measured at 8810 gausses were changed to 7640 gausses by multiplication by the experimentally determined factor 0.752, and all measurements made (usually four) averaged to give a mean value of Δw (in milligrams). Measurements were made in several tubes with slightly different diameters. Each tube was calibrated by the measurement of Δw for water against air, these values being close to 47. Experimentally determined corrections have been applied for dilution and for the diamagnetism of added reagents. All susceptibility measurements were made at temperatures between 22 and 26°.

The concentration of each hemoglobin solution used was determined by reducing to ferrohemoglobin with 0.3 or 0.6 g. of sodium hydrosulfite (for 30 ml. of solution), determining Δw, then saturating with carbon monoxide in the dark or in diffuse daylight and again determining Δw. The change in Δw corresponds to a change in molal susceptibility (per heme) of $12{,}430 \times 10^{-6}$ c. g. s. u. at 24°, the effective magnetic moment of ferrohemoglobin per heme being taken[2] as 5.46 Bohr magnetons. Representing Δw for ferrohemoglobin solution by Δw_{Hb}, that for carbonmonoxyhemoglobin solution by Δw_{COHb}, and that for the solution being studied by Δw (these Δw's corresponding to the same molal concentration), the molal susceptibility (per heme) at 24° for the solution being studied is given by the equation

$$\chi_{\mathrm{molal}} = \frac{\Delta w - \Delta w_{\mathrm{COHb}}}{\Delta w_{\mathrm{Hb}} - \Delta w_{\mathrm{COHb}}} \cdot 12{,}430 \cdot 10^{-6}\ \text{c. g. s. u.} \quad (1)$$

and the effective magnetic moment per heme by the equation

$$\mu = \left(\frac{\Delta w - \Delta w_{\mathrm{COHb}}}{\Delta w_{\mathrm{Hb}} - \Delta w_{\mathrm{COHb}}}\right)^{1/2} \cdot 5.46\ \text{Bohr magnetons} \quad (2)$$

Measurements of *p*H values were made with a Beckman glass-electrode *p*H meter generously furnished by Professor A. O. Beckman of these Laboratories. As a standard 0.05 *M* potassium hydrogen phthalate solution, *p*H 3.97, was used. Corrections were applied in alkaline solutions for error due to sodium and potassium ions present as recommended by the manufacturer of the instrument.

Preparation of Ferrihemoglobin Solutions.—In the preliminary work ferrihemoglobin solutions were prepared by oxidation of oxyhemoglobin solutions with potassium ferricyanide, the excess of this reagent then being converted to ferrocyanide (which has zero magnetic moment) by the addition of sodium sulfite, which does not reduce ferrihemoglobin. It was also found that on addition to oxyhemoglobin solutions of sodium hydrosulfite and then of potassium ferricyanide the hemoglobin is oxidized to ferrihemoglobin and the excess ferricyanide reduced to ferrocyanide by the sulfite formed earlier by oxidation of the hydrosulfite by oxyhemoglobin.

In order to avoid the possibility of magnetic effects of the added iron the following very satisfactory method of preparing ferrihemoglobin solutions was developed, involving the auto-oxidation of oxyhemoglobin solutions at *p*H 4.8 to 5.3. Corpuscles obtained by centrifuging bovine blood are washed three times with 0.14 *M* potassium chloride solution, laked with ether, and centrifuged, the ether then being removed by bubbling air through the solution. To this solution enough 6 *N* lactic acid solution is added (about 16 ml. per liter) to bring the *p*H to about 4.9. At this *p*H the auto-oxidation reaction is complete after the solution has remained for forty-eight hours at room temperature. The solution, now at *p*H about 5.15, is brought to *p*H 7.0 by the addition with vigorous stirring of 1 *N* potassium hydroxide solution and centrifuged to remove the small amount of denatured protein formed during acidification. For different preparations used in this work the concentration of ferrihemoglobin, determined magnetically as described above, ranged from 0.0113 to 0.0176 formal in heme-iron.

By following the concentration of oxyhemoglobin magnetically it was found that in twenty-four hours at *p*H 5.2 and temperature 22° the amount of oxyhemoglobin present had fallen to 1% of its initial value;[5] in order to ensure completion of the reaction it is recommended that the solution stand for another twenty-four hours.

Ferrihemoglobin and Ferrihemoglobin Hydroxide.—The pronounced change in spectrum of ferrihemoglobin solutions accompanying change in *p*H from the acid to the alkaline range corresponds to the addition of hydroxyl ion to ferrihemoglobin (acid methemoglobin) to form ferrihemoglobin hydroxide (alkaline methemoglobin). Spectrophotometric studies of this reaction made by Austin and Drabkin[6] have shown that the amounts of substances present at intermediate *p*H values correspond closely to the equilibrium[7]

$$\mathrm{HbOH} \rightleftharpoons \mathrm{Hb}^+ + \mathrm{OH}^- \quad (3)$$

In Table I there are given magnetic data for ferrihemoglobin solutions of low ionic strength

(4) S. Freed and C. Kasper, *Phys. Rev.*, **36**, 1002 (1930).

(5) The reaction is approximately first-order in oxyhemoglobin and first-order in hydrogen ion; the rate-determining step may be

$$\mathrm{HbO_2} + \mathrm{H}^+ \longrightarrow \mathrm{Hb}^+ + \mathrm{HO_2}$$

with the production of hydrogen superoxide. The temperature coefficient of the rate is high, about 5 for 10°.

(6) J. H. Austin and D. L. Drabkin, *J. Biol. Chem.*, **112**, 67 (1935); see also F. Haurowitz, *Z. physiol. Chem.*, **138**, 68 (1924).

(7) We use the symbol Hb^+ to represent the amount of ferrihemoglobin containing one heme, and HbOH, Hb, $\mathrm{HbO_2}$, etc., to represent corresponding amounts of ferrihemoglobin hydroxide, ferrohemoglobin, oxyhemoglobin, etc., respectively.

in the pH range 6.7 to 12.0. The data represent experiments made with three solutions, A, B and C. Two of these, A and B, were prepared by the lactic acid auto-oxidation method described above, and the third by a similar method, hydrochloric acid being used for acidification in place of lactic acid. The results for the three solutions show no differences greater than the experimental error. The hemoglobin concentration of each solution was determined by the ferrohemoglobin–carbonmonoxyhemoglobin magnetic method.

TABLE I

MAGNETIC SUSCEPTIBILITY OF FERRIHEMOGLOBIN SOLUTIONS OF LOW IONIC STRENGTH

Solution	pH	$\chi_{molal}\cdot 10^{6}$ [a]	Solution	pH	$\chi_{molal}\cdot 10^{6}$
B	6.73	13,975	A	8.32	10,790
B	6.73	13,800	A	8.32	10,840
B	6.73	14,010	A	8.34	10,930
B	6.73	13,980	A	8.49	9,950
A	6.86	13,900	A	8.57	9,835
A	6.86	13,760	A	8.60	9,890
A	6.86	13,620	A	8.61	9,620
A	6.86	13,725	A	8.63	9,660
A	6.86	13,630	C	8.97	9,080
A	6.86	14,000	A	9.02	8,960
A	6.86	13,770	A	9.06	9,250
A	6.86	13,820	A	9.54	8,775
A	6.86	13,900	A	9.61	8,340
C	6.88	13,730	A	9.82	8,730
C	6.88	13,630	C	10.01	8,420
C	6.88	13,450	A	10.02	8,020
C	6.88	13,630	A	10.30	8,040
C	6.88	13,730	B	10.78	8,765
A	7.05	13,785	A	10.81	8,380
A	7.34	13,220	B	10.82	8,310
C	7.69	12,190	B	10.90	8,470
A	7.87	12,080	A	10.92	8,355
A	7.87	12,200	A	11.73	8,150
A	8.12	11,320			

$\chi_{Hb^+}\cdot 10^6 = 14{,}060 \pm 50.$

$\chi_{HbOH}\cdot 10^6 = 8350 \pm 90.$

$pK_{HbOH} = 8.15 \pm 0.02.$

Ionic strength at pH 8.15 = 0.20.

[a] Paramagnetic contribution to susceptibility per mole of heme.

A portion of ferrihemoglobin solution was placed in one compartment of a tube and Δw determined. Successive small portions of potassium hydroxide solution (0.87 N) were then added with vigorous stirring, the values of pH and Δw being determined after the addition of each portion. The values of Δw were corrected for dilution and the diamagnetism of reagents, and converted into χ_{molal}, the paramagnetic contribution to the susceptibility per mole of heme, by means of equation 1. The data of Table I are shown by the open circles in Fig. 1.

The ionic strength of the hemoglobin solutions, due in part to the salts originally in solution in the erythrocytes and in part to added acid and base, changes somewhat with change in pH, the values at pH 7, 8 and 9 being about 0.17, 0.20 and 0.23, respectively. The contribution of hemoglobin to the ionic strength was ignored.

The data for these solutions represent a typical titration curve, the molal susceptibility changing rapidly over the range pH 7 to 9.5 from an asymptotic value of about $14{,}000\cdot 10^{-6}$, representing ferrihemoglobin, Hb^+, to an asymptotic value of about $8300\cdot 10^{-6}$, representing ferrihemoglobin hydroxide, HbOH. Using these estimated asymptotes to calculate concentrations of Hb^+ and HbOH, it is found that the experimental points when plotted on a graph of log ($[Hb^+]/[HbOH]$) against pH lie close to a straight line, with slope unity to within about five per cent., showing that the reaction is first-order in hydroxyl ion, as was reported by Austin and Drabkin. A theoretical susceptibility curve was fitted to the experimental points of Fig. 1 by the following procedure. The equilibrium constant K_{HbOH} for the reaction

$$HbOH \rightleftharpoons Hb^+ + OH^-$$

is related to the mole fraction x of total ferrihemoglobin in the form Hb^+ by the equation

$$\log x/(1 - x) = pK_{HbOH} - pH \qquad (4)$$

in which pK_{HbOH} is log (K_{HbOH}/K_W), with K_W the equilibrium constant for the ionization of water. The value of the molal susceptibility χ as a function of pH is then given by the expression

$$\chi = x\chi_{Hb} + (1 - x)\chi_{HbOH} \qquad (5)$$

in which x is determined by equation 4. With χ_{HbOH}, χ_{Hb^+} and pK_{HbOH} as variable parameters, this equation was fitted to the experimental points in such a way as to minimize the mean deviation from the curve. The final curve was found to have $\chi_{HbOH} = 8350 \pm 90\cdot 10^{-6}$, $\chi_{Hb^+} = 14{,}060 \pm 50\cdot 10^{-6}$, and $pK_{HbOH} = 8.15 \pm 0.02$, the indicated probable errors for the susceptibility values being half the mean deviations and that for pK_{HbOH} being an estimated value.

The data given in Table II and represented by the solid circles in Fig. 1 were obtained with a solution, D, to which potassium chloride had been added, the ionic strength being increased in this way to the value 1.3 (at pH 8.56). These data also are well represented by a curve corre-

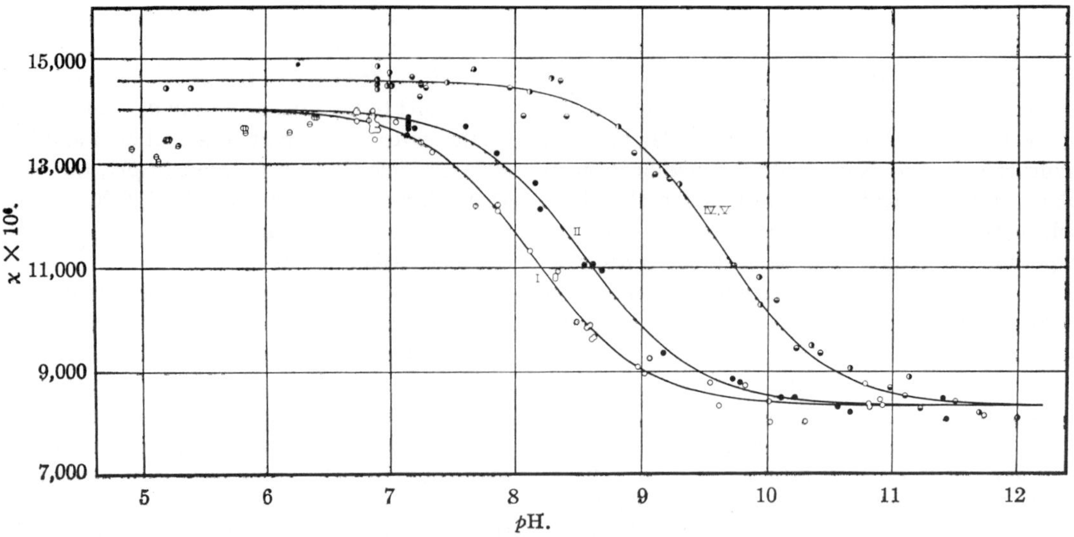

Fig. 1.—The dependence of the magnetic susceptibility of ferrihemoglobin solutions on *p*H: O, solutions of low ionic strength, Table I; ●, solutions of high ionic strength, Table II; ⊖ ⦶, solutions in low *p*H range, Table III; ◑, solutions of low ionic strength with added fluoride, Table IV; ◒, solutions of high ionic strength, with added fluoride, Table V.

sponding to equations 5 and 4, the values of the parameters for the curve giving the best fit being $\chi_{HbOH} = 8320 \pm 60 \cdot 10^{-6}$, $\chi_{Hb^+} = 14{,}000 \pm 70 \cdot 10^{-6}$, and $pK_{HbOH} = 8.56 \pm 0.02$.

TABLE II

MAGNETIC SUSCEPTIBILITY OF FERRIHEMOGLOBIN SOLUTIONS OF HIGH IONIC STRENGTH

Solution D

*p*H	$\chi_{molal} \cdot 10^6$	*p*H	$\chi_{molal} \cdot 10^6$
7.14	13,530	8.62	11,070
7.15	13,880	8.69	10,940
7.15	13,650	9.17	9,350
7.15	13,760	9.72	8,850
7.15	13,790	9.78	8,780
7.15	13,720	10.11	8,500
7.20	13,670	10.22	8,500
7.61	13,700	10.56	8,325
7.86	13,190	10.66	8,220
8.16	12,620	11.40	8,480
8.20	12,120	11.43	8,080
8.55	11,050		

$\chi_{Hb^+} \cdot 10^6 = 14{,}000 \pm 70.$
$\chi_{HbOH} \cdot 10^6 = 8320 \pm 60.$
$pK_B = 8.56 \pm 0.02.$
Ionic strength at *p*H 8.56 = 1.3.

The values found for χ_{HbOH} and χ_{Hb^+} in solutions at high and low ionic strength are in excellent agreement. For χ_{Hb^+} we accept the average value $14{,}040 \cdot 10^{-6}$, giving double weight to the asymptote of curve I. For χ_{HbOH} we obtain the mean value $8340 \cdot 10^{-6}$ by considering in addition to the two values $8350 \cdot 10^{-6}$ and $8320 \cdot 10^{-6}$ discussed above the values $8370 \cdot 10^{-6}$ and $8320 \cdot 10^{-6}$ obtained from solutions containing fluoride ion, discussed in the following section. The structural interpretation of the susceptibility values will be considered in the concluding section of the paper.

The equilibrium constant K_{HbOH} is dependent on the ionic strength of the solution, pK_{HbOH} changing from 8.15 to 8.56 with increase in ionic strength from 0.20 to 1.3. This dependence was observed by Austin and Drabkin, who found that pK_{HbOH} could be represented approximately by the equation

$$pK_{HbOH} = \text{constant} + \alpha\sqrt{\mu} \qquad (6)$$

in which μ is the ionic strength, the value for α being about 0.6. Our measurements support this,[8] the change for curves I and II corresponding to $\alpha = 0.59$.

The value $pK_{HbOH} = 8.12 \pm 0.01$ reported by Austin and Drabkin for canine hemoglobin at ionic strength 0.10 agrees reasonably well with our value 8.15 ± 0.02 at ionic strength 0.20 (corresponding to 8.07 at ionic strength 0.10) for bovine hemoglobin; complete agreement would, of course, not be expected for hemoglobins from different species.

It has been reported by Drabkin and Austin

(8) An experiment carried out involving the determination of change in Δw and *p*H for a portion of solution A, initially at *p*H 8.57, on the addition of successive portions of potassium chloride solution led to rough verification of the form of equation 6.

that ferrihemoglobin solutions become turbid at *p*H values less than 6. Two series of magnetic experiments were made in this *p*H range. In the first series a portion of solution A was made more and more acid by the addition, with rapid stirring, of successive small portions of 1 *N* hydrochloric acid solution, values of *p*H and Δw being determined after each acidification. These data are given in Table III and shown by the horizontally barred circles in Fig. 1. The formation of a small amount of coagulum in the acid solution was noticed; in order to eliminate possible error due to this coagulation, the following series of measurements was made. A portion of solution C was brought to *p*H 5.2 by the addition of hydrochloric acid, and centrifuged to remove the coagulum formed on acidification. The solution obtained in this way (solution E) was then made more and more alkaline by the addition of small portions of potassium hydroxide solution, values of *p*H and Δw being determined at each step. No coagulum was formed during this treatment. The ferrihemoglobin concentration, made uncertain by the loss of the coagulum resulting from the initial acidification, was determined by identifying the measured susceptibility at *p*H 7.25 (the most alkaline point) with that corresponding to the theoretical curve I. The data obtained in this way, given in Table III and represented in Fig. 1 by vertically barred circles, are in reasonably good agreement with those obtained by acidification. The measurements correspond to a decrease of about 5% in molal susceptibility in very acid solutions.

TABLE III

MAGNETIC SUSCEPTIBILITY OF FERRIHEMOGLOBIN SOLUTIONS OF LOW IONIC STRENGTH IN ACID SOLUTIONS

Solution	*p*H	$\chi_{molal}\cdot10^6$	Solution	*p*H	$\chi_{molal}\cdot10^6$
A	4.92	13,280	A	5.84	13,580
A	5.12	13,130	A	6.20	13,595
A	5.14	13,050	A	6.36	13,750
E	5.20	13,470	E	6.40	13,870
E	5.23	13,470	E	6.41	13,870
A	5.30	13,340	A	6.83	13,815
E	5.82	13,670	E	6.88	13,600
E	5.83	13,670	E	7.25	(13,410)

Ferrihemoglobin Fluoride.—The absorption spectrum of an acid ferrihemoglobin solution is changed in a pronounced manner by the addition of fluoride, indicating the formation of a compound. A crystalline compound was prepared and analyzed by Haurowitz,[9] who reported the substance to contain one atom of fluorine per atom of iron. Our magnetic studies have verified this, and have led to the evaluation of the equilibrium constants for the reactions

$$\mathrm{HbF} + \mathrm{OH^-} \rightleftarrows \mathrm{HbOH} + \mathrm{F^-}, \text{ and} \qquad (7)$$
$$\mathrm{HbF} \rightleftarrows \mathrm{Hb^+} + \mathrm{F^-}$$

The data given in Tables IV (for solutions A, B, C) and V (for solution D) were obtained by adding 0.5 g. of sodium fluoride to a 32-ml. portion of ferrihemoglobin solution and adding small portions of 0.87 *N* potassium hydroxide solution, with vigorous stirring, *p*H and Δw being determined after the addition of each portion. The

TABLE IV

MAGNETIC SUSCEPTIBILITY OF FERRIHEMOGLOBIN SOLUTIONS OF LOW IONIC STRENGTH WITH ADDED FLUORIDE

Solution	*p*H	$\chi_{molal}\cdot10^6$	Solution	*p*H	$\chi_{molal}\cdot10^6$
A	5.2	14,430	A	8.12	14,355
C	5.4	14,430	A	8.29	14,620
A	6.90	14,620	A	8.82	13,700
A	6.90	14,510	A	9.30	12,590
A	6.90	14,410	A	9.93	10,820
A	6.90	14,845	A	9.94	10,280
B	7.0	14,730	A	10.34	9,500
B	7.0	14,520	A	10.66	9,060
C	7.0	14,480	A	11.13	8,900
C	7.0	14,480	A	11.69	8,210
A	7.46	14,530	A	12.0	8,115
A	7.68	14,790			

$\chi_{HbF}\cdot10^6$ = 14,630 ± 70 (uncorr.), 14,660 ± 70 (corr.).
$\chi_{HbOH}\cdot10^6$ = 8370 ± 140.
$pK_{OH,F}$ = 9.63 ± 0.01 (uncorr.), 9.65 ± 0.01 (corr.).
Ionic strength at *p*H 9.63 = 0.54.
$[F^-]$ = 0.34 at *p*H 9.63.

TABLE V

MAGNETIC SUSCEPTIBILITY OF FERRIHEMOGLOBIN SOLUTIONS OF HIGH IONIC STRENGTH WITH ADDED FLUORIDE

Solution D

*p*H	$\chi_{molal}\cdot10^6$	*p*H	$\chi_{molal}\cdot10^6$
7.18	14,640	9.11	12,780
7.24[a]	14,270	9.22	12,680
7.25	14,490	9.73	11,030
7.25	14,520	10.07	10,360
7.29	14,430	10.23	9,440
7.96	14,430	10.42	9,350
8.07	13,900	10.98	8,690
8.36[b]	14,570	11.10	8,530
8.41	13,880	11.22	8,290
8.94[a]	13,190	11.50	8,425

$\chi_{HbF}\cdot10^6$ = 14,500 ± 80 (uncorr.), 14,550 ± 80 (corr.).
$\chi_{HbOH}\cdot10^6$ = 8320 ± 40.
$pK_{OH,F}$ = 9.60 ± 0.01 (uncorr.), 9.64 ± 0.01 (corr.).
Ionic strength at *p*H 9.60 = 1.6.
$[F^-]$ = 0.34 at *p*H 9.60.

[a] By adding 1 *N* HCl to solution of *p*H 10.98. [b] By adding 1 *N* HCl to solution of *p*H 9.22.

(9) F. Haurowitz, *Z. physiol. Chem.*, **138**, 68 (1924).

points for solutions A, B and C, with low ionic strength (0.54 at pH 9.6), and those for solution D, with high ionic strength (1.6 at pH 9.6), lie close to the same curve, showing that there is no appreciable salt effect for the reaction. This provides support for the postulate that the reaction consists essentially in the conversion of ferrihemoglobin fluoride into ferrihemoglobin hydroxide (eq. 7), for which no appreciable salt effect would be expected. Further evidence for equation 7 will be mentioned later.

The molal susceptibility values of Tables IV and V can be approximated closely by curves of the type used above for the ferrihemoglobin–ferrihemoglobin hydroxide equilibrium. For reaction 7 the equilibrium constant $K_{OH,F}$ has the value

$$K_{OH,F} = [HbOH][F^-]/[HbF][OH^-] \qquad (9)$$

and the molal susceptibility is given by the equation

$$\chi = y\chi_{HbF} + (1 - y)\chi_{HbOH} \qquad (10)$$

in which y is the mole fraction of total ferrihemoglobin in the form of ferrihemoglobin fluoride, given by the equation

$$\log y/(1 - y) = pK_{OH,F} - pH \qquad (11)$$

in which $pK_{OH,F}$ is $-\log (K_{OH,F}K_W/[F^-])$. The asymptotes of the curves which fit the data most closely are $14{,}630 \cdot 10^{-6}$ and $8{,}370 \cdot 10^{-6}$ for solutions A, B and C, and $14{,}500 \cdot 10^{-6}$ and $8320 \cdot 10^{-6}$ for solution D. The values $8370 \cdot 10^{-6}$ and $8320 \cdot 10^{-6}$, representing χ_{HbOH}, are in excellent agreement with the values $8350 \cdot 10^{-6}$ and $8320 \cdot 10^{-6}$ found for solutions without added fluoride.

The values for the other asymptote represent molal susceptibility not of ferrihemoglobin fluoride itself but of ferrihemoglobin fluoride containing a small fraction of ferrihemoglobin (Hb^+). From the equilibrium constants given below the values of the concentration ratio $[Hb^+]/[HbF]$ are found to be 0.047/0.953 for solutions A, B and C, and 0.094/0.906 for solution D. Using these ratios and the known value of χ_{Hb^+} to correct for the ferrihemoglobin present, the values $\chi_{HbF} = 14{,}660 \pm 70 \cdot 10^{-6}$ and $14{,}550 \pm 80 \cdot 10^{-6}$ are obtained; we accept for χ_{HbF} the mean of these closely agreeing values, $14{,}610 \cdot 10^{-6}$.

The values 9.63 and 9.60 for $pK_{OH,F}$ given by the curves lead on similar consideration of the $[Hb^+]/[HbF]$ ratio to the corrected values 9.65 and 9.64, which agree to within their estimated probable error of 0.01. Introducing the values of $[F^-]$ (0.34 at pH 9.6) and K_W ($10^{-14.01}$), we obtain for the equilibrium constant $K_{OH,F}$ at 24° the value $0.78 \cdot 10^4$.

The equilibrium constant K_{HbOH} for reaction 3 is given by the equation

$$\log K_{HbOH} = -6.12 + 0.59\sqrt{\mu} \qquad (12)$$

Combining with this the value of $K_{OH,F}$, we obtain for K_{HbF}, the equilibrium constant for the reaction

$$HbF \rightleftarrows Hb^+ + F^- \qquad (13)$$

the expression

$$\log K_{HbF} = -2.23 + 0.59\sqrt{\mu} \qquad (14)$$

Three sets of measurements were made to test these equilibrium constants by direct titration with potassium fluoride solution. In the first run a portion of solution A was brought to pH 8.60, and values of pH and Δw were measured after the addition of successive small portions of known volume of 2.00 f potassium fluoride solution. The average of nine values of K_{HbF} corresponding to the nine added portions of fluoride solution is 0.0145, with a mean deviation of 0.0012; this value is in good agreement with the value 0.013 given by equation 14 for μ = 0.34. When the formality of fluoride in the solution had reached 0.28, the pH was observed to have changed to the value 8.89, an increase of 0.29; this agrees well with the calculated change in pH expected because of the replacement of hydroxyl in ferrihemoglobin hydroxide by fluorine, 0.30. A less reliable set of six measurements made on solution D at pH 8.6 gave the average value 0.049 for K_{HbF}. This is somewhat larger than the value 0.029 given by equation 14 for μ = 1.4, perhaps because of experimental error, the total change in Δw values during the run being less than 1 mg. A set of five measurements made on solution A at pH 6.2 gave for K_{HbF} the average value 0.008, somewhat lower than the value 0.010 expected for μ = 0.22; the difference may be due to error in the measurement of Δw, which changed by only 0.40 mg. during the run.

A determination of the value of K_{HbF} by a spectrophotometric method has been reported by Lipmann.[10] The value found, about 0.015, for a solution of swine hemoglobin of uncertain ionic strength agrees with that calculated from equation 14 with μ given the reasonable value 0.5. Lipmann also reported a decrease of the constant in very acid solutions.

Ferrihemoglobin Cyanide.—Crystalline ferrihemoglobin cyanide was first obtained by Zey-

(10) F. Lipmann, *Biochem. Z.*, **206**, 171 (1929).

nek,[11] who by analysis showed the substance to contain one cyanide per heme. We have verified this by titration of a ferrihemoglobin solution with cyanide. The data represented in Fig. 2 were obtained with a solution containing equal volumes

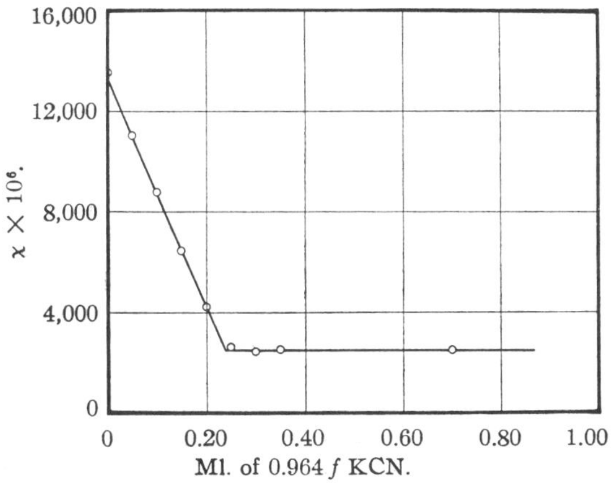

Fig. 2.—The magnetic titration of ferrihemoglobin with potassium cyanide at *p*H 6.75.

of ferrihemoglobin solution B and a phosphate buffer, the *p*H of the solution being 6.75. To 31.25 ml. of this solution there was added from a 1-ml. glass syringe graduated in hundredths successive measured volumes (0.050 ml.) of 0.964 *f* potassium cyanide solution. When the molal susceptibility is plotted against the amount of cyanide added the first five points fall close to a straight line and the last four to a horizontal straight line. The intersection of these lines occurs at 0.235 ml., which agrees to within the experimental error with the value 0.236 ml. calculated for one cyanide per heme. The value found for χ_{HbCN} in this experiment is $2520 \cdot 10^{-6}$. A similar experiment performed at *p*H 10.8 gave similar results.

In order to determine the molal susceptibility of ferrihemoglobin cyanide accurately sets of duplicate measurements were made with unbuffered solution B at *p*H 6.7 and with solution B brought to *p*H 10.9 by the addition of potassium hydroxide solution, an excess of potassium cyanide solution being added in each case. The values found for $\chi_{HbCN} \cdot 10^6$ are 2590 and 2620 at *p*H 6.7 and 2630 and 2590 at *p*H 10.9. There is accordingly no dependence of χ_{HbCN} on *p*H. The average of the measured values is $2610 \cdot 10^{-6}$. Values approximating this were also obtained in several preliminary measurements and in measurements made incidental to other experiments.

In order to obtain an approximate value for the equilibrium constant K_{HbCN} for the reaction

$$HbCN \rightleftarrows Hb^+ + CN^- \quad (15)$$

two cyanide titration experiments were made with ferrihemoglobin solution B brought to *p*H 4.77 by the addition of an equal volume of an acetate buffer solution 2 *M* in acetic acid and 2 *M* in sodium acetate. A portion of the solution was placed in one compartment of a tube, which was then closed with a thin rubber stopper. Portions of 0.964 *f* potassium cyanide solution were then added by means of a syringe, the needle being inserted through the rubber stopper, and Δw values were measured. The corresponding values of χ_{molal} for the two runs are shown in Fig. 3. When about one-half of the stoichiometric amount of cyanide has been added it is almost entirely in the form of HbCN. With larger amounts of cyanide the formation of HbCN is incomplete, the amount of cyanide present as HCN becoming appreciable. The ratio $[CN^-]/[HCN]$ is about 10^{-4}, so that the equilibrium measured is essentially

$$Hb^+ + HCN \rightleftarrows HbCN + H^+ \quad (16)$$

The experimental points lie close to the theoretical equilibrium expression, represented by the curve in Fig. 3; the value found for the corresponding

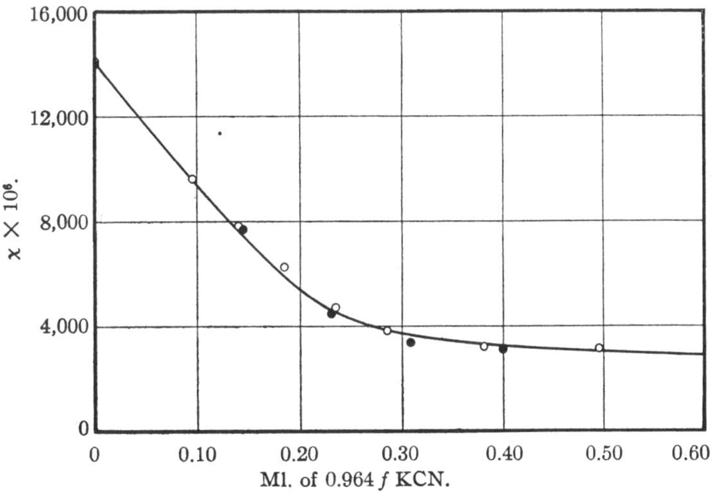

Fig. 3.—The magnetic titration of ferrihemoglobin with potassium cyanide at *p*H 4.77.

(11) R. v. Zeynek, *Z. physiol. Chem.*, **33**, 426 (1901).

equilibrium constant (presumably essentially independent of ionic strength) is 18. This value multiplied by the ionization constant of HCN, $2.0 \cdot 10^{-9}$, gives for K_{HbCN} the value $3.6 \cdot 10^{-8}$, at zero ionic strength.[12] The dependence on ionic strength is probably nearly the same as for K_{HbOH} and K_{HbF}; we hence write for K_{HbCN} the equation

$$\log K_{HbCN} = -7.44 + 0.59\sqrt{\mu} \qquad (17)$$

Ferrihemoglobin Hydrosulfide.—It was discovered by Keilin[13] that ferrihemoglobin forms a compound with hydrogen sulfide,[14] containing one sulfur atom per heme. By analogy with other compounds of ferrihemoglobin we consider it likely that this compound is ferrihemoglobin hydrosulfide, HbSH.

Solutions of ferrihemoglobin hydrosulfide made by addition of sodium hydrosulfide to ferrihemoglobin solution buffered to *p*H values in the range 5 to 7 were found to decompose rapidly, apparently undergoing auto-reduction to ferrohemoglobin, as shown by its spectrum and molal susceptibility and by the spectrum (that of oxyhemoglobin) observed after air is admitted to the solution. In order to evaluate χ_{HbSH} the following experiments were performed. To a portion of ferrihemoglobin solution B an equal volume of acetate or phosphate buffer was added. About 30 ml. of the solution was then placed in a compartment of a tube, which was closed with a thin rubber stopper, and 0.5 ml. (or, in one experiment, 0.1 ml.) of 4 *f* sodium hydrosulfide solution was added through the stopper with a syringe. The tube was then placed in position, and readings of Δw were taken at intervals of about two minutes, the first being made about six minutes after the addition of the hydrosulfide. The measured values of Δw correspond to a rapid increase of the molal susceptibility with time, χ being at first linear in t and then approaching an asymptote. The data are represented by the theoretical expression for a reaction first order with respect to HbSH, with the asymptotic value of χ_{molal} equal to that for ferrohemoglobin. To evaluate χ_{HbSH} the susceptibility values over the nearly linear portion of the curve were extrapolated to zero time; for four solutions, with *p*H 5.1, 5.7, 5.7, and 7.0, respectively, the values 2240, 2110, 2260 and 1930.10^{-6}, average $2140 \cdot 10^{-6}$, were obtained. A value for the dissociation constant of the substance is not provided by our data.[15]

The rate constant $k = -d(\ln[HbSH])/dt$ has the approximate value $5 \cdot 10^{-3}$ (with t measured in minutes), the observed values of $k \cdot 10^3$ being 5.0 at *p*H 5.08, 12.0 at *p*H 5.73, 5.2 at *p*H 7.02, and 3.0 at *p*H 5.73. (The third experiment was made with phosphate buffer, the others with acetate buffers; the fourth was made with 0.1 ml., the others with 0.5 ml. of 4 *f* sodium hydrosulfide solution added.) It is seen that no more than about 2-fold variation was observed over the *p*H range 5.1 to 7.0, and that the rate seems to increase with increase in the concentration of hydrosulfuric acid.

The Interpretation of the Molal Susceptibility Values

The values of the paramagnetic part of the molal susceptibility (per heme) at 24° of the substances studied in this investigation are collected in Table VI. If it be assumed that these values result from the independent orientation of the magnetic moments of the hemes and that Curie's law is applicable, they correspond to the values of the magnetic moment μ, in Bohr magnetons, shown in the last column of the table, calculated with the equation

$$\mu = 2.84\sqrt{\chi_{molal}\, T}$$

in which T is the absolute temperature.

TABLE VI

VALUES OF THE PARAMAGNETIC MOLAL SUSCEPTIBILITY AND EFFECTIVE MAGNETIC MOMENT (PER HEME) OF FERRIHEMOGLOBIN AND SOME OF ITS COMPOUNDS

	$\chi_{molal} \cdot 10^{6a}$	μ^b
Ferrihemoglobin, Hb^+	14,040	5.80
Ferrihemoglobin hydroxide, HbOH	8,340	4.47
Ferrihemoglobin fluoride, HbF	14,610	5.92
Ferrihemoglobin cyanide, HbCN	2,610	2.50
Ferrihemoglobin hydrosulfide, HbSH	2,140	2.26

[a] At 24°. [b] In Bohr magnetons.

(12) An effort to check this result was made by determining magnetically the ratio ferrihemoglobin–ferrihemoglobin cyanide in a solution in equilibrium with solid silver cyanide and solid silver chloride. Three measurements, made with concentrations 0.20, 0.43 and 1.00 *f* of chloride ion, gave for the ratio $[Hb^+]/[HbCN]$ the values 0.20, 0.093 and 0.040, respectively. Taking $1.7 \cdot 10^{-10}$ as the solubility product of AgCl and $7 \cdot 10^{-15}$ as that of AgCN [as given by M. Randall and J. O. Halford, THIS JOURNAL, **52**, 178 (1930)], these measurements lead to $K_{HbCN} = 1.6 \cdot 10^{-6}$, in rather poor agreement with the value $1 \cdot 10^{-7}$ given by equation 17. It is possible that the disagreement is due to error in the solubility product of AgCN, inasmuch as an older value $4.5 \cdot 10^{-17}$ [Bodländer and Eberlein, *Z. anorg. Chem.*, **39**, 197 (1904)] leads to $K_{HbCN} = 1 \cdot 10^{-8}$.

(13) D. Keilin, *Proc. Roy. Soc.* (London), **B113**, 393 (1933).

(14) Keilin emphasized the fact that this compound is different from the green substance formed from ferrohemoglobin in the presence of hydrogen sulfide and oxygen.

(15) Keilin performed experiments to determine this dissociation constant. No mention of the auto-reduction reaction which we observed is made in his paper.

The possibility should be considered that the moments of the four hemes in a molecule of molecular weight 68,000 are not oriented independently, but instead are combined to a resultant constant moment for the molecule, with magnitude twice that given for μ in the table. This possibility was discussed in connection with ferrohemoglobin,[2] and reasons were advanced for rejecting it in favor of the alternative simple interpretation in terms of hemes which interact with one another only weakly; it was suggested that the observed effective moment per heme for ferrohemoglobin, 5.46, is somewhat larger than the expected value for four unpaired electrons (spin moment alone, 4.90; expected orbital contribution, about 0.1 to 0.4) because of a partial stabilization of parallel orientations of the heme moments through heme–heme interaction. Some of the arguments advanced in support of this interpretation for ferrohemoglobin are applicable to ferrihemoglobin and its compounds also, and we have, moreover, been able to interpret the susceptibility values in a reasonably satisfactory manner on this basis and not on the basis of constant moments for the four-heme molecules; hence we believe that in these substances too the heme moments are entirely or almost entirely independent of one another.[16]

The effective moment per heme observed for ferrihemoglobin fluoride is 5.92, which is identical with the theoretical spin moment for five unpaired electrons, 5.917. Moreover, for five unpaired *d* electrons the total moment is equal to the spin moment, the orbital contribution vanishing because of the occupancy of each orbit in the subgroup by one electron; there is actually very close agreement between the theoretical spin moment and the experimental values for iron-group ions and complexes with five unpaired electrons, representative observed moments being 5.94 for Mn^{++} and 5.86–5.98 for Fe^{+++} in aqueous solution, 5.88 for the fluoferriate complex $[FeF_6]^{\equiv}$, and 5.91 for the complex $[FeF_5 \cdot H_2O]^{=}$. The observed moment for ferrihemoglobin fluoride shows that in this substance, as in the fluoferriate complex, the bonds from iron to the surrounding atoms (fluorine, the four porphyrin nitrogens, one globin nitrogen atom) are essentially ionic in character.

The effective moment per heme for ferrihemoglobin itself, 5.80, shows that in this complex too the bonds from the iron atom to the surrounding atoms are essentially ionic. It is possible that the coördination number of iron is here only five; it seems to us probable, however, that the sixth octahedral position is occupied by a water molecule,[17] the complex being $[HbOH_2]^+$ rather than Hb^+. The transition to ferrihemoglobin hydroxide would then involve loss of a proton (with change in bond type—*v. infra*) rather than addition of an hydroxyl ion.

The difference[18] between the observed moment 5.80 and the theoretical value 5.92 we attribute to heme–heme interaction operating to stabilize configurations in which the heme moments are opposed. Why the interaction should decrease the effective moment of ferrihemoglobin, increase that of ferrohemoglobin, and leave that of ferrihemoglobin fluoride unchanged we do not know.

The observed decrease in susceptibility (by about 5%) of ferrihemoglobin solutions in the low *p*H range may be due to change in the heme–heme interactions. Further experimental data are needed before a reliable explanation of this phenomenon can be given.[19]

The susceptibilities of ferrihemoglobin cyanide and ferrihemoglobin hydrosulfide correspond to the effective heme moments 2.50 and 2.26, respectively. These are close to the value expected for one unpaired electron (1.732 plus an orbital contribution of 0.3 to 0.5), showing that in these molecules two *d* orbitals of each iron atom are involved in covalent bond formation. Without doubt the structures are similar to that of the ferricyanide ion[20] (with μ = 2.33), the iron atom being attached by essentially covalent bonds to six surrounding atoms arranged octahedrally, consisting of the four porphyrin nitrogens, one globin nitrogen, and the carbon of cyanide or sulfur of hydrosulfide. The observed moments

(16) The possibility that the moments of the hemes in pasri are combined to constant resultants also seems unlikely to us.

(17) It is of interest that one fluorine in $[FeF_6]^{\equiv}$ is easily replaced by a water molecule, forming $FeF_5 \cdot H_2O^{=}$.

(18) The reality of the difference is shown by the directly observed increase in susceptibility of ferrihemoglobin solution on the addition of fluoride.

(19) During the cyanide titrations it was observed that in solutions heavily buffered (with acetate) at *p*H 4.8 the molal susceptibility of ferrihemoglobin has the value $14{,}020 \cdot 10^{-6}$, which agrees well with the asymptotic value given by the curves rather than with the values observed for unbuffered solutions.

(20) We have made a determination of the molal susceptibility of potassium ferricyanide in solution by the differential method, the diamagnetic correction being made by using the susceptibility of the ferrocyanide solution obtained by reduction with sodium sulfite. Five measurements made at concentrations of 0.608 and 0.304 *f* gave for the paramagnetic molal susceptibility at 22.0° the value $2280 \pm 10 \cdot 10^{-6}$, corresponding to $\mu = 2.33 \pm 0.01$ Bohr magnetons.

are somewhat higher than expected, indicating some heme–heme interaction; it is possible, on the other hand, that the orbital contribution is greater than usual.

None, or one, or two of the $3d$ orbitals of trivalent iron atom in a complex may be involved in covalent bond formation, the number of unpaired electrons being five, three, or one, respectively, and the moment 5.92, about 4.2, or about 2.0 Bohr magnetons. Many complexes of the first type and many of the third type are known, whereas iron complexes of the intermediate type are very rare.[21] Ferrihemoglobin hydroxide apparently is of this type; the observed effective moment 4.47 is only slightly larger than that expected for three unpaired electrons (spin moment 3.88, orbital contribution about 0.4).

The nature of the bonds in this complex is somewhat uncertain, since, although the value of the magnetic moment is that which is associated with square coördination (as in nickel protoporphyrin[1]), there is little doubt that the configuration about the iron atom is octahedral. Four dsp^2 covalent bonds directed to the corners of a square would utilize one d orbital; in ferrihemoglobin hydroxide it is probable that these four covalent bonds resonate among the six adjacent atoms, each of which is then attached to the iron atom by a bond with roughly two-thirds covalent character (or perhaps somewhat less).

(21) Measurements of χ made by L. Cambi and A. Cagnasso, *Rend. Ist. Lombardo Sci.*, **67**, 741 (1934), for complexes of $Fe(CNS)_2$ and $Co(CN)_2$ with *o*-phenanthroline and 2,2′-bipyridyl indicate structures of this type.

Summary

Magnetic measurements at approximately 24° of solutions of ferrihemoglobin and some of its compounds have been made, leading to values of the paramagnetic part of the molal susceptibility which correspond to the following values of the effective magnetic moment per heme, in Bohr magnetons: ferrihemoglobin, 5.80; ferrihemoglobin hydroxide, 4.47; ferrihemoglobin fluoride, 5.92; ferrihemoglobin cyanide, 2.50; ferrihemoglobin hydrosulfide, 2.26. For ferrihemoglobin and its fluoride these correspond to five unpaired electrons per heme, indicating essentially ionic bonds; for the cyanide and hydrosulfide to one, indicating essentially covalent bonds; and for the hydroxide to three, indicating bonds of an intermediate type.

Values determined by magnetic titrations are reported for the dissociation constants of ferrihemoglobin hydroxide, fluoride and cyanide.

PASADENA, CALIF. RECEIVED FEBRUARY 8, 1937

Reprinted from SCIENCE, December 14, 1951, Vol. 114, No. 2972, pages 629–634.

The Combining Power of Hemoglobin for Alkyl Isocyanides, and the Nature of the Heme-Heme Interactions in Hemoglobin[1]

Robert C. C. St. George and Linus Pauling
Gates and Crellin Laboratories of Chemistry,[2] California Institute of Technology, Pasadena

THE OBSERVED SIGMOID CHARACTER of the oxygen equilibrium curve of hemoglobin was explained by Adair (*1*) as resulting from the successive addition of four oxygen molecules to the hemoglobin molecule, with four different equilibrium constants. The differences in equilibrium constants were later attributed to heme-heme interactions, with each heme interacting with two adjacent hemes, the four hemes being assumed to have a square configuration (*2*). For oxyhemoglobin and carbonmonoxyhemoglobin in buffered solutions the heme-heme interaction amounts to 1.5 kcal mole^{-1}, or 6 kcal mole^{-1} for the hemoglobin molecule.

A satisfactory structural basis for the heme-heme interactions in hemoglobin has not been previously developed. The simplest explanation is that the hemes are conjugated with one another, perhaps by way of the vinyl side chains (two per protoheme, permitting coupling along the two adjacent edges of a square), and that the conjugation energy is affected by the addition of oxygen or carbon monoxide to the iron atom of a heme, which causes the electronic structure of the iron atom to change (*3*) from a quintet state, with four unpaired electrons (in ferrohemoglobin), to a singlet state, with no unpaired electrons (in oxyhemoglobin and carbonmonoxyhemoglobin). This explanation is rendered unlikely by the fact that both the absorption spectrum (*4*) and the magnetic susceptibility (*5*) of partially oxygenated hemoglobin show no deviations from a linear dependence on the degree of oxygenation. An effort to investigate this hypothesis in a more direct manner was made by Harrison Davies (*6*) through the study of the oxygen equilibrium curves of hemoglobin prepared by attaching mesoheme (with ethyl groups in place of vinyl groups) to globin, but it was unsuccessful because of the failure to find conditions under which a reconstituted hemoglobin with effective heme-heme interactions could be made even from globin and protoheme.

Another mechanism of heme-heme interaction, recently proposed (*7*), is that of steric hindrance. It is postulated that the heme groups are not on the surface of the hemoglobin molecule, where an oxygen molecule or other ligand could approach the iron atom without hindrance, but, instead, are buried within the hemoglobin molecule in such a way that a pushing-apart of the molecule must occur, to make space for the ligated molecule. The amount of steric hindrance for a second ligand would be decreased because of the loosening of the structure caused by the conversion of the first heme to oxyheme or carbonmonoxyheme, and a similar cooperative effect of the second and third ligated molecules would further decrease the steric interference for combination with a third or fourth ligand. The observed identity of heme-heme interactions for oxyhemoglobin and carbonmonoxyhemoglobin would be explained on the basis of this postulate as resulting from the very close approximation in size of the oxygen molecule and the carbon monoxide molecule, which would result in similar steric effects.

It is evident that, if this picture of the hemoglobin molecule is correct, the equilibrium between a ligand and hemoglobin would be affected not only by the nature of the bond to the iron atom, but also by the size and shape of the ligated molecule, including parts so far removed from the iron atom as to have no direct effect on the bond. The alkyl isocyanides suggest themselves for the investigation of this point. It was reported by Warburg, Negelein, and Christian (*8*) that methyl isocyanide combines with horse hemoglobin with an affinity approximately one fortieth that of oxygen (the equilibrium constant being expressed in terms of concentration of dissolved material). It has also been found that ethyl isocyanide combines with ferrohemoglobin, and that the compound ethyl isocyanide-ferrohemoglobin is diamagnetic, the bonds between the iron atom and its surrounding atoms being essentially covalent (*9*), as in oxyhemoglobin. We have, accordingly, investigated the combination of ethyl isocyanide, isopropyl isocyanide, and tertiary butyl isocyanide with hemoglobin, and have found the three substances to differ greatly in their combining powers, the difference amounting to a factor of about 200 in equilibrium constant between tertiary butyl isocyanide and ethyl isocyanide, with isopropyl isocyanide having an intermediate value. Evidence that this large effect is a steric one, and not the result of a change in the strength of the iron-carbon bond through interaction (hyperconjugation) with the alkyl group, is provided by the observation that the three

[1] This work was carried out during the tenure of a Merck Postdoctoral Fellowship by one of us (St. George). The investigation was supported in part by a research grant from the National Institutes of Health, U. S. Public Health Service.
[2] Contribution No. 1593.

2

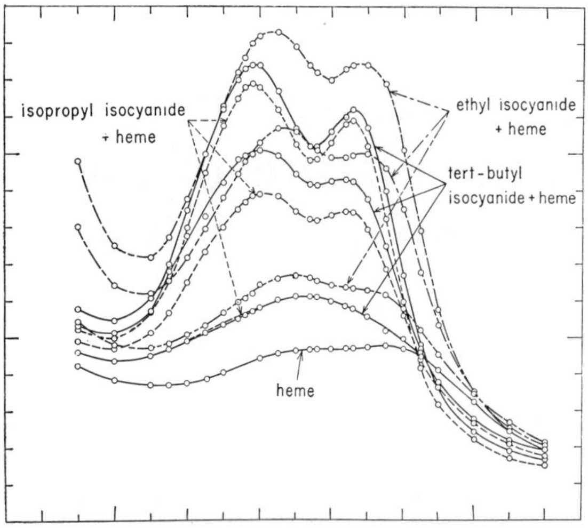

FIG. 1. The spectra of ferroheme and ferroheme partially and fully saturated with ethyl isocyanide, isopropyl isocyanide, and tertiary butyl isocyanide, at 25° C and pH 6.8. Total concentration of ferroheme, $5.87 \cdot 10^{-5}M$ in all cases. Total isocyanide concentrations $3.93 \cdot 10^{-5}$, $10.7 \cdot 10^{-5}$ ($13.8 \cdot 10^{-5}M$ for the second tert-butyl isocyanide curve), and $197 \cdot 10^{-5}M$. Ordinates, extinction 0.0 to 1.3; abscissae, wavelength 4800 A to 6200 A.

isocyanides have nearly the same power of combining with isolated ferroheme groups, not attached to protein.

The combination of isocyanides and ferroheme. In Fig. 1 there are given the absorption spectra of ferroprotoheme in aqueous solution at 25° C and pH 6.8 and of ferroprotoheme in the presence of different amounts of the three isocyanides, at the same temperatures and pH. The three curves for each isocyanide correspond to the presence in the solution of about 0.67, 1.8, and 34 moles of isocyanide per mole of heme. The curves show that two heme-isocyanide compounds are formed with each isocyanide; these are presumably the monoisocyanide and the diisocyanide. Spectrophotometric evidence that methyl isocyanide forms two compounds, the monoisocyanide and the diisocyanide, with both ferroprotoheme and ferrouroheme has been reported by Joan Keilin (*10*). Ferrohememonoisopropylisocyanide and ferrohememonotert-butylisocyanide have a broad absorption curve with a single maximum, at about 5520 A, and the corresponding diisocyanides have a double-maximum spectrum, the maxima being at 5380 A and 5660 A, and the intermediate minimum at 5550 A. The curve for ferrohemediethylisocyanide is similar to those for the other two diisocyanides, except that the extinction coefficient is somewhat larger, and the maxima and minimum are shifted to longer wavelengths. The maxima lie at 5440 A and 5760 A, and the minimum at 5590 A. The curve for ferroheme and the smallest amount of ethyl isocyanide used already shows two maxima, unlike the corresponding curves for the other isocyanides, possibly because of the presence of a smaller amount of the monoisocyanide and a larger amount of the diisocyanide in the solution.

The quantitative interpretation of the curves is rendered difficult by lack of knowledge of the extinction coefficient for the pure monoisocyanide compounds. If it be assumed, as a rough approximation, that the effect of one attached isocyanide group in changing the extinction coefficient is one half the effect of two attached isocyanide groups, the intermediate curves can be interpreted to give the number of bound isocyanide groups, and the concentration of free isocyanide can then be calculated by subtracting this quantity from the total amount of isocyanide added. Because of the change in shape of the curves from the monoisocyanide to the diisocyanide, values of the degree of saturation of the two combining sites per heme, calculated from the extinction coefficients at different wavelengths, are somewhat different, and the average values, which are given in Table 1, are reliable to only about 10 per cent.

TABLE 1

COMBINATION OF FERROHEME WITH ALKYL ISOCYANIDE
(Total ferroheme concentration $5.87 \cdot 10^{-5}M$)

Isocyanide	Degree of saturation, f	Free isocyanide $\cdot 10^{-5}M$	$K \cdot 10^{-10}M^{-2}$
A. Total isocyanide concentration $3.93 \cdot 10^{-5}M$			
Ethyl	0.23	1.23	0.20
Isopropyl	.21	1.47	.12
Tert-butyl	.20	1.58	.10
B. Total isocyanide concentration $10.7 \cdot 10^{-5}$			
Ethyl	0.69	2.6	0.33
Isopropyl	.63	3.3	.16
Tert-butyl	.74	5.1*	.11

* Total isocyanide concentration $13.8 \cdot 10^{-5}M$.

It is found that the data can be interpreted by neglecting the monoisocyanides. The last column of Table 1 contains values of the equilibrium constant K calculated by the equation

$$K = \frac{[\mathrm{HmI_2}]}{[\mathrm{Hm}][\mathrm{I}]^2} = \frac{f}{(1-f)[\mathrm{I}]^2}. \quad (1)$$

Here [Hm] is the concentration of uncombined heme, $[\mathrm{HmI_2}]$ the concentration of combined heme, [I] the concentration of free isocyanide, and f the fractional saturation. The spectrophotometric values of f are given in the second column of Table 1. The approximate constancy of K for each isocyanide indicates that the second isocyanide molecule adds to the heme more readily than the first.

The values of K in Table 1 are nearly the same (to within a factor of 2) for the three alkyl isocyanides, hence the affinity of ferroprotoheme for the isocyanide carbon atoms is nearly independent of the nature of the alkyl group.

The combination of alkyl isocyanides with ferrohemoglobin. The equilibria between reduced ferrohemoglobin and ethyl isocyanide, isopropyl isocyanide, and tertiary butyl isocyanide were studied spectrophotometrically at 25° C and pH 6.8. The ferrohemoglobin isocyanides have absorption spectra similar to the spectrum of carbonmonoxyhemoglobin, as shown in Table 2. Concentrations of free ferrohemoglobin

3

TABLE 2

Bovine Hemoglobin Isocyanides—Absorption Maxima of the Pure Substances and Isobestic Points in Mixtures with Ferrohemoglobin

Isocyanide	Absorption maxima (Wavelength, A)	Isobestic points (Wavelength, A)
Ethyl	5600, 5290	5430, 5500, 5690
Isopropyl	5590, 5290	5420, 5510, 5690
Tertiary butyl	? ?	5420, 5520, 5690

and of ferrohemoglobin isocyanide were measured spectrophotometrically. The results of the measurements for bovine ferrohemoglobin are given in Fig. 2, where the fractional saturation, [HbI]/([Hb] +

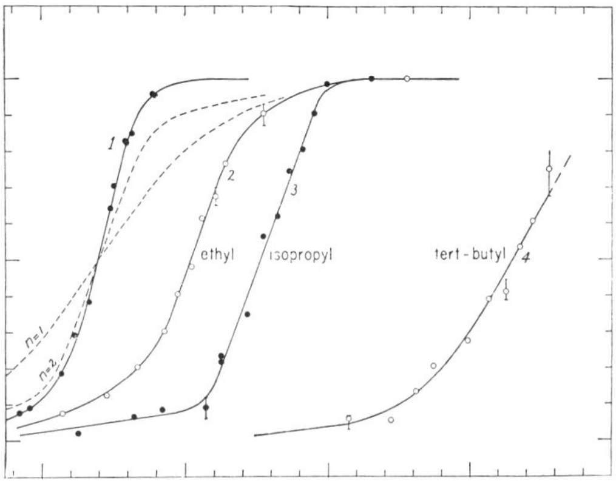

Fig. 2. Fractional saturation curves of ferrohemoglobin with oxygen, ethyl isocyanide, isopropyl isocyanide, and tertiary butyl isocyanide at 25° C and pH 6.8. The dashed curve $n = 1$ corresponds to no heme-heme interaction, and the dashed curve $n = 2$ to strong interactions between the hemes in pairs. Ordinates, fractional saturation 0 to 1 ; abscissae, log of concentration of unbound oxygen or isocyanide − 5.1 to − 1.0.

[HbI]), is plotted as a function of the logarithm of the concentration of free isocyanide in the solution. (Here the symbol Hb is used to refer to a quarter of a hemoglobin molecule, containing one heme.) The oxygen dissociation curve determined by Ferry and Green (*11*) at the same temperature and pH is given for comparison. Their data, which were reported in terms of the partial pressure of oxygen in the gas phase, have been replotted against the logarithm of the concentration of free oxygen in solution.

It is seen that there is a large difference in combining power of ethyl isocyanide, isopropyl isocyanide, and tertiary butyl isocyanide with hemoglobin. The combination constant of isopropyl isocyanide with hemoglobin is about one third as great as that of ethyl isocyanide, and that of tertiary butyl isocyanide is about one two hundredth as great as that of ethyl isocyanide. These differences are shown also by hemoglobins other than adult bovine hemoglobin. In Table 3 data are given for the combination constants (concentration of unbound isocyanide in the presence of hemoglobin half-saturated with isocyanide) with adult bovine hemoglobin, fetal bovine hemoglobin, normal human hemoglobin, and sickle cell anemia human hemoglobin. There is little change of the combination constant from one form of hemoglobin to another, but a very large dependence on the nature of the alkyl group in the isocyanide.

TABLE 3

Concentration of Unbound Isocyanide in the Presence of Half-saturated Hemoglobin

Type of hemoglobin	Days since blood was drawn	Ethyl isocyanide	Isopropyl isocyanide (Mm/l)	Tert-butyl isocyanide
Bovine (adult)	7	0.11	0.33	20
" "	1		.41	25
" (fetal)*	2		.27	13
Human (normal)	1	.075	.30	16
" (sickle cell)	1	.066	.29	14

* Cells lysed in the presence of toluene. (The values for adult bovine hemoglobin were found not to be changed by using toluene.)

It is very unlikely that the differences in combination of the different isocyanides with hemoglobin are due to a difference in the bond between the iron atom of the heme group and the carbon atom of the icocyanide. Only a very small inductive effect or hyperconjugation effect of the alkyl groups is observed in analogous compounds. Propionic acid, isobutyric acid, and trimethyl acetic acid have very nearly the same acid strengths. Similarly, the base strengths of ethylamine, isopropylamine, and tert-butylamine are very nearly the same (*12*, *13*). Moreover, our own studies (reported in Fig. 1) have shown that the three isocyanides have nearly the same combination constants with ferroheme itself.

The only pronounced difference between the molecules ethyl isocyanide, isopropyl isocyanide, and tertiary butyl isocyanide is the difference in size of the alkyl group. We conclude that it is this difference that determines the difference of affinity of the three isocyanides with hemoglobin. The difference in size of the alkyl group could make itself significant through the operation of steric hindrance, in case the structure of the hemoglobin molecule were such as to require a dilation on combination with the ligand. We accordingly conclude that the heme groups of hemoglobin are buried within the protein molecule in such a way that the ligated isocyanide molecules are required to push parts of the protein molecule away from one another, thus introducing a steric-hindrance effect.

Evidence that there is protein on both sides of the hemes in hemoglobin has been at hand for twenty years. There are two important heme-linked acid groups in the molecule, one of which is made more strongly acidic and the other less strongly acidic on oxygenation of the associated heme. It was suggested by Conant (*14*) that both of these acid groups are imidazole groups of histidine residues, and strong evidence for this assumption, from heats of ionization,

4

has been reported by Wyman (*15*). Coryell and Pauling (*16*), in discussing the mechanism of the heme-acid group interactions in terms of structure, found it necessary to assume that the two histidine residues are on the two sides of each heme, and that on oxygenation of the heme one imidazole ring is bonded more tightly than before, whereas the second is displaced by the oxygen molecule.

On this interpretation the differences in combination constants of the different isocyanides require that the energy of the steric-hindrance effect be 0.7 kcal $mole^{-1}$ per heme-isocyanide complex greater for isopropyl isocyanide than for ethyl isocyanide, and be 3.1 kcal $mole^{-1}$ greater for tertiary butyl isocyanide than for ethyl isocyanide.

It might be thought that the same additional steric-hindrance effect should result from the introduction of one methyl group, converting ethyl isocyanide into isopropyl isocyanide, as from the introduction of a second, converting isopropyl isocyanide into tertiary butyl isocyanide. However, it is possible that the shape of the protein layers in the neighborhood of the alkyl group may be such that the isopropyl group could fit into the protein with very little greater dilation than that caused by the ethyl group, whereas the tertiary butyl group, which has a third methyl group nearly at right angles to the plane formed by the isopropyl group, might exert a much greater steric effect. This situation would be described by saying that there is a hollow in the protein roughly complementary in configuration to the isopropyl group, and into which the ethyl and isopropyl groups fit almost equally well, but that the tertiary butyl group, with its larger dimensions, is not able to fit and, accordingly, causes a considerably larger separation of the protein layers.

It is of interest that Chance (*17*) concluded from the different velocities of combination of catalase with hydrogen peroxide, methyl hydrogen peroxide, and ethyl hydrogen peroxide (not observed for horseradish peroxidase and lactoperoxidase with these peroxides) that the heme group of catalase is embedded in the protein. Also, in a discussion of the heme-linked acid groups of hemoglobin, Wyman (*18*) has recently suggested that the heme-heme interactions are associated with an alteration in form of the hemoglobin molecule on addition of oxygen or carbon monoxide.

The nature of the heme-heme interactions in hemoglobin. The apparent existence of steric interactions of ligated groups in hemoglobin with the protein leads us to believe that the heme-heme interactions also have a steric origin. The energy of heme-heme interaction in the hemoglobin-isocyanide complexes, as given by the shape of the curves in Fig. 2, is seen to be essentially the same as in carbonmonoxyhemoglobin and oxyhemoglobin. Let us assume that, in addition to the protein below the groups, there is a layer of protein above the four heme groups of hemoglobin which needs to be separated to some extent from the rest of the molecule in order to permit the attachment of the isocyanide molecule to the iron atom of the heme. Let us also assume that the four parts of the protein layer covering the four hemes are connected with one another in such a way that, when the first part is lifted, it tends to carry a second and third along with it. This is equivalent to assuming heme-heme interactions along the edges of a square formed by the four hemes. If, now, the introduction of the first ligate causes a steric-hindrance effect greater than the average, and the introduction of the last one causes a steric-hindrance effect less than the average, the observed heme-heme interaction could be accounted for. Thus, for ethylisocyanide-hemoglobin, we might assume that the introduction of the first isocyanide group would require the energy 3.0 kcal $mole^{-1}$ to overcome steric hindrance, the introduction of the second and the third would require 1.6 and 1.0 kcal $mole^{-1}$, respectively, and the introduction of the fourth would require no steric-hindrance energy. This assumption would account for the observed amount of sigmoid character of the equilibrium curve.

The interaction between hemes through the steric-hindrance mechanism cannot be expected to be so simple as to correspond to constant heme-heme interactions along edges of a square. We think it likely that the four hemes in hemoglobin have, individually, the same configurations of protein around them. Precise determination of the values of the four successive equilibrium constants for oxyhemoglobin or carbonmonoxyhemoglobin, through methods such as those proposed by Roughton (*19*), might provide the basis for further conclusions about the steric relationships between these centers.

The steric-hindrance theory of heme-heme interaction in hemoglobin is in general compatible with our knowledge about its properties. It has been known for many years that on the removal of salt from a hemoglobin solution the oxygen affinity of the hemoglobin increases, and also the heme-heme interactions become smaller (*20–22*). Altschul and Hogness (*4*) reported that their dialyzed hemoglobin had an average combination constant with oxygen 37 times as great as that of undialyzed hemoglobin, corresponding to an average free energy difference of 2.1 kcal $mole^{-1}$ per heme-oxygen complex, and that the heme-heme interaction energy had decreased from 6.0 kcal $mole^{-1}$ to 2.4 kcal $mole^{-1}$. Both these changes are in the direction that would be predicted from the steric-hindrance theory. The removal of salt from a hemoglobin solution would cause a loosening of the structure of the protein, through the operation of electrostatic repulsive forces between the similarly charged portions of the molecule. In the presence of salt these repulsive forces are diminished by the ion atmospheres that surround the charged groups. The loosening of the protein structure would make it easier for a ligand to attach itself to the heme and thus would decrease both the steric-hindrance effect that influences the equilibrium constant and the heme-heme interaction through steric hindrance. Davies (*6*) found his reconstituted hemoglobin to combine more strongly with oxygen than the original material, the half-saturation pressure being only 0.2 times as great.

5

Wyman (*15*) has pointed out that there is a decrease of 60 per cent in the half-saturation pressure of horse hemoglobin on dissociation into half molecules in urea solution, and that this suggests heme-heme stabilization rather than oxyheme-oxyheme stabilization. From our point of view, it indicates a smaller steric-hindrance effect in the two-heme molecules than in the four-heme molecules, as is reasonable, particularly if the urea has caused a loosening of the molecular structure in addition to a separation into two parts. Wyman has also pointed out that the oxygen equilibrium data for horse hemoglobin in urea solution, in which it is split into molecules containing only two hemes, lead to a heme-heme interaction energy of about 3.5 kcal $mole^{-1}$, which we interpret as resulting from a steric-hindrance energy 3.5 kcal $mole^{-1}$ greater for the first oxygen molecule than for the second.

Crystallographic studies give evidence that the addition of oxygen to hemoglobin causes deep-seated structural changes in the molecule. Haurowitz (*23*) found that there is a significant difference in crystal form of horse ferrohemoglobin and horse oxyhemoglobin, which would hardly be expected from the small difference (0.2 per cent change in molecular weight) of the two molecules. He concluded that the addition of oxygen to hemoglobin must cause a significant change in the structure of the molecule, leading to a change in the nature of the intermolecular forces, and hence in the crystal structure. A similar conclusion was reached by Perutz (*24*). Such a change in intermolecular interactions could reasonably be expected on the basis of our postulate about the structure of hemoglobin, according to which a significant change in shape of the molecule occurs on oxygenation. The postulated change in shape is not so great, however, as to lead to a predicted observable change in viscosity or diffusion constant; this is compatible with the fact that no changes in these properties on oxygenation of hemoglobin were detected by Haurowitz or Gutfreund (*25*).

Our postulate provides an obvious explanation of the action of oxygen in preventing the sickling of sickle-cell-anemia erythrocytes. We have visualized the sickling process (*26*) as one in which complementary sites on adjacent hemoglobin molecules combine. It was suggested that erythrocytes containing oxyhemoglobin or carbonmonoxyhemoglobin do not sickle because of steric hindrance of the attached oxygen or carbon monoxide molecule. This steric-hindrance effect might be the distortion of the complementary sites through forcing apart of layers of protein, as is suggested by the isocyanide experiments.

Finally, a series of measurements on ferrihemoglobin lends support to the steric-hindrance theory of heme-heme interaction. This interaction appears not only when a ligand becomes attached to hemoglobin, but also in other reactions, such as the oxidation of ferrohemoglobin to ferrihemoglobin. Coryell (*27*) has pointed out that the heme-heme interaction energies for compounds of ferrihemoglobin are smaller than for compounds of ferrohemoglobin. The values found, per hemoglobin molecule, are 3.4 kcal $mole^{-1}$ for ferrihemoglobin hydrosulfide, 3.1 kcal $mole^{-1}$ for ferrihemoglobin azide, and 0 for ferrihemoglobin fluoride and ferrihemoglobin hydroxide, which are to be compared with 6.0 kcal $mole^{-1}$ for oxyhemoglobin and carbonmonoxyhemoglobin. In addition, the interaction energy of 2.6 kcal $mole^{-1}$ is reported for the oxidation of ferrohemoglobin to ferrihemoglobin (*27*). We believe that the latter is a steric effect resulting from the ligation of a water molecule to each iron atom in ferrihemoglobin. The water molecule is smaller than an oxygen molecule or carbon monoxide molecule and, accordingly, would be expected to produce a smaller steric effect. Moreover, the fluoride and hydroxide ions have essentially the same size as the water molecule, so that the replacement of a water molecule by either of these ions, in the formation of ferrihemoglobin fluoride or ferrihemoglobin hydroxide, would be expected to have no steric effect, and would thus lead to the observed absence of heme-heme interaction in the formation of these compounds. On the other hand, the hydrosulfide ion and the azide ion are larger—approximately as large as the oxygen molecule—so that a total steric effect about as great as in oxyhemoglobin, 6.0 kcal $mole^{-1}$, would be expected. On subtracting the interaction in ferrihemoglobin hydrate, 2.6 kcal $mole^{-1}$, we are led to 3.4 kcal $mole^{-1}$ as the predicted steric-hindrance effect in ferrihemoglobin hydrosulfide and ferrihemoglobin azide, in good agreement with observation. In this connection Perutz (*24*) has reported human ferrihemoglobin to form orthorhombic crystals isomorphous with those of oxyhemoglobin and carbonmonoxyhemoglobin, whereas crystals of human ferrohemoglobin itself (two modifications) are different; also, Jope and O'Brien (*28*) have found human ferrihemoglobin, oxyhemoglobin, and carbonmonoxyhemoglobin to have similar solubility curves, different in shape from the curve for ferrohemoglobin.

Experimental method. The isocyanides were prepared from the alkylhalides by the silver cyanide method: ethyl isocyanide and isopropyl isocyanide according to the directions of Gautier (*29*), and tertiary butyl isocyanide according to a modified method (*30*). The physical properties of the preparations used are as follows: ethyl isocyanide, bp, 77°–78° C at 750 mm; isopropyl isocyanide, bp, 86.5°–87° C at 750 mm; tertiary butyl isocyanide, bp, 91°–91.5° C at 750 mm, mp, 8°–10° C, density 0.72.

The hemoglobin was made from defibrinated bovine blood. The cells were washed three times with 1% NaCl solution by centrifuging, and were stored at 2° C. On the day on which the isocyanide titration was to be carried out the cells were lysed by adding three times their volume of distilled water. Ten minutes later six volumes of 0.133 M KH_2PO_4–Na_2HPO_4 buffer was added, the cell debris was centrifuged off, and the supernatant was diluted to the desired concentration (4 to $5 \cdot 10^{-5}$ gram atoms of iron per liter) with more buffer.

6

The cells used in obtaining the data shown in Fig. 2 had been stored one week. It was found in a set of check experiments reported in Table 3 that the affinity of hemoglobin for isopropyl isocyanide and for tertiary butyl isocyanide does not depend appreciably on whether the cells from which the hemoglobin was prepared had been stored for one day or one week.

The fetal bovine hemoglobin was obtained from cells that had been lysed with toluene. In check runs it was found that lysing with toluene leaves the isocyanide affinity of adult hemoglobin unchanged.

In measuring the degree of combination of isocyanides with hemoglobin the optical densities were measured at (*a*) 5260, 5280, and 5300 A, (*b*) 5580, 5600, and 5620 A, and (*c*) 5820, 5840, and 5860 A. The per cent saturation was calculated at each wavelength from the ratio of the change observed to the total change on saturation. The values within groups *a, b,* and *c* were then averaged. Finally the three values obtained were also averaged. The final averages are plotted in Fig. 2. Where the measured range is greater than ± 2 per cent, its magnitude is indicated in the graph.

Ethyl and isopropyl isocyanide were added in aqueous solution, and tertiary butyl isocyanide was added directly from a microburette. The hemoglobin was then reduced with dithionite solution. Because of the low solubility of tertiary butyl isocyanide it was not possible to build up a high enough concentration to saturate the hemoglobin completely and to obtain the spectrum of tertiary butyl isocyanide hemoglobin. The percentage saturation was therefore calculated on the assumption that the spectrum of tertiary butyl isocyanide hemoglobin is the same as that of the isopropyl isocyanide complex. This is probably a good assumption because (*a*) the spectra of ethyl and isopropyl isocyanide hemoglobin are nearly identical; (*b*) the isobestic points of reduced hemoglobin with isopropyl isocyanide hemoglobin are nearly the same as with tertiary butyl isocyanide hemoglobin; and (*c*) the spectra of the diisopropyl and of the ditertiary butyl isocyanide hemes are nearly the same. It turns out, then, that the highest degree of saturation of hemoglobin obtainable in a saturated aqueous solution of tertiary butyl isocyanide is about 70 per cent. With larger amounts of this isocyanide, present as a separate phase, the protein is denatured.

Ferriheme chloride was prepared by the method of Willstätter and Asahina (*31*), with use of a hemoglobin solution instead of blood as the starting material. Ferroheme solutions were made by dissolving the ferriheme chloride in a small amount of 0.2 *N* sodium hydroxide solution and then diluting with phosphate buffer to give a final concentration of $5.87 \cdot 10^{-5}$ *M*. Fresh centrifuged sodium dithionite solution was added to reduce the ferriheme after the desired amount of isocyanide dissolved in water had been added.

References

1. ADAIR, G. S. *Proc. Roy. Soc.*, **A109**, 292 (1925); *J. Biol. Chem.*, **63**, 529 (1925).
2. PAULING, L. *Proc. Natl. Acad. Sci., U. S.*, **21**, 186 (1935).
3. PAULING, L., and CORYELL, C. D. *Ibid.*, **22**, 210 (1936).
4. ALTSCHUL, A. M., and HOGNESS, T. R. *J. Biol. Chem.*, **129**, 315 (1939).
5. CORYELL, C. D., PAULING, L., and DODSON, R. W. *J. Phys. Chem.*, **43**, 825 (1939).
6. DAVIES, HARRISON. Unpublished investigation.
7. PAULING, L. In *Haemoglobin*, Barcroft Memorial Volume, New York: Interscience, 62 (1949).
8. WARBURG, O., NEGELEIN, E., and CHRISTIAN, W. *Biochem. Z.*, **214**, 26 (1929).
9. RUSSELL, C. D., and PAULING, L. *Proc. Natl. Acad. Sci., U. S.*, **25**, 517 (1939).
10. KEILIN, JOAN. *Biochem. J.*, **45**, 440, 448 (1949); *Nature*, **165**, 151 (1950).
11. FERRY, R. N., and GREEN, A. A. *J. Biol. Chem.*, **81**, 175 (1929).
12. MOORE, T. S., and WINMILL, T. F. *J. Chem. Soc.*, **101**, 1635 (1912).
13. BRUEHLMAN, R. J., and VERHOEK, F. H. *J. Am. Chem. Soc.*, **70**, 1401 (1948).
14. CONANT, J. B. *Harvey Lectures*, **28**, 159 (1932–33).
15. WYMAN, J., JR. In *Haemoglobin, loc. cit.*, 95.
16. CORYELL, C. D., and PAULING, L. *J. Biol. Chem.*, **132**, 769 (1940).
17. CHANCE, B. *Ibid.*, **179**, 1367 (1949).
18. WYMAN, J. Paper presented Sept. 5, 1951, at New York Meeting Am. Chem. Soc.
19. ROUGHTON, F. J. W. In *Haemoglobin, loc. cit.*, 83.
20. BARCROFT, J., and CAMIS, M. *J. Physiol.*, **39**, 118 (1909–10).
21. BARCROFT, J., and ROBERTS, F. *Ibid.*, **39**, 143 (1909–10).
22. SIDWELL, A. E., JR., *et al. J. Biol. Chem.*, **123**, 335 (1938).
23. HAUROWITZ, F. *Z. physiolog. Chem.*, **254**, 266 (1938).
24. PERUTZ, M. F. In *Haemoglobin, loc. cit.*, 146, 147.
25. GUTFREUND, H. Reported by M. F. Perutz in *Haemoglobin*, 146.
26. PAULING, L., *et al. Science*, **110**, 543 (1949).
27. CORYELL, C. D. *J. Phys. Chem.*, **43**, 841 (1939).
28. JOPE, H. M., and O'BRIEN, J. R. P. In *Haemoglobin, loc. cit.*, 274.
29. GAUTIER, A. *Ann. chim. et phys.*, **17**, (4), 215 (1869).
30. NEF, J. U. *Ann.*, **309**, 154 (1898).
31. WILLSTÄTTER, R., and ASAHINA, Y. *Ibid.*, **385**, 197 (1911).

(*Reprinted from Nature, Vol.* 203, *No.* 4941, *pp.* 182–183, *July* 11, 1964)

Nature of the Iron–Oxygen Bond in Oxyhæmoglobin

J. J. Weiss[1] has proposed that oxyhæmoglobin is a hæmoglobin peroxide, with the iron atom in the ferric state and the oxygen molecule present as an O_2^- ion, which is then taken up in the co-ordination shell of the ferric ion. He states that such an assumption accounts for the properties of oxyhæmoglobin.

Oxyhæmoglobin is, however, diamagnetic[2], and the generally accepted definition of oxidation number[3] requires that the iron atom in any diamagnetic molecule or complex containing one iron atom (no iron–iron bond) has an even oxidation number. It is accordingly not possible for an iron atom in oxyhæmoglobin to be described as ferric (oxidation number +3) unless a definition of oxidation number different from the usually accepted one is used.

The problem discussed by Weiss, the nature of the iron–oxygen bond in oxyhæmoglobin, was given a thorough discussion by me in two communications published about 15 years ago[4,5]. It is now possible to amplify the discussion somewhat, in part because of further development of the theory of the chemical bond[6] and in part because of the increased knowledge about the structure of oxyhæmoglobin and related molecules that has been obtained by the X-ray diffraction investigations of Kendrew[7] and Perutz[8].

Weiss has illustrated his proposal by writing the following equation:

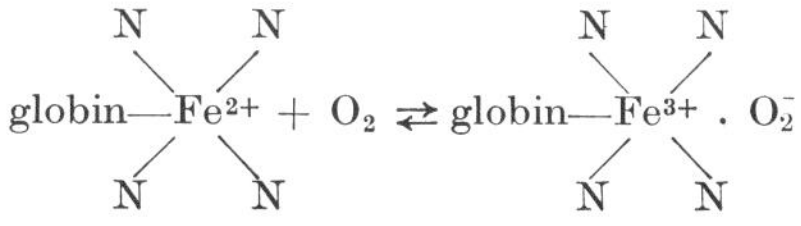

In the formula at the right, the dot represents not an electron but some sort of bond.

The structural formula which I proposed for oxyhæmoglobin[4,5] is the following:

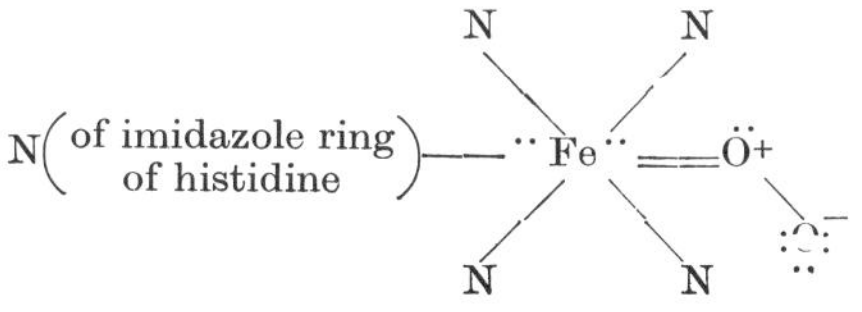

In this structural formula the iron atom is shown as forming a double bond with the first oxygen atom, which forms a single bond also with the second oxygen atom. (The iron atom then has two unshared electron pairs in its outer d^5sp^3 shell.) There is a formal charge $+1$ on the first oxygen atom and a formal charge -1 on the second oxygen atom. The difference in electronegativity of iron and oxygen is 1·7, which corresponds to 51 per cent of ionic character[9]. The electric charge assigned to the first oxygen atom is accordingly reduced to 0, whereas that of the outer oxygen atom remains -1. Except for the outer oxygen atom, the assigned structure is accordingly in good agreement with the principle of electroneutrality[10].

The negative charge assigned to the outer oxygen atom of oxyhæmoglobin can now be accounted for by the structure of myoglobin, as determined by Kendrew[7]. In the myoglobin molecule there is a residue of histidine adjacent to the iron atom on one side of the hæm group, and another residue of histidine constrained by the configuration of the polypeptide chain to a position a few Ångströms removed from the iron atom on the other side of the hæm. These two residues of histidine are those that were proposed originally by Conant[11], in order to account for the observed hæm-linked acid groups of hæmoglobin, and their positions relative to the iron atom are those postulated by Coryell and Pauling[12] in their detailed discussion of the mechanism of change in acid strength of hæmoglobin on oxygenation. It has been pointed out that the positive electric charge of the imidazolium side-chain of the second histidine residue probably serves, through its electrostatic interaction with the iron atom, to stabilize the bipositive state and assist in preventing oxidation of hæmoglobin to ferrihæmoglobin[13]. We see, also, that for oxyhæmoglobin the structure assigned above places an oxygen atom with a negative charge in close approximation to the positively charged imidazolium group and, indeed, in such a position as to permit hydrogen bonding with the negatively charged oxygen atom, thus further stabilizing the structure.

Similar structures, involving iron–carbon double bonds, were formerly assigned to carbonmonoxyhæmoglobin and the ferrohæmoglobin alkylisocyanides[4,5]. We may now recognize, however, that the arguments of the preceding paragraph strongly suggest that these molecules have structures in which the iron atom forms a triple bond with the carbon atom, and the outer oxygen atom or nitrogen atom has a negative electric charge:

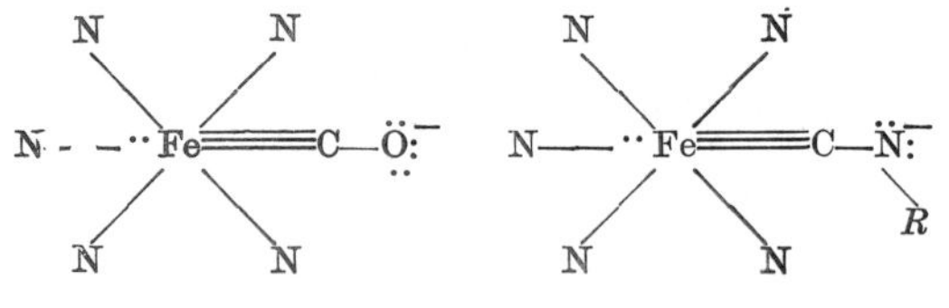

I conclude that oxyhæmoglobin and related hæmoglobin compounds are properly described as containing ferrous iron, rather than ferric iron, that their electronic structure involves essentially the formation of a double bond between the iron atom and the near-by oxygen atom in oxyhæmoglobin (with the iron–oxygen–oxygen bond angle about 120°) and a triple bond to the carbon atom in carbonmonoxyhæmoglobin and the hæmoglobin alkylisocyanides (with iron–carbon–oxygen or nitrogen bond angle 180°). The iron–oxygen and iron–carbon interatomic distances in these compounds are predicted to have the values corresponding to these multiple bonds.

LINUS PAULING

Big Sur, California.

1 Weiss, J. J., *Nature*, **202**, 83 (1964).
2 Pauling, L., and Coryell, C. D., *Proc. U.S. Nat. Acad. Sci.*, **22**, 210 (1936).
3 Pauling, L., *General Chemistry*, second ed. (W. H. Freeman and Co., San Francisco, 1954).
4 Pauling, L., *Stanford Med. Bull.*, **6**, 215 (1948).
5 Pauling, L., *Hemoglobin*, 57 (Butterworths Sci. Publ., London, 1949).
6 Pauling, L., *The Nature of the Chemical Bond* (Cornell Univ. Press, Ithaca, 1960).
7 Kendrew, J. C., "Myoglobin and the Structure of Proteins", in *Les Prix Nobel en 1962*, Stockholm, 1963, and references quoted there.
8 Perutz, M., "X-ray Analysis of Hæmoglobin", in *Les Prix Nobel en 1962*, and references quoted there.
9 Ref. 6, Chap. 3.
10 Pauling, L., *J. Chem. Soc.*, 1461 (1948); ref. 6, 172.
11 Conant, J. B., *Harvey Lectures*, **28**, 159 (1932–33).
12 Coryell, C. D., and Pauling, L., *Biol. Chem.*, **132**, 769 (1940).
13 Zuckerkandl, E., and Pauling, L., in Kasha, M., and Pullman, B., *Horizons in Biochemistry* (Academic Press, New York and London, 1962).

Printed in Great Britain by Fisher, Knight & Co., Ltd., St. Albans.

SECTION I: THE NORMAL HEMOGLOBINS 1

THE NORMAL HEMOGLOBINS AND THE HEMOGLOBINOPATHIES: BACKGROUND

Linus Pauling

Linus Pauling Institute of Science and Medicine, 440 Page Mill Road,
Palo Alto, California 94306

I am pleased that the editors of this review of work on human hemoglobin and hemoglobinopathies should have asked me to write on the background of our present knowledge, and especially of my own involvment in it.

In fact, during much of my scientific career I have worked on the problem of the structure and properties of the hemoglobin molecule and on other problems that have a bearing on this one. Our early work, beginning in 1922, dealt largely with the structure of inorganic compounds, including minerals, and of the simpler organic compounds, with emphasis on the experimental techniques of X-ray diffraction of crystals and electron diffraction of gas molecules and on the application of the theory of quantum mechanics to the problems. After about ten years, in the early 1930's, I began to think about the large molecules in the human body, and especially about proteins. The hemoglobin molecule, with its striking color and its property of combining reversibly with dioxygen, seemed to me to be especially interesting.

My first paper in this field dealt with the oxygen equilibrium of hemoglobin and its structural interpretation. Instead of four separate equilibrium constants for the addition of four successive dioxygen molecules to the hemoglobin molecule, with their corresponding four values of the standard free energy of combination, I simplified the problem to a two-parameter one. I assumed that the free energy of addition of dioxygen to each of the four heme groups would be the same and that an additional contribution to the free energy of the molecule would be made by each interacting pair of hemes with attached dioxygens. I found that the experimentally obtained oxygen equilibrium curve could be approximated reasonably well by either one of two simple assumptions: that the four hemes be at the corners of a regular tetrahedron or at the corners of a square. I suggested, however, that the square arrangement might be closer to the truth, largely on the basis of symmetry arguments (1).

The interaction energy for a pair of hemes was evaluated, for the square case (each heme interacting with two others), to be about 6.3 kJ $mole^{-1}$. During the next fifteen years I attempted to get experimental evidence about the nature of the interaction. One idea was that the interactions involve the double bonds of the vinyl chains of protoporphyrin. An effort to get experimental verification of this idea failed. Finally, in 1951, Robert C.C. St. George and I were able to advance the idea, with firm experimental support, that the interaction of adjacent hemes with ligated groups in hemoglobin has its basis in steric repulsion. We showed that the combination constants of ethylisocyanide, isopropylisocyanide, and tertiary butylisocyanide with hemoglobin fall off rapidly in that order, whereas those with free ferroheme groups are essentially constant.

We attributed the decrease in combination constant, as the alkyl group becomes larger, to a steric-hindrance effect: the fitting of the larger alkyl groups into a cavity in the hemoglobin molecule that is somewhat too small, thus decreasing the free energy of combination. We assumed that the loosening in structure of the polypeptide chain by the first ligand would lead to a change in structure of the adjacent polypeptide chains in such a way as to increase the size of the cavity into which the ligand fits, to such an extent as to decrease the energy required to fit it into the cavity (2).

At the beginning of my work on hemoglobin I recognized that there was uncertainty about the way in which oxygen molecules attach themselves to hemoglobin. Two possibilities were discussed: one, that the oxygen molecule forms chemical bonds with atoms in the hemoglobin; and the other, that the oxygen molecules occupy four sites on the surface of the hemoglobin molecule, where they are held by van der Waals attraction and other physical, rather than chemical, forces. Investigators in surface chemistry had evidence that adsorption of gas molecules on solids could occur in either one of these two ways. I thought that a decision could be made between the two by determining the magnetic susceptibility of oxyhemoglobin. The dioxygen molecule has a large magnetic moment, corresponding to its two unpaired electrons and its $^3\Sigma$ normal state.Presumably these electrons would remain unpaired if dioxygen were held to the hemoglobin molecule by physical forces, but would be paired if chemical bonds were formed. My student, Bright Wilson,Jr., had constructed a simple apparatus--based on an electromagnet that we had borrowed from George Ellery Hale, of the Mt. Wilson Observatory--to measure the magnetic susceptibility of nitroso compounds, for the experimental part of his doctoral thesis. A postdoctoral fellow, Charles D. Coryell, and I then adapted this apparatus to study blood and hemoglobin solutions. We began our paper on the results of this work with the following paragraph: "Over ninety years ago, on November 8, 1845, Michael Faraday investigated the magnetic properties of dried blood and made a note 'must try recent fluid blood.' If he had determined the magnetic susceptibilities of arterial and venous blood, he would have found them to differ by a large amount (as much as 20% for completely oxygenated and completely deoxygenated blood); this discovery, without doubt, would have excited much interest and would have influenced appreciably the course of research on blood and hemoglobin."

Coryell and I found that oxyhemoglobin is diamagnetic. This observation settled the question as to whether dioxygen is attached to hemoglobin by physical or chemical forces. It was clear that the dioxygen molecule forms chemical bonds with an atom or atoms in the hemoglobin molecule.

Moreover, we observed, much to our surprise, that the hemoglobin molecule, without attached dioxygen or other ligand such as carbon monoxide, had a large magnetic moment, corresponding to the spins of four unpaired electrons per iron atom. It was known, of course, that compounds of iron (II) are of two types: the low-spin type, with zero magnetic moment, and the high-spin type, with the magnetic moment corresponding to four unpaired electron spins. The idea had not occurred to us, however, that such closely related compounds as hemoglobin and oxyhemoglobin would differ in this way.

The high-spin complexes of ferrous iron, such as $[FeF_6]^{4-}$ and $[Fe(H_2O)_6]^{2+}$, involve bonds between the iron atom and the six ligated atoms that have a large amount of partial ionic character, whereas the low-spin complexes, such as the ferrocyanide ion $[Fe(CN)_6]^{4-}$, have bonds with high covalent character. This permitted Coryell and me, in our 1936 paper, to discuss the nature of the bonding of the iron atom with ligated carbon monoxide or dioxygen. We suggested that in carbonmonoxyhemoglobin the structure of the carbomonoxyheme group involves resonance between two structures, one with a single bond from iron to carbon and a triple bond from carbon to oxygen and the other with double bonds in each of these positions. For each of these structures the iron-carbon-oxygen group would be linear. For the oxyheme group, however, we suggested resonance between a structure with a single bond from iron to oxygen and a double bond between oxygen atoms, and another structure with the double bond from iron to oxygen and a single bond between the oxygen atoms. In each case the bond angle at the first oxygen is that between a single bond and a double bond, which is known to be about 117°, as in the ozone molecule (3,4). For some time there was controversy about the bonding of dioxygen in oxyhemoglobin, but recent studies have shown that the dioxygen molecule is attached to the iron atom by only one of the oxygen atoms, and that thw bond angle at the oxygen atom is close to 117°.

The discovery of the change in magnetic properties of hemoglobin accompanying various changes in chemical combination permitted many experimental studies to be made of chemical equilibrium and rates of reaction of hemoglobin and its derivatives, throwing additional light on the question of the structure of the hemoglobin molecule (5,6,7,8).

In 1936 Karl Landsteiner asked me to think about the nature of the forces involved in the interaction of antigens and antibodies. I was especially taken by the striking specificity of serological reactions, a specificity that permitted antisera to distinguish between closely related proteins such as hen ovalbumin and duck ovalbumin, and between azoproteins with closely similar attached groups, such as benzoic acid and the ortho, meta, and paratoluic acids. I felt that the mechanism of this specificity would probably be the mechanism of biological specificity in general, similar to that shown in heredity. Consideration of the observations reported by Landsteiner and others in the field of serology convinced me that the specificity of interaction of an antigen and its homologous antibody is the result of a detailed molecular complementariness in structure. The arguments leading to this conviction were presented in my paper on this subject in 1940 (9), in which some predictions were made: for example, that precipitating and agglutinating antibodies have two combining groups, complementary in structure to groups on the corresponding antigens. During the next eight years we carried out a great many experiments in this field. These studies led to the provision of such strong evidence for the idea that the specificity of the reactions is the result of detailed molecular complementariness that it is now impossible to reject this idea. The bivalence of precipitating and agglutinating antibodies and some other features of the theory of the structure of antibodies and the nature of serological reactions were also thoroughly checked (9). Many of our papers were published in the Journal of the American Chemical Society between 1942 and 1949. The principal investigators in this work were the late Professor Dan H. Campbell and the late Dr. David Pressman (10, 11, 12, 13, 14, 15).

Work on the amino-acid sequences in the polypeptide chains of human hemoglobin was begun in the California Institute of Technology by Dr. Walter Schroeder, and has been continued ever since by him and his associates. The first striking result of this work was the discovery that there are chains of two different kinds in normal adult human hemoglobin, the alpha chain and the beta chain, each doubly represented in the hemoglobin molecule (16).

Because of our interest in hemoglobin, Dr. Alfred E. Mirsky of the Rockefeller Foundation for Medical Research, who had done much research on hemoglobin, came for one year to the California Institute of Technology in the fall of 1945. He and I discussed many questions about hemoglobin and other proteins and developed a structural theory of native, denatured, and coagulated proteins, published in the Proceedings of the National Academy of Sciences in 1936 (17). We stated that native protein molecules, consisting of one polypeptide chain or sometimes two or more chains, have a structure in which the chain or chains are folded into a uniquely defined configuration, held by hydrogen bonds between the peptide nitrogen and oxygen atoms and also between the free amino and carboxyl groups of the diamino and dicarboxyl amino-acid residues. This structure is such that the molecules are able to remain in solution. If, however, the structure is broken up, as by thermal agitation, to produce a denatured protein molecule, some of these groups are freed, so that complementary groups of different molecules can combine with one another, leading to the formation of an insoluble coagulum.

The discovery of the abnormal human hemoglobins was the resultof my having been appointed to a committee. I have usually tried to avoid service on committees, but I was pleased to be a member of this one, because of the amount of information about medicine that I gathered in the course of my work on it. The committee, called the Medical Advisory Committee, was one of four set up to assist Dr. Vannevar Bush to reply to a letter in November, 1944, from President Franklin D. Roosevelt.The

President asked for advice about how the federal government could assist in scientific and medical research in the period following the Second World War. The members of the Medical Advisory Committee were, except for myself, physicians: Walter W. Palmer, Homer W. Smith, Kenneth B. Turner, William B. Castle, Edward A. Doisey, Ernest Goodpasture, Alton Ochsner, and James J. Waring.

During the next six months the committee interviewed 350 people: representatives from 73 of the 77 medical schools of the United States, from the armed services, the various medical research institutions, the pharmaceutical industry, and philanthropic foundations. Its report, Science, the Endless Frontier: Report to the President on a Program for Postwar Scientific Research by Vannevar Bush, Director of the Office of Scientific Research and Development, was published by the U.S. Government Printing Office in July, 1945, and was influential in the setting up of the National Science Foundation and the National Institutes of Health.

After a strenuous day listening to representatives of medical schools and medical research institutes in New York City, the members of the Medical Advisory Committee were having dinner together at a club in the city (19). Dr. William B. Castle, Professor of Medicine at Harvard University, began talking about the disease sickle-cell anemia, with which he had had some experience. I had only a mild interest in what he was saying, because at that time cells seemed to me to be far too complex to permit me to make any sort of attack on them. However, when Dr. Castle said that the red cells in the blood of a patient with this disease are sickled in the venous blood but not in the arterial blood, the idea occurred to me that sickle-cell anemia might be a disease of the hemoglobin molecule. Because of my background of knowledge and experience outlined in the preceding paragraphs, I thought at once of this possibility: that the hemoglobin molecules of these patients might have, as a result of a gene mutation, a structure formed so that one portion of the surface of the molecule would be sufficiently complementary to another portion to permit the molecules to aggregate into long chains. Further, these long chains would then line up side by side to form a needle-shaped crystal, which, as its length grew greater than the diameter of the red cell, would twist the red cell out of shape. These deformed cells would have properties sufficiently different from normal cells to give rise to the manifestation of the disease. The attachment of the dioxygen molecules to the iron atoms of the heme groups might interfere with the approach to one another of complementary regions of adjacent molecules close enough to permit a strong interaction to take place, so that oxygenation of the hemoglobin would reverse the sickling process.

In forming this idea, I drew first upon my knowledge about the interaction of complementary structures on antigen and antibody; second, upon the ideas that Mirsky and I had formulated about the structure of native protein molecules, such as to permit them to stay in solution, whereas after denaturation they attach themselves to one another to form an insoluble coagulum; third, my understanding of the change in structure of the polypeptide chains of hemoglobin when a ligand fits into its cavity, as discussed by St. George and me in our 1951 paper; and, finally, other information such as the knowledge that in a precipitate of silver cyanide there are long chains, with alternating silver atoms and cyanide groups, that line up side by side to form the needle-shaped crystal,

I was fortunate then in that Doisey recommended to me that one of his students, Dr. Harvey A. Itano, who had received his M.D. degree in 1945 and was then interning in Detroit, be accepted in the California Institute of Technology to work for a Ph.D. degree. He had been awarded the first American Chemical Society predoctoral fellowship in chemistry, for the three years 1946 to 1949. In a letter to Dr. Itano, I suggested that he investigate the hemoglobin from the red cells of patients with sickle-cell anemia in order to determine whether or not it is different in structure from normal adult hemoglobin.

I think that the rest of the story about the abnormal human hemoglobins is well known. Dr. Itano, with the assistance of Dr. S.J. Singer, succeeded in carrying out the electrophoresis of human hemoglobins from different sources, showing that sickle-cell-anemia hemoglobin differs from normal adult human hemoglobin. Hemoglobin C was then discovered by Itano and Neel (20), hemoglobin D by Itano (21), and hemoglobin E by Itano, Bergren, and Sturgeon (22). The later history of the abnormal human hemoglobins is discussed in other chapters in this book.

The study of sickle-cell-anemia hemoglobin has led to a number of developments in the fields of molecular disease and molecular evolution.The reader can learn more about these elsewhere (23,24,25,26,27).

REFERENCES

1. Pauling, L.: The oxygen equilibrium of hemoglobin and its structural interpretation. Proc Nat Acad Sci USA 21: 186, 1935.
2. St. George, R.C.C. and L. Pauling: The combining power of hemoglobin for alkyl isocyanides, and the nature of the heme-heme interactions in hemoglobin. Science 114: 629, 1951.
3. Pauling, L. and C.D. Coryell: Magnetic properties and structures of hemochromogens and related substances. Proc Nat Acad Sci USA 22: 210, 1936.
4. Coryell, C.D., F. Stitt, and L. Pauling: Magnetic properties and structure of ferrihemoglobin (methemoglobin) and some of its compounds. J Am Chem Soc 59:683, 1937.
5. Coryell, C.D., L. Pauling, and R.W. Dodson: Magnetic properties of intermediates in the reactions of hemoglobin. J Phys Chem 43:825, 1939.
6. Russell, C.D. and L. Pauling: Magnetic properties of the compounds ethylisocyanids-ferrohemoglobin and imidazole-ferrihemoglobin. Proc Nat Acad Sci USA, 25: 517, 1939.
7. Coryell, C.D. and L. Pauling: Structural interpretation of the acidity of groups associated with the hemes of hemoglobin and hemoglobin derivatives. J Biol Chem 132: 769,1940.
8. Pauling, L.: Electronic structure of haemoglobin. In:Haemoglobin , Butterworths Scientific Pub., London, pp. 57-65, 1949.
9. Pauling, L. A theory of the structure and process of formation of antibodies. J. Am. Chem Soc. 62: 2643, 1940.
10. Pauling, L., D.H. Campbell, and D. Pressman: Serological reactions with simple substances containing two or more haptenic groups. Proc Nat Acad Sci USA 27: 125, 1941.

Hemoglobins and Hemoglobinopathies: Background/Pauling 7

11. Pressman,D., D.H. Campbell, and L. Pauling: Agglutination of intact azo-erythrocytes by antisera homologous to the attached groups. J Immunology 44: 101, 1942.
12. Pauling, L., D. Pressman, D.H. Campbell, C. Ikeda and M. Ikawa: Serological properties of simple substances. I. Precipitation reactions betweem antibodies and substances containing two or more haptenic groups. J Am Chem Soc 64: 2994, 1942.
13. Pauling, L., D. Pressman and D.H. Campbell: Serological properties of simple substances. VI. Precipitation of a mixture of two specific antisera by a dihaptenic substance containing the two corresponding haptenic groups: evidence for the framework theory of serological precipitation. J Am Chem Soc 66: 330, 1944.
14. Pauling, L., D. Pressman and A.L. Grossberg: Serological properties of simple substance. VII. A quantitative theory of the inhibition by haptens of the precipitation of heterogeneous antisera with antigens, and comparison with experimental results for polyhaptenic simple substances and for azoproteins. J. Am Chem Soc 66: 784, 1944.
15. Pressman, D. and L. Pauling: Serological properties of simple substances. XV. The reactions of antiserum homologous to the 4-azophthalate ion. J Am Chem Soc 71: 2893, 1949.
16. Rhinesmith, H.S., W.A. Schroeder and L. Pauling: A quantitative study of the hydrolysis of human dinitrophenyl (DNP) globin: The number and kind of polypeptide chains in normal adult human hemoglobin. J Am Chem Soc 79: 4682, 1957.
17. Mirsky, A.E. and L. Pauling: On the structure of native, denatured, and coagulated proteins. Proc Nat Acad Sci USA 22: 439, 1936.
18. Pauling, L., H.A. Itano,S.J. Singer, and L.C. Wells: Sickle cell anemia, a molecular disease. Science 110: 543, 1949.
19. Pauling, L.: Abnormality of hemoglobin molecules in hereditary hemolytic anemias (Harvey Lecture, April 1954). In: The Harvey Lectures, 1953-1954, Series 49, Academic Press, New York, pp.216-241, 1955.
20. Itano, H.A. and J.V. Neel: A new inherited abnormality of human hemoglobin. Proc. Nat. Acad Sci USA 36: 613, 1950.
21. Itano, H.A.: A third abnormal hemoglobin associated with hereditary hemolytic anemia. Proc. Nat Acad Sci USA 37: 775, 1951.
22. Itano, H.A., W.R. Bergren and P. Sturgeon: Identification of a fourth abnormal human hemoglobin. J Am Chem Soc 76: 2278, 1954.
23. Pauling, L.: Molecular disease. Am J Orthopsychiat 29: 684, 1959.
24. Zuckerkandl, E., R.T. Jones, and L. Pauling: A comparison of animal hemoglobin by tryptic peptide pattern analysis. Proc Nat Acad Sci USA 46: 1349, 1960.
25. Zuckerkandl, E. and L. Pauling: Molecular disease, evolution and genic heterogeneity. In Horizons in Biochemistry, Szent-Gyorgyi Dedicatory volume. Academic Press, New York, pp. 189-225, 1962.
26. Pauling, L.: Chemical paleogenetics: Molecular "restoration studies" of extinct forms of life. Acta Chem Scand 17: suppl. No. 1, pp. S9-S16,1963.
27. Zuckerkandl, E. and L. Pauling: Molecules as documents of evolutionary history. J. Theoret Biol 8: 357, 1965.

Chapter 12

ANTIBODIES: STRUCTURE AND FUNCTION

Contents

SP 88 **A Theory of the Structure and Process of Formation of Antibodies** 890
J. Am. Chem. Soc. **62**, 2643–2657 (1940)

89 **The Serological Properties of Simple Substances. I. Precipitation Reactions Between Antibodies and Substances Containing Two or More Haptenic Groups** 905
(by Linus Pauling, David Pressman, Dan H. Campbell, Carol Ikeda, and Miyoshi Ikawa)
J. Am. Chem. Soc. **64**, 2994–3003 (1942)

90 **The Serological Properties of Simple Substances. II. The Effects of Changed Conditions and of Added Haptens on Precipitation Reactions of Polyhaptenic Simple Substances** 914
(by Linus Pauling, David Pressman, Dan H. Campbell, and Carol Ikeda)
J. Am. Chem. Soc. **64**, 3003–3009 (1942)

91 **The Nature of the Forces Between Antigen and Antibody and of the Precipitation Reaction** 921
(by Linus Pauling, Dan H. Campbell, and David Pressman)
Physiol. Rev. **23**, 203–219 (1943)

92 **The Serological Properties of Simple Substances. VI. The Precipitation of a Mixture of Two Specific Antisera by a Dihaptenic Substance Containing the Two Corresponding Haptenic Groups; Evidence for the Framework Theory of Serological Precipitation** 938
(by Linus Pauling, David Pressman, and Dan H. Campbell)
J. Am. Chem. Soc. **66**, 330–336 (1944)

93 **Antibodies and Specific Biological Forces** 945
Endeavour **7**(26), 43–53 (1948)

94 **The Reaction of Simple Antigens with Purified Antibody** 956
(by Arthur B. Pardee and Linus Pauling)
J. Am. Chem. Soc. **71**, 143–148 (1949)

[Reprinted from the Journal of the American Chemical Society, 62, 2643 (1940).]

[Contribution from the Gates and Crellin Laboratories of Chemistry, California Institute of Technology, No. 783]

A Theory of the Structure and Process of Formation of Antibodies*

By Linus Pauling

I. Introduction

During the past four years I have been making an effort to understand and interpret serological phenomena in terms of molecular structure and molecular interactions. The field of immunology is so extensive and the experimental observations are so complex (and occasionally contradictory) that no one has found it possible to induce a theory of the structure of antibodies from the observational material. As an alternative method of attack we may propound and attempt to answer the following questions: What is the simplest structure which can be suggested, on the basis of the extensive information now available about intramolecular and intermolecular forces,[1] for a molecule with the properties observed for antibodies, and what is the simplest reasonable process of formation of such a molecule? Proceeding in this way, I have developed a detailed theory of the structure and process of formation of antibodies and the nature of serological reactions which is more definite and more widely applicable than earlier theories, and which is compatible with our present knowledge of the structure and properties of simple molecules as well as with most of the direct empirical information about antibodies. This theory is described and discussed below.

II. The Proposed Theory of the Structure and Process of Formation of Antibodies

When an antigen is injected into an animal some of its molecules are captured and held in the region of antibody production.[2] An antibody to this antigen is a molecule with a configuration which is complementary to that of a portion of the antigen molecule.[3] This complementariness gives rise to specific forces of appreciable strength between the antibody molecule and the antigen molecule; we may describe this as a bond between the two molecules. I assume, with Marrack, Heidelberger, and other investigators,[4] that the precipitate obtained in the precipitin reaction is a framework,[5] and that to be effective in forming the framework an antibody molecule must have two or more distinct regions with surface configuration complementary to that of the antigen. The rule of parsimony (the use of the minimum effort to achieve the result) suggests that there are only two such regions, that is, that the antibody molecules are at the most bivalent. The proposed theory is based on this reasonable assumption. It would, of course, be possible to expand the theory in such a way as to provide a mechanism for the formation of antibody molecules with valence higher than two; but this would make the theory considerably more complex, and it is likely that antibodies with valence higher than two occur only rarely, if at all.

Antibodies are similar in amino-acid composition to one or another of the fractions of serum globulin of the animal producing the serum. It is known that there exist antibodies of different classes, with different molecular weights—the molecular weights of rabbit antibody and of monkey antibody (to pneumococcus polysaccharide) are about 157,000, whereas those of pig, cow, and horse antibodies are about 930,000.[6] The following discussion is for antibodies with molecular weight about 160,000, and similar in constitution to the γ fraction of serum globulin[7]; the changes to be made to cause it to apply to antibodies of other classes are obvious.

* Some of the material in this paper was presented on April 17th, 1940, at The Rockefeller Institute for Medical Research and on April 23rd, 1940, at the meeting of the National Academy of Sciences, Washington, D. C.

(1) See, for a summary of this information, L. Pauling, "The Nature of the Chemical Bond and the Structure of Molecules and Crystals," Cornell University Press, Ithaca, New York, second edition, 1940.

(2) There is some evidence that this is the cells of the reticuloendothelial system: see Florence R. Sabin, *J. Exptl. Med.*, **70**, 67 (1939), and references quoted by her.

(3) The idea of complementary structures for antibody and antigen was suggested by (a) F. Breinl and F. Haurowitz, *Z. physiol. Chem.*, **192**, 45 (1930); (b) Stuart Mudd, *J. Immunol.*, **23**, 423 (1932); (c) J. Alexander, *Protoplasma*, **14**, 296 (1931), and has come to be rather generally accepted. There is some intimation of it in the early work of Ehrlich and of Bordet.

(4) J. R. Marrack, "The Chemistry of Antigens and Antibodies," Report No. 230 of the Medical Research Council, His Majesty's Stationery Office, London, 1938; M. Heidelberger, *Chem. Rev.*, **24**, 323 (1939), and earlier papers.

(5) The framework is sometimes called a "lattice" by immunochemists; the use of this word in immunology is to be discouraged because of the implication of regularity associated with it through its application in crystallography.

(6) E. A. Kabat and K. O. Pedersen, *Science*, **87**, 372 (1938); E. A. Kabat, *J. Exptl. Med.*, **69**, 103 (1939).

(7) A. Tiselius, *Biochem. J.*, **31**, 1464 (1937); *Trans. Faraday Soc.*, **33**, 524 (1937); T. Svedberg, *Ind. Eng. Chem.*, **30**, 113 (1938).

The effect of an antigen in determining the structure of an antibody molecule might involve the ordering of the amino-acid residues in the polypeptide chains in a way different from that in the normal globulin, as suggested by Breinl and Haurowitz[3a] and Mudd.[3b] I assume, however, that this is not so, but that *all antibody molecules contain the same polypeptide chains as normal globulin, and differ from normal globulin only in the configuration of the chain; that is, in the way that the chain is coiled in the molecule.* There is at present no direct evidence supporting this assumption. The assumption is made because, although I have found it impossible to formulate in detail a reasonable mechanism whereby the order of amino-acid residues in the chain would be determined by the antigen, a simple and reasonable mechanism, described below, can be advanced whereby the antigen causes the polypeptide chain to assume a configuration complementary to the antigen. The number of configurations accessible to the polypeptide chain is so great as to provide an explanation of the ability of an animal to form antibodies with considerable specificity for an apparently unlimited number of different antigens,[8] without the necessity of invoking also a variation in the amino-acid composition or amino-acid order.[9]

The Postulated Process of Formation of Antibodies.—Let us assume that the globulin molecule consists of a single polypeptide chain, containing several hundred amino-acid residues, and that the order of amino-acid residues is such that for the center of the chain one of the accessible configurations is much more stable than any other, whereas the two end parts of the chain are of such a nature that there exist for them many configurations with nearly the same energy. (This point is discussed in detail in Section IV.) Four steps in our postulated process of formation of a normal globulin molecule are illustrated on the left side of Fig. 1. At stage I the polypeptide chain has been synthesized, the amino-acid residues having been marshalled into the proper order, presumably with the aid of polypeptidases and protein templates, and the two ends of the chain, A and C, each containing perhaps two hundred residues, have been liberated with the unstable extended configuration. (The horizontal line in each drawing separates the region, below the line, in which the polypeptide chain is not able to change its configuration from the region, above the line, where this is possible.) Each of these chain ends then coils up into the most stable or one of the most stable of the accessible configurations (stage II) and is tied into this configuration by the formation of hydrogen bonds and other weak bonds between parts of the chain. The central part B of the chain is then liberated (stage III) and assumes its stable folded configuration (stage IV) to give the completed globulin molecule.

There are also indicated in Fig. 1 six stages in the process of formation of an antibody molecule. In stage I there are shown an antigen molecule held at a place of globulin production and a globulin molecule with its two ends A and C liberated with the extended configuration. At stage II each of the ends has assumed a stable coiled configuration. These stable configurations A′ and C′ are not, however, identical with those A and C assumed in the absence of the antigen. The atoms and groups which form the surface of the antigen will attract certain complementary parts of the globulin chain (a negatively-charged group, for example, attracting a positively-charged group) and repel other parts; as a result of these interactions the configurations A′ and C′ of the chain ends which are stable in the presence of the antigen and which are accordingly assumed in the presence of the antigen will be such that there is attraction between the coiled globulin chain ends and the antigen, due to their complementarity in structure. The configuration assumed by the chain end may be any one of a large number, depending upon which part of the surface of the antigen happens to exert its influence on the chain end and how large a region of the surface happens to be covered by it.

When the central part B of the globulin chain is liberated from the place of its synthesis (stage III), one of two processes may occur. If the forces of attraction between the antigen and the portions A′ and C′ are extremely strong, they will remain bonded to the antigen for an indefinite time, and nothing further of interest will happen. If the forces are somewhat weaker, however, one will in time break away—dissociate from the antigen (stage IV). Then the portion B of the chain will fold up to achieve its normal stable configuration

(8) See K. Landsteiner, "The Specificity of Serological Reactions," Charles C. Thomas, Springfield, Ill., 1936.

(9) It has been pointed out by A. Rothen and K. Landsteiner, *Science*, **90**, 65 (1939), that the possibility of different ways of folding the same polypeptide chain to obtain different antibodies is worth considering.

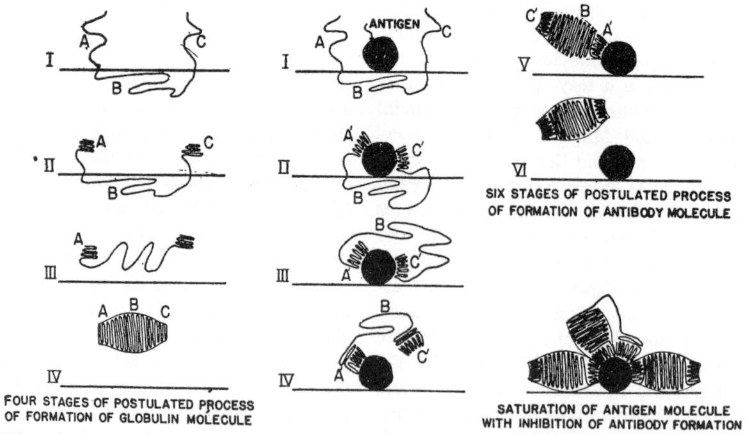

Fig. 1.—Diagrams representing four stages in the process of formation of a molecule of normal serum globulin (left side of figure) and six stages in the process of formation of an antibody molecule as the result of interaction of the globulin polypeptide chain with an antigen molecule. There is also shown (lower right) an antigen molecule surrounded by attached antibody molecules or parts of molecules and thus inhibited from further antibody formation.

(stage V), making a completed antibody molecule. In time this will dissociate from the antigen and float away (stage VI). It is possible that an auxiliary mechanism for freeing the active ends A′ and C′ from the antigen molecule comes into operation; this is discussed in Section VI.

The middle part of the antibody molecule thus produced would be like that of a normal globulin molecule, and the two ends would have configurations more or less complementary to parts of the surface of the antigen. These two active ends are effective in different directions, so that, after the antibody is completely formed, only one of them at a time can grasp a particular antigen molecule.

The antigen molecule, after its desertion by the newly-formed antibody molecule, may serve as the pattern for another, and continue to serve until its surface is covered by very strongly held antibodies or portions of antibodies or until the concentration of antibodies becomes so great that even with weak forces operating the antigen is combined with antibodies most of the time (as illustrated in Fig. 1), or until the antigen molecule is destroyed or escapes from the region of globulin formation.

III. Some Points of Comparison with Experiment

a. The Heterogeneity of Immune Sera.—The theory requires that the serum homologous to a given antigen be not homogeneous, but heterogeneous, containing antibody molecules of greatly varied configurations. Many of the antibody molecules will be bivalent, with two active ends with configuration complementary to portions of the surface of an antigen molecule. Great variety in this complementary configuration would be expected to result from the accidental approximation to one or another surface region, and further variety from variation in position of the antigen molecule relative to the point of liberation of the globulin chain end and from accidental coiling and linking of the chain end before it comes under the influence of the antigen. Some of the antibody molecules would be univalent, one of the chain ends having, because of its too great distance from the antigen, folded into a normal globulin configuration.

These predictions are verified by experimental results. It is well known that an immune serum to one antigen will, as a rule, react with a related heterologous antigen, and that after exhaustion with the latter there remains a fraction which will still react with the original antigen. Landsteiner and van der Scheer,[10] using as antigens azoproteins carrying various haptens containing the same active group, have shown that the antiserum for one antigen contains various fractions differing in the strength of their attraction for the haptens.

(10) K. Landsteiner and J. van der Scheer, *J. Exptl. Med.*, **63**, 325 (1936).

By the quantitative study of precipitin reactions Heidelberger and Kendall[11] reached the same conclusion, and showed in addition that even after prolonged immunization the antiserum studied (anti-egg albumin) contained much low-grade antibody, incapable by itself of forming a precipitate with the antigen, but with the property of being carried down in the precipitate formed with a more reactive fraction.[12]

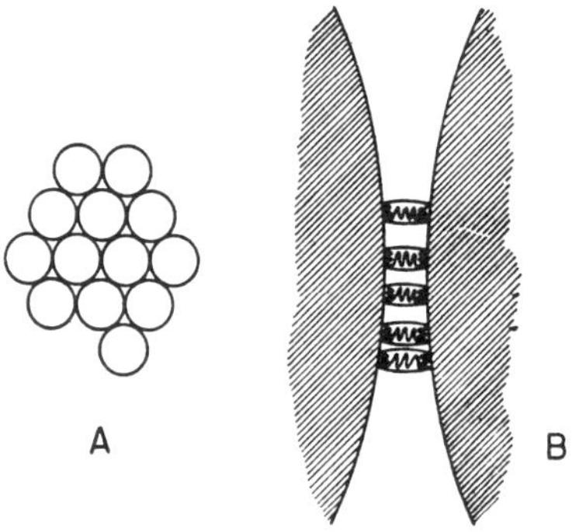

Fig. 2.—(A) Diagram representing agglutinated cells. (B) Diagram of the region of contact of two cells, showing the postulated structure and mode of action of agglutinin molecules.

b. The Bivalence of Antibodies and the Multivalence of Antigens.—Our theory is based on the idea that the precipitate formed in the precipitin reaction is a network of antibody and antigen molecules in which many or all of the antibody molecules grasp two antigen molecules apiece and the antigen molecules are grasped by several antibody molecules. The direct experimental evidence for this picture of the precipitate has been ably discussed by its propounders and supporters, Marrack and Heidelberger and Kendall, and need not be reviewed here. To the structural chemist it is clear that this picture of the precipitate must be correct. The great specificity of antibody–antigen interactions requires that a definite bond be formed between an antibody molecule and an antigen molecule. If antibodies or antigens were univalent, this would lead to complexes of one antigen molecule and one or more antibody molecules (or of one antibody molecule and one or more antigen molecules), and we know from experience with proteins that these aggregates would in general remain in solution. If both antibody and antigen are multivalent, however, the complex will grow to an aggregate of indefinite size, which is the precipitate.

This process is observed directly in the agglutination of cells. On the addition of an agglutinin to a cell suspension the cells are seen to clump together. It is obvious that the agglutinin molecules which are holding the cells together are bivalent[13]—each has two active ends, with configuration complementary to that of a portion of the surface of the cells; the agglutinin molecules hold the cells together at their regions of contact, as shown in Fig. 2.

It seems probable that all antibodies have this structure—that they are bivalent, with their two active regions oppositely directed. Heidelberger and his collaborators and Marrack have emphasized the multivalence of antibodies and antigens,[14] but limitation of the valence of antibodies to the maximum value two (ignoring the exceptional case of the attachment of two or more antigens or haptens to the same end region of an antibody) has not previously been made.

The maximum valence of an antigen molecule would be given by the ratio of its surface area to the area effectively occupied by one antibody molecule, if all regions of the antigen surface were active. In the special case that the antibody were able to combine only with one group (a hapten, say, with immunization effected by use of another antigen with the same hapten attached) the maximum valence of the antigen would be equal to the number of groups per molecule.

c. The Antibody–Antigen Molecular Ratio in Precipitates.—Our theory provides an immediate simple explanation of the observed antibody–antigen molecular ratios in precipitates. Under optimum conditions a precipitate will be formed in which all the valences of the antibody and antigen molecules are satisfied. An idealized repre-

(11) M. Heidelberger and F. E. Kendall, *J. Exptl. Med.*, **61**, 559 (1935); **62**, 467, 697 (1935).

(12) The older experimental results bearing on this question did not permit a clear distinction between antibody fractions differing in being complementary to different active groups in the antigen and fractions differing in the extent of their complementariness to the same group. The experiments are discussed in Marrack's monograph.

(13) Following Heidelberger and Kendall, I use the terminology of chemical valence theory in discussing the specific mutual attraction of antigen and antibody. The antibody–antigen "valence bonds" are not, of course, to be confused with ordinary covalent chemical bonds; they are due instead to the integrated weak forces discussed in Sec. IV.

(14) Professor Heidelberger has informed me that in their quantitative treatment of data on the precipitin reaction he and Dr. Kendall have found no incompatibility with this restriction; in their papers they discussed the general case of multivalence of antibody as well as of antigen.

sentation of a portion of such a precipitate is given in Fig. 3. The figure shows a part of a layer with each antigen molecule bonded to six surrounding antibody molecules; this structure represents the value $N = 12$ for the valence of the antigen, each antigen molecule being attached also to three antibody molecules above the layer represented and to three below. Each antibody molecule is bonded to two antigen molecules, one at each end. An ideal structure of the antibody–antigen precipitate for $N = 12$ may be described as having antigen molecules at the positions corresponding to closest packing, with the twelve antibody molecules which surround each antigen molecule lying along the lines connecting it with the twelve nearest antigen neighbors.

Similar ideal structures can be suggested for other values of the antigen valence. The antigen molecules might be arranged for $N = 8$ at the points of a body-centered cubic lattice, and for $N = 6$ at the points of a simple cubic lattice, with antibody molecules along the connecting lines. For $N = 4$ the antigen molecules, connected by antibody molecules, might lie at the points occupied by carbon atoms in diamond; or two such frameworks might interpenetrate, as in the cuprous oxide arrangement (copper and oxygen atoms being replaced by antibody and antigen molecules, respectively).

It is not to be inferred that the actual precipitates have the regularity of structure of these ideal arrangements. The nature of the process of antibody formation, involving the use of a portion of the antigen surface selected at random as the template for the molding of an active end of an antibody molecule, introduces so much irregularity in the framework that a regular structure analogous to that of a crystal is probably never formed. The precipitate is to be compared rather with a glass such as silica glass, in which each silicon atom is surrounded tetrahedrally by four oxygen atoms and each oxygen atom is bonded to two silicon atoms, but which lacks further orderliness of arrangement. Additional disorder is introduced in the precipitate by variation in the effective valence of the antigen molecules and by the inclusion of antibody molecules with only one active end.

The antibody-antigen molecular ratio R of a precipitate is given by the equation

$$R = N_{\text{eff.}}(\text{antigen})/N_{\text{eff.}}(\text{antibody}) \quad (1)$$

in which $N_{\text{eff.}}$(antigen) and $N_{\text{eff.}}$(antibody) are the

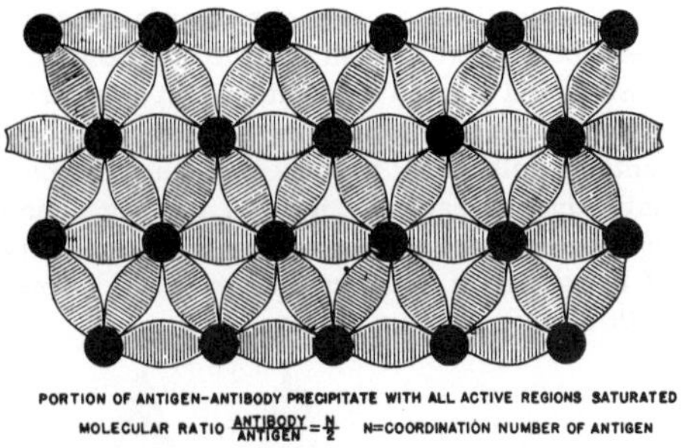

Fig. 3.—A portion of an ideal antibody–antigen framework. One plane of the structure corresponding to the value twelve for the valence of the antigen molecules is shown.

average effective valences of antigen and antibody molecules, respectively. The maximum value of $N_{\text{eff.}}$(antibody) is 2 (ignoring the exceptional possibility that two small haptens can attach themselves to the same combining region at one end of the antibody; steric repulsion of antigen molecules would usually prevent this occurrence), and under optimum conditions for formation of the most stable precipitate we may expect this maximum value to be closely approached. The antibody–antigen molecular ratio then becomes

$$R = N/2 \quad (2)$$

in which N is $N_{\text{eff.}}$(antigen). Now a sphere can be brought into contact with twelve surrounding spheres equal to it in size; hence a spherical antigen molecule with molecular weight equal to that of the antibody (157,000) might have the valence $N = 12$, if all regions of the antigen surface were active and if the antibody molecules were spherical; the assumption of elongated antibody molecules would permit the valence to be somewhat larger. The value 12 of N corresponds to the value 6 for the ratio R. For larger antigens larger values of R would be expected, and for smaller ones smaller values. Even for antigens with molecular weight as small as 11,000 the predicted maximum value of R is 4 (for spherical antibodies) or larger. In fact, a simple calculation based on the packing of spheres[15] leads to the results given in Table I.[16] It is seen

(15) See L. Pauling, ref. 1, Sec. 48a.

(16) It may be noted that values of R calculated in the text change with molecular weight in about the same way as those calculated by W. C. Boyd and S. B. Hooker, *J. Gen. Physiol.*, **17**, 341 (1934), on the assumption that each antigen molecule is surrounded by a close-packed layer of (univalent) antibody molecules. The Boyd–Hooker values agree roughly with experiment (Marrack, *loc. cit.*, p. 161).

TABLE I

COÖRDINATION OF SPHERICAL ANTIBODY MOLECULES ABOUT SPHERICAL ANTIGEN MOLECULES

No. of antibody molecules about antigen	Minimum ratio of antigen radius to antibody radius	Minimum mol. wt. of antigen (antibody 160,000)	Maximum mol. ratio Antibody/Antigen in satd. ppt.	Maximum mass ratio Antibody/Antigen
12	1.000	160,000	6	6
8	0.732	63,000	4	10
6	.414	11,000	3	44
4	.225	1,800	2	178

that our theory provides a simple explanation of the fact that for antigens of molecular weight equal to or less than that of the antibody the precipitate contains considerably more antibody than antigen. The values given in Table I are not to be considered as having rigorous quantitative significance. The calculated maximum molecular ratio would be larger for elongated antibody molecules than for spherical antibody molecules, and larger for non-spherical than for spherical antigen molecules, and, moreover, in many sera the antibodies might be complementary in the main only to certain surface regions of the antigen, the number of these determining the valence of the antigen. That this is so is indicated by the observation[17] that after long immunization of a rabbit with egg albumin serum was obtained giving a precipitate with a considerably larger molecular ratio than that for earlier bleedings.[18]

Observed values of R for precipitates formed in the equivalence zone (with amounts of antigen solution and serum so chosen that neither excess antibody nor excess antigen can be detected in the supernate) for antigens with molecular weights between about 4000 and 700,000 lie between about 2.5 for the smaller antigens and 15 for the larger ones.[19] It is seen that the values of R are somewhat less than the corresponding values from Table I, which indicates that not all of the surface regions of the antigens are effective.

(17) M. Heidelberger and F. E. Kendall, *J. Exptl. Med.*, **62**, 697 (1935).

(18) The discussion of the nature of this phenomenon of change in the serum on continued immunization must await the detailed treatment of intracellular antibody–antigen interactions. The phenomenon may involve the masking of the more effective surface regions of the injected antigen molecules by serum antibodies produced by earlier inoculations, leaving only the less effective regions available for template action.

(19) It is our restriction of the valence of the antibody to the maximum value two which leads to our explanation of the antibody–antigen ratios. The general observation of values of R considerably greater than 1 is not accounted for by a framework theory in which antibody and antigen molecules are both multivalent, unless some auxiliary postulate is invoked to make the effective valence of antigen considerably greater than that of antibody.

The data given in Table II are those of Heidelberger and his collaborators; the values reported by other investigators are similar in magnitude.

TABLE II

VALUES OF ANTIBODY–ANTIGEN MOLECULAR RATIOS FOR PRECIPITATES FROM RABBIT ANTISERA[a]

Antigen	Mol. wt.	R Equivalence zone	R Extreme antibody excess	R Antigen excess	R Soluble compound
Egg albumin	42,000	2.5–3	5	2	1
Dye egg albumin[b]	46,000	2.5–3	5	3/4	1/2
Serum albumin	67,000	3–4	6	2	1
Thyroglobulin	700,000	10–14	40	2	1

[a] The experimental values are those obtained by Heidelberger and collaborators, and quoted by M. Heidelberger, THIS JOURNAL, **60**, 242 (1938). [b] *R*-salt-azobiphenylazo egg albumin.

In a precipitate formed from a solution containing an excess of antibody not all of the antibody valences will be saturated. At the limit of antibody excess the precipitate will be a network of linear aggregates with a structure such as that represented in Fig. 4. Here each antigen molecule (with an occasional exception) is surrounded by N antibody molecules, only two of which bond it to neighboring antigen molecules. The padded strings formed in this way are tied together by an occasional cross-link to form the precipitate. The antibody-antigen molecular ratio is seen to be close to $N - 1$, which is one less than twice the value $N/2$ for the valence-saturated precipitate. The predicted relation between these ratios

$$R_{\text{antibody excess}} = 2R_{\text{equivalence zone}} - 1$$

is seen from the data in Table II to be verified approximately by experiment for the antigens other than thyroglobulin.

The discrepancy shown by thyroglobulin is, indeed, to be expected for an antigen with molecular weight greater than that of the antibody. The requirements of geometry are such that an arrangement in which each antigen is bonded equivalently by antibodies to more than twelve surrounding antigens is impossible. Hence for large antigen molecules the molecular ratio can exceed 6 in the valence-saturated precipitate only if two or more antibody molecules are shared between the same pair of antigen molecules, whereas in the antibody-excess region the entire surface of the large antigen may be covered by antibody molecules.

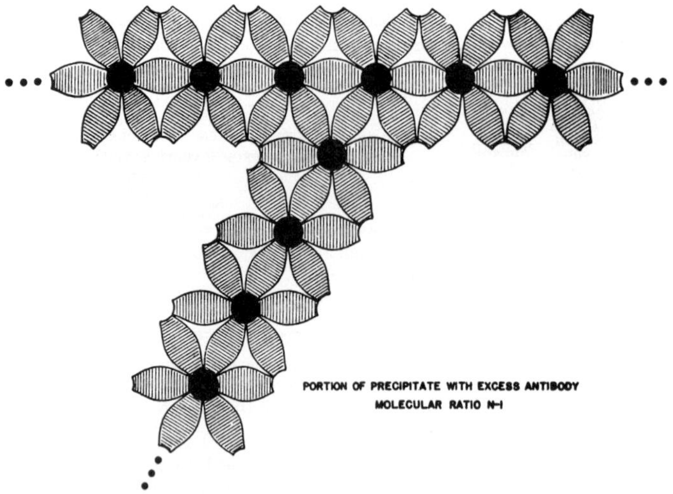

Fig. 4.—A portion of an antibody–antigen network formed in the region of antibody excess.

With antigen excess the precipitate formed will have the limiting structure shown in Fig. 5, in which (with an occasional exception) both antigen and antibody are bivalent, the molecular ratio approaching unity. The reported experimental values for this ratio (Table II) lie between 2 and $^3/_4$.

With great excess of antigen finite complexes which remain in solution are formed, with structures such as shown in Fig. 6. For these the molecular ratio varies between 1 and the minimum value $^1/_2$. It is observed that in general no precipitate forms in the region of great antigen excess, and Heidelberger and his collaborators have in fact assigned values 1 and $^1/_2$ to the molecular ratios for the complexes in solution.

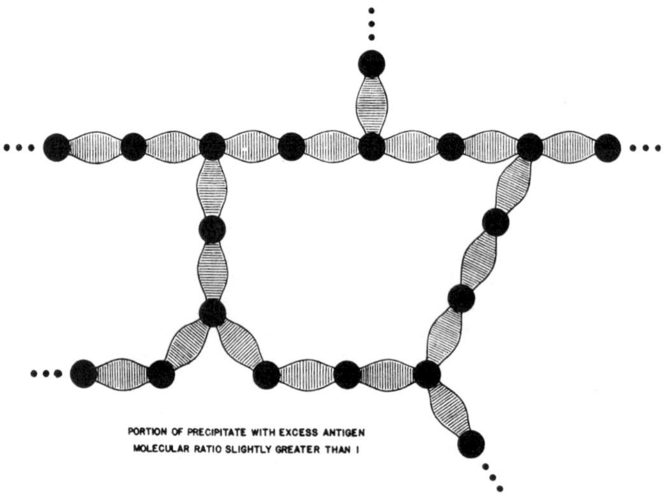

Fig. 5.—A portion of the network formed in the region of antigen excess.

Whereas precipitation is inhibited by antigen excess, it usually occurs even with great antibody excess, although soluble complexes with molecular ratio N and the structure shown in Fig. 7 are expected to exist. It seems probable that the difference in behavior of systems in the excess antigen region and excess antibody region is to be attributed to the fact that the molecular ratio for precipitate and soluble complex differs by a factor as great as two for the former case, and by only $N/(N-1)$ for the latter.

d. The Use of a Single Antigen Molecule as the Template for an Antibody Molecule.—There are two ways in which an antibody molecule with two opposed active regions complementary to the antigen might be produced. One is the way described in Section II. The other would involve the manufacture of the antibody molecule in its final configuration between two antigen molecules, one of which would serve as the pattern for one antibody end and the other for the second. No attempt, to decide between these alternatives seems to have been made before; there exists

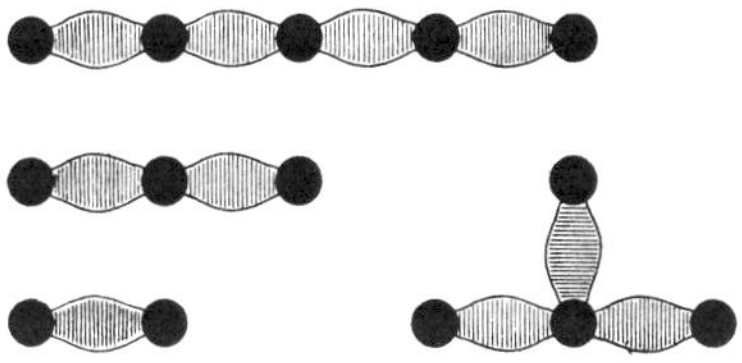

Fig. 6.—Representative soluble complexes formed with excess antigen.

evidence, however, some of which is mentioned below, to indicate that the first method of antibody production, involving only one antigen molecule, occurs predominantly. It is for this reason that I have developed the rather complicated theory described above, with the two end portions of the antibody forming first, one (or both) then separating from the antigen, and the central part of the antibody then assuming its shape and holding the active ends in position for attachment to two antigen molecules.

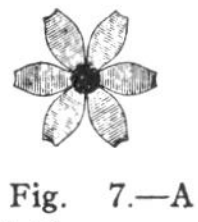

Fig. 7.—A soluble complex formed with excess antibody.

This theory requires that the formation of antibody be a reaction of the first order with respect to the antigen, whereas the other alternative would require it to be of the second order. There exists very little evidence as to whether on immunization with small amounts of antigen the antibody production is proportional to the amount of antigen injected or to its square. Some support for the one-antigen-molecule theory is provided by the experiments dealing with the injection of a mixture of antigens. If two antigen molecules were required for antibody formation, it would be expected that antibodies A′–B′, A′–C′, B′–C′, · · · complementary to two different antigens A and B, A and C, B and C, · · · as well as those A′–A′, B′–B′, C′–C′, · · · complementary to a single antigen would be formed. The available evidence speaks strongly against this. Thus Dean, Taylor, and Adair[20] have reported that the serum produced by immunization with a mixture of egg albumin and serum albumin contains distinct antibodies homologous to the two antigens, and that precipitation with one antigen leaves the amount of the heterologous antibody unaltered. An even more rigorous demonstration was furnished by Heidelberger and Kabat,[21] who, from the serum of a cow which had been injected with types I, II, and III pneumococci, isolated in succession, with the corresponding specific polysaccharide, the three anticarbohydrates, each in an apparently pure state and with no appreciable cross-reactivity as to pneumococcus type. In another striking experiment Hektoen and Boor[22] found that a serum obtained on injecting a rabbit with a mixture of 35 antigens reacted with 34 of the antigens, and that absorption with any one had in the main little effect on subsequent reaction with another. Since on the two-antigen-molecule theory the amount of antibodies A′–A′, B′–B′, · · · capable of causing precipitation with a single antigen would be small compared with the total amount of antibody (of the order of $1/n$, for n antigens—about 3% in this case), these qualitative observations provide significant evidence in favor of the alternative theory.

e. Criteria for Antigenic Power.—There has been extensive discussion of the question of what makes a substance an antigen, but no generally accepted conclusions have been reached. Our theory permits the formulation of the following reasonable criteria for antigenic activity:

1. The antigen molecule must contain active groups, capable of sufficiently strong interaction with the globulin chain to influence its configuration.

2. The configuration of the antigen molecule must be well-defined over surface regions large enough to give rise to an integrated antibody–antigen force sufficient to hold the molecules together.

3. The antigen molecule must be large enough to have two or more such surface regions, and in case that the antigenic activity depends upon a particular group the molecule must contain at least two of these groups. (This criterion applies to antibodies effective in the precipitin and agglutinin reactions and in anaphylaxis.)

These criteria are satisfied by substances known to have antigenic action. Many proteins, some carbohydrates with high molecular weight (bacterial polysaccharides, invertebrate glycogen[23]), and some lipids and carbohydrate–lipid complexes are antigenic. The simple chemical substances so far studied have been found to be inactive, except those which are capable of combining with proteins in the body. Non-antigenic

(20) H. R. Dean, G. L. Taylor and M. E. Adair, *J. Hyg.*, **35**, 69 (1935).

(21) M. Heidelberger and E. A. Kabat, *J. Exptl. Med.*, **67**, 181 (1938).

(22) L. Hektoen and A. K. Boor, *J. Infect. Diseases*, **48**, 588 (1931).

(23) D. H. Campbell, *Proc. Soc. Exptl. Biol. Med.*, **36**, 511 (1937); *J. Parasitol.*, **23**, 348 (1937).

substances have been reported to become antigenic when adsorbed on particles (Forssman antigen on kaolin[24]); in this case the particle with adsorbed hapten is to be considered the antigen "molecule" of our theory. I predict that relatively simple molecules containing two or more haptens will be found to be antigenic; experiments to test this prediction are now under way.

IV. A More Detailed Discussion of the Structure of Antibodies and Other Proteins

There has been gathered so far very little direct evidence regarding the detailed structure of protein molecules. Chemical information is compatible with the polypeptide-chain theory of protein structure, and this theory is also supported by the rather small amount of pertinent X-ray evidence.[25] It was pointed out some years ago[26] that the well-defined properties of native proteins require that their molecules have definite configurations, the polypeptide chain or chains in a molecule being coiled in a definite way and held in position by forces acting between parts of the chains. The phenomenon of denaturation involves the loss of configuration through the partial or complete uncoiling of the chains. Of the forces involved in the retention of the native configuration those described as hydrogen bonds are probably the most important. Our knowledge of the properties of the hydrogen bond has increased to such an extent during the past five years as to justify some speculation as to the nature of the stable configurations of protein molecules.

Hydrogen bonds can be formed by the peptide carbonyl and imino groups of a polypeptide chain, and also by the carboxy, amino, hydroxy, and other oxygen- and nitrogen-containing groups in the side chains of the amino-acid residues. In a stable configuration as many strong hydrogen bonds as possible will be present. One configuration in which all of the peptide carbonyl and imino groups are forming strong hydrogen bonds is that shown in Fig. 8. Here extended chains are bonded together to form a compact layer, with the side chains extending alternately above and below the plane of the layer (provided that the *levo* configuration is the only one represented by the amino-acid residues). This configuration has been assigned to β-keratin and other fibrous proteins by Astbury[27] on the basis of X-ray data. Although the structure has not been verified in detail by the analysis of the X-ray data, the agreement in the dimensions found experimentally for the pseudo-unit cell of β-keratin and those predicted from the complete structure determinations of glycine[28] and diketopiperazine[29] makes it very probable that the structure is essentially correct, with, however, the chains somewhat distorted from the completely extended configuration.[30]

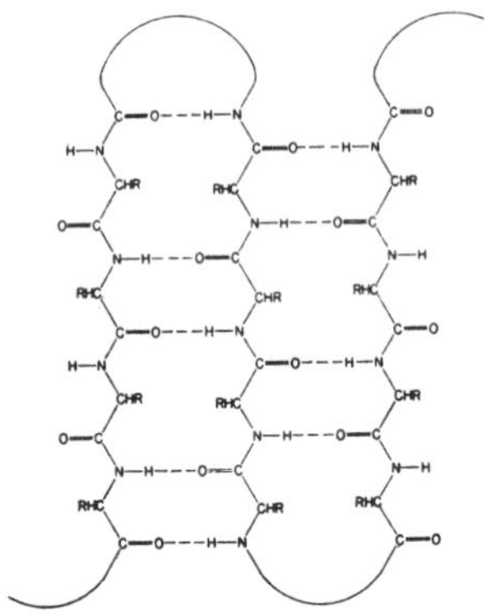

Fig. 8.—The folding of polypeptide chains into a layer held together by imino-carbonyl hydrogen bonds.

It is to be noted that the —NH—CO—CHR— sequence alternates in alternate lines in such a layer, so that a layer of finite size could be constructed by running a single polypeptide chain back and forth. A globular protein could then be made by building several such layers parallel to one another and in contact, like a stack of pancakes, the layers being held together by side-chain interactions as well as by the polypeptide chain itself. A protein molecule with, for example, roughly the shape and size of a cube 40 Å. on edge might contain four layers, each with about eight strings of about twelve residues each.

A few years ago I noticed, by studying molecular models, that a proline or hydroxyproline residue in the chain would interfere with the

(24) P. Gonzales and M. Armangué, *Compt. rend. soc. biol.*, **106**, 1006 (1931); K. Landsteiner and J. Jacobs, *Proc. Soc. Exptl. Biol. Med.*, **30**, 1055 (1933).

(25) A brief statement of the situation has been made by L. Pauling and C. Niemann, THIS JOURNAL, **61**, 1860 (1939); see also R. B. Corey, *Chem. Rev.*, **26**, 227 (1940).

(26) A. E. Mirsky and L. Pauling, *Proc. Nat. Acad. Sci.*, **22**, 439 (1936); H. Wu, *Chinese J. Physiol.*, **5**, 321 (1931).

(27) W. T. Astbury, *Trans. Faraday Soc.*, **29**, 193 (1933), and other papers.

(28) G. Albrecht and R. B. Corey, THIS JOURNAL, **61**, 1087 (1939).

(29) R. B. Corey, *ibid.*, **60**, 1598 (1938).

(30) R. B. Corey, ref. 25.

structure shown in Fig. 8 in such a way as to cause the chain to tend to turn through 180°; hence if these residues were suitably distributed along the chain during its synthesis the chain would tend to assume the configuration discussed above.

Layer structures other than this one might also be assumed, in which the chains are not extended. Some fibrous proteins, such as α-keratin, are known to have structures of this general type, but the nature of the folding of the chains has not yet been determined.

We have postulated the existence of an extremely large number of accessible configurations with nearly the same energy for the end parts of the globulin polypeptide chain. A layer structure, with variety in the type of folding in the layer, would not, it seems to me, give enough configurational possibilities to explain the great observed versatility of the antibody precursor in adjusting itself to the antigen, and I think that skew configurations must be invoked. But simple considerations show that it would be difficult for the chain to assume a skew configuration in which most of the peptide carbonyl and imino groups take part in forming hydrogen bonds, as they do in the layer structures; and in consequence the skew configurations would be much less stable than the layer configurations. The way out of this difficulty is provided by the postulate that *the end parts of the globulin polypeptide chains contain a very large proportion (perhaps one-third or one-half) of proline and hydroxyproline residues and other residues which prevent the assumption of a stable layer configuration.*

We may, indeed, anticipate that globular proteins may be divided into two main classes, comprising, respectively, those in which there is a layer structure and those in which the stable configuration of the chains is more complex; the latter may show the high proline and hydroxyproline content postulated for the globulin chain ends.

About one-third of the residues in gelatin are proline and hydroxyproline. We expect accordingly that there are many configurations with nearly the same energy accessible to a gelatin molecule, and that gelatin is not characterized by a single well-defined native molecular configuration.[31] Since the substance contains no strong antigenic groups, a definite configuration is a requisite for antigenic activity. These considerations thus provide a possible explanation of the well-known fact that gelatin is not effective as an antigen.[32]

Serological experiments with artificial conjugated antigens,[33] especially azoproteins, have provided results of great significance to the theory of antibody structure. Many of the arguments based on these results are presented in the books of Landsteiner and Marrack. The data obtained regarding cross-reactions of azoprotein sera with related azoproteins show that electrically charged groups (carboxyl, sulfonate, arsenate) interact strongly with homologous antibodies, and that somewhat weaker interactions are produced by hydrogen-bond-forming groups and groups with large electric dipole moments (hydroxy, nitro). The principal action of a weak group such as alkyl, phenyl or halogen is steric; this is shown clearly by the strong cross-reactions between similar chloro and methyl haptens. The data on specificity of antibodies with respect to haptens indicate strongly that the hapten group fits into a pocket in the antibody, and that the fit is a close one. It should not be concluded that all antibody–antigen bonds are of just this type; for example, the fitting of an antibody group into a pocket in the antigen may also be often of importance. Extensive work will be needed to determine the detailed nature of the antibody structures complementary to particular haptens and antigens.

V. Further Comparison of the Theory with Experiment. Possible Experimental Tests of Predictions

a. Methods of Determining the Valence of Antibodies.—The following methods may be proposed to determine the valence of antibodies. First, let a serum be produced by injection of an azoprotein of the following type: its hapten is to be sufficiently strong (that is, to interact sufficiently strongly with the homologous antibody) that one hapten group forms a satisfactory antibody–antigen bond, and the number of hapten groups per molecule is to be small enough so that in the main only one group will be present in the area serving as a pattern for an antibody end.

(31) Collagen has a definite fiber structure, as shown by X-ray photographs. A possible atomic arrangement has been suggested by W. T. Astbury and F. O. Bell, *Nature*, **145**, 421 (1940).

(32) Gelatin with haptens attached is antigenic (for references see Landsteiner, *loc. cit.*, p. 102, and C. R. Harington, *J. Chem. Soc.*, 119 (1940)). The presumption is that the haptens interact with the antibody so effectively that one hapten forms the antigen–antibody bond, and the lack of definite configuration of the gelatin is of no significance.

(33) K. Landsteiner and H. Lampl, *Biochem. Z.*, **86**, 343 (1918); K. Landsteiner, *ibid.*, **104**, 280 (1920).

The same hapten is then attached to another protein, and this azoprotein is precipitated with the serum. If it be assumed that the precipitate is valence-saturated, the ratio of hapten groups to antibody molecules in the precipitate gives the average valence of the antibody.

Data of essentially this sort, obtained with arsanilic acid as the hapten and the rabbit as the experimental animal, have been published by Haurowitz and his collaborators[34] in a paper reporting many interesting experiments. The equivalent weight of antibody per arsanilic acid residue was found to vary from 23,000 for antigens with very many attached haptens to 51,000 for those with only a few. The expected value for saturation of all haptens by bivalent antibody molecules is one-half the antibody molecular weight, that is, about 79,000. The low experimental values are probably due to failure of some haptens to combine with antibodies. In particular, if two haptens are attached to the same tyrosine or histidine residue steric interactions of antibodies may permit only one of the haptens to be effective.[35]

The bivalence of antibodies and our postulate that only one antigen molecule is involved in the formation of an antibody molecule require that a precipitin-effective antihapten be produced only if the injected antigen contain at least two hapten groups. Pertinent data have been obtained on this point by Haurowitz and his collaborators, who found that effective antihapten precipitin serum was produced by an azoprotein, made from arsanilic acid and horse globulin, containing 0.24% arsenic (4.8 haptens per average molecule of molecular weight 157,000), and a trace by one containing 0.13% arsenic (2.6 haptens per molecule).

b. The Possible Antigenic Activity of Simple Substances.—The criteria given above for antigenic activity would be satisfied by a substance of relatively low molecular weight in which several hapten groups (such as several arsanilic acid residues) are present in the molecule. A substance of this sort would be expected also to show the precipitin reaction with its own serum or with serum homologous to an azoprotein containing the same hapten, and to be capable also of producing anaphylaxis.

A substance with only two hapten groups per molecule might be expected not to give a precipitate with the homologous antibody, but rather to form long strings with antibody and antigen molecules alternating. These strings would remain in solution, and would confer on the solution the property of pronounced birefringence of flow. If, however, there were in one end of some of the antibody molecules complementary regions for two hapten groups, making these molecules effectively trivalent, the strings would be tied together and a precipitate would be formed. It is probably significant in this connection that Landsteiner and van der Scheer[36] have observed both the precipitin reaction and anaphylactic shock (in guinea pigs sensitized with the corresponding azoproteins) with azo dyes, such as resorcinoldiazo-*p*-suberanilic acid, $(OH)_2C_6H_2$-$(NNC_6H_4NHCO(CH_2)_6COOH)_2$, formed by coupling two anilic acid molecules with resorcinol. Landsteiner himself has explained these observations as resulting from the low solubility of the dyes, but the explanation advanced above seems more probable.

c. Experiments with Two or More Different Haptens in the Same Antigen.—We would predict that if there were used as an antigen a molecule to which several hapten groups A and several different hapten groups B were attached, the serum obtained would contain three kinds of bivalent antibodies, A′–A′, B′–B′, and A′–B′ (as well as the univalent antibodies A′– and B′–). Our picture of the process of formation of antibodies permits little or no correlation between the two chain ends of the globulin in the selection of surface regions of the antigen molecule to serve as templates, except that the two regions must not overlap. We accordingly predict that for n_A and n_B groups A and B, respectively, per antigen molecule the numbers of bivalent antibody molecules of different kinds in the serum would be

$$N_{A'-A'} = \alpha^2 n_A(n_A - 1) \quad (3a)$$

$$N_{A'-B'} = 2\alpha\beta n_A n_B \quad (3b)$$

$$N_{B'-B'} = \beta^2 n_B(n_B - 1) \quad (3c)$$

(34) F. Haurowitz, F. Kraus and F. Marx, *Z. physiol. Chem.*, **245**, 23 (1936).

(35) From the data discussed above, which in our opinion indicate that two hapten groups combine with a bivalent antibody molecule, Haurowitz drew the different conclusion that one arsenic-containing group in the antigen combines with one antibody molecule. He reached this result by assuming 100,000 (rather than 157,000) for the molecular weight of the antibody and by assuming that the active group in the antigen consists of two haptens attached to a tyrosine or histidine residue. It is, of course, likely that this occurs in antigens with high arsenic content, but it seems probable that the haptens are mainly attached to separate residues in the antigens containing only a few haptens per molecule.

(36) K. Landsteiner and J. van der Scheer, *Proc. Soc. Exptl. Biol. Med.*, **29**, 747 (1932); *J. Exptl. Med.*, **56**, 399 (1932).

in which α and β are coefficients which give the probabilities that the groups serve as templates. It is seen that *the amount of A'–B' antibodies is predicted to be equal to or slightly greater than twice the geometric mean of the amounts of A'–A' and B'–B'*. The only data permitting a quantitative test of this relation which have come to my attention are those obtained by Haurowitz and collaborators (*loc. cit.*) by use of azoproteins made from arsanilic acid. A serum produced by injecting rabbits with an antigen made from arsanilic acid and horse globulin was found to contain antibodies to the arsanilic hapten (precipitating an azoprotein made from arsanilic acid and rabbit globulin), others to horse globulin itself, and still others to the homologous antigen, these last presumably being complementary in structure both to the hapten and to the active groups of horse globulin. The quantitative results obtained are the following: 10 ml. of immune serum gave 17–18 mg. of precipitate with the maximal precipitating amount of azoprotein from arsanilic acid and rabbit globulin, and 8 mg. of precipitate with horse globulin, these amounts being independent of the order of the two precipitations. After exhaustion with these two antigens the serum gave 17 mg. of precipitate with the homologous antigen. If we assume that the conditions of each precipitation were such that only suitable bivalent antibody molecules were incorporated in the precipitate, the equations above would require the third precipitate to weigh about 24 mg.; the experiment accordingly provides some support for the theory. The quantitative discrepancy may possibly be due to the incorporation of some effectively univalent antibody molecules in the first two precipitates.

The qualitative experimental results which have been reported are in part compatible and in part incompatible with the theory. The most interesting of the experiments are those of Landsteiner and van der Scheer,[37] who prepared azoproteins containing two different kinds of haptens and studied the antibodies produced by them. In some cases only one of the haptens was effective in antibody formation. With the azoprotein made from 3-amino-5-succinylaminobenzoyl-*p*-aminophenylarsenic acid, however, a serum was obtained which would combine not only with the homologous antigen but also with azoproteins formed either from *m*-aminosuccinanilic acid or from *p*-aminophenylarsenic acid. After the serum was exhausted by interaction with stromata coupled with either one of these two simple haptens, it reacted as strongly (as measured by the estimated amount of precipitate) with the azoprotein containing the other simple hapten as before exhaustion. From this experiment and others the investigators concluded that there appeared to be present in the sera, if any, only small amounts of antibodies with two combining groups capable of interaction with the two different haptens in the antigen. It seems possible that the conclusion is not justified by the data, and, with the kind coöperation of Dr. Landsteiner, we are continuing this investigation.

d. The Antigenic Activity of Antibodies.— Our picture of an antibody molecule requires that the configuration of its middle portion be the same as that of normal serum globulin. Hence antibodies should have antigenic activity, with essentially complete cross-reactions with normal globulin. This is in agreement with experiment; Landsteiner and Prasek[38] found that precipitins which precipitated the serum of an animal precipitated also the agglutinins in the serum, and Eisler[39] showed that a precipitin to horse serum would precipitate tetanus antitoxin. On the other hand, according to our picture the active end regions of the antibody molecules would not have effective antigenic power, since their configurations would be different from molecule to molecule (depending on the accidentally selected template region and accidental way of coiling), and an antibody complementary to one antibody end would as a rule not combine with another. The antibody ends would hence in the main be left free in the precipitate formed by an antibody and its precipitin, as well as in that formed by an antibody and the precipitin to normal globulin. In agreement with this, Smith and Marrack[40] found that a precipitate formed by a precipitin and a serum containing diphtheria antitoxin has the power of combining with diphtheria toxin; and similar results have also been obtained recently by Heidelberger and Treffers[41] in the case of specific precipitates formed by pneumococcus antibody with homologous antiserum.

(37) K. Landsteiner and J. van der Scheer, *J. Exptl. Med.*, **67**, 709 (1938).

(38) K. Landsteiner and E. Prasek, *Z. Immunitäts*, **10**, 68 (1911).

(39) M. Eisler, *Zentr. Bakt. Parasitenk. Infekt.*, **84**, 46 (1920).

(40) F. C. Smith and J. Marrack, *Brit. J. Exptl. Path.*, **11**, 494 (1930).

(41) Personal communication of unpublished material by Professor M. Heidelberger and Dr. H. P. Treffers.

e. Factors Affecting the Rate of Antibody Production and the Specificity of Antibodies.— In order that an antibody be effective, the surface region of the antigen covered by an antibody end must be large enough so that the integrated attractive forces constitute an antibody–antigen bond of significant strength. With a bond of average strength the antibody molecule will be dissociated from its antigen template, perhaps with the aid of the auxiliary mechanism mentioned in Section VI, until an equilibrium or steady-state concentration is built up in the serum. Now if the antigen surface contains a large number of strong groups, capable of interacting strongly with complementary structures in the antibody, the antibody–antigen bond may be so strong that the antibody is not able to separate itself from the antigen, and only a small antibody concentration can be built up in the serum. We hence conclude that an *antigen containing weak groups will in general be a good antigen, whereas one containing many strong groups will be a poor antigen, with respect to antibody production.*

This prediction, which at first thought seems paradoxical, is in fact borne out by experiment. Thus a bland protein such as egg albumin is a good antigen, as are also conjugated proteins with weak groups attached. An azoprotein with many strong groups attached (arsenic acid, sulfonic acid, nitro, etc.; the azo group itself is a rather strong group, capable of forming hydrogen bonds) is a poor antigen; in order to obtain serum to such haptens an azoprotein containing a limited number of the groups must be used. I am told by Dr. Landsteiner that this observation was made in the early days of the study of azoproteins.[42] Pertinent data have been reported in recent years by Haurowitz and his collaborators (*loc. cit.*), who found the optimum arsenic content for the production of antihapten by azoprotein made from arsanilic acid to be between 0.5 and 1.0%; very little antibody is produced by antigens with over 2% arsenic, although strong precipitin reaction is shown by azoprotein with an arsenic content as great as 10%.

A second deduction, relating to specificity, can also be made. To achieve a sufficiently strong antibody–antigen bond with an antigen containing only weak groups a large surface region of the antigen must come into play, whereas with an antigen containing strong groups only a small region (in the limit one group) is needed. Hence *antibodies to antigens containing strong groups show low specificity, and those to antigens containing weak groups show high specificity.* This prediction is substantiated by many observations. Egg albumin, hemoglobin, and similar proteins give highly specific sera, whereas azoproteins produce sera which are less specific, strong cross-reactions being observed among various proteins with the same hapten attached. This shows, indeed, that a single hapten group gives a sufficiently strong bond to hold antibody and antigen together. In such a case the approximation of the antibody to a strong hapten is very close, and great specificity is shown with regard to the hapten itself, this specificity being the greater the stronger the hapten. Many examples of these effects are to be found in Landsteiner's work.

f. The Effect of Denaturing Agents.—We made the fundamental postulate that the end parts of the polypeptide chains of the globulin molecule are characterized by having a very large number of accessible configurations with nearly the same energy, whereas there is only one stable configuration for the central part. It is accordingly probable that the end configurations, giving characteristic properties to the antibodies, would be destroyed before the central part of the molecule is affected and, moreover, that the sensitivity to denaturing agents or conditions of antibodies to different antigens would be different. The available meager experimental information seems to be compatible with these ideas.[43]

Some remarks may be made regarding the difference in behavior of antibodies and antigens in the presence of denaturing agents. An antigen molecule may undergo a considerable change in configuration without losing completely its power of reacting with the homologous serum; if some of the surface regions remain essentially unchanged after partial denaturation of the protein, the antibody molecules complementary to these regions will retain the power of combining with them, whereas the antibody molecules complementary to the regions which have been greatly changed by denaturation will no longer be effective. In particular some native proteins may be built of superimposed layers, as described in Section IV. The antigenic regions on top of the top layer and on the bottom of the bottom layer would still be effective after the partial denatura-

(42) See K. Landsteiner and H. Lampl, *loc. cit.*

(43) See Marrack, *loc. cit.*, pp. 48–53.

tion of the molecule by the unleafing of the layers, whereas the antigenic regions at the sides of the original molecule would in large part lose their effectiveness by this unleafing. The observation by Rothen and Landsteiner[44] that egg albumin spread into surface films 10 Å. thick retains the ability to combine with anti-egg-albumin rabbit serum is most simply explained by the assumptions that the native egg albumin molecule has the layer structure suggested above and that the process of surface denaturation of this molecule involves the unleafing of the layers without the loss of their structure.

As mentioned above, it is probable that for most antibodies the end regions are affected by denaturing agents more easily than the central region, and that the first step in denaturation of an antibody involves these end regions and leads to loss of their specific properties. It has been shown by Danielli, Danielli and Marrack[45] that the reactivity of antibodies is destroyed by surface denaturation.[46]

An interesting possible method of producing antibodies from serum or globulin solution outside of the animal is suggested by the theory. The globulin would be treated with a denaturing agent or condition sufficiently strong to cause the chain ends to uncoil; after which this agent or condition would be removed slowly while antigen or hapten is present in the solution in considerable concentration. The chain ends would then coil up to assume the configurations stable under these conditions, which would be configurations complementary to those of the antigen or hapten.

Many of the experiments suggested above are being undertaken in our Laboratories, with the collaboration of Dr. Dan Campbell.

VI. Processes Auxiliary to Antibody Formation

It seems not unlikely that certain processes auxiliary to antibody formation occur. The reported increase in globulin (aside from the antibody fraction) after immunization suggests the operation of a mechanism whereby the presence of antigen molecules accelerates the synthesis of the globulin polypeptide chains. There is little basis for suggesting possible mechanisms for this process at present.

The occurrence of the anamnestic reaction—the renewed production of antibodies to an antigen caused by injection of a second antigen—may be explained by the assumption that following the synthesis of an antibody a mechanism comes into operation in the cell to facilitate the removal of the antibody from the antigen, perhaps by changing the hydrogen-ion or salt concentration or dielectric constant. This would assist in removing antibodies not only from the second antigen but also from those molecules of the first antigen which had remained, covered with homologous antibody attached too firmly for spontaneous removal, in the cell. The evidence indicates that the anamnestic reaction is not in general strong. At a time after inoculation with typhoid bacillus or erythrocytes long enough that the corresponding agglutinins are no longer detectable in the serum injection of another antigen gives rise to the presence of these agglutinins in amounts detectable by the very sensitive agglutination test; but Kabat and Heidelberger[47] found that the amount of additional antibody to serum albumin produced by injection of egg albumin or typhoid toxin was too small to be detected by their method of analysis.

The mechanism for catching the antibody molecule and holding it in the region of globulin synthesis may be closely related to that of antibody production—possibly a partially liberated globulin chain which forms a bond or two bonds with an antigen molecule directly above it is prevented from freeing its central part from the cell wall, and so serves as an anchor.

The renewed production of antibody in the serum after bleeding is to be attributed to the presence of trapped antigen molecules in the cells. The greater duration of active than of passive immunization may be attributed to this or to the presence of complexes of antigen and surrounding antibodies, the outer ends of which could combine with additional antigen.

Acknowledgments.—My interest in immunology was awakened by conversations with Dr. Karl Landsteiner; I am glad to express my gratitude to him, and to acknowledge my indebtedness to him for ideas as well as for facts. I wish also to thank Professors Michael

(44) A. Rothen and K. Landsteiner, *Science*, **90**, 65 (1939).

(45) J. F. Danielli, M. Danielli and J. R. Marrack, *British J. Exptl. Path.*, **19**, 393 (1938).

(46) Rothen and Landsteiner (*loc. cit.*) have pointed out that from these facts regarding surface denaturation the conclusion can be drawn that "the specific reactivity of antibodies is to a large extent dependent upon structures different from those which mainly determine the specificity of antigens."

(47) E. A. Kabat and M. Heidelberger, *J. Exptl. Med.*, **66**, 229 (1937).

Heidelberger and Dan Campbell for advice and assistance.

Summary

It is assumed that antibodies differ from normal serum globulin only in the way in which the two end parts of the globulin polypeptide chain are coiled, these parts, as a result of their amino-acid composition and order, having accessible a very great many configurations with nearly the same stability; under the influence of an antigen molecule they assume configurations complementary to surface regions of the antigen, thus forming two active ends. After the freeing of one end and the liberation of the central part of the chain this part of the chain folds up to form the central part of the antibody molecule, with two oppositely-directed ends able to attach themselves to two antigen molecules.

Among the points of comparison of the theory and experiment are the following: the heterogeneity of sera, the bivalence of antibodies and multivalence of antigens, the framework structure and molecular ratio of antibody–antigen precipitates, the use of a single antigen molecule as template for an antibody molecule, criteria for antigenic activity, the behavior of antigens containing two different haptens, the antigenic activity of antibodies, factors affecting the rate of antibody production and the specificity of antibodies, and the effect of denaturing agents. It is shown that most of the reported experimental results are compatible with the theory. Some new experiments suggested by the theory are mentioned.

PASADENA, CALIFORNIA RECEIVED JUNE 25, 1940

[Reprinted from the Journal of the American Chemical Society, **64**, 2994 (1942).]

[CONTRIBUTION FROM THE GATES AND CRELLIN LABORATORIES OF CHEMISTRY, CALIFORNIA INSTITUTE OF TECHNOLOGY, No. 885]

The Serological Properties of Simple Substances. I. Precipitation Reactions between Antibodies and Substances Containing Two or More Haptenic Groups

BY LINUS PAULING, DAVID PRESSMAN, DAN H. CAMPBELL, CAROL IKEDA, AND MIYOSHI IKAWA

The study of serological precipitation reactions is complicated by the fact that ordinarily these reactions involve two proteins, the antigen and the antibody. The understanding of these reactions was greatly advanced by the introduction into their study of precise microanalytical methods and a further simplification involving the use of a nitrogen-free multivalent hapten of pneumococcus polysaccharide.[1] A few years ago it was reported by Landsteiner and van der Scheer[2a] that the precipitin reaction and anaphylaxis could be produced by simple substances formed by coupling two haptenic groups with resorcinol or tyrosine, in place of the azoprotein (containing the same haptenic group) which has been used as the antibody-producing antigen. Landsteiner[2b] suggested that "the ready precipitability of these

(1) M. Heidelberger and F. E. Kendall, *J. Exptl. Med.*, **50**, 809 (1929): **61**, 559, 563 (1935).

(2) (a) K. Landsteiner and J. van der Scheer, *Proc. Soc. Exptl. Biol. Med.*, **29**, 747 (1932); *J. Exptl. Med.*, **56**, 399 (1932); **57**, 633 (1933); **67**, 79 (1938). (b) K. Landsteiner, "The Specificity of Serological Reactions," Charles C Thomas, Baltimore, Md., 1936, p. 120.

dyes is dependent upon peculiarities in constitution which, like those of fatty groups, diminish solubility in water and favor the formation of colloidal solutions." However, it seemed probable to us, on the basis of a theory of the structure of antibodies and the nature of the precipitin reaction,[3] that simple substances containing two or more haptenic groups would react with antibodies in essentially the same way as the homologous protein antigens containing the same haptenic groups; we accordingly prepared a half-dozen simple substances of this sort, each with two or more phenylarsonic acid groups per molecule, and observed each to precipitate antisera homologous to phenylarsonic acid azoprotein,[4] thus obtaining evidence for the generality of the phenomenon discovered by Landsteiner and van der Scheer.

The problem of obtaining from precipitation experiments evidence about the structure of antibodies and the nature of serological reactions is obviously greatly simplified by the replacement of protein antigens by simple substances of known structure. For this reason we began and are carrying on an extensive program of investigation of the reactions of simple substances with antisera. In this paper we report the quantitative study of the precipitin reaction for twenty simple substances containing two or more haptenic groups, and the results of tests of seven substances containing one group. It is found that the observations support the framework theory of serological precipitates.[5]

Discussion of Experimental Methods

Simple Antigens.—The simple antigens and haptens used in the investigation are listed in Table I; methods of preparations of these substances and the intermediates used are described in the following section.

Protein Antigens.—The immunizing antigens used for inoculations were made from diazotized arsanilic acid and sheep serum by the method described by Landsteiner and van der Scheer.[6] The ratio of arsenic to protein in these antigens ranged from 2 to 3%.

Test antigens were similarly made from purified ovalbumin by treatment with diazotized arsanilic acid, diazotized *p*-(*p*-aminophenylazo)-phenylarsonic acid, or diazotized *p*-aminobenzanilide-*p*'-arsonic acid. The azo-ovalbumins contained, respectively, 0.16, 4.0, and 2.0% arsenic.

Arsanilic acid was diazotized in hydrochloric acid solution by the addition of sodium nitrite solution at 0° to the starch–iodide end-point. The diazotizations of *p*-(*p*-aminophenylazo)-phenylarsonic acid and of *p*-aminobenzanilide-*p*'-arsonic acid were similarly carried out at 10° with end-point the disappearance of the slightly soluble amine hydrochlorides.

Preparation of Antisera.—Twenty-five rabbits were injected intraperitoneally or intravenously with 1- or 2-ml. portions of the atoxylazo-sheep-serum antigen described above, containing 0.5% protein. Several weekly courses of 3 to 5 injections were given, with intervening rest periods of a week or more. The rabbits were bled from the ear on the eighth, ninth, and tenth days after the last injection, 40 ml. of blood being taken from each rabbit each day The blood was permitted to clot, and the antisera were pooled according to titer. The courses of injections and subsequent bleedings were repeated to obtain more pools of serum.

A measure of the total amount of antibody homologous to the atoxyl hapten (the phenylarsonic acid group) was made by determining the maximum amounts of antibody precipitated by the azo-ovalbumin test antigens. The most effective test antigen, that made from *p*-(*p*-aminophenylazo)-phenylarsonic acid, precipitated 2 mg. of antibody per ml. of antiserum A, 4 mg. per ml. of B, 1.5 mg. per ml. of C, and 4 mg. per ml. of D.

The Reaction of Antigen and Antiserum.—The precipitation tests were carried out by mixing portions of undiluted antiserum, usually 2 ml., with equal volumes of saline solution containing dye; usually four to six dye concentrations were tested, differing by powers of 2. The tubes were allowed to stand for one hour at room temperature and then overnight in the refrigerator. The precipitates were then centrifuged down, washed with three or four 10-ml. portions of normal saline, and analyzed for nitrogen. In some experiments colorimetric determinations were made of the amount of dye in the redissolved precipitates. In the tests with serum A the customary visual estimates of cloudiness were made one-half hour after the solutions were mixed.

Methods of Analysis.—Analyses for nitrogen were made by the semimicro Kjeldahl method, using the apparatus described by Redemann.[7] Sulfuric acid, copper sulfate, and potassium sulfate were used in the digestion mixture; hydrogen peroxide was found not to be needed.

The arsenic determinations were carried out by the method of Haurowitz and Breinl.[8] Carbon and hydrogen analyses were made with the usual semimicro technique. Colorimetric determinations of dye and azoprotein were made with a Klett photoelectric colorimeter after dissolving the precipitates in a few drops of 2 *N* sodium carbonate solution.

The reported values of antibody in precipitates are the values of antibody nitrogen multiplied by the factor 6.25, the antibody nitrogen being the difference between total nitrogen in the precipitate and antigen nitrogen calculated

(3) L. Pauling, THIS JOURNAL, **62**, 2643 (1940). J. R. Marrack and F. C. Smith, *Brit. J. Exptl. Path.*, **13**, 394 (1932), had made the tentative suggestion that precipitation by azohaptens depends upon the presence in the molecule of two or more haptenic groups.

(4) L. Pauling, Dan H. Campbell and D. Pressman, *Proc. Nat. Acad. Sci.*, **27**, 125 (1941).

(5) R. J. Marrack, "The Chemistry of Antigens and Antibodies," His Majesty's Stationery Office, London, 1938; M. Heidelberger, *Chem. Rev.*, **24**, 323 (1939); *Bact. Rev.*, **3**, 49 (1939); L. Pauling, ref. 3.

(6) K. Landsteiner and J. van der Scheer, *J. Exptl. Med.*, **55**, 781 (1932).

(7) C. E. Redemann, *Ind. Eng. Chem., Anal. Ed.*, **11**, 635 (1939).

(8) F. Haurowitz and F. Breinl, *Z. physiol. Chem.*, **205**, 259 (1932).

TABLE I

SUBSTANCES USED IN PRECIPITATION TESTS

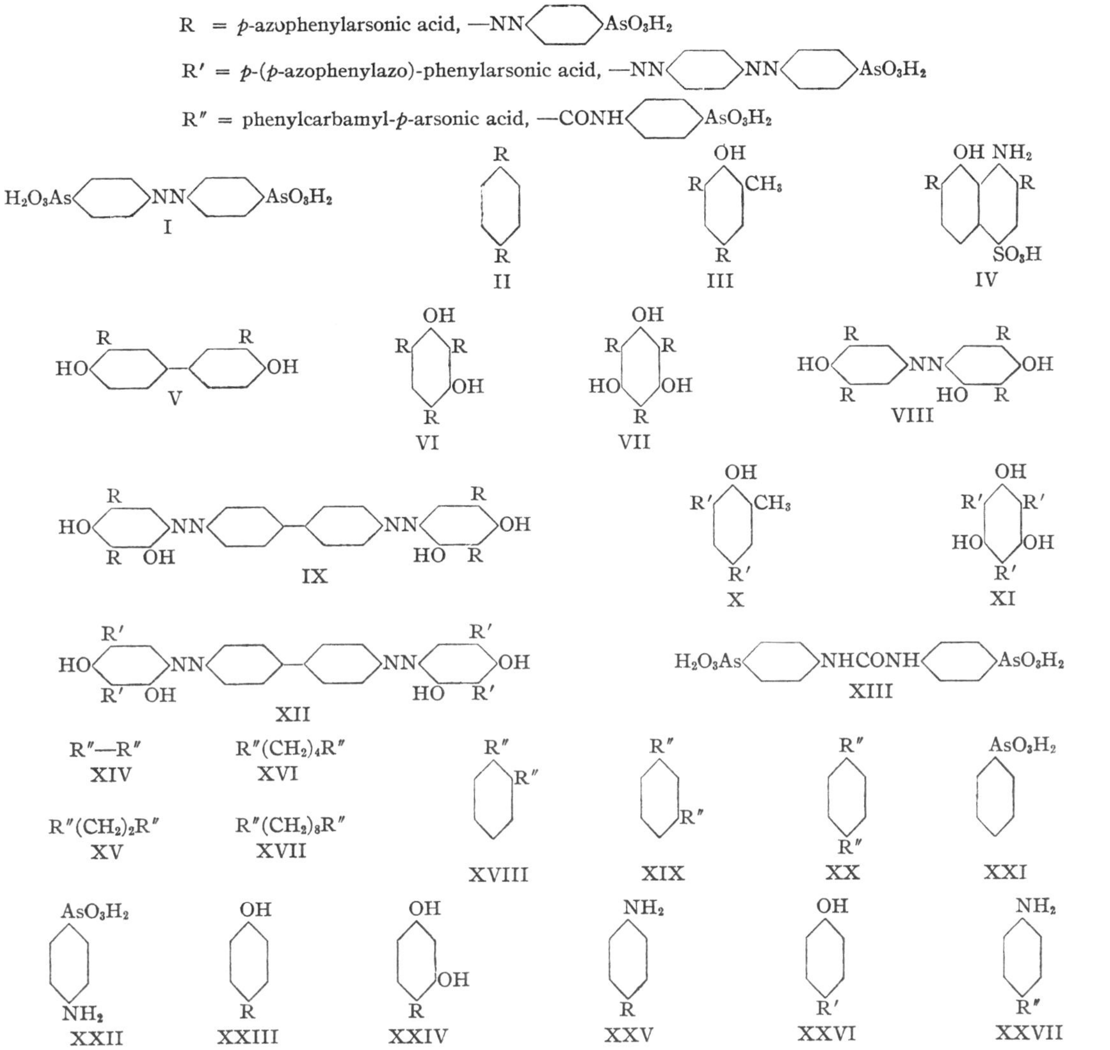

from the amount of antigen as determined colorimetrically (for simple antigens this correction is very small).

The Preparation of Compounds

XXI, Phenylarsonic acid was prepared by Mr. David Brown by the Bart reaction.[9]

XXII, Arsanilic acid was prepared by the method of Bechamp.[10]

XXV, *p*-(*p*-Aminophenylazo)-phenylarsonic acid was made (a) by the hydrolysis in 2 *N* sodium hydroxide of the acetyl derivative made by condensing *p*-nitrosophenylarsonic acid with *p*-aminoacetanilide (20% excess) in glacial acetic acid by refluxing for three hours, and (b) by the hydrolysis for twenty minutes in boiling 1 *N* sodium hydroxide of the ω-methylsulfonate formed by reaction in 0.3 *N* sodium carbonate of diazotized arsanilic acid and aniline-ω-methylsulfonate (20% excess).[11] The two products, purified as the sodium salts, appeared to be identical.

Anal. Calcd. for $C_{12}H_{11}O_3N_3AsNa$: C, 42.04; H, 3.12. Found: (a) C, 41.98; H, 3.23; (b) C, 42.12; H, 3.40.

Nitrosophenylarsonic acid was made by the method of Karrer[12] from arsanilic acid and Caro's acid.

I, Azobenzene-*p*,*p*′-diarsonic acid was prepared by the method of Karrer[12] from *p*-nitrosophenylarsonic acid and arsanilic acid and was purified by repeated precipitation with acid from alkaline solution.

Anal. Calcd. for $C_{12}H_{12}O_6N_2As_2$: C, 33.50; H, 2.79 Found: C, 33.50, 33.56; H, 2.83, 2.97.

(9) H. Gilman, "Organic Syntheses," John Wiley and Sons, New York, N. Y., 1935, Vol. XV, p. 59.

(10) *Ibid.*, 1932, Coll. Vol. I, p. 63.

(11) F. G. Pope and W. I. Willett, *J. Chem. Soc.*, 1259 (1913); H. Bucherer and A. Schwalbe, *Ber.*, **39**, 2798 (1906).

(12) S. Karrer, *ibid.*, **45**, 2066, 2376 (1912).

TABLE II

Compound	Reactant	Excess of diazo compound	Formula	Analyses, % Calcd. C	Calcd. H	Found C	Found H	Color in alkali	Color in H_2SO_4
III	*o*-Cresol	50%	$C_{19}H_{18}O_7N_4As_2$	40.07	3.09	40.17	3.20	Orange	Purple
IV	1-Amino-8-naphthol-4-sulfonic acid	100%	$C_{22}H_{19}O_{11}N_5As_2S$					Purple	Green
V	*p*,*p*-Dihydroxybiphenyl	150	$C_{24}H_{20}O_8N_4As_2$	44.85	2.74	45.43	2.93	Light yellow	Light yellow
						45.47	2.83		
VI	Resorcinol	33	$C_{24}H_{21}O_{11}N_6As_3$	36.27	2.67	36.52	2.62	Orange	Pink
						36.67	2.62		
VII	Phloroglucinol	33	$C_{24}H_{21}O_{12}N_6As_3$	35.55	2.61	35.58	2.91	Yellow	Pink
						35.65	2.91		
VIII	2,4,4′-Trihydroxyazobenzene	30	$C_{36}H_{30}O_{15}N_{10}As_4$	37.85	2.64	38.62	2.98	Brown	Pink
						38.80	3.18		
IX	4,4′-Bis-(azo-2,4-dihydroxy)-biphenyl	25	$C_{48}H_{38}O_{16}N_{12}As_4$	43.07	2.86	43.17	3.40	Brown	Purple
X	*o*-Cresol	50	$C_{31}H_{26}O_7N_8As_2$	48.21	3.37	48.42	3.64	Violet-red	Purple
						48.27	3.64		
XI	Phloroglucinol	33	$C_{42}H_{33}O_{12}N_{12}As_3$	44.92	2.96	44.71	2.96	Violet	Blue
XII	4,4′-Bis-(azo-2,4-dihydroxy)-biphenyl	10	$C_{72}H_{54}O_{16}N_{30}As_4$	49.27	3.10	49.85	3.66	Brown	Violet
XXIII	Phenol	20[a]	$C_{12}H_{11}O_4N_2As$	44.60	3.42	44.68	3.42	Yellow	Yellow
XXIV	Resorcinol	100[a]	$C_{12}H_{11}O_5N_2As$	42.59	3.28	42.65	3.26	Orange	Yellow
XXVI[b]	Phenol	20[a]	$C_{18}H_{14}O_4N_4As$-Na	48.20	3.16	48.20	3.42	Red-orange	Blue-violet

[a] Excess of phenol (per cent.). [b] Sodium salt.

II, *p*-Di-(*p*-azophenylarsonic acid)-benzene was similarly made from *p*-nitrosophenylarsonic acid and *p*-phenylenediamine and similarly purified.

Anal. Calcd. for $C_{18}H_{16}O_6N_4As_2$: C, 40.46; H, 2.94. Found: C, 38.85, 38.81; H, 3.12, 3.19.

III, 2-Methyl-4,6-di-(*p*-azophenylarsonic acid)-phenol; IV, 1-Amino-2,7-di-(*p*-azophenylarsonic acid)-4-sulfo-8-naphthol; V, 3,3′-Di-(*p*-azophenylarsonic acid)-4,4′-dihydroxybiphenyl; VI, 1,3-Dihydroxy-2,4,6-tri-(*p*-azophenylarsonic acid)-benzene; VII, 1,3,5-Trihydroxy-2,4,6-tri-(*p*-azophenylarsonic acid)-benzene; VIII, 2,4,4′-Trihydroxy-3,5,3′,5′-tetra-(*p*-azophenylarsonic acid)-azobenzene; IX, 4,4′-Bis-(azo-2,4-dihydroxy)-3,5-di-(*p*-azophenylarsonic acid)-biphenyl; X, 2-Methyl-4,6-di(*p*-(*p*-azophenylazo)-phenylarsonic acid)-phenol; XI, 1,3,5-Trihydroxy-2,4,6-tri-(*p*-(*p*-azophenylazo)-phenylarsonic acid)-benzene; XII, 4,4′-Bis-(azo-2,4-dihydroxy)-3,5-di-(*p*-(*p*-azophenylazo)-phenylarsonic acid)-biphenyl; XXIII, *p*-(*p*-Azophenylarsonic acid)-phenol; XXIV, 1,3-Dihydroxy-4-(*p*-azophenylarsonic acid)-benzene; XXVI, *p*-(*p*-(*p*-azophenylazo)-phenylarsonic acid)-phenol.—Compounds III to XII, XXIII, XXIV, and XXVI were made by coupling diazotized arsanilic acid or diazotized *p*-(*p*-aminophenylazo)-phenylarsonic acid with the appropriate phenolic nucleus in dilute sodium carbonate solution or (for XXVI) in sodium acetate–acetic acid solution. For III, X, and XI pyridine was added[13] in amount about 10% of the volume of the reaction mixture. The reaction mixtures were allowed to stand for a few days (1 to 4) and the products were then precipitated with hydrochloric acid and purified by repeated solution in dilute sodium hydroxide and reprecipitation with acid. Compounds X, XI, and XII were in addition purified by dialysis through Visking sausage casing against dilute borax solution with *p*H 10; this membrane permits passage of molecules containing only one haptenic group R′ but not those containing two or more of these groups. The compounds were washed free of sodium chloride and dried *in vacuo*. Experimental details, analytical results, and colors of solutions in alkali and in concentrated sulfuric acid are given in Table II.

A chromatographic method of analysis of the purity of these compounds was developed and applied by Mr. A. Pardee. The column packing used was a mixture of 30% Celite and 70% Neutrol Filtrol. A dilute solution of the dye was poured into the packed column and the chromatogram was developed with phosphate buffer of *p*H varying from 8 to 12 in different cases. The rate of passage of the dyes was greater the higher the *p*H. Compounds VI, VII, X, XI, XXIII, XXIV, and XXVI appeared to be quite pure, having at most only a slight trace of a second band. On the other hand, compounds I, II, III, V, VIII, IX, and XII contain appreciable colored impurity, as shown by the appearance of more than one band on the column. Compounds V, VIII, IX, and XII might be expected to be impure from the large number of steps in their preparation. The impurity in compound II was identified as XXV from a mixed chromatogram.

2,4,4′-Trihydroxyazobenzene was prepared by coupling diazotized *p*-aminophenol with resorcinol (100% excess) in the presence of sodium hydroxide. The product was purified by dissolving it in sodium hydroxide and reprecipitating with acid and finally by two crystallizations from 70% alcohol.

Anal. Calcd. for $C_{12}H_{10}O_3N_2$; C, 62.60; H, 4.38. Found: C, 62.39; H, 4.46.

4,4′-Bis-(azo-2,4-dihydroxy)-biphenyl was prepared by adding bisdiazotized benzidine to resorcinol (100% excess) in a sodium acetate buffered solution. The mixture finally was made alkaline with sodium hydroxide. The product was purified by washing.

Anal. Calcd. for $C_{24}H_{18}O_4N_4$; C, 67.60; H, 4.25. Found: C, 66.37; H, 4.62.

XXVII, *p*-Aminobenzanilide-*p*′-arsonic acid was prepared by the method of King and Murch[14] by the hydrolysis of the carbethoxyamino compound obtained by coupling *p*-arsanilic acid with *p*-carbethoxyaminobenzoyl chloride.

(13) K. H. Saunders, "The Aromatic Diazo-Compounds," Edward Arnold and Company, London, 1936, p. 115.

(14) H. King and W. O. Murch, *J. Chem. Soc.*, 2595 (1924).

TABLE III

Compound	Reactant	Formula	Analyses, % Calcd C	Calcd H	Found C	Found H
XIII	Phosgene	$C_{13}H_{14}O_7N_2As_2$	33.93	3.07	33.83	3.09
XIV	Oxalyl chloride	$C_{14}H_{14}O_8N_2As_2$	34.45	2.89	34.30	3.05
XV	Succinic anhydride	$C_{16}H_{18}O_8N_2As_2$	37.23	3.52	37.15	3.70
XVI	Adipyl chloride	$C_{18}H_{22}O_8N_2As_2$	39.72	4.08	39.69	4.24
XVII	Sebacyl chloride	$C_{22}H_{30}O_8N_2As_2$	44.01	5.04	43.97	5.03
XVIII	*o*-Phthalyl chloride	$C_{20}H_{18}O_8N_2As_2$	42.57	3.22	42.55	3.53
XIX	Isophthalyl chloride	$C_{20}H_{18}O_8N_2As_2$	42.57	3.22	42.42	3.28
XX	Terephthalyl chloride	$C_{20}H_{18}O_8N_2As_2$	42.57	3.22	42.76	3.43

Anal. Calcd. for $C_{13}H_{13}O_4N_2As$; C, 46.44; H, 3.90. Found: C, 46.91, 46.74; H, 4.08, 3.97.

XIII, Carbanilide-*p,p'*-diarsonic Acid; XIV, Oxanilide-*p,p'*-diarsonic Acid; XV, Succinanilide-*p,p'*-diarsonic Acid; XVI, Adipanilide-*p,p'*-diarsonic Acid; XVII, Sebacanilide-*p,p'*-diarsonic Acid; XVIII, Phthalanilide-*p,p'*-diarsonic Acid; XIX, Isophthalanilide-*p,p'*-diarsonic Acid; XX, Terephthalanilide-*p,p'*-diarsonic Acid.—These dianilide compounds were prepared by essentially the methods of King and Murch[14] and Morgan and Walton,[15] involving the reaction of arsanilic acid with the required acid chloride or anhydride in a basic or buffered aqueous solution. The compounds were purified by repeated solution in sodium hydroxide and precipitation with hydrochloric acid followed by a thorough washing with hot water. The reactants used and the results of analyses are given in Table III.

Results of the Precipitation Experiments

The Precipitation Reactions between Multivalent Compounds and Antisera.—Precipitation tests were carried out between the antisera A, B, C, and D homologous to atoxylazoprotein and the twenty compounds I to XX, each of which contains in its molecule two or more haptenic groups R, R', or R". *In every case precipitation occurred.* The amount of precipitate was found to vary with concentration of the antigen in the same way as for ordinary antigens. There is an optimum antigen concentration (or amount while the volume of antigen solution is held constant) at which the amount of precipitate reaches a maximum.

The amounts of precipitated antibody found by analysis are given in Tables IV, V, VI, VII, and VIII and are represented graphically in Figs. 1, 2, 3, 4, and 5.

Errors in the determination of the amounts of precipitated antibody may arise in the separation and washing of the precipitate or in the nitrogen analysis. Examination of the reported results of duplicate determinations indicates a probable error of about 5% for individual determinations.

(15) G. T. Morgan and E. Walton, *J. Chem. Soc.*, 615 (1931); 91 (1933), 902 (1936).

TABLE IV

AMOUNTS OF ANTIBODY PRECIPITATED FROM ANTISERUM A BY ANTIGENS I TO XII

Amounts of solutions used: 2 ml. antiserum, 3 ml. saline-antigen. The *p*H of the supernatant solutions was between 8.3 and 8.5. In Tables IV to VIII, inclusive, the amount of antigen used, in micrograms per ml. of antiserum, is given at the top of each column. The numbers in the columns are the amounts of precipitated antibody, in micrograms per ml. of antiserum. The meaning of the other symbols is given in the text.

Amount of antigen	3.13	6.25	12.5	25	50
I	−+	−+	−++	−+	−−
II	−++	−84	++134	+119	−53
III	−47	±103	++150	+100	−22
IV	−69	+138	++225	++213	±75
V	±106	+222	++322	+213	−63
VI	±100	+188	++222	+135	−↓
VII	−48	+116	++128	+94	−↓
VIII	+141	++185	+++338	++250	±110
IX	±106	+141	++291	+++347	−69
X	±66	+144	++188	+++397	↓ 660
XI	++232	↓ 425	↓ 920	↓ 1560	↓ 1690
XII	+128	+197	+++456	↓ 890	↓ 1210

TABLE V

AMOUNTS OF ANTIBODY PRECIPITATED FROM ANTISERUM B BY ANTIGENS II TO IX

Amounts of solutions used: 2 ml. antiserum, 3 ml. saline-antigen. The *p*H of the supernate was in each test 8.3.

Amount of antigen	6.25	12.5	25	50	100
II	270	520	660	470	300
	250	490	670	460	280
III	120	290	470	120	0
	110	350	490	140	9
IV	100	440	950	330	80
	200	490	850	330	100
V	330	760	1330	520	80
	270	840	1120	460	
VI	330	920	(1600)[a]	230	60
	310		(1490)	180	20
VII	180	350	770	300	100
	180	370	710	190	50
VIII	330	840	1590	340	20
	310	860	1400	280	10
IX	110	160	650	920	260
	110	130	690	670	260

[a] There is strong evidence that these values are in error.

TABLE VI

AMOUNTS OF ANTIBODY PRECIPITATED FROM ANTISERUM C BY ANTIGENS VI, IX, X, XI, AND XII

Amounts of solutions used: 2 ml. antiserum, 2 ml. saline-antigen for VI and IX; 1 ml. antiserum, 1 ml. saline-antigen for X, XI, and XII. The *p*H of the supernate was 8.4 for VI and IX, 8.5 for X, XI, and XII.

Amount of antigen	6.25	12.5	25	50	100	200
VI	290	460	270	160		
	290	380	270	150		
IX	220	440	400	230		
	180	430	390	220		
X	110	220	380	540	640	600
	110	220	350	530	650	710
XI	310	710	1360	1330	1000	660
	320	610	1350	1360	1190	750
XII	310	440	900	910	640	350
	320	440	840	990	680	360

TABLE VII

AMOUNTS OF ANTIBODY PRECIPITATED FROM ANTISERUM B BY ANTIGENS XIII TO XX

Amounts of solutions used: 3 ml. antiserum, 3 ml. saline-antigen. The *p*H of the supernate was 8.2 for each antigen.

Amount of antigen	6.25	12.5	25	50	100
XIII	50	75			10
	25	75			50
XIV	180	385	440	340	265
	165	355	475	340	270
XV	130	345	315	200	150
	150	255	325	225	140
XVI	15	35	55	35	15
	35	50		35	
XVII	35	35	50	45	30
	35	50	50	40	40
XVIII	10	10	60	85	50
	0	0	55	65	55
XIX	65	305	465	200	75
	65	300		215	
XX	120	505	695	440	195
	120	435	735	500	175

Test of the Customary Visual Estimation Method.—For the series reported in Table IV we took advantage of the opportunity of testing the customary visual-estimation method of studying antigen-antibody precipitation reactions. The tubes were inspected one-half hour after the solutions were mixed, and a record made of their appearance; this is given by the symbols in the table, which have the following meanings: −, no apparent cloudiness; ±, slight cloudiness; +, ++, +++, increasingly pronounced cloudiness; ↓, formation of clumps of precipitate. Comparison of these with the amounts of antibody precipitated in twenty-four hours shows that the

TABLE VIII

AMOUNTS OF ANTIBODY PRECIPITATED FROM ANTISERUM D BY ANTIGENS II AND XIII TO XX

Amounts of solutions used: 2 ml. antiserum, 3 ml. saline-antigen.

Amount of antigen	3.13	6.25	12.5	25	50	100
II	140	280	780	1160	1150	930
XIII			20	20		
XIV			570	690	700	570
			460	640	640	510
				740	760	
XV	60	190	440	600	480	
				700	660	
XVI	10	10	50	110	90	
XVII	10	60	80	100	120	100
			80	90	110	90
			80	90	110	
XVIII	20	30		50	40	
XIX	10	180		1060	420	
XX	120	250	610	1170	800	

correlation is reasonably good, and suggests that visual estimates of the amount of precipitate may be trusted to within about ± 30%. In this case each symbol represented about 60% more precipitate than the preceding one, as follows: −, 75; ±, 100; +, 150; ++, 250; +++, 400; ↓, 500 μg/ml.

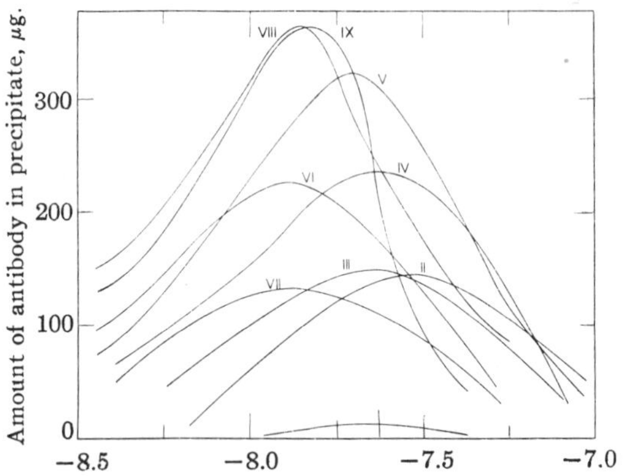

Fig. 1.—Amounts of antibody precipitated from antiserum A (per ml.) by antigens I to IX as functions of log of molal antigen concentration (Table IV). The curve for I represents an estimate. Smooth curves have been drawn through the experimental points, which are not shown.

Comparison Tests with Normal Serum.—Similar tubes were set up for each of the twenty antigens with normal rabbit serum in place of antiserum. In none did precipitation occur.

The Failure of Monohaptenic Compounds to Produce Precipitates with Antisera.—The seven

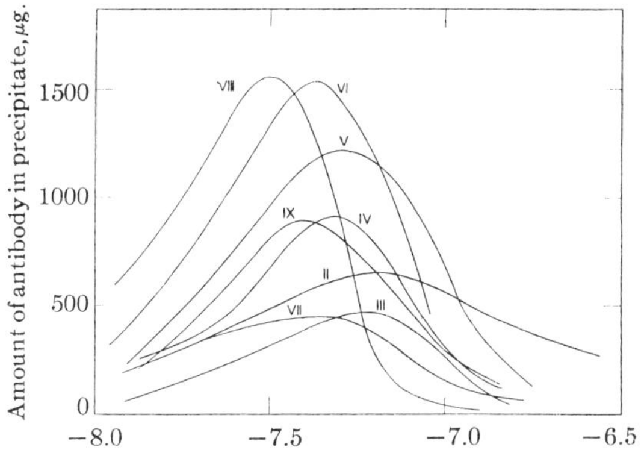

Fig. 2.—Amounts of antibody precipitated from antiserum B (per ml.) by antigens II to IX (Table V).

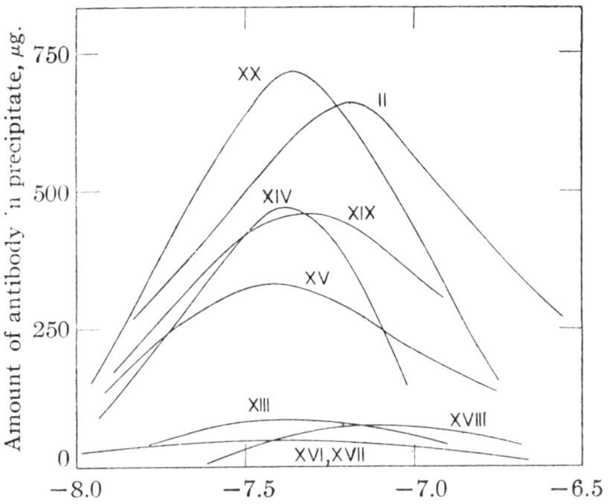

Fig. 4.—Amounts of antibody precipitated from antiserum B (per ml.) by antigens XIII to XX (Table VII) with antigen II for comparison (Table V).

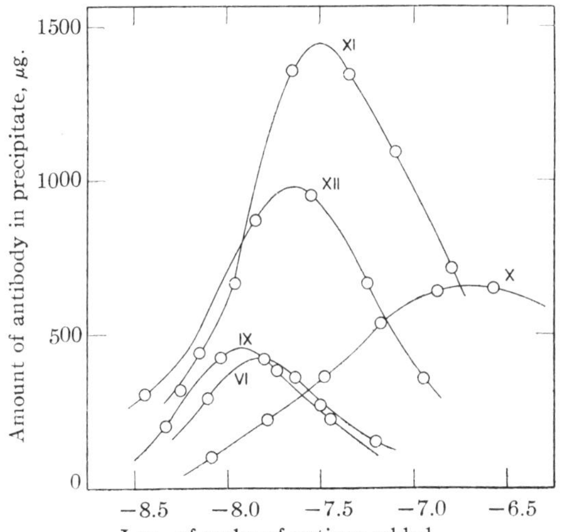

Fig. 3.—Amounts of antibody precipitated from antiserum C (per ml.) by antigens VI, IX, X, XI, and XII (Table VI).

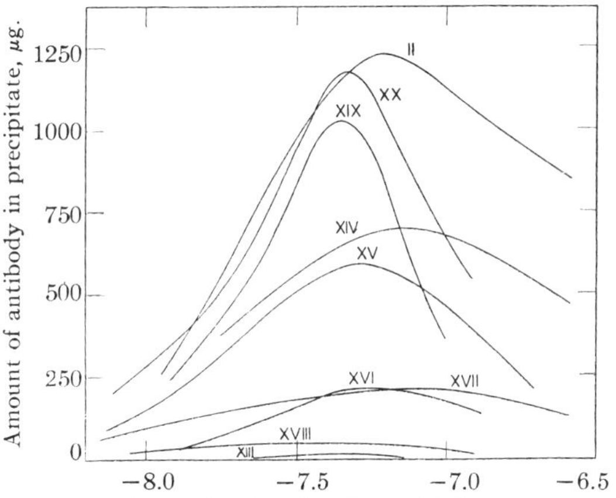

Fig. 5.—Amounts of antibody precipitated from antiserum D (per ml.) by antigens II and XIII to XX (Table VIII).

compounds XXI to XXVII, each of which contains one haptenic group per molecule, were tested with the antisera in the same way as the twenty compounds I to XX; no precipitate was formed by any of these "univalent" substances. This result is, of course, to be expected, in view of the failure of Landsteiner[16] to obtain precipitates between haptens and homologous antisera during his extensive experiments on the inhibition by homologous haptens of the azoprotein-antibody precipitation reaction.

Discussion

The fact that each of the twenty polyhaptenic substances precipitates hapten-homologous antibody, whereas substances containing only one haptenic group do not have this effect, provides strong support for the Marrack–Heidelberger framework theory (lattice theory) of the structure of serological precipitates, inasmuch as the framework theory gives a simple explanation of the phenomenon, and no other reasonable explanation has been proposed. Multivalent antibody molecules, whose existence is postulated in the framework theory, would be expected to combine with polyhaptenic molecules to form infinite aggregates, which would grow into visible precipitates, whereas with monohaptenic molecules they would form soluble small complexes containing one antibody molecule.

The general shape of all the curves of Figs. 1 to

(16) K. Landsteiner, "The Specificity of Serological Reactions," Charles C. Thomas, Baltimore, 1936, p. 118.

5, showing the dependence of amount of antibody precipitated on the amount of added antigen, is the same; the amount of precipitated antibody first increases, then reaches a maximum, and finally decreases, reaching zero for large amounts of antigen. This inhibition of precipitation by excess antigen, which occurs also for protein antigens, is explained as resulting from the formation of soluble complexes, such as A–B–A (A_2B) for a bivalent antibody (A = antigen, B = antibody). There may also be formed soluble complexes AB_2, A_2B_3, etc., which contain an excess of antibody, saturating the effective valences of the antigen molecules. The interpretation of the precipitation data in terms of the strength of the antigen-antibody bond is complicated not only by the necessity of considering these soluble complexes but also by the heterogeneity of antibody molecules.

The data for nine compounds I to IX which contain the group  used in the immunizing azoprotein are reproduced in Figs. 1 and 2. It is seen that there is agreement between the results obtained with the two antisera in most respects. From Fig. 1 we obtain the following sequence of precipitating ability of the haptens: VIII = IX > V > VI = IV > VII = III = II > I. Nearly the same sequence is given by Fig. 2. The principal difference is the interchange of VI and IX, and other data (as in Fig. 3) indicate that the order given by Fig. 1 is to be preferred. It is possible that some gross error was made in the work; on the other hand, the difference between Figs. 1 and 2 may well be real, resulting from difference between the two antisera.

The smallness of the amount of precipitate given by I  we attribute to two causes. First, the molecule contains only one azo group, so that in holding two antibody molecules the molecule cannot exert toward each the influence of the complete haptenic group  (present in the immunizing azoprotein); and second, two large antibody molecules clasping the two end halves of this very small antigen would be expected to interfere sterically with each other, and be prevented by this steric interference from forming as strong an antigen-antibody bond as would otherwise be possible.

Of the five dihaptenic compounds I to V the two most effective, IV and V, are those with the greatest hapten-hapten distance. We attribute their relative effectiveness to the resultant diminution in steric interaction of the attached antibodies.

Since in the immunizing azoprotein the azophenylarsonic acid groups are attached in part to tyrosine residues, a very effective haptenic group is expected to be 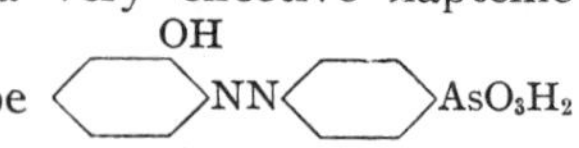 with one hydroxyl adjacent to the azo group. By considering this we explain the superiority of V

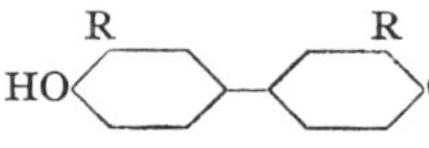 to IV 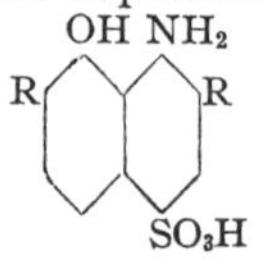 and of

VI 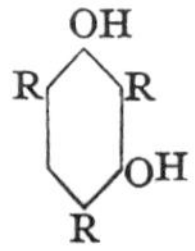 to VII 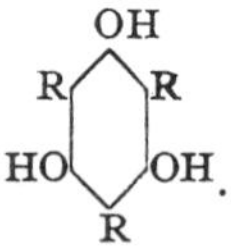.

It is interesting that the trihaptenic (VI and VII) and tetrahaptenic (VIII and IX) compounds are superior but not greatly superior to the dihaptenic compounds of similar size and structure. It was expected[3] that dihaptenic compounds would be inferior in precipitating ability because of the formation with bivalent antibody of soluble strings A–B–A–B–A–B– rather than insoluble frameworks. The observation that good precipitates are obtained with dihaptenic compounds indicates either that the long strings themselves precipitate easily or that enough trivalent antibody molecules are present to link the strings together.

The compounds X, XI, and XII, which contain the long haptenic group

are much more effective than the corresponding compounds III, VII, and IX; this increased effectiveness we attribute to decreased steric interference between the attached antibodies. The comparison of X

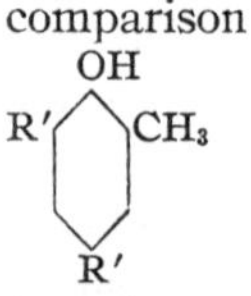 and XI 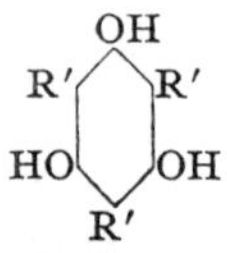 is especially interesting. The superiority of the latter is clearly to be attributed to its having three haptenic groups instead of two.

Why, then, does not XII

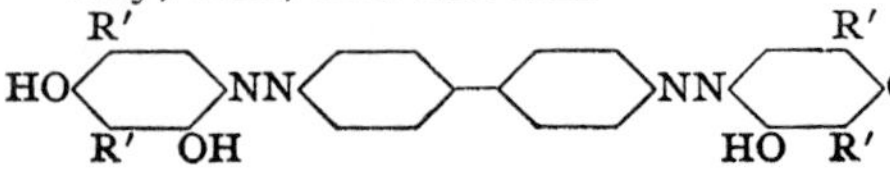, with

four, precipitate still more antibody? The answer may be that little gain is to be expected from increasing the number of haptenic groups from three to four, since three are enough to tie the strings together into a framework. In fact, XII is as effective as XI at low concentrations, and its later inferiority may be due to greater ability to form soluble complexes.

There is a striking relation between the number of haptenic groups per antigen molecule and the optimum antigen concentration for precipitation; a great change occurs from dihaptenic to trihaptenic antigens, and a much smaller change on addition of a fourth haptenic group. This is seen clearly in Fig. 3; the logarithm of the optimum molal concentration of X is −6.8, of XI −7.5, and of XII −7.65. Similarly the five dihaptenic antigens of Fig. 1 are grouped together at −7.6, and the others at −7.9.

The data represented in Figs. 4 and 5 show that the amide haptenic group 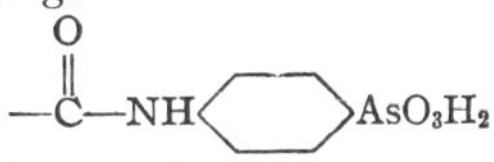 is essentially equivalent to the azo haptenic group  in its power to combine with antibodies homologous to the azo haptenic group. A straightforward comparison can be made between the amide compound XX  and the azo compound II 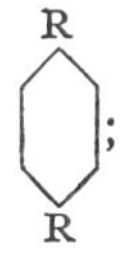; it is seen from the corresponding curves in Figs. 4 and 5 that these compounds are nearly equal in precipitating power.

The compound XIII 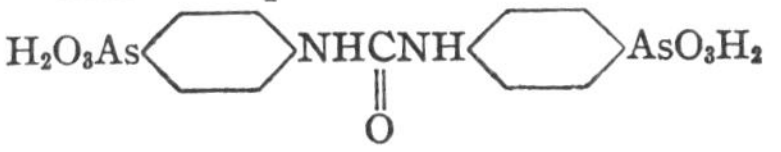 would be expected from steric considerations to be somewhat more effective than I , because of the added CO group; this is borne out by experiment. The compound XIV 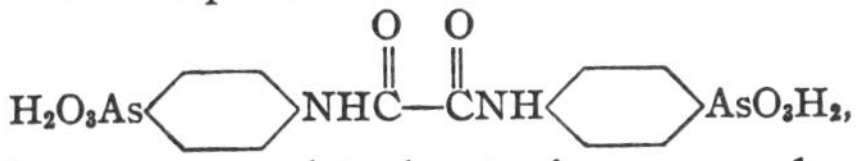, containing two complete haptenic groups, should be still more effective. It is in fact surprisingly effective, with nearly the precipitating power of XX. The compounds XVIII 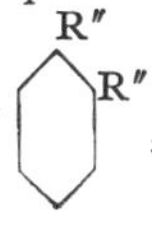, XIX  and XX 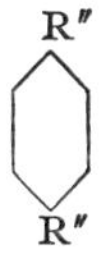 are increasingly effective in this order, with XVIII by far the weakest of the three. This is reasonably interpreted as due to steric interference of the attached antibody molecules.

This explanation cannot be extended, however, to the sequence XIV R″—R″, XV R″$(CH_2)_2$R″, XVI R″$(CH_2)_4$R″, XVII R″$(CH_2)_8$R″, since these compounds decrease in precipitating power in this order. A structural interpretation of this result in terms of the lack of rigidity of the polymethylene chain might be developed, but it is not very convincing.

The curves for compounds XIV and XV differ in shape from those for XIX and XX; no reasonable explanation of this has occurred to us.

It is important to note that the polyhaptenic simple antigens are not greatly inferior in precipitating power to azoproteins. Indeed, the best of the simple antigens, XI, was found to precipitate as much antibody as the test azoprotein.

Boyd[17] has recently reported failure to obtain precipitates between antisera and a number of simple substances containing two or more haptenic groups (including several substances also studied by us), and on the basis of this negative evidence he has drawn conclusions contrary to those which we have reached. In view of our observations, we consider it likely that his experiments were carried out under conditions unfavorable to precipitation—his antisera may have been weak, or his antigens may have contained monohaptenic impurities.

We are grateful to the Rockefeller Foundation for financial support of this work. We wish to thank Mr. Paul Faust and Mr. Shelton Steinle for their assistance in carrying out analyses, and Mr. David Brown for the preparation of phenylarsonic acid.

Summary

Twenty-seven simple substances containing the phenylarsonic acid group as haptenic group were prepared and used in precipitin tests with antisera made by injecting rabbits with azophenylarsonic acid sheep serum.

The twenty simple antigens containing two or more haptenic groups per molecule were found to give precipitates with the antisera, whereas

(17) W. C. Boyd, *J. Exptl. Med.*, **75**, 407 (1942); S. B. Hooker and W. C. Boyd, *J. Immunol.*, **42**, 419 (1941).

the seven monohaptenic substances failed to precipitate. It is pointed out that these results provide strong support of the framework theory of the precipitin reaction.

Data on the amounts of precipitate formed are discussed in relation to the structure of the simple antigens.

PASADENA, CALIFORNIA RECEIVED JULY 6, 1942

[CONTRIBUTION FROM THE GATES AND CRELLIN LABORATORIES OF CHEMISTRY, CALIFORNIA INSTITUTE OF TECHNOLOGY, No. 886]

The Serological Properties of Simple Substances. II. The Effects of Changed Conditions and of Added Haptens on Precipitation Reactions of Polyhaptenic Simple Substances

BY LINUS PAULING, DAVID PRESSMAN, DAN H. CAMPBELL, AND CAROL IKEDA

During the course of the investigation of precipitation reactions of polyhaptenic simple substances reported in the preceding paper of this series[1] we found it desirable to carry out a study of the effects of changed conditions of precipitation and washing on the amount of residual precipitate. We also made some experiments on the inhibition of precipitation by added haptens, in order to see how great would be the effects of monohaptenic impurities possibly present in the substances studied. The results obtained are presented and discussed in this paper.

Experimental Methods.—The experiments were carried out in the way described in the preceding paper (I). In addition to antisera C and D mentioned in paper I, three antisera, E, F, and G, were used. E and G contained amounts 0.6 and 3.2 mg. per ml., respectively, of antibody precipitable by azo-ovalbumin test antigen; the strength of F was not determined.

The borate buffer solutions were made by adding suitable amounts of 0.16 *N* sodium hydroxide solution to 0.2 *M* boric acid solution containing 0.9% sodium chloride.

The Effect of Changed Conditions of Precipitation and Washing on the Amount of Precipitate.—It is seen from the data reported in Tables I, II, and III, obtained with two antigens (VI and X) and three antisera (C, E, and F), that the antigen-antibody precipitate is either dissolved slightly or carried away mechanically by the saline or borate buffer solutions with which it is washed. The loss in this way is, however, small, amounting to about 5 to 15% for eight or ten extra washings with 10-ml. portions of solution.

A few experiments were made (Tables II and III) to test the effects of changing the time and temperature of precipitation. It was found that increasing the precipitation time from one day to two days increases the amount of precipitate by about 10%. The amount of precipitate formed seems to increase with increase in temperature of

TABLE I

EFFECT OF NUMBER OF WASHINGS ON AMOUNT OF RESIDUAL PRECIPITATE

3 ml. antigen VI, 25 μg./ml. in saline solution, plus 3 ml. antiserum C						2 ml. antigen VI plus 2 ml. antiserum E		
	Composition of precipitate			Composition of precipitate			Composition of precipitate	
Washings	Antibody[a]	Antigen[a]	Washings	Antibody[a]	Antigen[a]	Washings	Antibody[a]	Antigen[a]
2	1665	8.9	6	1600	8.7	3	1110	5.7
	1660	8.4		1540	8.7		1125	6.2
	1660	8.9		1590	8.7	10	970	5.7
3	1600	8.7	7	1590	8.7		930	6.0
	1600	8.9		1530	8.9	15	960	6.0
	1545	8.7		1510	8.9			
4	1545	8.7	8	1510	9.2			
	1570	8.9		1530	8.7			
	1660	8.7		1480	8.1			
5	1570	8.9	10	1520	9.5			
	1570	8.9		1510	7.6			
	1570	9.5		1320	9.5			

[a] Amounts of precipitated antibody and antigen in micrograms. *p*H of all supernates 8.1.

(1) Linus Pauling, David Pressman, Dan H. Campbell, Carol Ikeda, and M. Ikawa, THIS JOURNAL, **64**, 2994 (1942). We shall refer to this as paper I.

TABLE II

EFFECT OF CHANGED CONDITIONS ON AMOUNT OF PRECIPITATE

3 ml. antigen VI, 12.5 μg./ml., in saline solution, plus 3 ml. antiserum E. *p*H of supernates 8.0 to 8.3.

Conditions of precipitation	Washings	Composition of precipitate Antibody, μg.	Antigen, μg.
Room 1 hr., refrigerator 24 hr.	3	1600	9.2
		1610	8.9
	10	1190	8.4
		1310	8.1
		1140	8.7
	15	1140	9.2
Room 1 hr., refrigerator 48 hr.		1780	9.2
	3	1590	8.9
Refrigerator 24 hr.	3	1590	8.9
		1640	9.2
Refrigerator 48 hr.	3	1610	8.9
		1690	8.9
Room 24 hr.	3	1080	7.0
		980	7.0

TABLE III

EFFECT OF CHANGED CONDITIONS ON AMOUNT OF PRECIPITATE

5 ml. antigen X, 20 μg./ml. in saline solution, 2.5 ml. antiserum F, and 7.5 ml. borate buffer of *p*H 9.0.

Conditions of precipitation	Washing solution	Washings	Composition of precipitate Antibody, μg.	Antigen, μg.
Room 1 hr., refrigerator 24 hr.	Saline	3	520	3.3
			520	3.3
	Saline	10	525	3.1
			530	3.1
	Saline, iced	3	585	3.3
			520	2.9
	Buffer,	3	600	3.3
	Buffer, iced	3	630	2.9
			600	2.9
Room 1 hr., refrigerator 48 hr.	Saline	3	580	2.9
			595	2.9
Room 3 hr., refrigerator 22 hr.	Saline	3	610	3.1
			550	3.1
Room 24 hr.	Saline	3	675	3.1
			605	3.1
35° 1 hr., refrigerator 24 hr.	Saline	3	560	2.9
			555	2.9

the tube during the first few hours of the precipitation period (Table III). However, the evidence is inconclusive as to whether the final amount of precipitate is increased or not by refrigeration during the later part of the precipitation period.

The data reported in Table IV show that the amount of precipitated antibody is decreased by the addition of buffer solution to the antigen-antibody mixture. The decrease is not proportional to the volume of buffer added, so that the phenomenon is not analogous to the solution of a

TABLE IV

THE EFFECT OF DILUTION WITH BUFFER SOLUTION ON AMOUNT OF PRECIPITATED ANTIBODY

2.5 ml. antigen VI, varying volume of borate buffer solution with *p*H 8.0, 2.5 ml. antiserum C.

Volume of buffer solution	Amount of antigen used (μg.) 15.6	31.3	62.5
	Amount of precipitated antibody (μg.)		
0	750	1375	775
2.5	600	1030	990
5	480	845	525
10	470	655	450

well-defined compound. These observations provide further evidence of the heterogeneity of the antibodies in immune sera.

The Effect of Hydrogen-ion Concentration on Amount of Precipitate.—The results of experiments to test the effect of change in *p*H on the amount of precipitate are given in Table V. It is seen that for this antigen-antibody system the optimum *p*H is about 8.1, the amounts of precipitate in this region being greater than for either more acidic or more basic solutions. Evidence that the buffering substances do not have a large direct influence on the reaction is given by the agreement of the values at *p*H 8.1 for added saline solution and added borate buffer.

The optimum antigen concentration is seen to be changed only slightly by change in *p*H.

Less extensive experiments were also carried out with the amide antigen XIV (R″–R″) in place of the azo antigen; the results obtained, given in Table VI, are similar.

The Effect of Dilution with Normal Serum or Buffer Solution.—In order to determine the effect of change in the strength of a serum of fixed antibody composition on the position of the optimum zone, identical experiments were made with sera obtained by mixing antiserum G and normal serum. The results are given in Table VII. It is seen that for both antigens (III and VI) the maximum precipitation occurs at an amount of antigen about midway between 25 and 50 μg. (for 2 ml. of antiserum), and that on two-fold dilution of the antiserum, decreasing its antibody concentration by the divisor 2, the optimum amount of antigen is also decreased by about the divisor 2. The effect of diluting the antiserum is hence to cause the optimum zone to shift in such a way as to keep constant the ratio of antigen to antibody.

The same result is given by experiments on the effect of dilution with buffer solution, reported

TABLE V

EFFECT OF HYDROGEN-ION CONCENTRATION ON AMOUNT OF PRECIPITATE

3 ml. saline solution VI, 3 ml. antiserum E, and 3 ml. saline or buffer solution; 48 hrs. in refrigerator. B = antibody in precipitate, micrograms. A = antigen in precipitate, micrograms.

Amount of antigen (μg.)		22.2		33.3		50		75	
Added solution	*p*H of supernate	B	A	B	A	B	A	B	A
Boric acid	7.6–7.7	640	4.9	820	5.9	820	5.7	610	4.1
		640	4.6	820	5.7	820	5.7	620	5.4
Saline	8.1–8.2	890	4.6	1140	6.5	1020	6.2	610	3.0
		840	4.6	1140	6.5		6.2	550	3.8
Borate buffer,	8.1	820	5.7	1080	8.4	940	7.0	650	4.9
*p*H 8.0		790	5.7	1080	7.3	920	6.0	590	5.4
Borate buffer,	8.8–8.9	680	5.1	780	5.9	640	6.2	520	5.4
*p*H 9.0		650	4.6	820	6.5	720	6.2	520	4.3

TABLE VI

EFFECT OF HYDROGEN-ION CONCENTRATION ON AMOUNT OF PRECIPITATED ANTIBODY

1 ml. antigen XIV, 1 ml. antiserum G, 2 ml. saline or buffer solution.

Amount of added antigen (μg.)		12.5	25	50
Added solution	*p*H of supernate	Amount of precipitated antibody (μg.)		
Boric acid	7.6	200	275	260
		250	290	240
Saline	8.0–8.2	270	305	270
		290	345	265
Borate buffer,	8.1	270	290	225
*p*H 8.0		290	375	325
Borate buffer,	8.8–9.0	215	290	240
*p*H 9.0		280	295	270

TABLE VII

THE EFFECT OF DILUTION WITH NORMAL SERUM ON AMOUNT OF PRECIPITATED ANTIBODY

3 ml. of antigen III or VI plus 2 ml. of mixture of normal serum and antiserum G.

Amount of antigen used (μg.)			12.5	25	50	100
Antigen	Antiserum	Normal serum	Amount of precipitated antibody (μg.)			
III	2	0	230	495	495	265
			215		555	230
	1	1	305	340	200	105
			305	280	215	55
	0.5	1.5	225	170	105	10
			230	170	105	0
VI	2	0	395	1150	1070	170
			400	1100	1240	170
	1	1	495	455	195	65
			425	510	205	70
	0.5	1.5	290	120	40	0
			330	200	45	0
*p*H of supernate	antigen III		8.1	8.1	8.1	8.1
	antigen VI		8.2	8.3	8.4	8.5

in Table IV. In these experiments, in which the amount of antibody is kept constant, the position of the optimum zone does not change significantly.

These conclusions agree with those reached by many earlier investigators with protein antigens.

The Inhibition of Precipitation by Hapten.—It was discovered by Landsteiner[2] that a hapten such as arsanilic acid present in reasonable concentration in a mixture containing an azoprotein (made from this hapten) and the hapten-homologous antiserum decreases the amount or inhibits the formation of the antigen-antibody precipitate. It has also been found[3] that this phenomenon occurs with polyhaptenic simple substances replacing the azoprotein. Quantitative results were obtained by us with antigens VI, X, and XX, containing the groups R, R′, and R″, respectively, and haptens XXI, XXII, XXIII, and XXVII. These are given in Tables VIII to XI and Figs. 1 and 6. It is seen (Table VIII) that addition of hapten decreases the amount of precipitate without noticeable shift in the equivalence zone, and (Tables X, XI) that haptens differ in their inhibiting power.

TABLE VIII

THE INHIBITION OF PRECIPITATION BY HAPTEN

3 ml. antigen VI, 1 ml. hapten XXI, 3 ml. antiserum E. *p*H of supernates 8.2.

Amount of antigen (μg.)	15	22.2	31.3	50	75
Amount of hapten (μg.)	Amount of precipitated antibody (μg.)				
0	540	790	1280	1020	560
12.5	520	740	910	850	540
25	520	580		780	530
50	340	560	630	640	530
100	330	360	610	550	450
200	290	340	430	440	360
400	200	270	290		270

TABLE IX

INHIBITION BY HAPTEN

1 ml. antigen VI (12.5 μg.), 0.5 ml. hapten XXIII, 1 ml. antiserum D. *p*H of supernates 8.1.

Amount of hapten, μg.	Amount of precipitated antibody, μg.
0	505 470
3.13	505
6.25	480
12.5	355
25	220
50	70

(2) K. Landsteiner, *Biochem. Z.*, **93**, 117 (1919); **104**, 280 (1920); K. Landsteiner and J. van der Scheer, *J. Exptl. Med.*, **48**, 315 (1928); **50**, 407 (1929); **54**, 295 (1931); **55**, 781 (1932).

(3) K. Landsteiner and J. van der Scheer, *ibid.*, **56**, 399 (1932).

TABLE X

INHIBITION BY HAPTEN

0.5 ml. antigen X (25 μg.), 0.2 ml. hapten XXII or XXIII, 0.5 ml. antiserum D.

Amount of hapten, μg.	Amount of precipitated antibody, μg. Hapten XXII	XXIII
0	570	565
1.25	505	565
2.50	495	530
5	420	390
10	455	315
20	295	125
40	210	65
*p*H of supernates	8.5	8.8

TABLE XI

INHIBITION BY HAPTEN

1 ml. antigen XX (25 μg.), 0.5 ml. hapten XXII or XXVII, 1 ml. antiserum D.

Amount of hapten, μg.	Amount of precipitated antibody, μg. Hapten XXII	XXVII
0	1305	1200
3.13	1295	1120
6.25	1225	845
12.5	1075	590
25	960	215
50	650	
100	475	40
*p*H of supernates	8.2	8.3

The results obtained show that no significant error would be introduced in the experiments by the presence of haptens as impurities in the polyhaptenic antigens in amounts as great as 5%. It is improbable that any of the substances used contained this much monohaptenic impurity.

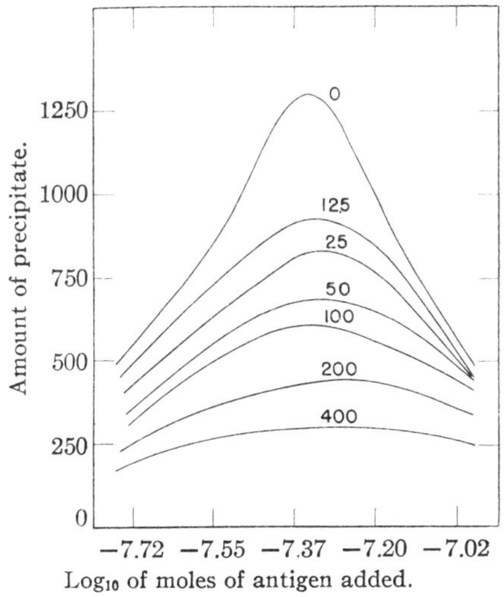

Fig. 1.—Effect of added hapten (in amounts given) on amount of antigen-antibody precipitate (Table VIII).

The data reported in Table XII indicate that the same final equilibrium is reached by the system antigen-antibody-hapten when the order of combining the reactants is changed.

TABLE XII

EFFECT OF ORDER OF COMBINATION OF REACTANTS

Reactants combined as indicated and allowed to stand indicated times at room temperature, then 24 hours in refrigerator: A, 3 ml. solution of antigen VI (37.5 μg.); H, 1 ml. solution of hapten XXI (100 μg.); S, 3 ml. antiserum E. *p*H of supernates 8.1.

	Amount of precipitated antibody, μg.	
1. A + S + 1 ml. saline solution, $1^1/_2$ hours	1200	1090
2. A + S + H, $1^1/_2$ hours	520	520
3. S + H, $^1/_2$ hour, + A, 1 hour	550	525
4. A + S, $^1/_2$ hour, + H, 1 hour[a]	550	520
5. Same as 4, with shaking.[a]	550	525

[a] A precipitate formed before hapten was added.

Discussion

A reasonable interpretation of these results and those of the preceding paper can be given in terms of the multivalent-antibody theory. This interpretation is conveniently presented with the aid of a simplified model susceptible to easy mathematical treatment.[4]

Let us assume that our idealized antigen–antibody system consists of a solution containing antigen molecules A, antibody molecules B, soluble complex molecules A_2B, and molecules AB in equilibrium with a precipitate AB. We ignore other complexes A_3B_2, A_4B_3, AB_2, etc., and the known heterogeneity of antibody molecules in a serum.

For simplicity we assume that each of the two bonds in A–B–A is equal in strength to the bond in A–B, and that the equilibrium constants for the two reactions

$$A + B = AB$$

and

$$A + AB = A_2B$$

differ only by the entropy factor 4. We represent these by $4K$ and K, respectively, with K the equilibrium constant for combination of a single haptenic group of an antigen molecule and a single complementary region of an antibody molecule, and derive the equation

$$AB(\text{pp}) = A_{\text{total}} - s - \frac{1 + 2Ks}{2 + 2Ks}\{A_{\text{total}} - B_{\text{total}} + [s(1 + Ks)/K + (A_{\text{total}} - B_{\text{total}})^2]^{1/2}\} \quad (1)$$

(4) Somewhat similar quantitative theories of the precipitin reaction have been published by M. Heidelberger and F. E. Kendall, *J. Exptl. Med.*, **61**, 563, **62**, 467, 697 (1935); **66**, 229 (1937); F. E. Kendall, *Annals N. Y. Acad. Sci.*, 153, 85 (1942); and A. D. Hershey, *J. Immunol.*, **42**, 455 (1941). These theories are designed to apply more broadly than ours, which is based on postulates suited to the special antigens and haptens which we are studying.

in which AB (pp) is the amount of precipitated compound, with solubility s, and A_{total} and B_{total} are the total amounts (per unit volume) of antigen and antibody in all molecular species, including the precipitate.

The curves of amount of precipitate for a given antiserum with varying amounts of antigen calculated with this equation have the general shape indicated by the experimental points.

Curves for the arbitrary values $s = 1$, $K = {}^1/_2$ are plotted against A_{total} in Figure 2 for each of several values of B_{total}, corresponding to the strength of the serum. It is seen that in each case the maximum amount of precipitate is produced by an amount of antigen approximately equal to the amount of antibody. This is in agreement with the results obtained with diluted serum (Tables IV and VII).

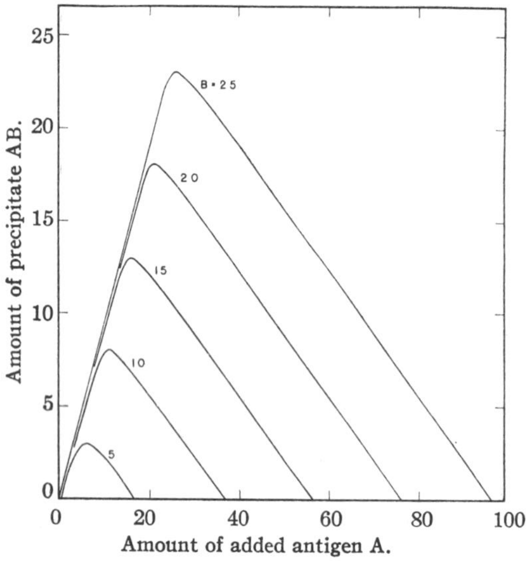

Fig. 2.—Theoretical curves showing amount of precipitate AB as function of amount of antigen A for antisera with varying antibody concentration B = 5 to 25. Values of constants used are $s = 1$, $K = {}^1/_2$.

The observation, reported in the preceding paper, that different polyhaptenic antigens containing the same haptenic group have the same molal concentration for maximum precipitation with a given serum, although the amount of precipitable antibody in the serum varies with the antigen, requires explanation, since it might well be expected that the optimum antigen concentration would be proportional to the amount of precipitate. Let us assume that for antigens containing the same haptenic group the A–B bond constant K has the same value, but that the solubility s of the precipitate may vary. This might reasonably result from steric interference of the large antibody molecules in the chains –A–B–A–B–A–B– in the precipitate, which might cause a second bond formed by a bivalent antigen molecule to be much weaker than the first bond. The curves in Fig. 3, with K constant and s varying, represent this situation. We see that, as the result of the consecutive equilibria A + B = AB and AB + A = A_2B, the position of the maximum is constant when K is constant and the solubility s varies. It is found from the equation, in fact, that the maximum occurs at the point $A_{total} - B_{total} = 1/2K$, and is independent of s.

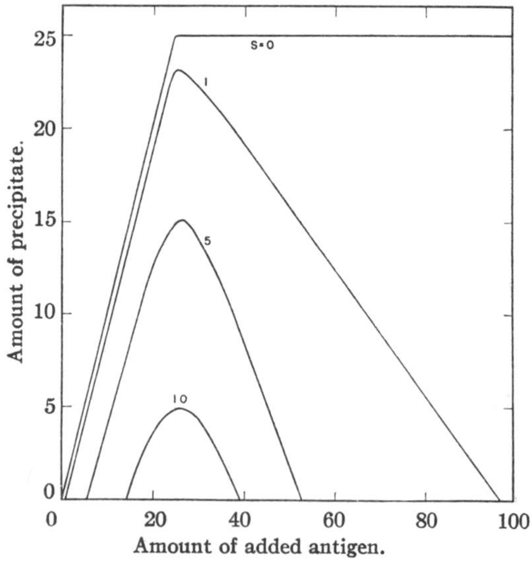

Fig. 3.—Calculated effect of variation of solubility of antigen-antibody precipitate on amount of precipitate; all curves for initial antibody concentration B = 25 and $K = {}^1/_2$.

It is also found (Fig. 4) that variation in K produces little change in the optimum antigen concentration, except when the value of K becomes very small. It accordingly seems probable that the difference of the optimum concentration for antigens containing the haptenic group R and those containing the longer group R′, as shown in Fig. 3 of paper I, is evidence that the effective strength of the serum for groups R′ is greater than that for R, because of the presence of antibodies capable of combining with R′ and not with R.

The maximum amount of precipitate is independent of K; its value, as found from Equation 1, is $B_{total} - 2s$.

The fact that change in effective strength of a serum with change in pH is not accompanied by

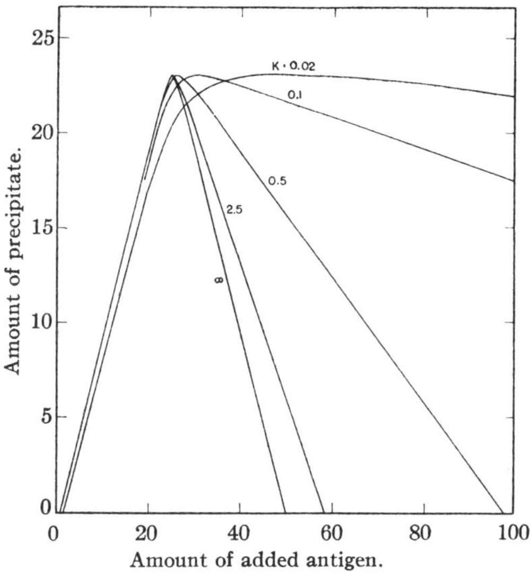

Fig. 4.—Calculated effect of variation of A–B bond-strength constant K on amount of precipitate; all curves for initial antibody concentration $B = 25$ and AB solubility $s = 1$.

shift of the optimum antigen concentration indicates that the effect is not due simply to change in the concentration of effective antibody molecules. Further experiments on the pH effect are under way.

The phenomenon of hapten inhibition has been explained by Landsteiner as resulting from combination of hapten and antibody to form a soluble complex, thus effectively neutralizing the antibody. The formation of soluble complex instead of a precipitate by antibody and hapten is explained by the framework theory as the result of the univalence of the hapten. It might be expected that as the maximum amount of precipitate which can be obtained from a serum is decreased by addition of hapten there would occur a corresponding decrease in the optimum antigen concentration, as was observed in the dilution experiments.

It is seen from Fig. 1, however, that the optimum antigen concentration is not shifted very much by addition of hapten. A small shift can be predicted by an extension of our simple theory. If we consider a system containing in solution the molecular species H, BH, BH_2, and ABH (H = hapten) as well as A, B, AB, and A_2B, and assume the B—H bond strength to be such that the equilibrium constants for the reactions

$$B + H = BH$$
$$BH + H = BH_2$$

and

$$AB + H = ABH$$

are $2K'$, $K'/2$, and K', respectively, we obtain the set of equations

$$AB(\text{pp}) = A_{\text{total}} - s - \alpha\{1 + 2K(s + z)\} \quad (2)$$

$$A_{\text{total}} - B_{\text{total}} = \alpha - \frac{s}{4K\alpha} + \alpha K s - z - \frac{\alpha K z^2}{s} \quad (3)$$

$$z = -\frac{1}{2}\left(\frac{s}{2K\alpha} + s + \frac{1}{K'}\right) \pm \left\{\frac{1}{4}\left(\frac{s}{2K\alpha} + s + \frac{1}{K'}\right)^2 + \frac{sH_{\text{total}}}{2K\alpha}\right\}^{1/2} \quad (4)$$

in which H_{total} is the total hapten concentration. The auxiliary variables α (the concentration of the molecular species A in solution) and z (the concentration of the molecular species HB in solution) are related by Equation 4, by means of which z can be calculated for an assumed value of α. Then by use of Equation 2 the amount of precipitate can be found, and by Equation 3 the variable α can be replaced by A_{total} and B_{total}.

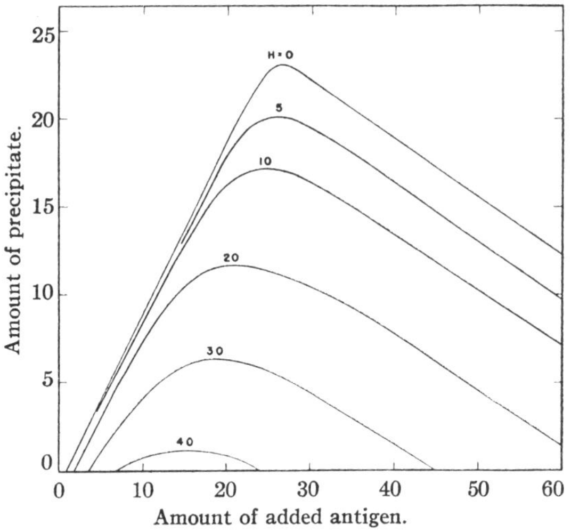

Fig. 5.—Calculated effect of addition of hapten on amount of antigen-antibody precipitate, for $B_{\text{total}} = 25$, $K = 1/2$, and $K' = s = 1$.

In Fig. 5 there are shown curves calculated in this way for $K = 1/2$, $K' = s = 1$, $B_{\text{total}} = 25$, and $H_{\text{total}} = 0$, 5, 10, 20, 30, and 40. It is seen that there is a small shift of the maxima toward lower antigen concentrations. In the region of the equivalence zone, where A_{total} equals B_{total}, the amount of precipitate formed is proportional to the hapten concentration, as is given by Equation 5, which is derived from Equations 2, 3, and 4.

$$\frac{dAB(\text{pp})}{dH_{\text{total}}} = -\frac{1/2 + Ks + (K^2s^2 + Ks)^{1/2}}{1 + Ks + \left(1 + \frac{1}{K's}\right)(K^2s^2 + Ks)^{1/2}} \quad (5)$$

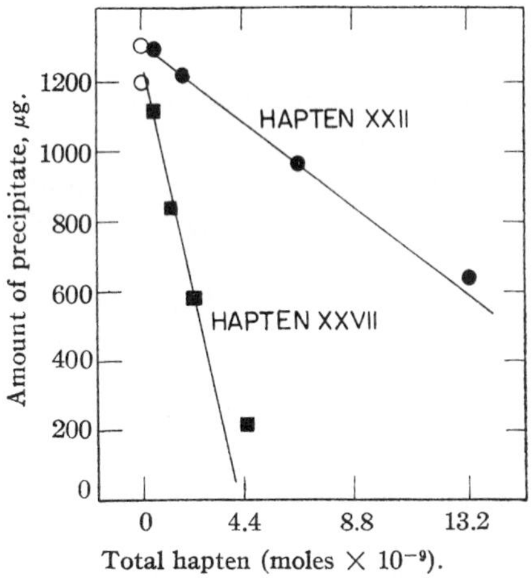

Fig. 6.—Observed effect of haptens XXII and XXVII on amount of precipitate between antigen XX and antiserum (Table XI).

The experimental data at low hapten concentrations are in rough agreement with the straight-line relation, as shown in Fig. 6 for the data of Table XI.

In Fig. 7 there is shown the predicted effect of variation of the hapten-antibody bond-strength constant K'.

It might be expected from its similarity in structure with the antigen that hapten XXVII, 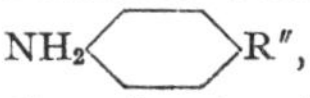, would be much more effective than hapten XXII, arsanilic acid, in inhibiting precipitation by antigen XX, 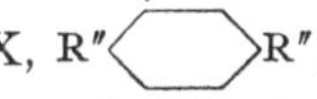; that this expectation is borne out can be seen from the slopes of the curves of Fig. 6. The data of Table X show that hapten XXIII, , has greater inhibiting effect for antigen X, 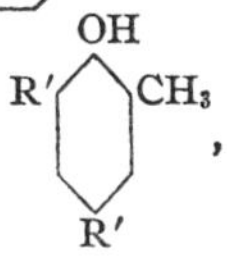, than has hapten XXII.

The data of Tables VIII and IX cannot be interpreted so reliably, since the antigen used (VI) is trihaptenic. The observation that hapten XXIII is very much more effective than hapten XXI, phenylarsonic acid, is however to be expected from the structures. We are planning to continue work on hapten inhibition, with the hope of obtaining quantitative information about the relative bond strengths of different haptenic groups with antibody.

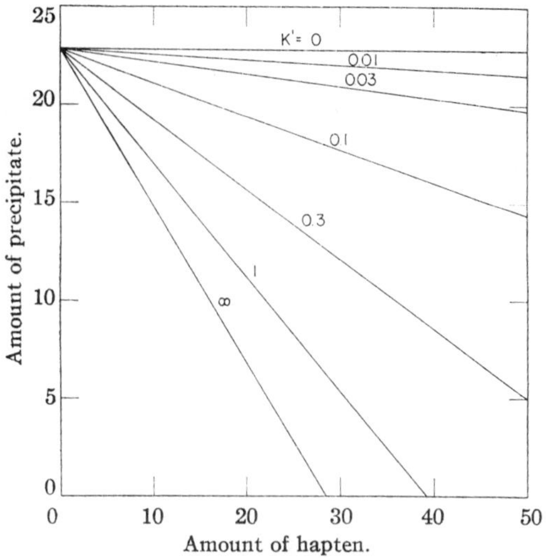

Fig. 7.—Calculated dependence of amount of antigen-antibody precipitate on hapten-antibody bond-strength constant K', for $A_{total} = B_{total} = 25$, $K = {}^1/_2$, $s = 1$.

We thank the Rockefeller Foundation for financial support of this work. We are indebted also to Dr. Verner Schomaker for helping with the theoretical treatment of antibody-antigen-hapten interactions, and to Mr. Shelton Steinle for carrying out analyses.

Summary

The results are reported of experimental studies of the effect of changed conditions, including time and temperature of precipitation, washing, addition of buffer solution and normal serum, hydrogen-ion concentration, and addition of hapten, on amount of precipitate formed by antisera and simple polyhaptenic antigens of known structure.

A simple theory of antibody-antigen-hapten interaction is formulated on the assumption of the bivalence of antibodies. It is found that this theory provides a reasonable interpretation of the experiments.

PASADENA, CALIFORNIA RECEIVED JULY 6, 1942

Reprinted from PHYSIOLOGICAL REVIEWS
Vol. 23, No. 3, July, 1943

THE NATURE OF THE FORCES BETWEEN ANTIGEN AND ANTIBODY AND OF THE PRECIPITATION REACTION

LINUS PAULING, DAN H. CAMPBELL AND DAVID PRESSMAN

Contribution no. 925 from the Gates and Crellin Laboratories of Chemistry, California Institute of Technology, Pasadena

In one of his lectures on immunochemistry at the University of California in the summer of 1904 Svante Arrhenius said (1) that Ehrlich and other investigators, because of incomplete knowledge of the phenomenon of chemical equilibrium, had been led to invent artificial hypotheses in order to explain their observations in the field of immunology. Since that time, and especially during the last few years, workers in this field have made greater and greater use of the concepts and methods of physical chemistry, and in consequence many previously puzzling observations have been reasonably interpreted.

Another branch of chemistry which is of importance to immunology is modern structural chemistry, which deals with the detailed structure of molecules and with the nature of interatomic and intermolecular interactions (2). Our present knowledge of this subject, in large part won during the past dozen years, is now so firmly founded and so extensive that it can be confidently used as the basis for a more penetrating interpretation of immunological observations than would be provided by the observations alone.

In this paper we present, after a brief historical introduction, a discussion of the nature of the specific forces between antigen and antibody and of the precipitation reaction from the point of view of modern chemistry. Only the simpler aspects of the phenomena are discussed; such complicating factors as the rôles of complement, lipids, etc., in the reactions are disregarded in our discussion.

The history of the precipitation reaction began in 1897, when Rudolf Kraus (3) reported the results of his work with anticholera and antityphoid sera. His observations were soon verified and extended by Nicolle, Tchistovich, Bordet, Myers and other workers, who prepared precipitating antisera against a great number of antigens of varied nature. We shall not review this early work here, nor the later studies of the methods of preparing antisera and carrying out the precipitation reaction, since these topics and others dealing with special phases of the reaction have been very well covered in earlier reviews (4, 5, 6).

Two most important advances in the attack on the problem of the nature of immunological reactions were the discovery that the specific precipitate contains both antigen and antibody (7) and the discovery that antibodies, which give antisera their characteristic properties, are proteins. The verification of these facts was provided by the work of many investigators over a score of years. This work, which is summarized in Marrack's monograph (6, chap. II), culminated in the preparation of purified antibody by Felton and Bailey (8), Heidelberger and collaborators (9), and others, and the determination of its properties, including amino-acid composition and molecular weight, which show that it is very closely related to normal serum globulin (6, chap. II).

The work of Landsteiner (10) and other investigators on artificial conjugated antigens provided a great body of qualitative information on the specificity of antibodies, which, together with the experimental results for natural antigens, led to the independent proposal by Breinl and Haurowitz (11), Alexander (12), and Mudd (13) in 1930–32 of the theory of structural complementariness of antigen and specific antibody. The framework theory of precipitation was then developed by Marrack (6) and Heidelberger (14). These and other theories are discussed in some detail in the following sections of this paper.

A new period in the study of the precipitation reaction was initiated by the careful quantitative studies of Heidelberger and his collaborators (15) who determined the amounts of antibody and antigen in precipitates, and the similar work of Haurowitz (16) and others. Very recently, in order to test certain aspects of his detailed theory of the structure of antibodies (17), Pauling and his collaborators have carried out many quantitative experiments on the precipitation of antisera by polyhaptenic simple substances (18, 19, 20), a phenomenon first observed by Landsteiner and Van der Scheer (21).

THE NATURE OF THE SPECIFIC FORCES BETWEEN ANTIGEN MOLECULES AND ANTIBODY MOLECULES. The detailed information which has been gathered in recent years regarding the nature of the chemical bonds which hold atoms together into stable molecules has been summarized in monographs (2, 22). Instead of interacting strongly with one another, with interaction energy of 20 kilocalories per mole or more, to produce a chemical bond, two atoms may interact more weakly. The nature of these weak interactions is now well understood, and a brief discussion of it is given in the following paragraphs. The properties of antigen-antibody systems, especially the reversibility of complex formation, are such as to indicate that the antigen-antibody attraction is due to these weaker interactions and not to the formation of ordinary chemical bonds.

The weak interactions between two molecules may be classified as electronic van der Waals attraction, Coulomb attraction, attraction of electric dipoles or multipoles, hydrogen-bond formation, etc. The forces increase rapidly in magnitude as the molecules approach one another more and more closely, and the attraction between the molecules reaches its maximum when the molecules are as close together as they can come. The molecular property which determines the distance of closest approach of two molecules is the electronic spatial extension of the atoms in the molecules. It is possible to assign to each atom a *van der Waals radius*, which describes its effective size with respect to intermolecular interactions. These radii vary in value from 1.2 Å for hydrogen through 1.4–1.6 Å for light atoms (fluorine, oxygen, nitrogen, carbon) to 1.8–2.2 Å for heavy atoms (chlorine, sulfur, bromine, iodine, etc.). The shape of a molecule can be predicted by locating the atoms within the molecule with use of bond distances and bond angles and then circumscribing about each atom a spherical surface corresponding to its van der Waals radius. This shape determines the ways in which the molecule can be packed together with other molecules (2, sec. 24).

The most general force of intermolecular attraction, which operates between every pair of molecules, is *electronic van der Waals attraction.* This type of electronic interaction between molecules was first recognized by London (23). A molecule (of methane, for example) which has no permanent average electric dipole moment may have an instantaneous electric dipole moment, as the center of charge of the electrons, in their rapid motion in the molecule, swings to one side or the other of the center of charge of the nuclei. This instantaneous dipole moment produces an instantaneous electric field, by which any other molecule in the neighborhood would be polarized; the electrons of the second molecule would move relative to its nuclei in such a way as to give rise to a force of attraction toward the first molecule.

This electronic van der Waals attraction operates between every atom in a molecule and every atom in other molecules in the near neighborhood. The force increases very rapidly with decreasing interatomic distance, being inversely proportional to the seventh power of the interatomic distance. Hence the electronic van der Waals attraction between two molecules in contact is due practically entirely to interactions of pairs of atoms (in the two molecules) which are themselves in contact; and the magnitude of the attraction is determined by the number of pairs of atoms which can be brought into contact. In consequence, two molecules which can bring large portions of their surfaces into close-fitting juxtaposition will in general show much stronger mutual attraction than two molecules with less extensive complementariness of surface topography.

Other types of molecular interactions result from the possession of a permanent electric charge, electric dipole moment, or electric moment of higher order by one or both of the interacting molecules. The effects of these charges and moments have been classified in various ways, as ion-ion forces, dipole-dipole forces, forces of electronic polarization of one molecule in the dipole field of another, etc. All electrostatic interactions are very much smaller in water than in a medium of low dielectric constant, and it can be shown by calculation, making use of known values of the effective dielectric constant of water for charges a given distance apart (24), that in general these electric forces are of minor importance, except when an isolated or essentially isolated electric charge is involved. The electrostatic attraction of a positive group such as a substituted ammonium ion and a negative group such as a carboxyl ion becomes significantly strong, with bond energy 5 kilocalories per mole or more, if the structure of the molecules containing the groups is such that they can come into juxtaposition.

A type of intermolecular attractive force which ranks in importance with the electronic van der Waals attraction and the attraction of oppositely charged groups is that associated with the structural feature called the *hydrogen bond.* The importance and generality of occurrence of the hydrogen bond were first pointed out in 1920 by Latimer and Rodebush (25) and summaries of the properties of the bond are given in the monographs quoted above. A hydrogen bond results from the attraction of a hydrogen atom attached to one electronegative atom for an unshared electron pair of another electronegative atom. The strength of a hydrogen bond depends on the electronegativity of the two atoms which are bonded together by hydrogen; fluorine, oxygen, and nitrogen, the most electronegative of all atoms, are the atoms which form the strongest hydrogen bonds. The energy of a hydrogen bond between two of these atoms is of the order of magnitude of 5 kcal. per mole. This is so large as to have a very important effect on the intermolecular interactions of molecules capable of forming hydrogen bonds and on the properties of the substances consisting of these molecules.

In synthesizing our knowledge of intermolecular forces and of immunological phenomena into a definite picture of the antigen-antibody bond the immunological property of greatest significance is the specificity of the combining power of antibody for the immunizing antigen.

The forces of van der Waals attraction, hydrogen-bond formation, and interaction of electrically charged groups are in themselves not specific; each atom of a molecule attracts every other atom of another molecule by van der Waals attraction, each hydrogen atom attached to an electronegative atom attracts every other electronegative atom with an unshared electron pair which comes near it, and each electrically charged group attracts every other oppositely charged group in its neighborhood. The van der Waals repulsive forces which determine the van der Waals radii of atoms also are not specific; each atom in a molecule repels every other atom of another molecule, holding it at a distance corresponding to the sum of the pertinent van der Waals radii. We see, however, that specificity can arise in the interaction of large molecules as a result of the shapes of the molecules. Two large molecules may have such spatial configurations that the surface of one cannot be brought into contact with the surface of the other except at a few isolated points. In such a case the total electronic van der Waals attraction between the two molecules would be small, because only the pairs of atoms near these few isolated points of contact would contribute appreciably to this interaction, and, moreover, the distribution of hydrogen-bond forming groups and of positively and negatively charged groups of two molecules might be such that only a small fraction of these groups could be brought into effective interaction with one another for any position and orientation of one molecule with respect to the other; the energy of attraction of these two molecules would then be small. If, on the other hand, the two molecules possessed such mutually complementary configurations that the surface of one conformed closely to the surface of the other, if, moreover, the electrically charged groups of one molecule and those of the other were so located that oppositely charged groups were brought close together as the molecules came into conformation with one another, and if the hydrogen-bond forming groups were also so placed as to form the maximum number of hydrogen bonds, the total energy of interaction would be very great, and the two molecules would attract one another very strongly. We see that this strong attraction might be highly specific in the case of large molecules which could bring large areas of their surfaces into close contact. A molecule would hence show strong attraction for another molecule which possessed complete complementariness in surface configuration and distribution of active electrically charged and hydrogen-bond forming groups, somewhat weaker attraction for those molecules with approximate but not complete complementariness to it, and only very weak attraction for all other molecules.

This specificity through complementariness of structure of the two interacting molecules would be more or less complete, depending on the greater or smaller surface area of the two molecules involved in the interaction. It may be emphasized that this explanation of specificity as due to a complementariness in structure which permits non-specific intermolecular forces to come into fuller operation than would be possible for non-complementary structures is the only explanation which the present knowledge of molecular structure and intermolecular forces provides.

This theory of structural complementariness of antigen and antibody was first suggested, in less detailed form than above, by Breinl and Haurowitz (11), Alexander (12), and Mudd (13). A detailed discussion of the structure of antibodies and of a postulated method of their formation has been presented by Pauling (17), who has also reviewed the evidence supporting the theory of complementariness.

It was suggested by Breinl and Haurowitz and by Mudd that the effect of an antigen in determining the structure of an antibody might involve the ordering of the amino-acid residues in the polypeptide chains in a way different from that in the normal globulin. Rothen and Landsteiner (26) then pointed out that the possibility of different ways of folding the same polypeptide chain is worth considering, and this postulate was amplified by Pauling (17), who assumed that all antibody molecules contain the same polypeptide chains as normal globulin, and differ from normal globulin only in the configuration of the chains. This assumption was made because it permits the formulation of a simple proposed mechanism of manufacture of specific antibodies. An antibody molecule, capable of existing in any one of a great number of configurations with nearly the same energy, is synthesized, except for the final folding step, in the same way as normal globulin. If no foreign substance is present, the chain then folds into a stable configuration, characteristic of normal globulin; but if an antigen molecule is present, the chain folds into a configuration stable in the presence of the antigen, that is, into a configuration complementary to that of a portion of the surface of the antigen molecule. This explanation of the ability of an animal to form antibodies with considerable specificity for an apparently unlimited number of different antigens (27), as shown especially by the work of Landsteiner (10), is compatible with the principles of structural chemistry and thermodynamics as well as with the immunological evidence.

To illustrate the way in which the complementariness theory accounts for many reported observations we shall mention only one point, taken from the great body of results on azoproteins obtained by Landsteiner. He observed a pronounced cross reaction between an azoprotein made from *m*-aminobenzoic acid and an antiserum to an azoprotein made from 4-chloro-3-aminobenzoic acid and a different protein, but no reaction with the haptenic groups reversed. The explanation of this is that the antibody to the 3-azo-4-chlorobenzoic acid group conforms closely to this haptenic group, allowing either this group or the

3-azobenzoic acid group, which differs in the replacement of the chlorine atom by a smaller atom, hydrogen, to fit into the complementary cavity in the antibody; but the 3-azo-4-chlorobenzoic acid group cannot fit into a cavity designed for the smaller haptenic group, and so the reverse cross reaction does not occur. In a quantitative extension of Landsteiner's work on hapten inhibition of precipitation reactions of simple polyhaptenic substances (10), Pauling and collaborators (28) have recently reported a great deal of evidence in support of the complementariness theory. They interpreted their results on the inhibition of the precipitation reaction between dyes containing *p*-azophenylarsonic acid groups and antisera to hapten-homologous azoproteins to obtain numerical values of the strength of the bonds formed by these antibodies with over twenty-five different haptens. The observed correlation between the bond strengths and the structure of the haptens is that which would be expected from the complementariness theory.

This theory is not greatly different from some earlier proposals, such as Ehrlich's lock-and-key analogy (29), but it differs greatly from others. For example, Buchner (30) considered that antigen molecules are split up and incorporated into the antibody molecules, thus imparting specificity to them. This theory or a closely related theory has been supported by many people, including Burnet (31), who proposed a mechanism for the manufacture of antibodies in the image of the antigens: the antigens act as templates for the manufacture by the body of specific enzymes, which then serve as the molds for the production of antibodies similar to the original antigens. Until recently there has been no suggestion as to why antibodies similar in structure to an antigen should combine specifically with it. Recently, however, Jordan (32) has stated that a strong attraction would occur between such identical or nearly identical molecules because of the quantum-mechanical resonance phenomenon; this has been denied by Pauling and Delbrück (33), who pointed out that the resonance energy would be so small as to be ineffective. Chemical evidence against the identity of antibodies and specific antigens has been presented by many authors, of whom the most recent are Haurowitz, Vardar, and Schwerin (34).

Forty years ago there was under way a keen controversy between Ehrlich and Bordet, and their respective supporters, as to whether the bonds between antibodies and antigens are chemical bonds or are physical forces of the sort producing surface phenomena such as adsorption. The modern point of view resolves this argument, but not in favor of either side; in fact, as in recent years an understanding has been obtained of the forces responsible for surface phenomena it has been found that these forces are the same as those which are operative in chemical reactions, so that the old distinction between chemical and physical forces has lost most of its meaning.

The nature of the precipitate. Under suitable conditions (salt concentration, antibody-antigen ratio, etc.) the first stage of combination of antibody and antigen, which may make itself evident in change in toxicity or other properties of the antigen, is followed by precipitation. There has been much discussion as to whether or not this second stage is specific, like the first stage, or is non-specific. Direct experimental evidence on this point, while not conclusive, favors the view that the reaction is specific. The most pertinent observations are those on the agglutination of mixed cellular antigens by mixed antisera; Topley and collaborators (35) noted the formation of separate clumps by the different cells, whereas Abramson (36) observed mixed clumping. Hooker and Boyd (37) found that mixed human and chicken erythrocytes gave separate clumps under some conditions and mixed clumps under other conditions. The fact that separate clumping is observed at all strongly favors the concept of a specific second stage, since mixed clumping might result from mechanical intertwining of specific clumps, whereas separate clumping would hardly be expected to result from non-specific interaction. Heidelberger has pointed out that in those cases where mixed clumping takes place the cells used were either very large or of greatly different sizes (38).

A reasonable theory of agglutination and precipitation, the framework theory (lattice theory[1]), was proposed in 1934 by Marrack (6), and has received strong support from the theoretical considerations and experiments of Heidelberger (9, 14, 15) and Pauling (18, 19, 20, 28) and their collaborators.

It is clear that, after we have accepted a mechanism for the specific attachment of antibody molecules to a cellular antigen, the simplest possible explanation of the agglutination of the cells is that it results from the same mechanism; if an antibody molecule had the power of specific attachment to two cells, it could form specific bonds with the two cells and thus hold them together, and the repetition of this process would lead to the formation of larger and larger clumps. Specific precipitation of antibodies and molecular antigens would result from the same mechanism if both antibody molecules and antigen molecules were multivalent (capable of forming two or more antigen-antibody bonds); larger and larger complexes A—B, A—B—A, A—B—A—B, etc., would form until the aggregates became macroscopic in size. The evidence supporting the framework theory has been reviewed by Marrack (6), Heidelberger (14), and Pauling (17); some of it, including that provided by recent work, is presented in the following section in connection with a discussion of the valence of antibody molecules.

The first of the theories of non-specific precipitation is the theory of neutralization of electrical charges. This theory was supported by many early investigators, who were attracted by the analogy with the well-known phenomenon of the mutual precipitation of oppositely charged colloids. Teague and Field (39) investigated the charges of agglutinins and bacteria and concluded that the former are positively and the latter negatively charged; it is now known, however, as the result of the application of improved experimental methods, that under ordinary conditions (normal hydrogen-ion and salt concentrations) antibody molecules and most antigens are negatively charged, and the theory of neutralization has in consequence been abandoned.

(The failure of precipitation or agglutination to occur in antigen-antibody

[1] We have adopted the name "framework theory" instead of "lattice theory" because of our belief that the framework of antibody-antigen precipitates does not usually have the regularity of structure which would be indicated by use of the latter expression.

systems with low salt concentration is, indeed, attributed to the electrostatic repulsion of the negatively-charged complexes in solution, which prevents the formation of large aggregates; agglutination or precipitation may occur in the presence of salt, the cations of which neutralize the negative charges of the complexes.)

Another theory of non-specific precipitation which was proposed soon after the discovery of the precipitation reaction is that the reaction results from the formation of a hydrophobic colloid, which precipitates in the presence of electrolytes. This theory has been revived recently by Eagle (40) and by Hooker and Boyd (37). Eagle's suggestion that the polar groups of the antigen which are assumed to be responsible for its solubility are masked by a layer of antibody molecules, which themselves turn their polar groups inward and present only non-polar groups toward the solvent, has been discussed by Marrack (6), who has marshalled some arguments against it. An important argument is that particles can be agglutinated by an amount of antibody very much smaller than the amount required to coat their surface; the most recently reported experiments of this sort (41) indicate that azoerythrocytes can be agglutinated by less than 0.02 per cent as much antibody as would cover their surface with a layer 3.5 Å thick.

Hooker and Boyd (37) have presented several arguments in support of the thesis that "... particles grow to visible size by the indiscriminate and non-specific accretion of other related or unrelated, small or large, aggregates whose primary nuclei are molecules or particles of antigen coated with antibody-globulin." As additional evidence for this concept and against the framework theory Boyd and Hooker (42) reported their failure to inhibit the agglutination of erythrocytes by use of an excess of hemagglutinin. In our opinion the fact that inhibition of agglutination of particles (43) as well as of precipitation of molecular antigens (44) by excess antibody has been observed gives strong support to the framework theory. The failure of inhibition to occur under ordinary circumstances may be due to the difficulty in saturating the multivalent antigens, especially cellular antigens with thousands of combining groups, as is indicated by the theories of Hershey (45) and Pauling (17). In particular, the experiments of Heidelberger and Kabat (43) on the agglutination by untreated pneumococci of pneumococci coated with antibody and then thoroughly washed are most easily explained by the framework theory. Hooker and Boyd (46, 47) have recently proposed the theory that precipitation of antibody by polyhaptenic dyes may result from the action of the dye molecules in pulling the antibody molecules to which they are bonded so tightly together as to prevent the solvent from reaching the polar groups. This theory seems to be incompatible with our observation (18) that in general the dyes of smaller molecular size, which according to Boyd's theory should pull the molecules more closely together, in fact precipitate less completely than those of larger molecular size.

COMPOSITION OF ANTIBODY-ANTIGEN PRECIPITATES AND VALENCE OF ANTIBODY. An essential requirement for agglutination or precipitation according to the framework theory is that both antigen and antibody be multivalent. The experimental observations which indicate multivalence of antigens and of agglutinins and precipitins have been summarized by Marrack (6), Heidelberger (14), and Pauling (17).

The most straightforward evidence for the necessary multivalence of antigen is given by experiments on the reactions of antibodies with simple substances of known structure. Subsequent to Landsteiner's discovery (10) that simple haptens inhibit the precipitation and agglutination reactions by forming soluble complexes with antibody, it was found by Landsteiner and Van der Scheer (21) that simple substances containing two or more haptenic groups form precipitates with hapten-homologous antibodies. We have shown that of the twenty-seven simple substances containing phenylarsonic acid groups which were tested with antisera made by injecting rabbits with azoprotein made from *p*-arsanilic acid each of those (twenty in number) which contained two or more of the haptenic groups gave the precipitation reaction, whereas none of the monohaptenic substances formed precipitates. These facts support the framework theory strongly.

(The failure to obtain precipitates with some polyhaptenic substances reported by Hooker and Boyd (46) and Boyd (47) may have been due to their failure to work under conditions favorable to precipitation. We have obtained precipitates with some of the same substances, and have observed that substances which give precipitates with strong antisera may fail to do so with weak antisera.)

Experiments which have been reported on the number of haptenic groups per azoprotein molecule necessary for precipitation with hapten-homologous antibody (48, 49) and some of our unpublished results indicate that a few groups are needed, but so far they have not been precise enough to distinguish between 1 and 2 as the minimum.

Direct proof of the bivalence of diphtheria antitoxin is given by the studies of the antitoxin in presence of an excess of toxin with use of the ultracentrifuge, which showed that the complexes ToxAntitox and Tox_2Antitox exist in the solution (50).

The fact that slides can be coated with alternate unimolecular layers of antigen and antibody in specific combination (51, 52) indicates effective bivalence of antibody molecules as well as antigen molecules.

It has long been known that in antigen-antibody precipitates molecules of antibody are present in larger numbers than those of antigen, the antibody-antigen molecular ratio being considerably greater than unity for nearly all systems (6, p. 161; 53). This was shown convincingly by the accurate quantitative investigations of Heidelberger and his collaborators (54). A simple explanation of this fact, which does not follow directly from the framework theory in its original form (6, 14), is given by the theory as modified by Pauling (17), who made the assumption that antibodies in general are at the most bivalent. (This assumption was made because his proposed structural theory of the process of formation of antibodies is such as to make unlikely the occurrence of antibodies of higher valence.) The maximum valence N of antigens toward homologous antibodies is assumed to be determined by the sizes and shapes of the antigen and antibody molecules, being equal to the number of antibody molecules which, when bonded to the antigen molecule, can be packed around it. If the antigen and antibody molecules were spheres of the same size, this number would be $N = 12$; for smaller antigen molecules it would be smaller, and for larger ones it would be larger.

The predicted antibody-antigen molecular ratio for small antigen molecules would be $N/2$ at the equivalence zone, with the maximum valences of both antibody and antigen effective; the limiting values of the ratio for antigen excess would be 1, and for antibody excess $N - 1$. For very large antigens the expected ratio at the equivalence zone would be less than $N/2$. These predictions are in reasonably good agreement with Heidelberger's observations (54, 17), which correspond to the following values of N: ovalbumin and R-salt-azobiphenylazoovalbumin (molecular weight 40,000–46,000), $N = 6$; serum albumin (m.w. 67,000), $N = 6$ to 8; thyroglobulin (m.w. 700,000), $N = 30$ to 40. Data of other investigators (55, 56, 57, 58, 59, 60, 61) correspond to similar values of N, with assumed bivalence of antigen.

If the valence of the antigen were known, measurement of the antibody-antigen molecular ratio could be interpreted to give the valence of antibody molecules. The only reliable experiments of this sort which have been reported so far are those of Pauling, Pressman, and Ikeda (20) with simple antigens of known structure. They found that the dihaptenic antigens 2-methyl-4,6-di(*p*-azophenylarsonic acid)phenol and 2-methyl-4,6-di(*p*-azobenzene(p-azophenylarsonic acid))phenol gave with antisera homologous to the *p*-azophenylarsonic acid group precipitates with the same molecular ratio throughout the range of relative concentrations from antibody excess to antigen excess. This is what would be expected if both antibody and dihaptenic antigen were bivalent, the predicted molecular ratio under all conditions then being 1, corresponding to the structure —A—B—A—B—A—B—A—B— for the precipitate. The observed molecular ratio (average of 119 analyses) was 0.75. The same independence of molecular ratio on relative concentration was found also for the four trihaptenic and tetrahaptenic substances studied; this was interpreted as resulting from the effective bivalence of these molecules also, as the result of their small size in comparison with the antibody molecules. The average molecular ratios found for the trihaptenic and tetrahaptenic substances, 0.85 and 0.83, respectively, are only slightly less than unity. These results, with assumed effective bivalence of the antigens, indicate for antibody molecules the effective valence 2.3.

Substantiating evidence for the multivalence of precipitating antibodies is provided by the observations which have been interpreted as resulting from the presence in antisera of antibodies with a valence of one. Heidelberger and Kendall (62) demonstrated the presence in rabbit antisera of univalent antibodies, which are able to combine specifically with antigen but are not able, in the absence of multivalent antibodies, to form precipitates. These univalent antibodies also occur in considerable amount in horse antisera; they seem to be produced in large amount by the first injections of ovalbumin into horses, precipitating (multivalent) antibody being formed only on repeated injection (63, 64). It is probable that univalent antibody confers on horse antitoxins the peculiar properties which they show, in particular the pronounced prezone (region of antitoxin excess in which precipitation does not occur). The change in properties of antisera on heat treatment (65, 66, 67) or treatment with formaldehyde (9) or other denaturing agent is probably due to the conversion of bivalent antibody to univalent antibody by destruction of one of the combining regions.

Quantitative theories of the precipitation reaction. Although the quantitative physico-chemical treatment of immunological phenomena was begun early in the present century, by Arrhenius and Madsen (1, 68), it is only during the last decade that significant progress has been made. The reason for the delay is not far to seek—it lies in the fact that the mathematical treatment of numerical data of low accuracy has little significance so long as a sound qualitative understanding of the phenomenon has not been developed. We may mention in illustration Arrhenius' discussion (1, p. 147) of the formula $C = K\ B^{2/3}$ which he found to express the relation between the amount C of agglutinin bound by bacterial cells and the amount B of free agglutinin; Arrhenius interpreted this equation (whose validity for the complex system we would ascribe to the accidental distribution of the heterogeneous antiserum) as showing that the agglutinin molecules are divided between two solvents, one within and one without the bacterial cells, and that three molecules of the bound agglutinin are formed from two of the free substance.

Recent theories are of two kinds: those based on thermodynamic equilibrium among the reacting substances, and those based on the rates of reactions under non-equilibrium conditions. There has been considerable discussion as to whether or not immunological reactions are reversible—whether, for example, an antigen-antibody precipitate is soluble, and is in equilibrium with free antigen and antibody in solution. We know from general principles, however, that, given time enough, every system reaches equilibrium, and every material is more or less soluble; the questions of interest deal rather with such quantitative points as the length of time required for the system to reach equilibrium, and the magnitude of the solute concentrations in equilibrium with the precipitate.

That the precipitation reaction in some cases reaches equilibrium in the hours or days usually allowed it is shown by various experiments on solution of the precipitate by salt (69), acid (70), alkali (71), and excess antigen, even after ageing for several months (72), including experiments in which there was variation in the method of approaching equilibrium (19).

Experiments on the Danysz phenomenon (73) and other related experiments indicate that a long time—many days—is needed for equilibrium to be approached for reactions involving change in composition of antigen-antibody precipitates.

The first quantitative theory which we shall discuss, that of Heidelberger and Kendall (14), was based on consideration of the rate of antigen-antibody combination under non-equilibrium conditions. The authors assumed that antigen A and antibody B first react completely and rapidly to form the complex AB, which uses up all the A (B being assumed present in excess). There then occur two competing slow reactions:

$$B + AB \rightarrow AB_2$$
$$AB + AB \rightarrow A_2B_2$$

The rates of formation of AB_2 (total number of units α) and A_2B_2 (total number of units β) are

$$\frac{d\alpha}{dt} = K[\mathrm{B}][\mathrm{AB}]$$

and

$$\frac{d\beta}{dt} = K'[\mathrm{AB}]^2$$

in accordance with the laws of chemical kinetics. It is assumed arbitrarily that $K = K'$ and that the reactions are not reversible; by integration over the course of the reaction until the solution is exhausted of AB there is obtained to represent the composition of the precipitate, which consists of all the AB_2 and A_2B_2 formed, the equation

$$\frac{y}{a} = 2R - \frac{R^2a}{b_0} \tag{1}$$

in which

y = milligrams of antibody precipitated
a = milligrams of antigen added
b_0 = total milligrams of antibody
R = antibody/antigen weight ratio at equivalence point

This equation for the composition of the precipitate, which corresponds to change from $2R$ at large antibody excess to R at the equivalence point, has been shown (15) to be in satisfactory agreement with the excellent experimental data obtained by the authors. In view of the arbitrary and unlikely assumptions originally used for its derivation, it is gratifying that Kendall himself (74) has recently derived the equation in another way, and that we have found (unpublished work) that the equation is obtained as an approximation from general considerations of chemical equilibrium when the assumption is made that the ratio may vary between the limits $2R$ and R and an expansion is made in powers of a/b.

Kendall's derivation is essentially the following. (He considers also some more general cases.)

Let antibody and antigen both be bivalent. For B_0 and A_0 molecules of antibody and antigen, respectively, there are $2B_0$ and $2A_0$ combining groups. Assume, for antibody excess, that all of the $2A_0$ antigen groups are bonded to antibody groups, and that they are distributed at random among the $2B_0$ antibody groups, without regard to whether or not the antibody molecule is already bonded at the other end. Since the chance that an antibody group is bound is $2A_0/2B_0$, the chance that it is free is $1—A_0/B_0$, and the fraction of antibody molecules free at both ends is $(1—A_0/B_0)^2$. The fraction not free at both ends is $1—(1—A_0/B_0)^2 = 2A_0/B_0—(A_0/B_0)^2$, and the number not free at both ends is this multiplied by the total number, B_0. If it be assumed that all antibody molecules not free at both ends are carried down in the precipitate with the antigen molecules the molecular ratio for the precipitate becomes

$$\frac{B_{\text{pp}}}{A_{\text{pp}}} = 2 - A_0/B_0 \tag{2}$$

and, introducing the ratio R of molecular weights, the weight ratio is found to be

$$\frac{y}{a} = 2R - R^2 \frac{a}{b_0} \tag{3}$$

This is identical with equation 1.

Similar considerations have also been used by Ghosh (75) for the derivation of related equations.

An involved theory of antigen-antibody equilibria based in part on probability considerations has been extensively developed by Hershey (45). The theory, in common with others based on multivalent antigen and antibody, is in qualitative and rough quantitative accord with experiment.

The only theory of the precipitation reaction which, following the program begun by Arrhenius, has been developed by straightforward application of the principles of chemical equilibrium is that of Pauling, Pressman, Campbell and Ikeda (19). This theory applies only to relatively simple systems, namely, those composed of bivalent antigen and bivalent antibody, univalent hapten, certain soluble complexes, and precipitate with invariant composition AB.

In order to show the nature of the treatment, we present here the derivation of the equation for the amount of precipitate formed in absence of hapten, generalized over the earlier treatment by consideration also of the complex AB_2.

Let the molecular species A, B, AB, A_2B, and AB_2 in solution be in equilibrium with each other and with solid AB_{pp}. We represent the concentrations of the five solutes by the symbols

$$\begin{aligned} [A] &= \alpha \\ [B] &= \beta \\ [AB] &= s \\ [A_2B] &= a \\ [AB_2] &= b \end{aligned}$$

The quantities α, β, a, and b are variable, whereas s is constant for a given system with precipitate present; it is the solubility of the precipitate. The equilibrium expressions for the three reactions

$$\begin{aligned} A + B &= AB \\ A + AB &= A_2B \\ B + AB &= AB_2 \end{aligned}$$

are respectively

$$\frac{s}{\alpha\beta} = 4K \tag{4}$$

$$\frac{a}{\alpha s} = K \tag{5}$$

$$\frac{b}{\beta s} = K'' \tag{6}$$

For simplicity we have assumed that each of the bonds in the complex A—B—A has the same strength as the bond in AB; this is probably a good approximation, in view of the fact that the two bonding regions are probably far apart on the large antibody molecules. (The theory can be carried through without this assumption.) The constant K corresponds to equilibrium for one antibody valence and one antigen valence, and the factor 4 is an entropy factor or symmetry factor. We use K'' rather than K for the second bond in AB_2 because steric repulsion between the two antibody molecules attached to the same small antigen would be expected to decrease the stability of this complex.

The expressions for the total amounts of antigen and antibody in the system (per unit volume of solution) are

$$AB_{\text{pp}} + s + \alpha + 2a + b = A_{\text{total}} \tag{7}$$

and

$$AB_{\text{pp}} + s + \beta + a + 2b = B_{\text{total}} \tag{8}$$

Subtracting equation 8 from 7 we obtain

$$\alpha - \beta + a - b = A_{\text{total}} - B_{\text{total}}$$

Eliminating β, a, and b with the use of equations 4, 5, and 6 we obtain

$$\alpha - \frac{s}{4K\alpha} + Ks\alpha - \frac{K''s^2}{4K\alpha} = A_{\text{total}} - B_{\text{total}}$$

This quadratic equation in α gives on solution

$$\alpha = \frac{1}{2(1 + Ks)}\{(A_{\text{total}} - B_{\text{total}} + [s(1 + K''s)(1 + Ks)/K + (A_{\text{total}} - B_{\text{total}})^2]^{\frac{1}{2}} \tag{9}$$

(The positive rather than the negative sign before the radical is seen to be correct by the consideration of limiting cases.) From equation 7 we find on eliminating a and b the expression

$$AB_{\text{pp}} = A_{\text{total}} - s - (1 + Ks)\alpha - \frac{K''s^2}{4K\alpha} \tag{10}$$

Equations 9 and 10 give the solution to our problem; from 9 the value of α is to be found in terms of A_{total} and B_{total} and the parameters of the system s, K, and K'', and this on substitution in 10 gives the amount of precipitate.

It is shown in the original paper how this equation (with $K'' = 0$) accounts for many observed properties of antigen-antibody systems.

Future progress which may be anticipated involves the extension of this straightforward thermodynamic treatment to include the case of variable composition and randomness of structure of the solid phase. This will require the development of satisfactory approximate expressions for the free energy of such a solid phase, by application of the methods of statistical mechanics on the basis of sound structural concepts. This problem is not an easy one; but fortunately promising methods for attacking it have been developed in recent years by able theoretical physicists interested in the problem of the stability of alloys with greater or smaller degree of randomness of atomic arrangement, and we may be confident that great progress will soon be made in the formulation of a satisfactory quantitative theory of the precipitation reaction.

SUMMARY

The forces responsible for combination and attraction of antigen and antibody molecules may be classified as electronic van der Waals attraction, Coulomb attraction, attraction of electric dipoles or multipoles, formation of hydrogen bonds, etc. The specificity of interaction of antigen and antibody molecules arises from their structural complementariness, which permits close contact of the molecules over sufficient area for these weak forces to co-operate in forming a strong antigen-antibody bond.

The weight of evidence indicates that further combination of the initial antigen-antibody complexes to form a precipitate is a specific rather than a nonspecific reaction and is due to a continuation of the primary combination step to form a framework structure of alternate antigen and antibody molecules.

Furthermore it appears that both precipitating antigen and precipitating antibody must be multivalent, at least bivalent.

The more recent quantitative theories of the precipitation reaction are discussed.

REFERENCES

(1) ARRHENIUS, S. Immunochemistry. Macmillan and Co., 1907, p. 31.

(2) PAULING, L. The nature of the chemical bond and the structure of molecules and crystals. Cornell University Press, Ithaca, N. Y., 2nd ed., 1940.

(3) KRAUS, R. Wien. klin. Wchnschr. **10:** 736, 1897.

(4) KRAUS, R. In KOLLE and WASSERMANN. Handbuch path. Microörganismen **4:** 592, 1904.

(5) DEAN, H. R. A system of bacteriology, (V. 6—Immunity p. 424). Medical Research Council, His Majesty's Stationery Office, London, 1931.

(6) MARRACK, J. R. The chemistry of antigens and antibodies. Report no. 230 of the Medical Research Council, His Majesty's Stationery Office, London, 1934; second edition, 1938.

(7) HARTLEY, P. Brit. J. Exper. Path. **7:** 55, 1926.

(8) FELTON, L. D. AND G. H. BAILEY. J. Infect. Dis. **38:** 131, 1926.
(9) HEIDELBERGER, M. ET AL. J. Exper. Med. **64:** 161, 1936; **67:** 181, 1938; **68:** 913, 1938.
(10) LANDSTEINER, K. The specificity of serological reactions. C. C. Thomas, Springfield, Ill., 1936.
(11) BREINL, F. AND F. HAUROWITZ. Ztschr. physiol. Chem. **192:** 45, 1930.
(12) ALEXANDER, J. Protoplasma **14:** 296, 1931.
(13) MUDD, S. J. Immunol. **23:** 423, 1932.
(14) HEIDELBERGER, M. Chem. Rev. **24:** 323, 1939; J. Exper. Med. **61:** 563, 1935.
(15) HEIDELBERGER, M. AND F. E. KENDALL. J. Exper. Med. **50:** 809, 1929.
(16) HAUROWITZ, F. AND F. BREINL. Ztschr. physiol. Chem. **214:** 111, 1933.
(17) PAULING, L. J. Am. Chem. Soc. **62:** 2643, 1940.
(18) PAULING, L., D. H. CAMPBELL AND D. PRESSMAN. Proc. Nat. Acad. Sci. **27:** 125, 1941.
PAULING, L., D. PRESSMAN, D. H. CAMPBELL, C. IKEDA AND M. IKAWA. J. Am. Chem. Soc. **64:** 2994, 1942.
(19) PAULING, L., D. PRESSMAN, D. H. CAMPBELL AND C. IKEDA. J. Am. Chem. Soc. **64:** 3003, 1942.
(20) PAULING, L., D. PRESSMAN AND C. IKEDA. J. Am. Chem. Soc. **64:** 3010, 1942.
(21) LANDSTEINER, K. AND J. VAN DER SCHEER. Proc. Soc. Exper. Biol. and Med. **29:** 747, 1932; J. Exper. Med., **56:** 399, 1932.
(22) RICE, O. K. Electronic structure and chemical binding. McGraw-Hill Book Co., New York, 1940.
(23) LONDON, F. Ztschr. Physik **63:** 245, 1930.
(24) SCHWARZENBACH, G. Ztschr. physik. Chem. **A176:** 133, 1936.
(25) LATIMER, W. M. AND W. H. RODEBUSH. J. Am. Chem. Soc. **42:** 1419, 1920.
(26) ROTHEN, A. AND K. LANDSTEINER. Science **90:** 65, 1939.
(27) HEKTOEN, L. AND A. K. BOOR. J. Infect. Dis. **48:** 588, 1931.
(28) PRESSMAN, D., D. H. BROWN AND L. PAULING. J. Am. Chem. Soc. **62:** 3015, 1942.
(29) EHRLICH, P. Studies on immunity. John Wiley and Sons, New York, 1906.
(30) BUCHNER, H. Münch. med. Wchnschr. **40:** 449, 1893; **47:** 277, 1900.
(31) BURNET, F. M. The production of antibodies. Macmillan & Co., Ltd., Melbourne, Australia, 1941, Chap. X.
(32) JORDAN, P. Ztschr. Physik **113:** 431, 1939; Ztschr. f. Immunitätsforsch. **97:** 330, 1940; Naturwissenschaften **29:** 89, 1941.
(33) PAULING, L. AND M. DELBRÜCK. Science **92:** 77, 1940.
(34) HAUROWITZ, F., M. VARDAR AND P. SCHWERIN. J. Immunol. **43:** 327, 1942.
(35) TOPLEY, W. C. C., J. WILSON AND J. T. DUNCAN. Brit. J. Exper. Path. **16:** 116, 1935.
(36) ABRAMSON, H. A. Nature **135:** 995, 1935.
(37) HOOKER, S. B. AND W. C. BOYD. J. Immunol. **33:** 337, 1937.
(38) HEIDELBERGER, M. Bact. Rev. **3:** 49, 1939.
(39) TEAGUE, O. AND C. W. FIELD. J. Exper. Med. **9:** 86, 1907.
(40) EAGLE, H. J. Immunol. **18:** 393, 1930.
(41) PRESSMAN, D., D. H. CAMPBELL AND L. PAULING. J. Immunol. **44:** 101, 1942.
(42) BOYD, W. C. AND S. B. HOOKER. Proc. Soc. Exper. Biol. and Med. **39:** 491, 1938.
(43) HEIDELBERGER, M. AND E. A. KABAT. J. Exper. Med. **65:** 885, 1937.
(44) DEAN, H. R. Proc. Roy. Soc. Med. **5:** 62, 1911.
(45) HERSHEY, A. D. J. Immunol. **42:** 455, 1941.
(46) HOOKER, S. B. AND W. C. BOYD. J. Immunol. **42:** 419, 1941.
(47) BOYD, W. C. J. Exper. Med. **75:** 407, 1942.
(48) HOOKER, S. B. AND W. C. BOYD. J. Immunol. **23:** 446, 1932.
(49) HAUROWITZ, F., F. KRAUS AND F. MARX. Ztschr. physiol. Chem. **245:** 23, 1936.
(50) PAPPENHEIMER, A. M., JR., H. P. LUNDGREN AND J. W. WILLIAMS. J. Exper. Med. **71:** 247, 1940.
(51) PORTER, E. F. AND A. M. PAPPENHEIMER, JR. J. Exper. Med. **69:** 755, 1939.

(52) Fourt, W., W. D. Harkins and G. G. Wright. Abstract 98th Meeting Am. Chem. Soc., 1939.
(53) Marrack, J. R. Ann. Rev. Biochem. **11:** 629, 1942.
(54) Heidelberger, M. J. Am. Chem. Soc. **60:** 242, 1938.
(55) Boyd, W. C. and S. B. Hooker. J. Gen. Physiol. **17:** 341, 1934; **22:** 281, 1939.
(56) Kleczkowski, A. Brit. J. Exper. Path. **22:** 44, 1941.
(57) Liu, S. C. and H. Wu. Chinese J. Physiol. **15:** 237, 1940.
(58) Wu, H., L. H. Cheng and C. P. Li. Proc. Soc. Exper. Biol. **25:** 407, 1928.
(59) Culbertson, J. T. J. Immunol. **23:** 439, 1932.
(60) Marrack, J. R. and F. C. Smith. Brit. J. Exper. Med. **12:** 30, 1931.
(61) Pappenheimer, A. M., Jr. and E. S. Robinson. J. Immunol. **32:** 291, 1937.
(62) Heidelberger, M. and F. E. Kendall. J. Exper. Med. **62:** 697, 1935.
(63) Pappenheimer, A. M., Jr. J. Exper. Med. **71:** 263, 1940.
(64) Heidelberger, M., H. P. Treffers and M. Mayer. J. Exper. Med. **71:** 271, 1940.
(65) Hooker, S. B. and W. C. Boyd. Ann. N. Y. Acad. Sci. **43:** 107, 1942.
(66) Shibley, C. S. J. Exper. Med. **50:** 825, 1929.
(67) Tyler, A. Biol. Bull. **81:** 190, 1941.
(68) Madsen, T. Comm. de l'institut serotherapique de l'Etat Danois, 1907–13.
(69) Heidelberger, M. and F. E. Kendall. J. Exper. Med. **64:** 161, 1936.
(70) Liu, S. and H. Wu. Chinese J. Physiol. **15:** 465, 1940.
(71) Felton, L. D. J. Immunol. **22:** 453, 1932.
(72) Boyd, W. C. J. Immunol. **38:** 143, 1940.
(73) Healey, M. and S. Pinfield. Brit. J. Exper. Path. **16:** 535, 1935.
(74) Kendall, F. E. Ann. N. Y. Acad. Sci. **43:** 85, 1942.
(75) Ghosh, B. N. Ind. J. Med. Research **23:** 285, 1935.

[Reprinted from the Journal of the American Chemical Society, **66**, 330 (1944).]

[CONTRIBUTION FROM THE GATES AND CRELLIN LABORATORIES OF CHEMISTRY, CALIFORNIA INSTITUTE OF TECHNOLOGY, No. 947]

The Serological Properties of Simple Substances. VI. The Precipitation of a Mixture of Two Specific Antisera by a Dihaptenic Substance Containing the Two Corresponding Haptenic Groups; Evidence for the Framework Theory of Serological Precipitation

BY LINUS PAULING, DAVID PRESSMAN AND DAN H. CAMPBELL

The framework theory (lattice theory) of serological precipitation and agglutination, first proposed by Marrack,[1] was shown by Marrack and by Heidelberger and Kendall[2] to account for many experimental observations. Because of its simplicity and its compatibility with the available information about intermolecular forces, this theory was incorporated in his general theory of the structure and process of formation of antibodies by one of the present authors.[3]

Strong support of the framework theory has been provided during the past two years by the results of extensive studies of the reactions of antibodies and simple substances,[4] based upon the observations by Landsteiner and Van der Scheer[5] of the precipitation of antibody by certain simple substances containing two haptenic groups. It was found[4] from experiments with about fifty substances that all of those (about twenty) containing two or more haptenic groups (azophenylarsonic acid groups) per molecule gave precipitates with antiserum homologous to this haptenic group, and that none of the monohaptenic substances gave a precipitate. This fact is most readily accounted for by the framework theory.

The argument might be made, however, that no more than one of the haptenic groups of a molecule of a polyhaptenic substance is involved in interaction with antibody molecules, and that the difference in precipitability of polyhaptenic and monohaptenic substances with antiserum is due to some difference in properties of these two classes of substances, such as a tendency to asso-

(1) J. R. Marrack, "The Chemistry of Antigens and Antibodies," Report No. 194 of the Medical Research Council, His Majesty's Stationery Office, London, 1934; Second Edition, Report No. 230, 1938.

(2) M. Heidelberger and F. E. Kendall, *J. Exptl. Med.*, **61**, 559, 563; **62**, 467, 697 (1935); M. Heidelberger, *Chem. Rev.*, **24**, 323 (1939).

(3) Linus Pauling, THIS JOURNAL, **62**, 2643 (1940).

(4) Linus Pauling, David Pressman, Dan H. Campbell, and collaborators, THIS JOURNAL, **64**, 2994, 3003, 3010, 3015 (1942); **65**, 728 (1943).

(5) K. Landsteiner and J. Van der Scheer, *Proc. Soc. Exptl. Biol. Med.*, **29**, 747 (1932).

ciate into colloidal particles.[6] In order to test this point, we have carried out a new experiment, the results of which show that each of the two haptenic groups of a dihaptenic substance enters into specific combination with antibody in the formation of the precipitate.

The experiment[7] was made with use of two different antisera, one of which (anti-R serum) was prepared by injecting rabbits with azoprotein containing R groups (*p*-azophenylarsonic acid groups) and the other (anti-X serum) by injecting with azoprotein containing X groups (*p*-azobenzoic acid groups). A substance was used as precipitating antigen which had one R group and one X group per molecule. It was found that this RX substance does not give a precipitate with either anti-R serum or anti-X serum alone, but does give a precipitate with a mixture of these two antisera. Similar results were obtained also with another RX substance.

This striking experimental result corresponds exactly to the behavior predicted by the framework theory: according to this theory the RX substance cannot give a precipitate with anti-R serum because its molecules contain only one R group, and are hence univalent with respect to anti-R antibodies and so are unable to form a framework with them; and similarly it can not give a precipitate with anti-X serum. (Indeed, it is predicted, and verified by experiment, that this effectively monohaptenic substance can inhibit the precipitation of either anti-R serum or anti-X serum by corresponding polyhaptenic simple substances or azoproteins.) But the RX substance is effectively dihaptenic and bivalent with respect to a mixture of anti-R serum and anti-X serum, since each molecule can form two bonds, one with an anti-R antibody molecule and one with an anti-X antibody molecule; and accordingly a specific framework precipitate, containing equal numbers of the two kinds of antibodies, can be formed with the mixed antiserum.

Let us now ask to what extent this experimental result supports the framework theory—whether it might not be equally compatible with some other theory. The answer is that our experiment shows that both of the two haptenic groups R and X of the RX molecule enter into specific combination with antibody, that the molecule of precipitating antigen is hence truly bivalent, and that this bivalence is necessary for precipitation. This, however, is common to two theories—the framework theory, which requires that the antibody molecules, as well as the antigen molecules, be bivalent or multivalent, and an alternative theory, which is based on the assumption that a complex of a multivalent antigen molecule and two or more univalent antibody molecules is for some reason or other insoluble, and constitutes the precipitate.[8]

Basis for decision between these theories is provided by the results of two other experiments, one relating to the antibody–antigen molecular ratio in the precipitate and the other to the solubility of the precipitate in excess antibody. Since we know that the RX precipitating antigen is truly bivalent, the effective valence of the antibody molecules can be calculated from the antibody–antigen molecular ratio. The value found by analysis for this ratio is 0.7, which corresponds to an average effective antibody valence of $2/0.7 = 2.8$; for univalent antibody the molecular ratio would be 2. (The assumption is made here that the antigen molecules are not associated into complexes, which might then form a precipitate with antibody with use of only some of the haptenic groups. Evidence against association of this sort is furnished by the failure of the RX substances to precipitate with anti-R serum or anti-X serum alone.)

The second experiment is based on the following argument. If antibody molecules are univalent, and the precipitate consists of RX antigen molecules each of which has two attached antibody molecules, increase in antibody concentration, with amount of antigen held constant, would necessarily increase the amount of precipitate; decrease in the amount of precipitate, which could occur only by formation of a soluble complex, would not occur because the antigen molecules in the precipitate would be already saturated with antibody, and so could not increase their valence, and the antibody molecules could not decrease their valence (below the value 1) and remain attached to antigen. But a framework precipitate with multivalent antibody could dissolve in excess antibody by formation of soluble complexes, the effective valence of the antibody molecules in these complexes being less than that in the precipitate (1 instead of 2 or 3), and the antibody–antigen ratio being greater than that for the precipitate; according to the principles of chemical equilibrium, increase in the antibody concentration would then lead to solution of the precipitate. The observation that increase in the amount of mixed antiserum results in pronounced decrease in the amount of precipitate formed with the RX substance accordingly eliminates the theory of univalent antibody and provides further proof of the framework theory.

A detailed discussion of these experiments and of other experiments with the RX substance (hapten inhibition, etc.) is given below, following the section on experimental methods and results.

Experimental Methods and Results

The following substances containing groups R =

(6) K. Landsteiner, "The Specificity of Serological Reactions," Charles C. Thomas, Baltimore, Md., 1936, p. 120.

(7) A brief statement about this work has been published in *Science*, **98**, 263 (1943).

(8) The "occlusion" theory proposed by W. C. Boyd, *J. Exptl. Med.*, **75**, 407 (1942), is a theory of this sort. See also F. Haurowitz and P. Schwerin, *British J. Exptl. Path.*, **23**, 146 (1942).

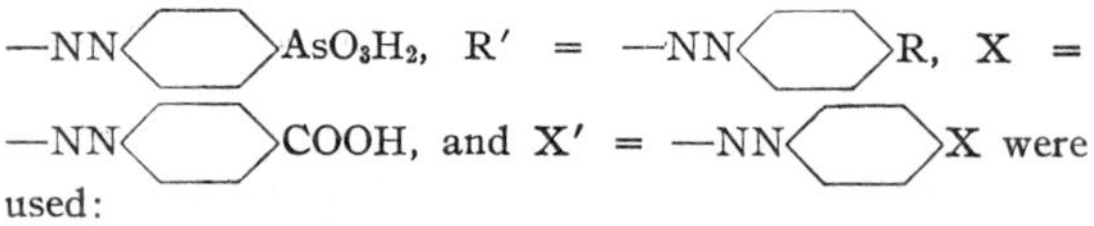
—NN AsO3H2, R' = —NN R, X =
—NN COOH, and X' = —NN X were

used:

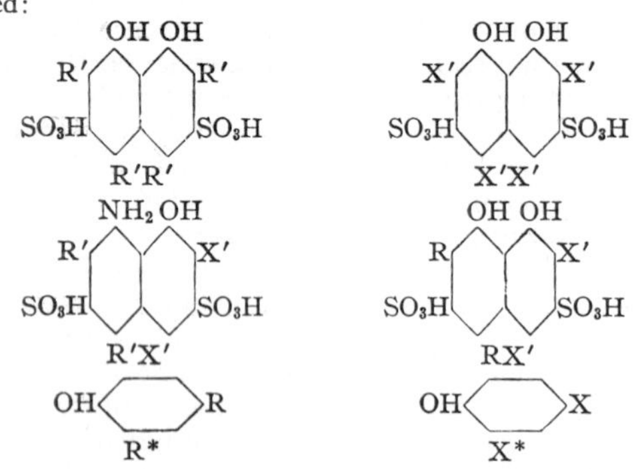

The substances R'X' and RX' are the "RX substances" referred to in the preceding discussion.

Preparation.—The substances R'R' and R* have already been described,[4] and X'X' and X* are described in a paper to be published soon.[9] The substance R'X' was prepared by diazotizing 1.3 g. (0.0038 mole) of *p*-(*p*-aminophenylazo)-phenylarsonic acid, removing excess nitrite with urea, and coupling with 1.35 g. (0.0040 mole) of recrystallized "H-acid" in acetate buffer. The intermediate red compound was precipitated by addition of hydrochloric acid and washed twice with 500 ml. of water containing 10 ml. of 6 *N* hydrochloric acid and 3 g. of sodium chloride. The substance was then dissolved in sodium carbonate solution and added to a diazotized solution of 0.312 g. (0.0012 mole) of *p*-(*p*-aminophenylazo)-benzoic acid.[9] The product was precipitated with hydrochloric acid, dissolved in 500 ml. of sodium hydroxide solution, and reprecipitated with hydrochloric acid and sodium chloride and an equal volume of ethanol, leaving most of the remaining intermediate in solution. The process of solution and precipitation was repeated until the solution showed constant color, and sodium chloride was then removed from the product by extraction with 90% ethanol. Although the product was not crystallized, there is little doubt of its identity and purity; the method of preparation is based on the fact that at low *p*H substitution occurs in the 7 position of H-acid, ortho to an amino group, and not in the 2 position, ortho to a hydroxyl group, reaction at the latter position occurring in basic solution.

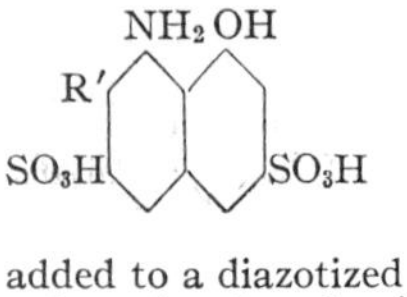

The RX' compound was made by diazotizing 4.0 g. (0.020 mole) of *p*-arsanilic acid and adding it very slowly to 6.0 g. (0.019 mole) of chromotropic acid in solution with sodium carbonate (final *p*H 9). After one hour a diazotized and neutralized solution of 1.3 g. (0.005 mole) of *p*-(*p*-aminophenylazo)-benzoic acid was added and the *p*H was adjusted to 9. The product was purified in the way described above; this process removes any RR compound which is formed.

Anti-R, anti-R', and anti-X sera were prepared as described elsewhere,[4,9] and the precipitation experiments were carried out in the usual way.[4] Several pools of antisera were tested; the same pools of anti-R and anti-X antisera were used for all of the quantitative experiments reported in this paper.

The substance RX' was found not to precipitate with anti-X serum, anti-R serum, or anti-R' serum, or with a mixture of anti-R' serum and anti-X serum; it formed heavy precipitates with a mixture of anti-R serum and anti-X serum.

(9) David Pressman, Stanley M. Swingle, Allan L. Grossberg and Linus Pauling, paper to be submitted for publication in THIS JOURNAL.

The substance R'X' gave no precipitates or only very slight precipitates with various pools of anti-X serum, anti-R serum, or anti-R' serum; heavy precipitates were formed with mixtures of anti-R serum and anti-X serum and with mixtures of anti-R' serum and anti-X serum. The quantitative experiments reported in Tables I to V were carried out with pools of sera which separately gave no precipitate with R'X'.

The failure of the substance RX' to precipitate with a mixture of anti-R' serum and anti-X serum is related to the fact[4] that anti-R' serum in general does not precipitate with polyhaptenic substances containing R groups.

It was found that the substances RX' and R'X' act as monohaptenic substances in inhibiting the specific precipitation of R'R' with anti-R' serum or anti-R serum and of X'X' with anti-X serum. Data showing the effect of haptens R* and X* on some precipitation reactions are given in Table II.

TABLE I

PRECIPITATION OF ANTIGENS R'R', X'X', R'R' + X'X', AND R'X' WITH MIXTURES OF ANTI-R SERUM AND ANTI-X SERUM

Antigen solution, 2 ml.; mixed antiserum, 1 ml.; 1 hour at room temperature and 2 nights in refrigerator; *p*H of supernates 8.1.

Ratio anti-R serum: anti-X serum[a]	Antigen	Moles of antigen × 10⁹: 6	18	54	167	500
		Amount of precipitated antibody, μg.[b]				
93:7	R'R'	52	522	1134	721	162
	X'X'	0	0	0	0	0
	R'R' + X'X'[c]	14	136	913	1070	564
	R'X'	48	60	72	60	60
62:38	R'R'	44	(261)	(446)	(169)	25
	X'X'	4	22	80	50	23
	R'R' + X'X'[c]	8	104	439	481	236
	R'X'	22	158	792	821	623
17:83	R'R'	(12)	16	9	5	4
	X'X'	9	23	706	756	280
	R'R' + X'X'[c]	9	12	137	1024	906
	R'X'	12	16	25	66	90

[a] The ratio 62:38 was selected as that for which the optimum concentrations of the antigens R'R' and X'X' are the same; the other ratios are 8 and 1/8 times as great, respectively. [b] Average of triplicate analyses, with mean deviation ±2%; duplicate analyses in parentheses. [c] In equimolal amounts, with totals as given above.

Discussion of Results and Comparison with Theory

The qualitative behavior of the substances RX' and R'X' in precipitation and in hapten inhibition, as described above, is just that predicted by the framework theory. The quantitative data given in Tables I and II are also in accord with this theory. In particular, there is some verification of the prediction of the framework theory that the precipitate with R'X' should contain equal numbers of anti-R molecules and anti-X molecules; this would require that the amount of precipitate be small in case that either anti-R or anti-X serum be present in small amount, in agreement with observation.

It is also predicted, and verified by experiment, that precipitation of mixed antiserum with R'X' can be completely inhibited by the presence of either hapten R* or hapten X*, whereas precipitation with a mixture of R'R' and X'X' is only partially inhibited by either of these haptens.

TABLE II

INHIBITION BY HAPTENS R* AND X* OF PRECIPITATION OF ANTIGENS WITH MIXTURES OF ANTI-R SERUM AND ANTI-X SERUM

Antigen solution, 1 ml.; mixed antiserum, 1 ml.; hapten solution, 1 ml.; 1 hour at room temperature and 2 nights in refrigerator; *p*H of supernates 8.1.

Ratio anti-R serum: anti-X serum	Antigen[a]	0	Moles of hapten R* × 10⁹ 37	111	333	1000	Moles of hapten X* × 10⁹ 37	111	333	1000
		Amount of precipitated antibody, μg.[b]								
93:7	R'R' + X'X'[c]	985	441	102	20	0	956	949	944	1004
	R'X'	64	41	(21)	(10)	0	50	27	19	5
62:38	R'R'	321	85	19	5	0	315	311	305	292
	X'X'	(71)	(61)	(59)	57	56	(43)	(15)	0	0
	R'R' + X'X'[c]	344	132	61	52	51	316	342	302	300
	R'X'	827	527	290	88	15	553	219	48	15
17:83	R'R' + X'X'[c]	496	502	470	434	319	223	48	11	11
	R'X'	49	39	31	21	11	34	17	9	7

[a] Total moles of antigen used: R'R', 50×10^{-9}; X'X', 50×10^{-9}; R'R' + X'X', 100×10^{-9}; R'X', 100×10^{-9}.
[b] Average of triplicate analyses, with mean deviation ±2%; duplicate analyses in parentheses. [c] In equimolal amounts.

TABLE III

ANTIBODY-ANTIGEN MOLECULAR RATIO IN PRECIPITATES FORMED BY R'X' ANTIGEN AND A MIXTURE OF ANTI-R SERUM AND ANTI-X SERUM

Ratio of antisera 62:38; equal volumes of antigen solution and mixed antiserum: 1.5 ml. of each for first series, 2.0 ml. for second; 1 hour at room temperature and 2 nights in refrigerator; *p*H of supernates 8.1.

Moles of antigen × 10⁹	Amount of precipitated antibody[a] μg.	moles	Amount of precipitated antigen[a] μg.	moles	Molecular ratio, antibody/antigen[a]
100	1421	8.89×10^{-9}	10.8	11.7×10^{-9}	0.76
33	456	2.85	4.2	4.5	.63

[a] Average of triplicate analyses, mean deviation ±2% for antibody, ±2% for antigen, ±2% for ratio. Assumed molecular weight of antibody 160,000, of antigen 927.

TABLE IV

PRECIPITATION OF MIXTURES OF NORMAL SERUM AND ANTISERUM AT CONSTANT AMOUNT OF ANTIGEN

Antigen solution, 2 ml., containing either 5×10^{-9} moles of antigen R'R' (series A) or 6.7×10^{-9} moles of antigen R'X' (series B); mixture of antiserum and normal serum, 2 ml.; 1 hour at room temperature and 2 nights in refrigerator; *p*H of all supernates 8.1.

Antiserum concentration in mixture[a]	Amount of antibody precipitated, μg.[b] Series A[c]	Series B[d]
1.00	39	39
0.67	35	106
.45	93	120
.30	128	109
.20	133	81
.13	120	47
.088	90	23
.059	51	6

[a] Anti-R serum (Series A) or 62:38 mixture of anti-R serum and anti-X serum (series B) diluted with normal rabbit serum to the extent indicated in this column. [b] Averages of triplicate analyses, with mean deviation ±3%. [c] Blank of anti-R serum and buffer, 5 μg.; blank of normal serum and antigen R'R', 5 μg. [d] Blank of 62:38 antiserum mixture, 6 μg.; blank of normal serum and antigen R'X', 5 μg.

An interesting feature of the data in Tables I and II is that the 62:38 mixture of antisera gives a larger amount of precipitate with the R'X' antigen than with R'R' and X'X' together. This is presumably related to the fact that the amount of precipitate formed with antigen R'R' (or X'X') alone falls off very rapidly with decrease in the fraction of anti-R (or anti-X) antiserum in the mixture; for example, the amount of precipitate formed by X'X' with the 62:38 mixture is only about one-ninth of that formed with the 17:83 mixture, although the first mixture contains nearly one-half as much anti-X antibody as the second. The amount of precipitate formed by antigens R'R' and X'X' together would be expected to be about equal to the sum of the amounts formed by these antigens separately, since the two precipitates are mutually independent. But a single precipitate is formed by R'X', containing equal amounts of the two kinds of antibody, and accordingly the amount of precipitate would be expected to be about equal to that formed by R'R' or X'X' with a mixture containing double the amount of anti-R serum or anti-X serum as is, indeed, observed.

It has been noted in general that simple antigens, in contradistinction to protein antigens, are far from completely precipitated by antisera, and that the fraction of the antigen remaining in solution at the optimum region increases on dilution of the antiserum with normal serum or heterologous antiserum; this effect, as discussed above, is to be noted in Table I. We have not yet carried out a sufficiently detailed experimental study of these phenomena to justify an extended discussion of them.

A mathematical discussion based on the simplified model treated previously[4,10] can be carried out for the RX precipitation reaction, leading to results in general correspondence with experiment. Let A represent the R'X' antigen, $B^{(1)}$ and $B^{(2)}$ the anti-R and anti-X antibodies, respectively, which are assumed to be bivalent and homogeneous, and $H^{(1)}$ and $H^{(2)}$ the haptens R* and X*, respectively. It is assumed that, in addition to the precipitate $A_2B^{(1)}B^{(2)}$, certain soluble complexes exist, and that their equilibrium constants of formation have the values given below; these values have been assigned on the basis of structural considerations, with inclusion of the entropy term arising from the symmetry numbers of the molecules.

$$\begin{array}{rll}
A + B^{(1)} = AB^{(1)} & & 2K_1 \\
A + B^{(2)} = AB^{(2)} & & 2K_2 \\
2A + B^{(1)} = AB^{(1)}A & & K_1^2 \\
2A + B^{(2)} = AB^{(2)}A & & K_2^2 \\
2A + B^{(1)} + B^{(2)} = AB^{(1)}AB^{(2)} & & 4K_1^2K_2'' \\
2A + B^{(1)} + B^{(2)} = AB^{(2)}AB^{(1)} & & 4K_1''K_2^2
\end{array}$$

(10) Linus Pauling, Dan H. Campbell, and David Pressman, *Physiol. Rev.*, **23**, 203 (1943).

$$H^{(1)} + B^{(1)} = H^{(1)}B^{(1)} \qquad 2K_1'$$
$$H^{(2)} + B^{(2)} = H^{(2)}B^{(2)} \qquad 2K_2'$$
$$2H^{(1)} + B^{(1)} = H^{(1)}B^{(1)}H^{(1)} \qquad K_1'^2$$
$$2H^{(2)} + B^{(2)} = H^{(2)}B^{(2)}H^{(2)} \qquad K_2'^2$$
$$A + B^{(1)} + H^{(1)} = AB^{(1)}H^{(1)} \qquad 2K_1K_1'$$
$$A + B^{(2)} + H^{(2)} = AB^{(2)}H^{(2)} \qquad 2K_2K_2'$$

By straightforward solution of the equilibrium equations there is obtained for the case that no hapten is present a set of simultaneous equations

$$\beta_1^2(1 + K_1\alpha)^2 - \beta_1(B^{(1)}_{\text{total}} - B^{(2)}_{\text{total}}) - S(1 + K_2\alpha)^2/\alpha^2 = 0 \quad (1)$$

$$A_{\text{total}} - B^{(1)}_{\text{total}} - B^{(2)}_{\text{total}} - \alpha + \beta_1(1 - K_1^2\alpha^2) + S(1 - K_2^2\alpha^2)/\alpha^2\beta_1 = 0 \quad (2)$$

$$A_2B^{(1)}B^{(2)}(\text{pp}) = B^{(1)}_{\text{total}} - \beta_1(1 + K_1\alpha)^2 - s \quad (3)$$

These three equations are to be solved for three unknown quantities, the amount of precipitate $A_2B^{(1)}B^{(2)}$(pp) and the two auxiliary variables α, the concentration of molecular species A in solution, and β_1, the concentration of molecular species $B^{(1)}$; the other quantities are the equilibrium constants K_1 and K_2, the solubility product $S = [A]^2[B^{(1)}][B^{(2)}]$ of the precipitate, and the solubility s of the precipitate as the complexes $AB^{(1)}AB^{(2)}$ and $AB^{(2)}AB^{(1)}$, with $s = 4(K_1''K_2^2 + K_1^2K_2'')S$. A convenient method of solving the equations is to leave A_{total} undetermined, and to introduce numerical values of the other parameters; introduction in Equation 1 of a trial value of α permits solution for β_1, after which A_{total} may be evaluated from Equation 2 and $A_2B^{(1)}B^{(2)}$(pp) from Equation 3.

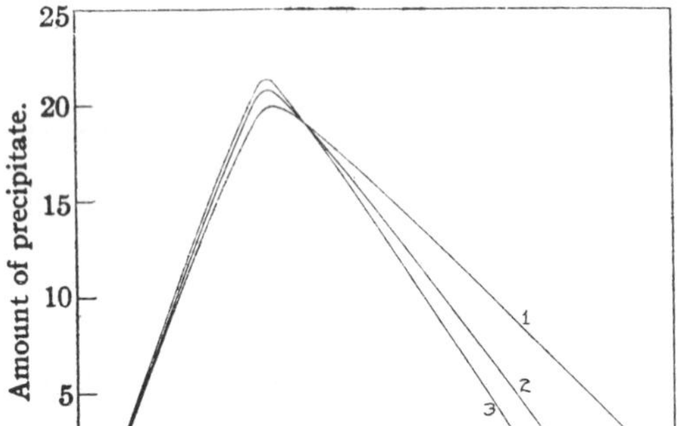

Fig. 1.—Curves showing results of theoretical calculations of amount of precipitate formed by antigen RX with an equimolal mixture of anti-R serum and anti-X serum, as function of amount of antigen: $B^{(1)}_{\text{total}} = B^{(2)}_{\text{total}} = 25$, $s = 2$, and $K_1 = K_1'' = 1$ for all curves; $K_2 = K_2'' = 0.2$ for curve 1, 1 for curve 2, and 5 for curve 3.

Curves showing dependence of amount of precipitate on amount of antigen with equal amounts of the two kinds of antibody present (Fig. 1) are similar to those for the precipitation of one kind of antibody by homologous dihaptenic antigen.[4] As the ratio of the amounts

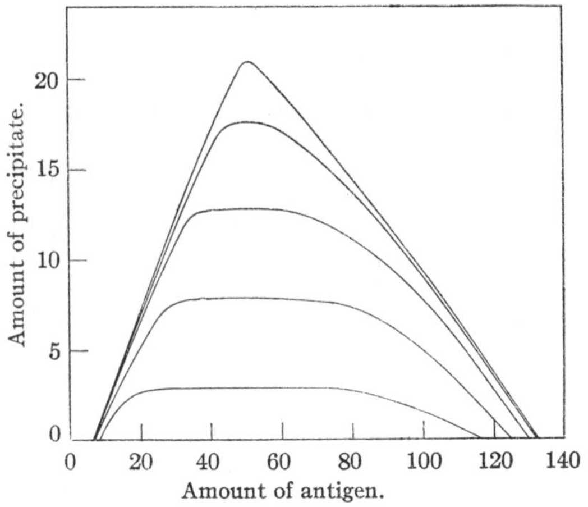

Fig. 2.—Curves showing calculated amount of precipitate as functions of amount of antigen RX for various relative amounts of anti-R and anti-X antibody. All curves are for $K_1 = K_1'' = K_2 = K_2'' = 1$ and $s = 2$. From top to bottom the curves correspond to the following pairs of values of $B^{(1)}_{\text{total}}$ and $B^{(2)}_{\text{total}}$: 25, 25; 20, 30; 15, 35; 10, 40; 5, 45.

of the two kinds of antibody is varied more and more from unity, the calculated curves of the amount of precipitate show broader and broader regions of approximate constancy (Fig. 2), resulting from the buffering action of the antibody present in excess, which combines with excess antigen to form a soluble complex, and thus interferes with the reaction of solution of the precipitate in excess of antigen; it is seen that the antigen concentration at which the amount of precipitate reaches its approximate maximum is determined by the amount of that kind of antibody which is present in smaller quantity, and that the antigen concentration at which the amount of precipitate begins to decrease is determined by the amount of the other kind of antibody. Experimental verification of these details of prediction cannot be expected until essentially homogeneous antibody solutions have been made by fractionation of antisera; however, the predicted general difference in nature is shown by the R′X′ data given in Table I for the 93:7 mixture (broad region with nearly constant amount of precipitate) and the 62:38 mixture (rather narrow optimum region). Verification of the existence of the broad region of nearly constant amount of precipitate was obtained by the repetition of the experiment with the 93:7 mixture, with inclusion of points for smaller and larger amounts of the antigen R′X′; the amount of precipitated antibody was found to be constant (80 ± 10 μg.) over the range 18 to 500 × 10^{-9} mole of antigen, and to decrease for smaller and for larger amounts of antigen (34 μg. at 6.2 × 10^{-9} mole of antigen, 60 μg. at 1500 × 10^{-9}).

When hapten of one kind is also present in the system, the equilibrium expressions may be combined to form the set of simultaneous equations 4 to 7.

$$\beta_1^2(1 + K_1\alpha + K_1'\gamma_1)^2 - \beta_1(B_{\text{total}}^{(1)} - B_{\text{total}}^{(2)}) - S(1 + K_2\alpha)^2/\alpha^2 = 0 \quad (4)$$

$$\beta_1^2\{(1 + K_1'\gamma_1)^2 - K_1^2\alpha^2\} + \beta_1\{A_{\text{total}} - B_{\text{total}}^{(1)} - B_{\text{total}}^{(2)} - \alpha\} + S(1 - \Lambda_2^2\alpha^2)/\alpha^2 = 0 \quad (5)$$

$$H_{\text{total}}^{(1)} = \gamma_1 + 2K_1'\beta_1\gamma_1 + 2K_1'^2\beta_1\gamma_1^2 + 2K_1K_1'\alpha\beta_1\gamma_1 \quad (6)$$

$$A_2B^{(1)}B^{(2)}(\text{pp}) = B_{\text{total}}^{(1)} - \beta_1(1 + K_1\alpha + K_1'\gamma_1)^2 - s \quad (7)$$

Here $H_{\text{total}}^{(1)}$ is the total amount of hapten (of the first kind) added and $\gamma_1 = [H^{(1)}]$ is another auxiliary variable. A similar set of equations can be derived for the case that hapten of the second kind is present. Solution of these equations may be made by selecting a value of γ_1 and finding by successive approximations the values of α and β_1 which satisfy Equations 4 and 5, and then determining $H_{\text{total}}^{(1)}$ and $A_2B^{(1)}B^{(2)}(\text{pp})$ from Equations 6 and 7, respectively.

The calculated curves for hapten inhibition by either kind of hapten when equal amounts of antibodies of the two kinds are present are closely similar to those for the simpler systems previously treated.[4] However, an interesting effect is predicted to occur when one antiserum predominates in the mixture and a small amount of antigen is present: a normal inhibition curve, linear in amount of hapten added, is found from the equations for one hapten, whereas the other hapten, that homologous to the antibody present in the larger amount, is predicted to be ineffective in small amounts, and to produce inhibition of precipitation only after enough of the hapten has been added to neutralize the excess of antibody in solution; calculated curves are given in Fig. 3. To test this prediction experiments were carried out, in triplicate, for the 93:7 mixture with the R′X′ antigen (antigen solution, 2 ml., containing 200 × 10^{-9} moles of R′X′ antigen; mixed antiserum, 2 ml.; hapten solution, 2 ml.; one hour at room temperature and two nights in refrigerator; *p*H of supernates 8.1; mean deviation of triplicate analyses ± 2%); the results are shown in Fig. 4. It is seen that there is a small initial region for which inhibition by the R* hapten does not occur. (In the comparison of Figs. 3 and 4 it should be remembered that the curves for Fig. 3 are calculated for haptens of equal bond-strength constant K', whereas K' for hapten R* is considerably larger than that for X*.) It is likely that the predicted effect would be shown more strikingly by fractionated antibody solutions than by the heterogeneous antisera used in these experiments.

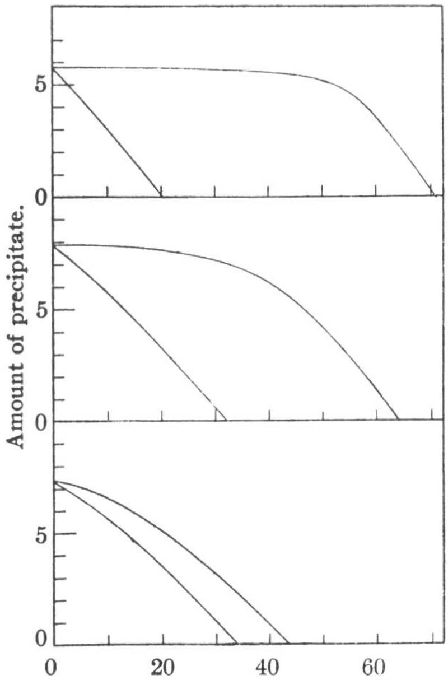

Fig. 3.—Calculated effects of haptens $H^{(1)}$ and $H^{(2)}$ on precipitation of antigen RX with an antiserum mixture containing four times as much anti-R antibody as anti-X antibody ($B_{\text{total}}^{(1)} = 10$, $B_{\text{total}}^{(2)} = 40$). From top to bottom the curves are for $A_{\text{total}} = 20$, 50, and 100; of each pair the curve at the left is for hapten $H^{(1)}$ and the other for hapten $H^{(2)}$.

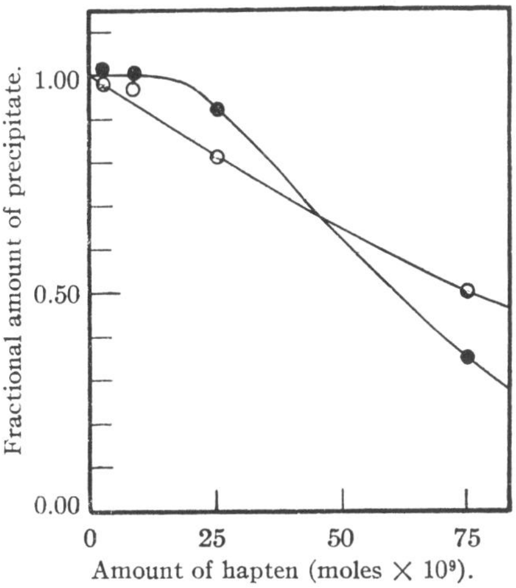

Fig. 4.—Experimental values of amount of precipitate formed by antigen R′X′ with 93:7 mixture of anti-R serum and anti-X serum in presence of hapten R* (solid circles) or hapten X* (open circles).

Experimental data bearing on the valence of the antibody molecules are given in Tables III and IV. The observed antibody/antigen molecular ratio of about 0.7 in both the equivalence zone and the region of antibody excess (Table III) corresponds, with antigen known to be bivalent, to 2/0.7 = 2.8 for the average valence of the antibody molecules in the precipitate.

The second of our arguments for the multivalence of antibody molecules is based on the solubility of antigen–antibody precipitates in excess antibody. This phenomenon has been reported for many antigen–antibody systems, especially with horse antibody. We have observed it to occur with rabbit antibody and simple polyhaptenic antigens; the results of an experiment of this sort, with anti-R serum and the substance R′R′, are given in Table IV. In this table there are given also data for a similar experiment with the substance R′X′ and a mixture of anti-R serum and anti-X serum; the observed decrease in the amount of precipitate with increase in amount of antiserum can be accounted for only with the theory of multivalent antibody.

It is known that excess of one component causes decrease in the rate of precipitation of antibody and antigen, and the question may be asked as to whether the effect of excess antibody reported in Table IV might be due to failure of the reaction to be completed in the time allowed. An experimental test of this possibility was made, with the results given in Table V. Tubes were set up corresponding to the first and fifth experiments of Series A of Table IV, representing, respectively, the region of antibody excess (Series C) and the region of optimum precipitation (Series D). These tubes were allowed to stand for times ranging from two to fifteen days, and then were analyzed. The values obtained after two days are close to the corresponding values of Table IV. In both series the amount of precipitate was observed to increase for six days, and then to remain essentially constant. The difference in amount of precipitate for Series C and Series D remained constant throughout the period; from these results we conclude that the effect of excess antibody in decreasing the amount of precipitate is real, and is not due to a difference in the rate of precipitation.

TABLE V

EFFECT OF TIME OF STANDING ON AMOUNT OF PRECIPITATE OBTAINED IN THE REGIONS OF ANTIBODY EXCESS AND OPTIMUM PRECIPITATION

Solution of antigen XXX, 2 ml. (5×10^{-9} mole); anti-R serum, 2 ml. for series C and 0.4 ml. plus 1.6 ml. of normal serum for series D; 1 hour at room temperature and indicated times in refrigerator.

Time of standing in refrigerator, days	Amount of antibody precipitated (μg.)[a] Series C	Series D
2	34	110
4	37	138
6	78	150
8	59	156
10	27	138
15	62	131

[a] Averages of triplicate analyses with mean deviation ±2 μg. Values are corrected by subtraction of the blanks for serum and buffer, which ranged between 14 and 23 μg. The large values of the blank in comparison with those of Table IV (5 μg.) may be due to change in the antiserum during the period of three months between the experiments.

Solubility of the precipitate in excess of antiserum has been reported for a few systems, such as diphtheria toxin and horse antitoxin, but not for antigen–antibody systems in general. Theoretical considerations[3,11] indicate that solubility in antibody excess should occur for antigens with small valence (such as the dihaptenic substances of Table IV) but not for antigens with large valence, which would only with difficulty be saturated with antibody to form a soluble complex.

This work was carried out with the aid of a grant from The Rockefeller Foundation. Dr. Stanley Swingle and Mr. Allan L. Grossberg helped with the experimental work.

Summary

It has been found by experiment that substances of the type RX, containing two different haptenic groups, do not form precipitates with either anti-R serum or anti-X serum alone, but do form precipitates with a mixture of the two specific antisera. This provides proof of the effective bivalence of the dihaptenic precipitating antigen, and thus furnishes further evidence for the framework theory of antigen–antibody precipitation. In these experiments the anti-R serum and anti-X serum were made by injecting rabbits with sheep serum coupled with diazotized *p*-arsanilic acid and diazotized *p*-aminobenzoic acid, respectively, and the RX substances used were 1-amino-2-*p*-(*p*-azophenylazo)-phenylarsonic acid-3,6-disulfonic acid-7-*p*-(*p*-azophenylazo)-benzoic acid-8-hydroxynaphthalene and 1,8-dihydroxy-2-*p*-azophenylarsonic acid-3,6-disulfonic acid-7-*p*-(*p*-azophenylazo)-benzoic acid-naphthalene.

The antibody–antigen molecular ratio in the precipitate was found by analysis to be 0.7, which, with antigen known to be bivalent, leads to the average valence 2.8 for the antibody molecules in the precipitate. Further evidence that the antibody valence is greater than 1 is given by the observation that the precipitate is soluble in excess of antiserum.

A simple physicochemical theory of the precipitation of RX antigen with mixed antiserum and of its inhibition by haptens is developed and compared with experiment.

PASADENA 4, CALIFORNIA RECEIVED SEPTEMBER 7, 1943

(11) A. D. Hershey, *J. Immunol.*, **42**, 455 (1941).

Antibodies and specific biological forces

LINUS PAULING

Advances in organic chemistry during the past century have made it possible to identify the structure of many of the simple substances found in living cells and to understand such chemical changes as respiration and excretion. Only recently, however, have there seemed sure grounds for believing that knowledge and technique have advanced far enough to permit an eventual elucidation of such highly complex problems as the relationship between antibody and antigen, and other fundamental phenomena of life.

Significant progress has recently been made in the attack on the problem of the nature of specific biological forces, and also on the related problem of the mechanism of manufacture of complex biological molecules with specific properties. For example, animals are able to synthesize various proteins fulfilling special functions—such as haemoglobin, which carries oxygen from the lungs to the tissues. The molecule of haemoglobin is large and complex. It contains about 10,000 atoms, and its molecular weight is 68,000. It has the property of combining reversibly with oxygen, and will react to the presence of carbon dioxide by tending to liberate the combined oxygen. The properties of haemoglobin are not exactly the same for animals of different species, but vary from species to species. The variation is sometimes of such a nature as to be obviously useful to the animal—thus the haemoglobin of cold-water fishes liberates its oxygen at lower temperatures than does that of warm-blooded animals.

The problem of the way in which an animal is able to manufacture the special molecule of haemoglobin that it needs is part of the general problem of the manufacture of specific biological substances. For example, a virus molecule in the proper environment (that provided by its host) is able to cause the production of replicas of itself, and the phenomena of heredity depend upon the similar autocatalytic action of molecules of genes present in the chromosomes and also in the cytoplasm of cells.

The specificity shown by the synthesizing system in cells in manufacturing haemoglobin molecules of a particular type, or other complex substances with specific biological properties, is observed in many other phenomena. Plants and animals form enzymes which have the special powers of catalysing certain chemical reactions, such as the hydrolysis of a polypeptide chain at the link between two definite amino-acid residues or the successive stages in the oxidation of foods. A striking example of specificity in properties is shown by antibodies, substances produced by an animal after the injection of a foreign material, the antigen. In general, antibodies produced in response to the injection of a particular antigen have the power of combination with the homologous antigen used in their production, but not with other substances, except those that are very closely related in molecular structure. Thus the problem of the structure of antibodies, in relation to their power of specific combination with the homologous antigen and to the mechanism of their production, bears on both of the basic problems of biology mentioned above.

The work of Karl Landsteiner contributed greatly to the development of the present understanding of the nature of serological reactions. After he had discovered the human blood-groups, Landsteiner began a penetrating series of experimental studies, designed to throw light on the problem of the nature of serological reactions. Most important was his discovery that it is possible to cause an animal to manufacture antibodies with the power of specific combination with various chemical groups of known structure. Beginning in 1917, he and his collaborators prepared artificially conjugated antigens by coupling relatively simple chemical substances to proteins, and then injecting these artificial antigens (usually azoproteins, with structure protein—N=N—R, made by coupling a diazotized amine to the protein molecule) into animals. Landsteiner thus produced antisera which were found to contain antibodies with the power of combining with the protein used in manufacturing the azoprotein, and also antibodies with the specific power of combining with the attached group of known structure, which he called the haptenic group. The most useful method for studying the combining power between antibodies and homologous

Reprinted from ENDEAVOUR *Volume VII, Number 26*
April 1948

antigens is the precipitation method—an antiserum mixed in proper proportions with a solution of the homologous antigen forms a precipitate with it, whereas it is not able, in general, to form a precipitate with other substances. For example, Landsteiner prepared an azoprotein by diazotizing *p*-aminobenzoic acid, 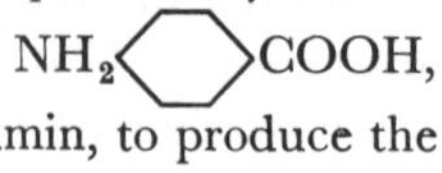

and coupling it with egg albumin, to produce the azoprotein

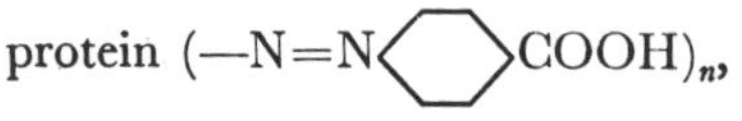

containing several haptenic groups attached to each molecule of protein. He found that the antiserum made by injecting a rabbit with this azoprotein not only had the power of forming a precipitate with a solution of the original azoprotein, made from egg albumin, but could form a precipitate with a solution of an azoprotein made with any other protein, such as bovine serum albumin, by coupling it with diazotized *p*-aminobenzoic acid. This property of precipitation with any azoprotein containing the *p*-aminobenzoate ion group shows that the antibody has developed the power of combining with this haptenic group. The combining power is, moreover, highly (but not completely) specific. The antiserum does not form a precipitate with an azoprotein made by coupling some unrelated substance, such as diazotized *p*-aminosuccinanilic acid, with a protein, but has the power of forming a small amount of precipitate with azoproteins made with closely related substances, such as a substituted *p*-aminobenzoic acid with a group or atom such as methyl or chlorine also attached to the benzene ring. The picture of the structure of antibodies described below had its origin in large part in an attempt to interpret the nature of the serological cross-reactions observed by Landsteiner in terms of the molecular structure of the haptenic groups. Landsteiner's work is summarized in his book *The Specificity of Serological Reactions*, the first edition of which was published in 1936 and the second edition (after his death) in 1945 [1].

The following picture of the process of formation of antibodies under the influence of a molecule of antigen was developed during a vigorous effort to imagine the simplest structure that could be suggested, on the basis of the information available about intramolecular and intermolecular forces, for a molecule with the properties observed for antibodies, and also to imagine the simplest reasonable process of formation of such a molecule [2]. This theory of the structure and process of formation of antibodies is based upon the concepts that the forces between an antibody and its homologous antigen are the ordinary short-range forces known to exist between simpler molecules, and that the great specificity results from a detailed 'complementariness' in configuration extending over a considerable surface of the antigen molecule and the corresponding combining region of the antibody molecule. The concept of specificity of serological reactions as resulting from complementariness in structure was originally suggested by Breinl and Haurowitz [3], and was then independently presented by Jerome Alexander [4] and Stuart Mudd [5]. There is some intimation of it in the early work of Ehrlich (the lock-and-key theory), and of Bordet. The concept of the multivalency of antibody molecules and antigen molecules, now incorporated in the theory, is due to J. R. Marrack [6], and support for it was provided by the work of Michael Heidelberger. Many experimental studies carried out at Pasadena since 1940 have provided evidence in favour of the concept of complementariness and also of the multivalency of antibodies.

Let us describe the immediate precursor of a molecule of normal γ-globulin as a polypeptide chain, containing a thousand or more amino-acid residues arranged in a sequence determined by the nature of the system of enzymes and reticular structures constituting the cell in which the γ-globulin is being synthesized. We assume that this polypeptide chain can become either a molecule of normal γ-globulin or a molecule of antibody with specific combining power for an arbitrary antigen, according to the way in which the long and complex polypeptide chain is folded. It seems likely that some native proteins are of such a nature—i.e. have such a sequence of amino-acid residues—that of the many ways in which the chain can be folded one is characterized by being specially stable. This specially stable configuration might be that of the native protein; it is known that certain proteins, such as trypsin and haemoglobin, can lose their specific properties under the action of a denaturing agent, and can then regain them, presumably by re-coiling to the normal, stable configuration as the denaturing agent is slowly removed. However, we assume that the polypeptide chain of γ-globulin is of such a nature that a very great number of alternate ways of coiling the polypeptide chain have nearly equal stabilities. When the polypeptide chain is synthesized in the cell, in the absence of any foreign molecules the chain will tend to coil

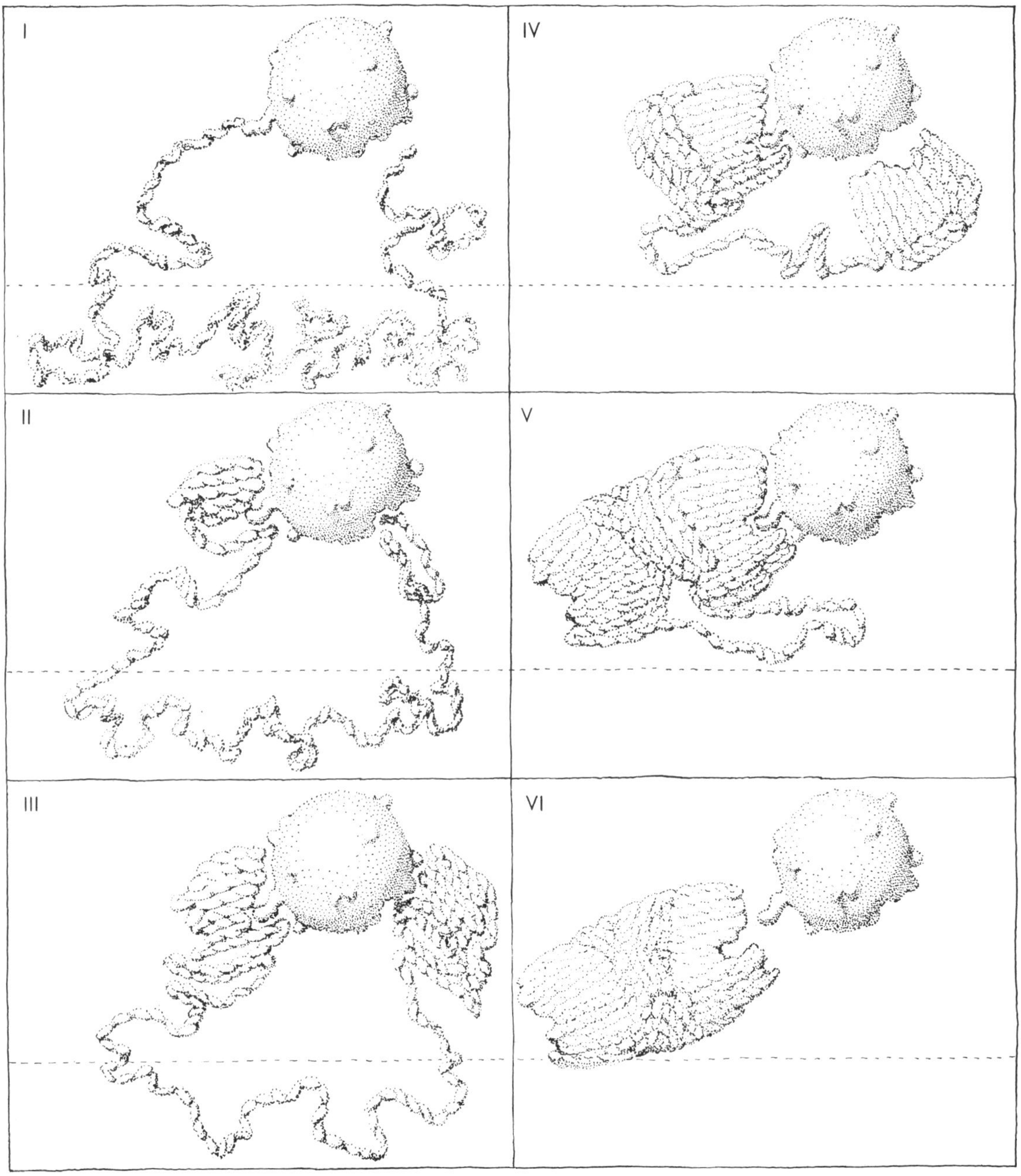

FIGURE I – *A postulated process of formation of antibody molecules. The polypeptide chain of the antibody precursor is folded, in the presence of an antigen molecule, into configurations complementary to the antigen.* (*See page* 49.)

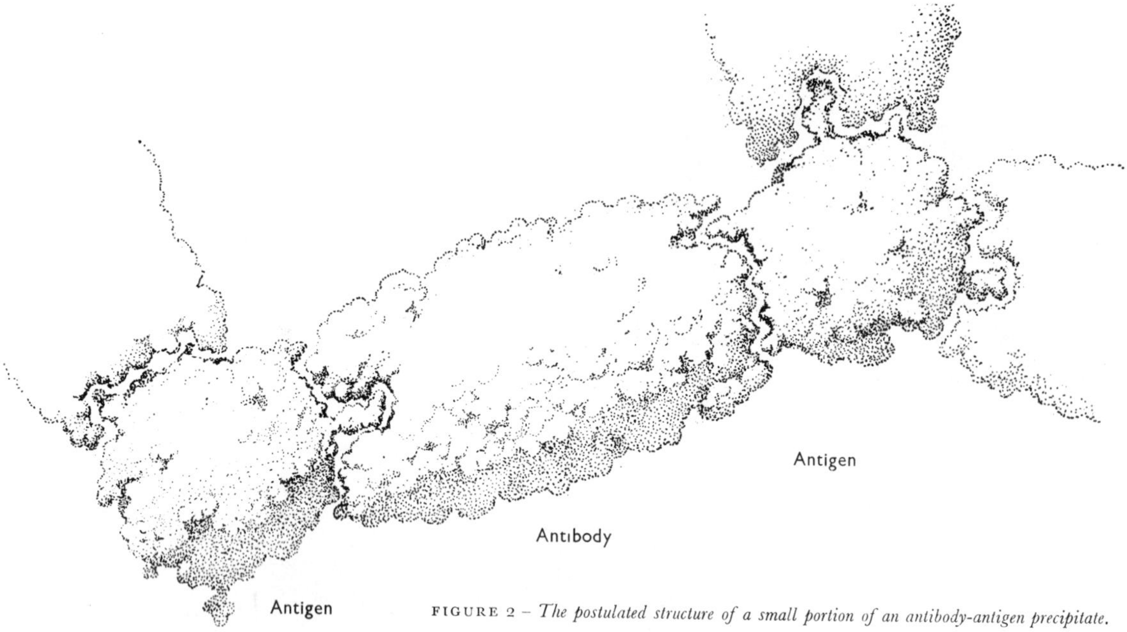

FIGURE 2 – *The postulated structure of a small portion of an antibody-antigen precipitate.*

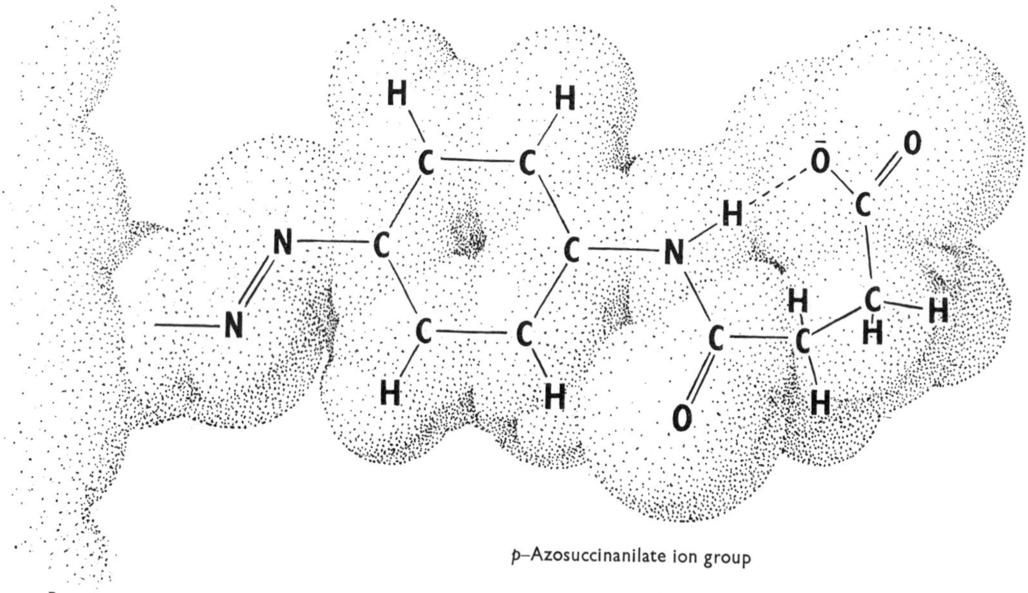

FIGURE 3 – *The structure of the haptenic group of an azoprotein, ovalbumin-*p*-azosuccinanilate ion.*

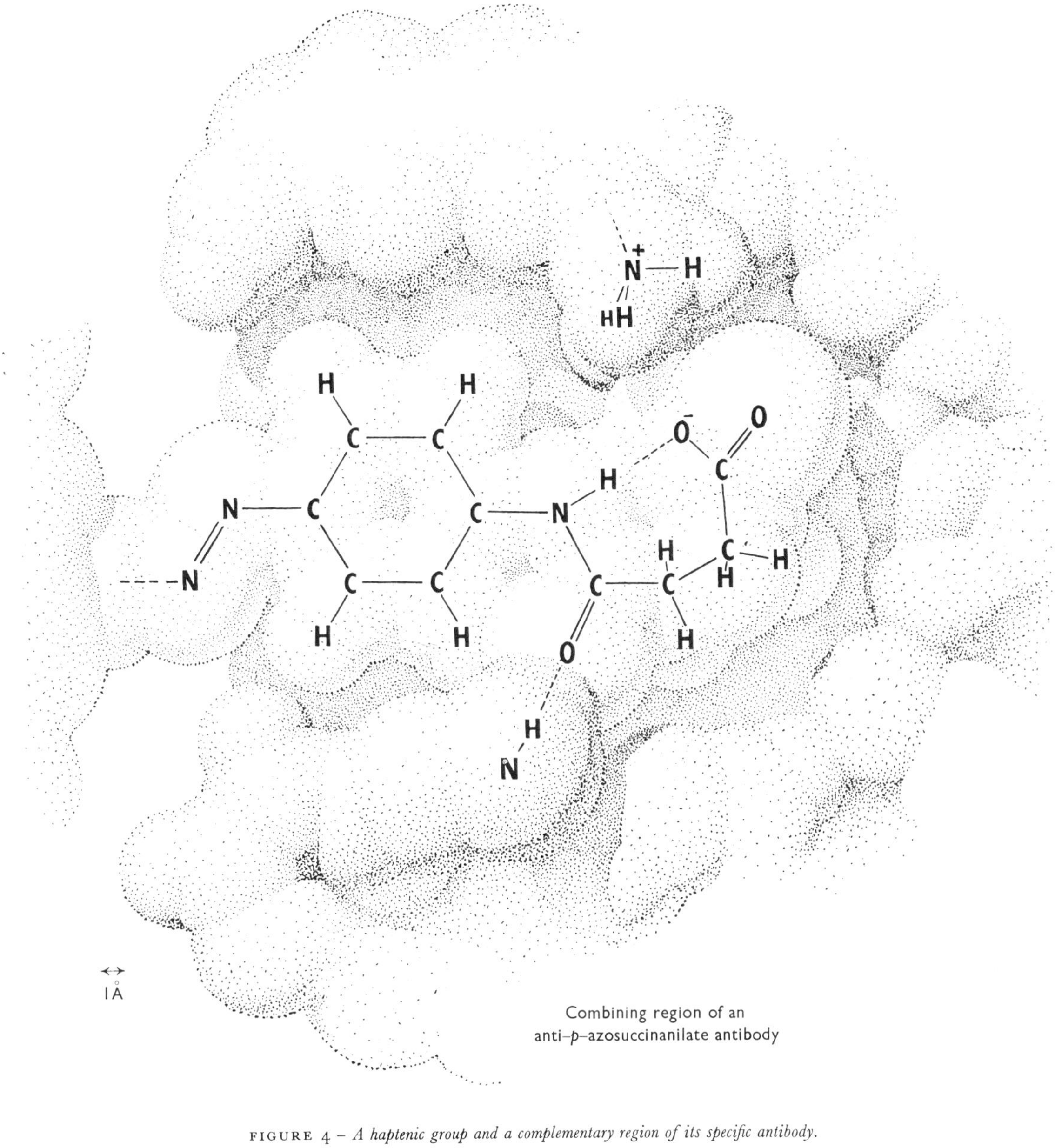

FIGURE 4 – *A haptenic group and a complementary region of its specific antibody.*

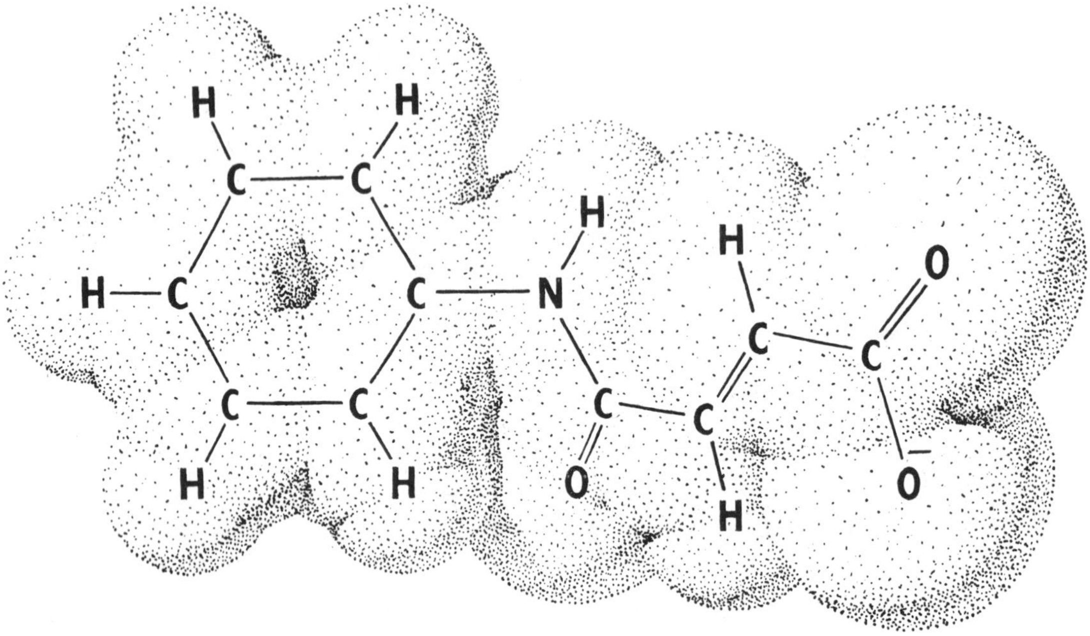

FIGURE 5 – *The fumaranilate ion, which has very small combining power with the antibody shown in figure 4.*

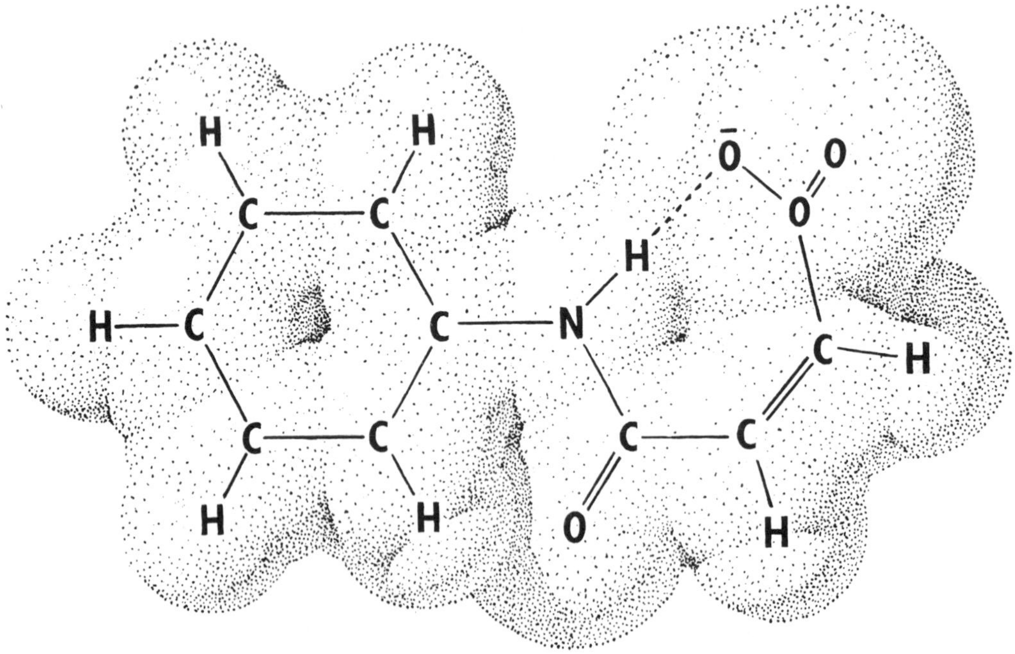

FIGURE 6 – *The maleanilate ion, which combines strongly with the antibody shown in figure 4.*

into that configuration which is most stable under the normal circumstances within the cell, producing a molecule of normal γ-globulin. If, however, there is present in the cell a foreign molecule, a molecule of antigen, the environment in which the γ-globulin polypeptide chain finds itself is different, and the difference in environment will be such as to stabilize others of the configurations accessible to the polypeptide chain, namely those which have the greatest power of attraction for the antigen molecule. This hypothetical process provides an automatic method by which a molecule is produced that is complementary in structure to a portion of the surface of an antigen molecule. The phenomenon is the same as the production of a coin by a die, or in general of a replica by the process of pressing a plastic material against a mould and permitting it to harden. The polypeptide chain, with its power of assuming alternative configurations, is the plastic material, and the surface of the antigen serves as the die or mould. The process of hardening is the result of the operation of the weak forces between different portions of the polypeptide chain that find themselves in juxtaposition; these weak forces, which individually could not withstand the disrupting effect of thermal agitation, co-operate, after the very large protein molecule has assumed its final configuration, to hold the molecule in that configuration.

In figure 1 the antigen molecule is represented as a roughly spherical aggregate of atoms, with a surface structure shown by protuberances and hollows. One of these protuberances might represent, for example, a *p*-azobenzoate ion haptenic group. The precursor of γ-globulin is shown as being synthesized in a region of the cell separate from that occupied by the antigen (this region being indicated by the line of dashes). In the first section of figure 1 two ends of a polypeptide chain are shown as liberated first from this region, and permitted to fold into the most stable of the configurations accessible to them. The ends of these two polypeptide chains are seen in the second section of the figure to be attracted to the surface of the antigen, and to assume configurations that make the forces of attraction between them and the molecule of antigen as great as possible. Inasmuch as most of the forces that are effective between molecules fall off very rapidly with increasing distance, and are strong only over a range of a few Ångström units, the stabilized configurations of the end of the polypeptide chain will be those that bring as large a portion as possible of each chain-end into immediate juxtaposition with a surface of the antigen molecule; that is, the structures of the combining regions will have as great a complementariness as possible with the surface structure of the antigen molecule. This complementariness includes not only the similarity in the surface configuration but the juxtaposition of special combining groups, such as a negatively charged group in the antibody with a positively charged group in the antigen, and a hydrogen-bond-forming group carrying the proton with a similar group presenting an electron pair.

The third section of figure 1 shows the antigen with two combining regions of the antibody formed. The central section of the polypeptide chain has not yet been folded. In the fourth drawing one of the combining regions of the antibody is represented as breaking away from the antigen, under the influence of thermal agitation; the central section of the polypeptide chain then folds into its stable configuration, fastening the two combining regions of the antibody together into the completed antibody molecule. In the fifth drawing this molecule is shown still attached to the antigen by one of its combining groups; in the sixth drawing the completed antibody molecule is represented as breaking away from the antigen under the influence of thermal agitation, forming a free antibody which may build up the antibody titre of the blood-plasma.

The assumption that most antibody molecules have two regions able to combine with antigen (i.e. are bivalent) accounts for some of the properties of antisera. For example, antibodies homologous to cellular antigens, such as red blood-cells, are able to cause these cells to agglutinate. The serum from humans with blood of type B or O contains antibodies capable of combining with the A antigen, which is present on human red blood-cells of types A or AB. When this serum is added to a suspension of these cells, the red cells clump together. The simplest explanation of this clumping is that an antibody molecule uses one of its combining groups to attach itself to the outer surface of one red blood-cell by combining with the A haptenic group present on this cell wall, and that then, on collision with another red blood-cell, it uses its other combining region to form a bond with the second cell, thus holding the two cells together. This process may continue until all of the cells are agglutinated into clumps. A similar explanation leads immediately to an understanding of the phenomenon of serological precipitation, as was pointed out by Marrack. The antigen molecule may use various portions of its surface to attach itself to several antibody molecules, as indicated in figure 2, and each of these antibody molecules may then use its second combining group to bind further molecules of antigen. As this process continues, the aggregates of antigen molecules and antibody molecules become larger and larger until finally they constitute a visible precipitate. Thus the phenomena of agglutination and serological precipitation do not require any further explanation, once the power of formation of a specific bond between antibody and antigen has been explained and a mechanism of providing an antibody with two combining groups has been revealed. Not all antisera have the power of serological precipitation; the non-precipitating antisera presumably contain univalent antibodies with only one combining group.

Landsteiner discovered a specially interesting way of studying the degree of specificity of antibodies. He found that when benzoic acid itself was added to an anti-*p*-aminobenzoic acid serum no precipitate was formed. However, a reaction had occurred between the benzoate ions and the antiserum, because when the homologous azoprotein was added to the solution it failed to precipitate with its antiserum, which in the absence of the benzoate ion would have given a precipitate with it. This inhibition of precipitation by the simple haptenic benzoate ion was explained by Landsteiner as resulting from the formation of a soluble complex between the antibody and the benzoate ion. The framework theory of serological precipitation and the assumed bivalency of antibodies immediately provide an explanation of the phenomenon: the bivalent antibody is able to combine with two molecules of benzoate ion, one of which attaches itself to each of its combining groups, but it is not able to form a framework precipitate with benzoate ion, because the benzoate ion ('antigen') itself has the power of combining with only one antibody molecule. The azoprotein antigen, with several *p*-azobenzoate ion haptenic groups, can form the framework in the absence of benzoate ion, but in the presence of benzoate ion all the antibody is tied up in soluble complexes with this ion, and so is prevented from precipitating.

Continuing the work of Landsteiner, Professor Dan H. Campbell, Dr David Pressman, and I, with a number of students, have carried out quantitative studies of the relative inhibiting powers of a great number of haptens, and have in this way obtained detailed information about the closeness of fit of the combining region of the antibody to haptenic groups of known structure. It has been found that, in general, the replacement of one group by another group that differs in shape from it by as much as one or two Ångström units leads to a significant decrease in combining power with the antibody. For example, the introduction of a methyl group in place of a *meta* hydrogen atom in the benzene ring of the benzoate ion decreases the combining power with an anti-*p*-aminobenzoic acid serum to about one-tenth of its original value. This is explained as the result of a resistance to fitting the larger methyl group, which has an effective radius of about 2·0 A, into the region which in the process of synthesis of the antibody was occupied by a hydrogen atom of the haptenic group of the immunizing antigen, this hydrogen atom having an effective radius of about 1·2 A. It has also been shown that a negatively charged group is present in the antibody at very nearly the minimum distance of approach to a positively charged group in the haptenic group of an azoprotein used in producing the antiserum [7].

An example of the degree of the effect of molecular shape on the power of combining with an antibody is illustrated in figures 3–6. Figure 3 gives a representation of the *p*-azosuccinanilate ion group, present with many other similar groups in an azoprotein molecule, used in the production of an anti-*p*-azosuccinanilate antiserum by injection into rabbits. Figure 4 shows a schematic drawing of the combining regions of such an antiserum, surrounding the haptenic group. It will be noticed that the atoms that constitute the surface of the combining region of the antibody are shown as being in approximate (about 1 A) contact with the surface of the atoms of the haptenic group. It seems likely that there is still closer approximation of the surface atoms of the antibody and the antigen in many cases, and that the antibody has enough elasticity to permit the introduction of haptens that differ in shape by 1 or 2 A in linear dimensions from the haptenic group of the immunizing antigen. An ammonium ion group, with a positive charge, is indicated in the drawing as at the minimum distance of approach to the negatively charged carboxyl group of the succinanilate hapten, and an $>$NH group, capable of forming a hydrogen bond with the carbonyl group of the succinanilate, is shown in the proper position to form this hydrogen bond.

The maleanilate ion:

$$C_6H_5.NH.CO.CH{=}CH.COO^-$$

and the fumaranilate ion (with the same formula) are closely similar substances, differing only in their spatial configuration, the maleanilate ion having the *cis* configuration about the carbon–carbon double bond and the fumaranilate ion having the *trans* configuration about this bond. These ions were found to differ very greatly in their inhibiting power for anti-*p*-azosuccinanilate serum. The maleanilate ion has nearly as great an inhibiting power as the homologous hapten succinanilate ion itself, whereas the fumaranilate ion has a very slight effect, only about 1 per cent. of that of the maleanilate ion. This power of the antibody to distinguish between two ions of such closely related structure was discovered by Landsteiner, and has been verified through the study of hapten inhibition by hundreds of substances. The relatively small difference in structure of the fumaranilate ion and the maleanilate ion is shown by the drawings in figures 5 and 6.

The importance of a positive charge in the antibody is indicated by the fact that no haptens except those with a negatively charged group—either a carboxyl ion group or a closely similar ion group—have any power of combination with antibodies homologous to the *p*-azosuccinanilate ion. The presence of the indicated hydrogen-bond-forming group in the antibody molecule, shown as combining with the carbonyl group, is verified by the fact that only haptens that contain this carbonyl group have a significant power of combining with the antibody.

It is interesting that a study of serological reactions can be used to provide information about the molecular structure and configuration of simple substances [8]. Thus it might be thought that the *p*-azosuccinanilate ion group, indicated in figure 3, would have an extended configuration, somewhat similar to that shown for the fumaranilate ion, inasmuch as there is considerable freedom of rotation about the carbon–carbon bond. However, from the fact that the maleanilate ion has a large combining power with anti-*p*-azosuccinanilate ion antibodies, while the fumaranilate ion has a small combining power, it can be deduced that these antibodies are complementary in configuration to the maleanilate ion, and that consequently the *p*-azosuccinanilate ion group, which serves as the template for the manufacture of these antibodies, itself has a configuration closely similar to that of the maleanilate ion. This configuration, shown in figure 3, is presumably stabilized, relative to the *trans* configuration, by the formation of an N—H . . . O hydrogen bond between the imino group and an oxygen of the carboxyl group. The fact that the succinanilate ion itself, other amides of succinic acid, and the mono-alkyl esters of succinic acid also combine strongly with the antiserum shows that these ions tend to assume a similar *cis* configuration. However, the succinate ion ($^{-}OOC.CH_2.CH_2.COO^{-}$) has a surprisingly small power of combination with the antiserum—approximately that of the fumarate ion, and much smaller than that of the maleate ion—from which it can be inferred that the succinate ion has essentially a *trans* configuration about the carbon–carbon single bond. An obvious explanation of the stability of the *trans* configuration for this ion is the Coulomb repulsion of the two negatively charged carboxyl groups.

In a preceding paragraph a rather complicated mechanism for the manufacture of antibodies with two combining groups has been described. The existence of these bivalent antibodies is presumably explained by their usefulness in causing the agglutination of cells or the precipitation of molecular antigens. However, the lysis and phagocytosis of cells are aided by the attachment of antibodies to the cells, but seem not to require agglutination; hence this part of the mechanism of protection against disease might be just as well done by univalent as by bivalent antibodies. Similarly the neutralization of toxins is effectively achieved by univalent antitoxins, and the reason for the manufacture by animals of bivalent antitoxins is here also not clear.

Nevertheless the phenomena of agglutination and serological precipitation are striking ones, and it is of interest to see whether they can be explained by the same forces that produce specific combination of antibody and antigen, or whether some additional explanation must be invoked. In suggesting that the first of these alternatives is correct, and that agglutination and precipitation result from multivalency of the antibody, Marrack [6] adduced various experimental observations in support of his proposal. Further support for the framework theory was provided by the work of Heidelberger and Kendall [9].

Recently a striking experiment was carried out, the results of which leave little doubt that the framework theory of serological precipitation and agglutination is correct [10]. Landsteiner and van der Scheer [11] had observed that certain simple substances containing two haptenic groups were able to produce precipitates with the hapten-homologous antisera. It seemed to us likely that the formation of a precipitate with these substances, containing two haptenic groups, and the failure to form precipitates with substances containing only one haptenic group, which in general have only an inhibiting power, due to the formation of soluble complexes, could be explained directly by the framework theory, and that accordingly a test of the theory could be made by investigating a large number of monohaptenic and polyhaptenic substances. This was done with the use of antisera homologous to the *p*-azobenzenearsonate ion group [12]. It was found that of the substances tested those containing two or more *p*-azobenzenearsonate ion groups in the molecule produced precipitates with the antisera, there being about twenty of these substances, whereas the thirty substances tested which contained only one benzenearsonate ion group in the molecule did not form precipitates but rather inhibited precipitation with a precipitating antigen.

The argument might be made, however, that the tendency of the polyhaptenic molecules to form precipitates with the antisera is due to some property other than their polyhaptenic character, and that only one of the haptenic groups may be actually involved in combination with antibody. A test was made by carrying out a special experiment [10], the results of which showed that each of the two haptenic groups of a dihaptenic substance enters into specific combination with antibody in the formation of the serological precipitate.

It was found that antisera could be obtained from two rabbits, and a substance could be synthesized of such nature that neither of the two antisera alone forms a precipitate with the substance, but that a mixture of the two antisera, from the two rabbits, has the power of precipitating with the substance. This serological reaction is, then, one involving a substance of known structure, which serves as precipitating antigen, and two different antisera, which must co-operate in producing the precipitate with the substance.

One of the rabbits was injected with an azoprotein containing R groups (*p*-azobenzenearsonate ion groups), and the other rabbit was injected with a protein containing X groups (*p*-azobenzoate ion groups). The two antisera are accordingly an anti-R serum and an anti-X serum. Each of these sera is able to form a precipitate with substances containing two or more of the homologous haptenic groups. The two RX substances used in the experiment were made by coupling one R group and one X group to a molecule of known structure; the substances were 1-amino-2-*p*-(*p*-azobenzeneazo)-benzenearsonic acid-3,6-disulphonic acid-7-*p*-(*p*-azobenzeneazo)-benzoic acid-8-hydroxynaphthalene and 1,8-dihydroxy-2-*p*-azobenzenearsonic acid-3,6-disulphonic acid-7-*p*-(*p*-azobenzeneazo)-benzoic acid-naphthalene. Each of these substances was found to form no precipitate (or only a very slight precipitate) with either anti-X or anti-R serum alone, but to form a large amount of precipitate with a mixture of these two antisera. This observation provides strong evidence that the precipitation of antibody and antigen involves both the haptenic groups of the dihaptenic antigen. When mixed with anti-R serum alone, a substance RX uses its R groups to combine with the anti-R antibodies, two molecules of RX thus forming a soluble complex with one molecule of bivalent antibody; but because the molecules RX are effectively monohaptenic when only anti-R antibodies are present, a framework precipitate cannot be formed. In the presence of a mixture of anti-R and anti-X antibodies, however, the molecule RX is effectively bivalent, and can form a framework in which the anti-R and anti-X antibodies alternate.

There is now very convincing evidence that the specificity of combining power of antibodies can be explained in terms of short-range forces of known nature, the specificity itself resulting from complementariness in structure of the combining region of the antibody and the surface of the homologous antigen. It seems not unlikely that biological specificity in general is to be accounted for in a similar manner, as resulting from the ordinary non-specific, short-range forces that operate between all molecules, with the specificity of the forces in the biological systems due to the complex surface configuration of the large molecules present in these systems. The evidence as to the nature of the biological forces in systems other than serological systems is, however, not so extensive. Phenomena such as competition between the sulphonamide drugs and *p*-aminobenzoic acid, and the inhibition of enzyme systems generally by substances related in structure to those that take part in the catalysed reaction, suggest, by their similarity to hapten-inhibition of serological precipitation, that the same forces are acting as in the serological systems. For example, the malonate ion serves to inhibit the catalytic activity of the enzyme dehydrosuccinase, presumably by competing with the succinate ion for the position of attachment to the active region of the enzyme. The fact that the malonate ion is especially effective as an inhibitor of this catalyst may be significant with respect to the mechanism of catalytic activity. If the catalytically active region of the enzyme were closely complementary in structure to the succinate ion it would presumably not accelerate the reaction of dehydrogenation of the succinate ion, but would instead retard it, since by combining with the succinate ion it would stabilize this ion, and thus increase the amount of energy necessary to convert the ion into the activated complex for the dehydrogenation reaction. If, however, the enzyme were more closely complementary in structure to the activated complex itself than to the succinate ion, it would decrease the activation energy for the reaction, and in this way speed the reaction up. The effectiveness of the malonate ion in inhibiting the reaction suggests that this ion is more closely similar in structure to the activated complex of the dehydrogenation reaction for succinate ion than is the succinate ion itself, and that it is for this reason that the malonate ion is able to compete effectively with the succinate ion for the position on the active region of the enzyme.

It seems unlikely that the specific differences between related enzymes produced by animals of different species, or between other proteins, such as haemoglobin, are due to so simple a difference in structure as a different way of coiling the polypeptide chains. It is probable that antibodies are unique in using this mechanism alone for the assumption of specific differences, and that other biological macromolecules depend upon more deep-seated changes in structure to produce these differences in specific properties. In particular, I think it likely that in general a mutation is due not simply to a changed way of coiling the filamentous structural unit of the gene nucleoprotein, but to a change in the composition of this unit which affects its catalytic activity in such a way as to permit the controlled manufacture of duplicates of itself, with the changed structure. It seems likely that a gene that is damaged by ultraviolet light or X-radiation or other catastrophe in such a way as ultimately to produce a mutant is not itself the mutated gene, but is simply a damaged gene. It might be damaged so badly as to be unable to carry on an essential function in the development of the organism, the damage then being lethal. Or the damage might be of such a nature as to permit the gene to serve as the template and to exercise its other functions so as to produce replicas of itself as it was before it was damaged. Or again it might happen that the damaged gene could carry out its functions and produce new genes of somewhat changed structure, which themselves could serve as templates, until ultimately a steady state was achieved, at which there would be no difference between the gene that served as the template and the gene produced with it is as a model. This would be the phenomenon of mutation.

There is at present little reliable evidence as to the detailed nature of the process of production of replicas of complex biological molecules, such as viruses or genes, but it is clear that the phenomenon of production of complementary structures, as in antibody formation, provides a possible mechanism. The manufacture of a replica, of a gene for example, might require the production of an intermediate, the gene producing a molecule complementary in structure to itself, which in turn serves as the template for the reproduction of a replica of the original gene. Some support for this concept is given by the known existence of complementary structures in living organisms, such as the blood-group antigens and their homologous agglutinins, and the mutually complementary substances found in eggs and sperm [13].

REFERENCES

[1] LANDSTEINER, K. *The Specificity of Serological Reactions* (Charles C. Thomas, Springfield, Ill., 1936).

[2] PAULING, L. *J. Am. Chem. Soc.*, **62**, 2643 (1940).

[3] BREINL, S., and HAUROWITZ, F. *Zeit. physiol. Chem.*, **192**, 45 (1930).

[4] ALEXANDER, J. *J. Protoplasma*, **14**, 296 (1931).

[5] MUDD, STUART. *J. Immunol.*, **23**, 423 (1932).

[6] MARRACK, J. R. 'The Chemistry of Antigens and Antibodies,' Report No. 230 of the Medical Research Council (2nd ed., 1938) (H.M. Stationery Office, London, 1934).

[7] PRESSMAN, D., GROSSBERG, A. L., PENCE, L. H., and PAULING, L. *J. Am. Chem. Soc.*, **68**, 250 (1946).

[8] PRESSMAN, D., BRYDEN, J., and PAULING, L. *J. Am. Chem. Soc.* (to appear shortly).

[9] HEIDELBERGER, M., and KENDALL, F. E. *J. Exp. Med.*, **61**, 559, 563; **62**, 467, 697 (1935).

[10] PAULING, L., PRESSMAN, D., and CAMPBELL, D. H. *J. Am. Chem. Soc.*, **66**, 330 (1944).

[11] LANDSTEINER, K., and VAN DER SCHEER, J. *Proc. Soc. Exp. Biol. Med.*, **29**, 747 (1932).

[12] PAULING, L., PRESSMAN, D., CAMPBELL, D. H., and collaborators. *J. Am. Chem. Soc.*, **64**, 2994, 3015 (1942); **65**, 728 (1943).

[13] TYLER, A. *Proc. Nat. Acad. Sci.*, **25**, 317 (1939); **26**, 249 (1940); *Growth*, **10** (supplement), 7 (1947).

[Reprinted from the Journal of the American Chemical Society, **71**, 143 (1949).]

[CONTRIBUTION NO. 1135 FROM THE GATES AND CRELLIN LABORATORIES OF CHEMISTRY CALIFORNIA INSTITUTE OF TECHNOLOGY]

The Reaction of Simple Antigens with Purified Antibody[1]

BY ARTHUR B. PARDEE[2] AND LINUS PAULING

Many biological reactions, including the interaction of antibodies and antigens, the effect of enzymes on their substrates, and the self-reproducing behavior of genes, are characterized by a high degree of specificity. The great problem of the nature of the forces responsible for this biological specificity is being attacked vigorously in many ways at the present time. During the past six years we have carried out a series of investigations[1] on the reactions of antisera with simple substances, extending and refining the work of Landsteiner.[3] The results obtained provide strong support for the concept that biological specificity is due to a detailed complementariness in surface configuration of the molecules involved (antigen and antibody) and that the forces which contribute to specific attraction of two molecules—van der Waals electronic forces, hydrogen bond forces, etc.—are in general short-range forces, effective over distances of a few Ångström units. The conclusion has been reached that the surface approximation of antibody and the haptenic groups of antigens is to within about 1 Å.

Much of our work has consisted of studies of the precipitation of an antiserum by a simple polyhaptenic substance. The observation that certain simple substances containing two or more haptenic groups would form precipitates with the homologous antiserum was made by Landsteiner and van der Scheer.[4] It was suggested by Landsteiner that the forces between dye molecules which favor the formation of colloidal solutions, that is, of polymerized aggregates, are responsible for the ready precipitability of these substances, many of which are dyes. We, however, have presented evidence that the presence of two or more haptenic groups in each molecule, making the formation of a framework possible, is responsible for their precipitability.

The suggestion that it is polymerization of these simple precipitating antigens that gives them their precipitating power has been revived by Boyd and Behnke,[5] who reported that they had found one of the simple antigens used by us to be highly (11-fold) polymerized in saline solution, and who stated that accordingly the results of our earlier investigations might not justify the

(1) The Serological Properties of Simple Substances. XIV. For number XIII of this series see D. Pressman, J. H. Bryden, and L. Pauling, THIS JOURNAL, **70**, 1352 (1948).

(2) Present address: McArdle Memorial Laboratory, The Medical School, The University of Wisconsin, Madison 6, Wisconsin.

(3) See K. Landsteiner, "The Specificity of Serological Reactions," Harvard University Press, Cambridge, Massachusetts, 1945.

(4) K. Landsteiner and J. van der Scheer, *Proc. Soc. Exptl. Biol. Med.*, **29**, 747 (1932); *J. Exptl. Med.*, **56**, 399 (1932); **57**, 633 (1933); **67**, 79 (1938).

(5) W. C. Boyd and J. Behnke, *Science*, **100**, 13 (1944). In this preliminary note about their work these authors wrote that details would be published elsewhere; their detailed paper has not yet appeared.

interpretation that we had given them. This criticism has been repeated by Kabat.[6]

In the present paper we report the results of experiments on the interaction of simple precipitating antigens with purified antibody solutions. It has been found that the behavior of the simple antigens with purified antibody solutions is somewhat different from that with antiserum, the difference apparently being due to a rather strong non-specific combination of the dye molecules with the constituents of serum not present in purified antibody, presumably mainly albumin.[7,8] The effect of this combination seems to be to reduce the concentration of the free dye molecules to such an extent that in serum solution they react mainly in the monomeric form.

Experimental

Synthetic Antigens.—The compounds used as synthetic antigens and their preparation have been described in previous papers except resorcinol-R_3', which was prepared by Mr. A. L. Grossberg, by the method used for chloroglucinol-R_3' (XI) with use of a different phenolic nucleus.[9,10]

R represents the *p*-azobenzenearsonic acid group, R′ the *p*-(*p*-azobenzeneazo)-benzenearsonic group, and X′ the *p*-(*p*-azobenzeneazo)-benzoic acid group.

Resorcinol-R_3' (R_3')

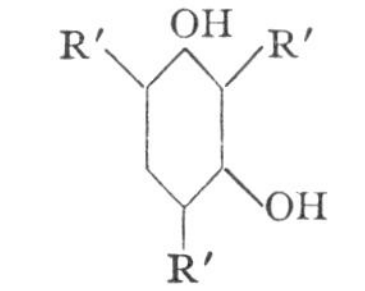

H-acid-R′X′ (R′X′)

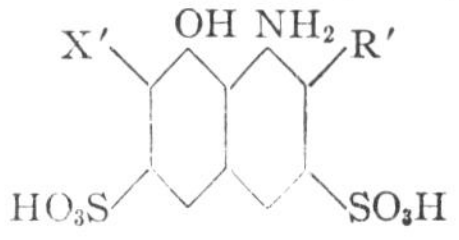

Chromatropic acid-R_2' (XXX)

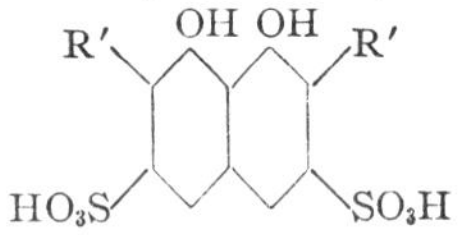

Antibodies and Other Substances.—The anti-R serum was obtained from rabbits by the method previously described.[9]

Specifically purified anti-R antibody (referred to in this paper as purified antibody) was prepared by Prof. D. H. Campbell.[11] Various preparations using resorcinol-R_3' as precipitant were pooled and used for the succeeding experiments.

The pneumococcus type I antibody used in some of the experiments was a commercial preparation, Lederle Refined and Concentrated Rabbit Globulin. The purified polysaccharide was prepared by Dr. J. E. Cushing, Jr.

The rabbit serum albumin was prepared by Dr. George A. Feigen, by a single salt precipitation.

Salt and Buffer.—All experiments were done in 0.9% sodium chloride solution, without buffer present unless otherwise noted. When a buffer is referred to as xF it is meant that the sum of all forms of the buffering material is x formula weights per liter.

Precipitation Reaction; General Methods.—Antigen and antibody at approximately the final *p*H were mixed and allowed to stand for one hour at room temperature and then for two to five days at 5°. (It was found that resorcinol-R_3' gave the same amount of precipitate in two hours as in three days, and that the maximum turbidity under the conditions used was reached in ten minutes at room temperature.) After having been washed,[9] the precipitates were analyzed for antigen by adding 2.5 ml. of 1 *N* sodium hydroxide, making up to 5.5 ml. in a centrifuge tube, and reading the light absorption on a Beckman spectrophotometer at an appropriate wave length and slit width. Compound resorcinol-R_3' was read at 500 mμ and 0.03 mm. slit width, and antigens XXX and R′X′ at 600 mμ and 0.04 mm. slit width. The first two compounds showed no tendency to fade over a period of several days in alkali or over several months in approximately neutral solution, but R′X′ faded about 20% in a day in alkali. The compounds reacted slowly with the soft glass bottles. The error in the colorimetric procedure is estimated at ±3% or less for samples containing more than 4×10^{-9} moles of antigen, and is somewhat greater for smaller samples.

The same sample was then analyzed for protein by the modified Folin–Ciocalteu method.[12] A correction was made for the color of the antigen. To test the method ten triplicate analyses were run on purified antibody samples by both the modified Folin–Ciocalteu and the Nessler method by two analysts. The first method gave values which averaged 1.035 ± 0.035 times the second.

In some of the experiments with purified antibody the supernatant liquids were decanted from the centrifuged precipitates, both parts were analyzed, and the precipitates were corrected for the small amount of antigen and protein in the remaining supernatant liquid. In some of the later experiments only the precipitates were analyzed and the correction was calculated from the amount of reactants added.

The molecular weight of the antibody was taken as 160,000 for calculations.

Results

Precipitates were analyzed both with and without washing, as described in the previous section. The amounts of precipitate and the mole ratios in the precipitate for the two procedures are compared in Table I and also in Table II. It can be seen that for the antigen used (XXX) the amount of precipitate washed away is 30 to 45% of the original amount; the ratio of antigen to antibody in the precipitate remains about the same, however. Only 10 to 15% of the precipitate with antigen resorcinol-R_3' (at *p*H 6.9) was lost on washing.

It was found that the amount of precipitate obtained with antigen XXX and purified antibody at *p*H 7.6 and 8.1 was only slightly dependent on the volume of the reaction mixture, the amount of precipitate varying over about a 20% range for any given amounts of antigen and antibody on three-fold dilution with buffer solution. The mole ratio in the precipitate was essentially independent of the volume of the mixture. In contrast, a three-fold increase in the volume of a system containing antiserum and a simple poly-

(6) E. A. Kabat, "Annual Review of Biochemistry," Annual Reviews, Inc., Stanford University, California, 1946, p. 528.

(7) B. D. Davis, *Am. Scientist*, **34**, 611 (1946).

(8) I. Klotz, THIS JOURNAL, **68**, 2299 (1946); **69**, 1609 (1947).

(9) L. Pauling, D. Pressman, D. H. Campbell, C. Ikeda, and M. Ikawa, *ibid.*, **64**, 2994 (1942).

(10) L. Pauling, D. Pressman, and D. H. Campbell, *ibid.*, **66**, 330 (1944).

(11) D. H. Campbell, R. H. Blaker, and A. B. Pardee, *ibid.*, **0**, (2496 1948).

(12) D. Pressman, *Ind. Eng. Chem., Anal. Ed.*, **15**, 357 (1943).

TABLE I

PRECIPITATION OF PURIFIED ANTIBODY BY ANTIGEN XXX

Varying amounts of antigen XXX in 2 ml. of saline were added to 2 ml. of saline containing 4.2×10^{-9} mole of antibody. Final pH 8.1, no buffer. Values are averages of triplicate analyses

Antigen added, moles $\times 10^9$	Precipitated antibody, moles $\times 10^{9a}$		Mole ratio antigen/antibody in ppt.[a]	
264	0.45	0.81	9.2	11.6
132	.53	.91	5.9	8.2
66	.62	1.03	4.9	5.5
33	.97	1.53	2.6	3.4
16	1.28	1.84	2.0	2.7
8	1.51	2.40	1.9	1.8
4	1.44	2.06	1.5	1.6
2	1.00	1.46	1.2	1.3
1	0.59	1.03	(1.0)	(0.9)

[a] The first column is for experiments in which the precipitate was washed in the conventional way; the second is for experiments in which the supernate was decanted from the precipitate and both were analyzed as described in the text.

TABLE II

PRECIPITATION OF PURIFIED ANTIBODY BY ANTIGEN XXX

Varying amounts of purified antibody in 2 ml. of saline were added to 4.1×10^{-9} mole of antigen XXX in 2 ml. of saline. Final pH 8.1, no buffer. Analyses in triplicate

Antibody added, moles $\times 10^9$	Precipitated antibody, moles $\times 10^{9a}$		Mole ratio antigen/antibody in ppt.[a]	
16.9	2.0	2.5	1.1	1.1
12.5	2.0	2.8	1.4	1.1
8.5	2.1	2.8	1.4	1.2
4.2	1.4	2.1	1.5	1.6
2.1	0.7	1.2	1.8	1.9
1.0	.2	0.4	(1.8)	(2.7)

[a] See note of Table I.

haptenic antigen[13] caused the amount of precipitate to decrease by 50%.

A number of experiments other than those reported in Table II were carried out in which varying amounts of purified antibody were added to a constant amount of antigen in a constant volume of solution. A 5 to 10% decrease in precipitated protein was noted at the largest amounts of antibody in every experiment. This effect of inhibition of precipitation by antibody excess is similar to but not nearly so marked as the effect in serum,[10] in which a four-fold increase in amount of antiserum above the optimum was found to decrease the amount of precipitate to a third of the maximum value.

Many precipitation experiments in which constant amounts of purified antibody were added to various amounts of antigen were carried out. The detailed results were in all cases similar to those reported in Table I, and their important features are summarized in Table III. Some experiments with added serum are also reported at the end of Table III for comparison. The first two columns give the antigen used and the pH of the system. A stands for antigen and B stands for antibody. Columns 3 to 5 have to do with conditions at the optimum, *i. e.*, the conditions which give the maximum amount of precipitate. Column 3 gives the per cent. of total antibody precipitated. The values in this column depend on the antigen, and, for resorcinol-R'_3, decrease with increasing pH. In the presence of serum an optimum is obtained at about pH 9.0.[14] The presence of serum greatly reduces the amount of precipitable antibody, especially in the case of antigen R'X', which gives no precipitate in serum.

TABLE III

SUMMARY OF PRECIPITATION EXPERIMENTS WITH PURIFIED ANTIBODY

Antigen	pH	Per cent of antibody pptd. at optimum	Mole ratio A/B in system at optimum	Mole ratio A/B in precipitate at optimum	A/B in precipitate at A = 0
XXX	8.1	73	1.8	1.3	0.9
XXX	8.1	58	1.5	1.8	1.1
Resorcinol-R'_3	6.6	82	18	19	2.5
Resorcinol-R'	6.9	84	7.0	10	5.0
Resorcinol-R'_3	7.6[a]	76	5.1	5.0	1.1
Resorcinol-R'_3	7.9	68	5.7	4.5	2.0
Resorcinol-R'_3	8.7[b]	63	6.0	7.0	1.5
R'X'	7.9[b]	50	12	4.6	2.8
XXX (in serum)	8.1[b]	19	7	..	...
Resorcinol-R'_3 (in serum)	8.5	36	5.1	2.2	2.0
R'X' (in serum)	8.3	0	..	..	...

[a] This experiment was started at pH 9.9; when no precipitate appeared, the mixtures were acidified to pH 7.6. [b] 0.01 F Veronal buffer present. Mr. Leonard Lerman found no non-specific precipitation between resorcinol-R'_3 and normal rabbit γ-globulin under conditions similar to the above.

Column 4 gives the number of moles of antigen required per mole of antibody to form the maximum amount of precipitate. The antigen resorcinol-R'_3, which is more highly associated than the others,[15] requires a higher ratio, and the ratio is greater at lower pH. Antigen R'X' has a strikingly different ratio than the structurally similar compound XXX. The presence of serum has little effect on this ratio for resorcinol-R'_3 but a large effect for XXX.

The fifth column gives the mole ratio of antigen to antibody in the precipitate at the optimum. The values are similar to those in column 4 in the absence of serum, and decrease with increasing pH. The antigen R'X' is an exception, requiring a much higher ratio in the system than in the precipitate. Experiments in serum also show a lower antigen–antibody ratio in the precipitate than in the system, probably because much of the antigen is non-specifically combined with serum proteins.

(13) L. Pauling, D. Pressman, D. H. Campbell, and C. Ikeda, THIS JOURNAL, 64, 3003 (1942).

(14) D. Pressman, unpublished work.

(15) See the following paper, THIS JOURNAL, 71, 148 (1949).

The sixth column gives the mole ratio of antigen to antibody in the precipitate extrapolated to zero antigen–antibody ratio. This value can be obtained from either an experiment with variable antigen and constant antibody or an experiment with constant antigen and variable antibody. The values for the two types of experiment are the same within experimental error for antigen resorcinol-R_3' and for XXX but are 4.0 and 2.8 for R'X'. The ratio is of interest because one would expect non-specific reactions of antigen with proteins, and also association, to be least important when there is a large excess of antibody in the system. Ideally the ratio should be unity according to the framework theory.[16] This is the value found for antigen XXX and approached by resorcinol-R_3' as the *p*H is increased. In the experiment which was started at *p*H 9.9 and finished at *p*H 7.6 it appeared as if the antigen combined as a small aggregate at the higher *p*H and remained unassociated as the *p*H was decreased. Earlier experiments by Pressman[14] showed that if resorcinol-R_3' were aged for even one hour at *p*H values below 9 before addition to antibody, the amount of precipitate was appreciably decreased below the optimum amount obtained with antigen kept above *p*H 9. R'X' has a higher limiting ratio than the structurally similar antigen XXX, indicating that R'X' is associated in the precipitate even at antigen–antibody ratios less than unity. If the values in the fifth and sixth columns of Table III agree, the non-specific effect in the purified antibody experiments is probably small. Experiments in serum all led to fairly low ratios.[10]

The shapes of the plots of amount of precipitate *vs.* amount of antigen added were generally similar One quarter to one sixth of the optimum amount of antigen was required to give half the maximum amount of precipitate with purified antibody, and slightly more was required with antiserum. Generally about eight times the optimum amount of antigen was required to reduce the amount of precipitate to a half by antigen-excess inhibition An exception was R'X', which required more than a fifteen-fold excess.

The Effect of Addition of Serum or Serum Albumin.—Normal rabbit serum was added to the purified antibody and precipitation tests were made, with the results reported in Table III. The effect of the serum was to reduce the amount of precipitate with antigens resorcinol-R_3' and XXX, and to prevent precipitation with antigen R'X'. Serum albumin had the same effect as normal rabbit serum: it completely inhibited precipitation with R'X', and reduced the amount of precipitate obtained with antigen XXX. The tests in serum albumin were made at *p*H 6.2.

(16) L. Pauling, THIS JOURNAL, **62**, 2643 (1940); L. Pauling, D. Pressman, and C. Ikeda, *ibid.*, **64**, 3010 (1942).

A Study of Non-specific Adsorption.—To see how much antigen was held non-specifically to the precipitates in the purified antibody experiments, precipitations were carried out between purified pneumococcus polysaccharide type I and a commercial globulin preparation from rabbit serum, containing about 10% antibody precipitable by the polysaccharide. In the first four tubes of the series reported in Table IV the compound resorcinol-R_3' was present at the time of precipitation of approximately optimal amounts of the reactants. In tubes 5 to 8 the precipitation was completed and then the supernatant liquid was replaced by solutions of resorcinol-R_3'. The results given in the table show that there is a very strong non-specific affinity of the precipitate for the simple antigen. That the antigen is not merely trapped in the precipitate is shown by the fact that the second four experiments gave ratios very similar to the first four. Although there is ten times as much protein in the solution as in the precipitate, the precipitate can hold 80% of the antigen. That the increase in precipitate with increasing amounts of antigen is a non-specific effect is shown by the ninth and tenth tubes of the table, in which no polysaccharide was present. Experiments with compound XXX and the pneumococcus polysaccharide system showed that the protein in the precipitate is more effective for non-specific combination than is the protein in solution for this compound also, and that the effect is not due to a difference in the type of protein, because in some experiments only a fraction of the antibody was precipitated and there was still preferential

TABLE IV

NON-SPECIFIC ADSORPTION OF COMPOUND RESORCINOL-R_3'

To 10^{-4} g. pneumococcus polysaccharide type I in 1 ml. saline there was added 1 ml. of saline containing 4.5 mg. of commercial anti-polysaccharide type I. The precipitation was done in the presence of 1 ml. of saline containing compound resorcinol-R_3' in tubes 1–4, and the supernate of the precipitate was replaced by 3 ml. of saline containing compound resorcinol-R_3' in tubes 5–8. Tubes 9 and 10 were controls, containing no polysaccharide. Final *p*H 8.0, no buffer, duplicate analyses

	Resorcinol-R_3' added moles × 10^9	Precipitated protein moles × 10^9	Mole ratio R_3' / protein in precipitate	% of total resorcinol-R_3' in precipitate
1	400	3.7	12.0	11
2	85	3.3	8.8	34
3	22	2.9	4.8	63
4	7	2.6	1.9	71
5	400	2.4	7.1	4
6	85	2.6	9.5	29
7	22	2.8	5.7	72
8	7	2.6	2.1	78
9	85	0.3	30	11
10	22	.2	14	14

A Referee suggests that it would be preferable to carry out this experiment with antibody prepared by more gentle methods because the Lederle product may have an altered affinity for haptens due to its method of preparation.

combination with the precipitate. Compound XXX was not bound as strongly as was resorcinol-R$'_3$. R'X' was bound to about the same extent as XXX.

These observations can perhaps best be explained by considering the precipitate to have a few locations where a dye-polymer molecule can attach itself very firmly, perhaps by combining with several protein molecules. Such sites would be absent in the protein in solution. Since only a few polymer molecules could go into these locations, the fraction of dye combined with the precipitate would be large when there was little dye in the system, and would become smaller when the dye was increased and the stronger locations were filled. Then the protein in the solution would compete successfully with the smaller amount of precipitate.

Hapten Inhibition.—Varying amounts of the strong hapten *p*-(*p*-hydroxybenzeneazo)-benzenearsonic acid were added to an optimal mixture of purified antibody and antigen XXX at *p*H 7.6 before precipitation had begun. This hapten, in 20-fold excess over the added antigen, was able to inhibit the precipitation almost completely, and gave a heterogeneity index[17] of σ = 1.5. A 50-fold excess of the weaker hapten benzenearsonic acid had almost no effect on the amount of precipitate obtained with purified antibody and resorcinol-R$'_3$. These results on hapten inhibition with purified antibody are closely similar to those obtained with antiserum.

Discussion of Results

In our earlier work on the molecular ratio of antigen to antibody in the precipitate formed by dye antigens with hapten-homologous antiserum, values of about 1.2 for this ratio were found. These values indicate a valence of approximately 2 (2.4) for antibody if the dyes are assumed to be bivalent. The equality of the values found for trihaptenic and tetrahaptenic dyes with those for dihaptenic dyes was explained as resulting from steric hindrance of the large antibody molecules, which prevents more than two of the haptenic groups of the small dye molecules from being effective. A constant antigen-antibody ratio was obtained (to within about $\pm 20\%$) even when the amount of dye added to a constant amount of antiserum was varied over a wide range, from antigen excess to antigen deficiency, and when the *p*H was varied from 7.6 to 9.2. Essentially the same results were obtained for seven dyes. The new results, obtained with different dyes and with purified antibody, are much different. A strong dependence of the ratio on the amount of antigen is observed, and the ratios are much higher than those found before, except in the antibody-excess region. The results found in the present investigation are of the nature to be expected in case that the dyes form aggregates that are taken up by the precipitate through non-specific adsorption. The amount of non-specific adsorption of resorcinol-R$'_3$ shown in Table IV is enough to account for the high values of the molecular ratio of dye and antibody reported in Tables I, II and III.

The limiting values of the molecular ratio extrapolated to zero dye concentration are approximately 1.0 for antigen XXX and somewhat larger for resorcinol-R$'_3$.

We believe that the difference between the results obtained with purified antibody and those obtained earlier with antiserum is due to the non-specific combination of the dye molecules with serum albumin or other constituents of serum not present in the purified antiserum,[7,8] and that in the presence of serum this effect causes the concentration of uncombined dye molecules to be so small as to keep the amount of aggregation down to a low value. It seems likely that the high values for the amount of dye in the precipitates formed with purified antibody are due largely to the inclusion of the aggregates in the precipitate through non-specific adsorption, and possibly also to some extent to their incorporation in the framework as effective precipitating antigens.

The results obtained with the antigen R'X' indicate that dimers or larger aggregates of this molecule are able to form a precipitate with anti-R serum, the precipitation being inhibited by the presence of serum. In our earlier investigation[10] we observed that a small amount of precipitate was formed by R'X' with some pools of anti-R serum; it is likely that these precipitates were due, not to cross-reaction, but to the presence of aggregates, these pools of antiserum perhaps containing smaller amounts of the complexing material than are usually present.

The conclusion that may be reached from the experimental results obtained is that the earlier work, carried out with serum, is presumably reliable, despite the tendency of the dyes to aggregate, because this tendency was counteracted by the presence of complexing materials in the serum. Moreover, the limiting molecular ratio found by extrapolation to antibody excess of the results obtained in the present investigation, with purified antibody, supports the molecular ratio of about 1 given by the earlier work. It would, however, be desirable, in order to carry conviction, for additional experiments to be made, with use of non-aggregating polyhaptenic precipitating antigens.

In the earlier studies it was found that non-aggregating polyhaptenic compounds (amides, rather than azo dyes) form precipitates with antisera, whereas the monohaptenic substances do not; however, analyses of the precipitates were not made because of the difficulties of determining the amount of the colorless precipitating antigens.

It may be pointed out that the quantitative hapten inhibition experiments reported in earlier

(17) L. Pauling, D. Pressman, and A. L. Grossberg, THIS JOURNAL, **66**, 784 (1944).

papers of this series are not made unreliable by the phenomenon of aggregation of the precipitating antigen, for two reasons: first, the experiments were carried out in the presence of serum, which presumably inhibited the aggregation of the antigens; and second, the experiments all involve the comparison of the concentration of hapten required for a standard amount of inhibition (50%) with the concentration of a standard hapten producing the same effect, and this concentration ratio of the haptens, which do not themselves form aggregates, would be expected to be the same whether the test substance (the precipitating antigen) was aggregated or not.

The investigation reported in this paper was carried out with the aid of a grant from The Rockefeller Foundation. We wish to thank Professors Dan H. Campbell and Verner Schomaker, and Drs. David Pressman, Frank Lanni, J. E. Cushing, Jr., George Feigen and Stanley Swingle for advice in connection with the work and for providing some of the materials used. Mr. Dan Rice carried out many of the protein analyses.

Summary

Experiments have been carried out on the precipitation under various conditions of specifically purified anti-azobenzenearsonic acid antibodies by three azo dyes, serving as precipitating antigens. The non-specific combination of the dyes with pneumococcus polysaccharide:antipolysaccharide precipitates has also been investigated.

In solution these dyes exist as monomeric molecules and as aggregates. The molecular ratio of dye to antibody in the precipitate was found in general to be much larger than in earlier experiments involving antiserum, and to depend greatly upon conditions of the precipitation, whereas the earlier ratio (approximately 1.2) was essentially independent of these conditions. The dye is also carried down in non-specific serological precipitates (pneumococcus polysaccharide:antipolysaccharide antibody), apparently through non-specific adsorption or entrapment in the precipitate. The large molecular ratio in the precipitate formed with the homologous antibody is probably due mainly to the non-specific adsorption of dye aggregates. It is suggested that the results obtained with antiserum, which seem not to be affected by aggregation of the dye molecules, are to be explained as resulting from the combination of the dye molecules with constituents of serum, mainly albumin, which are not present in the purified antibody, this combination keeping the concentration of free antibody low, and preventing the formation of appreciable amounts of the aggregates.

PASADENA, CALIFORNIA RECEIVED JUNE 7, 1948

Chapter 13

The Alpha Helix and The Structure of Proteins

Contents

SP 95 **On the Structure of Native, Denatured, and Coagulated Proteins** **965**
(by A. E. Mirsky and Linus Pauling)
Proc. Natl. Acad. Sci. USA **22**, 439–447 (1936)

96 **The Structure of Proteins** **974**
(by Linus Pauling and Carl Niemann)
J. Am. Chem. Soc. **61**, 1860–1867 (1939)

97 **The Structure of Proteins: Two Hydrogen-Bonded Helical Configurations of the Polypeptide Chain** **982**
(by Linus Pauling, Robert B. Corey, and H. R. Branson)
Proc. Natl. Acad. Sci. USA **37**, 205–211 (1951)

98 **Atomic Coordinates and Structure Factors for Two Helical Configurations of Polypeptide Chains** **989**
(by Linus Pauling and Robert B. Corey)
Proc. Natl. Acad. Sci. USA **37**, 235–240 (1951)

99 **The Structure of Synthetic Polypeptides** **995**
(by Linus Pauling and Robert B. Corey)
Proc. Natl. Acad. Sci. USA **37**, 241–250 (1951)

100 **The Pleated Sheet, a New Layer Configuration of Polypeptide Chains** **1005**
(by Linus Pauling and Robert B. Corey)
Proc. Natl. Acad. Sci. USA **37**, 251–256 (1951)

101 **The Structure of Feather Rachis Keratin** **1010**
(by Linus Pauling and Robert B. Corey)
Proc. Natl. Acad. Sci. USA **37**, 256–261 (1951)

SP 102 **The Structure of Hair, Muscle, and Related Proteins** **1015**
(by Linus Pauling and Robert B. Corey)
Proc. Natl. Acad. Sci. USA **37**, 261–271 (1951)

103 **The Structure of Fibrous Proteins of the Collagen-Gelatin Group** **1026**
(by Linus Pauling and Robert B. Corey)
Proc. Natl. Acad. Sci. USA **37**, 272–281 (1951)

104 **The Polypeptide-Chain Configuration in Hemoglobin and Other Globular Proteins** **1036**
Proc. Natl. Acad. Sci. USA **37**, 282–285 (1951)

105 **The Planarity of the Amide Group in Polypeptides** **1040**
(by Robert B. Corey and Linus Pauling)
J. Am. Chem. Soc. **74**, 3964 (1952)

106 **Fundamental Dimensions of Polypeptide Chains** **1041**
(by R. B. Corey and Linus Pauling)
Proc. Roy. Soc. Lond. B **141**, 10–20 (1953)

107 **Stable Configurations of Polypeptide Chains** **1053**
(by Linus Pauling and R. B. Corey)
Proc. Roy. Soc. Lond. B **141**, 21–33 (1953)

108 **Compound Helical Configurations of Polypeptide Chains: Structure of Proteins of the α-Keratin Type** **1066**
(by Linus Pauling and R. B. Corey)
Nature **171**, 59–61 (1953)

109 **Two Rippled-Sheet Configurations of Polypeptide Chains, and a Note About the Pleated Sheets** **1074**
(by Linus Pauling and R. B. Corey)
Proc. Natl. Acad. Sci. USA **39**, 253–256 (1953)

110 **The Structure of Tussah Silk Fibroin** **1078**
(with a note on the structure of β-poly-l-alanine)
(by Richard E. Marsh, Robert B. Corey, and Linus Pauling)
Acta Cryst. **8**, 710–715 (1955)

111 **The Discovery of the α Helix** **1084**
The Chemical Intelligencer **2**(1), 32–38 (1996)

Reprinted from the Proceedings of the NATIONAL ACADEMY OF SCIENCES
Vol. 22, No. 7, pp. 439–447. July, 1936.

ON THE STRUCTURE OF NATIVE, DENATURED, AND COAGULATED PROTEINS

BY A. E. MIRSKY* AND LINUS PAULING

GATES CHEMICAL LABORATORY, CALIFORNIA INSTITUTE OF TECHNOLOGY, PASADENA, CALIFORNIA

Communicated June 1, 1936

In this paper a structural theory of protein denaturation and coagulation is presented. Since denaturation is a fundamental property of a large group of proteins, a theory of denaturation is essentially a general theory of the structure of native and denatured proteins. In its present form our theory is definite and detailed in some respects and vague in others; refinement in regard to the latter could be achieved on the basis of the results of experiments which the theory suggests. The theory (some features of which have been proposed by other investigators) provides a simple structural interpretation not only of the phenomena connected with denaturation and coagulation which are usually discussed (specificity, solubility, etc.) but also of others, such as the availability of groups, the entropy of denaturation, the effect of ultra-violet light, the heat of activation and its dependence on pH, coagulation through dehydration, etc.

I. The experimental basis upon which the present theory rests will be briefly described.

1. The most significant change that occurs in denaturation is the loss of certain highly specific properties by the native protein. Specific differences between members of a series of related native proteins and specific enzymatic activities of native proteins disappear on denaturation, as the following observations demonstrate:

(*a*) Many native proteins can be crystallized and the crystal form is characteristic of each protein. No denatured protein has been crystallized.

(*b*) Most native proteins manifest in their immunological properties a high degree of specificity, which is diminished by denaturation.[1]

(*c*) The native hemoglobins of closely related animal species can be distinguished from each other by differences in crystal form, solubility,[2] gas affinities, positions of absorption bands, and other properties.[3] On the other hand, in the denatured hemoglobins some of these properties, the positions of the absorption bands, for example, can be subjected to precise measurement, and it is found that differences between the various hemoglobins have disappeared.[4]

(*d*) A number of enzymes have recently been isolated as crystalline proteins. When these proteins are denatured their enzymatic activity vanishes. In pepsin and trypsin, where an especially careful study has been made of this phenomenon, there is a close correlation between loss of activity and formation of denatured protein.[5]

2. Striking changes in the physical properties of a protein take place during denaturation. At its isoelectric point a denatured protein is insoluble, although the corresponding native protein may be quite soluble. It was the loss of solubility that first drew attention to the phenomenon of denaturation, and denaturation is now usually defined by the change in solubility. The denatured protein after precipitation has taken place is called a coagulated protein, the process of coagulation being considered to include both denaturation and aggregation of denatured protein in the form of a coagulum.[6] If the denatured protein is dissolved, by acid, alkali, or urea, the solution is found to be far more viscous than a solution of native protein of the same concentration.[7]

3. Changes in the availability of sulfhydryl, disulfide, and phenol groups appear as a consequence of denaturation. All of the SH and S-S groups found in a protein after hydrolysis can be detected in a denatured protein even before hydrolysis, while in the corresponding native protein only a fraction of these groups is detectable. In native egg albumin no SH or phenol groups are detectable. In other native proteins (hemoglobin, myosin, proteins of the crystalline lens, for example) some groups can be detected; new groups appear when the protein is made more alkaline, although not alkaline enough to cause denaturation, and then disappear when the original pH is restored. In all the different ways of coagulating a protein a close correlation between appearance of groups and loss of solubility has until recently been observed.[8] Now, however, it has been found that when myosin is rendered insoluble by drying (or when water is removed by freezing) there is no change in availability of its SH groups.[9] Furthermore, if the insoluble myosin is treated with a typical denaturing agent, such as heat or acid, all of the SH groups in the protein become available, although no change in solubility is observed. Hitherto protein coagulation due to dehydration has not been distinguished from coagulation

caused by other agents, but the tests for availability show that in coagulation by dehydration the change in protein constitution is distinctly different from that caused by any of the known protein coagulating agents. In our theory of coagulation both types of change will be considered. Probably both types occur biologically. It has been suggested that when light converts visual purple, a conjugated protein, into visual yellow the former is denatured in much the same way as it is by heat, alcohol, or acid.[10] And it has been shown that in the course of fertilization and in the rigor of muscle a coagulation of protein similar to that caused by dehydration takes place.[11,12]

4. In a list of the large number of different agents, with apparently little in common, that cause denaturation are heat, acid, alkali, alcohol, urea, salicylate, surface action, ultra-violet light, high pressure. The temperature coefficient of heat denaturation of many proteins (egg albumin, ferrihemoglobin, trypsin, etc.) is about 600 for a rise in temperature of ten degrees, and from this the energy of activation can be calculated. On either side of a point between the isoelectric point and neutrality the temperature coefficient of denaturation by heat is diminished.[13] A dry preparation of egg albumin is not readily denatured by heat.[6]

5. The denaturation of certain proteins, notably hemoglobin, serum albumin, and trypsin, is reversible.[14] From the effect of temperature on the equilibrium constant, the heat of reaction and the entropy change can be calculated. In the denaturation of hemoglobin by salicylate and in the denaturation of trypsin by heat or acid, the equilibrium between native and denatured protein is not affected by changes in the total concentration of protein present,[15] from which it can be inferred that in the denaturation of hemoglobin and trypsin by these agents no change in molecular weight takes place. Since typical denaturation can occur without change in the molecular weight, it is important to distinguish between denaturation and the changes in particle size observed by Svedberg. Although denaturation can occur without change in molecular weight, under certain conditions, as in the denaturation of myosin by urea, denaturation may be accompanied by depolymerization. Weber found the molecular weight of myosin to be of the order of a million and that of myosin in urea to be thirty-five thousand.[16] On the other hand, it is unlikely that depolymerization is always accompanied by denaturation, for under some of the conditions of depolymerization described by Svedberg it is improbable that denaturation (loss of specificity, loss of solubility, or appearance of previously inaccessible groups) takes place. In the case of hemocyanin, for example, decomposition into products $^1/_2$ and $^1/_{16}$ of the size of the original molecule readily occurs, and these smaller particles seem to have the properties of native hemocyanin.[17]

6. The shape of the native protein molecule appears to have little sig-

nificance for an understanding of denaturation. Denaturation occurs in the spherical molecules of egg albumin and hemoglobin, in the elongated particles of soluble myosin[18] and (as indicated by SH groups becoming detectable) even in myosin that has been formed into insoluble fibres by drying.[9]

7. In certain conjugated proteins stability of the protein and presence of the prosthetic group are related. In hemoglobin, the yellow oxidizing ferment of Warburg, and visual purple, it is necessary to denature the protein in order to detach the prosthetic group with the use of present methods.[10,19] After removal of the prosthetic group it is possible to reverse the denaturation of globin and the protein of the oxidizing ferment, but these native proteins are more unstable (with respect to denaturation) than they are when conjugated with their prosthetic groups. In hemoglobin the ease of denaturation depends upon the state of the prosthetic group. Carbon monoxide hemoglobin, for example, is less readily denatured by heat, acid, or alkali than are oxyhemoglobin and ferrihemoglobin.[20] In visual purple presence of the prosthetic group causes the protein to be denatured by visible light. Denaturation in this case appears to be reversible.[10]

II. Our conception of a native protein molecule (showing specific properties) is the following. The molecule consists of one polypeptide chain which continues without interruption throughout the molecule (or, in certain cases, of two or more such chains); this chain is folded into a uniquely defined configuration, in which it is held by hydrogen bonds[21] between the peptide nitrogen and oxygen atoms and also between the free amino and carboxyl groups of the diamino and dicarboxyl amino acid residues.

We shall not enter into a long discussion of the precise configurations of native proteins, about which, indeed, little reliable information is available. From the x-ray investigations of Astbury[22] and his collaborators it seems probable that in most native proteins the polypeptide chain, with the extended or one of the contracted configurations discussed by Astbury, folds back on itself in such a way as to form a layer in which peptide nitrogen and oxygen atoms of adjacent chains are held together by hydrogen bonds; several of these layers are then superposed to form the complete molecule, the bonds between layers (aside from the continuation of the polypeptide chain from one layer to the next) being hydrogen bonds between side-chain amino and carboxyl groups. In general not all of the side chain groups will be used in forming bonds within the molecule; some will be free on the surface of the molecule.

The importance of the hydrogen bond in protein structure can hardly be overemphasized. No complete review of the large amount of recent work on this bond is available; we shall mention only the most striking of its properties.[21] The hydrogen bond consists of a hydrogen atom which bonds two electronegative atoms together (F, O, N), the hydrogen atom lying between the two bonded atoms. The bond is essentially electrostatic in nature. The bonded atoms are held more closely together than non-bonded atoms, the N-H-O distance being about 2.8 Å. The energy of a strong hydrogen bond is 5000 to 8000 cal. per mole, the lower value being approximately correct for an N-H-O bond as in proteins. Side-chain bonds in proteins we consider to involve usually an amino and a carboxyl group, the nitrogen atom forming a hydrogen bond with each of two oxygen atoms and holding also one unshared hydrogen atom. In acid solutions hydrogen bonds may be formed between two carboxyl groups, as in the double molecules of formic acid.[23]

The characteristic specific properties of native proteins we attribute to their uniquely defined configurations.

The denatured protein molecule we consider to be characterized by the absence of a uniquely defined configuration. As the result of increase in temperature or of attack by reagents (as discussed below) the side-chain hydrogen bonds are broken, leaving the molecule free to assume any one of a very large number of configurations. It is evident that with loss of the uniquely defined configuration there would be loss of the specific properties of the native protein; it would not be possible to grow crystals from molecules of varying shapes, for example, nor to distinguish between closely related proteins when the molecules of each protein show a variability in configuration large compared with the differences in configuration of the different proteins.

Strong support of this view of the phenomenon of denaturation is provided by the known difference in entropy of native and denatured proteins, which is about 100 E. U. for trypsin, the entropy of the denatured form being the greater.[5] This very large entropy difference cannot be ascribed to a difference in the translational, vibrational, or rotational motion, but must be due to a difference in the number of accessible configurations. It corresponds to about 10^{20} accessible configurations for a denatured protein molecule. The large entropy of denaturation thus shows clearly that the phenomenon of denaturation consists in the change of the molecules of the native protein to a much less completely specified state.

The magnitude of the heat change, entropy change, and activation energy of denaturation (about 30,000 cal./mole, 100 E. U., and 150,000 cal./mole, respectively, for trypsin) can be interpreted in terms of our hydrogen-bond picture of the protein molecule. We consider the native protein molecule to be held in its definite configuration by side-chain hydrogen bonds, about fifty in number for this protein (corresponding to about twenty-five amino and twenty-five carboxyl side chains), each with a bond energy of about 5000 cal./mole, as in simpler systems. The activation energy of 150,000 cal./mole shows that in order for the molecule to lose its native configuration about thirty of the bonds must be broken. Some of the side-chain groups then again form hydrogen bonds; the heat of denaturation shows that on the average there are about six fewer such bonds in the denatured molecule than in the native molecule. (The activation energy for transition of a denatured protein molecule from one configuration to another is without doubt smaller than the activation energy of denaturation, as it involves breaking a smaller number of hydrogen bonds, so that at a temperature at which the rate of denaturation is appreciable the denatured molecule would run rapidly through its various configurations.) The magnitude of the entropy of denaturation also fits into our picture; it corresponds to the number of configurations obtained by forming hydrogen bonds at random between about twenty amino side chains and twenty carboxyl side chains.

The reagents which cause denaturation are all substances which affect hydrogen-bond formation. Alcohol, urea, and salicylate are well-known hydrogen-bond-forming substances; they form hydrogen bonds with the protein side chains, which are thus prevented from combining with each other and holding the protein in its native configuration. Acids act by supplying protons individually to the electronegative atoms which would otherwise share protons, and bases by removing from the molecule the protons needed for hydrogen-bond formation. This conception provides an explanation of the facts that the isoelectric point of a protein shifts toward the neutral point on denaturation[24] and that the pH at which the activation energy for denaturation has its maximum value is in general not at the isoelectric point of the native protein, but between this point and the neutral point.[25] In the native protein molecule of the usual type some amino and carboxyl side-chain groups are paired together by forming hydrogen bonds. The acid-base properties of the molecule are in the main determined by the groups which are left free. On denaturation some of the paired groups are freed, amino and carboxyl in equal numbers, and in consequence the isoelectric point of the denatured protein is shifted toward neutrality. We have pictured the process of denaturation as involving the rupture of a large number of hydrogen bonds, to form a labile activated molecule. This labile molecule is stabilized by action of base on its free carboxyl groups or of acid on its free amino groups, the activation energy thus having a maximum at a pH value between the neutral point and the isoelectric point of the native protein.

The action of ultra-violet light must be different. It is not possible to formulate a reasonable mechanism whereby a quantum of light can break twenty or thirty hydrogen bonds. Instead the light must attack the molecule in a different place, probably breaking the main polypeptide chain after absorption in a tyrosine or other phenolic residue, as suggested by Mitchell.[26] That this occurs is indicated by the observation[27] that after illumination with ultra-violet light in the cold denaturation occurs only on warming, though then at a lower temperature than without illumination; it is clear that illumination in the cold causes a break in the molecule, which, however, is restrained to configurations near to its native configuration by the side-chain hydrogen bonds. (That some loosening of the molecule occurs is shown by the observation of an increase in the available groups in egg albumin after illumination in the cold.) On warming, these bonds are broken; because of the break in the molecule, however, it can be denatured "in parts," and hence at a lower temperature than before illumination. We predict that it will be found that denaturation by illumination with ultra-violet light is in general not reversible.

In a conjugated protein the prosthetic group plays a part in holding the molecule in the native configuration (hemoglobin, yellow oxidizing ferment, visual purple). It is possible in such a protein for reversible denaturation to take place after absorption of light by the prosthetic group, no permanent damage being done to the molecule.

In a protein coagulum side-chain hydrogen bonds hold adjacent molecules together. Native proteins do not coagulate because most of the side chains are in protected positions inside of the molecule; denatured proteins (at the isolectric point) do coagulate because they have a larger number of free side chains and because in the course of time, as the molecule assumes various configurations, all of the side chains become free. The increase of viscosity of protein solutions on denaturation we attribute to the change from the compact configuration of the native protein molecules to more extended configurations.

A native protein molecule of small molecular weight may have free side chains so arranged as to permit it to combine with similar molecules to form a polymer with properties differing little from those of the small molecules. The observations of Svedberg and his collaborators indicate that this is the case for hemocyanin, casein, and certain other proteins.

Our theory of denaturation leads to definite predictions regarding the availability of groups to attack by reagents. In the large compact molecule of a native protein, of the order of magnitude of 50 Å in diameter and having the same structure as every other molecule of the protein, all groups would be protected from attack by reagents except those on the surface or near the surface. After denaturation of the protein, however, the molecule (in solution) would in the course of time assume various configurations, and every group in the molecule would become available to attack. These statements are in complete agreement with the experimental results regarding sulfhydryl, disulfide, and phenol groups mentioned above. The observation that the number of available groups is increased when the solution is made alkaline, though not alkaline enough to cause denaturation, shows that under these conditions some of the hydrogen bonds in the protein molecule are broken, causing it to assume a configuration somewhat more open than its original configuration.

Although a denatured protein molecule in a coagulum is not free to assume all configurations, being restrained by bonds to its neighbors, in general its configuration will be so open as to make all groups accessible to attack. However, as mentioned above, there is no change in the availability of groups in myosin when it is rendered insoluble by drying or when water is removed by freezing. We interpret this as showing that *coagulation of this type is not accompanied by denaturation; that is, the molecules do not lose their uniquely defined configurations.* On dehydration of the native protein the surface side chains of adjacent molecules form bonds sufficiently strong to produce a coagulum of native protein molecules. It would be expected that the mechanical bolstering effect of adjacent molecules in this coagulum would aid the molecules to retain their native configurations; this is observed to be the case for egg albumin and some other proteins, which in this state are denatured by heat only at a temperature considerably higher than before dehydration.[6]

Some features of the theory of protein structure discussed above have been suggested before. Many investigators have correlated the specific properties of native proteins with definitely specified molecular configurations and the loss of specific properties on denaturation with change in configuration. Astbury and his collaborators in particular, in discussing their x-ray investigations, which have provided so much valuable information on protein structure, have stated[22] that dehydration and temperature denaturation lead to destruction of the original special configuration of the native protein, giving a débris consisting simply of peptide chains. Our picture agrees with theirs except in regard to the effect of dehydration, which we believe to consist primarily in the coagulation of molecules with essentially unchanged structure. The hydrogen bonds which we postulate to exist between side-chain carboxyl and amino groups (each nitrogen atom attached by hydrogen to two oxygen atoms, with the distances N-H-O equal to 2.8 Å) are a refinement of the side-chain salt-like linkages discussed by Astbury. Our picture of the spreading of protein films on water, involving the unfolding of compact molecules to layers one amino-acid-residue thick, is not essentially different from that of Gorter and Neurath, who have suggested unfolding in connection with film formation, though not with denaturation in general.[28]

In this paper we have discussed in some detail a very drastic change in configuration of protein molecules, that connected with denaturation. We have pointed out that the large entropy of denaturation provides strong support for our suggested structures of native and denatured protein molecules, and that many other phenomena can also be interpreted in a simple way from this point of view. There also has been put forth recently some evidence that small changes in configuration of proteins occur, which play an important part in protein behavior. Thus Northrop and his collaborators have prepared an enzymatically inactive protein which on slight hydrolysis by trypsin is transformed into native trypsin,[29] and another which is similarly transformed by pepsin into native pepsin,[30] and Gorter[31] has shown that myosin spreads on water only after slight hydrolysis. The structural interpretation of these phenomena can be made only after further experimental information is available.

* On leave of absence from the Hospital of the Rockefeller Institute.

[1] Zinsser, H., and Ostenberg, Z., *Proc. New York Path. Soc.*, **14,** 78 (1914).

[2] Landsteiner, K., and Heidelberger, M., *Jour. Gen. Physiol.*, **6,** 131 (1933).

[3] Barcroft, J., *The Respiratory Function of the Blood,* Part II, Cambridge Univ. Press, 1928.

[4] Anson, M. L., and Mirsky, A. E., *Jour. Physiol.*, **60,** 50 (1925).

[5] Northrop, J. H., *Jour. Gen. Physiol.*, **13,** 756 (1930); **16,** 33, 323 (1932).

[6] Chick, H., and Martin, C. J., *Jour. Physiol.*, **40,** 404 (1910); **43,** 1 (1911–12); **45,** 61, 261 (1912–13).

[7] Anson, M. L., and Mirsky, A. E., *Jour. Gen. Physiol.*, **15,** 341 (1932).

[8] Heffter, A., *Mediz. nat. Arch.*, **1,** 81 (1907); Arnold, V., *Zeit. Physiol. Chem.*, **70,** 300 (1911); Mirsky, A. E., and Anson, M. L., *Jour. Gen. Physiol.*, **18,** 307 (1934–35); **19,** 427, 439 (1936).

[9] Mirsky, A. E., unpublished experiments.

[10] Wald, G., *Jour. Gen. Physiol.*, **19,** 351 (1935); Mirsky, A. E., *Proc. Nat. Acad. Sci.*, **22,** 147 (1936).

[11] Mirsky, A. E., *Jour. Gen. Physiol.*, **19** (1936), in press.

[12] Mirsky, A. E., unpublished experiments.

[13] Lewis, P. S., *Biochem. Jour.*, **20,** 978, 984 (1927); Loughlin, W. J., *Biochem. Jour.* **27,** 1779 (1933).

[14] Anson, M. L., and Mirsky, A. E., *Jour. Gen. Physiol.*, **9,** 169 (1925); *Jour. Phys. Chem.*, **35,** 185 (1931).

[15] Anson, M. L., and Mirsky, A. E., *Jour. Gen. Physiol.*, **17,** 393, 399 (1934).

[16] Weber, H. H., *Ergeb. d. Physiol.*, **36,** 109 (1934).

[17] Svedberg, T., *Science,* **79,** 327 (1934).

[18] von Muralt, A., and Edsall, J. T., *Jour. Biol. Chem.*, **89,** 315 (1930).

[19] Theorell, H., *Biochem. Zeitschr.*, **278,** 263 (1935).

[20] Hartridge, H., *Jour. Physiol.*, **44,** 34 (1912).

[21] Latimer, W. M., and Rodebush, W. H., *Jour. Am. Chem. Soc.*, **42,** 1419 (1920); Pauling, L., *Proc. Nat. Acad. Sci.*, **14,** 359 (1928); Sidgwick, N. V., *The Electronic Theory of Valency,* Oxford University Press, 1929; Bernal, J. D., and Megaw, H. D., *Proc. Roy. Soc.*, **A151,** 384 (1935).

[22] Astbury, W. T., and Street, A., *Phil. Trans. Roy. Soc.*, **A230,** 75 (1931); Astbury and Woods, H. J., *Ibid.*, **A232,** 333 (1933); Astbury and Sisson, W. A., *Proc. Roy. Soc.*, **A150,** 533 (1935); Astbury and Lomax, R., *Jour. Chem. Soc.*, **1935,** 846; Astbury, Dickinson, S., and Bailey, K., *Biochem. Jour.* **29,** 2351 (1935); see also Meyer, K. H., and Mark, H., *Der Aufbau der Hochpolymeren Organischen Naturstoffe,* Akademische Verlagsgesellschaft, M. B. H., Leipzig, 1930.

[23] Pauling, L., and Brockway, L. O., *Proc. Nat. Acad. Sci.*, **20,** 336 (1934).

[24] Heidelberger, M., and Pederson, K. O., *Jour. Gen. Physiol.*, **19,** 95 (1935); Pederson, K. O., *Nature,* **128,** 150 (1931).

[25] Loughlin, W. J., *Biochem. Jour.*, **27,** 1779 (1933).

[26] Mitchell, J. S., *Nature,* **137,** 509 (1936).

[27] Bovie, W. T., *Science,* **37,** 373 (1913).

[28] Gorter, E., and Grendel, F., *Proc. Acad. Sci. Amsterdam,* **29,** 371 (1926); Neurath, H., *J. Phys. Chem.*, **40,** 361 (1936).

[29] Kunitz, M., and Northrop, J. H., *Science,* **80,** 505 (1934).

[30] Herriott, R. M., and Northrop, J. H., *Ibid.*, **83,** 469 (1936).

[31] Gorter, E., *Nature,* **137,** 502 (1936).

[Reprinted from the Journal of the American Chemical Society, **61**, 1860 (1939).]

[CONTRIBUTION FROM THE GATES AND CRELLIN LABORATORIES OF CHEMISTRY, CALIFORNIA INSTITUTE OF TECHNOLOGY, No. 708]

The Structure of Proteins

BY LINUS PAULING AND CARL NIEMANN

1. Introduction

It is our opinion that the polypeptide chain structure of proteins,[1] with hydrogen bonds and other interatomic forces (weaker than those corresponding to covalent bond formation) acting between polypeptide chains, parts of chains, and side-chains, is compatible not only with the chemical and physical properties of proteins but also with the detailed information about molecular structure in general which has been provided by the experimental and theoretical researches of the last decade. Some of the evidence substantiating this opinion is mentioned in Section 6 of this paper.

Some time ago the alternative suggestion was made by Frank[2] that hexagonal rings occur in proteins, resulting from the transfer of hydrogen atoms from secondary amino to carbonyl groups with the formation of carbon–nitrogen single bonds. This *cyclol hypothesis* has been developed extensively by Wrinch,[3] who has considered the geometry of cyclol molecules and has given discussions of the qualitative correlations of the hypothesis and the known properties of proteins.

It has been recognized by workers in the field of modern structural chemistry that the lack of conformity of the cyclol structures with the rules found to hold for simple molecules makes it very improbable that any protein molecules contain structural elements of the cyclol type. Until recently no evidence worthy of consideration had been adduced in favor of the cyclol hypothesis. Now, however, there has been published[4] an interpretation of Crowfoot's valuable X-ray data on crystalline insulin[5] which is considered by the authors to provide proof[6] that the insulin molecule actually has the structure of the space-enclosing cyclol C_2. Because of the great and widespread interest in the question of the structure of proteins, it is important that this claim that insulin has been proved to have the cyclol structure be investigated thoroughly. We have carefully examined the X-ray arguments and other arguments which have been advanced in support of the cyclol hypothesis, and have reached the conclusions that there exists no evidence whatever in support of this hypothesis and that instead strong evidence can be advanced in support of the contention that bonds of the cyclol type do not occur at all in any protein. A detailed discussion of the more important pro-cyclol and anti-cyclol arguments is given in the following paragraphs.

2. X-Ray Evidence Regarding Protein Structure

It has not yet been possible to make a complete determination with X-rays of the positions of the atoms in any protein crystal; and the great complexity of proteins makes it unlikely that a complete structure determination for a protein will ever be made by X-ray methods alone.[7] Nevertheless the X-ray studies of silk fibroin by Herzog and Jancke,[8] Brill,[9] and Meyer and Mark[10] and of β-keratin and certain other proteins by Astbury and his collaborators[11] have provided strong (but

(1) E. Fischer, "Untersuchungen über Aminosäuren, Polypeptide und Protein," J. Springer, Berlin, 1906 and 1923.

(2) F. C. Frank, *Nature*, **138**, 242 (1936); this idea was first proposed by Frank in 1933: see W. T. Astbury, *J. Textile Inst.*, **27**, 282 (1936).

(3) D. M. Wrinch, (a) *Nature*, **137**, 411 (1936); (b) **138**, 241 (1936); (c) **139**, 651, 972 (1937); (d) *Proc. Roy. Soc.* (London), **A160**, 59 (1937); (e) **A161**, 505 (1937); (f) *Trans. Faraday Soc.*, **33**, 1368 (1937); (g) *Phil. Mag.*, **26**, 313 (1938); (h) *Nature*, **143**, 482 (1939); etc.

(4) (a) D. M. Wrinch, *Science*, **88**, 148 (1938); (b) THIS JOURNAL, **60**, 2005 (1938); (c) D. M. Wrinch and I. Langmuir, *ibid.*, **60**, 2247 (1938); (d) I. Langmuir and D. M. Wrinch, *Nature*, **142**, 581 (1938).

(5) D. Crowfoot, *Proc. Roy. Soc.* (London), **A164**, 580 (1938).

(6) In ref. 4d, for example, the authors write "The superposability of these two sets of points represented the first stage in the proof of the correctness of the C_2 structure proposed for insulin. ... These investigations, showing that it is possible to deduce that the insulin molecule is a polyhedral cage structure of the shape and size predicted, give some indication of the powerful weapon which the geometrical method puts at our disposal."

(7) A protein molecule, containing hundreds of amino acid residues, is immensely more complicated than a molecule of an amino acid or of diketopiperazine. Yet despite attacks by numerous investigators no complete structure determination for any amino acid had been made until within the last year, when Albrecht and Corey succeeded, by use of the Patterson method, in accurately locating the atoms in crystalline glycine [G. A. Albrecht and R. B. Corey, THIS JOURNAL, **61**, 1087 (1939)]. The only other crystal with a close structural relation to proteins for which a complete structure determination has been made is diketopiperazine [R. B. Corey, *ibid.*, **60**, 1598 (1938)]. The investigation of the structure of crystals of relatively simple substances related to proteins is being continued in these Laboratories.

(8) R. O. Herzog and W. Jancke, *Ber.*, **53**, 2162 (1920).

(9) R. Brill, *Ann.*, **434**, 204 (1923).

(10) K. H. Meyer and H. Mark, *Ber.*, **61**, 1932 (1928).

(11) W. T. Astbury, *J Soc. Chem. Ind.*, **49**, 441 (1930); W. T. Astbury and A. Street, *Phil. Trans. Roy. Soc.*, **A230**, 75 (1931); W. T. Astbury and H. J. Woods, *ibid.*, **A232**, 333 (1933); etc.

not rigorous) evidence that these fibrous proteins contain polypeptide chains in the extended configuration. This evidence has been strengthened by the fact that the observed identity distances correspond closely to those calculated with the covalent bond lengths, bond angles, and N-H ··· O hydrogen bond lengths found by Corey in diketopiperazine.

The X-ray work of Astbury also provides evidence that α-keratin and certain other fibrous proteins contain polypeptide chains with a folded rather than an extended configuration. The X-ray data have not led to the determination of the atomic arrangement, however, and there exists no reliable evidence regarding the detailed nature of the folding.

X-Ray studies of crystalline globular proteins have provided values of the dimensions of the units of structure, from which some qualitative conclusions might be drawn regarding the shapes of the protein molecules. An interesting attempt to go farther was made by Crowfoot,[5] who used her X-ray data for crystalline insulin to calculate Patterson and Patterson–Harker diagrams.[12] Crowfoot discussed these diagrams in a sensible way, and pointed out that since the X-ray data correspond to effective interplanar distances not less than 7 Å. they do not permit the determination of the positions of individual atoms;[13] the diagrams instead give some information about large-scale fluctuations in scattering power within the crystal. Crowfoot also stated that the diagrams provide no reliable evidence regarding either a polypeptide chain or a cyclol structure for insulin.

Wrinch and Langmuir[4] have, however, contended that Crowfoot's X-ray data correspond in great detail to the structure predicted for the insulin molecule on the basis of the cyclol theory, and thus provide the experimental proof of the theory. We wish to point out that the evidence adduced by Wrinch and Langmuir has very little value, because their comparison of the X-ray data and the cyclol structure involves so many arbitrary assumptions as to remove all significance from the agreement obtained. In order to attempt to account for the maxima and minima appearing on Crowfoot's diagrams, Wrinch and Langmuir made the assumption that certain regions of the crystal (center of molecule, center of lacunae) have an electron density less than the average, and others (slits, zinc atoms) have an electron density greater than the average. The positions of these regions are predicted by the cyclol theory, but the magnitudes of the electron density are not predicted quantitatively by the theory. Accordingly the authors had at their disposal seven parameters, to which arbitrary values could be assigned in order to give agreement with the data. Despite the numbers of these parameters, however, it was necessary to introduce additional arbitrary parameters, bearing no predicted relation whatever to the cyclol structure, before rough agreement with the Crowfoot diagrams could be obtained. Thus the peak B″, which is the most pronounced peak in the $P(xy0)$ section (Fig. 2 of Wrinch and Langmuir's paper) and is one of the four well-defined isolated maxima reported, is accounted for by use of a region (V) of very large negative deviation located at a completely arbitrary position in the crystal; and this region is not used by the authors in interpreting any other features of the diagrams. This introduction of four arbitrary parameters (the three coördinates and the intensity of the region V) to account for one feature of the experimental diagrams would in itself make the argument advanced by Wrinch and Langmuir unconvincing; the fact that many other parameters were also assigned arbitrary values removes all significance from their argument.

It has been pointed out by Bernal,[14] moreover, that the authors did not make the comparison of their suggested structure and the experimental diagrams correctly. They compared only a fraction of the vectors defined by their regions with the Crowfoot diagrams, and neglected the rest of the vectors. Bernal reports that he has made the complete calculation on the basis of their structure, and has found that the resultant diagrams show no relation whatever to the experimental diagrams. He states also that with seven density values at closest-packed positions as arbitrary parameters he has found that a large number of structures which give rough agreement with the experimental diagrams can be formulated.

We accordingly conclude that there exists no satisfactory X-ray evidence for the cyclol structure for insulin.

(12) A. L. Patterson, *Z. Krist.*, **90**, 517 (1935); D. Harker, *J. Chem. Phys.*, **4**, 825 (1936).

(13) It has also been pointed out by J. M. Robertson, *Nature*, **143**, 75 (1939), that the intensities of 60 planes could not provide sufficient information to locate the several thousand atoms in the insulin molecule.

(14) J. D. Bernal, *Nature*, **143**, 74 (1939); see also D. P. Riley and I. Fankuchen, *ibid.*, **143**, 648 (1939).

3. Thermochemical Evidence Regarding Protein Structure

It is, moreover, possible to advance a strong argument in support of the contention that the cyclol structure does not occur to any extent in any protein.

X-Ray photographs of denatured globular proteins are similar to those of β-keratin, and thus indicate strongly that these denatured proteins contain extended polypeptide chains.[15] Astbury[16] has also obtained evidence that in protein films on surfaces the protein molecules have the extended-chain configuration, and this view is shared by Langmuir, who has obtained independent evidence in support of it.[17] Now the heat of denaturation of a protein is small—less than one hundred kilogram calories per mole of protein molecules for denaturation in solution,[18] that is, only a fraction of a kilogram calorie per mole of amino acid residues. Consequently the structure of native proteins must be such that only a very small energy change is involved in conversion to the polypeptide chain configuration.

It is unfortunate that there exist no substances known to have the cyclol structure; otherwise their heats of formation could be found experimentally for comparison with those of substances such as diketopiperazine which are known to contain polypeptide chains or rings. It is possible, however, to make this comparison indirectly in various ways. A system of values of bond energies and resonance energies has been formulated[19] which permits the total energy of a molecule of known structure to be predicted with an average uncertainty of only about 1 kcal./mole for a molecule the size of the average amino acid residue. The polypeptide chain (amide form) and cyclol can be represented by the following diagrams

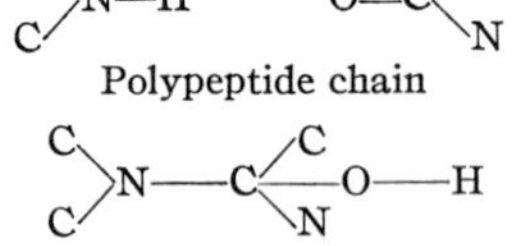

Polypeptide chain

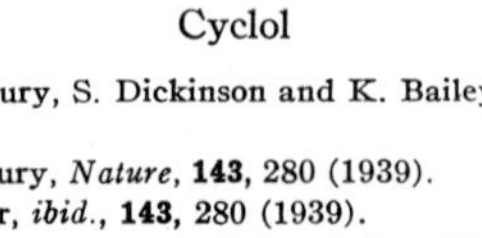

Cyclol

The change in bonds from polypeptide chain to cyclol is N–H + C=O ⟶ N–C + C–O + O–H. With N–H = 83.3, C=O = 152.0, N–C = 48.6, C–O = 70.0, and O–H = 110.2 kcal./mole, the bonds of an amino acid residue are found to be 6.5 kcal./mole less stable for the cyclol configuration than for the chain configuration. This must further be corrected for resonance of the double bond (resonance of the type

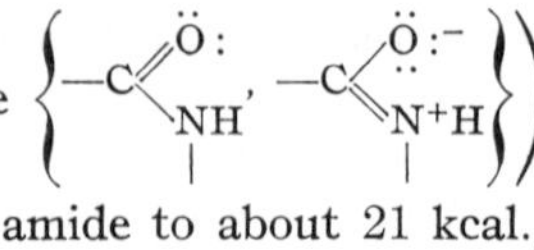

), which amounts for an amide to about 21 kcal./mole[19]; there is no corresponding resonance for the cyclol, which involves only single bonds. We conclude that the cyclol structure is less stable than the polypeptide chain structure by 27.5 kcal./mole per amino acid residue.

This value relates to gaseous molecules, containing no hydrogen bonds, and with the ordinary van der Waals forces also neglected. It is probable that the ordinary van der Waals forces would have nearly the same value for a cyclol as for a polypeptide chain; and the available evidence[19,20] indicates that the polypeptide hydrogen bonds would be slightly stronger than the hydrogen bonds for the cyclol structure. Moreover, the observed small values (about 2 kcal./mole) for the heat of solution of amides and alcohols show that the stability relations in solution are little different from those of the crystalline substances. We accordingly conclude that the polypeptide chain structure for a protein is more stable than the cyclol structure by about 28 kcal./mole per amino acid residue, either for a solid protein or a protein in solution (with the active groups hydrated[21]).

The comparison of the polypeptide chain and cyclol can also be made without the use of bond energy values. The heat of combustion of crystalline diketopiperazine, which contains two glycine residues forming a polypeptide chain,[22] is known;[23] from its value, 474.6 kcal./mole, the heat of formation of crystalline diketopiperazine (from elements in their standard states) is calculated to be 128.4 kcal./mole, or 64.2 kcal./mole per glycine residue. A similar calculation cannot be made directly for the cyclol structure, because no sub-

(15) W. T. Astbury, S. Dickinson and K. Bailey, *Biochem. J.*, **29**, 2351 (1935).

(16) W. T. Astbury, *Nature*, **143**, 280 (1939).

(17) I. Langmuir, *ibid.*, **143**, 280 (1939).

(18) M. L. Anson and A. E. Mirsky, *J. Gen. Physiol.*, **17**, 393, 399 (1934).

(19) (a) L. Pauling and J. Sherman, *J. Chem. Phys.*, **1**, 606 (1933); (b) L. Pauling, "The Nature of the Chemical Bond," Cornell University Press, Ithaca, N. Y., 1939. The values quoted above are from the latter source; they involve no significant change from the earlier set.

(20) M. L. Huggins, *J. Org. Chem.*, **1**, 407 (1936).

(21) The suggestion has been made [F. C. Frank, *Nature*, **138**, 242 (1936)] that the energy of hydration of hydroxyl groups might be very much greater than that of the carbonyl and secondary amino groups of a polypeptide chain; there exists, however, no evidence indicating that this is so.

(22) R. B. Corey, ref. 7.

(23) M. S. Kharasch, *Bur. Standards J. Research*, **2**, 359 (1929).

stance is known to have the cyclol structure; but an indirect calculation can be made in many ways, such as the following. One hexamethylenetetramine molecule and one pentaerythritol molecule contain the same bonds as four cyclized glycine residues and three methane molecules; hence the heat of formation of a glycine cyclol per residue is predicted to have the value 32.2 kcal./mole found experimentally[24] for $\frac{1}{4}C_6H_{12}N_4(c) + \frac{1}{4}C(CH_2OH)_4(c) - \frac{3}{4}CH_4(c)$. Similarly the value for $N(C_2H_5)_3(c) + C_2H_5OH(c) - 3C_2H_6(c)$ is 40.2 kcal./mole. The average of several calculations of this type, 36 kcal./mole, differs from the experimental value of the heat of formation of diketopiperazine per residue, 64 kcal./mole, by 28 kcal./mole. This agrees closely with the value 27.5 kcal./mole found by the use of bond energies, and we can be sure that the suggested cyclol structure for proteins is less stable than the polypeptide chain structure by about this amount per amino acid residue. Since denatured proteins are known to consist of polypeptide chains, and native proteins differ in energy from denatured proteins by only a very small amount (less than 1 kcal./mole per residue), we draw the rigorous conclusion that *the cyclol structure cannot be of primary importance for proteins; if it occurs at all (which is unlikely because of its great energetic disadvantage relative to polypeptide chains) not more than about three per cent. of the amino acid residues could possess this configuration.*

The above conclusion is not changed if the assumption be made that polypeptide chains are in the imide rather than the amide form,[3b] since this would occur only if the imide form were the more stable. In this case the experimental values of heats of formation (such as that of diketopiperazine) would still be used as the basis for comparison with the predicted value for the cyclol structure, and the same energy difference would result from the calculation.

It has been recognized[25-27] that energy relations present some difficulty for the cyclol theory (although the seriousness of the difficulty seems not to have been appreciated), and various suggestions have been made in the attempt to avoid the difficulty. In her latest communication[3h] Wrinch writes, "The stability of the globular proteins, under special conditions, in solution and in the crystal, we attribute to definite stabilizing factors;[26,27] namely, (1) hydrogen bonds between the oxygens of certain of the triazine rings, (2) the multiple paths of linkage between atoms in the fabric, (3) the closing of the fabric into a polyhedral surface which eliminates boundaries of the fabric and greatly increases the symmetry, and (4) the coalescence of the hydrophobic groups in the interior of the cage." These factors are, however, far from sufficient to stabilize the cyclol structure relative to the polypeptide chain structure. (1) The hydrogen bonds between hydroxyl groups in the cyclol structure would have nearly the same energy (about 5 kcal./mole) as those involving the secondary amino and carbonyl groups of the polypeptide chain. The suggestion[26] that resonance of the protons between oxygen atoms would provide further stabilization is not acceptable, since the frequencies of nuclear motion are so small compared with electronic frequencies that no appreciable resonance energy can be obtained by resonance involving the motion of nuclei. (2) We are unable to find any aspects of the bond distribution in cyclols which are not taken into consideration in our energy calculation given above. (3) There is no type of interatomic interaction known to us which would lead to additional stability of a cage cyclol as the result of eliminating boundaries and increasing the symmetry. (4) The stabilizing effect of the coalescence of the hydrophobic groups has been estimated[26] to be about 2 kcal./mole per CH_2 group, and to amount to a total for the insulin molecule of about 600 kcal./mole. It seems improbable to us that the van der Waals interactions of these groups are much less than this for polypeptides. The maximum of 600 kcal./mole from this source is still negligibly small compared with the total energy difference to be overcome, amounting to about 8000 kcal./mole for a protein containing about 288 residues.[28]

We accordingly conclude that the cyclol structure is so unstable relative to the polypeptide structure that it cannot be of significance for proteins.

It may be pointed out that a number of experiments[29-31] have added the weight of their evi-

(24) The values of heats of combustion used are $C_6H_{12}N_4(c)$, 1006.7; $C(CH_2OH)_4(c)$, 661.2; $CH_4(c)$, 210.6; $N(C_2H_5)_3(c)$, 1035.5; $C_2H_5OH(c)$, 325.7; $C_2H_6(c)$, 370.0 kcal./mole.

(25) F. C. Frank, *Nature*, **138**, 242 (1936).

(26) I. Langmuir and D. Wrinch, *ibid.*, **143**, 49 (1939).

(27) D. Wrinch, *Symposia on Quant. Biol.*, **6**, 122 (1938).

(28) Other suggestions regarding the source of stabilizing energy which have been made hardly merit discussion. "Foreign molecules" (Wrinch, ref. 27), for example, cannot be discussed until we have some information as to their nature.

(29) G. I. Jenkins and T. W. J. Taylor, *J. Chem. Soc.*, 495 (1937).

(30) L. Kellner, *Nature*, **140**, 193 (1937).

(31) H. Meyer and W. Hohenemser, *ibid.*, **141**, 1138 (1938).

dence to the general conclusion reached in this communication that the cyclol bond and the cyclol fabric are energetically impossible.

4. Further Arguments Indicating the Non-existence of the Cyclol Structure

There are many additional arguments which indicate more or less strongly that the cyclol structure does not exist. Of these we shall mention only a few.

It has been found experimentally that two atoms in adjacent molecules or in the same molecule but not bonded directly to one another reach equilibrium at a distance which can be represented approximately as the sum of certain van der Waals radii for the atoms.[19,32] Two carbon atoms of methyl or methylene groups not bonded to the same atom never approach one another more closely than about 4.0 Å., and two hydrogen atoms not bonded to the same atom are always at least 2.0 Å. apart. It has been pointed out by Huggins[33] that the cyclol structure places the carbon atoms of side chains only 2.45 Å. apart, and that in the C_2 structure for insulin there are hydrogen atoms only 0.67 Å. apart. We agree with Huggins that this difficulty alone makes the cyclol hypothesis unacceptable.

A closely related argument, dealing with the small area available for the side chains of a cyclol fabric, has been advanced by Neurath and Bull.[34] The area provided per side chain by the cyclol fabric, about 10 sq. Å., is far smaller than that required; and, as Neurath and Bull point out, the suggestion[3e,26] that some of the side chains pass through the lacunae of the fabric to the other side cannot be accepted, because this would require non-bonded interatomic distances much less than the minimum values found in crystals.

One of the most striking features of the cyclol fabric is the presence of great numbers of hydroxyl groups: in the case of cyclol C_2 there are 288 hydroxyl groups exclusive of those present in the side chains. Recently Haurowitz[35,36] has subjected the cyclol hypothesis to experimental tests on the basis of the existence or non-existence of cage hydroxyl groups. In the first communication[35] Haurowitz concludes on the basis of his and previous experiments[37–39] on the acylation and alkylation of proteins that the experimental evidence is in decided opposition to the conception that proteins possess great numbers of hydroxyl groups and therefore to the cyclol hypothesis. It seems to us that the objection raised by Haurowitz[36] is worthy of consideration and it certainly cannot be disposed of on the grounds that the original structure has been destroyed unless some concrete evidence can be submitted to indicate that this is the case. In a second communication Haurowitz and Astrup[36] write that "According to the classical theory of protein structure the carboxyl and amino groups found after hydrolytic splitting of a protein come from —CO—NH— bonds. According to the cyclol hypothesis, however, the free carboxyl and amino groups must be formed, during the splitting, from bonds of the structure =C(OH)—N=. The classical theory would predict on hydrolysis no great change in the absorption spectrum below 2400 Å. because the CO groups of the amino acids and of the peptide bonds both are strongly absorbing in this region.[40] On the other hand, the cyclol hypothesis would predict a greatly increased absorption because of the formation of new CO groups. . . . The absorption for genuine and for hydrolyzed protein is about equal. This seems to be in greater accordance with the classical theory of the structure of proteins than with the cyclol theory."

Mention may also be made of the facts that no simple substances with the cyclol structure have ever been synthesized[29] and that in general chemical reactions involving the breaking of covalent bonds are slow, whereas rapid interconversion of polypeptide and cyclol structure must be assumed to occur in, for example, surface denaturation. These chemical arguments indicate strongly that the cyclol theory is not acceptable.[41]

5. A Discussion of Arguments Advanced in Support of the Cyclol Theory

Although a great number of papers dealing with the cyclol theory have been published, we have

(32) N. V. Sidgwick, "The Covalent Link in Chemistry," Cornell University Press, Ithaca, N. Y., 1933; E. Mack, Jr., THIS JOURNAL, **54**, 2141 (1932); S. B. Hendricks, *Chem. Rev.*, **7**, 431 (1930); M. L. Huggins, *ibid.*, **10**, 427 (1932).

(33) M. L. Huggins, THIS JOURNAL, **61**, 755 (1939).

(34) H. Neurath and H. D. Bull, *Chem. Rev.*, **23**, 427 (1938).

(35) T. Haurowitz, *Z. physiol. Chem.*, **256**, 28 (1938).

(36) T. Haurowitz and T. Astrup, *Nature*, **143**, 118 (1939).

(37) J. Herzig and K. Landsteiner, *Biochem. Z.*, **61**, 458 (1914).

(38) B. M. Hendrix and F. Paquin, Jr., *J. Biol. Chem.*, **124**, 135 (1938).

(39) K. G. Stern and A. White, *ibid.*, **122**, 371 (1938).

(40) M. A. Magill, R. E. Steiger and A. J. Allen, *Biochem. J.*, **31**, 188 (1937).

(41) Another argument against cyclols of the C_2 type can be based on the results reported by J. L. Oncley, J. D. Ferry and J. Shack, *Symposia on Quant. Biol.*, **6**, 21 (1938), H. Neurath, *ibid.*, **6**, 196 (1938), and J. W. Williams and C. C. Watson, *ibid.*, **6**, 208 (1938), who have shown that dielectric constant measurements and diffusion measurements indicate that the molecules of many proteins are far from spherical in shape.

had difficulty in finding in them many points of comparison with experiment (aside from the X-ray work mentioned above) which were put forth as definite arguments in support of the structure.

One argument which has been advanced is that the cyclol theory "readily interprets the total number of amino acid residues per molecule, without the introduction of any *ad hoc* hypothesis"[3e] and that "The group of proteins with molecular weights ranging rom 33,600 to 40,500 are closed cyclols of the type C_2 containing 288 amino acid residues."[3e] Now the presence of imino acids (proline, oxyproline) in a protein prevents its formation of a complete cyclol such as C_2, and many proteins in this molecular weight range are known to contain significant amounts of proline: for insulin 10% is reported,[3f] for egg albumin 4%,[42] for zein 9%,[43] for Bence-Jones protein 3%,[44] and for pepsin 5%.[45] Wrinch has stated that "a future modification" (in regard to the number of residues) "is also introduced if imino acids are present"[3c]; "these numbers perhaps being modified if imino acids are present";[3e] and "if certain numbers of imino acid residues are present, these numbers" (of residues) "may be correspondingly modified."[3g] This uncertainty regarding the effect of the presence of imino acids in cyclols on the expected number of residues leaves the argument little force. In fact, even the qualitative claim that the cyclol hypothesis implies the existence of polyhedral structures containing certain numbers of amino acid residues and so predicts that globular proteins have molecular weights which fall into a sequence of separated classes can be doubted for the same reason.

It has been claimed[36] that the cyclol hypothesis explains the facts that proteins contain certain numbers of various particular amino acid residues and that these numbers are frequently powers of 2 and 3,[46] and it is proper that we inquire into the nature of the argument. Wrinch states[27] "An individual R group" (side chain) "is presumably attached, not to just any α-carbon atom, but only to those whose environment makes them appropriate in view of its specific nature. As an example of different environments, we may refer to the cyclol cages; here the pairs of residues at a slit have 'different environments' and the residues not at a slit fall into sets which again have 'different environments.' We therefore expect characteristic proportions to be associated with aromatic, basic, acidic, and hydrocarbon R groups, respectively, even perhaps with individual R groups. In any case a non-random distribution of the proportions of each residue in proteins in general is to be expected on any fabric hypothesis. On the cyclol hypothesis, for example, α-carbons having equivalent environments occur in powers of 2 and 3. . . . It is difficult to avoid interpreting the many cases which have recently been summarized in which the proportions of many types of residue are powers of 2 and 3 as further direct evidence in favor of the cyclol fabric. This fabric consists of an alternation of diazine and triazine hexagons, with symmetries respectively 2 and 3." Also it has been said by Langmuir[47] that "The occurrence of these factors, 2 and 3, furnishes a powerful argument for a geometrical interpretation such as that given by the cyclol theory. In fact, the hexagonal arrangement of atoms in the cyclol fabric gives directly and automatically a reason for the existence of the factors 2 and 3 and the non-occurrence of such factors as 5 and 7."

On examining the cyclol C_2, however, we find that these statements are not justified. The only factors of 288 are of the form $2^n\ 3^m$; moreover, the framework of the cyclol C_2 has the tetrahedral symmetry T, so that if the distribution of side chains conforms to the symmetry of the framework the amino acid residues would occur in equivalent groups of twelve. But in view of the rapid decrease in magnitude of interatomic forces with distance there would seem to be little reason for the distribution of side chains over a large protein molecule to conform to the symmetry T; it is accordingly evident that any residue numbers might occur for the cyclol C_2. We conclude that the cyclol hypothesis does not provide an explanation of the occurrence of amino acid residues in numbers equal to products of powers of 2 and 3.

Although there is little reason to expect that the distribution of side chains would correspond to the symmetry of the framework, it is interesting to note that the logical application of the methods of argument used by Wrinch suggests strongly that sixty residues of each of two amino acids should be present in a C_2 cyclol. This cyclol contains twenty lacunae of a particular type—each surrounded by a nearly coplanar border of twelve diazine and triazine

(42) H. O. Calvery, *J. Biol. Chem.*, **94**, 613 (1931).

(43) T. B. Osborne and L. M. Liddle, *Am. J. Physiol.*, **26**, 304 (1910).

(44) C. L. A. Schmidt, "Chemistry of Amino Acids and Proteins," C. C. Thomas, Springfield, Ill., 1938

(45) Unpublished determination by one of the authors.

(46) M. Bergmann and C. Niemann, *J. Biol. Chem.*, **115**, 77 (1936); **118** 307 (1937).

(47) I. Langmuir, *Symposia on Quant. Biol.*, **6**, 135 (1938).

rings. Each of these has trigonal symmetry so far as this near environment is concerned. Hence it might well be expected that a particular amino acid would be represented by three residues about each of these twenty lacunae, giving a total of sixty residues. But the number 60 cannot be expressed in the form $2^n 3^m$, it is not a factor of 288, and the integer nearest the quotient 288/60, 5, also cannot be expressed in the form $2^n 3^m$.

One of the most straightforward arguments advanced by Wrinch[3c,d] is that a protein surface film must have all its side chains on the same side, which would be the case for a cyclol fabric but not for an extended polypeptide chain. This argument now has lost its significance through the recently obtained strong evidence that proteins in films have the polypeptide structure,[16,17] and not the cyclol structure.

There can be found in the papers by Wrinch many additional statements which might be construed as arguments in support of the cyclol structure. None of these seems to us to have enough significance to justify discussion.

6. The Present State of the Protein Problem

The amount of experimental information about proteins is very great, but in general the processes of deducing conclusions regarding the structure of proteins from the experimental results are so involved, the arguments are so lacking in rigor, and the conclusions are so indefinite that it would not be possible to present the experimental evidence at the basis of our ideas of protein structure[48] in a brief discussion. In the following paragraphs we outline our present opinions regarding the structure of protein molecules, without attempting to do more than indicate the general nature of the evidence supporting them. These opinions were formed by the consideration not only of the experimental evidence obtained from proteins themselves but also of the information regarding interatomic interactions and molecular structure in general which has been gathered by the study of simpler molecules.

We are interested here only in the role of amino acids in proteins—that is, in the simple proteins (consisting only of α-amino and α-imino acids) and the corresponding parts of conjugated proteins; the structure and linkages of prosthetic groups will be ignored.

The great body of evidence indicating strongly that the amino acids in proteins are linked together by peptide bonds need not be reviewed here.

The question now arises as to whether the polypeptide chains or rings contain many or few amino acid residues. We believe that the chains or rings contain many residues—usually several hundred. The fact that in general proteins in solution retain molecular weights of the order of 17,000 or more until they are subjected to conditions under which peptide hydrolysis occurs gives strong support to this view. It seems to us highly unlikely that any protein consist of peptide rings containing a small number of residues (two to six) held together by hydrogen bonds or similar relatively weak forces, since, contrary to fact, in acid or basic solution a protein molecule of this type would be decomposed at once into its constituent small molecules.

There exists little evidence as to whether a long peptide chain in a protein has free ends or forms one more peptide bond to become a ring. This is, in fact, a relatively unimportant question with respect to the structure, as it involves only one peptide bond in hundreds, but it may be of considerable importance with respect to enzymatic attack and biological behavior in general.

A native protein molecule with specific properties must possess a definite configuration, involving the coiling of the polypeptide chain or chains in a rather well-defined way.[49] The forces holding the molecule in this configuration may arise in part from peptide bonds between side-chain amino and carboxyl groups or from side-chain ester bonds or S–S bonds; in the main, however, they are probably due to hydrogen bonds and similar interatomic interactions. Interactions of this type, while individually weak, can by combining their forces stabilize a particular structure for a molecule as large as that of a protein. In some cases (trypsin, hemoglobin) the structure of the native protein is the most stable of those accessible to the polypeptide chain; the structure can then be reassumed by the molecule after denaturation. In other cases (antibodies) the native configuration is not the most stable of those accessible, but is an unstable configuration impressed on the molecule by its environment (the influence of the antigen) during its synthesis; denaturation is not reversible for such a protein.

Crystal structure investigations have shown

(48) We believe that our views regarding the structure of protein molecules are essentially the same as those of many other investigators interested in this problem.

(49) H. Wu, *Chinese J. Physiol.*, **5**, 321 (1931); A. E. Mirsky and L. Pauling, *Proc. Nat. Acad. Sci.*, **22**, 439 (1936).

that in general the distribution of matter in a molecule is rather uniform. A protein layer in which the peptide backbones are essentially coplanar (as in the β-keratin structure) has a thickness of about 10 Å. If these layers were arranged as surfaces of a polyhedron, forming a cage molecule, there would occur great steric interactions of the side chains at the edges and corners. (This has been used above as one of the arguments against the C_2 cyclol structure.) We accordingly believe that *proteins do not have such cage structures*.[50] A compact structure for a globular protein might involve the superposition of several parallel layers, as suggested by Astbury, or the folding of the polypeptide chain in a more complex way.

One feature of the cyclol hypothesis—the restriction of the molecule to one of a few configurations, such as C_2—seems to us unsatisfactory rather than desirable. The great versatility of antibodies in complementing antigens of the most varied nature must be the reflection of a correspondingly wide choice of configuration by the antibody precursor. We feel that the biological significance of proteins is the result in large part of their versatility, of the ability of the polypeptide chain to accept and retain that configuration which is suited to a special purpose from among the very great number of possible configurations accessible to it.

Proteins are known to contain the residues of some twenty-five amino acids and it is not unlikely that this number will be increased in the future. A great problem in protein chemistry is that of the order of the constituent amino acid residues in the peptide chains. Considerable evidence has been accumulated[46] suggesting strongly that the stoichiometry of the polypeptide framework of protein molecules can be interpreted in terms of a simple basic principle. This principle states that the number of each individual amino acid residue and the total number of all amino acid residues contained in a protein molecule can be expressed as the product of powers of the integers two and three. Although there is no direct and unambiguous experimental evidence confirming the idea that the constituent amino acid residues are arranged in a periodic manner along the peptide chain, there is also no experimental evidence which would deny such a possibility, and it seems probable that steric factors might well cause every second or third residue in a chain to be a glycine residue, for example.

(50) Wrinch recently has suggested[3h] that even if proteins are not cyclols the cage structure might be significant.

The evidence regarding frequencies of residues involving powers of two and three leads to the conclusion that there are 288 residues in the molecules of some simple proteins. It is not to be expected that this number will be adhered to rigorously. Some variation in structure at the ends of a peptide chain might be anticipated; moreover, amino acids might enter into the structure of proteins in some other way than the cyclic sequence along the main chain.[51] The structural significance of the number 288 is not clear at present. It seems to us, however, very unlikely that the existence of favored molecular weights (or residue numbers) of proteins is the result of greater thermodynamic stability of these molecules than of similar molecules which are somewhat smaller or larger, since there are no interatomic forces known which could effect this additional stabilization of molecules of certain sizes. It seems probable that the phenomenon is to be given a biological rather than a chemical explanation—we believe that the existence of molecular-weight classes of proteins is due to the retention of this protein property through the long process of the evolution of species.

We wish to express our thanks to Dr. R. B. Corey for his continued assistance and advice in the preparation of this paper, and also to other colleagues who have discussed these questions with us.

Summary

It is concluded from a critical examination of the X-ray evidence and other arguments which have been proposed in support of the cyclol hypothesis of the structure of proteins that these arguments have little force. Bond energy values and heats of combustion of substances are shown to lead to the prediction that a protein with the cyclol structure would be less stable than with the polypeptide chain structure by a very large amount, about 28 kcal./mole of amino acid residues; and the conclusion is drawn that proteins do not have the cyclol structure. Other arguments leading to the same conclusion are also presented. A brief discussion is given summarizing the present state of the protein problem, with especial reference to polypeptide chain structures.

PASADENA, CALIF. RECEIVED APRIL 22, 1939

(51) H. Jensen and E. A. Evans, Jr., *J. Biol. Chem.*, **108**, 1 (1935), have shown that insulin probably contains several phenylalanine groups attached only by side-chain bonds to the main peptide chain.

Reprinted from the Proceedings of the NATIONAL ACADEMY OF SCIENCES,
Vol. 37, No. 4, pp. 205–211. April, 1951

THE STRUCTURE OF PROTEINS: TWO HYDROGEN-BONDED HELICAL CONFIGURATIONS OF THE POLYPEPTIDE CHAIN

BY LINUS PAULING, ROBERT B. COREY, AND H. R. BRANSON*

GATES AND CRELLIN LABORATORIES OF CHEMISTRY,
CALIFORNIA INSTITUTE OF TECHNOLOGY, PASADENA, CALIFORNIA†

Communicated February 28, 1951

During the past fifteen years we have been attacking the problem of the structure of proteins in several ways. One of these ways is the complete and accurate determination of the crystal structure of amino acids, peptides, and other simple substances related to proteins, in order that information about interatomic distances, bond angles, and other configurational parameters might be obtained that would permit the reliable prediction of reasonable configurations for the polypeptide chain. We have now used this information to construct two reasonable hydrogen-bonded helical configurations for the polypeptide chain; we think that it is likely that these configurations constitute an important part of the structure of both fibrous and globular proteins, as well as of synthetic polypeptides. A letter announcing their discovery was published last year.[1]

The problem that we have set ourselves is that of finding all hydrogen-bonded structures for a single polypeptide chain, in which the residues are equivalent (except for the differences in the side chain R). An amino acid residue (other than glycine) has no symmetry elements. The general operation of conversion of one residue of a single chain into a second residue equivalent to the first is accordingly a rotation about an axis accompanied by translation along the axis. Hence the only configurations for a chain compatible with our postulate of equivalence of the residues are helical configurations. For rotational angle 180° the helical configurations may degenerate to a simple chain with all of the principal atoms, C, C′ (the carbonyl carbon), N, and O, in the same plane.

We assume that, because of the resonance of the double bond between the carbon-oxygen and carbon-nitrogen positions, the configuration of each residue

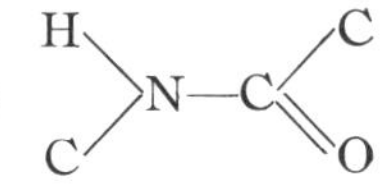

is planar. This structural feature has been verified for each of the amides that we have studied. Moreover, the resonance theory is now so well grounded and its experimental substantiation so extensive that there can be no doubt whatever about its application to the amide group. The observed C—N distance, 1.32 Å, corresponds to nearly 50 per cent double-bond character, and we may conclude that rotation by as much as 10° from the planar configuration would result in instability by about 1 kcal. mole^{-1}. The interatomic distances and bond angles within the residue are assumed to have the values shown in figure 1. These values have been formulated[2] by consideration of the experimental values found in the crystal structure studies of DL-alanine,[3] L-threonine,[4] N-acetylglycine,[5] and β-glycylglycine[6] that have been made in our Laboratories. It is further assumed that each nitrogen atom forms a hydrogen bond with an oxygen atom of another residue, with the nitrogen-oxygen distance equal to 2.72 Å, and that the vector from the nitrogen atom to the hydrogen-bonded oxygen atom lies not more than 30° from the N—H direction. The energy of an N—H · · · O=C hydrogen bond is of the order

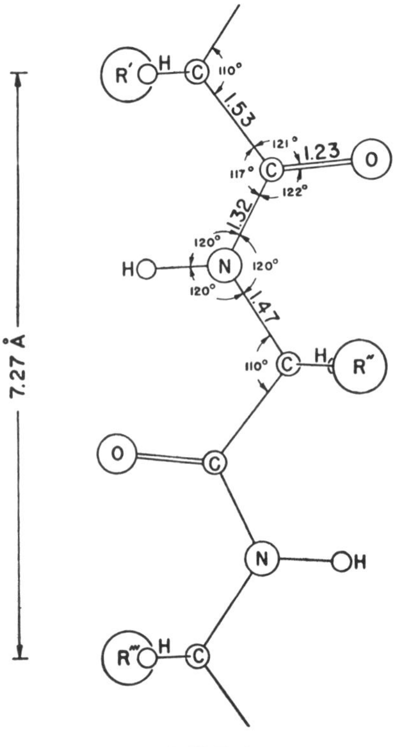

FIGURE 1

Dimensions of the polypeptide chain.

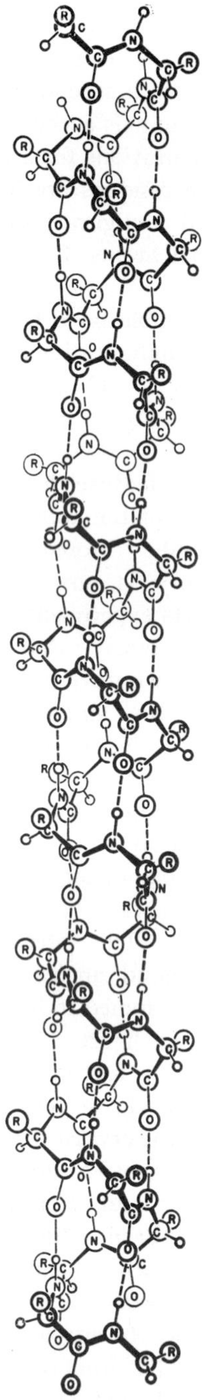

FIGURE 2
The helix with 3.7 residues per turn

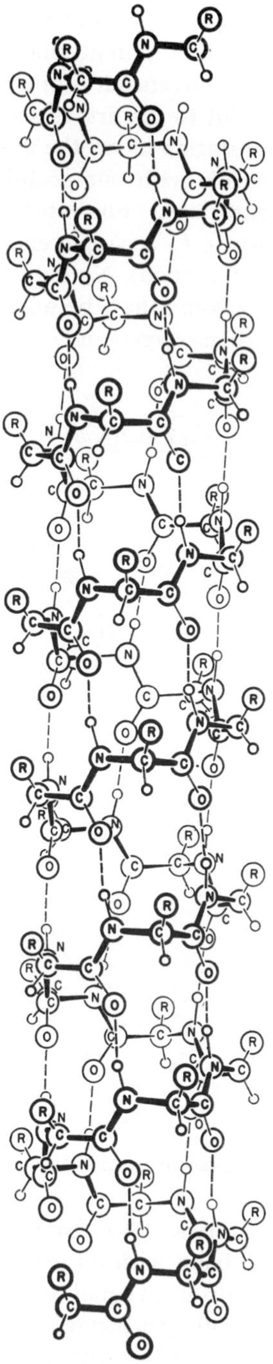

FIGURE 3
The helix with 5.1 residues per turn

of 8 kcal. mole^{-1}, and such great instability would result from the failure to form these bonds that we may be confident of their presence. The N—H · · · O distance cannot be expected to be exactly 2.72 Å, but might deviate somewhat from this value.

Solution of this problem shows that there are five and only five configurations for the chain that satisfy the conditions other than that of direction of the hydrogen bond relative to the N—H direction. These correspond to the values 165°, 120°, 108°, 97.2° and 70.1° for the rotational angle. In the first, third, and fifth of these structures the >CO group is negatively and the >N—H group positively directed along the helical axis, taken as the direction corresponding to the sequence—CHR—CO—NH—CHR— of atoms in the peptide chain, and in the other two their directions are reversed. The first three of the structures are unsatisfactory, in that the

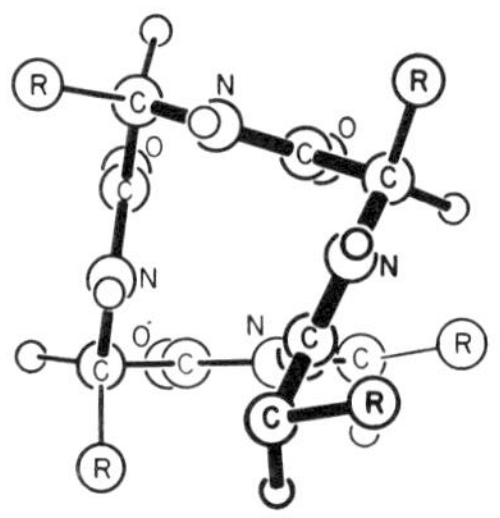

FIGURE 4

Plan of the 3.7-residue helix.

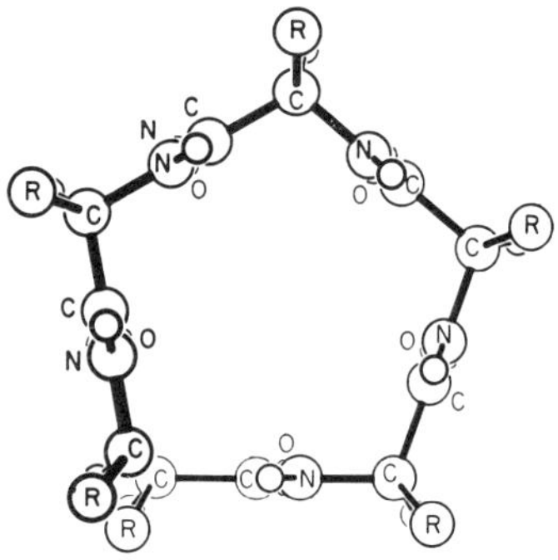

FIGURE 5

Plan of the 5.1-residue helix.

N—H group does not extend in the direction of the oxygen atom at 2.72 Å; the fourth and fifth are satisfactory, the angle between the N—H vector and N—O vector being about 10° and 25° for these two structures respectively. The fourth structure has 3.69 amino acid residues per turn in the helix, and the fifth structure has 5.13 residues per turn. In the fourth structure each amide group is hydrogen-bonded to the third amide group beyond it along the helix, and in the fifth structure each is bonded to the fifth amide group beyond it; we shall call these structures either the 3.7-residue structure and the 5.1-residue structure, respectively, or the third-amide hydrogen-bonded structure and the fifth-amide hydrogen-bonded structure.

Drawings of the two structures are shown in figures 2, 3, 4, and 5.

For glycine both the 3.7-residue helix and the 5.1-residue helix could occur with either a positive or a negative rotational translation; that is, as either a positive or a negative helix, relative to the positive direction of the helical axis given by the sequence of atoms in the peptide chain. For other amino acids with the L configuration, however, the positive helix and the negative helix would differ in the position of the side chains, and it might well be expected that in each case one sense of the helix would be more stable than the other. An arbitrary assignment of the R groups has been made in the figures.

The translation along the helical axis in the 3.7-residue helix is 1.47 Å, and that in the 5.1-residue helix is 0.99 Å. The values for one complete turn are 5.44 Å and 5.03 Å, respectively. These values are calculated for the hydrogen-bond distance 2.72 Å; they would have to be increased by a few per cent, in case that a larger hydrogen-bond distance (2.80 Å, say) were present.

The stability of our helical structures in a non-crystalline phase depends solely on interactions between adjacent residues, and does not require that the number of residues per turn be a ratio of small integers. The value 3.69 residues per turn, for the third-amide hydrogen-bonded helix, is most closely approximated by 48 residues in thirteen turns (3.693 residues per turn), and the value 5.13 for the other helix is most closely approximated by 41 residues in eight turns. It is to be expected that the number of residues per turn would be affected somewhat by change in the hydrogen-bond distance, and also that the interaction of helical molecules with neighboring similar molecules in a crystal would cause small torques in the helixes, deforming them slightly into configurations with a rational number of residues per turn. For the third-amide hydrogen-bonded helix the simplest structures of this sort that we would predict are the 11-residue, 3-turn helix (3.67 residues per turn), the 15-residue, 4-turn helix (3.75), and the 18-residue, 5-turn helix (3.60). We have found some evidence indicating that the first and third of these slight variants of this helix exist in crystalline polypeptides.

These helical structures have not previously been described. In addition to the extended polypeptide chain configuration, which for nearly thirty years has been assumed to be present in stretched hair and other proteins with the β-keratin structure, configurations for the polypeptide chain have been proposed by Astbury and Bell,[7] and especially by Huggins[8] and by Bragg, Kendrew, and Perutz.[9] Huggins discussed a number of structures involving intramolecular hydrogen bonds, and Bragg, Kendrew, and Perutz extended the discussion to include additional structures, and investigated the compatibility of the structures with x-ray diffraction data for hemoglobin and myoglobin. None of these authors proposed either our 3.7-residue helix or our 5.1-residue helix. On the other hand, we would eliminate, by our basic postulates, all of the structures proposed by them. The reason for the difference in results obtained by other investigators and by us through essentially similar arguments is that both Bragg and his collaborators and Huggins discussed in detail only helical structures with an integral number of residues per turn, and moreover assumed only a rough approximation to the requirements about interatomic distances, bond angles, and planarity of the conjugated amide group, as given by our investigations of simpler substances. We contend that these stereochemical features must be very closely retained in stable configurations of polypeptide chains in proteins, and that there is no special stability associated with an integral number of residues per turn in the helical molecule. Bragg, Kendrew, and Perutz have described a structure topologically similar to our 3.7-residue helix as a hydrogen-bonded helix with 4 residues per turn. In their thorough comparison of their models with Patterson projections for hemoglobin and myoglobin they eliminated this structure, and drew the cautious conclusion that the evidence favors the non-helical 3-residue folded α-keratin configuration of Astbury and Bell, in which only one-third of the carbonyl and amino groups are involved in intramolecular hydrogen-bond formation.

It is our opinion that the structure of α-keratin, α-myosin, and similar fibrous proteins is closely represented by our 3.7-residue helix, and that this helix also constitutes an important structural feature in hemoglobin, myoglobin, and other globular proteins, as well as of synthetic polypeptides. We think that the 5.1-residue helix may be represented in nature by supercontracted keratin and supercontracted myosin. The evidence leading us to these conclusions will be presented in later papers.

Our work has been aided by grants from The Rockefeller Foundation, The National Foundation for Infantile Paralysis, and The U. S. Public Health Service. Many calculations were carried out by Dr. S. Weinbaum.

Summary.—Two hydrogen-bonded helical structures for a polypeptide chain have been found in which the residues are stereochemically equivalent, the interatomic distances and bond angles have values found in amino acids, peptides, and other simple substances related to proteins, and the conjugated amide system is planar. In one structure, with 3.7 residues per turn, each carbonyl and imino group is attached by a hydrogen bond to the complementary group in the third amide group removed from it in the polypeptide chain, and in the other structure, with 5.1 residues per turn. each is bonded to the fifth amide group.

* Present address, Howard University, Washington, D. C.

† Contribution No. 1538.

[1] Pauling, L., and Corey, R. B., *J. Am. Chem. Soc.*, **72**, 5349 (1950)

[2] Corey, R. B., and Donohue, J., *Ibid.*, **72**, 2899 (1950).

[3] Lévy, H. A., and Corey, R. B., *Ibid.*, **63,** 2095 (1941). Donohue, J., *Ibid.*, **72,** 949 (1950).

[4] Shoemaker, D. P., Donohue, J., Schomaker, V., and Corey, R. B., *Ibid.*, **72,** 2328 (1950).

[5] Carpenter, G. B., and Donohue, J., *Ibid.*, **72,** 2315 (1950).

[6] Hughes, E. W., and Moore, W. J., *Ibid.*, **71,** 2618 (1949).

[7] Astbury, W. T., and Bell, F. O., *Nature,* **147,** 696 (1941).

[8] Huggins, M. L., *Chem. Rev.*, **32,** 195 (1943).

[9] Bragg, L., Kendrew, J. C., and Perutz, M. F., *Proc. Roy. Soc.*, **A203,** 321 (1950).

Reprinted from the Proceedings of the NATIONAL ACADEMY OF SCIENCES,
Vol. 37, No. 5, pp. 235–285. May, 1951

ATOMIC COORDINATES AND STRUCTURE FACTORS FOR TWO HELICAL CONFIGURATIONS OF POLYPEPTIDE CHAINS

BY LINUS PAULING AND ROBERT B. COREY

GATES AND CRELLIN LABORATORIES OF CHEMISTRY,* CALIFORNIA INSTITUTE OF TECHNOLOGY, PASADENA, CALIFORNIA

Communicated March 31, 1951

During recent years we have been gathering information about interatomic distances, bond angles, and other properties of simple substances related to proteins, and have been attempting to formulate configurations of the polypeptide chain that are compatible with this information and that might constitute a structural feature of proteins. We have reported the discovery of two helical configurations that satisfy these conditions.[1,2] In the following paragraphs we discuss the atomic positions for these configurations, and their form factors for diffraction of x-rays in the equatorial direction.

The γ *Helix.*—Let us first discuss the 5.1-residue helix. This configuration is obtained by coiling a polypeptide chain into a helical form, in such a way that the planar amide groups,

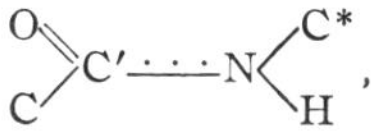

are in the trans configuration (the carbonyl group being almost directly opposed to the imino group), and each amide group forms hydrogen bonds with the fifth more distant group in each direction along the chain. The structure is represented diagrammatically in figure 1, and a drawing of it has been recently published.[2] We base our discussion on the values of interatomic distances and bond angles given in figure 4; these differ from those described earlier only in the change from 120° to 123° for the angle C′—N—C*.

The fifth amide group beyond a given group in the helix is nearly directly above it, and if the hydrogen bonds determine the orientation of the plane of the amide groups there seems to be no reason for this plane not to be parallel to the axis of the helix. The following calculations are made with this assumption. It is found that the rise per residue—the magnitude of the translation associated with the rotatory translation that converts one residue into the adjacent one along the chain—is determined nearly exactly by the length of the hydrogen bond. For N—H···O distance 2.72 A the rise per residue is 0.97 A, for 2.78 A it is 0.98 A, and it continues to increase linearly by about 0.002 A for each 0.010 A increase in length of the hydrogen-bond distance. The number of residues per turn, on the other hand, is determined essentially by the N—C—C′ angle of the α carbon

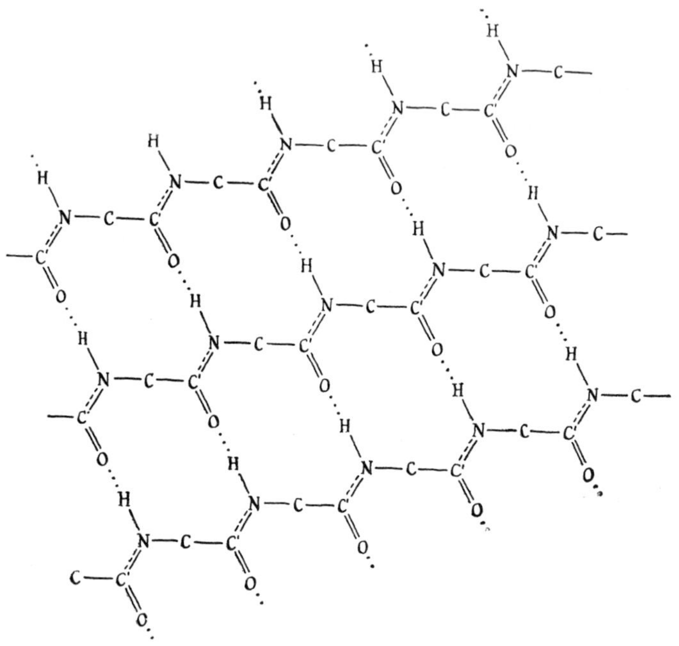

FIGURE 1

Diagrammatic representation of the 5.1-residue helical configuration of the polypeptide chain.

atom. For 21 residues in 4 turns (5.25 residues per turn) this angle has the value 111.2°, for 26 residues in 5 turns (5.20) its value is 110.6°, for 31 residues in 6 turns (5.17) 110.1°, and for 36 residues in 7 turns (5.14 residues per turn) 109.8°. These values are all within a reasonable range for this angle; the existing experimental evidence, for scores of molecules containing a tetrahedral carbon atom, suggests 110° as the best value for the angle, but a change by as much as 1° would introduce such a small amount of strain energy, in any case, that it should be allowed. There is, of course, no reason in an isolated molecule of a polypeptide or protein for the number of residues per turn to be rational. In a crystal, however, the intermolecular forces might constrain the molecule to assume the symmetry of a position in the crystal. The most likely eventuality is that the cylindrical molecules would take up a hexagonal close-packed arrangement, and that the molecules would assume a sixfold screw axis. This is provided from among the foregoing possibilities only by the configuration with 36 residues in 7 turns, which has a 36-fold screw axis, with a sixfold screw axis contained within its symmetry group. We present in table 1 calculated atomic coordinates for this case, for which the angle of rotation of the fundamental rotatory-translational operation is exactly 70°. The coordinates have been calculated for a rise per residue 0.98 A, corresponding to the

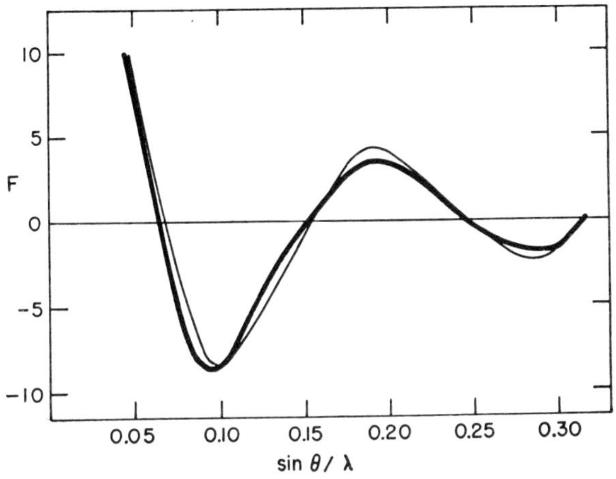

FIGURE 2

X-ray form factor for the 5.1-residue helix, calculated for equatorial reflections, and for cylindrical symmetry. The light line represents the scattering due to the four main-chain atoms of the amide group, and the heavy line includes the scattering of one β carbon atom per residue.

distance N—H···O equal to 2.78 A. The rise per turn is then 5.04 A, and the identity distance along the helical axis is 35.28 A.

The form factor for x-ray scattering for this molecule could, of course, be calculated from the coordinates of the atoms. A very good approximation for the equatorial reflections can be made by assuming cylindrical symmetry about the axis. The structure factor is then $F = \sum_i f_i J_0 (4\pi\rho_i \sin \theta/\lambda)$, in which J_0 is the Bessel function of order zero with the indicated argument, and ρ_i is the radius of the ith atom, in cylindrical coordinates, as given in table 1. (In this table there are given both cartesian coordinates, x, y, z, and cylindrical coordinates, ρ, ϕ, z, the latter being relative to the axis of the helix.) The form factor F as calculated by this method with use of the atomic form factors given in the International Tables for the

Determination of Crystal Structures is shown in figure 2, both for the residue alone, corresponding to polyglycine, and for a molecule with a β carbon atom in each residue. The two form factors do not differ very

TABLE 1

ATOMIC COORDINATES FOR THE 36-RESIDUE 7-TURN γ HELIX

x, y, z, ρ IN A

ATOM	x	y	z	ρ	ϕ
C	0.00	0.00	0.00	3.22	0.0°
C′	1.52	0.00	−0.15	2.66	27.8°
O	2.05	0.00	−1.26	2.65	39.3°
N	2.23	0.00	0.97	2.67	43.2°
C*	3.70	0.00	0.98	3.22	70.0°
βC	−0.50	−0.74	1.26	4.11	0.7°
or					
βC	−0.52	−0.72	−1.26	4.11	−0.7°
Axis	1.85	2.64	...	0.00	...

much. The square of F, which determines the intensity of reflection of x-rays, is shown in figure 3. It is seen that we predict that the equatorial

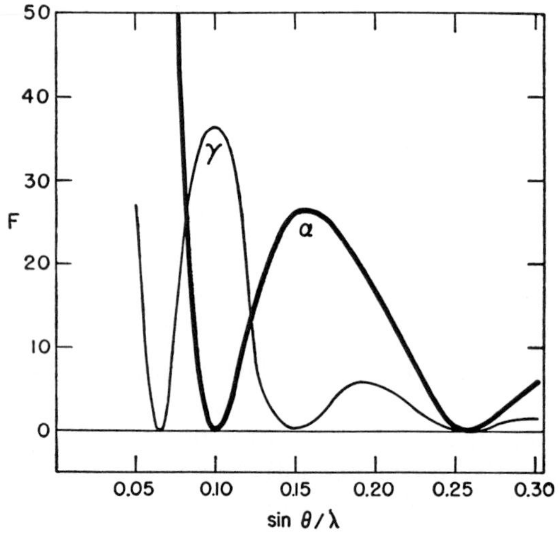

FIGURE 3

The square of the form factor for the 5.1-residue helix (γ) and the 3.7-residue helix (α), for equatorial reflections.

reflections from an array of molecules with this structure would be weak in the regions of interplanar distance 7.7 A, 3.3 A, and 2.0 A, and strong at about 5.0 A and 2.4 A. A discussion of the question of possible existence of protein molecules with this structure will be given in a later paper.

The α Helix.—The α helix is a configuration of the polypeptide chain in which each amide group forms hydrogen bonds with the amide groups removed by three from it in either direction along the chain. The α helix and the γ helix are the only helical configurations in which the residues are all equivalent and in which the stereochemical requirements, including formation of intramolecular hydrogen bonds, are satisfied. The structure is represented diagrammatically in figure 4. With the structural parameters described above, we find that a translation per residue of 1.47 A corresponds to a hydrogen-bond distance 2.75 A, and that the translation increases by 0.01 A for every 0.03 A increase in the hydrogen-bond distance. The number of residues per turn is fixed primarily by the bond angle at the α carbon atom; it varies from 3.60 for bond angle 108.9° to 3.67 for bond angle 110.8°. α-Helixes with these numbers of residues per turn have been found (see the following paper[3]) to explain the x-ray reflections[4] from highly oriented fibers of poly-γ-methyl-L-glutamate and poly-γ-benzyl-L-glutamate, respectively. The first ratio corresponds to 18 residues in 5 turns and the second to 11 residues in 3 turns.

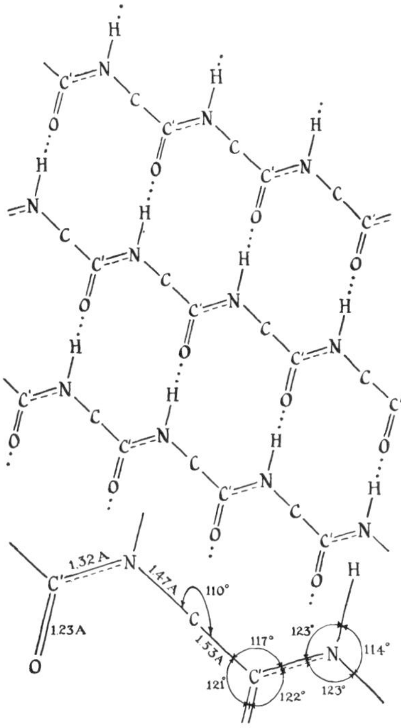

FIGURE 4

Diagrammatic representation of the 3.7-residue helical configuration of the polypeptide chain.

We have chosen to present parameters for the 18-residue 5-turn helix, which has a sixfold screw axis that would be stabilized in hexagonal packing. The parameters given in table 2 correspond to this helix, with a translation of 1.50 A per residue (5.40 A per turn, 27.0 A identity distance along the axis), these being the dimensions found for the polymethylglutamate. The corresponding hydrogen-bond distance is 2.86 A.

The structure factor for equatorial reflections calculated for cylindrical symmetry is shown in figure 5, and the corresponding intensity curve in figure 3. It is seen that the equatorial reflections should be very weak at about 5.0 A and 2.0 A, and strong at about 3.4 A. The correlation between theory and experiment for synthetic polypeptides and for proteins with the α-keratin structure is discussed in following papers of this series.

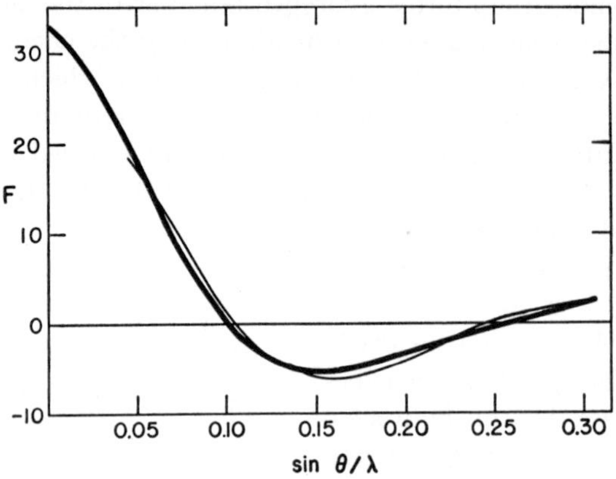

FIGURE 5

X-ray form factor for the 3.7-residue helix, calculated for equatorial reflections, and for cylindrical symmetry. The light line represents the scattering due to the main-chain atoms only, and the heavy line includes also the scattering of one β carbon per residue.

TABLE 2

ATOMIC COORDINATES FOR THE 18-RESIDUE 5-TURN α HELIX

x, y, z, ρ IN A

ATOM	x	y	z	ρ	ϕ
C	0.00	0.00	0.00	2.29	0.0°
N	1.16	0.00	0.89	1.59	27.8°
C′	2.42	0.00	0.44	1.61	73.8°
O	2.69	0.00	−0.76	1.74	82.0°
C*	3.52	0.00	1.50	2.29	100.0°
βC	−1.33	0.20	0.76	3.34	−17.6°
or					
βC	−0.03	−1.34	−0.76	3.34	17.6°
Axis	1.76	1.47	...	0.00	...

This investigation was aided by grants from The Rockefeller Foundation, The National Foundation for Infantile Paralysis, and The United States Public Health Service.

* Contribution No. 1550.

[1] Pauling, L., and Corey, R. B., *J. Am. Chem. Soc.*, **72,** 5349 (1950).

[2] Pauling, L., Corey, R. B., and Branson, H. R., these PROCEEDINGS, **37,** 205 (1951).

[3] Pauling, L., and Corey, R. B., *Ibid.*, **37,** 241 (1951).

[4] Bamford, C. H., Hanby, W. E., and Happey, F., *Proc. Roy. Soc.*, **A205,** 30 (1951).

VOL. 37, 1951 CHEMISTRY: PAULING AND COREY 241

THE STRUCTURE OF SYNTHETIC POLYPEPTIDES

BY LINUS PAULING AND ROBERT B. COREY

GATES AND CRELLIN LABORATORIES OF CHEMISTRY,* CALIFORNIA INSTITUTE OF TECHNOLOGY, PASADENA, CALIFORNIA

Communicated March 31, 1951

In a preliminary communication last year[1] we stated that there are only two helical configurations of polypeptide chains in which the residues are all equivalent and intramolecular hydrogen bonds are formed, and in which the interatomic distances, bond angles, and other structural features, especially the coplanarity of the conjugated amide system, are as required by earlier work in these Laboratories on amino acids, simple peptides, and other substances related to proteins. These two helical configurations were described in detail in a later paper[2] and it was mentioned that there is evidence that they occur in α keratin, α myosin, supercontracted keratin and myosin, and other fibrous proteins, and also constitute an important structural feature of hemoglobin and other globular proteins.[3] In the following paragraphs we discuss evidence that one of the helical structures is assumed also by synthetic polypeptides.

Oriented films and fibers of several synthetic polypeptides have been prepared and examined by x-ray diffraction by Bamford, Hanby, and Happey,[4] and the oriented films have been investigated with polarized infrared spectroscopy by Ambrose and Elliott.[5] The best photographs were given by poly-γ-methyl-L-glutamate and poly-γ-benzyl-L-glutamate. The x-ray data indicate an identity distance in the direction of the fibers of 5.50 A for the first substance, and a slightly larger value for the second substance. The dichroism observed for the N—H stretching and bending infrared absorption bands and for the C=O stretching band indicates that these groups are oriented nearly parallel to the fiber axis, and the conclusion is accordingly drawn that intramolecular hydrogen bonds are formed nearly parallel to this axis. These authors have interpreted all of their data as providing strong support for the α_{II} structure shown in figure 1. This structure was first proposed by Huggins[6] and has been discussed by Zahn,[7] Simanouti and Mizushima,[8] and Ambrose and Hanby.[9]

The α_{II} structure must, however, be rejected. In the structure as discussed by Bamford and coworkers the amide group is not assigned a planar configuration. There is, in fact, very strong theoretical and experimental evidence that the carbon-nitrogen bond has a large amount of double-bond character, and that the four atoms adjacent to these two atoms must form a coplanar system with them. The carbon-nitrogen distance is 1.32 A, which is 0.15 A less than the single-bond distance between these atoms. This shortening corresponds to about 50 per cent double-bond character, which is great enough to lead to effective coplanarity of the system. (Ambrose and Elliott[5] say "It remains to be shown whether resonance in polypeptides and proteins effectively limits the rotation about the C—N bond, and we regard the matter as by no means settled." In fact, however, our present knowledge of structural chemistry is such that there can be no doubt on this point: non-planar configurations of this group must surely be accompanied by pronounced instability.) In our investigation of helical configurations of the polypeptide chain[2] we found it impossible to construct an acceptable configuration resembling α_{II}. The closest approximation to an acceptable configuration that can be constructed, in which the hydrogen atom of the N—H group is brought to within 1.8 A of the carbonyl oxygen atom, is unsatisfactory because the N—H···O angle differs from a straight angle by about 70°, and thus this configuration would be presumed

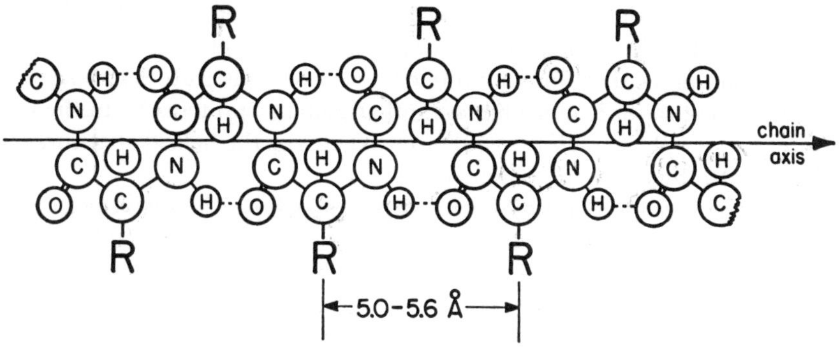

FIGURE 1

The α_{II} structure for the polypeptide chain, discussed by Bamford, Hanby, and Happey, and by Ambrose and Elliott.

not to correspond to a reasonably strong hydrogen bond. It was for this reason that the α_{II} structure was eliminated in our earlier considerations. There are also several other arguments against the α_{II} structure. Ambrose and Elliott mention that their model gives a dichroic ratio of 1.41:1 for the C=O stretching calculated for perfectly oriented molecules, and that the observed dichroism in poly-γ-benzyl-L-glutamate, 2.6:1, is considerably greater than this value. Moreover, the α_{II} structure corresponds to an extension of only approximately 40 per cent during the $\alpha \rightarrow \beta$ transformation. Ambrose and Elliott suggest that the observed reversible extension of wool, 100 per cent or more, is not to be interpreted as necessarily requiring a similar reversible extension from the α to the β configuration of the polypeptide chain, because there is present some amorphous material in wool; but it seems to us, as also to Astbury,[10] very unlikely that such a great discrepancy could exist. In addition, it may be pointed out that the synthetic polypeptides seem to be hexagonal or closely pseudohexagonal in structure, and the flat α_{II} configuration provides no explanation of this fact.

All of these difficulties are overcome if it is assumed that the synthetic polypeptides have the configuration of the third-amide hydrogen-bonded helix which we have described. In this helix the distance per residue along the helical axis is predicted to be approximately 1.50 A and there are approximately 3.7 residues per turn. The translation along the axis per turn is thus predicted to be about 5.5 A. This is in excellent agreement with the fiber axis translations reported by Bamford, Hanby, and Happey,

TABLE 1

X-RAY DATA FOR POLY-γ-METHYL-L-GLUTAMATE
HEXAGONAL UNIT WITH a_0 = 11.96 A, c_0 = 27.5 A

Equator:

$HI \cdot L$	$d_{obs.}$	$d_{calc.}$	$F_1{}^a$	F_2	$I_{2_{calc.}}$	$I_{obs.}$
10·0	10.35 A	10.35 A	18.5	16.2	360	VVS
11·0	5.98	5.98	5.1	6.2	29	S
20·0	5.22	5.18	1.3	3.2	7	M
21·0	3.89	3.91	−4.0	−4.8	22	S
30·0	3.45	3.45	−5.0	−7.1	21	S
22·0	3.00	2.99	−5.0	−6.1	14	S
31·0	2.87	2.87	−4.9	−1.8	2	W

Second hyperbola

$HI \cdot L$	$d_{obs.}$	$d_{calc.}$
11·2	5.51 A	5.48 A
20·2	4.88	4.87

Fifth hyperbola

$HI \cdot L$	$d_{obs.}$	$d_{calc.}$
10·5	4.82 A	4.86 A
11·5	4.05	4.05
20·5	3.75	3.77

Sixth hyperbola: a polar arc with $d_{obs.}$ = 4.43 A; $d_{calc.}$ for $\{10.6\}$, 4.20 A.

[a] Calculated for cylindrically symmetrical distributions of atomic centers, with radii from helical axis 2.29 A for C, 1.59 A for N, 1.61 A for C′, 1.74 A for O, and 3.34 A for βC.

5.50 A for poly-γ-methyl-L-glutamate and slightly larger for the benzyl ester. Moreover, the helical molecules would be expected to pack together essentially as would cylinders, in a hexagonal packing, and the predicted interplanar distances for the hexagonal lattice agree well with those observed, and shown in tables 1 and 2.

Bamford, Hanby, and Happey interpreted the data for the benzyl ester in terms of an orthorhombic unit with axes 10.35 A, 5.98 A, and 5.50 A. The number of molecules in this unit calculated from the density 1.34 is 1.9. The authors note that the first two axial lengths are in the ratio $\sqrt{3}:1$, which is compatible with hexagonal symmetry, and they ask to what extent the x-ray results are consistent with a helical configuration of the molecules. They point out that a hexagonal unit with $a_0 = 11.96$ A and $c_0 = 5.50$ A and containing three amino-acid residues, corresponding to a three-fold helix, would have a density 1.06 g cm^{-3}, which is too small. Our helix, with about 3.7 residues per turn, would lead, however, to the density 1.29, which is acceptable. A similarly satisfactory value of the density, 1.3, is also calculated for the corresponding hexagonal unit for the benzyl ester, with interplanar distances given in table 2.

Although most of the reflections observed by Bamford, Hanby, and Happey are accounted for by the orthorhombic unit or the hexagonal unit

TABLE 2

X-RAY DATA FOR POLY-γ-BENZYL-L-GLUTAMATE

PSEUDO-ORTHORHOMBIC UNIT WITH $a_0 = 25.0$ A, $b_0 = 17.3$ A, $c_0 = 14.42$ A

Equator:

hkl	$d_{calc.}$	$d_{obs.}$	F_1	F_2	$I_{2calc.}$	$I_{obs.}$
100	25.0	...	0	1.3	0.5	W
001	14.4	...	0	1.1	0.4	W
101, 200	12.5	12.6	21.5	17.8	165	VVS
002, 301	7.21	7.21	9.5	11.1	35	VS
202, 400	6.25	6.21	6.0	9.0	20	S
103, 402, 501	4.72	4.87	−0.8	−2.2	2	M (diffuse)
303, 600	4.17	4.07	−3.8	−7.4	9 }	S (diffuse)
004, 602	3.61	3.85	−4.4	−4.8	4 }	
204, 503, 701	3.49	...	−4.8	+0.8	0.2	...

First hyperbola: faint reflections at 16 A.
Second hyperbola: faint reflections at 8 A.
Third hyperbola:

hkl	$d_{calc.}$	$d_{obs.}$	$I_{obs.}$
151, 250	5.24 A	5.27 A	VS
052, 351	4.50	4.48 }	S (diffuse)
252, 450	4.24	4.08 }	
153, 452, 551	3.65	3.60 }	

with $c_0 = 5.50$ A, some reflections which they could not explain were reported by them. They concluded that these reflections, observed for both of the esters, are due to a second phase in each specimen, because the reflections seemed not to have any obvious relation to those that had been indexed. The second phase would have to be assumed to have parallel orientation to the main phase. It has seemed to us likely that these weak reflections indicate the presence of a larger unit for each crystal. A larger unit of structure would, of course, be predicted on the basis of our 3.7-residue helix, inasmuch as a true identity distance along the helical axis could occur only after a number of turns. Even though the interatomic interactions that are operative in stabilizing the helix would not be expected, in a non-crystalline phase, to lead to a rational number of residues per turn, the forces operating during crystallization might well be great enough to produce a small torque in the helix, such as to cause it to assume a configuration with a rational number of residues per turn. The three simplest rational helixes of the 3.7-residue hydrogen-bonded type are those with 11 residues in 3 turns (3.67 residues per turn), 15 residues in 4 turns (3.75 residues per turn), and 18 residues in 5 turns (3.60 residues per turn). We have found that the weak reflections for the methyl ester can be accounted for on the basis of the 18-residue 5-turn helix, and those for the benzyl ester on the basis of the 11-residue 3-turn helix. Plans of these helixes are shown as figures 2 and 3.

The 18-residue 5-turn helix has a sixfold screw axis of symmetry, and it would accordingly be expected that these helical molecules would arrange themselves side by side in hexagonal packing in such a way that the unit

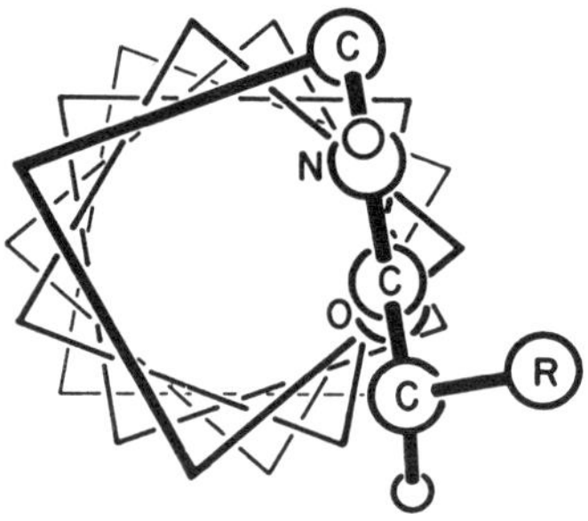

FIGURE 2

Plan of the 18-residue 5-turn helix.

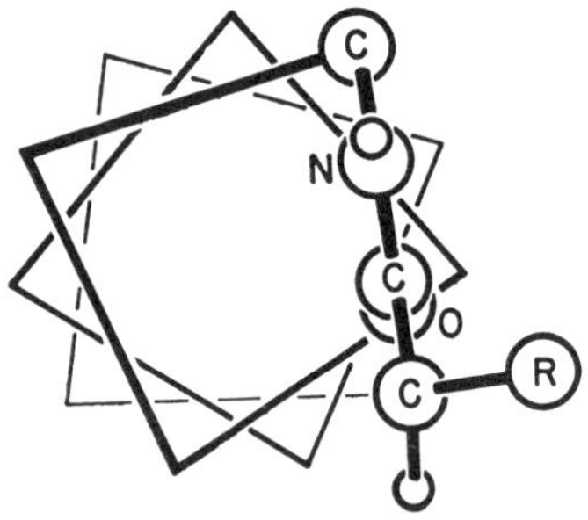

FIGURE 3

Plan of the 11-residue 3-turn helix.

of structure would contain only one 18-residue segment of one helix. The predicted dimensions of the hexagonal unit of structure would, then, be $a_0 = 11.96$ A, $c_0 = 27.5$ A. All of the data reported by Bamford, Hanby, and Happey are accounted for by this unit, as shown in table 1. The two reflections in the second hyperbola and the one in the sixth hyperbola are those which are not accounted for by the smaller unit. The earlier authors described the two reflections in the second hyperbola as corresponding to a c-axial length of about 13.3 A, which is close to half of our value of c_0. The reflection at 4.43 A is a polar arc, which probably could not be measured very accurately.

The determination of the structure of the crystals of the methyl ester would involve only the determination of the orientation of one of the helixes and of the positions of the atoms in the side chains. Simple considerations show that the observed intensities are approximately those predicted for a structure of this sort.

We have calculated intensities of the equatorial reflections for this hexagonal structure, with the four peptide atoms CNCO and the β carbon atom distributed over cylindrical surfaces, the structure factor then being $F_{HI.L} = \sum_i f_i J_0(2\pi\rho_i/d_{HI\cdot L})$; here f_i is the atomic structure factor for the ith atom, ρ_i is the distance of the atom from the helical axis, as given by our calculations,[2] and J_0 is the Bessel function of order zero. The values found are included in table 1, as F_1. It is seen that there is rough general agreement. Our calculations have included only five of the ten heavy atoms per residue; and although it may be predicted that the contribution of the

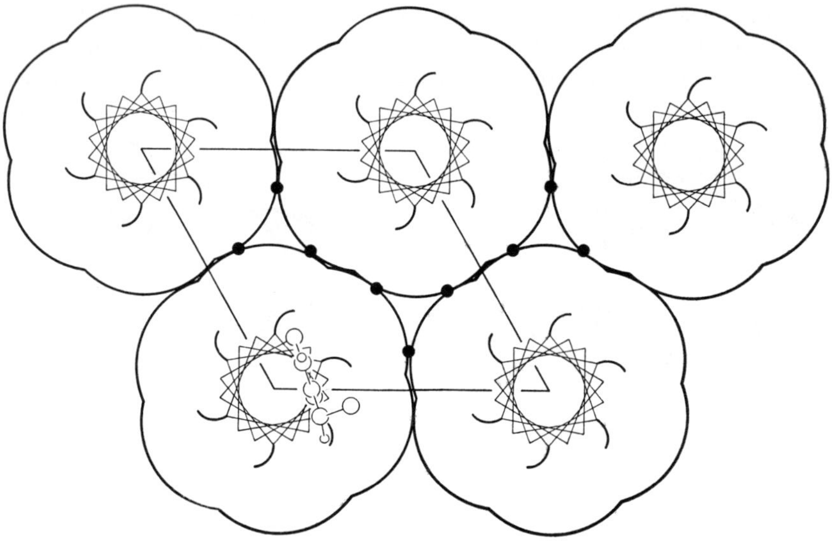

FIGURE 4

Proposed structure of poly-γ-methyl-L-glutamate.

other five (side-chain) atoms would be less, because of greater mutual interference, than that of the five main-chain atoms, it should be significant, and can explain the observed medium intensity of $\{20.0\}$ and weak intensity of $\{31.0\}$. That this can be done is seen by comparison with F_2, and with I_2, the corresponding intensities $pLPF_2^2$, in which p is the frequency factor, L the Lorentz factor, and P the polarization factor. In this calculation it is assumed that side-chain atoms are grouped about the 3-fold screw axes, in the positions xyz, with $xy = {}^3/_7\,{}^6/_7, {}^3/_7\,{}^4/_7, {}^1/_7\,{}^4/_7, {}^4/_7\,{}^1/_7, {}^6/_7\,{}^3/_7$, and ${}^4/_7\,{}^3/_7$. These positions are indicated by solid circles in figure 4, which represents the proposed structure for the methyl ester polymer. One atom per residue has been placed in these positions; the calculated intensities correspond to one oxygen atom and two carbon atoms for each asymmetric unit of three residues.

It is interesting to note that a reasonable explanation of the observed intensities of the equatorial reflections can be obtained with neglect of 40 per cent of the heavy atoms in the crystal. We believe that this phenomenon, which seems to be rather general for proteins and related substances, is due to the larger mutual interference of the rays scattered by the side-chain atoms than of those scattered by the main-chain atoms. The side-chain atoms, of which there are 90 in the unit, are located in 15 sets of 6-fold positions, presumably at 15 different values of the radius from the helical axis; whereas the 90 main-chain atoms (including the β carbon atoms) are more regularly arranged, corresponding to the pseudo 18-fold screw axis of the helix, and only five values of the radius are represented.

The anomalous reflections reported for the benzyl ester are described as lying on diffuse hyperbolas at 16 A and 8 A. We interpret these reflections as resulting from the presence of the 11-residue 3-turn helix, for which the value 16.8 A for c_0 would be predicted, on the assumption that the residue distance is the same as in the methyl ester, 1.53 A. This corresponds to 5.60 A per turn; Bamford, Hanby, and Happey suggested 5.76 A or a somewhat smaller value for the translation along the fiber axis, and we have found the data to correspond to $3 \times 5.76 = 17.3$ A, or 1.57 A per residue. The increase of 0.04 A over the methyl ester may well be due to van der Waals repulsion of the side chains.

The 11-residue 3-turn helix does not have a 6-fold axis or 3-fold axis, and accordingly it would not be expected to form hexagonal crystals with one helix per unit. The larger residue weight of the benzyl ester (219 in place of 143 for the methyl ester) would lead to a larger pseudo-hexagonal unit, with a_0 about 14.4 A, rather than 10.96 A (table 2). (The values given in table 2 are those obtained with the x-ray beam through the edge of an oriented film; closely similar values were also obtained with the x-ray beam perpendicular to the film, and with oriented fibers.)

Let us consider ways in which an asymmetric helix might be surrounded by others. We assume that the helixes are all equivalent, and that they are in contact with one another in a simple way. The six vectors from each asymmetric helix to its six neighbors are of six different kinds, representing six kinds of interaction. These might consist of three pairs, with the two of each pair differing only in polarity. It is found that there are only three arrangements of the three pairs in which the helixes are equivalent, and that none of them explains the occurrence of the 25-A reflection and the 14.4-A reflection for the benzyl polymer. It is possible, however, that two adjacent helixes interact in a way without polarity, to form a doublet. This would occur, for example, if one helix were related to the other by a 2-fold screw axis. There are many ways of arranging these doublets in a crystal; one of the simplest, shown in figure 5, has monoclinic (pseudo-hexagonal) symmetry, and for the benzyl ester corresponds to axes a_0 = 25.0 A, b_0 = 17.3 A, c_0 = 14.42 A, and β = 90°. This unit accounts for all of the reflections observed for this polymer.

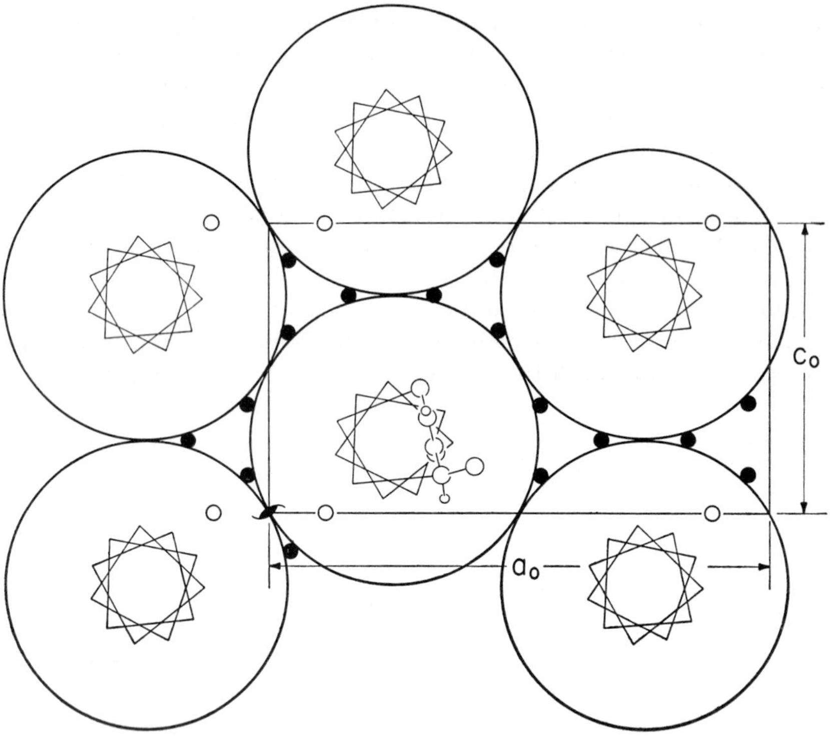

FIGURE 5

Proposed structure of poly-γ-benzyl-L-glutamate.

The intensities of all equatorial reflections can be roughly accounted for by consideration of only the five main-chain atoms, cylindrically distributed, as described above, about helical axes with coordinates x, z, and $\bar{x}$, $\bar{z}$, with $x = {}^1/_4$ and $z = {}^1/_4$, and one additional atom per residue, in the same positions as assumed for the methyl ester. The weak reflections {100}and{001} can be accounted for by any one of various slight distortions from the ideal hexagonal structure. The values of F_1 in table 2 are the calculated form factors for the five main-chain atoms per residue, and those of F_2 include the contributions of additional atoms in the positions assumed for the methyl ester (shown by full circles in figure 5). The scattering power in this position has been taken as 50 per cent greater than for the methyl ester. An additional concentration of electrons equivalent to one oxygen atom per unit of eleven residues has also been placed at each of the positions $x, z = \pm {}^1/_8, 0$ (shown by open circles in the figure). This small additional concentration of electrons, representing the deviation from the hexagonal symmetry of the methyl ester, accounts for the faint reflections $\{100\}$ and $\{001\}$. The close approximation of the structure to hexagonal packing of circular cylinders is made evident not only by the close approximation of the axial ratio a/c to $\sqrt{3}$, but also by the fact that the intensities of the two reflections requiring the larger unit can be explained by the scattering power of two heavy atoms, out of the 330 in the unit cell. Any one of many alternative small deviations of the distribution of scattering power from the ideal distribution with hexagonal symmetry would, of course, account for these observed weak reflections.

It seems likely that a poly-L-glutamate helix would be more stable with one screw sense (right-handed or left-handed) than with the other, and that helixes of only one kind are formed in significant number in the process of folding. A hexagonal array of helixes might then consist of equal numbers with positive and with negative orientation and essentially random distribution over the lattice points, equal numbers in an ordered array, only helixes with one orientation, or (much less likely) an ordered array in a ratio other than 1:1. For poly-γ-methyl-L-glutamate the detail of the photographs and the presence of only one helix per unit cell strongly suggest that the crystals contain helixes with only one orientation, and that segregation has occurred in crystallization. For the poly-benzyl ester the two-molecule unit also suggests that the helixes all have the same orientation; the strong reflections all have $h + k + l$ even, as required for an approximately body-centered unit. The possibility that segregation of positively and negatively oriented molecules occurs during crystallization suggests that annealing the specimens might improve the photographs.

Bamford, Hanby, and Happey also investigated several copolymers, and found from x-ray investigation that some tended to crystallize with the α configuration and some, especially those containing a large fraction of glycine residues, with the β configuration, involving sheets of extended polypeptide chains with lateral hydrogen bonds. They found that the α phase was oriented much more readily by stretching than the β phase, and attributed this difference to the presence of intramolecular hydrogen bonds in the α phase molecules. This feature remains unchanged by our attribution of our helical structure to the α phase. We explain the stability of the β phase for the glycine polymers by the stability of the lateral hydrogen bonds that can be formed; we have noted that with other residues there is serious steric interference between side chains, for the β configuration, leading to increase in the hydrogen-bond distance by 0.2 A, and a corresponding decrease in stability, relative to the α helix.

Bamford, Hanby, and Happey report interplanar distances leading to fiber-axis pseudo identity distances of 5.75 A for each of the three copolymers poly-DL-β-phenylalanine, poly-(DL-β-phenylalanine:γ-methyl-L-glutamate), and poly-(DL-β-phenylalanine:L-leucine). Larger steric repulsion of side chains would be expected for these copolymers than for poly-γ-methyl-L-glutamate.

The infrared dichroism observed by Ambrose and Elliott is accounted for very satisfactorily by our helical structure. In the 3.7-residue hydrogen-bonded helix the N—H and C=O bonds are oriented nearly parallel to the helical axis, the angle of deviation being only about 12°. This leads to a predicted dichroic ratio of about 44:1 for the N—H and C=O stretching vibrations. The largest observed dichroism, 14:1 for N—H stretching, is well within the predicted limit; it corresponds to an average angle of deviation of about 20° in the partially oriented specimen.

We conclude that there is strong evidence that crystals of poly-γ-methyl-L-glutamate and poly-γ-benzyl-L-glutamate, and also of the peptide copolymers, contain molecules with the third-amide hydrogen-bonded helical configuration, with about 3.7 residues per turn; and that the intermolecular forces operating in the crystals have caused the helixes in the first substance to assume the 18-residue 5-turn configuration, and those in the second to assume the 11-residue 3-turn configuration, these configurations having 3.60 and 3.67 residues per turn, respectively.

This investigation was aided by grants from The Rockefeller Foundation, The National Foundation for Infantile Paralysis, and The U. S. Public Health Service.

* Contribution No. 1551.

[1] Pauling, L., and Corey, R. B., *J. Am. Chem. Soc.*, **72,** 5349 (1950).

[2] Pauling, L., Corey, R. B., and Branson, H. R., these PROCEEDINGS, 205–211 (1951).

[3] Pauling, L., and Corey, R. B., *Ibid.*, to be published later.

[4] Bamford, C. H., Hanby, W. E., and Happey, F., *Proc. Roy. Soc.*, **A205,** 30 (1951).

[5] Ambrose, E. J., and Elliott, A., *Ibid.*, **A205,** 47 (1951).

[6] Huggins, M. L., *Chem. Rev.*, **32,** 195 (1943).

[7] Zahn, H., *Z. Naturforsch.*, **2B,** 104 (1947).

[8] Simanouti, T., and Mizushima, S., *Bull. Chem. Soc. Japan*, **21,** 1 (1948).

[9] Ambrose, E. J., and Hanby, W. E., *Nature*, **163,** 483 (1949).

[10] Astbury, W. T., *Ibid.*, **164,** 439 (1949).

VOL. 37, 1951 *CHEMISTRY: PAULING AND COREY* 251

THE PLEATED SHEET, A NEW LAYER CONFIGURATION OF POLYPEPTIDE CHAINS

BY LINUS PAULING AND ROBERT B. COREY

GATES AND CRELLIN LABORATORIES OF CHEMISTRY,* CALIFORNIA INSTITUTE OF TECHNOLOGY, PASADENA, CALIFORNIA

Communicated March 31, 1951

For many years it has been assumed that in silk fibroin, stretched hair and muscle, and other proteins with the β-keratin structure the polypeptide chains are extended to nearly their maximum length, about 3.6 A per residue, and during the last decade it has been assumed also that the chains form lateral hydrogen bonds with adjacent chains, which have the opposite orientation. A hydrogen-bonded layer of this sort is represented diagrammatically in figure 1.[1–4]

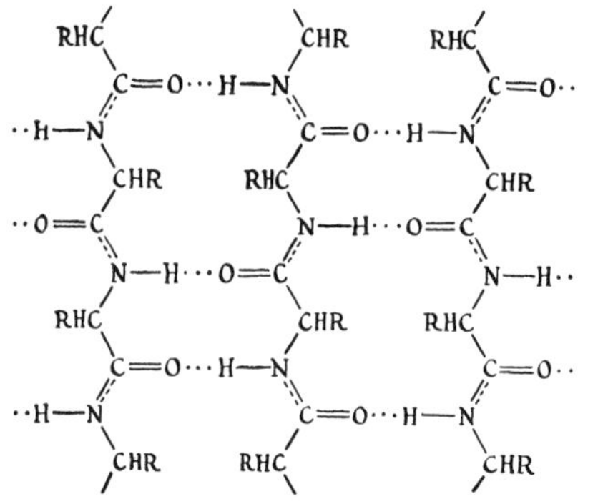

FIGURE 1

Diagrammatic representation of a hydrogen-bonded layer structure of polypeptide chains with alternate chains oppositely oriented.

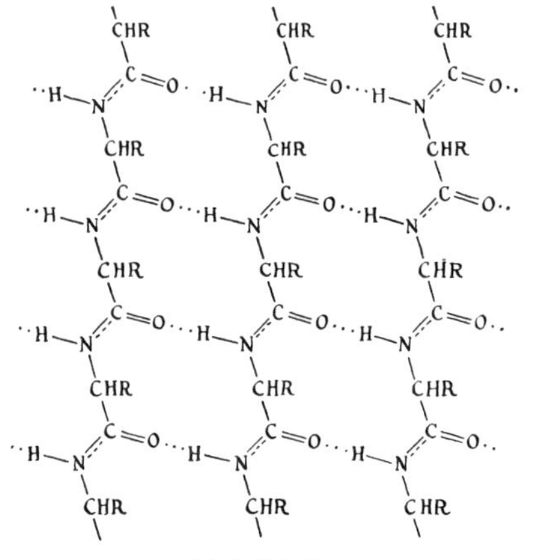

FIGURE 2

Diagrammatic representation of a hydrogen-bonded layer structure of polypeptide chains with all chains similarly oriented (the pleated sheet).

We have now discovered that there is another, rather similar hydrogen-bonded layer configuration of polypeptide chains, which differs from that of figure 1 in several ways. In the new configuration, which we shall call the pleated-sheet configuration, the plane formed by the two chain bonds of the α carbon atom is perpendicular to the plane of the sheet, as shown in figures 2 and 3, rather than being coincident with it. In this structure the successive residues in a chain are similarly oriented, directing their carbonyl groups in one direction and their imino groups in the opposite direction, and all of the chains are oriented in the same way, instead of adjacent chains being opposed in direction.

Let us assume that a polypeptide chain with the configuration indicated diagrammatically in figure 2 is bent in such a way that the planes of the successive amide groups form dihedral angles whose edges are perpendicular to the plane formed by the axes of the groups (the lines connecting successive α carbon atoms). It is found that if the bond distances and bond angles are given the values that we have used in our recent considerations of protein configurations the dihedral angle has the value 106.5°, and the vertical component of the axis of each residue is 3.07 A. It is also found that the carbonyl and imino groups are oriented in such a way that they can form satisfactory hydrogen bonds with corresponding groups in chains

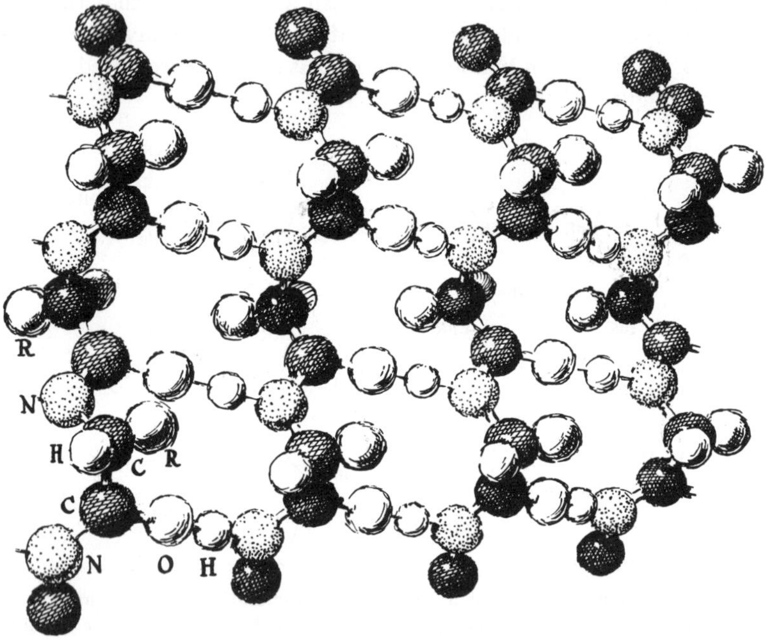

FIGURE 3

Drawing representing the pleated-sheet configuration of polypeptide chains.

obtained by lateral translation. If the lateral translation is given the value 4.75 A the N—H···O distance is 2.75 A; this is a normal hydrogen-bond distance. The N—H axis lies within 6° of the N···O axis, indicating that a stable hydrogen bond should be formed. The coordinates of atoms for the pleated-sheet configuration are given in table 1, and a drawing of the configuration is shown as figure 3.

It is to be noted that each amide group in the chain (neglecting the side chains) may be described as obtained from the preceding one by the operation of a glide plane of symmetry. Because of this, side chains of L-amino acid residues are related differently to the structure when attached to one α carbon atom than when attached to the α carbon atom of an adjacent residue. The pleated-sheet configuration can accordingly be described as involving only one kind of glycine residue, in case that it were to be assumed by a polyglycine, but two kinds of residues for all optically active amino-acid polymers. These two kinds differ in that, for the L configuration, a residue of one kind points its β carbon atom in the C=O direction, and a residue of the other kind points its β carbon atom in the N—H direction.

We have found some evidence to support the belief that the pleated-sheet configuration is present in stretched muscle, stretched hair, feather keratin, and some other fibrous proteins that have been assigned the β-keratin structure. These proteins give x-ray diagrams on which there is a strong meridional reflection corresponding to spacing about 3.3 A, which is a few per cent larger than the fiber-axis distance per residue for the undistorted

TABLE 1

Coordinates of Atoms in the Polypeptide Pleated-Sheet Configuration (in A)

	Unrotated			7° Rotation			20° Rotation		
Atom	x	y	z	x	y	z	x	y	z
C_1	0.00	1.15	0.00	0.00	1.09	0.00	0.00	0.96	0.00
N_1	−0.36	0.30	1.14	−0.36	0.46	1.17	−0.36	0.35	1.29
C_1'	0.53	−0.28	1.91	0.53	−0.31	1.96	0.50	−0.40	1.98
O_1	1.74	−0.14	1.73	1.72	−0.31	1.75	1.63	−0.64	1.58
C_2	0.00	−1.15	3.07	0.00	−1.09	3.15	0.00	−0.96	3.32
N_2	−0.36	−0.30	4.21	−0.36	−0.39	4.31	−0.34	−0.14	4.49
C_2'	0.53	0.28	4.98	0.53	0.22	5.12	0.50	0.08	5.49
O_2	1.74	0.14	4.80	1.72	−0.04	4.95	1.63	−0.39	5.50
C_1^*	0.00	1.15	6.14	0.00	1.09	6.30	0.00	0.96	6.64

pleated sheet, but much smaller than the value 3.6 A for fully extended polypeptide chains. We have noticed that the pleated sheet can be subjected, without rupturing the hydrogen bonds, to a considerable distortion, in such a way as to increase the fiber-axis distance. This distortion is effected by rotating each amide group about its C—C* axis through a small angle. The rotation moves one of the two β positions of each carbon atom farther from the median plane and the other nearer, and the effective rotations for the two non-equivalent kinds of optically active residues are such as to permit each to be an L residue with its side chain farther from the median plane than in the undistorted structure. Presumably the van der Waals repulsion of the side chain atoms and the main chain atoms would be operating in proteins of normal chemical composition with the pleated-sheet configuration, and this would cause some distortion of the chain-lengthening sort. (It is to be noted that two kinds of pleated sheets can be constructed of L-amino-acid residues, of which for one the deformation that relieves the strain of side chain van der Waals repulsion increases the fiber-axis length, and for the other it decreases it.) It might well occur that the magnitude of the deformation would be such as to give the fiber-axis residue length observed for the β-keratin proteins, about 3.3 A. This deformation results from a 20° rotation of the amide groups, which gives 3.32 A as the residue length. Coordinates for the structure with 20° rotation and also for a less deformed structure, with 7° rotation, are given in table 1.

The deformed structures require some distortion of the hydrogen bonds, in that if the hydrogen atom is kept coplanar with the amide group the N—H direction deviates from the N···O axis by an angle somewhat greater than the distorting angle of rotation. The nature of the distortion is such, however, as to suggest that not much strain energy is involved. Let us consider the effect on the stability of the amide group of moving the hydrogen atom onto (or nearly onto) the N···O axis. This motion would keep the hydrogen atom nearly in a plane normal to the 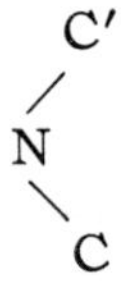 plane; that is, it involves moving the hydrogen atom toward one of the tetrahedral corners of the nitrogen atom. If the nitrogen atom were forming a pure double bond with the carbonyl carbon C′ there would be strong resistance to this motion of the hydrogen atom. However, it forms a bond with about one-half double-bond character and one-half single-bond character, corresponding to the resonance

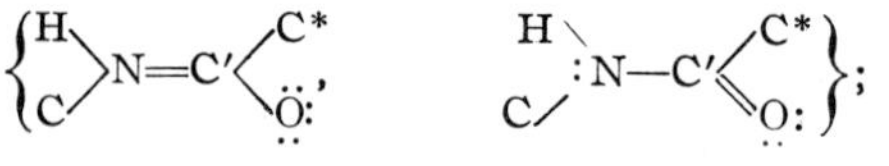

and for the second of the structures the tetrahedral position for the hydrogen atom would be the normal one, whereas for the first the planar position is stable. According we would predict that this rotational distortion of the pleated sheet would not involve so much strain as if the bonds were double bonds.

We may now ask to what extent distortion of the amide group from the planar configuration, through rotation of the two ends 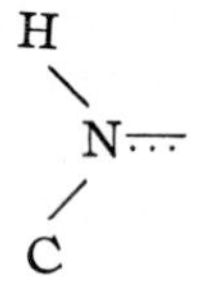 and 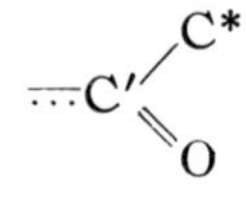 in opposite directions about the N┄C′ axis, might be expected to occur. The strain energy of this distortion, which is essentially also the strain energy of distortion of the hydrogen atom out of the plane, can be estimated in the following way. With δ the dihedral angle formed by the planes of the two end groups, the amide resonance energy may be taken equal to $-A \sin^2 (\delta - \pi/2)$, and the strain energy to $A \sin^2 \delta$. The factor A is the amide resonance energy for the planar configuration. This may be estimated as about 30 kcal mole^{-1}. (The experimental value for the carboxylate ion, in which each of the two C⋯O bonds has 50 per cent double-bond character, is 36 kcal mole^{-1}, and somewhat smaller values are found for gas molecules of amides, esters, and related substances.[5]) We thus find about 0.9 kcal mole^{-1} strain energy for 10° distortion of the amide group, 3.5 kcal mole^{-1} for 20° distortion, and so on, and we may predict that distortions as large as 20° might well occur in structures in

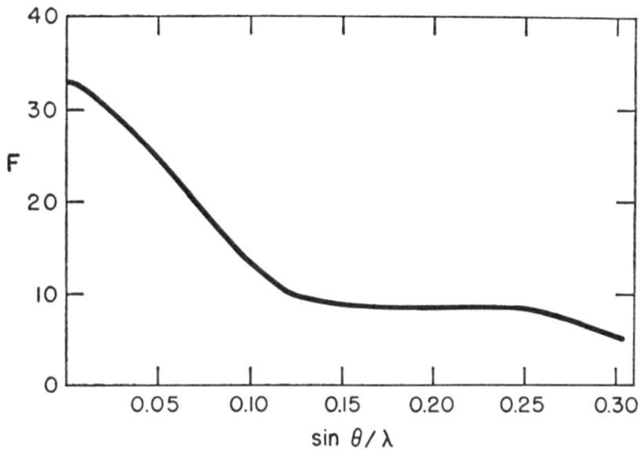

FIGURE 4

Calculated x-ray form factors for the pleated sheet, for planes parallel to the plane of the sheet.

which these distortions would relieve a larger strain, but that in general the polypeptide chain would avoid structures involving such strains. In any case, we would expect the distortion to be divided between the amide residue and the hydrogen bond. In calculating the coordinates of table 1 we have not taken account of these distortions.

The discussion of the pleated sheet in β keratin and other proteins will be presented in following papers. In this discussion we make use of the x-ray scattering form factor for the sheet. The form factor, calculated for reflections from planes parallel to the median plane of the undistorted sheet, is for convenient later reference given here, in figure 4, as calculated from the equation $F = \sum_i f_i \cos (2\pi y_i \sin \theta/\lambda)$, with f_i values as given in the International Tables for the Determination of Crystal Structures. The sum has been taken over the atoms of one residue of the undistorted structure, including also a β carbon atom, with $y = 2.04$.

This investigation was aided by grants from The Rockefeller Foundation, The National Foundation for Infantile Paralysis, and The United States Public Health Service.

* Contribution No. 1552.

[1] Astbury, W. T., *Trans. Faraday Soc.*, **29**, 193 (1933).

[2] Huggins, M. L., *J. Org. Chem.*, **1**, 407 (1936); *Chem. Rev.*, **32**, 195 (1943).

[3] Pauling, L., *J. Am. Chem. Soc.*, **62**, 2643 (1940).

[4] Lotmar, W., and Picken, L. E. R., *Helv. Chim. Acta*, **25**, 538 (1942).

[5] Pauling, L., *The Nature of the Chemical Bond*, Cornell University Press, Ithaca, N. Y., 1939.

THE STRUCTURE OF FEATHER RACHIS KERATIN

BY LINUS PAULING AND ROBERT B. COREY

GATES AND CRELLIN LABORATORIES OF CHEMISTRY,* CALIFORNIA INSTITUTE OF TECHNOLOGY, PASADENA, CALIFORNIA

Communicated March 31, 1951

The rachis of feathers gives rise to x-ray diffraction patterns of great complexity—they have been described as the most complex known for the naturally occurring fibrous substances. For their interpretation there is required a unit of structure with dimensions at least 9.5 A $\times$ 34 A $\times$ 94.6 A. In the following paragraphs we propose a structure for this protein that accounts for the principal features of the x-ray pattern and for some physical properties of the substance.

Astbury and other workers in the field have mentioned that the pattern somewhat resembles that of stretched hair, stretched muscle, and other proteins with the β-keratin structure, but that the x-ray diagram indicates that the length per residue is only 3.07 A, somewhat shorter than expected for an extended polypeptide chain, about 3.6 A, and than observed for silk fibroin, about 3.5 A, and for the β-keratin proteins, about 3.3 A. Astbury suggested that the chains might be in a somewhat collapsed β-keratin configuration, and pointed out that the reversible extensibility of feather keratin through about 7 per cent supported this assumption.[1] We were struck by the identity of the indicated fiber-axis residue length, 3.07 A, and the corresponding length predicted for the undistorted pleated-sheet configuration of hydrogen-bonded polypeptide chains, described in the preceding paper, and we investigated the possibility that feather keratin is composed of these pleated sheets in parallel orientation. This can be ruled out as unsatisfactory, however, in that, although the predicted distance between chains in the direction of the hydrogen bonds, 4.75 A, agrees closely with that indicated by the x-ray diagram, about 4.68 A, the other equatorial reflections on the diagram cannot be accounted for by such a structure. The diagram[3] shows four equatorial reflections, with spacings 34 A, 17 A, 11 A, and 8.56 A, that seem to be the first four orders from a unit with edge about 34 A. The magnitude of this dimension suggests three layers of protein, each about 11 A thick, and these layers could not be identical, inasmuch as all four orders of reflections are observed, rather than only the third order. Consideration of alternative possibilities led us to the conclusion that the layers consist of a pleated sheet and two layers of α helixes, these helixes having the configuration described in recent papers.[4]

A plan of the proposed structure is shown as figure 1, and a schematic drawing as figure 2. The chains of the pleated sheet are at the positions $x = 0$, $y = 0$, and $x = {}^1/_2$, $y = 0$, the base of the unit of structure having the dimensions $a_0 = 9.5$ A and $b_0 = 34.2$ A. The centers of the α helixes

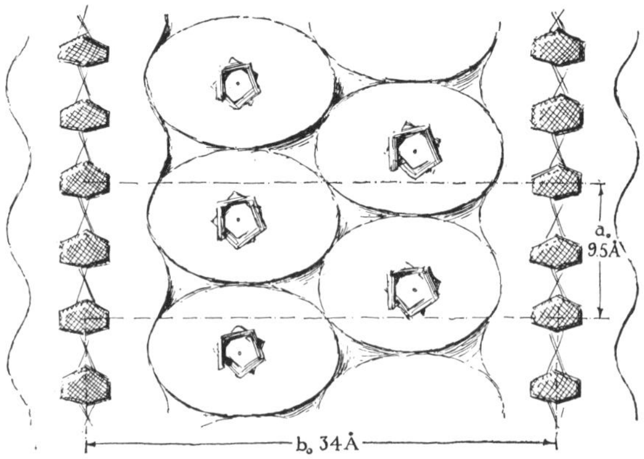

FIGURE 1

Plan of the proposed structure for feather rachis keratin. The structure consists of pleated-sheet layers, between which there are double layers of 3.7-residue helixes.

are at ${}^1/_4$, y and ${}^3/_4$, $\bar{y}$, with y approximately ${}^1/_3$. The α helixes are indicated to be in the close-packed arrangement given by these parameters by the absence of an equatorial reflection at 9.5 A. A diffuse reflection corresponding to the second order is observed at about 4.68 A.

We have found that the intensities of the equatorial reflections can be rather well explained by this structure, with consideration only of the atoms of known fixed position in the pleated sheet and in the α helixes. The atomic parameters and also the form factors for the pleated sheet have been reported in a preceding paper,[2] and those for the α helix are given in the following paper. The value of the parameter y indicated by the data is 0.275. The structure factors calculated with use only of the atoms of known position, including the β carbon atom for each residue in both the pleated she t and the α helixes, and with the approximation of cylindrical symmetry for the α helixes, are given in table 1 under the heading F_1.

Although these structure factors are in rough agreement with the observed intensities, somewhat improved agreement is obtained by making a correction for the remaining atoms of the side chains of the pleated-sheet residues. We have carried out this calculation by assuming two carbon atoms, or atoms of equivalent scattering power, per residue, arranged about the positions $y = +{}^1/_{12}$ and $-{}^1/_{12}$. These positions, at 2.85 A from the center of the pleated sheet, are those expected for the pleated-sheet side chains. The distribution of the atoms about these positions has been approximated by using an F-factor for these atoms proportional to that calculated for the pleated sheet.[2] The values F_2 in table 1 were calculated with inclusion of these side-chain atoms, and the quantity $I_{\text{calc.}}$ given in the following column was obtained by multiplying the square of F_2 by the Lor-

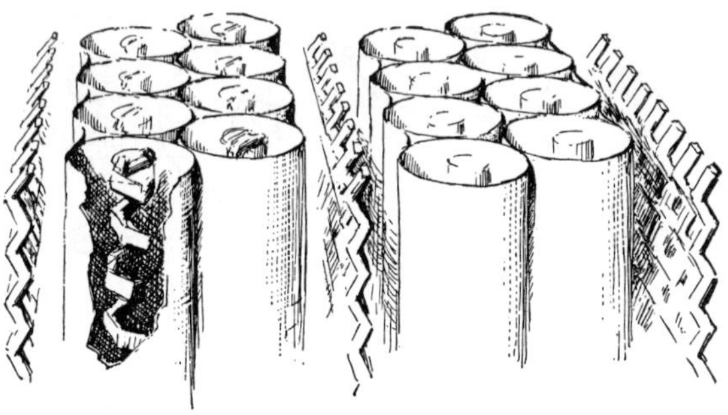

FIGURE 2

Drawing representing the proposed structure for feather rachis keratin.

entz, polarization, and frequency factors. The calculated intensity for (040) seems to be small; however, this calculated intensity is increased to the value 330 by inclusion of the intensity, 175, calculated for the form $\{110\}$, which is the only important diagonal reflection in this region of the equatorial plane. The structure factor for $\{120\}$ is small. The strong, diffuse reflection at 4.68 A is due to $\{200\}$, $\{210\}$, and $\{220\}$, with calculated spacings 4.75 A, 4.71 A, and 4.58 A. The structure factor for the α helix is very nearly 0 for this interplanar distance, and the form factor for these reflections is that of the chain in the β sheet, with contributions from the side-chain atoms. The value of F_2 has been obtained by assuming that the main-chain and side-chain atoms have the same effective distribution about their central axis as has the pleated sheet in the y direction. The general agreement of observed and calculated intensities is seen to be satisfactory.

In addition to these equatorial reflections, Corey and Wyckoff[3] reported three others, at 51.0 A (faint), 81.8 A (medium), and 115 A (medium). Bear[5] has pointed out that the central regions of the x-ray pattern given by feather rachis are confused by radiation artifacts related to the strong reflections, and that this is probably the source of these three large equatorial spacings. If we accept this interpretation, the equatorial reflections of feather keratin are all accounted for by our unit.

The meridional reflections observed by Corey and Wyckoff[3] and by Bear[5] are given in table 2. They can nearly all be accounted for as orders of a large identity distance, 94.6 A. In addition to the meridional reflections given in table 2, Bear reported a number of small-angle near-meridional reflections, corresponding to other orders of the identity distance 94.6 A.

TABLE 1

Comparison of Calculated and Observed Intensities of Equatorial X-Ray Reflections for Feather Rachis Keratin

Pseudo-orthorhombic, triclinic unit with a_0 = 9.50 A, b_0 = 34.2 A, c_0 = 94.6 A, $\alpha \cong 90°$, $\beta \cong 90°$, $\gamma \cong 90°$

hkl	$d_{calc.}$	F_1	F_2	$I_{calc.}$	$I_{obs.}$[a]	$d_{obs.}$[a]
010	(34.2 A)	22.8	42.6	730	Strong	33.3 A
020	17.1	−20.8	−10.4	24	Faint	17.1
030	11.4	45.5	45.5	295	Medium	11.0
110	9.14	−30.0	−30.0	210 }	Medium	8.56
040	8.56	46.1	37.6	155 }		
050	6.84	9.4	−3.2	1	...	...
060	5.70	12.3	0.4	0	...	...
070	4.89	11.4	3.1	1	...	...
200	4.75	...	25	38 }		
210	4.71	...	25	75 }	Strong	4.68
220	4.58	...	25	75 }	(diffuse)	
080	4.28	10.6	6.6	2		
090	3.80	16.4	16.4	12		
0·10·0	3.42	9.0	12.2	6		

[a] Corey and Wyckoff.[3]

For the ideal pleated sheet we have calculated a fiber-axis distance of about 3.07 A per residue, which agrees in numerical value with the spacing of the outermost meridional reflection. We think it likely, however, that the fiber-axis length per pleat in the pleated sheet in this substance is in fact not 6.14 A, but 6.30 A, and that there are fifteen of these units (thirty amino-acid residues) in the unit with c_0 = 94.6 A. The corresponding fiber-axis distance per residue, 3.15 A, can be achieved by a small deformation of the ideal pleated sheet, amounting to rotation by 7° around the C—C* axes of the amide groups, as described in the preceding paper.[2] A possibility that seems to us less likely is that there are $15^1/_2$ units of the pleated sheet in the distance 94.6 A, in which case the length per residue would be 3.05 A, and the value of c_0 would be twice as great, 189.2 A.

The fiber-axis length per residue for the α helix in substances so far investigated ranges from 1.53 to 1.56 A. These values correspond to the integers 62 and 61, respectively, as the number of residues in unit length 94.6 A along the c axis, the value 62 giving 1.525 A and 61 giving 1.551 A as the length per residue. The number of residues per unit turn is predicted to be close to 3.69, and values between 3.6 and 3.67 have so far been reported. With 17 turns in c_0, 61 residues would give 3.59 residues per turn, and 62 residues would give 3.65 residues per turn; no other value than 17 turns seems likely, inasmuch as it is improbable that the α-carbon bond angle would be strained enough to give 3.81 residues per turn (61 residues in 16 turns) or 3.45 residues per turn (62 residues in 18 turns). The most likely possibility is thus the 62-residue 17-turn helix, with 3.65 residues per turn. The distribution of side chains along the polypeptide chain presumably is such as to stabilize this helix, and the scattering of x-rays by the different side chains, in positions that remain to be determined, gives rise to the true meridional reflections, with contributions from the atoms of the main chains in the helix and the pleated sheet for some reflections.

TABLE 2

MERIDONAL REFLECTIONS FROM FEATHER RACHIS KERATIN

ORDER OF REFLECTION	BEAR		COREY AND WYCKOFF	
	d	INTENSITY	d	INTENSITY
4	23.6 A	10	23.1 A	Strong
			17.2	?
8	11.90	3	...	...
9	10.46	3	...	...
?	...	..	9.08	Faint
15	6.30	6	6.20	Strong
17	5.53	3	...	...
19	4.98	6	4.90	Strong
21	4.45	4	4.37	Medium
24	...	..	3.95	Faint
27	...	..	3.52	Faint
?	...	..	3.22	Faint
31	...	..	3.07	Medium

It is interesting to note that the reflection (0·0·17) is reported by Bear, at 5.35 A; this would correspond to one turn of the α helix. A strong reflection at 6.20 A (Corey and Wyckoff) or 6.30 A (Bear) is interpreted by Bear as the fifteenth basal plane reflection, (0·0·15). We interpret this as involving collaboration of the 15 units of the pleated sheet.

The identity distance 94.6 A receives a rational explanation as resulting from the presence of two structures, the slightly distorted pleated sheet with identity distance along the a axis of 6.30 A, and the α helix with 5.57 A per turn. The simplest ratio of integers approximating the ratio of these numbers is 17:15, corresponding to the mutual identity distance

94.6 A. These two structures accordingly might well be expected to form a protein such as feather keratin, with a triclinic unit with a_0 = 9.50 A, b_0 = 34.2 A, c_0 = 94.6 A, $\alpha \cong 90°$, $\beta \cong 90°$, $\gamma \cong 90°$.

It seems likely that the pleated sheets are all oriented similarly in the structure—there is no significant indication of a unit with b_0 = 68 A, corresponding to two kinds of pleated sheets, with opposite orientations. A pleated sheet is polar: all of the C=O groups point in one direction, and the N—H groups in the opposite direction, and in addition the side chains on one side of the sheet are arranged differently with respect to the residues than are those on the other side of the sheet, so that an isolated sheet would be curved. It is interesting to speculate that this curvature of the pleated sheets may be related to the natural curvature of the feather rachis. An example of a polar sheet in the inorganic field is the kaolin sheet.[6] Curved crystals of the clay minerals have been recently observed with use of the electron microscope, and their curvature has been assumed to result from the polar nature of the kaolin sheets.[7,8]

This investigation was aided by grants from The Rockefeller Foundation, The National Foundation for Infantile Paralysis, and The United States Public Health Service.

* Contribution No. 1553.

[1] Astbury, W. T., Cold Spring Harbor Symposia on Quantitative Biology, **2,** 15 (1934).

[2] Pauling, L., and Corey, R. B., these PROCEEDINGS, **37,** 251 (1951).

[3] Corey, R. B., and Wyckoff, R. W. G., *J. Biol. Chem.*, **114,** 407 (1936).

[4] Pauling, L., Corey, R. B., and Branson, H. R., these PROCEEDINGS, **37,** 205 (1951); Pauling, L., and Corey, R. B., *Ibid.* **37,** 235 (1951).

[5] Bear, R. S., *J. Am. Chem. Soc.*, **66,** 2043 (1944).

[6] Pauling, L., these PROCEEDINGS, **16,** 123 (1930).

[7] Davis, D. W., Rochow, T. G., Rowe, F. G., Fuller, M. L., Kerr, T. S., and Hamilton, T. K., Electron Micrographs of Reference Clay Minerals, Preliminary Report No. 6 of American Petroleum Institute Project 49 (1950).

[8] Bates, T. F., Hildebrand, S. A., and Swineford, A., Morphology and Structure of Endelite and Halloysite, *American Mineralogist*, **35,** 463 (1950).

THE STRUCTURE OF HAIR, MUSCLE, AND RELATED PROTEINS

BY LINUS PAULING AND ROBERT B. COREY

GATES AND CRELLIN LABORATORIES OF CHEMISTRY,* CALIFORNIA INSTITUTE OF TECHNOLOGY, PASADENA, CALIFORNIA

Communicated March 31, 1951

It is thirty years since x-ray photographs were first made of hair, muscle, nerve, and sinew, by Herzog and Jancke.[1] During this period, despite the efforts of many investigators, the photographs have eluded detailed interpretation, and the molecular structures of the proteins have remained undetermined. In the present paper we propose structures for hair, muscle, and related proteins in the extended state (β keratin and β myosin) and the contracted state (α keratin and α myosin), and discuss the extent to which the diffraction data are accounted for by the proposed structures.

The α-Keratin Structure.—It seems not unlikely that the polypeptide chains in unstretched hair, contracted muscle, horn, nail, quill, and other proteins that give the α-keratin x-ray pattern have the 3.7-residue helical configuration[2] (which for convenience we shall call the α helix).

Let us consider the structure expected for an aggregate of α helixes. The molecules, with the approximate form of circular cylinders, would be expected to pack in a hexagonal or pseudohexagonal array, as do the synthetic polypeptides poly-γ-methyl-L-glutamate and poly-γ-benzyl-L-glutamate.[3] The average residue weight 120 and approximate density 1.30 g cm^{-3} lead to 11 A as the value of a_0 for the hexagonal unit. The predicted equatorial reflections are $\{10\cdot0\}$ at 9.5 A, $\{11\cdot0\}$ at 5.5 A, $\{20\cdot0\}$ at 4.8 A, $\{21\cdot0\}$ at 3.6 A, $\{30\cdot0\}$ at 3.2 A, etc. The observed pattern of α keratin, as described by Astbury and Street,[4] has a strong equatorial reflection at 27 A, a very strong reflection at about 9.8 A, and a vague region of darkening around 3.5 A. We would attribute the 27-A reflection to a long-range order that can be elucidated only through further study. The 9.8-A reflection is described as covering a range of spacings of about 3 A centered at 9.8 A, which suggests that the packing is only pseudo-hexagonal, and that the hexagonal form $\{10\cdot0\}$ is split into several forms with different spacings. The failure to observe the reflections $\{11\cdot0\}$ and $\{20\cdot0\}$ can be attributed to the smallness of the x-ray form factor for equatorial scattering by the α helix,[5] which has a node at 5 A; indeed, the α-keratin x-ray photographs show a light band which is centered at about 5 A. The form factor then has a maximum at 3.4 A, which corresponds to the vague region of darkening, with center around 3.5 A, reported by Astbury and Street.

The principal meridional feature of the α-keratin x-ray pattern is a strong arc at 5.15 A. This reflection has been accepted as indicating that the c-axis identity distance is 5.15 A or a simple multiple of it, and it has usually been assumed that the c-axis length per residue is either $^1/_3\cdot5\cdot15 = 1.72$ A or $^1/_2\cdot5\cdot15 = 2.58$ A. The 5.15-A arc seems on first consideration to rule out the α helix, for which the c-axis period must be a multiple of the axis distance per turn, which is about 5.6 A. However, it was noted by Bamford, Hanby, and Happey[6] that the "meridional arc" observed on photographs of partially oriented fibers of poly-γ-benzyl-L-glutamate is on other photographs resolved into off-meridional spots in positions corresponding to the value 5.76 A for the c-axis translation. It seems probable that the 5.15-A arc seen on the α-keratin photographs is to be interpreted in a similar way.

A very significant contribution has been made by Herzog and Jancke[7] in 1926 and Lotmar and Picken[8] in 1942. These investigators obtained, by non-reproducible procedures, preparations of rather well crystallized muscle. Herzog and Jancke observed eight forms, and Lotmar and Picken eighteen. Lotmar and Picken's preparation was a piece of posterior-valve-closing muscle (of the mussel *Mytilus edulis*) that had been dried for 48 hours under 10-g tension and then allowed to stand in a can for a year. They indexed their photograph and (also Herzog and Jancke's) with a monoclinic unit with a_0 = 11.70 A, b_0 = 5.65 A (fiber axis), c_0 = 9.85 A, and β = 70.5°. The fiber-axis translation 5.65 A is in fine accordance with prediction for the 3.7-residue helix.

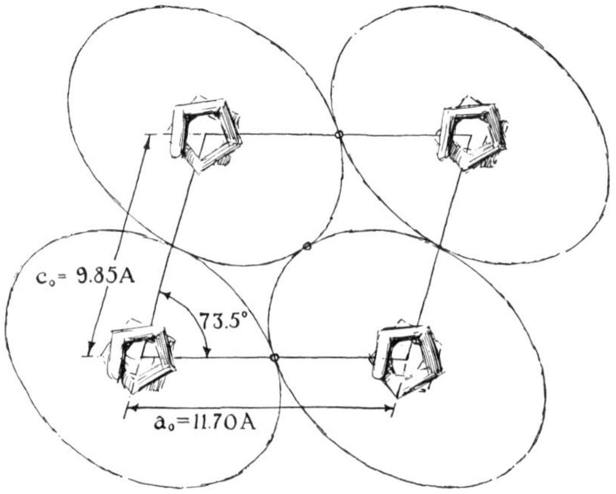

FIGURE 1

Plan of the monoclinic unit proposed for crystalline muscle by Lotmar and Picken, with the 3.7-residue helixes, suggested in the present paper, represented by elliptical cylinders. The dots indicate the positions assumed for side-chain carbon atoms.

Lotmar and Picken stated that their excellent photograph presumably shows the x-ray diagram of crystallized myosin, and that there is no sign of the 5.15-A meridional arc. They considered their preparation to represent a new molecular modification of myosin, with two residues per 5.65-A length along the *c*-axis. We think it likely that the process of crystallization has involved only the ordering of the α helixes, as shown in figures 1 and 2, in such a way that the larger side chains are grouped into layers with $x \cong {}^1/_2$, the region near $x = 0$, $z = {}^1/_2$ being free of side chains, thus permitting the helixes to come into close contact at these points.

The occurrence of the 5.15-A arc on the x-ray photographs of poorly ordered aggregates of α-helical molecules can be explained by the consideration illustrated in figure 3. Here a helical curve with pitch 5.65 A is shown on a cylinder with radius 1.81 A, the average radius for the peptide atoms C, N, C′, and O. The angle of inclination of the helix is 26°, and the perpendicular distance between adjacent turns of the curve is 5.65 cos 26° = 5.1 A. We would thus predict strong reenforcement of x-rays scattered by an α molecule in directions about 26° from the fiber axis. (A related pertinent fact is that the maximum for the calculated radial distribution function for the α helix comes at 5.0 A; this function will be reproduced in a later paper.) In the case of a somewhat poorly ordered fibrous aggregate of the molecules or of a crystalline phase with very large unit the intermolecular interference would in general permit strong diffraction maxima with spacing 5.1 A to occur in the near-meridional region, whereas

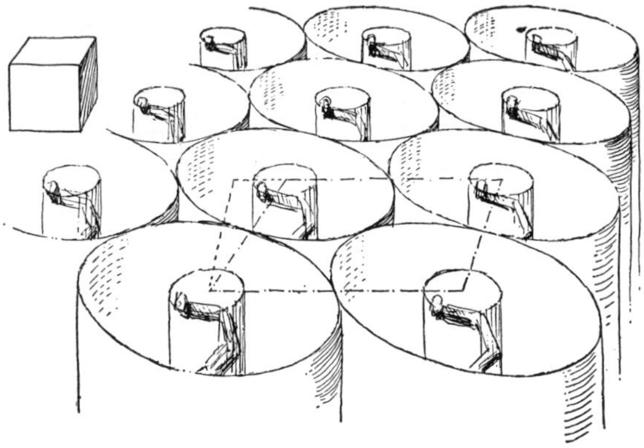

FIGURE 2

Drawing of the proposed structure of α keratin.

for a well-crystallized specimen intermolecular interference could cause these reflections to fail to appear, despite their large molecular form factor.

(Added April 10, 1951: It has been pointed out to us by Professor Verner Schomaker that the foregoing argument is not reliable. He has evaluated the form factor for x-ray scattering by a uniform helix, and has found that the maximum scattering by the helix with the dimensions given above occurs at angles considerably larger than 26° from the meridional direction, and at a Bragg angle corresponding to a spacing of about 4.2 A, rather than 5.1 A. He has further shown that if four main-chain atoms and the β carbon atom are represented by helixes with the correct radii for the 3.7-residue helical structure, and four side-chain atoms are represented by another helix, with the same pitch and with radius 4.0 A, a pronounced maximum is predicted to occur at 26° from the meridional direction and at a Bragg angle corresponding to a spacing of 5.1 A.)

The interplanar distances and estimated intensities of the basal-plane reflections on the x-ray photographs of crystalline muscle reported by Lotmar and Picken and by Herzog and Jancke are given in table 1, together with calculated values of the interplanar distance d, structure factor F, and intensity $I = LPF^2$, with L the Lorentz factor and P the polarization factor. The structure factor is that for the α helix, including a β carbon atom for each residue, evaluated for the case of cylindrical symmetry,[5] plus a side-chain carbon atom per residue at $x = {}^1/_2$, $z = 0$ and another one at $x = {}^1/_2$, $z = {}^1/_2$. If the crystal has monoclinic symmetry, as assumed by Lotmar and Picken, there are 2-fold screw axes passing through these points, and a greater-than-average density of atoms might be expected near these axes; we have accordingly tried to approximate the effect of the side-chain atoms on the structure factor by placing atoms in these positions.

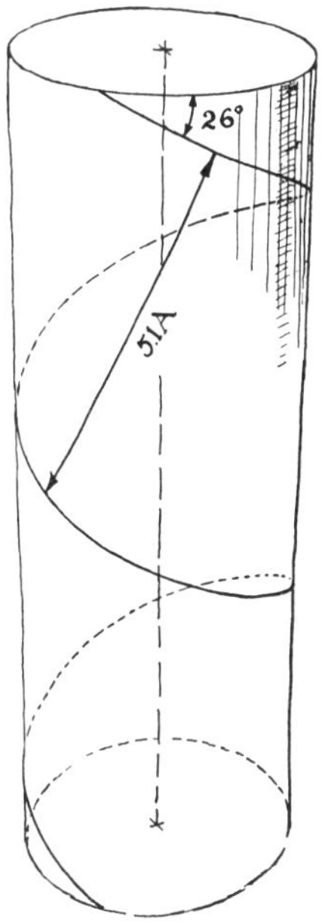

FIGURE 3

Diagrammatic representation of the 3.7-residue helix, indicating the origin of the meridional arc at 5.15 A observed on x-ray photographs.

It is seen that the calculated intensity pattern corresponds surprisingly well with the observed pattern, when it is considered that no variable parameter is involved in the calculation. The only arbitrary decision involved in the calculation is that the side-chain scattering, for about four heavy atoms per residue, can be approximated by placing a carbon atom near each of two 2-fold axes.

The first reflection, A_1, is seen on the reproduction of their photograph in Lotmar and Picken's paper to be very strong—we estimate it to be perhaps ten times as strong as A_2 or A_4. Its breadth is about 3 A, which is enough to include the forms {100}, {001}, and {101}. This spot closely resembles the corresponding spot given by ordinary preparations of α keratin, and described by Astbury and Street as covering a range of about 3 A, centered at 9.8 A. We accordingly think that it is likely that the ordinary preparations of α keratin have in fact a monoclinic structure closely resembling that of crystalline muscle, and only rather slightly disordered. The 27-A reflection seen on photographs of hair and ordinary muscle seems not to be present on Lotmar and Picken's photograph.

The second reflection, A_2, is observed to be strong, and calculated as weak. An additional side-chain carbon atom in phase for this reflection (near the line $x + z = 1$) would bring the value of $I_{\text{calc.}}$ to 103. (We have not attempted to find a distribution of side-chain atoms that would give

TABLE 1

COMPARISON OF CALCULATED AND OBSERVED INTENSITIES OF EQUATORIAL REFLECTIONS FOR CRYSTALLINE MUSCLE

TRICLINIC UNIT WITH a_0 = 11.70 A, b_0 = 5.65 A, c_0 = 9.85 A, $\alpha \cong 90°$, β = 73.5°, $\gamma \cong 90°$

RE-FLEC-TION	$h0l$	$d_{calc.}$	$F_{calc.}$	$I_{calc.}$	$I_{obs.}LP$[a]	$I_{obs.}HJ$[a]	$d_{obs.}LP$[a]	$d_{obs.}HJ$[a]
A_1	100	11.21 A	10.0	150				
	001	9.44	16.8	353	S(broad)	S(broad)	9.3 A	10.0 A
	101	8.51	14.0	217				
A_2	$10\bar{1}$	6.38	6.1	32	S	M	6.6	6.4
A_3	201	5.56	2.1	3	?[b]	M	6.0	5.8
	200	5.60	9.7	52				
A_4	102	4.87	−9.8	60	S	...	4.78	...
	002	4.72	7.7	35				
	$20\bar{1}$	4.31	−3.1	5				
	202	4.25	5.3	15				
A_5	$10\bar{2}$	3.97	−12.1	73	W	W	3.97	3.94
	301	3.87	−4.3	9				
A_6	300	3.74	−12.7	74	M	...	3.73	...
A_7	302	3.44	−12.4	65	M	...	3.50	...
	103	3.28	−5.0	10				
	$20\bar{2}$	3.19	2.1	2				
	$30\bar{1}$	3.18	−4.9	9				
	003	3.15	−4.9	9				
	203	3.15	−4.9	9				
A_8	401	2.92	−4.4	7	?[b]	...	2.94	
	303	2.84	−4.2	6				
	$10\bar{3}$	2.83	−4.2	6				
	400	2.80	2.9	3				
A_9	402	2.78	2.6	2	W	...	2.75	
	$30\bar{2}$	2.59	−9.6	28				
	$40\bar{1}$	2.50	−3.0	3				
	403	2.47	−2.9	2				
	$20\bar{3}$	2.46	−2.9	2				
A_{10}	104	2.45	−8.8	22	W	...	2.48	
	204	2.44	3.1	3				
	004	2.36	3.2	3				
	501	2.34	−2.4	2				
	304	2.31	−7.8	16				
	502	2.29	−7.7	15	M	...	2.18	
A_{11}	500	2.24	−7.3	13				
	$10\bar{4}$	2.19	−6.9	11				
	$40\bar{2}$	2.16	3.8	5				
	503	2.14	−1.3	0.4				
	$30\bar{3}$	2.13	−1.3	0.4				
	404	2.13	3.9	4				

[a] LP = Lotmar and Picken, HJ = Herzog and Jancke.

[b] Presumably the symbol means that Lotmar and Picken were doubtful as to the presence of these two spots.

better general agreement with the entire observed pattern, inasmuch as the approximate agreement given by the less arbitrary calculation that we have made seems to us to be more significant.)

The questionable reflections A_3 and A_8 cannot be seen on the reproduced photograph. A_5, which is described by Lotmar and Picken as weak (whereas $I_{calc.}$ is as large for it as for A_6 and A_7, described as strong), lies in a region of general blackening, which may have caused its intensity to be underestimated—it is interesting that Herzog and Jancke report A_5, but not A_6 nor A_7. Whether A_9 can be assigned to the form $\{30\bar{2}\}$ is doubtful; but in any case the appearance of the three reflections A_9, A_{10}, and A_{11} in nearly the calculated positions and with the calculated intensities can hardly be the result of coincidence. The sequence of forms with very small values of $I_{calc.}$ in the regions for which no reflections are reported by Lotmar and Picken and also the rather striking agreement found for the observed reflections strongly favor the conclusion that the assumed structure is not greatly different from the actual structure.

Some evidence for the 3.7-residue helix is provided also by the meridional reflections reported for muscle fibers and porcupine quill by Corey and Wyckoff,[9] MacArthur,[10] and Bear.[11] These reflections correspond to large *c*-axis identity distances, about 726 A for muscle fiber and 198 A for porcupine quill. The 726-A unit for muscle is shown also in electron micrographs of muscle fibrils treated with osmic acid.[12] We would expect that side chains of different kinds on the α helix would repeat after an integral number of residues, corresponding to an integral multiple of the residue length along the helix axis, about 1.53 A, and that accordingly those orders of basal plane reflection for which the spacing approximated closely to certain multiples of 1.53 A would be enhanced in intensity. It is in fact found that about 80 per cent of the meridional reflections are of this type; for both Venus clam muscle and porcupine quill they are multiples of 1.51 A. The reflections at 1.49 A, 3.05 A, 4.50 A, and 6.19 A for porcupine quill, which represent the first four orders of enhancement, are the strongest features of the wide-angle meridional pattern, except for the 5.2-A arc.

The β-Keratin Structure.—Hair and muscle can be reversibly stretched to about 100 per cent elongation.[13] Some authors have expressed doubt as to whether this elongation is to be attributed to the polypeptide chains, but it seems to us that Astbury's contention that it should be is justified. With a fiber-axis length of 1.53 A per residue for the α helix, an extended chain in the β-keratin structure would be predicted, on this assumption, to have a fiber-axis residue length of about 3.1 A. The principal meridional x-ray reflection of stretched hair, stretched muscle, and other proteins with the β-keratin structure[4] has in fact a spacing reported by Astbury as about 3.32 A, which is presumably the fiber-axis residue length, and would thus correspond to 117 per cent extension of the α helix. That the β-keratin structure involves extended polypeptide chains was first suggested by Brill[14] in 1923, and for the past fifteen years it seems to have been rather generally assumed that the chains are essentially coplanar, and that they alternate in direction in forming hydrogen-bonded non-polar sheets.[15–17, 8] Another hydrogen-bonded layer structure, the pleated sheet, which avoids the difficulty of large steric interference of side-chain groups predicted for the planar sheet, has recently been described.[18] The pleated sheet can easily assume a configuration corresponding to the residue-length 3.32 A, and it seems to us likely that it represents the β-keratin structure.

At present there is not much direct evidence to support this view. The observed equatorial x-ray reflections[4] at 9.8 A (strong) and 4.65 A (very strong) have been shown by Astbury and Sisson[19] to correspond to the plane of the β-keratin layers and the lateral direction in the layers, respectively. The lack of other equatorial reflections, except one weak reflection at 2.4 A, and lack of knowledge of positions of side-chain atoms make a calculation of intensities of reflection of little value. It may be pointed out, however, that if the one-molecule unit that accounts for the x-ray pattern of crystalline muscle obtained by Lotmar and Picken is correct the α helixes must all be oriented in the same sense, and accordingly this muscle on stretching could be transformed into the pleated sheet, but not into the planar sheet.

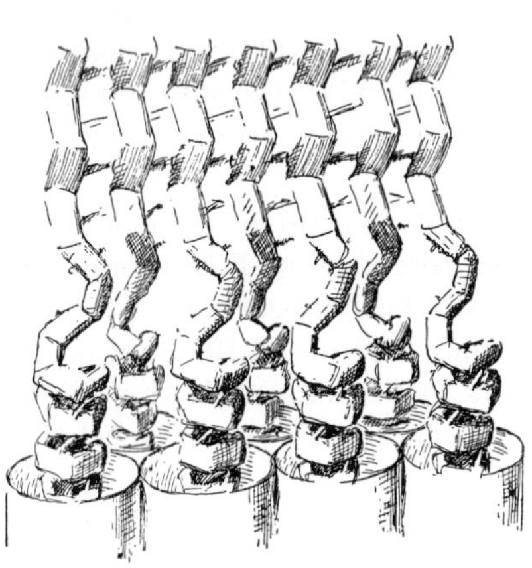

FIGURE 4

Drawing illustrating the proposed mechanism of conversion of a pleated sheet into a double row of 3.7-residue helixes; this is proposed as the process involved in the contraction of muscle.

When hair or muscle is treated with hot water or steam it shortens in the direction of the fiber axis, and swells laterally. The resultant material is called supercontracted keratin. The contraction from the α state is about 30 per cent for both hair[20] and myosin.[21] It is possible that supercontracted keratin has the configuration of the 5.1-residue helix,[2, 5] but there is very little evidence to support this suggestion. The fiber-axis residue length for this helix is about 0.99 A, which corresponds to 35 per cent contraction from the α helix, with residue length 1.53 A. The agreement of this value with the experimental value for the amount of supercontraction provides some support for the suggestion that the γ helix is present in supercontracted keratin. A careful study of the x-ray diagram should permit a decision on this point to be made.

The Mechanism of Contraction of Muscle.—The assignment of the pleated-sheet configuration to extended muscle and of the α-helical configuration to contracted muscle suggests that a discussion be made of the mechanism of contraction of muscle.

We have noticed that in order for a pleated sheet to be converted into a double layer of α helixes it is not necessary that all of the hydrogen bonds in the pleated sheet be initially broken. Instead, it is necessary to break only enough hydrogen bonds, four or five, to liberate four or five residues in each polypeptide chain, these being at about the same horizontal level in the pleated sheet. The liberated chains can then coil into the α-helical configuration, to produce the double layer of packed cylindrical α helixes, as shown in figure 4.

The sheet configuration is, for a normal protein, somewhat unstable relative to the α helix. This is indicated to be the case for synthetic polypeptides involving residues other than glycine by the fact that these polypeptides have been observed to form crystals of the α type.[6] Indication that the instability of the pleated sheet is due to steric repulsion between side chains is given by the fact that copolymers containing a large fraction of glycine residues assume the β configuration.[6] The steric-hindrance explanation of the instability of the pleated sheet is made reasonable by a consideration of the area available per side chain. In a normal β keratin, with fiber-axis length per residue 3.32 A and side-chain spacing 4.75 A and both sides of the sheet available for side chains, the area per side chain is 31.6 A^2. For the α helix we may take the radius of the side-chain median cylindrical surface to be 4.16 A, which is midway between the centers of the β carbon atoms (radius 3.34 A) and the point of contact with adjacent molecules ($^1/_2c_0$ = 4.98 A); with the fiber-axis residue length 1.53 A, the area per side chain is calculated to be 40.0 A^2, which is 25 per cent larger than for the pleated sheet. Moreover, in the discussion of the pleated sheet we have pointed out that the ideal pleated sheet has fiber-axis length per residue 3.07 A, and that the extension to 3.32 A, as observed in myosin, seems to be associated with a steric interference with the side chains, which causes the residues to rotate out of the position most favorable to hydrogen-bond formation, and thus introduces a strain, essentially of bending, in the hydrogen bonds. The normal hydrogen-bond energy for peptides may be estimated at 7.5 kcal $mole^{-1}$; it is expected to be large, because of the negative charge conferred on the carbonyl oxygen atom and the positive charge conferred on the imino nitrogen atom by the amide resonance. The energy of the strain introduced may be estimated to be of the order of magnitude of about 3 per cent of the hydrogen-bond energy, or 0.22 kcal $mole^{-1}$, which is about 1.8 cal per gram of myosin. Since muscle is about 10 per cent myosin, we estimate (very roughly) the work that could be done by 1 g of muscle to be about 0.18 cal, in a single complete twitch. Muscle has density 1.06 g cm^{-3}, and 1 g of muscle in the extended state with cross-section 1 cm^2 would be 0.94 cm long. If the entire muscle were to shorten proportionately to the myosin molecules, and these molecules were to shorten from the residue length 3.32 A (for the pleated sheet) to 1.53 A (for the α helix), the contracted length would be 0.43 cm, the contraction being by 54 per cent. The contracting pleated sheet would be predicted to exercise the same force throughout its contraction; for initial cross-section 1 cm^2 (in the extended state) this force is calculated from the assumed strain energy 0.18 cal per g of muscle, assuming it to be free energy, to have the value 1.53 kg.

The foregoing rough calculation agrees well with experiment in some respects. A. V. Hill has reported[22] that the maximum force exerted in a twitch by frog's muscle at 0°C is 1 to 2 kg cm^{-2}, which agrees well with the value 1.5 kg cm^{-2} calculated above. It is interesting also that the observed maximum shortening,[23] by 50 to 60 per cent for toad muscle, is close to the predicted shortening, 54 per cent, for the transition from the pleated sheet to the α helix. The heat liberated by frog's sartorius muscle on tetanic contraction has been measured,[24] and found to be about 400 g cm per cm of shortening and per cm^2 cross-section, or 0.05 cal per g of muscle, assuming complete contraction (by 54 per cent). Somewhat smaller values were found in a later study.[23] These values are considerably smaller than the estimated energy difference of the extended and the contracted forms of the myosin of muscle, as given above (0.18 cal g^{-1}).

In order to account for the observed mechanical properties of hair (great extension under a load of 500 to 2000 kg cm^{-2}, depending on humidity) in the same way, it must be assumed that the strain of the β configuration is very much greater, about 10 cal $mole^{-1}$ per residue. Moreover, the great dependence on humidity shows that side-chain interactions, which are changed by hydration, are involved.

The pleated-sheet configuration of extended muscle is metastable. In order for the polypeptide chains to contract an excitation energy would be needed, the energy of breaking four or five hydrogen bonds, to liberate four or five residues, in each chain. We suggest that the mechanism whereby the reaction of contraction is initiated may involve the production in or transfer to this region of the muscle of a number of hydrogen-bond-forming molecules, which can attack the hydrogen bonds of the pleated sheet, and through the formation of hydrogen bonds with the carbonyl and imino groups of the chains decrease the energy of activation of the sheet-disrupting process. As the α helixes are formed these molecules are liberated, and might continue to attack hydrogen bonds in the pleated sheet. It is, however, not necessary that they do so, in order for the process of formation of α helixes to continue, once that it is started: as the freed residues coil into the α helixes, and form more stable hydrogen bonds within these helixes, they would exercise a mechanical strain, communicated along the polypeptide chain, on the adjacent residues that are still held by hydrogen bonds in the pleated sheet, and this strain would have the effect of reducing the activation energy for the liberation of further residues. Accordingly once that the reaction were initiated, it would be expected to continue until all of the pleated sheet had been converted into a double row of α helixes.

It is not so easy to suggest a single reasonable way in which the muscle can be reconverted to the stretched state. There are many conceivable ways in which this could be done, with the use of chemical reactions of various sorts—especially the change in the nature of the environment of the α helixes. One possibility, suggested over twenty years ago by Meyer and Mark,[25] is that through a change in the ionic environment in the muscle the polypeptide chain is provided with a sequence of similarly charged side chains, not paired with neutralizing charges of the opposite sign. The electrostatic repulsion of these side chains would then tend to cause the chains to stretch out into the pleated-sheet configuration. It is of interest that Meyer and Mark illustrated this mechanism with use of a simple helical curve for the molecule of contracted muscle.

This investigation was aided by grants from The Rockefeller Foundation, The National Foundation for Infantile Paralysis, and The United States Public Health Service.

* Contribution No. 1554.

[1] Herzog, R. O., and Jancke, W., "Festschrift der Kaiser Wilhelm Gesellschaft," 1921.

[2] Pauling, L., Corey, R. B., and Branson, H. R., these PROCEEDINGS, **37**, 205 (1951).

[3] Pauling, L., and Corey, R. B., *Ibid.*, **37**, 241 (1951).

[4] Astbury, W. T., and Street, A., *Phil. Trans. Roy. Soc.*, **A230**, 75 (1931).

[5] Pauling, L., and Corey, R. B., these PROCEEDINGS **37**, 235 (1951).

[6] Bamford, C. H., Hanby, W. E., and Happey, F., *Proc. Roy. Soc.*, **A205**, 30 (1951).

[7] Herzog, R. O., and Jancke, W., *Naturwiss.*, **14**, 1223 (1926).

[8] Lotmar, W., and Picken, L. E. R., *Helv. Chim. Acta*, **25**, 538 (1942).

[9] Corey, R. B., and Wyckoff, R. W. G., *J. Biol. Chem.*, **114**, 407 (1936).

[10] MacArthur, I., *Nature*, **152**, 38 (1943).

[11] Bear, R. S., *J. Am. Chem. Soc.*, **66**, 2043 (1944).

[12] Jakus, M. A., Hall, C. E., and Schmitt, F. O., *Ibid.*, **66**, 313 (1944).

[13] Astbury, W. T., and Woods, H. J., *Nature*, **126**, 913 (1930).

[14] Brill, R., *Ann. d. Chem.*, **434**, 204 (1923).

[15] Astbury, W. T., *Trans. Faraday Soc.*, **29**, 193 (1933).

[16] Huggins, M. L., *J. Org. Chem.*, **1**, 407 (1936).

[17] Pauling, L., *J. Am. Chem. Soc.*, **62**, 2643 (1940).

[18] Pauling, L., and Corey, R. B., these PROCEEDINGS, **37**, 251 (1951).

[19] Astbury, W. T., and Sisson, W. A., *Proc. Roy. Soc.*, **A150**, 533 (1935).

[20] Astbury, W. T., and Woods, H. J., *Phil. Trans. Roy. Soc.*, **A232**, 333 (1933).

[21] Astbury, W. T., and Dickinson, S., *Proc. Roy. Soc.*, **B129**, 307 (1940).

[22] Hill, A. V., *Proc. Roy. Soc.*, **B136**, 243 (1949).

[23] Hill, A. V., *Ibid.*, **B136**, 195 (1949).

[24] Hill, A. V., *Ibid.*, **B126**, 136 (1938).

[25] Meyer, K. H., and Mark, H., "Der Aufbau der hochpolymeren organischen Naturstoffe" Akademische Verlagsgesellschaft M.B.H., Leipzig, 1930.

272 CHEMISTRY: PAULING AND COREY PROC. N. A. S.

THE STRUCTURE OF FIBROUS PROTEINS OF THE COLLAGEN-GELATIN GROUP

By Linus Pauling and Robert B. Corey

Gates and Crellin Laboratories of Chemistry,* California Institute of Technology, Pasadena, California

Communicated March 31, 1951

Collagen is a very interesting protein. It has well-defined mechanical properties (great strength, reversible extensibility through only a small range) that make it suited to the special purposes to which it is put in the animal body, as in tendon, bone, tusk, skin, the cornea of the eye, intestinal tissue, and probably rather extensively in reticular structures of cells. During the last thirty years, following the pioneer work of Herzog and Jancke,[1] a number of investigators have attempted to find the structure of collagen (and of gelatin, which gives similar x-ray photographs), but no one has previously proposed any precisely described configuration, nor has attempted to account for the positions and intensities of the x-ray diffraction maxima.

The diffraction pattern of collagen and gelatin is characterized by a meridional arc at 2.86 A. (Good reproductions of x-ray photographs have been published by Astbury.[2]) This arc remains essentially uninfluenced by a change in the source of material or its previous treatment; Bear[3] found that it varied only between the limits 2.82 A and 2.90 A for 26 samples, ranging from demineralized mammoth tusk to plain surgical gut (sheep intestinal submucosa). On the other hand, the principal equatorial reflection, which for thoroughly dried tendon[2] corresponds to the spacing 10.4 A, varies greatly in spacing with source and treatment of the material. At ordinary humidity it is about 11.5 A, and Bear reported the value 15.5 A for kangaroo tail tendon treated with water. It is evident that collagen consists of molecules (polypeptide chains) extending along the fiber axis, and rather loosely packed in parallel orientation. It will be pointed out below that the equatorial reflections correspond to a hexagonal packing of circular cylinders.

The 2.86-A fiber-axis spacing suggests that the amide groups of the polypeptide chain are in the cis configuration, as has been mentioned by Astbury.[2] (Astbury has suggested[2, 4] a structure for collagen which bears little resemblance to our structure.) If we accept the configuration of the amide group in a peptide as predicted from x-ray investigations of related simple substances, and described in our earlier paper[5] (with a single change —we have replaced the value 120° for the C′—N—C angle by the value 123°, which is suggested by general considerations as a reasonable value for the angle between a single bond and a bond with about 50 per cent double-bond character), we predict that the length of an amide group with the cis configuration is 2.83 A, which is close to the observed fiber-axis spacing. Although the principal meridional arc at 5.1 A observed for proteins with the α-keratin structure, which had always been accepted as representing a fiber-axis distance, has been found[6, 7] to be in fact a diagonal spacing, there seems to be little reason to doubt that the 2.86 A spacing of collagen represents a fiber-axis distance.

We found it impossible to formulate a satisfactory structure for collagen

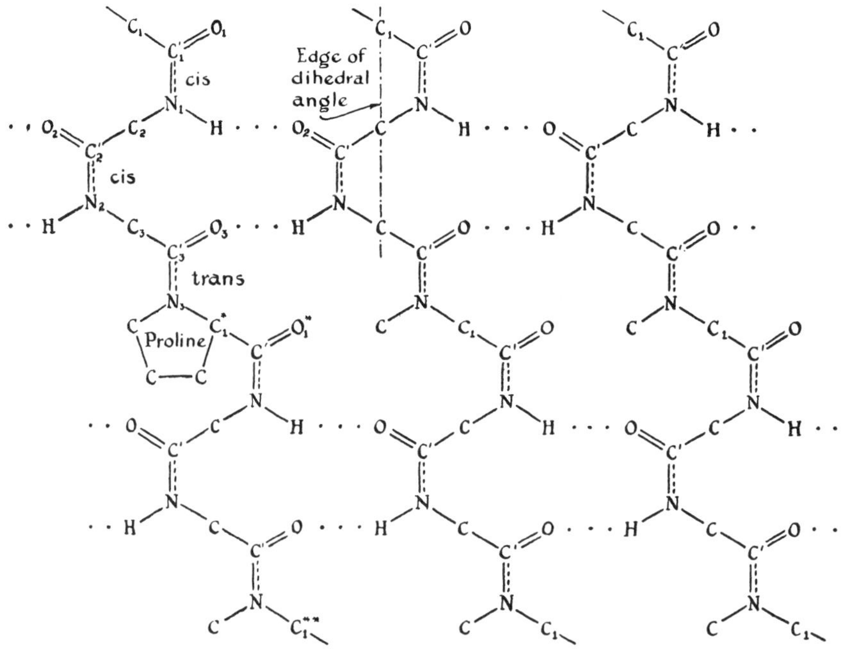

FIGURE 1

Diagrammatic representation of the configuration of polypeptide chains in the collagen-gelatin three-chain helix.

from cis amide groups alone. However, a satisfactory structure, described in detail in the following paragraphs, has been formulated with use of polypeptide chains in which there is an alternation of two cis groups and one trans group, as shown in figures 1 and 2. The angular orientation which the trans group is required to assume by its bonds with the contiguous cis groups is such as to cause its component of length along the fiber axis to be about 2.92 A, the average fiber-axis length per residue for the cis-cis-trans chain thus being 2.86 A, as observed for the collagen fibers.

Let us now consider the folding of the cis-cis-trans chain in such a way as to form satisfactory hydrogen bonds. The ease of lateral swelling of collagen indicates that hydrogen bonds between molecules of the protein are not present, and hence that a structure involving intramolecular hydrogen bonds is to be sought.

The principal equatorial spacing for thoroughly dried collagen is 10.4 A. For cylindrical molecules in hexagonal packing this value for $d_{10\cdot0}$ leads to 125 A^2 as the basal-plane area per molecule, and with 2.86 A as the fiber-axis length to 358 A^3 as the volume per unit, or 215 cm^3 per mole of units. If the density is taken as 1.35 g cm^{-3} (the reported density of dry gelatin) the mass per unit is 291 g. The average residue weight for collagen and gelatin is found by analysis to be slightly less than 100; it is hence evident that there are about three residues in the unit of the molecule (2.86 A length). We accordingly reach the conclusion that the molecule of collagen and gelatin, essentially cylindrical in shape, is not a single polypeptide chain, but consists of three polypeptide chains. This is pleasing, inasmuch as there is no way in which a single polypeptide chain with a cis-cis-trans sequence of amide groups can be folded to give intramolecular hydrogen bonds and a fiber-axis residue length of 2.86 A. The intramolecular hydrogen bonds in the collagen-gelatin molecule must be lateral bonds between the three polypeptide chains that constitute the molecule.

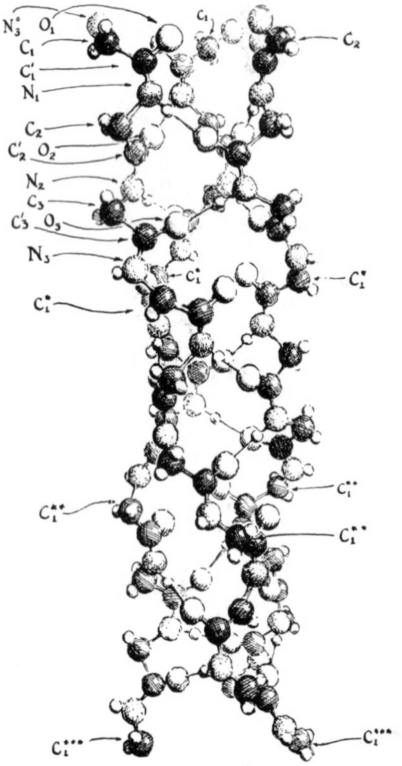

FIGURE 2

Drawing representing the proposed structure of the collagen-gelatin molecule.

A satisfactory structure can be built in which each of the three polypeptide chains is coiled into a helix, the coiling being achieved through the bending of the chain at the positions of the α carbon atoms, and the three helixes having a common axis. For the ideal configuration, in which the three α carbon atoms are similarly oriented directly above and below one another along a line parallel to the fiber axis, the dihedral angle at these carbon atoms is 97°. Some distortion of the structure is required in order that satisfactory hydrogen bonds be formed. This distortion consists in a rotation around the two single bonds formed by the α carbon atoms, so as to draw the α carbon atom C_2 closer to the axis of the helix, by 0.34 A, than the α carbon atoms C_1 and C_3, as shown in figure 2. The hydrogen bonds are introduced in such a way that the three chains of the molecule are related to one another by a three-fold axis of symmetry. The α carbon atoms C_2 of the three chains lie at the corners of an equilateral triangle of edge 5.60 A, and the atoms C_1 and C_3 at the corners of an equilateral triangle with edge 6.20 A. The two cis amide residues are rotated by about 9° out of the orientation parallel with the fiber axis. The trans residue is rotated through 30° about its $C_3 - C_1^*$ axis. The trace of this axis on the basal plane is 2.44 A. The relation of this trace to the 6.20-A triangle of the C_3 and C_1 atoms is such that the angle of rotation about the fiber axis that converts one three-residue element of a polypeptide chain into the element following it in the chain is 40°. We have found that this angle of rotation can hardly be varied by more than 3° without introducing unsatisfactory structural features. The helix formed by a single chain of the collagen molecule is thus found to have very nearly a 9-fold screw axis of symmetry.

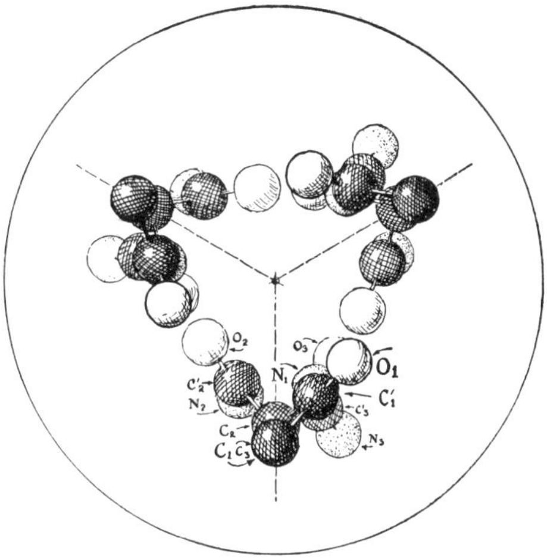

FIGURE 3

Plan of the three-chain configuration proposed for collagen and gelatin.

There is, of course, no reason for an isolated gelatin molecule to have exactly the angle 40° for the rotation of its rotatory translation, or to have exactly a 9-fold screw axis. In an aggregate of molecules in hexagonal packing, however, the influence of adjacent molecules on a given molecule might well be such as to introduce a small torque that would cause it to assume exactly the value 40° for this angle, and thus to assume a true 9-fold screw axis, in addition to the 3-fold symmetry axis that converts one chain into another, as shown in basal-plane projection in figure 3. We have found this phenomenon to occur in the hexagonal crystals of poly-γ-methyl-L-glutamate,[6] in which the α helix, normally with approximately 3.69 residues per turn, is constrained to the value $18/5 = 3.60$ in order to achieve a 6-fold screw axis. In the following discussion we describe a molecule with a 9-fold screw axis.

The translation of the operation that converts one structural element into the following one in the chain is given for our model as 8.58 A, which corresponds to 2.86 A per residue, in exact agreement with the x-ray value.

The coordinates of atoms in a structural unit of the molecule are given in Table 1. The coordinates x, y, and z refer to cartesian axes centered at the α carbon atom C_2, and the coordinates ρ, θ, and z are cylindrical coordinates referred to the axis of the helix. These atomic positions correspond closely to the assumed bond angles, bond distances, and planarity of the

TABLE 1

ATOMIC PARAMETERS FOR THE COLLAGEN THREE-CHAIN HELIX

x, y, z, AND ρ IN A

ATOM	x	y	z	ρ	ϕ
C_1	−0.25	−0.25	2.83	3.59	0.0°
C_1'	1.16	−0.09	2.19	2.63	19.6°
O_1	2.18	−0.09	2.90	2.38	42.3°
N_1	1.17	0.00	0.88	2.55	18.9°
C_2	0.00	0.00	0.00	3.25	0.0°
C_2'	0.05	1.24	−0.90	2.47	−19.9°
O_2	0.22	2.38	−0.41	2.07	−47.5°
N_2	−0.04	1.06	−2.21	2.63	−17.2°
C_3	−0.25	−0.25	−2.83	3.59	0.0°
C_3'	1.01	−0.37	−3.73	2.94	19.3°
O_3	2.03	0.21	−3.41	2.10	37.9°
N_3	0.89	−1.09	−4.80	3.66	22.5°
C_1^*	1.98	−1.32	−5.75	3.59	40.0°
Axis	2.29	2.29	...	0.00	...

TABLE 2

COMPARISON OF EQUATORIAL FEATURES OF COLLAGEN X-RAY PHOTOGRAPHS AND CALCULATED MAXIMA AND MINIMA OF THE FORM FACTOR

CALCULATED SPACINGS		OBSERVED SPACINGS	
7.14 A,	minimum intensity	7.8 A,	center of light band
4.72	maximum	4.37	center of diffuse dark band
2.89	minimum	2.74	center of light band
2.25	maximum	2.18	center of dark band
1.64	minimum		

amide groups. Because of the complexity of the structure, we are not sure that a considerable displacement of the chains from the proposed positions might not exist, in such a way as to retain these structural features.

The nature of the inter-chain interactions is shown in figures 1 and 3. Two hydrogen bonds are formed per unit. The N_1—H···O_2 bond has the length 2.63 A. This is slightly shorter (by about 0.06 A) than any N—H···O hydrogen bond so far reported, but it seems to us to be possible that the bond in this molecule is indeed this short. The N_2—H···O_3 hydrogen bond has the length 2.83 A. If our model were to be changed by a small additional rotation of the trans amide group the distance would be significantly shortened.

There is an interesting correlation between the chemical composition of the collagen-gelatin proteins and the existence in the structure of only two, rather than three, hydrogen bonds per element. It is found by chemical analysis that about one-third of the amino acids obtained by hydrolysis of collagen and gelatin are proline or hydroxyproline. In residues of these amino acids the nitrogen atom does not have an attached hydrogen atom and so does not enter into hydrogen-bond formation. The proposed structure may explain this aspect of the chemical composition of these proteins. It seems likely that if long polypeptide chains were to be synthesized with every third residue a proline or hydroxyproline residue and every third residue a glycine residue they would spontaneously aggregate into complexes of three, with the collagen structure, the other principal known structures for fibrous proteins, the α-helix structure, pleated-sheet structure, and γ-helix structure, being rendered unstable by the steric interference of the 5-membered rings that are held in a fixed orientation.

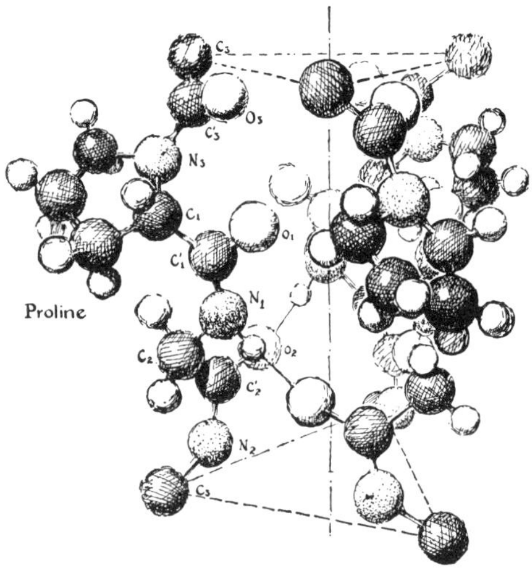

FIGURE 4

Drawing of a portion of the proposed structure of the collagen-gelatin molecule, showing the positions of the proline residues.

It is of interest that the sense of the collagen helix is uniquely related to the configuration of the amino acid residues. It is possible for only one of the two alternative positions of a β carbon atom on the α carbon atom C_1 of the proline residue to be connected with the nitrogen atom to form a 5-membered ring, for a given sense of the helix. The acceptable configuration is shown in figure 4.

The structure of the collagen-gelatin molecule can be represented diagrammatically as shown in figure 5. The three helical polypeptide chains are shown projected onto the surface of a cylinder. It is interesting to note that the structure of the molecule provides an immediate explanation of the principal mechanical property of collagen, its extensibility over only a limited range. The effective fiber-axis length of the trans residue in the molecule as shown is 2.93 A. If, on the application of force, this residue were to be twisted into a parallel orientation, its effective length would be

3.83 A, the C_3—C_1* distance. This maximum increase in length corresponds to a 10 per cent extension for the molecule. It is likely, however,

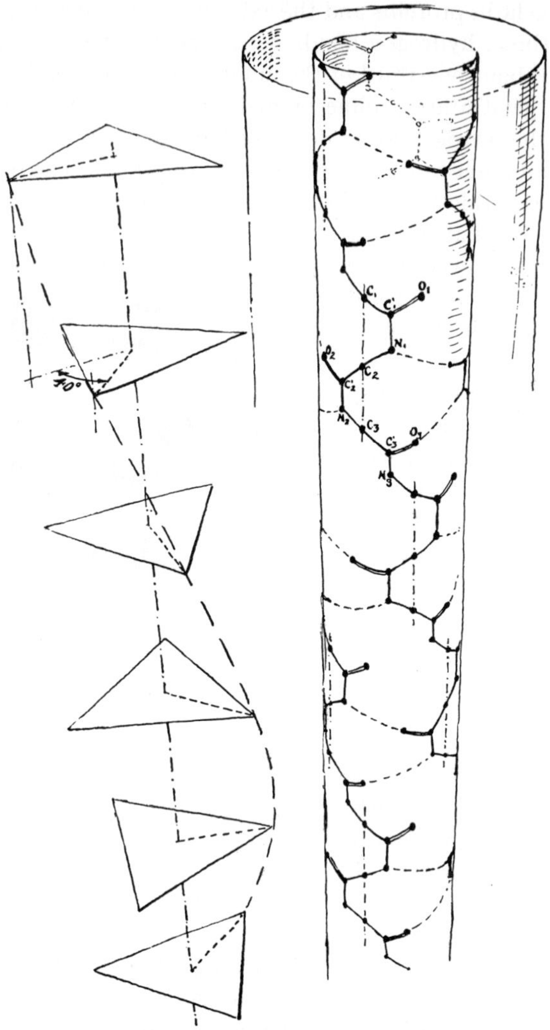

FIGURE 5

Diagrammatic representation of the collagen three-chain helix.

that the bond angles and planarity of amide groups would prevent complete parallel orientation, and that only a somewhat smaller extension could be achieved.

The way in which the molecules are packed together in a fiber of tendon or connective tissue is shown in figure 6. The cylindrical molecules are arranged in hexagonal packing. As discussed above, they are all helical, with the same sense, either right-handed or left-handed. It cannot be predicted whether in tendon or other collagenous material the molecules would all be oriented similarly, or whether some would be oriented in one direction and others in the opposite direction. It seems likely that the distribution of side chains reflects the direction of the polypeptide chains in the molecules, in such a way that the packing would be better in case that the molecules are all similarly oriented than if they were to alternate in

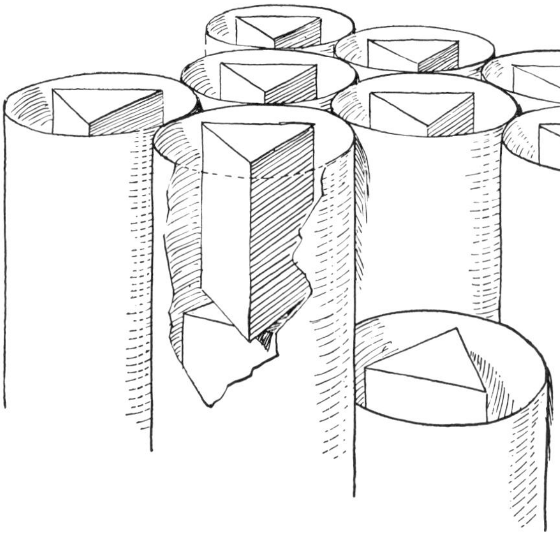

FIGURE 6

Diagrammatic representation of an aggregate of collagen molecules in a collagen fiber.

orientation. A preliminary study of the packing of side chains indicates that the glycine residues lie immediately below the proline and hydroxyproline residues (in the molecule oriented as shown in figure 2); that is, that the carbon atom C_2 is a methylene carbon atom of a glycine residue. It has been found by analysis that about one-third of the residues are glycine residues.

The proposed structure of the collagen-gelatin molecule accounts in a striking way for the principal features of the x-ray diffraction pattern of collagen and gelatin. A fiber consisting of these cylindrical molecules in hexagonal packing, with a_0 = 12.50 A (at normal humidity), would be predicted to produce equatorial reflections $\{10\cdot0\}$, with d = 10.83 A, $\{11\cdot0\}$, with d = 6.25 A, {20·0}, with d = 5.42 A, and {22·0}, with d = 3.13 A. The lines reported by Corey and Wyckoff,[8] Astbury,[2] and other investigators are a strong reflection at 10.9 A and a medium reflection at 5.42 A. We may ask why these reflections appear, and {20·0} and {22·0} do not. The answer is given by the calculation of the form factor for the collagen molecule, with the parameters of table 1, and the assumption of cylindrical symmetry. The equation $F = \sum_i f_i J_0 (4\pi\rho_i \sin \theta/\lambda)$ leads to the function F shown in figure 7. This function has nodes and maxima as given in table 2, where comparison is made with the corresponding features as measured by us on the photographs of raw kangaroo tendon made by Corey and

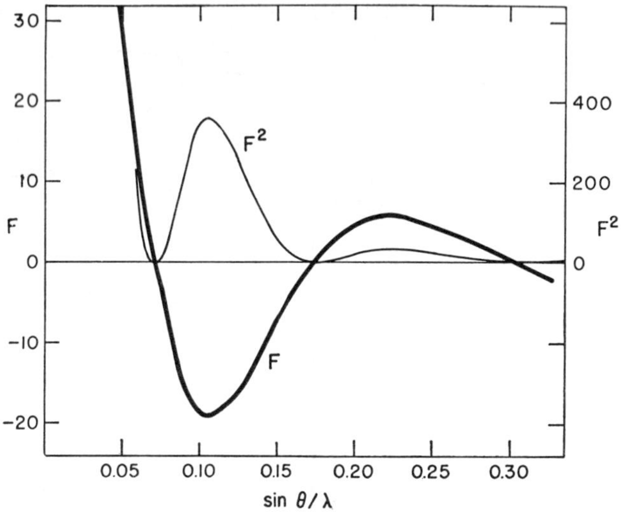

FIGURE 7

Calculated form factor and square of form factor for equatorial reflections of the collagen three-chain helix, calculated for cylindrical symmetry.

Wyckoff.[8] We see that it is predicted that a minimum in intensities of reflections would occur at about d = 7.14 A and another minimum at d = 2.89 A; these values are close to those for {20·0} and {22·0}, 6.25 A and 3.13 A, respectively. Moreover, there is observable on the photographs, and on published photographs of collagen, such as those of Corey and Wyckoff and of Astbury, a general light band, at about the azimuthal angle predicted for the minimum at 7.14 A equatorial spacing, followed by a region of general darkening, for which Astbury has located the center at 4.4 A, and which we have also measured at 4.37 A. This band is followed by a lighter band, and then a faint darker ring, the center of which we measure at 2.18 A, in excellent agreement with the predicted maximum of intensity, 2.25 A. It seems likely that the presence of the 3-chain collagen helix can be recognized in a fibrous material more easily from this general blackening, arising from disordered molecules, than from the crystallographic diffraction maxima.

The predicted unit of structure of crystalline collagen, except for perturbations due to the distribution of side chains, is a hexagonal unit with a_0 = 12.5 A (for collagen at normal humidity) and c_0 = 25.74 A, the fiber-axis distance for nine residues. We have noted that the wide-angle pattern of collagen can be completely or nearly completely indexed in terms of this unit. Except for the side chains, the suggested hexagonal structure involves only one undetermined parameter, the azimuthal angle fixing the orientation of the molecules relative to the crystal axes. The meridional small-angle reflections reported by Corey and Wyckoff,[8] Bear,[3] and other workers are also seen to be related to the unit. These reflections seem to represent orders from the large spacing of about 640 A (which appears not only in x-ray photographs but also in electron micrographs[9]), apparently enhanced when they bear a nearly rational relation to c_0.

(Added May 5, 1951.) An interesting paper on infrared spectra and structure of fibrous proteins, by E. J. Ambrose and A. Elliott, *Proc. Roy. Soc.*, **A 206,** 206 (1951), has just appeared. The results obtained agree well with our proposed structures. In particular, Ambrose and Elliott conclude that in collagen "the N—H bond must be nearly if not exactly normal to the chain axis, and this applies also to the C=O bond," in excellent substantiation of the prediction on the basis of our three-chain helical structure.

This work was aided by grants from The Rockefeller Foundation, The National Foundation for Infantile Paralysis, and The U. S. Public Health Service.

* Communication No. 1555.

[1] Herzog, R. O., and Jancke, W., *Ber.*, **53,** 2162 (1920).
[2] Astbury, W. T., *J. Int. Soc. Leather Trades' Chem.*, **24,** 69 (1940).
[3] Bear, R. S., *J. Am. Chem. Soc.*, **66,** 1297 (1944).
[4] Astbury, W. T., and Bell, F. O., *Nature*, **145,** 421 (1940).
[5] Pauling, L., Corey, R. B., and Branson, H. R., these PROCEEDINGS, **37,** 205 (1951).
[6] Pauling, L., and Corey, R. B., *Ibid.*, **37,** 241 (1951).
[7] Pauling, L., and Corey, R. B., *Ibid.*, **37,** 261 (1951).
[8] Corey, R. B., and Wyckoff, R. W. G., *J. Biol. Chem.*, **114,** 407 (1936).
[9] Schmitt, F. O., Hall, C. E., and Jakus, M. A., *J. Cell. Comp. Physiol.*, **20,** 11 (1942); *J. Am. Chem. Soc.*, **64,** 1234 (1942).

282 CHEMISTRY: PAULING AND COREY PROC. N. A. S.

THE POLYPEPTIDE-CHAIN CONFIGURATION IN HEMOGLOBIN AND OTHER GLOBULAR PROTEINS

BY LINUS PAULING AND ROBERT B. COREY

GATES AND CRELLIN LABORATORIES OF CHEMISTRY,* CALIFORNIA INSTITUTE OF TECHNOLOGY, PASADENA, CALIFORNIA

Communicated March 31, 1951

In the immediately preceding papers we have described several hydrogen-bonded planar-amide configurations of polypeptide chains, and have discussed the evidence bearing on the question of their presence in fibrous proteins. It seems worth while to consider the possibility that these configurations—the pleated sheet, the 3.7-residue α helix, the 5.1-residue γ helix, and the three-chain collagen helix—are represented in molecules of the globular proteins.

It may first be noted that many globular proteins, such as ovalbumin, can on denaturation be converted into a form showing the β-keratin x-ray pattern.[1] The fiber-axis residue distance that is observed, about 3.3 A, is the same as for β keratin, for which we have suggested the pleated-sheet configuration,[2] and it seems reasonable that the same structure should be represented by these denatured proteins. It is, of course, to be expected that a layer structure, such as the pleated sheet, would be assumed by a protein when pressed flat, and the extension of the chains in the pleated-sheet structure makes it reasonable that such a structure should also be assumed by a protein when drawn into a fiber.

The most significant published data bearing on the structure of globular proteins are those on horse carbonmonoxyhemoglobin that have been obtained through the well-planned and diligent efforts of Perutz and his coworkers.[3, 4] These data have been published mainly as a set of sections of a three-dimensional Patterson diagram. We have observed that the data provide some support for the idea that the 3.7-residue helix is a principal feature of the structure of this protein.

Perutz has pointed out that his data indicate that the hemoglobin molecule is about 57 A long, and between 34 A and 57 A in other dimensions, and that there are present rods extending in the 57-A direction, and packed in a pseudohexagonal array, with the centers of the rods about 10.5 A apart. He concluded that the rods probably have the same structure as the molecules in α keratin, for which we have recently suggested the 3.7-residue helical configuration.[5]

There are several facts that favor the view that the 3.7-residue helix is represented in hemoglobin. First, there is the similarity to α keratin, pointed out by Perutz, and the evidence supporting the 3.7-residue helical configuration for the fibrous proteins with the α-keratin structure.[5] Closely related is the fact that from the density and the average residue weight for hemoglobin one would predict that molecules with this helical configuration would be spaced about 11 A apart (from center to center), in agreement with Perutz's conclusion that the rods in hemoglobin are about 10.5 A apart. (A calculation of this sort at once eliminates the 5.1-residue helix, for which the predicted average spacing of the rods is 14 A.)

Another bit of supporting evidence is provided by the integrated vector density in a strip of the xz Patterson section through the origin of the 3-dimensional diagram and in the direction of the axes of the rods. Bragg, Kendrew, and Perutz[6] have reproduced this quantity, plotted as a function of the distance from the origin, in connection with their painstaking analysis

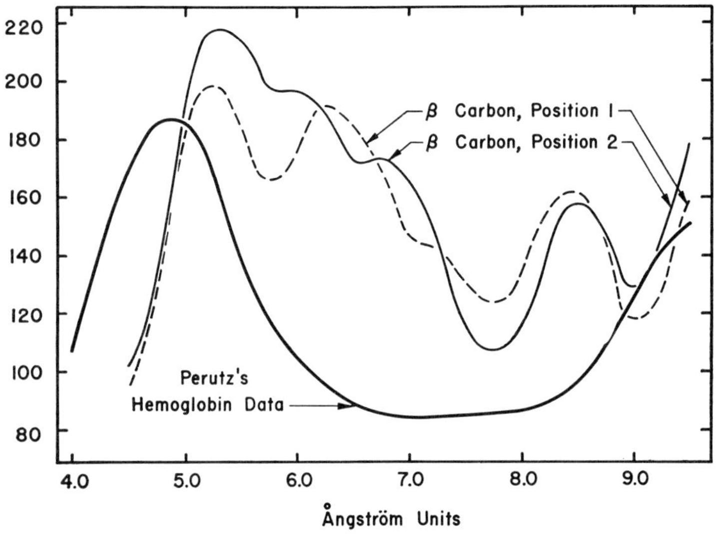

FIGURE 1

Comparison of the radial distribution function calculated for the 5.1-residue helical configuration, with inclusion of a β carbon atom per residue, and the experimental radial distribution function for carbonmonoxyhemoglobin, as calculated from the three-dimensional Patterson function given by Perutz.

of the data for hemoglobin and also for myoglobin[7] and discussion of the correlation of the data with alternative polypeptide configurations. The function has peaks at about 5 A, 11.5 A, 16.5 A, 21.5 A, 27 A, 32 A, etc. We have evaluated a corresponding function for the 3.7-residue helix by including interatomic vectors deviating by not more than 2 A from the direction of the helical axis, and weighting the vectors proportionately to the product of the atomic numbers of the two atoms. The function obtained in this way for an 18-residue 5-turn helix with fiber-axis residue length 1.53 A has maxima at 5.1 A, 10.6 A, 16.7 A, 21.4 A, 27.5 A, 32.6 A, etc., in excellent agreement with the experimental points.

Another test of the proposed configuration can be made by comparison of the calculated and observed radial distribution functions. Perutz pointed out that the Patterson diagram shows a strong shell at about 5 A from the origin. We have obtained a radial distribution function corresponding to his data for hemoglobin by numerical integration over the contoured Patterson sections published in his paper; this function is shown in figures 1 and 2. It is seen that it has a maximum at about 4.8 A. The calculated radial distribution functions for the 5.1-residue helix are also shown in figure 1. The two curves represent respectively the function for

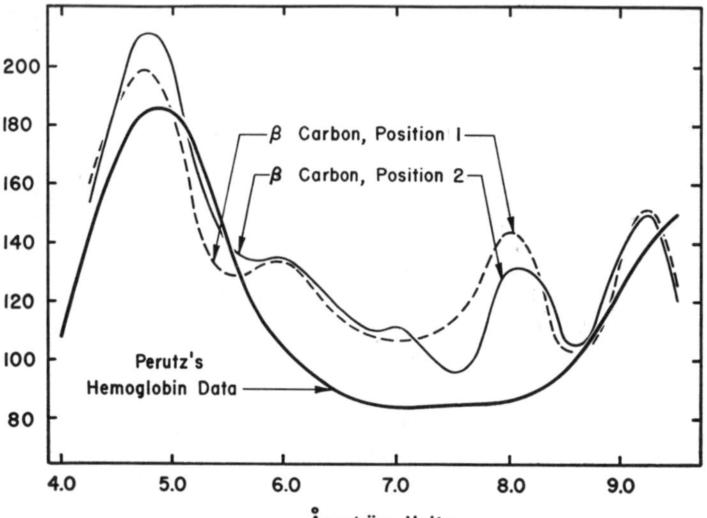

FIGURE 2

Comparison of the radial distribution function calculated for the 3.7-residue helical configuration of the polypeptide chain, with inclusion of a β carbon atom for each residue, and the experimental radial distribution function for hemoglobin.

the four main-chain atoms C, C′, O, and N and a β carbon atom in one of the two alternative positions, and the function for the four main-chain atoms and a β carbon atom in the other position. It is seen that there is no agreement with the hemoglobin curve. The same two calculated radial distribution functions for the 3.7-residue helix are given in figure 2. We think that the rough agreement with the hemoglobin curve is to be considered as significant; it is to be remembered that even with inclusion of the β carbon atom only about 60 per cent of the heavy atoms in the molecule have been taken into consideration in the calculation. The neglected side-chain atoms are, of course, far more randomly arranged than the main-chain atoms of the helix, and would for this reason tend to distribute their vectors rather uniformly, and thus not to mask the characteristic features of the function due to the main-chain and β carbon atoms.

The comparison of radial distribution functions may thus be construed as giving additional evidence in favor of the suggestion that the rods that Perutz has reported to be present in the hemoglobin molecule have the 3.7-residue helical configuration.

We think that it is not unlikely that this polypeptide configuration is represented in other globular proteins also. In particular, its presence in myoglobin, which is closely related to hemoglobin, would not be surprising; however, it must be pointed out that the Patterson projection for myoglobin on a plane perpendicular to the axis of the rods, given by Bragg, Kendrew, and Perutz,[6] seems hardly to be compatible with this structure. It is possible, of course, that side-chain atoms happen to cooperate effectively in changing the aspect of this projection, or that the axes of the rods do not lie exactly along the direction of projection. The evidence favoring the 3.7-residue helix for myoglobin is contained in Kendrew's description of the myoglobin molecule, as deduced from his data, as consisting of a layer of four rods about 9.5 A apart and with vector maxima spaced 5 A apart in the direction of the axes of the rods. The layers themselves are about 15 A apart, which suggests that if the structure does involve the 3.7-residue helix the side chains are distributed as in crystalline muscle,[5] in which the molecules have an effectively elliptical cross-section, with major and minor diameters 13.1 A and 9.8 A, respectively.

This investigation was aided by grants from The Rockefeller Foundation, The National Foundation for Infantile Paralysis, and The United States Public Health Service. We acknowledge with gratitude the assistance and encouragement of our colleagues in The Gates and Crellin Laboratories of Chemistry throughout the period during which the studies reported in this series of papers and also the investigations on which this work is based were made. We are especially grateful to Professor Verner Schomaker, who has helped by giving us the benefit of both his deep understanding of structural chemistry and his profound critical insight.

* Contribution No. 1556.

[1] Astbury, W. T., and Lomax, R., *J. Chem. Soc.*, **1935**, 846; Astbury, W. T., Dickinson, S., and Bailey, K., *Biochem. J.*, **29**, 2351 (1935).

[2] Pauling, L., and Corey, R. B., these PROCEEDINGS, **37**, 251 (1951).

[3] Boyes-Watson, J., Davidson, E., and Perutz, M. F., *Proc. Roy. Soc.*, **A191**, 83 (1947).

[4] Perutz, M. F., *Ibid.*, **A195**, 474 (1949).

[5] Pauling, L., and Corey, R. B., these PROCEEDINGS, **37**, 261 (1951).

[6] Bragg, W. L., Kendrew, J. C., and Perutz, M. F., *Proc. Roy. Soc.*, **A203**, 321 (1950).

[7] Kendrew, J. C., *Ibid.*, **A201**, 62 (1950).

[Reprinted from the Journal of the American Chemical Society, **74**, 3964 (1952).]

THE PLANARITY OF THE AMIDE GROUP IN POLYPEPTIDES

Sir:

Dr. M. L. Huggins has kindly sent us copies of his Letters,[1,2] in which he has proposed a helical configuration of polypeptide chains as an alternative to the α helix described in our earlier publications.[3,4,5] In his configuration the amide group is not planar. The deformation of the amide group from the planar configuration can be described as a rotation of 17.5° of the NHC* plane about the C′–N axis plus a bending of 15° of the N–C* bond and the N–H bond out of the rotated plane, to the same side. The part of the strain energy due to the rotation of the π orbital of the nitrogen atom can be calculated by the formula[6,7] $A \sin^2 \delta$ with $A = 30$ kcal. mole^{-1} and $\delta = 17.5°$; this calculation gives 2.7 kcal. mole^{-1}. The strain energy of deformation of the N–C* bond and the N–H bond can be calculated by the assumption that the bond energy is proportional to the strength of the bond orbital of the nitrogen atom in the bond direction, which is for these bonds 15° from the the direction of maximum strength. With use of the bond energies of the bonds (48.6 and 83.7 kcal. mole^{-1}, respectively), this calculation leads to 3.3 kcal. mole^{-1} for the bending energy of the two bonds. The total strain energy for the distorted amide group is thus found to be 6 kcal. mole^{-1}. This strain energy, which in the structure proposed by Huggins applies to every residue, is so great as to make the structure unacceptable in comparison with the α helix, which is just as satisfactory in every other respect, so far as we are aware, and which involves planar amide groups.

(1) M. L. Huggins, THIS JOURNAL, **74**, 3963 (1952).
(2) M. L. Huggins, *ibid.*, **74**, 3963 (1952).
(3) L. Pauling and R. B. Corey, *ibid.*, **72**, 5349 (1950).
(4) L. Pauling, R. B. Corey and H. R. Branson, *Proc. Nat. Acad. Sci.*, **37**, 205 (1951).
(5) L. Pauling and R. B. Corey, *ibid.*, **37**, 235 (1951).
(6) L. Pauling and R. B. Corey, *ibid.*, **37**, 251 (1951).
(7) R. B. Corey and L. Pauling, *Proc. Roy. Soc.* (*London*), to be published; presented at the Discussion Conference of the Royal Society of London, May 1, 1952.

GATES AND CRELLIN LABORATORIES OF CHEMISTRY
CALIFORNIA INSTITUTE OF TECHNOLOGY LINUS PAULING
PASADENA 4, CALIFORNIA ROBERT B. COREY

RECEIVED JULY 7, 1952

FUNDAMENTAL DIMENSIONS OF POLYPEPTIDE CHAINS

R. B. COREY AND L. PAULING

STABLE CONFIGURATIONS OF POLYPEPTIDE CHAINS

L. PAULING AND R. B. COREY

Gates and Crellin Laboratories of Chemistry
California Institute of Technology, Pasadena 4, California

Reprinted from
Proceedings of the Royal Society, B, 141, 10-33 (1953)

10 R. B. Corey and L. Pauling (Discussion Meeting)

Fundamental dimensions of polypeptide chains

By R. B. Corey and L. Pauling, For.Mem.R.S.

Gates and Crellin Laboratories of Chemistry,
California Institute of Technology, Pasadena, California

The dimensions of the polypeptide chain and of its associated N—H···O hydrogen bonds can be inferred with confidence. Data from many sources—X-ray diffraction analyses of crystals of organic acids, amides, peptides and related compounds; polarized infra-red studies of crystals—together with fundamental concepts of structural chemistry, now provide a basis for satisfactory knowledge and understanding of the dimensions and configurations of the amide group, 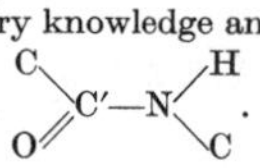. The normal coplanarity of the atoms of this group is the result of resonance which gives rise to partial double-bond character of the N—C′ peptide bond. Rotation about this bond is, in general, severely restricted. The *trans* configuration of the amide group appears to be preferred. Interatomic distances and bond angles of the amide group which may be derived from the present data are:

N—C = 1·47 Å	C—N—C′ = 123°
N—C′ = 1·32	C—N—H = 114
C′—O = 1·24	H—N—C′ = 123
C′—C = 1·53	N—C′—O = 125
	N—C′—C = 114
	O—C′—C = 121
	C′—C—N = 110

The dimensions and directional characteristics of N—H···O hydrogen bonds have been determined in a great variety of crystals. In these crystals the N···O distances vary, falling generally within the limits 2·79 ± 0·12 Å. The H···O vector rarely makes an angle greater than 20° with the probable direction of the N—H bond.

1. *Introduction*

More than fifteen years ago Albrecht (Albrecht & Corey 1939) began a detailed X-ray analysis of crystals of glycine, the simplest of the amino-acids. This work was the first of a series of crystal-structure studies of amino-acids and simple peptides made at the California Institute of Technology as part of a basic attack on the problems of protein structure. These crystal-structure analyses provided direct experimental information about the interatomic distances and bond angles in molecules of amino-acids, about the packing of the molecules, and about the dimensions and directional characteristics of the N—H···O hydrogen bonds which largely determine the molecular arrangement in the crystals. This information, together with data from crystals of other organic compounds, was used to derive probable dimensions for the polypeptide chain of proteins (Corey 1940). As new and more accurate data were accumulated these dimensions were progressively revised (Corey 1948; Corey & Donohue 1950).

During the last few years the precision which is attainable in practical crystal-structure analysis has been greatly increased through the use of high-speed calculating equipment, especially of the punched-card type. With crystals of the complexity of an amino-acid or simple peptide, we formerly had to rely on the intensities of perhaps one or two hundred prism zone reflexions for establishing the positions of the atoms. Atomic co-ordinates were refined by the computation from these intensities of two-dimensional projections of electron density parallel to particular directions in the crystal, or by other equivalent methods of two-dimensional refinement. Now, with modern computing equipment, use can be made of all, perhaps a thousand or more, X-ray reflexions obtainable from the crystal. The three-dimensional distribution of electron density can be calculated to give plots in which all atoms are resolved. Three-dimensional refinement, based on all intensity data, is capable of establishing interatomic distances within a limit of error (three times the probable error) of about 0·02 Å and bond angles within 1·0°. The re-determinations by Robertson and his co-workers of the crystal structures of naphthalene (Abrahams, Robertson & White 1949) and anthracene (Mathieson, Robertson & Sinclair 1950) are outstanding examples of the resolution and accuracy which can be attained by three-dimensional Fourier methods.

Crystals of several amino-acids and of a few simple peptides and other compounds related to proteins have been analyzed recently by three-dimensional methods. These and other accurate determinations of structure constitute the best experimental sources of information about the dimensions and configuration of the polypeptide chain. The present paper comprises a critical summary of this information—in particular, of the principal structural features of the amide group and the N—H···O hydrogen bonds, as derived from X-ray diffraction analyses of crystals of amino-acids, peptides, and other organic compounds.

2. *The amide group*

(*a*) *Dimensions of the amide group*

Reference to figure 2 shows that the polypeptide chain may be regarded as composed of tetrahedral carbon atoms (the α-carbon atoms of the constituent amino-acids) joined by amide groups. The amide group, 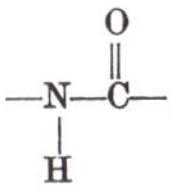, is therefore a fundamental structural component of the polypeptide chain. The structure of the amide group can be defined in terms of bond lengths and bond angles associated with its carbon and nitrogen atoms and of its configuration with respect to rotation around the C—N peptide bond.

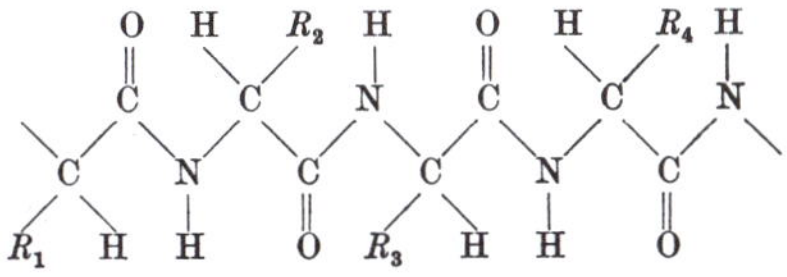

FIGURE 2. The polypeptide chain.

Table 2 lists interatomic distances found in X-ray diffraction analyses of crystals of amino-acids, peptides, and compounds of urea. These structure determinations, all but two of which are based upon three-dimensional refinement of the atomic parameters, are probably the most reliable sources of the distances which characterize the amide group. We are indebted to the authors for permission to present throughout this paper the results of the analyses of DL-serine, hydroxy-L-proline, α-glycylglycine, *N*, *N'*-diglycylcystine, urea and urea oxalate in advance of publication. Work on the structures of α-glycylglycine and *N*, *N'*-diglycylcystine is not quite completed, but the final values of the interatomic distances are not expected to differ by more than 0·01 Å from those given here. An earlier and less accurate determination of the structure of hydroxy-L-proline has been published by Zussman (1951). A structure determined by Dyer (1951) for cysteylglycine-sodium iodide is not established with certainty.

TABLE 2. BOND LENGTHS OF THE AMIDE GROUP AS DERIVED FROM THOSE FOUND IN CRYSTALS OF AMINO-ACIDS, PEPTIDES AND RELATED COMPOUNDS

An asterisk indicates three-dimensional refinement of the atomic positions.

	N—αC (Å)	C′—αC (Å)	C′—O (Å)	C′—N (Å)	reference
DL-alanine*	1·50	1·54	—	—	Donohue (1950)
L-threonine*	1·49	1·52	—	—	Shoemaker, Donohue, Schomaker & Corey (1950)
DL-serine*	1·49	1·53	—	—	Shoemaker, Donohue, Barieau & Lu (unpubl.)
hydroxy-L-proline*	1·50	1·52	—	—	Donohue & Trueblood (1952)
N-acetylglycine*	1·45	1·50 1·51	1·24	1·32	Carpenter & Donohue (1950)
β-glycylglycine	1·48	1·53 1·53	1·23	1·29	Hughes & Moore (1949)
α-glycylglycine*	1·47	1·56	1·24	1·32	Hughes & Biswas (unpubl.)
N, *N'*-diglycylcystine*	1·48	1·56 1·52	1·21	1·35	Hughes & Yakel (unpubl.)
urea*	—	—	1·26	1·34	Vaughan & Donohue (1952)
urea · H_2O_2	—	—	1·24	1·34	Lu, Hughes & Giguere (1941)
urea oxalate*	—	—	1·26	1·34 1·35	Schuch, Merritt & Sturdivant (unpubl.)
selected value	1·47	1·53	1·24	1·32	—

The length of the αC—N single bond is about 1·49 Å in amino-acid crystals, and is somewhat shorter in the four peptides, β-glycylglycine, *N*-acetylglycine, α-glycylglycine, and *N*, *N'*-diglycylcystine. The selected value, 1·47 Å, is identical with that given by the sum of the single-bond radii (Pauling 1940, p. 164). The length of the C—C bond, about 1·53 Å in the amino-acids and surprisingly short in *N*-acetylglycine, is probably slightly, but significantly, shorter than the C—C distance found in diamond, 1·5445 Å (Lonsdale 1947). The selected value, 1·53 Å, represents this probable shortening.

The lengths of the C—O bond found in the different crystals are all in good agreement; the selected value, 1·24 Å, is significantly greater than the sum of the double-bond covalent radii, 1·215 Å (Pauling 1940, p. 164). The length (1·32 Å) of the peptide bond between the nitrogen atom and the carbonyl carbon atom is much less than that of the typical single bond (1·47 Å). In *N*, *N'*-diglycylcystine the C′—O bond is somewhat shorter and the C′—N bond somewhat longer than in the other peptides, but these apparent differences may not be significant.

There is still considerable uncertainty about the bond angles of the amide group. Data from the linear peptides for which accurate analyses have been made are given in table 3. Except for *N*, *N′*-diglycylcystine, they are restricted to glycine peptides, so that no observations can be made regarding the possible effect of side-chain atoms. Fortunately, precise three-dimensional analyses of other peptides are now in progress and will soon supply additional information.

TABLE 3. BOND ANGLES OF THE AMIDE GROUP AS DERIVED FROM THOSE FOUND IN CRYSTALS OF PEPTIDES

	around carbonyl carbon atom			around amide nitrogen atom		
	αC—C′—O	N—C′—O	αC—C′—N	C′—N—αC	C—N—H	αC—N—H
N-acetylglycine*	121·0°	121·3°	117·7°	119·6°	—	100° ± 10°†
β-glycylglycine	121	125	114	122	—	—
α-glycylglycine*	121·1	124·2	114·4	119·3	—	—
N, *N′*-diglycylcystine*	120·6	125·3	113·2	121·6	—	—
selected values	121	125	114	123	123	114

* Three-dimensional refinement.

† Polarized infra-red study (Newman & Badger 1951).

In the last three peptides listed in table 3, the angles around the carbonyl carbon atom are in good agreement. They are selected as the most probable values. It is interesting to note that they differ significantly from the angles found in *N*-acetylglycine, which contains only one amino-acid residue.

Around the nitrogen atom the bond angles are known with much less definiteness. In particular, the X-ray data have essentially nothing to say about the position of the hydrogen atom. On the other hand, the direction of the N—H bond in crystals of *N*-acetylglycine has been the subject of a recent infra-red study by Newman & Badger (1951). Using very thin sections, these authors have measured the change in the intensity of the absorption band at 3340 cm^{-1} with change in angle of polarization of the incident radiation. They concluded that the direction of the N—H bond is probably close to that of the plane of the molecule and that the angle between the N—H bond and the bond joining the nitrogen atom to the α-carbon atom is approximately 100°. They further state that 'the error in this estimate may be as large as 10°, but is not so large as to admit the possibility that the hydrogen atom lies on a line connecting nitrogen and oxygen atoms, which requires an angle of 132°'. The selected values for the bond angles around the nitrogen atom are shown in table 3. The value, 123°, for the angles adjacent to the peptide C—N bond is a compromise between the values found in the crystal structures and the angle, 125°, predicted for a 100 % double bond. The corresponding angle, 114°, between the N—H bond and the αC—N bond is not far from the upper limit of that indicated by the infra-red measurements.

(b) *Planarity of the amide group*

The shortening of the C′—N peptide bond from the normal single bond length 1·47 to 1·32 Å and the slight lengthening of the C′—O bond from the normal double bond length 1·215 to 1·24 Å are to be ascribed primarily to resonance between the two structures

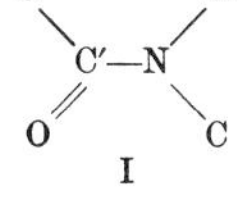

I

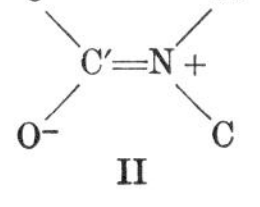

II

It is therefore to be expected that the atoms of the amide group are coplanar. Upon assumption of 60 and 40 % contributions from structures I and II respectively, the bond lengths calculated (Pauling 1940, p. 175) are C′—O = 1·25 Å and C′—N = 1·33 Å. Both are in excellent agreement with those observed. About 40 % double-bond character appears to be associated with the C′—N peptide bond. We may estimate the strain energy involved in rotation around this bond. If the planes of the two ends of the amide group form a dihedral angle δ, and if A is the amide resonance energy for the planar configuration, the strain energy may be taken equal to $A \sin^2 \delta$. A reasonable value for A is about 30 kcal/mole. From this we can calculate strain energies of about 0·9 kcal/mole for 10° distortion of the amide group, 3·5 kcal/mole for 20° distortion, and so on.

Experimental confirmation of the planarity of the amide group is supplied by structural studies already cited. In β-glycylglycine (Hughes & Moore 1949) all

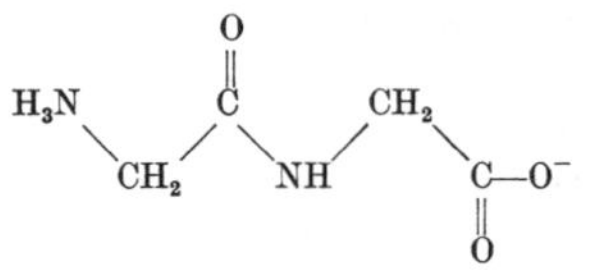

carbon, nitrogen and oxygen atoms lie in the same plane within the limit of error of the determination, with the single exception of the terminal charged nitrogen atom. In the crystal of acetylglycine (Carpenter & Donohue 1950) the parameters

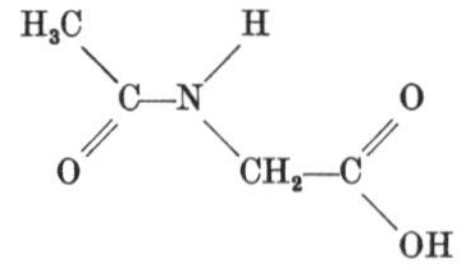

found for the carbon, nitrogen and oxygen atoms of the amide group (CH_3—CO—N) place these atoms within 0·005 Å of a common plane. The α-carbon atom is 0·03 Å from this plane, corresponding to a rotation around the C′—N peptide bond of 1·3°. The entire molecule is coplanar within 0·1 Å. Newman & Badger (1951) concluded from their polarized infra-red studies that the direction of the N—H bond is close to the plane of the molecule.

In α-glycylglycine the carbonyl carbon atom and the three atoms bonded to it are within 0·03 Å of a common plane, but the position of the α-carbon atom bonded to the nitrogen atom indicates that there is a rotation of 5 to 6° around the C′—N peptide bond. A rotation of about the same amount is found in N,N'-diglycylcystine. It corresponds to a strain energy of approximately 0·2 kcal/mole.

In the structures found for the sodium and potassium benzylpenicillins (Crowfoot, Bunn, Rogers-Low & Turner-Jones 1949) the amide group is planar within the rather wide limits of error of the determination. In a recently published structure of cysteylglycine-sodium iodide (Dyer 1951) the position of the α-carbon atom indicates a rotation of more than 40° around the C′—N bond of the amide group. Because of this exceptional configuration we feel that this structure should not be considered definitely established until it has been confirmed by additional study.

The planarity of the amide group in simple amides, 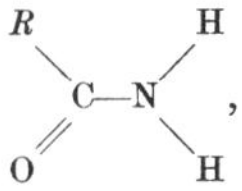 , has not been established directly by X-ray diffraction analysis because of the relatively small contribution made by hydrogen atoms to X-ray scattering. Nevertheless, there is strong evidence of planarity in simple amides. The dimensions of the molecule of acetamide as determined by Senti & Harker (1940) are shown in figure 3, p. 16, together with the relative positions of the oxygen atoms of two adjacent molecules to which the nitrogen atom is hydrogen-bonded. If probable positions are assigned to the hydrogen atoms in the plane of the molecule with each N—H bond making an angle of 125° with the N—C bond, we find that these positions lie close to lines joining the nitrogen atom and the two hydrogen-bonded oxygen atoms. The latter are seen to be only 0·24 and 0·39 Å from the plane of the acetamide molecule. From these geometrical relationships we may infer that the hydrogen atoms do indeed lie in or very close to the plane of the amide group. It can be argued that the position of the hydrogen atoms in the molecular plane, imposed upon them by the resonance characteristic of the amide group, is an important factor in determining the steric distribution of the molecules of acetamide in the crystal (Corey 1948).

Another simple amide is carbamide, urea. The configuration of its molecule will be expected to correspond to contributions from the three resonance structures

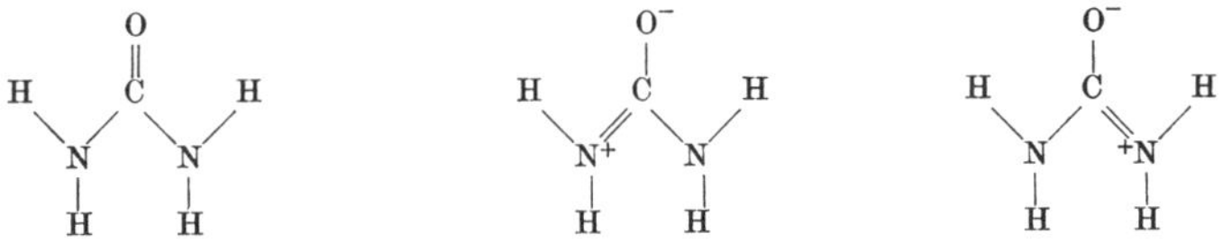

I II III

Like acetamide, we should expect the amide groups to be planar, and the hydrogen atoms to occupy positions in the plane of the molecule. The crystal structure of urea was determined by Hendricks in 1928, and has been refined by other workers (Wyckoff 1930, 1932; Wyckoff & Corey 1934; Vaughan & Donohue 1952). If the positions of the hydrogen atoms conform to the space group of the crystal as determined by X-ray analysis, they are strictly coplanar with the other atoms of the molecule, since the molecule occupies a plane of symmetry in the crystal. The possibility that the hydrogen atoms may not lie in the molecular plane has been investigated by Waldron & Badger (1950), using polarized infra-red radiation and single micro-crystals of urea. From their excellent absorption spectra with very high resolution in the 2·9μ region they conclude that 'the complete planarity of the urea molecule in the crystal is now reasonably well established by spectroscopic evidence'.

In view of the experimental evidence from X-ray crystal analysis and from other sources, and of the theoretical arguments, we accept the planarity of the amide group as a sound structural principle, and we believe that a significant departure from the planar configuration in a postulated structure must be justified as a highly unusual steric relationship. We have estimated that rotation around the C′—N peptide bond by as much as 10° from the planar configuration would result in instability by about 1 kcal/mole.

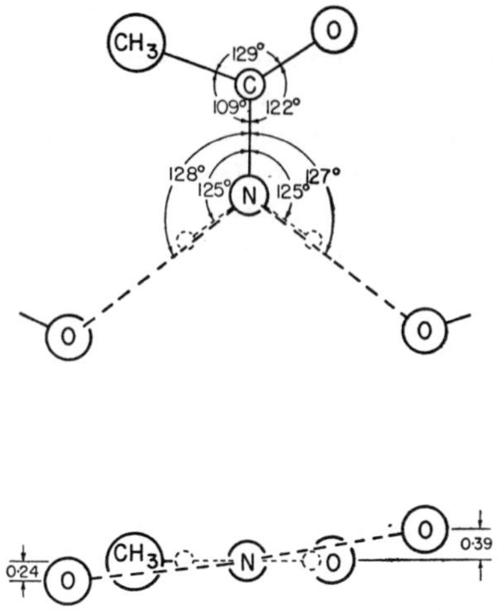

FIGURE 3. A diagrammatic representation of the relative positions of the planar acetamide molecule and the oxygen atoms of two adjacent molecules hydrogen-bonded to the nitrogen atom. The small dotted circles show positions in the molecular plane assigned to the hydrogen atoms of the $—NH_2$ group.

(c) *Configuration of the amide group*

In most discussions of the polypeptide chain in proteins the tacit assumption seems to be made that it exists primarily in the all-*trans* orientation. X-ray data for most fibrous proteins have been interpreted as indicating that in these substances the amide groups are wholly in the *trans* configuration. Crystal-structure analysis has as yet provided only a single example of the *cis* modification, in the dipeptide diketopiperazine (Corey 1938), glycine anhydride,

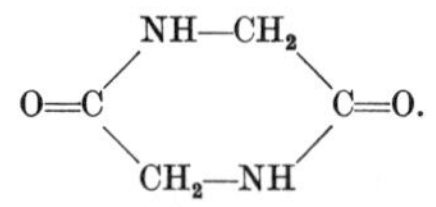

In a recent investigation of the seventy-two configurations of the polypeptide chain (with all residues equivalent) which can be derived from certain preferred orientations about the C—C and C—N single bonds, several sterically possible and plausible structures were discovered for the case that the amide groups have the *trans* configuration (Pauling & Corey 1951), but relatively few for the case that they have the *cis* configuration (Pauling & Corey 1952). In linear peptides a preference of the amide group for the *trans* configuration may be indicated by the published structures of β-glycylglycine and *N*-acetylglycine already discussed, and by incomplete investigations in these Laboratories of the structures of α-glycylglycine (Hughes & Biswas), diglycylcystine (Hughes & Yakel unpublished) and glycylasparagine (Katz, Pasternak & Corey 1952).

3. *The polypeptide chain*

Figure 4 is a diagrammatic representation of a completely extended polypeptide chain embodying the dimensions derived from the crystal-structure data and other experimental evidence discussed in the preceding paragraphs. We believe it to be unlikely that the structural parameters of polypeptide chains in proteins differ from those given in figure 4 by more than 0·03 Å in interatomic distances or 4° in bond angles. Some of this uncertainty may soon be removed as a result of additional accurate X-ray analyses of crystalline peptides. As in other organic compounds, small individual variations in dimensions will occur owing to the peculiarities of molecular configuration, the effects caused by side-chains, and the details of molecular packing.

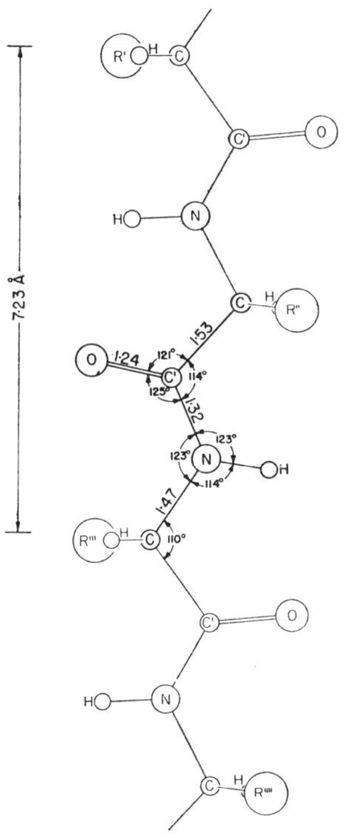

FIGURE 4. A diagrammatic representation of a fully extended polypeptide chain with the bond lengths and bond angles derived from crystal structures and other experimental evidence.

4. *Dimensions of* N—H···O *hydrogen bonds*

The important part which hydrogen bonds play in the structural chemistry of organic molecules has been recognized for many years. Mirsky & Pauling (1936) called attention to their significance in denaturation and other phenomena characteristic of protein molecules. Huggins (1936, 1943) discussed their probable role in stabilizing various possible configurations of the polypeptide chain in both fibrous and globular proteins. Since no data are yet available concerning the positions of individual atoms in proteins, the dimensions of the hydrogen bonds in these compounds cannot be determined by direct physical measurement. They can, however, be inferred from measurements made on amino-acids, peptides and related compounds. In table 4 are listed the N—H···O distances found in crystals of amino-acids and peptides. The corresponding angles which the N···O vector makes with the N—C single bond are taken from a recent review by Donohue (1953). Usually the length of the hydrogen bonds found in these crystals falls within the range $2{\cdot}79 \pm 0{\cdot}12$ Å, with an average value of 2·795 Å. There are, however, a few exceptionally long distances: L-threonine, 3·10 Å; hydroxy-L-proline, 3·17 Å; *N*-acetylglycine, 3·03 Å; *β*-glycylglycine, 3·07 Å. Donohue has called attention to the greater deviation of the longer bonds from a tetrahedral angle with the C—N covalent bond, corresponding perhaps to a greater deviation from the direction of the N—H bond in these crystals. We consider the shorter bonds to be normal and the longer bonds to be strained.

TABLE 4. DIMENSIONS OF N—H···O HYDROGEN BONDS FOUND IN CRYSTALS OF AMINO-ACIDS AND PEPTIDES

crystal	N—H···O (Å)	∠C—N···O angle found	∠C—N···O deviation from 110°
DL-alanine*	2·80	105°	5°
	2·84	103	7
	2·88	116	6
L-threonine*	2·80	98	12
	2·90	116	6
	3·10	132	22
DL-serine*	2·79	99	11
	2·81	98	12
	2·87	121	11
hydroxy-L-proline*	2·69	102, 113	8, 3,
	3·17	81, 133	29, 23
N-acetylglycine*	3·03	132	22
β-glycylglycine	2·68	100	10
	2·80	115	5
	2·81	99	11
	3·07	131	21
α-glycylglycine*	2·67	118	8
	2·77	114	4
	2·67	88	22
	2·75	117	7
N, N'-diglycylcystine*	2·75	84·5	25·5
	2·89	129·8	19·8
	2·75	111·8	1·8
most probable value	$2{\cdot}79 \pm 0{\cdot}12$		

* Three-dimensional analysis.

A factor which may influence the length of N—H···O bonds is the electrostatic charge on the bonding atoms. The amino-acids are dipolar ions, and in alanine, in which all bonds are short, all are between positively charged $—NH_3^+$ groups and negatively charged carboxyl oxygen atoms. In threonine the two short bonds are between —NH_3^+ and carboxyl oxygen atoms; the long bond joins NH_3^+ and the oxygen of the hydroxyl group. This reasoning breaks down with the next two amino-acids. In hydroxyproline the long and the short bond join the $\rangle NH_2^+$ group to crystallographically equivalent oxygen atoms. *N*-acetylglycine is not a dipolar molecule, and the single, and rather long, N—H···O bond joins the peptide nitrogen atom and the carboxyl oxygen atom without an attached hydrogen. In *β*-glycylglycine the three shorter bonds join the terminal —NH_3^+ group to two equivalent carboxyl oxygen atoms and one carbonyl oxygen atom; the longer bond connects the peptide nitrogen atom with a carboxyl oxygen atom. It may be that additional precise crystal analyses will provide data for establishing a relationship between electrostatic charge and the dimensions of N—H···O bonds, but the problem will always be complicated by the shapes of the molecules involved and the concessions which molecular packing must make to purely steric effects.

From his inspection of all types of hydrogen-bonded organic structures, Donohue (1953) concluded that 'only in very exceptional cases does a hydrogen atom bonded to a nitrogen atom or oxygen atom occupy a position such that hydrogen bond formation is impossible'; in other words, in crystals containing *X*—H groups (*X* is a nitrogen or oxygen atom) and oxygen atoms in appropriate relative numbers, one rarely finds an *X*—H group which is not involved in an *X*—H···O bond. From this observation we conclude that the polypeptide chain in proteins very probably assumes configurations which make maximum use of the hydrogen-bonding capacities of its N—H and C—O groups.

5. *Conclusions*

From this brief review it is apparent that the fundamental dimensions of the polypeptide chain are now known. There is therefore no reason why a specific configuration proposed for a polypeptide chain or a protein molecule cannot be precisely described in terms of co-ordinates of the carbon, nitrogen and oxygen atoms of the backbone structure, even including the *β*-carbon atoms. If a configuration is thus unambiguously described it can be criticized and evaluated, its X-ray diffraction effects can be calculated for comparison with observed X-ray patterns, its structural implications can be stated and tested by experiment. In view of our present knowledge of the dimensions of peptide chains, it seems reasonable to urge, and perhaps to demand, that a structure for which serious consideration is desired should be presented on a definite metrical basis—that co-ordinates for the atoms should be given, subject perhaps to stipulated limits of uncertainty.

The normal planarity of the amide group is established on both experimental and theoretical grounds as a sound structural principle. A structure in which the atoms of the amide group are not approximately coplanar should be regarded with scepticism until its relatively unstable configuration has been adequately confirmed.

20 R. B. Corey and L. Pauling (Discussion Meeting)

All of the experimental evidence indicates that stable structures for proteins and peptides involve the formation of close to the maximum possible number of N—H···O=C hydrogen bonds. Failure of a proposed configuration to conform to this structural principle should therefore raise grave doubts about its validity.

References (Corey & Pauling)

Abrahams, S. C., Robertson, J. M. & White, J. G. 1949 *Acta Cryst.* **2**, 233, 238.
Albrecht, G. & Corey, R. B. 1939 *J. Amer. Chem. Soc.* **61**, 1087.
Carpenter, G. B. & Donohue, J. 1950 *J. Amer. Chem. Soc.* **72**, 2315.
Corey, R. B. 1938 *J. Amer. Chem. Soc.* **60**, 1598.
Corey, R. B. 1940 *Chem. Rev.* **26**, 227.
Corey, R. B. 1948 *Advances in protein chemistry*, **4**, 385. New York: Academic Press, Inc.
Corey, R. B. & Donohue, J. 1950 *J. Amer. Chem. Soc.* **72**, 2899.
Crowfoot, D., Bunn, C. W., Rogers-Low, B. W. & Turner-Jones, A. 1949 *The chemistry of penicillin*, p. 310. Princeton University Press.
Donohue, J. 1950 *J. Amer. Chem. Soc.* **72**, 949.
Donohue, J. 1953 *J. Phys. Chem.* (in the Press).
Donohue, J. & Trueblood, K. N. 1952 *Acta Cryst.* **5**, 414, 419.
Dyer, H. B. 1951 *Acta Cryst.* **4**, 42.
Hendricks, S. B. 1928 *J. Amer. Chem. Soc.* **50**, 2455.
Huggins, M. L. 1936 *J. Org. Chem.* **1**, 407.
Huggins, M. L. 1943 *Chem. Rev.* **32**, 195.
Hughes, E. W. & Biswas, A. B. Unpublished.
Hughes, E. W. & Moore, W. J. 1949 *J. Amer. Chem. Soc.* **71**, 2618.
Hughes, E. W. & Yakel, H. L. Unpublished.
Katz, L., Pasternak, R. A. & Cory, R. B. 1952 *Nature, Lond.*, **170**, 1066.
Lonsdale, K. 1947 *Phil. Trans.* A, **240**, 244.
Lu, C.-S., Hughes, E. W. & Giguere, P. A. 1941 *J. Amer. Chem. Soc.* **63**, 1507.
Mathieson, A. M., Robertson, J. M. & Sinclair, V. C. 1950 *Acta Cryst.* **3**, 245, 251.
Mirsky, A. E. & Pauling, L. 1936 *Proc. Nat. Acad. Sci., Wash.*, **22**, 439.
Newman, R. & Badger, R. M. 1951 *J. Chem. Phys.* **19**, 1147.
Pauling, L. 1940 *The nature of the chemical bond.* Ithaca: Cornell University Press.
Pauling, L. & Corey, R. B. 1951 *Proc. Nat. Acad. Sci., Wash.*, **37**, 729.
Pauling, L. & Corey, R. B. 1952 *Proc. Nat. Acad. Sci., Wash.*, **38**, 86.
Schuch, A. F., Merritt, L. L. Jr. & Sturdivant, J. H. Unpublished.
Senti, F. & Harker, D. 1940 *J. Amer. Chem. Soc.* **62**, 2008.
Shoemaker, D. P., Donohue, J., Barieau, R. & Lu, C.-S. Unpublished.
Shoemaker, D. P., Donohue, J., Schomaker, V. & Corey, R. B. 1950 *J. Amer. Chem. Soc.* **72**, 2328.
Vaughan, P. & Donohue, J. 1952 *Acta Cryst.* **5**, 530.
Waldron, R. D. & Badger, R. M. 1950 *J. Chem. Phys.* **18**, 566.
Wyckoff, R. W. G. 1930 *Z. Krist.* **75**, 529.
Wyckoff, R. W. G. 1932 *Z. Krist.* **81**, 102.
Wyckoff, R. W. G. & Corey, R. B. 1934 *Z. Krist.* **89**, 462.
Zussman, J. 1951 *Acta Cryst.* **4**, 72, 493.

Stable configurations of polypeptide chains

By L. Pauling, For.Mem.R.S., and R. B. Corey

Gates and Crellin Laboratories of Chemistry, California Institute of Technology, Pasadena, California

Several configurations of polypeptide chains involving planar amide groups with the dimensions found by experiment for simple substances, and hydrogen bonds between the NH groups and the carbonyl oxygen atoms, have been discovered. One of these structures, the α-helix, with about 3·7 amino-acid residues per turn of the helix, has been assigned to synthetic polypeptides and proteins that give X-ray diagrams of the α-keratin type. The evidence supporting this assignment is reviewed. Other configurations of polypeptide chains, including the γ-helix, three pleated-sheet structures, and the three-chain helical structure proposed for collagen, are described, and evidence bearing on their possible presence in proteins is discussed.

1. *Introduction*

Fifteen years ago we attacked the problem of the configuration of polypeptide chains in proteins by assuming the amide groups to be planar, with reasonable values of interatomic distances and bond angles, and assuming that hydrogen bonds are formed between the imino groups and the oxygen atoms of the carbonyl groups. Although the assumed dimensions were essentially the same as those described in the preceding paper by Corey & Pauling, the attack was not a successful one, the lack of success being due in part to uncertainty as to the reliability of the assumed dimensions. We decided to gather information about interatomic distances and bond angles by making careful determinations of the structure of amino-acids, simple peptides and related substances. The results of this work are described in the preceding paper.

X-ray photographs of proteins were first made by Herzog & Jancke (1920), who studied muscle, nerve, sinew, hair and silk, and shortly thereafter the suggestion was made by Brill (1923), from analysis of the X-ray data, that silk contains long polypeptide chains. Meyer & Mark (1928) then pointed out that in silk the chains are essentially in the extended form, in which each residue has a length along the fibre axis of about 3·5 Å. Astbury & Street (1931) found that hair, wool and related fibres can exist in either a contracted form, which they named α-keratin, or an extended form, β-keratin. They suggested that in β-keratin the polypeptide chains are in the fully extended configuration, and that a chemical change involving the formation of rings of atoms takes place on contraction to α-keratin. Two years later Astbury & Woods (1933) described β-keratin as involving sheets of extended polypeptide chains, the chains being held together in the sheet by interaction of carbonyl groups and imino groups of the amide groups. This interaction was soon recognized as corresponding to the formation of N—H···O hydrogen bonds (Mirsky & Pauling 1936; Huggins 1936).

Several possible configurations were then suggested for the coiled polypeptide chain in α-keratin, by Astbury and other investigators. These structures are of two types: those with three residues in a fibre-axis length of 5·1 Å, such as that of Astbury (1942), and those with two residues in this fibre-axis length, such as that first suggested by Huggins (1943) and supported by other workers (Zahn 1947;

Simanouti & Mizushima 1948; Ambrose & Hanby 1949). Huggins (1943) formulated a number of configurations for polypeptide chains in which the coiling is such as to permit hydrogen bonds to be formed; most of his configurations are helical, with the number of amino-acid residues per turn of the helix an integer. A thorough survey of structures of this sort was made by Bragg, Kendrew & Perutz (1950), in a study of the possible types of folding of the polypeptide chain in molecules of haemoglobin and myoglobin. These investigators were the first to describe their structures precisely, by reporting atomic co-ordinates to 0·1 Å.

In all of this earlier work the investigators were handicapped by the fact that the assumptions made about acceptable configurations were not sufficiently precise to determine the configurations uniquely—the proposed structures could be distorted by rather large amounts without rendering them unacceptable. This vagueness of the suggested structures interfered seriously with their straightforward testing by comparison with the X-ray data. On examining the configurations that had previously been proposed for the polypeptide chain in proteins we found that every configuration that had been described with sufficient precision to permit its critical discussion is incompatible in one way or another with the structural features that we consider to be essential (preceding paper). The only structure not ruled out by the failure to conform to the values of interatomic distances, bond angles, hydrogen-bond distance, close linearity of the N—H···O group and planarity of the amide group is the planar sheet that is formed from fully extended polypeptide chains, alternating in direction, and with lateral hydrogen bonds. This structure, however, is eliminated for all substances except polyglycine by steric hindrance between the side-chain groups attached to the α-carbon atoms of adjacent chains, and so it, too, has to be considered as not acceptable for proteins.

In searching for acceptable configurations we first investigated configurations involving *trans* amide groups, with all of the amide groups structurally equivalent except for differences in the nature of the side-chain R, and with intramolecular hydrogen bonds. The general operation of conversion of a unit without symmetry element, such as an L-amino-acid residue, into an equivalent unit is the rotatory translation. The continued application of such an operation leads to a helix; accordingly, the structures that result from the foregoing assumptions are helical structures. It was found that there are only two such structures satisfying the assumed requirements (Pauling & Corey 1950; Pauling, Corey & Branson 1951). One of these structures, called the γ-helix or 5·1-residue helix, which has about 5·15 residues per turn of the helix, seems not to be an important one—no strong evidence has yet been found for its occurrence in proteins. The second structure, the α-helix, has about 3·7 residues per turn of the helix. The evidence for its presence in hair, muscle, horn and other fibrous proteins of the α-keratin type is very strong, and it seems also to be present in some synthetic polypeptides and some globular proteins, including haemoglobin, myoglobin and serum albumin.

The search for satisfactory structures was initially carried out analytically, with rather small use of simple molecular models. In the later stages, especially the systematic investigation of configurations corresponding to favoured orientations around the single bonds to the α-carbon atoms (Pauling & Corey 1951*i*,

1952*a*), there was nearly complete reliance on a set of molecular models, constructed from wood at the scale of 1 in. = 1 Å, and provided with metal pins with locking clamps that give rigidity to the model (Corey & Pauling 1953).

2. *The α-helix*

The α-helix is represented diagrammatically in figure 5, and in perspective in figure 6. Each amide group is attached by hydrogen bonds to the third amide group from it, in either direction, along the helix. The structure is formally equivalent to one described by Bragg *et al.* (1950) as having four residues per turn. Our assumptions about the dimensions of the chain do not, however, permit this number of residues per turn.

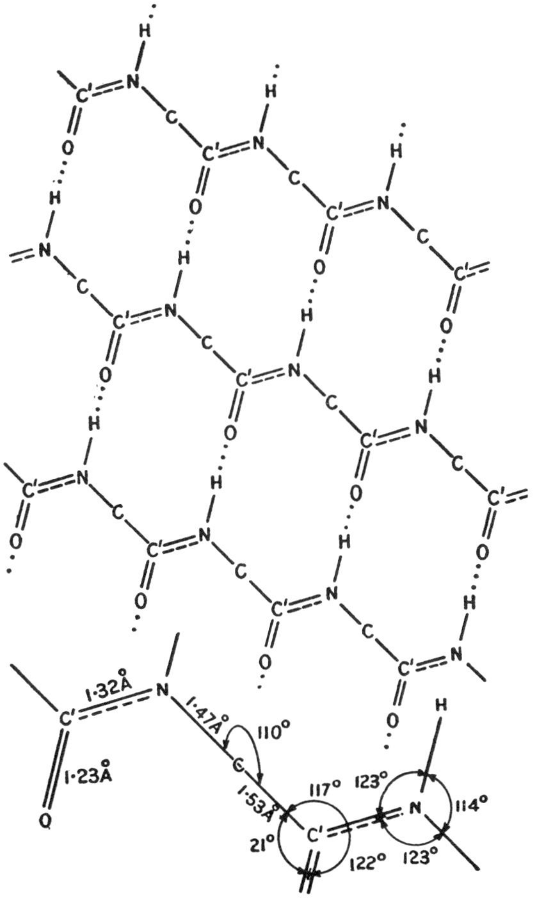

FIGURE 5. Diagrammatic representation of the 3·7-residue helical configuration of the polypeptide chain.

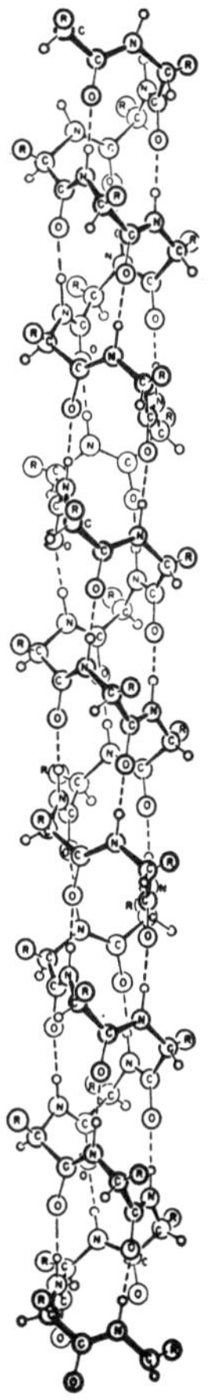

FIGURE 6. A drawing of the α-helix.

24 L. Pauling and R. B. Corey (Discussion Meeting)

The α-helix was first predicted to have 3·69 residues per turn, and length per residue of 1·47 Å along the fibre axis, leading to a fibre-axis length per turn of 5·44 Å. In a more detailed discussion (Pauling & Corey 1951*a*) it was pointed out that the fibre-axis length per residue 1·47 Å corresponds to the hydrogen-bond distance 2·75 Å, and that the length per residue increases by 0·01 Å for every 0·03 Å increase in the hydrogen-bond distance; the reasonable range 2·68 to 2·92 Å for the hydrogen-bond length would then correspond to the range 1·45 to 1·53 Å for the fibre-axis length per residue. The number of residues per turn is fixed primarily by the bond angle at the α-carbon atom; it varies from 3·60 for bond angle 108·9° to 3·67 for bond angle 110·8°. It was pointed out in the earlier discussion (Pauling & Corey 1951*a*) that these ratios correspond respectively to 18 residues in five turns and 11 residues in three turns. The minimum values and the maximum values just quoted lead to the values 5·22 and 5·62 Å, respectively, for the pitch of the helix.

The assignment of the α-helix to certain synthetic polypeptides (Pauling & Corey 1951*b*), proteins of the α-keratin type (Pauling & Corey 1951*e*) and globular proteins (Pauling & Corey 1951*g*) was made on the basis of comparison of observed and predicted dimensions of the unit of structure or pseudo-unit (especially the pitch of the helix), the intensity of reflexion of X-rays at right angles to the fibre axis, and the radial distribution function. The agreement between predicted and observed values of the pitch of the helix has been improved through further investigation during recent months. The pseudo-identity distance for α-keratin and related proteins, usually reported as 5·1 or 5·15 Å, lies below the predicted range for the pitch of the α-helix, 5·22 to 5·62 Å; it has been suggested by Professor V. Schomaker (Pauling & Corey 1951*e*) that the discrepancy is the result of an abnormality in background intensity of the poorly crystallized materials. For muscle giving a crystalline X-ray diagram Lotmar & Picken (1942) originally reported the value 5·65 Å for the fibre-axis identity distance. The smaller value 5·30 Å has now been reported by Bamford & Hanby (1951) from the re-analysis of the data of Lotmar & Picken, by Pauling & Corey (1952*b*) from re-measurement of the Lotmar-Picken X-ray negatives, and by Bear & Cannan (1951) from measurements of X-ray photographs of specimens that they had obtained. A similar change has taken place in the value of the pitch of the helix for poly-γ-methyl-L-glutamate. The pseudo-identity distance for this synthetic polypeptide was originally reported to be 5·50 Å (Bamford, Hanby & Happey 1951); it has since then been revised to 5·4 Å (Bamford & Hanby 1951; a redetermination of the value by Dr Harry Yakel, in our Laboratories, has led to the result 5·27 Å). It seems not unlikely that the value for poly-γ-benzyl-L-glutamate, given by Bamford *et al.* as 5·76 Å or somewhat smaller, is in fact considerably smaller, although a precise value has not yet been reported. There is accordingly at the present time no reliable experimental value for the pitch of the α-helix that lies outside of the predicted range 5·22 to 5·62 Å. The pitch 5·27 Å corresponds, for 3·60 residues per helix, to a fibre-axis length per residue of 1·47 Å, and to the hydrogen-bond length 2·75 Å, which is close to the mean of the values found in simple substances.

Strong evidence for the correctness of the assignment of the α-helix to porcupine quill tip was provided by the observation by MacArthur (1943) of an X-ray reflexion in the position corresponding to a spacing about 1·5 Å perpendicular to the fibre axis; this reflexion was also reported by Perutz (1951) for hair, poly-γ-benzyl-L-glutamate and haemoglobin. This spacing represents the fibre-axis length per residue, and results from the scattering in phase of the successive residues in the α-helix. It has been observed also for poly-γ-methyl-L-glutamate, with spacing 1·489 Å, by Dr Harry Yakel.

There now exists much evidence for the assignment of the configuration of the α-helix to the polypeptide chains of the synthetic polypeptide poly-γ-methyl-L-glutamate. Bamford *et al.* (1951) had reported for this substance an orthorhombic unit of structure with axes 10·35, 5·98 and 5·50 Å. They had noted that the first two axial lengths are in the ratio $\sqrt{3}:1$, which is compatible with hexagonal symmetry, but had eliminated a hexagonal structure as requiring three amino-acid residues per unit, which would lead to the density 1·06 $\text{g}\,\text{cm}^{-3}$, which is too small. We observed that the extra reflexions reported by Bamford *et al.*, and attributed by them to an oriented impurity, could be accounted for as intermediate layer-line reflexions for a unit with a five-fold increase in the fibre-axis identity distance. An α-helix repeating after five turns would contain 18 residues, and could accordingly be surrounded by similar molecules in a hexagonal array. We accordingly assigned a hexagonal structure to the substance, and showed that the intensities of equatorial reflexions could be satisfactorily explained on this basis. Verification of the repeat after 18 residues for one modification of the substance has been obtained by Dr Harry Yakel, who has measured the identity distance as 26·75 Å and the fibre-axis length per residue (the Perutz spacing) as 1·489 Å, the quotient of these numbers being 18·0. Strong support for the 18-residue 5-turn helix has been provided by the work of Cochran & Crick (1952), who have shown that consideration of the Bessel functions involved in the form factor for such a helix (Cochran, Crick & Vand 1952) leads to a direct explanation of the relative strengths of twenty-eight observed layer lines on the X-ray diagram of poly-γ-methyl-L-glutamate. Further progress has been made by Dr Yakel, who has shown that the intensities of the reflexions in the fifth layer line are explained satisfactorily by the α-helix with the β-carbon atom in position 2 (Pauling & Corey 1951*a*) and are in disagreement with prediction for β-carbon atom in position 1. In this work only five of the ten heavy atoms of the amino-acid residue are taken into consideration in the calculation of predicted intensities; the small contribution of the side-chain atoms is to be attributed to their having less regular distribution in space than the atoms of the amide group and the α- and β-carbon atoms, and hence having larger mutual interference of the scattered rays. In order to complete the determination of the structure of poly-γ-methyl-L-glutamate (aside from the side-chain atoms) there remains only a determination of the azimuthal angle for a helix, which might be achieved by the consideration of the intensities of layer-line reflexions.

The rough agreement between equatorial intensities of the X-ray diagram of proteins of the α-keratin type and those predicted for the α-helix supports the assignment of this configuration to the α-keratin proteins. Proof of the correctness of this assignment has not yet been obtained, but additional evidence is provided by the consideration of the nature of the Bessel functions involved in the form factor, as given by the equation of Cochran *et al.* Figure 7 is a rough representation of the Bessel-function pattern for the 18-residue 5-turn α-helix. The orders of Bessel functions, up to the fourth order, as calculated by the formula of Cochran *et al.*, are given in the figure for the first twenty-three layer lines, and regions of darkening are shown to represent the regions in which these Bessel functions have large values. The origin is indicated by the cross in the centre. It is seen that a

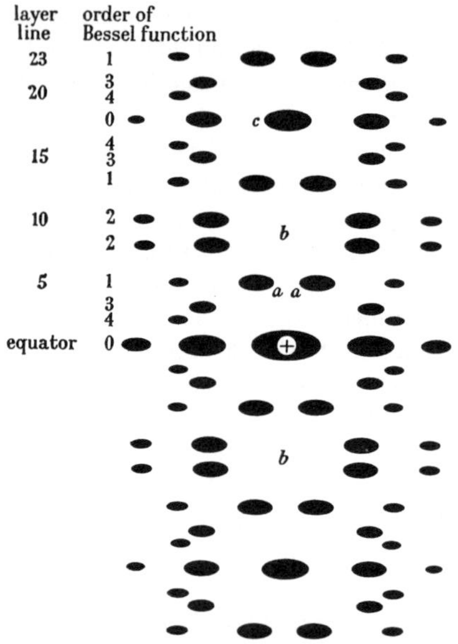

FIGURE 7. Diagram of the Bessel functions contributing to the form factor for the 18-residue 5-turn α-helix, as given by the formula of Cochran, Crick & Vand. Features *a* represent the 5·1 Å meridional arc, and *c* the 1·5 Å Perutz reflexion. The letter *b* shows the position of a sharp meridional reflexion observed on X-ray photographs of horse hair, and not accounted for by the simple theory of the α-helix.

somewhat smudged-out pattern of this sort, representing imperfect orientation, corresponds reasonably well with the familiar X-ray diagram of hair, horn and other α-keratin proteins. The features *a* represent the 5·1 Å arc, characteristic of these proteins. (A difficulty that remains unexplained is the observation of this reflexion on the meridian, instead of in the off-meridional positions.) These features, and the adjacent ones for the equator and layer lines 2 and 3, combine to form the ring of darkening that encloses a light ring about the central image. The 1·5 Å reflexion, resulting from a Bessel function of order zero, is feature *c*, in the eighteenth layer line. The general appearance of X-ray diagrams that we have made of descaled horse hair is similar to that shown in figure 7, except that a rather sharp meridional reflexion appears in positions *b* of the figure. Our measurement of this sharp meridional reflexion gives interplanar distance 2·96 Å, with reflexion *c* having the value 1·49 Å. Reflexion *b* is perhaps to be interpreted as a meridional reflexion involving co-operation of the residues in pairs.

The value 5·30 Å for the fibre-axis identity distance of the material giving the Lotmar-Picken X-ray diagram approximates the value 5·27 Å found for poly-γ-methyl-L-glutamate so closely as to indicate that this material is a large polypeptide or protein with the configuration of the α-helix (Pauling & Corey 1952*b*). The intensities of equatorial reflexions also provide evidence in favour of this assignment (Pauling & Corey 1951*e*).

The rods with diameter 10 Å running parallel to the 57 Å direction in the molecules of haemoglobin and myoglobin, discovered by Perutz and his collaborators (Boyes-Watson, Davidson & Perutz 1947; Perutz 1949) in haemoglobin, and by Kendrew (1950) in myoglobin, have been identified as α-helixes (Pauling & Corey 1951*g*) by comparison of calculated radial distribution functions with a rough experimental radial distribution function obtained by analysis of Perutz's data. The experimental radial distribution function did not show the fine structure that would enable a decision to be made between the positions 1 and 2 for the β-carbon atom. Riley & Arndt (1952) have reported a careful evaluation of the radial distribution function for bovine-serum albumin, and have concluded that the close approximation of the experimental function to the calculated function for the α-helix leaves little doubt of the correctness of the assignment of the configuration of the α-helix to polypeptide chains in this globular protein, and that, in fact, the β-carbon atoms can be rather confidently placed in position 2.

It seems not unlikely that the α-helix will be found to be the most generally prevalent mode of folding of polypeptide chains in proteins.

3. *The γ-helix*

In the γ-helix each amide group forms hydrogen bonds with the fifth more distant group in each direction along the polypeptide chain. The fibre-axis length per residue is 0·97 Å for hydrogen-bond distance 2·72 Å, and it increases by about 0·002 Å for each 0·010 Å increase in length of the hydrogen bond, to 1·01 Å for hydrogen-bond distance 2·92 Å. There are 5·14 residues per turn for bond angle 109·8° at the α-carbon atom, increasing to 5·25 residues per turn for bond angle 111·2°. The extremes of these expected values correspond to a pitch of the helix ranging from 5·00 to 5·30 Å (Pauling & Corey 1951*a*).

It was suggested (Pauling & Corey 1951*e*) that supercontracted keratin might have the configuration of the γ-helix, although it was pointed out that there is very little evidence to support the suggestion. A decision could be made by the calculation of the radial distribution function from a powder diagram of supercontracted keratin.

There is an evident defect in the γ-helix which suggests that it may not occur in proteins, or may occur only rarely. Whereas the α-helix is a tight helix, with van der Waals contact of atoms close to the helical axis, the γ-helix has a cylindrical hole along its axis, with diameter about 2·5 Å. This hole might be large enough to permit entry of a water molecule, but it seems unlikely that water molecules would occupy it, because their positions would be unfavourable to hydrogen-bond formation. If this cavity were to remain empty there would occur a serious loss in energy of van der Waals attraction of adjacent molecules. By use of the equation

$$E = -\sum_{A,B} 38 \frac{R_A R_B}{r^6_{AB}} \text{ kcal/mole}$$

for the energy of the London electronic dispersion interaction of groups A and B the distance r_{AB} apart (the equation is obtained from the approximate second-order perturbation theory with use of the average value 14 eV for excitation energy of the groups), and with the mole refraction R of the amide group and β-carbon atom taken as 13 cm³, it is calculated that the γ-helix is less stable than the α-helix, because of the decreased van der Waals attraction resulting from the presence of the hole, by about 4 kcal/mole per residue. This result indicates strongly that the γ-helix is not to be expected as a common constituent of proteins.

4. *The three pleated sheets*

The steric hindrance between side-chains that occurs for fully extended polypeptide chains forming lateral hydrogen bonds to the oppositely directed neighbouring chains can be avoided by rotating the residues about the bonds to the α-carbon atoms in such a way that the length per residue decreases below the maximum value, about 3·6 Å. The first structure of this sort to be discovered, the polar pleated sheet, involves parallel polypeptide chains, with all of the residues having their carbonyl groups oriented in the same way (Pauling & Corey 1951*c*). The length per residue for undistorted hydrogen bonds is 3·07 Å, and a rather large distortion is necessary to increase the fibre-axis residue length to 3·3 Å, the value observed for the β-keratin proteins. Also, a polar pleated sheet constructed of L-amino-acid residues is unsymmetrical, and would have a tendency to curve; for these reasons we think that it is likely that the two other pleated sheets occur in proteins, rather than the polar pleated sheet.

The parallel-chain pleated sheet (figure 8) and the anti-parallel-chain pleated sheet (figure 9) were discovered in the course of a systematic investigation of configurations of polypeptide chains with favoured orientations around single bonds, carried out with the use of molecular models (Pauling & Corey 1951*i*). For both of these structures the fibre-axis residue length of 3·34 Å, corresponding closely to the value usually observed for proteins with the β-keratin structure, is achieved with essentially unstrained hydrogen bonds, and with orientations around the single bonds to the α-carbon atom that are believed to be somewhat more stable than other orientations. An increase in fibre-axis residue length to 3·5 Å, the value observed in silk fibroin, can be achieved easily for the anti-parallel-chain pleated sheet, but not for the parallel-chain pleated sheet.

We think that it is likely that all polypeptides and proteins that give X-ray diagrams of the β-keratin type have the parallel-chain pleated-sheet structure, the anti-parallel-chain pleated-sheet structure, or a pleated-sheet structure in which the chains have positive and negative orientations at random.

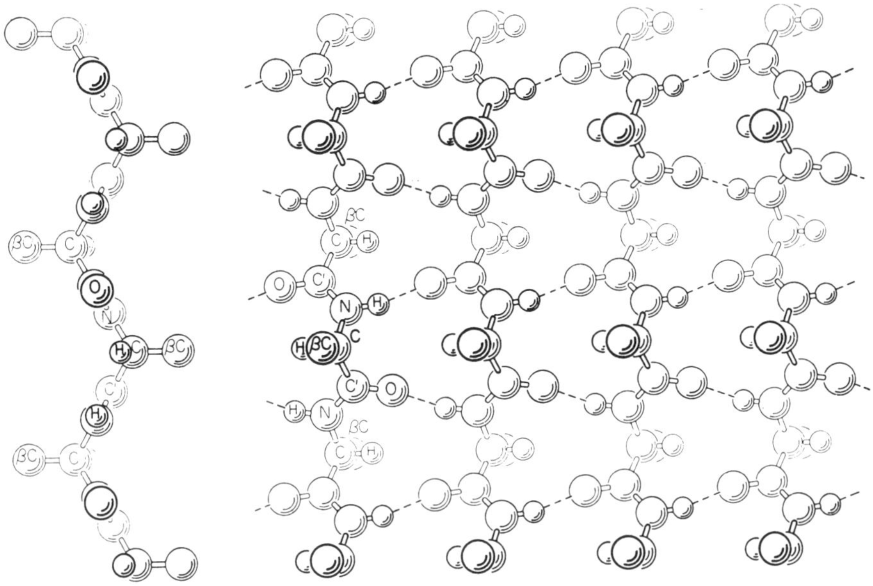

FIGURE 8. A drawing representing the parallel-chain pleated sheet of nearly extended polypeptide chains.

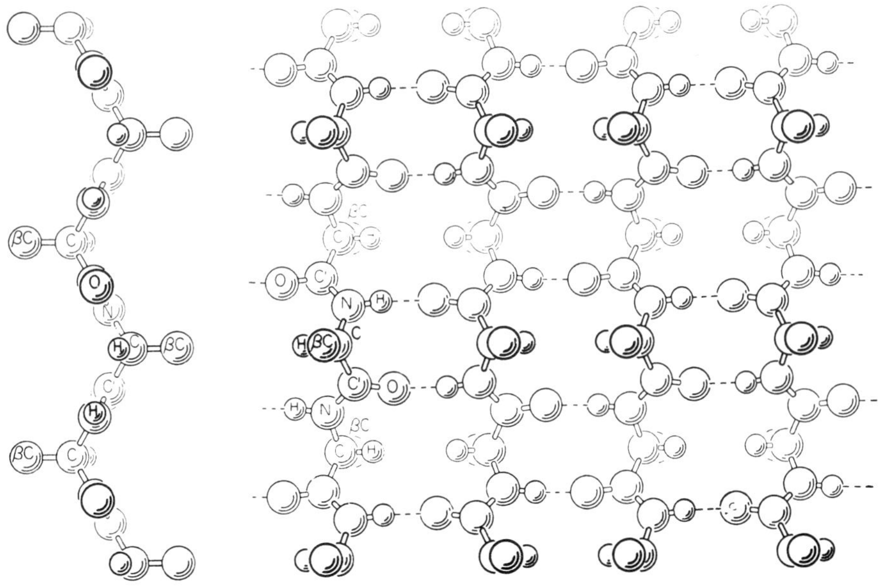

FIGURE 9. Drawing representing the anti-parallel-chain pleated-sheet structure.

30 L. Pauling and R. B. Corey (Discussion Meeting)

Together with Dr Richard E. Marsh, we have carried out a calculation of intensities to be expected on fibre diagrams of polyglycine, with an assumed anti-parallel-chain pleated-sheet structure, the unit of structure being essentially as suggested by Astbury (1949). The agreement with the observed powder pattern is sufficiently good to provide support for the assignment of this structure to the substance. We also have made a test of a pleated-sheet structure for silk fibroin, in collaboration with Dr Marsh and Dr Max Rogers, with promising results. This structure involves double pleated sheets with all side-chains contained in the space between them, the contacts with adjacent double pleated sheets being like those in polyglycine. Verification of the pleated-sheet configuration for other β-keratin proteins by the analysis of X-ray intensity data has not yet been carried out.

5. *The contraction of muscle*

We have suggested (Pauling & Corey 1951*e*, 1951*h*) that the process of contraction of muscle may involve the conversion of a contractile protein from a pleated sheet of nearly completely extended polypeptide chains to a double row of α-helices. This proposal seems to us to be a reasonable one which is compatible with our present knowledge of the structural chemistry of polypeptide chains, and which accounts for the usually observed extent of contraction of muscle, about 55 %. There is, however, little direct evidence in support of it at the present time.

6. *Complex protein structures*

It is not unlikely that two or more types of folding of polypeptide chains may occur together in proteins. We have proposed a structure of this sort for feather rachis protein. The X-ray diffraction pattern of feather rachis is very complex, and it requires for its interpretation a unit of structure with dimensions at least 9·5 by 34 by 94·6 Å. The identity distance 9·5 Å perpendicular to the fibre axis indicates the presence of pleated sheets. The proposed structure consists of a series of pleated sheets interleaved with double rows of α-helices. The structure can be made to account roughly for the observed intensities of equatorial reflexions, but the agreement is not sufficiently extensive to provide significant support for the structure. Evidence against the structure, though not decisive (Pauling & Corey 1951*h*), was reported by Perutz (1951), who stated that feather rachis keratin does not give rise to the 1·5 Å reflexion that is expected in general to occur for the α-helix. However, we have observed on Weissenberg photographs of sea-gull feather rachis a rather diffuse meridional reflexion with spacing $1{\cdot}52 \pm 0{\cdot}03$ Å, which may be the 1·5 Å reflexion of α-helices in this fibrous protein.

7. *The structure of collagen*

Collagen and gelatin give characteristic X-ray diagrams, showing a strong meridional reflexion for spacing 2·9 Å. This reflexion can be reasonably interpreted as an average fibre-axis length per residue in nearly extended polypeptide chains, and Astbury has suggested that some or all of the amide groups have the *cis* configuration (Astbury 1940). We noted that the X-ray diagrams are compatible with a hexagonal packing of cylindrical molecules, with diameter about 11 Å, and from a consideration of the lattice constants and density of collagen reached the conclusion that the molecule of collagen and gelatin is not a single polypeptide chain, but consists of three polypeptide chains (Pauling & Corey 1951*f*). We suggested a detailed structure for collagen, in which three polypeptide chains are intertwined about one another, and are held together by lateral hydrogen bonds. In each chain there is an alternation of two *cis* groups and one *trans* group, and the structure is of such a nature that every third residue may be a proline or hydroxyproline residue, with a glycine residue as one of its neighbours. This structure accounts in a striking way for the distribution of intensity along the equator of the X-ray fibre diagram of collagen and gelatin.

A few months ago it was pointed out by Bear (paper presented at the New York meeting of the American Chemical Society, September 1951) that the X-ray diagram of collagen shows layer lines at positions related to the spacing of the principal meridional reflexion, about 2·9 Å, in the approximate ratio of the integer 7 to other small integers, and that the diagram thus indicates an identity distance in the fibre-axis direction of 20 Å, seven times the average residue length, rather than 26 Å, as given by the three-chain structure with nine units per turn of each helical chain. He suggested that the structure might be that of the γ-helix, with amino-acid residues of different kinds arranged in units of three; the identity distance for the γ-helix with 21 residues in four turns is about 20 Å, and the seventh order of basal plane reflexion might appear as a strong reflexion, through reinforcement of side-chains.

This suggestion can be tested by application of the formula given by Cochran *et al.* (1952). With 21 residues in four turns the Bessel functions for successive layer lines are of orders 0, 5, 10, 6, 1, 4, 9, 7, 2, 3, 8, 8, 3, 2, 7, 9, 4, 1, 6, 10, 5, 0. The seventh layer line, which should be of order 0 to account for the 2·9 Å meridional reflexion, is in fact of order 7. The discrepancy can be avoided by assuming a 3-residue repeating unit. The sequence of orders of Bessel functions for a helix with seven 3-residue units in four turns is 0, 2, 3, 1, 1, 3, 2, 0. This sequence introduces J_0 at the seventh layer line, accounting for the 2·9 Å reflexion, but the layer lines represented by J_1, which should be the next strongest, are the third and fourth, rather than the second and fifth, as observed on the photographs. A structure based on the γ-helix must accordingly be considered improbable.

A study of the X-ray diagram of collagen has shown that it is unlikely that this diagram is completely compatible with the three-chain structure that we have proposed; there is indication, however, that the correct structure does not differ greatly from this three-chain structure. Instead of the regular sequence of *cis-cis-trans* units in each of the three chains, there may be inserted occasional units of four residues, presumably *cis-cis-trans-trans*, with only one of the four a proline or hydroxyproline residue. The presence of meridional reflexions with spacings about 9·73 and 2·91 Å, with ratio $0{\cdot}299 \pm 0{\cdot}002$, on the photographs indicates an approximate repetition of sequences of ten amino-acid residues, consisting of three groups of residues (*cis-cis-trans-cis-cis-trans-cis-cis-trans-trans*), or perhaps of 110 residues, consisting of seven *cis-cis-trans* groups and four *cis-cis-trans-trans* groups. The fine-structure of the collagen fibre shown in the electron microscope

32 L. Pauling and R. B. Corey (Discussion Meeting)

(Schmitt, Hall & Jakus 1942*a*, *b*) suggests that there is no simple repeating unit in collagen.

8. *Conclusion*

Two structures, the α-helix and the pleated sheet, that have been derived on the assumption that the amide group in polypeptides is planar and has interatomic distances and bond angles as found in simpler substances, and that hydrogen bonds are formed between the imino groups and the carbonyl oxygen atoms, have been found to be present in polypeptides and proteins with the α-keratin structure and the β-keratin structure respectively. The dimensions found in polypeptides and proteins substantiate the assumed values of the structural parameters; in particular the N—H···O hydrogen-bond distance between amide groups in proteins may be assumed to be close to 2·75 Å.

In view of the success that has so far been obtained by this method of attack, it seems justified to assume that proposed configurations of polypeptide chains in proteins that deviate largely from the structural principles that have now been formulated (planarity of amide group, correct interatomic distances and bond angles, formation of hydrogen bonds) may be ruled out of consideration.

Note added in proof, 9 January 1953. The problem of the origin of the 5·1 Å meridional reflexion on the X-ray photographs of the α-keratin proteins has now been overcome. It has been pointed out (Pauling & Corey 1953) that the presence of a repeating sequence of amino-acid residues should cause the axis of the α-helix itself to describe a helical course, and that there is evidence from the X-ray diagrams that seven-strand cables, each composed of six α-helices twisted about a central seventh α-helix, are present in hair, horn, muscle, and other proteins of the α-keratin type. (The suggestion that the α-helices present in α-keratin are twisted about one another was independently made by Crick (1952).) In hair, horn, and other α-keratin proteins there seem also to be individual α-helices, occupying the interstices between the seven-strand cables. It has also been suggested that feather rachis keratin consists of seven-strand cables, with the interstices occupied by three-strand ropes, each composed of three α-helices twisted about one another. The twist of the seven-strand cables is in the direction of a right-handed screw, and that of the three-strand ropes in the opposite direction.

The dimensions of the α-helix in synthetic polypeptides quoted in the paper as determined by Dr Harry Yakel differ by about 1 % from his earlier values, which have been found by him to be in error.

References (Pauling & Corey)

Ambrose, E. J. & Hanby, W. E. 1949 *Nature, Lond.*, **163**, 483.
Astbury, W. T. 1940 *J. Int. Soc. Leath. Chem.* **24**, 69.
Astbury, W. T. 1942 *J. Chem. Soc.* p. 337.
Astbury, W. T. 1949 *Nature, Lond.*, **163**, 722.
Astbury, W. T. & Street, A. 1931 *Phil. Trans.* A, **230**, 75.
Astbury, W. T. & Woods, H. J. 1933 *Phil. Trans.* A, **232**, 333.
Bamford, C. H. & Hanby, W. E. 1951 *Nature, Lond.*, **168**, 1085.
Bamford, C. H., Hanby, W. E. & Happey, F. 1951 *Proc. Roy. Soc.* A, **205**, 30.

Bear, R. S. 1951 Talk presented at meeting of American Chemical Society, New York City, September.
Bear, R. S. & Cannan, C. M. N. 1951 *Nature, Lond.*, **168**, 684.
Boyes-Watson, J., Davidson, E. & Perutz, M. F. 1947 *Proc. Roy. Soc.* A, **191**, 83.
Bragg, W. L., Kendrew, J. C. & Perutz, M. F. 1950 *Proc. Roy. Soc.* A, **203**, 321.
Brill, R. 1923 *Liebigs Ann.* **434**, 204.
Cochran, W. & Crick, F. H. C. 1952 *Nature, Lond.*, **169**, 234.
Cochran, W., Crick, F. H. C. & Vand, V. 1952 *Acta Cryst.* **5**, 581.
Corey, R. B. & Pauling, L. 1953 *Rev. Sci. Instrum.* (to be published).
Crick, F. H. C. 1952 *Nature, Lond.*, **170**, 882.
Herzog, R. O. & Jancke, W. 1920 *Ber. dtsch Chem. Ges.* **53**, 2162.
Huggins, M. L. 1936 *J. Org. Chem.* **1**, 407.
Huggins, M. L. 1943 *Chem. Rev.* **32**, 195.
Kendrew, J. C. 1950 *Proc. Roy. Soc.* A, **201**, 62.
Lotmar, W. & Picken, L. E. R. 1942 *Helv. chim. acta*, **25**, 538.
MacArthur, I. 1943 *Nature, Lond.*, **152**, 38.
Meyer, K. H. & Mark, H. 1928 *Ber. dtsch Chem. Ges.* **61**, 1932.
Mirsky, A. E. & Pauling, L. 1936 *Proc. Nat. Acad. Sci., Wash.*, **22**, 439.
Pauling, L. & Corey, R. B. 1950 *J. Amer. Chem. Soc.* **72**, 5349.
Pauling, L. & Corey, R. B. 1951*a* *Proc. Nat. Acad. Sci., Wash.*, **37**, 235.
Pauling, L. & Corey, R. B. 1951*b* *Proc. Nat. Acad. Sci., Wash.*, **37**, 241.
Pauling, L. & Corey, R. B. 1951*c* *Proc. Nat. Acad. Sci., Wash.*, **37**, 251.
Pauling, L. & Corey, R. B. 1951*d* *Proc. Nat. Acad. Sci., Wash.*, **37**, 256.
Pauling, L. & Corey, R. B. 1951*e* *Proc. Nat. Acad. Sci., Wash.*, **37**, 261.
Pauling, L. & Corey, R. B. 1951*f* *Proc. Nat. Acad. Sci., Wash.*, **37**, 272.
Pauling, L. & Corey, R. B. 1951*g* *Proc. Nat. Acad. Sci., Wash.*, **37**, 282.
Pauling, L. & Corey, R. B. 1951*h* *Nature, Lond.*, **168**, 550.
Pauling, L. & Corey, R. B. 1951*i* *Proc. Nat. Acad. Sci., Wash.*, **37**, 729.
Pauling, L. & Corey, R. B. 1952*a* *Proc. Nat. Acad. Sci., Wash.*, **38**, 86.
Pauling, L. & Corey, R. B. 1952*b* *Nature, Lond.*, **169**, 494.
Pauling, L. & Corey, R. B. 1953 *Nature, Lond.*, **171**, 59.
Pauling, L., Corey, R. B. & Branson, H. R. 1951 *Proc. Nat. Acad. Sci., Wash.*, **37**, 205.
Perutz, M. F. 1949 *Proc. Roy. Soc.* A, **195**, 474.
Perutz, M. F. 1951 *Nature, Lond.*, **167**, 1053.
Riley, D. P. & Arndt, U. W. 1952 *Nature, Lond.*, **169**, 138.
Schmitt, F. O., Hall, C. E. & Jakus, M. A. 1942*a* *J. Cell. Comp. Physiol.* **20**, 11.
Schmitt, F. O., Hall, C. E. & Jakus, M. A. 1942*b* *J. Amer. Chem. Soc.* **64**, 1234.
Simanouti, T. & Mizushima, S. 1948 *Bull. Chem. Soc. Japan*, **21**, 1.
Zahn, H. 1947 *Z. Naturf.* **2B**, 104.

(*Reprinted from Nature, Vol.* 171, *p.* 59, *January* 10, 1953)

COMPOUND HELICAL CONFIGURATIONS OF POLYPEPTIDE CHAINS: STRUCTURE OF PROTEINS OF THE α-KERATIN TYPE*

By Prof. LINUS PAULING, For.Mem.R.S., and Prof. ROBERT B. COREY

Gates and Crellin Laboratories of Chemistry, California Institute of Technology, Pasadena 4

LAST year we described several configurations for polypeptide chains, and suggested that one of them, the α-helix, which has about 3·6 amino-acid residues per turn of the helix, is present not only in synthetic polypeptides but also in proteins of the α-keratin type, and in hæmoglobin and other globular proteins[1]. We pointed out that the dimensions of the unit of structure, as indicated by the X-ray photographs, and the distribution of intensities in the equatorial direction are roughly accounted for by the α-helix. In addition, we mentioned that the structure is supported also by the meridional reflexions observed for muscle fibres and porcupine quill, in particular the reflexions with spacings 1·5 A. (the length per residue along the axis of the α-helix) and multiples of this value ; the value of this evidence was emphasized by Perutz[2], who observed the 1·5-A. reflexion for hair, horn, and other α-keratin proteins.

We also pointed out that the X-ray pattern of α-keratin, as described by Astbury and Street[3], has a strong equatorial reflexion at 27 A., which is not accounted for by a structure involving α-helixes in regular parallel orientation, and we suggested that this reflexion is due to a long-range order to be elucidated through further study. Other difficulties for the proposed simple structure of the α-keratin proteins have been emphasized to us by several workers in the field. The most important of these difficulties are a discrepancy between the observed and calculated density, and the failure of the α-helix to explain the 5·2-A. arc on the X-ray photographs as a true meridional reflexion.

We have recently noticed that an α-helix for a polypeptide chain involving repeating sequences of amino-acid residues of different kinds would be expected not to have a straight axis ; instead, the

* Contribution No. 1745.

axis of the helix would itself be predicted to pursue a helical course. A protein might consist of a single polypeptide chain with the configuration of a compound helix; three such helixes might well be expected to twist about one another, to form a three-strand rope, and six might be expected to twist about a seventh, to form a seven-strand cable; still more complex structures may also be formed. There is good reason for believing that hair, horn, quill, and other proteins of the α-keratin type consist of seven-strand protein cables in parallel orientation, with single compound α-helixes occupying interstices between them.

The study of simple substances related to proteins has not only provided reliable values of interatomic distances and bond-angles, but has also indicated that the N—H···O hydrogen-bond distance may be expected to vary by about $\pm$ 0·12 A. about its average value[4] in compounds of this sort, 2·80 A. Variation in the hydrogen-bond distance might be caused directly by the interaction of side-chains of amino-acid residues with the carbonyl and imino groups of the adjacent amide groups, or indirectly by steric hindrance (especially adjacent to a proline or hydroxyproline residue) or by attraction between side chains. Let us consider an α-helix composed of a polypeptide in which a unit of four amino-acid residues of different sorts is continually repeated. Two of the hydrogen bonds might be longer than the other two, by about 0·2 A. This difference in length would cause a curvature of the axis of the α-helix. If the α-helix has 3·6 residues per turn, the normal to the curved helical axis would be rotated by 0·09 revolution by progressing from one unit to the next unit of four residues along the chain, which corresponds to a complete revolution in about eleven units. The axis of the α-helix would itself accordingly describe a larger helix, with pitch approximately 66 A., the axial length of 44 residues. The radius of the larger helix would be about 1·5 A., and the sense of the larger helix would be the same as that of the α-helix. A compound helix of this sort is represented in Figs. 1 and 2.

Another simple case is that of the compound helix with a repeating unit of seven amino-acid residues. An α-helix with 3·60 residues per turn executes 97·2 per cent of two turns in seven residues, and would accordingly be expected to complete a turn of the larger helix in about thirty-five turns of the α-helix, or 126 residues, corresponding to about 190 A. for the pitch of the helix. However, the prediction of the pitch of this compound helix is rather uncertain; decrease by 1·5 per cent of the number of residues per turn would cause the predicted pitch to be doubled, to the value about 400 A. The radius

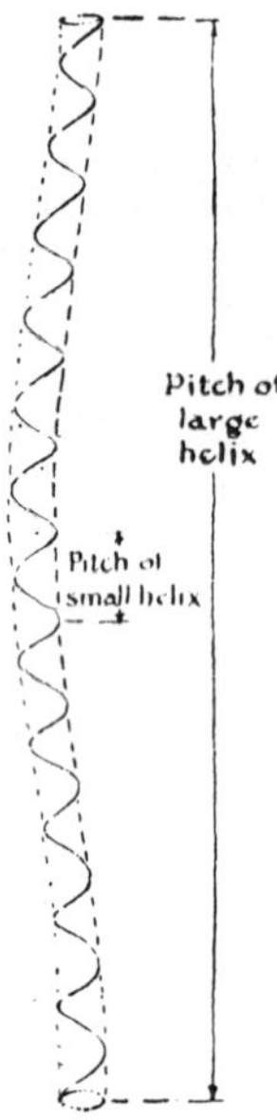

Fig. 1

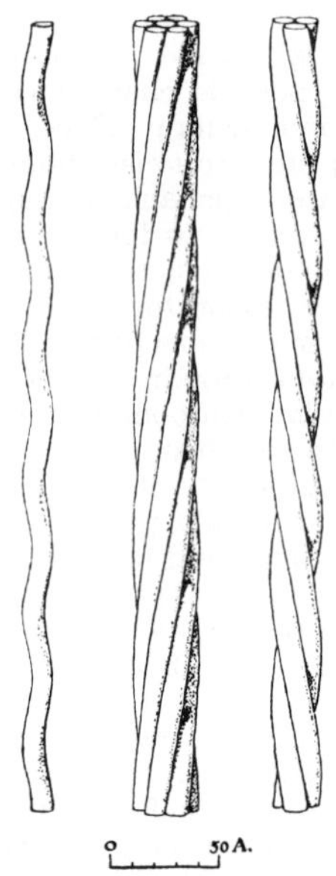

Fig. 2

Fig. 1. A compound helix with pitch of the large superimposed helix equal to 12·5 times the pitch of the small helix. For the compound α-helix the values might be 68 A. and 5·44 A. respectively

Fig. 2. At the left, a compound α-helix with pitch about 67 A. The diameter, shown as about 10 A., includes the volume occupied by side chains as well as the main chains of the protein. Centre, a seven-strand α-cable, with lead of about 400 A. In the proposed structures of proteins of the α-keratin type these cables are packed together, with compound helixes as shown at the left in the interstices. At the right, a three-strand rope, with lead of about 200 A., and with sense opposite to that of the seven-strand cable. In the suggested structure for feather rachis keratin, these ropes are packed together with seven-strand α-cables, in the ratio of two ropes to one α-cable

of the large helix might easily be as great as 10 A., with the variation in hydrogen-bond length mentioned above. The sense of the large helix of the seven-residue compound α-helix is opposite to that of the α-helix itself.

A radius of 6 A. for the large helix would permit three compound helixes to twist about one another, to form a three-strand rope (Fig. 2). Such a rope with the sense of twist of the strands opposite to that of the rope would be formed by the seven-residue compound helix, or with the sense of the strands the same as that of the rope by, for example, the fifteen-residue compound helix (with a repeating unit of fifteen residues, comprising nearly four turns of the α-helix).

Six compound helixes with radius of the large helix equal to 10 A. could twist about a central straight α-helix, to form a seven-strand cable. The repeating unit of seven amino-acid residues, comprising nearly two turns of the α-helix, is the simplest one that would give rise to compound helixes suitable to a seven-strand cable; the next simplest is the repeating unit of fourteen (perhaps containing one proline residue). A drawing of the seven-strand α-cable is shown in Fig. 2. We suggest that the symbol AB_6 be used for this cable.

The seven-strand α-cable is about 30 A. in diameter. A fibre containing these cables in parallel orientation would have a hexagonal unit of structure or pseudo-unit with a equal approximately to 30 A. The observed equatorial reflexion with spacing 27 A. could then be explained as the 10·0 reflexion corresponding to this unit. The X-ray patterns of hair, horn, porcupine quill, and other α-keratin proteins are reasonably well accounted for by such a unit, with $a = 32{\cdot}4$ A. (for porcupine quill a multiple of this unit, indicating further superstructure, is needed to explain weak reflexions with larger spacing).

A plan of the packing of the seven-strand α-cable is shown in Fig. 3. It is seen that there is room enough at the positions $\frac{1}{3}\,\frac{2}{3}$ and $\frac{2}{3}\,\frac{1}{3}$ for other polypeptide chains to be introduced. If the adjacent seven-strand cables are staggered in azimuthal orientation, at a given level, α-helixes may be introduced in the positions $\frac{1}{3}\,\frac{2}{3}$ and $\frac{2}{3}\,\frac{1}{3}$. Because of the rotation of the cables, for which the value of the

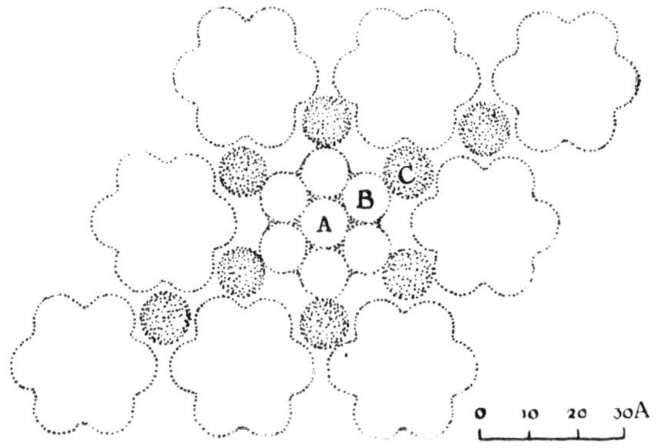

Fig. 3. A cross-section of the α-keratin structure, showing the α-cables AB_6 and the interstitial compound helixes C. The orientation of the cross-section of the cable changes with co-ordinate along the fibre axis. The central cable is shown in the most unfavourable orientation for the interstitial α-helixes. The protein chains are not so nearly circular in cross-section as indicated in the drawing, and space is filled more effectively than is indicated

lead (vertical distance between a point on one strand and an equivalent point on the same strand after one turn about the cable) is about 400 A., and that of the pitch (vertical distance from one strand to an adjacent strand) is about 66 A., the inserted α-helixes must be compound helixes, with radius of the large helix about 1·5 A., and with pitch equal to the pitch of the cable, about 66 A. The sense of these compound helixes must be opposite to that of the cables, in order that they may fit between the strands of an adjacent cable. The four-residue compound helix, which has the proper sense and approximately the correct pitch, is accordingly satisfactory. We suggest the symbol C for this compound helix.

The volume of a portion of the hexagonal unit with $a = 32{\cdot}4$ A., and with height 1·50 A., the axial length per residue, is 136·3 A.^{3}. The proposed structure involves nine amino-acid residues in this portion of the unit. With the reported range 107–118 for the average residue weight, the density of the protein is calculated to be 1·17–1·30 gm.cm.$^{-3}$. The indicated disagreement with the experimental value, about 1·32 gm.cm.$^{-3}$, may be due to the presence of a few per cent of water of hydration in the fibrous proteins. The much larger discrepancy between calculated and observed density that results from assuming the centre of the strong but broad equatorial reflexion, at 9·8 A., to correspond to the spacing of the reflexion 10·0 for the small hexagonal unit containing one α-helix is eliminated by the new indexing, which assigns the strong reflexion with centre at 9·8 A. to the overlapping of 21·0 at 10·6 A., and 33·0 at 9·3 A.

The strong 5·2-A. meridional arc, characteristic of the α-keratin proteins and previously given only rather unsatisfactory explanation in terms of the α-helix, is now explained in a straightforward manner as resulting from the co-operation of the seven residues, in a repeat involving approximately two turns of the α-helix, in the B compound helixes of the AB_6 cable. With 1·50 A. as residue length of the α-helix, the unit of seven residues has length 10·5 A. The component of this distance in the direction of the fibre axis, assuming the cable to have lead 400 A. and radius 10 A., is calculated from the angle of inclination, 9·0°, to be 10·36 A. A weak meridional reflexion would be expected to occur with this spacing; the second order of this reflexion, with spacing 5·18 A., should be strong, because it involves reinforcement rather than extinction by the two turns of the α-helix in the repeating unit of seven residues. The 5·2-A. meridional arc, for which Astbury and MacArthur have recently reported spacings of about 5·15 A. or 5·18 A., is accordingly well explained by the proposed structure.

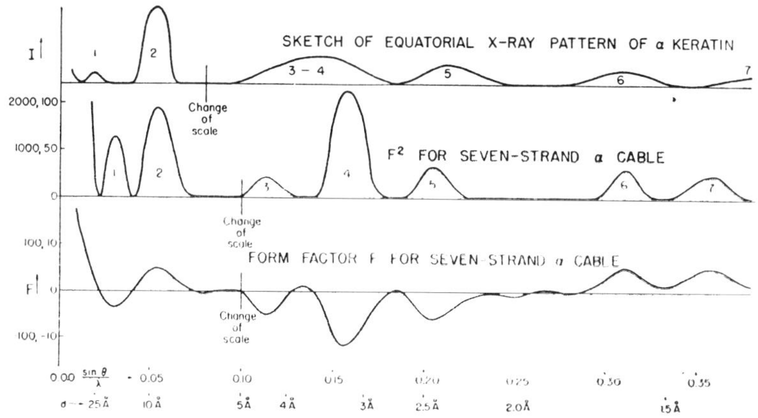

Fig. 4. Comparison of observed and calculated equatorial intensities on X-ray photographs of α-keratin. At the top there is a sketch of the observed intensity of reflexion by hair and horn. The decrease in intensity with increase in angle is more rapid than indicated. At the bottom, there is given the calculated form factor for the AB_6 cable, and in the centre the square of this form factor, which may be compared with the observed intensity curve

Strong support of the proposed structure is also given by many other features of the X-ray patterns of various proteins of the α-keratin type. One comparison of observed and calculated features is given in Fig. 4. At the top of this figure there is a sketch of the equatorial X-ray pattern of α-keratin. Astbury and Street[3] described the α-keratin pattern as having a strong reflexion at 27 A., a very strong and broad reflexion at about 9·8 A., and a vague region of darkening around 3·5 A. Examination of heavily exposed photographs shows that the last of these features is, in fact, a broad maximum, extending from about 5 A. to about 3·2 A. (This band may be due in part to the very strong 4·7-A. reflexion of β-keratin; it is known that specimens of α-keratin usually contain some β-keratin, and some of our photographs show a rather strong 4·7-A. ring.) There is then a minimum at about 2·8 A., a maximum at about 2·4 A., a minimum at about 1·8 A., a maximum at about 1·6 A., a minimum at 1·4 A., and a further maximum. The form factor F for the α-cable shown at the bottom of the figure is calculated by multiplying the form factor for the α-helix, including the β-carbon atom, as reported previously[1], by the factor $1 + 6J_0(4\pi\rho \sin \theta/\lambda)$, in which J_0 is the Bessel function of order 0, and ρ, the radius of the B helixes of the cable, is placed equal to 10 A. Above the curve for F there is shown the curve for F^2. It is seen that there is striking, although not complete, agreement between this curve and the sketch of the equatorial X-ray pattern. Inasmuch as the contribution of the two interstitial compound α-helixes and of the side chains, which together constitute just

50 per cent of the scattering matter in the fibres, have not been taken into consideration, the agreement must be considered to be excellent. It may be pointed out that, in the calculation of the form factor, no arbitrary parameters have been involved except the radius of the large helix; and that the value 10 A. for this quantity is required, to within a few per cent, by the dimensions of the unit of structure, if it is assumed that the three different kinds of polypeptide chains in the α-keratin proteins (the core of the cable, the six surrounding strands, and the interstitial compound helixes) have approximately the same average amino-acid residue weight.

The X-ray diffraction photographs of feather rachis keratin, which have previously been interpreted as involving β-keratin pleated sheets and α-keratin helixes in molecular distribution, may show a superimposed β-keratin pattern and α-keratin pattern. If this is correct, the α-keratin is different in nature from hair and horn α-keratin, probably consisting of AB_6 cables with D_3 ropes (three α-helixes coiled about one another) in the interstices. This structure accounts in a striking way for the characteristic features of the X-ray photographs.

The proposed structure of α-keratin involves three kinds of α-helixes, which presumably differ in amino-acid composition. We propose to examine these materials by fractionation. It is interesting that Goddard and Michælis have reported that wool put into solution by reduction with thioglycollic acid or other reducing agent and treated with iodoacetic acid to form a carboxymethyl derivative can be converted into fractions with different chemical composition by ammonium sulphate precipitation[5]. We suggest that a solution of wool might be fractionated into the AB_6 cables and the protein keratin-C, and that the AB_6 cables might be further separated into keratin-A and keratin-B.

We think it not unlikely that actomyosin has the structure shown in Fig. 3, and that its fractionation into myosin and actin is a separation of the seven-strand cables AB_6 (myosin) from the single α-helixes C (actin). The amounts of myosin and actin obtained by the fractionation are approximately in the predicted ratio 7 : 2, and, as predicted, the 5·2-A. meridional X-ray reflexion is observed for actomyosin and myosin but not for actin.

We are indebted to Prof. W. T. Astbury, Dr. I. MacArthur, Mr. F. H. C. Crick, and other workers in the field of the structure of proteins for having discussed with us the question of the structure of α-keratin, and especially for having emphasized the necessity for refinement of our suggestions. The detailed description of compound α-helixes and aggregates of them, and discussion of fibrous proteins which they compose, will be presented in later papers, to be published in the *Proceedings of the National Academy of Sciences* of the United States of America. This investigation was supported in part by a research grant from the National Institutes of Health, Public Health Service, and by a contract (Nonr-220(05)) with the Office of Naval Research. [Oct. 14.

[1] Pauling, L., and Corey, R. B., *J. Amer. Chem. Soc.*, **72**, 5349 (1950). Pauling, L. Corey, R. B., and Branson, H. R., *Proc. U.S. Nat. Acad. Sci.*, **37**, 205 (1951). Pauling, L., and Corey, R. B., *ibid.*, **37**, 235, 241, 251, 256, 261, 272, 282, 729 (1951); *ibid.*, **38**, 86 (1952); *Nature*, **168**, 550 (1951); **169**, 494, 920 (1952).

[2] Perutz, M. F., *Nature*, **167**, 1053 (1951).

[3] Astbury, W. T., and Street, A., *Phil. Trans. Roy. Soc.*, A, **230**, 75 (1951).

[4] Donohue, J., *J. Phys. Chem.*, **56**, 502 (1952). Corey, R. B., and Pauling, L., *Proc. Roy. Soc.* (in the press).

[5] Goddard, D. R., and Michaelis, L., *J. Biol. Chem.*, **106**, 605 (1934); **112**, 361 (1935).

Printed in Great Britain by Fisher, Knight & Co., Ltd., St. Albans.

Reprinted from the Proceedings of the NATIONAL ACADEMY OF SCIENCES,
Vol. 39, No. 4, pp. 253–256. April, 1953

TWO RIPPLED-SHEET CONFIGURATIONS OF POLYPEPTIDE CHAINS, AND A NOTE ABOUT THE PLEATED SHEETS

BY LINUS PAULING AND ROBERT B. COREY

GATES AND CRELLIN LABORATORIES OF CHEMISTRY,* CALIFORNIA INSTITUTE OF TECHNOLOGY

Communicated January 30, 1953

About a year ago, in the course of the consideration of configurations of polypeptide chains with favored orientations around single bonds, we described two pleated-sheet structures.[1] These structures are suited to polypeptide chains constructed entirely of L amino-acid residues or of D amino-acid residues. In one pleated sheet alternate polypeptide chains are antiparallel, and in the other they are parallel. The amide groups have the trans configuration.

We have observed that closely related structures can be constructed in which polypeptide chains of D and L amino-acid residues alternate. The configuration of these layer structures is such as to make it appropriate to call them rippled sheets.

The Antiparallel-Chain Rippled Sheet.—The antiparallel-chain rippled sheet, represented in figure 1, is closely similar to the antiparallel-chain

TABLE 1

ATOMIC COORDINATES FOR THE ANTIPARALLEL-CHAIN RIPPLED SHEET

a_0 = 9.44 A, b_0 = 7.00 A, c_0 = 1.00 A (assumed). Four atoms in x, y, z; $^1/_2 - x, ^1/_2 + y, \bar{z}$; $\bar{x}, \bar{y}, \bar{z}$; $^1/_2 + x, ^1/_2 - y, z$

	x	y	z
C	0.287	0.040	0.64
C′	0.200	0.216	0.28
O	0.068	0.220	0.45
N	0.275	0.359	−0.22

pleated sheet, and the diagrammatic representation given in figure 4 of the earlier paper[1] applies to both structures. One structure is converted into the other by the reflection of alternate chains into their enantiomers, in the plane of the sheet. The unit of structure was determined by measurement of a model constructed of units precisely built on the scale 10 cm. = 1A. The value of the lateral translation was found to be 9.44 A, and that of the translation in the direction of the polypeptide chains 7.00 A. Atomic coordinates are given in table 1.

The Parallel-Chain Rippled Sheet.—The parallel-chain rippled sheet, shown in figure 2, is closely similar to the parallel-chain pleated sheet, and has the same diagrammatic representation, shown in figure 5 of the previous paper.[1] The unit of structure was found by measurement of a model to have lateral identity distance a_0 = 9.60 A, and identity distance along the fiber axis b_0 = 6.50 A. Atomic coordinates are given in table 2.

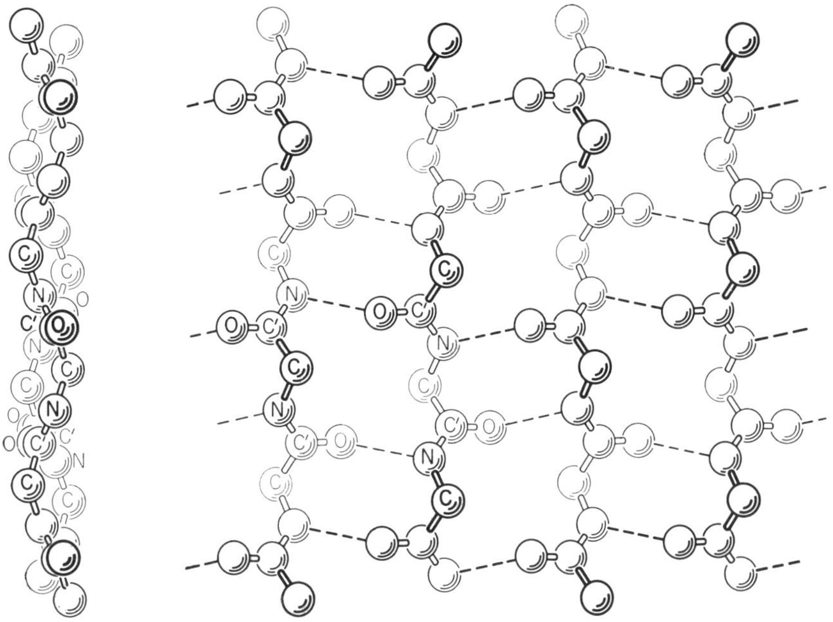

FIGURE 1

The antiparallel-chain rippled sheet of hydrogen-bonded polypeptide chains.

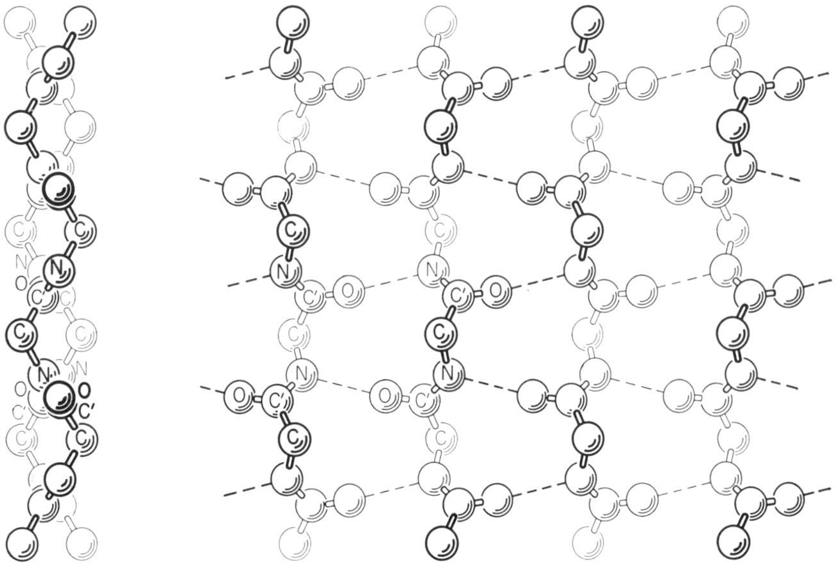

FIGURE 2

The parallel-chain rippled sheet of hydrogen-bonded polypeptide chains.

The antiparallel-chain rippled sheet and the parallel-chain rippled sheet are satisfactory structures, in that they involve linear hydrogen bonds, the interatomic distances and bond angles have the accepted values, and the amide groups are planar. The orientations of the bonds around the α carbon atoms are such that there is room for a side chain, projecting out nearly perpendicularly to the sheet, if alternate chains are constructed of D amino-acid residues and L amino-acid residues. These structures are accordingly satisfactory ones for equimolal mixtures of D polypeptides and L polypeptides.

Moreover, mixtures of D polypeptides and L polypeptides in which the enantiomeric polypeptide chains are present in unequal numbers can assume sheet structures which represent a mixture of a pleated-sheet configuration and a rippled-sheet configuration. Irregular sequences of polypeptide chains with positive and negative orientations can also lead to reasonably satisfactory hydrogen-bonded layer structures. The value of the identity distance in the fiber-axis direction would be intermediate between the value

TABLE 2

ATOMIC COORDINATES FOR THE PARALLEL-CHAIN RIPPLED SHEET

a_0 = 9.60 A, b_0 = 6.50 A, c_0 = 1.00 A (assumed). Four atoms in x, y, z; $\bar{x}, {}^1/_2 + y, \bar{z}$; ${}^1/_2 + x, y, \bar{z}$; ${}^1/_2 - x, {}^1/_2 + y, z$

	x	y	z
C	−0.006	0.000	0.98
C′	0.059	0.185	0.28
O	0.188	0.205	0.25
N	−0.027	0.314	−0.26

7.00 A corresponding to linear hydrogen bonds for the antiparallel-chain sheets and the value 6.50 A corresponding to linear hydrogen bonds for the parallel-chain sheets. An intermediate value of this identity distance would require that all hydrogen bonds be somewhat bent, and presumably also somewhat strained.

A Note on the Pleated Sheets.—In the earlier discussion of the pleated-sheet configurations of the polypeptide chains[1] the assumption was made that certain orientations around the bonds between the α carbon atom and adjacent atoms in the amide group are favored over other orientations. The assumed favored orientations around these single bonds led to the predicted value 6.68 A for the identity distance in the fiber-axis direction (the direction of the polypeptide chains). Although the existence of a potential function causing certain orientations around these bonds to be favored is likely from *a priori* considerations, the magnitude of the effect is uncertain, and it may well be that very little strain (less than 0.1 kcal. mole^{-1} per residue) is involved in rotating the chain from favored to less favored configurations.

We have now found through the construction of large-scale models that other values of the identity distance in the direction of the polypeptide chains are indicated for the two pleated sheets, corresponding to slightly different orientations about the single bonds to the α carbon atom. The new configurations, described by the coordinates in tables 3 and 4, involve linear hydrogen bonds—that is, the angle N—H · · · O is equal to 180°. The fiber-axis identity distance for the antiparallel-chain pleated sheet is 7.00 A, and that for the parallel-chain pleated sheet is 6.50 A; these values are the same as for the corresponding rippled sheets.

The two pleated sheets provide satisfactory structures for proteins and for polypeptides composed exclusively of L amino-acid residues (or D amino-acid residues). We think that it is likely that silk fibroin, for which the

TABLE 3

ATOMIC COORDINATES FOR THE ANTIPARALLEL-CHAIN PLEATED SHEET

a_0 = 9.50 A, b_0 = 7.00 A, c_0 = 1.00 A (assumed). Four atoms in x, y, z; $\bar{x}, {}^1/_2 + y, \bar{z}$; ${}^1/_2 - x, \bar{y}, z$; ${}^1/_2 + x, {}^1/_2 - y, \bar{z}$

	x	y	z
C	0.034	−0.005	−0.70
N	−0.030	0.173	−0.20
C′	0.051	0.320	0.21
O	0.180	0.326	0.22
βC	0.024	−0.005	−2.24

TABLE 4

ATOMIC COORDINATES FOR THE PARALLEL-CHAIN PLEATED SHEET

a_0 = 4.85 A, b_0 = 6.50 A, c_0 = 1.00 A (assumed). Two atoms in x, y, z; $\bar{x}, {}^1/_2 + y\ \bar{z}$,

	x	y	z
C	0.012	0.000	−0.98
N	−0.066	0.186	−0.26
C′	0.118	0.315	0.28
O	0.371	0.295	0.25
βC	−0.093	0.014	−2.45

observed fiber-axis identity distance is 7.0 A, has the antiparallel-chain pleated-sheet structure, and that the β-keratin proteins, for which the observed fiber-axis identity distance is about 6.6 A, have the parallel-chain pleated-sheet structure, or a structure in which a considerable number of adjacent polypeptide chains have parallel orientations.

The atomic coordinates given in tables 3 and 4 differ only slightly from those previously reported.[1]

Acknowledgments.—This investigation was aided by grants from The Rockefeller Foundation and The National Institutes of Health, Public Health Service.

* Contribution No. 1771.

[1] Pauling L., and Corey, R. B., these PROCEEDINGS, **37**, 729 (1951).

Reprinted from *Acta Crystallographica*, Vol. 8, Part 11, November 1955

PRINTED IN DENMARK

Acta Cryst. (1955). **8**, 710

The Structure of Tussah Silk Fibroin*

(with a note on the structure of β-poly-L-alanine)

BY RICHARD E. MARSH, ROBERT B. COREY AND LINUS PAULING

Gates and Crellin Laboratories of Chemistry, California Institute of Technology, Pasadena, California, U.S.A

(*Received* 11 *March* 1955)

A detailed structure for Tussah silk fibroin has been derived which is in agreement with the X-ray diffraction data. The structure is similar to that of *Bombyx mori* fibroin in that it is based on anti-parallel-chain pleated sheets; the method of packing of the sheets, however, is quite different. This difference in packing can be explained on the basis of the chemical compositions of the two silks.

It seems highly probable that the structure of the β (stretched) form of poly-L-alanine is essentially that derived for Tussah silk.

Introduction

A detailed structure for commercial silk fibroin (*Bombyx mori*) has recently been formulated in these Laboratories (Marsh, Corey & Pauling, 1955). A prominent feature of the structure of *Bombyx mori* silk fibroin is the occurrence of glycine as alternate residues along the polypeptide chains.

Another form of silk fibroin is that derived from Tussah silk (commonly called wild silk). Previous investigators (Kratky & Kuriyama, 1931; Trogus & Hess, 1933) have shown that the X-ray diffraction pattern of Tussah silk fibroin, although having many features in common with the pattern obtained from *Bombyx mori*, is significantly different in several respects. Its chemical composition also differs from that of *Bombyx mori* in a very significant way (Table 1). The most striking differences are in the relative amounts of glycine and alanine. In particular, the amount of glycine in Tussah silk (26·6 residue %) is insufficient to permit the occurrence of glycine a alternate residues along the polypeptide chains. I

* Contribution No. 1978 from the Gates and Crellin Laboratories of Chemistry. The work described in this article was carried out under a contract (Nonr-220(05)) between the Office of Naval Research and the California Institute of Technology.

Table 1. *Composition of the fibroins of* Bombyx mor *and Tussah silks**

Amino-acid residue	*Bombyx mori* (residue %)	Tussah silk (residue %)
Glycine	44·4	26·6
Alanine	30·2	44·2
Serine	11·9	11·8
Tyrosine	4·9	4·9
Aspartic acid	1·4	4·7
Arginine	0·4	2·6
Valine	2·1	0·6
Glutamic acid	0·9	0·8
Tryptophan	0·2	1·1
Phenylalanine	0·6	0·5
Isoleucine	0·5	0·4
Leucine	0·5	0·4
Histidine	0·2	0·8
Proline	0·4	0·3
Threonine	1·0	0·1
Lysine	0·3	0·1
Cystine	0·1	—
Mean residue weight	78·3	83·5

* Calculated from the data of Schroeder & Kay (1955).

contrast to the smaller percentage of glycine, the percentage of alanine is much larger in Tussah silk than in *Bombyx mori*. It seems reasonable to expect that the structural differences between the two silks, as evidenced by the differences in their X-ray patterns, are related in a simple way to these striking differences in their chemical compositions.

In the current investigation, a detailed structure for Tussah silk fibroin has been derived which is compatible with both chemical and X-ray data. This structure also has strong implications concerning the structure of the β (stretched) form of poly-L-alanine.

Experimental

Samples of Tussah silk (*Antherea pernyi*) were kindly supplied by Prof. S. Mizushima of the University of Tokyo. For X-ray photography, degummed fibers (Schroeder & Kay, 1955) were arranged in the form of a bundle about 0·5 mm. in diameter. Diffraction photographs were taken in an evacuated 3-cm.-radius camera and in a helium-filled 10-cm.-radius camera with nickel-filtered copper $K\alpha$ radiation ($\lambda = 1{\cdot}5418$ Å).

Standard fiber photographs were taken with the X-ray beam perpendicular to the axis of the fibers; a typical photograph is reproduced in Fig. 1, together

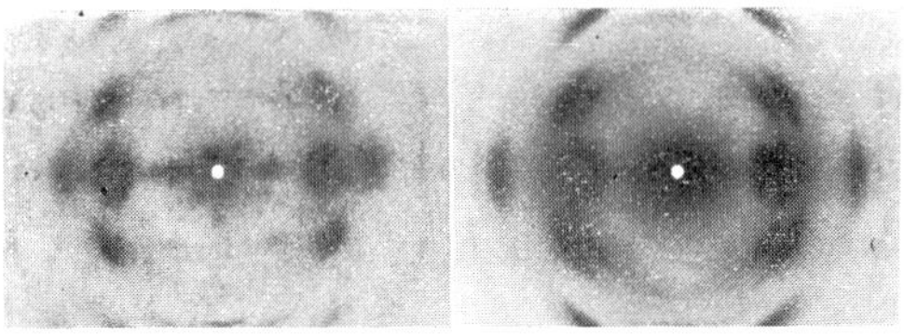

Fig. 1. X-ray diffraction photographs of *Bombyx mori* (left) and Tussah (right) silk fibroins, prepared with Cu $K\alpha$ radiation in a cylindrical camera. The fiber axes are vertical.

with a photograph of *Bombyx mori* fibroin. In addition, 5° oscillation photographs were prepared in which the axis of oscillation was perpendicular to the fiber axis; at the center of oscillation, the X-ray beam made an angle of 48° with the fiber axis. By this means the strong sixth-order meridional reflection at 1·16 Å was brought into reflecting position. For calibration purposes, the powder spectrum of sodium fluoride was superimposed on this pattern; the lattice parameter of the NaF sample was obtained separately from measurements of Straumanis-type powder photographs.

It was found that, within the limits of visual observation, the 400 reflection of NaF was exactly coincident with the center of the sixth-order meridional reflection of Tussah silk. The spacing of the 400 reflection of NaF, as measured on the Straumanis photographs, was 1·158 Å. In confirmation of this spacing, the lattice parameter a_0 was obtained from a least-squares treatment of all of the observed powder lines; it was found to be 4·6327 Å with a probable error of 0·0003 Å, in satisfactory agreement with the value 4·620±0·004 kX. (= 4·629±0·004 Å) obtained by Davey (1923). Accordingly, the fiber-axis identity distance of Tussah silk is 6·949 Å, with an estimated limit of error of about 0·010 Å.

The value 6·95 Å for the fiber-axis identity distance is slightly smaller than that found for *Bombyx mori*—6·97±0·03 Å (Marsh *et al.*, 1955). As a confirmation of this shortening, a corresponding oscillation photograph of *Bombyx mori* was prepared with the powder pattern of NaF superimposed. On this photograph, the sixth-order meridional reflection occurred at a significantly larger spacing than the 400 reflection from NaF; the fiber-axis identity distance for *Bombyx mori* was calculated to be 6·988±0·015 Å. Thus there is a real, though small, difference in the fiber-axis identity distances of the two types of silk fibroin. This difference was first reported by Bamford, Brown, Elliott, Hanby & Trotter (1953); the identity distances given by them are 6·92 Å and 6·94 Å for Tussah and *Bombyx mori*, respectively.

The measured spacings and visually-estimated intensities of the equatorial reflections from Tussah silk are listed in Table 2; these values are in good agreement with those reported by Trogus & Hess (1933). For comparison, the spacing and intensity data for *Bombyx mori* are also listed in Table 2. Intensity and spacing data for non-equatorial reflections are given in Table 5, together with the intensities calculated on the basis of the proposed structure.

Derivation of the structure

A comparison of data in Table 2 shows that the reflections which, in the case of *Bombyx mori*, have been

Table 2. *Spacings and relative intensities for equatorial reflections for* Bombyx mori *and Tussah silk fibroins*

Tussah					*Bombyx mori*			
No.	d_o (Å)	I	$h0l$	d_c (Å)	No.	d_o (Å)	I	$h0l$
					1	9·70±0·25	90	001
1	5·35±0·10	*vvvs*	002	5·30	2	4·70±0·20	450	002
2	4·35±0·08	*vvs*	201	4·31	3	4·25±0·15	900	201, 20$\bar{1}$
3	2·64±0·05	*s*	004	2·65	4	3·05±0·02	180	003
4	2·36±0·04	*mw*	400	2·36	5	2·35±0·01	20	400
					6	2·10±0·05	*vw*	402, 40$\bar{2}$
					7	1·80±0·05	*vvw*	005
5	1·57±0·02	*w*	601	1·56	8	1·56±0·01	18	601, 60$\bar{1}$
					9	1·20±0·05	*vw*	800

indexed as $h00$ or $h01$ occur with very similar spacings and intensities in the two types of silk; reflections of the type $00l$, however, do not compare. The indexing of the *Bombyx mori* pattern was accomplished unambiguously from data obtained from doubly-oriented specimens, and it has been shown (Kratky & Kuriyama, 1931; Marsh *et al.*, 1955) that the $h00$ reflections arise from sets of diffraction planes oriented perpendicular to the plane of rolling of the specimens. The plane of rolling has been identified with the plane of the antiparallel-chain pleated sheets which are the basic structural feature of *Bombyx mori*, and the a axis lies within the plane of the sheets. On the other hand, the $00l$ reflections arise from sets of diffraction planes oriented perpendicular to the plane of rolling of the doubly-oriented specimens, and the c axis is an identity distance perpendicular to the pleated sheets.

Thus, the two identity distances which, in the case of *Bombyx mori*, have been shown to lie within the plane of the pleated sheets—the b-axis (fiber axis) identity distance, which represents the distance between alternate residues along the polypeptide chains, and the a-axis identity distance, which represents the distance between alternate polypeptide chains in the hydrogen-bonded sheet—appear to have close counterparts in Tussah silk. On the other hand, the methods of packing of the sheets, as represented by the $00l$ reflections, are apparently different in the two silks.

It should be pointed out that in both silk fibroins the $00l$ reflections are quite diffuse, whereas the $h00$ and $h01$ reflections are relatively sharp. This is readily understandable in terms of a pleated-sheet structure: the strong hydrogen bonding within the sheets would be expected to lead to well-defined repeat distances and, hence, sharp reflections in the direction of the a axis, whereas any disorder in the packing of adjacent sheets would result in diffuse reflections in the direction of the c axis.

With the aid of the *Bombyx mori* pattern, we have been able to index the diffraction pattern of Tussah silk on the basis of an orthogonal unit cell with

$$a_0 = 9{\cdot}44,\quad b_0\ (\text{fiber axis}) = 6{\cdot}95,\quad c_0 = 10{\cdot}60\ \text{Å};$$

the corresponding values for the pseudo unit of *Bombyx mori* are 9·40, 6·97 and 9·20 Å, respectively. The indexes and calculated spacings of the equatorial reflections listed in Table 2 are based on this unit cell. This cell will account for all of the reflections observed for Tussah silk, both equatorial and non-equatorial, as is shown subsequently in Table 5. Accordingly, we have chosen it as a unit of structure, bearing in mind that, as was the case for *Bombyx mori*, it is too small to accommodate the larger amino-acid residues known to be present in the protein, and hence must be regarded only as a pseudo unit cell.

Prof. R. Brill has called to our attention that the dimensions of this orthorhombic unit cell of Tussah silk are essentially equivalent to those proposed by him (Brill, 1943) for a monoclinic unit cell for Satonia-type silk.

In the formulation of positional parameters for al of the atoms within this pseudo unit cell, we have assumed that the basic structural component is the antiparallel-chain pleated sheet (Pauling & Corey 1953). The reasons for this choice have been discussed in connection with *Bombyx mori*; they are founded principally on the confidence we place in our knowledge of the geometry of polypeptide chains and hydrogen bonds. In particular, the values for the a- and b-axis identity distances—9·44 and 6·95 Å—are almost exactly those calculated for the antiparallel-chain pleated sheet—9·5 and 7·0 Å (Pauling & Corey, 1953).

The formulation of the detailed structure of Tussah silk was arrived at by a consideration of the ways in which adjacent pleated sheets might pack together. The absence of odd orders of reflections of the type $00l$ and the relatively high intensities of the 002 and 004 reflections indicate that the pleated sheets are spaced at approximately equal intervals of 5·3 Å along the c axis; the relative position of adjacent pleated sheets was determined by packing considerations.

The pseudo unit of structure of Tussah silk is shown diagrammatically in Fig. 2; the positional parameters

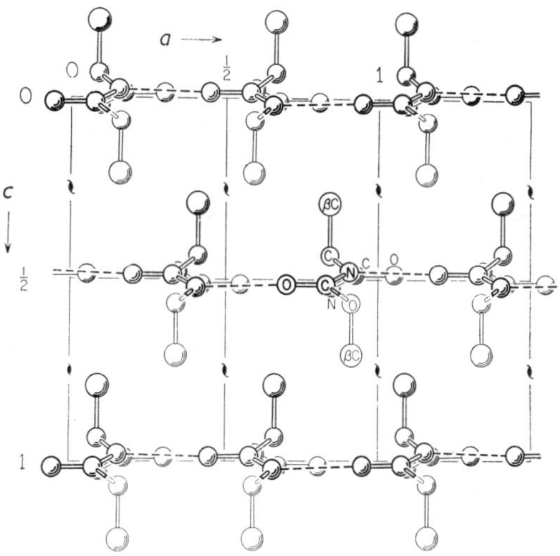

Fig. 2. A representation of the pseudo unit of structure of Tussah silk, viewed along the fiber axis.

for all of the main-chain atoms and for the β-carbon atoms of the side chains are listed in Table 3. The parameters are consistent with a planar peptide group and linear hydrogen bonds; they lead to the bond distances and bond angles listed in Table 4, where the accepted values (Corey & Pauling, 1953) are also listed for comparison. The shortest distance between non-bonded atoms in neighboring sheets is 3·9 Å.

The symmetry of the pseudo structure is that of the space group D_2^4-$P2_12_12_1$; there are eight amino-acid

Table 3. *Atomic parameters for the pseudo unit of structure of Tussah silk*

Space group: $P2_12_12_1$.
Equivalent positions: (x, y, z); $(\frac{1}{2}-x, \bar{y}, \frac{1}{2}+z)$; $(\bar{x}, \frac{1}{2}+y, \frac{1}{2}-z)$; $(\frac{1}{2}+x, \frac{1}{2}-y, \bar{z})$.
$a_0 = 9{\cdot}44$, $b_0 = 6{\cdot}95$, $c_0 = 10{\cdot}60$ Å.

	Residue I					Residue II				
Atom	C′	O	N	C	β-C	C′	O	N	C	β-C
x	0·328	0·197	0·404	0·340	0·340	0·422	0·553	0·346	0·410	0·410
y	0·676	0·676	0·824	0·002	0·002	0·176	0·176	0·324	0·502	0·502
z	0·514	0·514	0·478	0·434	0·291	0·486	0·486	0·522	0·566	0·709

Table 4. *Interatomic distances and bond angles for the pseudo structure*

Distances	Calculated from parameters in Table 3	Accepted values*	Angle	Calculated from parameters in Table 3	Accepted values*
C′=O	1·24 Å	1·24 Å	O=C′−N	123°	123°
C′−N	1·31	1·32	O=C′−C	120	121
N−C	1·45	1·47	C−C′−N	116	114
C−C′	1·54	1·53	C′−N−C	122	123
C−βC	1·52	1·54	C′−N···O	122	123
O···N	2·77	2·79	C−N···O	115	114
			N−C−C′	110	110
			N−C−βC	109	109½
			C′−C−βC	111	109½

* Corey & Pauling, 1953.

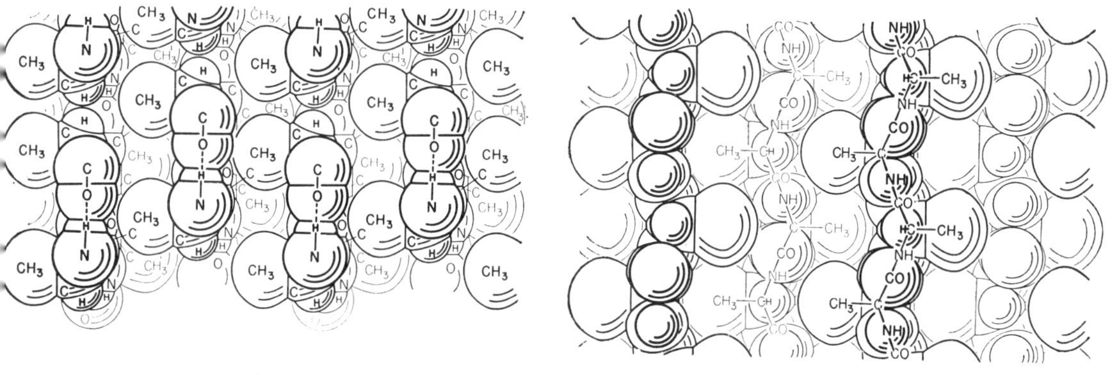

(a) (b)

Fig. 3. A packing drawing of the pseudo structure of Tussah silk (*a*) viewed along the fiber axis, (*b*) viewed perpendicular to the fiber axis and parallel to the plane of the pleated sheets.

esidues within the pseudo unit cell, and hence two I and II in Table 3) within each asymmetric unit. They are successive residues of a single polypeptide hain. Thus, the space group differs from that derived or the pseudo unit cell of *Bombyx mori*—$P2_1$. The difference is due to the method of packing of adjacent pleated sheets; in *Bombyx mori* adjacent sheets are eparated by distances alternately 3·5 and 5·7 Å, whereas in Tussah silk the sheets are spaced at equal ntervals (5·3 Å).

The parameters of Table 3 were used in the calcula-ion of structure factors for the equatorial reflections nd those occurring on the first three layer lines. Atomic form factors of McWeeny (1951) were used; no temperature factor was applied. The resulting values of F^2 are listed in Table 5. The agreement with the observed intensities is seen to be quite satisfactory; it could be improved by the application of an anisotropic temperature factor to take account of the apparent disorder in the c direction which is manifested by the diffuseness of the $00l$ reflections. It should be pointed out that the calculations included one β-carbon atom for each residue of the structure, whereas the results of chemical analysis of Tussah silk (Schroeder & Kay, 1955) indicate the presence of about 27% glycine, and hence approximately one-quarter of the β-carbon atoms are in fact replaced by hydrogen atoms.

Discussion of the structure

Packing drawings of the pseudo structure of Tussah silk fibroin are shown in Fig. 3. Within each pleated

Table 5. *Observed intensities for Tussah silk compared with values of F^2 calculated for the pseudo unit of structure*

		$k = 0$		$k = 1$		$k = 2$		$k = 3$	
$h\ l$	d_{h0l}	I_o	F_c^2	I_o	F_c^2	I_o	F_c^2	I_o	F_c^2
0 0	—	—	—	—	0		5	—	0
0 1	10·60 Å	—	0	—	0	*ms*	0		0
1 0	9·44	—	0	—	15		193	*m*	132
1 1	7·05	—	1	—	16	—	10		106
0 2	5·30	*vvvs*	1248	*vs*	239	*s*	390	—	7
2 0	4·72	—	0	—	0	—	0	*ms*	0
1 2	4·62	—	0	—	33	—	10		69
2 1	4·31	*vvs*	444	*vs*	260	—	12	—	45
0 3	3·53	—	0	—	0	—	0	—	0
2 2	3·53	—	0	—	0	—	0	—	0
1 3	3·31	—	8	—	28	—	6		55
3 0	3·15	—	0	—	18	—	0		10
3 1	3·02	—	18	—	3	*w*	79	**mw*	48
2 3	2·83	—	24	*w*	89	—	36		3
3 2	2·71	—	7	—	14	*vvw*	71	—	12
0 4	2·65	*s*	900	—	37		120	—	57
1 4	2·55	—	13	—	13	—	7		50
4 0	2·36	*mw*	233	*mw*	87	*w*	16	**vw*	1
3 3	2·35		1		19		47		7
2 4	2·31	—	0	—	0	—	0		0
4 1	2·30	—	0	—	0	—	0	—	0
4 2	2·16	—	26	*w*	102	—	17	—	8
0 5	2·12	—	0		0	—	0	—	0
1 5	2·07	—	8	—	9	—	12	—	38
3 4	2·03	—	3	—	13	—	34	—	15
4 3	1·96	—	0	*vvw*	0	—	0		0
2 5	1·93	—	47		100	—	4	*vvw*	4
5 0	1·89	—	0	—	27	—	37		72
5 1	1·86	—	18	—	5	—	22	—	35
5 2	1·78	—	9		4		20	—	4
0 6	1·77	—	161		182		164	—	7
3 5	1·76	—	0		13	*vvw*	34	—	12
4 4	1·76	—	33	*vvw*	44		8	—	4
1 6	1·74	—	7	—	11		13	—	19
5 3	1·67	—	0	—	6	—	4	—	2
2 6	1·65	—	0	—	0	—	0	—	0
6 0	1·57	*w*	0	—	0	*vvw*	0	*w*	0
6 1	1·56		103	—	8		13		1

* These two reflections appear to be double.

The brackets around the F_c^2 values include all reflections having scattering angles within the range imposed by the uncertainties of the measurements of the observed maxima.

sheet, adjacent polypeptide chains are held together in an antiparallel sense by linear hydrogen bonds (Pauling & Corey, 1953). Adjacent pleated sheets are separated by a distance of 5·3 Å ($= \frac{1}{2}c_0$), the space between sheets being occupied by the side-chain atoms of the various amino-acid residues. In Fig. 3, all of the side-chains are represented as equivalent; their size as shown is approximately that expected for the methyl groups of alanine residues. The side chains of adjacent pleated sheets interlock in a highly efficient manner, a side chain of one sheet being surrounded by four side chains of the next sheet. Furthermore, as can be seen in Fig. 3(*b*), each side chain falls in the concavity created by the 'pleating' of the adjacent sheet.

We can now see how the differences in the structures of *Bombyx mori* and Tussah silk fibroins are related to the striking differences in their chemical compositions. The principal difference between the pseudo structures of the two silks is in the method of packing of the pleated sheets. The predominant feature of the structure of *Bombyx mori* is the alternation of intersheet distances between the values 3·5 and 5·7 Å; these distances correspond to back-to-back and front-to-front packing, respectively, between pleated sheets having the methyl-group side chains of alanine (or serine) residues protruding from the back side. Such pleated sheets can be formed from the polypeptide chains of *Bombyx mori* in which the residues are alternately glycine and alanine (or serine); the sequence *-G-X-G-X-G-* (*G* = glycine, *X* = alanine or serine) occurs frequently in this silk. In Tussah silk, however, adjacent pleated sheets are spaced regularly at a distance of 5·3 Å; thus, both sides of the pleated sheets appear to be structurally equivalent and the X-ray data give no information concerning the sequence of residues.

These structural differences can be readily explained in light of the chemical compositions of the two silk ibroins (see Table 1). In *Bombyx mori*, glycine accounts or about 44% of the amino-acid residues, and alanine nd serine together make up an additional 42%. Thus, sequence of the type *–G–X–G–X–G–* will account for bout 85% of the residues. In Tussah silk, however, he amount of glycine is quite insufficient to allow uch a sequence to predominate in the structure. Accordingly, the pleated sheets in Tussah silk cannot rrange themselves in the manner found in *Bombyx mori*; instead, they adopt another simple structure which is particularly appropriate in view of the high lanine content.

It should be emphasized that, as in the case of *Bombyx mori*, the structure we have derived for Tussah silk must be regarded as only a pseudo structure. On he basis of the proposed unit cell, containing eight mino-acid residues, and an assumed density of about 1·35 g.cm.$^{-3}$, the average residue weight is calculated o be about 71. This value is exactly the weight of n alanine residue, but is considerably smaller than he value 83·5 calculated for Tussah silk from the data of Schroeder & Kay (1955). Furthermore, the distance between adjacent pleated sheets, 5·3 Å, is insufficient o accommodate the side chains of the larger amino-acid residues, such as tyrosine. There seems to be no valid reason for presuming that the crystalline portion of Tussah silk contains only the smaller amino-acid esidues glycine, alanine and serine; rather, it is probable that there are regions in the structure where adjacent pleated sheets are separated by distances greater than 5·3 Å.

A note on the structure of β-poly-L-alanine

Bamford, Brown, Elliott, Hanby & Trotter (1954) have reported that the X-ray diffraction pattern of the β (stretched) form of poly-L-alanine is almost identical o that of Tussah silk; they have proposed a unit cell for polyalanine which, except for a halving of the *a* axis, has dimensions in good agreement with our pseudo unit cell of Tussah silk. In view of the large amount of alanine in Tussah silk and the efficiency of packing of alanine residues in the pseudo structure, it seems likely that the structure of β-poly-L-alanine is very closely related to the pseudo structure which we have derived for Tussah silk fibroin. As pointed out previously in this paper, the average residue weight calculated for the pseudo structure of Tussah silk, assuming a density of 1·35 g.cm.$^{-3}$, is 71, the weight of an alanine residue.

Bamford *et al.* report two significant differences between the diffraction patterns of polyalanine and Tussah silk. First, they find (Bamford *et al.*, 1953) that the fiber-axis identity distance in poly-L-alanine is slightly shorter than in Tussah silk fibroin; this shortening, which may connote a slightly greater twist in the polypeptide chains, is too small to require significant revision of the atomic positional parameters listed in Table 3. Second, they report (Bamford *et al.*, 1954) that the intensity of the 5·3 Å equatorial reflection is distinctly lower in Tussah silk than in poly-L-alanine. This reflection, which has been indexed as 002 on the basis of our pseudo unit cell, is the first order of the spacing between adjacent pleated sheets. Thus, any distortions in the packing of the sheets, as might be required by the large amino-acid residues in Tussah silk, would be expected to lower the intensity of this reflection; no such distortions would be expected in poly-L-alanine.

In view of the experimental evidence, it is difficult to escape the conclusion that the structure of β-poly-L-alanine is essentially that of the pseudo unit of Tussah silk formulated in Table 3.

The authors are grateful to Prof. S. Mizushima for a generous supply of Tussah silk, and to Dr W. A. Schroeder for chemical assistance and for permitting the use of the analytical data of Table 1 prior to their publication elsewhere.

References

Bamford, C. H., Brown, L., Elliott, A., Hanby, W. E. & Trotter, I. F. (1953). *Nature, Lond.* **171**, 1149.

Bamford, C. H., Brown, L., Elliott, A., Hanby, W. E. & Trotter, I. F. (1954). *Nature, Lond.* **173**, 27.

Brill, R. (1943). *Z. phys. Chem.* B, **53**, 61.

Corey, R. B. & Pauling, L. (1953). *Proc. Roy. Soc.* B, **141**, 10.

Davey, W. P. (1923). *Phys. Rev.* **21**, 143.

Kratky, O. & Kuriyama, S. (1931). *Z. phys. Chem.* B, **11**, 363.

McWeeny, R. (1951). *Acta Cryst.* **4**, 513.

Marsh, R. E., Corey, R. B. & Pauling, L. (1955). *Biochim. Biophys. Acta*, **16**, 1.

Pauling, L. & Corey, R. B. (1953). *Proc. Nat. Acad. Sci., Wash.* **39**, 253.

Schroeder, W. A. & Kay, L. M. (1955). *J. Amer. Chem. Soc.* To be published.

Trogus, C. & Hess, K. (1933). *Biochem. Z.* **260**, 376.

The Discovery of the Alpha Helix

LINUS PAULING

There are many of Dr. Linus Pauling's writings that remain unpublished—among them, for instance, one about Dutch elm disease, another about the development of the hen's egg, quite a few about vitamin C, not so many about quasicrystals—still, a few—and, of course, several about nuclear structure, rejected in his later years by the editors of Physical Review and Physical Review Letters.

Here at the Linus Pauling Institute, Dr. Pauling's research assistant, Dr. Zelek Herman, and I have for many years been compiling and updating Dr. Pauling's list of publications, and a few months after Dr. Pauling's death in August of 1994, I realized with sudden shock that I would no longer be able, happily and proudly (because I had helped in the preparation), to add another just-published paper to the list, which numbered at that time 1069 publications.

Dorothy Munro, secretary/assistant to Linus Pauling at the Linus Pauling Institute from 1973 to his death in 1994. She continues there as Coordinator of Public Information.

I remembered, however, one manuscript in the files on which I had at one time scribbled the notation "Was this ever published?" Its title was "The Discovery of the Alpha Helix," dictated in 1982 by Dr. Pauling on his ancient dictaphone and transcribed by me from the resulting now-ancient dictabelts. It had been written at the request of the editor-in-chief of W.H. Freeman and Company, as a chapter for a book. The paper had historical significance; had it ever been published? I decided to find out.

I communicated with publishers W.H. Freeman and Company and John Wiley & Sons, and finally with author Dr. Donald Voet, who told me by phone on April 25, 1995, that the essay had not been published by either Freeman or Wiley. Meanwhile, we had obtained a copy of Dr. Voet's book, for which the essay had been intended, and had determined that indeed it had not been included.

So I was free to find a publisher and get the paper published. At the suggestion of Dr. Robert Paradowski, Dr. Pauling's authorized biographer, I approached Dr. Hargittai, who kindly accepted the paper for publication in his lively new magazine, The Chemical Intelligencer.

Thus I shall once again, happily and proudly, be able to add a newly published paper to the list of publications of Professor Linus Pauling, and regain—for a little while—my lost sense of continuity.

DOROTHY MUNRO
Linus Pauling Institute of Science and Medicine
440 Page Mill Road, Palo Alto, CA 94306

I am still astonished to think that I have carried on research on proteins. I have never thought of myself as a biochemist. In 1922, when I began my career in research, the problem of understanding the simplest branches of chemistry, both inorganic and organic, seemed to me to be difficult and challenging, to such an extent that I was not able to envisage the progress that would be made during the next 15 years. The theory of the chemical bond was still in a very primitive stage. Gilbert Newton Lewis in Berkeley had in 1916 defined the chemical bond as a pair of electrons held jointly by two atoms, occupying the space between them, but how to make this compatible with the Bohr theory of the atom was a question that was not answered until 1927, when the quantum-mechanical theory of the chemical bond began to be developed.

Linus Pauling and Robert B. Corey. (Courtesy of California Institute of Technology.)

My first work as a graduate student, under the tutelage of Roscoe Gilkey Dickinson, who had in 1920 been given the first Ph.D. degree ever awarded by the California Institute of Technology, was the study of the structure of minerals and other inorganic crystals by the X-ray diffraction tech-

nique. At that time this technique, which had been developed eight years earlier by W.L. Bragg and his father W.H. Bragg, had been applied in the determination of the structures of some hundreds of elements and inorganic compounds. I immediately became interested in the precise values of the interatomic distances—bond lengths—in the crystals. I tried to develop a system of atomic radii that would permit prediction of these bond lengths. The effort to understand interatomic distances was largely successful within a period of 10 years, but in fact I still continue to work on the problem of correlating interatomic distances and bond angles with the electronic structures of molecules and crystals.

By 1932 I felt reasonably well satisfied with my understanding of inorganic compounds, including such complicated ones as the silicate minerals. The possibility of getting a better understanding also of organic compounds then presented itself. There was as yet not any large amount of experimental information about bond lengths and bond angles in molecules of organic compounds. The first organic compound to have its structure determined, hexamethylene tetramine, had been investigated by Dickinson, together with an undergraduate student named Albert Raymond, in 1922. The carbon–nitrogen bond length had been found to be 1.47 Å, and the bond angles at both carbon and nitrogen were about 109.5°, the tetrahedral angle. By 1932, structure determinations had been made also of a few other crystals containing molecules of organic substances, but not a great many. In 1930, however, I had learned about a new method of determining the structure of molecules that had been invented by Dr. Herman Mark, in Germany. It was the electron diffraction method of studying gas molecules. The determination of the structure of a crystal of an organic compound, even with rather simple molecules, at that time was often difficult because the molecules tended to be packed together in the crystal in a complicated way. The method of electron diffraction of molecules had the advantage that a simple molecule always gave a simple electron diffraction pattern, so that one could be almost certain of success in determining the structure by this method. My student Lawrence Brockway began in 1930 to construct the first electron diffraction apparatus for studying gas molecules that had been built anywhere but in Mark's laboratory in Germany. Herman Mark had been good enough to say that he was not planning to continue work along this line and that he would be glad to see it done in the California Institute of Technology. He also gave me the drawings showing how the instrument could be constructed.

Within a few years we and other investigators had amassed a large amount of information about bond lengths and bond angles in organic compounds. This information had great value in permitting new ideas in structural chemistry, such as the theory of resonance, to be checked against experiment and even to be refined. For example, it was observed that in organic compounds many bonds between carbon atoms or a carbon atom and a nitrogen or oxygen atom were intermediate in length between a single bond and a double bond. This fact was interpreted as showing that the bonds were covalent single bonds with a certain amount of double-bond character. The observations were generally in accord with the results of quantum-mechanical calculations, and it became clear by 1935 that a far more extensive, precise, and detailed understanding of organic compounds had been developed than had been available to chemists in the earlier decades.

It was just at this time that I began to think about proteins. The first protein to attract my interest was hemoglobin. I had read that the equilibrium curve for hemoglobin, oxygen, and oxyhemoglobin was not represented by any simple theoretical expression of the sort that physical chemists had devised for chemical equilibria. I also knew that some eight years earlier it had been shown by Adair in Cambridge that the hemoglobin molecule contains four iron atoms—that is, four heme groups, each being a porphyrin with an iron atom linked to it—and that the molecule could combine with as many as four oxygen atoms. I formulated a theory, published in 1935, on the oxygen equilibrium of hemoglobin and its structural interpretation. The theory was that each iron atom can attach one oxygen molecule to itself, by forming a chemical bond with it. There is an interaction, however, between each heme group and the adjacent heme groups such that addition of the oxygen molecule to one iron atom changes the equilibrium constant for the combination of the other iron atoms with oxygen molecules. I had several ideas as to the nature of the heme–heme interaction, and somewhat later my student Robert C.C. St. George and I published a paper showing that the addition of a group such as the oxygen molecule to one of the iron atoms deforms the molecule through a steric hindrance effect in such a way as to make it easier for oxygen molecules to attach themselves to other iron atoms in the molecule (1951). While I was thinking about the oxygen equilibrium curve in 1935, it occurred to me that measurement of the magnetic properties of hemoglobin, carbon monoxyhemoglobin, and oxyhemoglobin should provide information about the nature of the bonds formed by the iron atoms with the surrounding groups (two distinct kinds of compounds of bipositive iron were known) and the electronic structure

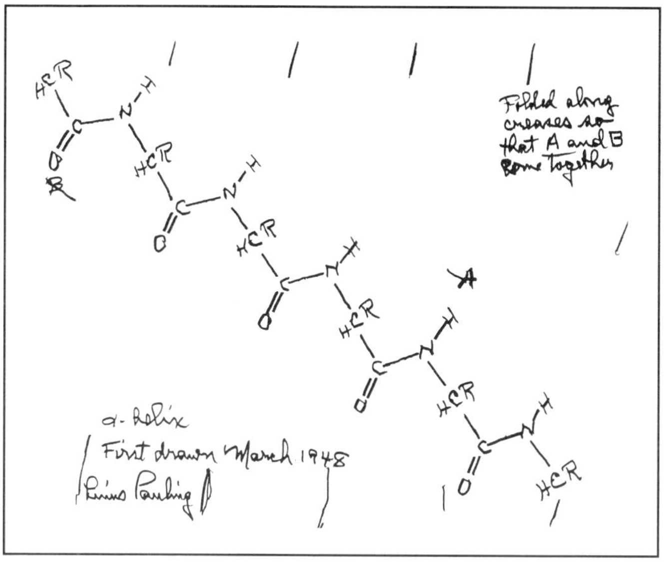

Fig. 1. α-Helix. First drawn in March 1948 by Linus Pauling.

of the oxygen molecule in oxyhemoglobin. Charles Coryell and I carried out measurements of the magnetic properties of these compounds, showing that the iron atoms change their electronic structure when the oxygen molecule is attached, and also that the oxygen molecule changes from having two unpaired electron spins to having none. My first work on proteins accordingly dealt essentially with the physical chemistry and structural chemistry of the heme group and the attached ligand, rather than with the apoprotein, the globin.

The measurement of the magnetic susceptibility of solutions of hemoglobin and related substances turned out to be a valuable technique, and we immediately began applying it to determine equilibrium constants, rates of reaction, and other properties. A leading protein chemist, Dr. Alfred Mirsky, was sent to Pasadena by the Rockefeller Institute of Medical Research to work with us during the year 1935–36. He had been especially interested in the phenomenon of the denaturation of proteins by heat or chemical substances, such as hydrogen ion, hydroxide ion, urea, etc. After many discussions, he and I formulated a general theory of the denaturation of proteins. The theory involved the statement that a native protein consists of polypeptide chains that are folded in a regular way, with the type of folding determined and stabilized by the weaker interactions, especially hydrogen-bond formation. Denaturation, we said in our 1936 paper, is incomplete or complete unfolding of the polypeptide chains, producing molecules that can assume a large number of conformations, giving increased entropy and increased intermolecular interaction.

These considerations about the folding of the polypeptide chains in denatured protein molecules immediately raised the question, of course, as to the nature of the folding. It was a question to which I applied myself during the next 15 years.

Shortly after X-ray diffraction had been discovered, several investigators had made X-ray diffraction photographs of protein fibers. These photographs for the most part showed only rather diffuse diffraction maxima, insufficient to permit structure determinations to be deduced from them. There were two principal types, one shown by keratin fibers such as hair, horn, porcupine quill, and fingernail, and the other shown by silk. William T. Astbury and his collaborators in the early 1930s had reported that the diffraction pattern of a hair changes when the hair is stretched. He called the normal pattern alpha keratin and the stretched-hair pattern, which is somewhat like that of silk, beta keratin. In the early summer of 1937, when I was free of my teaching duties, I decided to try to determine the alpha-keratin structure. My plan was to use my knowledge of structural chemistry to predict the dimensions and other properties of a polypeptide chain and then to examine possible conformations of the chain, to find one that would agree with the X-ray diffraction data. The principal piece of information supplied by the rather fuzzy diffraction photographs of hair and other alpha-keratin proteins came from a rather diffuse arc on the meridian, above and below (that is, in the direction of the axis of the hair). The measured position of this reflection indicated that the structural unit in the direction along the axis of the hair would repeat in 5.10 Å. This fact required that there be at least two amino acid residues for this apparent repeat distance of the alpha-keratin structure.

Because of the large amount of theoretical and experimental progress that had been made, I felt that I could predict the dimensions of the peptide group with reliability. This group is shown in Fig. 1. The alpha-carbon atom forms a single bond with a hydrogen atom, a single bond with the group R characteristic of the amino acid, a single bond to an adjacent main-chain carbon atom, and a single bond to the main-chain nitrogen atom. The single-bond lengths were known to within about 0.01 Å: 1.54 Å for C–C and 1.47 Å for C–N (as determined by Dickinson and Raymond as early as 1922 and verified in many compounds). However, for the other bond between carbon and nitrogen we have to consider the theory of resonance. According to this theory, there are two structures that can be written for a peptide group: in one the carbon–oxygen bond is a double bond, and the carbon–nitrogen bond is a single bond, and in the other the carbon–oxygen bond is a single bond (one of the electron pairs in the double bond having shifted out onto the oxygen atom, giving it a negative charge), and the car-

Linus Pauling through the decades: 1950s, 1960s, 1970s, and 1980s on the following pages. (Photographs courtesy of Linus Pauling Institute of Science and Medicine.)

bon–nitrogen main-chain bond is a double bond (with the nitrogen atom having a positive charge). Because of the separation of charges, the second structure is less stable than the first, and the estimate that could be made is that it should contribute about 40%, so that this bond has 40% double-bond character. The expected bond length is then 1.32 Å, rather than 1.47 Å. Moreover, because of the 40% double-bond character for this bond, these two atoms and the four adjacent atoms should all lie in the same plane, this quality of planarity being characteristic of compounds of molecules in which there are double bonds. In this way I reached the conclusion that these peptide groups in the molecule would have a well-defined rigid structure, with bond lengths and bond angles as shown in Fig. 1, and that there would be two degrees of freedom for the chain, rotation around the single bonds from carbon and nitrogen to the alpha carbon atom. Accordingly, the conclusion, on the basis of the theory of resonance, that the peptide group should be planar greatly restricts the possible structures.

Despite this restriction, I was unable to find a way of folding the polypeptide chain to give a repeat in 5.10 Å along the fiber axis. After working for several weeks on this problem I stopped, having reached the conclusion that there probably was some aspect of structural chemistry characteristic of proteins and remaining to be discovered. This conclusion was, in fact, wrong, but it led to a large amount of experimental work.

Dr. Robert B. Corey was a chemist who, after getting his Ph.D. in chemistry in Cornell University and teaching analytical chemistry there for five years, had joined a leading X-ray crystallographer, Ralph W. G. Wyckoff, in the Rockefeller Institute for Medical Research. He worked with him on crystallographic problems for nine years and then came, in 1937, to spend a year as research fellow in the California Institute of Technology. He and Wyckoff had made some X-ray photographs of proteins, and he was interested in the problem of determining the structure of proteins. I told him about my failure to find the way of folding the polypeptide chains in alpha keratin and my conclusion that there might be some structural feature that we had ignored. I had assumed that the poly-peptide chain should be folded in such a way as to permit the NH group to form a hydrogen bond with the oxygen atom of the carbonyl group of an adjacent peptide group, with the N–H···O distance 2.90 Å, as indicated by structure measurements on compounds other than the amino acids. At that time there had been no correct structure determination made for any amino acid or any peptide. The state of X-ray crystallography was such that a year's work, at least, would be needed to make such a structure determination, even for such a simple compound as glycine, and the efforts of several investigators in other institutions to do such a job had resulted in failure. I suggested to Dr. Corey that he, together with graduate students, attack the problem of determining the structure of some simple amino acid crystals and simple peptides. He agreed, and within little more than a year he and two graduate students (Gustav Albrecht and Henri Levy) had succeeded in making completely satisfactory determinations of the structures of glycine, alanine, and diketopiperazine. This work was continued with vigor, with many students and postdoctoral fellows in chemistry in the California Institute of Technology involved in it, during the following years, interrupted to a considerable extent, however, by the Second World War.

In the spring of 1948 I was in Oxford, England, serving as George Eastman Professor for the year and as a fellow of Balliol College. I caught cold and was required to stay in bed for about three days. After two days I had got tired of reading detective stories and science fiction, and I began thinking about the problem of the structure of proteins. By this time Dr. Corey and the other workers back in Pasadena had determined with high reliability and accuracy the structures of a dozen amino acids and simple peptides, by X-ray diffraction. No other structure determinations of substances of this sort had been reported by any other investigators. I realized, on thinking about the structures, that there had been no surprises whatever: every structure conformed to the dimensions—bond lengths and bond angles and planarity of the peptide group—

that I had already formulated in 1937. The N–H···O hydrogen bonds, present in many crystals, were all close to 2.90 Å in length. I thought that I would attack the alpha-keratin problem again. As I lay there in bed, I had an idea about a new way of attacking the problem. Back in 1937 I had been so impressed by the fact that the amino acid residues in any position in the polypeptide chain may be any of 20 different kinds that the idea that with respect to folding they might be nearly equivalent had not occurred to me. I accordingly thought to myself, what would be the consequences of the assumption that all of the amino acid residues are structurally equivalent, with respect to the folding of the polypeptide chain? I remembered a theorem that had turned up in a course in mathematics that I had attended, with Professor Harry Bateman as the teacher, in Pasadena 25 years before. This theorem states that the most general operation that converts an asymmetric object into an equivalent asymmetric object (such as an L-amino acid into another molecule of the same L-amino acid) is a rotation–translation—that is, a rotation around an axis combined with a translation along the axis—and that repetition of this operation produces a helix. Accordingly, the problem became that of taking the polypeptide chain, rotating around the two single bonds to the alpha carbon atoms, with the amounts of rotation being the same from one peptide group to the next, and on and on, keeping the peptide groups planar and with the proper dimensions and searching for a structure in which each NH group performs a 2.90-Å hydrogen bond with a carbonyl group. I asked my wife to bring me pencil and paper and a ruler. By sketching a polypeptide chain on a piece of paper and folding it along parallel lines, I succeeded in finding two structures that satisfied the assumptions. One of these structures was the alpha helix, with 4.6 residues per turn, and the other was the gamma helix. The gamma helix has a hole down its center that is too small to be occupied by other molecules but is large enough to decrease the van der Waals stabilizing interactions, relative to those in the alpha helix. It seems to me to be a satisfactory structure in every respect but this one, but, so far as I am aware, it has not been observed in any of the protein structures that have been determined so far, and it has been generally forgotten.

I got my wife to bring me my slide rule, so that I could calculate the repeat distance along the fiber axis. The structure does not repeat until after 18 residues in 5 turns, the calculated repeat distance being 27.0 Å, which corresponds to 5.4 Å per turn. This value did not agree with the experimental value, given by the meridional arcs on the X-ray diffraction patterns, 5.10 Å. I tried to find some way of adjusting the bond lengths or bond angles so as to decrease the calculated distance from 5.4 Å to 5.1 Å, but I was not able to do so.

I was so pleased with the alpha helix that I felt sure that it was an acceptable way of folding polypeptide chains and that it would show up in the structures of some proteins when it finally became possible to determine them experimentally. I was disturbed, however, by the discrepancy with the experimental value 5.10 Å, and I decided that I should not publish an account of the alpha helix until I understood the reason for the discrepancy. I had been invited to give three lectures on molecular structure and biological specificity in Cambridge University, and while I was there I talked with Perutz about his experimental electron density distribution functions for the hemoglobin crystal that he had been studying. It seemed to me that I could see in his diagrams evidence for the presence of the alpha helix, but I was troubled so much by the 5.1-Å value that I did not say anything to him about the alpha helix.

On my return to Pasadena in the fall of 1948 I talked with Pro-fessor Corey about the alpha helix and the gamma helix, and also with Dr. Herman Branson, who had come for a year as a visiting professor. I asked Dr. Branson to go over my calculations and, in particular, to see if he could find any third helical structure. He reported that the calculations were all right and that he could not find a third structure. More than a year went by, and then a long paper on ways of folding the polypeptide chain, including helical structures, was published by W. Lawrence Bragg, John Kendrew, and Max Perutz, in *Proceedings of the Royal Society of London.* They

described about 20 structures, and they reached the conclusion that none of them seemed to be satisfactory for alpha keratin. Moreover, none of them agreed with my assumptions, in particular, the assumption of planarity of the peptide group. Lord Todd has told the story of his having told Bragg, when they were just beginning their work, that the main-chain carbon–nitrogen bond has some double-bond character but that Bragg did not understand that that meant that the peptide group should be planar. My efforts during a year and a half to understand the 5.1-Å discrepancy had failed, but Dr. Corey and I decided that we should publish a description of the alpha helix and the gamma helix. It appeared in the *Journal of the American Chemical Society* in the fall of 1950. It was followed in 1951 by a more detailed paper, with Branson as coauthor, and a number of other papers on the folding of polypeptide chains. An important development had been the publication of X-ray photographs of fibers of synthetic polypeptides, in particular of poly-gamma-methyl-L-glutamate, by investigators at Courtaulds. These striking diffraction photographs showed clearly that the pseudo repeat distance along the fiber axis is 5.4 Å rather than 5.1 Å. There are strong reflections near the meridional line, corresponding to 5.1 Å, but they are not true meridional reflections. On the X-ray photographs of hair, the reflections overlap to produce the arc that seems to be a meridional reflection. It was this misinterpretation that had misled all of the investigators in this field. It was accordingly clear that the alpha helix is the way in which polypeptide chains are folded in the alpha-keratin proteins.

Moreover, we reached the conclusion, as did Crick, that in the alpha-keratin proteins the alpha helices are twisted together into ropes or cables. This idea essentially completed our understanding of the alpha-keratin diffraction patterns.

The apparent identity distance in the fiber X-ray diagrams of silk is somewhat smaller than corresponds to a completely extended polypeptide chain. We accordingly concluded that the polypeptide chains have a zigzag conformation in silk and the beta-keratin structure. We reported in detail three proposed sheet structures. The first one, which we called the rippled sheet, involves amino acid residues of two different kinds, one of which cannot be an L-amino acid residue, but can be a residue of glycine. It was known that Bombyx mori silk fibroin has glycine in 50% of its positions, with L-alanine or some other L-amino acid residue (such as L-serine) in the alternate positions, so that the rippled sheet seemed to be a possibility for Bombyx mori silk fibroin. It turned out, however, that Bombyx mori silk fibroin has the structure of the antiparallel-chain pleated sheet. The third pleated sheet structure, the parallel-chain pleated sheet, is also an important one.

About 85% of the amino acid residues in myoglobin and hemoglobin are in alpha-helix segments, with the others involved in the turns around the corners. In other globular proteins the alpha helix, the parallel-chain pleated sheet, and the antiparallel-chain pleated sheet all are important structural features. These three ways of folding polypeptide chains have turned out to constitute the most important secondary structures of all proteins. Dr. Corey, to some extent with my inspiration, designed molecular models of several different kinds that were of much use in the later effort to study other methods of folding polypeptide chains. I used these units to make about 100 different possible structures for folding polypeptide chains. For example, if the hydrogen bonds are made alternately a little shorter and a little longer than 2.90 Å in a repeated sequence, an additional helical twist is imposed upon the alpha helix. Some of the models that I constructed related to ways of changing the direction of the axis of the alpha helix. I reported on all of this work at a protein conference in Pasadena in 1952, but then I became interested in other investigations and stopped working in this field.

It pleases me to think that our work in Pasadena in the Division of Chemistry and Chemical Engineering, first in collecting experimental information about the structure of molecules, then in developing structural principles, and then in applying these principles to discover the alpha helix and the pleated sheets, has shown how important structural chemistry can be in the field of molecular biology.

Chapter 14

Molecular Biology: The Role of Large Molecules in Life and Evolution

Contents

SP 112 **The Nature of the Intermolecular Forces Operative in Biological Processes** **1093**
(by Linus Pauling and Max Delbrück)
Science **92**, 77–79 (1940)

113 **Molecular Architecture and Biological Reactions** **1097**
The Journal of NIH Research **2**(6), 59–64 (1990)

114 **The Nature of Forces Between Large Molecules of Biological Interest** **1103**
Nature **161**, 707–709 (1948)

115 **Molecular Architecture and the Processes of Life** **1109**
Sir Jesse Boot Foundation, Nottingham, England, 13 pp. (1948)

116 **A Proposed Structure for the Nucleic Acids** **1123**
(by Linus Pauling and Robert B. Corey)
Proc. Natl. Acad. Sci. USA **39**, 84–97 (1953)

117 **Specific Hydrogen-Bond Formation Between Pyrimidines and Purines in Deoxyribonucleic Acids** **1137**
(by Linus Pauling and R. B. Corey)
Arch. Biochem. Biophys. **56**, 164–181 (1956)

118 **A Comparison of Animal Hemoglobins by Triptic Peptide Pattern Analysis** **1156**
(by Emile Zuckerkandl, Richard T. Jones, and Linus Pauling)
Proc. Natl. Acad. Sci. USA **46**, 1349–1360 (1960)

SP 119 **Molecular Disease, Evolution, and Genic Heterogeneity** 1168
(by Emile Zuckerkandl and Linus Pauling)
In *Horizons in Biochemistry*, eds. Michael Kasha and Bernard Pullman, Academic Press, NY, pp. 189–225 (1962)

120 **Chemical Paleogenetics: Molecular "Restoration Studies" of Extinct Forms of Life** 1205
(by Linus Pauling and Emile Zuckerkandl)
Acta Chem. Scand. **17** (Suppl. no. 1), S9–S16 (1963)

121 **Molecules as Documents of Evolutionary History** 1213
(by Emile Zuckerkandl and Linus Pauling)
J. Theoret. Biol. **8**, 357–366 (1965)

Reprinted from SCIENCE, July 26, 1940, Vol. 92,
No. 2378, pages 77–79.

THE NATURE OF THE INTERMOLECULAR FORCES OPERATIVE IN BIOLOGICAL PROCESSES

IN recent papers P. Jordan[1] has advanced the idea that there exists a quantum-mechanical stabilizing interaction, operating preferentially between identical or nearly identical molecules or parts of molecules, which is of great importance for biological processes; in particular, he has suggested that this interaction might be able to influence the process of biological molecular synthesis in such a way that replicas of molecules present in the cell are formed. He has used the idea in connection with suggested explanations of the reproduction of genes, the growth of bacteriophage, the formation of antibodies, and other biological phenomena. The novelty in Jordan's work lies in his suggestion that the well-known quantum-mechanical resonance phenomenon would lead to attraction between molecules containing identical groups and to autocatalytic reproduction of molecules. Jordan himself expressed some doubt as to whether resonance attraction could really be operative in this way; after studying the question, we have reached the conclusion that the theory can not be applied in the ways indicated by him, and that his explanations of biological phenomena on this basis can not be accepted. In this note we wish to state our objections to Jordan's hypothesis and to formulate briefly our view of the present status of the chemical problems involved in these phenomena. We shall not discuss here Jordan's biological arguments for the occurrence of autocatalytic reactions, as distinct from the arguments concerning their mechanism.

Let us consider two identical molecules or parts of molecules, A and B, which interact with each other, the interaction being perhaps the electrostatic inter-

[1] P. Jordan, *Phys. Zeits.*, 39: 711, 1938; *Zeits. f. Phys.*, 113: 431, 1939; *Fundam. Radiol.*, 5: 43, 1939; *Zeits. f. Immun. forsch. u. exp. Ther.*, 97: 330, 1940.

2

action of electric dipoles in the molecules, as considered by Jordan. If both molecules are in their lowest states the interaction is normal. If, however, one is in its lowest state and the other in an excited state there occurs a resonance phenomenon; the wave function is either the symmetric or the antisymmetric combination of two functions, one representing molecule A normal and molecule B excited, and the other the reverse. For one of these symmetry types there is a resonance stabilization and attraction between the two molecules, and for the other a resonance repulsion.

It is this resonance stabilization between identical molecules which Jordan invokes as the cause of his postulated process of synthesis of molecules similar to a molecule present in the cell. His argument requires, of course, that the stabilization occur only between identical molecules, or, at any rate, between molecules which differ only slightly. Moreover, since the phenomenon does not occur for two molecules in their lowest energy states, he assumes that there exist excited energy states differing from the lowest state by a small amount of energy ($\sim kT$), so that thermal excitation raises the molecules to these excited states.

Now let us examine Jordan's argument. In addition to those objections which the author himself has pointed out we may advance the following ones.

(1) The resonance stabilization between two identical molecules is equal to the resonance integral $H'_{n'_A n''_B,\, n''_A n'_B}$ (here n' and n'' are quantum numbers for the states involved in the resonance), which to be effective in determining the behavior of the system must be at least of the order of kT in magnitude. But *under the assumed circumstances the resonance between unlike molecules would be nearly as great as that between like molecules.*[2] If the two molecules A and B are unlike, but have excited states with nearly the same energy difference ($\sim kT$) from their normal states, then the resonance stabilization will be equal to the resonance integral; even if the energy values of the excited states for the two molecules differed by as

[2] This is found to be so by evaluating the roots of the secular equation corresponding to the perturbation.

3

much as one half the resonance integral ($\sim {}^1/_2 kT$), the resonance stabilization would still be about three quarters as great as for identical molecules.[3]

(2) For large molecules in solution, such as protein molecules, the complexity of the molecules and the perturbing influence of the environment would be such as to make the energy spectrum of a molecule effectively a continuum,. and would wipe out the distinction between unlike and like molecules.

(3) The reasonance stabilization is equal to the resonance integral, and must be of magnitude kT or greater to be effective; *i.e.,* it must be as great as the energy difference between ground state and excited state. On the other hand, the theory requires the excited state which is involved to be non-degenerate. But by definition a "non-degenerate" state in perturbation problems is one for which the energy differences with all other states are greater than the resonance integral; hence Jordan's argument is inconsistent.

Summing up, then, we find that under the conditions of excitation and perturbation prevailing in aqueous solutions the resonance interaction could not cause a specific attraction between like molecules and therefore could not be effective in bringing about autocatalytic reactions.

It is our opinion that the processes of synthesis and folding of highly complex molecules in the living cell involve, in addition to covalent-bond formation, only the intermolecular interactions of van der Waals attraction and repulsion, electrostatic interactions, hydrogen-bond formation, etc., which are now rather well understood. These interactions are such as to give stability to a system of two molecules with *complementary* structures in juxtaposition, rather than of two molecules with necessarily identical structures; we accordingly feel that complementariness should be given primary consideration in the discussion of the

[3] The relation between the amount of resonance stabilization and the energy difference of the resonating states is given, for example, by L. Pauling, ''The Nature of the Chemical Bond,'' p. 18. Cornell University Press, Ithaca, N. Y. 1940.

4

specific attraction between molecules and the enzymatic synthesis of molecules.

A general argument regarding complementariness may be given. Attractive forces between molecules vary inversely with a power of the distance, and maximum stability of a complex is achieved by bringing the molecules as close together as possible, in such a way that positively charged groups are brought near to negatively charged groups, electric dipoles are brought into suitable mutual orientations, etc. The minimum distances of approach of atoms are determined by their repulsive potentials, which may be expressed in terms of van der Waals radii; in order to achieve the maximum stability, the two molecules must have complementary surfaces, like die and coin, and also a complementary distribution of active groups.

The case might occur in which the two complementary structures happened to be identical; however, in this case also the stability of the complex of two molecules would be due to their complementariness rather than their identity. When speculating about possible mechanisms of autocatalysis it would therefore seem to be most rational from the point of view of the structural chemist to analyze the conditions under which complementariness and identity might coincide.

From the biological side it would seem most rational to postulate the possibility of both processes; *viz.*, formation of complementary non-identical structures and formation of complementary identical structures, and to proceed by analyzing experimental data for clear-cut evidence as to their occurrence.

LINUS PAULING

GATES AND CRELLIN LABORATORIES
OF CHEMISTRY,
CALIFORNIA INSTITUTE OF TECHNOLOGY

MAX DELBRÜCK

PHYSICS DEPARTMENT,
VANDERBILT UNIVERSITY

LANDMARKS

Molecular Architecture And Biological Reactions

Linus Pauling*

Reprinted with permission from *Chemical & Engineering News* May 25, 1946, **24**(10), pp. 1375–1377. Copyright 1946 American Chemical Society.

Answers to many basic problems of biology—nature of growth, mechanism of duplication of viruses and genes, action of enzymes, mechanism of physiological activity of drugs, hormones, and vitamins, structure and action of nerve and brain tissue—may lie in knowledge of molecular structure and intermolecular reactions.

There are two subjects that I am deeply interested in—structure, the detailed nature of molecules, crystals, and cells, described in terms of their constituent atoms, with interatomic distances determined to within 0.01 Å, an interest that began in my youth and has received most of my attention until recent years; and the basis of the physiological activity of substances, an interest that is more recent but just as keen. It is with a deep feeling of satisfaction that I have reached the firm conclusion in recent years that these two fields are most intimately related.

Why have we still so little understanding of the structural basis of the physiological activity of chemical substances, despite the interest and effort of many able physiologists and chemists during recent decades? I believe that it is because the problem has been examined, in the main, from one point of view only—not the wrong point of view, but one which, unaided, gives a vista insufficient to reveal its true complex nature. This point of view is that which surveys the chemical reactivity of molecules—their tendency to break their chemical bonds, the very strong bonds between atoms, and to form new chemical bonds. The other point of view which is needed is that which directs the mind's eye to the detailed size and shape of the molecules and the nature of the weak interactions of molecules with other molecules, in particular with the macromolecules and macromolecular stromatic structures which characterize the living organism. Until very recently physiologists and pharmacologists have barely thought of this aspect of their great problem—and I am convinced that once they begin to use this new idea seriously a period of the greatest development will have started. I believe that the next twenty years will be as great years for biology and medicine as the past twenty have been for physics and chemistry.

Eddington has said that the study of the physical world is a search for structure rather than a search for substance. If we ignore the philosophical implications of the words, we may say that the chemist and biologist in their study of living organisms must carry on both a search for structure and a search for substance, and that the second of these must precede the first. Investigators have had great success in isolating chemical substances from living organisms, and in determining the chemical composition of the simpler of these substances. The chemical composition is also known of many substances of external origin which exert physiological activity on living organisms. We may consider this work of isolation and identification of active chemical substances as the search for substance in biology.

◆ The Landmark Interviews ◆

The First Molecular Biologist

Forty-four years ago, when **Linus Pauling** published the landmark paper "Molecular architecture and biological reactions," the words "molecular" and "biology" were rarely uttered in the same breath. The DNA double helix had yet to be discovered. It was not even clear that DNA existed as elongated strands of covalently linked nucleotides. No proteins had been sequenced, and the idea that they fold into discrete three-dimensional structures had never been proposed. The basis of enzyme action was still an unchartered frontier. Pauling himself had just begun to study the nature of interactions between antibodies and antigens . . .

interview continued on page 61

*At the time of publication of this article, Linus Pauling was chairman of the division of chemistry and chemical engineering at the California Institute of Technology in Pasadena. Today he is research professor at the Linus Pauling Institute of Science and Medicine in Palo Alto, Calif.

The Search for Structure

The search for structure has also made great progress. From the one side biologists have, by visual observation with the microscope, made thorough studies of the apparent structure of aggregates of cells, of cells themselves, and of certain constituents of cells, such as chromosomes. This visual observation has provided information about structures in size extending down to 10^{-4} cm., 10,000 Å. Forty years ago the dark forest of the dimensional unknown stretched from this limit of the visible microscope back indefinitely into the region of smaller dimensions. In recent years the region from 10^{-7} down to 10^{-12} cm., containing atoms and simple molecules, has been thoroughly explored by an expedition outfitted with x-rays and similar tools, and the physicists are strongly pushing back into the region of the structure of atomic nuclei, below 10^{-12} cm. Another detailed exploration is being carried out with the electron microscope. This has pushed the nearer boundary of the unknown back from 10^{-4} to 10^{-6} cm., although the major portion of this region has been only sketchily explored during the few years since the development of the electron microscope, and a very great amount of work still remains to be done.

The answers to many of the basic problems of biology—the nature of the process of growth, the mechanism of duplication of viruses, genes, and cells, the basis for the highly specific interactions of these structural constituents, the mode of action of enzymes, the mechanism of physiological activity of drugs, hormones, vitamins, and other chemical substances, the structure and action of nerve and brain tissue—the answers to all these problems are hiding in the remaining unknown region of the dimensional forest, mostly in the strip between 10 and 100 Å., 10^{-7} and 10^{-6} cm.; and it is only by penetrating into this region that we can track them down.

There are many ways of investigating this region—by x-rays, ultracentrifuges, light-scattering techniques, the study of chemical equilibria, the techniques of degradation, isolation, identification, and synthesis used by the organic chemist, serological methods, chemical genetics, the use of both radioactive and nonradioactive tracers, the use of electron microscopes of improved resolving power—but no one method is good enough to solve the problem, and all these methods must be applied as effectively as possible if the problem is to be solved.

At the present time we know in complete detail the atomic structure of many simple molecules, including a few amino acids; but we do not know in detail how the amino acids are combined to form proteins. We do not know, except very roughly, even the shapes of such important molecules as serum proteins, enzymes, genes, the substances which make up protoplasm—and if we are to obtain a thorough understanding of the structure of living organisms detailed information about the atomic arrangement of these substances must be obtained.

Let us imagine ourselves increased in size by the linear factor 250,000,000—the commonly used factor in molecular models, which makes 1 Å., 10^{-8} cm., become approximately 1 inch, atoms on this scale being 2 or 3 inches in diameter. With this magnification we would become about equal in height to the distance from the earth to the moon. Let us consider ourselves examining the earth, which would appear to us to be about the size of a billiard ball; and let us concentrate our attention on a small organism on the surface of the earth—New York City—which would appear as a spot about 0.01 inch in diameter, barely visible to the naked eye, and showing itself to be living by slow changes in shape and size.

To obtain a better view of this organism we could use a microscope, the resolving power of which would be about 1,000 feet; we could distinguish Central Park, the rivers, and such aggregates of skyscrapers as Rockefeller Center, but the individual skyscrapers would not be clearly defined. By "chemical" methods we would know that, running through the veins and arteries of this organism, there were substances such as street cars, busses, automobiles, ships, and people; and we might, by the use of membranes of known pore size or by some similar method, obtain the molecular weight of these. In addition, we would have obtained, through the application of a strange method of experimental investigation, the diffraction of x-rays and electron waves, complete information about the structure of objects smaller than about 1 foot in diameter, such as a storage battery, a small electric motor, a piece of cable, a small gear wheel, a bolt or rivet.

The use of the electron microscope, with resolving power about 10 feet, would give us very much additional information. We would know exactly—that is, to within 10 feet—the shape of the Empire State Building, though we might not be sure about the separate smaller rooms into which it is divided, and we could not obtain by the electron microscope information about the elevators and the machinery for operating them, the steel girders of which the building is constructed, and other structural features of similar size. We would be able to see, with the electron microscope, an automobile only as a particle, barely discernible, and roughly spherical in shape, and the human beings in the city would not be visible. We could get complete information about a storage battery, a ring gear, a brake pedal—but not about the automobile built up of these and many other parts; and it is clear that to obtain an understanding of the structure of this city we would still need to find a method of exploring objects in the range 1 to 10 feet.

Our hope for achieving precise knowledge about biological structures and reactions is based largely on the electron microscope and on diffraction methods. The diffraction studies of simple molecules have been carried out in sufficient number to permit the formulation of generalizations about atomic radii, bond angles, and other features of molecular configuration; it is still very important that the exact structure be determined of vitamins, bacteriostatic agents, and other physiologically active substances—the complete crystal structure determination of the rubidium salt of penicillin so ably made by Dorothy Crowfoot and Barbara Rogers-Low (*6*) has provided not only decisive information about the chemical formula of the substance but also the structural basis for later consideration of the detailed mechanism of its bacteriostatic activity.

continued from page 59

...but questions concerning antibody diversity and specificity were still unanswered. Before anybody knew much about the structure of biological macromolecules, Pauling began to speculate about the nature of the forces that would bring them to life—to fold into three-dimensional forms and to interact with other molecules to carry out the business of the cell.

Pauling thought that molecular complementarity—how molecules fit together—was the key to understanding the actions of the movers and shakers of physiology: nucleic acids, proteins, enzymes, and antibodies. Chemists had been concerned with the making and breaking of strong covalent bonds between molecules, but Pauling thought that the weaker bonds were just as important, because such interactions would dictate the size and shape of the cell's molecules. In his 1946 paper, Pauling prophesied about the future of biology and medicine and why understanding the nature of complementarity is so important to the future of the field.

Pauling's ideas are still being exploited today. One of the hottest new developments in biochemistry is the production of catalytic antibodies—antibodies that behave like enzymes (see "Catalytic Antibodies: Reinventing The Enzyme," page 77). Pauling proposed in 1946 that enzymes work by binding in a complementary fashion to the "activated complex," or transition state, of a chemical reaction, thus stabilizing the transient species. He suggested that antibodies work in the same way—by specifically interacting with molecules of a complementary shape, like a lock and key. The new research on catalytic antibodies takes advantage of these concepts.

Pauling also speculated that the same forces of molecular complementarity would be responsible for gene duplication. Indeed, it is difficult to imagine any area of contemporary molecular biology that does not depend upon the principles espoused by Pauling nearly 45 years ago.

Pauling's contributions have not gone unnoticed. In 1954 he won the Nobel Prize in chemistry for his theory on the nature of the chemical bond, and in 1962 he won the Nobel Peace Prize for his stance against nuclear arms testing. He has received scores of honorary degrees and awards and is still an active and much sought-after lecturer. He continues his research on vitamins at the Linus Pauling Institute in Palo Alto, Calif., and is busy breaking ground in another area of science—the nature of quasi-crystals.

Although it is difficult to identify Pauling's greatest contribution to science, most scientists, including Alexander Rich of the Massachusetts Institute of Technology in Cambridge, agree that Pauling's paper, published in *Chemical & Engineering News* in 1946, is truly a landmark paper.

Alexander Rich

"Reading this paper, published in 1946," says Rich, "is rather like being in a time warp. Linus writes in a language that makes complete sense with the way we think about biological problems at the molecular level today. You could take this literature out and insert it into any textbook today, and the words don't need to be changed. It's as if the paper were written today instead of 45 years ago."

Rich, who now studies the molecular biology of Z-DNA, was a postdoc with Pauling at California Institute of Technology in Pasadena from 1949 until 1954, where he worked on the structures of several biological macromolecules. "It was a very exciting period," recalls Rich. "Linus would frequently think about things, then come into the lab to discuss his ideas. There was an exceptionally able group of young postdocs in the lab. Most of us regarded it as a kind of high point in our careers, because of all the excitement generated by the things going on in the lab.

"Linus was the first to really understand the true nature of interactions in biological systems... the fact that the important interactions are going to be those on the surfaces of macromolecules, where individual surface atoms interact either with other macromolecules or with small molecules. This is a very powerful idea, and this is the first clear statement of it. It is applicable to catalysis, for example in the case of catalytic antibodies, relevant to the idea of how large molecules bind to each other, and to how factors bind to receptors—all of the phenomena of biology that we now look at. Today when we look at structures of antibodies binding to large-molecule substrates, it's simply a graphic and explicit demonstration of that which Linus was predicting and describing 45 years ago. That's a landmark paper."

But what Rich thinks is even more remarkable than Pauling's contemporary relevance is his "sense of looking into the future." By 1945, a fair amount of work had been done with small molecules, smaller than 10 Å in diameter. Larger structures, greater than 100 Å, were also being studied by light—and even electron—microscopy. "But Linus knew that in the range between 10 and 100 angstroms, which is the space occupied by proteins, that's where the thrust has to come in the future," says Rich. "Furthermore, he makes the statement, which is pretty prophetic in 1946, 'I believe that the next 20 years will be as great for biology and medicine as the past 20 years have been for physics and chemistry.' Mind you, between 1925 and 1945, the ground work of physics and chemistry was set—the application of quantum mechanics to physical and chemical phenomena. He thinks the next wave of exciting discovery will occur within the next 20 years. Now, by 1965, we had already solved the genetic code and we knew the great outlines of what goes on in biological systems."

Rich counts Pauling as "among the first molecular biologists, if not the first. Why? Because he described the essence of molecular biology in a way that all the rest of us are now following," says Rich. His unearthing of the cause of sickle cell anemia "was the first discovery of a molecular mutation that gives rise to disease. It's very near the beginning of molecular biology as we know it."

Not everyone immediately appreciated Pauling's ideas at the time, says Rich. "Linus was a man who was looked up to enormously, then, as he is now. His comments were always taken seriously." But according to Rich, there was an educational problem. "The average biochemist of that day didn't know what a van der Waals interaction was, didn't know about hydrogen bonds or electrostatic potentials," says Rich. "All of that was not understood by the biochemists of that day. He was not a part of their world. For the people who were knowledgeable, it had an impact, but I suspect that the group was very small."

continued on page 63

Structure of Protein

The most important of all structural problems is the problem of the structure of proteins: until this problem is solved all discussions of the exact molecular basis of biological reactions remain in some degree speculative. The polypeptide-chain structure of proteins proposed by Fischer is now generally accepted, and there is little doubt that the picture of folded chains held by hydrogen bonds, van der Waals forces, and related weak interactions in more or less well-defined configurations, as discussed eleven years ago by Mirsky and me (*9*) is essentially correct. But this whole picture remains very vague—for only a few proteins (such as β-lactoglobulin (*3*) do we have nearly complete knowledge of the numbers of residues of the different amino acids in the molecule, and for no protein does there exist more than fragmentary information either about the sequence of the different residues in the polypeptide chain or about the way in which the chain is folded. Only for fibrous proteins in the completely extended state do we have knowledge (still very rough) of the configuration and relative orientation of the polypeptide chains (as originally determined by Astbury), and this knowledge applies only to the backbone of the chains and not to the side groups. There is urgent need for complete and accurate structure determinations of proteins and related substances. So far these determinations have been reported for only four such substances (*5*)—two amino acids and two simple polypeptides—all made in our Pasadena laboratories; and it is my hope that, now that the war is over, precise information will rapidly accrue, including ultimately detailed structures of fibrous proteins, respiratory pigments, antibodies, enzymes, reticular proteins of protoplasm, and others.

Importance of Shape

Despite the lack of detailed knowledge of the structure of proteins, there is now very strong evidence that the specificity of the physiological activity of substances is determined by the size and shape of molecules, rather than primarily by their chemical properties, and that the size and shape find expression by determining the extent to which certain surface regions of two molecules (at least one of which is usually a protein) can be brought into juxtaposition—that is, the extent to which these regions of the two molecules are complementary in structure. This explanation of specificity in terms of "lock-and-key" complementariness is due to Paul Ehrlich, who expressed it often, in words such as "only such substances can be anchored at a particular part of the organism which fit into the molecule of the recipient combination as a piece of mosaic fits into a certain pattern".

In recent years the concept of complementariness of surface structure of antigen and antibody was emphasized by Breinl and Haurowitz (*4*), Mudd (*10*), and Alexander (*1*), and then was strongly supported by me (*11*) in the course of an effort to understand and interpret serological phenomena in terms of molecular structure and molecular interactions. Since 1940 my collaborators (Dan H. Campbell, David Pressman, Carol Ikede, L. H. Pence, G. G. Wright, S. M. Swingle, D. H. Brown, J. H. Bryden, A. L. Grossberg, L. A. R. Hall, Miyoshi Ikawa, Frank Lanni, J. T. Maynard, and A. B. Pardee) and I have gathered a great amount of experimental evidence about antigen-antibody interaction (*12*), which not only supports the general thesis that serological specificity is the consequence of structural complementariness, but provides information about extent of complementariness.

It has been verified that the closeness of fit of an antibody molecule to its homologous haptenic group is to within better than 1 Å.—that a methyl group (van der Waals radius 2.0 Å.) can replace a chlorine atom (radius 1.8 Å.) in a haptenic group with little interference with its combination with antibody (as was first shown by Landsteiner), but that interference is caused by replacing a hydrogen atom (radius 1.2 Å.) by a methyl group. The complementariness in structure with respect to proton-donating and proton-accepting hydrogen bond–forming groups has been found to be very important in determining the strength of attraction of antibody and haptenic group; and the complementary electrical charge in antibody homologous to the *p*-azophenyltrimethylammonium group has been shown to be within about 2 Å. of the minimum possible distance from the charge of opposite sign in the haptenic group. The great amount of quantitative data which has been gathered for scores of different haptens and antigens and successfully interpreted in terms of molecular structure and the concept of complementariness leaves no doubt that this structural explanation of serological specificity is correct.

The phenomenon of specificity, so common in biology, is rare in chemistry (with the sole general exception mentioned below). Only very occasionally does there occur a unique representative of a class of compounds, such as the ion $W_2Cl_9^{---}$, which owes its special stability to the ratio of radii of the atoms of chlorine and tripositive tungsten, which permits a covalent bond to be formed between the two tungsten atoms in the complex. The one general chemical phenomenon with high specificity is closely analogous in both its nature and its structural basis to biological specificity: this phenomenon is crystallization. There can be grown from a solution containing molecules of hundreds of different species, crystals of one substance which are essentially pure. The reason for the great specificity of the phenomenon of crystallization is that a crystal from which one molecule has been removed is very closely complementary in structure to that molecule, and molecules of other kinds cannot in general fit into the cavity in the crystal or are attracted to the cavity less strongly than a molecule of the substance itself. Only if the foreign molecule is closely similar in size and shape and the location and nature of active (hydrogen bond–forming) groups to the molecule it is replacing will it fit into the crystal; and it is indeed found that the tendency to solid-solution formation depends upon the same structural features (such as replacement of a chlorine atom by a methyl group) as the tendency to serological cross reaction.

Examples of Biological Specificity

Many isolated examples of biological specificity and biological similarity determined by molecular size and shape and the detailed nature of intermolecular forces *continued from page 61*

Rich finds it difficult to pinpoint Pauling's greatest scientific contribution. "The nature of Linus is that he's a person who has worked in several fields and has made monumental contributions in all of them. You might say his most significant contribution was the discovery of the first molecular disease, due to sickle hemoglobin. But how would you rank that versus the discovery that carbon atoms have tetrahedral chemical bonds, by hybridizing orbitals? That's a profound and basic discovery that underlies everything in chemistry and biology. Or take his theory that made it possible to predict, based on the size of ions, the kind of lattice and structures they would form—very powerful."

However, Rich does not hesitate to acknowledge Pauling's influence on his own career. "The main thing I learned from Linus," says Rich, "is, put your feet up on the table and think. Think about the way the thing ought to work. Get a clear idea and follow it through. He was a great inspiration."

Linus Pauling

"I had been working for about 10 years on the molecular basis of biological specificity," recalls Pauling. "In 1940, I published a paper on the structure of antibodies, which included a rather general statement that biological specificity is the result of a detailed complementariness in structure between one molecule and another. In particular, we carried out a great many experiments with antibodies and were able to prove, essentially, that [molecular complementarity] is the mechanism of specificity of antibodies."

Pauling also proposed that proteins fold into discrete structures and, in 1949, discovered the alpha helix and the beta sheet. "I was thinking about these problems around 1946—how the polypeptide chains in the proteins such as enzymes are folded," recounts Pauling. "And I was thinking about complementariness as the basis for specificity. So all of these ideas were mulling around together in my head. I began to think more about enzyme specificity.

"Some enzymes are so specific in their action that there's only one reaction that they catalyze," he says. "I could explain this specificity by assuming that it involved complementary structures. The question was, what is the structure that the enzyme is complementary to? I knew about chemicals and thermodynamics. I knew that if an enzyme catalyzes a reaction of a substance to form a product, such as methylmalonic acid to form succinic acid, it had to catalyze the reverse reaction. So the fact that it has to speed up both the forward and back reaction indicated to me that the structure that the enzyme is complementary to had to be midway between the reactants and the products. That meant the activated complex. Then it was obvious—to me, at any rate—what the answer was, to why an enzyme is able to speed up a chemical reaction by as much as 10 million times. It had to do this by lowering the energy of activation—the energy of forming the activated complex. It could do this by forming strong bonds with the activated complex, but only weak bonds with the reactants or products."

Pauling says that the unique arrangement of atoms on the surfaces of molecules is the key to understanding complementarity. "The active region of the enzyme consists of atoms arranged in a structure, such that there is a cavity into which a haptenic group—in this case, the activated complex—fits very closely. We were able to show that the fit is very good, to within a fraction of the diameter of an atom. I also proposed that the gene consists of two strands of molecules that are mutually complementary. I published that in detail some years before the double helix was discovered." He adds, "There's no example of biological specificity that doesn't fit this generalization."

Pauling also tinkered with the idea of turning antibodies into enzymes. "Around this time, 1946, I tried to make an artificial enzyme by using an inhibitor for an enzyme-catalyzed reaction as the template," he says. "I thought I would use malonic acid because it inhibits catalase." Catalase is an enzyme that breaks down hydrogen peroxide. "I carried out one experiment, and I had hoped to get an antibody preparation to malonic acid that had catalatic activity—that would decompose hydrogen peroxide. I'm sure it would have worked, but I just got busy with other things and never repeated the experiment. It was a very busy time for me," says Pauling.

Pauling was also starting to think about the molecular basis of disease. His musings led to what Pauling now considers one of his most creative ideas. "It was known that sickle cell anemia was hereditary, but it wasn't quite clear yet as to what the hereditary nature was that caused the red cells to be deformed into sort of sickle-shaped cells," says Pauling. "I heard that in arterial blood, they regained their normal shape. They were deformed only in the venous blood. That was what made me think it was a disease of the hemoglobin molecule. Because in the venous blood, you have hemoglobin molecules and in the arterial blood you have oxyhemoglobin molecules in the red cells. So I thought, it's only the hemoglobin molecules, not oxyhemoglobin, that cause the red cells to change their shape."

With the idea that sickle cells contain a mutant form of hemoglobin that crystallizes in the absence of oxygen and deforms the red cell, Pauling convinced one of his students, Harvey Itano, now retired from the University of California at San Diego, to carry out the experiment. "We were able to show that the sickle cell hemoglobin differs from normal hemoglobin as the result of a mutation that changes one of the amino acid residues in the β-chains," says Pauling. "Well, I thought that was a pretty nice idea that I had in 1945, about molecular diseases."

Most scientists would be ecstatic to have but one of Pauling's "nice ideas." One of his students once asked him how he goes about coming up with good ideas. Pauling told him, "Well, you have to have a lot of ideas and throw out the bad ones." Pauling says there are two parts to that. "First, if you want to make discoveries, it's a good thing to have good ideas. It's a good thing to have a lot of ideas. And second, you have to have sort of a sixth sense—the result of judgment and experience—which ideas are worth following up. I seem to have the first thing, a lot of ideas, and I also seem to have good judgment as to which are the bad ideas that I should just ignore, and the good ones, that I'd better follow up," he says.

But there was one idea that he wishes he had followed up. "I should have taken the time to repeat my experiments with malonic acid as an antigen to produce antibodies with catalytic activity. I should have done that back in the 1940s."

—NANCY TOUCHETTE

might be mentioned, such as the similarity in physiological (antipyretic-antineuralgic) activity of 4-isopropylantipyrine and 4-dimethylaminoantipyrene (pyramidon), which is clearly the result of the similarity in size and shape of the isopropyl group and the dimethylamino group. I shall, however, discuss in detail only the specificity of enzymatic reactions.

From the standpoint of molecular structure and the quantum mechanical theory of chemical reaction, the only reasonable picture of the catalytic activity of enzymes is that which involves an active region of the surface of the enzyme which is closely complementary in structure not to the substrate molecule itself, in its normal configuration, but rather to the substrate molecule in a strained configuration, corresponding to the "activated complex" for the reaction catalyzed by the enzyme: the substrate molecule is attracted to the enzyme, and caused by the forces of attraction to assume the strained state which favors the chemical reaction—that is, the activation energy of the reaction is decreased by the enzyme to such an extent as to cause the reaction to proceed at an appreciably greater rate than it would in the absence of the enzyme. This is, I believe, the picture of enzyme activity which is usually accepted.

Experimental data have not been gathered which permit the induction of so precise a representation of the structure and configuration of the active region of any enzyme as for the antibodies discussed above, but there do exist some data which support the general concept. If the enzyme were completely complementary in structure to the substrate, then no other molecule would be expected to compete successfully with the substrate in combining with the enzyme, which in this respect would be similar in behavior to antibodies; but an enzyme complementary to a strained substrate molecule would attract more strongly to itself a molecule resembling the strained substrate molecule than it would the substrate molecule. Examples of this behavior have been found: the hydrolysis of benzoyl-*l*-tyrosylglycine amide by either chymotrypsin or papain was found by Bergmann and Fruton (*2*) to be practically completely inhibited by an equal amount of benzoyl-*d*-tyrosylglycine amide. This suggests that the strained configuration of the *l*-isomer during the enzymatic hydrolysis is somewhat similar to the normal configuration of the *d*-isomer.

More extensive quantitative studies of inhibition of enzyme activity might well provide very interesting information about the configuration of the enzyme molecules. Carl Niemann and I have studies of this kind under way.

It is highly probable that many chemotherapeutic agents exercise their activity by acting as inhibitors to an enzymatic reaction through competition with an essential metabolite of similar structure. It was shown by Woods (*16*) in 1940 that the bacteriostatic action of sulfanilamide results from an inhibitory competition with *p*-aminobenzoic acid, and can be overcome by increasing the concentration of the latter substance. The metabolite and its inhibitor are closely related in molecular shape, differing in the replacement of a carboxyl group by a sulfonamide group. Other pairs in which a carboxyl group is replaced by a sulfonic acid or a sulfonamide group are nicotinic acid and pyridine-3-sulfonic acid or its amide (*7*), pantothenic acid and pantoyltaurine (*14*) and the α-aminocarboxylic acids and the corresponding α-aminosulfonic acids (*8*).

An interesting case of inhibition is that of thiamine by pyrithiamine (*13*), the corresponding substance with the 6-membered pyridine ring in place of the 5-membered thiazole ring. The effective competition of pyrithiamine with thiamine for combination with the enzyme or other macromolecule involved might well have been predicted from the known cross reactivity of aromatic 5-membered rings containing sulfur and 6-membered rings not containing sulfur, as is strikingly shown by the formation of solid solutions by thiophene and benzene. An analogous situation has been reported (*15*) by D. S. Tarbell of the University of Rochester. He has found that any substitution in the benzenoid ring of 2-methylnaphthoquinone destroys its vitamin K activity, but that the substance with a sulfur atom in place of —CH=CH— in the benzenoid ring retains this activity. These facts indicate that in the process of exerting vitamin K activity the benzenoid end of the molecule must fit into a pocket carefully tailored to it; that the other end is not so surrounded is shown by the retention of activity on changing the alkyl group in the 2-position. On the other hand, the failure of pyrithiamine to replace thiamine as a metabolite indicates that the sulfur atom of the thiazole ring in thiamine not only is effective in binding the molecule into its seat of action but also takes part in some way in the subsequent chemical reactions involved in the metabolic process.

Many Sciences Cooperate

The complete understanding of physiological activity will require consideration not only of molecular structure and weak intermolecular forces, but also of the chemical reactivity of the substances and of such other properties as solubility in different phases and degree of ionization, as well as of those properties of living organisms which may long defy simplification to chemical description; the importance of the problem for practical medicine as well as for fundamental biology is so great as to justify the attention and effort of many workers, in various fields of science, through whose cooperative effort the solution will some day be found.

Literature Cited

(1) Alexander, J., *Protoplasma*, **14**, 296 (1931).
(2) Bergmann, M., and Fruton, J. S., *J. Biol. Chem*, **138**, 124, 321 (1941).
(3) Brand, E., Saidell, L. J., Goldwater, W. H., Kassell, B., and Ryan, F. H., *J. Am. Chem. Soc.*, **67**, 1524 (1945).
(4) Breinl, F., and Haurowitz, F., *Z. physiol. Chem.*, **192**, 45 (1930).
(5) Corey, R. B., *Chem. Rev.*, **26**, 227 (1940).
(6) Crowfoot, D., and Rogers-Low, B., mentioned in *Science*, **102**, 627 (1946).
(7) McIlwain, H., *Brit. J. Exptl. Path.*, **21**, 136 (1940).
(8) McIlwain, H., *J. Chem. Soc*, **1941**, 75; *Brit. J. Exptl. Path.*,**22**, 148 (1941).
(9) Mirsky, A. E., and Pauling, L., *Proc. Nat. Acad. Sci.*, **22**, 439 (1936).
(10) Mudd, Stuart, *J. Immunol.*, **23**, 423 (1932).
(11) Pauling, L., *J. Am. Chem. Soc.*, **62**, 2643 (1940).
(12) Pauling, L., and collaborators, *J. Am. Chem. Soc.*, **68**, 250 (1946), and earlier papers.
(13) Robbins, W. J., *Proc. Nat. Acad. Sci.*, **27**, 419 (1941); Woolley, D. W., and White, A. G. C., *J. Exptl. Med.*, **78**, 489 (1943).
(14) Snell, E. E., *J. Biol. Chem.*, **139**, 975,**141**, 121 (1941).
(15) Tarbell, D. S., private communication to author.
(16) Woods, D. D., *Brit. J. Exptl. Path.*, **21**, 74 (1940).

(*Reprinted from* NATURE, *Vol.* 161, *page* 707, *May* 8, 1948.)

NATURE OF FORCES BETWEEN LARGE MOLECULES OF BIOLOGICAL INTEREST*

By PROF. LINUS PAULING
California Institute of Technology

AS I look at a living organism, I see reminders of many questions that need to be answered. Not all these questions are obviously important, nor would their answers be useful—but we want them answered. Thomas Wright in 1601 said, "Nothing is so curious and thirsty after knowledge of dark and obscure matters as the nature of man"—of scientific men especially, he might have said.

What is skin, fingernail ? How do fingernails grow ? How do I feel things ? How are nerves built and how do they function ? How do I see things ? How can I smell things, and why does benzene have one smell and *iso*-octane another ? Why is sugar sweet and vinegar sour ? How does the hæmoglobin in my blood do its job of carrying oxygen from the lungs to the tissues ? How do the enzymes in my body break up the food that I eat, burn it to keep me warm and to permit me to do work, and build new tissues for me from the food fragments ? Why do I catch cold when exposed through contact with an ailing person, get pneumonia, and then recover after treatment with a specific antiserum or a sulpha drug ? How does penicillin carry out its wonderful function of fighting disease ? Why am I immune to measles, whooping cough, poliomyelitis, smallpox, whereas some other people are not ? And finally, why is it that my children, as they grow and develop, become human beings, and show characteristics similar to mine, and their mother's—how have these characters been transmitted to them ?

The basic answers to all these questions are not to be found in books. Even though Chaucer said

"For out of olde feldes, as men seith,
Cometh al this newe corn fro yeer to yere ;
And out of olde bokes, in good feith,
Cometh al this newe science that men lere",

he was before long corrected by Francis Bacon :

"Books must follow sciences, and not sciences books".

To understand all these great biological phenomena we need to understand atoms, and the molecules that

* Friday Evening Discourse at the Royal Institution, delivered on February 27.

they form by bonding together ; and we must not be satisfied with an understanding of simple molecules—nitrous oxide, NNO, that Davy used to produce hysteria ; benzene, C_6H_6, discovered by Faraday ; penicillin, containing forty atoms in its molecule. We must also learn about the structure of the giant molecules in living organisms, such as insulin, with 2,000 atoms in its molecule, hæmoglobin, with about 10,000, the disease-producing viruses, a thousand times larger still.

Just two thousand years ago the Roman poet Lucretius wrote

"Wine flows easily because its particles are smooth and round and roll easily over one another, whereas the sluggish olive oil hangs back because it is composed of particles more hooked and entangled one with another".

This is essentially the modern point of view, expressed by Lucretius, of course, only as a surmise, whereas to us it is a fact, of which we have a detailed understanding. Lucretius went beyond us in interpreting the taste also of substances in terms of the shape of their molecules, writing

"There is this, too, that the liquids of honey and milk give a pleasant sensation of the tongue, when rolled in the mouth ; but on the other hand the loathsome nature of wormwood and biting centaury set the mouth awry by their noisome taste ; so that you may easily know that those things which can touch the senses pleasantly are made of smooth and round bodies, but that on the other hand all things which seem to be bitter and harsh, these are held together with particles more hooked".

I shall now present a detailed discussion of the shapes of molecules, and the forces between molecules, in order to show the extent to which this surmise, too, of Lucretius is substantiated.

Even though hæmoglobin is spoken of as a large molecule, the molecule is very small in comparison with our usual scale. If I look at my hand or watch a few feet away, the atoms subtend the same angle that oranges would on the surface of the moon. The molecules might thus be described as an aggregate of oranges, held together by bonds which in actual molecules are electrons circulating between the atoms. On this scale of 250,000,000-fold linear magnification, the nitrous oxide molecule of Davy's laughing gas consists of three atoms in line, N–N–O, the distance between the two nitrogen atoms being about 1·12 in. and between the nitrogen and oxygen atoms 1·19 in. (in the molecules themselves 1·12 A. and 1·19 A., where 1 A. is 1/100,000,000 cm.). Faraday's benzene molecule consists of twelve atoms: six carbon atoms arranged at the corners of a regular hexagon, with the carbon-carbon distance 1·39 A., and six hydrogen atoms, in the same plane, at the corners of a larger hexagon, the distance of each hydrogen atom to the carbon atom to which it is attached being 1·06 A. The structure of penicillin, determined by Mrs. Hodgkin and her collaborators, is also known, as are the structures of many hundreds of simpler molecules.

As yet the detailed atomic structure of no protein molecule has been determined; but the vigour of the attack that is being made and the increasing power of the marvellous methods of investigation used in structural study permit the confident hope that the next decade or two will bring the solution of this great problem.

Much information is already available, however, about the general sizes and shapes of large molecules of biological importance, and also about their function. The most striking general property of these molecules is the specificity of their activity. For example, hæmoglobin, the red protein in the red cells of the blood, has the specific property of combining with oxygen in the lungs, and then releasing this oxygen in the tissues, thus making it available for the oxidation of foods. There is no other substance known that is able to combine with oxygen and release so large a fraction of the combined oxygen reversibly with the small change in partial pressure of oxygen existing between the lungs and the tissues. Then there are the enzymes, such as those involved in the catalysis of the processes of degradation and oxidation of foodstuffs. These substances all show a high degree of specificity in their action: in general, a particular enzyme serves the purpose of speeding up a particular single chemical reaction in the body.

Specificity is shown in a striking way by antibodies. These are substances produced in the living body in response to its invasion by foreign organic material, such as viruses or bacteria. They have the power of combining specifically with the material (the antigen) that led to their production, and of neutralizing or incapacitating the antigen. Thus an antitoxin is able to neutralize its homologous toxin; an antibody against a virus or bacterium is able to prevent the virus or bacterium from reproducing, and thus to aid in the control of disease. An unfortunate incidental effect of this protective mechanism of the body is the occasional production of protein sensitization, leading to asthma and hay fever. Sufferers from protein sensitization, such as those people who are sensitive to egg white (ovalbumin), can testify as to the specificity of the reaction and to its delicacy.

The specificity of serological reactions can be

shown by simple serological tests carried out in the laboratory. Thus if a minute amount of ovalbumin is injected into a rabbit, it is found that after a week or two the serum of the rabbit's blood will produce a precipitate when mixed with a solution of ovalbumin, whereas the serum of an uninoculated rabbit will not produce such a precipitate. The precipitate contains ovalbumin from the ovalbumin solution that is added to the serum, and antibody molecules from the serum of the rabbit. These antibody molecules are protein molecules, with molecular weight about 160,000, corresponding to the presence in the molecule of about 20,000 atoms. These anti-ovalbumin molecules have the power of combining with molecules of ovalbumin, but in general not with molecules of other substances. Only if another substance is extremely closely similar in its structure to the ovalbumin with which the rabbit was originally injected can the antibody combine with it. Thus a precipitate, somewhat smaller in amount than with hen ovalbumin, is formed by antibodies produced by hen ovalbumin when mixed with a solution of duck ovalbumin ; but no precipitate is formed when these antibodies are mixed with a solution of ovalbumin from the eggs of birds of other species.

Dr. Karl Landsteiner showed by his experiments that animals are able to manufacture antibodies specific to chemical groups that do not exist in Nature. The power of formation of antibodies able to combine only with an injected antigen is hence not a power that was developed in the course of evolution, in response to interaction with this antigen at some earlier age, and in a particular individual called into operation by the inoculating injection ; but is instead a general power, which enables the animal to cope with a foreign material, even of completely foreign nature to the animal and its ancestors throughout its course of evolution.

I became interested in the problem of the structure of antibodies in 1936, as the result of conversations with Dr. Landsteiner. I found that the complex phenomena of immunology could be clarified and brought into order by a theory of the structure of antibodies based upon the idea of the folding of the basic polypeptide chain of the protein precursor of the antibody into the most stable of the configurations accessible to it. This theory of the structure and process of formation of antibodies involves the acceptance of the suggestion that antibody and antigen have molecular structures that are mutually complementary, originally made by Breinl and Haurowitz, Jerome Alexander, and Stuart Mudd. If the molecule of gamma globulin that might become an antibody is considered to consist of a chain that might fold up into any one of a large number of configurations, with essentially the same stability, then in the presence of a molecule of antigen that configuration would be selected which is stabilized by interaction with the antigen. The configuration that would be stabilized is the one in which there is the greatest force of attraction between the folded polypeptide chain and the molecule of antigen ; and this greatest force of attraction would result from the assumption by the folding polypeptide chain of a configuration which permits the parts of the chain to approach as closely as possible to a portion of the surface of the antigen molecule, and which also brings positive charges in the chain in close proximity to negative charges in the antigen, and brings hydrogen-bond-forming groups in the chain into juxtaposition with their complementary hydrogen-bond-forming groups in the antigen. A very high degree of specificity can be obtained if the surface area over which the complementariness of structure is exercised is great enough to include a good number of interacting structural units.

Experiments carried out with antibodies and antigens have shown that the configuration of the antibody molecules is very closely complementary to that of the surface of the homologous antigen ; the antibody reflects or reproduces, in a negative way, the shape of the surface of the immunizing antigen to within about 1 A., that is, to within about one-half or one-quarter of the diameter of an atom. The forces that lead to the production of the specific bonds between antibody and antigen are thus interatomic forces that operate between atoms essentially in contact with one another, and are not long-range specific forces of attraction operating through great distances (on the atomic scale) through space.

There is a highly specific phenomenon of the chemistry of simpler substances that is closely analogous in its nature and its cause to the highly specific phenomenon of serological interaction ; namely, the phenomenon of crystallization. Chemists are accustomed to using the process of crystallization as a method of purification : a crystal growing in a complex mixture of molecules is able to select from the mixture just the molecules of one kind, rejecting all others. Thus pure crystals of sugar may deposit from a jam in which there are molecules of thousands of different substances. The specificity of crystallization is the result of the same striving toward complementariness and the operation of the same interatomic and intermolecular forces that are responsible for the specificity of antibodies. A molecular crystal has the structure that gives it the greatest stability, which would result from the maximum amount of attraction for each molecule in the crystal and the surrounding molecules. Each molecule in the crystal is then in a cavity that conforms in shape to the shape of the molecule itself. The molecule may be described as complementary in structure to the remainder of the crystal ; and other molecules, with different shape and structure, would not fit into this cavity nearly so well, and in general would not be incorporated in the growing crystal. We may hence say that life has borrowed from inanimate processes the same basic mechanism used in producing those striking structures that are crystals, with their beautiful plane faces, their unfailingly constant interfacial angles, and their wonderfully complex geometrical forms.

I believe that the same mechanism, dependent on a detailed complementariness in molecular structure, is responsible for all biological specificity. I think that enzymes are molecules that are complementary in structure to the activated complexes of the reactions that they catalyse, that is, to the molecular configuration that is intermediate between the reacting substances and the products of reaction for these catalysed processes. The attraction of the enzyme molecule for the activated complex would thus lead to a decrease in its energy, and hence to a decrease in the energy of activation of the reaction, and to an increase in the rate of the reaction. Although convincing evidence is not yet at hand, I believe that it will be found that the highly specific powers of self-duplication shown by genes and viruses are due to the same intermolecular forces, dependent upon atomic contact, and the same processes of replica formation through complementariness in structure as are operative in the formation of antibodies under the influence of an antigen. I believe that it is molecular size and shape, on the atomic scale, that are of primary importance in these phenomena, rather than the ordinary chemical properties of the substances, involving their power of entering into reactions in which ordinary chemical bonds are broken and formed.

Even though the general picture of some important biological processes is becoming clear, our present knowledge of the detailed structure of the complex substances of biological importance is vague. We may expect that as more precise information about the structure of these molecules is obtained in the future, a more penetrating understanding of biological reactions will develop, and that this understanding will lead to great progress in the fields of biology and medicine.

PRINTED IN GREAT BRITAIN BY FISHER, KNIGHT AND CO., LTD., ST. ALBANS

THE UNIVERSITY OF NOTTINGHAM

Sir Jesse Boot
Foundation Lecture
1948

Molecular Architecture and the Processes of Life

BY

Professor LINUS PAULING, B.S., M.A., Ph.D., D.Sc.,
Hon. Mem. R.I., For. Mem. R.S.

MOLECULAR ARCHITECTURE AND THE PROCESSES OF LIFE.

The last twenty-five years have seen great progress made in our understanding of the nature of life. This progress has been along two lines : first the chemical substances that make up the living body have been isolated, and information has been obtained about their properties and about the work that they do, and second, great insight has been obtained into the structure of the molecules of chemical substances generally, in terms of atoms and electrons, and this understanding of the properties of chemical substances in terms of their molecular architecture is now being extended to include the very complicated substances responsible for life.

What are the features that are characteristic of a living organism ? As we look about us we see such organisms everywhere—human beings, other animals, plants ; and we know that there are very many forms of life that we do not see so easily, such as the bacteria that do their beneficial work in the soil converting waste organic material into substances that can be used by plants in their growth, and that also, in some cases, work for harmful ends, as when they produce illnesses, such as pneumonia and typhoid fever. These bacteria, when they are examined under the microscope, are seen to be far simpler in structure than the larger organisms ; they consist perhaps of a single cell, whereas the larger organisms may consist of many millions of cells, with specialized functions, working in co-operation with one another. But we recognize that the bacteria are living : they are able to grow in size, and to reproduce themselves, and they hand on to their progeny the specific characters that they themselves possess.

It is these properties that differentiate living matter from non-living matter—the possession of specific characters, and the ability to produce progeny, to which these specific characters are passed on. Can we obtain an understanding of these properties ? Do we know what the nature of life is ? I believe that we can understand these properties of living matter, and that we do know what the nature of life is (aside from consciousness), in terms of molecular architecture, the atomic structure of the molecules that constitute living organisms.

2

Let us first consider how the body works. It does its work by use of special molecules, molecules that have specific properties suitable to the use to which they are put. A few of these molecules are simple ones, representing ordinary chemical substances, such as water, oxygen, carbon dioxide. Others are more complicated, with ten or twenty or thirty atoms per molecule. These include the necessary food substances called the vitamins—vitamin A, vitamin B, vitamin C, and so on, with formulas such as $C_{20}H_{30}0$ (Vitamin A), and special foods such as sugar, $C_{12}H_{22}0_{11}$. And then there are the giant molecules, containing tens of thousands or hundreds of thousands of atoms, and with certain well-defined properties—the ability to do very special jobs that serve the purposes of the organism. Thus in living we make use of oxygen of the air, to burn certain materials in our tissues, and in this way to obtain heat and energy to keep our bodies warm and to permit us to do work, and also at the same time to get rid of some unwanted materials by burning them to water and carbon dioxide. The oxygen that we inhale goes into the blood in the lungs. It does this by combining with a special substance in the blood, the red substance called hemoglobin. Hemoglobin is an extremely interesting substance. It is practically the only substance known that has the property of combining easily with oxygen from the air, and then of giving up the oxygen under slightly changed conditions. A molecule of hemoglobin contains about 10,000 atoms—atoms of carbon, nitrogen, oxygen, sulfur, and four atoms of iron. It is these four iron atoms that are directly concerned in the job of carrying oxygen from the lungs to the tissues; each of the iron atoms attaches to itself one molecule of oxygen, 0_2, and carries it along in the arterial bloodstream to the tissues, where it gives it up. Ordinary iron atoms, in ordinary compounds of iron, do not have this property. *It is the special structure of the hemoglobin molecule, the special arrangement of other atoms, of nitrogen, carbon, hydrogen, and sulfur, in the neighbourhood of the iron atom in the hemoglobin molecule, that gives to this molecule this special property.* Then in the muscles and other tissues another protein, myoglobin, only one-quarter as large as hemoglobin, takes the oxygen from the hemoglobin and carries it around within the muscle. Here still other special substances, giant molecules, begin to work. Under ordinary circumstances the presence of oxygen molecules, in the air, in the neighbourhood of a food or other combustible material does not lead to the oxidation of this material. It is usually necessary to light a fire, in order to raise the temperature high enough for the process of oxidation to go on. The body operates at body temperature, 98.6°F., by having developed some ways of causing this burning to go on even at this relatively low temperature. These ways involve giant molecules, enzymes, with 10,000 or 20,000 or more atoms per molecule, which have the ability of causing chemical reactions to go on at lower temperatures than in the absence of the enzymes. The body contains many special enzymes connected with the oxidation of the breakdown products of foodstuffs in the body, and with the oxidation of the parts of the body itself that are no longer needed. Just as hemoglobin has the specific ability of combining with oxygen and carrying it from the lungs to the tissues, so do the various enzymes have the specific ability of oxidizing certain materials. In general each enzyme has one use : it catalyzes one reaction of the many thousands of reactions that take place in the living body.

Moreover, it is giant molecules, presumably molecules of nucleoprotein, that determine the characters of individual living organisms and that are involved in the transmission of these characters to their progeny. These giant molecules are the genes, which are usually present in structures in the cell called chromosomes.

The Gregorian monk Mendel noted that the inheritance of characters by pea plants, such as the character of tallness or of dwarfness, or the character of having purple flowers or white flowers, could be understood on the basis of hereditary units transmitted from the parent to the offspring. Thomas Hunt Morgan and his collaborators identified these units with genes arranged in a linear array in the chromosomes. They have determined the arrangement of several hundreds of these genes in the chromosomes of the vinegar fly, *Drosophila melanogaster*, and it is likely that even this very small organism has many more than this number of genes, probably as many as 20,000.

A wonderful step forward has been made in recent years by Professor George W. Beadle and his collaborators[1], who have developed techniques for investigating chemical genes, the genes that are responsible for the ability of organisms to carry out the chemical reactions that are fundamental to their nature. This has introduced a great simplification into genetics. The character of having short wings rather than long wings is clearly not a very simple one—it is not easily interpreted in terms of chemistry and of molecular structure. On the other hand, the colour of the eye might be easily interpreted, inasmuch as the presence or absence of an enzyme capable of catalyzing the reaction of formation of a special pigment producing eye colour might determine the colour of the eye, and the corresponding chemical reaction would be catalyzed by a special enzyme. Accordingly the fundamental nature of this character would be the presence or absence of this enzyme, and the basic property of the corresponding gene would be that of being able to cause the manufacture of this specific enzyme in the body. Beadle and his collaborators have carried out the investigation of hundreds of these chemical genes that are responsible for the production of hundreds of different specific enzymes, catalyzing different chemical reactions. The organism that they have used is the red bread mold, *Neurospora.*

I may mention some examples of chemical characters[1]. One of our chemical senses is the sense of taste—the taste of a substance depends upon the structure of its molecules. There are two kinds of people with respect to ability to taste a certain substance, phenylthiourea. This substance seems to some people to be very bitter, whereas to others it is quite tasteless. The inheritance of this character is determined by a single gene, transmitted by the pure Mendelian mechanism.

Beadle has pointed out that it would be difficult to determine this character for a fruit fly, because of the difficulty of finding out whether the material seemed to be bitter or tasteless to this small animal. It is interesting in this connection to mention, however, that some investigators in Maryland thought that they would try to find out whether the substance was also either bitter or tasteless to rats, and they began a series of experiments with this in view. They discovered, instead, that the substance, while harmless to human beings, is very poisonous for rats, and, following up this line, they found that a closely related substance, α thiourea, is still more poisonous to rats. This substance is now finding extensive practical use as a rat poison.

Another simple chemical gene in man is the one involved in the oxidation of phenylpyruvic acid. The chemical reaction of oxidizing this substance

5

is determined by an enzyme, the production of which depends upon a single gene. The ability to carry out this oxidation is a character that is inherited phenylketonurea, an invariable symptom of which is idiocy or imbecility.

The viruses are vectors of disease that are still smaller than bacteria. Measles, smallpox, and many other diseases are caused by viruses. The study of their nature was difficult before the invention of the electron microscope, because the particles of viruses are too small to be seen in the ordinary microscope, using visible light. There is now strong evidence that the viruses, at any rate the simplest ones, may well be considered as being molecules rather than organisms that have the power of growth. That is, this form of life seems to consist of individual groups of atoms that do not change their nature, once that they are formed, but remain in existence for an indefinite period without change, in the same way that molecules may. This conclusion was suggested by the discovery by Wendell Stanley that viruses can be crystallized. In general the reason that a substance can form crystals is that its molecules are identical in shape and size, and so can be piled together in a regular arrangement. The crystallization of viruses by Stanley accordingly indicated that all of the individual particles of a virus are essentially identical with one another in shape and size. This conclusion has been verified, for the simpler viruses, by electron micrographs. The electron microscope has such a large resolving power that it is possible to see extremely small particles, containing only a few thousand atoms. The fundamental particles of, for example, the virus that causes mosaic disease in beans, are seen under the electron microscope to be nearly spherical in shape, and to be all of the same size. They can arrange themselves in a regular array, like a pile of marbles, producing a crystal. Each of these virus molecules contains about one million atoms. Although these molecules may not ordinarily carry out the processes of respiration of air and metabolism of foodstuffs that we usually associate with life, they have one important property that causes us to regard them as living, the property of producing progeny. When the virus invades the tissues of its host, it so influences the chemical reactions that are going on there as to cause the production of many molecules that are replicas of itself, and that are not produced in its absence.

6

This effect is shown in a striking way by the bacterial viruses, which prey upon the bacteria themselves. One molecule of a bacterial virus may infect a single bacterium, and cause it, after the lapse of a certain time, of the order of an hour, to burst open, liberating from its interior some hundreds of molecules that are identical with the molecule of bacterial virus that originally infected the bacterium. The bacterial virus molecule thus has the ability to influence the chemical reactions in the bacterium in a specific way. This is a property very similar to the property associated with genes themselves, and it has indeed become difficult to make a sharp distinction between viruses and genes—the virus may be described as a gene that has escaped from the control of the parent organism.

I have been especially interested in that aspect of biological specificity that is involved in the mechanism of protection of the body against disease. When we first become infected with the measles virus, we become ill. After a few days, however, the protective mechanism of the body has come into operation, and we recover from the disease—the symptoms of the disease were the result of the multiplication of the molecules of measles virus within our body, and the recovery is the result of the development of a police force that stops this multiplication. This police force consists of special molecules, formed in the cells of the body, and circulating in the blood stream. These molecules are antibodies, anti-measles antibodies, with the specific property of being able to combine with the measles virus and to prevent its reduplication. Other diseases also lead to the development of special antibodies to counteract them, and even foreign proteins generally, whether they are harmful or not, cause this process of production of special antibodies to become operative.

The antibodies are closely related to a protein, normal gamma-globulin, which is present to the extent of about one per cent in the blood stream. The molecules of antibody against measles are not, however, identical with the molecules of normal gamma-globulin. They have instead a certain definite structure that permits them to combine with the molecules of measles virus. The specificity of antibodies is great—the antibodies that protect us against measles, causing us to be immune to this disease after a first attack, do not protect us against any other disease. In general we must build up an immunity against each disease, either through exposure to the disease, which may cause a mild attack or a severe attack, or by some special process. Inoculation against diphtheria, typhoid fever, and other diseases, and vaccination against smallpox have in recent decades led to the avoidance of an immense amount of human suffering that would otherwise have occurred.

The process of vaccination against smallpox consists of the introduction into the body not of the virus of smallpox itself, but instead of a few molecules that are closely similar in nature to smallpox virus. The similarity between these molecules, of vaccinia virus, and the molecules of smallpox virus is so great that the antibodies that are produced in response to the injection of vaccinia virus molecules have the power also of combining with smallpox virus, and preventing this virus from reduplicating itself. This very simple and sensible method of combating smallpox has reduced it from the terrible scourge that it once was to a rare disease, that can become important again in the civilized parts of the world only if we forget that vaccination is necessary to keep the disease under control.

The example of the vaccinia virus molecules and smallpox molecules shows that the specificity of antibody molecules is not quite complete. However, it is very great—in general the antibodies that are produced in response to the inoculation by a certain antigen, such as the molecules of a virus, have the power of combining with only that particular antigen, and with no others. Sometimes, if the molecules of two substances are very closely similar, the antibody against one substance may show some power of combining with the molecules of the other substance. For example, if a minute amount of hen albumin, the protein in egg white, is injected into a rabbit, the animal will in a few days produce a large number of molecules of antibody specific to hen egg albumin, and capable of forming a precipitate with it. This antibody does not have the power of combining with any other substances, any of the thousands of the different plant and animal proteins that are known, except the egg albumin of birds of very closely related species, such as the duck—but not even with the egg albumin of birds of more distant species, such as pigeon or ostrich. The antibodies that are produced by a rabbit in response to an injection of human hemoglobin have the power of combining with hemoglobin, but not with the hemoglobin molecules from any other animal, except the monkey.

8

When we think about the mechanism of manufacture of these specific antibodies, with the power of combining essentially only with the molecules of the antigen that were originally injected into the animal, we may consider two reasonable possibilities : first, that the power of producing these antibodies specific to the particular antigen, say, egg albumin, is a power that was developed in the course of evolution of the animal, as a means of protecting itself against hen egg albumin, and that this power is called into play by the first injection of the egg albumin ; and second, that the power is not a special one, developed in the course of evolution as a response to the particular antigen, but is a general power, namely, the power of moulding the precursor of the antibody into a form suitable to any foreign substance, antigen, that may get into the animal body. Dr. Karl Landsteiner, of The Rockefeller Institute for Medical Research, carried out experiments to show that the second of these statements is the correct one. Dr. Landsteiner was the man who discovered the human blood groups, and made it possible for blood transfusions to be carried out safely. He found that he could cause animals to produce antibodies specific to chemical substances that were surely different from any that the animal or any of its ancestors in the course of evolution could have had contact with. One of the substances with which he carried out a large amount of work is a compound of arsenic, para-aminobenzenearsonic acid. Landsteiner caused the molecules of this substance to be attached to an ordinary protein, obtained from sheep serum, and injected this azoprotein into rabbits. In response to this injection rabbits produced antibodies with the specific power of combining with the benzenearsonic acid molecules. This power is shown by the ability of these antibodies to form a precipitate with any azoprotein containing benzenearsonic acid groups, attached to a protein that might be different from that used in the original injection, or even to combine with benzene-arsonic acid itself, and simple derivatives of benzenearsonic acid.

The explanation of the power of specific combination of antibodies with benzenearsonic acid and other antigens was proposed, in a general way, about eighteen years ago, by Breinl and Haurowitz[5], Jerome Alexander[6], and Stuart Mudd[7]. This explanation is that the antibody moulds itself to a portion of the antigen molecule, producing a combining region of the antibody that is complementary in structure to a portion of the antigen molecule.

9

All atoms attract all other atoms, with the general van der Waals electronic forces of attraction. These forces are weak in case that the atoms are far from one another, and strong only when they are essentially in contact. The forces between a few atoms in an antigen molecule and a few atoms in an antibody molecule would not be enough to produce a bond between the two molecules strong enough to resist the disrupting influence of thermal agitation of the molecules. If, however, the combining region of the antibody molecule is complementary in configuration to a portion of the surface of the antigen molecule, so that a large number of atoms of the antibody molecule are able to bring themselves into contact with corresponding atoms in the antigen molecule, then the integrated forces of attraction become large, enough to constitute a significant bond between the two molecules. Other types of intermolecular interaction—the formation of hydrogen bonds, and the forces of attraction between a positive charge and a complementary negative charge—may also contribute significantly, if the structures are complementary with respect to them also. It is clear that a good approximation of the combining region of the antibody to the surface of the antigen can be achieved only by having complementary structures, and if the surface of the antigen were changed by adding onto it a group of atoms even as much as one or two atomic diameters in size, this change might effectively prevent a large part of the combining region of the antibody from getting into satisfactory contact with the surface of the changed antigen, and thus prevent the formation of a significant bond. It is in fact found that even so small a change as the replacement of one of the hydrogen atoms of the benzearsonic acid group by a chlorine atom, which is roughly 50% larger in size (diameter 1.8 Å for chlorine, 1.2 Å for hydrogen), is enough to interfere with the combination of the molecules with the antibody homologous to the unsubstituted benzenearsonic acid.

It is possible to propose a reasonable mechanism for the manufacture of antibody. This mechanism is based upon the concept that a protein molecule is a very long chain of atoms, perhaps a thousand times as great in length as in diameter, which may coil up into a well-defined configuration. The precursor of gamma-globulin, as it is formed in the cells of the body, may be such a long chain, of such a nature that its end portions may coil up into any one of a large number of alternative configurations. In general

10

that configuration will be assumed by the molecule that is the most stable under the conditions of its coiling. If there is an antigen molecule present in the cell, the interaction of the atoms in the surface of the antigen molecule with the atoms of the chain of the gamma-globulin precursor will tend to cause this chain to coil into the configuration that is stabilized to the greatest amount by the forces of interaction with the antigen. This will be just the configuration that is complementary in structure to a portion of the surface of the antigen molecule ; that is, just the configuration that corresponds to the formation of a bond with the antigen.[12,13]

This concept thus gives us an automatic method of producing a substance with a specific biological property, that of combining with the molecules of the antigen. The mechanism of obtaining this property is one of moulding a plastic material, the coiling chain, into a die or mould, the surface of the antigen molecule. I believe that the same process of moulding of plastic materials into a configuration complementary to that of another molecule, which serves as a template, is responsible for all biological specificity. I believe that the genes serve as the templates on which are moulded the enzymes that are responsible for the chemical characters of the organisms, and that they also serve as templates for the production of replicas of themselves.

The detailed mechanism by means of which a gene or a virus molecule produces replicas of itself is not yet known. In general the use of a gene or virus as a template would lead to the formation of a molecule not with identical structure but with complementary structure. It might happen, of course, that a molecule could be at the same time identical with and complementary to the template on which it is moulded. However, this case seems to me to be too unlikely to be valid in general, except in the following way. If the structure that serves as a template (the gene or virus molecule) consists of, say, two parts, which are themselves complementary in structure, then each of these parts can serve as the mould for the production of a replica of the other part, and the complex of two complementary parts thus can serve as the mould for the production of duplicates of itself. In some cases the two complementary parts might be very close together in space, and in other cases more distant from one another—they might constitute individual molecules, able to move about within the cell.

11

Evidence supporting the idea of the existence of mutually complementary structures in living organisms is provided by reported experimental results. The late Professor F. R. Lillie of the University of Chicago found that there is contained in the gelatinous coats of eggs of the sea-urchin and other marine animals a material, called fertilizin, that has the property of agglutinating the sperm of the same species of animal.[11] Thus the animals of this organism produce two substances, fertilizin and the material in the outside of the sperm, that have the power of combining with one another, and presumably are mutually complementary in structure. My colleague Professor Albert Tyler, at the California Institute of Technology, has also obtained from the interior of the egg of sea-urchins a substance, antifertilizin, that has the property of combining specifically with the fertilizin that is obtained from the gelatinous coat of the same eggs[4]. Antifertilizin has the power to agglutinate the egg from the interior of which it can be obtained, in the same way that an antibody specific to cells of a particular sort is able to agglutinate these cells. The presence of this pair of complementary substances in the same cell suggests strongly that other such pairs may be present, and that the presence of pairs of complementary substances may indeed be of fundamental importance to life.

Professor Tyler has in fact found that there is present in the blood serum and in the liver of the Gila monster a substance that is able to combine with the toxic substance present in the venom secreted by this reptile. He has suggested that the poisonous substance, venom, and its complementary substance, antivenin, in combination, are produced by an organ, probably the liver, and are liberated into the blood stream ; and that the venom gland then effects a separation of the two, the venom accumulating in the gland and the antivenin being released into the blood stream. Professor Tyler has summarized the significance of these observations in the following words :

> " The view that cells are made up of constituents that bear the same sort of relation to one another as antigen and antibody leads to the inference that their origin is the result of the operation of the same kind of processes as are involved in antibody formation. For the formation of immune antibodies there is now fairly general acceptance of the views of Breinl and Haurowitz (5), Alexander (6), and Mudd (7) which have been

extended and experimentally supported by Pauling (3,8). These views involve the incorporation of antigen in the site of synthesis of serum globulin so that, under an orienting influence of the antigen, the new globulin that is formed bears regional structural configurations that are complementary to specific regions of the antigen. We add to this the inference that, in the absence of foreign, introduced antigen, the normal globulin is complementary to structures normally present at the site of synthesis. Similarly, then, any of the macromolecular constituents synthesized in a cell would be complementary to the substances comprising the sites of synthesis. Since growth consists primarily in the formation of such substances that comprise the integral structure of the cell, we may regard the mechanism of the process of growth to be essentially analogous to that manifested in antibody formation. For the increase of self-duplicating bodies such as genes, one may invoke the formation of substances that are both complementary and identical, as Pauling and Delbruck (9) have done, or the formation of an intermediate template which Emerson (10) lists as an alternative possibility."

Progress in the understanding of the molecular basis of serological reaction and related biological phenomena in which molecules interact specifically with one another has been very rapid in recent years, and we may hope confidently that a great deepening of our understanding will be obtained as the result of the work of scientists in the near future. This progress will not be of value only in satisfying our curiosity ; it will, instead, surely lead to significant practical results, especially in the battle against disease. During the first half of the present century the workers in the field of medical research have won out in the battle against most of the infectious diseases. The diseases of childhood — diphtheria, measles, scarlet fever, whooping cough—are no longer the scourges they once were, the death rate from them being as low as one-twentieth of those of three decades ago. It is the degenerative diseases—heart disease and related diseases of the kidney and periphero-vascular system, and cancer—that now are the leading causes of death of human beings. The problem of attacking these diseases is made most difficult because of the lack of complete understanding of their nature, and even of the nature of the human organism

13

itself, in terms of molecular structure. When once we know what the molecular architecture of the proteins and other large molecules that carry out the physiological activity of the human body is, what the relation of the structure of these molecules is to that of the vectors of disease, and of the drugs, such as penicillin and the sulpha drugs, that serve effectively in protecting us against infectious disease, what changes in molecular architecture are associated with the degenerative diseases—then we can attack the problem of the degenerative diseases in an effective way, using the methods of attack that are suggested by this knowledge. The study of molecular structure is as important a part of medical research as is the work of the clinical investigator in the hospital. We may have confidence that, through the joint efforts of these research men, working in different fields, further great progress will be made, leading to a great increase in the well-being and happiness of man.

BIBLIOGRAPHY

1. G. W. Beadle, *Chem. Rev.*, *37*, 15 (1945).
2. K. Landsteiner, The Specificity of Serological Reactions, Harvard University Press, Cambridge, Mass., 1945.
3. L. Pauling, *J. Am. Chem. Soc.*, *62*, 2643 (1940).
4. A. Tyler, *Proc. Nat. Acad. Sci. America*, *26*, 249 (1940).
5. F. Breinl and F. Haurowitz, *Z. physiol. Chem.*, *192*, 45 (1930).
6. J. Alexander, *Protoplasma*, *74*, 296, (1932).
7. S. Mudd, *J. Immunol.*, *23*, 423 (1932).
8. L. Pauling and D. H. Campbell, *J. Exp. Med.*, *76*, 211 (1942).
9. L. Pauling and M. Delbruck, *Science*, *92*, 77 (1940).
10. S. Emerson, *Ann. Missouri Bot. Garden*, *32*, 243 (1945).
11. F. R. Lillie, *J. Exp. Zool.*, *14*, 515 (1913).
12. L. Pauling, *Endeavour*, *VII*, 43 (1948).
13. L. Pauling, *Nature*, *161*, 707 (1948).

Reprinted from the Proceedings of the NATIONAL ACADEMY OF SCIENCES,
Vol. 39, No. 2, pp. 84–97. February, 1953

A PROPOSED STRUCTURE FOR THE NUCLEIC ACIDS

BY LINUS PAULING AND ROBERT B. COREY

GATES AND CRELLIN LABORATORIES OF CHEMISTRY,* CALIFORNIA INSTITUTE OF TECHNOLOGY

Communicated December 31, 1952

The nucleic acids, as constituents of living organisms, are comparable in importance to the proteins. There is evidence that they are involved in the processes of cell division and growth, that they participate in the transmission of hereditary characters, and that they are important constituents of viruses. An understanding of the molecular structure of the nucleic acids should be of value in the effort to understand the fundamental phenomena of life.

We have now formulated a promising structure for the nucleic acids, by making use of the general principles of molecular structure and the available information about the nucleic acids themselves. The structure is not a vague one, but is precisely predicted; atomic coordinates for the principal atoms are given in table 1. This is the first precisely described structure for the nucleic acids that has been suggested by any investigator. The structure accounts for some of the features of the x-ray photographs; but detailed intensity calculations have not yet been made, and the structure cannot be considered to have been proved to be correct.

The Formulation of the Structure.—Only recently has reasonably complete information been gathered about the chemical nature of the nucleic acids. The nucleic acids are giant molecules, composed of complex units. Each unit consists of a phosphate ion, HPO_4^{--}, a sugar (ribose in the ribonucleic acids, deoxyribose in the deoxyribonucleic acids), and a purine or pyrimidine side chain (adenine, guanine, thymine, cytosine, uracil, 5-methylcytosine). The purine or pyrimidine group is attached to carbon atom 1′ of the sugar, through the ring nitrogen atom 3 in the case of the pyrimidine nucleotides,[1] and the ring nitrogen atom 9 in the case of the purine nucleotides.[2] Good evidence has recently been obtained as to the nature of the linkage between the sugar and the phosphate, through the investigations of Todd and his collaborators;[3] it seems likely that the phosphate ester links involve carbon atoms 3′ and 5′ of the ribose or deoxyribose. New chemical evidence that the natural ribonucleosides have the β-D-ribofuranose configuration has also been reported by Todd and his collaborators,[4] and spectroscopic evidence indicating that the deoxyribonucleosides have the same configuration as the ribonucleosides has been obtained.[5] The β-D-ribofuranose configuration has been verified for cytidine by the determination of the structure of

TABLE 1

ATOMIC COORDINATES FOR NUCLEIC ACID

ATOM	ρ	ϕ	z	ATOM	ρ	ϕ	z
P	2.65 Å	0.0°	0.00 Å	O_1'	4.4 Å	45.4°	2.65 Å
O_I	2.00	28.3°	−0.67	O_2'	6.1	81.0°	2.1
O_{II}	2.00	−28.3°	0.67	N_3	6.7	52.8°	2.8
$O_{III} = O_5'$	3.72	13.5°	0.93	C_4	7.85	59.3°	2.8
$O_{IV} = O_3'$	3.72	−13.5°	−0.93	C_5	9.1	55.2°	2.8
C_5'	3.4	35.3°	0.7	C_6	9.35	46.9°	2.8
C_4'	3.2	51.6°	1.9	N_6	10.7	44.9°	2.8
C_3'	3.8	74.6°	1.55	N_1	8.45	39.9°	2.8
C_2'	5.3	70.3°	1.75	C_2	7.05	41.5°	2.8
C_1'	5.3	58.2°	2.8	O_2	6.35	32.4°	2.8

Identity distance along z axis = 27.2 Å.

Twenty-four atoms of each kind, with cylindrical coordinates (right-handed axes).

$\rho, \phi + n\cdot105.0°, n\cdot3.40 + z;\ \rho, \phi + n\cdot105.0° + 120°, n\cdot3.40 + z;\ \rho, \phi + n\cdot105.0° + 240°, n\cdot3.40 + z;\ n = 0, 1, 2, 3, 4, 5, 6, 7.$

the crystal by x-ray diffraction; cytidine is the only nucleoside for which a complete x-ray structure determination has been reported.[6]

X-ray photographs have been made of sodium thymonucleate and other preparations of the nucleic acids by Astbury and Bell.[7, 8] It has recently been reported by Wilkins, Gosling, and Seeds[9] that highly oriented fibers of sodium thymonucleate have been prepared, which give sharper x-ray photographs than those of Astbury and Bell. Our own preparations have given photographs somewhat inferior to those of Astbury and Bell. In the present work we have made use of data from our own photographs and from reproductions of the photographs of Astbury and Bell, especially those published by Astbury.[10] Astbury has pointed out that some information about the nature of the nucleic acid structure can be obtained from the x-ray photographs, but it has not been found possible to derive the structure from x-ray data alone.

A configuration of polypeptide chains in many proteins is the α helix.[11] In this structure the amino-acid residues are equivalent (except for differences in the side chains); there is only one type of relation between a residue and neighboring residues, one operation which converts a residue into a following residue. Through the continued application of this operation, a rotation-translation, the α helix is built up. It seems not unlikely that a single general operation is also involved in the construction of nucleic acids, polynucleotides, from their asymmetric fundamental units, the nucleotide residues. The general operation involved would be a rotation-reflection, and its application would lead to a helical structure. We assume, accordingly, that the structure to be formulated is a helix. The giant molecule would thus be cylindrical, with approximately circular cross section.

Some evidence in support of this assumption is provided by the electron micrographs of preparations of sodium thymonucleate described by Williams.[12] The preparation seen in the shadowed electron micrograph is clearly fibrous in nature. The small fibrils or molecules seem to be circular in cross-section, and their diameter is apparently constant; there is no evidence that the molecules are ribbon-like. The diameter as estimated from the length of the shadow is 15 or 20 Å. Similar electron micrographs, leading to the estimated molecular diameter 15 ± 5 Å, have been obtained by Kahler and Lloyd.[13] Also, estimates of the diameter of the molecules of native thymonucleic acid in the range 18 to 20 Å have been made[14, 15] on the basis of sedimentation velocity in the ultracentrifuge and other physicochemical data. The molecular weights reported are in the range 1 million to 4 million.

The x-ray photographs of sodium thymonucleate show a series of equatorial reflections compatible with a hexagonal lattice. The principal equatorial reflection, corresponding to the form 10·0, has spacing 16.2 Å or larger, the larger values corresponding to a higher degree of hydration of the substance. The minimum value,[7] 16.2 Å, corresponds to the molecular diameter 18.7 Å. From the average residue weight of sodium thymonucleate, about 330, and the density, about 1.62 g. cm.$^{-3}$, we calculate that the volume per residue is 338 Å.^{3} The cross-sectional area per residue is 303 Å^2; hence the length per residue along the fiber axis is about 1.12 Å.

The x-ray photographs show a very strong meridional reflection, with spacing about 3.40 Å. This reflection corresponds to a distance along the fiber axis equal to three times the distance per residue. Accordingly, the reflection is to be attributed to a unit consisting of three residues.

If the molecule of nucleic acid were a single helix, the reflection at 3.4 Å. would have to be attributed to a regularity in the purine-pyrimidine sequence, or to some other structural feature causing the three nucleotides in the structural unit to be different from one another. It seems unlikely that there is a structural unit composed of three non-equivalent nucleotides.

The alternative explanation of the x-ray data is that the cylindrical molecule is formed of three chains, which are coiled about one another. The structure that we propose is a three-chain structure, each chain being a helix with fundamental translation equal to 3.4 Å, and the three chains being related to one another (except for differences in the nitrogen bases) by the operations of a threefold axis.

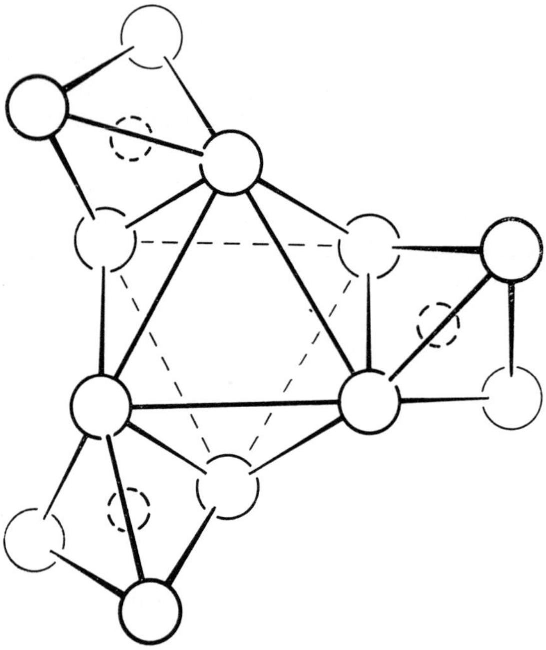

FIGURE 1

A group of three phosphate tetrahedra near the axis of the nucleic acid molecule. Oxygen atoms are indicated by full circles and phosphorus atoms by dashed circles.

The first question to be answered is that as to the nature of the core of the three-chain helical molecule—the part of the molecule closest to the axis. It is important for stability of the molecule that atoms be well packed together, and the problem of packing atoms together is a more difficult one to solve in the neighborhood of the axis than at a distance away from the axis, where there is a larger distance between an atom and the equivalent atom in the next unit. (An example of a helical structure which seems to satisfy all of the structural requirements except that of close packing of atoms in the region near the helical axis is the 5.2-residue helix of polypeptide chains. This structure seems not to be represented in proteins, whereas

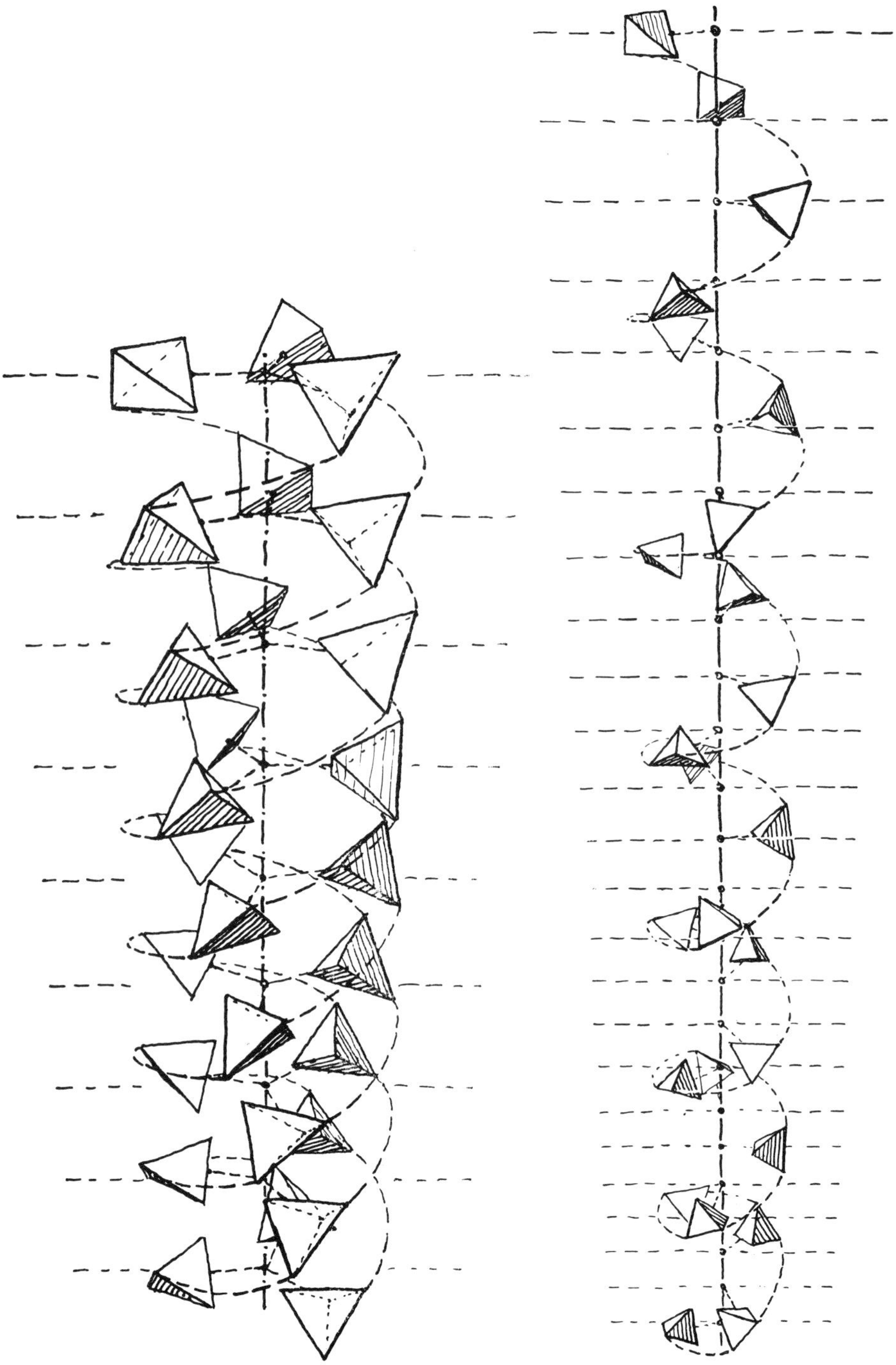

the similar α helix, in which the atoms are packed in a satisfactorily close manner about the axis, is an important protein structure.) There are three possibilities as to the composition of the core: it may consist of the purine-pyrimidine groups, the sugar residues, or the phosphate groups. It is found by trial that, because of their varied nature, the purine-pyrimidine groups cannot be packed along the axis of the helix in such a way that suitable bonds can be formed between the sugar residues and the phosphate groups; this choice is accordingly eliminated. It is also unlikely that the sugar groups constitute the core of the molecule; the shape of the ribofuranose group and the deoxyribofuranose group is such that close packing of these groups along a helical axis is difficult, and no satisfactory way of packing them has been found. An example that shows the difficulty of achieving close packing is provided by the polysaccharide starch, which forms helixes with a hole along the axis, into which iodine molecules can fit. We conclude that the core of the molecule is probably formed of the phosphate groups.

A close-packed core of phosphoric acid residues, HPO_4^{--}, can easily be constructed. At each level along the fiber axis there are three phosphate groups. These are packed together in the way shown in figure 1. Six oxygen atoms, two from each tetrahedral phosphate group, form an octahedron, the trigonal axis of which is the axis of the three-chain helical molecule. A similar complex of three phosphate tetrahedra can be superimposed on this one, with translation by 3.4 Å along the fiber axis, and only a small change in azimuth. The neighborhood of the axis of the molecule is then filled with oxygen atoms, arranged in groups of three, which change their azimuthal orientation by about 60° from layer to layer, in such a way as to produce approximate closest packing of these atoms.

The height (between two opposite edges) of a phosphate tetrahedron is about 1.7 Å. If the same distance were preserved between the next oxygen layers, the basal-plane distance along the fiber axis would be 3.4 Å. This value is the spacing observed for the principal meridional reflection.

It is to be expected that the outer oxygen atoms of the complex of three phosphate groups would be attached to the ribofuranose or deoxyribofuranose residues, and that the hydrogen atom of the HPO_4^{--} residues

---

FIGURE 2

Figure 2 (*left*). A 24-residue 7-turn helix representing a single polynucleotide chain in the proposed structure for nucleic acid. The phosphate groups are represented by tetrahedra, and the ribofuranose groups by dashed arcs connecting them.

FIGURE 3

Figure 3 (*right*). One unit of the 3-chain nucleic acid structure. Eight nucleotide residues of each of the three chains are included within this unit. Each chain executes $3^1/_3$ turns in this unit.

would be attached to one of the two inner oxygen atoms, and presumably would be involved in hydrogen-bond formation with another of the inner oxygen atoms, of an adjoining phosphate group. The length of the O–H ···O bond should be close to that observed in potassium dihydrogen phosphate, 2.55 Å. The angle P—O—H should be approximately the tetrahedral angle. It is found that the spacing 3.4 Å is not compatible with this bond angle, if the hydrogen bonds are formed between one phosphate group

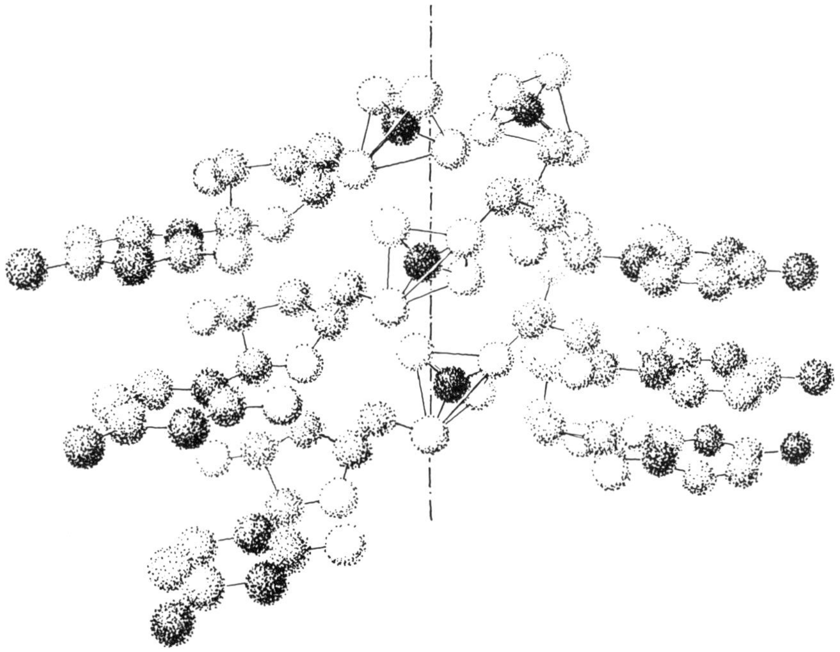

FIGURE 4

Perspective drawing of a portion of the nucleic acid structure, showing the phosphate tetrahedra near the axis of the molecule, the β-D-ribofuranose rings connecting the tetrahedra into chains, and the attached purine and pyrimidine rings (represented as purine rings in this drawing). The molecule is inverted with respect to the coordinates given in table 1.

and a group in the layer above or below it. Accordingly we assume that hydrogen bonds are formed between the oxygen atoms of the phosphate groups in the same basal plane, along outer edges of the octahedron in figure 1.

The maximum distance between the oxygen atoms 3′ and 5′ of a ribofuranose or deoxyribofuranose residue permitted by the accepted structural parameters (C—C = 1.54 Å, C—O = 2.43 Å, bond angles tetrahedral, with the minimum distortion required by the five-membered ring, one atom of the five-membered ring 0.5 Å from the plane of the other four, as reported by Furberg[6] for cytidine) is 4.95 Å. It is found that it is very difficult to assign atomic positions in such a way that the residues can form a bridge between an outer oxygen atom of one phosphate group and an outer oxygen atom of a phosphate group in the layer above, without bringing some atoms into closer contact than is normal. The atomic parameters given in Table

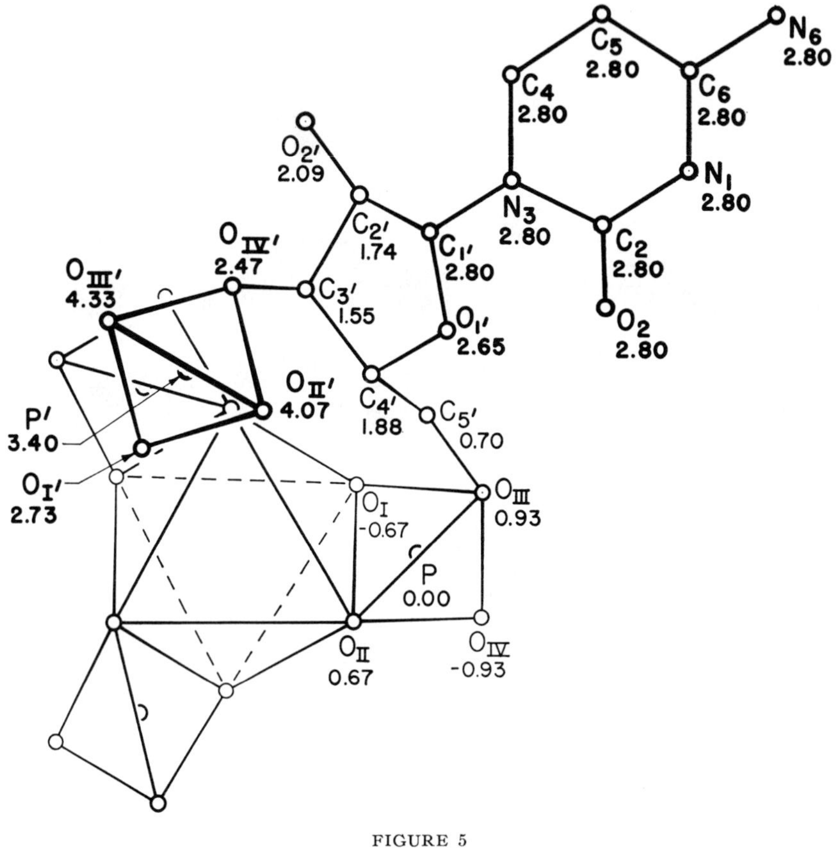

FIGURE 5

A plan of the nucleic acid structure, showing four of the phosphate groups, one ribofuranose group, and one pyrimidine group.

1 represent the best solution of this problem that we have found; these parameters, however, probably are capable of further refinement. The structure is an extraordinarily tight one, with little opportunity for change in position of the atoms.

The phosphate groups are unsymmetrical: the P—O distance is 1.45 Å for the two inner oxygen atoms, and 1.60 Å for the two outer oxygen atoms.

which are involved in ester linkages. This distortion of the phosphate group from the regular tetrahedral configuration is not supported by direct experimental evidence; unfortunately no precise structure determinations have been made of any phosphate di-esters. The distortion, which corresponds to a larger amount of double bond character for the inner oxygen atoms than for the oxygen atoms involved in the ester linkages, is a reason-

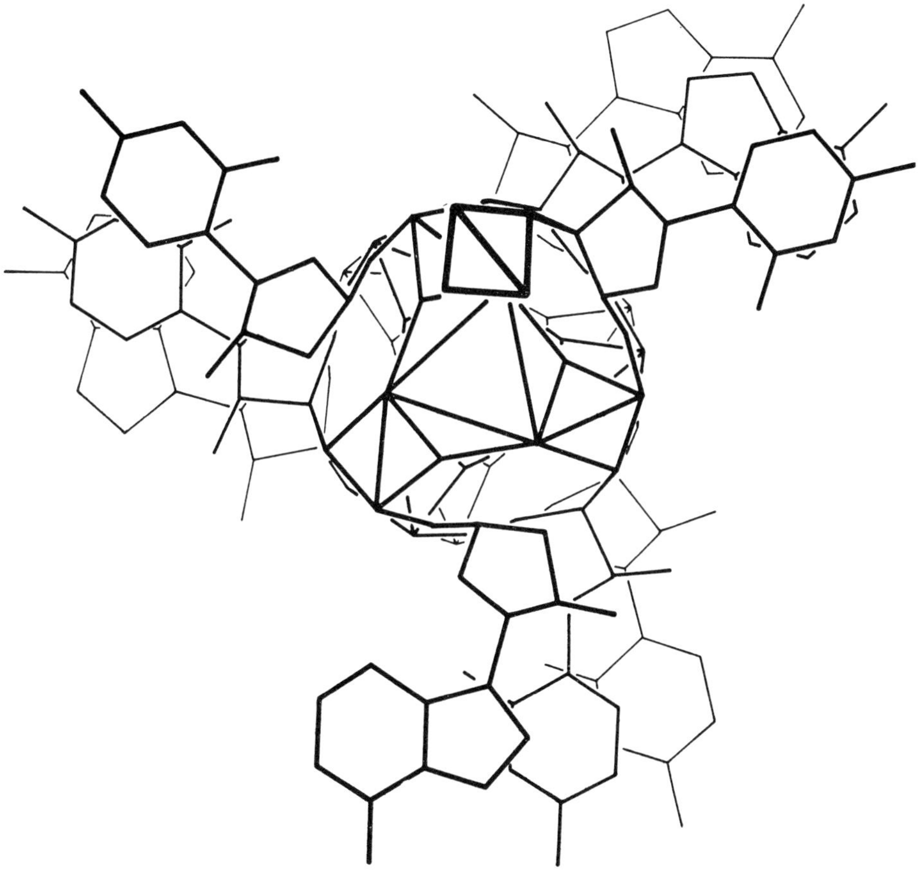

FIGURE 6

Plan of the nucleic acid structure, showing several nucleotide residues.

able one, and the assumed distances are those indicated by the observed values for somewhat similar substances, especially the ring compound S_3O_9, in which each sulfur atom is surrounded by a tetrahedron of four oxygen atoms, two of which are shared with adjacent tetrahedra, and two unshared. The O—O distances within the phosphate tetrahedron are 2.32 Å (between the two inner oxygen atoms), 2.46 Å, 2.55 Å, and 2.60 Å. The hydrogen-bond distance is 2.50 Å, and each phosphate tetrahedron has two O—O contacts at 2.50 Å, with tetrahedra in the layer above. The group of three phosphate tetrahedra in each layer is obtained from that in the layer below by translation upward by 3.40 Å, and rotation in the direction corresponding to a left-handed screw by the azimuthal angle 15°. Thus there are strings of phosphate tetrahedra that are nearly superimposed, and execute a slow twist to the left. These strings are not connected together into a single polynucleotide chain, however. The sugar residues connect each phosphate group with the phosphate group in the layer above that is obtained from it by the translation by 3.40 Å and rotation through the azimuthal angle 105°, in the direction corresponding to a right-handed screw, as shown in figure 2. This gives rise to a helical chain, with pitch 11.65 Å, and with 3.43 residues per turn of the helix. The chain has an identity distance or approximate identity distance of 81.5 Å, corresponding to 24 nucleotide residues in seven turns, as shown in figure 3. The three chains of the molecule interpenetrate in such a way that the pitch of the triple helix is 3.88 Å, and the identity distance or approximate identity distance is 27.2 Å, corresponding to eight layers (see also Figs. 4, 5, and 6).

The structure requires that the sugar residues have the β-furanose configuration; steric hindrance would prevent the introduction of purine or pyrimidine groups in the positions corresponding to the α configuration. The planes of the purine and pyrimidine residues may be perpendicular or nearly perpendicular to the axis of the molecule. This causes these groups to be superimposed in layers that execute a slow left-handed turn about the molecule, the distance between the planes of successive groups being 3.4 Å. The orientation of the groups is accordingly that required by the observed strong negative birefringence of the nucleic acid fibers. The assignment of the sense of the helical molecules corresponding to the right-handed screw is required by the nature of the structure (the packing of the atoms near the axis, and the absolute configuration of the sugar, as given by the recent experimental determination[16] that absolute configurations are correctly given by the Fischer convention).

The structure bears some resemblance to the structures that have been suggested earlier, and described in a general way, without atomic coordinates. Astbury and Bell[7] suggested that the nucleic acid molecule consists of a column of nucleotide residues, with the purine and pyrimidine groups arranged directly above one another, in planes 3.4 Å apart. Astbury[10] considered the possibility that the nucleotides are arranged in a spiral around the long axis of the molecule, and rejected it, on the grounds that it does not lead to a sufficiently close packing of the groups, as is required by the high density of the substance. He pointed out that it is unlikely that adjacent molecules could interleave their purine and pyrimidine residues in such a way as to lead to the high density. Our structure solves this problem by the device of intertwining three helical polynucleotide chains, in such a way that there are three nearly vertical purine-pyrimidine columns, consisting of purine and pyrimidine residues from the three chains in alternation. Furberg[17] suggested two single helical configurations, each resembling in a general way one of our helical polynucleotide chains, but his structures involve orientations of phosphate tetrahedra and the ribofuranose rings that are quite different from ours, and it is doubtful that three chains with either of the configurations indicated in his drawing could be intertwined.

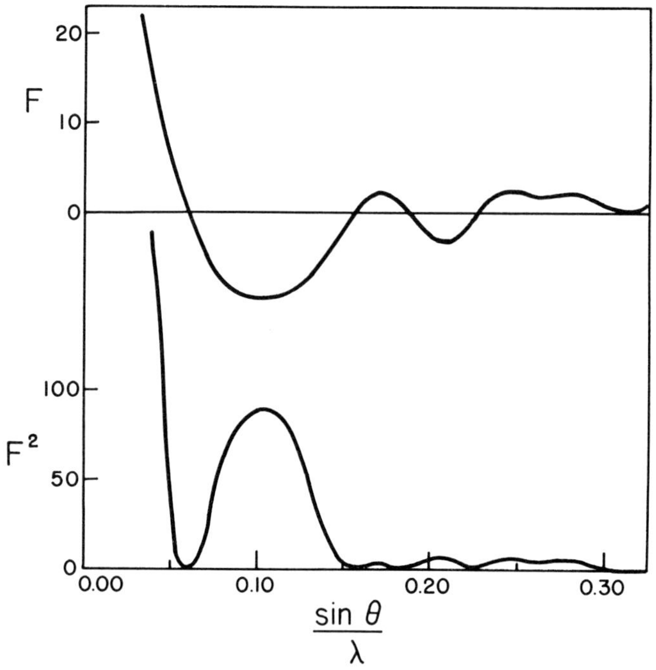

FIGURE 7

The calculated x-ray form factor F and its square F^2 for equatorial reflections of nucleic acid.

The proposed structure accounts moderately well for the principal features of the x-ray patterns of sodium thymonucleate and other nucleic acid derivatives. The spacing 3.40 Å between successive layers of three nucleotides along the molecular axis is required to within about 0.10 Å by the structural parameters of the nucleotides. The prediction that the helixes have 24 nucleotide residues per turn, corresponding to identity distance 8×3.4 Å in the direction of the fiber axis, is in good agreement with the fact that the x-ray diagrams can be reasonably well indexed by placing the 3.4 Å meridional reflection on the eighth layer line. The formula of Cochran,

Crick, and Vand[18] for the form factor for helical structures requires that the orders of Bessel functions for the successive layer lines from 0 to 8 be 0, 3, 6, 9, 12, 9, 6, 3, and 0. The layer-line intensities agree satisfactorily with this prediction, in the region from layer line 4 to layer line 8. There is an unexplained blackening near the meridian for layer lines 2 to 4, which, however, differs in nature for sodium thymonucleate and clupein thymonucleate, and which probably is to be attributed to material between the polynucleotide chains.

The distribution of intensity along the equator can be accounted for satisfactorily. In figure 7 there are shown the calculated form factor in the

TABLE 2

CALCULATED AND OBSERVED EQUATORIAL X-RAY REFLECTIONS FOR SODIUM THYMONUCLEATE. HEXAGONAL UNIT WITH a_0 = 22.1 Å

hkl	$d_{calc.}$	F_1	F_2	pF_2^2	$I_{obs.}$[a]	$d_{obs.}$[a]
10.0	19.1 Å	55	47	6600	m	18.1 Å
11·0	11.0	9.6	21	1350	m	11.2
20·0	9.5	4.7	−1.0	3		
21·0	7.22	−3.4	−8.9	480	w	7.16
30·0	6.37	−7.7	3.1	29		
22·0	5.52	−9.2	1.3	5		
31·0	5.30	−9.4	−14.6	1280	m	5.30
40.0	4.78	−9.3	−14.4	620		
32·0	4.38	−9.1	−14.1	1200 }	m	19[b]
41.0	4.17	−8.8	1.0	6 }		
50·0	3.83	−6.1	−10.8	350		
33.0	3.68	−5.1	4.2	53		
42·0	3.61	−4.3	−8.9	480 }	vw	3.57
51·0	3.43	−2.6	−7.1	300 }		

The symbol *p* in column 5 is the frequency factor for the form.

[a] The observed intensity values and interplanar distances are those reported by Astbury and Bell.

[b] The reflection covers the angular range corresponding to interplanar distances 4.0 to 4.4 Å, and may arise in part from overlapping from the adjacent layer lines.

equatorial direction, and the square of the form factor. It is seen that the form factor vanishes at a spacing of about 8 Å, and has a maximum in the region near 5 Å. Calculated intensities, given in table 2, are obtained by making a correction for interstitial material, at the coordinates 1/3 2/3 and 2/3 1/3, the amount of this material being taken as corresponding in scattering power to 1.5 oxygen atoms per nucleotide residue. There is reasonably satisfactory agreement with the experimental values; on the other hand, similar agreement might be given by any cylindrical molecule with approximately the same diameter. A comparison of observed and calculated radial distribution functions would provide a more reliable test of the structure; this comparison has not yet been carried out.

It is interesting to note that the purine and pyrimidine groups, on the periphery of the molecule, occupy positions such that their hydrogen-bond forming groups are directed radially. This would permit the nucleic acid molecule to interact vigorously with other molecules. Moreover, there is enough room in the region of each nitrogen base to permit the arbitrary choice of any one of the alternative groups; steric hindrance would not interfere with the arbitrary ordering of the residues. The proposed structure accordingly permits the maximum number of nucleic acids to be constructed, providing the possibility of high specificity. As Astbury has pointed out, the 3.4-Å x-ray reflection, indicating a similar distance along the axis of the molecule, is approximately the length per residue in a nearly extended polypeptide chain, and accordingly the nucleic acids are, with respect to this dimension, well suited to the ordering of amino-acid residues in a protein. The positions of the amino-acid residues might well be at the centers of the parallelograms of which the corners are occupied by four nitrogen bases. The 256 different kinds of parallelograms (neglecting the possibility of two different orientations of each nitrogen base) would permit considerable power of selection for each position.

(Added in proof.) Support of the assumed phosphorus-oxygen distances in the phosphate di-ester group is provided by the results of the determination of the structure of ammonium tetrametaphosphate.[19, 20] In this crystal there are P_4O_{12} complexes, consisting of four tetrahedra each of which shares two oxygen atoms with other tetrahedra. The phosphorus-oxygen distance is 1.46 Å for the oxygen atoms that are not shared, and 1.62 Å for those that are shared. These values are to be compared with the values that we have assumed, 1.45 Å for the inner oxygen atoms (which are not shared), and 1.60 Å for the outer ones, which have bonds to carbon atoms.

This investigation was aided by grants from The National Foundation for Infantile Paralysis and The Rockefeller Foundation.

* Contribution No. 1766.

[1] Levene, P. A., and Tipson, R. S., *J. Biol. Chem.*, **94**, 809 (1932); **97**, 491 (1932); **101**, 529 (1933).

[2] Gulland, J. M., and Story, L. F., *J. Chem. Soc.*, **1938**, 259.

[3] Brown, D. M., and Todd, A. R., *Ibid.*, **1952**, 52.

[4] Clark, V. M., Todd, A. R., and Zussman, J., *Ibid.*, **1951**, 2952.

[5] Manson, L. A., and Lampen, J. P., *J. Biol. Chem.*, **191**, 87 (1951).

[6] Furberg, S., *Acta Cryst.*, **3**, 325 (1950).

[7] Astbury, W. T., and Bell, F. O., *Nature*, **141**, 747 (1938); *Cold Spring Harbor Symp Quant. Biol.*, **6**, 109 (1938).

[8] Astbury, W. T., and Bell, F. O., *Tabulae Biologicae*, **17**, 90 (1939).

[9] Wilkins, N. H. F., Gosling, R. J., and Seeds, W. E., *Nature*, **167**, 759 (1951).

[10] Astbury, W. T., in *Nucleic Acids, Symposia of the Society for Experimental Biology, No. 1*, Cambridge University Press (1947).

[11] Pauling, L., and Corey, R. B., these PROCEEDINGS, **37,** 235 (1951).
[12] Williams, R. C., *Biochimica et Biophysica Acta,* **9,** 237 (1952).
[13] Kahler, H., and Lloyd, B. J., Jr., *Biochim. et Biophys. Acta,* in press.
[14] Cecil, R., and Ogston, A. G., *J. Chem. Soc.,* **1948,** 1382.
[15] Kahler, H., *J. Phys. Colloid Chem.,* **52,** 207 (1948).
[16] Bijvoet, J. M., Peerdeman, A. F., and van Bommel, A. J., *Ibid.,* **168,** 271 (1951).
[17] Furberg, S., *Acta Chemica Scand.,* **6,** 634 (1952).
[18] Cochran, W., Crick, F. H. C., and Vand, V., *Acta Cryst.,* **5,** 581 (1952).
[19] Romers, C., Ketelaar, J. A. A., and MacGillavry, C. H., *Nature* **164,** 960 (1949).
[20] Romers, C., dissertation, Amsterdam. 1948.

Reprinted from ARCHIVES OF BIOCHEMISTRY AND BIOPHYSICS, Volume 65, No. 1, November, 1956
Academic Press Inc. *Printed in U.S.A.*

Specific Hydrogen-Bond Formation between Pyrimidines and Purines in Deoxyribonucleic Acids

Linus Pauling and Robert B. Corey

ARCHIVES OF BIOCHEMISTRY AND BIOPHYSICS 65, 164–181 (1956)

Specific Hydrogen-Bond Formation between Pyrimidines and Purines in Deoxyribonucleic Acids

Linus Pauling and Robert B. Corey

From the Gates and Crellin Laboratories of Chemistry, California Institute of Technology, Pasadena, California

Received June 7, 1956

Introduction

Three years ago a novel structure was suggested for deoxyribonucleic acids (DNA) by Watson and Crick (1–3). The proposed structure is of extraordinary interest because it involves a detailed complementariness of two intertwined polynucleotide chains. The possibility that a detailed complementariness in structure is operative in the process of duplication of genes had been suggested earlier by Pauling and Delbrück (4), as a result of the consideration of the strong evidence supporting the idea that an antigen molecule and a homologous antibody molecule have mutually complementary structures.

In the Watson-Crick structure there is assumed to be at every level of the polynucleotide chain (every 3.4 A. along the molecular axis) a nitrogen base (adenine, guanine, thymine, or cytosine) for each of the two helical polynucleotide chains. The two nitrogen bases at each level form a complementary pair, assumed by Watson and Crick to be held together by two hydrogen bonds. Watson and Crick pointed out that the structures of the four nitrogen bases are such as to permit complementary pairs to be formed in four ways: adenine–thymine, thymine–adenine, guanine–cytosine, and cytosine-guanine; in each case they assumed that two hydrogen bonds are formed between the two residues. Adenine and guanine are purines, and thymine and cytosine are pyrimidines; there is one purine and one pyrimidine in each complementary pair. The authors pointed out (2, 3) that a postulate about the duplication of genes is provided by this model of DNA: the two polynucleotide chains of the double helix could separate from one another, and each could then serve as the template for the synthesis of a duplicate of the other.

These considerations are so attractive, and the problem is so important, that we have thought it worth while to analyze the available experimental information about the detailed molecular structure of the pyrimidines and purines. Not many reliable x-ray determinations of the structure of crystals of these substances have been made, and at the present time only tentative conclusions can be reached as to the probable interatomic distances and bond angles in the residues of adenine, thymine, guanine, and cytosine in the polynucleotide chains of DNA. Consideration of the available evidence has, however, permitted us to reach the conclusion that cytosine and guanine should form three hydrogen bonds with one another, rather than two, as suggested by Watson and Crick. Adenine and thymine are able to form only two hydrogen bonds with one another; hence our suggestion of a change in the detailed nature of the hydrogen bonds that can be formed by these pairs of molecules is such as to correspond to a higher degree of specificity than indicated by the considerations of Watson and Crick, rather than a lower degree of specificity, and it thus strengthens the arguments presented by these authors as to the possible role of complementariness in structure of two DNA polynucleotide chains in the process of duplication of the gene.

Interatomic Distances and Bond Angles in Pyrimidines and Purines

Reasonably reliable structure determinations have been reported for crystals of four pyrimidines. The intramolecular and interatomic distances and bond angles for the four molecules are given in Fig. 1, 2, 3, and 4.

In Fig. 1 the molecule of 2-amino-4,6-dichloropyrimidine is represented, as determined by Clews and Cochran (5). These authors also studied the isomorphous substance 2-amino-4-methyl-6-chloropyrimidine; the molecule of this substance has essentially the same dimensions as those of the dichloropyrimidine. The interatomic distances shown in Fig. 1 are not those reported by the authors, but are the values calculated from the atomic parameters that they give in their paper; the distances as calculated from the parameters differ by as much as 0.05 A. from those reported by the authors, and the bond angles differ by as much as 6°. Except for those in Fig. 12, the distances and bond angles reported for the other substances discussed below are those given by the respective authors; some of them have been checked by us.

The molecule of 4-amino-2,6-dichloropyrimidine is represented in

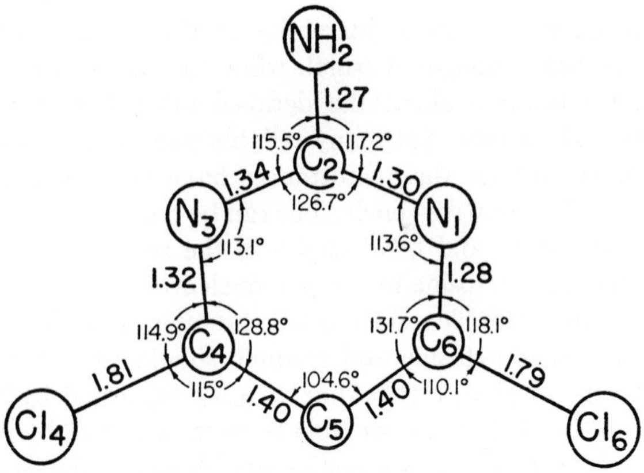

FIG. 1. A drawing showing the dimensions of the molecule of 2-amino-4,6-dichloropyrimidine as determined by x-ray crystal analysis.

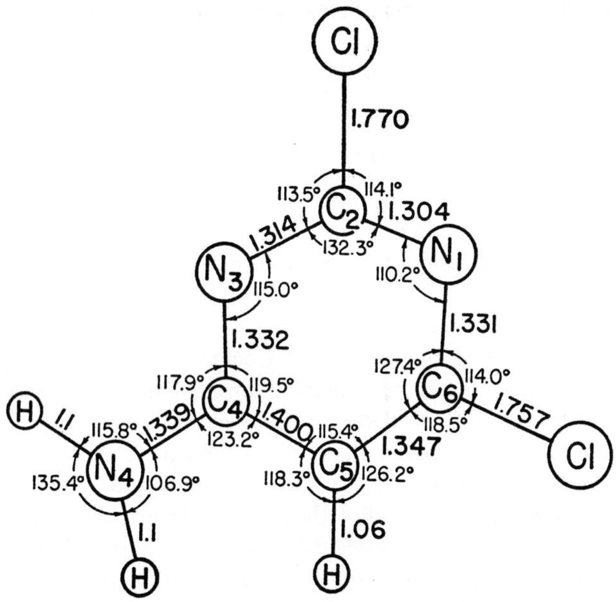

FIG. 2. A drawing showing the dimensions of the molecule of 4-amino-2,6-dichloropyrimidine as determined by x-ray crystal analysis.

Fig. 2, and that of 5-bromo-4,6-diaminopyrimidine in Fig. 3, both as determined by Clews and Cochran (6). The structure of the molecule of uracil is shown in Fig. 4, as determined by Parry (7).

As was pointed out by the investigators, these studies provide important information about the location of the hydrogen atoms. For each

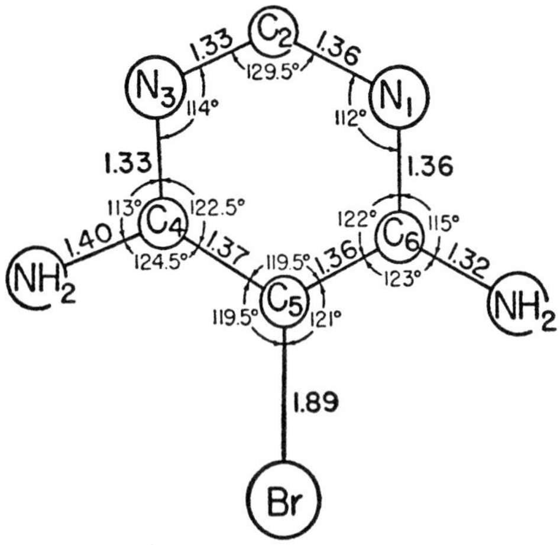

FIG. 3. A drawing showing the dimensions of the molecule of 5-bromo-4,6-diaminopyrimidine as determined by x-ray crystal analysis.

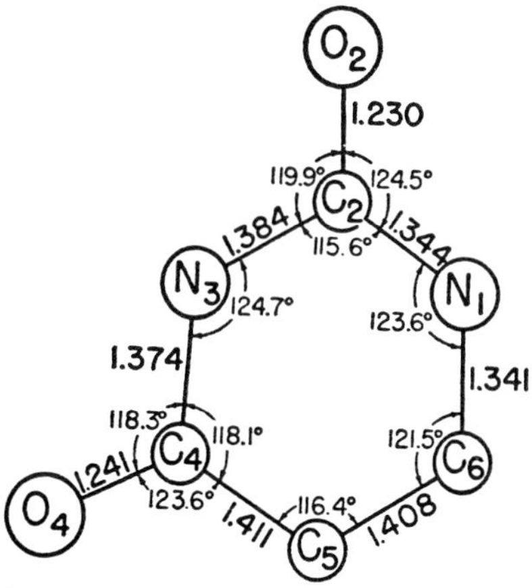

FIG. 4. A drawing showing the dimensions of the molecule of uracil as determined by x-ray crystal analysis.

substance there are alternative structures, involving no hydrogen atoms attached to the two ring nitrogen atoms, a hydrogen atom attached to one of these nitrogen atoms, or in some cases hydrogen atoms attached to both of these nitrogen atoms. If hydrogen atoms are not attached to either of the ring nitrogen atoms, the Kekulé structures, with three double bonds in the ring, would be expected to make a larger contribution to the electronic structure of the molecule than if hydrogen atoms were attached to these atoms. Clews and Cochran (6) reported that they were able to locate the hydrogen atoms in 4-amino-2,6-dichloropyrimidine by means of a three-dimensional Fourier synthesis, and that they found the hydrogen atoms to be attached to the amino nitrogen atom, and not to the ring nitrogen atoms. There is additional evidence supporting the assumption that in the substances represented in Figs. 1, 2, and 3 the ring nitrogen atoms are free of attached hydrogen atoms, and the Kekulé structures make a large contribution to the electronic structures of the molecules. The expected C—N distance in the ring is 1.32 A., in case that the Kekulé structures predominate; it would be somewhat large if the double bonds tend to be formed between ring carbon atoms and side-chain atoms. The average value of the four distances in the ring for the three molecules represented in Figs. 1, 2, and 3 is 1.32 A., as expected for an aromatic ring.

On the other hand, the average value of the four C—N distances in the ring of uracil, Fig. 4, is 1.36 A., which is significantly larger, and suggests that there is a large amount of double-bond character between the carbon atoms 2 and 6 in the ring and the attached oxygen atoms. Moreover, the C—O interatomic distances in uracil, 1.241 and 1.230 A., correspond to a large amount of double-bond character, approximately 60%.

We reach the conclusion, accordingly, that in uracil the hydrogen atoms are attached to the ring nitrogen atoms; that is, that uracil has a double keto structure:

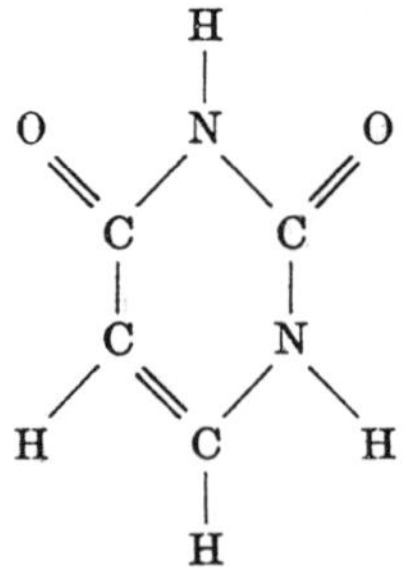

The bond angles provide interesting information. In the substances of Figs. 1, 2, and 3 the bond angles for the ring nitrogen atoms average 113°, the range being 110° to 115°. Inasmuch as the six bond angles of the ring must average 120°, the bond angles for carbon are greater, approximately 124°. We may ask why there is such a great difference, about 10°, in the bond angles for the nitrogen atoms and the carbon atoms in an aromatic ring. The answer can be given by a consideration of bond orbitals. Each of the carbon atoms is forming four bonds: two σ-bonds in the ring, a π-bond in the ring, and a σ-bond outside of the ring. The orbitals for these bonds are sp^3 hybrid orbitals, and the bond angles may be expected to be the normal ones for a tetrahedral atom. The normal value of the angle between a single bond and a double bond for a tetrahedral atom is 125°16′, which is 5° greater than the value 120° for benzene. The nitrogen atoms in the ring, on the other hand, form three bonds —two σ-bonds and a π-bond in the ring—and have an unshared electron pair in the valence shell. The greater stability of a $2s$ orbital occupied by an electron than of a $2p$ orbital will cause the orbital occupied by the unshared pair to have a large amount of s character, leaving only a small amount of s character for the bond orbitals. There is no independent information about the value of a bond angle to be expected between a single bond and a double bond when the bond orbitals have largely p character, but in general it is found that bond angles are determined mainly by the nature of the σ-bonds. We can accordingly make a comparison of the bond angles between single bonds, in order to make a prediction. The bond angle between two sp^3 tetrahedral single bonds is 109°28′, and that between two pure p bonds is 90°. We accordingly expect that the angle between a single bond and a double bond of a ring nitrogen atom in an aromatic molecule (without a hydrogen atom attached to the nitrogen atom) would tend to be approximately 19° less than the value of the tetrahedral bond angle between a single bond and a double bond; that is, about 106°, which is 19° less than 125°16′. The observed difference between the nitrogen bond angles and the carbon bond angles in these molecules is not so great, but clearly indicates a tendency for the bond angle of the carbon atoms to approach the tetrahedral value and for that of the nitrogen atoms to approach a value considerably less than 120°.

The foregoing argument is supported by recently reported experimental values for bond angles in *s*-triazine (8) and *s*-tetrazine (9). In *s*-triazine, $C_3N_3H_3$, the molecule is planar and has trigonal symmetry. The bond angle NCN is found to have the value 126.8 ± 0.4°, and the bond angle CNC to have the value 113.2 ± 0.4°. The C—N distance is reported to be 1.319 ± 0.005 A., in agreement with the value 1.32 A. mentioned above. The bond angles N—C—N and C—N—N in *s*-tetrazine are reported to be 127.4 ± 0.7° and 115.9 ± 0.7°, respectively.

On the other hand, the situation is much different in uracil, as shown in Fig. 4: here the bond angle at the two nitrogen atoms is 124°, and at the carbon atoms 118°, on the average. The diketo structure written above for uracil provides an explanation of the fact that in this molecule it is the nitrogen atoms that have larger bond angles than 120°, and the carbon atoms smaller, whereas in the other pyrimidines the carbon atoms have the larger angles and the nitrogen atoms the smaller. Each of the nitrogen atoms in uracil has a hydrogen atom attached to it; its bond orbitals are accordingly sp^3 orbitals, and the angle in the ring would be expected to have the tetrahedral value, 125°16′, for the angle between a single bond and a double bond. On the other hand, the carbon atoms 2 and 6 are to a large extent forming single bonds within the ring, and their ring bond angles would be expected to approach the single-bond value 109°28′.

We can understand that the hydrogen atoms in uracil are attached not to oxygen, but rather to the ring nitrogen atoms, in the light of the considerable acidity of aromatic hydroxy groups and basicity of ring nitrogen atoms of aromatic heterocycles. On the other hand, the aromatic amino group has only very low acidity, and can be expected to retain

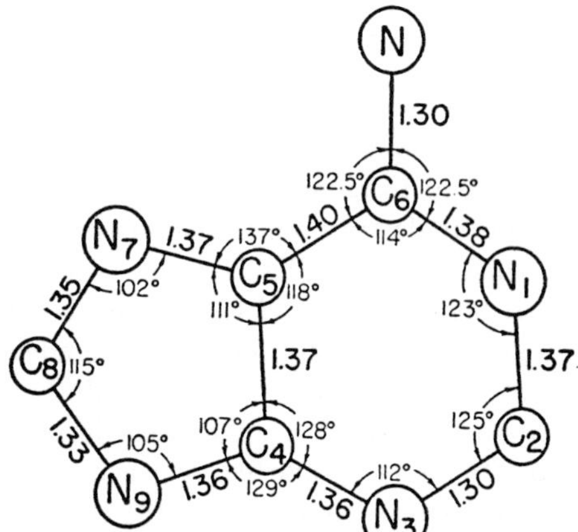

FIG. 5. A drawing showing the dimensions of the molecule of adenine as determined by x-ray analysis of crystals of adenine hydrochloride hemihydrate.

its hydrogen atoms. The general conclusion may be reached that amino groups attached to a pyrimidine ring retain their hydrogen atoms, whereas one or two oxygen atoms attached to the ring are held by double bonds.

Another crystal structure determination of a pyrimidine has been reported (10). We have chosen not to present any discussion of this substance, 4,6-dimethyl-2-hydroxypyrimidine, because values of the atomic parameters are not given in the paper.

The dimensions of the molecules of two purines are given in Figs. 5 and 6. Figure 5 is from the investigation of adenine hydrochloride hemihydrate by Cochran (11), following an earlier study by Broomhead (12), and Fig. 6 from that of guanine hydrochloride monohydrate (13), by Broomhead.

These molecules, like those of the pyrimidines, are planar to within the experimental error of the structure determinations. The bond distances and bond angles are reasonable. The oxygen atom of guanine is indicated not to have a hydrogen atom attached to it both by the C—O distance, 1.20 A., and the ring bond angle of the carbon atom, 108°. The nitrogen atom N_1 has bond angle 126° or 127° in each of the two substances, whereas the other nitrogen atom in the six-membered ring, N_3, has bond angle 113° or 114°. We might interpret these values as indicating clearly that there is a hydrogen atom attached to N_1, but no hydrogen atom attached to N_3.

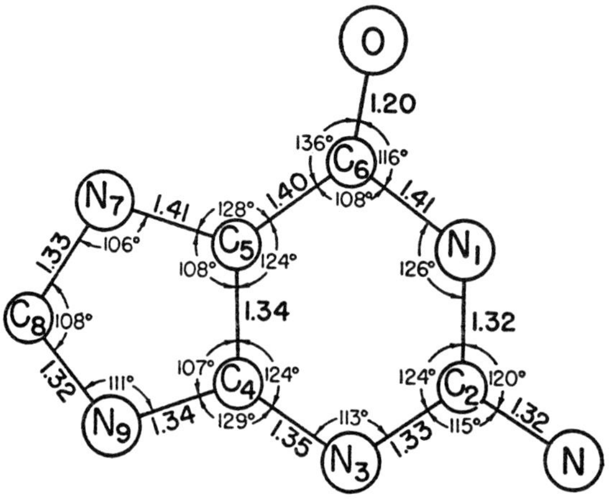

FIG. 6. A drawing showing the dimensions of the molecule of guanine as determined by x-ray analysis of crystals of guanine hydrochloride monohydrate.

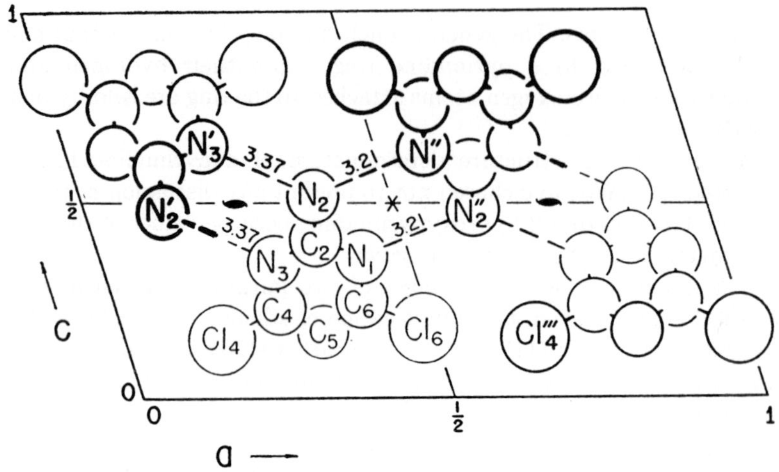

FIG. 7. A drawing of the crystal structure of 2-amino-4,6-dichloropyrimidine. Hydrogen bonds are indicated by dashed lines.

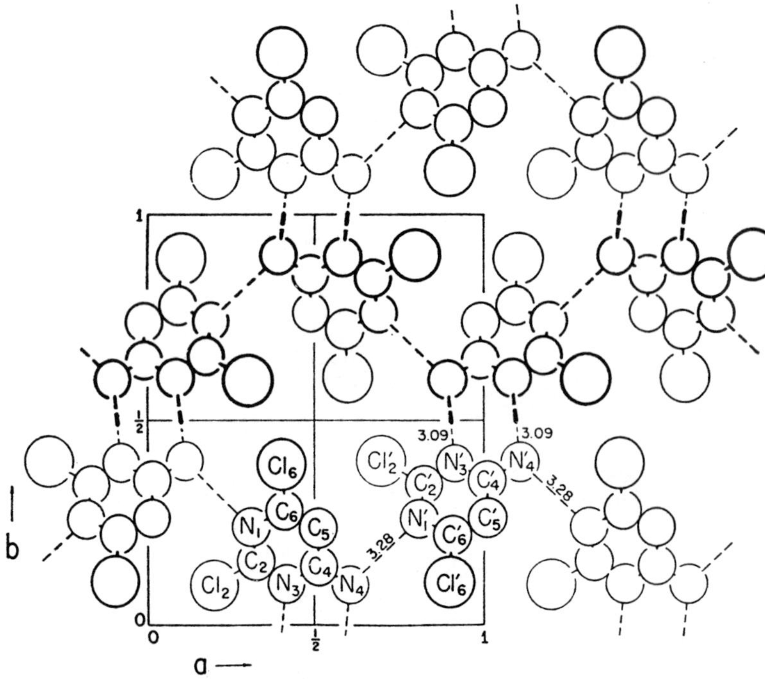

FIG. 8. A drawing of the crystal structure of 4-amino-2,6-dichloropyrimidine. Hydrogen bonds are indicated by dashed lines.

It is probable that there is a small error in the parameters determining the position of the oxygen atom; it is unlikely that the two bond angles of the C—O bond and the adjacent bonds in the ring should differ so much as the reported values, 136° and 116°.

Drawings showing the arrangement of molecules in crystals and giving the values of hydrogen-bond distances for these four pyrimidines and two purines are shown in Figs. 7–13.

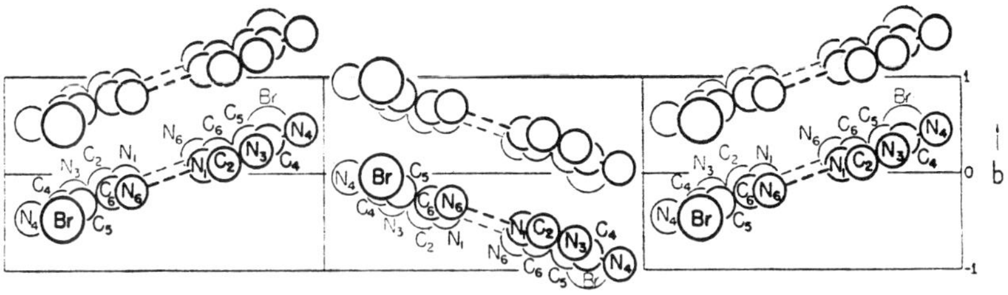

FIG. 9. A drawing of the crystal structure of 5-bromo-4,6-diaminopyrimidine viewed parallel to the molecular layers. Hydrogen bonds are indicated by dashed lines.

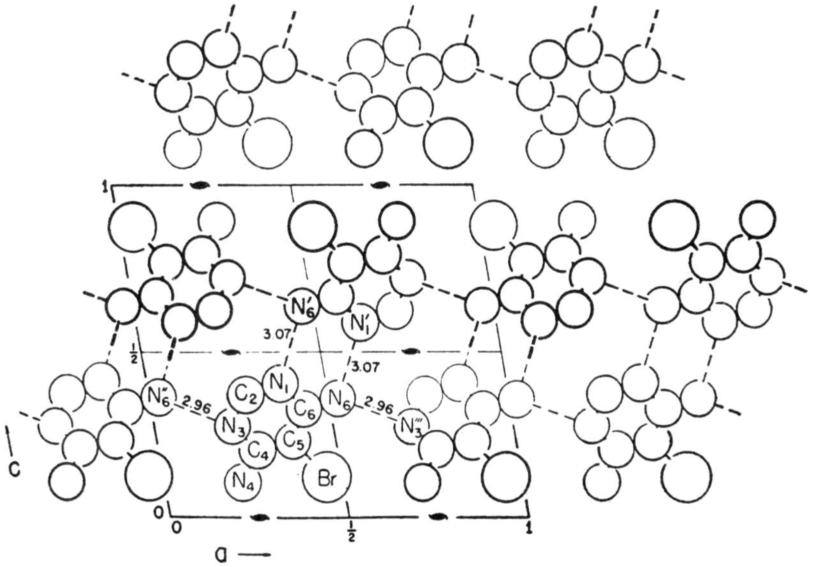

FIG. 10. A drawing of the crystal structure of 5-bromo-4,6-diaminopyrimidine viewed perpendicular to the molecular layers. Hydrogen bonds are indicated by dashed lines.

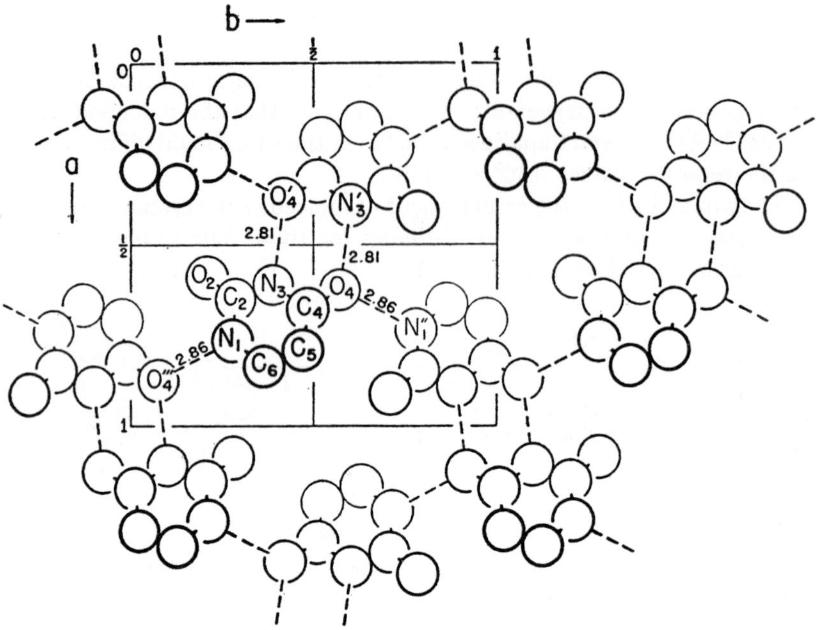

FIG. 11. A drawing of the crystal structure of uracil. Hydrogen bonds are indicated by dashed lines.

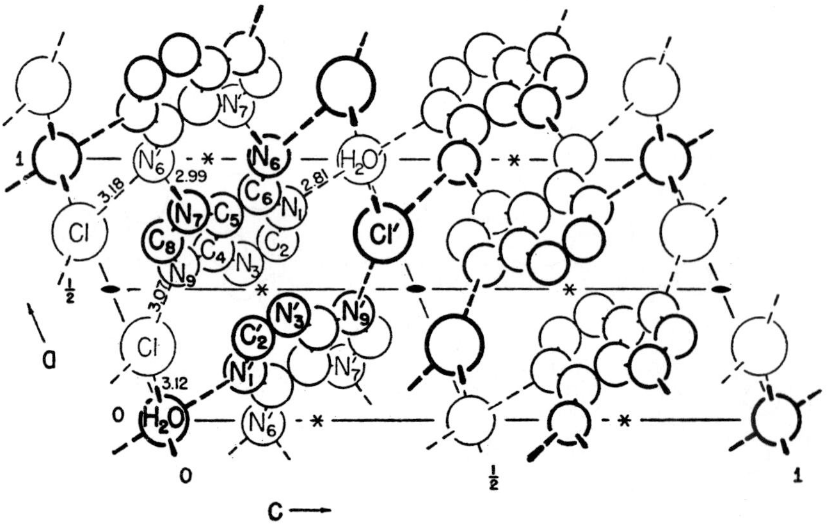

FIG. 12. A drawing of the crystal structure of adenine hydrochloride hemihydrate. Hydrogen bonds are indicated by dashed lines. The dimensions were calculated from the coordinates given by Cochran (11).

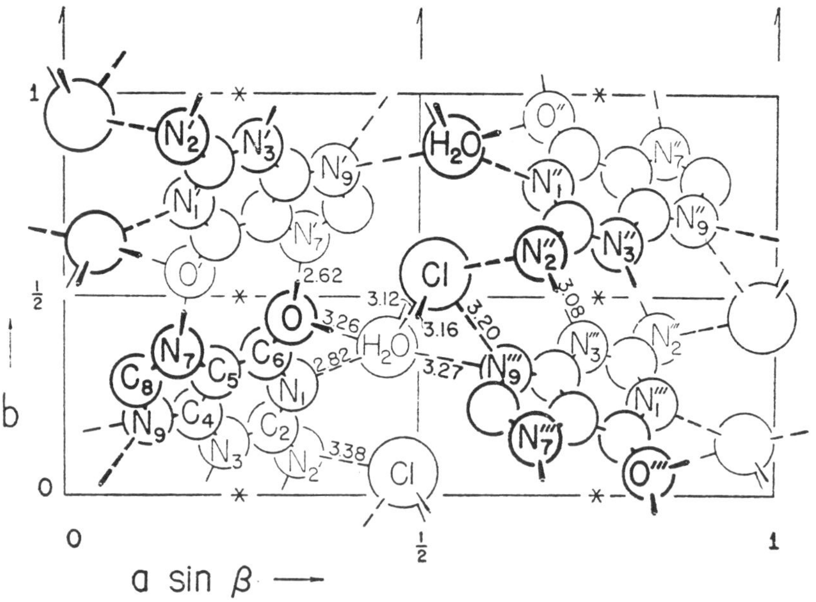

FIG. 13. A drawing of the crystal structure of guanine hydrochloride monohydrate. Hydrogen bonds are indicated by dashed lines.

The hydrogen bonds formed by the predominantly aromatic molecules represented in Figs. 7, 8, and 10 are those between a ring nitrogen atom without a hydrogen atom and an amino group. The observed distances range from 2.96 to 3.37 A. Similar hydrogen bonds for purines, with values 2.99 and 3.08 A., are shown in Figs. 12 and 13.

The N—H···O hydrogen bonds in uracil (Fig. 11) have lengths 2.81 and 2.86 A. A similar bond in guanine (Fig. 13) is reported to be 2.62 A. long. The other hydrogen bonds represented in Figs. 12 and 13 are those with water molecules; their lengths are between 2.82 and 3.27 A.

Complementary Purine–Pyrimidine Structures

The information provided by these investigations and by reasonable arguments based upon considerations of electronic structure permit the discussion of complementary structures between adenine and thymine and between guanine and cytosine. Drawings of reasonable structures of the four molecules are given in Figs. 14–17. In making these drawings the C—C distances in the rings have been taken uniformly as 1.40 A. The C—N distances in the ring have been taken as 1.32 A. if the nitrogen atom

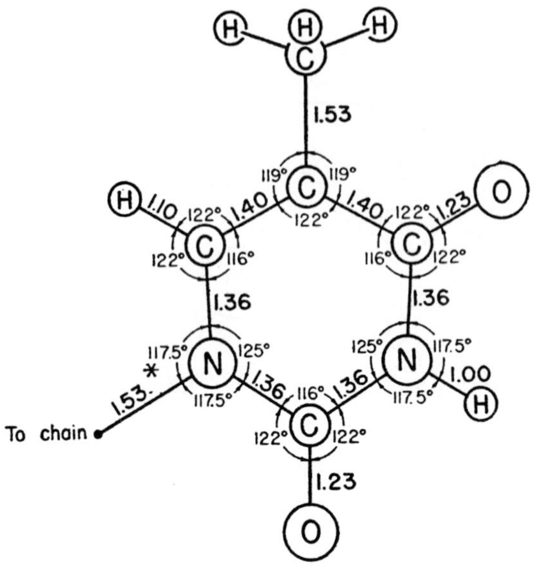

FIG. 14. A drawing showing the molecule of thymine with dimensions derived from x-ray studies of purines and pyrimidines. The point of attachment to a polynucleotide chain is indicated.

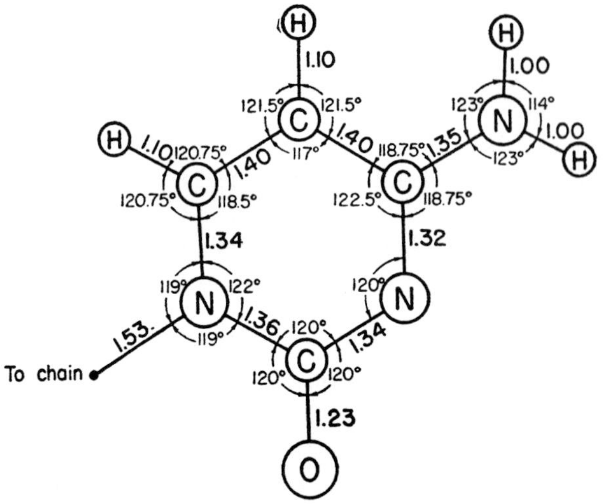

FIG. 15. A drawing showing the molecule of cytosine with dimensions derived from x-ray studies of purines and pyrimidines. The point of attachment to a polynucleotide chain is indicated.

*The value 1.53 A. was given, through an error, as 1.59 A. in the published paper.

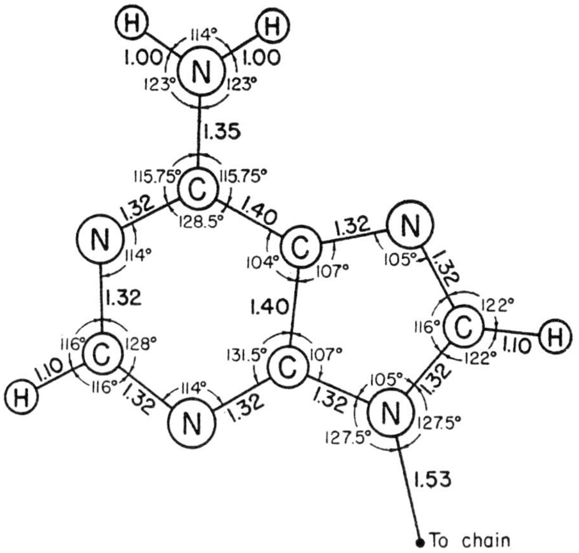

Fig. 16. A drawing showing the molecule of adenine with dimensions derived from x-ray studies of purines and pyrimidines. The point of attachment to a polynucleotide chain is indicated.

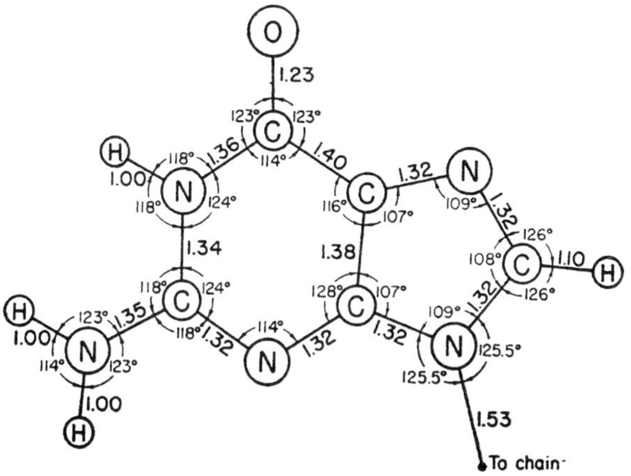

Fig. 17. A drawing showing the molecule of guanine with dimensions derived from x-ray studies of purines and pyrimidines. The point of attachment to a polynucleotide chain is indicated.

has an unshared pair, and 1.36 A. if it forms a bond outside the ring; some small variations from these values have been made, as indicated by consideration of the resonating bond structures. The bond angle for a nitrogen atom with an unshared pair in a six-membered ring has been taken as 113°, and that for a nitrogen atom with a bond outside the ring as 123°, with corresponding values for the carbon atoms; in some cases the requirements of geometry have permitted only a rough approximation to these values. The bond angles in the five-membered rings have been taken as having minimum deviations from 108°. The two bond angles of a bond outside a ring and the two bonds in a ring have been taken as equal.

Structures of the two complementary pairs are shown in Figs. 18 and 19. The hydrogen-bond distances have been selected to approximate those discussed in the preceding paragraphs. It has been found that, as shown in the figures, adenine and thymine can form two hydrogen bonds, with, of course, exactly the assumed distances, and guanine and cytosine can form three hydrogen bonds, involving only a small deviation from the expected distances. The two N—H···O distances in

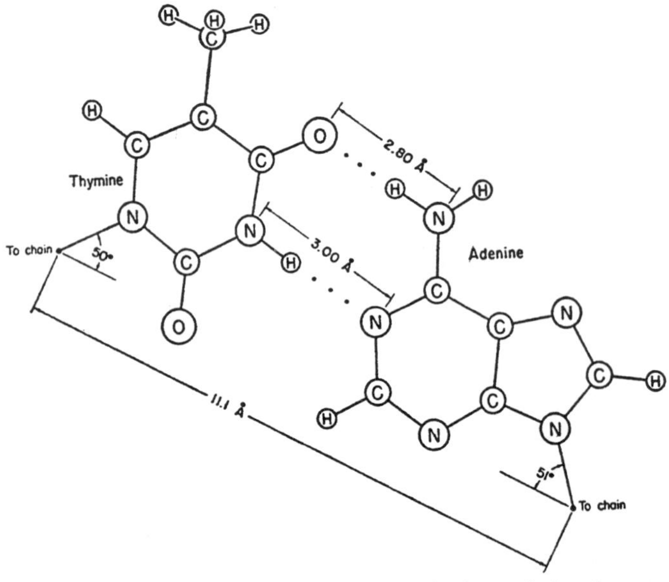

FIG. 18. A drawing showing how molecules of adenine and thymine may form a complementary pair held together by two hydrogen bonds.

Fig. 19 have been taken as 2.93 A.* This requires the N—H···N hydrogen bonds to have the length 2.96 A., which is within the range 2.96–3.37 A. observed for pyrimidine crystals.

We think that there is no doubt that the guanine–cytosine pair involves three hydrogen bonds, and we may conclude that the difference between this pair and the adenine–thymine pair, with two hydrogen bonds, is such as to introduce greater specificity in the action of a polynucleotide as a template than was indicated by the considerations of Watson and Crick, who had assumed that two hydrogen bonds are formed in each case.

An important requirement of the Watson-Crick structure is that the distance between the two deoxyribose carbon atoms, one in each of the two chains, be the same for the two complementary pairs, permitting replacement of one pair by the other, and also that the angle formed by the line connecting these two carbon atoms with the bond between one of the carbon atoms and the purine nitrogen atom be equal to the corresponding angle formed with the bond between the other carbon atom and the pyrimidine nitrogen atom, and that these angles be the same

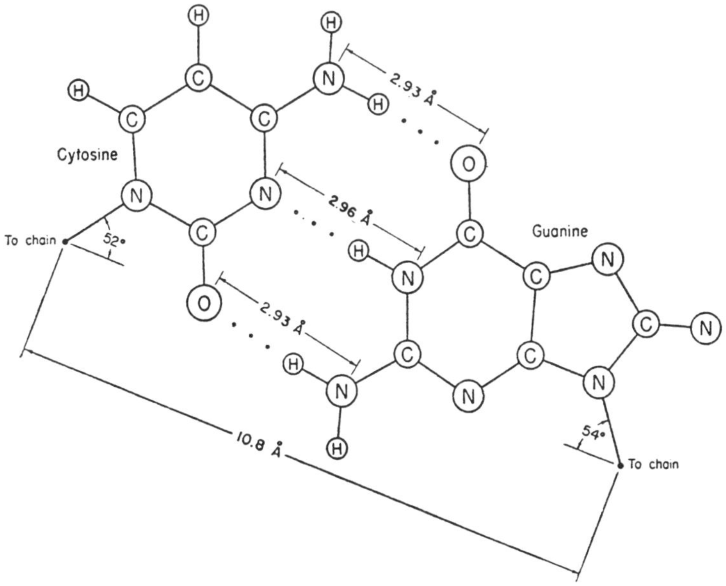

FIG. 19. A drawing showing how cytosine and guanine may form a complementary pair held together by three hydrogen bonds.*

*The distances 2.93 A. and 2.96 A. appeared in the published paper as 2.90 A. and 3.00 A., respectively; the changed values were found to be correct on remeasurement.

for the two pairs. The structural considerations illustrated in Figs. 18 and 19 provide satisfactory substantiation of the statement made by Watson and Crick that these qualifications are satisfied. The distance between the two deoxyribose carbon atoms connected by the purine–pyrimidine pair is found from the drawing of Fig. 18 to be 11.1 A., and the corresponding distance from Fig. 19 is 10.8 A. The four angles between these lines and the bonds are 50°, 51°, 52°, and 54°. The distances are close enough to one another to permit their being made equal by small deformation, and similarly the angles are nearly equal to the average angle; we estimate that the strain energy involved would be less than 50 cal./mole. If the N—H···O distance in Fig. 18 is increased to 2.90 A., both angles become 52° and the distance between the two deoxyribose carbon atoms decreases to 11.0 A.

A number of other types of hydrogen-bonded interactions between purines and pyrimidines have been discussed recently by Donohue (14). In the absence of evidence indicating that these structures occur in nature and of thoroughly reliable information about the dimensions of the molecules, we have not thought it worth while to carry out a detailed metrical study of these structures.

The uncertainties in the structure determinations of purines and pyrimidines described in this paper are such as to indicate that further careful studies of crystals of these substances should be made.

Summary

From crystal structure data for purines and pyrimidines it is concluded that in Watson and Crick's structure for DNA cytosine and guanine should form three hydrogen bonds. This conclusion strengthens the arguments of Watson and Crick as to the role of complementariness of structure of two DNA polynucleotide chains in the duplication of the gene.

References

1. Watson, J. D., and Crick, F. H. C., *Nature* **171,** 737–8 (1953).
2. Watson, J. D., and Crick, F. H. C., *Nature* **171,** 964–7 (1953).
3. Watson, J. D., and Crick, F. H. C., *Cold Spring Harbor Symposia Quant. Biol.* **18,** 123–31 (1953).
4. Pauling, L., and Delbrück, M., *Science* **92,** 77–9 (1940).
5. Clews, C. J. B., and Cochran, W., *Acta Cryst.* **1,** 4–11 (1948).
6. Clews, C. J. B., and Cochran, W., *Acta Cryst.* **2,** 46–57 (1949).
7. Parry, G. S., *Acta Cryst.* **7,** 313–20 (1954).
8. Wheatley, P. J., *Acta Cryst.* **8,** 224–6 (1955).

9. PERTINOTTI, F., GIACOMELLO, G., AND LIQUORI, A. M. The Structure of Heterocyclic Compounds Containing Nitrogen. I. Crystal and Molecular Structure of *s*-Tetrazine. *Acta Cryst.* **9,** 510 (1956).
10. PITT, G. J., *Acta Cryst.* **1,** 168–74 (1948).
11. COCHRAN, W., *Acta Cryst.* **4,** 81–92 (1951).
12. BROOMHEAD, J. M., *Acta Cryst.* **1,** 324–9 (1948).
13. BROOMHEAD, J. M., *Acta Cryst.* **4,** 92–100 (1951).
14. DONOHUE, J., *Proc. Natl. Acad. Sci. U. S.* **42,** 60–5 (1956).

Reprinted from the Proceedings of the NATIONAL ACADEMY OF SCIENCES
Vol. 46, No. 10, pp. 1349–1360. October, 1960.

A COMPARISON OF ANIMAL HEMOGLOBINS BY TRYPTIC PEPTIDE PATTERN ANALYSIS*

BY EMILE ZUCKERKANDL,† RICHARD T. JONES, AND LINUS PAULING

DIVISION OF CHEMISTRY AND CHEMICAL ENGINEERING,‡ CALIFORNIA INSTITUTE OF TECHNOLOGY

Communicated August 22, 1960

The complete amino acid sequences (primary structure) of hemoglobins can, in principle, be determined by methods currently available. Although detailed studies of the primary structure of human and horse hemoglobins are in progress in several laboratories,[1] the methods are so laborious that complete sequences have not yet been established. Important questions in the realm of genetics and evolution require the immediate examination of the structure, primary and other, of many different hemoglobins. The application of methods that are quicker, though less informative and reliable, than the techniques required for complete sequence determination is therefore in order as a provisional means of securing useful information. Such a method is the analysis of peptide patterns obtained by combined paper electrophoresis and chromatography of tryptic hydrolysates of denatured hemoglobin.[2] Of particular interest are comparisons between hemoglobin components present in (*a*) organisms of one animal species at a given time in development, (*b*) organisms of one species at different stages of development, and (*c*) organisms of different species. The present paper is concerned exclusively with the last type of comparison. In order to scan the range of variation of hemoglobin structure throughout evolution, hemoglobins from a number of animals both closely and distantly related to man have been selected and compared as to tryptic peptide patterns with human hemoglobin A. Whole hemoglobin preparations from adult animals have been studied throughout. The problem of individual heterogeneity will be treated elsewhere.

Materials and Methods.—Ape bloods[3a] were anticoagulated and transported in Alsever's solution. The apes studied included two lowland gorillas (*Gorilla gorilla*), two chimpanzees (*Pan troglodytes*), and three orangutans (*Pongo pygmaeus*). Erythrocytes from Rhesus monkeys (*Macaca mulatta*) were obtained from clotted blood.[3b] Heparinized porcine and bovine bloods were secured during bleedings at Los Angeles slaughter houses. Heparinized bloods were obtained from the marine lungfish, *Pimelometopon pulcher* (sheepshead), and from the cartilaginous fish *Cephaloscyllium uter* (swell shark).[3c] Blood was also obtained from a live specimen of *Lepidosiren paradoxa* (Dipneust, South American fresh water lungfish).[3c, d] Contamination of the latter blood by tissue fluid was unavoidable. Blood of the Pacific hagfish, *Polistotrema stouti*, was also examined.[4] Coelomic cells (hemoglobin cells) were obtained from *Urechis caupo* (Echiurid marine "worm")[5] collected at low tide from mud flats near the Kerckhoff Marine Laboratory at Corona del Mar, California.[3e]

In general, the hemoglobin preparations examined were from single individuals. However, in the case of *Urechis* single as well as pooled samples of coelomic cells were used without apparent differences. Only a pooled sample of blood from twelve hagfish was examined. All of the animals studied were judged to be adults either from their size or their known age. The youngest apes were one orangutan 2 years old and one gorilla 2 years old. Peptide patterns of their hemoglobins were indistinguishable from those of older individuals of the same species. The *Lepidosiren*, 14 inches long, was at a minimum 8 months of age, and might be considered a subadult.

The erythrocytes were washed four times with cold 0.9% NaCl with the exception of *Pimelometopon* and *Cephaloscyllium* cells, which were washed with 1.2% NaCl.[6] The red cells from hagfish were washed with chilled 3% NaCl and the coelomic cells of *Urechis* with 3.2% NaCl. Washing with 0.9% NaCl led to considerable hemolysis in the case of the fresh water fish *Dipneust.*

The washed cells were hemolyzed in a standard fashion with distilled water and toluene[7] with the exception of *Urechis*, where ether was used instead of toluene. The washed red cells of hagfish were stored for one day at 4°C before lysing. A major portion of the hemoglobin obtained was insoluble possibly due to acidification during storage. The peptide pattern shown below was obtained from the remaining soluble fraction. After lysis, each sample was centrifuged at high speeds in order to remove solid debris and the toluene phase. The resulting hemoglobin solutions were saturated with carbon monoxide and dialyzed for a week against distilled water saturated with carbon monoxide (three changes of water with the ratio of hemoglobin solution to water of the order of 1:100). This prolonged dialysis was employed to permit the flocculation of non-hemoglobin proteins.[6] A crystal of thymol was added to one *Urechis* preparation in order to eliminate the possibility of the formation of peptides by bacterial contamination. No differences were found in the peptide patterns obtained with and without thymol.

With species closely related to man the hemoglobin concentration in the final preparation could be assayed approximately by the use of spectrophotometric constants established for human hemoglobins. In other species the quantities of material to be used on peptide patterns were estimated roughly from the consumption of sodium hydroxide during tryptic digestion.

Hydrolysis with trypsin of heat-denatured hemoglobin preparations was performed at constant pH using a Radiometer automatic titrator as a pH-stat. The hydrolyses were carried out at 40°C at pH 8.0 in the presence of Ca ion (0.01 *M*) with an enzyme to hemoglobin ratio of between 1 and 2% by weight (Worthington twice crystallized trypsin). Ninety minutes was allowed for the hydrolysis except in the case of the apes, where the digestion was stopped after sixty minutes.

Those tryptic peptides which are soluble at pH 6.5 were analyzed by a combination of electrophoresis and chromatography on paper. These peptide patterns were obtained essentially as specified by Ingram[2] except that the voltage rate was 900–1,000 V for a duration of $3^1/_2$ hours. In general, a human control sample was run as a pair mate with every animal hemoglobin in order to have a reference pattern for each unknown. Even with such parallel determinations, the spreading occurring during electrophoresis was not always identical. The polarity of the electrophoretic field is marked on each pattern and the point of application of the sample is designated by a cross (+). The peptide spots resulting from tryptic hydrolysis of human hemoglobin have been assigned arbitrary numbers by Ingram.[2] This convention has been employed in the present study. The neutral band region is comprised of peptide spots 1 through 7. These peptides do not migrate to any extent in the electric field, but do move some from the point of application due to fluid movement. Phenylalanine and lysine have been added as reference spots at least on one pattern in the case of every species. Chromatography was ascending in direction and the peptides were detected mainly by their reaction with ninhydrin. The Sakaguchi[8] test for arginine and the cinnamaldehyde[9] reaction for trypophan were also employed. Patterns from each species were obtained at least in triplicate.

Results.—Figure 1 is a peptide pattern of human hemoglobin A as currently obtained in our laboratories. A diagram indicating the peptide spot numbering system is also included.

Figure 2 shows peptide patterns from three different apes (gorilla, chimpanzee, and orangutan) and one monkey (Rhesus). The gorilla, chimpanzee, and human patterns are almost identical in appearance. In the case of the gorilla peptide patterns, spot numbers 12 and 24 both appear to be double. In humans the peptides that make up these two spots are from the beta chain. No difference was

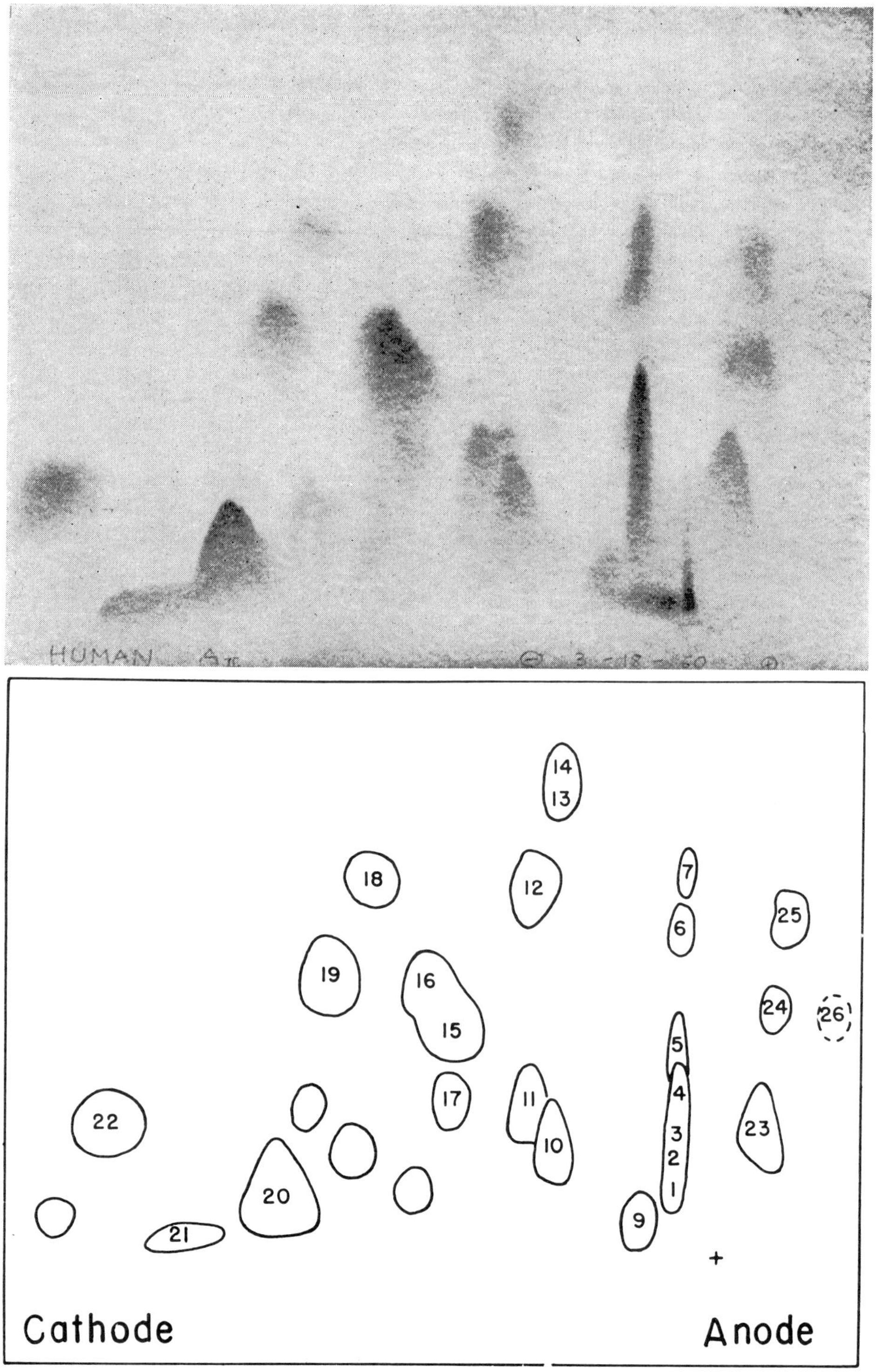

FIG. 1.—A tryptic peptide pattern of adult human hemoglobin and its schematic representation. The spots are numbered according to the convention introduced by Ingram.[2] The site of spot 26 is shown on the diagram, although not seen on the pattern represented. The locations of four hitherto undescribed spots usually found on patterns obtained in our laboratories are indicated.

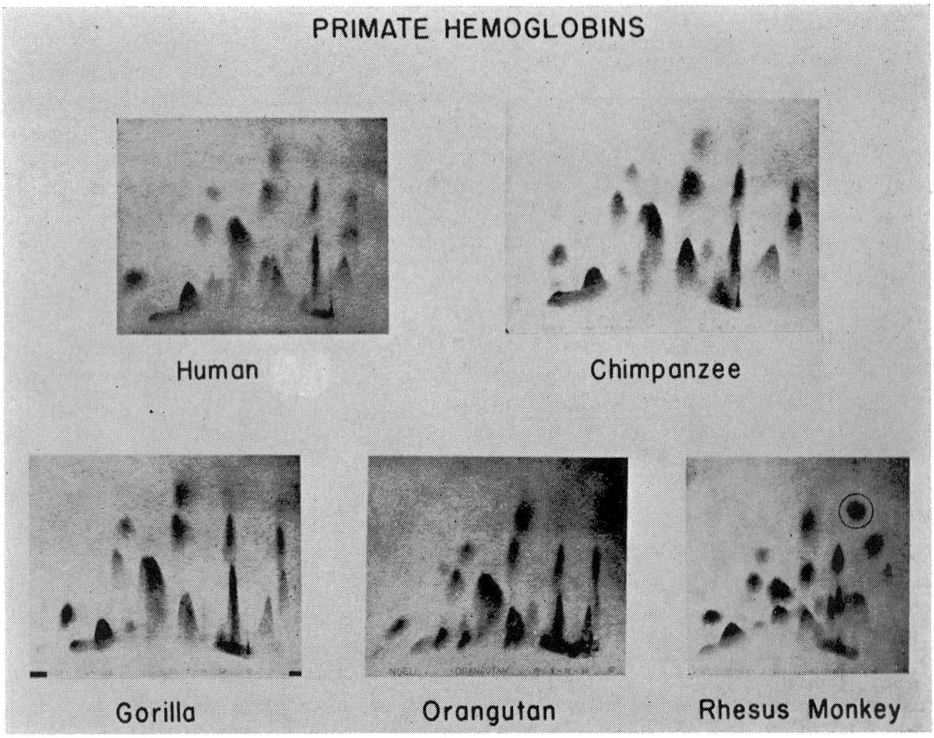

Fig. 2.—Tryptic peptide patterns of primate hemoglobins. The circled spot on the Rhesus monkey pattern represents phenylalanine added two and a half inches to the anodal side of the point of application of the peptide mixture.

observed between two gorillas, one male and one female. Spots 12 and 24 also seem to be double in the peptide patterns of chimpanzee. In addition, extra material may be present in the 15–16 region. In humans at least four major peptides are present in this region.[1] Two chimpanzees, both females, were found to have identical patterns. Of course further differences between these two types of apes and humans may be discovered upon the analysis of the individual peptides or the study of the protein residue which remains insoluble after hydrolysis with trypsin.

The difference from human patterns is somewhat greater for orangutan peptide patterns than for the patterns of the two apes just mentioned. In two orangutans (one male and one female) spots 12 and 24 are double. The third orangutan (female), whose pattern is illustrated in Figure 2, was observed to have single 12 and 24 spots. Further differences between the patterns of the individual orangutans are not apparent; however, each differs from human patterns by an apparent increase in peptides in the 15–16 region. The orangutans also differ from humans and the other two apes studied by the appearance of two new spots, one anodal to spot 20 and the other cathodal and below spot 10. The latter does not appear to be the same as spot 9 of human. All of the spots containing arginine and tryptophan are the same as in human.

Spot 12 is absent in the peptide patterns of Rhesus monkey hemoglobin. A new spot, possibly representing a modification of the peptide of spot 12, is present just below the 12 region. The new spot, like 12 from human, contains tryptophan. In another individual Rhesus monkey, whose pattern is not presented, spot 12 and the new spot are both present simultaneously. Patterns of both monkeys also reveal another spot on the cathodal side of the 2, 3, and 4 region. Spot 7 is absent or modified (almost fused with 6). Other differences may be present in the neutral band region, which is poorly resolved.

The human spot 26, which contains arginine and is derived from the beta chain, is not always seen because of poor color production with ninhydrin. The existence of this spot can, however, be ascertained by the Sakaguchi test for arginine. The primate patterns presented do not show spot 26, either because it did not react with ninhydrin or because it migrated off the paper during electrophoresis. By undimensional electrophoresis and the application of the Sakaguchi test the presence of arginine in the zone corresponding to human peptide 26 has been ascertained in all the primates examined.

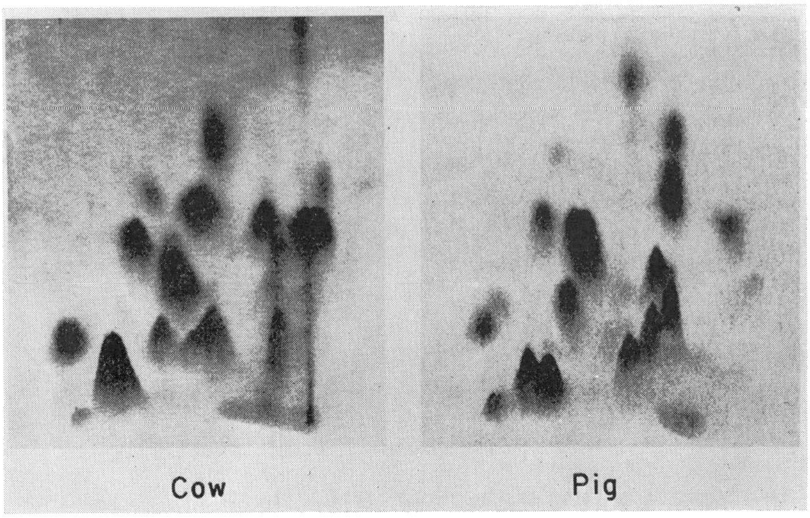

FIG. 3.—Tryptic peptide patterns of bovine and porcine hemoglobins.

Tryptic peptide patterns of bovine and porcine hemoglobins are presented in Figure 3. Cow and pig belong to the same order. In many respects these patterns appear similar to human patterns. However, the differences are numerous and striking.

In bovine and porcine hemoglobin, sequence val-leu has been found to be N-terminal in one of the polypeptide chains.[10] Both hemoglobins therefore contain alpha chains, according to the terminology introduced by Rhinesmith, Schroeder, and Martin.[11] However, the cow alpha chain is different in several respects from the human alpha chain. Although the chains have not yet been separated, the peptide patterns indicate that the alpha chain peptides 5, 10, and 17 are absent from bovine patterns, while the alpha chain peptides 12, 13, 18 and an alpha component of spot 16, detected by a positive Sakaguchi test, are present.[1,12] Although the second chain by definition is not a beta chain,[10, 11] the human beta chain spots

12 and 19 are seen clearly. The human pure beta spots 2, 4, 14 (as ascertained by a negative reaction with cinnamaldehyde) and 25 are absent.

The porcine alpha chain may be somewhat more similar to the human alpha chain since the alpha spots 10, 13, and 18 are seen, and the presence of the alpha spots 11, 17, and 23 is suspected. The presence of an alpha-component in the composite spot 15 is likely because of a positive cinnamaldehyde test, and the presence of an alpha-component in the composite spot 16 is indicated by a positive Sakaguchi test. The alpha spot 5 is absent. Among the beta spots, 14 (cinnamaldehyde positive), 26 (Sakaguchi positive), 4, 19, 24, and possibly 25 are present; while 12 is absent. Like bovine hemoglobin, porcine hemoglobin does not contain beta chains; however, there are a number of tryptic peptides similar to those from the human beta chain observed.

The composite human spots 13–14, 15–16, 20 and 21 as well as the lysine spot 22 (arising from both α and β chains[1]) are seen in pig as well as in cow. Thus the

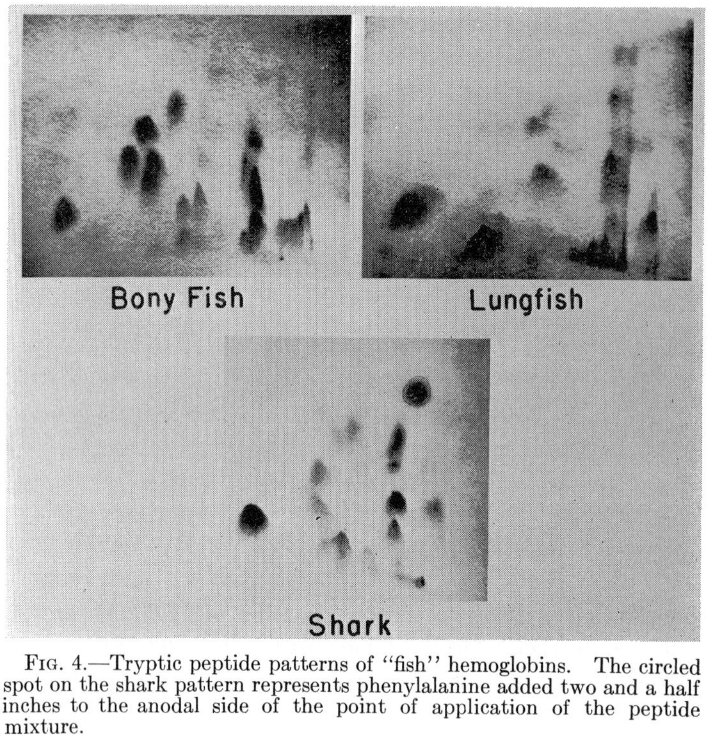

FIG. 4.—Tryptic peptide patterns of "fish" hemoglobins. The circled spot on the shark pattern represents phenylalanine added two and a half inches to the anodal side of the point of application of the peptide mixture.

apparent similarities between bovine and human patterns are in many respects the same as those noted between porcine and human; however, the differences are not the same. From three independent pattern studies of the same bovine sample it appears that many of the new spots are slightly on the anodal side of the neutral band region, whereas in porcine patterns most of the new spots are on the cathodal side. Arginine is present in at least two of the new porcine peptides, one neutral and one acidic.

Spot 12, as stated, and as confirmed by cinnamaldehyde tests, is absent from pig, but present in the cow. The decapeptide[1] that accounts for this spot in humans may have undergone independent alteration (mutation?) at different times. In pig, a novel tryptophan peptide is found in the upper neutral region. It may correspond to an altered peptide No. 12.

The three fish from which peptide patterns have been obtained (Fig. 4) belong to entirely different groups which modern systematicians do not unite under the heading of a single class. None of the three patterns is similar to the patterns of human hemoglobin. In the bony fish (*Pimelometopon*), spot 22 is present as are spots 18 and 19, which in humans are dipeptides.[1] Possibly spots 24 and 25 and some material in the regions of 4 and 15 are common to human and *Pimelometopon.* The remaining spots appear to be different from human. Three spots containing tryptophan are seen with no analogy in human hemoglobin. One other spot at the base of the neutral band resembles a similarly located spot on pig patterns in that it gives a blue rather than purple color with ninhydrin and a positive Sakaguchi test. In the shark (*Cephaloscyllium*), spot 22 is again present as are probably spots 4, 5, and 7. A spot is present in the region of peptide 12, but differs from spot 12 as shown by the absence of tryptophan. Spots 18 and 19 are absent as well as spots 20 and 21. Four peptides containing tryptophan are seen with no analogy in human or in *Pimelometopon.*

In the case of the peptide patterns of lungfish (*Lepidosiren*) hemoglobin, spot 22, (lysine, identified by position and shade of color) is present. Spots 19 and 23 may also be present. The absence of spots 20 and 21 appears to be characteristic of all three types of fish. Three Sakaguchi-positive peptides are seen in the middle and upper neutral band region and one cinnamaldehyde positive peptide adjacent to the lower part of the neutral band. None of these are present in human. Among the vertebrates examined, *Lepidosiren* hemoglobin may be the one with the least number of primary structural traits in common with human hemoglobin.

There may be some peptides in common between *Cephaloscyllium* and *Pimelometopon,* notably one Sakaguchi-positive spot in the upper neutral band and some between *Cephaloscyllium* and *Lepidosiren. Pimelometopon* and *Lepidosiren* appear to have only one ninhydrin spot in common beyond the lysine spot, No. 22.

The Cyclostomes are the most primitive group of living vertebrates, although they are very specialized. The molecular weight of those Cyclostome hemoglobins that have been studied is reported to be one quarter[13, 14] of the molecular weight of other vertebrate hemoglobins. Therefore it might be presumed that these Cyclostome hemoglobins are composed of single polypeptide chains. Because direct estimates of the molecular weight of *Polistotrema* hemoglobin are not available, an ultracentrifuge study was made.[15] The soluble fraction of our *Polistotrema* hemoglobin (see *Materials and Methods*) appeared to sediment as a single component with an $s_{20,\,w}$ of 1.9. This value, in the case of hemoglobin from the related *Petromyzon,* has been interpreted by others[14] to correspond to a molecular weight of approximately 23.600. However, the peptide pattern of *Polistotrema* hemoglobin shown in Figure 5 may not represent a single kind of polypeptide chain because two major and possibly some minor hemoglobin components were shown to be present by starch gel electrophoresis.[16]

It is not possible to make definite statements about the similarities between

Polistotrema and other patterns in the region of the neutral band because of incomplete resolution of the *Polistotrema* spots. Except possibly for the neutral band region and the lysine spot, the *Polistotrema* peptide pattern has no feature in common with human, lungfish, or shark patterns. One or two spots in the bony fish pattern appear similar to spots in the *Polistotrema* pattern.

In order to test the range of variation in hemoglobins in general, a peptide pattern of a single invertebrate, the Echiurid "worm" *Urechis* was studied (Fig. 6). In this group the hemoglobins are intracellular, as in the vertebrates. The heavily

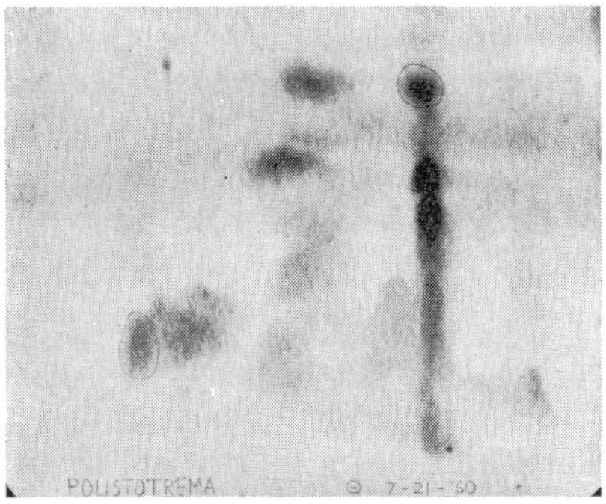

FIG. 5.—Tryptic peptide pattern of hagfish hemoglobin. The circled spots represent phenylalanine and lysine respectively added at the point of application of the peptide mixture.

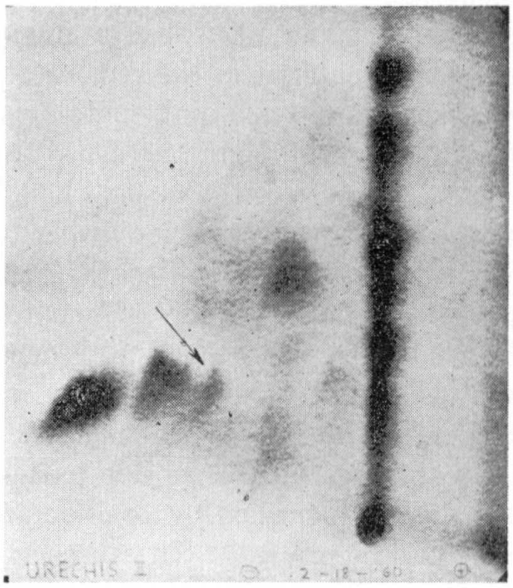

FIG. 6.—Tryptic peptide pattern of hemoglobin from an Echiurid "worm." Most of the spots in the neutral band, as well as the one indicated by the arrow, are not tryptic peptides (see text).

loaded neutral band region obtained has been shown to contain many amino acids[17] that are not the result of tryptic hydrolysis. The spots on the cathodal side of the neutral band, with the exception of the one indicated by the arrow, were found to be from *Urechis* hemoglobin. The latter conclusion was obtained by comparing, peptide patterns of the total undigested hemoglobin preparation with the undigested supernatant obtained after heat denaturation and centrifugation of the precipitate. The only spot identified with any certainty is spot 22. All other *Urechis* spots appeared to be without a match in human.

Discussion.—Peptide patterns as obtained by the present technique do not yield unequivocal information for several reasons: (*a*) spots found in identical locations may not represent identical peptides, (*b*) single spots may contain more than one peptide, (*c*) a single peptide, through partial secondary alteration or because of incomplete splitting into two peptides, may give rise to more than one spot, and (*d*) a significant part of the hemoglobin molecule (about one third in human A, S, and F[18]) is not broken down by trypsin sufficiently to become soluble, and accordingly is not represented on tryptic peptide patterns.

Although, in view of (*d*), all findings must be substantiated by the examination of the undigested core, observations (*a*), (*b*), and (*c*) do not seriously impair the conclusion that Primate and human hemoglobins are very similar, especially gorilla, chimpanzee and human. There is no doubt that the identification of individual spots needs to be based on data more specific than position, shape, and (sometimes) color shade of the ninhydrin-developed spots, or even reaction with reagents specific for certain amino acid residues. However, when two complex peptide patterns look similar as a whole the probability that most of the spots actually represent identical or highly similar sequences becomes high.

That chimpanzee should be extremely close to man while orangutan and Rhesus monkey are somewhat more distant when examined by hemoglobin peptide patterns is in accord with serological data.[19] On the other hand, while human hemoglobin strongly crossreacts with Rhesus monkey hemoglobin, it has been observed not to do so with porcine hemoglobin and only exceptionally crossreacts with cattle hemoglobin.[20] The similarity between human and Primate hemoglobins suggests that these hemoglobins probably have not been modified extensively since the times of their common ancestor. It is possible that the genes for the α and β chains of a normal adult human hemoglobin are more stable than the mutated genes formed from them by simple mutations, and that the mutated genes often undergo back-mutation to the more stable normal genes. In addition to natural selection, the thermodynamic stability of the genes themselves may be an important factor in determining the distribution of alleles in a population. Our results indicate that stable hemoglobin genes had been developed before the separation of humans and anthropoids from their common stock.

The idea that the normal genes have greater thermodynamic stability than their mutant alleles, and that in consequence the mutant alleles have a greater mutation rate than the normal genes, provides an explanation of the fact that some of the abnormal hemoglobins have their abnormalities at the same site.[21] The alleles corresponding to one site may represent one mutant from the normal gene and other alleles formed by second mutations from this mutant, and involving its unstable site.

Human adult hemoglobin peptide patterns actually differ more from human fetal patterns than from adult gorilla and chimpanzee patterns. In view of the theory of recapitulation it would not appear paradoxical that an adult organism should be in a sense a more distant "relative" of its own embryo than of other closely related adult organisms. Adaptative factors in response to the environment may play an equally important role in establishing the differences. In the only animal examined (sheep) fetal hemoglobin differs from adult hemoglobin in molecular shape.[22]

As one gets further away from the group of Primates, the amount of primary structure that is shared with human hemoglobin decreases. Different primary structures may be compatible with rather constant tertiary structures.[23] The limits to this statement remain to be investigated.

The observation that pig and human on one hand and steer and human on the other show to a large extent apparent identity of the same peptides but diverging differences for the others may again indicate the existence in the hemoglobin molecule of zones more prone to mutation than others, or of zones where the occurring mutations are more liable to be preserved by natural selection. Part of the Primate pattern may have evolved relatively early in mammalian history, if not before, while another part may have varied more frequently throughout the groups.

Sections of both the alpha and beta chains, or of chains taking their place, appear to be involved. So far there is no evidence that one of the chains remains more stable through evolution than the other.

Firm conclusions about the partial stability of chains are premature before the sequence of each of the tryptic peptides that are common to different hemoglobins is determined. On the basis of a chance distribution of amino acids in the hemoglobin molecule, it is evident that the smaller the peptide, the greater the chance of finding it in different species. The presence of spot 22 seems to be the only trait shared by all hemoglobins. Because 22 is pure lysine, this only means that in all of the hemoglobins examined there is, in at least one place, a lysine next to another lysine or an arginine. In human hemoglobin a lys-lys sequence has been found to occur in both the alpha and beta chains.[1] The frequently seen spots 18 and 19 are dipeptides.[1] Spots 20 and 21 are a tetrapeptide and a pentapeptide, respectively, which are closely related.[1] It is true that some of the spots that may be common to pig, steer, and human (spots 10, 11, 24, and 25) contain much higher peptides.

However, the significance of the similarities observed between pig, steer, and human hemoglobins is increased by the observation that many of these similarities are not shared by the three fish and none of them (except the lysine spot) by the Cyclostome and the *Urechis* "worm." The three "fish" patterns differ among themselves considerably more than the mammalian patterns examined. Although shark and bony fish as well as shark and lungfish hemoglobins may share a small number of tryptic peptides, bony fish patterns appear about as dissimilar from lungfish as from human patterns. The findings are consistent with the view that these "fish" belong to widely divergent evolutionary lines.

Unless a greater constancy were found in the undigested core than in the rest of the molecule, no large tryptic peptide would be constant throughout the vertebrate series. Thus, most of the hemoglobin molecule, at least in the portion solubilized by trypsin, has been subject to successful mutation at some time during vertebrate evolution. Successful mutations in each tryptic peptide region have probably occurred more than once considering the differences between groups of fish and between Cyclostomes and fish. This observation, to be sure, does not exclude the existence of preferential sites of (successful) mutation. But such sites, if they exist, must be distributed throughout the molecule and not be concentrated in any one part of it, in contradistinction to what has so far been observed with insulin.[24]

* This work was supported in part by Grant No. H3136 from the National Institutes of Health, U.S. Public Health Service, and was presented in part at the 138th meeting of the American Chemical Society.

† On leave from Centre National de la Recherche Scientifique, Paris.

‡ Contribution No. 2618.

[1] Groups investigating the amino acid sequence of human hemoglobin include Dr. G. Braunitzer and co-workers, Drs. R. J. Hill and W. Koingsberg, Drs. V. M. Ingram and J. A. Hunt, and Dr. W. A. Schroeder and co-workers. Information concerning the tryptic peptides is from unpublished sequence studies of W. A. Schroeder and co-workers.

[2] Ingram, V. M., *Biochim. Biophys. Acta*, **28**, 539–545 (1958).

[3] Various hemoglobin specimens have been supplied for this research. The authors wish to acknowledge the very valuable help in this respect of (*a*) Dr. W. P. Henschele, Head Veterinarian, San Diego Zoo; (*b*) Dr Renato Dulbecco, Division of Biology, California Institute of Tech-

nology; (*c*) Dr. K. S. Norris, Curator of Marineland of the Pacific, Palos Verdes, California; (*d*) Dr. W. H. Hildemann, Department of Zoology, University of California at Los Angeles, who drew the lungfish blood; and (*e*) Dr. A. Tyler, Division of Biology, California Institute of Technology.

[4] Specimens of the Pacific hagfish were caught and bled by Dr. David Jensen of the Oceanography Institute, La Jolla, California. The special methods employed were devised by him.

[5] Redfield, A. C., and M. Florkin, *Biol. Bull.*, **61**, 185–210 (1931).

[6] Roche, J., Y. Derrien, and M. S. Chouaiech, *Ann. Inst. Oceanograph.*, **20**, 97–113 (1940).

[7] Drabkin, D. L., *Arch Biochem.*, **21**, 224 (1949).

[8] Acher, R., and C. Crocker, *Biochim. Biophys. Acta*, **9**, 704 (1952).

[9] Method of J. Harley–Mason, *Biochem. J.*, **69**, 60 P (1958), modified by Kenneth N. F. Shaw (personal communication).

[10] Ozawa, H., and K. Satake, *J. Biochem.*, **42**, 641–647 (1955).

[11] Rhinesmith, Herbert S., W. A. Schroeder, and Nancy Martin, *J. Am. Chem. Soc.*, **80**, 3358 (1958).

[12] Hunt, J. A., *Nature* **183**, 1373–1375 (1959).

[13] Svedberg, The, *J. Biol. Chem.*, **103**, 311–325 (1933).

[14] Lenhert, P. G., W. E. Love, and F. D. Carlson, *Biol. Bull.*, **111**, 293–294 (1956).

[15] Made under the direction of Mr. John E. Hearst in the laboratory of Dr. J. Vinograd.

[16] Performed by Mr. Donald Sheffler in the laboratory of Dr. Ray Owen.

[17] The identification of these amino acids was made by Dr. Thomas L. Perry and Dr. Kenneth Shaw.

[18] Barrett, H. W., and W. A. Schroeder, unpublished.

[19] Mollison, Th., quoted by P. Kramp in *Primatologia* I, ed. H. Hofer, A. Schultz, and D. Starck (New York: S. Karker, Basel, 1956), pp. 1015–1034.

[20] Hektoen, L., and A. K. Boor, *J. Infect. Dis.*, **49**, 29–36 (1931).

[21] Ingram, V. M., *Brit. Med. Bull.*, **15**, 27–32 (1959), and unpublished work from this laboratory.

[22] Bragg, W. L., and M. F. Perutz, *Acta Cristall.*, **5**, 323–328 (1952).

[23] Perutz, M. F., M. G. Rossmann, A. F. Cullis, H. Muirhead, G. Will, and A. C. T. North, *Nature*, **185**, 416–422 (1960).

[24] Harris, J. I., F. Sanger, and M. A. Naughton, *Arch. Biochem. Biophys.*, **65**, 427–438 (1956).

Reprinted from
HORIZONS IN BIOCHEMISTRY
Copyright 1962
Academic Press Inc.. New York, N. Y.

Molecular Disease, Evolution, and Genic Heterogeneity[1]

EMILE ZUCKERKANDL[2] AND LINUS PAULING

Division of Chemistry and Chemical Engineering, California Institute of Technology, Pasadena, California

I. Generalities

A. The Notion of Molecular Disease and Its Field of Application

Life is a relationship between molecules, not a property of any one molecule. So is therefore disease, which endangers life. While there are molecular diseases, there are no diseased molecules. At the level of the molecules we find only variations in structure and physicochemical properties. Likewise, at that level we rarely detect any criterion by virtue of which to place a given molecule "higher" or "lower" on the evolutionary scale. Human hemoglobin, although different to some extent from that of the horse (Braunitzer and Matsuda, 1961), appears in no way more highly organized. Molecular disease and evolution are realities belonging to superior levels

[1] Contribution No. 2774 from the Gates and Crellin Laboratories of Chemistry, California Institute of Technology, Pasadena, California.

[2] On leave from the Centre National de la Recherche Scientifique, Paris.

of biological integration. There they are found to be closely linked, with no sharp borderline between them. The mechanism of molecular disease represents one element of the mechanism of evolution. Even subjectively the two phenomena of disease and evolution may at times lead to identical experiences. The appearance of the concept of good and evil, interpreted by man as his painful expulsion from Paradise, was probably a molecular disease that turned out to be evolution. Subjectively, to evolve must most often have amounted to suffering from a disease. And these diseases were of course molecular.

Relationships between molecules, which define states of health and disease, may be altered by environmental factors, or by factors of aging, or by inherited internal factors. The two last types of factors are partly the same, inasmuch as aging is itself determined by genetic factors. The two first types of factors are also partly the same, inasmuch as aging is due to the cumulative effect of external agents. The term molecular disease in its more restricted sense, the only useful one, relates to the third type of factors, to altered relationships between molecules traceable to altered genes.

To the extent to which we have grounds to believe today that inheritance is linked to nucleic acids and that the primary products of nucleic acids, beside perhaps other nucleic acids, are proteins and no other types of molecules, the notion of molecular disease relates exclusively to the inheritance of altered protein and nucleic acid molecules. Abnormal glycogens, for instance (Cori, 1954), are to be traced to abnormal enzymes, proteins responsible for their production.

The abnormal nucleic acid and the abnormal protein produced under its control represent two aspects of the same reality. A great many more protein molecules are present at a given time than nucleic acid molecules responsible for their production, and this is one of the reasons why it is much easier at the moment to study the phenomenon of molecular disease from the protein end.

An abnormal protein causing molecular disease has abnormal enzymatic or other physicochemical properties. Changes in such properties are necessarily linked to changes in structure. It seems unwarranted at the present time to draw a basic distinction between two types of structural changes as causes of molecular disease, change in folding of the polypeptide chains and changes in the sequence of amino acids, the building stones of the chains. It becomes increasingly probable that the changes in spatial configuration (conformation) are direct expressions of changes in sequence (Crick, 1958; Mizushima and Shimanouchi, 1961). Therefore a molecular disease can probably be defined at the molecular level in a way that is potentially complete by determining the alteration of the amino-acid sequence of a protein (or the nucleotide sequence in the corresponding nucleic acid). This statement applies of course to a given intracellular and extracellular environment. If there are changes in the environment, the spatial configuration of a protein may be altered without any change in amino-acid sequence and a pathological condition may ensue. Such a change in environment results either from external influences, which do not concern us here, or ultimately from the change in amino-acid sequence in other proteins, to which the molecular disease must then be traced.

B. Molecular Disease, Evolution, and Environment

The study of molecular diseases leads back to the study of mutations, most of which are known to be detrimental. All loss mutations in a broad sense of the word—involving either the total loss of a protein or the loss of protein function through a structural alteration of the protein—are molecular diseases. Loss mutations, on the other hand, are among the conditions of adaptation of the organism to changes in its environment and adaptation, the conditions of evolution. A loss of function, when compatible with survival thanks to the nature of the environment, may make cellular energy and genic raw materials available for the acquisition of new functions. More highly evolved organisms have lost powers of synthesis that more primitive organisms possess (Lwoff, 1943). It thus appears possible that there would be no evolution without molecular disease. A maintenance of molecular health, although in the interest of the individual, is opposed to evolution. However, only a small fraction of the molecular diseases that occur are used by and turned into evolution.

A bacterium that loses by mutation the ability to synthesize a given enzyme has a molecular disease. The first heterotrophic organisms suffered from molecular diseases, of which they cured themselves by feeding on their fellow creatures. At the limit, life itself is a molecular disease, which it overcomes temporarily by depending on its environment. Every vitamin we need today bears testimony to a molecular disease our ancestors contracted sometimes hundreds of millions of years ago. These molecular diseases are not experienced as such under normal circumstances, because our environment constantly supplies palliative drugs. Conversely, if phenylalanine happened to be present only in low amounts in our usual diet, the mutation leading to phenylketonuria, characterized by the inability to convert toxic amounts of phenylalanine into tyrosine, would also not be experienced as a molecular disease, whereas it actually is one under the prevailing circumstances. We might say that evolution is based in part on the appearance of molecular diseases whereof the environment can cure the symptoms. Since our remote ancestors must have been autotrophic, we may consider ourselves as degenerate autotrophic organisms. Whereas, in order to achieve superiority, it is not sufficient to be degenerate, it is however necessary.

In many cases the notion of molecular disease is thus closely linked to the nature of the environment. It is not so in other cases, such as a structural change of the hemoglobin molecule leading to the loss of its ability to combine reversibly with oxygen. Both types of cases are similar in that chemicals available in the environment either cannot be used (oxygen or vitamin precursors from which vitamins are built) or cannot be disposed of (phenylalanine). The difference between the two types of molecular diseases resides in the fact that in one case, that of vitamin need and of phenylketonuria, the environment can make up for the lost biochemical reaction either by furnishing its product (the vitamin) or by ceasing to furnish its starting material (phenylalanine), and in the other case it is not that products are needed or that toxic substances must be excluded, but the process itself of making the product is essential to the organism. Thus when oxygen cannot be carried to the tissues efficiently it would be of no avail to furnish the tissues with oxidation products. The oxidation must be carried out by the organism itself, mainly because living matter requires chemical energy to be set free at the right time in the right place. Thus molecular diseases are defined in relation to the environment when the requirement involved is that of a substance or of less specific forms of energy such as heat, and they are not so defined when the requirement involved is that of a process fundamental to the existence of living matter, that is, of a high degree of specificity in the release of energy in relation to time and space. Life can get everything out of the environment except a degree of specificity approaching its own.

Considering molecular disease and the environment in relation to evolution, we are faced with a two-way relationship. Evolution has probably been influenced not only by heritable changes in the organism that the environment could prevent from being deleterious, but also by changes in the environment that molecular changes in the organism could prevent from being deleterious. Molecular disease can be selected for as a defense against diseases caused by external agents. For example, the past incidence of malaria has been shown to be positively correlated with the abnormal hemoglobin, HbS, present in sickle-cell anemia; with thalassemia, another type of genetically controlled hemoglobin disease; with glucose-6-phosphate dehydrogenase deficiency; and with color blindness (Allison, 1957; Siniscalco *et al.*, 1961). Apparently the presence in the environment of the agent of a highly dangerous disease, *Plasmodium falciparum*, favors the conservation and spreading in the human species of molecular diseases that afford protection against the infectious agent by unknown mechanisms. The molecular diseases, at least in the heterozygous ("trait") condition, are less lethal than the infectious disease. Observations of this kind extend the interaction pattern between molecular disease and environment. A "molecular disease" may be maintained in the species because certain agents in the environment render it innocuous; or it may, on the contrary, be maintained because it renders relatively innocuous certain agents present in the environment. On account of this latter effect it seems possible that external disease-causing agents, notably infectious agents, have played a role in evolutionary sequences of noncompensated degenerative nature, such as those leading to parasitism.

The sickle-cell gene increases the life expectancy of the individual in the heterozygous state, while in the homozygous state it decreases it probably at least as radically as does malaria. When two carriers of sickle-cell trait marry, on the average half of their offspring will again be heterozygotes. The other half will have a decreased life expectancy, because of either sickle-cell disease (sickle-cell gene homozygotes) or malaria ("wild-type" homozygotes). In malaria-infested countries the sickle-cell gene will thus have a tendency to spread in the population. This would hardly be the case if the mutant gene were advantageous in the homozygous instead of the heterozygous condition. A newly appearing mutation that would be retained only in the homozygote would usually have no chance of establishing itself in the population. We may point out that the replacement in a population of a given gene by a mutant gene may often require two successive mutations, except when the population is very small (close inbreeding). After the first mutation the mutant is selected for in the heterozygous state. This allows the mutant gene to establish itself, but at the same time the corresponding wild-type gene is preserved. To eliminate the wild-type gene a second mutation must now occur, such that the doubly mutated gene is most advantageous in the homozygous state. This double mutant would not cause what would appear as a molecular disease, while the single mutant that precedes it might. In this sense molecular disease may be a frequent intermediary step in those evolutionary sequences that lead to the total replacement of a gene by a mutant allele, and that require that the heterozygous condition be at first advantageous. During this phase the homozygous condition may indeed often be deleterious.

II. Hemoglobin, Its Multiplicity and Evolution

A. Introduction to the Hemoglobin Molecule

If one examines molecular disease in relation to evolution it is unavoidable at the present time to center the discussion on the hemoglobin molecule. So far this molecule is the only one that has been studied in many pertinent respects: amino-acid sequence, structure of the site directly involved in function, structural changes leading to molecular disease, normal structural multiplicity, different rates of synthesis of structurally distinct "editions" of the molecule, and their change in the course of time.

All vertebrates save the most primitive ones seem to have hemoglobins composed of four polypeptide chains, linked to each other by bonds much weaker than the peptide bonds that are instrumental in lining the amino acids up unidimensionally within the chains. Each of these chains is composed of slightly less than 150 amino acids and carries one heme group that contains the iron atom capable of binding oxygen reversibly. The string of amino acids winds about in space in a highly specific fashion that may be common to all vertebrate oxygen-carrying pigments, since even sperm whale myoglobin (muscle hemoglobin) shows a similar conformation in spite of the fact that its amino-acid sequence differs very considerably from the sequences so far found in blood hemoglobins (Watson and Kendrew, 1961). Nature has produced a great many such hemoglobin and myoglobin chains that differ in their amino-acid sequence and therefore in various physicochemical properties and yet apparently remain similar in their over-all conformation and in the fundamental characteristics of their relation with the heme group. Not only does the amino-acid sequence of these chains always vary in different animal species, except perhaps in some extremely closely related ones, but any one individual of any given species produces a number of different hemoglobin chains, in part successively and in part simultaneously.

So far the tetrahemic hemoglobin molecule of higher vertebrates has always been found to be normally made up of two kinds of chains that combine two by two. Thus human adult hemoglobin, HbA, contains two so-called α- and two so-called β-chains (Rhinesmith *et al.*, 1957, 1958). This type of hemoglobin is predominant only from the time of birth on, while during intrauterine life by far the greatest proportion of the hemoglobins produced is represented by fetal hemoglobin, HbF, composed of two α- and two γ-chains (Schroeder and Matsuda, 1958; Hunt, 1959, Schroeder *et al.*, 1959b, Shelton and Schroeder, 1960). There exists another human hemoglobin chain during postnatal life, but normally never in more than small amounts, the δ-chain. Two α-chains combine with two δ-chains to form the minor component known as HbA_2 (Kunkel and Wallenius, 1955; Kunkel *et al.*, 1957; Ingram and Stretton, 1961). This is as far as the list of structurally distinct *normal* human hemoglobin chains goes at present. Occasionally, in the diseases called the thalassemias, one or two of the chains are present in subnormal amounts. Sometimes a relative excess of the partner chain is produced, which then associates with its own kind to form tetramers. This leads to the formation of abnormal hemoglobin such as HbH, composed of four β-chains (Jones *et al.*, 1959), or Hb "Bart's," composed of four γ-chains (Lehmann, 1959). Thus the association of two different types of chains is not an absolute requirement for the formation of chain tetramers but probably a matter of preferential affinity. All the known normal human hemoglobins have one type of chain in common, the α-chain, but in other species this is not necessarily so, as has been shown for chicken (C. J. Muller, 1961).

B. Hemoglobin Heterogeneity

It is remarkable that hemoglobin-chain heterogeneity has been found in all species. So far only vertebrates have been examined, but of widely different classes, ranging from mammals to fish and Cyclostomes (reviewed by Gratzer and Allison, 1960; also Huisman *et al.*, 1960, and unpublished results from this laboratory). The

Cyclostomes belong to the most primitive group of vertebrates, whose living representatives, hagfish and lampreys, seem to have hemoglobins composed of single, unassociated polypeptide chains (Svedberg, 1933; Roche and Fontaine, 1940; Lenhert *et al.*, 1956), presumably of the same general type as those which in higher forms associate into tetramers. In this respect Cyclostome hemoglobins resemble the myoglobins. Even though lamprey hemoglobin chains do not normally associate into higher molecular units, these animals also possess several distinct types of chains (Andinolfi *et al.*, 1959). Apparently the multiplicity of hemoglobin chains is not an evolutionary consequence of their association, but their association is an evolutionary consequence of their multiplicity. This conclusion is confirmed by the observation that there exist also several distinct types of myoglobins in all individuals (Rossi-Fanelli and Antonini, 1956; Rumen, 1959; Rossi-Fanelli *et al.*, 1960; Edmundson and Hirs, 1961), while myoglobin polypeptide chains usually do not associate to yield molecular units of a higher order. [Evidence of the presence of dimers in solutions of some invertebrate myoglobins has recently been reported by Manwell (1958b, 1960)]. It appears justified at the present time to extrapolate from these and other observations to polypeptides and proteins in general and to state that proteins and polypeptides of all kinds may usually be expected to coexist within every individual in structurally distinct "editions." Several of these may be synthesized in the same cells, as fetal and adult hemoglobins often are (Kleihauer *et al.*, 1957; Itano, 1956); or they may be produced in different tissues. This generalization rests now on a number of investigations, among which are those of Markert and Møller (1959) and of Kaplan *et al.* (1960). The latter authors showed that in vertebrates as well as invertebrates lactic dehydrogenases extracted from different tissues of one animal are different from one another. By analogy with hemoglobin we may suppose that for most kinds of proteins there will be found in every organism, in addition to major components that succeed each other in time or coexist in different tissues, structurally distinct minor components. The importance of these minor components is probably negligible from the point of view of function, but not of evolution. Molecular diseases will of course relate to the quantitatively important "major" components only. From what precedes it will be recognized that the greater the role of minor components in evolution, the smaller that of molecular disease.

If this picture is correct and it probably is, the number of distinct proteins in the human organism that has been estimated by one of us (L.P.) to be of the order of 100,000 will have to be multiplied by a factor of presently unknown magnitude.

It must be pointed out that a number of minor hemoglobin components apparently do not differ in amino-acid sequence from one of the major ones (Jones, 1961). Some arise apparently through a secondary combination with some other molecule, such as glutathione (C. J. Muller, 1961). Others may be oxidation or denaturation products or chromatographic artifacts, or unusual combinations of hemoglobin chain dimers or monomers, or they may be hemoglobin polymers. Finally, it remains probable, though the contention still awaits experimental confirmation, that some minor components might have an altered amino-acid sequence without being produced under the control of an altered gene. Such components would express "errors" in the synthesis of normal hemoglobin chains (Pauling, 1957a).

If we consider structurally different hemoglobin chains found in one given species, man, we may divide them into two groups. The chains in one group differ from each other by more than one amino-acid substitution. Thus the number of changes in sequence, when the α-, β-, γ-, and δ-chains are compared with each other, varies from about 6 to a little less than 80 (computed from Braunitzer *et al.*, 1960a,b; G. Braunitzer, 1961, personal communication; Konigsberg *et al.*, 1961; Schroeder *et al.*, 1961; W. A. Schroeder and R. Shelton, personal communication; Ingram and Stretton, 1961). These chains which show marked differences have all been found to be controlled by distinct genetic loci (Itano, 1957; Smith and Thorbert, 1958; Cepellini, 1959a,b, Ingram and Stretton, 1961), and they are present in all normal human individuals. In the second group we find chains that differ from one of the others by only one amino-acid substitution. In all cases that have been examined, the gene that controls such a chain has been found to be an allele of the gene in control of the nearly identical chain. Each of these different alleles occurs only in small proportions of the population in different areas of the world. They are the abnormal hemoglobins. While some go unnoticed by their carriers, others lead to characterized molecular diseases in the homozygous condition (see for instance Itano and Pauling, 1957; Itano, 1957; Neel, 1959; Ingram, 1961a). Of course chains differing by more than one amino-acid substitution and controlled by allelic genes may be discovered, but they will presumably remain a small minority, and, for reasons that will become clear presently, we may expect that chains differing by more than a very small number of changes in amino-acid sequence or, more accurately, chains that have been affected by more than a very small number of mutational events will generally be traceable to distinct genetic loci.

On the other hand, when we compare hemoglobin chains from different species, chains controlled by corresponding genetic loci may differ considerably in amino-acid sequence. This is of course only a presumption, since we have no means of matching genetic loci of different species. For instance, there seem to be two differences between the human and gorilla α-chains (Zuckerkandl and Schroeder, 1961), yet there is no reason to suppose that the genic loci controlling their production are not homologous.

We may venture the following generalization. While in different species markedly different hemoglobin chains may conceivably be and probably quite often are controlled by homologous loci (by genes that would be shown to be allelic if fertile crosses between the species were possible), within one species a greater difference between chains is associated with greater independence in their genetic control. In this respect it is suggestive that the α- and β-chain genes, among the most different within the species, have been shown to be on separate chromosomes, or at least not to be closely linked (Smith and Thorbert, 1958), while the β- and the δ-chain genes, which resemble each other most, appear to be linked (Cepellini, 1959b).

C. The Evolution of the Hemoglobin Chains

The foregoing observations can be understood at once if it is assumed that in the course of time the hemoglobin-chain genes duplicate, that the descendants of the duplicate genes "mutate away" from each other, and that the duplicates eventually become distributed through translocations over different parts of the genome. Different non-allelic genes are thus thought to have arisen from an original mother gene. Since it seems justified to consider effective (i.e., viable) translocations as phenomena that occur more rarely than effective amino-acid substitutions, one would expect that genes related to a common ancestor but not closely linked differ from each other by a number of mutational changes, in accordance with observation. Ideas of this kind have been evolved by Bridges (1935), Metz (1947), and notably Lewis (1951), and have been applied to hemoglobin evolution by Itano (1957), ably developed by Ingram (1961b), and elaborated quite independently by ourselves in 1960. It seems likely that the *intraspecific* multiplicity of proteins of a given type is to be explained in these terms. Considerable *interspecific* differences between proteins of a certain type may, on the other hand, as stated before, be compatible with homology of genic loci and not require the intervention of gene duplication. As species gradually get to be more different from each other, so presumably do the genes at the homologous loci.

All we can check at present are homologies of chain structure as expressed by correspondences between amino-acid sequences in hemoglobin chains, and such homologies, whether inter- or intraspecific, suggest a common evolutionary origin. An alternate hypothesis would be convergence by selection for functionally adaptive hemoglobin-chain structures. While convergence may play a significant role, this role is most likely confined to a relatively small number of features of amino-acid sequence. The over-all similarity must be an expression of evolutionary history. This is indicated by the gradually increased amount of differences found when human hemoglobin is compared with hemoglobins from progressively more distant species (Zuckerkandl *et al.*, 1960; C. J. Muller, 1961). The difference between human and fish hemoglobins is such that no common features, except the presence of free lysine, could be detected by the comparison of peptide patterns obtained by spreading the products of a tryptic digestion two-dimensionally over filter paper by successive electrophoresis and chromatography. The absence of common features in these patterns in no way implies the absence of significant stretches of similar amino-acid sequences, but nevertheless expresses qualitatively a degree of difference. A comparable result has been obtained by comparing mammalian and fish insulin (Wilson and Dixon, 1961). Insulin, on the whole, seems less variable than hemoglobin, even taking into account its smaller molecular weight.

At the other extreme we may compare human and gorilla adult hemoglobins. From the amino-acid analysis of separate gorilla α- and β-chains it appears that there are only two differences in the α-chain and one in the β-chain. The amino-acid analysis of isolated tryptic peptides from gorilla hemoglobin, two thirds of which has been completed (unpublished), has so far furnished no evidence of further changes. It is therefore possible that the gorilla β-chain and the human normal and abnormal β-chains form one single population. (As mentioned before, the abnormal human chains differ from the normal ones by only one amino-acid substitution.) Since gorillas get along well with their hemoglobin, as they prove by existing, it is not likely that the gorilla β-chain, if it were present in humans, would cause molecular disease. The required oxygenation properties of hemoglobin must be rather similar in the two species that are otherwise so much alike. Thus, if the gorilla β-chain occurred in a human family the physician's attention would probably not be attracted to it. Moreover, it would probably go unnoticed in general surveys, because the nature of its difference with the human β-chain —probably a substitution of a lysyl for an arginyl residue—seems to be of the kind that current scanning techniques do not detect. Conversely, it is also possible that the human β-chain occurs in some gorillas.

Some of the hemoglobin chains coexisting within one individual differ from each other as much as or more than corresponding chains may be expected to differ in the most distantly related vertebrates. While human β- and γ-chains are only moderately different, they are much more different than gorilla and human β-chains. Therefore with respect to hemoglobin an adult man resembles an adult gorilla much more than his own human embryo. Morphological observation also suggests this relationship, which is now confirmed at the biochemical level. The human α-chain differs much more from the β-, γ-, and δ-chains than the latter from each other. Nevertheless even in the case of the α-chain the remaining similarities with the others are striking. When the sequence of the first 30 amino acids of the β-chain became known from G. Braunitzer's laboratory, our own knowledge of the α-chain was limited to amino-acid sequences in peptides isolated from the chain by tryptic digestion by W. A. Schroeder, R. T. Jones, J. R. Shelton, and their collaborators. The succession of these peptides along the α-chain was unknown. By fitting them into the β-chain according to the principle of maximum homologies, a sequence of the first 31 residues of the α-chain was predicted that was later confirmed by Braunitzer *et al.* (1961). This showed for the first time that the homology principle could be put to work effectively, even between chains that differ considerably.

It is possible to evaluate very roughly and tentatively the time that has elapsed since any two of the hemoglobin chains present in a given species and controlled by non-allelic genes diverged from a common chain ancestor. The figures used in this evaluation are the number of differences between these chains, the number of differences between corresponding chains in different animal species, and the geological age at which the common ancestor of the different species in question may be considered to have lived. Braunitzer and Matsuda (1961) have recently found that there are a minimum of 15 differences in sequence between the horse and human α-chains. (Only one of the two main horse hemoglobin components was analyzed.) This number is not likely to be increased very much by subsequent work. If we estimate that the real number of differences in sequence is between 15 and 20, we may take 18 as a probable mean. From paleontological evidence it may be estimated that the common ancestor of man and horse lived in the Cretaceous or possibly in the Jurassic period, say between 100 and 160 millions of years ago (Piveteau, 1955; Dodson, 1960). For the sake of the calculation it is assumed that most effective mutations result in single amino-acid substitutions, as evidence from abnormal human hemoglobins indicates, and that the evolutionarily effective mutation rate, i.e., the rate of the mutations that have not been eliminated by natural selection, fluctuated during the time of evolution of hemoglobin around a mean without showing a predominant trend to increase or to decrease. Under these conditions the presence of 18 differences between the human and horse α-chains would indicate that each chain averages 9 evolutionarily effective mutations in 100 to 160 millions of years. This yields the figure of 11 to 18 million years per amino-acid substitution in a chain of about 150 amino acids, with a medium figure of 14.5 million years. Our results for the gorilla hemoglobin chains yield somewhat different figures. Because of considerable fluctuations that may be expected in cases where the number of evolutionarily effective mutations has been very small, it seems advisable to use the figure derived from the horse α-chain alone. As the amino-acid sequences of more animal hemoglobin chains become known and paleontological dating is improved, the calculation will have to be revised. Also the number of differences between the human chains is subject to moderate revision, especially the comparisons involving the γ- and δ-chains, based on the results of Schroeder *et al.* (1961), W. A. Schroeder and J. R. Shelton (personal communication, 1961), and Ingram and Stretton (1961).

As Table I shows, the evaluation of the time elapsed since the β- and δ-chains differentiated places their common ancestor at the time of origin of the Primates or somewhat earlier. This checks with the fact that so far δ-chains have been found only in Primates (Kunkel *et al.*, 1957) and furnishes evidence that, at least with respect to more recent evolution, the present evaluation is not unreasonable. Also the time of derivation of man and gorilla from their common ancestor as calculated on the basis of the figures derived from man and horse, 11 million years, falls on the lower limit of the range estimated on paleontological grounds, 11 to 35 million years.

TABLE I

THE APPROXIMATE TIME OF DERIVATION OF DIFFERENT HEMOGLOBIN CHAINS FROM THEIR COMMON ANCESTOR

Chains being compared	Number of differences[a]	Estimated time of derivation from common chain ancestor	Corresponding geological period
β and δ	~6	44×10^6 years	Eocene
β and γ	~36	260	Beginning of Carboniferous
α and β	78	565	Toward end of Pre-Cambrian
α and γ	~83	600	Toward end of Pre-Cambrian
Gorilla α and human α	2	14.5 } Mean 11	Pliocene
Gorilla β and human β	1	7.3 }	

[a] The presence or absence of one to several contiguous amino-acid residues in one of the chains is counted as one mutational change.

Of course, the uncertainty increases as we go further back in time. The common ancestry of the β- and γ-chains is placed at the beginning of the Carboniferous period, that is, about at the time of the first amphibians. Differences between fetal and adult hemoglobins have however been found also in contemporary fish (Manwell, 1957, 1958a). It is conceivable that these have arisen from a gene duplication independent of the one that led to the differentiation of β- and γ-chains.

The α- and β-chains are so different that the present evaluation places their common chain ancestor in the Pre-Cambrian, before the apparent onset of vertebrate evolution. The differences between the α- and γ-chains check reasonably well with those between the α- and β-chains. If the figures were taken at face value, it would seem that vertebrate hemoglobin with its differentiation into hemoglobin polypeptide chains derives from an invertebrate hemoglobin. Thus the ancestors of the present α- and β-chains would have to have been already present as differentiated chains in those primitive vertebrates in which the hemoglobin chains presumably did not associate into tetramers, as in the contemporary Cyclostomes, mentioned earlier. The finding of hemoglobin heterogeneity in the lamprey (Andinolfi *et al.*, 1959) is suggestive in this respect.

The figures in Table I also make it appear unlikely that corresponding chains, say α-chains and their homologs in animals, when the most distantly related vertebrates are compared, will be found to differ from each other more than human α- and β-chains differ. The figures also strengthen the presumption that hemoglobin has not been evolved independently more than once during vertebrate evolution and suggest, as stated, that even the most primitive among the ancestral vertebrates had already inherited their hemoglobin from other forms.

Polypeptide chains that are clearly not homologous, such as horse-heart cytochrome c (Margoliash and Tuppy, 1960) and mammalian hemoglobin chains, may still have a common molecular ancestor, in the sense in which all protein molecules of a given organism may conceivably have one, but such an ancestor would have existed so far back in Pre-Cambrian times that comparative studies on contemporary organisms have no significant chance of revealing a kinship. For all practical purposes it is therefore correct to say that horse-heart cytochrome c and horse hemoglobin chains have evolved independently.

Our best excuse for making the present evaluation is that it affords us the opportunity to point out why it is probably wrong. The sources of error involved are factors in gene evolution that deserve to be mentioned here.

We do not know whether the present "major" hemoglobin components have once been derived from "minor" components. The contribution of minor components as oxygen carriers is mostly negligible. Unless they have other unknown functions, natural selection will not be expected to act upon them. Thus all mutations will probably be preserved in a minor component until one of the three following possible events occurs: a mutation that makes it unrecognizable as a hemoglobin chain; a mutation that brings about a total inhibition of its synthesis, or a mutational change that transforms it into a major component. If the ancestors of the human hemoglobin chains that are important quantitatively (the α-, β-, and γ-chains) have started out as minor components, they will during that remote period have retained many more mutations per unit time than we have assumed, and from this point of view the figures given in the table would be overestimates.

They tend to be underestimates for other reasons. In the comparison between the chains, possible back-mutations, of which we have no knowledge, had to be neglected, as well as successive different effective substitutions at the same amino-acid residue. The likelihood of these events increases with the increase in the number of amino acids affected by change in a given chain. Thus the number of effective mutational events that have actually occurred since the α- and β-chains have evolved from their common ancestor may be significantly greater than is presently apparent.

Furthermore, even if we assume the intrinsic mutation rate of the hemoglobin chain genes to have remained fairly constant throughout the geological periods, disregarding the probable effects of changes in temperature and in intensity of ionizing radiations, the effective mutation rate may have varied widely according to the "ecological" conditions of hemoglobin within the organism and of the organism within its environment. In particular, during evolutionary transition periods such as afforded by a change from aquatic to terrestrial habits the effective mutation rates may have been much higher than at other times. The size of the populations at every stage is also of paramount importance in determining the evolutionarily effective mutation rate. Some other factors, the impact of which is equally difficult to evaluate, should also be taken into account. Fortunately the over-all result of the interplay of all factors is expressed in the speed of evolution, which has been evaluated. The general finding is that in the course of time evolution has become accelerated (Rensch, 1954). The more recent terrestrial groups of animals, on the average, have evolved faster than the more ancient aquatic groups. We may expect that this generalization, based on morphological characteristics, has its counterpart in the speed of evolution of deoxyribonucleic acid (DNA) and of the proteins. On the average, a lesser number of evolutionarily effective mutations per unit time may thus have affected the hemoglobin molecules during the initial phases of vertebrate evolution than in later periods. Our guess is that the numbers given in Table I are more likely to be underestimates than overestimates.

In the preceding evaluations we have equated one mutational event to one amino-acid substitution in the polypeptide chain. As mentioned, present evidence tends to show that this is the most frequent type of evolutionarily effective mutation in hemoglobin genes. It is too early to generalize to structural genes at large. Work in progress in several laboratories has a direct bearing on this question, such as the work of Yanofsky's group on *Escherichia coli* tryptophan synthetase at Stanford University, that of Fraenkel-Conrat's group on tobacco mosaic virus (TMV) mutants at Berkeley, that of Levinthal's group on *E. coli* alkaline phosphatase at M.I.T., and that of Brenner's group on phage-head protein at the Cavendish laboratory. There is ample evidence that mutational events other than single amino-acid substitutions exist. Some such evidence is derived from the comparison of the hemoglobin α- and β-chains themselves. In several regions of both chains sequences of contiguous amino acids numbering from 1 to 5 (or 6, if we include in the comparison the C-terminal sequence of sperm whale myoglobin, to which no counterpart is as yet known in hemoglobin chains) are missing. Three types of mechanisms might account for such observations: terminal growth of the chains, in the case of terminal differences; deletions; or insertions. Insertions may be duplications of chain segments, associated or not with a reversion of the segments. Although the events at the level of the chromosome and of a single DNA molecule may be qualitatively quite different, one must not discount the possibility that events of the type first described by Sturtevant (1925)—the insertion into a chromosome of a duplicate of a chromosomal region—may have its counterpart within one structural gene.

There are two one-amino-acid "holes" in the α as compared to the β-chain and one two-amino-acid "hole" and one five-amino-acid "hole" in the β as compared to the α-chain. If each of these is tentatively considered to be attributable to a single mutational event, then of a total of 78 mutational events that have led to the present differentiation of the α- and β-chains four events, i.e., 5% of the total, represent deletions, insertions, or terminal chain growth. It is possible that the actual proportion of mutations that result in events other than the substitution of a single amino acid in the polypeptide chain is much higher, if such mutations are more often lethal than "substitution mutations," as seems likely indeed.

"Substitution mutations," such as have so far always been found in mutant alleles of hemoglobin chain genes, if occurring in a major chain gene and advantageous either will be eliminated rather quickly or will eventually replace entirely the wild-type gene. Likewise, "substitution mutations" that cause molecular disease either will be eliminated before there is time for a second mutation to occur in the mutant or, if a selective advantage exists under special circumstances for the heterozygote, as in the case of sickle-cell hemoglobin, will be confined to a small enough number of individuals so that the appearance of a second mutation in the same gene will be improbable. Considering a given species, there are only two types of cases where the appearance, by repeated single substitutions, of more than one difference between originally identical genes would be favored: (1) when the heterozygote is universally favored over the wild-type homozygote; this would apply to sickle-cell hemoglobin, if humans were universally exposed to malaria; (2) when a gene has duplicated and the conformity of the duplicate gene to the original model is not selected for. Assuming that the duplicate gene contributes to protein production, this conformity will be selected for only if an increase in output of a given polypeptide chain is advantageous.

D. The Destiny of Duplicate Genes and the Function of Genic Multiplicity

If gene duplication is one of the means of increasing the output of a given protein, one may distinguish two phases in this respect. Up to an optimum number of duplications, the duplicate gene will be selectively retained with a structure identical to and a position rather near to that of the mother gene. Beyond this point, duplicate genes will be progressively more strongly selected against. During this latter phase they will be in part eliminated with their carriers, and in part subjected to progressive change. When they are preserved and changed, their destiny may be of three types. They might evolve new useful functional properties. In this case they will be retained as active genes, and to the extent to which polypeptide output depends on gene duplication their own duplicates will be kept unchanged by natural selection. Secondly, functionless or unfavorable duplicates will not maintain duplicates to their own likeness and may themselves be translocated to other chromosome parts and be reduced to minor-component genes by a position effect. Some such minor-component genes, more or less profoundly changed in the meantime, may be selected for later in evolution and be changed into major-component genes. Thirdly, the activity of the duplicate may be reduced to zero.

This elimination of gene activity may again take three forms. The changed structural gene may be bodily eliminated through the loss of the part of the chromosome that carries it; or it may be modified to such an extent that its products, although significant in amount, are no longer recognizable in terms of the original protein; or it may be preserved in a modified state, but totally or subtotally deprived of the power of expression.

The existence of such "dormant genes," although difficult to verify, is a plausible inference from two types of observations: firstly, that within a given tissue, say the hematopoietic tissue, major and minor structurally distinct components of a given type of polypeptide chain are found. Since at a given time the relative quantities of hemoglobin chains vary between 100 and 1%, other genetically distinct minor components may be present in such small amounts that they are practically undetectable. Secondly, there are numerous examples of proteins that are produced exclusively in one type of tissue, and of which no trace is found with presently available analytical means in other tissues. The nonproduction of hemoglobin in muscle cells and of myoglobin in reticulocytes is one example. This example shows that some among even relatively closely related structural genes may, within a given tissue, be the ones strongly expressed, the others unexpressed. We must assume that all the structural genes have during embryological development been communicated to all cell lineages. It is therefore quite likely that there exist in every organism numerous structural genes that do not find in any of the existing tissues conditions favorable to their expression and thus remain permanently dormant.

Furthermore the relative structural similarity of minor hemoglobin components to one of the major components affords yet another argument in favor of the existence of dormant genes. Indeed, as we have seen, the human δ- and β-chains are quite similar. Likewise structurally distinct components of orangutan hemoglobin have been found to be quite similar, and the same holds for pig hemoglobin components (E. Zuckerkandl, R. T. Jones, Y. Nishiwaki and L. Pauling, 1959–1961, unpublished). If duplicate genes remained usually expressed, one would expect to find a series of minor-component chains differing from all the other chains as much as the human α- from the human β-chain. This is, however, not the case. One is led to think that, in the long run, duplicate minor-component genes most often cease to be expressed. There is no apparent reason why one should assume that they have most times been bodily eliminated. Other possibilities are that they have been implied in a further translocation, that a mutation or transposition of a controlling element (repressor) has occurred, or that the structurally modified mutated gene possesses specificity characteristics that fail to comply with the specificity requirements for polypeptide production under the conditions prevailing in the cell.

Dormant genes of course are conceived as dormant only as far as their expression, and not as far as their mutability goes. Mutations in dormant genes and in minor-component genes will never be lethal, unless the latter have some distinct specific function, which would then lead us to consider them as "major components" of another protein type. Minor-component genes and, mainly, dormant genes may thus furnish an important and perhaps the principal part of the genic raw material for macro-evolutionary experiments of nature. A new translocation, or the transposition of a controlling element such as those described by McClintock (1956), or some other genetic modification may reactivate the dormant gene after a very long period of time during which mutations have changed it enough so that it now controls the production of a new kind of protein. In this fashion new enzymes, new functions can arise without the corresponding loss of old enzymes, old functions. We have recalled earlier the importance to evolution of the loss of functions through the mutations of active genes. But it is evident that evolution, while it makes the best of such losses and of molecular disease, could not be based on them alone. There are a great many more different functions to be carried out by a great many more different types of enzymes than we are allowed to suppose can have existed in early evolutionary times. Primordial living matter must have been limited to a few simple functions. Therefore the notion of evolution by gain is a necessary complement to the notion of evolution by loss.

Horowitz (1945) made a lasting contribution to our thinking about evolutionary gain at the enzymatic level. He described how new reaction chains might arise in certain circumstances through the chance combination of the necessary genes and furthermore proposed a general mechanism for the stepwise building up of complex enzyme-systems, presenting us with a plausible scheme of macro-evolution at the molecular level. Obviously, as enzyme systems become more complex, more different enzyme molecules are needed. There are reasons to think that the same molecules cannot usually be expected to carry out several different enzymatic functions. Therefore, new genes are needed for the building up of new functions. This is where the concept of the mutational reactivation of dormant genes complements Horowitz's picture. Minor-component genes are not to be excluded from a similar role, but may usually not yet be different enough from their parental genes to be fit for carrying out novel functions. Most minor-component genes, as stated above, are liable to be eventually turned into dormant genes because natural selection will not prevent their transfer to synthetically inactive chromosome regions, nor their coming otherwise under the influence of a repressor gene, nor structural changes that might place the genes outside the range of specificity requirements of the available macromolecules that collaborate with the structural gene in protein synthesis. As dormant genes become reactivated after periods of cryptic existence corresponding perhaps to geological ages, they may produce potential enzymes that do not disrupt existing chains of reactions, but are able to add new processes to the old ones, perhaps in the ways described by Horowitz. One of the possible mechanisms of the reactivation of dormant genes is the reactivation of the chromosomal region where it is located through a change in intracellular environment. The hope of demonstrating the existence of dormant genes rests on this possibility. Such a change in intracellular environment could result from the adaptation of the organism to changes in the external milieu. In adaptational changes, during an initial phase, gain and loss mutations may be balanced, or loss mutations only may occur, so that the total complexity of the organism either remains constant or tends to decrease. In the process, however, the environment of the chromosomes may be altered in some tissue and, on account of this alteration, genes be activated in some regions of the chromosomes, inactivated in others. Inactivations of this kind will be mostly lethal and genomes with corresponding inactivation-resistant chromosome regions will be selected for. The newly activated genes on the other hand will now be available to respond to further adaptational needs and will furnish a series of gain mutations without corresponding losses. Parts of this concept are supported by observation. Changes in conformation of the genic DNA molecules are presumably related to changes in the activity of the genes, and Schmitt (1956) has shown that the state of chromosomal DNA depends on the chemical environment of the chromosomes. Furthermore, genes in mice that display the same activity characteristics during the individual's life time have been found to be closely linked on the same chromosome (Paigen and Noell, 1961).

Through the reactivation of dormant genes by a modification of the intracellular environment an initial adaptational stress of great magnitude would appear instrumental in producing a rise in complexity of the organism. Thus would be solved the old paradox expressed in this question: why should organisms get to be more complex, since simpler organisms are evidently adapted as well to their environment? Once more Biology will show that it can do without any "élan vital" or "entelechy."

The present concept leads one to predict that the fastest evolution toward more highly organized forms should take place after the occurrence of major environmental stresses. Evolutionary history bears out this expectation. For a long time paleontologists have noted that the initial phases in the development of new types of forms has an "explosive" character (consult, for instance, Rensch, 1954). According to the present theory, we assume that at the onset of such an explosive evolutionary phase a change in intracellular environment has brought about the reactivation of a number of previously dormant genes. Rensch believes that the phenomenon of explosive evolutionary phases is adequately accounted for by increased natural selection accompanying the conquest of new biotopes. He thus considers as instrumental a change in *external* environment only. This however does not explain the trend toward increased complexity of the forms.

In the sense that has been laid out here a marked change in environmental conditions may lead to what a layman would call a shake-up of the genome, and this may be the part of reality behind the poorly documented and falsely interpreted observations of the Soviet anti-geneticist A. Lysenko.

To sum up, mutations of active genes controlling "major" protein components suffice to explain how an organism can adapt to changes in its environment. Mutations of minor-component genes and dormant genes, however, seem to be able to furnish the organism with the genic raw materials that eventually allow it not only to adapt to a new environment but also, in the process, to become more highly organized. Minor-component genes and dormant genes may thus prepare the major steps in evolution.

Beside a minor-component multiplicity, there is also a major-component multiplicity in protein production, especially the one characterized by a succession of major components in time. Why are hemoglobin β-chains substituted for γ-chains? The reason may not be that β-chains would not be fit to meet the respiratory prerequisities of intrauterine life and the γ-chains those of adult life, but, rather, that the structural genes corresponding to the two chains are located in two distinct chromosomal regions, one of which is activated in the particular intracellular environment determined by the organism at early developmental stages, and will not be active under other conditions, while the second chromosomal region, on the contrary, will be activated only at the end of embryonic development due to the presence or absence of some particular factors in the cell at that time.

To carry out a given function it is thus not sufficient for a cell to possess a favorable structural gene. A further prerequisite is that this gene either be located in a chromosome region that remains active in protein synthesis in spite of the changes of intracellular environment that occur during development, as is the case of the hemoglobin α-chain gene; or that there exist several duplicates of the gene, each of which is located in a chromosomal region that is active during certain phases of development, and that these duplicates are distributed in such a way over the genome that at any time of development at least one of them remains "on duty." These different genes will never be identical because, although they have supposedly arisen by duplication of an original gene, translocation is apparently a rarer event than amino-acid substitution, so that translocation genes will be expected to differ from each other by more than one amino-acid substitution.

Constant vital functions thus frequently need to have at their disposal several editions of a given type of genes in several regions of the genome that are successively activated and inactivated, or vice versa, with respect to protein-synthesizing ability. This view is advanced as an explanation of the generality of the most important types of "major-component" multiplicity in proteins. These types are on the one hand the successive embryonic and adult editions of a protein; and on the other hand the different editions found at any one time in different tissues of the same animal. The latter type of protein heterogeneity has been referred to earlier and may be interpreted in the same terms. In each tissue the particular intracellular environment provides different conditions for the distribution of active and inactive regions throughout the genome, and thus different duplicates of a given gene that have undergone different translocations will be "on duty" in different tissues.

It thus appears that the unavoidable change in intracellular environment during embryonic development is a great challenge to embryonic development itself, because of the obligation that many critically important proteins be produced throughout and in spite of this change. One may therefore venture to say that there could be no embryonic development without gene duplication followed by gene translocation. In this theory the same events that furnish the genic raw materials for evolution also furnish the genic raw materials for ontogeny.

Other types of genes, of course, are active in protein synthesis during one or the other phase of ontogenetic development only. Certain functions, for instance, are developed only in the adult organism. When their expression is delayed or hastened, paedomorphosis or palingenesis ensues (see for instance Rensch, 1954). Chromosomal events of the type described may be postulated to cause these phenomena also.

After having insisted on the cause and on the function of genic multiplicity, in particular in relation to hemoglobin structural heterogeneity, we must here point out that there is a limit to hemoglobin heterogeneity, at least so far as phenotypic expression goes. A universal intraspecific structural multiplicity of all types of proteins is on the way to being established and this finding may be an important step forward in the recognition of biological reality. Yet part of this step is the rejection of older speculations about protein heterogeneity. These postulated a continuous spectrum of structural variants in the case of each protein. As a result, the apparent amino-acid composition of a protein would be only a mean composition, and proteins would not exist as strictly defined chemical species. Work of recent years has shown that, except for a possible low percentage of "errors" in synthesis, the chemical formula of proteins is as rigorously defined as that of simple molecules. We can no longer agree with J. S. Haldane and J. G. Priestley, who wrote in 1935 "It does, in fact, appear to be fairly certain that each individual has a specific kind of hemoglobin, just as he has a specific nose." Concordant amino-acid sequence analyses in three laboratories (Braunitzer in the Max Planck Institute in Munich; Schroeder in the California

Institute of Technology; Hill and Konigsberg in the Rockfeller Institute) of human hemoglobin chains obtained from different individuals, as well as the comparison of tryptic peptide patterns (Ingram, 1958) obtained with different human hemoglobins in many more instances, shows that this is not so. Of course a greater number of structurally distinct human hemoglobin chains are probably produced than we presently know about. This applies in particular to the abnormal human hemoglobins, about thirty of which have so far been described, and whose inventory one will probably never be able to consider as complete. This also applies to the normal human hemoglobin chains controlled by non-allelic genes, of which a few more may eventually be found.

Some workers go further and believe that a number of undetected hemoglobin alleles may be present in the population. They may escape notice if they behave identically in electrophoresis, but differ in uncharged amino acids. One of us (L.P.) and Itano (1957) have proposed this idea as a possible explanation for the inhibition of hemoglobin synthesis in the molecular diseases known under the name of the thalassemias. Changes in structural genes may indeed lead to an inhibition and even to a total loss of the ability to synthesize a protein. Ingram and Stretton (1959) have developed this idea as one of several possible explanations of the thalassemias. It has ceased to be likely that such cryptic mutants are of very general occurrence in normal populations. The great similarity between gorilla and human hemoglobin chains is a piece of evidence against such a view. With the exception of one difference relating to a serine residue in one of the chains, the uncharged amino acids may all be the same. This similarity suggests that nonpathogenic undetected structural variants of hemoglobin chains must be rather rare, unless a human is more often like a gorilla than like a human. It is true that this hypothesis is supported by more observations than most biological theories.

III. Three Types of Molecular Diseases

After having considered molecular disease in its relation to evolution and genic variability in its relation to evolution, we may comment on genic variability in relation to molecular disease. Three types of molecular disease may be distinguished. Mutations may (1) interfere with molecular function, (2) interfere with the adaptation of the molecule to the intracellular environment, and (3) interfere with the rate of synthesis of a molecule that is functionally fit. Hemoglobin mutants offer examples of all three types.

A. Interference with Function

The only known alterations of amino-acid sequence that lead directly to an interference with the oxyphoric function of hemoglobin are those of the pigments collectively known under the name of HbM. They are all characterized by the formation of methemoglobin, in which the iron is oxidized to the tripositive state (Gerald, 1960). It was recently shown that several distinct mutations give rise to abnormal methemoglobin formation (Gerald, 1958, and personal communication, 1960). It is interesting that structural studies have revealed that the amino-acid substitutions leading to HbM formation all affect a certain region of the hemoglobin polypeptide chains. This region has been called the "basic center," because it comprises a relatively large amount of basic amino acids, and we may call it "basic center I" to distinguish it from a second basic center further along the chain. The two basic centers have in common the property of containing a histidine that seems to be in relation with the heme iron. While the main linkage of the heme iron to the globin is supposed to be to a histidine of basic center II, the histidine of basic center I is, according to present evidence (Watson and Kendrew, 1961), placed opposite the sixth coordination position of the iron, the one that binds oxygen in the oxygenated state and water in the deoxygenated state (Haurowitz, 1949). It may be that the second histidine is the one mentioned by Conant (1934) and Coryell and Pauling (1940). Table II shows the variability of some features of sequence and the constancy of others of the peptide region around basic center I in different animal species. Abnormal human mutants in which this region is affected are also listed. Corresponding results relative to basic center II have not yet been forthcoming.

The seven consecutive amino-acid residues shown in the table comprise residues numbers 56 to 62 in the α-chain and numbers 61 to 67 in the β-chain, counting from the amino end. Three of the seven residues are shown to be substituted in some normal respiratory pigments and the changes, as in the ovine α-chain and in one of the orangutan chains, may involve the substitution of an acid for a neutral amino acid without any fundamental interference with hemoglobin function. Of course such substitutions would be expected to affect one or several of the physical parameters of oxygenation.

TABLE II
"BASIC CENTER I"

LIST OF HEMOGLOBIN CHAINS DIFFERING IN THIS REGION FROM THE HUMAN α- OR β-CHAIN

Chain	Sequence	Reference
Human α, normal	–lys–gly–his–gly–lys–lys–val–	Schroeder *et al.* (1961)
Human β, normal	–lys–ala–his–gly–lys–lys–val–	Braunitzer *et al.* (1961)
Human α, Norfolk	–lys–*asp*–his–gly–lys–lys–val–	Baglioni (1961)
Human β, $M_{Milwaukee}$	–lys–ala–his–gly–lys–lys–*glu*–	P. S. Gerald *et al.* (personal communication, 1960)
Human α, M_{Boston}	–lys–gly–*tyr*–gly–lys–lys–val–	P. S. Gerald *et al.* (personal communication, 1960)
Human β, M_{Emory}	–lys–ala–*tyr*–gly–lys–lys–val–	P. S. Gerald *et al.* (personal communication, 1960)
Human β, Zürich	–lys–ala–*arg*–gly–lys–lys–val–	C. J. Muller and Kingma (1961)
Bovine α	–lys–gly–his–gly–*ala*–lys– (or arg)	C. J. Muller (1961)
Ovine α	–lys–gly–his–gly–*glu*–lys– (or arg)	C. J. Muller (1961)
Goat α	–lys–gly–his–gly–*glu*–lys– (or arg)	C. J. Muller (1961)
Horse α	–lys–*ala*–his–gly–lys–lys–	Inferred from Braunitzer and Matsuda (1961)
Orangutan α or β	–lys–*asp*–his–gly–lys–lys– (or *glu*)	C. Baglioni (personal communication, 1961)
Sperm whale myoglobin	–lys–*val*–his–gly–*ileu*–glu–val– (or glu)	Watson and Kendrew (1961); Edmundson and Hirs (1961)

The reduction of the basicity of this peptide region is still more marked in sperm whale myoglobin. While both types of chains of human, gorilla, chimpanzee, beef, and horse hemoglobins, one of the chains of orangutan hemoglobin, and the β-chains of sheep and goat possess four basic amino acids in the stretch of seven shown in the table, sperm whale myoglobin probably retains only the histidine opposite the heme group as well as the initial lysine. In various types of fish the corresponding peptide must also be less basic than in man, as tryptic peptide patterns indicate (Zuckerkandl *et al.*, 1960). However, four amino-acid residues out of the seven have so far not been shown to vary in any of the normal respiratory pigments, and may be essential to the oxyphoric function.

Methemoglobin formation in man has so far been shown to be caused by substitutions at two different residues, and one of these substitutions has been observed in both α- and β-chains. In $HbM_{Milwaukee}$ there is a substitution at the valine in the fourth position C-terminally with respect to the histidine that is supposedly in relation with the heme iron. In the two other kinds of HbM that have been structurally analyzed, HbM_{Boston} and HbM_{Emory}, it is this histidine that is affected; in both cases it is replaced by tyrosine. A replacement of this histidine by another amino acid does however not necessarily lead to methemoglobinemia, since methemoglobinemia has not been reported as a feature normally observed in the family possessing $Hb_{Zürich}$ (Hitzig *et al.*, 1960), although oxydation may be facilitated. In $Hb_{Zürich}$ the histidine in question is replaced by arginine. Thus the basic character of histidine seems to be more essential in protecting the heme iron from oxidation than its particular configuration, and the affinity for heavy metals, which is much greater in histidine than in arginine (Albert, 1952), seems also not to be involved.

Since the "basic center I" region is probably part of an α-helix, as is the corresponding region in sperm whale myoglobin (helical region E, Watson and Kendrew, 1961), the fourth residue after the critical histidine should lie next to it on the helix. It is plausible that the acid residue found in this position in the case of $HbM_{Milwaukee}$ interferes with function. Thus all types of HbM that have so far been analyzed seem characterized by a change in state or in kind of the critical histidine residue. On the other hand, the aspartic acid that replaces a neutral amino acid in $Hb_{Norfolk}$ does not seem to interfere basically with the function of the neighboring histidine, and this is probably due to the fact that because of the different orientations of neighboring residues in a helix the side chains of neighboring residues can be further apart than the side chains of residues four removed along the helix.

A further interesting observation related to the basic center I is that the change in the orangutan peptide is apparently the same as in the abnormal $Hb_{Norfolk}$ (C. Baglioni, 1961, and personal communication, 1961). In this respect human carriers of $Hb_{Norfolk}$ have orangutan hemoglobin. The disease is mild; but this hemoglobin has not yet been observed in the homozygous state. The case shows that what may appear as a "molecular disease" in one species may be the norm in another.

B. Interference with Normal Intermolecular Relations

Mutations that do not significantly affect the oxyphoric function of hemoglobin may nevertheless lead to severe molecular diseases if they alter the physicochemical properties of the molecules that are of importance in its relation to sister molecules and to other constituents of the red cell. Among the most important of these properties is solubility. Considering the great proportion of nonpolar groups usually found in proteins, the building of a soluble protein molecule appears to be a difficult accomplishment. The readiness with which solubility is lost upon changing spatial comformation (denaturation) demonstrates that the solubility of most proteins is very sensitive to the distribution of their side-chains in space. Mutations that result in the substitution of a polar by a nonpolar amino-acid residue, as well as other mutations that weaken the conformational stability, may thus be expected to interfere frequently with the functionally required solubility characteristics of a protein. One might surmise that many molecular diseases should involve such losses of solubility. If it is justified to consider aging as a multiple molecular disease arising through somatic mutations, then aging probably also expresses in part the loss of solubility of certain proteins.

Hemoglobin is expected to be particularly sensitive in this respect to mutational change because it is in solution in the red cell at a concentration of about 30%, a concentration that not many molecules are able to achieve. Even a slight change in the properties of molecular interaction may under such circumstances lead to a drastic effect. Under certain conditions the abnormal human hemoglobins H [composed of four β-chains (Jones *et al.*, 1959a)] and Zürich (Hitzig *et al.*, 1960) tend to precipitate in the red cell. Among the cases in point, the most well known, and the one that typified molecular diseases in general (Pauling *et al.*, 1949), is that of HbS, the hemoglobin of sickle-cell anemia chemically characterized by the substitution of a valyl for a glutamyl residue in the sixth position from the N-terminus of the β-chain (Ingram, 1959). This substitution does not lead to a decrease in solubility of the oxyhemoglobin (Perutz and Mitchison, 1950), but upon deoxygenation the molecules, when in sufficiently concentrated solution, interact and align along fibers that seem to form a network reminiscent of gelification (Bessis *et al.*, 1958), and at the level of cellular dimensions the alignment of the fibers is expressed by the formation of tactoids (Harris, 1950), which deform the cell membrane and lead to an early destruction of the red cells and to interference with blood flow in capillaries.

The oxygen-dissociation curve, its position, shape, and dependence on pH, and thus the fitness of the hemoglobin with respect to a given set of circumstances may be affected indirectly by amino-acid substitutions that alter the relation between the respiratory pigment and its cellular environment. The oxygen affinity within the erythrocytes of HbS is lower than that of HbA (Becklake *et al.*, 1955), whereas very similar affinities have been reported for these two pigments in cell-free solution (Allen and Wyman, 1954). Recent evidence indicates however that also in cell-free solution the oxygen affinity of HbS is lower than that of HbA (Riggs and Wells, 1961). As to HbF, it has a greater oxygen affinity than HbA within the red cell, yet in cell-free dialysed solution the affinities of HbA and HbF appear to be nearly identical (Allen *et al.*, 1953), except at pH values below 7 (Manwell, 1960). Red cells containing fetal or adult hemoglobin, respectively, differ in certain parameters such as surface, volume, and thickness (Riegel *et al.*, 1959). This difference may possibly express the influence of structural characteristics of the hemoglobin molecule on structural characteristics of the red cell. Admittedly other factors, such as the relative rate of synthesis of different hemoglobins, will also intervene here. This leads us to consider briefly the third type of molecular diseases, associated with the hampering of protein synthesis.

C. Interference with Synthesis

Hemoglobin synthesis may be interfered with by various mechanisms. One of the most obvious is the absence of red-cell production. This apparently occurs as a normal feature in a family of antarctic bony fish, the *Chaenichthydae* (Ruud, 1954). When red cells are present a decrease in the output per cell of individual hemoglobin chains is known as a heritable character in many human families. Most of the known abnormal human hemoglobins, while not unfit as oxygen carriers, are present in the red cells in subnormal amounts.

HbJ, which is present in higher quantities per red cell than HbA, is so far the only exception to this statement (Thorup *et al.*, 1956). The decreased ratio of abnormal to normal hemoglobins may be assumed to express a decreased relative rate of synthesis.

Itano (1953, 1957) was the first to point out that the structure of a gene may have a direct influence on its synthetic activity. The decrease in rate of synthesis of most abnormal hemoglobins may be visualized in at least two different ways. It may be due to an interaction between the structural gene and its substrate on the chromosome; or it may be due to an interaction between the gene and extrachromosomal macromolecules, which might be called co-determinant factors. The collaboration of co-determinant factors with the genes or the primary gene products is assumed to be necessary for protein synthesis. An alteration of the gene structure would upset the balance of attractive and repulsive forces (of electrostatic, van der Waals and steric nature) between these molecules. A decreased rate of synthesis would then result from a decreased degree of fitting between the two entities. We have referred earlier to the probable change of state in the genes during various phases of the development of an organism. Co-determinant factors also may change in tertiary structure as a result of altered intracellular conditions, or they may combine with smaller molecules, perhaps hormones, that change their specificity and activity. The rate of protein synthesis as a function of the intracellular environment may thus be determined by changes of state of the genes on the chromosomes as well as by changes of state of other macromolecules in solution.

Evidence is accumulating to the effect that large parts of protein molecules are not connected with any of their known specific functions. Various hypotheses can be used as partial tentative explanations of this fact such as the stabilization of the tertiary structure in larger proteins, the random survival of structures that were useful during earlier evolutionary stages, and the resistance to diffusion of large molecules (Pauling, 1957b). Consideration should also be given to the possibility that many parts of the protein molecule that appear functionally neutral may have a function in connection with the rate of synthesis of the protein. Natural selection may well act not only at the level of the finished protein product, but also at the level of its production.

The decrease in the rate of synthesis of hemoglobin chains is often more considerable in thalassemias than in the case of abnormal hemoglobins. While a single genetic event probably causes both the amino-acid substitution and the change in rate of synthesis in abnormal hemoglobins (Itano, 1959), there is evidence that more than one type of genetic event may be involved in thalassemias. The mechanism of these diseases is being investigated in many quarters and has been recently discussed by Ingram and Stretton (1959) as well as other authors. None of the theories so far proposed is completely satisfactory. Whatever the theory, it should be kept in mind that the inhibition of protein synthesis is a nonspecific effect that may have different causes and that there may be several types of thalassemia, not only, as is commonly recognized, with respect to the particular hemoglobin chain that is most severely affected, but also with respect to the underlying mechanism.

IV. Fighting Molecular Disease

In the more or less distant future an enzyme deficiency such as the one that causes phenylketonuria may be met by endowing the organism permanently with a certain quantity of a stable artificial enzyme (Pauling, 1956). Of course, such a solution could hardly be considered in the case of a respiratory pigment, where the quantities required with respect to the "substrate" are not catalytic, but stoichiometric.

Another conceivable means of fighting molecular disease that cannot at present be theoretically excluded is the activation in a certain tissue through drug action of a minor-component gene or a dormant gene, representing a functionally fit duplicate of the functionally unfit "major-component" mutant. Whether or not such a treatment is possible depends on the nature of the control of gene activity. If this activity is in part controlled by position effects in relation to factors present in the intracellular environment, medical research might take such a course, although of course the problems of cell permeability and of the toxicity to the activity of other important genes of the agents capable of modifying specifically the intracellular environment might create major obstacles. The fact that this question can be raised shows however that research on the mechanisms of control of gene activity is not only of fundamental importance to biology, but of great interest to medicine as well.

None of the means of meeting the challenge of molecular disease will, however, be as satisfying as the elimination of the disease-causing mutant genes from the human populations. This is theoretically the best, and it is at present the only concrete efficient measure that can be proposed. H. J. Muller (1959, 1961) who has given considerable attention to eugenics, has recently proposed the creation of germ-cell banks, from which prospective parents could draw the choicest human genomes. Such discussions and proposals are of paramount importance, even though one might not share Muller's optimism, which leads him to believe that in a process of free choice of genomes on the part of the populations the greatest human values would all get their fair share. It may be anticipated that governments would advocate and propagandize the choice of the socially minded, the active, the efficient, at the expense of the contemplative person; and the "good fellow" who represents the majority of humans would tend to procreate by choice a "super good fellow," a super corporation man, more able in conforming than in intellectual accomplishments. Without the contemplative, endowed with refined perceptive abilities of the qualitative and of the significance of forms, the human race would deteriorate. They are indeed humanity's greatest asset, both in the realm of doing and in the realm of being. It is true that the present development of the world toward a huge hive where nothing can stand in the path of technology and mass production promises to individuals so endowed little in the future except frustration and unhappiness, and one might contend that to relieve their suffering it might be a good deed to eliminate them genetically.

On the other hand, no objection can be legitimately raised, it seems to us, against the ambition to eliminate from human heredity those genes that lead to clearly pathological manifestations and great human suffering. The means of achieving this goal need to be discussed. We know now that in the United States about 10% of the Negro population carry one HbS or HbC gene. Therefore about one 400th of the children born to Negro parents have the deadly disease sickle-cell anemia. A simple test permitting the detection of the heterozygous carriers of a sickle-cell-hemoglobin gene exists, and as a first protective step there should be a law requiring all persons within a population in which this gene is present to any significant extent to submit themselves to this test.

If people carrying the mutant gene were to refrain from marrying one another, but married normal individuals, the incidence of the gene would remain constant in the population, and the problem of eliminating the gene would not be solved. To eliminate the mutant gene the following rules may be proposed. If two heterozygotes marry they should have no children of their own. If a heterozygote marries a normal person they should have a number of progeny smaller than the average. In this way the mutant gene would be eliminated in the course of time in a way not involving the suffering caused by the birth of the defective children. Similar measures should be taken in the case of phenylketonuria and other molecular diseases.

In a marriage of heterozygotes, who do not suffer from the diseases caused by recessive genes, the chance that each child is homozygous for the defective gene is 25%. This percentage is much too high to let private enterprise in love combined with ignorance take care of the matter. And although interference of law in questions that are to a great extent of a very personal nature is to be avoided whenever possible, it would be clearly unethical to oppose such an interference in the case of molecular disease, at the very least in those cases, such as sickle-cell anemia, where presently available palliative measures are inadequate.

We may accordingly have hope that the increase in knowledge about molecular disease will in the course of time lead to a significant decrease in the amount of human suffering in the world.

Acknowledgment

We thank Professors E. B. Lewis and N. H. Horowitz for stimulating discussions about some of the topics referred to in this article.

References

Albert, A. (1952). *Biochem. J.* **50,** 690.

Allen, D. W., and Wyman, J., Jr. (1954). *Rev. hématol.* **9,** 155.

Allen, D. W., Wyman, J., Jr., and Smith, C. A. (1953). *J. Biol. Chem.* **203,** 81.

Allison, A. C., (1957). *Exptl. Parasitol.* **6,** 418.

Andinolfi, M., Chieffi, G., and Siniscalco, M. (1959). *Nature* **184,** 1325.

Baglioni, C. (1961). *Federation Proc.* **20,** 254.

Becklake, M. R., Griffith, S. B., McGregor, M., Goldman, H. I., and Schreve, J. P. (1955). *J. Clin. Invest.* **34,** 751.

Bessis, M., Nomarski, G., Thiery, J. P., Breton-Gorini, J. (1958). *Rev. hématol.* **13,** 249.

Braunitzer, G., and Matsuda, G. (1961). *Z. physiol. Chem.* **324,** 91.

Braunitzer, G., Rudloff, V., Hilse, K., Liebold, B., and Müller, R. (1960a). *Z. physiol. Chem.* **320,** 283.

Braunitzer, G., Hilschmann, N., Hilse, K., Liebold, B., and Müller, R. (1960b). *Z. physiol. Chem.* **322,** 96.

Braunitzer, G., Hilschmann, N., Rudloff, V., Hilse, K., Liebold, B., and Müller, R. (1961). *Nature* **190,** 480.

Bridges, C. B. (1935). *J. Heredity* **26,** 60.

Cepellini, R. (1959a). *Acta Genet. Med. et Gemellol.* **8,** Suppl. II, p. 47.

Cepellini, R. (1959b). *In* "Biochemistry of Human Genetics" (Ciba Foundation Symposium), p. 133. Churchill, London.

Conant, J. B. (1934). *Harvey Lectures Ser.* **28,** 159.

Cori, G. T. (1954). *Harvey Lectures Ser.* **48,** 145.

Coryell, C. D., and Pauling, L. (1940). *J. Biol. Chem.* **132,** 769.

Crick, F. H. C. (1958). *Symposia Soc. Exptl. Biol.* **12,** 138.

Dodson, E. O. (1960). "Evolution: Process and Product," 352 pp. Reinhold, New York.

Edmundson, A. B., and Hirs, C. H. W. (1961). *Nature* **190,** 663.

Gerald, P. S. (1958). *Blood* **13,** 936.

Gerald, P. S. (1960). *In* "The Metabolic Basis of Inherited Disease" (J. B. Stanbury, J. B. Wyngaarden, and D. S. Fredrickson, eds.), pp. 1068–1085. McGraw-Hill, New York.

Gratzer, W. B., and Allison, A. C. (1960). *Biol. Revs.* **35,** 459.

Haldane, J. S., and Priestley, J. G. (1935). "Respiration," New ed., 493 pp. Clarendon Press, Oxford.

Harris, J. W. (1950). *Proc. Soc. Exptl. Biol. Med.* **75,** 197.

Haurowitz, F. (1949). *In* "Haemoglobin" (P. J. W. Roughton and J. C. Kendrew, eds.), pp. 53–56. Interscience, New York.

Hitzig, W. H., Frick, P. G., Betke, K., and Huisman, T. H. J. (1960). *Helv. Paediat. Acta.* **15,** 499.

Horowitz, N. H. (1945). *Proc. Natl. Acad. Sci. U.S.* **31,** 153.

Huisman, T. H. J., van de Brande, J., and Meyering, C. A. (1960). *Clin. Chim. Acta* **5,** 375.

Hunt, J. A. (1959). *Nature* **183,** 1373.

Ingram, V. M. (1958). *Biochim. et Biophys. Acta* **28,** 539.

Ingram, V. M. (1959). *Biochim. et Biophys. Acta* **36,** 402.

Ingram, V. M. (1961a). "Hemoglobin and its Abnormalities," 153 pp. Charles C Thomas, Springfield, Illinois.

Ingram, V. M. (1961b). *Nature* **189,** 704.

Ingram, V. M., and Stretton, A. O. W. (1959). *Nature* **184,** 1903.

Ingram, V. M., and Stretton, A. O. W. (1961). *Nature* **190,** 1079.

Itano, H. A. (1953). *Am. J. Human Genet.* **5,** 34.

Itano, H. A. (1956). *Ann. Rev. Biochem.* **25,** 331.

Itano, H. A. (1957). *Advances in Protein Chem.* **12,** 215.

Itano, H. A. (1959). *In* "Abnormal Haemoglobins" (J. H. P. Jonxis and J. F. Delafresnaye, eds.), pp. 1–17. Charles C Thomas, Springfield, Illinois.

Itano, H. A., and Pauling, L. (1957). *Svensk Kem. Tidskr.* **69,** 509.

Jones, R. T. (1961). Chromatographic and Chemical Studies of Some Abnormal Human Hemoglobins and Some Minor Hemoglobin Components. Ph.D. Thesis, California Institute of Technology.

Jones, R. T., Schroeder, W. A., Balog, J. E., and Vinograd, J. R. (1959a). *J. Am. Chem. Soc.* **81,** 3161.

Jones, R. T., Schroeder, W. A., and Vinograd, J. R. (1959b). *J. Amer. Chem. Soc.* **81,** 4749.
Kaplan, N. O., Ciotti, M. M., Hamolsky, M., and Bieber, R. E. (1960). *Science* **131,** 392.
Kleihauer, E., Braun, H., and Betke, K. (1957). *Klin. Wochschr.* **35,** 637.
Konigsberg, W., Guidotti, G., and Hill, R. J. (1961). *J. Biol. Chem.* **236,** PC55.
Kunkel, H. G., and Wallenius, G. (1955). *Science* **122,** 288.
Kunkel, H. G., Ceppellini, R., Muller-Eberhard, U., and Wolf, J. (1957). *J. Clin. Invest.* **36,** 1615.
Lehmann, H. (1959). *Nature* **184,** 872.
Lenhert, P. G., Love, W. E., and Carlson, F. D. (1956). *Biol. Bull.* **111,** 293.
Lewis, E. B. (1951). *Cold Spring Harbor Symposia Quant. Biol.* **16,** 159.
Lwoff, A. (1943). "L'évolution physiologique. Etude des pertes de fonctions chez les microorganismes," 308 pp. Hermann, Paris.
McClintock, B. (1956). *Cold Spring Harbor Symposia Quant. Biol.* **21,** 197.
Manwell, C. (1957). *Science* **126,** 1175.
Manwell, C. (1958a). *Physiol. Zool.* **31,** 93.
Manwell, C. (1958b). *J. Cellular Comp. Physiol.* **52,** 341.
Manwell, C. (1960). *Comp. Biochem. Physiol.* **1,** 267.
Margoliash, E., and Tuppy, H. (1960). Presented at the 138th Annual Meeting of the American Chemical Society, New York, September, 1960.
Markert, C. L., and Møller, F. (1959). *Proc. Natl. Acad. Sci. U.S.* **45,** 753.
Metz, C. W. (1947). *Am. Naturalist* **81,** 81.
Mizushima, S. I., and Shimanouchi, T. (1961). *Advances in Enzymol.* **23,** 1.
Muller, C. J. (1961). A Comparative Study on the Structure of Mammalian and Avian Hemoglobins. Ph.D. Thesis, Groningen, Netherlands.
Muller, C. J., and Kingma, S. (1961). *Biochim. et Biophys. Acta* **50,** 595.
Muller, H. J. (1956). *Acta genet. statist. med.* **6,** 157.
Muller, H. J. (1959). *Prespectives in Biol. and Med.* **3,** 1.
Muller, H. J. (1961). *Science* **134,** 643.
Neel, J. V. (1959). *In* "Abnormal Haemoglobins" (J. H. P. Jonxis and J. F. Delafresnaye, eds.), p. 158. Charles C Thomas, Springfield, Illinois.
Paigen, K., and Noell, W. K. (1961). *Nature* **190,** 148.
Pauling, L. (1956). *Am. J. Psychiat.* **113,** 492.
Pauling, L. (1957a). *In* "Arbeiten aus dem Gebiet der Naturstoffchemie. Festschrift Arthur Stoll," p. 597. Birkhäuser, Basel.
Pauling, L. (1957b). *Am. Inst. Biol. Sci. Publ. No.* **2,** 186.
Pauling, L., Itano, H. A., Singer, S., and Wells, I. C. (1949). *Science* **110,** 543.
Perutz, M. F., and Mitchison, J. M. (1950). *Nature* **166,** 677.
Piveteau, J. (1955). *In* "Traité de Zoologie" (P. P. Grassé, ed.), vol. 17, p. 1. Masson, Paris.
Rensch, B. (1954). "Neuere Probleme der Abstammungslehre. Die transspezifische Evolution," 2nd ed., 346 pp. Ferdinand Enke, Stuttgart.
Rhinesmith, H. S., Schroeder, W. A., and Pauling, L. (1957). *J. Am. Chem. Soc.* **79,** 4682.
Rhinesmith, H. S., Schroeder, W. A., and Martin, N. (1958). J. *Am. Chem. Soc.* **80,** 3358.
Riegel, K., Bartels, H., and Schneider, J. (1959). *Z. Kinderheilk.* **83,** 209.

Riggs, A., and Wells, M. (1961). *Biochim. Biophys. Acta* **50,** 243.
Roche, J., and Fontaine, M. (1940). *Ann. inst. oceanog.* **20,** 77.
Rossi-Fanelli, A., and Antonini, E. (1956). *Arch. Biochem. Biophys.* **65,** 587.
Rossi-Fanelli, A., Antonini, E., and Giuffrè, R. (1960). *Nature* **186,** 896.
Rumen, N. M. (1959). *Acta Chem. Scand.* **13,** 1542.
Ruud, J. T. (1954). *Nature* **173,** 848.
Schmitt, F. O. (1956). *Proc. Natl. Acad. Sci. U.S.* **42,** 806.
Schroeder, W. A., and Matsuda, G. (1958). *J. Am. Chem. Soc.* **80,** 1521.
Schroeder, W. A., Jones, R. T., Shelton, J. R., Shelton, J. B., Cormick, J., and McCalla, K. (1961). *Proc. Natl. Acad. Sci. U.S.* **47,** 811.
Shelton, J. R., and Schroeder, W. A. (1960). *J. Am. Chem. Soc.* **82,** 3342.
Siniscalco, M., Bernini, L., Latte, B., and Motulsky, A. G. (1961). *Nature* **190,** 1179.
Smith, E. W., and Thorbert, J. V. (1958). *Bull. Johns Hopkins Hosp.* **101,** 38.
Sturtevant, A. H. (1925). *Genetics* **10,** 117.
Svedberg, T. (1933). *J. Biol. Chem.* **103,** 311.
Thorup, O. A., Itano, H. A., Wheby, M., and Leavell, B. S. (1956). *Science* **123,** 889.
Watson, H. C., and Kendrew, J. C. (1961). *Nature* **190,** 670.
Wilson, S., and Dixon, G. H. (1961). *Nature* **191,** 876.
Zuckerkandl, E., and Schroeder, W. A. (1961). *Nature* **192,** 984.
Zuckerkandl, E., Jones, R. T., and Pauling, L. (1960). *Proc. Natl. Acad. Sci. U.S.* **46,** 1349.

ACTA CHEMICA SCANDINAVICA 17 (1963) S9–S16

Chemical Paleogenetics

Molecular "Restoration Studies" of Extinct Forms of Life

LINUS PAULING and EMILE ZUCKERKANDL*

*Division of Chemistry and Chemical Engineering, California Institute of Technology, Pasadena, California, USA***

Attention is attracted to the possibility of reconstructing the amino-acid sequence of ancestral polypeptide chains by virtue of a comparison between the amino-acid sequences of related polypeptide chains found in contemporary organisms. A tentative partial structure is proposed for two ancestral hemoglobin polypeptide chains. Some perspectives of paleobiochemistry are outlined.

In different hemoglobin polypeptide chains, derived either from one individual organism (man) or from different vertebrate species, identical amino-acid residues are often found in corresponding positions along the chains (*cf.* Braunitzer[1]). This occurs too frequently to be due to chance, and it appears to be too constant a feature over long periods of evolutionary time to be attributable to convergence by natural selection from primitively heterologous polypeptide starting materials. Consequently, the homology is plausibly interpreted by the assumption of the past existence of common polypeptide-chain ancestors, controlled by common ancestral genes. At least a few times during evolution an evolutionarily effective duplication, *i. e.*, a duplication that has been spread at least temporarily by natural selection, of either a hemoglobin gene or a chromosome carrying such a gene is thought to have occurred. The resulting daughter genes are considered to have differentiated by independent mutation (Itano[2]; Ingram[3]; Zuckerkandl and Pauling[4]).

On the basis of this hypothesis, the degree of difference between two homologous polypeptide chains is a measure of the relative time at which the common ancestor of the structural genes controlling these chains existed, and it is also, within large limits of error, a measure of this time in absolute units (Zuckerkandl and Pauling[4]). We now direct attention to two further types of information that can be derived from the comparison of different homologous polypeptide chains. First, it is possible to determine, with some probability, the amino-acid sequence of their presumed common polypeptide-chain ancestor. Second, when two polypeptide chains do not possess the same amino-acid residue at a certain

* On leave from Centre National de la Recherche Scientifique, Paris.

** Contribution No. 2957.

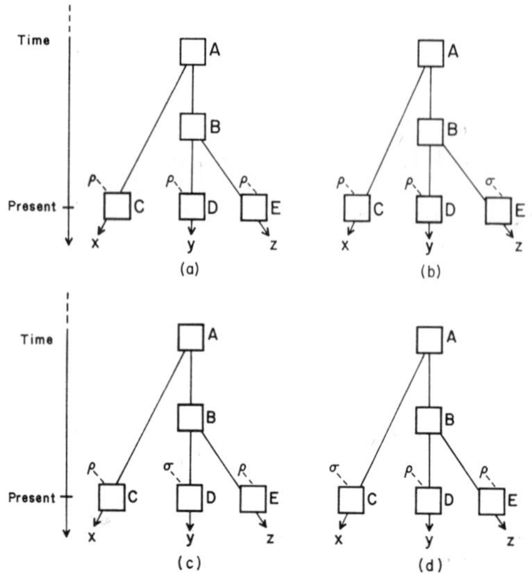

Fig. 1. Different evolutionary relationships of the amino-acid residues ϱ and σ found at corresponding sites in the homologous polypeptide chains C, D, and E. See text.

molecular site, it is possible to determine in which line of descent the mutation responsible for this difference has occurred since the epoch of the common ancestor.

If the amino-acid residue is the same in two chains at a given molecular site, there is a certain likelihood that this residue was also present in the common ancestor of the chains*. If three or more chains are compared and the same residue is found at corresponding sites in two or more chains whereas it differs in one chain, there is a certain likelihood that the mutation responsible for the difference occurred in the line of descent leading to the chain showing this difference, and that it occurred since the time of the most recent molecular ancestor that can be assumed to relate the variant chain to one of the others. The further apart the organisms from which homologous polypeptide chains are derived, the more significant are the identities found, not only because they then suggest that the residue at the molecular site considered has remained the same for a long evolutionary time and therefore must have an important function within the molecule, but also because in widely different forms the occurrence of convergence at the molecular level appears to us to be a negligible possibility.

In Figure 1 three lines of molecular descent, x, y, and z, are represented, and the respective common molecular ancestors at two branching points are A and B. Squares stand for related polypeptide chains. Two different amino-acid residues, ϱ and σ, are considered for one given molecular site in the homologous polypeptide chains, C, D, and E. These chains are located on the time scale at a

* Convergence effects may possibly intervene under some circumstances even at the molecular level, but are not considered to be likely to affect an important proportion of amino-acid residues in polypeptides.

level representing present time. The following probable statements can be made about the four situations pictured: Situation (a): the common chain-ancestors A and B had residue ϱ at the molecular site under consideration. Situation (b): the chain ancestors A and B had residue ϱ; a mutation to residue σ has occurred in evolutionary lineage z after it became distinct from evolutionary lineage y. Situation (c): the chain ancestors A and B had residue ϱ; a mutation to residue σ occurred in evolutionary lineage y after lineage z branched off from lineage y. Situation (d): the chain-ancestor B had residue ϱ, but on the basis of the evidence the nature of the residue in chain-ancestor A is undetermined; further related polypeptide chains have to be drawn into the comparison to permit a conclusion to be reached. The probability of correctness of the other deductions can be increased by use of information about other chains.

Three out of the four types of chains that make up human hemoglobin molecules (there are four chains of two types per molecule), namely the α-, β-, and γ-chains, have been defined by their N-terminal sequence (Schroeder[5]). This sequence may change partially or totally during evolution, and yet the structural genes that control it may still be homologous. We need to be able to refer to homologous structural genes irrespective of the actual amino-acid sequence of their polypeptide products. Let us propose, therefore, to speak of the I^{α}, II^{β}, III^{γ}, and IV^{δ}-hemoglobin-chain genes and the corresponding polypeptide chains, the superscript representing in each case the reference for homology considerations. The common ancestor of two or more of these genes is then designated by juxtaposition of the symbols referring to the genes derived by duplication. Thus, the II^{β}–IV^{δ}-gene is the common ancestor of the II^{β}-gene and the IV^{δ}-gene; the II^{β}–III^{γ}–IV^{δ}-gene is the common ancestor of the III^{δ}-gene and the II^{β}–IV^{δ}-gene; and the I^{α}–II^{β}–III^{δ}–IV^{δ}-gene is the common ancestor of the I^{α}-gene and the II^{β}–III^{γ}–IV^{δ}-gene (Fig. 2).

Tables 1a, 1b, and 1c illustrate the procedure by presenting two partially and tentatively reconstructed ancestral hemoglobin polypeptide chains. The evidence on which these reconstructions are based are the human α-, β-, γ-, and δ-chains, the sperm-whale myoglobin chain (see Schroeder[7] for references and latest information on the structure of these chains), the horse α-chain (Braunitzer and Matsuda, in: Cullis *et al.*[8]), and the human myoglobin chain (Hill[9]). One of us (L. P.[10]) has listed earlier probable residues for 70 loci in the common precursor

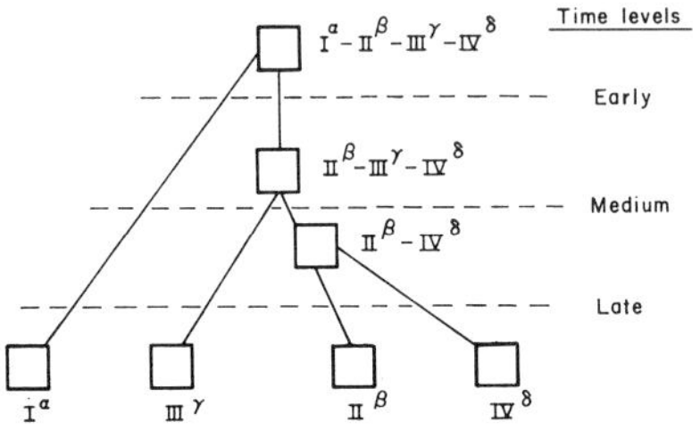

Fig 2. Schematic representation of the successive gene duplications that are presumed to have lead to the hemoglobin genes found in man. *Cf.* Ingram[3]. See text.

Table 1 a. Tentative partial structure of two chain-ancestors of the human hemoglobin polypeptide chains. The numbering of the residues is the one usually applied to the human α-chain. The abbreviations for the amino acids are those commonly employed, except for asparagin (asg) and glutamine (glm). Other abbreviations: E = early epoch; M = medium epoch; L = late epoch; abs = absent. "None", in column (e), means: probably no evolutionrily effective mutation occurred at the site under consideration in the line of descent leading from the ancestral genes to the human genes. The residues and comments are placed in parentheses when the conclusion reached is partly based on the consideration of human and sperm whale myoglobins.

(a) Residue number
(b) Partial sequence of the II^{β}–III^{γ}–IV^{δ}-chain (late form)
(c) Partial sequence of the I^{α}–II^{β}–III^{γ}–IV^{δ}-chain (late form)
(d) Chain(s) in whose direct ancestry the mutation(s) seem to have occurred
(e) Nature of the substitution
(f) Qualitative evaluation of the time of mutation
(g) From evidence relating to a number of different animals (cf. Gratzer and Allison[6])
(h) Not after the time of ancestor common to horse and man
(i) After the time of ancestor common to horse and man
(j) Amino-acid residue x present at a homologous site in the two myoglobins
(k) Polypeptide chain I^{α} or II^{β}–III^{γ}–IV^{δ}

(a)	1	1a	2			5					10		12
(b)	val—	his—	leu—	thr—	pro—	glu—	asp—	lys—	?—	?—	val—	thr—	ala—
(c)	val—	?—	leu—	?—	pro—	?—	asp—	lys—	?—	?—	val—	?—	ala—
(d)	III^{γ}	(k)	III^{γ}	(k)	III^{γ} and horse I^{α}	(k)	II^{β}—IV^{δ}				III^{γ}	(k)	III^{γ}
(e)	val→gly		leu→phe		III^{γ}: pro→glu; horse I^{α}: pro→ala		asp→glu	none			val→ileu		ala→ser
(f)	M or L	E(g)	M or L	E or M(h)	Man: M or L Horse: (i)	E or M(h)	M					E or M(h)	M or L

(a)	13		15					20					25			28
(b)	—leu—	try—	gly—	lys—	val—	asg—	val—	abs—	abs—	?—	glu—	?—	gly—	gly—	glu—	ala—
(c)	—?—	try—	gly—	lys—	val—	?—	?—	?—	?—	?—	glu—	?—	gly—	?—	glu—	ala—
(d)	(k)		I^{α} of horse			(k)	(k)				III^{γ}			(k)		III^{γ}
(e)		none	gly→ser	none	none						glu→asp		none		none	ala→thr
(f)	E or M(h)		(i)			E or M(h)	E or M(h)				M or L			E or M(h)		M or L

Table 1 b. Continuation of Table 1 a. See Table 1 a for legend.

(a)	29	30					35					40	41
(b)	—leu—	gly—	arg—	leu —	leu —	val—	val—	tyr—	pro—	try—	thr—	glm—	arg—
(c)	—leu—	? —	arg—	(leu)—	(phe)—	? —	? —	? —	pro—	? —	thr—	? —	? —
(d)		(k)		(I^{α}) (leu→met)	(II^{β}-III^{γ}-IV^{δ}) (phe→leu)	(k)	(k)	(k)		(k)		(k)	(k)
(e)	none		none	x=leu (j)	x=phe(j)				none		none		
(f)		E or M(h)		(E or M)	(E)	E or M(h)	E or M(h)	E or M(h)		E or M(h)		E or M(h)	E or M(h)

(a)	42			45		46a 47			50					54a	54b
(b)	—phe—	phe—	(asp)—	ser—	phe—	gly-asp—	leu—	ser—	ser—	? —	ser—	ala—	? —	met—	gly—
(c)	— ? —	phe—	(asp)—	? —	phe—	? -asp—	leu—	ser—	? —	? —	ser—	ala—	? —	? —	? —
(d)	(k)		(I^{c}) and (II^{β}-IV^{δ}) (brace over phe—(asp))	(k)		III^{γ}			(k) (brace over ser—?)	II^{β}	II^{β}-IV^{δ}				
(e)		none	↓ (?→pro) x=asp(j) ↓; ↓ (asp→glu) ↓		none	asp→asg	none	none	↓	ser→thr ↓	ser→asp	none			
(f)	E or M(h)		(E or M(h)) (M)	E or M(h)		M or L			E or M(h)	L	M				

(a)	54c	54d	54e	55					60					65			68
(b)	—asg—	pro—	lys—	val—	lys—	ala—	his—	gly—	lys—	lys—	val—	leu—	? —	ala—	leu—	(gly)—	asp—
(c)	— ? —	? —	? —	val—	lys—	ala—	his—	gly—	lys—	lys—	val—	? —	? —	ala—	leu—	(gly)—	? —
(d)						I^{α}						(k)		III^{γ}	II^{β}-IV^{δ}	(I^{α}) and (II^{β}-IV^{δ}) (brace over (gly))	(k)
(e)				none	none	ala→gly	none	none	none	none	none			ala→ser	leu→phe	↓ (gly→thr) x=gly(j) ↓; ↓ (gly→ser) ↓	
(f)						M or L(i)						E or M(h)		M or L	M	E or M(h); M	E or M(h)

(a)	69		70				75					80			83
(b)	—ala—	?	ala—	his—	leu—	asp—	asp—	leu—	lys—	gly—	thr—	phe—	ala—	? —	leu—
(c)	—ala—	?	ala—	his—	leu—	asp—	asp—	? —	? —	? —	? —	? —	? —	? —	leu—
(d)	II^{β}-IV^{δ}		III^{γ}		I^{α}		II^{β}-IV^{δ} or II^{β}	(k)	(k)	(k)	(k)	(k)	(k)		
(e)	ala→gly		ala→lys	none	leu→val	none	asp→asg								none
(f)	M		M or L		L		M or L	E or M(h)	E or M(h)	E or M(h)	E or M(h)	E or M(h)	E or M(h)		

Table 1 C. Continuation of Table 1 a. See Table 1 a for legend.

(a)	84	85					90					95				99
(b)	—ser—	glu—	leu—	his—	cys—	asp—	lys—	leu—	his—	val—	asp—	pro—	glu—	asg—	phe—	lys—
(c)	—ser—	? —	leu—	his—	(ala)—	? —	lys—	leu—	? —	val—	asp—	pro—	? —	asg—	phe—	lys—
(d)		(k)			(II^{β}-III^{γ}-IV^{δ})	(k)			(k)				(k)			II^{β}-IV^{δ} or II^{β}
(e)	none		none	none	(ala→cys) (x=ala (j))		none	none		none	none	none		none	none	lys→arg
(f)		E or M(h)			(E)	E or M(h)			E or M(h)				E or M(h)			M or L

(a)	100					105					110					115
(b)	—leu—	leu—	gly—	asg—	val—	leu—	val—	? —	val—	leu—	ala—	? —	his—	phe—	gly—	lys—
(c)	—leu—	leu—	(ser)—	? —	? —	leu—	? —	? —	? —	leu—	ala—	? —	his—	? —	? —	? —
(d)			(II^{β}-III^{γ}-IV^{δ})	(k)	(k)		(k)		(k)					(k)	(k)	(k)
(e)	none	none	(ser→gly) (x = gly (j))			none				none	none					
(f)			(E)	E or M(h)	E or M(h)		E or M(h)		E or M(h)					E or M(h)	E or M(h)	E or M(h)

(a)	116				120					125					130	
(b)	—glu—	phe—	thr—	pro—	? —	val—	glm—	ala—	ser—	? —	glm—	lys—	? —	val—	ala—	gly—
(c)	— ? —	phe—	thr—	pro—	? —	val—	? —	ala—	ser—	? —	(asp)—	lys—	? —	? —	ala—	? —
(d)	(k)						(k)		II^{β}-IV^{δ} or II^{β}		(II^{β}-III^{γ}-IV^{δ})			(k)	III^{γ}	(k)
(e)		none	none	none				none	ser→ala		asp→glm (x = asp (j))	none			ala→thr	
(f)	E or M(h)						E or M(h)		M or L		(E)			E or M(h)	M or L	E or M(h)

(a)	132	133		135					140	
(b)	—val—	ala—	? —	ala—	leu—	? —	ser—	lys—	tyr—	his
(c)	—val—	? —	? —	? —	leu—	? —	ser—	lys—	tyr—	?
(d)		(k)		(k)			II^{β}-IV^{δ}	III^{γ}		(k)
(e)	none				none		ser→his	lys→arg	none	
(f)		E or M(h)		E or M(h)			M	M or L		E or M(h)

of α-chains and non-α-chains. A qualitative estimate is made of the time at which a given mutation supposedly occurred. (In a majority of cases the changes probably involve more than one substitution, according to the genetic codes proposed by Yukes[11] and by Smith[12]). Three periods are distinguished: early, medium, and late, according to the time off the mutation or mutations in relation to the successive gene duplications (Fig. 2).

Fossil remains no doubt express the activity of only a fraction of the genes of a given organism (although perhaps a significant fraction) and this fraction cannot be analyzed into its components. Paleobiochemistry, through molecular restoration studies on the basis of existing related polypeptide chains, provides the means of investigating the structure of such components for any part of the genome of extinct organisms. This holds, however, only in relation to structural genes, as long as the object of such studies is confined to the polypeptide products rather than extended to the genic material itself. Yet, once the structures of ancestral polypeptide chains are known, it will in the future be possible to synthesize these presumed components of extinct organisms. Thus one will be able to study the physico-chemical properties of these molecules and to make inferences about their functions. For instance, the oxygen affinity and its dependence on pH of ancestral hemoglobins might be studied as well as the affinity of ancestral enzymes for various substrates and the probable nature of these substrates in past evolutionary history. As information about various paleogenes belonging to a given group of extinct organisms will accumulate, some deductions concerning these organisms will be possible in relation to levels of biological integration higher than the level of individual macromolecules. When a fossil record is available, knowledge about the organisms concerned will go far beyond what has so far been believed possible. Important information will also be provided about forms that have left no fossil record whatsoever, such as many soft-bodied animals.

When a gene-duplication has occurred and one of the duplicate genes is not needed for carrying out the function for which its partner suffices, the polypeptide chain synthesized under its control may undergo a change in structure, including a change in spatial conformation, and the latter change especially may sometimes lead to the appearance of a new function. If the new function is retained by natural selection, it appears likely that during the subsequent evolutionary period many more mutational changes will be preserved by the corresponding gene than by genes whose functions have long been established. The latter genes are submitted to forces of selection that are conservative most of the time. Two reasons may be advanced for a more rapid alteration of the amino-acid sequence in a conformationally and functionally altered polypeptide: First, a newly evolved function is likely to be perfectible through further structural changes; second, while over very long periods of evolution the amino-acid sequence of homologous polypeptide chains can be deeply transformed even when the spatial conformation is kept nearly constant, as the comparison of myoglobin and hemoglobin suggests, a faster transformation is to be expected when the spatial conformations differ, since in that case the residues required to remain unchanged in order to preserve a given spatial conformation are not any more the same in the precursor polypeptide and in the polypeptide derived from it. Can paleobiochemical studies be anticipated to trace the filiation between genes that control polypeptides endowed with different functions? In a polypeptide whose function has been replaced by another one the amino-acid sequence may eventually be modified to such an extent that it becomes unrecognizable in terms of the molecular ancestor. Thus the comparison of the amino-acid sequences in contemporary polypeptide chains endowed with different functions probably will lead only in a minority of cases to the detection of the evolutionary relationships that actually obtain, *viz.*, mainly when the acquisition of a different molecular function by a polypeptide has been relatively recent. In such cases convincing homologies in amino-acid sequence will still be found, although the spatial conformations of the proteins may differ. Even then the discovery of polypeptides related by evolution but functionally different might appear to involve either an unlikely coincidence or a formidable task. This task could, however, be substantially reduced, if a systematic investigation were made of the amino-acid sequences in polypeptides that have been shown by genetic studies to be controlled by closely linked genes, notably by genes controlled by a common operator gene (Jacob and Monod[13]). The functions carried out by the genes included in a given operon (the unit controlled by an operator gene) can be quite diversified, and it is to be expected that the spatial conformations of the corresponding polypeptides are diverse also. Yet, since at least some genes derived through the duplication of a mother gene will probably remain closely linked during a considerable evolutionary time and since some of them may change in their function, the chances of discovering homologies between apparently unrelated proteins are likely to be greatest in a survey of the amino-acid sequences of polypeptides controlled by closely linked genes. An effort in that direction may result in a worthy contribution to the theory of evolution, in that it might show that even apparently unrelated proteins can indeed have a common molecular ancestor. Thus suggestive evidence would be furnished in support of the view that most or all apparently heterologous genes derive ultimately from a common gene-ancestor.

REFERENCES

1. Braunitzer, G., Hilschmann, N., Rudloff, V., Hilse, K., Liebold, B. and Müller, R. *Nature* **190** (1961) 480.
2. Itano, H. A. *Adv. Protein Chem.* **12** (1957) 216.
3. Ingram, V. M. *Nature* **189** (1961) 704.
4. Zuckerkandl, E. and Pauling, L. In Kasha, M. and Pullman, B. *Horizons in Biochemistry*, Academic Press, New York and London, 1962, p. 189.
5. Schroeder, W. A. *Fortschr. der Chem. Org. Naturstoffe* **17** (1959) 322.
6. Gratzer, W. B. and Allison, A. C. *Biol. Rev. Cambridge Phil. Soc.* **35** (1960) 459.
7. Schroeder, W. A. *Ann. Rev. Biochem.* **32** (1963). *In press.*
8. Braunitzer, G. and Matsuda, G. *Cf.* Cullis, A. F., Muirhead, H., Perutz, M. F. and Rossmann, M. G. *Proc. Roy. Soc. London* **A 265** (1962) 161.
9. Hill, R. *Personal communication* (1962).
10. Pauling, L., communicated at Hemoglobin Workshop, Arden House, Columbia University, Nov. 1962. *Cf.* Ingram, V. M., Richards, D. W. and Fishman, A. P. *Science* **138** (1962) 996.
11. Jukes, T. H. *Proc. Natl. Acad. Sci. U. S.* **48** (1962) 1809.
12. Smith, E. *Proc. Natl. Acad. Sci. U. S.* **48** (1962) 859.
13. Jacob, F. and Monod, J. *J. Mol. Biol.* **3** (1961) 318.

Received March 29, 1963.

H. Halls Boktr. A.-B., Jönköping, Sweden
48130 - 1963

J. Theoret. Biol. (1965) **8**, 357–366

Molecules as Documents of Evolutionary History

EMILE ZUCKERKANDL AND LINUS PAULING

Gates and Crellin Laboratories of Chemistry,
California Institute of Technology, Pasadena, California, U.S.A.

(*Received* 17 *September* 1964)

Different types of molecules are discussed in relation to their fitness for providing the basis for a molecular phylogeny. Best fit are the "semantides", i.e. the different types of macromolecules that carry the genetic information or a very extensive translation thereof. The fact that more than one coding triplet may code for a given amino acid residue in a polypeptide leads to the notion of "isosemantic substitutions" in genic and messenger polynucleotides. Such substitutions lead to differences in nucleotide sequence that are not expressed by differences in amino acid sequence. Some possible consequences of isosemanticism are discussed.

1. The Chemical Basis for a Molecular Phylogeny

Of all natural systems, living matter is the one which, in the face of great transformations, preserves inscribed in its organization the largest amount of its own past history. Using Hegel's expression, we may say that there is no other system that is better *aufgehoben* (constantly abolished and simultaneously preserved). We may ask the questions where in the now living systems the greatest amount of their past history has survived and how it can be extracted.

At any level of integration, the amount of history preserved will be the greater, the greater the complexity of the elements at that level and the smaller the parts of the elements that have to be affected to bring about a significant change. Under favorable conditions of this kind, a recognition of many differences between two elements does not preclude the recognition of their similarity.

One may classify molecules that occur in living matter into three categories, designated by new terms, according to the degree to which the specific information contained in an organism is reflected in them:

(1) Semantophoretic molecules or semantides—molecules that carry the information of the genes or a transcript thereof. The genes themselves are the

primary semantides (linear "sense-carrying" units). Messenger-RNA molecules are secondary semantides. Polypeptides, at least most of them, are tertiary semantides.

(2) Episemantic molecules—molecules that are synthesized under the control of tertiary semantides. All molecules built by enzymes in the absence of a template are in this class. They are called episemantic because, although they do not express extensively the information contained in the semantides, they are a product of this information.

(3) Asemantic molecules—molecules that are not produced by the organism and therefore do not express, either directly or indirectly (except by their presence, to the extent that this presence reveals a specific mechanism of absorption), any of the information that this organism contains. However, the organism may often use them, and may often modify them anabolically and thus change them into episemantic molecules to the extent of this modification. The same molecular species may be episemantic in one organism and asemantic in another. Vitamins constitute examples. Simple molecules such as phosphate ion and oxygen also fall into this category. Macromolecules found in an organism for any length of time are never asemantic, viruses excepted. Viruses and other "episomes", i.e., particles of external origin that may be integrated into a genome (Wollman & Jacob, 1959) are asemantic when present in the host cell in the vegetative, autonomous state: they are semantophoretic when integrated into the genome of the host.

Products of catabolism are not included in this classification. During the enzymatic breakdown of molecules, information contained in enzymes is expressed, but instead of being manifested in both the reaction and the product, this information is manifested in the reaction only. Since we are considering products, catabolites as such are non-existent with respect to the proposed classification.

The relevance of molecules to evolutionary history decreases as one passes from semantides to asemantic molecules, although the latter may represent quantitative or qualitative characteristics of groups. As such they are, however, unreliable and uninformative. It is plain that asemantic molecules are not worthy of consideration in inquiries about phylogenetic relationships.

Neither can episemantic molecules furnish the basis for a universal phylogeny, for such molecules, if universal, are not variable (ATP), and, if variable, are not universal (starches). It appears however possible *a priori* that parts of the phylogenetic tree could be defined in terms of episemantic molecules. An attempt in this direction has been made for instance on the basis of carotenoids in different groups of bacteria (cf. Goodwin, 1962). It is characteristic of such studies that they need independent confirmation. Such independent confirmation may be obtained by direct or indirect studies of semantides. In relation to a number of organic molecules, such as vitamin B_{12}, organisms as far apart on the evolutionary scale as bacteria, flagellates, and higher vertebrates differ, not in that the compound is present or absent, required or not required, but in the prevalent "pattern of specificity" (Hutner, 1955). By this is meant the measure of functional effectiveness of compounds closely similar to but not identical to the one that is actually present, say, vitamin B_{12}. Thereby the difference of organisms in relation to the organic molecules under consideration is reduced to differences in enzymes and, in the last analysis, to the difference in primary structure of polypeptide chains. Because of this relationship to studies of semantides, it is possible that the establishment of different patterns of specificity is one of the best uses to which episemantic and asemantic molecules may be put in phylogenetic studies.

Whereas semantides are of three types only (DNA, RNA, polypeptides), episemantic molecules are of a great variety of types. Their interest for phylogeny is proportional to their degree of complexity. Polysaccharides such as cellulose are large molecules, but their complexity is small because of the monotonous repeat of the same subunits. In fact, not only episemantic molecules, but also semantides vary in their degree of complexity. The complexity of semantides is largest in the case of large globular polypeptide chains and smallest in the case of structural proteins characterized by numerous repeats of simple sequences. There may be a region of overlap of semantides with the lowest degree of complexity and of episemantic molecules with the highest degree of complexity. The former, however, will still contain more information than the latter about the present and the past of the organism. Indeed, episemantic molecules are mostly polygenic characters, in that enzymes controlled by several distinct structural genes have to collaborate in their synthesis; moreover, they express the information contained in the active centers of enzymes only, and in no other enzymatic region; and even then express this information ambiguously; i.e., probably with considerable "degeneracy". There is thus a great loss of information as one passes from semantides to episemantic molecules. Incidentally, one cannot yet be sure that all polypeptides are semantides. Some, especially among the small ones, but also among the large structural ones, may be episemantic. Thus, it has not been possible to split glutenin into subunits of reproducible molecular weight (Taylor & Cluskey, 1962). This raises the suspicion that glutenin might not be produced as the transcript of a template.

Because tertiary semantides (enzymes) with different primary structures can lead to the synthesis of identical episemantic molecules as long as the active enzymatic sites are similar, wrong inferences about phylogenetic relationships may be drawn from the presence of identical or similar episemantic molecules in different organisms. Amylopectins in plant starches and animal glycogens are very similar, yet we may expect (a point it will be of interest to verify) that the amino-acid sequence of the enzymes responsible for the synthesis of these polysaccharides in animal and plant kingdoms is very different. Moreover a similar end-product, in the case of episemantic molecules, may be obtained by different pathways, so that not even the active sites of the enzymes involved need to be similar. The synthesis of nicotinic acid and that of tyrosine are carried out via different pathways in bacteria and in other organisms (cf. Cohen, 1963). Therefore, the presence of these molecules in no way points to a phylogenetic relationship between bacteria and these other organisms. The number of possible historical backgrounds to the presence of a molecule synthesized by an organism will tend toward unity only as the number of enzymes involved in the synthesis of this molecule increases significantly. It is not likely that a whole pyramid of enzymatic actions has been built more than once or twice during evolution. This consideration implies that the best phylogenetic characters among episemantic molecules are not just the most complex molecules but, among these, the ones that are built from the least complex asemantic molecules.

The preceding discussion suggests that the most rational, universal, and informative molecular phylogeny will be built on semantophoretic molecules alone. Evolution, in these molecules, seems to proceed most frequently by the substitution of one single building stone out of, say 50 to 300 for polypeptides or, on the basis of a triplet code, 150 to 900 for the corresponding nucleic acids. Even these small changes can have profound consequences at higher levels of organic integration, through an alteration of the established pattern of molecular interaction. Therefore, in macromolecules of these types there is more history in the making and more history preserved than at any other single level of biological integration.

In previous communications (Zuckerkandl & Pauling, 1962; Pauling & Zuckerkandl, 1963) we have discussed ways of gaining information about evolutionary history through the comparison of homologous polypeptide chains. This information is threefold: (1) the approximate time of existence of a molecular ancestor common to the chains that are being compared; (2) the probable amino-acid sequence of this ancestral chain; and (3) the lines of descent along which given changes in amino-acid sequence occurred. The first type of information is obtained in part through an assessment of the overall differences between homologous polypeptide chains. The second and third types of information are obtained through a comparison of individual amino-acid residues as found at homologous molecular sites. Our purpose was to spell out principles of how to extract evolutionary history from molecules, rather than to write any part thereof in its final form—an attempt that would require more information than is presently available even in the case of hemogloblins.

Beside the analysis of amino-acid sequence of a greater number of homologous polypeptide chains, two other sources of knowledge will help in retracing the evolutionary history of molecules. One is a consideration of the genetic code, to assess whether the passage of one character of sequence to another could have occurred in one step, to discover possible intermediary states of sequence, and to evaluate to a better approximation than by the simple comparison of the amino-acid sequence of two homologous polypeptide chains the minimum number of mutational events that separate these two chains on the evolutionary scale.

A second is the study of three dimensional molecular models, to permit one to make predictions about the effects of particular substitutions and, on the basis of the transitions allowed by the genetic code, to exclude some substituents, as incompatible with the preservation of molecular function, from the list of possible evolutionary intermediates.

This cursory outline of methodology in chemical paleogenetics applies directly to the analysis of polypeptide chains only. Although techniques are not yet available for a thorough investigation of sequence in other types of semantides, it is of interest to examine the relationship between the different types of semantides with respect to the information they contain.

2. Cryptic Genetic Polymorphism through Isosemantic Substitution

For any one corresponding set of molecules, the three scripts used by nature in semantophoretic molecules, the DNA-, RNA- and polypeptide scripts, represent largely, but presumably not exactly, the same message. One major reason for this is probably the following. In view of the "degeneracy" of the genetic code, many amino acids appearing to be coded for by more than one type of codon (Weisblum, Benzer & Holley, 1962; Jones & Nirenberg, 1962), one must assume that information is lost in the passage from secondary (RNA) to tertiary (polypeptide) semantides. Moreover many primary semantides may not be transcribed: there are significant stretches of DNA that are apparently not expressed in polypeptide products, and the base sequence along these stretches may represent important documents about the history of the organism as well as its present organization and potentialities.

The degeneracy of the genetic code, then, leads one to predict the existence of isosemantic heterozygosity, namely of differences in base sequence in allelic stretches of DNA that do not lead to differences in amino-acid sequence in the corresponding polypeptide chains. The base sequence of a codon may be changed, but the "sense" of the "word", in terms of amino acids, may remain the same. The same inference has been drawn independently by Jones (1946, personal communication).†

Eck (1963) proposes that one of the three letters of each codon, perhaps the middle letter, is recognized by transfer-RNA's only as "purine" or "pyrimidine"; that is, according to the bulk of the molecule rather than to its exact species. Thus shifts between adenine and guanine or between cytosin and uracil in the middle letter of messenger-RNA codons will not be heeded by transfer-RNA. If this is so, one must distinguish two levels of crypticity. Some base substitutions will remain cryptic, unexpressed at the level of the polypeptide chains, but will be recognized at the level of transfer-RNA (secondary crypticity). Other base substitutions will remain cryptic at the level of both the polypeptide chain and the transfer-RNA (primary crypticity). (A third, more superficial level of crypticity was often referred to in the past, namely cryptic amino-acid substitutions in polypeptides, substitutions that were supposed to actually exist, but not to be detected by available chemical means.) According to Eck's code, primary crypticity should exist for every amino acid, because of the peculiar role tentatively attributed to the middle letter of the codon, and secondary crypticity should exist for eleven amino acids out of twenty. Amino acids that occur with high frequencies usually seem to have degenerate codes. The opportunities for isosemantic substitution and cryptic genetic polymorphism, even of the secondary type of crypticity, should therefore be very widespread indeed.

As is well known, the abnormal human hemoglobins HbS and HbC differ from HbA in that a valyl residue replaces a glutamyl residue in the β-chain of HbS at the sixth position from the amino-end, whereas a lysyl residue replaces the same glutamyl residue in HbC (Ingram, 1958; Hunt & Ingram, 1958, 1959). According to the proposals for the genetic code made by Jukes (1962), by Wahba *et al.* (1963) and by Eck (1963), the shift from valine to lysine is one of the rare ones that require three base pair substitutions in DNA and therefore, presumably, three mutational steps. If correct, this conclusion would render unlikely the hypothesis, previously formulated by one of us (Pauling, 1961), that HbC is derived from HbS rather than from HbA. The three genetic codes, on the other hand, are compatible with a one-step transition between HbA and HbS as well as between HbA and HbC. According to Eck's code—not according to the other proposals for a genetic code mentioned above—the valine of HbS and the lysine of HbC must

† *Note added in proof.* Since the present article was written this inference has also been drawn independently or made use of in various connections by H. A. Itano (Symposium on Abnormal Haemoglobins, Ibadan, Nigeria, 1963), C. Baglioni (personal communication), B. N. Ames & P. Hartman (*Cold Spr. Harb. Symp. quant. Biol.* (1963), **28**, 349), G. S. Stent (*Science* (1964), **144**, 816) and T. Sonneborn (in "Evolving Genes and Proteins", H. J. Vogel *et al.*, eds., Academic Press, New York, in the press).

however have derived from two distinct isosemantic codons for glutamic acid in HbA (Fig. 1). Whether or not Eck's code will, in the end, be shown to be correct in this respect, we may accept it provisionally for the sake of this discussion. Indeed, even if HbS and HbC are not the products of mutations in two isosemantic codons, other cases of this type are likely to be found in the future.

FIG. 1. The relation between coding triplets in hemoglobins A, S and C according to Eck's code. Possible one-step transitions are marked by double arrows. Middle letters of codons: P = purine, Y = pyrimidine. Other symbols as usual (cf. Eck, 1963).

If the situation is as represented in Fig. 1, the two isosemantic codons for glutamic acid (which are actually, according to Eck, resolvable into four isosemantic triplets, AAG, AGG, UAG and UGG) must be thought to have at one time been widespread in the human population, and may even today constitute a case of perhaps widely occurring cryptic genetic polymorphism. A search for it might be made in particular among individuals who appear to possess different isoalleles of HbA, in the nomenclature of Itano (1957), namely different heritable relative levels of HbA production. Because all transfer-RNAs that would correspond to all possibilities of a degenerate code might conceivably not be available in excess in certain organisms or in certain tissues of an organism, isosemantic substitutions may lead to increased or decreased rates of polypeptide synthesis. Thus there could exist an operator-independent change in rate of protein synthesis through base-pair substitutions in structural genes. One of us (L. P.) and Itano (1957) used to think that the existence of isoalleles is probably linked to cryptic substitutions. In terms of cryptic amino-acid substitutions, this hypothesis has ceased to be as likely as it appeared to be, at least as a very generally applicable explanation. Yet it may be reintroduced on the basis of cryptic base substitutions in DNA and RNA.

If the scarceness of some species of isosemantic transfer-RNA's, can affect the rate of synthesis of polypeptides, the synthesis of a single polypeptide chain should not proceed at constant speed along the chain, but so to speak in jerks, with sudden decelerations at molecular sites where the codons happen not to correspond to species of transfer-RNA that are present in excess. No evidence for or against this effect is available to our knowledge.

One may ask the question whether cryptic isosemantic substitutions may offer a possible alternate explanation, beside those already proposed, of the thalassemic inhibition of certain hemoglobin chain genes. A single isosemantic substitution may form a bottle-neck in synthesis, but an additive affect of such substitutions is also possible. The human δ-chain may be one that is universally "thalassemic" in this sense. This interpretation of thalassemia or of the low rate of synthesis of δ-chains would imply that normal amounts of the corresponding messenger-RNA's are produced and occupy a significant percentage of the available ribosomal sites without leading to much synthesis. Such an effect, is not very likely, especially in thalassemia. Indeed, an extensive survey of the literature has shown in heterozygotes for α-thalassemia or β-thalassemia a "compensatory" increase of mean *absolute* amounts per cell of the chain synthesized under the control of the allelic gene, whether this allelic gene be normal or structurally abnormal (Zuckerkandl, 1964). This observation suggests that more ribosomal sites have become available to messenger-RNA produced by a single allele than is the case when both alleles are normally active. If correct, this interpretation would imply that the output of messenger-RNA by thalassemic hemoglobin genes is actually reduced, and that the block of polypeptide synthesis in thalassemia is at the genic level rather than at the level that involves the action of transfer-RNA. On the other hand, one may surmise that the low rate of synthesis of δ-chains is correlated with a low genic output of messenger-RNA more probably than with low synthetic efficiency at the ribosomal level. Moreover, the hypothetical effect on rate of synthesis of isosemantic substitutions is not supported by Eck's code for hemoglobin S, if it is assumed that HbS has arisen from HbA. According to Eck, there is indeed only one codon for valine recognizable by transfer-RNA, namely UYG (i.e., UCG and UUG; Y stands for "pyrimidine"). One can therefore not say, without resorting to an auxiliary hypothesis, that the apparent slower relative rate of HbS synthesis as compared to HbA synthesis in HbA/HbS heterozygotes is perhaps due to the appearance of a codon whose corresponding transfer-RNA is present in limiting amounts. The auxiliary hypothesis is that a given kind of transfer-RNA is not entirely indifferent to the exact chemical species of the central purine or pyrimidine in a codon. It is possible that the exact chemical species of the presumed middle "letter", while without action on the coding, influences the rate of synthesis of the polypeptide. However, very recent evidence (Levere & Lichtman, 1963) suggests that the rate of synthesis of HbS may in reality not be inferior to that of HbA. The present status of these problems is uncertainty.

Although there is therefore no evidence in favor of considering isosemantic substitutions as a significant factor in the regulation of the rate of polypeptide synthesis, the possibility is not ruled out and should be kept in mind as furnishing a basis for Itano's idea, expressed here in a slightly modified way, that rate of synthesis and structure, at the level of a given structural gene, are intimately linked (Itano, 1957).

If isosemantic substitutions recognized by transfer-RNA actually exerted an effect on rate of polypeptide synthesis, one would expect natural selection to act quite strongly on such substitutions. If natural selection did not act on the other postulated type of isosemantic substitutions, those of "primary crypticity", not recognized by transfer RNA, the occurrence of such substitutions would be random. It would be more probable that some effect is present and that natural selection acts here also. A possible effect of the exact chemical species of the presumed middle letter of a codon on rate of synthesis has just been mentioned. Other effects might be considered. In particular, it has been shown that the frequency of crossing over is inversely related to the degree of heterozygosity in a chromosome pair (Stadler & Towe, 1962). Via this mechanism isosemantic substitutions of primary crypticity as well as those of secondary crypticity may have far reaching effects on population genetics.

One may also examine the possibility that isosemantic substitutions have some effect on evolutionary stability. Benzer (1961) has pointed out that, since AT base pairs are held together much less strongly than GC base pairs, a genetic region rich in AT pairs will tend to be more subject to substitution. By selecting for isosemantic triplets rich in GC and low in AT content, an organism might reduce its mutation rate without changing the structure of any of its proteins. Whether such an effect would be significant enough to influence the rate of evolution remains to be seen. It will be of interest to compare the base composition of DNA from "living fossils" such as *Lingula* or *Limulus* with base composition in more rapidly evolving animals.

Finally, isosemantic substitutions in those regions of DNA that carry out the function of operators (Jacob & Monod, 1961) might well lead to a modification of the stereochemical relationship between the operators and the repressor molecules. Thereby such isosemantic substitutions would have an effect on rate of polypeptide synthesis, distinct from the operator-independent effect discussed earlier.

Due to isosemantic substitutions, there probably is more evolutionary history inscribed in the base sequence of nucleic acids than in the amino-acid sequence of corresponding polypeptide chains. By its implications, a degenerate code thus emphasizes the role of nucleic acids as "master molecules" over polypeptides, (a role still doubted by some (Commoner, 1962)), even though polypeptides may interact with nucleic acids to regulate the rate of synthesis of both polypeptides and nucleic acids. All the potentialities of an individual may be assumed to be inscribed in polypeptide chains that are actually synthesized, or could be synthesized, by the cells under certain circumstances, and in the structures that control the actual and potential rates of this synthesis. Yet it appears conceivable, since equal rates of synthesis under the control of distinct but isosemantic codons are possible, that the individual contains information, not only, as we know, beyond that which it actually uses for its realization, but even beyond that which defines its potentialities. This part of its "being", necessarily cryptic in terms of the phenotype, would at best be expressed only in relation to the evolution of the species.

One of the authors (E. Z.) is greatly indebted to Professor Joshua Lederberg for discussions about the topics treated in this paper.

This article, written in 1963, was published early in 1964, in a Russian translation, in the volume *Problems of Evolutionary and Technical Biochemistry*, dedicated to the Academician A. I. Oparin (Science Press, Academy of Sciences of the U.S.S.R., pp. 54–62). The original text of the article is published here, as contribution No. 3041 of the Division of Chemistry and Chemical Engineering of the California Institute of Technology. This work was supported by a grant GM 04276–99 from the National Institutes of Health.

REFERENCES

BENZER, S. (1961). *Proc. nat. Acad. Sci., Wash.* **47,** 403.

COHEN, S. S. (1963). *Science,* **139,** 1017.

COMMONER, B. (1962). *In* "Horizons in Biochemistry", pp. 319 (M. Kasha & B. Pullman, eds.). New York: Academic Press.

ECK, R. V. (1963). *Science,* **140,** 477.

GOODWIN, T. W. (1962). *In* "Comparative Biochemistry", Vol. IV, pp. 643 (M. Florkin and H. S. Mason, eds.). London: Academic Press.

HUNT, J. A. & INGRAM, V. M. (1958). *Nature,* **181,** 1062.

HUNT, J. A. & INGRAM ,V. M. (1959). *Nature,* **184,** 870.

HUTNER, S. H. (1955). *In* "Protozoa", Vol. II (S. H. Hutner and A. Lwoff, eds.). New York: Academic Press.

INGRAM, V. A. (1958). *Nature,* **181,** 1062.

ITANO, H. A. (1957). *Advanc. Protein Chem.* **12,** 215.

JACOB, F. & MONOD, J. (1961). *Cold Spr. Harb. Symp. quant. Biol.* **26,** 193.

JONES, O. W. & NIRENBERG, N. W. (1962). *Proc. nat. Acad. Sci., Wash.* **48,** 2115.

JONES, R. T. (1964). *In* "Symposium on Foods: Proteins and their Reactions" (H. W. Shultz and A. F. Anglemier, eds.). The Avi Publishing Co..

JUKES, T. H. (1962). *Proc. nat. Acad. Sci., Wash,* **48,** 1809.

LEVERE, R. D. & LICHTMAN, H. C. (1963). *Blood,* **22,** 334.

PAULING, L. (1963). Rudolf Virchow Memorial Lecture, New York, 1961; Proceedings of the Rudolf Virchow Medical Society, **21,** (S. Karger and A. G. Basel).

PAULING, L. & ZUCKERKANDL, E. (1963). *Acta Chem. Scand.,* **17,** 59.

STADLER, D. R. & TOWE, A. M. (1962). *Genetics,* **47,** 839.

TAYLOR, N. W. & CLUSKEY, J. E. (1962). *Arch. biochem. Biophys.* **97,** 399.

WAHBA, A. J., GARDNER, R. S., BASILIO, C., MILLER, R. S., SPEYER, J. F. & LENGYEL, P. (1963). *Proc. nat. Acad. Sci., Wash.* **49,** 116.

WEISBLUM, B., BENZER, S. & HOLLEY, R. W. (1962). *Proc. nat. Acad. Sci., Wash.* **48,** 1449.

WOLLMAN, E. L. & JACOB, F. (1959). "La sexualité des bactéries". Paris: Masson.

ZUCKERKANDL, E. (1964). *Cold Spr. Harb. Symp. quant. Biol.* **29,** 357.

ZUCKERKANDL, E. & PAULING, L. (1962). *In* "Horizons in Biochemistry", pp. 189 (M. Kasha & B. Pullman, eds.). New York: Academic Press.

Photos for Part III

Photo 17 Linus Pauling and Prof. Dan H. Campbell (left) with laboratory rabbits used in serological research at Caltech in the early 1940's (Chapter 12).

Photo 18 (right) Linus Pauling in the early 1950's, studying a polyhedral model that may be related to structures of intermetallic compounds.

Photo 19 (left) Linus Pauling (middle row to right of center) in 1950 with co-authors Dr. David Shoemaker (lower left), Prof. László Zechmeister (middle row left), and (top row left to right) Prof. Dan Campbell, Dr. Joseph B. Koepfli, and Prof. Carl Nieman. Shoemaker (SP 54) was an expert in the structures of metals and alloys. Nieman (SP 96) was professor of organic chemistry. Campbell (several SPs in Chapter 12) appears also in Photo 17. Koepfli collaborated with Pauling and Campbell in developing an artificial substitute for blood serum, called oxypolygelatin (paper [51-2]). Zechmeister co-authored two papers on prolycopene ([41-5] and [43-3]). Other Caltech faculty in the picture are Prof. John Kirkwood (top row right) and Dr. David M. Mason (lower right).

Photo 20 Linus Pauling with Dr. Fred J. Ewing (right) in the mid 1950's. Ewing worked with Pauling on metals and intermetallic compounds (SP 30, SP 54). In this photo the writing on the blackboard shows that Pauling has just given a lecture on hydrogen bonding.

Photo 21 Linus Pauling with Prof. Youqi Tang, co-author of SP 53. Tang received his Ph.D. at Caltech in 1950, and then returned to his family in the P.R.C. (China), becoming professor of chemistry in Peking University. His laboratory did extensive x-ray studies of the structure of insulin. Photo was taken in 1991. Here and in Photos 23, 35, 41 and 46, Pauling sports his characteristic beret.

Photo 22 Prof. J. Holmes Sturdivant (left) and Prof. Robert B. Corey (right) in 1948. These men were Pauling's close and steadfast colleagues. Sturdivant was Pauling's first graduate student, and worked with him over the period 1926 to 1964. Corey worked with him from 1937 to 1964. Sturdivant specialized in x-ray diffraction study of the structure of ionic crystals (SP 42), and in diffraction instrumentation; he also contributed to the development of the Pauling oxygen meter (SP 62). Corey collaborated extensively with Pauling in the study of protein structure, and co-authored many papers (Chapter 13). See also Photo 23.

Photo 23 Linus Pauling with Robert B. Corey at about the time the α-helix and other protein structures were discovered (Chapter 13), in the early 1950's. Corey's specialty played a crucial role in these discoveries. He determined by x-ray diffraction the structures of amino acids, peptides, and other molecules derived from or related to proteins (SP 99, SP 105, SP 106).

ıoto 24 Linus Pauling working with a protein molecular model in the early 1950's. The rectangular ɜces of flat sheet metal represent the planar amide groups, so important in the structure of proteins P 105). The model represents Pauling and Corey's proposed structure for collagen (SP 103).

Photo 25 Linus Pauling and Fred J. Ewing carrying α-helix-type molecular models into the Athenaeum at Caltech, for a lecture. Mid 1950's.

Photo 26 Linus Pauling and Prof. George Beadle (Nobel Prize in 1958) examining a molecular model of the α-helix at an American Cancer Society exhibit in the early 1950's.

Photo 27 (left) Linus Pauling with Dr. Richard E. Marsh (left), in 1975. Marsh is a distinguished crystallographer who collaborated with Pauling in the structural study of hydrogen bonding by water molecules in crystals (SP 37). He also collaborated in studies of intermetallic compounds (SP 54) and proteins (SP 110).

Photo 28 Linus Pauling with models of the structure of water based on the pentagonal dodecahedron, ca. 1960 (SP 37, SP 38, and SP 129).

Photo 29 Dr. Sterling Hendricks, co-author of SP 39 and several other papers on mineral structures. He came to Caltech in 1924, but is here shown in his later years. He became a renowned expert on the clay minerals and other sheet silicates. Pauling wrote a memorial of him in 1982 (paper [82-13], in Group 18, Appendix III).

Part IV

HEALTH AND MEDICINE

Introduction To Part IV

Health and Medicine

Growing out of the first phase of Pauling's biomolecular research, covered in Part III, there emerged, beginning in the 1950's, a second phase of activity, in which Pauling championed the belief that basic principles of structural chemistry could shed light on human disease and provide a route to better health. This is the subject of Part IV. In Pauling's publications we can trace his efforts to connect the fundamental elements of scientific knowledge with their practical applications in the medical realm.

Part IV begins with what is without doubt one of Linus Pauling's greatest contributions to medical science—the concept of *molecular disease*, developed in the five papers in Chapter 15. The discovery was made in human hemoglobin, and thus has roots in the papers in Chapter 11, as described in SP 87. But important as hemoglobin is, the concept of molecular disease extends far beyond it and has become one of the foundations of modern medicine. Yet there is only a slight distinction between this concept as a basis for medical treatment and as a key to biochemical understanding, as noted in the Introduction to Part III.

Chapter 16 gathers together several health-related subjects. Pauling's concern with the *health hazards from nuclear radiation* arose from the great human devastation caused by the atomic bombs dropped on Hiroshima and Nagasaki, and was an important component of his involvement in advocacy of humanitarian causes, especially peace (see Chapter 18). Two samples of his scientific work on radiation damage to health are SP 127 and SP 128. They were written in the context of his suspicion, probably well founded, that the then immensely powerful Atomic Energy Commission was withholding or distorting scientific information on the health effects of radiation. The *"hydrate microcrystal" theory of general anesthesia* (SP 129) sprang from Pauling's concept of the hydrogen-bonded structure of liquid water (SP 38, in Chapter 4), which in turn was rooted in the appearance of pentagonal dodecahedra of hydrogen-bonded water molecules enclosing "guest" molecules in the crystal structure of clathrate crystalline hydrates (SP 37). He and his colleagues endeavored to test the theory by laboratory experiments on brine shrimp and goldfish, but definitive confirmation was not achieved.

SP 131 describes Pauling's 1933 discovery and verification of the superoxide free radical, which now, 67 years later—has become of much interest as an active agent in oxidative damage to tissues and in the aging process. The control of this substance in living organisms, by means of the superoxide dismutase enzyme that catalyzes its decomposition, is an example of a potential task of orthomolecular medicine (see below).

SP 132 and SP 133 serve to represent several of Pauling's papers devoted to biomedical statistics and their use in interpreting the results of clinical trials of drugs or nutrients. He became interested in this subject in endeavoring to evaluate reported results of tests of the medical effectiveness of vitamins. SP 128 addresses similar statistical issues in radiation-dose testing of the survival of mice.

Orthomolecular medicine, the subject of Chapter 17, was Pauling's main scientific work in his later years—aside from the interests represented in Chapter 10 and the ever-present interest in the chemical bond, which he pursued "for recreation". Pauling invented the term "orthomolecular psychiatry" (SP 134) and then enlarged it to "orthomolecular medicine" (SP 135). It refers to the maintenance or restoration of good mental or physical health by control of the concentrations of substances normally present in the human body, this control being exerted generally by nutritional control of the intake of such substances or their precursors or inhibitors. These substances are nowadays called "micronutrients" because they are normally ingested in small amounts, such as vitamins. Pauling's interest in nutritional medicine was probably stimulated by his life-threatening encounter in 1941 with glomerular nephritis, which was cured by a nutritional/dietary regime.

SP 130 and SP 142 are examples of results of experimental work from the Institute of Orthomolecular Medicine, which Pauling founded in 1973 to carry out scientific studies in support of the principles of orthomolecular medicine. It was later renamed the Linus Pauling Institute of Science and Medicine, and is now located at Oregon State University (Corvallis), Pauling's alma mater.

Pauling's interest was initially in vitamin C (ascorbic acid) as a preventative or palliative for the common cold (SP 137), but his interest later extended to other micronutrients (SP 131, SP 139) and to the control of cancer (SP 139, SP 140, SP 141), AIDS (SP 142) and cardiovascular disease (SP 143). He wrote several lay-oriented books on these subjects. Although controversial among doctors and nutritional scientists, Pauling's ideas were and still are widely read and quoted, and they substantially stimulated popular awareness of micronutrients. Part IV ends with a conclusionary section from the book *Cancer and Vitamin C* (SP 144).

As noted in the Introduction to Part III, Linus Pauling can rightfully be regarded as the father of modern molecular biology, and many consider this to be his most important contribution to biology and medical science. Most of the landmark biomolecular papers that led to this renown are in Part III, but some, including especially those on the discovery of molecular genetic disease, are in Part IV (Chapter 15). "Pauling's grand vision of molecular biology and medicine has been realized to an extent he could never have forseen" (—B. J. Strasser, *Science*, Vol. 286, p. 1490, 1999). Or could he?

Chapter 15

Molecular Disease

Contents

SP 122 **Sickle Cell Anemia, a Molecular Disease** **1240**
(by Linus Pauling, Harvey A. Itano, S. J. Singer, and Ibert C. Wells)
Science **110**, 543–548 (1949)

123 **The Hemoglobin Molecule in Health and Disease** **1246**
Proc. Am. Phil. Soc. **96**, 556–565 (1952)

124 **Abnormality of Hemoglobin Molecules in Hereditary Hemolytic Anemias** **1256**
In *The Harvey Lectures 1953-1954*, Series 49,
Academic Press, N.Y., pp. 216–241 (1955)

125 **Abnormal Hemoglobin Molecules in Relation to Disease** **1283**
(by Harvey A. Itano and Linus Pauling)
Svensk Kemisk Tidskrift **69**, 509–523 (1957)

126 **Molecular Disease and Evolution** **1298**
Proc. Rudolf Virchow Med. Soc. N.Y. **21**, 131–140 (1963)

Reprinted from SCIENCE, November 25, 1949, Vol. 110, No. 2865, pages 543–548.

Sickle Cell Anemia, a Molecular Disease[1]

Linus Pauling, Harvey A. Itano,[2] S. J. Singer,[2] and Ibert C. Wells[3]

Gates and Crellin Laboratories of Chemistry, California Institute of Technology, Pasadena, California[4]

THE ERYTHROCYTES of certain individuals possess the capacity to undergo reversible changes in shape in response to changes in the partial pressure of oxygen. When the oxygen pressure is lowered, these cells change their forms from the normal biconcave disk to crescent, holly wreath, and other forms. This process is known as sickling. About 8 percent of American Negroes possess this characteristic; usually they exhibit no pathological consequences ascribable to it. These people are said to have sicklemia, or sickle cell trait. However, about 1 in 40 (*4*) of these individuals whose cells are capable of sickling suffer from a severe chronic anemia resulting from excessive destruction of their erythrocytes; the term sickle cell anemia is applied to their condition.

The main observable difference between the erythrocytes of sickle cell trait and sickle cell anemia has been that a considerably greater reduction in the partial pressure of oxygen is required for a major fraction of the trait cells to sickle than for the anemia cells (*11*). Tests *in vivo* have demonstrated that between 30 and 60 percent of the erythrocytes in the venous circulation of sickle cell anemic individuals, but less than 1 percent of those in the venous circulation of sicklemic individuals, are normally sickled. Experiments *in vitro* indicate that under sufficiently low oxygen pressure, however, all the cells of both types assume the sickled form.

The evidence available at the time that our investigation was begun indicated that the process of sickling might be intimately associated with the state and the nature of the hemoglobin within the erythrocyte. Sickle cell erythrocytes in which the hemoglobin is combined with oxygen or carbon monoxide have the biconcave disk contour and are indistinguishable in that form from normal erythrocytes. In this condition they are termed promeniscocytes. The hemoglobin appears to be uniformly distributed and randomly oriented within normal cells and promeniscocytes, and no birefringence is observed. Both types of cells are very flexible. If the oxygen or carbon monoxide is removed, however, transforming the hemoglobin to the uncombined state, the promeniscocytes undergo sickling. The hemoglobin within the sickled cells appears to aggregate into one or more foci, and the cell membranes collapse. The cells become birefringent (*11*) and quite rigid. The addition of oxygen or carbon monoxide to these cells reverses these phenomena. Thus the physical effects just described depend on the state of combination of the hemoglobin, and only secondarily, if at all, on the cell membrane. This conclusion is supported by the observation that sickled cells when lysed with water produce discoidal, rather than sickle-shaped, ghosts (*10*).

It was decided, therefore, to examine the physical and chemical properties of the hemoglobins of individuals with sicklemia and sickle cell anemia, and to compare them with the hemoglobin of normal individuals to determine whether any significant differences might be observed.

Experimental Methods

The experimental work reported in this paper deals largely with an electrophoretic study of these hemoglobins. In the first phase of the investigation, which concerned the comparison of normal and sickle cell anemia hemoglobins, three types of experiments were performed: 1) with carbonmonoxyhemoglobins; 2) with uncombined ferrohemoglobins in the presence of dithionite ion, to prevent oxidation to methemoglobins; and 3) with carbonmonoxyhemoglobins in the presence of dithionite ion. The experiments of type 3 were performed and compared with those of type 1 in order to ascertain whether the dithionite ion itself causes any specific electrophoretic effect.

Samples of blood were obtained from sickle cell anemic individuals who had not been transfused within three months prior to the time of sampling. Stroma-free concentrated solutions of human adult hemoglobin were prepared by the method used by Drabkin (*3*). These solutions were diluted just before use with the

[1] This research was carried out with the aid of a grant from the United States Public Health Service. The authors are grateful to Professor Ray D. Owen, of the Biology Division of this Institute, for his helpful suggestions. We are indebted to Dr. Edward R. Evans, of Pasadena, Dr. Travis Winsor, of Los Angeles, and Dr. G. E. Burch, of the Tulane University School of Medicine, New Orleans, for their aid in obtaining the blood used in these experiments.

[2] U. S. Public Health Service postdoctoral fellow of the National Institutes of Health.

[3] Postdoctoral fellow of the Division of Medical Sciences of the National Research Council.

[4] Contribution No. 1333.

2

appropriate buffer until the hemoglobin concentrations were close to 0.5 grams per 100 milliliters, and then were dialyzed against large volumes of these buffers for 12 to 24 hours at 4° C. The buffers for the experiments of types 2 and 3 were prepared by adding 300 ml of 0.1 ionic strength sodium dithionite solution to 3.5 liters of 0.1 ionic strength buffer. About 100 ml of 0.1 molar NaOH was then added to bring the pH of the buffer back to its original value. Ferrohemoglobin solutions were prepared by diluting the concentrated solutions with this dithionite-containing buffer and dialyzing against it under a nitrogen atmosphere. The hemoglobin solutions for the experiments of type 3 were made up similarly, except that they were saturated with carbon monoxide after dilution and were dialyzed under a carbon monoxide atmosphere. The dialysis bags were kept in continuous motion in the buffers by means of a stirrer with a mercury seal to prevent the escape of the nitrogen and carbon monoxide gases.

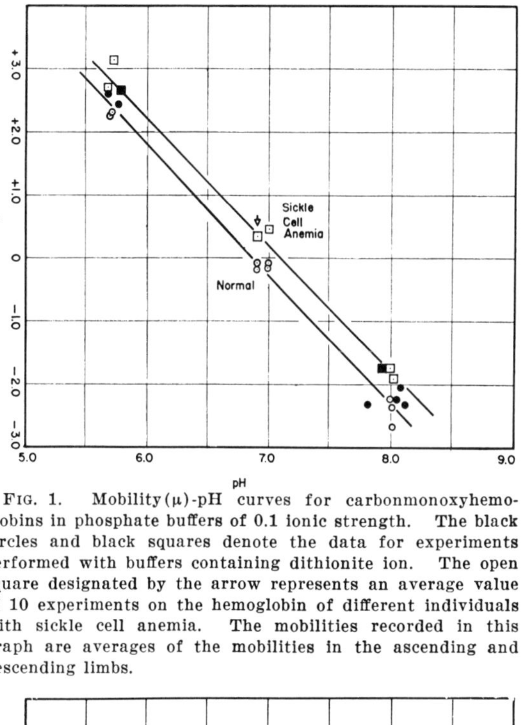

FIG. 1. Mobility (μ)-pH curves for carbonmonoxyhemoglobins in phosphate buffers of 0.1 ionic strength. The black circles and black squares denote the data for experiments performed with buffers containing dithionite ion. The open square designated by the arrow represents an average value of 10 experiments on the hemoglobin of different individuals with sickle cell anemia. The mobilities recorded in this graph are averages of the mobilities in the ascending and descending limbs.

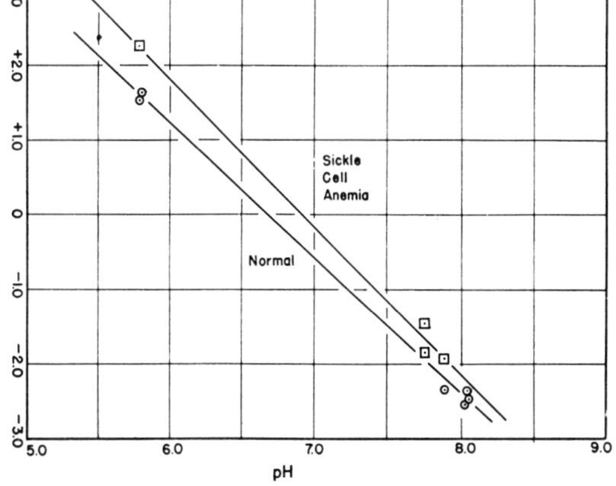

FIG. 2. Mobility (μ)-pH curves for ferrohemoglobins in phosphate buffers of 0.1 ionic strength containing dithionite ion. The mobilities recorded in the graph are averages of the mobilities in the ascending and descending limbs.

The experiments were carried out in the modified Tiselius electrophoresis apparatus described by Swingle (*14*). Potential gradients of 4.8 to 8.4 volts per centimeter were employed, and the duration of the runs varied from 6 to 20 hours. The pH values of the buffers were measured after dialysis on samples which had come to room temperature.

RESULTS

The results indicate that a significant difference exists between the electrophoretic mobilities of hemoglobin derived from erythrocytes of normal individuals and from those of sickle cell anemic individuals. The two types of hemoglobin are particularly easily distinguished as the carbonmonoxy compounds at pH 6.9 in phosphate buffer of 0.1 ionic strength. In this buffer the sickle cell anemia carbonmonoxyhemoglobin moves as a positive ion, while the normal compound moves as a negative ion, and there is no detectable amount of one type present in the other.[4] The hemoglobin derived from erythrocytes of individuals with sicklemia, however, appears to be a mixture of the normal hemoglobin and sickle cell anemia hemoglobin in roughly equal proportions. Up to the present time the hemoglobins of 15 persons with sickle cell anemia, 8 persons with sicklemia, and 7 normal adults have been examined. The hemoglobins of normal adult white and negro individuals were found to be indistinguishable.

The mobility data obtained in phosphate buffers of 0.1 ionic strength and various values of pH are summarized in Figs. 1 and 2.[5]

[4] Occasionally small amounts (less than 5 percent of the total protein) of material with mobilities different from that of either kind of hemoglobin were observed in these uncrystallized hemoglobin preparations. According to the observations of Stern, Reiner, and Silber (*12*) a small amount of a component with a mobility smaller than that of oxyhemoglobin is present in human erythrocyte hemolyzates.

[5] The results obtained with carbonmonoxyhemoglobins with and without dithionite ion in the buffers indicate that the dithionite ion plays no significant role in the electrophoretic properties of the proteins. It is therefore of interest that ferrohemoglobin was found to have a lower isoelectric point in phosphate buffer than carbonmonoxyhemoglobin. Titration studies have indicated (*5*, *6*) that oxyhemoglobin (similar in electrophoretic properties to the carbonmonoxy compound) has a lower isoelectric point than ferrohemoglobin in

3

The isoelectric points are listed in Table 1. These results prove that the electrophoretic difference between normal hemoglobin and sickle cell anemia hemoglobin

TABLE 1

ISOELECTRIC POINTS IN PHOSPHATE BUFFER, $\mu = 0.1$

Compound	Normal	Sickle cell anemia	Difference
Carbonmonoxyhemoglobin	6.87	7.09	0.22
Ferrohemoglobin	6.68	6.91	0.23

exists in both ferrohemoglobin and carbonmonoxyhemoglobin. We have also performed several experiments in a buffer of 0.1 ionic strength and pH 6.52 containing 0.08 M NaCl, 0.02 M sodium cacodylate, and 0.0083 M cacodylic acid. In this buffer the average mobility of sickle cell anemia carbonmonoxyhemoglobin is 2.63×10^{-5}, and that of normal carbonmonoxyhemoglobin is 2.23×10^{-5} cm/sec per volt/cm.[6]

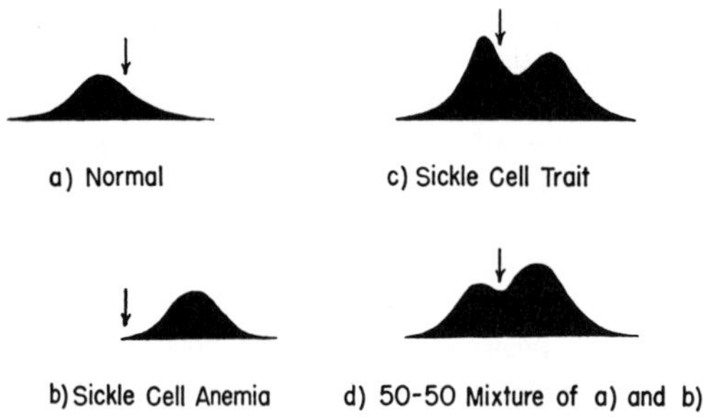

FIG. 3. Longsworth scanning diagrams of carbonmonoxyhemoglobins in phosphate buffer of 0.1 ionic strength and pH 6.90 taken after 20 hours' electrophoresis at a potential gradient of 4.73 volts/cm.

These experiments with a buffer quite different from phosphate buffer demonstrate that the difference between the hemoglobins is essentially independent of the buffer ions.

Typical Longsworth scanning diagrams of experiments with normal, sickle cell anemia, and sicklemia carbonmonoxyhemoglobins, and with a mixture of the first two compounds, all in phosphate buffer of pH 6.90 and ionic strength 0.1, are reproduced in Fig. 3. It is apparent from this figure that the sicklemia material contains less than 50 percent of the anemia component. In order to determine this quantity accurately some experiments at a total protein concentration of 1 percent were performed with known mixtures of sickle cell anemia and normal carbonmonoxyhemoglobins in the cacodylate-sodium chloride buffer of 0.1 ionic strength and pH 6.52 described above. This buffer was chosen in order to minimize the anomalous electrophoretic effects observed in phosphate buffers (*7*). Since the two hemoglobins were incompletely resolved after 15 hours of electrophoresis under a potential gradient of 2.79 volts/cm, the method of Tiselius and Kabat (*16*) was employed to allocate the

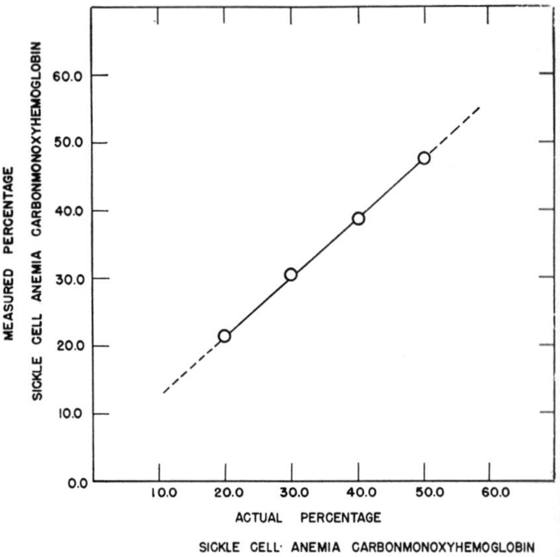

FIG. 4. The determination of the percent of sickle cell anemia carbonmonoxyhemoglobin in known mixtures of the protein with normal carbonmonoxyhemoglobin by means of electrophoretic analysis. The experiments were performed in a cacodylate sodium chloride buffer described in the text.

areas under the peaks in the electrophoresis diagrams to the two components. In Fig. 4 there is plotted the percent of the anemia component calculated from the areas so obtained against the percent of that component in the known mixtures. Similar experiments were performed with a solution in which the hemoglobins of 5 sicklemic individuals were pooled. The relative concentrations of the two hemoglobins were calculated from the electrophoresis diagrams, and the actual proportions were then determined from the plot of Fig. 4. A value of 39 percent for the amount of the sickle cell anemia component in the sicklemia hemoglobin was arrived at in this manner. From the experiments we have performed thus far it appears that this value does not vary greatly from one sicklemic individual to another, but a more extensive study of this point is required.

Up to this stage we have assumed that one of the two components of sicklemia hemoglobin is identical with sickle cell anemia hemoglobin and the other is identical with the normal compound. Aside from the

the absence of other ions. These results might be reconciled by assuming that the ferrous iron of ferrohemoglobin forms complexes with phosphate ions which cannot be formed when the iron is combined with oxygen or carbon monoxide. We propose to continue the study of this phenomenon.

[6] The mobility data show that in 0.1 ionic strength cacodylate buffers the isoelectric points of the hemoglobins are increased about 0.5 pH unit over their values in 0.1 ionic strength phosphate buffers. This effect is similar to that observed by Longsworth in his study of ovalbumin (7).

4

genetic evidence which makes this assumption very probable (see the discussion section), electrophoresis experiments afford direct evidence that the assumption is valid. The experiments on the pooled sicklemia carbonmonoxyhemoglobin and the mixture containing 40 percent sickle cell anemia carbonmonoxyhemoglobin and 60 percent normal carbonmonoxyhemoglobin in the cacodylate-sodium chloride buffer described above were compared, and it was found that the mobilities of the respective components were essentially identical.[7] Furthermore, we have performed experiments in which normal hemoglobin was added to a sicklemia preparation and the mixture was then subjected to electrophoretic analysis. Upon examining the Longsworth scanning diagrams we found that the area under the peak corresponding to the normal component had increased by the amount expected, and that no indication of a new component could be discerned. Similar experiments on mixtures of sickle cell anemia hemoglobin and sicklemia preparations yielded similar results. These sensitive tests reveal that, at least electrophoretically, the two components in sicklemia hemoglobin are identifiable with sickle cell anemia hemoglobin and normal hemoglobin.

Discussion

1) *On the Nature of the Difference between Sickle Cell Anemia Hemoglobin and Normal Hemoglobin*: Having found that the electrophoretic mobilities of sickle cell anemia hemoglobin and normal hemoglobin differ, we are left with the considerable problem of locating the cause of the difference. It is impossible to ascribe the difference to dissimilarities in the particle weights or shapes of the two hemoglobins in solution: a purely frictional effect would cause one species to move more slowly than the other throughout the entire pH range and would not produce a shift in the isoelectric point. Moreover, preliminary velocity ultracentrifuge[8] and free diffusion measurements indicate that the two hemoglobins have the same sedimentation and diffusion constants.

The most plausible hypothesis is that there is a difference in the number or kind of ionizable groups in the two hemoglobins. Let us assume that the only groups capable of forming ions which are present in carbonmonoxyhemoglobin are the carboxyl groups in the heme, and the carboxyl, imidazole, amino, phenolic hydroxyl, and guanidino groups in the globin. The number of ions nonspecifically adsorbed on the two proteins should be the same for the two hemoglobins under comparable conditions, and they may be neglected for our purposes. Our experiments indicate that the net number of positive charges (the total number of cationic groups minus the number of anionic groups) is greater for sickle cell anemia hemoglobin than for normal hemoglobin in the pH region near their isoelectric points.

According to titration data obtained by us, the acid-base titration curve of normal human carbonmonoxyhemoglobin is nearly linear in the neighborhood of the isoelectric point of the protein, and a change of one pH unit in the hemoglobin solution in this region is associated with a change in net charge on the hemoglobin molecule of about 13 charges per molecule. The same value was obtained by German and Wyman (*5*) with horse oxyhemoglobin. The difference in isoelectric points of the two hemoglobins under the conditions of our experiments is 0.23 for ferrohemoglobin and 0.22 for the carbonmonoxy compound. This difference corresponds to about 3 charges per molecule. With consideration of our experimental error, sickle cell anemia hemoglobin therefore has 2–4 more net positive charges per molecule than normal hemoglobin.

Studies have been initiated to elucidate the nature of this charge difference more precisely. Samples of porphyrin dimethyl esters have been prepared from normal hemoglobin and sickle cell anemia hemoglobin. These samples were shown to be identical by their x-ray powder photographs and by identity of their melting points and mixed melting point. A sample made from sicklemia hemoglobin was also found to have the same melting point. It is accordingly probable that normal and sickle cell anemia hemoglobin have different globins. Titration studies and amino acid analyses on the hemoglobins are also in progress.

2) *On the Nature of the Sickling Process*: In the introductory paragraphs we outlined the evidence which suggested that the hemoglobins in sickle cell anemia and sicklemia erythrocytes might be responsible for the sickling process. The fact that the hemoglobins in these cells have now been found to be different from that present in normal red blood cells makes it appear very probable that this is indeed so.

We can picture the mechanism of the sickling process in the following way. It is likely that it is the globins rather than the hemes of the two hemoglobins that are different. Let us propose that there is a surface region on the globin of the sickle cell anemia, hemoglobin molecule which is absent in the normal molecule and which has a configuration complementary to a different region of the surface of the hemoglobin molecule. This situation would be somewhat analogous to that which very probably exists in antigen-antibody reactions (*9*). The fact that sick-

[7] The patterns were very slightly different in that the known mixture contained 1 percent more of the sickle cell anemia component than did the sickle cell trait material.

[8] We are indebted to Dr. M. Moskowitz, of the Chemistry Department, University of California at Berkeley, for performing the ultracentrifuge experiments for us.

5

ling occurs only when the partial pressures of oxygen and carbon monoxide are low suggests that one of these sites is very near to the iron atom of one or more of the hemes, and that when the iron atom is combined with either one of these gases, the complementariness of the two structures is considerably diminished. Under the appropriate conditions, then, the sickle cell anemia hemoglobin molecules might be capable of interacting with one another at these sites sufficiently to cause at least a partial alignment of the molecules within the cell, resulting in the erythrocyte's becoming birefringent, and the cell membrane's being distorted to accommodate the now relatively rigid structures within its confines. The addition of oxygen or carbon monoxide to the cell might reverse these effects by disrupting some of the weak bonds between the hemoglobin molecules in favor of the bonds formed between gas molecules and iron atoms of the hemes.

Since all sicklemia erythrocytes behave more or less similarly, and all sickle at a sufficiently low oxygen pressure (*11*), it appears quite certain that normal hemoglobin and sickle cell anemia hemoglobin coexist within each sicklemia cell; otherwise there would be a mixture of normal and sickle cell anemia erythrocytes in sicklemia blood. We might expect that the normal hemoglobin molecules, lacking at least one type of complementary site present on the sickle cell anemia molecules, and so being incapable of entering into the chains or three-dimensional frameworks formed by the latter, would interfere with the alignment of these molecules within the sicklemia erythrocyte. Lower oxygen pressures, freeing more of the complementary sites near the hemes, might be required before sufficiently large aggregates of sickle cell anemia hemoglobin molecules could form to cause sickling of the erythrocytes.

This is in accord with the observations of Sherman (*11*), which were mentioned in the introduction, that a large proportion of erythrocytes in the venous circulation of persons with sickle cell anemia are sickled, but that very few have assumed the sickle forms in the venous circulation of individuals with sicklemia. Presumably, then, the sickled cells in the blood of persons with sickle cell anemia cause thromboses, and their increased fragility exposes them to the action of reticulo-endothelial cells which break them down, resulting in the anemia (*1*).

It appears, therefore, that while some of the details of this picture of the sickling process are as yet conjectural, the proposed mechanism is consistent with experimental observations at hand and offers a chemical and physical basis for many of them. Furthermore, if it is correct, it supplies a direct link between the existence of "defective" hemoglobin molecules and the pathological consequences of sickle cell disease.

3) *On the Genetics of Sickle Cell Disease*: A genetic basis for the capacity of erythrocytes to sickle was recognized early in the study of this disease (*4*). Taliaferro and Huck (*15*) suggested that a single dominant gene was involved, but the distinction between sicklemia and sickle cell anemia was not clearly understood at the time. The literature contains conflicting statements concerning the nature of the genetic mechanisms involved, but recently Neel (*8*) has reported an investigation which strongly indicates that the gene responsible for the sickling characteristic is in heterozygous condition in individuals with sicklemia, and homozygous in those with sickle cell anemia.

Our results had caused us to draw this inference before Neel's paper was published. The existence of normal hemoglobin and sickle cell anemia hemoglobin in roughly equal proportions in sicklemia hemoglobin preparations is obviously in complete accord with this hypothesis. In fact, if the mechanism proposed above to account for the sickling process is correct, we can identify the gene responsible for the sickling process with one of an alternative pair of alleles capable through some series of reactions of introducing the modification into the hemoglobin molecule that distinguishes sickle cell anemia hemoglobin from the normal protein.

The results of our investigation are compatible with a direct quantitative effect of this gene pair; in the chromosomes of a single nucleus of a normal adult somatic cell there is a complete absence of the sickle cell gene, while two doses of its allele are present; in the sicklemia somatic cell there exists one dose of each allele; and in the sickle cell anemia somatic cell there are two doses of the sickle cell gene, and a complete absence of its normal allele. Correspondingly, the erythrocytes of these individuals contain 100 percent normal hemoglobin, 40 percent sickle cell anemia hemoglobin and 60 percent normal hemoglobin, and 100 percent sickle cell anemia hemoglobin, respectively. This investigation reveals, therefore, a clear case of a change produced in a protein molecule by an allelic change in a single gene involved in synthesis.

The fact that sicklemia erythrocytes contain the two hemoglobins in the ratio 40 : 60 rather than 50 : 50 might be accounted for by a number of hypothetical schemes. For example, the two genes might compete for a common substrate in the synthesis of two different enzymes essential to the production of the two different hemoglobins. In this reaction, the sickle cell gene would be less efficient than its normal allele. Or, competition for a common substrate might occur at some later stage in the series of reactions leading to the synthesis of the two hemoglobins. Mechanisms of this sort are discussed in more elaborate detail by Stern (*13*).

6

The results obtained in the present study suggest that the erythrocytes of other hereditary hemolytic anemias be examined for the presence of abnormal hemoglobins. This we propose to do.

Based on a paper presented at the meeting of the National Academy of Sciences in Washington, D. C., in April, 1949, and at the meeting of the American Society of Biological Chemists in Detroit in April, 1949.

References

1. Boyd, W. *Textbook of pathology.* (3rd Ed.) Philadelphia: Lea and Febiger, 1938. P. 864.
2. Diggs, L. W., Ahmann, C. F., and Bibb, J. *Ann. int. Med.*, 1933, **7**, 769.
3. Drabkin, D. L. *J. biol. Chem.*, 1946, **164**, 703.
4. Emmel, V. E. *Arch. int. Med.*, 1917, **20**, 586.
5. German, B. and Wyman, J., Jr. *J. biol. Chem.*, 1937, **117**, 533.
6. Hastings, A. B. *et al.* *J. biol. Chem.*, 1924, **60**, 89.
7. Longsworth, L. G. *Ann. N. Y. Acad. Sci.*, 1941, **41**, 267.
8. Neel, J. V. *Science*, 1949, **110**, 64.
9. Pauling, L., Pressman, D., and Campbell, D. H. *Physiol. Rev.*, 1943, **23**, 203.
10. Ponder, E. *Ann. N. Y. Acad. Sci.*, 1947, **48**, 579.
11. Sherman, I. J. *Bull. Johns Hopk. Hosp.*, 1940, **67**, 309.
12. Stern, K. G., Reiner, M. and Silber, R. H. *J. biol. Chem.*, 1945, **161**, 731.
13. Stern, C. *Science*, 1948, **108**, 615.
14. Swingle, S. M. *Rev. sci. Inst.*, 1947, **18**, 128.
15. Taliaferro, W. H. and Huck, J. G. *Genetics*, 1923, **8**, 594.
16. Tiselius, A. and Kabat, E. *J. exp. Med.*, 1939, **69**, 119.

THE HEMOGLOBIN MOLECULE IN HEALTH AND DISEASE

LINUS PAULING

Professor of Chemistry, California Institute of Technology

(*Read April 19, 1951*)

I. THE HEMOGLOBIN MOLECULE IN HEALTH

HEMOGLOBIN is one of the most interesting chemical substances in the world—to me it is the most interesting of all. Each of us carries around with him his own supply, amounting to a pound or two, approximately one per cent of the body weight. This supply is in the red corpuscles of the blood. Hemoglobin is the pigment of blood: it has a beautiful red color in arterial blood, and a purple color in venous blood. It is hemoglobin that gives a pink flush to our skin; we are pale when there is a deficiency of hemoglobin in the skin, either because of a general deficiency of the substance in the body, an anemia, or because blood is driven from the skin to the interior of the body by the contraction of the blood vessels in the skin.

The red corpuscles in man are flattened disks about 70,000 A in diameter and 10,000 A thick. In an ordinary microscope they have the appearance shown in figure 1. These red cells are suspended in the plasma of the blood, and they constitute about one third of the blood. They are full of hemoglobin, which makes up about 35 per cent of each red cell. There are about 100 million hemoglobin molecules in each red cell; this number is small in part because the red cell itself is small, and in part because the molecules of hemoglobin are large. Their molecular weight is 68,000, which may be compared with 18 for water, 46 for ethyl alcohol, and 342 for sucrose. The molecule contains about 10,000 atoms, of carbon, nitrogen, hydrogen, oxygen, sulfur, and other elements. There are four atoms of iron in the molecule, which play a special part in the principal function of hemoglobin, that of combination with oxygen.

The main work done by the blood is that of carrying oxygen from the lungs to the tissues, and carbon dioxide and other products of breakdown of tissues and foods to the lungs and excretory organs. The hemoglobin molecule is involved in carrying oxygen from the lungs to the tissues and in carrying carbon dioxide from the tissues to the lungs. The hemoglobin molecule can combine with four molecules of oxygen; the resultant oxyhemoglobin is bright red in color. In the tissues, where the partial pressure of oxygen is less than in the lungs, it gives up part of its load of oxygen, which then is used in oxidation reactions of various sorts. The carbon dioxide produced by oxidation of compounds containing carbon is then carried by the blood back to the lungs, and released in the exhaled air.

The four oxygen molecules that can be taken up by the hemoglobin molecule attach themselves to the four iron atoms that are present in the molecule. These iron atoms are present as the central atoms in complexes called hemes, with the structure shown in figure 2. These four flat groups of atoms are present in the hemoglobin combined with the rest of the molecule, a protein called globin. It is the hemes that are responsible for the color of hemoglobin. The nature of the bonds in the hemoglobin molecule has been elucidated in considerable part by the study of the magnetic properties of hemoglobin. It was discovered over fifteen years ago [1] that venous blood is paramagnetic—that is, it is attracted into a magnetic field—whereas arterial blood is diamagnetic, and these magnetic properties have been found to be closely correlated with the bonding of the iron atoms.

Although some of the carbon dioxide that is carried by the blood from the tissues to the lungs is in chemical combination with hemoglobin, most of it is carried in solution in the blood. The hemoglobin contributes to this mechanism of transport of carbon dioxide in a very ingenious manner. There are in the hemoglobin molecule four acid groups which are coupled with the heme groups in such a way that their acidity is greater for an oxygenated heme than for a deoxygenated heme. Accordingly when the blood reaches the

[1] Pauling, L., and C. D. Coryell, The magnetic properties and structure of hemoglobin, oxyhemoglobin, and carbonmonoxyhemoglobin, *Proc. Nat. Acad. Sci.* **22**: 210–216, 1936.

PROCEEDINGS OF THE AMERICAN PHILOSOPHICAL SOCIETY, VOL. 96, NO. 5, OCTOBER, 1952
Reprint *Printed in U. S. A.*

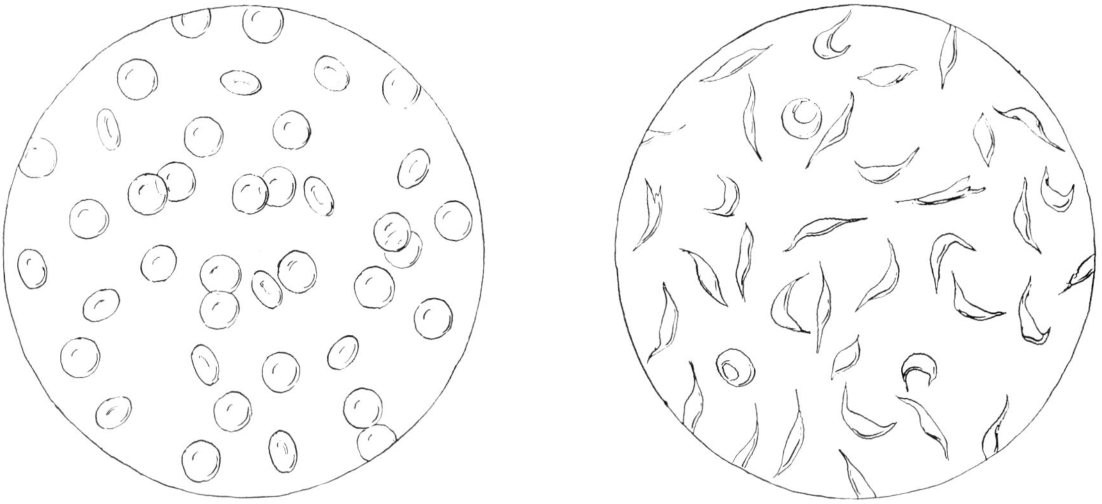

FIG. 1. At the left a drawing of normal human red cells, as seen through the ordinary microscope; at the right, sickled red cells, present in the venous blood of patients with sickle-cell anemia.

lungs, and oxygen molecules attach themselves to the iron atoms of the four heme groups in the hemoglobin molecule, the acid groups coupled with these heme groups become stronger, and liberate hydrogen ions. This increase in acidity of the blood causes some of the bicarbonate ion dissolved in the blood to change to carbonic acid, H_2CO_3, which then breaks down to water and carbon dioxide. The heme-linked acid groups

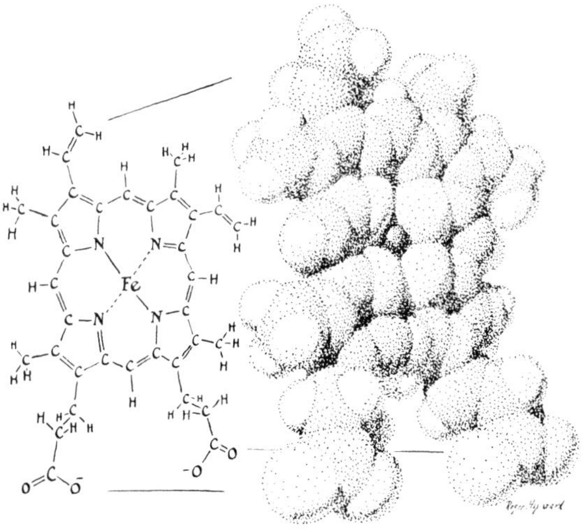

FIG. 2. A heme group, the compound of an iron atom and a protoporphyrin molecule. Four of these hemes are present in the hemoglobin molecule; they are responsible for the red color of hemoglobin, and also are involved in the combination of hemoglobin with oxygen.

assist in this way in the liberation of carbon dioxide in the lungs. Similarly when the blood containing oxyhemoglobin reaches the tissues, and the oxygen is liberated from the hemoglobin, the acid groups become weaker, and the blood becomes more basic, thus increasing the solubility of carbon dioxide in the blood, and assisting in removing it from the tissues.

It may well be that the hemoglobin molecule carries out other functions, but not so much is known about them as about these functions of assisting in the transport of oxygen from the lungs to the tissues and of carbon dioxide from the tissues to the lungs.

There are many different kinds of hemoglobin. All vertebrate animals and many invertebrate animals use hemoglobin as an oxygen carrier, and the hemoglobin molecule is, so far as is known, different for every animal from that of every other animal. The differences may be small, but they are detectable by the sensitive methods of examination of crystals of the substances, and testing with antisera that are produced by injection of hemoglobin of different sorts into animals. The differences are due entirely to the protein part of the molecule, the globin; the heme is the same in all hemoglobins that have yet been investigated.

In some animals, including man, a different sort of hemoglobin is present in the blood of the fetus from that in the blood of the mature animal. Human fetal hemoglobin makes up all of the blood in the human fetus until about two months before birth. At this epoch there usually begins to appear some adult hemoglobin, which normally has completely replaced the fetal hemoglobin by two months after birth.

Our information about the nature of hemoglobin is due to many investigators. The striking sigmoid shape of the oxygen equilibrium curve was discovered by Barcroft, and the first attempts to explain it were made by A. V. Hill. The molecular weight of hemoglobin was determined by Adair, through the measurement of the osmotic pressure of a hemoglobin solution. The effects of the heme-linked acid groups were discovered by Bohr and Hasselbalch. The structure of heme was elucidated by Hans Fischer, and the nature of the bonds between the iron atom and the surrounding atoms was determined by magnetic investigations.[1] In recent years much information about the hemoglobin molecule has been obtained through the x-ray and optical investigations of Perutz and his co-workers. The identification of the groups in the globin that are adjacent to the iron atoms of the hemes, as imidazole rings of histidine side chains, was made by J. B. Conant. Measurements of the heat of oxygenation of hemoglobin solutions, made by J. Wyman, Jr., have been valuable in this identification.

II. THE HEMOGLOBIN MOLECULE IN DISEASE

Until recently it was thought that all adult human beings had the same kind of hemoglobin molecules in their red cells. Then it was discovered [2] that an abnormal form of hemoglobin is present in the red cells of people suffering from the disease sickle-cell anemia, and more recently still other abnormalities have come to light.

Sickle-cell anemia is a hereditary disease that is prevalent among Negroes. It is characterized by the extraordinary aspects of the red cells in the venous blood. The red cells in fresh arterial blood seem to be normal; those in venous blood, or in arterial blood kept for some time away from contact with air, or to which an agent that removes oxygen has been added, have an abnormal form, as shown in figure 1. They are twisted into crescent or sickle-like shapes, with longest dimension considerably greater than that of the normal cell. They become pleochroic, indicating that the hemoglobin molecules have been oriented, and they are quite rigid—the normal cell is almost jelly-like in its flexibility, but when sickling occurs the cell loses this flexibility, so that it has been described as appearing to be as rigid as a crystal of ice as it moves about and abuts against fixed objects. These distorted cells, which seem also to be sticky, have difficulty in passing through capillaries, many of which are so small as just to allow passage of normal erythrocytes in single file. When sickling becomes enhanced, in a crisis of the disease, the capillaries become jammed with red cells, and the flow of blood is prevented. The interference with the flow of blood leads to anoxia, and consequent damage to the tissues. All of the clinical manifestations of the disease seem to be due to this effect. These clinical manifestations include pains in the bones and joints, kidney damage, damage to other organs, poor circulation in the extremities leading to chronic indolent skin ulcers, and poor development of the extremities. The malformed red cells tend to be removed from the circulation by the spleen and leucocytes, and this removal of the red cells leads to the characteristic anemia. The spleen becomes small and fibrotic because of numerous thromboses, so that after several crises of the disease there is little circulation of the blood through it.

From this description of the disease it would seem that it involves a pathology of the red cell, and is to be considered, like other diseases, to be a cellular disease. However, the extraordinary fact that sickling occurs in the venous blood and not in the arterial blood suggested strongly that the hemoglobin molecule is involved. This conclusion was given greater probability by the fact that sickle-cell-anemia blood saturated with carbon monoxide does not contain sickled cells, even in the absence of oxygen; carbon monoxide combines with hemoglobin, to form carbonmonoxyhemoglobin, which is closely similar to oxyhemoglobin in nature, whereas the other properties of carbon monoxide are much different from those of oxygen. As a result of these considerations a careful study was made of the contents of red cells from sickle-cell-anemia patients, in order to see whether or not differences in properties of the hemoglobin present in these red cell contents and normal hemoglobin could be detected. This investigation led to the discovery that the red cells of patients with sickle-cell anemia contain an abnormal hemoglobin, and no normal adult human hemoglobin.[2] Sickle-cell anemia was in this way

[2] Pauling, L., H. A. Itano, S. J. Singer, and I. C. Wells, Sickle cell anemia, a molecular disease, *Science* **110**: 543–548, 1949.

found to be a molecular disease, involving a pathological hemoglobin molecule; it is the first disease to be clearly characterized as a molecular disease.

The property used to show the difference between sickle-cell-anemia hemoglobin and normal adult hemoglobin was its electrophoretic mobility—the motion of molecules, in aqueous solution, in an applied electrical field, as determined with use of the Tiselius electrophoresis apparatus. The electrophoretic patterns for normal adult human hemoglobin, sickle-cell-anemia hemoglobin, sickle-cell-trait hemoglobin, and a mixture of normal adult human hemoglobin and sickle-cell-anemia hemoglobin are shown in figure 3. Under the conditions of this study (phosphate buffer of 0.1 ionic strength and *p*H 6.90), the molecules of normal adult hemoglobin have a negative charge, and move toward the anode, whereas those of sickle-cell-anemia hemoglobin have a positive charge, and move toward the cathode. The difference in electrical charge amounts to about three electronic units per molecule, and corresponds to a difference in isoelectric point of 0.2 *p*H units.

The third pattern in figure 3 is that of the red-cell contents of a person with sickle-cell trait, a carrier of the disease sickle-cell anemia. These people are not ill—they do not show the symptoms of sickle-cell anemia, nor do they have in their venous circulation any large number of sickled cells. Their red cells can be made to sickle, however, though not so easily as those of patients with sickle-cell anemia. It is seen from the electrophoresis pattern that their hemoglobin is a mixture of normal adult human hemoglobin and sickle-cell-anemia hemoglobin. Usually there is about 60 per cent normal adult hemoglobin, and 40 per cent sickle-cell-anemia hemoglobin, but the ratios vary rather widely.[3, 4] Both parents of sickle-cell-anemia patients are in general found to have sickle-cell trait.

The results of the investigation of the hemoglobin of individuals with sickle-cell trait and sickle-cell anemia clarify the genetics of the disease, and lead to conclusions identical with those reached by Neel [5] by direct genetic studies; namely, that the gene responsible for the sickling characteristic is in heterozygous condition in individuals with sickle-cell trait, and in homozygous condition in those with sickle-cell anemia. The existence of normal hemoglobin and sickle-cell-anemia hemoglobin in individuals with sickle-cell trait is, according to this postulate, a result of the presence in the cells of these individuals of an allele for normal hemoglobin and an allele for sickle-cell-anemia hemoglobin. In the cells of patients with sickle-cell anemia there are two doses of the sickle-cell allele and a complete absence of the normal hemoglobin allele, whereas in the cells of normal individuals there are two doses of the normal hemoglobin allele.

The fact that the blood of individuals with sickle-cell trait usually contains normal hemoglobin in somewhat larger amount than sickle-cell-anemia hemoglobin, the ratio of the two being somewhat different for different individuals, has been ascribed recently by Itano [6] to a genetic difference of rate of manufacture of normal hemoglobin, as compared with the rate of manufacture of sickle-cell-anemia hemoglobin.

After the discovery of the existence of an abnormal form of adult human hemoglobin in sickle-

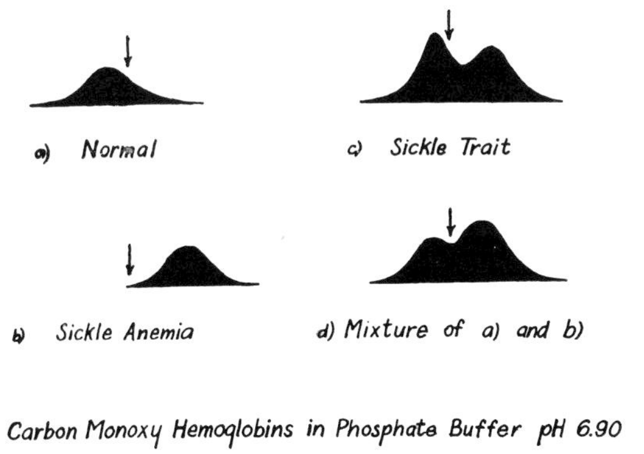

FIG. 3. The electrophoretic patterns for (*a*) normal adult human hemoglobin, (*b*) sickle-cell-anemia hemoglobin, from the red cells of patients with sickle-cell anemia, (*c*) sickle-cell-trait hemoglobin, which is indicated to be a mixture of normal adult human hemoglobin and sickle-cell-anemia hemoglobin, and (*d*) a mixture of normal adult human hemoglobin and sickle-cell-anemia hemoglobin, prepared by mixing the red-cell contents from normal blood and sickle-cell-anemia blood.

[3] Wells, I. C., and H. A. Itano, Ratio of sickle-cell-anemia hemoglobin to normal hemoglobin in sicklemics, *Jour. Biol. Chem.* **188**: 65–74, 1951.

[4] Neel, J. V., I. C. Wells, and H. A. Itano, Familial differences in the proportion of abnormal hemoglobin present in the sickle cell trait, *Jour. Clin. Invest.* **30**: 1120–1124, 1951.

[5] Neel, J. V., The inheritance of sickle cell anemia, *Science* **110**: 64–66, 1949; The inheritance of the sickling phenomenon, with particular reference to sickle cell disease, *Blood* **5**: 389–412, 1951.

[6] Itano, H. A., The inheritance of three molecular species of adult human hemoglobin; a paper submitted for publication.

cell-anemia patients and individuals with sickle-cell trait, two more types of abnormal adult human hemoglobin were discovered. The second abnormal hemoglobin, hemoglobin *c* (the letters *a* and *b* being used to represent normal adult human hemoglobin and sickle-cell-anemia hemoglobin, respectively), was discovered by Itano and Neel.[7] It differs in its isoelectric point from normal adult human hemoglobin by twice as much as does sickle-cell anemia hemoglobin. Its presence in blood is easily shown by an electrophoretic experiment. Four patients, suffering with a disease that had been diagnosed as sickle-cell anemia, were found to contain in their red cells roughly equal amounts of sickle-cell-anemia hemoglobin and abnormal hemoglobin *c*. Investigation of the parents showed one parent to be an individual with sickle-cell trait: his red cells were found to contain roughly equal amounts of normal human hemoglobin and sickle-cell-anemia hemoglobin; whereas the other parent was found to be a carrier of the new abnormal hemoglobin *c*, with red cells containing roughly equal amounts of normal human hemoglobin and hemoglobin *c*. The rules of Mendelian genetics would lead to the prediction that about one quarter of the children should be of genetic type *bc,* and should contain in their red cells approximately equal amounts of sickle-cell-anemia hemoglobin *b* and the new abnormal hemoglobin *c*. The type of anemia resulting from this genetic constitution must be considered a new disease. It is similar to sickle-cell anemia in that the red cells sickle nearly as readily as those of a sickle-cell-anemia patient, and it is presumably this phenomenon that causes the clinical manifestations of the disease. The carriers of the new abnormality, like the carriers of sickle-cell anemia, are not anemic. Moreover, their red cells cannot be made to sickle.

Another abnormal form of hemoglobin, hemoglobin *d,* has also been recognized by Itano.[8] The electrophoretic properties of hemoglobin *d* are very closely similar to those of sickle-cell-anemia hemoglobin. However, the solubility characteristics of hemoglobin *d* are different from those of sickle-cell-anemia hemoglobin, and, moreover, the cells of carriers of hemoglobin *d* cannot be made to sickle. The new disease, shown by individuals with sickle-cell-anemia hemoglobin and hemoglobin *d* in their erythrocytes, is similar in nature to sickle-cell anemia.

So far individuals of seven genetic types, involving hemoglobins *a, b, c,* and *d,* have been discovered: the types represented are *aa* (normal individuals), *ab* (individuals carrying sickle-cell trait), *bb* (patients with sickle-cell anemia), *ac* (carriers of the second abnormal hemoglobin *c*), *bc* (patients with the first new disease, involving the inheritance of a sickle-cell-anemia allele and an allele for the second abnormal hemoglobin, *c*), *ad* (carriers of the third abnormal hemoglobin, *d*), and *bd* (patients with the second new disease, resulting from the inheritance of a sickle-cell-anemia allele and an allele of the third abnormal hemoglobin, *d*). Individuals of types *cc,* homozygous in abnormal hemoglobin *c,* and *dd,* homozygous in abnormal hemoglobin *d,* have not yet been discovered, nor have individuals of type *cd,* carrying both of these two abnormal alleles.

Hematological abnormalities involving human fetal hemoglobin have recently been discovered. Last year it was reported by Liquori[9] that the hemoglobin of some individuals with thalassemia major (Cooley's anemia, Mediterranean anemia) contained approximately 50 per cent normal hemoglobin and 50 per cent human fetal hemoglobin. Further studies by Alexander Rich[10] led to the discovery of two patients with thalassemia major whose red cells contained 100 per cent (to within 5 per cent) of fetal hemoglobin, although these patients were past the fetal stage (age two years). Itano[11] has found that human fetal hemoglobin is present in small amount in the red cells of sickle-cell-anemia patients, and, together with sickle-cell-anemia hemoglobin and normal hemoglobin, in the blood of individuals who have inherited the thalassemia gene and the sickle-cell-anemia gene. There thus exists strong indication that the presence of severe anemia can cause the continued manufacture of fetal hemoglobin, in an effort to counteract the anemia. Fetal hemoglobin has also been reported by Singer, Chernoff, and Singer[12]

[7] Itano, H. A., and J. V. Neel, A new inherited abnormality of human hemoglobin, *Proc. Nat. Acad. Sci.* **36**: 613–617, 1950.

[8] Itano, H. A., A third abnormal hemoglobin associated with hereditary hemolytic anemia, *Proc. Nat. Acad. Sci.* **37**: 775–784, 1951.

[9] Liquori, A. N., Presence of foetal haemoglobin in Cooley's anemia, *Nature* **167**: 950–051, 1951.

[10] Rich, A., Studies on the hemoglobin of Cooley's anemia and Cooley's trait, *Proc. Nat. Acad. Sci.* **38**: 187–196, 1952.

[11] Itano, H. A., The identification of fetal hemoglobin in sickle-cell anemia by electrophoretic, spectrophotometric, and solubility studies; unpublished investigation.

[12] Singer, K., A. I. Chernoff, and L. Singer, Studies on abnormal hemoglobins, *Blood* **5**: 413–435, 1951.

to be present in the blood of adult patients suffering from anemias secondary to leukemia or carcinoma.

III. THE STRUCTURE OF THE HEMOGLOBIN MOLECULE

Although it has not been found possible as yet to make a complete structure determination for any hemoglobin molecule, a large amount of evidence bearing on the problem has been obtained, and many features of the structure can now be discussed with confidence.

Much of our knowledge about this molecule has been obtained through the vigorous efforts of M. Perutz and his collaborators at Cambridge University. Their x-ray studies [13, 14] have led to the conclusion that the hemoglobin molecule (horse hemoglobin) is about 57 A long, and that its dimensions in the other two directions are about 40 or 50 A. Moreover, strong evidence has been obtained by Perutz that the molecule consists of rods of polypeptide chains extending in the 57-A direction. These rods are about 10.5 A in diameter, and they are packed together in approximate hexagonal packing. There is considerable experimental evidence supporting the suggestion [15] that the rods are based upon a helical configuration of polypeptide chains, the configuration being that of the α helix, in which the coiling, with formation of hydrogen bonds between planar amide groups, is such as to correspond to about 3.7 residues per turn of the helix.[16, 17] Strong support of this suggestion was then obtained by Perutz [18] through the discovery that suitably oriented crystals of hemoglobin give an x-ray reflection with spacing 1.50 A, representing the collaboration of successive residues in the helix, which are spaced 1.50 A apart along the helical axis. These helical rods are indicated in figure 4, the details of their arrangement being, however, hypothetical. There are only five or six polypeptide chains in the molecule, and accordingly some of the rods must be connected with one another, as a single polypeptide chain.

In addition to the polypeptide chains of the protein part of the molecule, globin, the hemoglobin molecule contains four heme groups, the structure of which is completely known. These molecules are conjugated systems, and are essentially planar in configuration. Their orientation in a crystal can be determined by measurement of the pleochroism of the crystal, since the light is absorbed only when the electric vector of the light wave has a component in the plane of the molecule. It was found by Perutz [19] that in crystals of horse carbonmonoxyhemoglobin all of the heme groups lie in parallel orientations, their planes being perpendicular to the 57-A axis of the hemoglobin molecule, which is also an axis of the crystal. We thus know that the four hemes are to be attached to the globin in such a way that their planes are normal to the direction of the polypeptide rods.

In figure 4 the hemes are shown with this orientation, but not attached at the ends of the hemoglobin molecule; instead they are shown inserted in slits between layers of globin. The most direct evidence in support of this position is provided by measurements of the combining power of hemoglobin with alkyl isocyanides.[20] It has been assumed for twenty years, since the suggestion by Conant, that a heme group is attached by the iron atom, on one side of the plane of the group, to an imidazole nitrogen atom of a histidine side chain, and that the oxygen molecule or other ligand attaches itself to the iron atom on the other side of the plane of the heme group. Detailed information about the nature of the bonds formed by the iron atom has been obtained through the investigation of the magnetic properties of hemoglobin and oxyhemoglobin,[1] and it is known that in hemoglobin itself the iron atom forms bonds of essentially ionic nature with adjacent atoms, whereas in oxyhemoglobin and similar compounds the iron atom forms six covalent bonds, which are directed towards the corners of an octahedron. These six bonds are formed with the four nitro-

13 Boyes-Watson, J., E. Davidson, and M. F. Perutz, An x-ray study of horse methaemoglobin, *Proc. Roy. Soc.* **A191**: 83–132, 1947.

14 Perutz, M. F., An x-ray study of horse methaemoglobin. II, *Proc. Roy. Soc.* **A195**: 474–499, 1949.

15 Pauling, L., and R. B. Corey, The polypeptide chain configuration in hemoglobin and other globular proteins, *Proc. Nat. Acad. Sci.* **37**: 282–285, 1951.

16 Pauling, L., R. B. Corey, and H. R. Branson, The structure of proteins: two hydrogen-bonded helical configurations of the polypeptide chain, *Proc. Nat. Acad. Sci.* **37**: 205–211, 1951.

17 Pauling, L., and R. B. Corey, Atomic coordinates and structure factors for two helical configurations of polypeptide chains, *ibid.* **37**: 235-240, 1951.

18 Perutz, M. F., New x-ray evidence on the configuration of polypeptide chains, *Nature* **167**: 1053–1054, 1951.

19 Perutz, M. F., Absorption spectra of single crystals of haemoglobin in polarized light, *Nature* **143**: 731, 1939.

20 St. George, R. C. C., and L. Pauling, The combining power of hemoglobin for alkyl isocyanides, and the nature of the heme-heme interactions in hemoglobin, *Science* **114**: 629–634, 1951.

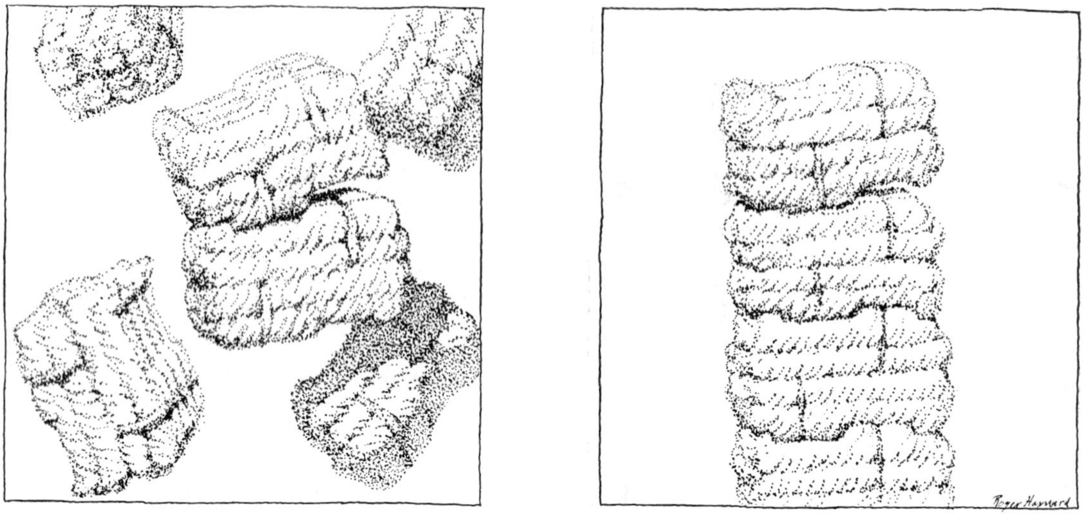

FIG. 4. A drawing indicating some of the features of the structure of the hemoglobin molecule, and the postulated mechanism of sickling of sickle-cell-anemia erythrocytes. The four hemes are indicated to be contained within slits in the hemoglobin molecule, their planes being perpendicular to the axes of the helical rods in the protein. At the right the molecules of sickle-cell-anemia hemoglobin, without oxygen attached to the hemes, are shown as having self-complementary configurations, which permit them to aggregate into long strings of molecules. At the left, the addition of oxygen or other ligand to the hemes is shown as swelling them enough to destroy the self-complementariness of the molecules, thus interfering with the formation of the aggregates.

gen atoms of the porphyrin molecule, which lie in the plane of the molecule, and with the nitrogen atom of the imidazole ring, to one side of the plane, and the oxygen molecule or other ligand, to the other side of the plane. The fact that the compounds ethyl isocyanide, isopropyl isocyanide, and tertiary butyl isocyanide show successively smaller combining powers with hemoglobin, although their combining powers with free heme groups are essentially the same, provides strong evidence that there is steric interference with the attachment of the isocyanide molecule, and this steric interference can be produced only by a part of the globin molecule. Accordingly the conclusion is reached that there is part of the globin molecule on each side of the heme group, as sketched in figure 4. Many other pieces of information are compatible with this structure, and difficult to interpret in terms of a structure in which the heme groups are attached to the surface of the globin. In particular, this model provides an explanation of the nature of the oxygen combining curve of hemoglobin—the fact that a second oxygen molecule, and a third and fourth, attach themselves to the hemoglobin molecule more readily than does the first.

We may now ask about the nature of the difference in structure of the abnormal human hemoglobins and normal adult human hemoglobin. First, it is found that the heme groups in sickle-cell-anemia hemoglobin, which has been investigated more than hemoglobins *c* and *d*, are identical with protoheme, the heme present in normal hemoglobin. The abnormality is thus to be attributed to the globin part of the hemoglobin.

An obvious suggestion is that there is a difference in amino-acid composition of sickle-cell-anemia hemoglobin and normal adult human hemoglobin. Determination of the amino-acid composition of these two hemoglobins has, however, led to the discovery of no abnormality.[21] A difference of one or two residues of one or another amino acid might be permitted by the analyses, but no difference is required by them. In addition, studies of the end groups have shown that there are present in sickle-cell-anemia hemoglobin, as well as in normal adult human hemoglobin, about five or six end groups with free amino groups, and that the amino acid represented, valine, is the same for sickle-cell-anemia hemoglobin as for normal adult human hemoglobin.[22] The

[21] Schroeder, W. A., L. M. Kay, and I. C. Wells, Amino acid composition of hemoglobins of normal Negroes and sickle cell anemics, *Jour. Biol. Chem.* **187**: 221–240, 1950.

[22] Havinga, E., and F. C. Green, End-group analyses of sickle-cell hemoglobin; unpublished investigation.

uncertainty in this investigation is about one residue.

Electrophoretic studies on the globins obtained from sickle-cell-anemia hemoglobin and normal adult human hemoglobin by careful removal of the hemes have shown that there is a difference in electrical charge on the globin molecules approximately the same as that on the two kinds of hemoglobin.[23] If the globin is treated with guanidinium chloride in solution, and the guanidinium chloride is then removed by dialysis, the resultant globins, which might be described as denatured globins, have a changed mobility, this mobility being identical to within the experimental error for sickle-cell-anemia denatured globin and normal adult human denatured globin.[23] There is thus indication that the polypeptide chains involved in sickle-cell-anemia globin are the same as those involved in normal adult human globin, and that the difference in structure between these molecules is simply a difference in the way in which the polypeptide chains are folded. In order to account for the difference in electrical charge it is necessary to assume that the difference in folding changes the acid strength of some of the groups in the molecule. If this hypothesis is correct, we shall have to conclude that there is a gene in the cells of the human body that is responsible for the folding of the polypeptide chains, in the proper way, in the manufacture of adult hemoglobin.

In the paper announcing the discovery of sickle-cell-anemia hemoglobin [2] the mechanism of the sickling process was discussed in the following way.

> It is likely that it is the globins rather than the hemes of the two hemoglobins that are different. Let us propose that there is a surface region on the globin of the sickle-cell-anemia hemoglobin molecule which is absent in the normal molecule and which has a configuration complementary to a different region of the surface of the hemoglobin molecule. This situation would be somewhat analogous to that which probably exists in antigen-antibody reactions.[24] The fact that sickling occurs only when the partial pressures of oxygen and carbon monoxide are low suggests that one of these sites is very near to the iron atom of one or more of the hemes, and that when the iron atom is combined with either one of these gases, the complementariness of the two structures is considerably diminished. Under the appropriate conditions, then, the sickle-cell-anemia hemoglobin molecules might be capable of interacting with one another at these sites sufficiently to cause at least a partial alignment of the molecules within the cell, resulting in the erythrocyte's becoming birefringent, and the cell membrane's being distorted to accommodate the now relatively rigid structures within its confines. The addition of oxygen or carbon monoxide to the cell might reverse these effects by disrupting some of the weak bonds between the hemoglobin molecules in favor of the bonds formed between gas molecules and iron atoms of the hemes.

In the discussion of the combining power of hemoglobin with alkali isocyanides, and the postulate that the heme groups are buried within the globin of the hemoglobin molecule,[20] a further discussion of the nature of the process of sickling was given, as follows:

> Our postulate provides an obvious explanation of the action of oxygen in preventing the sickling of sickle-cell-anemia erythrocytes. We have visualized the sickling process as one in which complementary sites on adjacent hemoglobin molecules combine. It was suggested that erythrocytes containing oxyhemoglobin or carbonmonoxyhemoglobin do not sickle because of steric hindrance of the attached oxygen or carbon monoxide molecule. This steric-hindrance effect might be the distortion of the complementary sites through the forcing apart of layers of protein, as is suggested by the isocyanide experiments.

In figures 4 and 5 the postulated mechanism of interaction of sickle-cell-anemia hemoglobin molecules is illustrated. Because of the assumed complementariness in structure, the molecules of sickle-cell-anemia hemoglobin (without oxygen molecules or other molecules attached) could interact to form long chains of molecules. These long chains of molecules could attract one another into parallel orientation, causing the formation of a crystal or liquid crystal. Evidence has recently been obtained by Harris [25] in support of this picture, through the observation that solutions of sickle-cell-anemia hemoglobin, containing over 10 per cent of the protein, form liquid crystals of the nematic type, with the shape of double circular cones. Also, Perutz and Mitchison [26] have made

[23] Havinga E., and H. A. Itano, Electrophoretic studies on the globins of sickle-cell-anemia hemoglobin and normal adult human hemoglobin; unpublished investigation.

[24] Pauling, L., A theory of the structure and process of formation of antibodies, *Jour. Am. Chem. Soc.* **62**: 2643–2657, 1940.

[25] Harris, J. W., Studies on the destruction of red blood cells. VIII. Molecular orientation in sickle cell hemoglobin solutions, *Proc. Soc. Exptl. Biol. Med.* **75**: 197–201, 1950.

[26] Perutz, M. F., and J. M. Mitchison, State of haemoglobin in sickle-cell anemia, *Nature* **166**: 677–679, 1950.

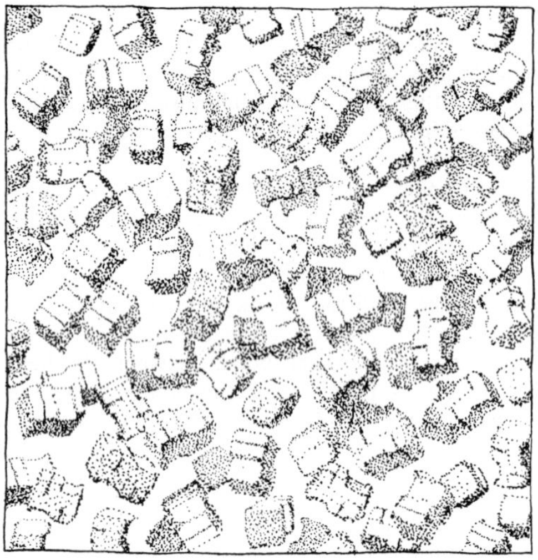

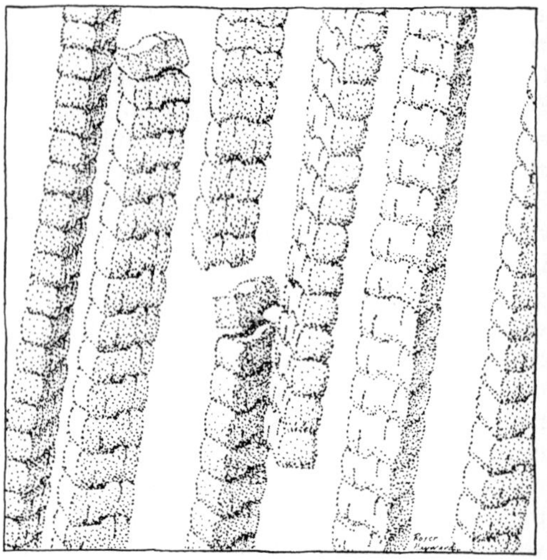

FIG. 5. At the left, molecules of normal hemoglobin or of oxygenated sickle-cell-anemia hemoglobin are shown, with random orientations, and at about the average distance apart characteristic of red-cell contents. At the right long strings of molecules of deoxygenated sickle-cell-anemia hemoglobin are shown, assuming the parallel orientation characteristic of the nematic liquid crystals that presumably form within the red cells in the venous blood of patients with sickle-cell anemia, and twist the red cells into the abnormal shape characteristic of the disease.

a quantitative study of the pleochroism of sickled cells, and have shown that the pleochroism is compatible with this postulate, the orientation of the heme groups being as indicated in figure 4, namely, the planes of the heme groups being parallel to the long axis of the sickled cell.

In the postulated mechanism, the introduction of an oxygen molecule or carbon monoxide molecule causes an effective increase in thickness of the heme groups, and, as shown in figure 4, destroys the complementariness in configuration of the surfaces of the molecule, and thus prevents the formation of linear aggregates, and the sickling of the cells.

On the basis of the available information we may surmise that the folding of the polypeptide chains in the globin of normal adult human hemoglobin is such that this complementariness in structure is not present, or at any rate is not so pronounced. The abnormal hemoglobins *c* and *d* seem to be intermediate in nature; we assume that there is an approximate complementariness shown by these molecules, which permits them to fit into the aggregates, together with sickle-cell-anemia hemoglobin molecules, although their own tendency to form aggregates is not sufficiently great to cause cells containing either of these abnormal hemoglobins together with normal adult human hemoglobin to sickle.

Thus at the present time we have a considerable amount of knowledge of the structure of the normal adult human hemoglobin molecule, and of the abnormal forms of this molecule that are responsible for three known molecular diseases. Although the knowledge of the structure of these molecules is as yet far from complete, it has led to the suggestion of possible methods of chemotherapeutic treatment of the diseases, which are now under investigation. In the course of time, through continued attack on the problem, the complete structure of the hemoglobin molecule will be discovered, and the precise nature of the abnormalities that are present in the molecules of sickle-cell-anemia hemoglobin, abnormal hemoglobin *c*, and abnormal hemoglobin *d*; we may feel confident that this knowledge will permit the deduction of improved therapeutic methods, and that in the future a similar attack on other diseases, through the determination of the structure of the molecules that are involved, can also be made.

I am indebted to Dr. Harvey A. Itano for his collaboration in work on hemoglobin and for assistance in the preparation of this paper.

IV. SUMMARY

It has been discovered that, in addition to fetal human hemoglobin and normal adult human hemoglobin, three abnormal forms of hemoglobin occur, in the red cells of certain individuals. One of these abnormal hemoglobins is associated with the disease sickle-cell anemia, and the two others are associated with two newly recognized hereditary anemias, resembling sickle-cell anemia. These diseases are to be considered as not diseases of the red cell itself, but rather diseases involving molecular abnormalities.

Although it has not yet been found possible to determine completely the structure of the molecules of any form of hemoglobin, a considerable amount of information about the structure of these molecules has been obtained. On the basis of this information it is possible to suggest a plausible mechanism whereby the abnormal hemoglobin molecules produce the clinical manifestations of the diseases.

Abnormality of Hemoglobin Molecules in Hereditary Hemolytic Anemias

LINUS PAULING
Professor of Chemistry, California Institute of Technology, Pasadena, California

Reprinted from
THE HARVEY LECTURES
Series XIL, 1953–54
ACADEMIC PRESS INC., 1955

ABNORMALITY OF HEMOGLOBIN MOLECULES IN HEREDITARY HEMOLYTIC ANEMIAS*

LINUS PAULING

Professor of Chemistry, California Institute of Technology, Pasadena, California

TWENTY years ago, after having worked for a decade on the determination of the structure of relatively simple inorganic and organic molecules, I became interested in hemoglobin. This interest arose from the consideration of the structural origin of the sigmoid oxygen equilibrium curve.[1] It was soon extended to include the denaturation of hemoglobin and other proteins[2] and the magnetic properties of hemoglobin and its derivatives.[3-11] The study of magnetic properties has been especially fruitful in providing information about the nature of the bonds formed by the iron atoms in hemoglobin with the neighboring atoms of the porphyrin ring system, the globin, and attached molecules such as the oxygen molecule.[3,12-15]

The discovery of the abnormal hemoglobins was the result of the consideration of hypothetical molecular mechanisms of the disease. In the spring of 1945 I, together with eight men from medical schools of the country, was serving as a member of the Medical Advisory Committee which assisted in the preparation of the Bush Report.[16] One evening Dr. William B. Castle, Professor of Medicine in Harvard University, mentioned to the other members of the Committee the disease sickle-cell anemia, with which he had had some experience. He told about the discovery of the disease by Dr. J. B. Herrick, in 1910,[17] and described the characteristic change in shape of the red corpuscles and the effect of oxygen in preventing the sickling and of carbon dioxide in accelerating it. I suggested that the action of carbon dioxide was to accelerate the dissociation of oxygen from oxyhemoglobin, through the Bohr-Hasselbalch effect (it had in fact been clearly stated by Hahn and Gillespie[18] in 1927 that sickling occurs only when the partial pres-

* Lecture delivered April 29, 1954.

sure of oxygen is small), and I pointed out that the relation of sickling to the presence of oxygen clearly indicated that the hemoglobin molecules in the red cell are involved in the phenomenon of sickling, and that the difference between sickle-cell-anemia red corpuscles and normal red corpuscles could be explained by postulating that the former contain an abnormal kind of hemoglobin, which when deoxygenated has the power of combining with itself into long rigid rods, which then twist the red cell out of shape. The opportunity to test this idea arose when Dr. Harvey A. Itano came to the California Institute of Technology, in the fall of 1946. He had been a student of Professor Edward A. Doisy, of St. Louis University School of Medicine, where Dr. Itano had received his M.D. degree in 1945. Dr. Doisy suggested that he work with me, and the opportunity for doing so arose in the course of his year as an intern, when he was awarded an American Chemical Society Predoctoral Fellowship in Chemistry, for the three years 1946 to 1949. In a letter to Dr. Itano I suggested that he investigate the hemoglobin from the red cells of sickle-cell-anemia patients, in order to see whether it was different from normal adult human hemoglobin. On his arrival in Pasadena in September, 1946, he began this investigation. He verified the published reports[18] that carbonmonoxyhemoglobin, like oxyhemoglobin, prevents sickling of the red cells, and found that some other hemoglobin derivatives, including alkyl isocyanide-ferrohemoglobin, ferrihemoglobin, ferrihemoglobin azide, and ferrihemoglobin cyanide similarly prevent sickling. He developed a rapid diagnostic test for sickle-cell anemia and sickle-cell trait, based on the use of a chemical reducing agent.[19] Most of the properties of the hemoglobin from the blood of sickle-cell-anemia patients were found to be the same, to within the error of determination, as those of hemoglobin from normal individuals, but it was finally clearly shown, by careful measurement of electrophoretic mobility, that the blood of the patients contains nearly 100 per cent of an abnormal hemoglobin, differing from normal adult human hemoglobin, and that the blood of the parents of patients contains an approximately half-and-half mixture of the abnormal hemoglobin and normal adult human hemoglobin.[20] This electrophoretic work was carried out with the collaboration of Dr. S. J. Singer and Dr. Ibert C. Wells.

THE INHERITANCE OF SICKLE-CELL ANEMIA

The electrophoretic patterns reported in the first publication[20] are shown in Fig. 1. They were made by electrophoresis for 20 hours at a potential gradient of 4.73 volts per centimeter of solutions of carbonmonoxyhemoglobins in phosphate buffer of 0.1 ionic strength and pH 6.90. The peaks *a* and *b*, representing normal hemoglobin and hemoglobin from the red cells of patients with sickle-cell anemia, are single peaks, corresponding in each

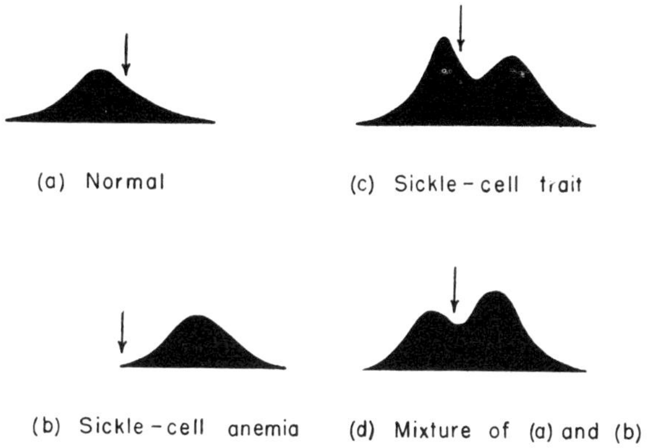

FIG. 1. Longsworth scanning diagrams of carbonmonoxyhemoglobin in phosphate buffer of 0.1 ionic strength and pH 6.90, taken after 20 hours electrophoresis at a potential gradient of 4.73 volts per centimeter.

case to an electrophoretically homogeneous material. The electrophoretic mobilities are different for the two hemoglobins; in fact, at this pH the molecules of normal hemoglobin move toward the anode, showing that they have a negative electric charge, and those of sickle-cell-anemia hemoglobin move toward the cathode, showing that they have a positive charge. The isoelectric points in phosphate buffer of ionic strength 0.1 were found to be 6.87 (in pH units) for normal adult human carbonmonoxyhemoglobin and 7.09 for sickle-cell-anemia hemoglobin. The difference between these values is nearly the same as that between the observed values 6.68 for normal ferrohemoglobin and 6.91 for sickle-cell-anemia ferrohemoglobin.

The electrophoretic diagram of a solution containing a mixture of normal carbonmonoxyhemoglobin and sickle-cell-anemia carbonmonoxyhemoglobin, in equal amounts, is shown as *d* in Fig. 1, and that of hemoglobin from the red cells of a parent of a patient is shown as *c*. The blood from which this hemoglobin was obtained showed the characteristic properties of sickle-cell trait (sicklemia); the cells could be made to sickle, on removal of oxygen, but less readily than the cells of a sickle-cell-anemia patient. It is seen that the sicklemic hemoglobin is a mixture of two hemoglobins, presumed to be normal adult human hemoglobin and sickle-cell-anemia hemoglobin, with the normal hemoglobin present in an amount somewhat greater than 50 per cent.

The indication of a genetic basis for the sickling of erythrocytes had been recognized by Emmel[21] in 1917; and Taliaferro and Huck,[22] at a time when the distinction between sicklemia and sickle-cell anemia was not clearly understood, suggested that a single dominant gene was involved. The inheritance of sickle-cell disease was then clarified by Neel, who in 1947[23] had suggested "that there is present in the colored population a certain factor which, when heterozygous, may have no discernible effect, but usually results in sickling, and, when homozygous, tends to result in sickle cell anemia." In 1949 he reported[24] that every one of 42 tested parents of children with sickle-cell anemia was found to be sicklemic, their blood containing red cells which could be made to sickle, though less readily than that of the sickle-cell-anemia patients. He concluded that sickle-cell anemia is the result of the homozygous condition of the sickle-cell gene, and sicklemia the result of the heterozygous condition. Beet[25] arrived at the same conclusion independently and almost simultaneously. The electrophoretic patterns shown in Fig. 1 had permitted this inference to be drawn before Neel's paper and Beet's paper were published. Moreover, the gene responsible for the sickling process could be identified with an alternative pair of alleles of which neither one is recessive or dominant, one allele being responsible for a part of the process of manufacture of normal adult human hemoglobin, and the other for the manufacture of sickle-cell-anemia hemoglobin. The fact that all the red cells of a sicklemic individual can be made to sickle by removal of oxygen shows that the cells are not of two classes, one containing normal hemoglobin and the other abnormal hemoglobin, but that each cell contains a mixture of the two kinds of hemoglobin. The presence of a larger amount of normal than of abnormal hemoglobin in the blood of sicklemic individuals indicates that the process of manufacture of the abnormal hemoglobin is somewhat less efficient than that of normal hemoglobin. It was suggested in the first paper on sickle-cell-anemia hemoglobin[20] that the two genes in the heterozygous individual might compete for a common substrate in the synthesis of two different enzymes essential to the production of the two different hemoglobins, or that competition for a common substrate might occur at a later stage in the series of reactions leading to the synthesis of the two hemoglobins themselves.

An investigation of the amount of abnormal hemoglobin in the blood of sicklemic individuals was carried out by Wells and Itano,[26] who found, using the electrophoretic method, that the amount of sickle-cell-anemia hemoglobin varied from 24 per cent to 45 per cent in 42 individuals with sicklemia. Neel, Wells, and Itano[27] reported a study of 32 sicklemic individuals who were members of 7 Negro families, comprising 74 individuals altogether. The amounts of abnormal hemoglobin, ranging from 22.3 per cent to 45.2 per cent, showed significant differences between family means. A postulate to explain the apparent inheritance of a factor determining the amount of abnormal hemoglobin in sicklemic blood was made by Itano.[28] He suggested that the differences can be attributed to differences in the rate of synthesis of normal hemoglobin, and that the evidence requires that there be at least three rate-determining modifications of the mechanism of synthesis of normal hemoglobin.

Additional contributions to the problem of the genetics of normal and abnormal hemoglobins have been made by Neel[29–31] and other workers.[32,33]

The Properties of Sickle-Cell-Anemia Hemoglobin and the Mechanism of Sickling

Sickle-cell-anemia hemoglobin is closely similar to normal adult human hemoglobin in most of its properties.[20] The two proteins have approximately the same sedimentation and diffusion constants, and hence nearly the same molecular weights. The acid-base titration curves of both hemoglobins in the neighborhood of neutrality are linear, a change of 1 pH unit of the solution being associated with the change in charge of the hemoglobin of about 13 electronic charges per molecule. The normal molecule has about three more negative charges than the abnormal molecule in this region. In the search for the structural basis for this difference samples of porphyrin dimethyl esters were prepared from the two hemoglobins, and the samples were shown by their x-ray powder photographs and by identity of their melting points and mixed melting points to be identical. The difference in structure was hence attributed to a difference in the globins.

An investigation by Schroeder, Kay, and Wells[34] of the amino acid composition of normal adult human hemoglobin (from normal Negro individuals) and sickle-cell-anemia hemoglobin gave results indicating that the hemoglobins do not differ with respect to their content of basic and acidic amino acids; the investigators concluded that sickle-cell-anemia hemoglobin probably contains slightly less leucine and more serine than normal hemoglobin, and possibly less valine and more threonine. These amino acids do not contribute directly to the net charge of the proteins, but they might affect the folding or coiling of the polypeptide chains in such a way as to change the acid or basic constants of other groups. Havinga[35] investigated the phosphorus content, optical rotation, ease of separation of hemes and globin, and number of terminal amino acid residues of normal adult human hemoglobin and sickle-cell-anemia hemoglobin, and found no significant differences between the two proteins. Globins carefully prepared from the two hemoglobins were investigated electrophoretically by Havinga and Itano[36] and found to have the same difference in electrophoretic mobility as the hemoglobins themselves. On denaturation by treatment with 4 *N* guanidinium chloride for 1 hour at 4° C. and removal of the guanidinium chloride by dialysis, the globins were found to have increased markedly in heterogeneity, and to have essentially the same electrophoretic properties. These results indicate that the normal and abnormal hemoglobin molecules might be composed of the same polypeptide chains, folded, however, in different ways, and that on denaturation with guanidinium ion the resulting denatured proteins have the same complex of configurations. The interesting possibility exists that the gene responsible for the sickle-cell abnormality is one that determines the nature of the folding of polypeptide chains, rather than their composition.

It was pointed out by Sherman[37] in 1940 that sickled red cells are observed under the polarizing microscope to be birefringent, whereas normal cells are optically isotropic. Ponder[38] suggested, on the basis of this observation, that in sickled cells the hemoglobin molecules assume an orderly or paracrystalline arrangement, which is responsible for the sickling. A detailed mechanism of the sickling process was suggested in the first paper on sickle-cell-anemia hemoglobin,[20] as follows: "We can picture the mechanism of the sickling process in the following way. It is likely that it is the globins rather than the hemes of the two hemoglobins that are different. Let us propose that there is a surface region on the globin of the sickle-cell-anemia hemoglobin molecule which is absent in the normal molecule and which has a configuration complementary to a different region of the surface of the hemoglobin molecule. This situation would be somewhat analogous to that which very probably exists in antigen-antibody reactions.[39] The fact that sickling occurs only when the partial pressures of oxygen and carbon monoxide are low suggests that one of these sites is very near to the iron atom of one or more of the hemes, and that when the iron atom is combined with either one of these gases, the complementariness of the two structures is considerably diminished. Under the appropriate conditions, then, the sickle-cell-anemia hemoglobin molecules might be capable of interacting with one another at these sites sufficiently to cause at least a partial alignment of the molecules within the cell, resulting in the erythrocyte's becoming birefringent, and the cell membrane's being distorted to accommodate the now relatively rigid structure within its confines. The addition of oxygen or carbon monoxide to the cell might reverse these effects by disrupting some of the weak bonds between the hemoglobin molecules in favor of the bonds formed between gas molecules and iron atoms of the hemes."

A more detailed discussion of the effect of oxygen was made possible by the results of an investigation of the combination of hemoglobin with alkyl isocyanides.[40] It was found that ethyl isocyanide, isopropyl isocyanide, and tertiary butyl isocyanide differ greatly in their combining powers with hemoglobin, although they have essentially the same combining power with heme; and this fact was interpreted as showing that the four hemes in the hemoglobin molecule are buried within the globin: "Our postulate provides an obvious explanation of the action of oxygen in preventing the sickling of sickle-cell-anemia erythrocytes. We have visualized the sickling process[20] as one in which complementary sites on adjacent hemoglobin molecules combine. It was suggested that erythrocytes containing oxyhemoglobin or carbonmonoxyhemoglobin do not sickle because of steric hindrance of the attached oxygen or carbon monoxide molecule. This steric-hindrance effect might be distortion of the complementary sites through forcing apart of layers of protein, as is suggested by the isocyanide experiments."

Substantiation of this picture was soon obtained through microscopic investigations. Rebuck, Sturrock, and Monaghan,[41] substantiating the work of Sherman,[37] observed that in the early stages of sickling the intracellular hemoglobin forms anisotropic aggregates, suggestive of incipient crystallization. Perutz and Mitchison,[42] at the suggestion of Dr. C. A. Stetson of the Rockefeller Institute, compared the dichroism of sickled cells and hemoglobin crystals, and additional studies of the same sort were reported by Perutz, Liquori, and Eirich.[43] These investigators found that the dichroism of the sickled cells corresponds to an orientation of the hemoglobin molecules such that the normal to the plane of the heme groups is perpendicular to the long axis of the crystal needles and of the sickled cells. (This statement is based on the paper of Perutz, Liquori, and Eirich;[43] there is some conflict with the earlier paper.[42]) They also pointed out[42,43] that the solubility of sickle-cell-anemia hemoglobin is much smaller than that of normal hemoglobin or of either normal oxyhemoglobin or sickle-cell-anemia oxyhemoglobin. A detailed study of the solubilities of mixtures of sickle-cell-anemia hemoglobin and other hemoglobins has been made by Itano,[44] who has shown that a solubility measurement provides a simple way of determining roughly the amount of sickle-cell-anemia hemoglobin present in a mixture of hemoglobins. A most significant investigation was then reported by

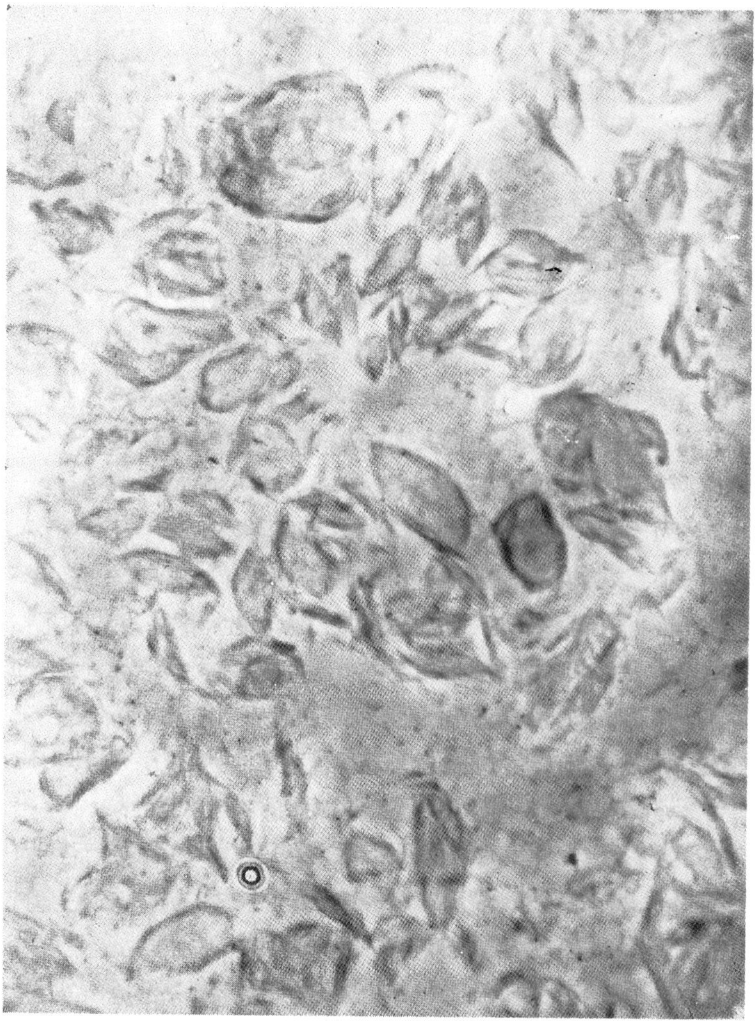

FIG. 2. Hemoglobin formed in stroma-free solutions of deoxygenated sickle-cell-anemia hemoglobin. Phase photomicrography, × 375. From John W. Harris, *Proc. Soc. Exptl. Biol. and Med.* **75**, 197 (1950).

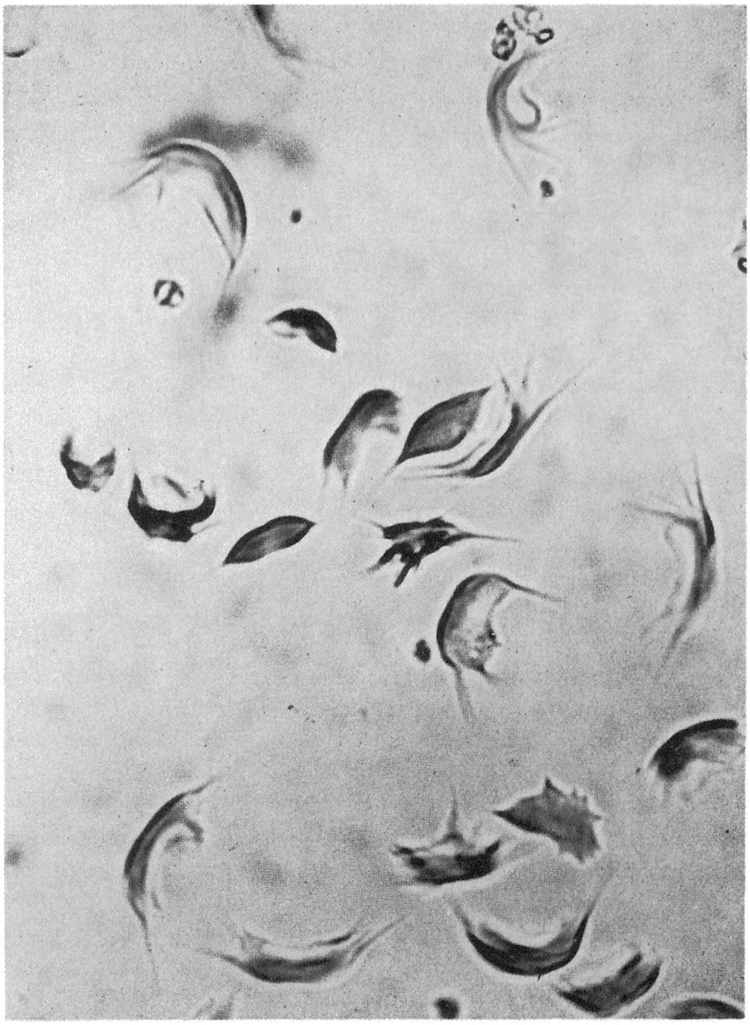

Fig. 3. Sickled erythrocytes in oxygen-depleted whole blood from a patient with sickle-cell anemia, demonstrating the similarities in shape to tactoids formed in stroma-free solutions of their deoxygenated hemoglobin. Phase photomicrography, × 375. From John W. Harris, *Proc. Soc. Exptl. Biol. and Med.* **75**, 197 (1950).

Harris.[45] He showed that a stroma-free solution of sickle-cell-anemia hemoglobin with concentration 15.2 or 23.5 g. per 100 ml. on deoxygenation forms birefringent spindle-shaped bodies varying in length from 1 to 15 μ. He identified these bodies as tactoids (liquid crystals) of the nematic type. His photomicrograph is shown as Fig. 2, which may be compared with Fig. 3, a similar photomicrograph (375 $\times$ magnification) of sickled erythrocytes.

It hence seems probable that sickle-cell anemia can be described as a molecular disease, resulting from the difference in molecular structure of sickle-cell-anemia hemoglobin and normal adult human hemoglobin. The properties of the abnormal hemoglobin are such that when deoxygenated the molecules combine with one another to form long molecular strings, which, through intermolecular attraction, aggregate into tactoids. These tactoids have enough mechanical strength to distort the red cell, changing the viscosity of the blood, and causing the clinical and pathological manifestations of the disease.

The close approximation in structure of sickle-cell-anemia hemoglobin to normal adult human hemoglobin is strikingly shown by the antigenic properties, investigated by Goodman and Campbell.[46] They found that, whereas large differences in serological specificity are shown by human fetal hemoglobin and human adult hemoglobin, suggesting that only a few antigenic groups are shared in common by these two hemoglobins, only a small difference in antigenic specificity could be demonstrated between sickle-cell-anemia hemoglobin and normal adult human hemoglobin. Antiserums obtained by injection of rabbits with these two forms of adult human hemoglobin showed no differences in antigenic specificity. The injection of chickens with these two forms of adult hemoglobin produced antiserums with a significant, though small, difference in properties, indicating that the two hemoglobins have a predominance of antigenic groups in common, but that a small number are different.

The results of an investigation by Ingbar and Kass[47] of the number of titratable sulfhydryl groups (two groups per molecule in normal hemoglobin, and three in sickle-cell-anemia hemoglobin) may provide an additional clue as to the difference in structure of the molecules.

Other Abnormal Hemoglobins, Fetal Hemoglobin, Synergy with Thalassemia

During the five years since the discovery of sickle-cell-anemia hemoglobin three other abnormal varieties of adult human hemoglobin (named hemoglobin C, D, and E, respectively; A represents normal adult human hemoglobin, and S sickle-cell-anemia hemoglobin) have been discovered. These abnormal hemoglobins, either alone or in synergistic interaction with sickle-cell-anemia hemoglobin or with thalassemia, have been observed in association with five diseases, which, together with a sixth disease resulting from the simultaneous existence in the individual of sicklemia and thalassemia minor, had not previously been recognized as clinical entities. Another interesting complication in the constitution of the blood has also been recognized, the continued manufacture of fetal hemoglobin by anemic individuals.

Hemoglobin C was discovered by Itano and Neel[48] and has been further investigated by Neel,[49–51] Spaet *et al.*,[52] Ranney *et al.*,[53,54] and Smith and Conley.[55] The difference in electrophoretic mobility between hemoglobin C and normal adult human hemoglobin is nearly twice as great as the difference between sickle-cell-anemia hemoglobin and normal hemoglobin: the electric charge of hemoglobin C differs by about 5 electronic charges from that of A. Hemoglobin C was first found to be present together with S in the blood of some anemia patients. These patients may be described as carrying one allele for C and one for S: they are heterozygous in hemoglobin C and also in hemoglobin S, having inherited one of the two abnormalities from each parent. On investigation, the red cells of one parent were found to contain a mixture of A and C, and of the other parent to contain a mixture of A and S. Dilution of S with C does not inhibit the sickling tendency so much as dilution of S with A, and in consequence the individuals of type SC are anemic, their disease being called sickle-cell:hemoglobin-C disease. The heterozygous condition in C does not lead to a pathologic state. Sickle-cell:hemoglobin-C disease can be readily differentiated from sickle-cell anemia on clinical grounds.[49–55] The characteristic differences in electrophoretic behavior of A, S, and C are shown in the paper electrophoresis patterns of Fig. 4, reproduced from the work of Smith and Conley.[55] Under the conditions of this investigation, which was carried out in veronal buffer at pH 8.6 and ionic strength 0.06, all three hemoglobins are negatively charged, with A having the largest negative charge and C the smallest charge.

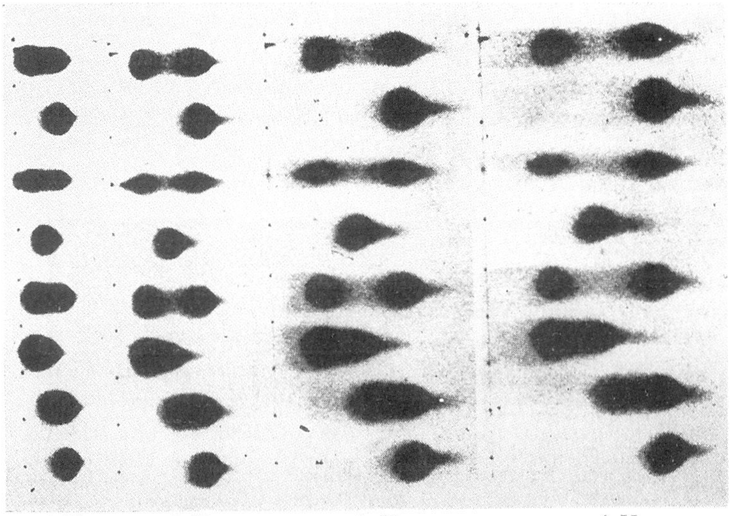

FIG. 4. Paper electrophoresis patterns of hemoglobins. Veronal buffer, 0.06 ionic strength, pH 8.6; Whatman paper No. 3, 6 × 6 inches, 15 mils, 380 volts. The migration begins at the dotted line; hemoglobin A is the fastest moving, hemoglobin S has intermediate mobility, and hemoglobin C migrates most slowly. From Ernest W. Smith and T. Lockard Conley, *Bull. Johns Hopkins Hosp.* **93,** 94 (1953).

Using paper electrophoresis, Smith and Conley[55] made an investigation of the hemoglobins of 500 white persons and 500 Negroes. In the 500 white persons they found no evidence of the presence of any hemoglobin differing electrophoretically from normal adult human hemoglobin. Hemoglobin S was found to occur in 8.4 per cent of the 500 Negroes surveyed, and hemoglobin C in 2 per cent. Seven patients with sickle-cell:hemoglobin-C disease were found among the 500 Negroes.

If the incidence of the heterozygous state AC is 2 per cent, the occurrence of the C allele is 0.01, and it can be predicted that homozygous hemoglobin-C disease should occur among Negroes with frequency 0.01^2, or one in 10,000. It was reported by Spaet and co-workers[52] and Ranney and co-workers[53] that patients with a mild hemolytic anemia have been found with only hemoglobin C in their red cells, and it is probable that they are homozygous in hemoglobin C, and that their disease is homozygous hemoglobin-C disease.

An interesting complication in the hemoglobin pattern has been resolved through the discovery that anemic individuals may continue to manufacture fetal hemoglobin long after they have passed the fetal stage of development. Wells and Itano, in the course of

FIG. 5. Electrophoretic pattern of the hemoglobin of a patient with sickle-cell:hemoglobin-C disease, showing three hemoglobins, C, S, and F, from left to right.

their investigation of the ratio of S to A in sicklemic individuals,[26] discovered that there is present in the blood of patients with an abnormally mild form of sickle-cell anemia a small amount, 5 per cent to 20 per cent of the total, of a protein other than S. They identified this protein as A, on the basis of its electrophoretic pattern, but they pointed out that the patients could be assumed, from the facts that both parents were sicklemic and that hemoglobin S predominated, to be homozygous in S, and that the presence of A was anomalous. Itano and Neel[48] reported that one patient with sickle-cell:hemoglobin-C disease gave a hemoglobin electrophoretic pattern showing S and C in large amounts (39 per cent and 48 per cent, respectively) and an additional hemoglobin, 13 per cent, with the electrophoretic mobility, at the pH used (pH 6.50), of A. The electrophoretic pattern of this patient is given in Fig. 5. It was pointed out by Singer, Chernoff, and Singer[56] that there is present in the blood of many patients with sickle-cell anemia and other hematologic disorders a hemoglobin fraction that is resistant to alkali denaturation, and that the properties of this fraction permit it to be identified with fetal hemoglobin, F; in particular, these authors suggested that the hemoglobin reported by Wells and Itano to be present in small amounts in the blood of some sickle-cell-anemia patients is hemoglobin F, and not A. At about the same time Liquori[57] reported that he had found that as much as 50 per cent of the hemoglobin in the blood of patients with thalassemia is fetal hemoglobin, and Rich[58] found that two patients with thalassemia major had in their red cells no hemoglobin other than F. It is now recognized that anemic patients may continue to manufacture significant amounts of F throughout their lives, and that the presence of F ameliorates their disease. Hemoglobin F is identified, in relation to A, only with difficulty by electrophoretic methods, but it is easily identified by its resistance to alkali denaturation and by its characteristic ultraviolet absorption spectrum.[59-64] Goodman and Campbell[46] verified the identification of the minor hemoglobin in patients with sickle-cell anemia by studying its antigenic properties, which were found to be either identical with or closely similar to those of fetal hemoglobin.

Thalassemia (Cooley's anemia, Mediterranean anemia) is a hereditary anemia that results from a gene-controlled interference with the process of synthesis of adult human hemoglobin. It was first clearly recognized by Cooley and Lee.[65] The disease is largely confined to persons derived from the northern shores of the Mediterranean. The thalassemia allele in the homozygous form produces a very serious anemia, thalassemia major, which usually terminates fatally in childhood. In heterozygous form the allele produces a mild anemia, called thalassemia minor. In some parts of Italy, notably the region around Ferrara, the incidence of thalassemia minor is about 10 per cent, corresponding to a probability of the allele of 5 per cent.[30]

In 1953 Powell, Rodarte, and Neel[66] reported one case of a disease due to the combination of thalassemia minor and sicklemia. The resulting sickle-cell:thalassemia disease is a more serious disease than thalassemia minor. Other patients with the disease have been described by Sturgeon, Itano, and Valentine[67] and by Neel, Itano, and Lawrence.[68] A case of thalassemia:hemoglobin-C disease is mentioned in a footnote at the end of a paper by Kaplan, Zuelzer, and Neel.[51]

The third abnormal hemoglobin to be discovered, hemoglobin D, was found by Itano[69] in five members of a single family. Two of the family were patients with a disease resembling sickle-cell anemia but somewhat milder than the normal form of this disease. The electrophoretic patterns of the blood of these patients were closely similar to those of hemoglobin S. The hemoglobin from the red cells of the three other individuals gave an electrophoretic pattern like that obtained from sicklemics. It was found, however, by measurement of the solubility of the hemoglobins that an abnormal hemoglobin is present, with electrophoretic properties like that of S, and solubility like that of A. This hemoglobin, hemoglobin D, seems not to interfere with the process of sickling so much as does A, so that the state of double heterozygosity SD leads to a moderately serious anemia. The blood of each of the two patients with sickle-cell: hemoglobin-D disease was found to contain a small amount (6 per cent to 12 per cent) of F.

Two individuals containing the fourth abnormal hemoglobin, E, have been discovered by Itano, Bergren, and Sturgeon. One of them is an anemic patient with thalassemia: hemoglobin-E disease.[70] The mother of the patient has thalassemia minor, and no hemoglobin other than A in her red cells.[70] The red cells of the patient contain 41 per cent F and 59 per cent E. The father of the patient is of type AE, with 72 per cent A and 28 per cent E (no F); he is a carrier of E (personal communication from Itano, Bergren, and Sturgeon). The patient has inherited thalassemia minor from the mother and the allele for E from the father. The thalassemia allele seems to have completely suppressed the manufacture of A.

Hemoglobin E shows a striking change in electrophoretic mobility with change in pH. At pH 6.5, in cacodylate buffer of ionic strength 0.1, its mobility is greater than that of A and slightly less than that of S, and at pH 8.8, in 0.01 F disodium hydrogen phosphate, its mobility is nearly identical with that of C. The absorption spectrum, solubility, and lability to alkali denaturation of E are similar to those of A.

A summary of the hereditary hemolytic anemias related to abnormal hemoglobins is presented in Fig. 6. There are five forms of adult human hemoglobin: A, S, C, D, and E. Assuming that these five correspond to five alleles occupying the same locus in a chromosome, there are fifteen possible combinations, the five homozygous states AA, SS, CC, DD, and EE, and the ten heterozygous states AS, AC, AD, AE, SC, SD, SE, CD, CE, and DE. Of these fifteen states nine have been found to occur. In addition to sickle-cell anemia, six other types of hereditary anemia involving an abnormality of hemoglobin have been recognized.[71]

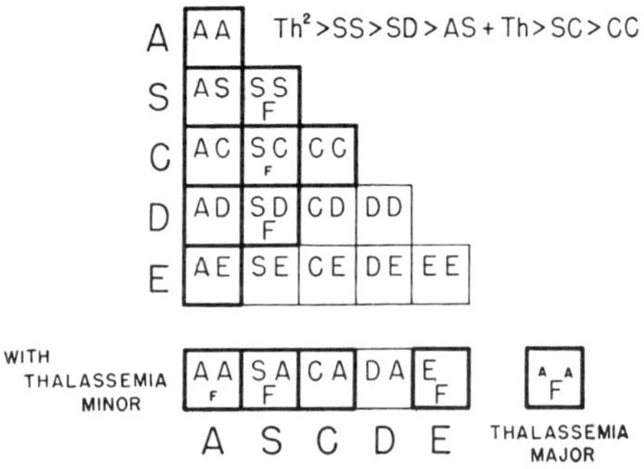

FIG. 6. A chart representing possible combinations of the alleles A, S, C, D, and E. The horizontal row at the bottom represents simultaneous occurrence with thalassemia minor. The amount of fetal hemoglobin usually present is also indicated: F means a few per cent, F means 10 per cent or more. Observed conditions are shown within heavy borders. At the top the seriousness of different kinds of anemia is indicated.

The seriousness of these diseases (not including AC + Th and E + Th, for which the number of patients is too small to permit an estimate) is indicated at the top of Fig. 6. There is some variability in the seriousness of the diseases, in part as the result of the extent to which compensation of abnormal hemoglobins is achieved through the manufacture of fetal hemoglobin. The mildest anemias are SC and CC. Not all SC and CC individuals are anemic. Although they probably have a greater than normal rate of hemolysis, they may be able to compensate completely with a greater than normal rate of production of red blood cells.[52,72]

The conditions CD, DD, SE, CE, DE, EE, and DA + Th have not yet been observed; it may be found that they are associated with anemia.

Except for the fetal-hemoglobin fraction, the hemoglobin of patients with thalassemia major has properties reported to be the same as those of normal adult human hemoglobin,[58] and it is usually considered that thalassemia is not associated with the production of an abnormal hemoglobin. There are some facts, however, that indicate that thalassemia hemoglobin is a fifth abnormal kind of adult human hemoglobin, with properties so closely similar to those of normal adult human hemoglobin that the differences have escaped detection. It has been observed that thalassemia minor involves a greater interference with the synthesis of the normal hemoglobin than of abnormal hemoglobin. For example, whereas in sicklemic individuals the ratio A/S is always greater than 1, this ratio is much less than 1 in thalassemia:sickle-cell disease. Two patients reported by Neel, Itano, and Lawrence[68] were found to have the ratios A:S:F equal to 20:61:19 and 11:84:5, respectively. The hemoglobin of the patient with thalassemia:hemoglobin-E disease contains no detectable amount of normal adult hemoglobin, but only E and F. The thalassemia allele interferes with the manufacture of normal adult hemoglobin, and seems not to interfere seriously with the manufacture of the abnormal hemoglobins. The simplest explanation of this fact is that the thalassemia allele occupies the same locus in the chromosome as the alleles for the other abnormal hemoglobins, and is itself responsible for an abnormal globin in which the abnormality is of such a nature as to interfere with the step of inclusion of the hemes in the molecule. If this postulate is correct it should be possible to show a difference between thalassemia hemoglobin and normal adult human hemoglobin.

Under this circumstance there would be six alleles occupying the same locus, A, S, C, D, E, and Th (more than six if there is more than one thalassemia allele). The six states at the bottom of the triangle in Fig. 6 should then be written ATh, STh, CTh, DTh, ETh, and ThTh. Of the possible twenty-one combinations of the six alleles, fourteen would be ascribed to known individuals.

The possibility that the thalassemia gene is allelomorphic with the sickle-cell gene has been discussed by Neel.[30] He pointed out that both of the children of a patient with thalassemia:sickle-cell disease and married to a normal woman exhibited thalassemia minor, and mentioned that if either of these children had been normal (without either thalassemia minor or sicklemia) the hypothesis of allelomorphism would have to be abandoned. Silvestroni and Bianco, who a number of years ago described the disease resulting from simultaneous inheritance of thalassemia minor and sicklemia,[73] have also discussed the genetic aspects of thalassemia and sickling, and have concluded that the two are not allelomorphic.[74]

Sickle-Cell-Anemia Hemoglobin and Malaria

A number of interesting questions about the origin and heredity of the abnormal hemoglobins remain to be answered. It has, for example, been pointed out by Lehmann[75] that sickle-cell anemia seems to be a far less serious disease in Africa than in America. A possible explanation may be that the patients in Africa for some reason manufacture a larger amount of fetal hemoglobin than those in America; the question can presumably be answered by a thorough investigation of the hemoglobins of Africans, which has not yet been carried out.

The question of the continued high incidence of the sickle-cell allele, despite its continued loss because of the lethal character of the homozygous condition, has been raised by Neel,[30] who has suggested three alternative explanations: (1) continued production of the allele through mutation; (2) the existence of an abnormal genetic mechanism that favors the heterozygous condition, AS, over the normal condition, AA; (3) a positive selection of the heterozygote, perhaps through increased fertility. The first explanation has to be rejected because the rate of mutation that would be required is far greater than any that has ever been observed for any organism. There now exists evidence indicating that the third alternative provides the correct explanation, and that malaria is involved. It was first suggested by Brain[76] that the presence of S in the red cells might give protection against malaria parasites, and thus confer an advantage to the sicklemic individual that would balance the disadvantage of the lethal homozygosity. Lehmann[77] wrote that "The lethal tendency of a gene potentially causing sickle cell anemia may thus be counteracted by its conferring a resistance to malaria similar to that found in early infancy." A test of the hypothesis was carried out recently by Allison,[78] who infected 15 healthy adult Africans with the sickle-cell trait and 15 similar healthy adult Africans without the sickle-cell trait with *P. falciparum,* by subinoculation with 15 ml. of blood containing a large number of trophozoites or by being bitten by heavily infected anopheles mosquitoes, in which the presence of sporozoites was confirmed by dissection of the mosquito. The infection was established in 14 out of the 15 Africans without the sickle-cell trait, and in only 2 of the 15 with this trait. It was concluded by Allison that the abnormal erythrocytes of individuals with the sickle-cell trait are less easily parasitized by *P. falciparum* than are normal erythrocytes, and that accordingly those who are heterozygous for the S allele have a selective advantage over normal individuals in regions where malaria is hyperendemic. It is, of course, not unreasonable that the abnormal hemoglobin may be less effective than normal hemoglobin in nourishing the parasites.

The question of the origin of sickle-cell anemia has been investigated especially by Lehmann, who has studied the incidence of the ability of red cells to sickle in several parts of the world.[79–81] In Africa the incidence of the sickle-cell allele is highest in the north and east, diminishing somewhat toward the west and south, and being virtually zero in South Africa itself. In India the major groups of the population, classed as Dravidians, show no sickling, and among the pre-Dravidians living in the hills of Southern India the Veddoids alone have a high incidence of sickling. A high incidence of sickling was found also in a Veddoid community of Southern Arabia, although sickling is absent among the Semitic Arabs. Some sicklemic individuals are found in Italy, and in Greece there are a few communities, each of a few hundred individuals, with a high incidence of sickling. Lehmann concluded[77] that it is not unlikely that the sickle-cell gene originated, presumably through a mutation, in a Veddoid community in Southern Arabia, and that it spread from this point to India, the Mediterranean region, and especially to Africa.

Molecular Diseases

Sickle-cell anemia has been described as a molecular disease.[20] It may be that all diseases can be described as molecular diseases, inasmuch as the human body and the vectors of disease are all composed of molecules. For example, carbon monoxide poisoning is the result of combination of molecules of carbon monoxide with molecules of hemoglobin. Erythroblastosis fetalis involves the interaction of the molecules or haptenic groups of an Rh antigen on red cells with molecules of the homologous antibody. An inborn error of metabolism such as alcaptonuria results from the failure of the body to manufacture molecules of the enzyme that, if present, would catalyze the oxidation of homogentisic acid in the body. Any hereditary disease may be said to be a molecular disease, if the genes are described as molecules, in that it involves an abnormality of a gene.

There is, however, one sense of the expression molecular disease that permits it to be applied to sickle-cell anemia and the other diseases associated with the abnormal hemoglobins. These diseases have been shown to result from the manufacture by the patient of abnormal molecules, in the place of normal molecules which are manufactured by normal individuals, and the abnormal molecules have been characterized. I think that it is not unlikely that many diseases will in the course of time be found to be molecular diseases in this sense. The discovery of the abnormal molecules responsible for other molecular diseases may be far more difficult than the discovery of the abnormal hemoglobins. Hemoglobin is unique among the proteins of the human body because of its presence in very large amount, about 1 per cent by weight of the body, and because of the ease with which it can be obtained from the individual and characterized. Other substances in general are present in far smaller amounts. I have estimated that there are of the order of 100,000 different kinds of proteins in the human body. If the number of substances making up the human body is about 100,000, the average amount of each substance present (aside from water) is about 100 mg., and some important substances may be present in far smaller amounts. The isolation and identification of abnormal forms of molecules of these substances might be extremely difficult. Because of the insight into the nature of disease that would be provided by the discovery of additional molecular diseases, however, it seems important to prosecute investigations along this line with much vigor.

ACKNOWLEDGMENTS

It is a pleasure for me to acknowledge the financial support of the work on hemoglobin received from The Rockefeller Foundation during the entire score of years that the work has been carried on in our laboratories, and from the Public Health Service of the National Institutes of Health (Research Grant 257, on Investigations of the Chemistry of Blood) during the period 1946 to 1953. I am glad also to have an opportunity to express my thanks to my co-workers, Alfred E. Mirsky, Charles D. Coryell, Fred Stitt, Donald S. Taylor, Richard W. Dodson, Charles D. Russell, S. J. Singer, Ibert C. Wells, Robert B. Corey, Ray D. Owen, W. A. Schroeder, Lois M. Kay, Alexander Rich, E. Havinga, D. H. Campbell, Morris Goodman, Robert C. C. St. George, William R. Bergren, Phillip Sturgeon, and especially Harvey A. Itano, who has been the discoverer or co-discoverer of each of the four known abnormal forms of adult human hemoglobin.

REFERENCES

1. Pauling, L. 1935. The oxygen equilibrium of hemoglobin and its structural interpretation. *Proc. Natl. Acad. Sci. U.S.* **21**, 186.
2. Mirsky, A. E., and Pauling, L. 1936. On the structure of native, denatured, and coagulated proteins. *Proc. Natl. Acad. Sci. U.S.* **22**, 439.
3. Pauling, L., and Coryell, C. D. 1936. The magnetic properties and structure of the hemochromogens and related substances. *Proc. Natl. Acad. Sci. U.S.* **22**, 159.
4. Pauling, L., and Coryell, C. D. 1936. The magnetic properties and structure of hemoglobin, oxyhemoglobin, and carbonmonoxyhemoglobin. *Proc. Natl. Acad. Sci. U.S.* **22**, 210.
5. Coryell, C. D., Stitt, F., and Pauling, L. 1937. The magnetic properties and structure of ferrihemoglobin (methemoglobin) and some of its compounds. *J. Am. Chem. Soc.* **59**, 633.
6. Taylor, D. S., and Coryell, C. D. 1938. The magnetic susceptibility of the iron in ferrohemoglobin. *J. Am. Chem. Soc.* **60**, 1177.
7. Stitt, F., and Coryell, C. D. 1939. The magnetic study of the equilibrium between ferrohemoglobin, cyanide ion, and cyanide ferrohemoglobin. *J. Am. Chem. Soc.* **61**, 1263.
8. Taylor, D. S. 1939. The magnetic properties of myoglobin and ferrimyoglobin, and their bearing on the problem of the existence of magnetic interactions in hemoglobin. *J. Am. Chem. Soc.* **61**, 2150.
9. Coryell, C. D., and Stitt, F. 1940. Magnetic studies of ferrihemoglobin

reactions. II. Equilibria and compounds with azide ion, ammonia, and ethanol. *J. Am. Chem. Soc.* **62,** 2942.
10. Coryell, C. D., Pauling, L., and Dodson, R. W. 1939. The magnetic properties of intermediates in the reactions of hemoglobin. *J. Phys. Chem.* **43,** 825.
11. Russell, C. D., and Pauling, L. 1939. The magnetic properties of the compounds ethylisocyanide-ferrohemoglobin and imidazole-ferrihemoglobin. *Proc. Natl. Acad. Sci. U.S.* **25,** 517.
12. Coryell, C. D. 1939. The existence of chemical interactions between the hemes in ferrihemoglobin (methemoglobin) and the role of interactions in the interpretation of ferro-ferrihemoglobin electropotential measurements. *J. Phys. Chem.* **43,** 841.
13. Coryell, C. D., and Pauling, L. 1940. A structural interpretation of the acidity of groups associated with the hemes of hemoglobin and hemoglobin derivatives. *J. Biol. Chem.* **132,** 769.
14. Pauling, L. 1948. The interpretation of some chemical properties of hemoglobin in terms of its molecular structure. *Stanford Med. Bull.* **6,** 215.
15. Pauling, L. 1949. The electronic structure of haemoglobin, *in* "Haemoglobin," pp. 57–65, Butterworths Scientific Publications, London.
16. Bush, V. 1945. Science, the Endless Frontier: A Report to the President, U.S. Government Printing Office, Washington, D.C.
17. Herrick, J. B. 1910. Peculiar elongated and sickle shaped red corpuscles in a case of severe anemia. *Arch. Internal Med.* **6,** 517.
18. Hahn, E. V., and Gillespie, E. B. 1927. Sickle cell anemia: Report of a case greatly improved by splenectomy. Experimental study of sickle cell formation. *Arch. Internal Med.* **39,** 233.
19. Itano, H. A., and Pauling, L. 1949. A rapid diagnostic test for sickle cell anemia. *Blood* **4,** 66.
20. Pauling, L., Itano, H. A., Singer, S. J., and Wells, I. C. 1949. Sickle cell anemia, a molecular disease. *Science* **110,** 543.
21. Emmel, V. E. 1917. A study of the erythrocytes in a case of severe anemia with elongated and sickle-shaped red blood cells. *Arch. Internal Med.* **20,** 586.
22. Taliaferro, W. H., and Huck, J. G. 1923. The inheritance of sickle cell anemia in man. *Genetics* **8,** 594.
23. Neel, J. V. 1947. The clinical detection of the genetic carriers of inherited disease. *Medicine* **26,** 115.
24. Neel, J. V. 1949. The inheritance of sickle cell anemia. *Science* **110,** 64.
25. Beet, E. A. 1949. The genetics of the sickle cell trait in a Bantu tribe. *Ann. Eugen.* **14,** 279.
26. Wells, I. C., and Itano, H. A. 1951. Ratio of sickle-cell anemia hemoglobin to normal hemoglobin in sicklemics. *J. Biol. Chem.* **188,** 65.
27. Neel, J. V., Wells, I. C., and Itano, H. A. 1951. Familial differences in the proportion of abnormal hemoglobin present in the sickle cell trait. *J. Clin. Invest.* **30,** 1120.

28. Itano, H. A. 1953. Qualitative and quantitative control of adult hemoglobin synthesis—a multiple allele hypothesis. *Am. J. Human Genet.* **5**, 34.
29. Neel, J. V. 1951. The inheritance of the sickling phenomenon, with particular reference to sickle cell disease. *Blood* **6**, 389.
30. Neel, J. V. 1951. The population genetics of two inherited blood dyscrasias in man. *Cold Spring Harbor Symposia Quant. Biol.* **15**, 141.
31. Neel, J. V. 1933. Data pertaining to the population dynamics of sickle cell disease. *Am. J. Human Genet.* **5**, 154.
32. Singer K. 1951. The pathogenesis of sickle cell anemia: A review. *Am. J. Clin. Pathol.* **21**, 858.
33. Silvestroni, B., and Bianco, I. 1952. Genetic aspects of sickle cell anemia and microdrepanocytic disease. *Blood* **7**, 429.
34. Schroeder, W. A., Kay, L. M., and Wells, I. C. 1950. Amino acid composition of hemoglobins of normal negroes and sickle-cell anemics. *J. Biol. Chem.* **187**, 221.
35. Havinga, E. 1953. Comparison of the phosphorus content, optical rotation, separation of hemes and globin, and terminal amino acid residues of normal adult human hemoglobin and sickle cell anemia hemoglobin. *Proc. Natl. Acad. Sci. U.S.* **39**, 59.
36. Havinga, E., and Itano, H. A. 1953. Electrophoretic studies of globins prepared from normal adult and sickle cell hemoglobins. *Proc. Natl. Acad. Sci. U.S.* **39**, 65.
37. Sherman, I. J. 1940. The sickling phenomenon, with special reference to the differentiation of sickle cell anemia from the sickle cell trait. *Bull. Johns Hopkins Hosp.* **67**, 309.
38. Ponder, E. 1948. "Hemolysis and Related Phenomena," p. 145, Grune and Stratton, New York.
39. Pauling, L., Pressman, D., and Campbell, D. H. 1943. The nature of the forces between antigen and antibody and of the precipitation reaction. *Physiol. Revs.* **23**, 203.
40. St. George, R. C. C., and Pauling, L. 1951. The combining power of hemoglobin for alkyl isocyanides, and the nature of the heme-heme interactions in hemoglobin. *Science* **114**, 629.
41. Rebuck, J. W., Sturrock, R. M., and Monaghan, D. A. 1950. Sickling processes in anemia and trait erythrocytes with the electron microscopy of their incipient crystallization. *Federation Proc.* **9**, 340.
42. Perutz, M. F., and Mitchison, J. M. 1950. State of haemoglobin in sickle-cell anemia. *Nature* **166**, 677.
43. Perutz, M. F., Liquori, A. M., and Eirich, F. 1951. X-ray and solubility studies of the haemoglobin of sickle-cell anaemia patients. *Nature* **167**, 929.
44. Itano, H. A. 1953. Solubilities of naturally occurring mixtures of human hemoglobin. *Arch. Biochem. and Biophys.* **47**, 148.
45. Harris, J. W. 1950. Studies on the destruction of red blood cells. VIII. Molecular orientation in sickle cell hemoglobin solutions. *Proc. Soc. Exptl. Biol. Med.* **75**, 197.

46. Goodman, M., and Campbell, D. H. 1953. Differences in antigenic specificity of human normal adult, fetal, and sickle cell anemia hemoglobin. *Blood* **8**, 422.
47. Ingbar S. H., and Kass, E. H. 1951. Sulfhydryl content of normal hemoglobin and hemoglobin in sickle cell anemia. *Proc. Soc. Exptl. Biol. Med.* **77**, 74.
48. Itano, H. A., and Neel, J. V. 1950. A new inherited abnormality of human hemoglobin. *Proc. Natl. Acad. Sci. U.S.* **36**, 613.
49. Kaplan, E., Zuelzer, W. W., and Neel, J. V. 1951. A new inherited abnormality of hemoglobin and its interaction with sickle cell hemoglobin. *Blood* **6**, 1240.
50. Neel, J. V., Kaplan, E., and Zuelzer, W. W. 1953. Further studies on hemoglobin C. I. A description of three additional families segregating for hemoglobin C and sickle cell hemoglobin. *Blood* **8**, 724.
51. Kaplan, E., Zuelzer, W. W., and Neel, J. V. 1953. Further studies on hemoglobin C. II. The hematologic effects of hemoglobin C alone and in combination with sickle cell hemoglobin. *Blood* **8**, 735.
52. Spaet, T. H., Alway, R. H., and Ward, G. 1953. Homozygous type "C" hemoglobin. *Pediatrics* **12**, 483.
53. Ranney, H. M., Larson, D. L., and McCormack, G. H., Jr. 1953. Some clinical, biochemical, and genetic observations on hemoglobin C. *J. Clin. Invest.* **32**, 1277.
54. Larson, D. L., and Ranney, H. M. 1953. Filter paper electrophoresis of human hemoglobin. *J. Clin. Invest.* **32**, 1070.
55. Smith, E. W., and Conley, C. L. 1953. Filter paper electrophoresis of human hemoglobins with special reference to the incidence and clinical significance of hemoglobin C. *Bull. Johns Hopkins Hosp.* **93**, 94.
56. Singer, K., Chernoff, A. I., and Singer, L. 1951. Studies on abnormal hemoglobins. I. Their demonstration in sickle cell anemia and other hematologic disorders by means of alkali denaturation. II. Their identification by means of the method of fractional denaturation. *Blood* **6**, 413, 429.
57. Liquori, A. M. 1951. Presence of foetal haemoglobin in Cooley's anemia. *Nature* **167**, 950.
58. Rich, A. 1952. Studies on the hemoglobin of Cooley's anemia and Cooley's trait. *Proc. Natl. Acad. Sci. U.S.* **38**, 187.
59. Beaven, G. H., Hoch, H., and Holiday, E. R. 1951. The haemoglobins of the human foetus and infant. Electrophoretic and spectroscopic differentiation of adult and foetal types. *Biochem. J.* **49**, 374.
60. Zinsser, H. H. 1952. Electrophoretic studies on human hemoglobin in the premature and new-born. *Arch. Biochem. and Biophys.* **38**, 195.
61. Beaven, G. H., and White, J. C. 1953. Detection of foetal and sickle-cell haemoglobins in human anaemias. *Nature* **172**, 1006.
62. Singer, K., and Chernoff, A. I. 1952. Studies on abnormal hemoglobins. III. The inter-relationship of type S (sickle cell) hemoglobin and type F (alkali-resistant) hemoglobin in sickle cell anemia. *Blood* **7**, 47.

63. Chernoff, A. I., and Singer, K. 1952. Studies on abnormal hemoglobins. IV. Persistence of fetal hemoglobin in the erythrocytes of normal children. *Pediatrics* **9**, 469.
64. Singer, K., and Fisher, B. 1952. Studies on abnormal hemoglobins. V. The distribution of type S (sickle cell) hemoglobin and type F (alkali-resistant) hemoglobin within the red cell population in sickle cell anemia. *Blood* **7**, 1216.
65. Cooley, T. B., and Lee, P. 1925. Series of cases of splenomegaly in children with anemia and peculiar bone changes. *Trans. Am. Pediat. Soc.* **37**, 29.
66. Powell, W. N., Rodarte, J. G., and Neel, J. V. 1950. The occurrence in a family of Silician ancestry of the traits for both sickling and thalassemia. *Blood* **5**, 887.
67. Sturgeon, P., Itano, H. A., and Valentine, W. N. 1952. Chronic hemolytic anemia associated with thalassemia and sickling traits. *Blood* **7**, 350.
68. Neel, J. V., Itano, H. A., and Lawrence, J. S. 1953. Two cases of sickle cell disease presumably due to the combination of the genes for thalassemia and sickle cell hemoglobin. *Blood* **8**, 434.
69. Itano, H. A. 1951. A third abnormal hemoglobin associated with hereditary hemolytic anemia. *Proc. Natl. Acad. Sci. U.S.* **37**, 775.
70. Itano, H. A., Bergren, W. R., and Sturgeon, P. 1954. Identification of a fourth abnormal human hemoglobin. *J. Am. Chem. Soc.* **76**, 2278.
71. Itano, H. A. 1953. Human hemoglobin. *Science* **117**, 89.
72. Crosby, W. H., and Akeroyd, J. H. 1952. The limit of hemoglobin synthesis in hereditary hemolytic anemia. *Am. J. Med.* **13**, 273.
73. Silvestroni E., and Bianco, I. 1946. Una nuova entitá nosologica: La malatia microdrepanocitica. *Haematologica* **29**, 455.
74. Silvestroni, E., and Bianco, I. 1952. Genetic aspects of sickle cell anemia and microdrepanocytic disease. *Blood* **7**, 429.
75. Lehmann, H. 1953. Sickle-cell anaemia. *Brit. Med. J.* **ii**, 1217.
76. Brain, P. 1952. Sickle-cell anaemia in Africa. *Brit. Med J.* **ii**, 880.
77. Lehmann, H. 1953. The distribution of the sickle cell trait. *J. Clin. Pathol.* **6**, 329.
78. Allison, A. C. 1954. Protection afforded by sickle cell trait against subtertian malarial infection. *Brit. Med. J.* **i**, 290.
79. Lehmann, H. 1951. Sickle-cell anaemia and sickle-cell trait as homo- and heterozygous gene-combinations. *Nature* **167**, 931.
80. Lehmann, H., and Cutbush, M. 1952. Sickle-cell trait in Southern India. *Brit. Med. J.* **i**, 404.
81. Choremis, C., Ikin, E. W., Lehmann, H., Mourant, A. E., and Zannos, L. 1953. Sickle-cell trait and blood-groups in Greece. *Lancet* 911.

Särtryck ur Svensk Kemisk Tidskrift nr 69, 509, 1957

Abnormal hemoglobin molecules in relation to disease*

By *Harvey A. Itano* and *Linus Pauling*

The National Institute of Arthritis and Metabolic Diseases of the National Institutes of Health, Public Health Service, U.S. Department of Health, Education, and Welfare, Bethesda, Md. and the Gates and Crellin Laboratories of Chemistry, Pasadena, Calif., U.S.A.

The human body is made up of molecules, as are also bacteria and other vectors of disease. We might accordingly say that all diseases are molecular diseases, involving molecules in one way or another. For example, phenylketonuria, which causes feeble-mindedness or more serious mental impairment, is an inborn error of metabolism such that the patient is not able to carry out the oxidation of phenylalanine to tyrosine. This disease is due to an abnormal gene, present in double dose; either the gene is not able to manufacture the enzyme catalysing the oxidation reaction, or it manufactures abnormal enzyme molecules, with decreased effectiveness.

A few years ago it was discovered[1] that the disease *sickle-cell anemia* is a disease of the hemoglobin molecule; the patients with this disease manufacture abnormal molecules of hemoglobin, rather than normal adult human hemoglobin, and there is strong indication that the abnormal molecules of hemoglobin are directly responsible for the symptoms of the disease. Since then it has been discovered that there are a number of other hereditary hemolytic anemias that are caused by abnormal kinds of human hemoglobin. Several review articles in this field have been published recently.[2, 3, 4, 5, 6, 7, 8, 9, 10, 11]

The disease sickle-cell anemia was first described by Dr. J. B. Herrick in 1910.[12] The red cells of patients with this disease when partially deoxygenated (as in venous blood) undergo a change in shape, as shown in Fig. 1. This change in shape, called sickling (deformation to resemble a sickle), can also be caused to occur in a specimen of blood from a parent of a sickle-cell-anemia patient, by making the partial pressure of oxygen small enough; sickling does not occur to significant extent in the venous blood of these individuals, who are not anemic. It was suggested by Emmel[13] in 1917 and Taliaferro and Huck[14] in 1923 that the disease is hereditary, and the Mendelian inheritance was made clear by Neel,[15, 16] who presented evidence that there is a sickle-cell gene that causes some tendency to sickling, but not sickle-cell anemia, when it is present in the heterozygous condition (one sickle-cell gene and one normal gene), and causes sickle-cell anemia when it is present in the homozygous condition (two sickle-cell genes).

* Contribution No. 2174, California Institute of Technology, Pasadena. Calif., U.S.A.

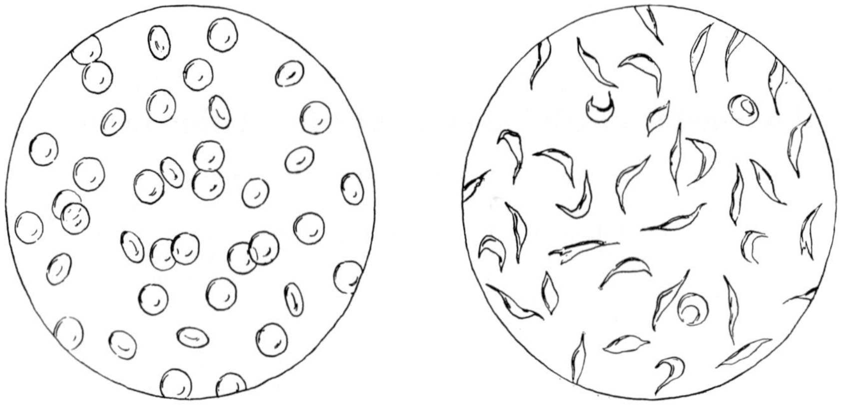

Fig. 1. At the left a drawing of normal human red cells, as seen through the ordinary microscope; at the right, sickled red cells, present in the venous blood of patients with sickle-cell anemia.

The discovery that sickle-cell anemia is a molecular disease[1] was made by the study of *electrophoretic patterns* of hemoglobin from normal individuals, sickle-cell-anemia patients, and the carriers of the disease (parents of patients), as shown in Fig. 2. At pH 6.90 normal adult human hemoglobin has a negative electric charge, and moves toward the anode, whereas sickle-cell-anemia hemoglobin moves toward the cathode. The difference in electric charge of the two kinds of molecules is about three electronic charges. The hemoglobin from the red cells of sickle-cell heterozygotes is a mixture of the two kinds, usually with normal adult hemoglobin present in amount somewhat greater than 50 per cent.[17] A postulate to explain the amounts of the two hemoglobins, A and S, in heterozygotes was made by Itano,[18] who suggested that the differences can be attributed to differences in the rate of synthesis of normal hemoglobin, and that the evidence requires that there be at least three rate-determining modifications of the mechanism of synthesis of normal hemoglobin.

Sickle-cell-anemia hemoglobin in closely similar to normal adult human hemoglobin in most of its properties.[1] The difference in structure seems to be a difference in the nature of the globins, inasmuch as the porphyrin from both hemoglobins has been identified as protoporphyrin. Schroeder, Kay, and Wells[19] found that there is little difference in *amino acid composition* of the two hemoglobins, and there is a possibility that they differ in the nature of the folding or amino acid sequence of the polypeptide chains, rather than in their composition. Amino acid compositions of A, F, S, and C (C is another abnormal hemoglobin, discussed in later paragraphs in this paper) have been reported by Huisman, Jonxis, and van der Schaaf,[20] who verified the work of Schroeder *et al.* for A and S, and found C to contain about three more residues of lysine and one or two more of histidine than are present in A. The amino acid

Fig. 2. The electrophoretic patterns for (a) normal adult human hemoglobin, (b) sickle-cell-anemia hemoglobin, from the red cells of patients with sickle-cell anemia, (c) sickle-cell-trait hemoglobin, which is indicated to be a mixture of normal adult human hemoglobin and sickle-cell-anemia hemoglobin, and (d) a mixture of normal adult human hemoglobin and sickle-cell-anemia hemoglobin, prepared by mixing the red-cell contents from normal blood and sickle-cell-anemia blood. The two hemoglobins migrate in opposite directions in this buffer, normal adult hemoglobin as an anion and sickle-cell-anemia hemoglobin as a cation.

composition of another abnormal form, hemoglobin E, is similar to that of A and S.[21] Valine is the only N-terminal amino acid in the human hemoglobins. Hemoglobins A and F have five and two residues, respectively, according to Porter and Sanger.[22] Although different investigators agree that the abnormal hemoglobins do not differ from A in their content of N-terminal valine, they do not agree on the number per molecule. Havinga[23] reported four or five each in A and S in different experiments, Masri and Singer[24] reported an average of four in A, S, and C, Huisman and Drinkwaard[25] reported five in A, S, C and E, Brown[26] reported five in A, S, and hemoglobins from patients with several diseases, and Rhinesmith, Schroeder, and Pauling[27] reported 3.6 for A (all on the basis 66,700 for the molecular weight).

It was pointed out by Sherman[28] in 1940 that sickled red cells are observed under the polarizing microscope to be *birefringent*, whereas normal cells are optically isotropic. Ponder[29] suggested, on the basis of this observation, that in sickled cells the hemoglobin molecules assume an orderly or paracrystalline arrangement, which is responsible for the sickling. A detailed mechanism of the sickling process was formulated in the first paper on sickle-cell-anemia hemoglobin,[1] as follows: "We can picture the mechanism of the sickling process in the following way. It is likely that it is the globins rather than the hemes of the two hemoglobins that are different. Let us propose that there is a surface region on the globin of the sickle-cell-anemia hemoglobin molecules which is absent in the normal molecule and which has a configuration complementary to a different region of the surface of the hemoglobin molecule. This situation would be somewhat analogous to that which very probably exists in antigen-antibody reactions.[30] The fact that sickling occurs only when the partial pressures of oxygen and carbon monoxide are low suggests that one of these sites is very near to the iron atom of one or more of the hemes, and that when the iron atom is combined with either one of these gases, the complementariness of the two structures is considerably diminished. Under the appropriate conditions, then, the sickle-cell-anemia hemoglobin molecules might be capable of interacting with one another at these sites sufficiently to cause at least a partial alignment of the molecules within the cell, resulting in the erythrocytes's becoming birefringent, and the cell membrane's being distorted to accommodate the now relatively rigid structure within its confines. The addition of oxygen or carbon monoxide to the cell might reverse these effects by disrupting some of the weak bonds between the hemoglobin molecules in favor of the bonds formed between gas molecules and iron atoms of the hemes."

Substantiation of this picture was soon obtained through *microscopic* investigations. Rebuck, Sturrock, and Monaghan,[31] substantiating the work of Sherman,[26] observed that in the early stages of sickling the intracellular hemoglobin forms anisotropic aggregates, suggestive of incipient crystallization. Perutz and Mitchison,[32] (also Perutz, Liquori, and Eirich[33]) found a pronounced dichroism of sickle cells, similar to that of hemoglobin crystals. It was then reported by Harris[34] that a stromafree solution of sickle-cell-anemia hemoglobin with concentration 10 g or more per 100 ml on deoxygenation forms birefringent spindle-shaped bodies varying in length from 1 to 15 μ, which he identified as tactoids (liquid crystals) of the nematic type. It accordingly is probable that sickle-cell anemia can be described as a molecular disease, resulting from the difference in molecular structure of sickle-cell-anemia hemoglobin and normal adult human hemoglobin, the properties of the abnormal hemoglobin being such that when deoxygenated the molecules combine with one another to form long molecular strings, which, through intermolecular attraction, aggregate into tactoids, which have enough mechanical strength to distort the red cell, changing the viscosity of the blood, and causing the clinical and pathological manifestations of the disease.

Perutz and his collaborators[33, 34] found that the *solubility* of sickle-cell-anemia hemoglobin is much smaller than that of normal hemoglobin or of either normal oxyhemoglobin or sickle-cell-anemia oxyhemoglobin. A detailed study of the solubilities of mixtures of sickle-cell-anemia hemoglobin and other hemoglobins has been made by Itano,[35] who has shown that a solubility measurement provides a simple way of determining roughly the amount of sickle-cell-anemia hemoglobin present in a mixture of hemoglobins.

During the six years since the discovery of sickle-cell-anemia hemoglobin several other *abnormal varieties of human hemoglobin* (named hemoglobin C, D, E, G, H, I, and J respectively) have been discovered. The abnormal hemoglobins, either alone or in synergistic interaction with one another or with thalassemia, have been observed in association with several diseases that had not previously been recognized as clinical entities. Another interesting complication in the constitution of the blood has also been discovered, the continued manufacture of fetal hemoglobin (F) by anemic individuals.

Hemoglobin C was discovered by Itano and Neel.[36] The difference in electrophoretic mobility between hemoglobin C and hemoglobin A is nearly twice as great as the difference between S and A: the electric charge of C differs by about 5 electronic charges from that of A. Hemoglobin C was first found to be present together with S in the blood of some anemic patients. These patients may be described as carrying one gene for C and one for S: they are heterozygous in hemoglobin C and also in hemoglobin S, having inherited one of the two abnormalities from each parent. On investigation, the red cells of one parent were found to contain a mixture of A and C, and of the other parent to contain a mixture of A and S. Dilution of S with C does not inhibit the sickling tendency so much as dilution of S with A, and in consequence the individuals of type SC are anemic, their disease being called *sickle-cell:hemoglobin-C disease*. The heterozygous condition in C does not lead to a pathologic state. An interesting relationship has been observed among the three factors, ease of sickling, altitude, and splenic infarction, the last being a probable consequence of intravascular sickling. Smith and Conley[37] have reviewed 16 cases of splenic infarct during aerial flight among individuals with sickling cells. Four had sickle-cell:hemoglobin-C disease and the others were sickle-cell heterozygotes. All of the sickle-cell heterozygote infarcts occurred at 10,000 to 15,000 feet altitude, whereas two of the sickle-cell:hemoglobin-C disease cases were at only 4000 to 6000 feet. Infarction of the spleen is a common occurrence in sickle-cell anemia, so that atrophy of the spleen commonly occurs during childhood. Atrophy is less common in sickle-cell:hemoglobin-C disease, and hypertrophy is frequently observed.[38]

Sickle-cell: hemoglobin-C disease can be distinguished from sickle-cell anemia on clinical grounds.[37, 38, 39, 40, 41, 42, 43, 44, 45] Smith and Conley,[45] using paper electrophoresis, found hemoglobin S to occur in 8.4 per cent of 500 Negroes investigated and hemoglobin C in 2 per cent. These correspond to incidence of the S gene of 4.2 per cent, and of the C gene of 1 per cent. Edington and Lehmann[46] found the incidence of the S gene to be 20.5 per cent and of the C gene to be 12 per cent in West Africa, and Schneider[47] found the incidence of the C gene to be 1.5 per cent among 505 Negroes in Texas. A gene incidence of 1 per cent would correspond to the occurrence of homozygous hemoglobin-C individuals with frequency 0.01^2 or 1 in 10,000. Several patients with a mild hemolytic anemia and with only hemoglobin C in their red cells, who presumably have homozygous hemoglobin-C disease, have been reported.[42, 43, 46, 47, 48, 49, 50, 51, 52, 53, 54] Hemoglobin crystals have been observed in some of the erythrocytes of homozygous hemoglobin-C disease patients following splenectomy.[50, 54]

An interesting complication in the hemoglobin pattern has been resolved through the discovery that anemic individuals may continue to manufacture fetal hemoglobin long after they have passed the fetal stage of development. Wells and Itano[17] discovered that there is present in the blood of patients with a mild form of sickle-cell anemia a small amount, 5 per cent to 20 per cent of the total, of a protein other than S, and Itano and Neel[34] reported that one patient with sickle-cell:hemoglobin-C disease gave a hemoglobin electrophoretic pattern showing S and C in large amounts (39 per cent and 48 per cent, respectively) and an additional hemoglobin, 13 per cent, with the electrophoretic mobility, at the pH used (pH 6.50), of A. It was then pointed out by Singer, Chernoff, and

Singer[55] that there is present in the blood of many patients with sickle-cell anemia and other hematologic disorders a hemoglobin fraction that is resistant to alkali denaturation, and that the properties of this fraction permit it to be identified with *fetal hemoglobin, F;* they suggested that the hemoglobin reported by Wells and Itano is hemoglobin F, and not A. Liquori[56] found as much as 50 per cent of the hemoglobin in the blood of patients with thalassemia to be fetal hemoglobin, and Rich[57] found that two patients with thalassemia major had in their red cells no hemoglobin other than F. It is now recognized that anemic patients may continue to manufacture significant amounts of F throughout their lives, and that the presence of F ameliorates their disease.

Thalassemia (Cooley's anemia, Mediterranean anemia) is a hereditary anemia resulting from a gene-controlled interference with the process of synthesis of adult human hemoglobin.[58] The disease was thought to be largely confined to persons derived from the northern shores of the Mediterranean; however, recent reports indicate that it occurs among natives of Southeast Asia.[59, 60, 61] The thalassemia gene in the homozygous form produces a very serious anemia, thalassemia major, which usually terminates fatally in childhood. In heterozygous form the gene produces a mild anemia, called thalassemia minor. In some parts of Italy, notably the region around Ferrara, the incidence of thalassemia minor is about 10 per cent, corresponding to a gene probability of 5 per cent.[62]

The interaction of the sickle cell and thalassemia genes to produce a hemolytic anemia was first reported by Silvestroni and Bianco in 1945. This observation and later ones by the same authors have been summarized in their review.[63] Powell, Rodarte, and Neel[64] reported a case in 1950 and discussed the possible genetic mechanisms. Although a single dose of either gene does not result in hemolytic anemia, their simultaneous presence causes a mild to severe anemia. Electrophoretic studies by Sturgeon, Itano, and Valentine[65] and by Neel, Itano, and Lawrence[66] revealed that the erythrocytes in this condition contained a high proportion of sickle-cell hemoglobin. It was concluded that the thalassemia gene interferes with the synthesis of hemoglobin A but not of S, and that the hemolytic process, as in sickle cell anemia, is due to cells which sickle more readily than sickle cell trait cells. Other cases of this disease in which hemoglobin compositions were determined have been reported.[38, 67, 68, 69]

Several patients with another new disease, *thalassemia:hemoglobin-C disease*, have also been reported.[70, 71, 72] The patient mentioned in a footnote in reference 41 is the same one studied in reference 71.

A third abnormal hemoglobin to be discovered, *hemoglobin D*, was found by Itano [73] in five members of a single family. Two of the family were patients with a disease resembling sickle-cell anemia but somewhat milder than the normal form of this disease. The blood of each of these two patients with sickle-cell:hemoglobin-D disease was found to contain a small amount (6 per cent to 12 per cent) of F, in addition to S and D. Hemoglobin D has electrophoretic properties like those of S, and solubility like that of A. A mother of this family, who is a carrier of hemoglobin D, is of English, Irish, and American Indian ancestry.[74] Additional occurrences of hemoglobin D have been reported from England,[4, 75] India,[76, 77] Algeria,[78] Turkey,[79] and Uganda (in Indian immigrants).[80] These have included cases of sickle-cell:hemoglobin D disease,[4, 75] and apparent homozygous hemoglobin D disease.[77]

Three individuals containing the fourth abnormal *hemoglobin E*, were discovered by Itano, Bergren, and Sturgeon.[74, 81] One of them is an anemic patient with thalassemia:hemoglobin-E disease. The mother of the patient has thalassemia minor, and no hemoglobin other than A in her red cells. The red cells of the patient contain 41 per cent F and 59 per cent E. Both the father of the patient[7] and his half-sister[82] are of type AE; their common parent, presumed to be the carrier of the disease for E, came from India. The patient has inherited thalassemia minor from her mother and the allele for E from the father; her red cells contain only hemoglobin E and hemoglobin F (41 per cent—the thalassemia gene seems to have completely suppressed the manufacture of A).

Homozygous hemoglobin-E disease and thalassemia:hemoglobin-E disease have been observed in Thailand.[83, 84, 85] The hemoglobin components in six individuals with homozygous hemoglobin-E disease were E, 94 to 99 per cent, and F, 6 to 1 per cent, and the components in thirty-two individuals with thalassemia:hemoglobin-E disease were E, 15 to 95 per cent, and F, 85 to 5 per cent.[84] Normal adult hemoglobin was not detectable in any patient. Of 1,006 individuals surveyed in Thailand, 137 had hemoglobin-E trait and 3 had homozygous hemoglobin-E disease.[85] Hemoglobin E has also been observed in Ceylon,[86] Indonesia,[87] and among Burmese residing in England.[88]

Hemoglobin E shows a striking change in electrophoretic mobility with change in pH. At pH 6.5 its mobility is greater than that of A and slightly less than that of S, and at pH 8.8 it is nearly identical with that of C. The absorption spectrum, solubility, and lability to alkali denaturation of E are similar to those of A.[81]

Several new hemoglobins have been observed during the past year. Edington and Lehmann[89] observed a specimen (from West Africa) which contained hemoglobin A and a component migrating between the A and S positions on paper at pH 8.6; hemoglobin F migrates in this position, but the new component is unlike F in that it is labile to alkali denaturation. They named the component *hemoglobin G*. Edington, Lehmann, and Schneider[90] have reported that nine siblings of this individual are also of type AG, and that the father is a G homozygote, the mother being normal.

Hemoglobin H is the first abnormal hemoglobin to be found with lower isoelectric point than A. It was found, together with hemoglobin A, in two siblings in a family of Chinese descent.[91] Hemoglobin H is not found in the red cells of either parent. One parent has thalassemia minor, and the children with hemoglobin H have anemias indistinguishable from thalassemia major. Bergren, Itano, and Sturgeon[82] and Motulsky[92] have observed abnormal hemoglobins with the same electrophoretic behavior as hemoglobin H in individuals of Scottish descent and of Filipino descent.

A hemoglobin with isoelectric point between those of H and A was reported by Jensen. He was unaware of the work of Rigas, Koler, and Osgood, and used the name hemoglobin H in the abstract describing his discovery;[93] he has, however, agreed to change the name to *hemoglobin I*.[94] Hemoglobin I[94] was found in individuals of three generations of a Negro family in North Carolina. Only the trait condition has been seen, and the affected individuals contain hemoglobins A and I in the ratio 80: 20.

Another hemoglobin, named *hemoglobin J*,[95] has been found with isoelectric point between those of I and A. Hemoglobin J is present in the trait condition in two generations of a Negro family in Virginia, the two hemoglobins A and J occurring in the ratio 40:60. Neither hemoglobin-I trait nor hemoglobin-J trait is associated with clinical or hematologic abnormalities. An abnormal hemoglobin which behaves like H, I, or J in that it migrates more rapidly than A in electrophoresis at pH 8.6 has been reported by Cabannes.[96] Its presence is associated with mild anemia and morphologic abnormalities of erythrocytes.

A hemoglobin component associated with familial anemia similar to thalassemia has been reported by Battle and Lewis[97] and has been designated hemoglobin K.[8] Results obtained by Itano[9,10] on samples from the patients of Battle and Lewis suggest that this component does not differ from the minor component observed in small amounts in normal individuals (component X of Figure 3) and in slightly elevated amounts in thalassemia minor.[65, 66, 98, 99] Since it is a component of all normal samples, its genetic significance differs from that of the inherited abnormal

Table 1. Distinguishing properties of the human hemoglobins.

Physical property	Hemoglobin A	Fe	Sf	C	D	E	G	H	I	J
Isoelectric point[a]	6.87	6.98	7.09	7.2	7.09	—	—	—	—	—
Mobility $\times 10^5$ at pH 6.5[b]	2.4	2.2	2.9	3.2	2.9	2.8	between A and S	—0.1	1.7	2.0
Order of migration velocities at pH 8.6[c]	4	5	6	8	6	7	5	1	2	3
Solubility in 2.58 M phosphate at pH 6.8[d]	1.39	$> A$	$\ll A$	3.5	$\approx A$	$\approx A$	0.6	—	$\approx A$	$> A$

[a] As carbonmonoxyhemoglobin in phosphate buffers of ionic strength 0.1.[1, 109, 110]

[b] In $cm^2\ sec^{-1}\ volt^{-1}$ by the moving-boundary method. Cacodylate buffer of ionic strength 0.1.[36, 73, 81, 109]

[c] By paper electrophoresis in pH-8.6 barbital buffer of ionic strength 0.05—0.06.[43, 44, 81, 83, 89, 95, 111] Hemoglobin H is the fastest and C the slowest. The numbers have no quantitative significance.

[d] In grams per liter at 25° as amorphous ferrohemoglobin. Hemoglobin S has solubility 0.1 g/l in 2.24 M phosphate. The approximate solubilities of D, E, I, and J were deduced from those of their naturally occurring mixtures with A.[35, 90, 95, 109] Comparable information about crystalline ferrohemoglobin solubilities of A and S is available.[32, 33]

[e] Is resistant to denaturation at high pH[55] and has a maximum in its absorption spectrum at 2898 A.[112]

[f] Causes sickling when present as ferrohemoglobin in red cells[1] and forms tactoids when concentrated solution is deoxygenated.[34]

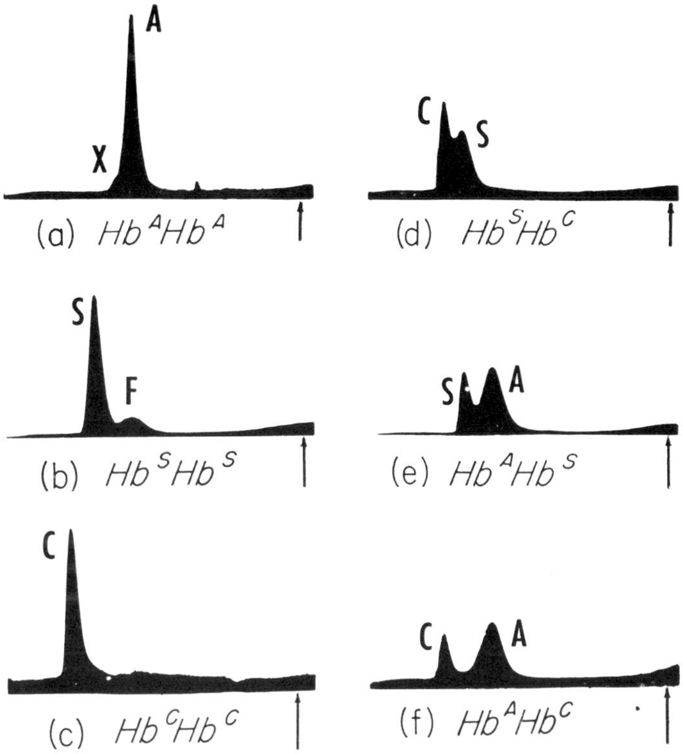

Fig. 3. The electrophoretic patterns of the hemoglobin mixtures found in several human genotypes, (a) normal adult, (b) sickle-cell anemia, (c) hemoglobin C disease, (d) sickle-cell:hemoglobin C disease, (e) sickle-cell trait, and (f) hemoglobin C trait. The electrophoretic analyses were conducted in cacodylate buffer of pH 6.5, in which all of the hemoglobins represented migrate as cations. The genetic significance of component X in normal adult hemoglobin is not known. The proportions of components present differ among individuals of the same genotype, and fetal hemoglobin (F) occurs in some cases of hemoglobin C disease, sickle-cell:hemoglobin C disease, and other severe anemias as well as in sickle-cell anemia.

hemoglobins. The use of the letter K to designate the component has, therefore, been withdrawn by mutual agreement of the investigators.

Table 1 lists some of the differences among the human hemoglobins. Only those techniques that have been applied to more than two or three of the hemoglobin types have been included. Table 2 lists the states known to be accompanied by the abnormal hemoglobins or the thalassemia allele. The sickle-cell-disease phenotype was defined early in the study of abnormal hemoglobins.[3] A recent review of the available information on other anemias associated with abnormal hemoglobins has revealed that these anemias are due to a combination of a mild hemolytic process and a relative inhibition of hemoglobin synthesis.[8] Thus, they resemble mild cases of thalassemia major[99] in their clinical and hematologic manifestations and, prior to the introduction of hemoglobin electrophoresis, have been mistaken for the latter condition in several instances. Data are available to support the postulate for allelism of the genes for A, S, C, and D,[18, 100] and for non-allelism of the thalassemia gene with the genes for S and C.[63, 70] On the other hand, the absence of hemoglobin A in hemoglobin-E:thalassemia disease[81] is best explained in terms of a thalassemia gene allelic with the gene for E.[5] It is not unlikely that genes both allelic and non-allelic with the abnormal hemoglobin alleles may produce the thalassemia effect.[8]

The postulated genotypes in Table 2 are based on the assumption that the genes which determine the type of hemoglobin are allelic with

Table 2. Observed hemoglobin contents and postulated genotypes for the abnormal hemoglobin states.

Phenotype	Condition	Hemoglobin[a] A	Hemoglobin[a] Abnormal	Hemoglobin[a] F	Genotype[b]
Asymptomatic state[c]	Sickle-cell trait	+	S	—	$Hb^A Hb^S Th^N Th^N$
	Hemoglobin-C trait	+	C	—	$Hb^A Hb^C Th^N Th^N$
	Hemoglobin-D trait	+	D	—	$Hb^A Hb^D Th^N Th^N$
	Hemoglobin-E trait	+	E	—	$Hb^A Hb^E Th^N Th^N$
	Hemoglobin-G trait	+	G	—	$Hb^A Hb^G Th^N Th^N$
	Hemoglobin-I trait	+	I	—	$Hb^A Hb^I Th^N Th^N$
	Hemoglobin-J trait	+	J	—	$Hb^A Hb^J Th^N Th^N$
	Homozygous G individual	—	G	—	$Hb^G Hb^G Th^N Th^N$
	Thalassemia minor	+	—	±	$Hb^A Hb^A Th^N Th^T$
Sickle-cell disease[d]	Sickle-cell anemia	—	S	±	$Hb^S Hb^S Th^N Th^N$
	Sickle-cell:hemoglobin-C disease	—	S, C	±	$Hb^S Hb^C Th^N Th^N$
	Sickle-cell:hemoglobin-D disease	—	S, D	+	$Hb^S Hb^D Th^N Th^N$
	Sickle-cell:thalassemia disease[e]	±	S	±	$Hb^A Hb^S Th^N Th^T$
Thalassemia disease[f]	Thalassemia major	±	—	+	$Hb^A Hb^A Th^T Th^T$
	Hemoglobin-C disease	—	C	±	$Hb^C Hb^C Th^N Th^N$
	Hemoglobin-C:thalassemia disease	+	C	±	$Hb^A Hb^C Th^N Th^T$
	Hemoglobin-E disease	—	E	+	$Hb^E Hb^E Th^N Th^N$
	Hemoglobin-E:thalassemia disease[g]	—	E	+	$Hb^A Hb^E Th^N Th^T$
	Hemoglobin-H:thalassemia disease	+	H	—	[h]

[a] +, present; —, absent; ±, present in some cases, absent in others.

[b] Hb^A, Hb^S, etc. — multiple allelic series of genes determining hemoglobin type; Th^T, Th^N — thalassemia gene and its normal allele, respectively.[100] (Science, in press). A thalassemia gene, Hb^T, allelic with the *Hb* locus, may also exist.[5, 8]

[c] Hemolytic disease absent although erythrocytes may show morphologic abnormalities.

[d] Hemolytic disease with sickling erythrocytes.[3]

[e] Individuals with hemoglobins S and F but no A be of genotype $Hb^S Hb^T Th^N Th^N$.[8]

[f] Hemolytic disease with relative inhibition of hemoglobin synthesis.[8]

[g] The genotype $Hb^E Hb^T Th^N Th^N$ would be consistent with the absence of hemoglobin A.[5]

[h] A reasonable postulate of the genotype with respect to the Hb and Th loci cannot be made on the basis of available data.

each other but not with the thalassemia allele. The gene notation follows that proposed by Allison.[101] The presence of fetal hemoglobin has been assumed not to be due to a direct action of any of the abnormal hemoglobin or thalassemia genes but to a compensatory mechanism, the chemical and genetic nature of which is still unknown. If we assume the eight types of hemoglobin A, S, C, D, E, G, I, and J to be allelomorphic, there are 36 hemoglobin types predicted: the eight homozygotes, AA, SS, etc., and the 28 heterozygotes, AS, AC, etc. Of the eight homozygotes six, AA, SS, CC, DD, EE, and GG, have been observed; those other than AA and GG are anemic (see Table 2). Ten of the heterozygotes (each involving either A or S) have also been observed, and in addition individuals representing the abnormal hemoglobins S, C, and E in synergy with thalassemia minor (Table 2).

The results of moving boundary *electrophoresis* experiments in cacodylate buffer[17] of pH 6.5 and ionic strength 0.1 are given in Fig. 3. Normal adult (AA), sickle-cell anemia (SS) homozygous hemoglobin C disease (CC), sickle-cell trait (AS), hemoglobin C trait (AC), and sickle-cell:hemoglobin C disease (SC) are shown. These are the most common types. In addition to the major components, hemoglobin F occurs in nearly all cases of sickle-cell anemia and in some cases of sickle-cell: hemoglobin C disease and homozygous hemoglobin C disease.

It is interesting that the thalassemia gene, which interferes with the manufacture of normal adult human hemoglobin, seems usually not to interfere seriously with the manufacture of the abnormal hemoglobins; one exception (involving hemoglobin C) has been reported.[71]

A number of interesting questions about the *origin* and *heredity* of the abnormal hemoglobins remain to be answered. The question of the continued high incidence of the sickle-cell gene, despite its continued loss because of the lethal character of the homozygous condition, has been raised by Neel,[62] who has suggested three alternative explanations: (1) continued production of the sickle-cell allele through mutation; (2) the existence of an abnormal genetic mechanism that favors the heterozygous condition, AS, over the normal condition, AA; (3) a positive selection of the heterozygote, perhaps through increased fertility. The first explanation has to be rejected because the rate of mutation that would be required is far greater than any that has ever been observed for any organism. There now exists evidence indicating that the third alternative provides the correct explanation, and that malaria is involved. It was first suggested by Brain[102] that the presence of hemoglobin S in the red cells might give *protection against malaria* parasites, and thus confer an advantage to the sicklemic individual that would balance the disadvantage of the lethal homozygosity. Lehmann[103] wrote that "The lethal tendency of a gene potentially causing sickle-cell anemia may thus be counteracted by its conferring a resistance to malaria similar to that found in early infancy."

A test of the hypothesis was carried out recently by Allison,[104] who infected 15 healthy adult Africans with the sickle-cell trait and 15 similar healthy adult Africans without the sickle-cell trait with *P. falciparum.*

The subjects were infected by subinoculation with 15 ml. of blood containing a large number of trophozoites or by being bitten by heavily infected anopheles mosquitoes, in which the presence of sporozoites was confirmed by dissection of the mosquito. The infection was established in 14 out of the 15 Africans without the sickle-cell trait, and in only 2 of the 15 with this trait. It was concluded by Allison that the abnormal erythrocytes of individuals with the sickle-cell trait are less easily parasitized by *P. falciparum* than are normal erythrocytes, and that accordingly those individuals who are heterozygous for the S allele have a selective advantage over normal individuals in regions where malaria is hyperendemic. It is, of course, not unreasonable that the abnormal hemoglobin may be less effective than the normal hemoglobin in nourishing the parasites.

Beutler, Dern, and Flanagan,[105] working with non-immune adult subjects, repeated Allison's experiment and failed to find a significant difference in the parasitemia of normal and sickle-cell trait subjects. A second inoculation in some of their original subjects resulted in somewhat less marked parasitemia among the sickle-cell trait subjects, but the authors considered the differences to be "unimpressive and of questionable significance". A more severe manifestation of malaria can place normal subjects at a selective disadvantage by decreasing their fertility and by increasing their mortality.

The role of malaria as a selecting agent is supported by the observation of Raper[106] that a relatively lower direct mortality from malaria occurs in sickling children. The viability of homozygous sicklers to reproductive age in Africa has been estimated by Allison[107] to be about 20 per cent of that of other genotypes. The relative fertility in adults with sickle cell anemia, which is the measure of their contribution to gene survival, was not, however, given. Lehmann and Raper,[108] in a study of a community with 39 per cent sickling, concluded that the survival of homozygotes plays no significant part in the maintenance of the high sickling incidence, but that 10.6—24.2 per cent of all normal homozygotes would have to die from malaria if all heterozygotes survived this disease. The possibility that the heterozygotes might have more children than normal homozygotes was discarded in their calculations.

Is is accordingly likely that the hemoglobin S gene was originally produced by a mutation in a region where normal individuals were at a disadvantage because of the protection that the presence of hemoglobin S in the erythrocytes of the AS heterozygotes provides against malaria. There is no evidence at the present time, except indirect evidence, such as the high incidence of hemoglobin C in Africa and of hemoglobin E in Thailand, Ceylon, and Indonesia, where malaria may have been an important evolutionary factor, that the other abnormal hemoglobins have, in the heterozygous condition, protective power against malaria. It is possible that some of the abnormal hemoglobins may have some other evolutionary advantage, as yet undiscovered.

There remains the problem of the nature of the difference in molecular structure among the hemoglobins. Despite several attacks on this problem, no detailed information on this point has been obtained, and it may be that the question will remain unanswered until the general attack on the problem of the structure of globular proteins has been crowned with success.

References

1. Pauling, L., Itano, H. A., Singer, S. J., and Wells, I. C., *Science* **110**, 543 (1949).
2. Harris, H., An Introduction to Human Biochemical Genetics. Cambridge University Press, p. 46 (1953).
3. Itano, H. A., *Science* **117**, 89 (1953).
4. White, J. C., and Beaven, G. H., A review of the varieties of human haemoglobin in health and disease. *J. Clin. Pathol.* **7**, 175 (1954).
5. Pauling, L., *The Harvey Lectures* **41**, 216 (1955).
6. Singer, K., *Am. J. Med.* **18**, 633 (1955).
7. Chernoff, A. I., *New England J. Med.* **253**, 322, 365, 416 (1955).
8. Itano, H. A., Am. Med. Ass. *Arch. Internal Med.* **96**, 287 (1955).
9. Itano, H. A., *Ann. Rev. Biochem.* **25**, 331 (1956).
10. Itano, H. A., Sturgeon, P., and Bergren, W. R., The abnormal hemoglobins. *Medicine* **35**, 121 (1956).
11. Zuelzer, W. W., Neel, J. V., and Robinson, A. R., *Progr. Hematol.* **1**, 91 (1956).
12. Herrick, J. B., *Arch. Internal Med.* **6**, 517 (1910).
13. Emmel, V. E., *Arch. Internal Med.* **20**, 586 (1917).
14. Taliaferro, W. H., and Huck, J. G., *Genetics* **8**, 594 (1923).
15. Neel, J. V., *Medicine* **26**, 115 (1947).
16. Neel, J. V., *Science* **110**, 64 (1949).
17. Wells, I. C., and Itano, H. A., *J. Biol. Chem.* **188**, 65 (1951).
18. Itano, H. A., *Am. J. Human Genet.* **5**, 34 (1953).
19. Schroeder, W. A., Kay, L. M., and Wells, I. C., *J. Biol. Chem.* **187**, 221 (1950).
20. Huisman, T. H. J., Jonxis, J. H. P., and van der Schaaf, P. C., *Nature* **175**, 902 (1955).
21. Jonxis, J. H. P., Huisman, T. H. J., van der Schaaf, P. C., and Prins, K. H., *Nature* **177**, 627 (1956).
22. Porter, R. R., and Sanger, F., *Biochem. J.* **42**, 287 (1948).
23. Havinga, E., *Proc. Natl. Acad. Sci. U.S.* **39**. 59 (1953).
24. Masri, M. S., and Singer, K., *Arch. Biochem. Biophys.* **58**, 414 (1955).
25. Huisman, T. H. J., and Drinkwaard, J., *Biochim. et Biophys. Acta.* **18**, 588 (1955).
26. Brown, H., *Arch Biochem. Biophys.* **61**, 241 (1956).
27. Rhinesmith, H. S., Schroeder, W. A., and Pauling, L., *J. Am. Chem. Soc.* In press. (1957).
28. Sherman, I. J., *Bull. Johns Hopkins Hosp.* **67**, 309 (1940).
29. Ponder, E., "Hemolysis and Related Phenomena", p. 145, Grune and Stratton, New York, (1948).
30. Pauling, L., Pressman, D., and Campbell, D. H., *Physiol. Revs.* **23**, 203 (1943).
31. Rebuck, J. W., Sturrock, R. M., and Monaghan, D. A., *Federation Proc.* **9**, 340 (1950).
32. Perutz, M. F., and Mitchison, J. M., *Nature* **166**, 677 (1950).
33. Perutz, M. F., Liquori, A. M., and Eirich, F., *Nature* **167**, 929 (1951).
34. Harris, J. W., *Proc. Soc. Exptl. Biol. Med.* **75**, 197 (1950).
35. Itano, H. A., *Arch. Biochem. Biophys.* **47**, 148 (1953).
36. Itano, H. A., and Neel, J. V., *Proc. Natl. Acad. Sci. U.S.* **36**, 614 (1950).
37. Smith, E. W., and Conley, C. L., *Bull. Johns Hopkins Hosp.* **96**, 35 (1955).
38. Smith, E. W., and Conley, C. L., *Bull. Johns Hopkins Hosp.* **94**, 289 (1954).
39. Kaplan, E., Zuelzer, W. W., and Neel, J. V., *Blood* **6**, 1240 (1951).
40. Neel, J. V., Kaplan, E., and Zuelzer, W. W., *Blood* **8**, 724 (1953).
41. Kaplan, E., Zuelzer, W. W., and Neel, J. V., *Blood* **8**, 724 (1953).
42. Spaet, T. H., Alway, R. H., and Ward, G., *Pediatrics* **12**, 483 (1953).

43. Ranney, H. M., Larson, D. L., and McCormack, G. H., Jr., *J. Clin. Invest*, **32**, 1277 (1953).
44. Larson, D. L., and Ranney, H. M., *J. Clin. Invest.* **32**, 1070 (1953).
45. Smith, E. W., and Conley, C. L., *Bull. Johns Hopkins Hosp.* **93**, 94 (1953).
46. Edington, G. M., and Lehmann, H., *Trans. Roy. Soc. Trop. Med. Hyg.* **48**, 332 (1954).
47. Schneider, R. G., *J. Lab. Clin. Med.* **44**, 133 (1945).
48. Terry, D. W., Motulsky, A. C., and Rath, C. E., *New England J. Med.* **251**, 365 (1954).
49. Singer, K., Chapman, A. Z., Goldberg, S. R., Rubenstein, H. M., and Rosenblum, S. A., *Blood* **9**, 1023 (1954).
50. Diggs, L. W., Kraus, A. P., Morrison, D. B., and Rudnicki, R. P. T., *Blood* **9**, 1172 (1954).
51. Hartz, W. H., and Schwartz, S. O., *Blood* **10**, 235 (1955).
52. Thomas, E. D., Motulsky, A. C., and Walters, D. H., *Am. J. Med.* **18**, 832 (1955).
53. Lange, R. D., and Hagen, P. S., *Am. J. Med. Sci.* **229**, 655 (1955).
54. Wheby, M. S., Thorup, O. A., and Leavell, B. S., *Blood* **11**, 266 (1956).
55. Singer, K., Chernoff, A. I., and Singer, L., *Blood* **6**, 413 (1951).
56. Liquori, A. M., *Nature* **167**, 950 (1951).
57. Rich, A., *Proc. Natl. Acad. Sci. U.S.* **38**, 187 (1952).
58. Cooley, T. B., and Lee, P., *Trans. Am. Pediat. Soc.* **37**, 29 (1925).
59. Minnich, V., Na-Nakorn, S., Chongchareonsuk, S., and Kochasemi, S., *Blood* **9**, 1 (1954).
60. Lie-Injo Luan Eng and Jo Kian Tjay, *Doc. Med. Geogr. Trop.* **7**, 273 (1955).
61. Siddoo, J. K., Siddoo, S. K., Chase, W. H., Morgan-Dean, L., and Perry, W. H., *Blood* **11**, 197 (1956).
62. Neel, J. V., *Cold Spring Harbor Symposia Quant. Biol.* **15**, 141 (1951).
63. Silvestroni, E., and Bianco, I., *Blood* **7**, 429 (1952).
64. Powell, W. N., Rodarte, J. G., and Neel, J. V., *Blood* **5**, 887 (1950).
65. Sturgeon, P., Itano, H. A., and Valentine, W. N. *Blood* **7**, 350 (1952).
66. Neel, J. V., Itano, H. A., and Lawrence, J. S., *Blood* **8**, 434 (1953).
67. Banks, L. O., and Scott, R. B., *Am. Med. Ass. Am. J. Dis. Child.* **84**, 601 (1952).
68. Humble, J. G., Anderson, I., White, J. C., and Freeman, T., *J. Clin. Pathol.* **7**, 201 (1954).
69. Singer, K., Singer, L., and Goldberg, S. R., *Blood* **10**, 405 (1955).
70. Singer, K., Kraus, A. P., Singer, L., Rubenstein, H. M., and Goldberg, S. R., *Blood* **9**, 1032 (1954).
71. Zuelzer, W. W., and Kaplan, E., *Blood* **9**, 1047 (1955).
72. Erlandson, M., Smith, C. H., and Schulman, I., *Pediatrics* **740** (1956).
73. Itano, H. A., *Proc. Natl. Acad. Sci. U.S.* **37**, 775 (1951).
74. Sturgeon, P., Itano, H. A., and Bergren, W. R., *Blood* **10**, 389 (1955).
75. Stewart, J. W., and MacIver, J. E., *Lancet* **i**, 23 (1956).
76. Bird, G. W. G., Lehmann, H., and Mourant, A. E., *Trans. Roy. Soc. Trop. Med. Hyg.* **49**, 399 (1955).
77. Bird, G. W. G., and Lehmann, H., *Man* **55**, 1 (1956).
78. Cabanes, R., Sendra, L., and Dalaut, *Algérie Médicale* **59**, 387 (1955).
79. Aksoy, M., and Lehmann, H., *Trans. Roy. Soc. Trop. Med. Hyg.* **50**, 178 (1956).
80. Jacob, G. F., Lehmann, H., and Raper, A. B., *East African Med. J.* **33**, 135 (1956).
81. Itano, H. A., Bergren, W. R., and Sturgeon, P., *J. Am. Chem. Soc.* **76**, 2278 (1954).
82. Bergren, W. R., Itano, H. A., and Sturgeon, P., Unpublished observations.
83. Chernoff, A. I., Minnich, V., and Chongchareonsuk, S., *Science* **120**, 605 (1954).
84. Chernoff, A. I., Minnich, V., Na-Nakorn, S., Tuchinda, S., Kashemsant, C., and Chernoff, R. R., *J. Lab. Clin. Med.* **47**, 455 (1956).
85. Na-Nakorn, S., Minnich, V., and Chernoff, A. I., *J. Lab. Clin. Med.* **47**, 490 (1956).
86. Graff, J. A. E., Ikin, E. W., Lehmann, H., Mourant, A. E., Parkin, D. M., and Wickremasinghe, R. L., *J. Physiol.* **127**, 41P (1955).
87. Lie-Injo Luan Eng, *Nature* **176**, 469 (1955).

88. Lehmann, H., Story, P., and Thein, H., *Brit. Med. J.* **i**, 544 (1956).
89. Edington, G. M., and Lehmann, H., *Lancet* **267**, 173 (1954).
90. Edington, G. M., Lehmann, H., and Schneider, R. G., *Nature* **175**, 850 (1955).
91. Rigas, D. A., Koler, R. D., and Osgood, E. E., *Science* **121**, 372 (1955).
92. Motulsky, A. G., Personal communication.
93. Jensen, W. N., Page, E. B., and Rucknagel, D. L., *Clin. Research Proc.* **3**, 93 (1955).
94. Rucknagel, D. L., Page, E. B., and Jensen, W. N., *Blood* **10**, 999 (1955).
95. Thorup, O. A., Itano, H. A., Wheby, M., and Leavell, B. S., *Science* **123**, 889 (1956).
96. Cabannes, R., Sendra, L., and Dalaut *Compt. rend soc. biol.* **149**, 914 (1955).
97. Battle, J. D., and Lerois, L., *J. Lab. Clin. Med.* **44**, 764 (1954).
98. Kunkel, H. G., and Wallenius, G., *Science* **122**, 288 (1955).
99. Sturgeon, P., Itano, H. A., and Bergren, W. R. *Brit. J. Haemat.* **1**, 264 (1955).
100. Ranney, H. M., *J. Clin. Invest.* **33**, 1634 (1954).
101. Allison, A. C., *Science* **122**, 640 (1955).
102. Brain, P., *Brit. Med. J.* **ii**, 880 (1952).
103. Lehmann, H., *J. Clin. Path.* **6**, 329 (1953).
104. Allison, A. C., *Brit. Med. J.* **i**, 290 (1954).
105. Beutler, E. Dern, R. J., and Flanagan, C. L., *Brit. Med. J.* **i**, 1189 (1955).
106. Raper, A. B., *Brit. Med. J.* **i**, **965** (1956).
107. Allison, A. C., *Trans. Roy. Soc. Trop. Med. Hyg.* **50**, 185 (1956).
108. Lehmann, R., and Raper, A. B., *Brit. Med. J.* **ii**, **333** (1956).
109. Itano, H. A., Unpublished observations.
110. Zinsser, H. H., *Arch. Biochem. Biophys.* **38**, 195 (1952).
111. Bergren, W. R., Sturgeon, P., and Itano, H. A., *Acta Haemat.* **12**, 160 (1954).
112. Beaven, G. H., Hoch, H., and Holiday, E. R., *Biochem. J.* **49**, 374 (1951).

MOLECULAR DISEASE AND EVOLUTION*

LINUS PAULING

Professor of Chemistry
California Institute of Technology
Pasadena, California

THE universe is made up of matter and radiant energy. The human body is made up of molecules—molecules of all sorts; little molecules, such as the water molecule, consisting of only three atoms—a very important molecule, which is present in larger numbers than any other in the human body; larger molecules, of medium size, such as those that constitute the vitamins; and many very large molecules, protein molecules, polysaccharide molecules, nucleic acid molecules.

I believe that it is likely that a human being manufactures 50,000 or 100,000 different kinds of protein molecules. A representative protein molecule, such as hemoglobin, is built of about 10,000 atoms. It has a well-defined structure; for most of the protein molecules not a single atom is out of place.

The protein molecules of different kinds are manufactured by genes, which are themselves molecules of deoxyribonucleic acid. Each one of us inherits half of his complement of genes, approximately 50,000, from his father, and the other half, approximately 50,000, from his mother. It is these molecules, 100,000 molecules of DNA, that make the human being what he is, that confer his character upon him.

These are the most important molecules in the world. The pool of human germ plasm is a precious heritage of the human race.

A few years ago it was discovered that some diseases are molecular diseases, diseases of protein molecules. A gene, a molecule of deoxyribonucleic acid, may be damaged by cosmic radiation or some other mutagenic agent in such a way that a few atoms are out of place. This gene then duplicates itself in its new, mutated, form. Moreover, when it

* This article is a revision of a lecture given before a meeting of The Rudolf Virchow Medical Society in the City of New York, held at The New York Academy of Medicine, November 5, 1961. It is published here by permission of S. Karger, Basel/New York, having originally appeared in the *Karger Gazette* No. 7-8 (1963) and in the *Proceedings of the Rudolf Virchow Medical Society in the City of New York, 1962,* v. 21, 1963.

serves its other function, the function other than self-duplication, it determines the nature of a protein molecule, which it has the responsibility of manufacturing. A mutated gene produces an altered protein molecule, with a few atoms different from the corresponding normal protein molecule.

Molecular disease is closely connected with evolution. The appearance of the concept of good and evil that was interpreted by Man as his painful expulsion from Paradise probably was a molecular disease that turned out to be evolution.

Among the molecular diseases there are many that involve enzymes. For example, the disease phenylketonuria, which is responsible for 1 per cent of the institutionalized mentally defective individuals in the United States, is a simple molecular disease that is reasonably well understood. One person in eighty has an abnormal gene that is called the gene for phenylketonuria. A normal person has two genes that manufacture, independently of one another, an enzyme in the liver that catalyzes the oxidation of phenylalanine to tyrosine. This is a mechanism for converting part of the phenylalanine in our food, which is present in excess of our need, into another amino acid, tyrosine, which is then used in various ways in the human body. One person in eighty has one normal gene, which manufactures this enzyme, and one abnormal gene (the gene for phenylketonuria), which does not manufacture the enzyme, or which manufactures an abnormal enzyme molecule that is lacking in enzyme activity.

These people, the carriers of a single gene for phenylketonuria, manufacture only 50 per cent as much of the enzyme as normal individuals; but this 50 per cent is enough to take care of the phenylalanine that they ingest. They are called phenylketonuric heterozygotes. They are not damaged significantly by carrying the gene in single dose.

However, when two of these heterozygotes marry one another there occurs the great lottery, the greatest of all lotteries in the world, in which the prospective child, the fertilized ovum, carries out the selection of one or the other of the pair of genes that the father has and of one or the other of the pair of genes that the mother has. On average, a quarter of the children inherit the defective gene from the father and also the defective gene from the mother. They have the defective gene in double dose, and they manufacture none of the enzyme that catalyzes the oxidation of phenylalanine to tyrosine. When such a homozygote eats his food, containing ordinary protein, the phenylalanine builds up in his blood stream and cerebrospinal fluid to concentrations as great as fifty times that in normal individuals. This high concentration of phenylalanine and of other substances made from it interferes with the growth and function of the brain in such a way as to cause him to be mentally defective, perhaps with an I.Q. as low as 20. In addition, the phenylketonuria genes in double dose cause him to have severe eczema and other somatic difficulties.

It has been recognized in recent years that it is possible to treat this disease, phenylketonuria. A diagnosis of the disease may be made at an age as early as one month, and the infant then may be fed a diet of protein hydrolysate from which most of the phenylalanine has been removed. Children treated in this way seem to develop in an essentially normal manner.

Many molecular diseases that have arisen in the course of evolution have been controlled in a somewhat similar manner. Human beings require many vitamins. Pellagra is an example of a vitamin deficiency disease—a molecular disease that originated through a mutation, perhaps millions of years ago, and was then cured by the heterotrophic process of eating other organisms that manufacture the vitamin. Scurvy and other avitaminoses are also diseases of this sort. It is not customary for us to admit that we have these diseases, because we treat them as a matter of habit by eating what is called a proper diet.

Organisms such as the red bread mould are able to manufacture not only all of the vitamins, but also all of the amino acids. At some time in our evolutionary history we suffered mutations that resulted in the loss of our power to manufacture the various enzymes involved in these syntheses. Each of these mutations produced in our predecessors a disease—one disease for each vitamin that we now require, and one disease for each of the nine amino acids that are essential for man. Most of us keep these diseases under control by ingesting the proper food.

I have been especially interested in the hemoglobinopathies, which are the diseases, including sickle-cell anemia, to which the name molecular disease was first applied. I remember very well the time, some fifteen years ago, when three of my students—Dr. Harvey Itano, Dr. S. J. Singer, and Dr. I. C. Wells, carried out the crucial experiment that showed that sickle-cell anemia is a disease of the hemoglobin molecule. I had made this prediction three years earlier, and Dr. Itano had worked for three years, toward the end with Drs. Singer and Wells, to test it.

Patients with sickle-cell anemia are anemic because their red cells tend to twist out of shape. These deformed cells are then recognized by the spleen as abnormal, and are destroyed so rapidly as to make it impossible for the patient to manufacture new erythrocytes fast enough to prevent anemia from developing. Moreover, the deformed cells are sticky; they clamp on to one another and clog up the capillaries in such a way as to interfere with the flow of blood and thus to cause different organs of the body to be damaged by anoxia. This disease, involving deformation of the red cell, might seem to be a classic disease of cells, as described by Rudolf Virchow. However, the fact that the cells sickle only in the venous circulation and regain their normal shape in the arterial circulation seemed to me, in 1945, to provide very strong indication that the disease is in fact a disease of the hemoglobin molecule, which is present as hemoglobin in the venous blood and as a different molecule, oxyhemoglobin, in arterial blood.

We all know that protein molecules tend to be sticky—it is hard for a protein molecule to keep from being sticky. If a solution of protein molecules, manufactured by some living organism and selected by the evolutionary process of trial and error so as not to be sticky, but to remain in solution instead of forming an insoluble coagulum, is disturbed a bit by warming, even to as low a temperature as 60°C., so that the molecules become slightly unfolded (denatured), then the characteristic property of stickiness makes itself evident; the denatured protein molecules clamp on to one another, to form an insoluble coagulum of denatured protein. It need not surprise us that, although the normal hemoglobin molecules, selected by the evolutionary process, are able to remain separated from one another even in the concentrated solution (30% protein) that is inside the red cell, a change in structure resulting from a gene mutation may cause the altered hemoglobin molecule to have a sticky region on its surface, such as to make it tend to clamp on to another one, which would clamp on to a third one, a fourth one, and so on, to form a long rod of these molecules. These rods would then line up side by side, attracted by the Van der Waals forces of attraction, to form a sort of needle-like crystal that would grow longer and longer until, as it became longer than the diameter of the red cell, it would twist the red cell out of shape, and would deform the red cell membrane, making it sticky, causing the red cells to get tangled up with one another in the capillaries and causing the spleen to destroy these red cells, and thus produce the manifestation of the disease. We accordingly have a molecular explanation of the manifestations of the disease, based upon the hypothesis that it is a disease of the hemoglobin molecule, a molecular disease in which the abnormal molecule is manufactured by a mutated gene. We can also understand that in the molecules of oxyhemoglobin, molecules of hemoglobin to which molecules of oxygen are attached, the attached oxygen molecules may interfere, by steric hindrance, with the Van der Waals forces of attraction and thus prevent the sickling of the red cells in the arterial circulation.

The incidence of the gene for phenylketonuria is small enough to permit it to be explained as the result of a steady state determined by the rate at which new genes for phenylketonuria are produced by mutation and the rate at which the phenylketonuria genes are removed from the pool of human germ plasm by the death without progeny of the phenylketonuria homozygotes. But the incidence of the gene for sickle-cell hemoglobin is much too great to be explained in this way. It was recognized that the sickle-cell gene must carry some advantageous character, to compensate the disadvantage of death of the sickle-cell homozygotes without progeny. The suggestion was made by Dr. Russell Brain that the heterozygotes, carrying one sickle-cell gene, are protected against malaria—he had noticed that there is a higher incidence of sickling in villages in Africa where malaria is endemic than in other villages, where malaria is not endemic. Dr. Anthony Allison, of Oxford, then carried out an experiment that provided good evidence that the sickle-cell heterozygotes are protected against malignant subtertian malaria *(Plasmodium falciparum)*. We can accordingly understand why the sickle-cell gene spread in the African population. A heterozygote, carrying a sickle-cell gene newly formed through mutation, was protected against malaria. Half of his children inherited the sickle-cell gene, and, because of their protection against malaria, they helped in rapidly spreading the gene through the population. Finally, the incidence of the gene approached the steady-state value. In marriages between heterozygotes, who would then make up a large fraction of the population, one quarter of the children would inherit two normal genes for hemoglobin, and would, in large part, die of malaria; one quarter would inherit two sickle-cell genes, and would die of sickle-cell anemia; but one half would be heterozygotes, like their parents, and would be protected against malaria and would not have the disease sickle-cell anemia. This process gives a yield of only fifty per cent in children, but only recently has this yield been thought to be unsatisfactory.

The next step in the process should be a mutation that would manufacture a kind of hemoglobin, such that in the homozygous state it would provide protection against malaria and would not produce a disease such as sickle-cell anemia. This newly mutated gene could spread rapidly through the population, provided that the double heterozygotes, in the new gene and in the sickle-cell gene, were also protected against malaria and did not have a serious disease. It seems not unlikely that another known form of abnormal hemoglobin, hemoglobin C, represents a step in this direction.

Since the discovery of sickle-cell anemia hemoglobin 14 years ago some scores of other abnormal human hemoglobins have been discovered. These abnormal hemoglobins are associated with many different diseases.

The nature of the difference between sickle-cell anemia hemoglobin (hemoglobin S) and normal adult human hemoglobin (hemoglobin A) has now been discovered, principally through the efforts of Vernon M. Ingram and his collaborators. Immediately after the discovery of hemoglobin S, Dr. Walter A. Schroeder and his associates in the California Institute of Technology made an amino acid analysis of hemoglobin S and hemoglobin A. They were able to report that the amino acid composition of the two hemoglobins is closely similar, with no amino acid represented by a difference of more than two residues. Ingram then developed a new and powerful way of investigating the structure of hemoglobin molecules, which he called the fingerprint method—it is also called the peptide-pattern method. The sample of hemoglobin is split into peptides by the proteolytic action of an enzyme, such as trypsin. About twenty-six peptides, containing on average about twelve amino-acid residues each, are obtained in the mixture produced in this way. The mixture is then separated into the constituent peptides by a two-dimensional process carried out on a sheet of filter paper. The separation on the basis of mobility in an electric field is carried out along the horizontal axis of the sheet of filter paper, and then separation by the chromatographic method, involving a flowing solvent, is carried out in the vertical direction. In this way Ingram was able to show that hemoglobin S differs from hemoglobin A only in the replacement of a single amino acid residue in one-half of the hemoglobin molecule by the residue of another amino acid.

Schroeder and his associates in Pasadena found that hemoglobin molecules usually consist of four polypeptide chains, two of one kind and two of another kind. The normal adult human hemoglobin molecule contains two alpha chains, which have the sequence val-leu-ser-pro-ala...(total 141 residues), measured from the free-amino end, and two beta chains, which have the sequence val-his-leu-thr-pro-glu-... (total 146 residues). Ingram and Schroeder found that the alpha chains of hemoglobin S are the same as those of hemoglobin A, but the beta chains are different: the beta chain of hemoglobin S has valine in the sixth position, in place of glutamate; the other 145 residues are the same.

The amino acid sequence has been determined for many other abnormal hemoglobins. For every one of those studied so far, the mutation involves only a single amino acid residue. Thus, for hemoglobin C there is lysine in the sixth residue of the beta chain, in place of the glutamate of hemoglobin A or the valine of hemoglobin S. For hemoglobin E there is lysine in the 26th position, in place of glutamate.

Other abnormal hemoglobins involve an abnormality in the alpha chain rather than the beta chain. An interesting example is hemoglobin M_{Boston}. In the alpha chain of normal adult hemoglobin the 58th residue is histidine. This residue of histidine is known to be close to the iron atom of the heme group. Histidine usually carries a positive charge, because of the attachment of a proton to the imidazole ring. In hemoglobin M_{Boston} the alpha chain has tyrosine in the 58th position, in place of histidine. Because the tyrosine residue does not carry a positive charge, we may expect that it would be easier for the iron atom of the heme group to assume an extra positive charge, leading to a ferriheme group, containing tri-positive iron, in place of the normal ferroheme, containing bi-positive iron. The presence of tripositive iron hemoglobin converts it into ferrihemoglobin (also called methemoglobin); and the carriers of the gene for hemoglobin M_{Boston} do in fact have a disease, a form of methemoglobinanemia.

Accordingly in this disease, as in sickle-cell anemia, the known difference in amino acid sequence of the abnormal hemoglobin provides a reasonable explanation of the manifestations of the disease produced by the molecular abnormality.

Some interesting conclusions about the process of evolution have been reached on the basis of the comparison of amino acid sequence of hemoglobin molecules of animals of different species, carried out especially by my collaborator Dr. Emile Zuckerkandl. It has been found that the peptide pattern of hemoglobins of animals of different species can be correlated reasonably well with the generally accepted ideas about evolutionary relationships between the species. For example, the peptide patterns of gorilla hemoglobin and chimpanzee hemoglobin are almost identical with those of human hemoglobin. The peptide pattern of Rhesus monkey hemoglobin is somewhat different from that of human hemoglobin. Still greater differences from human hemoglobin are shown by the patterns of cow hemoglobin, horse hemoglobin, pig hemoglobin, and the hemoglobins of other mammals. The differences are greater still for fish and worm hemoglobins.

A detailed study of horse hemoglobin has shown that the alpha chains differ from those of human hemoglobin by about 18 amino acid substitutions, as do also the beta chains of the two hemoglobins. If we accept 130,000,000 years as the time that has passed since the evolutionary lines of horse and human separated, as estimated by paleontologists, we conclude that each chain has on the average suffered an evolutionarily effective mutation every 14.5 million years. We may then use this value to discuss other evolutionary epochs.

The gorilla alpha chain and the human alpha chain differ in two residues, and the gorilla beta chain and human beta chain differ in one; the average, 1.5, indicates that about 11 million years has gone by since the derivation of these chains from their common chain ancestor—that is, that the evolutionary lines leading to the present-day gorillas and present-day human beings separated from one another about 11 million years ago. The estimates made by paleontologists for this epoch range from 10 million to 35 million years.

Another interesting question is that of the biochemical differences between adult human beings and human fetuses. The human fetus manufactures a special kind of hemoglobin, called hemoglobin F. In hemoglobin F the beta chains are abnormal. These abnormal beta chains, which are called gamma chains, differ from adult human beta in 36 of the 146 amino acid residues. Accordingly we calculate, assuming that there has been a constant rate of evolutionarily effective mutation, that the gamma chains and the beta chains separated from one another about 260 million years ago; that is, at the beginning of the

Carboniferous period.

This epoch was, of course, long before human beings had come into existence. Other mammals also have fetal hemoglobins differing from the adult hemoglobin for the species, and we might conclude that the fetal forms of different mammals separated from the adult forms some 260 million years ago, and in a sense constitute a group of species different from the adult group. With respect to hemoglobin, a human fetus resembles a fetal horse more closely than a human adult.

I believe that it will be possible, through the detailed determination of amino acid sequences of hemoglobin molecules and of other molecules, to obtain much information about the course of the evolutionary process, and to illuminate the question of the origin of species.

Moreover, I believe that the continued study of the molecular structure of the human body and the nature of molecular disease will provide information that will contribute to the control of disease and will significantly diminish the amount of human suffering. Molecular biology and molecular medicine are new fields of science that can be greatly developed for the benefit of mankind.

BIBLIOGRAPHY

Pauling, L., Itano, H. A., Singer, S. J. and Wells, I. C. Sickle cell anemia, a molecular disease, *Science 110*:543-48 (Nov. 25), 1949.

Pauling, L. Abnormality of hemoglobin molecules in hereditary hemolytic anemias, *The Harvey Lectures 49*:216-41, 1955. (New York, Academic Press, Inc.)

Zuckerkandl, E. and Pauling, L. Molecular disease, evolution, and genic heterogeneity. *Horizons in Biochemistry,* New York, Academic Press, Inc., 1962, pp. 189-225.

Ingram, V. M. *The Hemoglobins in Genetics and Evolution.* New York. Columbia University Press, 1963.

Chapter 16

Physiological Chemistry, Effects of Radiation, and Health Hazards

Contents

SP 127 **Genetic and Somatic Effects of Carbon-14** 1308
Science **128**, 1183–1186 (1958)

128 **The Effects of Strontium-90 on Mice** 1312
(by Barclay Kamb and Linus Pauling)
Proc. Natl. Acad. Sci. USA **45**, pp. 54–69 (1959)

129 **A Molecular Theory of General Anesthesia** 1328
Science **134**, 15–21 (1961)

130 **Results of a Loading Test of Ascorbic Acid, Niacinamide, and Pyridoxine in Schizophrenic Subjects and Controls** 1335
(by Linus Pauling, Arthur B. Robinson, Susanna S. Oxley, Maida Bergeson, Andrew Harris, Paul Cary, John Blethen, and Ian T. Keaveny)
In *Orthomolecular Psychiatry: Treatment of Schizophrenia*, eds. David Hawkins and Linus Pauling, W. H. Freeman, San Francisco, pp. 18–34 (1973)

131 **The Discovery of the Superoxide Radical** 1352
Trends in Biochemical Sciences **4**, 270–271 (1979)

132 **Biostatistical Analysis of Mortality Data for Cohorts of Cancer Patients** 1354
Proc. Natl. Acad. Sci. USA **86**, 3466–3468 (1989)

133 **Criteria for the Validity of Clinical Trials of Treatments of Cohorts of Cancer Patients Based on the Hardin Jones Principle** 1357
(by Linus Pauling and Zelek S. Herman)
Proc. Natl. Acad. Sci. USA **86**, 6835–6837 (1989)

Reprinted from SCIENCE, November 14, 1958, Vol. 128, No. 3333, pages 1183–1186.

Genetic and Somatic Effects of Carbon-14

This by-product of nuclear-weapon testing may do more genetic and somatic damage than has been supposed.

Linus Pauling

In his 1956 paper on radioactive fallout (*1*) Libby pointed out that neutrons released in the explosions of nuclear weapons react with nitrogen nuclei in the air to make carbon-14, which has a half-life of about 5600 years. In his discussion of bomb-test carbon-14 he said that "Fortunately, this radioactivity is essentially safe because of its long lifetime and the enormous amount of diluting carbon dioxide in the atmosphere." He pointed out that 5.2 tons of neutrons would be needed to "double the feeble natural radioactivity of living matter due to radiocarbon. Such an increase would have no significance from the standpoint of health." He mentioned that, for a given energy release, thermonuclear weapons produce more neutrons than fission weapons, and concluded that "the essential point is that the atmosphere is difficult to activate and the activities produced are safe."

Perhaps because of a feeling of reassurance engendered by these statements, I did not make any calculations of the genetic and somatic effects of the carbon-14 produced in the testing of nuclear weapons until April 1958. I was then surprised to find that these calculations, which form the subject of this article (*2*), lead to the conclusions that the genetic damage, as measured by the predicted number of children born with defects caused by the mutations induced by the radioactivity, may be greater for carbon-14 than for the fission products ordinarily classed as world-wide fallout, and that the somatic effects may be of the same order of magnitude.

The author is professor of chemistry at California Institute of Technology, Pasadena.

In his 1956 paper Libby stated that a 20-kiloton weapon, involving fission of 1 kg of plutonium or uranium, would produce 10 g of neutrons, of which 15 percent might reasonably be expected to escape and make carbon-14. The yield of carbon-14 would hence be 1.05 kg per megaton (the maximum would be 7 kg per megaton, if all neutrons were effective).

More information was given in his 27 March 1958 address on radioactive fallout, delivered at the symposium of the Swiss Academy of Medical Sciences in Lausanne and released on that day by the Atomic Energy Commission (*3*). In this address he said that 1 megaton with fusion and fission weighed as they have actually occurred would generate 3.2×10^{26} atoms of carbon-14, which is 7.4 kg. He pointed out that this estimate is higher than the earlier estimate based on an assumed 15-percent escape efficiency, and said that the new value is based on firmer information.

The old value was for fission alone. If we assume it to be valid, we might conclude that the sevenfold increase to the new value is to be attributed to a high yield for fusion. For example, if the energy yields for fusion and fission have been equal for past explosions the carbon-14 yield for fusion might be calculated to be 13.8 kg per megaton, about 13 times that given for fission.

(On 29 May 1958, after the calculations described in this article had been made, my attention was called by Ben Tucker to the paper "Radioactivity danger from the explosion of clean hydrogen bombs and ordinary atomic bombs," by O. I. Leipunskii, published in the December 1957 issue of the U.S.S.R. journal *Atomic Energy* (*4*). The values given there agree only very roughly with my values. Leipunskii gives 5.2 kg per megaton as the amount of carbon-14 produced by fission and 33 kg per megaton as the amount produced by fusion. The latter value represents a 96-percent effectiveness of the neutrons calculated to be released in the $H^2 + H^3$ reaction giving 1 megaton of energy, or a somewhat smaller effectiveness if some of the 12.5-Mev neutrons produce additional neutrons by $n,2n$ reactions. The Libby value 7.4 kg per megaton for fission and fusion in the ratio of past explosions is 39 percent of 19.1, the Leipunskii value for fission and fusion in 50:50 ratio.)

Libby gives 10^{28} as the best estimate of the number of carbon-14 atoms introduced into the atmosphere (mostly into the stratosphere) by the bomb tests so far, keeping in mind that a substantial amount falls back as calcium carbonate, especially in the case of ground shots over coral. The number 10^{28} atoms (232 kg) corresponds to 31 megatons of bombs. I assume that one-third of the generated carbon-14 is released to the atmosphere, two-thirds falling back as calcium carbonate. This estimate is based upon the statement by Libby (*5*) in December 1956 that total bomb tests up to the time his paper was written (it was submitted for publication on 17 October 1956) had liberated 30 megatons of fission products. It is my understanding, from the table of nuclear explosions given in *The Nature of Radioactive Fallout and Its Effects on Man* (*6*, pp. 2063–

2

2065), that fission products were first released in large amounts on 1 March 1954, the earlier explosions having been those of small bombs or of fission-fusion bombs with no large third stage. If the testing has continued at the same rate from October 1956 to January 1958 (reference date for the 1958 statement by Libby) as from 1954 to 1956, the value 232 kg of carbon-14 introduced into the atmosphere corresponds to 45 megatons of fission and, with the surmise that the fission-fusion ratio has been 1, to 90 megatons of total tests, and hence to the above estimate that one-third of the carbon-14 becomes atmospheric CO_2.

The 232 kg of carbon-14 (Libby's estimate) introduced into the atmosphere by the bomb tests had caused the carbon-14 concentration for atmospheric carbon dioxide in New Zealand to increase to 10 percent over its normal value by 1957 (7). The carbon-14 released into the atmosphere becomes mixed in a few years with the biosphere and the top layer (about 300 feet thick) of the ocean (*8, 9*). Mixing occurs more slowly with the deep layers of the ocean. Studies by several authors (*8, 9*) have led to closely similar conclusions about the rates of mixing. We shall make use of a simple model discussed by Arnold and Anderson (*9*); essentially the same conclusions would be reached with use of any model compatible with the value 600 years for the age of the dissolved carbon in the ocean.

Two Reservoirs of Carbon

In the simple model of Arnold and Anderson two reservoirs of carbon are considered. Reservoir A consists of the atmosphere (0.13 g of carbon per square centimeter), the land biosphere (0.05 g cm^{-2}), and humus (0.2 g cm^{-2}), totaling 2.0×10^{18} g of carbon, of which 3200 kg is carbon-14. Within this reservoir there is rapid equilibration of carbon-14. Reservoir C is the entire ocean, including the ocean biosphere; it contains 8.5 g cm^{-2} (44×10^{18} g) of carbon, 22 times as much as A.

The equilibrium between A and C can be expressed by a forward rate constant k and reverse rate constant k', with values $k = 0.035$ yr^{-1} and $k' = 0.0016$ yr^{-1}, respectively.

Let us consider N_0 atoms of carbon-14 released into A by 1 year's testing at a standard rate, which we assume to be 30 megatons per year, with 222 kg of carbon-14 made and 74 kg released into A. The later number (N_A) of these atoms in A is given by the equation

$$\frac{dN_A}{dt} = -kN_A + k'(N_0 - N_A) \qquad (1)$$

The solution of this equation is

$$N_A = \frac{k'}{k+k'}N_0 + \frac{k}{k+k'}N_0e^{-(k+k')t}$$

which with insertion of the values of k and k' becomes

$$N_A = 0.044N_0 + 0.956N_0e^{-0.0363t}$$

So far N_0 has been considered a constant. We replace it by $N_0e^{-0.000124t}$, corresponding to the radioactive decay of carbon-14 with mean life 8070 years (half-life 5586 years), to obtain

$$N_A = 0.044N_0e^{-0.000124t} + 0.956N_0e^{-0.0364t} \qquad (2)$$

Hence, the freshly made carbon-14 in reservoir A, which gives it access to the bodies of human beings, can be considered as consisting of a 4.4-percent fraction with mean life 8070 years and a 95.6-percent fraction with mean life 27.5 years (the reciprocal of 0.0364 yr^{-1}).

Genetic Effects at Present Population Levels

Let us first evaluate the genetic effect of the carbon-14 from bomb tests on the assumption that the population of the world will remain constant.

James F. Crow, a member of the National Academy of Sciences–National Research Council Committee on Genetic Effects of Atomic Radiation, presented an estimate of the genetic effects of a 0.1-roentgen exposure of the gonads in his testimony before the Special Subcommittee on Radiation of the Joint Congressional Committee on Atomic Energy on 4 June 1957 (*6*, p. 1021). He estimated that a 0.1-roentgen exposure of the gonads of the present world population would produce gene mutations that would in the course of many generations give rise to the birth of 80,000 children with gross physical or mental defect, 300,000 stillbirths and childhood deaths, and 700,000 embryonic and neonatal deaths. Of these, 8000, 20,000, and 40,000, respectively, were expected to occur in the first generation. In addition, he estimated that there would be produced a larger but unknown number of minor or intangible defects, which might represent the major part of the damage, because by virtue of their being milder they are less likely to cause the sterility or death of the person who possesses them and therefore are more likely to persist in the population and thus to affect a larger number of persons.

The estimates for the three categories were made in different ways, and the categories are not mutually exclusive. In particular, deaths at about the time of birth are included in both the second and the third category. Crow has told me that in his opinion there is little overlap between the first and the second category.

These estimates must be recognized as highly uncertain. Crow said that they might be 5 times too high or 5 times too low, or more, but that we are better off estimating even very crudely what the numbers involved are than not making any numerical estimates at all. I agree with this statement.

Uncertainty in these estimates does not affect the discussion of relative effects of carbon-14 and fission products given below.

It must be emphasized that, although large numbers are given below as the estimated effects of the testing of nuclear weapons at the recent rate, these numbers are very small in comparison with numbers representing the effects of natural radiation and other mutagenic agents. For example, it is stated in the National Academy of Sciences–National Research Council report that about 2 percent of total live births have tangible defects of simple genetic origin (this is roughly the first category of the three given above). With 75 million births per year, this corresponds to 1.5 million per year with gross physical and mental defect. The estimated effect of continued testing of nuclear weapons at the recent rate is an additional 15,000 per year (including the effect of carbon-14). Hence the bomb tests are expected to produce not more than a 1-percent increase in defective births (or between 0.2 percent and 5 percent, if we use Crow's suggestion about uncertainty in the estimates).

The estimate of the magnitude of the gonad exposure for the average rate of bomb testing for the 5 years preceding 1956, reported by the National Academy of Sciences–National Research Council committee, is 0.1 roentgen in 30 years. Hence 1 year of testing at that rate, it is estimated, will cause about 2700 children with gross physical or mental defect, 10,000 stillbirths and childhood deaths, and 23,000 embryonic and neo-

3

natal deaths. (This estimate ignores the effects of carbon-14.)

The *Twenty-third Semiannual Report of the Atomic Energy Commission* contains the statment that bomb testing at the present rate, it can be estimated, will cause between 2500 and 13,000 defective children to be born per year of testing. This statement is in the report of the Advisory Committee on Biology and Medicine. It seems to correspond to the above calculation, with recognition of the uncertainy of the amount of overlap between the first two categories.

The report of the National Academy of Sciences–National Research Council committee contains the sentence "With these understandings, it may be stated that U.S. residents have, on the average, been receiving from fallout over the past five years a dose which, if weapons testing were continued at the same rate, is estimated to produce a total 30-year dose of about *one tenth of a roentgen*; and since the accuracy involved is probably not better than a factor of 5, one could better say that the 30-year dose from weapons testing if maintained at the past level would probably be larger than 0.02 r and smaller than 0.50 r. The rate of fall-out over the past five years has not been uniform. If weapons testing were, in the future, continued at the largest rate which has so far occurred (in 1953 and 1955) then the 30-year fall-out dose would be about twice that stated above."

It is accordingly possible that a somewhat larger estimate than 0.1 roentgen in 30 years should be made for the average gonad exposure corresponding to the recent rate of testing of nuclear weapons. Little can be done to make the estimates of the effects of fission products more reliable in the absence of any published detailed discussion of the evidence upon which the estimates of gonad exposure are based.

Now let us consider the genetic effects of carbon-14. The gonad exposure due to natural carbon-14 has been given by Libby (*10*) as 0.0015 roentgen per year. This dosage was calculated on the basis of the assumptions that the body is 18 percent carbon, the specific activity of carbon is 15 disintegrations per minute per gram, and the mean energy of the beta radiation is 40 percent of the maximum energy, 167 kev.

If we take as the present rate of bomb testing the value 30 megatons per year (fission plus fusion), the initial activity of the carbon-14 from 1 year of bomb tests is 0.0015 roentgen per year multiplied by 74/3200, the ratio of the amount of carbon-14 released to reservoir A by the tests to the amount of natural (cosmic-ray produced) carbon-14. This initial activity is 35×10^{-6} roentgen per year. Of this amount, 1.46×10^{-6} roentgen is associated with the first term in Eq. 2 and 33×10^{-6} roentgen with the second term. The total gonad exposure is obtained by multiplying these quantities by the corresponding mean lives, 8070 and 27.5 years, respectively, to obtain 0.0118 and 0.0009, respectively, with sum 0.0127 roentgen.

We see that the second term (the nonequilibrium term with respect to mixing with the large ocean reservoir) contributes only about 8 percent as much as the first term to the total effect. On the other hand, it is the more important of the two with respect to the present generation and the next one.

The total gonad exposure due to carbon-14 over the entire life of the isotope (per person now living, world population assumed constant), 0.0127 roentgen, is 4 times that usually assumed for world-wide fallout (0.0033) roentgen, corresponding to 0.1 roentgen in 30 years). The estimated effects of carbon-14 from 1 year of bomb testing, from Crow's numbers, are 12,000 children with gross physical or mental defect, 38,000 stillbirths and childhood deaths, and 90,000 embryonic and neonatal deaths.

Genetic Effects at Predicted Population Levels

Now let us consider the effect of the increase in world population that can be reasonably anticipated. At the present time the world population is growing at a rate such as to double in about 50 years. If we assume that no catastrophe intervenes, this rate may continue for hundreds of years, and the population may then remain essentially constant, with a value for number of births per year 5 times the present value. The number of defective children corresponding to the first term of Eq. 2 would then be multiplied by a factor nearly equal to 5. If the world population were to increase in this way, the carbon-14 from 1 year of testing would cause an estimated total of about 55,000 children with gross physical and mental defect, 170,000 stillbirths and childhood deaths, and 425,000 embryonic and neonatal deaths. On this assumption about world population it is estimated that the bomb tests carried out so far (estimated total, including 1958, 150 megatons) will cause about 5 times these numbers of defective children and deaths.

Thus we see that the genetic effects of carbon-14 from bomb tests are estimated to be about 4 times as great as those of ordinary world-wide fallout (calculated for the customarily quoted value of gonad exposure) if the world population stays constant, and about 17 times as great if the world population increases as assumed.

There is a simpler way of making the calculation (*11*). Let us assume that there is very rapid mixing of the carbon-14 released in the bomb tests throughout the entire reservoir, including the depths of the ocean. With this assumption and the other assumptions given above, a straightforward calculation can be carried out, leading to nearly the some numbers.

These predicted effects of carbon-14, which over the period of thousands of years are greater than those of the fission products in the world-wide fallout, may be thought to have little significance because of uncertainty about the nature of the world of the rather distant future. It is accordingly of interest to calculate what the effects of 1 year of testing will be on the next generation.

Effects of One Year's Tests on the Next Generation

We may consider first the predicted numbers of seriously defective births in the next generation as a result of the ordinary fallout. From Crow's estimates and the gonad exposure 0.1 roentgen in 30 years, there are 270 children with gross physical or mental defect, 670 stillbirths and childhood deaths, and 1304 embryonic and neonatal deaths.

In calculating the number of seriously defective births expected to occur in the first generation as a result of the presence of added amounts of carbon-14 in the atmosphere we cannot neglect the rate of diffusion of carbon-14 into the depths of the ocean. The 74 kg of carbon-14 liberated into the atmosphere by 1 year of testing at the standard rate causes an initial increase of 2.3 percent of the carbon-14 concentration, the normal burden of the atmosphere, biosphere, humus, and upper part of the ocean being 3200 kg. This calculation agrees roughly with the statement by Libby that "the observed carbon-14 rise

4

might be as high as 3 percent per year as appears to have been observed." The rate of increase, reported from experiment, for carbon dioxide in the atmosphere is about 2.1 percent per year.

It may be pointed out that the observed rate of increase of carbon-14 in the atmosphere provides some justification of the assumed standard recent rate of testing, 30 megatons of fission plus fusion per year (together with the assumption that one-third of the carbon-14 that is produced is liberated to the atmosphere), as shown by the agreement of the calculated 2.3-percent increase per year and the observed 2.1-percent increase. The same rate of increase in the atmosphere and the same genetic and somatic effects would result from, say, 20 megatons per year with one-half escaping. The calculation of genetic and somatic effects could be based directly on the observed rate of increase of carbon-14 in the atmosphere.

The rate of diffusion of the carbon-14 into the depths of the ocean corresponds to a mean life of 27.5 years in the smaller reservoir. The gonad exposure for natural carbon-14 is 0.0015 roentgen per year and that for an amount 2.3 percent as much is 0.000035 roentgen per year. With a mean life for carbon-14 in the small reservoir of 27.5 years, the total gonad exposure for the first decades after the testing becomes 0.00096 roentgen. With world population at the present level, the estimated numbers in the three categories during the first generation due to carbon-14 from a single year of testing are 80, 200, and 400, respectively. These are smaller than estimates for the ordinary radioactive fallout. It is because of the very long life of carbon-14 that the total effect, throughout the life-times of the isotopes, becomes greater for carbon-14 than for the fission products.

The possibility must be considered of a special mutagenic action of carbon-14: the damage of a deoxyribonucleic acid molecule through the Szilard-Chalmers effect or the chemical effect of conversion to a nitrogen atom when a carbon-14 atom in the molecule undergoes radioactive decomposition. We assume 50,000 genes per individual, 200,000 carbon atoms per gene, 5×10^9 future world population up to 30 years of age, and a carbon-14 yield of 74 kg to the atmosphere per year of testing, and calculate 70,000 as the number of mutations by this mechanism per year of testing. This number, presumably an upper limit, is only about 10 percent of the numbers in the three categories expected to result from carbon-14 irradiation, and we conclude that the special mechanism involving carbon-14 atoms in the genes themselves is less important than irradiation in causing genetic damage.

The calculation of predicted somatic effects of bomb-test carbon-14 in comparison with those of fission products can be easily made. With the same assumptions as for the foregoing calculation of the genetic effects, including the assumption of a fivefold increase in world population, it is found that 1 year of testing of nuclear weapons produces carbon-14 irradiation, over the entire life of the radiocarbon, equivalent to the exposure of the present world population to a whole-body dose of 0.061 roentgen. This is much larger than the customarily quoted value of 0.0033 roentgen for whole-body irradiation by fission products from 1 year of testing, and somewhat larger than the estimated exposure of bone marrow and bone tissue by strontium-90 (given as 0.03 and 0.056 roentgen, respectively, per year of testing, as estimated by the Atomic Energy Commission's Advisory Committee on Biology and Medicine, in the *Twenty-third Semiannual Report of the Atomic Energy Commission,* 1958). Hence we calculate that the total number of cases of leukemia and bone cancer expected to be caused by carbon-14 is about equal to the number expected to be caused by fission products, including strontium-90, and that the number of cases of cancer of other sorts expected to result from radiation damage to tissues other than bone marrow and bone tissue is greater for bomb-test carbon-14 than for fission products.

Summary

On the basis of information about carbon-14 given by Libby, calculations are made of the predicted genetic and somatic effects of the carbon-14 produced by the testing of nuclear weapons. It is concluded that 1 year of testing (30 megatons of fission plus fusion) is expected to cause in the world (estimated future number of births per year 5 times the present number) an estimated total of about 55,000 children with gross physical or mental defects, 170,000 stillbirths and childhood deaths, and 425,000 embryonic and neonatal deaths. (There is an unknown amount of overlap of these three categories.) These numbers are about 17 times the numbers usually estimated as the probable effects of the fallout fission products from 1 year of testing. In addition, the somatic effects of bomb-test carbon-14 are expected to be about equal to those of fission products, including strontium-90, with respect to leukemia and bone cancer and greater than those of fission products with respect to diseases resulting from radiation damage to tissues other than bone tissue and bone marrow. All of the estimated numbers are subject to great uncertainty; they may be as much as 5 times too high or 5 times too low. The uncertainty in the estimation of the relative effects of carbon-14 and fission products in world-wide fallout is not so great.

References and Notes

1. W. F. Libby, *Science* **123**, 657 (1956).
2. This article is contribution No. 2396 from the Gates and Crellin Laboratories of Chemistry, California Institute of Technology, Pasadena.
3. Atomic Energy Commission release of 27 Mar. 1958; W. F. Libby, *Proc. Natl. Acad. Sci. U.S.* **44**, 800 (1958).
4. O. I. Leipunskii, *Atomic Energy (U.S.S.R.) English translation* **12**, 530 (1957).
5. W. F. Libby, *Proc. Natl. Acad. Sci. U.S.* **42**, 945 (1956).
6. *The Nature of Radioactive Fallout and Its Effects on Man* (U.S. Government Printing Office, Washington, D.C., 1957).
7. T. A. Rafter, *N.Z. J. Sci. Technol.* **37B**, 20 (1955); 18, 871 (1957); ——— and G. J. Fergusson, *Science* **126**, 557 (1957).
8. H. Craig, *Tellus* **9**, 1 (1957); R. Revelle and H. E. Suess, *ibid.* **9**, 18 (1957).
9. J. R. Arnold and E. C. Anderson, *ibid.* **9**, 28 (1957).
10. W. F. Libby, *Science* **122**, 57 (1955).
11. L. Pauling, letter to the editor, *New York Times* (16 May 1958).

Reprinted from the Proceedings of the NATIONAL ACADEMY OF SCIENCES
Vol. 45, No. 1, pp. 54–69. January, 1959.

*THE EFFECTS OF STRONTIUM-90 ON MICE**

BY BARCLAY KAMB AND LINUS PAULING

CALIFORNIA INSTITUTE OF TECHNOLOGY, PASADENA, CALIFORNIA

Communicated November 17, 1958

On Sept. 19, 1958 there was published in *Science* a paper by Dr. Miriam P. Finkel of Argonne National Laboratory in which she communicated her observations on the effects of strontium-90 injected into mice on life expectancy and on incidence of tumors of bone and blood-forming tissues.[1] She discussed the question of whether or not the effects are proportional to the amount of injected strontium-90 at low doses, and reached the conclusion that it is likely that there is a threshold with value for man between 5 and 15 μc. (as compared with the present average value from fallout, about 0.0002 μc., and the predicted steady-state value from fallout for testing of nuclear weapons at the average rate for the past five years, about 0.02 μc.). Her paper ends with the sentence "In any case, the present contamination with strontium-90 from fallout is so very much lower than any of these levels that it is extremely unlikely to induce even one bone tumor or one case of leukemia."

On the same day, Sept. 19, 1958, newspapers throughout the United States published accounts of this work. For example, the Pasadena (Calif.) *Star-News* contained an article with the headline "Tests on Mice Show Fallout Safe" and the first sentence, "A woman researcher says tests on mice show that the present fallout from nuclear weapons tests will not produce a single case of bone cancer or leukemia in humans." The *New York Times* published accounts of the work on both September 19 and September 28.

We have made an analysis of Dr. Finkel's data that shows that she had no justification whatever for her concluding statement. All of her data are compatible with a zero threshold for strontium-90. Moreover, the statistical analysis shows that in order for Dr. Finkel to have been justified with 90 per cent confidence (10 per cent type-II error) in making her concluding statement on the basis of her data she would have to have used over 1,000,000,000,000 mice in each of her groups, instead of the 150 or less that were used. It is hard for us to understand how such a serious error could be made in Dr. Finkel's argument, leading her to publish her seriously misleading statement about this matter of great importance.

The Mice Experiments.—In the studies described by Dr. Finkel young adult female mice (strain CF No. 1, about 70 days old) were given a single intravenous injection of an isotonic equilibrium mixture of strontium-90 chloride and yttrium-90 chloride. There were twelve injected groups, ranging in size from 150 mice for the group receiving the smallest amount (1.3 μc/kg body weight) to 15 for that receiving the largest amount (9330 μc/kg, an amount that caused death of about 50 per cent within 30 days). The control group contained 150 mice. The author states that there is 11 per cent retention (at 600 days) of the injected radioactive material. Report was made of the fractional decrease in average survival time, the incidence of animals with osteogenic sarcomas (among 150-day survivors), and the fractional decrease in time to a 20 per cent incidence of reticular tissue tumors compared with the 20 per cent incidence time of the controls.

Studies of this sort may be of great value in providing information about the probable amount of damage done to human beings by exposure to high-energy radiation, such as that from strontium-90 produced by nuclear weapons. It is important that the analysis of the experimental results be carried out correctly. We have found that in the treatment of problems of this sort the assumption that the probability of damage is strictly proportional to the amount of radiation exposure does not in general require that a response such as decrease in life expectancy be linear, except over a very small range. Moreover, we have found that this assumption together with the theory of statistics can be applied in a reasonably straightforward way in the discussion of data such as those obtained by Dr. Finkel, as shown in the following sections.

Analysis of the Experimental Data on Life Shortening.—Our analysis proceeds from the hypothesis, induced by Lewis[2] as a result of his study of the incidence of leukemia, that exposure of the bone marrow of an animal to radiation results in an increase in the probability per unit time that the animal will die at any time thereafter, the increase being proportional to the quantity of radiation absorbed. We shall suppose that this hypothesis applies to all of the radiation-induced effects in Dr. Finkel's experiments with mice.

Let N_0 be the number of animals at the beginning of a given experiment, taken to be at $t = 0$, and let $N(t)$ be the expected number (average for many experiments of the same kind) at the later time t. Further, let $N^0(t)$ be the expected number in a "control" experiment in which no strontium-90 is injected, so that

$$g(t) \equiv - \frac{1}{N_0} \frac{dN^0}{dt}$$

is the natural specific death-rate function. We denote by α the quantity of strontium-90, in μc/kg body weight, that is retained in the animals. Then our hypothesis yields the equation

$$\frac{dN}{dt} = -N\beta\alpha t - \frac{N}{N^0(t)}\cdot N_0 g(t) \tag{1}$$

where β is a constant of proportionality relating the quantity of strontium-90 retained to the increased probability per unit time that the animals will die, this probability of course increasing linearly with time owing to the nearly constant irradiation by the decaying strontium-90.

Let $n^0(t) = N^0(t)/N_0$, so that $g(t) = -\dot{n}^0(t)$. Then on integrating equation (1) we obtain

$$N = N_0 \exp\left[-\frac{1}{2}\beta\alpha t^2 - \int_0^t \frac{\dot{n}^0(t)}{n^0(t)} dt\right]$$
$$= N_0 n^0(t) e^{-(1/2)\beta\alpha t^2} = N^0(t) e^{-(1/2)\beta\alpha t^2} \tag{2}$$

To compare this result with the experimental data we calculate Δ, the fractional decrease in life expectancy (fractional decrease in average survival time after injection),

$$\Delta = \frac{t_0 - t_\alpha}{t_0}$$

where t_α is the life expectancy for a retained quantity α of strontium-90,

$$t_\alpha = \frac{1}{N_0}\int_0^\infty N(t)dt \tag{3}$$

The equation for Δ is

$$\Delta = 1 - \frac{1}{t_0}\int_0^\infty n^0(t) e^{-\gamma t^2} dt \tag{4a}$$

$$= \frac{1}{t_0}\int_0^\infty \left(t - \frac{1}{2}\sqrt{\frac{\pi}{\gamma}} \operatorname{Erf} t\sqrt{\gamma}\right) g(t)\, dt \tag{4b}$$

where for simplicity we put $\gamma = \frac{1}{2}\alpha\beta$. The result in equation (4b) is obtained by an integration by parts, and the error function is defined as

$$\operatorname{Erf} x = \frac{2}{\sqrt{\pi}}\int_0^x e^{-y^2} dy$$

If normally (for $\alpha = 0$) all animals lived to the age t_0 and then died, so that $g(t)$ were a delta function $\delta(t - t_0)$, then we would have simply

$$\Delta = 1 - \frac{\sqrt{\pi}}{2}\frac{\operatorname{Erf} t_0\sqrt{\gamma}}{t_0\sqrt{\gamma}} \tag{5}$$

However. the actual lifetimes scatter with sizable dispersion about t_0. The extent of this dispersion can be estimated from the acceptance region $\Delta \leq 0.07$ quoted by Dr. Finkel (Figs. 3 and 4) as appropriate to a test of the hypothesis of no difference between the responses of the control population and of a population injected with a given dose of strontium-90. If we assume (1) that the test of no difference applies to Curve A, the life-shortening data, (2) that the test was one-tailed, and (3) that the test accounted for the uncertainty in the mean lifetime of the control population and for the uncertainty in the mean lifetime of her group 10, the highest-dosed population to fall within the acceptance region (except for the "peculiar result" for group 8), then we find that the estimated standard deviation for $g(t)$ is $\hat{\sigma} = 258$ days.

These assumptions are somewhat uncertain, as explained later, but they are the best that can be made from the information given in Dr. Finkel's paper. The uncertainty in drawing any conclusions about $g(t)$ from Dr. Finkel's data lead us to take a more general approach. Gompertz discovered that for animal populations the logarithm of the "age-specific death rate" is closely a linearly increasing function of time. For man the age-specific death-rate doubling time is about 8 years. Jones[3] has pointed out that the doubling times for different animal species are approximately proportional to the mean life spans for the species. We shall use this information to derive a hypothetical death-rate function $g(t)$ for the mouse population used in Dr. Finkel's experiments.

The Gompertz law is

$$\ln\left(-\frac{1}{N^0}\frac{dN^0}{dt}\right) = C + Bt \tag{6}$$

which yields

$$n^0(t) = \exp\left[-A(e^{Bt} - 1)\right] \tag{7}$$

where A $(= e^C/B)$ is a constant and where B is related to the doubling time τ_D by

$$B = \frac{\ln 2}{\tau_D} \tag{8}$$

A is to be chosen so as to give the correct mean life span:

$$t_0 = \int_0^\infty \exp\left[-A(e^{Bt} - 1)\right]dt = \frac{e^A}{B}\left[-\mathrm{Ei}(-A)\right] \tag{9}$$

The exponential integral $\mathrm{Ei}(x)$ is defined by Jahnke and Emde.[4] If $t = 0$ is taken to be a time shortly after birth, but long enough after birth to exclude infant deaths (which are omitted in Gompertz' treatment), then t_0 is T, the mean life span from birth to death. If then T/τ_D is a constant for all animal species, we find from equations (8) and (9) that A is a constant, independent of species, given by the solution of the equation

$$e^A\left[-\mathrm{Ei}(-A)\right] = \frac{T}{\tau_D}\ln 2 \tag{10}$$

Assuming $T = 60$ years for man, with $\tau_D = 8$ years, we find $A = 0.0032$. The solution of equation (10) is obtained with the help of the expansion[4]

$$-\mathrm{Ei}(-x) = -\ln\Gamma x + x - \frac{x^2}{2!2} + \ldots \qquad (11)$$

where $\Gamma = 1.781$.

The death-rate function is

$$g(t) = AB \exp [Bt - A(e^{Bt} - 1)]$$

The dispersion of life spans is measured by the standard deviation σ of $g(t)$:

$$\sigma^2 = \int_0^\infty t^2 g(t)dt - t_0{}^2 = 2\int_0^\infty tn^0(t)dt - t_0{}^2$$

The second expression results from an integration by parts. A numerical integration is required to obtain σ^2, which can most easily be carried out with $n^0(t)$ values from equation (7). In this way we find from equations (9) and (10), with $t_0 = T$, that

$$\sigma = 0.24T$$

For Dr. Finkel's mice, reported to have $T = 670$ days, we have $\sigma = 161$ days, in rather poor agreement with the value $\hat{\sigma} = 258$ days inferred above from her paper.

In the calculations that follow we have used equations (7) and (10), with the assumption $T = t_0 = 600$ days, although actually the mice were about 70 days old at the beginning of the experiment. Thus we have used a doubling time τ_D of 80 days, and our $g(t)$ has standard deviation $\sigma = 141$ days. The assumption $\tau_D = 80$ days agrees with the value quoted by Jones[3] for mice. Values of $n^0(t)$ for these parameters are given in Table 1. The difference between assuming $T = 600$ days and assuming $T = 670$ days is not great; in fact, survival curves calculated from equation (5), which assumes $\sigma = 0$, do not differ greatly from curves obtained by the more refined procedure that we have used.

TABLE 1

t (Days)	$n^0(t)$	$e^{-\mu t^2}$
0	1.000	1.000
80	0.997	0.997
160	0.990	0.987
240	0.978	0.972
320	0.953	0.950
400	0.908	0.923
480	0.818	0.891
560	0.666	...
640	0.443	...
720	0.195	...
800	0.038	...
880	0.000	...

We proceed now to compare equation (4a), evaluated with the help of equations (7) and (10), with the experimental life-shortening data. We assume with Dr. Finkel that $\alpha = 0.11\alpha^*$ where α^* is the injected dose of strontium-90 in μc/kg body weight, and we attempt to choose the available parameters so as best to reproduce the observed life-shortening data $\Delta(\alpha^*)$. There are two parameters: the constant β, and the no-dose life-shortening Δ_0, the latter arising from the fact that we cannot give great weight to Dr. Finkel's zero point because of the statistical uncertainty in the observed mean life span of the control population. Thus the theoretical curve to be fitted to the data is

$$\Delta(\alpha^*) = \Delta_0 + 1 - \frac{1}{t_0}\int_0^\infty n^0(t)e^{-(0.11/2)\beta\alpha^* t^2}\,dt \qquad (12)$$

The inclusion of Δ_0 simply as a constant in equation (12) is not strictly correct: the additive term arising from an adjustment of the zero point should be written $\Delta_0(\alpha^*)$, where $\Delta_0(\alpha^*)$ is a somewhat complicated decreasing function of α^* that tends to zero as $\alpha^* \rightarrow \infty$. For simplicity, however, we ignore this complication, which proves to be unimportant for the lower radiation levels ($\alpha^* \lesssim 1000$ μc/kg), and which in any case does not much change the results obtained, because Δ_0 is small.

For very small values of α^* equation (12) reduces to

$$\begin{aligned}\Delta(\alpha^*) &= \Delta_0 + \alpha^*\left[\frac{1}{2}(0.11)\beta\int_0^\infty n^0(t)t^2dt\right]\\ &= \Delta_0 + \alpha^*\cdot\left.\frac{d\Delta}{d\alpha^*}\right|_{\alpha^*=0}\end{aligned} \qquad (13)$$

This is the linear response region. We can therefore choose preliminary values of Δ_0 and β by estimating a linear fit to the experimental points at low values of α^*. The integration in equation (13) is performed numerically, with use of equation (7).

We have calculated theoretical curves from equation (12) in three steps: (1) For $\alpha^* \leq 50$ μc/kg equation (13) applies; (2) for selected values of α^* in the range $50 \leq \alpha^* \leq 1000$ we carry out the integration in equation (12) numerically, using time intervals $\Delta t = 80$ days; (3) for $\alpha^* > 1000$ it is found that the asymptotic form of equation (12) is valid:

$$\Delta(\alpha^*) = \Delta_0 + 1 - \frac{1}{t_0}\sqrt{\frac{\pi}{0.11\beta\alpha^*}} \qquad (14)$$

Because we wish to examine the result statistically, we adjust the parameters by a weighted least squares procedure. We calculate two theoretical curves $y = f_0 + f(x, \beta_1)$ and $y = f_0 + f(x, \beta_2)$ (here y is fractional life shortening, Δ; x is injected dose, α^*; and f_0 is the constant Δ_0) for two nearly correct values β_1 and β_2 of the parameter β. We ask for values $\hat{f}_0 = f_0 + \Delta f_0$ and $\hat{\beta} = \beta_1 + (\beta_2 - \beta_1)\delta$ of the parameters such that the weighted sum of the squares of the differences between the experimental values y_i and the theoretical values $y(x_i)$ is a minimum:

$$\sum_i w_i(y_i - f(x_i, \hat{\beta}) - \hat{f}_0)^2 = \min \qquad (15)$$

Since $\beta_2 - \beta_1$ is small we can assume that

$$\begin{aligned}f(x_i, \hat{\beta}) &= f(x_i, \beta_1) + [f(x_i, \beta_2) - f(x_i, \beta_1)]\delta\\ &\equiv f(x_i, \beta_1) + \Delta f_i\delta\end{aligned} \qquad (16)$$

The parameter adjustments Δf_0 and δ are then given by

$$\Delta f_0 = \frac{(\sum w_i \Delta y_i)(\sum w_i \Delta f_i^2) - (\sum w_i \Delta y_i \Delta f_i)(\sum w_i \Delta f_i)}{(\sum w_i)(\sum w_i \Delta f_i^2) - (\sum w_i \Delta f_i)^2} \tag{17}$$

$$\delta = \frac{(\sum w_i \Delta y_i \Delta f_i)(\sum w_i) - (\sum w_i \Delta y_i)(\sum w_i \Delta f_i)}{(\sum w_i)(\sum w_i \Delta f_i^2) - (\sum w_i \Delta f_i)^2}$$

where $\Delta y_i = y_i - f_0 - f(x_i, \beta_1)$.

The weights w_i appearing in equations (15) and (17) should be inversely proportional to the *a priori* variances of the experimental values y_i. We take the variances to be inversely proportional to the number of animals in each experimental group. This ignores the effect of radiation in changing the dispersion of life spans, but a detailed examination shows it to be a not unreasonable procedure.

The theoretical curve obtained in the above manner is shown in Figure 1, with

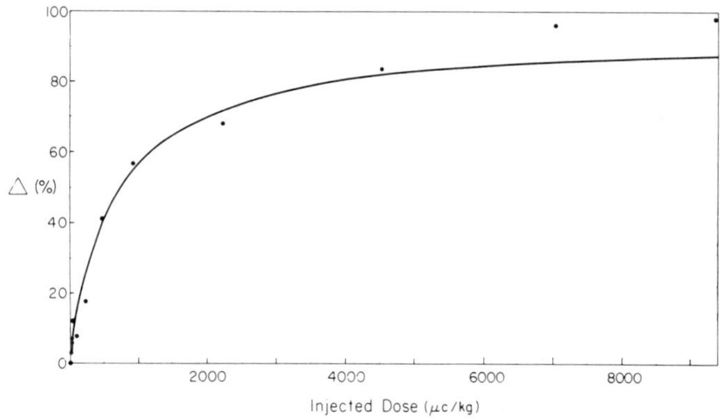

FIG. 1.—Percentage decrease in life expectancy, Δ, as a function of injected dose α^* of strontium-90. Solid curve is the theoretical curve calculated from equation (12). Solid circles are the experimental values reported by Dr. Finkel.

the experimental points for comparison. The two points at highest radiation levels lie well above the curve, doubtless because the mechanism of life shortening at the high radiation levels departs from what we have assumed, owing to the importance of subacute and acute irradiation disease, which Dr. Finkel reports to be the primary cause of death at injected doses above 2200 μc/kg. These points make little contribution to the least squares parameter adjustment, owing to their low weights, and can be omitted without sensibly changing the result. The parameters obtained are $\beta = 1.8 \times 10^{-7}$ day^{-2} (μc/kg retained)$^{-1}$, and $\Delta_0 = 2.5$ per cent.

The least-squares-fitted curve can be used to estimate the death-rate standard deviation σ: for experimental Δ values of unit weight (taken here to be for the control group and "group 12"), the estimated variance of the experimental Δ values is

$$\hat{\sigma}_\Delta^2 = \frac{1}{m-2} \sum_{i=1}^{m} w_i [y_i - \hat{f}_0 - f(x_i, \hat{\beta})]^2 \tag{18}$$

where m is the number of experimental points. Equation (18) takes into account the two-parameter adjustment. If M is the number of animals in groups having unit weight, then

$$\hat{\sigma}^2 = M\hat{\sigma}_\Delta{}^2 \tag{19}$$

From equations (18) and (19) we find $\hat{\sigma} = 191$ days in case the two highest points mentioned above are omitted ($\hat{\sigma} = 222$ days in case they are included). Comparing the value 191 days with the value $\sigma = 161$ days based on the Gompertz death-rate curve (670-day life span) and the value $\hat{\sigma} = 258$ days inferred from Dr. Finkel's data, we see that the theoretical curve fits the experimental values about as well as would be expected from the Gompertz curve, and somewhat better than would have been expected on the basis of Dr. Finkel's acceptance region.

In Figure 2 there is shown the portion of the theoretical curve for the lower radi-

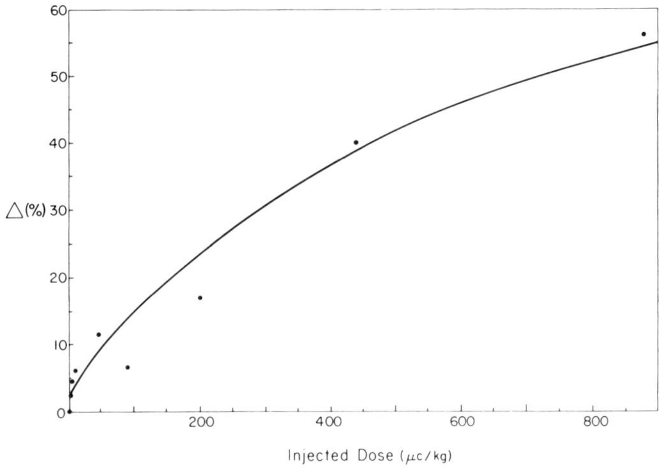

FIG. 2.—Percentage decrease in life expectancy in the low-dose region of Dr. Finkel's experiments. Theoretical curve and experimental points as in Fig. 1.

ation levels. The curvature is pronounced, and the linear response region is restricted to injected doses less than about 50 μc/kg. Most of Dr. Finkel's experiments were carried out in the nonlinear portion of the curve.

It is interesting to compare the above analysis with an alternative one based on the approach developed by Jones,[3] in which the effect of a given exposure of an animal to radiation is regarded as equivalent to an increase in physiological age of the animal by an amount proportional to the amount of radiation received. In terms of the Gompertz formulation of the natural death rate, this results in the case we are considering to the addition of a linear term in t to equation (6):

$$\ln\left(-\frac{1}{N}\frac{dN}{dt}\right) = C + Bt + B\eta\alpha t \tag{20}$$

The constant $B\eta$ in this treatment plays a role analogous to the constant β used previously.

From equations (3) and (20) we obtain

$$t_\alpha = \frac{-\mathrm{Ei}\left(-\dfrac{A}{1+\eta\alpha}\right)}{B(1+\eta\alpha)}\, e^{A/(1+\eta\alpha)} \tag{21}$$

where $A = e^C/B$ as before. In practice A is so small that the exponential integral can adequately be approximated by the logarithmic term in equation (11):

$$-\mathrm{Ei}\left(-\frac{A}{1+\eta\alpha}\right) = \ln\frac{1}{\Gamma A} + \ln(1+\eta\alpha) \tag{22}$$

Recognizing from equations (21) and (22) that

$$t_0 = \frac{1}{B}\ln\frac{1}{\Gamma A}$$

and making use of equation (8), we obtain

$$\Delta = 1 - \frac{1 + \epsilon\ln(1+\eta\alpha)}{1+\eta\alpha} \tag{23}$$

where $\epsilon = \tau_D/(t_0 \ln 2) \cong 1/5$.

By choosing $\eta = 0.014$ (μc/kg retained)$^{-1}$, and by adjusting the zero point slightly as done previously, we calculate from equation (23) a theoretical curve that matches closely the curve calculated from equation (12), which is the curve shown in Figures 1 and 2. The discrepancy in Δ between the two curves is less than 0.02 over the range $0 \leq \alpha^* \leq 3000$ μc/kg, and increases to 0.04 at $\alpha^* =$ 9000 μc/kg. The two curves fit the experimental data equally well, as shown by the estimates $\hat{\sigma} = 191$ days for the curve from equation (12) and $\hat{\sigma} = 189$ days for equation (23), calculated by the weighted sum-of-squares procedure described previously. Life-shortening data, at least of the accuracy involved here, are therefore unable to discriminate between the two analytical approaches.

Analysis of Incidence of Leukemia and Related Diseases.—The experimental data on the incidence of diseases of the blood and blood-forming tissues can be analyzed in the framework of the above treatment. However, because of the peculiar form in which the experimental results are presented ("Curve C: percentage decrease in time to a 20 per cent incidence of reticular tissue tumors compared with the 20 per cent incidence time of the controls"), the analysis is subject to greater uncertainties and difficulties and the data cannot so readily be evaluated statistically as those for the decreased life expectancy. We therefore content ourselves with a somewhat sketchy treatment, which should suffice to indicate the general nature of the problem.

Let $\lambda(t, \alpha)$ be the expected number of deaths due to these diseases that have occurred by the time t in a population having retained body burden α of strontium-90. We may then expect to find a death-rate probability parameter β_l for these diseases such that the death rate is

$$\frac{d\lambda}{dt} = N(t)\beta_l\alpha t + \frac{d\lambda_0}{dt}\frac{N(t)}{N^0(t)}$$

where $\lambda_0(t)$ is the number of deaths due to these diseases expected in the control population. To carry the analysis further we need to know the function $\lambda_0(t)$, but unfortunately Dr. Finkel presents no data that enable us to determine it. Of the various assumptions that could be made, we have chosen to assume that the natural deaths due to leukemia are distributed as though they were radiation-induced according to the same model as the deaths due to radiation from strontium-90. The natural leukemia death rate will then be equivalent to a "background" body burden α_0 of strontium-90, and equation (20) becomes

$$\frac{d\lambda}{dt} = N(t)\beta_l(\alpha_0 + \alpha)t$$

Obtaining $N(t)$ from equation (2), we have

$$\frac{\lambda(t)}{N_0} = \beta_l(\alpha_0 + \alpha) \int_0^t n^0(t)e^{-(1/2)\beta\alpha t^2}\, t\, dt$$

The expected 20 per cent incidence time τ is then the implicit solution of

$$0.11\beta_l(\alpha_0^* + \alpha^*) \int_0^\tau n^0(t)e^{-(0.11/2)\beta\alpha^* t^2}\, t\, dt = \frac{1}{5} \qquad (24)$$

and the expected no-dose 20 per cent incidence time τ_0 is given by

$$0.11\beta_l\alpha_0^* \int_0^{\tau_0} n^0(t)t\, dt = \frac{1}{5} \qquad (25)$$

τ_0 as given by equation (25) is not necessarily the same as the 20 per cent incidence time $\tau_0' = 565$ days observed for the control population.

To compare the theory with the experimental data we calculate from equation (24) the fractional decrease function $1 - \tau(\alpha^*)/\tau_0'$. An adequate approximate calculation for values of τ less than about 450 days ($1 - \tau/\tau_0' > 0.20$) can be made by approximating $n^0(t)$ by a Gaussian $e^{-\mu t^2}$, as shown in Table 1. In this case equation (24) becomes

$$0.11\beta_l(\alpha_0^* + \alpha^*) = \frac{\mu + \frac{1}{2}(0.11)\beta\alpha^*}{5\left[1 - \exp\left(-\tau^2\left(\mu + \frac{1}{2}0.11\beta\alpha^*\right)\right)\right]} \qquad (26)$$

To evaluate the parameters β_l and α_0^* we have fitted a smooth curve, by eye, to the experimental values of $1 - \tau/\tau_0'$, and used this curve to pick pairs of values (α^*, $\tau(\alpha^*)$) from which the quantity $0.11\beta_l(\alpha_0^* + \alpha^*)$ was calculated from equation (26). The quantity $0.11\ \beta_l\alpha_0^*$ was calculated from equation (25) by numerical integration, with the assumption $\tau_0 = \tau_0'$. When plotted against α^*, the values of $0.11\ \beta_l(\alpha_0^* + \alpha^*)$ calculated in this way lie nicely along a straight line, as required by the theory, for values of α^* in the range $0 \leq \alpha^* \leq 1000\ \mu c/kg$. Above 1000 $\mu c/kg$ the linear relation breaks down, reflecting the fact that the one experimental value in this higher range, at 2200 $\mu c/kg$, lies rather far from the theoretical curve. Ignoring this highest value we obtain in this way the parameters $\beta_l = 0.7 \times 10^{-7}\ day^{-2}\ (\mu c/kg)^{-1}$ and $\alpha_0^* = 200\ \mu c/kg$, from which the theoretical curve shown in Figure 3 is calculated. In addition to the point $\tau(0)$, and the points $\tau(\alpha^*)$ calculated from equation (24) over the range of validity of the Gaussian approximation, we have calculated the slope of the theoretical curve at $\alpha^* = 0$ from the following formula, which can be derived from equation (24):

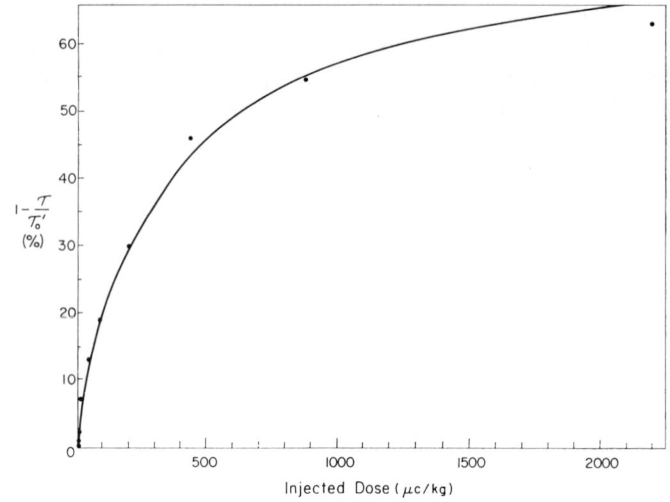

FIG. 3.—Percentage decrease in time to a 20 per cent incidence of bone tumors, as a function of injected dose of strontium-90. Solid curve is calculated from equation (26). Solid circles are experimental values reported by Dr. Finkel.

$$\frac{d}{d\alpha^*}\left(1-\frac{\tau}{\tau_0}\right)_{\alpha^*=0} = \frac{1-\dfrac{\beta}{10\beta_l}\dfrac{\int_0^{\tau_0} n^0(t)t^3dt}{\left[\int_0^{\tau_0} n^0(t)tdt\right]^2}}{10\gamma_0{}^2\tau_0{}^2n^0(\tau_0)}\cdot\frac{1}{2}(0.11)\beta_l$$

where $\gamma_0 = {}^1/_2(0.11)\beta_l\alpha_0{}^*$. The ratio of integrals appearing in the second term of the numerator in this equation can be shown to have a value close to unity (actually 1.06).

A comparison of the parameters $\beta_l = 0.7 \times 10^{-7}$ and $\beta = 1.8 \times 10^{-7}$ suggests that of the radiation-induced deaths the fraction due to leukemia and related diseases in Dr. Finkel's experiments on mice is rather larger than has been estimated for man. A particular sensitivity to these diseases on the part of this strain of mice is suggested also by the large "background dose level" $\alpha_0{}^*$, reflecting the relatively large number of deaths due to these diseases in the control population.

Statistical Examination of Dr. Finkel's Conclusions.—In searching for evidence for the existence of a threshold body burden of strontium-90, below which no harmful effects are caused, Dr. Finkel uses two methods: (1) statistical analysis of the experimental data, and (2) extrapolation of experimental curves. We now consider these two methods.

The statistical analysis consists of a *t*-test of the hypothesis of no difference in response between the control population and a population dosed with strontium-90. Dr. Finkel accepts the null hypothesis at the 10 per cent significance level ("10% probability level or higher") for the three lowest-dosed experimental groups, and considers that this acceptance constitutes "evidence that there might be a threshold" or that "a threshold . . . may lie between 4.5 and 44 μc/kg."

It constitutes nothing of the kind. It is clear that the width of the acceptance region for the null hypothesis (shaded region in Figs. 3 and 4 of Dr. Finkel's paper) should vary inversely as the square root of the number of animals in the experimental groups, assuming approximate normality of the death-rate curve $g(t)$, as Dr. Finkel must have done in applying the t-test. The threshold for which she finds "evidence" in the experiments is thus no threshold at all but simply a reflection of the statistical uncertainty of her information. It is clear that she could have found "evidence" of this sort for a threshold at any arbitrarily large radiation level (perhaps short of what would produce acute radiation sickness) by simply using few enough animals in her experiments.

The fallacy in Dr. Finkel's statistical argument is a failure to control the probability of type-II error of her test. Type-II error[5] is acceptance of the null hypothesis when it is in fact false. Consideration of the type-II error requires consideration of the alternative to the null hypothesis, which in this case is the theoretically likely linear response at low doses. If we use for the slope $d\Delta/d\alpha^*|_{\alpha^*=0}$ of the life-shortening response at low doses the value obtained above (eq. [13]) from a study of Dr. Finkel's results, namely, $d\Delta/d\alpha^* = 0.14\%\ (\mu c/kg)^{-1}$, and if we assume that Dr. Finkel's t-test acceptance region is appropriate to a one-tailed test at $\alpha^* = 8.9$ μc/kg, the highest experimental value for which the null hypothesis was accepted, then we can calculate the probability of type-II error. It is 85 per cent. This means that if there exists in fact no threshold at 8.9 μc/kg, Dr. Finkel's test would nevertheless have produced "evidence" for one in 85 experiments out of every 100 experiments performed. On the other hand, if there were in fact a threshold, the test would deny it in only 10 per cent of the experiments. Evidently the test is worthless as a proof of the existence of a threshold at this dose level (or lower, for which the probability of type-II error approaches the maximum that is possible, 90 per cent, for a 10 per cent probability of type-I error).

It is incumbent upon those who would extrapolate their threshold conclusions from 150 mice to 3×10^9 human beings that they demonstrate the existence of a significant experimental departure from the theoretically likely linear response, because although the existing burdens of strontium-90 are low, the number of individuals involved is very large, and the harmful consequences of proceeding on an unfounded assumption of a threshold are great.[6] As we have shown above, Dr. Finkel's results are in complete harmony with a linear law; in fact, the agreement between the linear law and the experimental results is better than could have been expected on the basis of the width of her null-hypothesis acceptance region.

As an alternative to the statistical tests, Dr. Finkel determines a threshold by extrapolating the experimental life-shortening curve. She states: "Since the [life-shortening] values for 1.3, 4.5, and 8.9 μc/kg do lie along a straight line when plotted semilogarithmically, it may be argued that they represent true departures from the control value. An extension of this straight line crosses the control value at 0.4 μc/kg." It is difficult to see why the semilog plot rather than some other should be used for the extrapolation. But in fact an extrapolation of any kind is groundless. The three response values lie within less than half the range of probable error (within $-1/2$ P.E. to $+1/2$ P.E.) of the difference d between experimental values of Δ, as determined from the width of the null-hypothesis acceptance region ($\sigma_d = 5.5$ per cent). If it is not obvious that no non-zero regression slope determined from these points can have any statistical significance, one can show[7] that the standard deviation of the regression slope estimator derived from the semilogarithmic plot is 2.2 times the estimated slope itself. If the semilogarithmically linear relation of the three points can be ascribed to anything but chance, then all of Dr. Finkel's statistical arguments are false. It is hard to imagine how two such mutually contradictory "proofs" could be advanced at one time.

There are other statistical points in Dr. Finkel's paper that merit scrutiny. In our discussion of her results we have had to rely on the correctness of the null-hypothesis acceptance region that she presents, but there are serious reasons for doubting its correctness. The width of the acceptance region corresponds to the estimate $\hat{\sigma} = 284$ days for the standard deviation of the death-rate function $g(t)$, if it applies to a one-tailed test on the difference between the life-shortening values obtained from two experimental groups of 150 animals each. On the other hand, Dr. Finkel's statement[8] that groups of 1393 animals would have been required to establish as significant at the 1 per cent level the difference observed (2.5 per cent) at the lowest dose corresponds to $\hat{\sigma} = 171$ days, a gross discrepancy. The latter value, we note, agrees reasonably with the values $\sigma = 161$ days from the Gompertz relation or $\hat{\sigma} = 191$ days from the agreement between experimental data and our theoretical curve.

It is clear that since the number of animals differs from one experimental group to another in Dr. Finkel's experiments, the null-hypothesis acceptance region cannot have width independent of injected dose α^*, as shown in her figures. From the information given there is no way to tell to which experimental groups the test appropriately applies.

More serious is the evident fact that Dr. Finkel applies the same acceptance region indiscriminately to the three very different sets of experimental data represented by her curves *A*, *B*, and *C*. It seems likely that the test was designed to handle the life-shortening data (curve *A*), because a *t*-test would not be inappropriate to life-span data, since the death-rate function $g(t)$ is (rather crudely) Gaussian. A statistical analysis of the curve-*C* data would be difficult, because the experimental statistic τ (20 per cent incidence time) is cumbersome to handle mathematically, as is evident in our discussion. But it is easy to show that Dr. Finkel's acceptance region is entirely inapplicable to the curve-*B* data ("proportion of animals that survived the latent period of 150 days and then died with osteogenic sarcomas").

The number of bone-cancer deaths in populations of a given size during a given time interval will be Poisson-distributed, if we neglect variations in population size due to deaths during the first 150 days, which is legitimate, as can be seen from Table 1 or from numbers given by Dr. Finkel, which show that the control group still contained close to 150 animals at $t = 150$ days. Whatever the low-dose regression function for curve *B*, it is clear from Dr. Finkel's Figure 4 that the expected number ξ of bone-cancer deaths is close to 3 for groups of 150 animals not dosed with strontium-90. To find acceptance regions for the null hypothesis of no significant difference in the number of such deaths between the control population and a dosed population of equal size we therefore find the value of the difference δ_P such that the probability of type-I error (one-tailed test) is P:

$$1 - P = e^{-2\xi} \sum_{\delta=-\infty}^{\delta_P} \xi^{\delta} \sum_{n=-\infty}^{+\infty} \frac{\xi^{2n}}{n!(n+\delta)!} \tag{27}$$

The results of a numerical evaluation of equation (27), in case $\xi = 3$, are $P = 0.10$, $\delta_P = 2.6$; $P = 0.01$, $\delta_P = 5$; $P = 0.001$, $\delta_P \cong 8.5$. Using the fact that the observed number of control deaths was 3 (2 per cent of 150), we find that the upper limit of the acceptance region at 10 per cent significance level is 3.7 per cent of 150. From Dr. Finkel's Figure 4 we therefore see that the highest strontium-90 dose that produced a "statistically non-significant increase" in the number of bone-cancer deaths is 8.9 μc/kg, not 200 μc/kg as stated by her. Her acceptance region represents for curve B a test having 0.1 per cent probability of type-I error. Her entire discussion of the statistical significance of the curve-B data is erroneous.

The Proper Testing of Evidence for a Threshold.—From the above discussion it is clear that a valid statistical test of the null hypothesis of "no response" at a given radiation level or a given dose of strontium-90 must use the type-II error as the basic parameter, rather than the type-I error, as is employed in the standard "cook book" tests, which are designed basically for application to the manufacture of goods. Alternatively stated, the null hypothesis that must be tested in the standard way is the hypothesis that the observed response values are in accord with a linear response curve at low doses.

Our analysis of Dr. Finkel's data for mice enables us to estimate reliably the linear decrease in life expectancy for low doses of strontium-90, and it is therefore possible for us to determine how many animals would have to be used in an experiment in which the mean lifetime of a control group is compared with the mean lifetime of a dosed group in order to establish the existence of a threshold at or above the dose used. We may consider two types of test: (A) the "minimal" test, that is, the test that requires as few animals as possible; (B) the "most powerful" test, which minimizes the probability both of type-I and of type-II error.

The null-hypothesis "no response" for test A is to be accepted if the dosed group exhibits no decrease in life expectancy, or an actual increase, when compared to the control group. Clearly this acceptance region makes the test minimal, because the probability of type-I error is 50 per cent, so that the test gives a neutral decision in case a threshold actually exists. If the number of animals used is greater than required for test A, then a decision as to the existence of a threshold will be more often right than wrong, in case that the threshold does actually exist. At the same time we can protect ourselves adequately against the serious alternative possibility by suitably choosing the type-II error.

Since the expected decrease in life expectancy for the dosed group is $\alpha \cdot d\Delta/d\alpha \big|_0$, the type-I and type-II errors are simultaneously minimized, and made equal, by choosing as the upper limit of the acceptance region for test B a decrease in life expectancy of $^1/_2\alpha \cdot d\Delta/d\alpha \big|_0 = 0.63\% \cdot \alpha$, where α is given in μc/kg retained in the body.

Since the expected decrease is proportional to α, the number of animals M required for the control group, if an equal number is used for the dosed group, is given by

$$M = \frac{\nu}{\alpha^2} \qquad (28)$$

where ν is a constant that depends on the type of test (A or B), on the probability of type-II error, and on the standard deviation σ of the natural death-rate function

$g(t)$. Because M proves in all cases of interest to be large, it is adequate to use the normal distribution in computing the constants ν in equation (28), owing to the Central Limit Theorem.

TABLE 2

VALUES OF THE CONSTANT ν IN EQUATION (28)

	TYPE OF TEST			
	A ("Minimal")		B ("Most Powerful")	
σ (days)	284	170	284	170
10 per cent probability of type-II error	4560	1630	18240	6520
1 per cent probability of type-II error	14950	5360	59800	21440

Using all of these principles, we have computed the coefficients ν for the various circumstances shown in Table 2. In particular we compare the results for σ = 284 days, derived from Dr. Finkel's acceptance region, with the results for $\sigma = 170$ days, which seems most reasonable on the basis of the previous discussion.

The numbers ν given in Table 2 are equal to the number of animals in the control and in the dosed groups required to establish the existence of a threshold at $\alpha^* =$ 9.1 μc/kg, just above the highest injected dose (8.9 μc/kg) for which Dr. Finkel accepted the hypothesis that a threshold exists. The numbers of animals used in her experiments were too small by factors of 10 to 400, for the conclusion that she reached. By solving equation (28) for α we may compute very simply the lowest threshold $\alpha_T{}^*$ that could have been recognized with statistical significance in her experiments, assuming that 150 animals were used both in the control group and in the dosed groups, which in general was not the case (fewer were used). These values of $\alpha_T{}^*$ are 91 and 54 for test A and 181 and 109 for test B (in each case for σ = 284 days and 170 days, respectively). It is clear from the experimental data that no threshold exists at any of these levels, and accordingly we are required to conclude that Dr. Finkel's data show that there is no threshold large enough to have been recognized with statistical significance from her data.

Conclusions about Effects on Man of Strontium-90 from Fallout.—We now turn to the discussion of Dr. Finkel's conclusion that the present contamination with strontium-90 from fallout is so very much lower than the "threshold" levels that it is extremely unlikely to induce even one bone tumor or one case of leukemia.

This statement by Dr. Finkel is shown by the argument given above to have no justification whatever from her experimental results, obtained with 150 mice or fewer in her control group and injected groups. We may ask how many mice would be needed in each group in order to permit Dr. Finkel's statement to be made with statistical significance (or to be shown to be false).

The present average body burden of strontium-90 in the world's population is about 0.0002 μc. per person. This corresponds, with Dr. Finkel's conversion factor (5 to 10 μc. per 70-kg man equivalent to 1 μc. retained per kg for mice) to a retained dose α = 0.00002 to 0.00004 μc/kg in mice. Hence in order to justify Dr. Finkel's statement evidence would be needed that the mouse threshold is as great as about 0.00004 μc/kg; that is, we must place α in equation (28) equal to 0.00004 μc/kg. From the values of the constant ν in Table 2 (we use the values for σ = 170 days, which we believe to be better than those for σ = 284 days) we find $M = 1 \times 10^{12}$ for the "minimal" test and 3.4×10^{12} for the "most powerful" test with 10 per cent type-II error, and 3.3×10^{12} and 13.5×10^{12}, respectively, with 1 per cent type-II error. We hence conclude that a study like that made by Dr. Finkel would have to use a much greater number of mice than the number of people in the world, in order to provide evidence that would justify her extreme statement to be made with statistical significance.

This conclusion is, of course, not at all unexpected. The difficulty of detecting by statistical methods an effect that causes a small increase in the annual number of deaths among the world's population is well known. For example, let us assume that the average number of deaths per year is 50 million. The statistical fluctuations from this average from year to year are measured roughly by the square root of this number, 7000; and accordingly a study of a larger population would be needed to show with statistical significance the existence of an effect resulting in an additional 1000 deaths per year (the rough estimate of the world-wide effect of the present body burden of strontium-90 from fallout, if there is no threshold). The same number of mice would be needed to test the equivalent effect in mice.

Summary.—We have developed methods of theoretical analysis of the results of experimental studies of the effects of injection of radioactive substances into animals on their life expectancy and on the incidence of tumors. These methods have been applied to the data reported for mice by Dr. Miriam P. Finkel, and it has been shown that her conclusion from these data that it is extremely unlikely that the strontium-90 from the fallout from nuclear weapons tests will induce even one bone tumor or one case of leukemia in human beings is completely unjustified.

* This paper is a contribution from the Division of the Geological Sciences (No. 908) and the Division of Chemistry and Chemical Engineering (No. 2421) of the Institute. A brief account of the work has been published by us (Letter to the Editor, *The New York Times*, Nov. 16, 1958), and a reply has been made by Dr. Finkel (*ibid.*, Nov., 30 1958).

[1] M. P. Finkel, *Science,* **128,** 637, 1958.

[2] E. B. Lewis, *Science,* **125,** 965, 1957.

[3] H. Jones, *The Nature of Radioactive Fallout and Its Effects on Man* (Washington, D. C.: Government Printing Office, 1957), p. 1109.

[4] Jahnke and Emde, *Tables of Functions* (New York: Dover Publications, 1945), pp. 1–9.

[5] A. M. Mood, *Introduction to the Theory of Statistics* (New York: McGraw-Hill Book Co. Inc., 1950), p. 247.

[6] L. Pauling, *No More War!* (Dodd-Mead & Co., 1958).

[7] Mood, *op. cit.*, pp. 292–294.

[8] Finkel, *op. cit.*, pp. 638–639.

Reprinted from Science, July 7, 1961, Vol. 134. No. 3471, pages 15-21

A Molecular Theory of General Anesthesia

Anesthesia is attributed to the formation in the brain of minute hydrate crystals of the clathrate type.

Linus Pauling

During the last twenty years much progress has been made in the determination of the molecular structure of living organisms and the understanding of biological phenomena in terms of the structure of molecules and their interaction with one another. The progress that has been made in the field of molecular biology during this period has related in the main to somatic and genetic aspects of physiology, rather than to psychic. We may now have reached the time when a successful molecular attack on psychobiology, including the nature of encephalonic mechanisms, consciousness, memory, narcosis, sedation, and similar phenomena, can be initiated. As one of the steps in this attack I have formulated a rather detailed theory of general anesthesia, which is described in the following paragraphs (*1*).

It is likely that consciousness and ephemeral memory (reverberatory memory) involve electric oscillations in the brain, and that permanent memory involves a material pattern in the brain, in part inherited by the organism (instinct) and in part transferred to the material brain from the electric pattern of the ephemeral memory (*2*). The detailed natures of the electric oscillations constituting consciousness and ephemeral memory, of the molecular patterns constituting permanent memory, and of the mechanism of their interaction are not known.

The electric oscillations of the brain make themselves evident in a crude way in electroencephalograms, which show patterns of electric oscillation that depend upon the state of consciousness and the nature of the encephalonic activity of the subject. Evidence that the ephemeral memory, with an effective life that is rarely longer than a few minutes, is electrical in nature is provided by a number of observations. It has been noted that unconsciousness produced by a blow to the head or electric shock often has caused complete loss of memory of the events experienced during the period of 10 or 15 minutes before the blow or shock to the brain. Moreover, when the formation of new permanent memories is interfered with by the decreased ability of the brain to carry on metabolic processes involving proteins, as in old age or Korsakoff's syndrome (alcoholism, protein starvation, thiamine deficiency), the memory continues for a period of 10 or 15 minutes, but usually not much longer; the memory seems to persist only so long as conscious attention is directed to it (*3*).

Consciousness and Ephemeral Memory

We may discuss the electric oscillations of consciousness and ephemeral memory in terms of the exciting mechanism and the supporting structure. The supporting structure is the brain, with its neuroglial cells, neurones, and synaptic interneuronal connections that determine the detailed nature of the oscillations. The average energy of the electric oscillations may be assumed to be determined by the activity of the exciting mechanism and the impedance of the neural network. Loss of consciousness such as occurs in sleep or in narcosis (general anesthesia) may be the result either of a decrease in activity of the exciting mechanism or of an increase in impedance of the supporting network of conductors, or of both. I think that it is likely that sleep results from a decrease in the activity of the exciting mechanism, and that many sedatives, such as the barbiturates, operate by a specific action on the exciting mechanism, such as to decrease its activity; similarly, stimulants such as caffeine may have a specific action on the exciting mechanism that increases its activity. I think that general anesthetics of the non-hydrogen-bonding type, such as cyclopropane, chloroform, nitrous oxide, and 1,1,1-trifluoro-2-chloro-2-bromoethane (halothane), operate by increasing the impedance of the encephalonic network of conductors, and that this increase in impedance results from the formation in the network, presumably mainly in the synaptic regions, of hydrate microcrystals formed by crystallization of the encephalonic fluid. These hydrate microcrystals trap some of the electrically charged side-chain groups of proteins and some of the ions of the encephalonic fluid, interfering with their freedom of motion and with their contribution to the electric oscillations in such a way as to increase the impedance offered by the network to the electric waves and thus to cause the level of electrical activity of the brain to be restricted to that characteristic of

The author is professor of chemistry, Gates and Crellin Laboratories of Chemistry, California Institute of Technology, Pasadena.

anesthesia and unconsciousness, despite the continued activity of the exciting mechanism. The formation of the hydrate microcrystals may also decrease the rate of chemical reactions by entrapping the reactant molecules and thus preventing them from coming close enough to one another to react; in particular, the catalytic activity of enzymes may be decreased by the formation of hydrate microcrystals in the neighborhood of their active sites.

Anesthetic Agents

This theory is forced upon us by the facts about anesthesia. Hundreds of substances are known to cause general anesthesia; among them are chloroform ($CHCl_3$), halothane ($CF_3CClBrH$), nitrous oxide (N_2O), carbon dioxide (CO_2), ethylene (C_2H_4), cyclopropane (C_3H_6), sulfur hexafluoride (SF_6) (*4*), nitrogen (N_2) and argon (Ar), which under high pressure cause narcosis (*5*), and xenon (Xe) (*6*). The substances given in this list have rather similar properties as general anesthetics; these properties show a rough correlation with their physical properties, such as the vapor pressure of the liquids. Ferguson (*7*) calls them the physical anesthetics. We may infer that they function in similar ways in causing narcosis. Their chemical properties are such that it is impossible to believe that they produce narcosis by taking part in chemical reactions involving the formation and breaking of ordinary chemical bonds (covalent bonds). Moreover, although it is known that in many physiological processes the formation and rupture of hydrogen bonds play an important part, these substances, with the exception of a few (nitrous oxide, carbon dioxide, chloroform), would not be expected to form even weak hydrogen bonds, and we may call them the non-hydrogen-bonding anesthetic agents. Other narcotics, such as ethanol, may be placed in the hydrogen-bonding class.

The most surprising anesthetic agents are the noble gases, such as xenon. Xenon is completely unreactive chemically. It has no ability whatever to form ordinary chemical compounds, involving covalent or ionic bonds. The only chemical property that it has is that of taking part in the formation of clathrate crystals. In these crystals the xenon atoms occupy chambers in a framework formed by molecules that interact with one another by the formation of hydrogen bonds. The crystal of this sort of greatest interest to us is xenon hydrate, $Xe \cdot 5\frac{3}{4} H_2O$. The crystals of xenon hydrate have been shown by x-ray examination to have the same structure as those of other hydrates of small molecules, such as methane hydrate and chlorine hydrate (*8, 9*). A thorough x-ray examination of chlorine hydrate has been made (*10*), showing that in the cubic unit of structure, with edge 11.88 A, there are 46 water molecules arranged in a framework such that each water molecule is surrounded tetrahedrally by four others, with which it forms hydrogen bonds with length 2.75 A, essentially the same as in ordinary ice (2.76 A). Whereas in ordinary ice the hydrogen-bonded framework does not contain any chambers large enough for occupancy by molecules other than those of helium or hydrogen, the framework for xenon hydrate and related hydrates contains eight chambers per cubic unit cell. Two of these chambers are defined by 20 molecules at the corners of a nearly regular pentagonal dodecahedron, and the other six are defined by 24 water molecules at the corners of a tetrakaidecahedron with 2 hexagonal faces and 12 pentagonal faces. These polyhedral chambers are illustrated in Figs. 1 and 2. The smaller chambers and the larger chambers may be occupied by the xenon atoms or methane molecules, but only the larger chambers permit occupancy by chlorine molecules, which are somewhat larger than the molecules of xenon or methane. In chlorine hydrate the dodecahedral chambers presumably are partially occupied by water molecules not forming hydrogen bonds, or, if air is present, by nitrogen molecules or oxygen molecules.

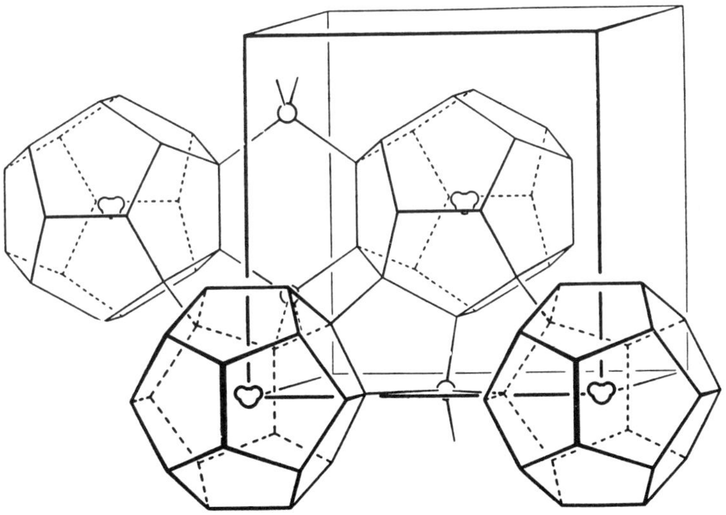

Fig. 1. The structure of the 12-A hydrate crystals of small molecules, such as xenon. The unit cube is about 12 A on an edge. The hydrogen-bonded framework of water molecules consists of 46 water molecules per unit cube. Of these, there are two sets of 20 at the corners of pentagonal dodecahedra, one about the corner of the cube and one about the center of the cube. Six more water molecules aid in holding the dodecahedra together by hydrogen bonds. All hydrogen bonds, indicated by lines in the figure, are about 2.76 A long, as in ordinary ice. There is room in each dodecahedron for a small molecule; a symbol suggesting a molecule of H_2O or H_2S is shown.

Hydrate crystals with somewhat similar structures are formed also by other anesthetic agents (*8, 9*). Chloroform, for example, forms the hydrate $CHCl_3 \cdot 17H_2O$, which has a cubic unit of structure with the cube edge 17.30 A. The hydrogen-bonded framework of 136 molecules per cube involves 16 small chambers per cube, with the pentagonal dodecahedron as the coordination polyhedron, and 8 large chambers, each formed by 28 water molecules at the corners of a hexakaidecahedron, with 4 hexagonal faces and 12 pentagonal faces (Fig. 3). Only the large chambers can accommodate a chloroform molecule. The smaller chambers may be occupied by smaller molecules, such as xenon, which with water and chloroform forms the crystal $CHCl_3 \cdot 2Xe \cdot 17H_2O$. The volume of the 17-A

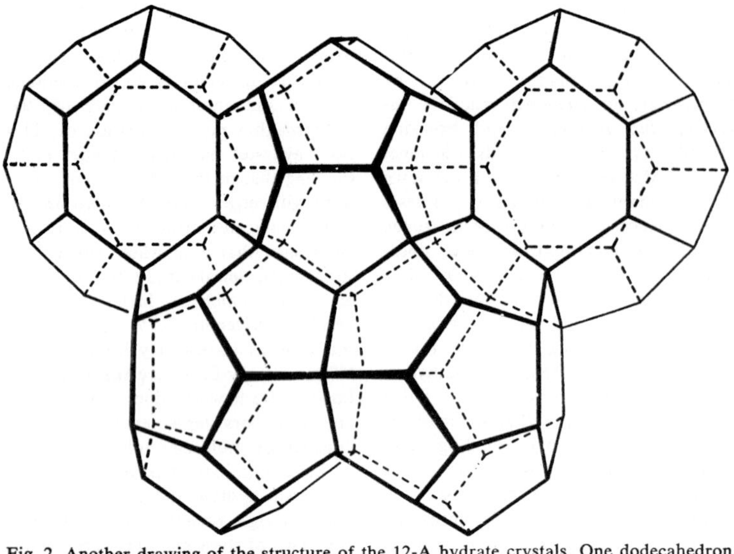

Fig. 2. Another drawing of the structure of the 12-A hydrate crystals. One dodecahedron is shown in the upper center. Around it are tetrakaidecahedra, which provide room for somewhat larger molecules than can fit in the dodecahedra. There are six tetrakaidecahedra and two dodecahedra per unit cube.

framework per water molecule is slightly larger than that of the 12-A framework—an 18-percent increase over ordinary ice (ice I), as compared with 16 percent for the 12-A framework.

The stability of the hydrate crystals results in part from the van der Waals interaction between the entrapped molecules and the water molecules of the framework and in part from the energy of the hydrogen bonds. So far as the energy of the hydrogen bonds is concerned, the stability of the framework alone would be expected to be the same as that of ordinary ice; however, the framework is more open for the hydrates than for ordinary ice, and in consequence the stabilization by van der Waals interaction of the water molecules with one another is less for the hydrate frameworks than for ordinary ice. A thorough study of experimental information about hydrate crystals by the methods of statistical mechanics, with the crystals treated as having variable occupancy of the chambers in the framework, has been carried out by van der Waals and Platteeuv; it shows that the free energy per water molecule of the empty framework is greater than that for ice I at 0°C by 0.167 kcal/mole for the 12-A framework and 0.19 kcal/mole for the 17-A framework (*11*).

The extent to which the crystals are stabilized by the van der Waals interaction of the entrapped molecules and the surrounding water molecules can be estimated by a simple calculation. The London equation for the energy of the electronic dispersion interaction between two molecules *A* and *B* is

$$W = -\frac{3}{2}\,\frac{\alpha_A\,\alpha_B\,E_A^*\,E_B^*}{r^6\,(E_A^* + E_B^*)} \qquad (1)$$

In this equation α_A and α_B are the electric polarizabilities of the two molecules, E_A^* and E_B^* are their effective energies of electronic excitation, and r is the distance between their centers.

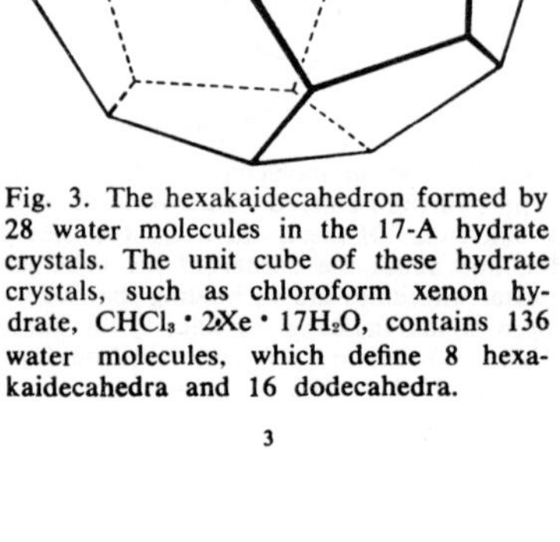

Fig. 3. The hexakaidecahedron formed by 28 water molecules in the 17-A hydrate crystals. The unit cube of these hydrate crystals, such as chloroform xenon hydrate, $CHCl_3 \cdot 2Xe \cdot 17H_2O$, contains 136 water molecules, which define 8 hexakaidecahedra and 16 dodecahedra.

It has been found that agreement between this equation and the observed enthalpies of sublimation of crystals of the noble gases is obtained by taking the effective excitation energy to be 1.57 times the first ionization energy (*12*). The first ionization energy of xenon is 280 kcal/mole, and the same value may be used for the water molecule. The interaction energy of two molecules then has the value $-aR_AR_B/r^6$, in which R_A and R_B are the mole refractions of *A* and *B*, in milliliters, and *a* is equal to 51 kcal/mole, with *r* measured in angstroms (the mole refraction is $4\pi N/3$ times the polarizability; *N* is Avogadro's number). In the crystal $8Xe \cdot 46H_2O$, two of the xenon molecules are in pentagonal dodecahedral chambers formed by 20 water molecules at the distance 3.85 A from the xenon atom, and the other six are in tetrakaidecahedral chambers formed by 24 water molecules, of which 12 are at 4.03 A and 12 at 4.46 A. The average energy of van der Waals attraction of a xenon atom (R = 10.16 ml) with its neighboring water molecules (R = 3.75 ml) is thus calculated to be −9.1 kcal/mole, which becomes −10.3 kcal/mole on addition of the similarly calculated values for the interaction with more distant water molecules and with other xenon atoms in the crystals.

The difference in enthalpy of the 12-A water framework and ordinary ice may be roughly evaluated by a similar calculation of the energy of van der Waals attraction between the water molecules (the nearest and next-nearest neighbors are at nearly the same distances in ordinary ice and the hydrate crystals, but the larger distances are different, corresponding to the more open structure of the hydrate framework). This calculation gives 0.16 kcal/mole for the 12-A framework and 0.20 kcal/mole for the 17-A framework; the close approximation of these values to the corresponding free-energy values indicates that there is little difference in entropy of the empty frameworks and ice I, as is expected from the similarity of the intermolecular forces that determine the vibrations of these hydrogen-bonded structures. The enthalpy of formation, at 0°C, of $Xe \cdot 5\frac{3}{4}H_2O$ from gaseous xenon and ice I is found by experiment (*8*) to be 8.4 kcal/mole. The value given by the foregoing calculations is $10.3 - 5.75 \times 0.16 = 9.4$ kcal/mole, minus a small correction for the van der Waals repulsion of the xenon atoms and the surrounding water molecules. The agreement shows the extent to which the stabilization of the hydrate crystals may be understood in terms of the van der Waals interactions of the molecules.

The relation between the logarithm of the equilibrium pressure (in millimeters of mercury) of hydrate crystals and water (and also ice I) at 0°C and the mole refraction of the molecules stabilizing the hydrate crystals is shown in Fig. 4, at the left. The energy of van der Waals attraction between the water framework and the entrapped molecules is directly proportional to the mole refraction of the entrapped molecules. Hence if no other interactions affected the free energy of the hydrate crystals the points for $X \cdot 5\frac{3}{4}H_2O$ would lie on a straight line and those for $X \cdot 17H_2O$ on another straight line. There is a general concordance with this expectation, and the deviations are reasonable. For example, the molecules acetylene, ethylene, and ethane increase in size in this order, and it is likely that the van der Waals repulsion between these molecules and the water molecules of their dodecahedral and tetrakaidecahedral cages increases rapidly in this sequence in such a way as to decrease the stability of the ethylene hydrate crystal and, still more, that of the ethane hydrate crystal, with corresponding increases in the equilibrium partial pressures.

Hydrate Microcrystal Theory

It is evident that the mechanism of narcosis cannot be simply the formation in the brain of the hydrate microcrystals $X \cdot 5\frac{3}{4}H_2O$ and $X \cdot 17H_2O$ that we have been discussing, because these crystals would not be stable under the conditions that lead to narcosis. For example, methyl chloride is narcotic for mammals at partial pressure about 0.14 atmosphere and temperature 37°C, but the crystals of its hydrate are not stable at 37° until the partial pressure reaches 40 atmospheres. In order to account for the formation of microcrystals of hydrates at body temperature we must assume that some stabilizing agent other than the anesthetic agent is also operating. I think that it is likely that the other stabilizing agents are side chains of protein molecules and solutes in the encephalonic fluid. It is known that substances resembling the charged side chains of proteins also interact with water to form hydrate crystals with a structure closely resembling that of the hydrates of the anesthetic agents. For example, tetra-*n*-butyl ammonium fluoride forms a hydrate with composition $(C_4H_9)_4NF \cdot 32H_2O$ and melting point 24.9°C. The crystals of this hydrate are tetragonal, with edge a = 23.78 A and edge c = 12.53 A, and with a structure that is believed to be closely similar in character to that of xenon hydrate and the related hydrates discussed above.

These crystals and similar crystals

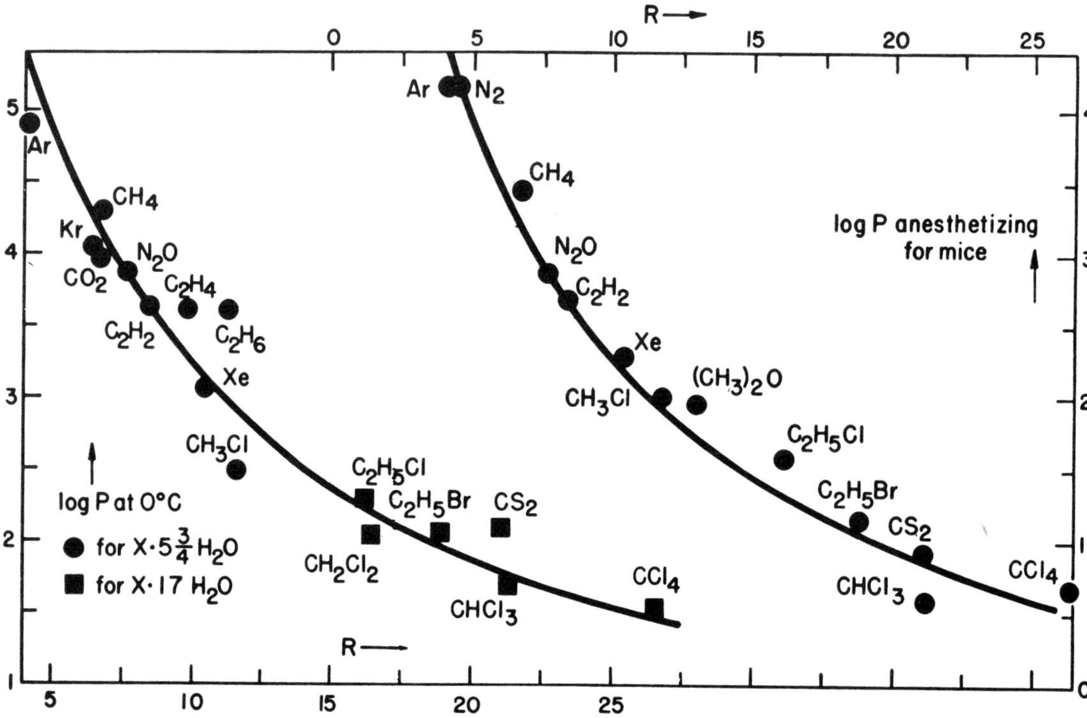

Fig. 4. At left are values of the partial pressure of anesthetic agents in equilibrium with their hydrate crystals and ordinary ice and water at 0°C, plotted against values of the mole refraction (shown by the scale at bottom). Circles correspond to the 12-A hydrate crystals; squares, to the 17-A hydrate crystals. The composition of the 12-A hydrate crystals is $X \cdot 5\frac{3}{4}H_2O$ for the smaller molecules, which can occupy both dodecahedra and tetrakaidecahedra, and $X \cdot 7\frac{2}{3}H_2O$ for the larger ones (ethane, methyl chloride), which occupy only the tetrakaidecahedra. At right the logarithm of the anesthetizing partial pressure for mice is plotted against the mole refraction of the anesthetizing agent (scale at top).

of tetra-*i*-amyl ammonium salts were first made by Fowler, Loebenstein, Pall, and Kraus (*13*). The determinations of the structure of these crystals and of related ones that they have prepared {$[(n\text{-}C_4H_9)_3S]F \cdot 20H_2O$, $[(i\text{-}C_5H_{11})_4N]F \cdot 38H_2O$, $[(n\text{-}C_4H_9)_4P]_2WO_4 \cdot 64H_2O$} are being carried out by Jeffrey and his co-workers (*14*).

It is known that two anesthetic agents can cooperate to increase the stability of a hydrate framework. For example, 1 atmosphere of xenon (*15*) increases the decomposition temperature of the 17-A hydrate of chloroform by a little over 14.7°C. In the absence of xenon the crystal has the composition $CHCl_3 \cdot 17H_2O$, and in its presence $CHCl_3 \cdot 2Xe \cdot 17H_2O$. The 17-A framework forms one hexakaidecahedron and two dodecahedra per $17H_2O$; the chloroform molecules are too large to enter the dodecahedra, which can, however, be occupied by atoms of xenon or other small molecules. Similar increases of 5° to 20°C in the decomposition temperatures of 17-A hydrate crystals of $CHCl_3$, CH_3CF_2Cl, $CHF{=}CF_2$, $CFCl_3$, SF_6, and some other substances by 1 atmosphere of krypton, H_2S, or H_2Se, as well as by xenon, have also been reported (*8, 15*). The 17-A hydrate $CHF_2CH_3 \cdot 2H_2S \cdot 17H_2O$ becomes stable in the presence of H_2S, whereas 1,1-difluorethane without other molecules forms a 12-A hydrate.

We may accordingly surmise that the stabilizing effect for hydrate crystals of amino acids and other solute molecules in encephalonic fluid and also of the alkyl ammonium side chains of lysyl residues and the alkyl carboxylate ion side chains of aspartate and glutamate residues, and perhaps also of certain other side chains of proteins, could operate effectively to stabilize hydrate crystals at temperatures not much lower than normal body temperature, perhaps about 25°C. The narcosis resulting from cooling of the brain, which is observed to take place at about 27°C in human beings, would then, according to our theory, be explained as resulting from the formation of these hydrate crystals in the synaptic regions of the brain and from the resultant increase in impedance of the neural network and correspondingly decreased energy of the electric oscillations. Hibernation may similarly involve the induction of unconsciousness by formation of hydrate crystals on decrease in temperature.

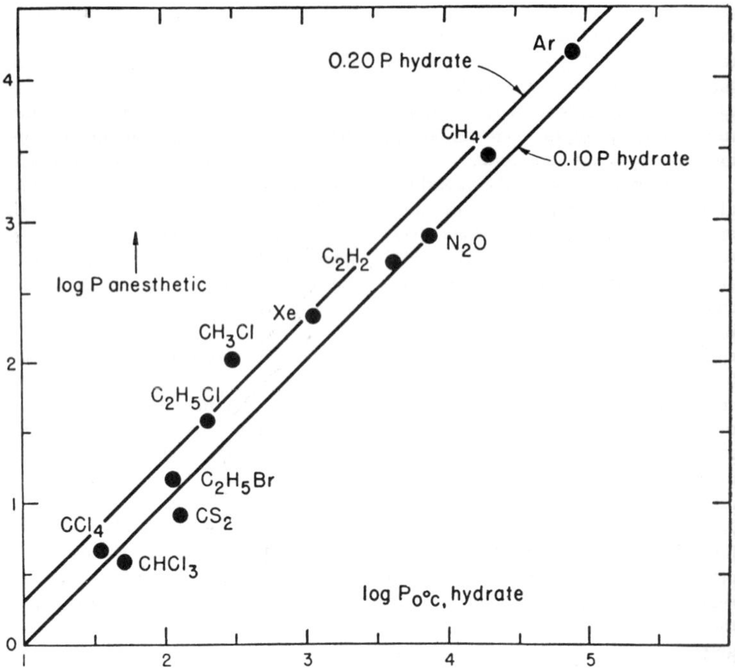

Fig. 5. A diagram showing the logarithm of the anesthetizing partial pressure of non-hydrogen-bonding anesthetic agents plotted against the equilibrium partial pressure of their hydrate crystals.

The molecules of the anesthetic agent, when present, would occupy some of the chambers in the hydrate crystal, with others occupied by the protein side chains and other groups normally present in the brain, in such a way as to give an increase in stability of the microcrystals such as to permit them to form at a temperature 10° or 15°C higher than that at which they are stable in the absence of the anesthetic agent. Through the formation of these microcrystals the conductance of the network would be decreased, with a consequent decrease in energy of the electric oscillations sufficient to cause unconsciousness. On decrease of the activity of the anesthetic agent in the encephalonic fluids, as elimination from the body takes place, the microcrystals would melt, the conductance of the synapses would be restored to its original level, and consciousness would be regained.

The logarithm of the anesthetizing partial pressure (in millimeters of mercury) for mice is shown as a function of the mole refraction of the non-hydrogen-bonding anesthetic agents in Fig. 4, at the right. The points lie close to a curve that resembles the curve for the equilibrium partial pressure of the hydrate crystals, shown at the left. The relation between the anesthetizing partial pressure and the partial pressure for the hydrate crystals at 0°C is shown in Fig. 5; the two pressures are proportional, the proportionality factor being about 0.14. The average deviation of the 11 points from the best line corresponds to the factor 1.4 (or 0.7) over a total pressure range of 4000 (for the logarithm, ± 0.15 over a range of 3.6).

Other Theories of Anesthesia

This agreement provides some support for the proposed theory, but not proof. Approximately the same correlation would be found between the anesthetizing partial pressure of the non-hydrogen-bonding anesthetic agents and any other property involving an energy of intermolecular interaction proportional to the mole refraction of the molecules. An example is the solubility in olive oil of the gaseous anesthetic agent at a standard pressure; another is the ratio of the solubility in olive oil to that in water (the oil-water distribution coefficient). The first depends largely on the energy of van der Waals attraction of the anesthetic molecules by the oil molecules, and the second on the difference between this energy and the energy of attraction by the water molecules, and each is proportional to the mole refraction of the anesthetic agent. These quantities are involved in the Meyer-Overton theory of narcosis (*16*). The thermodynamic activity theory of Ferguson (*7*) is based upon the observed rough constancy of the ratio of anesthetizing partial pressure of non-hydrogen-bonding anesthetic agents to the vapor pressure (thermodynamic activity) of the pure liquid at a standard temperature. This rough constancy is, of course, to be expected on any theory of anesthesia involving intermolecular forces, since the vapor pressure of a liquid is determined by the forces acting between its molecules.

The lipid theories of anesthesia seem to me to be less attractive than the hydrate microcrystal theory. First, brain, like other tissues of the human body, consists largely of water: about 78 percent, as compared with about 12 percent of lipids and 8 percent of proteins. The water contains ions and proteins with electrically charged side chains and is hence expected to be largely involved in the electric oscillations that constitute consciousness; the lipids probably function mainly as insulating materials, and their electrical properties are presumably changed only slightly by the presence of nonpolar solute molecules of the non-hydrogen-bonding anesthetic agents. Moreover, the postulated change in phase from liquid to hydrate microcrystal, with a correspondingly great change in properties, provides an explanation of the large change in encephalonic activity caused by a small amount of substance, and there is no evidence to cause us to expect such a change in phase for the lipids.

Agents That Function by Stabilization of Microcrystals

Anesthetic agents that function by the stabilization of hydrate microcrystals may be divided into several classes, determined by the sizes and shapes of their molecules. Those of the first class may be defined as having molecules sufficiently small to fit into a pentagonal dodecahedron formed by 20 hydrogen-bonded water molecules without serious van der Waals repulsion. Those of the second class include the larger molecules that are able to fit into the hexagonal tetrakaidecahedron without serious van der Waals hindrance. Those of the third class are the still larger molecules that fit into the hexakaidecahedron without serious steric hindrance. The molecules of other classes might fit into larger chambers in the hydrogen-bonded framework; for example, the tetra-*n*-butyl ammonium ion probably fits into the cavity formed by four contiguous tetrakaidecahedra about the tetrahedral position between four dodecahedra in the chlorine hydrate structure, with the elimination of the water molecule at this position, as found in the crystal-structure study of the trialkylsulfonium crystal carried out by Jeffrey and McMullan (*14*). It seems likely that several kinds of microcrystals are formed in brain tissue, and that they are variously stabilized by anesthetic agents of the several classes. It might accordingly be expected that the agents of different classes would act to some extent synergistically (and also to some extent competitively, in that molecules of an agent of one class can occupy the larger polyhedra corresponding to the succeeding classes, with, however, less stabilizing effect than for its own polyhedron because of the greater intermolecular distance). Hence it may be suggested that a mixture of agents of the dodecahedral, tetrakaidecahedral, and hexakaidecahedral classes, such as CF_4, CF_3Cl (or CF_3Br), and $CFCl_3$ (or $CF_3CClBrH$), would be a better anesthetic than any one substance.

It is not unlikely that magnesium ion, $Mg(OH_2)_6^{++}$, acts as an anesthetic agent by stabilizing hydrate microcrystals. This ion, with its attached water molecules, would become a part of the hydrogen-bonded framework. Molecules such as ethanol and tribromoethanol, CBr_3CH_2OH, may be expected to participate in the formation of microcrystals of hydrates in such a way that the molecule becomes a part of the hydrogen-bonded framework and also has a space-filling and van der Waals stabilizing effect. Other hydrogen-bond-forming narcotic agents may attach themselves by the formation of hydrogen bonds to protein molecules in a specific way so as to interfere specifically with certain encephalonic processes. The study of these specific effects will require the detailed investigation of the proteins and other substances present in brain and nerve tissue.

Related Studies

Many experiments by means of which evidence about the proposed molecular theory of anesthesia may be obtained are suggested by the theory; some of them are being carried out in our laboratories. Studies of crystalline hydrate phases formed in the presence of anesthetic agents, ions, and protein molecules or molecules and ions similar to protein side chains might yield interesting results.

The "iceberg" theory of ionic solutions (*17*) and of hydration of proteins (*18*) is closely related to the hydrate microcrystal theory of anesthesia; the only change suggested for these theories is that the ordered arrangement of water molecules about the solute ions and protein side chains has one or another of the clathrate structures rather than the more compact ice-I structure.

The hydrate microcrystal theory of anesthesia clearly suggests that the anesthetic agents should act on all tissues, and not just on brain and nerve tissue. It was pointed out nearly a century ago by Claude Bernard (*19*) that "an anesthetic agent is not just a special poison of the nervous system; it anesthesizes all elements, all tissues by numbing them, temporarily blocking their irritability." Many studies of the effects of anesthetic agents on physiological processes other than thinking have been reported (*20*).

At present there is little information available about the fraction of the aqueous phase in the brain that is changed into hydrate microcrystals during anesthesia, or about the dimensions of the microcrystals. Experiments now under way should provide some information. The results of density-gradient ultracentrifuge studies of solutions of deoxyribonucleic acid by Hearst and Vinograd (*21*) indicate that at 25°C the nucleic acid molecules have about 50 water molecules of hydration per nucleotide residue at water activity near unity. This suggests that the microcrystals have linear dimensions of about 20 A or 30 A (for nucleic acid, of course, they continue along the Watson-Crick double helix). A hydrate cube with edge 30A contains about 750 water molecules.

Conclusion

The hydrate-microcrystal theory of anesthesia by non-hydrogen-bonding agents differs from most earlier theories in that it involves primarily the interaction of the molecules of the anesthetic agent with water molecules in the brain, rather than with molecules of lipids. The postulated formation of hydrate microcrystals similar in structure to known hydrate crystals of chloroform, xenon, and other anesthetic agents as well as of the substances related to protein side chains, entrapping ions and electrically charged side chains of protein molecules in such a way as to decrease the energy of electric oscillations in the brain, provides a rational explanation of the effect of the anesthetic agents in causing loss of consciousness. The striking correlation between the narcotizing partial pressure of the anesthetic agents and the partial pressure necessary to cause formation of hydrate crystals provides some support for the proposed theory, but it is recognized that any theory based upon the van der Waals attraction of the molecules of the anesthetic agent for other molecules would show a similar correlation, inasmuch as the energy of intermolecular attraction is approximately proportional to the polarizability (mole refraction) of the molecules of the anesthetic agent. The proposed theory is sufficiently detailed to permit many predictions to be made about the effect of anesthetic agents in changing the properties of brain tissue and other substances, and it should be possible to carry out experiments that will disprove the theory or provide substantiation for it.

References and Notes

1. The work reported in this article (contribution No. 2697) is part of a program of investigation of the chemical basis of mental disease supported by grants to the California Institute of Technology made by the Ford Foundation and the National Institutes of Health. This theory has been presented in lectures at Pacific State Hospital, California State Department of Mental Hygiene, Spadra (23 May 1960); at a meeting of the Western Society of University Anesthetists, Stanford Medical School, Palo Alto, Calif. (21 Jan. 1961); at a meeting of the Hawaii section of the American Chemical Society and Sigma Pi Sigma, University of Hawaii, Honolulu (5 Apr. 1961); and at a meeting of the Mediterranean section of the Société de Chimie Physique, Toulouse, France (25 Apr. 1961).
2. L. A. Jeffress, Ed., *Cerebral Mechanisms in Behavior* (Wiley, New York, 1951), especially sections by W. S. McCulloch. McCulloch (p. 101) suggests that there are three kinds of memory: (i) reverberatory memory; (ii) a kind of alteration of the nervous net with use; and (iii) a storage memory with a bottleneck both in putting information in and in taking it out. If the second and third are to be differentiated at all, I think that they may be classed together as involving permanent or semipermanent molecular patterns.
3. McCulloch (2, p. 58) has reported an example of a man over 80 years old, and with no power of adding to his permanent memory, who held all the details of an important meeting of a board of directors in his mind during the 8 hours of its duration, so that he was able to summarize it brilliantly during the last half hour of the meeting; yet a few minutes after the meeting he had forgotten it completely and permanently.
4. R. W. Virtue and R. H. Weaver, *Anesthesiology* **13**, 605 (1952).
5. A. R. Behnke, R. M. Thompson, E. P. Motley, *Am. J. Physiol.* **112**, 554 (1935); A. R. Behnke and O. D. Yarborough, *U.S. Naval Med. Bull.* **36**, 542 (1938); E. M. Case and J. B. S. Haldane, *J. Hyg.* **41**, 225 (1941).
6. J. H. Lawrence, W. F. Loomis, C. A. Tobias, F. H. Turpin, *J. Physiol.* **105**, 197 (1946); S. C. Cullen and E. G. Gross, *Science* **113**, 580 (1951).
7. J. Ferguson, *Proc. Roy. Soc. London* **B127**, 387 (1939); "Mécanisme de la narcose," *Colloq. intern. centre natl. recherche sci. Paris* (1951), p. 25.
8. M. von Stackelberg *et al.*, *Fortschr. Mineral.* **26**, 122 (1947); ———, *Naturwissenschaften* **36**, 327, 359 (1949); ———, *ibid.* **38**, 456 (1951); ———, *ibid.* **39**, 20 (1952); ———, *J. Chem. Phys.* **19**, 1319 (1951); ———, *Z. Elektrochem.* **58**, 25, 40, 99, 104, 162 (1954).
9. W. F. Claussen, *J. Chem. Phys.* **19**, 259, 662, 1425 (1951).
10. L. Pauling and R. E. Marsh, *Proc. Natl. Acad. Sci. U.S.* **38**, 112 (1952).
11. J. H. van der Waals and J. C. Platteeuv, *Mol. Phys.* **1**, 91 (1958). Approximately the same value for the 12-A framework has been reported by R. M. Barrer and W. I. Stuart [*Proc. Roy. Soc. London* **A243**, 172 (1957)]. H. S. Frank and A. S. Quist [*J. Chem. Phys.* **34**, 604 (1961)] have reported about 0.200 kcal for the similar framework suggested in the theory of the structure of liquid water proposed by L. Pauling and P. Pauling [L. Pauling, in *Hydrogen Bonding*, D. Hadzi, Ed. (Pergamon Press, London, 1959), p. 1; *The Nature of the Chemical Bond* (Cornell Univ. Press, Ithaca, N.Y., ed. 3, 1960), p. 472].
12. L. Pauling and M. Simonetta, *J. Chem. Phys.* **20**, 29 (1952).
13. D. L. Fowler, W. V. Loebenstein, D. B. Pall, C. A. Kraus, *J. Am. Chem. Soc.* **62**, 1140 (1940).
14. R. K. McMullan and G. A. Jeffrey, *J. Chem. Phys.* **31**, 1231 (1959); G. A. Jeffrey and R. K. McMullan, American Crystallographic Association meeting, Washington, D.C., 24–27 Jan. 1960.
15. J. G. Waller, *Nature* **186**, 429 (1960).
16. H. H. Meyer, *Arch. exptl. Pathol. Pharmakol. Naunyn-Schmiedeberg's* **42**, 109 (1899); E. Overton, *Studien über die Narkose* (Jena, Germany, 1901).
17. H. S. Frank and M. W. Evans, *J. Chem. Phys.* **13**, 507 (1945); H. S. Frank and W. Y. Wen, *Discussions Faraday Soc. No.* **24** (1957), p. 133; H. S. Frank, *Proc. Roy Soc. London* **A247**, 481 (1958).
18. I. M. Klotz and S. W. Luborsky, *J. Am. Chem. Soc.* **81**, 5119 (1959); R. M. Featherstone, C. A. Muehlbaecher, J. A. Forsaith, F. L. DeBon, *Anesthesiology*, in press (studies of the solubility of anesthetic gases in protein solutions).
19. C. Bernard, *Leçons sur les anesthésiques et sur l'asphyxie* (Paris, 1875).
20. For example, L. V. Heilbroun, "Mecanisme de la narcose," *Colloq. intern. centre natl. recherche sci. Paris* (1951), p. 163; F. H. Johnson, H. Eyring, M. J. Polissar, *The Kinetic Basis of Molecular Biology* (Wiley, New York, 1954).
21. J. E. Hearst and J. Vinograd, in preparation.

Reprinted from *Orthomolecular Psychiatry: Treatment of Schizophrenia*, eds. David Hawkins and Linus Pauling, W.H. Freeman, San Francisco, pp. 18–34 (1973).

2

Results of a Loading Test of Ascorbic Acid, Niacinamide, and Pyridoxine in Schizophrenic Subjects and Controls

LINUS PAULING
ARTHUR B. ROBINSON
SUSANNA S. OXLEY
MAIDA BERGESON
ANDREW HARRIS
PAUL CARY
JOHN BLETHEN
IAN T. KEAVENY

INTRODUCTION

During the past fifteen years many studies of ascorbic acid in relation to schizophrenia have been published.[1] Our attention was brought to this field by the report by VanderKamp (1966) that a much larger intake of ascorbic acid is needed by hospitalized chronic schizophrenics than by other persons to reach a standard concentration of the acid in their urine (later verified by Herjanic and Herjanic, 1967). After making a preliminary quantitative study of the urinary excretion of ascorbic acid, we conducted a loading test (determination of the amount of ascorbic acid eliminated in the urine during the 6-hour period following oral ingestion of a sample) with schizophrenic subjects in University Hospital of San Diego County and in Mesa Vista Hospital (a private hospital in San Diego, California), in each case with a parallel control group of normal subjects. The technique of administering the vitamin was then improved somewhat, and the program was broadened to include two other

[1] See Chapter 14 for a summary of these studies.

TABLE 2-1.
Nature of the Vitamin Loading Tests

County, ascorbic acid:
46 County patients (23F, 23M), 14 controls (7F, 7M) from UCSD and County staff; vitamin in orange juice:*

County, niacinamide:
26 County patients, 20 controls from UCSD and County staff; vitamin in gelatin capsule, with one glass of water.

Mesa Vista, ascorbic acid:
16 Mesa Vista patients, 16 controls from Mesa Vista staff; vitamin in orange juice (after breakfast); vitamin administered on eight consecutive days, urine collected on first and eighth days (five patients failed to complete the experiment).

Patton-1, ascorbic acid, niacinamide, pyridoxine:
35 Patton patients† (18F, 17M), 15 controls‡ from UCSD staff; vitamins in gelatin capsules, with one can of Nutrament§ (instead of breakfast), two glasses of water.

Patton-2, ascorbic acid:
28 Patton patients (13F, 15M), of whom 18 (8F, 10M) had been used also in Patton-1; vitamins same as Patton-1, two months after Patton-1; analysis performed for ascorbic acid only.

Patton-(1 + 2), ascorbic acid:
44 Patton patients (22F, 22M); a combination of Patton-1 and Patton-2 where values for patients present in both Patton-1 and Patton-2 (8F, 10M) were averaged.

Stanford, ascorbic acid, niacinamide, pyridoxine:
44 controls (6F, 38M) from Stanford staff; vitamins in gelatin capsules, with one can of Nutrament (instead of breakfast), two glasses of water.

* All vitamins administered orally.
† One ascorbic acid sample and three niacinamide samples were lost during analysis.
‡ Only 13 controls were used for niacinamide and 11 controls for pyridoxine.
§ Nutrament breakfast drink is a registered brand name.

vitamins, niacinamide and pyridoxine, that had been reported to be of benefit to some schizophrenic patients. This program was then carried out with a group of schizophrenics in Patton State Hospital, Patton, California, with a control group there and a larger control group of normal subjects at Stanford University (Table 2-1).

In all our studies the schizophrenic groups consisted of persons who had been diagnosed as such by two physicians. We attempted no differential diagnosis.

THE LOADING TESTS

The vitamins were given orally to each patient in a dosage proportional to the two-thirds power of the body weight, with the constant such that a 200-pound subject received 0.01 mole of ascorbic acid and, if the other vitamins were being studied, 0.01 mole of niacinamide and 0.005 mole of pyridoxine. (An exception was the Stanford study, in which for convenience each subject received the 200-pound dose of each vitamin: 1.76 g of ascorbic acid, 1.22 g of niacinamide, and 1.03 g of pyridoxine hydrochloride.) Complete samples of urine were taken during the 2-hour period preceding the oral dose of vitamins and for each of three 2-hour periods following. The samples were immediately frozen in solid carbon dioxide (after addition of 1 ml of 1 N hydrochloric acid solution—see section on experimental methods) and taken to the laboratory for later analysis by the methods described in the section on experimental methods. No interesting results were obtained from the first samples (before the oral dose), and the analytical values for this period are not given in this paper. Of the three later 2-hour samples, the values for a subject were usually proportional, with the fraction of the dose appearing in the urine during the second period usually about twice that in either the first or third period. To simplify the discussion, the sum of the three values (the fraction of the dose appearing in the urine during the period of 6 hours following the oral administration of the dose) is discussed in this report. No significantly different conclusions were reached by consideration of the results of analysis of the separate 2-hour specimens.

THE TESTS WITH ASCORBIC ACID

The fractions of the oral dose of ascorbic acid present in the urine of the subjects during the next 6 hours are shown in Figure 2-1. The mean values for the five control groups are consistently higher than those for the five schizophrenic groups (the sixth, Mesa Vista 8th day, is discussed below).

Because the experimental techniques changed somewhat in the course of the work, the most reliable comparisons are those between each schizophrenic group and its control group, which was studied at the same time and by the same methods. Application of the Wilcoxon series test (described, for example, by Sokal and Rohlf, 1969) showed that in each case the null hypothesis that the schizophrenic group and the control group are randomly selected groups from a uniform population is rejected with statistical significance. The level of confidence ($1 - P$, one-tailed) with which the null hypothesis is rejected has the following values:

County	99.99%
Mesa Vista	99.6%
Patton–1	99.993%
Patton–2	99.97%
Patton–(1 + 2)	99.996%

We have formulated another method for statistical evaluation of the results. Examination of the distribution of the points for normal subjects in the earlier studies (County, Mesa Vista, Patton) showed that they were not distributed in accordance with a distribution function approximating the error function in either the fraction

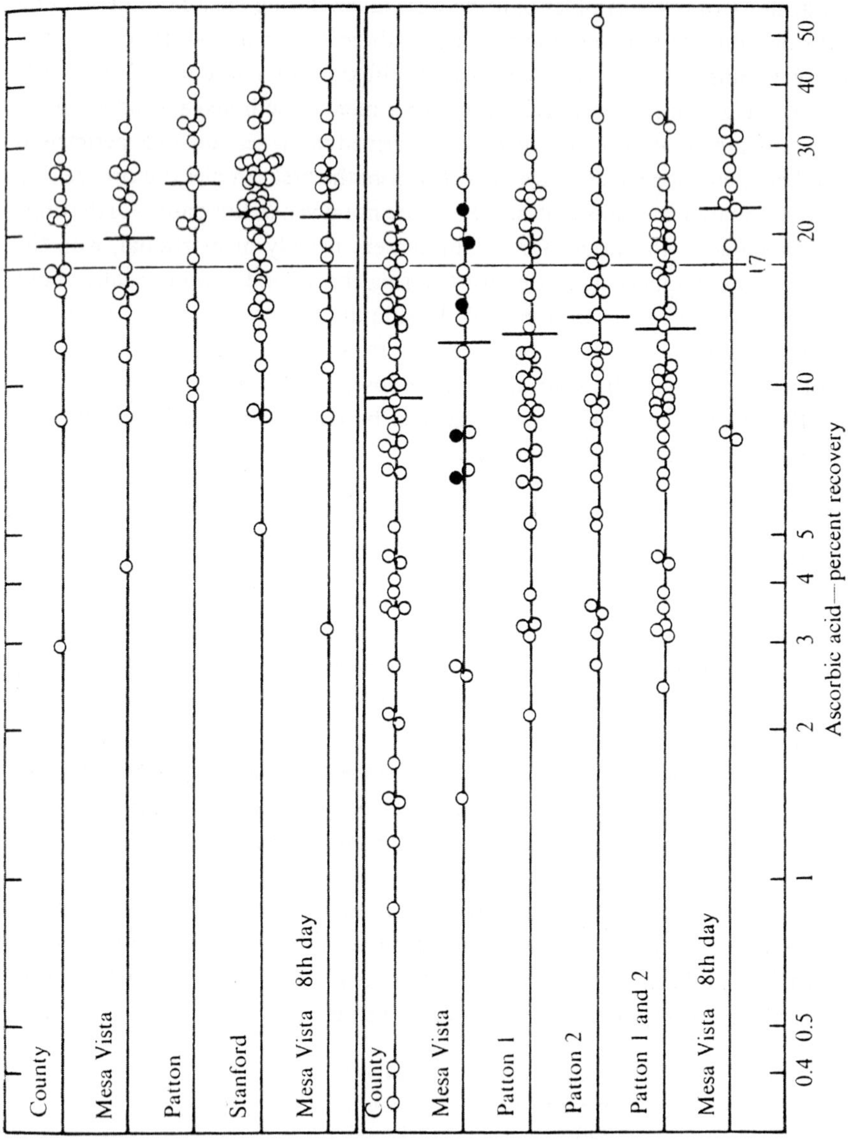

FIGURE 2-1.
Values of percentages of an oral dose of L-ascorbic acid found in the urine during the six-hour period after ingestion of the dose.

eliminated or its logarithm, and indicated instead a sum of three such functions. The Stanford study, with 44 normal subjects, was carried out to check this point and, as can be seen from Figure 2-1, gave results supporting the earlier indication. A distribution function derived from the Stanford points by a method formulated by us is shown as the upper curve in Figure 2-2. This function is the sum of 44 error functions, representing the 44 experimental values. Each error function is normalized to unit area, centered on its point, and has standard deviation 2.5 times the logarithmic distance to the third nearest neighboring point.

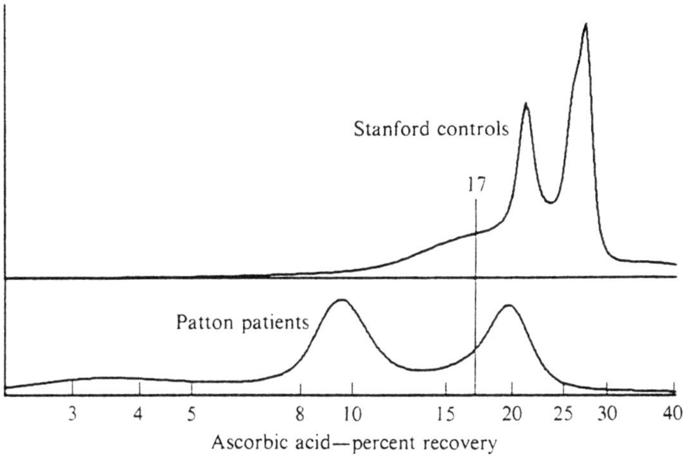

FIGURE 2-2.
Probability-distribution curves, calculated as described in the text, for percent of ascorbic acid recovered in the six-hour urine for 44 Stanford control subjects (above) and for 44 Patton schizophrenic subjects (below).

The distribution function shown in Figure 2-2 clearly indicates that with respect to the handling of ascorbic acid the 44 normal subjects are not essentially similar. The fraction of ascorbic acid eliminated in 6 hours differed from a normal (mean) value because of a number of randomly effective environmental and polygenic factors. The subjects are probably representative of three different populations: the high excretors (above 25 percent, 17/44), the medium excretors (17 to 25 percent, 16/44), and the low excretors (below 17 percent, 11/44). It seems to us unlikely that dietary differences would lead to such a distribution function. The reasonable hypothesis that the three groups result from two alleles *l* and *h* in a single gene locus, and represent the genotypes *ll* (low excretors), *lh* (medium), and *hh* (high) could, of course, be checked by familial studies, which we have not as yet undertaken.

The distribution curve for recovery of ascorbic acid by the Patton-(1 + 2) combined group of 44 schizophrenic subjects is shown in the lower part of Figure 2-2. There is clear indication of three groups, centered on 3.5 percent, 10 percent, and 20 percent, respectively. If the division into separate groups, as indicated in Figure 2-2, is accepted, there are two obvious alternative explanations of the differences between the upper curve and the lower curve. The first explanation is that the factor of decreased intake of ascorbic acid (poor nutrition of the schizophrenic subjects) operates to shift all values of percent recovery down. The normal peak around 27 percent would thus be shifted to 20 percent, that at around 22 percent to 10 percent, and that around 10 percent to 3.5 percent. The alternative is that the groups are not shifted, but instead the number of subjects in these groups is changed, with the top group, around 27 percent, much decreased in the schizophrenic subjects, and the groups around 10 percent and 3.5 percent much increased. Additional studies will be needed to check the reliability of this apparent division into groups and to distinguish between the alternative explanations of the differences between the distribution curves. There is no doubt, however, that there is a striking difference in the distribution of schizophrenic subjects and control subjects with respect to the recovery of an orally administered dose of ascorbic acid.

Let us assume that the upper limit of the low-excretor group is 17 percent elimination of the dose of ascorbic acid in 6 hours. The number of low excretors and the total group number are the following:

Controls: County 7/14, Mesa Vista 6/16, Patton 3/15, Stanford 11/44.
Schizophrenics: County 38/46, Mesa Vista 12/16, Patton–1 24/34, Patton–2 21/28, Patton–(1 + 2) 31/44.

Of the schizophrenic subjects 76 percent are low excretors, 2.5 times the incidence (30 percent) for the controls.

The numbers of low excretors can be used in a second test of the null hypothesis that the schizophrenic group and its control group are selected randomly from the same population. The values of χ^2 for a two-by-two test of these groups lead to the following values of the confidence level of rejection of the null hypothesis:

County	$1 - P$ (two-tailed) $> 99.5\%$
Mesa Vista	$> 96.5\%$
Patton–(1 + 2)	$> 99.93\%$

The four control groups show no statistically significant differences with one another, and may be combined into a single control group, with 27 low excretors in 89 subjects; similarly, the single combined schizophrenic group contains 81 low excretors in 106 subjects. These numbers lead to rejection of the null hypothesis at the confidence level $1 - P$ (two-tailed) > 99.9999 percent ($\chi^2 = 41.57$).

There is no doubt that the schizophrenic patients we studied differed significantly from the control groups in their elimination of ascorbic acid. It is known that cigarette smokers tend to have low blood levels of ascorbic acid, but among our subjects there were few smokers, and they did not show a significant difference from the nonsmokers. For example, the average excretion for the five Stanford smokers was 22.26 percent, as compared with 22.16 percent for all 44 Stanford subjects. (The Stanford subject with the lowest value, 5.16-percent excretion, was a cigarette smoker; the four others were not in the low group.) The sex of the subjects had no significant bearing on the results.

EFFECT OF INCREASED INGESTION OF ASCORBIC ACID

In the Mesa Vista series the loading test was repeated after extra ascorbic acid (1.23 to 2.04 g) had been administered each day for seven days. The results of the test on the eighth day are shown in Figure 2-1. For the control group there was essentially no change: average recovery increased from 20.0 percent to 21.5 percent, and low excretors decreased from 7/16 to 5/16. For the 11 members of the schizophrenic group who were tested on the eighth day,[2] however, there was a large change: average recovery increased from 11.3 percent for the 11 subjects to 21.7 percent, and low excretors decreased from 9/11 to 3/11. There is a striking difference between the schizophrenic subjects and the control subjects not only in the number of low excretors but also in the response to an increased intake of ascorbic acid. We surmise that many of the schizophrenic subjects had a high degree of "tissue unsaturation," which was rectified by the ingestion of about 10 g of extra ascorbic acid during the week of treatment.

THE TESTS WITH NIACINAMIDE

In Figure 2-3 the fractions of the oral dose of niacinamide found in the urine excreted by the subjects during the next 6 hours are shown. As was found for ascorbic acid, the mean values of the three control groups are consistently higher than those for the two schizophrenic groups by about 45 percent (as compared with 90 percent for ascorbic acid). Comparison of the County schizophrenic group with its control group and of the Patton schizophrenic group with its control group by the Wilcoxon sequence test showed that for each pair the null hypothesis that the two groups of the pair are randomly selected from a population uniform with respect to the handling of

---

[2] Five subjects failed to complete the 8-day study because of departure from the hospital.

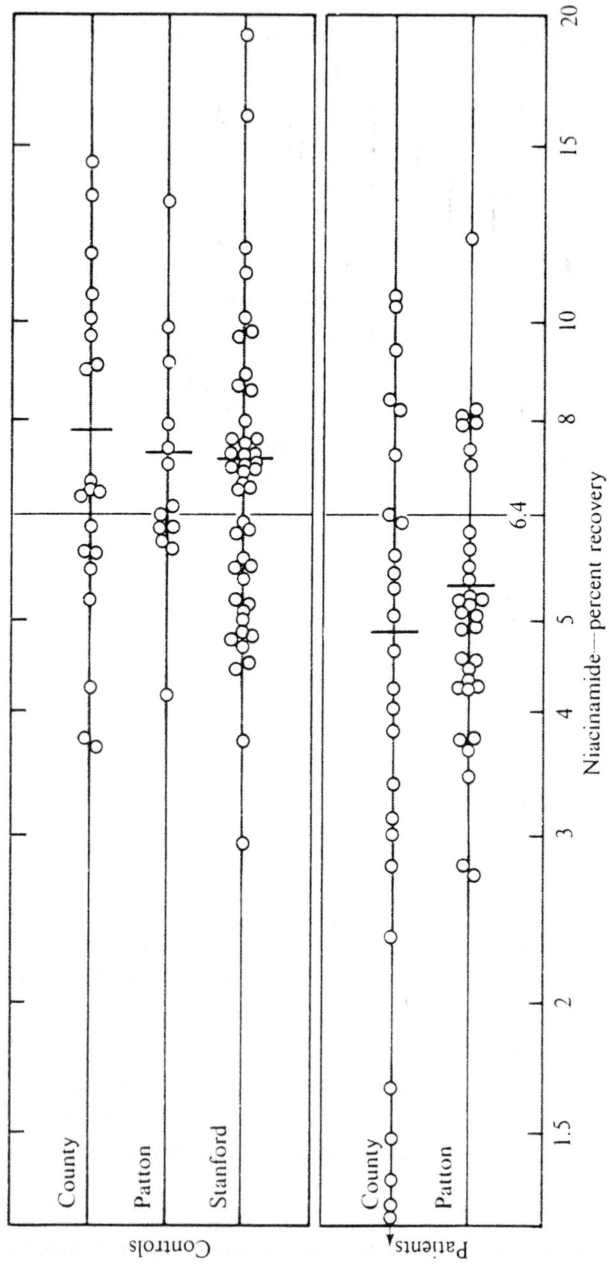

FIGURE 2-3.
Values of percentages of an oral dose of niacinamide found in the six-hour urine.

niacinamide is rejected with statistical significance, at the following values of the level of confidence (1 − *P*, one-tailed):

County	99.8%
Patton	99.98%

The niacinamide distribution function for the Stanford control group, calculated in the same way as for ascorbic acid (Figure 2-2), is shown in Figure 2-4. It is seen that a division into two or three groups is indicated, with one group consisting of the subjects with less than 6.4 percent of the dose of niacinamide, the low excretors. Of the 44

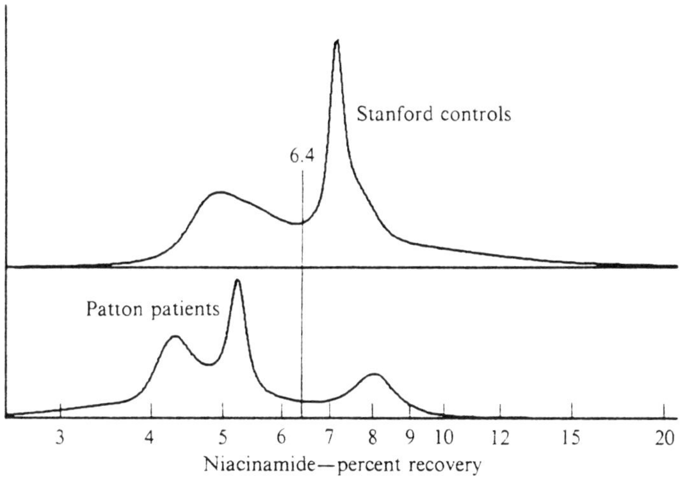

FIGURE 2-4.
Niacinamide-recovery probability-distribution curves for 44 Stanford control subjects (above) and for 32 Patton schizophrenic subjects (below).

Stanford subjects, 19 (43 percent) are low excretors. The second peak indicates the medium excretors, of whom there are 15 (34 percent) if the upper limit of their range is taken (rather arbitrarily) at 8.5 percent. The remaining 10 high excretors (23 percent) have values of recovery from 8.6 to 19 percent. Low excretors constitute 40 percent (8/20) of the County controls, 38 percent (5/13) of the Patton controls, 79 percent (20/26) of the County schizophrenics, and 78 percent (25/32) of the Patton schizophrenics. The values of χ^2 for a two-by-two test of pairs of numbers lead to the following values of the confidence level of rejection of the null hypothesis:

County	$1-P$ (two-tailed)	> 99.9%
Patton		> 99.9%

The three control groups and the two schizophrenic groups may be combined, giving 32 low excretors in 77 control subjects (42 percent) and 45 low excretors in 58

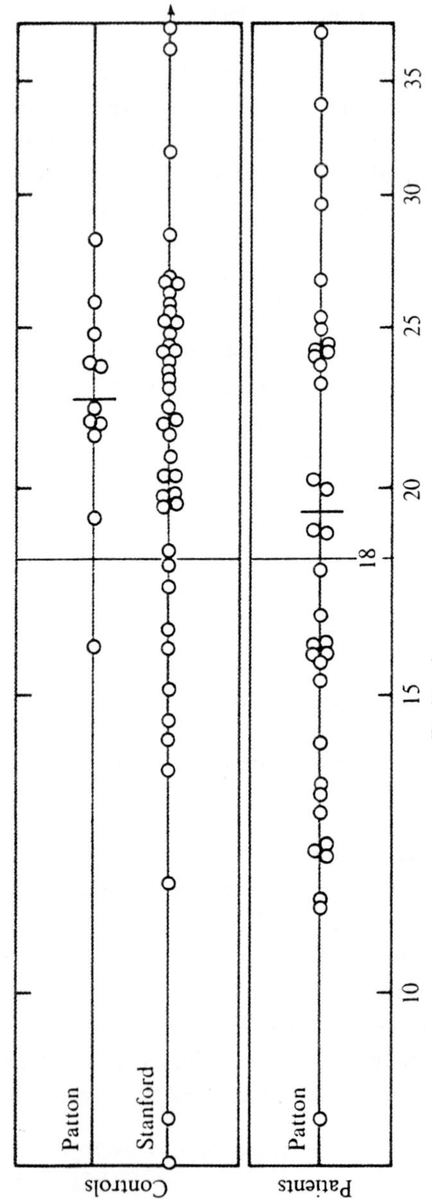

FIGURE 2-5.
Values of percentages of an oral dose of pyridoxine found in the six-hour urine.

schizophrenic subjects (78 percent). The statistical significance of this difference is very high, 99.997 percent ($\chi^2 = 17.52$). We conclude that with respect to their handling of niacinamide the schizophrenic groups in our study are significantly different from the control groups.

THE TESTS WITH PYRIDOXINE

The results of loading tests with pyridoxine for 34 schizophrenic subjects in Patton State Hospital and their 11 control subjects and for 44 normal subjects in the Stanford test are shown in Figure 2-5. The mean value for each of the two control groups is 17 percent higher than that for the schizophrenic group. Application of the Wilcoxon sequence test for the Patton schizophrenic group and its control group leads to rejection of the null hypothesis that the two groups are randomly selected from a population uniform with respect to the handling of pyridoxine at the confidence level $1 - P$ greater than 95.7 percent.

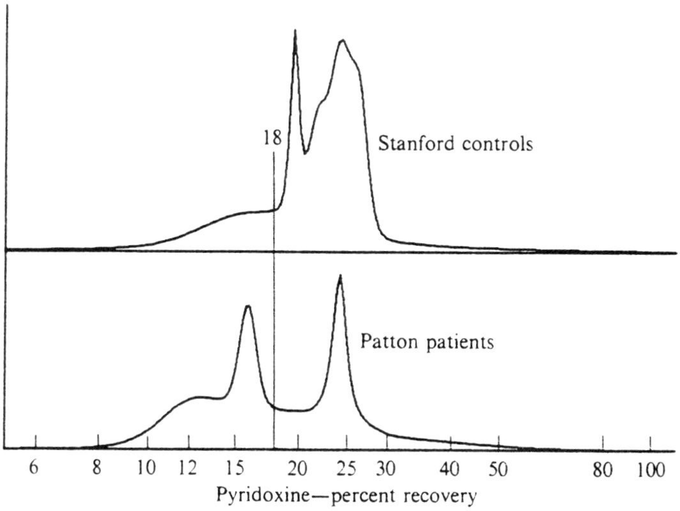

FIGURE 2-6.
Pyridoxine-recovery probability-distribution curves for 44 Stanford control subjects (above) and for 35 Patton schizophrenic subjects (below).

The pyridoxine distribution functions for the Stanford control group and the Patton patients, calculated in the same way as for ascorbic acid (Figure 2-2), are shown in Figure 2-6. It is seen that there are indications of a division into three groups. The group of low pyridoxine excretors, up to 18 percent (vertical line in Figure 2-6), includes 11 of the 44 Stanford control subjects (25 percent), 1 of the 11 Patton control subjects (9 percent), and 18 of the 35 Patton schizophrenic subjects (51 percent). The null hypothesis is rejected at the 99.2 percent confidence level by comparing Patton schizophrenics with the Patton control group, and at the 99.6 percent confidence level by comparing Patton schizophrenics with Patton controls plus Stanford controls.

CORRELATION OF LOW VALUES IN THE THREE VITAMINS

Loading tests with the three vitamins administered at the same time were carried out for 50 control subjects (Patton 6, Stanford 44) and for 31 schizophrenic subjects (Patton). Data on subjects with low excretion in one vitamin only, two vitamins, or all three are given in Table 2-2.

If low values were the result of poor nutrition there would be positive correlation of the incidence, and the number of subjects low in all three would increase. The numerals in parentheses in Table 2-2 are calculated from the values of the incidences, assuming them to be independent of one another. The differences between observed

TABLE 2-2.
Correlation of Low Excretion in Three Vitamins in Schizophrenic and Control Subjects

Characteristic	Controls[a]	Schizophrenics[b]
Number of subjects	50	31
Incidence of L(AA)[c]	0.320[d]	0.742[d]
Incidence of L(Ni)	0.460[d]	0.806[d]
Incidence of L(Py)	0.240[d]	0.516[d]
L(AA) only	3 (6.6)[e]	2 (2.2)[e]
L(Ni) only	10 (11.9)	3 (3.1)
L(Py) only	1 (4.4)	1 (0.8)
L(AA + Ni)	6 (5.6)	8 (9.0)
L(AA + Py)	4 (2.1)	1 (2.3)
L(Ni + Py)	4 (3.8)	2 (3.3)
L(AA + Ni + Py)	3 (1.8)	12 (9.6)
Not low in any	19 (14.0)	2 (0.7)

[a] Patton (6 subjects) plus Stanford (44 subjects).
[b] Patton.
[c] AA = ascorbic acid, Ni = niacinamide, Py = pyridoxine.
[d] Fraction of subjects with low excretion of the vitamin.
[e] Values in parentheses are calculated from values of incidence, rows 2 to 4, assuming independence in incidence from one another.

and calculated numbers are not statistically significant, but there is some indication of positive correlation, especially for the control group (15 observed low in two or all three vitamins, 11.7 calculated). The values for the schizophrenic group (22 observed, 23.7 calculated) indicate no correlation.

A reasonable possibility is that positive correlation, resulting from poor nutrition (low intake of all three vitamins or poor assimilation of all three vitamins), operates for both groups, and is effectively cancelled for the schizophrenic group by another effect, peculiar to the schizophrenic population, that introduces negative correlation. Let us assume that correlation introduces a factor in the frequency of subjects low in two vitamins and another factor in the frequency of subjects low in three vitamins. The value of the first factor is about 5, and that of the second is about 25 (incidences 0.10, 0.35, and 0.10 for ascorbic acid, niacinamide, and pyridoxine, respectively, then give calculated values within 1.5 of the observed values). An effect that increased the number of subjects low in one vitamin only in the schizophrenic population (relative to those low in two or three) could account for the differences between the correlations for controls and schizophrenics in Table 2-2. This effect might be a change in genotype affecting one vitamin only, thus increasing the probability of hospitalization for schizophrenia.

DISCUSSION AND SUMMARY

Our studies have shown that there is a pronounced difference, with high statistical significance, in the response of groups of schizophrenic patients and control groups to orally ingested doses of ascorbic acid, niacinamide, and pyridoxine, as shown by the fractions of the ingested doses eliminated in the urine during the period of 6 hours immediately following the doses. The observations for ascorbic acid are in general agreement with the results of many earlier studies, summarized by Herjanic in Chapter 14; those for niacinamide and pyridoxine are, we believe, new.

Our results indicate that for each of the three vitamins the population, either schizophrenic or normal, does not show a simple distribution function for the fraction of vitamin eliminated, but instead shows clustering into groups. The upper limit for the fraction excreted in 6 hours by low excretors has been taken as 17 percent, 6.4 percent, and 18 percent for ascorbic acid, niacinamide, and pyridoxine. The low excretors of the individual vitamins in the normal population constitute about 30, 40, and 20 percent, and in the schizophrenic population about 75, 80, and 50 percent, respectively. About 60 percent of the control population and 94 percent of the schizophrenic population are low excretors for one or more of the three vitamins.

The pattern of excretion of ascorbic acid is changed very little for the control subjects by the ingestion of 1 to 2 g of this vitamin per day for one week, but that for the schizophrenic subjects is shifted so that it becomes the same as for the controls. It is likely that the low excretors have a low body store of the vitamin, and that this store is brought up to the normal value by a sufficiently increased daily intake.

The observations reported in Table 2-2 make possible some statements about the probability of hospitalization for schizophrenia of persons in different categories of vitamin excretion (assuming that the low excretors in the schizophrenic group would remain low excretors if they had been tested at other times, and that the control subjects represent the population as a whole). The ratio of the probability of hospitalization for schizophrenia of the various kinds of excretors to that for persons not low in any of the three vitamins is: low in one vitamin, probability 5 times as great; low in two vitamins, 8 times as great; low in all three vitamins, 40 times as great.

Only 6 percent of our schizophrenic subjects were found by our tests not to be low in any of the three vitamins. A complete understanding of vitamins in relation to schizophrenia has not yet been obtained. There is uncertainty as to the relative contributions of genetic and environmental factors, and to the effect of the schizophrenic episodes and of hospitalization on the biochemical and physiological functioning of the patients. There is no uncertainty, however, about the fact that the great majority, 94 percent, of the hospitalized schizophrenic subjects studied by us show a low urinary excretion of one or more of the three vitamins which we have studied, and that this is an indication of a low content of vitamins in the body, which can be rectified by the methods of orthomolecular psychiatry—the increased daily intake of the vitamins, as discussed in other chapters of this book.

These vitamins are inexpensive and they are almost entirely nontoxic and free of undesirable side reactions, as compared with ordinary drugs. Studies of individual patients might indicate which vitamins are especially needed. We feel, however, that at the present time these vitamin studies of individual patients need not be carried out before megavitamin therapy is instituted, because almost all schizophrenics are low excretors for at least one of the three vitamins, and an increased intake of any of these vitamins has small probability of doing harm, and large probability of doing good.

This work may be summarized by the following statement: observations made by administration of an oral dose of ascorbic acid, niacinamide, and pyridoxine, and determination of the fractions excreted in the urine in the next 6 hours have shown that almost all of the schizophrenic patients studied are low excretors for one or more of the three vitamins. These results strongly support the use of megavitamin therapy for the prevention and treatment of schizophrenia.

EXPERIMENTAL METHODS

Analysis for Ascorbic Acid by Iodine Titration

All the urine samples were collected in 500-ml plastic bottles which contained 1 ml of 1 *N* HCl, to lower the *p*H of the urine and thus inhibit the oxidation of ascorbic acid. The samples were kept frozen at −20 °C until analysis. For the experiment using normal subjects at Stanford University, the samples were analyzed immediately after collection for ascorbic acid and were frozen in small glass bottles at −76 °C for later analysis for niacinamide and pyridoxine. The analysis was performed as follows: add 1 ml of 1 *N* HCl to a 50-ml conical flask; add 10 ml mixed urine to flask; add about 3 ml of chloroform; add a 1/4 inch magnetic stirring rod; place on magnetic stirrer and adjust to highest speed without splattering; titrate with I_2 solution (6.25 g of I_2 and 20.0 g of KI in 2 liters of water, 0.0123 M) from a 5-ml buret. In making the end-point determination, ignore the pink color that sometimes occurs immediately in the chloroform phase, titrate rapidly until the yellow color of iodine lingers at the point of addition, and then titrate very slowly until a lasting pink color is seen in the chloroform phase. The titrations were usually made in triplicate, and the iodine solutions were standardized every week by titrating weighed amounts of crystalline L-ascorbic acid.

Analysis for Niacinamide by Microbiological Assay

The analysis for niacinamide for all County subjects was done by the microbiological method described by Strohecker and Henning (1966) with use of a strain of *Lactobacillus arabinosus* (ATCC 8014) responding to niacin, niacinamide, and nicotinuric acid. The procedure was very time-consuming, and was discarded after the development of the chromatographic method described below. The measured amounts were converted by an empirical factor to correspond to the results of the chromatographic method.

Analysis of Niacinamide by Gas-Liquid Chromatography

Two 0.5-ml samples of urine for this analysis were stored at −76 °C. A Varian-2100 gas-liquid chromatograph with flame-ionization detectors was used. The column was 5 percent Carbowax 20M on Chromosorb W (regular), with oven at 240°C and injector and detector at 250°C. The internal standard was 1.00 mg methylniacinamide in 1.00 ml of water. Standardization was achieved by analyzing 0.5-ml samples of water containing various amounts of niacinamide between 10 and 1,000 mg, with standardization runs made before and after each set of urine-sample runs. The 0.5-ml sample of urine (or niacinamide solution) was thawed and 100 ml of the internal standard was added. Then 1 μl or 2 μl was injected into the chromatograph. The only large peaks are those of methylniacinamide and niacinamide, and the amount of methylniacinamide in urine is very small compared with the amount added.

Analysis of Pyridoxine by Gas-Liquid Chromatography

Two 0.1-ml samples of urine for this analysis were stored at −76°C in glass vials with teflon-lined caps. The chromatographic column was 1.5-percent OV-101 on Chromosorb G, with oven at 190°C, injector at 210°C, and detector at 220°C. The internal standard was heneicosane. Standardization runs before and after each set of urine analyses were made with 0.1-ml aqueous solutions containing between 10 and 1,000 mg of pyridoxine. The 0.1-ml sample of urine (or pyridoxine solution) was dried in high vacuum, capped, and frozen. Then 0.5 ml of 1 : 1 solution of bis-trimethylsilylacetamide in pyridine and 0.1 ml of pyridine containing 125 μg of heneicosane were added; the container was capped and shaken for 90 minutes, and 1 μl of the solution was injected into the chromatograph.

ACKNOWLEDGMENTS

We thank Dr. Robert Nichols and Mrs. Dorothy Marshall at University Hospital of San Diego County, Dr. L. Pratum of Mesa Vista Hospital, Dr. Halmuth Schaefer of Patton State Hospital, and the staffs of these hospitals for help in the collection of the urine samples.

This investigation was carried out with support of grants MH 18149 and MH 18149–02, National Institutes of Health.

REFERENCES

Herjanic, M., and Herjanic, B. L. M. (1967). *J. Schizophrenia* **1**, 257.

Sokal, R. R., and Rohlf, F. J. (1969). *Biometry*. San Francisco and London: W. H. Freeman and Co.

Strohecker, R., and Henning, H. M. (1966). *Vitamin Assay, Tested Methods*. Cleveland, Ohio: C.R.C. Press.

VanderKamp, H. (1966). *Internat. J. Neuropsychiat.* **2**, 204.

Reprinted from Trends in Biochemical Sciences – *November, 1979*

The discovery of the superoxide radical

Linus Pauling

Earlier this year Irwin Fridovich and H. Moustafa Hassan (May, 113) wrote of the toxic effects of the superoxide radical. Here Linus Pauling describes how this radical may have been the first important substance whose existence was predicted through arguments based on the theory of quantum mechanics.

During the last decade I have been mainly interested in the use of vitamin C in preventing and treating viral diseases and cancer. Vitamin C (ascorbate) has the power of inactivating viruses *in vitro*, and Murata and Kitagawa [1] in 1973 pointed out that oxygen is involved in this inactivation, and suggested that the inactivation may occur as a result of the formation of the superoxide radical, O_2^-, or the hydroxide radical, OH. Morgan, Cone and Elgert [2] have reported that superoxide dismutase has no effect on the reaction of superoxide with viruses, but catalase suppresses it. The cleavage of DNA induced by ascorbate is, however, unaffected by superoxide dismutase, although it is suppressed by catalase, so that it is probably the hydroxyl radical that is responsible for scission of DNA strands.

The enzyme superoxide dismutase was discovered only ten years ago [3]. It catalyses the reaction of two superoxide radicals to give oxygen and hydrogen peroxide. Superoxide dismutase is found in all aerobic organisms. Obligate anaerobes, which do not contain superoxide dismutase, are killed in air, and Morgan, Cone and Elgert have suggested that the enzyme played a crucial role in the evolution of life as the partial pressure of atmospheric oxygen increased. The superoxide radical may well be the principal cause of oxygen toxicity, and superoxide dismutase may be essential for survival.

Linus Pauling is associated with the Linus Pauling Institute of Science and Medicine, 2700 Sand Hill Road, California 94025, U.S.A.

During the last five years more papers have appeared on superoxide dismutase than on any other single enzyme (Malcolm and Coggins [4]) and a recent monograph on superoxide and superoxide dismutase [5] contains a history of the enzyme, but no history of its substrate. The superoxide radical may well be the first important substance to have been discovered through quantum mechanical arguments. In the early 1930s I wrote several papers about the nature of the chemical bond, the second of which dealt with the one-electron bond and the three-electron bond [6]. I pointed out that a stable one-electron bond can be formed only when there are two conceivable electronic states of the system with essentially the same energy, the states differing in that for one there is an unpaired electron attached to the second atom. A three-electron bond can be formed between two atoms, A and B, with A having an orbital occupied by a pair of electrons and B having an orbital occupied by a single electron, or the reverse, if the energies of these two structures are essentially the same. The resonance energy corresponding to the interaction of the two structures is the energy of the three-electron bond. It is usually about 60% as great as the energy of a shared-electron-pair bond between two atoms. I carried out a detailed calculation for the helium molecule ion, He_2^+, in which the three-electron bond corresponds to resonance between the two structures $He{:}\ {\cdot}He^+$ and $He^{\cdot+}\ {:}He$ [7]. At that time, 1931, I assigned to the oxygen molecule in its normal state the structure in which the two atoms are linked together by a single bond and two three-electron bonds.

I had then not yet thought about the possibility of the existence of the ion O_2^-, with a three-electron bond plus a single bond, corresponding to resonance between the two structures:

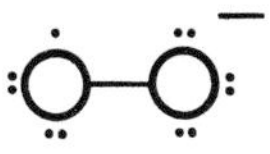 and 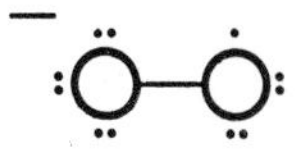

A few months later this idea occurred to me, and I realized that the ion O_2^-, intermediate between molecular oxygen and hydrogen peroxide, should have enough stability to exist. I knew that when potassium, rubidium and cesium burn in oxygen, higher oxides are formed which were called tetroxides and were assigned the formulas K_2O_4, Rb_2O_4, and Cs_2O in the reference books of inorganic chemistry. I also knew that the tetroxide ion, with the presumable structure of a chain containing three single oxygen–oxygen bonds, would be unstable, because of the well-known instability of the oxygen–oxygen single bond, as represented in hydrogen peroxide [8]. It seemed likely, accordingly, that these substances were in fact KO_2, RbO_2, CsO_2, containing a unipositive alkali ion and the anion O_2^-. A test of this hypothesis could be made by measuring the magnetic susceptibility of the substances, because the tetroxides would be diamagnetic and the compounds MO_2, containing an anion with an odd number of electrons, would have the paramagnetism corresponding to one unpaired electron spin. I accordingly asked a postdoctoral fellow working with me, Edward W. Neuman, to prepare a sample of this oxide of potassium and to measure its magnetic susceptibility. He carried out this measurement, and found the magnetic susceptibility to be that corresponding to

TIBS – November 1979

an odd electron for every pair of oxygen atoms.

I was at that time spending part of my time each year giving lectures in physics and chemistry in the University of California, Berkeley. On 22 April 1933 in Berkeley, in a discussion of the three-electron bond, I announced that Neuman had found the higher oxide of potassium to be paramagnetic, showing that it contains the radical O_2^-, in which the atoms are held together by a single bond and a three-electron bond. I pointed out that KO_2 should not be called potassium dioxide, because the term dioxide is usually reserved for compounds of quadrivalent metals, as in PbO_2, lead dioxide. Professor Wendell M. Latimer proposed the name potassium hypoperoxide, which I rejected. Professor E. D. Eastman and Professor W. C. Bray then both suggested that the substance be called potassium superoxide, with the radical O_2^+ called the superoxide ion or the superoxide radical. I accepted this suggestion, and Dr Neuman then published his paper (submitted for publication on 14 November 1933) with the title 'Potassium Superoxide and the Three-Electron Bond' [9].

I still retain an interest in structural chemistry, but in recent years I have also become increasingly interested in biochemistry and medicine, and it is a source of satisfaction to me that the superoxide radical, whose existence was predicted through arguments based upon the theory of quantum mechanics, should have turned out to be important in biology and medicine. Morgan, Cone and Elgert, in fact, have suggested that superoxide may have a role in phagocytosis, and have discussed other biological and medical implications of superoxide and superoxide dismutase.

References

1 Murata, A. and Kitagawa K. (1973) *Agric. Biol. Chem.* 37, 294–296

2 Morgan, A. R., Cone, R. L. and Elgert, T. M. (1976) *Nucleic Acids Res.* 3, 1139–1149

3 McCord, J. M. and Fridovich, I. (1969) *J. Biol. Chem.* 244, 6049

4 Malcolm, A. B and Coggins, J. R. (1976) *Annual Rep. Prog. Chem. Sect. B.* 73, 368–374

5 Michelson, A. M., McCord, J. M. and Fridovich, I. (1977) *Superoxide and Superoxide Dismutases*, Academic Press, London, New York and San Francisco

6 Pauling, L. (1931) *J. Am. Chem. Soc.* 53, 3225–3237

7 Pauling, L. *J. Chem. Phys.* 1, 56–59

8 Pauling, L. (1932) *J. Am. Chem. Soc.* 54, 3570–3582

9 Neuman, E. W. (1934) *J. Chem. Phys.* 2, 31–33

Proc. Natl. Acad. Sci. USA
Vol. 86, pp. 3466–3468, May 1989
Statistics

Biostatistical analysis of mortality data for cohorts of cancer patients

(Hardin Jones principle/Kaplan–Meier renormalization)

LINUS PAULING

Linus Pauling Institute of Science and Medicine, 440 Page Mill Road, Palo Alto, CA 94306

Contributed by Linus Pauling, February 13, 1989

ABSTRACT **The Hardin Jones principle states that for a homogeneous cohort of cancer patients the logarithm of the fraction surviving at time *t* has a constant slope. With use of this principle, the survival times of the members of a heterogeneous cohort can be analyzed to divide the cohort into subcohorts with different mortality rate constants. Probable values of the additional survival time can be estimated for members surviving at the closing date of a clinical trial, permitting them to be included in the biostatistical analysis of the results of the trial in a more significant way than through Kaplan–Meier renormalization.**

Cancer continues to constitute a great problem. Although there has been significant progress in the prevention and treatment of some kinds of cancer during recent decades, most kinds still cannot be prevented or successfully treated.

I have developed a powerful method of biostatistical analysis of the observed survival times of cancer patients on the basis of the Hardin Jones principle, as described in the following sections of this paper.

The Hardin Jones Principle

In 1956 Hardin Jones made a penetrating analysis of the demography of the cancer problem (1). An important conclusion that he formulated is that with a reasonably homogeneous cohort of cancer patients, such as those with the same kind of cancer who have reached the terminal or untreatable stage (for example, breast cancer patients with metastases who have not responded favorably to high-energy radiation or chemotherapy), the rate of death is given by the equation

$$\frac{dN}{N} = -\alpha t, \qquad [1]$$

where N is the number of survivors at time t and α is a constant, the probability of death in unit time for a member of the homogeneous cohort. Integration of this equation leads to

$$S = \frac{N}{N_0} = e^{-\alpha t}, \qquad [2]$$

in which N_0 is the number of patients in the cohort at $t = 0$ (the beginning of the study or the time of entrance of the patient into the study). This equation describes a first-order reaction; that is, the number of persons dying in unit time is a constant fraction of the number of survivors in the cohort, independent of the time.

If for some reason patients leave the study, renormalization is usually made by the Kaplan–Meier method (2).

Jones (1) reported the results of his analysis of about 50 sets of mortality data for cohorts of cancer patients on the basis of his principle, and Burch (3) reported similar results for 9 sets. An example is given in figure 11.4 of ref. 3, which shows as a good straight line the logarithm of the percentage survival for 9159 women of all ages with localized breast cancer (data were obtained from the 1963 California Tumor Registry). My associate Zelek S. Herman and I have made many similar analyses of the survival data for presumably homogeneous cohorts of cancer patients, verifying the general validity of the Hardin Jones principle (unpublished studies). This principle accordingly provides a sound basis for the formulation of a biostatistical theory of cancer mortality.

The Analysis of Trials Made on Cohorts Consisting of Two or More Significantly Different Subcohorts

Jones (1) found that the logarithm of the fraction of the cohort surviving at time t sometimes could be expressed as the sum of two or three exponential terms, rather than one,

$$\frac{N}{N_0} = \sum f_i e^{-\alpha_i t}. \qquad [3]$$

For example, for women with metastatic breast cancer who were treated by the Halsted operation, 67% had 50% survival time $t_{1/2}$ (= $0.693/\alpha$) of 0.69 years and 33% had $t_{1/2}$ of 4.25 years (1). For women seen initially without evidence of metastasis, there were three subcohorts: 36% with $t_{1/2} = 1.20$ years, 54% with $t_{1/2} = 5.37$ years, and 10% with $t_{1/2} = 35$ years (1). Burch (3) resolved the logarithm of the percent survival of 13,392 women in California with breast cancer at all stages into the sum of two terms: 30% with 50% survival time $t_{1/2}$ = 1.2 years and 70% with $t_{1/2}$ = 9.1 years.

A Theory of Modest Heterogeneity of a Cohort of Cancer Patients

The Hardin Jones plot for some presumably homogeneous cohorts of cancer patients shows some curvature, such as to suggest a moderate amount of heterogeneity. A reasonable assumption is that there is an error function distribution of the activation energy of the rate constant, α, about a mean value of the activation energy corresponding to an intermediate value α_0 of the rate constant. This assumption leads on expansion and integration to the introduction of a quadratic term:

$$\ln S = -\alpha_0 t + \beta t^2. \qquad [4]$$

The parameter β is related to the standard deviation σ in the activation energy error function by the equation

$$\beta = \frac{3}{4}\left(\frac{\sigma}{RT}\right)^2, \qquad [5]$$

The publication costs of this article were defrayed in part by page charge payment. This article must therefore be hereby marked "*advertisement*" in accordance with 18 U.S.C. §1734 solely to indicate this fact.

in which R is the molar gas constant and T is the absolute temperature.

Mean Values of Powers of Survival Times for a Homogeneous Cohort

The mean value $\langle t^n \rangle$ of the nth power of the survival time t for a homogeneous cohort is obtained by integrating the product of t^n and the fraction $-dS/dt$ dying between t and $t + dt$, which is $\alpha e^{-\alpha t}$:

$$\langle t^n \rangle = \int_0^\infty \alpha t^n e^{-\alpha t}\, dt. \quad [6]$$

The known value of the definite integral leads to the equation

$$\langle t^n \rangle = \frac{\Gamma(n+1)}{\alpha^n}. \quad [7]$$

Here the Γ function has values 0.9999422883, 0.886227, 1, and 2 for $n = 0.0001$, 0.5, 1, and 2, respectively. It is seen that the mean $\langle t \rangle$ is equal to $1/\alpha$. It is convenient to use the symbol τ for $1/\alpha$, the reciprocal of the rate constant.

From Eq. **7** we derive the following result:

$$\{\langle t^n \rangle / \Gamma(n+1)\}^{1/n} = \frac{1}{\alpha} = \tau. \quad [8]$$

This equation is valid for every positive value of n for a homogeneous cohort with the values of t distributed in accordance with Eq. **2**. Adherence to this equation accordingly provides a test for the homogeneity of a cohort.

Another convenient method for evaluating τ is to make use of the definite integral

$$\langle \ln t \rangle = \int_0^\infty \ln t e^{-t}\, dt = -\gamma, \quad [9]$$

in which γ is Euler's constant, with value 0.5772156649 . . .. In this equation τ has been taken to have the value 1, so that τ is equal to $e^\gamma \exp\langle \ln t \rangle$.

$$\tau = e^\gamma \exp\langle \ln t \rangle = 1.7810 \exp\langle \ln t \rangle. \quad [10]$$

Moreover, $\langle \ln t \rangle^{1/N_o}$, with N_o the number of terms, is equal to $\{\Pi(t_i)\}^{1/N_o}$, the N_oth root of the product of the values of t, giving the following equation (equivalent to Eq. **10**):

$$\tau = 1.7810\, \{\Pi(t_i)\}^{1/N_o}. \quad [11]$$

The surviving fraction corresponding to $\exp\langle \ln t \rangle = 1$ is $e^{-\gamma} = 0.56146$ . . ..

Still another method is to calculate the slope of the line connecting the points on the Hardin Jones plot with the origin; the reciprocal of this slope for each point is a value of τ:

$$\tau_i = -t_i/\ln S_i. \quad [12]$$

S_i is N_i, the number of survivors (half integral) on the day t_i when this number decreased by 1, divided by N_o, the number of members in the original cohort. The mean of τ_i is

$$\tau = \langle -t/\ln S_i \rangle. \quad [13]$$

The values of τ for t small may be in significant error because a change in t by 1 day changes τ by t^{-1}, and the values for t large may be in error because truncation of the cohort can make a large change in S.

Table 1 gives results for a representative example of a small cohort (10 patients). The mean of the four values of τ is 43.2 days, and the mean deviation from the mean is 2.7 days, which is close to the 10% error expected for τ determined by any one of the four methods for a cohort of 10. Since the causes of error have different effects for the four methods, the mean of the values provides a better approximation than any one value. Accordingly, I recommend that this mean be presented as the value of τ for a cohort. The value of $\langle t \rangle$ is usually close to this mean.

Table 1. Mean values of survival times for a cohort of 10 stomach cancer patients

Method	τ, days
$1.7810 \exp\langle \ln t \rangle$	48.9
$\{\langle t^{1/2} \rangle / \Gamma(3/2)\}^2$	43.7
$\langle t \rangle$	41.5
$\{\langle t^2 \rangle / 2\}^{1/2}$	38.7
Mean	43.2

The patients had reached the untreatable stage at time $t = 0$ and received no further treatment (group no. 1 in table 1 of ref. 4, all female, ages 56–66; t = 5, 8, 12, 21, 29, 36, 41, 54, 85, and 124 days).

For the 10-member cohort of Table 1, omitting one value of t leads to a mean deviation of ±9% for the value of τ determined by any one of the four methods. For N_o members, the mean deviation is somewhat less than N_o^{-1}.

An Alternative to the Kaplan–Meier Renormalization Procedure in the Biostatistical Analysis of Survival Times of a Cohort of Cancer Patients Some of Whom Are Alive at the Termination Time of the Study

In the Kaplan–Meier renormalization procedure (2), a member of the cohort who changes treatment, becomes unavailable, is alive at the termination of a mortality study, or for some other reason can no longer be considered to be a member of the cohort is removed from the study, decreasing the number at risk by 1. Valuable information may be lost by this procedure if the fraction of dropouts is large, especially if the dropouts tend to occur at larger values of t. In the case of mortality studies of cancer patients, there is an alternative procedure, which is to use the Hardin Jones principle to predict the probable survival time t for the member of the cohort surviving at time t^+ at the termination of the study. If the survivor is a member of a homogeneous cohort, the value of t for this survivor is given by the following equation, in which τ is the mean survival time of the cohort:

$$t = t^+ + \tau. \quad [14]$$

For a cohort consisting of N_o members with N_o^+ alive at the termination date of the trial or on withdrawal from the trial, a first approximation value of τ is τ_o, the mean value of t_i and t_i^+. The self-consistent value of τ is then given by the following equation:

$$\tau = \tau_o/(1 - N_o^+/N_o). \quad [15]$$

This value of τ is then to be added to each t_i^+ to obtain the estimated value of t_i.

For a cohort consisting of two subcohorts with fractions f_1 and f_2 and mean survival times τ_1 and τ_2, the mean expected additional survival time of a survivor is

$$\tau = \frac{\{f_1\tau_1 \exp(-t^+/\tau_1) + f_2\tau_2 \exp(-t^+/\tau_2)\}}{f_1 \exp(-t^+/\tau_1) + f_2 \exp(-t^+/\tau)}. \quad [16]$$

Values of f_1, f_2, τ_1, and τ_2 are to be obtained in a self-consistent manner by consideration of all values of t, including the predicted values for the survivors. For example, with a cohort of 15 patients with "untreatable" bronchial cancer who received daily doses of ascorbate (4) and who had values of t_i from 17 to 460 days, including one survivor with t^+ =

200^+ days, the value of τ increased from 137 to 146 days when t^+ was replaced by $t^+ + \tau$.

A Subcohort with a Single Member

With a single member, with survival time t, in a subcohort, the probability of survival at time t is $e^{-t/\tau}$. When t/τ is equal to 0.693, this probability is 1/2; this is accordingly the median of the values of τ leading to the value t. The mean survival time is equal to τ. It corresponds to the relation

$$\tau = t. \quad [17]$$

If the sole member of the subcohort is still surviving at time t^+ and the probable corresponding median value of the expected lifetime t is $t^+ + 0.6931\ \tau$, we obtain the equation

$$\tau = 3.2589\ t^+ \quad [18]$$

for τ and the expected lifetime, after entry into the study, for the survivor

$$t = 3.2589\ t^+. \quad [19]$$

Dividing a Cohort into Two Subcohorts

If a cohort can be represented as the sum of two homogeneous subcohorts, with mean survival times τ_1 and τ_2 and coefficients f_1 and f_2 ($f_1 + f_2 = 1$), the three parameters may be evaluated from three independent properties of the set of values of t. For example, the three properties $\langle t^{1/2}\rangle$, $\langle t\rangle$, and $\langle t^2\rangle$ provide a set of equations that can be solved for the unknowns. A large and truly representative cohort is needed for this analysis to be successful.

Another method makes use of three points, $S_1(t_1)$, $S_2(t_2)$, and $S_3(t_3)$, on a smoothed Hardin Jones plot of ln S vs. t. For one set of 130 breast cancer patients (4), this treatment with $S = 0.5$, 0.25, and 0.0625 gave $f_1 = 0.45$, $f_2 = 0.55$, $\tau_1 = 29$ days, and $\tau_2 = 83$ days. This method is unreliable for small cohorts. An alternative is to make a least-squares fit to all the points on a Hardin Jones plot with a two-term function or, if the cohort is very large, to carry out a Laplace transformation.

Treatment of a Cohort with Several Survivors

With the assumption that the cohort, with N_o members, consists of two subcohorts, the following treatment can be used. There are N_o^+ survivors, with survival times greater than t_i^+, and $N_o - N_o^+$ others, with known survival times t_i. The value of τ_1 for the latter is taken to be $\tau_1 = \langle t_i\rangle$. The quantity $\exp(-t_i^+/\tau_1)$ is then calculated for each t_i^+ and assumed to be the probability that this member is a member of the first subcohort. The value of τ_2 is evaluated by Eq. **15** with N_o^+ decreased by subtracting $\Sigma \exp(-t_i^+/\tau_1)$. The values of t_i for the surviving patients are found from the equation

$$t_i = \tau_2 - (\tau_2 - \tau_1)\exp(-t_i^+/\tau_1). \quad [20]$$

Outliers

An outlier is a member of a cohort with such a large value of t that it is likely that the member belongs to a separate subcohort. Many examples could be quoted. In one cohort the survivor had $t^+ = 857^+$ days (alive at the termination date of the study), with the other 16 members of the cohort of 17 untreatable patients with bronchial cancer having values of t from 16 days to 450 days. The value of τ_1 for these 16 members is 152 days (5). With this value of τ_1, the value of $N_o e^{-t^+/\tau}$ for $t^+ = 857$ days is 5.7%, and for $t = 2793$ days, the probable survival time of the outlier, it is about 10^{-7}. Accordingly this cohort consists of a 16-member subcohort with $\tau_1 = 152$ days and a one-member subcohort with $\tau_2 = 2793$ days.

Discussion

The Hardin Jones principle that the death rate of members of a homogeneous cohort of cancer patients is constant is supported by much empirical evidence. The use of this principle permits powerful methods of biostatistical analysis of cancer mortality data to be formulated. Heterogeneous cohorts can be resolved into two or more homogeneous subcohorts. Probable values of additional survival times of members of a cohort who have not yet died at the end of a study can be estimated, permitting a method of analysis to be carried out that provides more information than that given by Kaplan–Meier renormalization. Outliers, members of a subcohort with very large survival time, can be identified. These methods are especially useful in interpreting survival times for small cohorts.

I thank Ewan Cameron, Zelek S. Herman, and Dorothy Munro for their help. This investigation was supported by grants from the Japan Shipbuilding Industry Foundation and other donors to the Linus Pauling Institute of Science and Medicine.

1. Jones, H. B. (1956) *Trans. N.Y. Acad. Sci.* **18,** 298–333.
2. Kaplan, E. L. & Meier, P. (1958) *J. Am. Stat. Assoc.* **53,** 457–481.
3. Burch, P. R. J. (1976) *The Biology of Cancer, A New Approach* (University Park Press, Baltimore).
4. Cameron, E. & Pauling, L. (1976) *Proc. Natl. Acad. Sci. USA* **73,** 3685–3689.
5. Cameron, E. & Pauling, L. (1978) *Proc. Natl. Acad. Sci. USA* **75,** 4538–4542.

Proc. Natl. Acad. Sci. USA
Vol. 86, pp. 6835–6837, September 1989
Statistics

Criteria for the validity of clinical trials of treatments of cohorts of cancer patients based on the Hardin Jones principle

(vitamin C/ascorbic acid)

LINUS PAULING AND ZELEK S. HERMAN

Linus Pauling Institute of Science and Medicine, 440 Page Mill Road, Palo Alto, CA 94306

Contributed by Linus Pauling, May 8, 1989

ABSTRACT With the assumption of the validity of the Hardin Jones principle that the death rate of members of a homogeneous cohort of cancer patients is constant, three criteria for the validity of clinical trials of cancer treatments are formulated. These criteria are satisfied by most published clinical trials, but one trial was found to violate all three, rendering the validity of its reported results uncertain.

The Hardin Jones Principle

In 1956 Hardin Jones (1) pointed out that the death rate is constant for a presumably homogenous cohort of cancer patients, such as those with breast cancer with generalized advanced metastasis and classed inoperable when originally observed. For such a homogeneous cohort the fraction surviving, S, at time t is given by Eq. **1**, in which α is the death rate:

$$S = e^{-\alpha t}. \quad [1]$$

A plot of ln S against t is then a straight line, with negative slope equal to α. An example, from Burch (2), is shown in Fig. 1, and another example, from Cameron and Pauling (3), in Fig. 2. Sometimes, as has been pointed out in a recent discussion (4), resolution of the survival curve into the sum of two or more Hardin Jones exponentials, with death rates α_i and coefficients f_i, can be made:

$$S = \sum_i f_i e^{-\alpha_i t}. \quad [2]$$

An example is given in Fig. 3.

Some uses of the Hardin Jones principle, such as the identification and characterization of a long-lived subcohort in a heterogeneous cohort, have been discussed in earlier papers. Another use is in formulating criteria for the reliability of reported clinical trials of treatments of cancer patients, as discussed below.

The First Criterion of the Validity of a Trial of the Effectiveness of a Treatment of a Cohort of Cancer Patients

In order to increase the significance of the study, the cohort should be reasonably homogeneous, so that Eq. **1** or Eq. **2** can be applied in the analysis of the data. Moreover, the treatment of all the members of the cohort should be the same, and it should be continuous and unchanged from the time $t = 0$ when the patient enters the trial until the time t when the patient dies or t^+ when, without dying, is withdrawn from the set of survivors at risk. Any patient who stops the treatment or changes the treatment at any t^+ should be removed from the study and included in the analysis by the Kaplan–Meier renormalization procedure (5) or the alternative procedure developed in ref. 4. If the trial is to test the later response of a patient to a short-term course of treatment (with or without a following continuous treatment), the time $t = 0$ is to be taken as the time at which the short-term course was completed, with only those patients who survived the course included in the cohort.

The Second Criterion

For more than 200 studies of survival of cohorts of cancer patients examined by us, we have found that the Hardin Jones straight line passes through the 100% axis at time $t = 0$. No statistically significant lag period during which no deaths occur is observed. A trial producing a set of survival times with a significant lag period, during which the value of α would lead to the expectation that several deaths would occur, can be considered to be faulty.

The Third Criterion

If the cohort is heterogeneous and the study is properly carried out, with conditions constant during the period of the study, the semilogarithmic survival curve must bend away from the Hardin Jones initial straight line only in the direction of increased survival times for the longer-term survivors, as shown in Fig. 3. This is, of course, to be expected, since the subcohort of patients with shorter life expectancy is depleted, leaving the subcohort with longer life expectancy. The observation of a deviation of the curve in the opposite direction indicates faulty design or execution of the trial. For example, after the death of many or most of the patients in the subcohort with high death rate, some of the survivors might have had their treatment changed in such a way as to increase their death rate. Under our first criterion, they should at that time have been removed from the study.

An Example of a Study that Fails to Meet These Criteria

We have found a reported clinical trial that fails on each of the three criteria for validity discussed in the preceding sections (6).

This study, described as a randomized double-blind comparison of vitamin C (10 g per day) and a lactose placebo in 100 patients with advanced adenomatous colorectal cancer, gave survival times published only in a Kaplan–Meier figure. We have measured the published curves and produced the semilogarithmic plot shown as Fig. 4.

Each of the 51 vitamin-C patients received vitamin C for some time (median stated to be 2.5 months). The vitamin C was stopped for 19 patients with clearly measurable areas of malignant disease when there was a 50% increase in the product of the perpendicular diameters of any of these areas and for the other 32 when there was some other evidence of progression of the disease. Each of the 51 patients then entered a period of no treatment, during which some of them

The publication costs of this article were defrayed in part by page charge payment. This article must therefore be hereby marked "*advertisement*" in accordance with 18 U.S.C. §1734 solely to indicate this fact.

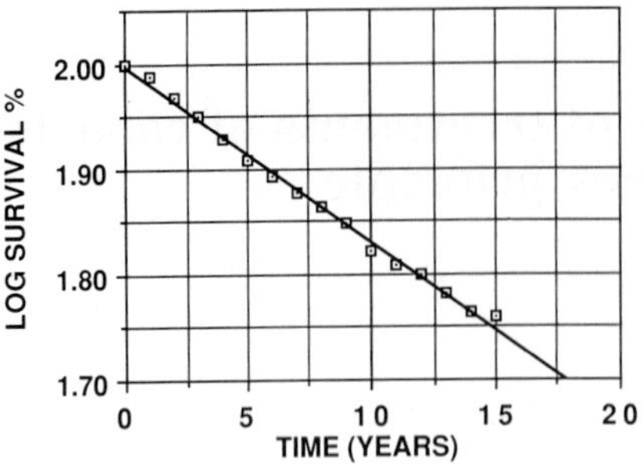

FIG. 1. Logarithm of percent survival of 5159 California women of all ages with localized breast cancer (after ref. 2); $t_{1/2}$ = 18.0 years.

died. Then 30 of them entered a third period, after, the authors say, they had "discontinued participation in this study." This was the period during which they received chemotherapy, mostly with fluorouracil, either alone or in combination with other agents. There were accordingly, by our first criterion, four separate periods for the patients: the period of regular intake of vitamin C, the following period of no treatment (with some possible aftereffects of the earlier vitamin C), the period of the chemotherapy treatment, and the period after the end of the course of chemotherapy.

The authors treat the data, however, as though it were a single clinical trial. This violates the first criterion of validity.

As shown in Fig. 4, the second criterion is also violated. The Hardin Jones straight line has an unexplained intercept.

The third criterion is also violated, in that after 350 days the death rate increases significantly.

All three criteria are also violated by the study of the 49 patients who began taking a placebo, then stopped it and entered the period of no treatment, then (28 of them) received chemotherapy, and then entered into the fourth period, with no treatment.

We are not able to explain the lag period of 70 days (Fig. 4), during which only one patient died, rather than about 25 expected from the Hardin Jones principle. Also, we are not able to make a good biostatistical analysis of the death rates under the four separate conditions (vitamin C or placebo being taken, no treatment, chemotherapy being given, then no treatment), because we could not obtain information about the times during which the individual patients were in each of these four periods. One conclusion, however, is certain: this study provides no information about the value of a continued intake of 10 g per day of vitamin C in extending survival time, because none of the patients died while taking the vitamin.

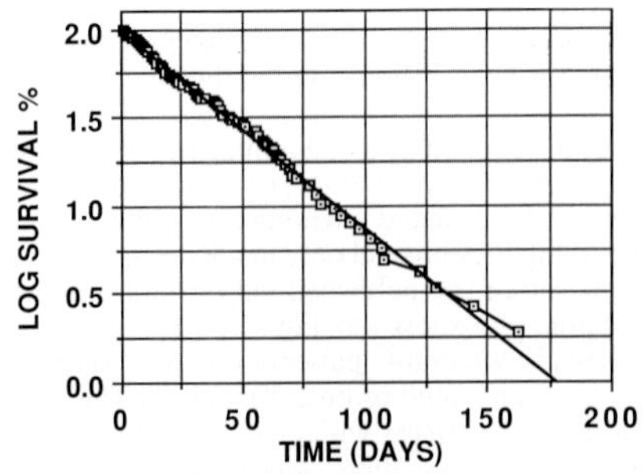

FIG. 2. Logarithm of percent survival of 130 Scottish patients with untreatable cancer of the colon (after ref. 3), compared with Eq. **1** (straight line) with average rate constant $\alpha = 2.58 \times 10^{-2}\ \mathrm{d}^{-1}$ ($t_{1/2}$ = 26.9 d).

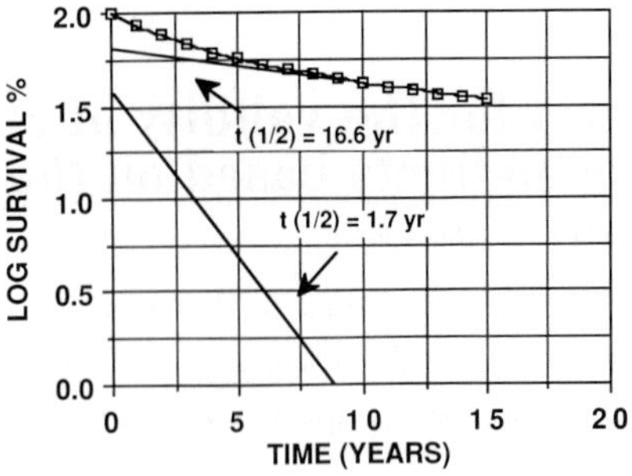

FIG. 3. Logarithm of percent survival of 13,392 California women of all ages and with breast cancer of all degrees of severity, roughly resolved into two negative exponential functions, one with $t_{1/2}$ = 1.7 years and one with $t_{1/2}$ = 16.6 years (after ref. 2).

Another conclusion can also be reached from the violation of the third criterion. It is that the life expectancy of these patients was decreased somewhat by chemotherapy.

It has been pointed out (ref. 7, page 117) that the sudden stopping of high doses of vitamin C leads to a rebound effect that may be dangerous for cancer patients. During the period from 70 to 120 days 10 vitamin-C patients died and only 4 placebo patients. Also, the reported death rate during this period for the vitamin-C patients was significantly greater than during the following 300 days. The fact that this increased death rate occurred immediately after the median withdrawal date of the vitamin C, 75 days, suggests that it may have been caused by the rebound effect.

The administration of chemotherapy to 58 patients who had withdrawn from the study had begun by 250 days, when 58 were still alive (Fig. 4). At about 350 days the death rate increased. It is certain, from the Hardin Jones principle, that something had changed in the treatment or environment of the patients, and the conclusion that this change was the administration of chemotherapy seems to be justified.

Conclusion

Three biostatistical criteria have been formulated for the validity of clinical trials of treatments of cancer patients. The first criterion is that the regimen of a patient in the cohort should not be changed during the period of the trial; if it is

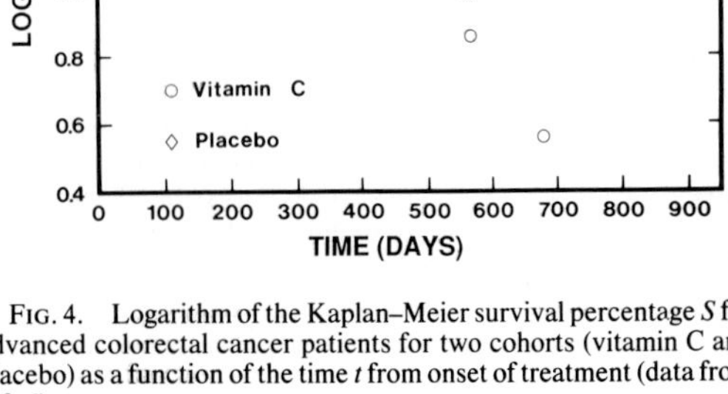

FIG. 4. Logarithm of the Kaplan–Meier survival percentage S for advanced colorectal cancer patients for two cohorts (vitamin C and placebo) as a function of the time t from onset of treatment (data from ref. 6).

changed, the patient should be withdrawn from the trial. The second criterion is that the Hardin Jones line on a semilogarithmic plot of the fraction surviving at time *t* should go to 100% at time 0; deviation by a significant amount indicates some sort of error in reporting the time at which patients entered the study. The third criterion is that the semilogarithmic curve should not deviate from the initial Hardin Jones straight line in the direction of increased mortality at later times. Such a deviation shows that the regimen has been changed for some members of the cohort, who should then have been removed from the study.

We have found that most of the reported results of clinical trials of cohorts of cancer patients satisfy these criteria of validity. One study, however, fails by all three. The conclusion reached in that study, that ascorbic acid has no more value than lactose in extending the survival of patients with advanced colorectal cancer, is not justified.

We thank Ewan Cameron and Dorothy Munro for their help. This investigation was supported in part by a grant from the Japan Shipbuilding Industry Foundation and other donors to the Linus Pauling Institute of Science and Medicine.

1. Jones, H. B. (1956) *Trans. N.Y. Acad. Sci.* **18,** 298–333.
2. Burch, P. R. J. (1976) *The Biology of Cancer, a New Approach* (University Park Press, Baltimore, MD).
3. Cameron, E. & Pauling, L. (1976) *Proc. Natl. Acad. Sci. USA* **73,** 3685–3689.
4. Pauling, L. (1989) *Proc. Natl. Acad. Sci. USA* **86,** 3466–3468.
5. Kaplan, E. L. & Meier, P. (1958) *J. Am. Stat. Assn.* **53,** 457–481.
6. Moertel, C. G., Fleming, T. R., Creagan, E. T., Rubin, J., O'Connell, M. J. & Ames, M. M. (1985) *N. Engl. J. Med.* **312,** 137–141.
7. Cameron, E. & Pauling, L. (1979) *Cancer and Vitamin C* (Pauling Inst. Sci. Med., Palo Alto, CA).

Chapter 17

Orthomolecular Medicine

Contents

SP 134 **Orthomolecular Psychiatry: Varying the Concentrations of Substances Normally Present in the Human Body May Control Mental Disease** **1363**
Science **160**, 265–271 (1968)

135 **Orthomolecular Somatic and Psychiatric Medicine** **1371**
Vitalstoffe-Zivilisationskrankheiten **1/68**, 3–5 (1968)

136 **Evolution and the Need for Ascorbic Acid** **1374**
Proc. Natl. Acad. Sci. USA **67**, 1643–1648 (1970)

137 **Ascorbic Acid and the Common Cold** **1380**
Am. J. Clin. Nutr. **24**, 1294–1299 (1971)

138 **Orthomolecular Psychiatry** **1386**
In *Orthomolecular Psychiatry: Treatment of Schizophrenia,* eds. David Hawkins and Linus Pauling, W. H. Freeman, San Francisco, pp. 1–17 (1973)

139 **Ascorbic Acid and the Glycosaminoglycans: An Orthomolecular Approach to Cancer and Other Diseases** **1403**
(by Ewan Cameron and Linus Pauling)
Oncology **27**, 181–192 (1973)

140 **Supplemental Ascorbate in the Supportive Treatment of Cancer: Prolongation of Survival Times in Terminal Human Cancer** **1415**
(by Ewan Cameron and Linus Pauling)
Proc. Natl. Acad. Sci. USA **73**, 3685–3689 (1976)

SP 141 **Ascorbic Acid and Cancer: A Review**
(by Ewan Cameron, Linus Pauling, and B. Leibovitz)
Cancer Research **39**, 663–681 (1979)

142 **Suppression of Human Immunodeficiency Virus Replication by Ascorbate in Chronically and Acutely Infected Cells**
(by Steve Harakeh, Raxit J. Jariwalla, and Linus Pauling)
Proc. Natl. Acad. Sci. USA **87**, 7245–7249 (1990)

143 **A Unified Theory of Human Cardiovascular Disease Leading the Way to the Abolition of this Disease as a Cause for Human Mortality**
(by Matthias Rath and Linus Pauling)
J. Orthomolecular Med. **7**(1), 17–23 (1992)

144 **Summary and Conclusions: The Role of Vitamin C in the Treatment of Cancer**
(by Ewan Cameron and Linus Pauling)
In *Cancer and Vitamin C* (Updated and Expanded Edition), Camino Books, Philadelphia, PA, Chapter 23, pp. 189–195 (1993)

Reprinted from
19 April 1968, Volume 160, pp. 265-271

SCIENCE

Orthomolecular Psychiatry

Linus Pauling

Orthomolecular Psychiatry

Varying the concentrations of substances normally present in the human body may control mental disease.

Linus Pauling

The methods principally used now for treating patients with mental disease are psychotherapy (psychoanalysis and related efforts to provide insight and to decrease environmental stress), chemotherapy (mainly with the use of powerful synthetic drugs, such as chlorpromazine, or powerful natural products from plants, such as reserpine), and convulsive or shock therapy (electroconvulsive therapy, insulin coma therapy, pentylenetetrazol shock therapy). I have reached the conclusion, through arguments summarized in the following paragraphs, that another general method of treatment, which may be called orthomolecular therapy, may be found to be of great value, and may turn out to be the best method of treatment for many patients.

Orthomolecular psychiatric therapy is the treatment of mental disease by the provision of the optimum molecular environment for the mind, especially the optimum concentrations of substances normally present in the human body (*1*). An example is the treatment of phenylketonuric children by use of a diet containing a smaller than normal amount of the amino acid phenylalanine. Phenylketonuria (*2*) results from a genetic defect that leads to a decreased amount or effectiveness of the enzyme catalyzing the oxidation of phenylalanine to tyrosine. The patients on a normal diet have in their tissues abnormally high concentrations of phenylalanine and some of its reaction products, which, possibly in conjunction with the decreased concentration of tyrosine, cause the mental and physical manifestations of the disease (mental deficiency, severe eczema, and others). A decrease in the amount of phenylalanine ingested results in an approximation to the normal or optimum concentrations and to the alleviation of the manifestations of the disease, both mental and physical.

The functioning of the brain is dependent on its composition and structure; that is, on the molecular environment of the mind. The presence in the brain of molecules of *N,N*-diethyl-D-lysergamide, mescaline, or some other schizophrenogenic substance is associated with profound psychic effects (*3*). Cherkin has recently pointed out (*4*) that in 1799 Humphry Davy described similar subjective reactions to the inhalation of nitrous oxide. The phenomenon of general anesthesia also illustrates the dependence of the mind (consciousness, ephemeral memory) on its molecular environment (*5*).

The proper functioning of the mind is known to require the presence in the brain of molecules of many different substances. For example, mental disease, usually associated with physical disease, results from a low concentration in the brain of any one of the following vitamins: thiamine (B_1), nicotinic acid or nicotinamide (B_3), pyridoxine (B_6), cyanocobalamin (B_{12}), biotin (H), ascorbic acid (C), and folic acid. There is evidence that mental function and behavior are also affected by changes in the concentration in the brain of any of a number of other substances that are normally present, such as L(+)-glutamic acid, uric acid, and γ-aminobutyric acid (*6*).

Optimum Molecular Concentrations

Several arguments may be advanced in support of the thesis that the optimum molecular concentrations of substances normally present in the body may be different from the concentrations provided by the diet and the gene-controlled synthetic mechanisms, and, for essential nutrilites (vitamins, essential amino acids, essential fatty acids) different from the minimum daily amounts required for life or the "recommended" (average) daily amounts suggested for good health. Some of these arguments are presented in the following paragraphs.

Evolution and Natural Selection

The process of evolution does not necessarily result in the normal provision of optimum molecular concentrations. Let us use ascorbic acid as an example. Of the mammals that have been studied in this respect, the only species that have lost the power to synthesize ascorbic acid and that accordingly require it in the diet are man, other Primates (rhesus monkey, Formosan long-tail monkey, and ring-tail or brown capuchin monkey), the guinea pig, and an Indian fruit-eating bat (*Pteropus medius*) (*7*). Presumably the loss of the gene or genes controlling the synthesis of the enzyme or enzymes involved in the conversion of glucose to ascorbic acid occurred some 20 million years ago in the common ancestor of man and other Primates, and occurred independently for the guinea pig and for one species of bat and one bird, in each case in an environment such that ascorbic acid was provided by the food. For a mutation rate of 1/20,000 per gene generation and for even a very small advantage for the mutant (0.01 percent more progeny) the mutant would replace the earlier genotype within about 1 million years. The advantage to the mutant of being rid of the ascorbic-acid-synthesis machinery (decrease in cell size and energy requirement, liberation of machinery for other purposes) might well be large, perhaps as much as 1 percent; a disadvantage nearly as large (less by 0.01 percent) resulting from a less than optimum supply of dietary ascorbic acid would not prevent the replacement of the earlier species by the mutant. Hence, even if the amount of the vitamin provided by the diet available at the time of the mutation were less than the optimum amount, the mutant might still be able to replace its predecessor. Moreover, it is possible that the environment has changed during the last 20 million years.

The author it professor of chemistry in residence at the University of California, San Diego, P.O. Box 109, La Jolla, California 92037.

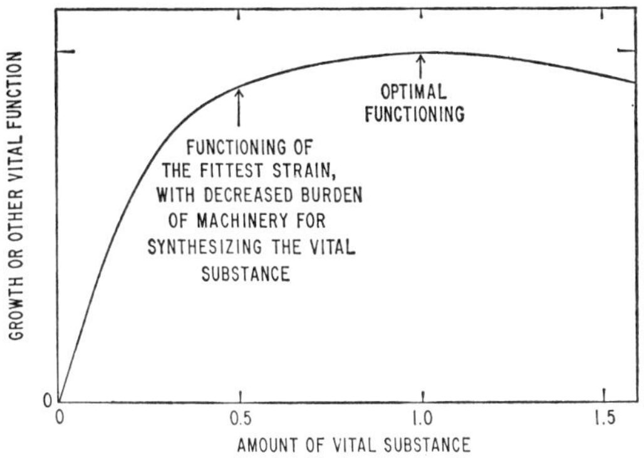

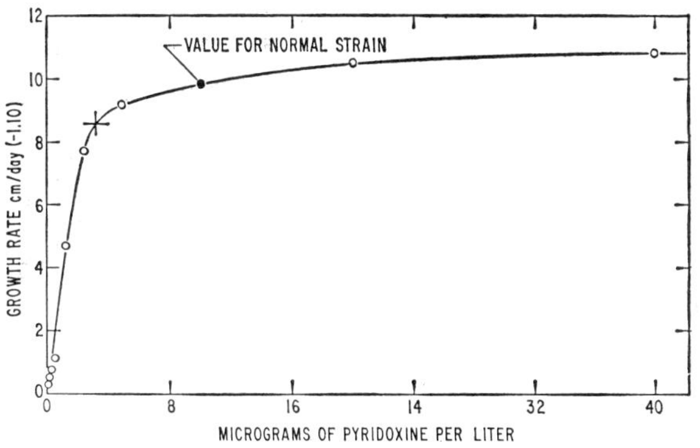

Fig. 1 (left). Diagrammatic representation of growth rate or other vital property of an organism as function of the concentration of vital substance in the organism, showing the concentration at which the differential advantage of an increased amount of vital substance is just balanced by the differential disadvantage resulting from an increased amount of machinery for synthesis, and the concentration that gives optimum functioning without consideration of the burden of the machinery for synthesis. Fig. 2 (right). The observed rate of growth of a pyridoxine-requiring *Neurospora* mutant (Beadle and Tatum, 1941), as function of the concentration of pyridoxine in the medium.

in such a way as to provide a decreased amount of the vitamin. Even a serious disadvantage of the changed environment would not lead to a mutation restoring the synthetic mechanism within a period of a few million years, because of the small probability of such mutations, far smaller than of those resulting in loss of function.

Moreover, the process of natural selection may be expected later on to lead to the survival of a species or strain that synthesizes somewhat less than the optimum amount of an autotrophic vital substance rather than of the species or strain that synthesizes the optimum amount. To synthesize the optimum amount requires about twice as much biological machinery as to synthesize half the optimum amount. As suggested in Fig. 1, the evolutionary disadvantage of synthesizing a less than optimum amount of the vital substance may be small, and may be outweighed by the advantage of requiring a smaller amount of biological machinery. Evidence from the study of microorganisms is discussed in the following paragraphs.

Evidence from Microbiological Genetics

Many mutant microorganisms are known to require, as a supplement to the medium in which they are grown, a substance that is synthesized by the corresponding wild-type organism (the normal strain). An example is the pyridoxine-requiring mutant of *Neurospora sitophila* reported by G. W. Beadle and E. L. Tatum in their first *Neurospora* paper, published in 1941 (*8*). Several species of *Neurospora* that have been extensively studied are known to be able to grow satisfactorily on synthetic media containing inorganic salts, an inorganic source of nitrogen, such as ammonium nitrate, a suitable source of carbon, such as sucrose, and the vitamin biotin. All other substances required by the organism are synthesized by it. Beadle and Tatum found that exposure to x-radiation produces mutant strains such that one substance must be added to the minimum medium in order to permit the growth at a rate approximating that of the normal strain. Their pyridoxine-requiring mutant was found to grow on the standard medium at a rate only 9 percent of that of the normal strain. When pyridoxine (vitamin B_6) is added to the medium, the rate of growth of this strain at first increases nearly linearly with the concentration of the added pyridoxine and then increases less rapidly, as shown in Fig. 2 (*9*). The growth rate of the normal strain without added pyridoxine is equal to that of the mutant with about 10 micrograms of the growth substance per liter in the medium. At a concentration about four times this value (40 micrograms per liter) the growth rate of the mutant strain reaches a value 7 percent greater than that of the normal strain without added pyridoxine.

The point of maximum curvature of the curve in Fig. 2, at about 3.2 micrograms of pyridoxine per liter (indicated by a cross), may be reasonably considered to mark the division between the region of vitamin deficiency (to the left) and the region of normal vitamin supply (to the right), such as might permit the mutant to compete with the wild type, which has the growth rate represented by the filled circle in Fig. 2. The point marked by the cross might well correspond to an "adequate" or "recommended" amount of the vitamin, in that the growth rate of the mutant is only 12 percent less than that of the wild strain, and that the amount of the vitamin would have to be increased threefold to make up this 12 percent (*10*).

As shown in Fig. 2, quadrupling the concentration of pyridoxine that gives the mutant a growth rate equal to that of the wild type causes a further increase in growth rate by nearly 10 percent. The growth rates of the mutant and the wild type at very large concentrations of the vitamin have not been measured, so far as I know, and the optimum concentration is not known. From the work of Beadle and Tatum the optimum concentration may be taken to be greater than 40 micrograms per liter; that is, more than ten times the "adequate" concentration for the mutant and more than four times the concentration equivalent to the synthesizing capability of the wild type. The growth rate of the mutant at the optimum concentration is more than 22 percent greater than that at the "adequate" concentration and more than 9 percent greater than that of the normal strain.

Similar results have been reported for other mutants of *Neurospora*. The values found by Tatum and Beadle (*11*) for a *p*-aminobenzoic-acid-requiring mutant of *Neurospora crassa* as a function of the concentration of *p*-aminobenzoic acid added to the standard medium are shown in Fig. 3. The growth-rate curve is similar in shape to that for the pyridoxine-requiring mutant. The value of the growth rate for the normal strain of *Neurospora crassa* with no added *p*-aminobenzoic acid is equal to that for the mutant at a concentration of added *p*-aminobenzoic acid of about 15 micrograms per liter. A value about 4 percent greater is found for the normal strain at 40 micrograms per liter and for the mutant strain at 80 micrograms per liter, as indicated in Fig. 3.

It is customary to plot values of the growth rate against the logarithm of the concentration of the growth substance, as shown in Fig. 4. The amount of increase accompanying a doubling in the concentration of the growth substance is a maximum at 1.25 to 2.5 micrograms per liter, and decreases thereafter to about half the value for each successive doubling.

From these two examples we see that there may be a significant increase in rate of growth of the normal strain through addition of some of the growth substance that it synthesizes to the medium in which it is grown; that is, that the amount of the growth substance that is synthesized by the normal strain is not the optimum amount, but is somewhat less—approximately 7 percent less in the case of pyridoxine (with the normal strain of *Neurospora sitophila*) and 4 percent less for *p*-aminobenzoic acid (with the normal strain of *Neurospora crassa*). Many other examples are known of microorganisms that grow more abundantly in a medium containing vitamins, amino acids, or other substances that they are able to synthesize than on a minimum medium.

Evidence supporting the above arguments has been presented recently by Zamenhof and Eichhorn (*11a*) in a paper entitled "Study of microbial evolution through loss of biosynthetic functions: Establishment of 'defective' mutants." These authors carried out experiments involving competitive growth in a chemostat of an auxotrophic mutant (a mutant requiring a nutrilite) and a prototrophic parent in a medium of constant composition containing the nutrilite. They found that the "defective" mutant has a selective advantage over the prototrophic parental strain under these conditions. For example, an indole-requiring mutant of *Bacillus subtilis* was found to show a strong selective advantage over the prototrophic back-mutant when the two were grown together in a medium containing tryptophan: the relative number of cells of the latter decreased 10^6-fold in 54 generations. They also found that greater advantage to the auxotroph accompanies a greater number of biosynthetic steps that have been dispensed with (earlier block in a series of reactions), with the final metabolite available. They point out that a mutant with a gene deletion would be at a distinct selective advantage over a point mutant, in that not only the synthesis of the metabolite, but also that of the structural gene, the messenger RNA, and perhaps the inactive enzyme itself would be dispensed with, and that accordingly the mutant with a deletion would replace the point mutant in competition. They mention evidence that some of the "defective" strains occurring in nature have lost one or more of their structural genes by deletions, rather than by point mutations.

Molecular Concentrations and Rate of Reaction

Most of the chemical reactions that take place in living organisms are catalyzed by enzymes. The mechanisms of enzyme-catalyzed reactions in general involve (i) the formation of a complex between the enzyme and a substrate molecule and (ii) the decomposition of this complex to form the enzyme and the products of the reaction. The rate-determining step is usually the decomposition of the complex to form the products, or, more precisely, the transition through an intermediate state of the complex, characterized by activation energy less than for the uncatalyzed reaction, to a complex of the enzyme and the products of reaction, with a rapid dissociation. Under conditions such that the concentration of the complex corresponds to equilibrium with the enzyme and the substrate, the rate of the reaction is given by the following equation [the Michaelis-Menten equation (*12*)]:

$$R = \frac{d[S]}{dt} = \frac{kE[S]}{[S] + (1/K)} \quad (1)$$

In this equation $[S]$ is the concentration of the substrate, E is the total con-

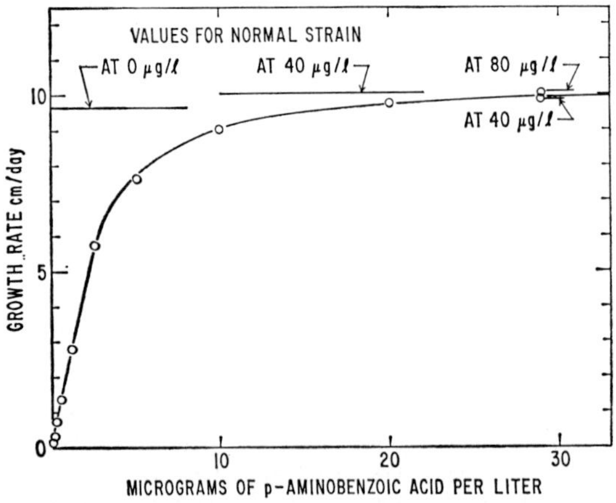

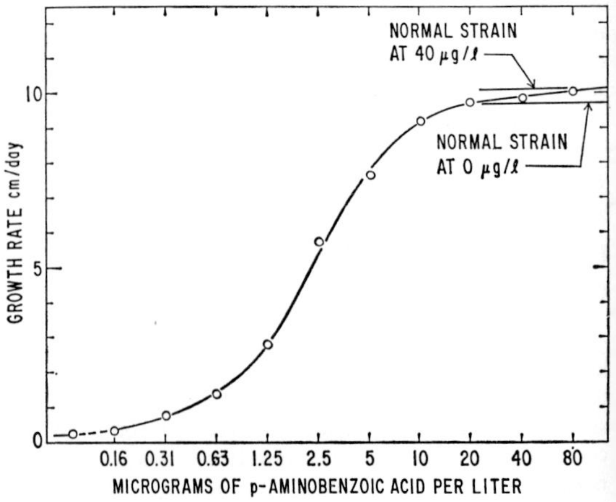

Fig. 3 (left). The observed rate of growth of a *p*-aminobenzoic-acid-requiring *Neurospora* mutant (Tatum and Beadle, 1942), as function of concentration of the growth substance in the medium. Fig. 4 (right). Observed rate of growth of a *p*-aminobenzoic-acid-requiring *Neurospora* mutant as function of the logarithm of the concentration of *p*-aminobenzoic acid.

centration of enzyme (present both as free enzyme and enzyme complex), K is the equilibrium constant for formation of the enzyme complex ES, and k is the reaction-rate constant for decomposition of the complex to form the enzyme and reaction products. This equation corresponds to the case in which there are no enzyme inhibitors present.

Values of the reaction rate calculated from this equation for different values of K are shown in Fig. 5. The curves are similar in shape to those of Figs. 2 and 3. At concentrations much smaller than K^{-1} the reaction rate is proportional to the concentration of substrate. At larger concentrations, as the amount of enzyme complex becomes comparable to the amount of free enzyme, the reaction rate changes from the linear dependence. At substrate concentration equal to K^{-1} the slope of the curve is one-quarter of the initial slope, and the value is one-half of the value corresponding to saturation of the enzyme by the substrate.

The similarity of the curves of Figs. 2 and 3 to appropriate curves in Fig. 5 suggests that the growth substance may be involved in an enzyme-catalyzed reaction in which it serves as the substrate. The normal strain of the organism manufactures an amount of the substrate such as to permit the reaction to take place at what may be considered a normal rate, 90 or 95 percent of the maximum rate, which corresponds to saturation of the enzyme. As described above, the gain in reaction rate associated with the manufacture of a larger amount of the substrate, with a corresponding advantage to the organism, might be balanced by the disadvantage to the organism associated with the upkeep of the larger amount of machinery required to manufacture the increased amount of substrate. An increase in rate of this reaction could also be achieved by an increase in the amount of the enzyme synthesized by the organism. Here, again, the advantage to the organism resulting from this increase may be overcome by the disadvantage associated with the increase in the amount of machinery required for the increased synthesis. During the process of evolution there has presumably been selection of genes determining the concentrations of the enzymes catalyzing successive reactions such as to achieve an approximation to the optimum reaction rate with the smallest amount of disadvantage to the organism.

The rate of an enzyme-catalyzed reaction is approximately proportional to the concentration of the reactant, until concentrations that largely saturate the enzyme are reached. The saturating concentration is larger for a defective enzyme with decreased combining power for the substrate than for the normal enzyme. For such a defective enzyme the catalyzed reaction could be made to take place at or near its normal rate by an increase in the substrate concentration, as indicated in Fig. 5. The short horizontal lines intersecting the curves indicate what may be called the "normal" reaction rate, 80 percent of the maximum. For $K = 2$ the "normal" rate is achieved at substrate concentration $[S] = 2$. At this substrate concentration the reaction rate is only 29 percent of the maximum and 35 percent of "normal" for a mutated enzyme with $K = 0.2$; it could be raised to the "normal" value by a tenfold increase in the substrate concentration, to $[S] = 20$. Similarly, the still greater disadvantage of low reaction rate for a mutated enzyme with K only 0.01 could be overcome by a 200-fold increase in substrate concentration, to $[S] = 400$. This mechanism of action of gene mutation is only one of several that lead to disadvantageous manifestations that could be overcome by an increase, perhaps a great increase, in the concentration of a vital substance in the body. These considerations obviously suggest a rationale for megavitamin therapy.

Molecular Concentrations and Mental Disease

The functioning of the brain and nervous tissue is more sensitively dependent on the rate of chemical reactions than the functioning of other organs and tissues. I believe that mental disease is for the most part caused by abnormal reaction rates, as determined by genetic constitution and diet, and by abnormal molecular concentrations of essential substances. The operation of chance in the selection for the child of half of the complement of genes of the father and mother leads to bad as well as to good genotypes, and the selection of foods (and drugs) in a world that is undergoing rapid scientific and technological change may often be far from the best. Significant improvement in the mental health of many persons might be

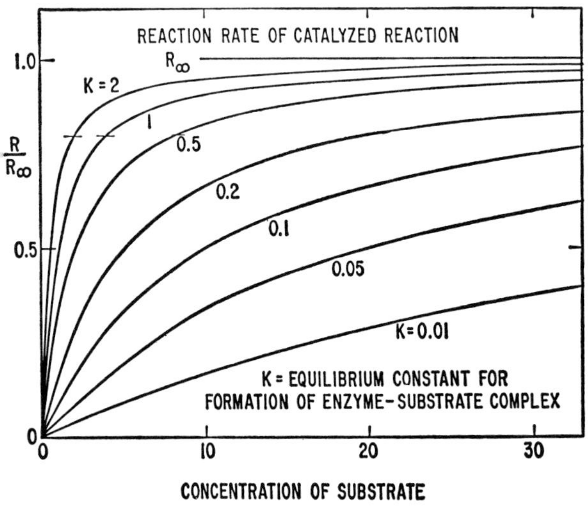

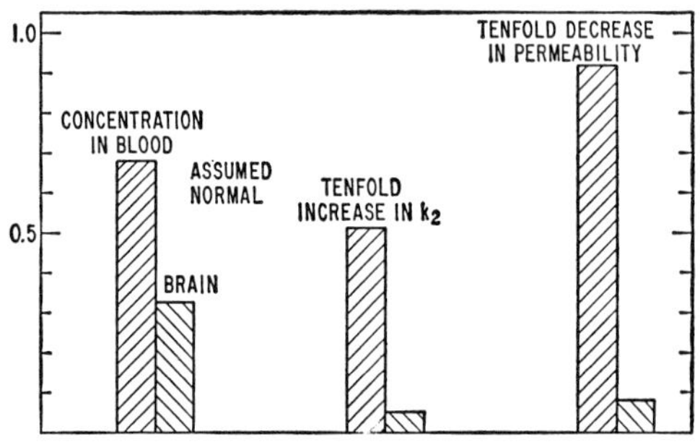

Fig. 5 (left). Curves showing calculated reaction rate R/R_∞ of catalyzed reaction as function of the concentration of the substrate, for different values of the equilibrium constant K for formation of the enzyme-substrate complex. Fig. 6 (above). Values of the concentration of a vital substance in the blood and in the cerebrospinal fluid for three different assumed sets of value of blood-brain barrier permeability and rate of destruction in the cerebrospinal fluid.

achieved by the provision of the optimum molecular concentrations of substances normally present in the human body. Among these substances, the essential nutrilites may be the most worthy of extensive research and more thorough clinical trial than they have yet received. One important example of an essential nutrilite that is required for mental health is vitamin B_{12}, cyanocobalamin. A deficiency of this vitamin, whatever its cause (pernicious anemia; infestation with the fish tapeworm *Diphyllobothrium*, whose high requirement for the vitamin results in deprivation for the host; excessive bacterial flora, also with a high vitamin requirement, as may develop in intestinal blind loops), leads to mental illness, often even more pronounced than the physical consequences. The mental illness associated with pernicious anemia [a genetic defect leading to deficiency of the intrinsic factor (a mucoprotein) in the gastric juice and the consequent decreased transport of cyanocobalamin into the blood] often is observed for several years in patients with this disease before any of the physical manifestations of the disease appear (*13*). A pathologically low concentration of cyanocobalamin in the serum of the blood has been reported to occur for a much larger fraction of patients with mental illness than for the general population. Edwin, Holten, Norum, Schrumpf, and Skaug (*14*) determined the amount of B_{12} in the serum of every patient over 30 years old admitted to a mental hospital in Norway during a period of 1 year. Of the 396 patients, 5.8 percent (23) had a pathologically low concentration, less than 101 picograms per milliliter, and the concentration in 9.6 percent (38) was subnormal (101 to 150 picograms per milliliter). The normal concentration is 150 to 1300 picograms per milliliter. The incidence of pathologically low and subnormal levels of B_{12} in the serums of these patients, 15.4 percent, is far greater than that in the general population, about 0.5 percent (estimated from the reported frequency of pernicious anemia in the area, 9.3 per 100,000 persons per year). Other investigators (*15*) have also reported a higher incidence of low B_{12} concentrations in the serums of mental patients than in the population as a whole, and have suggested that B_{12} deficiency, whatever its origin, may lead to mental illness.

Nicotinic acid (niacin), when its use was introduced, cured hundreds of thousands of pellagra patients of their psychoses, as well as of the physical manifestations of their disease. For this purpose only small doses are required; the recommended daily allowance (National Research Council) is 12 milligrams per day (for a 70-kilogram male). In 1939 Cleckley, Sydenstricker, and Geeslin (*16*) reported the successful treatment of 19 patients and in 1941 Sydenstricker and Cleckley (*17*) reported similarly successful treatment of 29 patients with severe psychiatric symptoms by use of moderately large doses of nicotinic acid (0.3 to 1.5 grams per day). None of these patients had physical symptoms of pellagra or any other avitaminosis. More recently many other investigators have reported on the use of nicotinic acid and nicotinamide for the treatment of mental disease. Outstanding among them are Hoffer and Osmond, who since 1952 have advocated and used nicotinic acid in large doses, in addition to the conventional therapy, for the treatment of schizophrenia (*18–20*). The dosage recommended by Hoffer is 3 to 18 grams per day, as determined by the response of the patient, of either nicotinic acid or nicotinamide, together with 3 grams per day of ascorbic acid. Nicotinic acid and nicotinamide are nontoxic [the lethal dose, 50 percent effective (LD_{50}), is not known for humans, but probably it is over 200 grams; the LD_{50} for rats is 7.0 grams per kilogram for nicotinic acid, and 1.7 grams per kilogram for nicotinamide], and their side effects, even in continued massive doses, seem not to be commonly serious. Among the advantages of nicotinic acid, summarized by Osmond and Hoffer (*19*), are the following: it is safe, cheap, and easy to administer, and it is a well-known substance that can be taken for years on end, if necessary, with only small probability of incidence of unfavorable side effects.

Another vitamin that has been used to some extent in the treatment of mental disease is ascorbic acid, vitamin C. A sometimes-recommended daily intake of ascorbic acid is 75 milligrams for healthy adults. Some investigators have estimated that the optimum intake is much larger (*21*): perhaps 3 to 15 grams per day, according to Stone (*22*). Williams and Deason (*23*) have emphasized the variability of individual members of a species of animals; they have reported their observation of a 20-fold range of required intake of ascorbic acid by guinea pigs, and have suggested that human beings, who are less homogeneous, have a larger range.

Mental symptoms (depression) accompany the physical symptoms of vitamin-C deficiency disease (scurvy). In 1957 Akerfeldt (*24*) reported that the serum of schizophrenics had been found to have greater power of oxidizing *N,N*-dimethyl-*p*-phenylenediamine than that of other persons. Several investigators then reported that this difference is due to a smaller concentration of ascorbic acid in the serum of schizophrenics than of other persons. This difference has been attributed to the poor diet and increased tendency to chronic infectious disease of the patients (*25*), and has also been interpreted as showing an increased rate of metabolism of ascorbic acid by the patients (*26*). It is my opinion, from the study of the literature, that many schizophrenics have an increased metabolism of ascorbic acid, presumably genetic in origin, and that the ingestion of massive amounts of ascorbic acid has some value in treating mental disease.

Other vitamins (thiamine, pyridoxine, folic acid) and other substances [zinc ion, magnesium ion, uric acid, tryptophan, L(+)-glutamic acid, and others] influence the functioning of the brain. I shall review work on L(+)-glutamic acid as a further example. L(+)-Glutamic acid is an amino acid that is present at rather high concentration in brain and nerve tissue and plays an essential role in the functioning of these tissues (*27*). It is normally ingested (in protein) in amounts of 5 to 10 grams per day. It is not toxic; large doses may cause increased motor activity and nausea. In 1944 Price, Waelsch, and Putnam (*28*) reported favorable results for glutamic acid therapy of convulsive disorders [benefit to one out of three or four patients with petit mal epilepsy (*29*)]. Zimmerman and Ross then reported an increase in maze-running learning ability of white rats given extra amounts of glutamic acid (*30*). Zimmerman and many other investigators then studied the effects of glutamic acid on the intelligence and behavior of persons with different degrees and kinds of mental retardation. L(+)-Glutamic is apparently more effective than its sodium or potassium salts. The effective dosage is usually between 10 and 20 grams per day (given in three doses with meals), and is adjusted to the patient as the amount somewhat less than that required to cause hyperactivity; improvement in personality and increase in intelligence (by 5 to 20 I.Q. points) have been reported for many patients with mild or moderate mental deficiency by several investigators (*31*).

Localized Cerebral Deficiency Diseases

The observation that the psychosis associated with pernicious anemia may manifest itself in a patient for several years before the other manifestations of this disease become noticeable has a reasonable explanation: the functioning of the brain and nervous tissue is probably more sensitively dependent on molecular composition than is that of other organs and tissues. The observed high incidence of cyanocobalamin deficiency in patients admitted to a mental hospital, mentioned above, suggests that mental disease may rather often be the result of this deficiency, and further suggests that other deficiencies in vital substances may be wholly or partly responsible for many cases of mental illness.

The foregoing arguments suggest the possibility that under certain circumstances a deficiency disease may be localized in the human body in such a way that only some of the manifestations usually associated with the disease are present. Let us consider, for example, an enzyme or other vital substance that is normally metabolized by the catalytic action of an enzyme normally present in the tissues and organs of the body. In a person of unusual genotype there might be an especially great concentration of this enzyme in one body organ, with essentially the normal amount in other organs. Through the action of this enzyme in especially great concentration the steady-state concentration of the vital substance in that organ might be decreased to a level much lower than that required for normal function. Under these circumstances there would be present a deficiency disease restricted to that organ.

An especially important case is that of the brain. We may, as a rough model of the human body, consider two reservoirs of fluid, the blood and lymph, with volume V_1, and cerebrospinal fluid, the extracellular fluid of the brain and spinal column, with volume V_2. We assume that a vital substance is destroyed in each of these reservoirs at a characteristic rate, corresponding to the rate constants k_1 and k_2, that it diffuses across the blood-brain barrier at a rate determined by the product of the permeability and area of the barrier and the difference $c_2 - c_1$ of the concentrations in the two reservoirs, and that it is introduced from the gastrointestinal tract into the first reservoir at a constant rate. The steady-state concentrations are then in the ratio

$$c_1/c_2 = 1 + (k_2V_2/PA)$$

where PA is the product of permeability and the area of the blood-brain barrier. The steady state corresponds to the following system:

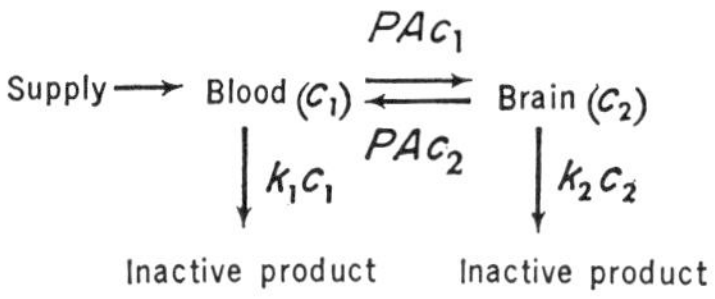

From this equation it is seen, as shown also in Fig. 6, that for small values of k_2V_2/PA the difference in steady-state concentrations in the cerebrospinal fluid and the blood is small, but that through either decrease in permeability of the barrier or increase in the metabolic rate constant k_2 the steady-state concentration in the brain becomes much less than that in the blood.

This simple argument leads us to the possibility of a localized cerebral avitaminosis or other localized cerebral deficiency disease. There is the possibility that some human beings have a sort of cerebral scurvy, without any of the other manifestations, or a sort of cerebral pellagra, or cerebral pernicious anemia. It was pointed out by Zuckerkandl and Pauling (*32*) that every vitamin, every essential amino acid, every other essential nutrilite represents a molecular disease (*33*) which our distant ancestors learned to control, when it began to afflict them, by selecting a therapeutic diet, and which has continued to be kept under control in this way. The localized deficiency diseases described above are also molecular diseases, compound molecular diseases, involving not only the original lesion, the loss of the ability to synthesize the vital substance, but also another lesion, one that causes a decreased rate of transfer across a membrane, such as the blood-brain barrier (*34*), to the affected organ, or an increased rate of destruction of the vital substance in the organ, or some other perturbing reaction.

It has been suggested by Huxley, Mayr, Osmond, and Hoffer (*35*), partially on the basis of the observations of Böök (*36*) and Slater (*37*) on the incidence of schizophrenia in relatives of schizophrenics, that schizophrenia is caused by a dominant gene with incomplete penetrance. They suggested that the penetrance, about 25 percent, may in some cases be determined by other genes and in some cases by the environment. I suggest that the other genes may in most cases be those that regulate the metabolism of vital substances, such as ascorbic acid, nicotinic acid or nicotinamide, pyridoxine, cyanocobalamin, and other substances mentioned above. The reported success in treating schizophrenia and other mental illnesses by use of massive doses of some of these vitamins may be the result of successful treatment of a localized cerebral deficiency disease involving the vital substances, leading to a decreased penetrance of the gene for schizophrenia. There is a possibility that the so-called gene for schizophrenia is itself a gene affecting the metabolism of one or another of these vital substances, or even of several vital substances, causing a multiple cerebral deficiency.

I suggest that the orthomolecular treatment of mental disease, to be successful, should involve the thorough study of and attention to the individual, such as is customary in psychotherapy but less customary in conventional chemotherapy. In the course of time it should be possible to develop a method of diagnosis (measurement of concentrations of vital substances) that could be used as the basis for determining the optimum molecular concentrations of vital substances for the individual patient and for indicating the appropriate therapeutic measures to be taken. My co-workers and I are carrying on some experimental studies suggested by the foregoing considerations, and hope to be able before long to communicate some of our results.

Summary

The functioning of the brain is affected by the molecular concentrations of many substances that are normally present in the brain. The optimum concentrations of these substances for a person may differ greatly from the concentrations provided by his normal diet and genetic machinery. Biochemical and genetic arguments support the idea that orthomolecular therapy, the provision for the individual person of the optimum concentrations of important normal constituents of the brain, may be the preferred treatment for many mentally ill patients. Mental symptoms of avitaminosis sometimes are observed long before any physical symptoms appear. It is likely that the brain is more sensitive to changes in concentration of vital substances than are other organs and tissues. Moreover, there is the possibility that for some persons the cerebrospinal concentration of a vital substance may be grossly low at the same time that the concentration in the blood and lymph is essentially normal. A physiological abnormality such as decreased permeability of the blood-brain barrier for the vital substance or increased rate of metabolism of the substance in the brain may lead to a cerebral deficiency and to a mental disease. Diseases of this sort may be called localized cerebral deficiency diseases. It is suggested that the genes responsible for abnormalities (deficiencies) in the concentration of vital substances in the brain may be responsible for increased penetrance of the postulated gene for schizophrenia, and that the so-called gene for schizophrenia may itself be a gene that leads to a localized cerebral deficiency in one or more vital substances.

References and Notes

1. I might have described this therapy as the provision of the optimum molecular composition of the brain. The brain provides the molecular environment of the mind. I use the word mind as a convenient synonym for the functioning of the brain. The word orthomolecular may be criticized as a Greek-Latin hybrid. I have not, however, found any other word that expresses as well the idea of the right molecules in the right amounts.
2. A. Følling, *Nord. Med. Tidskr.* **8**, 1054 (1934), *Z. Physiol. Chem.* **277**, 169 (1934).
3. See, for example, D. W. Woolley, *The Biochemical Bases of Psychoses* (Wiley, New York, 1962).
4. A. Cherkin, *Science* **155**, 266 (1967).
5. L. Pauling, *ibid.* **134**, 15 (1961); S. Miller, *Proc. Nat. Acad. Sci. U.S.* **47**, 1515 (1961).
6. The literature is so extensive that I refrain from giving references here.
7. For references see I. Stone, *Amer. J. Phys. Anthropol.* **23**, 83 (1965). The only other vertebrate known to require exogenous ascorbic acid is the red-vented bulbul *Pycnonotus cafer.*
8. G. W. Beadle and E. L. Tatum, *Proc. Nat. Acad. Sci. U.S.* **27**, 499 (1941).
9. The points in Fig. 2 represent my measurement of the slopes of the growth curves shown in fig. 1 of reference (*8*). They agree closely with the points of fig. 2 of reference (*8*) except for one point, that for 1.2 μg/liter, which may have been misplotted.
10. The reported growth rate for the normal strain in a medium with 40 μg of added pyridoxine per liter is 3 percent greater than that for the basic medium, as shown by the slopes of the lines in reference (*8*), fig. 1.
11. E. L. Tatum and G. W. Beadle, *Proc. Nat. Acad. Sci. U.S.* **28**, 234 (1942).

11a. S. Zamenhof and H. H. Eichhorn, *Nature* **216**, 465 (1967).

12. L. Michaelis and M. Menten, *Biochem. Z.* **49**, 333 (1913).
13. A. D. M. Smith, *Brit. Med. J.* **11**, 1840 (1950).
14. E. Edwin, K. Holten, K. R. Norum, A. Schrumpf, O. E. Skaug, *Acta Med. Scand.* **177**, 689 (1965).
15. T. Hansen, O. J. Rafaelson, P. Rødbro, *Lancet* **1966-II**, 965 (1966), report serum B_{12} concentration below 150 pg/ml in 13 of 1000 consecutive patients admitted to a Copenhagen psychiatric clinic; J. G. Henderson, R. W. Strachan, J. S. Beck, A. A. Dawson, M. Daniel, *ibid.*, p. 809, report that nine of 1012 unselected psychiatric patients in a region in Scotland were found to have B_{12} deficiency, in addition to five pernicious anemia patients in the group.
16. H. M. Cleckley, V. P. Sydenstricker, L. E. Geeslin, *J. Amer. Med. Ass.* **112**, 2107 (1939).
17. V. P. Sydenstricker and H. M. Cleckley, *Amer. J. Psychiat.* **99**, 83 (1941). References are given in this paper to some earlier work on nicotinic acid therapy.
18. A. Hoffer, H. Osmond, M. J. Callbeck, I. Kahan, *J. Clin. Exp. Psychopathol.* **18**, 131 (1957); A. Hoffer, *Niacin Therapy in Psychiatry* (Thomas, Springfield, Ill., 1962).
19. H. Osmond and A. Hoffer, *Lancet* **1962-I**, 316 (1962); review of a 9-year study.
20. A. Hoffer and H. Osmond, *Acta Psychiat. Scand.* **40**, 171 (1964); A. Hoffer, *Int. J. Neuropsychiat.* **2**, 234 (1966).
21. For example, E. D. Kyhos, E. L. Sevringhaus, D. R. Hagendorn, *Arch. Int. Med.* **75**, 407 (1945), found that for some subjects 1.5 to 2.8 grams per day were needed for saturation.
22. I. Stone, *Perspect. Biol. Med.* **10**, 135 (1967); *Acta Genet. Med. Gemell.* **15**, 345 (1966).
23. R. J. Williams and G. Deason, *Proc. Nat. Acad. Sci. U.S.* **57**, 1638 (1967).
24. S. A. Akerfeldt, *Science* **125**, 117 (1957).
25. J. D. Benjamin, *Psychosom. Med.* **20**, 427 (1958); S. S. Kety, *Science* **129**, 1528, 1590 (1959).
26. A. Hoffer and H. Osmond, *The Chemical Basis of Clinical Psychiatry* (Thomas, Springfield, Ill., 1960), p. 232; M. H. Briggs, *New Zealand Med. J.* **61**, 229 (1962).
27. H. Weil-Malherbe, *Biochem. J.* **30**, 665 (1936).
28. J. G. Price, H. Waelsch, T. J. Putnam, *J. Amer. Med. Ass.* **122** (1944).
29. H. Waelsch, *Amer. J. Ment. Defic.* **52**, 305 (1948).
30. F. T. Zimmerman and S. Ross, *Arch. Neurol. Psychiat.* **51**, 446 (1944).
31. A recent survey of the role of glutamic acid in cognitive behaviors has been published by W. Vogel, D. M. Broverman, J. G. Draguns, E. L. Klaiber, *Psychol. Bull.* **65**, 367 (1966). Many references to earlier work are given in this paper.
32. E. Zuckerkandl and L. Pauling, in *Horizons in Biochemistry*, M. Kasha and B. Pullman, Eds. (Academic Press, New York, 1962), p. 189.
33. L. Pauling, H. A. Itano, S. J. Singer, I. C. Wells, *Science* **110**, 543 (1949).
34. It has been suggested by B. Melander and S. Martens, *Dis. Nerv. Syst.* **19**, 478 (1958); *Acta Psychiat. Neurol. Scand.* **34**, 344 (1959), and by A. Hoffer and H. Osmond, *Int. J. Neuropsychiat.* **2**, 1 (1966), that the effects of taraxein [R. G. Heath, S. Martens, B. E. Leach, M. Cohen, C. A. Feigley, *Amer. J. Psychiat.* **114**, 917 (1958)] may result from changing the permeability of the blood-brain barrier.
35. J. Huxley, E. Mayr, H. Osmond, A. Hoffer, *Nature* **204**, 220 (1964).
36. J. A. Böök, *Acta Genet. Statist. Med.* **4** (I) (1953); *Proc. Int. Congr. Genet. 10th* **1**, 81 (1958).
37. I. E. Slater, *Acta Genet. Statist. Med.* **8**, 50 (1958).

Sonderdruck aus der Zeitschrift Vitalstoffe-Zivilisationskrankheiten Heft 1/68

Orthomolecular Somatic and Psychiatric Medicine

By Linus Pauling

A Communication to the Thirteenth International Convention on Vital Substances, Nutrition and the Diseases of Civilization at Luxembourg and Trier on 18–24 September 1967

I greatly appreciate the honor of being associated with the International Society for Research on Vital Substances, Nutrition, and the Diseases of Civilization, in succession to my friend Dr. Albert Schweitzer, and I take this opportunity to express my thanks to the members and officers of the Society.

Also, I express my regret that it has not been possible for me to be present at the Thirteenth International Convention of the Society. Some difficult problems associated with my move to La Jolla and the assumption of my Professorship in the University of California made it necessary for me to abandon my tentative plan of coming to Luxembourg to participate in the Convention.

In this communication to the Society I present some thoughts about molecular medicine, and especially about the use in the treatment of both physical and mental disease of substances that are normally present in the human body and are required for life.

I have introduced the expression orthomolecular medicine to describe one aspect of molecular medicine. In 1949 my co-workers Harvey Itano, S. J. Singer, and Ibert C. Wells and I published a paper with the title Sickle-cell Anemia, A Molecular Disease. In this paper we communicated our discovery that patients with the disease sickle-cell anemia have in their erythrocytes a form of hemoglobin differing from that manufactured by other people. We pointed out that the difference in molecular structure of the hemoglobin manufactured by persons suffering from this disease leads to a difference in properties of the hemoglobin molecules from those manufactured by other people, and that this difference in properties is responsible for the manifestations of the disease. The disease can properly be described as a disease of the hemoglobin molecule, rather than of the erythrocyte itself, and in consequence it may be called a molecular disease.

Later work in our laboratories in the California Institute of Technology and by other investigators elsewhere showed that the hemoglobin molecule contains four polypeptide chains, two of one kind, the alpha chains, and two of another kind, the beta chains, and also that sickle-cell-anemia hemoglobin differs from normal hemoglobin in having one of the 146 amino-acid residues in each beta chain different from that in normal human hemoglobin, the difference being a replacement of a glutamic-acid residue in the sixth position from the amino end of the beta chain in normal adult human hemoglobin by a residue of valine.

Many other abnormal forms of human hemoglobin have now been discovered, and many diseases have been recognized as diseases of the hemoglobin molecule. Other molecular diseases have also been identified. These diseases for the most part are genetic diseases, the result of a gene mutation.

The disease phenylketonuria, discovered about forty years ago by Følling in Norway, may be considered a molecular disease. This disease involves a gene mutation such that the patient fails to manufacture molecules of an enzyme normally present in the liver, which catalyzes the oxidation of phenylalanine to tyrosine, or produces an abnormal enzyme, with greatly reduced catalytic activity. The patients are homozygotes, who have inherited the gene in double dose, usually from a heterozygotic father and a heterozygotic mother. The patients on a normal diet have in their tissues abnormally high concentrations of phenylalanine and some of its reaction products, which cause the physical and the mental manifestations of the disease, severe eczema, mental deficiency, and so on. The treatment that has considerable success is to place the patients, from the age of one month or two months, on a diet of foods from which a considerable amount of phenylalanine has been removed. This decrease in the amount of phenylalanine ingested by the patients results in an approximation to the normal or optimal concentrations of phenylalanine and its reaction products in the body fluids, and to the alleviation of the physical and mental manifestations of the disease.

Another molecular disease for which a molecular treatment is available is diabetes. This disease results from a gene abnormality such that the patient does not manufacture the proper amount of the hormone insulin. The disease can be controlled by the injection of insulin, to bring the concentration to approximately the normal or optimal value.

Phenylketonuria involves the presence in the body of phenylalanine and its reaction products in amounts greater than normal. It is treated by reducing the intake of phenylalanine and in this way reducing the concentration of this substance and its reaction products to approximately the normal level. Diabetes involves a deficiency of insulin. It is treated by injecting insulin, and increasing the concentration to approximately the normal value. I have reached the conclusion, through arguments summarized in the following paragraphs, that a general method of treatment of disease, which may be called orthomolecular therapy, may be found to be of great value, and may turn out often to be the best method of treatment for many patients.

Orthomolecular therapy is the treatment of disease by the provision of the optimal molecular constitution of the body, especially the optimal concentration of substances that are normally present in the human body and are required for life. The adjective orthomolecular is used to express the idea of the right molecules in the right concentration. This word may be criticized as a Greek-Latin hybrid, but I have not thought of a better word.

I believe that orthomolecular therapy may have a special value in the treatment of mental disease. The functioning of the mind is dependent on its molecular environment, the molecular structure of the brain. The presence in the brain of molecules N,N-diethyl-D-lysergamide, mescaline, or some other schizophrenogenic substance often is associated with profound psychic effects. The phenomenon of general anesthesia also illustrates the dependence of the mind (consciousness, ephemeral memory) on its molecular environment.

The proper functioning of the mind is known to require the presence in the brain of molecules of many different substances. For example, mental disease, usually associated also with physical disease, results from a low concentration in the brain of thiamine (Vitamin B_1), nicotinic acid or nicotinamide (B_3), pyridoxine (B_6), cyanocobalamin (B_{12}), biotin (H), and ascorbic acid (Vitamin C).

Also, there is evidence that mental function and behavior are affected by changes in the concentration in the brain of any one of a number of other substances that are normally present, such as glutamic acid, uric acid, and gamma-aminobutyric acid.

Optimal Molecular Concentrations

Several arguments may be advanced in support of the thesis that the optimal molecular concentrations of substances normally present in the body may be different from the concentrations provided by the diet and by the gene-controlled synthetic mechanisms of the body, and also, for essential nutrilites, such as vitamins, essential amino acids, and essential fatty acids, different from the minimal daily amounts required for life for the average human being or the "recommended" daily amounts suggested for good health.

One argument can be developed through consideration of the process of evolution and natural selection. The process of evolution does not necessarily result in the normal provision of optimal molecular concentrations. Let us use ascorbic acid as an example. Dr. Irwin Stone has pointed out that of the mammals that have been studied in respect to their need for ascorbic acid, the only species that have lost the power to synthesize this substance and that accordingly require it in the diet are man, other primates (Rhesus monkey, Formosan long-tail monkey, ringtail or brown Capuchin monkey), the guinea pig, and an Indian fruit-eating bat (Pteropus medius); in addition one bird, the Red-vented Bulbul (Pycnonotus Cafer) is unable to synthesize its own ascorbic acid. Presumably the loss of the gene or genes controlling the synthesis of the enzyme or enzymes involved in the conversion of glucose to ascorbic acid occurred some twenty million years ago in the common ancestor of man and other primates, and occurred independently for the guinea pig and other isolated species mentioned above, in each case in an environment such that ascorbic acid was provided by the available food. The advantage to the mutant of being rid of the machinery for the synthesis of ascorbic acid (decrease in cell size and energy requirement, liberation of machinery for other purposes) might well be large. A disadvantage nearly as large resulting from a less than optimal supply of dietary ascorbic acid would not prevent the replacement of the earlier species by the mutant. Hence the amount of the vitamin provided by the diet available at the time of the mutation might be less than the optimal amount. Moreover, it is possible that the environment has changed during the last twenty million years in such a way as to provide a decreased amount of the vitamin for the average human being. Even a serious disadvantage of the changed environment would not lead to a mutation restoring the synthetic mechanism, because of the small probability of such mutations, far smaller than of those resulting in loss of function.

Individual Variation

The human race is characterized by large genetic heterogeneity. Enzyme concentrations in the tissues of different persons often differ by the factor of two or even a factor of ten or one hundred, as has been pointed out especially by Professor Roger J. Williams of the University of Texas. Heterozygosity in the gene for phenylketonuria halves the amount of the enzyme phenylalanine hydroxylase, and homozygosity in this gene reduces the amount or effectiveness of the enzyme by two or more orders of magnitude.

Molecular Concentrations and Rate of Reactions

The rate of an enzyme-catalysed reaction is approximately proportional to the concentration of reactant until concentrations are reached that largely saturate the enzyme. The saturating concentration is larger for a defective enzyme with decreased combining power for the substrate than for the normal enzyme. For such a defective enzyme the catalysed reaction could be made to take place at or near its normal rate by an increase in the concentration of the substrate. Moreover, an increase in concentration of an enzyme inhibitor can decrease the rate of reaction; for example, an increase in concentration of nicotinamide, with the consequent inhibition of the enzyme diphosphopyridine nucleotidase, decreases the rate of hydrolysis of diphosphopyridine nucleotide.

Evidence from Microbiological Genetics

Many mutant microorganisms are known to require, as a supplement to the medium on which they are grown, a substance that is synthesized by the corresponding wild-type organism. An example is the "pyridoxineless" mutant of Neurospora Sitophila reported by G. W. Beadle and E. L. Tatum in their first Neurospora paper, published in the Proceedings of the United States National Academy of Sciences in 1941. They found the rate of growth of this mutant on their standard medium to be only nine per cent of that of the wild type. When pyridoxine (Vitamin B_6) is added to the medium, the rate of growth at first increases nearly linearly with the concentration of the added pyridoxine, and then the growth-rate curve bends rather sharply, and continues to increase with a much smaller slope. The region of concentrations in which the growth increases rapidly with increase in concentration may be considered to be the region of vitamin deficiency, and the concentration at which the curve changes to much smaller slope may be considered to correspond to an "adequate" or "recommended" amount of the vitamin, in that the growth rate of the mutant is then only a few per cent less than that of the wild strain, and the amount of the vitamin would have to be increased three-fold to make up the difference.

Increasing the concentration of the growth substance to thirty-five times the adequate concentration for the mutant causes an increase in growth rate by about twenty-five per cent. The growth rate of the mutant is then ten per cent greater than that of the wild type. An increase in growth rate or in some other function of magnitude twenty-five per cent or ten per cent might under some circumstances be of great value, and might mean the difference between life and death or between good health and poor health, either physical or mental. Especially if the growth substance is non-toxic and free from side reactions, its therapeutic use in large amounts might well be justified.

Ascorbic acid, for example, is non-toxic for all or almost all human beings. It is required for life in amounts of a few milligrams per day. The ingestion of large amounts, three or five grams per day, seems to improve the general health of human beings, and to provide greater resistance to colds and other infectious diseases. I believe that it is likely that the optimal intake of ascorbic acid is far greater than the "approved" or "recommended" intake, and that a significant improvement in the health of human beings could be achieved by approximating this optimal intake.

Orthomolecular Psychiatry

The functioning of the brain is more sensitively dependent on the rate of chemical reaction than the functioning of other organs and tissues. I believe that mental disease is for the most part caused by abnormal molecular concentrations of essential substances. The operation of chance in the selection, for the child, of half of the complement of genes of the father and mother leads to bad as well as to good genotypes, and the selection of foods (and drugs) in a world that is undergoing rapid scientific and technological change may often be far from the best. Significant improvement in the mental health of many persons, especially those with borderline or mild mental illness or mental retardation, might be achieved by the provision of the optimal molecular concentrations of substances normally present in the human body, especially those that are not toxic or have low toxicity.

Among these substances, the essential nutrilites may be the most worthy of extensive research and more thorough clinical trial than they have yet received.

L(+)-Glutamic acid is a non-essential amino acid that is present at rather high concentration in brain and nerve tissue, and plays an essential role in the functioning of this tissue. It is normally ingested (in protein) in amounts of five to ten grams per day. It is not toxic; large doses may cause increased motor activity and nausea. In 1944 Price, Waelsch, and Putnam (Journal of the American Medical Association) reported favorable results for glutamic acid treatment of convulsive disorders (benefit to one out of three or four patients with petit mal epilepsy). Many investigators then studied the effects of glutamic acid on the intelligence and behavior of persons with different degrees and kinds of mental retardation. Increase in intelligence and improvement in personality have been reported for many patients with mild or moderate mental deficiency, when given doses of ten grams to twenty grams per day of L(+)-glutamic acid, adjusted to the patient as the amount somewhat less than required to cause hyperactivity.

Nicotinic acid, when its use was introduced, cured hundreds of thousands of pellagra patients of their psychoses, as well as of the physical manifestations of their disease. For this purpose only small doses are required; the recommended daily allowance (U.S. National Research Council) is twelve milligrams per day. In 1940 Streitwieser and his associates reported some success in the treatment of severe depression and other forms of mental disease by use of large doses of nicotinic acid, three grams or more per day. Other investigators, especially A. Hoffer of Saskatchewan and H. Osmond of New York, have advocated and used nicotinic acid in large doses for the treatment of schizophrenia. The dosage recommended by Hoffer is three grams per day, or more, up to eighteen grams per day, as determined by the response of the patient, of either nicotinic acid or nicotinamide, together with three grams per day of ascorbic acid. Nicotinic acid and nicotinamide are nontoxic (LD 50 not known for humans, but probably over 200 grams), and their side effects, even in continued massive doses, seem not to be commonly serious. The advantages of nicotinic acid therapy have been summarized by Osmond and Hoffer (The Lancet 10 Feb. 1962, 316) in the following words: "Niacin (nicotinic acid) has some though not all the qualities of an ideal treatment: it is safe, cheap (less than one cent per gram), and easy to administer, and it uses a known pharmaceutical substance which can be taken for years on end if necessary ... it does not seem to affect the more chronic illnesses, and even in acute illnesses its action is often less dramatic than that of some of the phenothiazenes. Its protective qualities, continuing long after patients have stopped taking it, are puzzling ... it has been proved to reduce the level of cholesterol in the blood. It seems to benefit some deliria not obviously associated with vitamin lack, and is claimed to improve many cases of intractable rheumatism. In our view it is a useful adjunct in the treatment of schizophrenia, both for acute cases and to reduce the chance of relapses."

It is my opinion, for reasons presented in the opening arguments of this paper, that for those patients for whom it is effective the control of mental disease by varying the concentrations in the brain of non-toxic substances that are normally present, such as nicotinic acid and ascorbic acid, is to be preferred to the use of phenylthiazenes and other means of therapy that involve a greater insult to the body and mind.

Varying the intake of other vitamins has also been used to some extent in psychiatry. The mental deterioration accompanying aging and cerebral vascular disease may be alleviated by supplementary biflavonoids (Vitamin P) and ascorbic acid. The use of cyancobalamin (B_{12}) in the treatment of mental disease has been reported, as well as the use of pyridoxine (B_6).

I believe that the study of the functioning of the brain in its relation to the concentrations and intake of the vitamins, essential amino acids, and other substances normally present in the brain constitutes a field of research in which much more work needs to be done. Biochemical and genetic arguments support the idea that orthomolecular therapy, the provision for the individual human being of the optimal concentrations of important normal constituents of the human body, may be the preferred treatment for many patients, especially those with mild mental retardation or mild psychosis. I suggest that this therapy, to be successful, should involve the thorough study of the individual, and continued attention to him, such as is customary in psychoanalysis but not in conventional chemotherapy. There is the possibility that analysis of body fluids and tests of the ability of the individual to utilize essential substances may indicate the types of orthomolecular therapy that would be most likely to be effective for the patient.

I express the hope that the members of the Society and other medical investigators will find the foregoing discussion of orthomolecular medicine interesting and suggestive, and that added attention will be given in the future to the possibility of improving the mental and physical health of human beings by varying the concentrations of substances normally present in the human body.

Author's address:

Linus Pauling,
Professor of Chemistry
Departments of Chemistry and Biology
University of California, San Diego
La Jolla, California 92037

Reprinted from the *Proceedings of the National Academy of Sciences*
Vol. 67, No. 4, pp. 1643-1648. December, 1970.

Evolution and the Need for Ascorbic Acid

Linus Pauling

DEPARTMENT OF CHEMISTRY, STANFORD UNIVERSTY, STANFORD, CALIFORNIA 94305

Communicated September 23, 1970

Abstract. Ascorbic acid differs from other vitamins in that an exogenous source is required by only a few animal species. It is pointed out that this fact indicates that the amount contained in a diet of raw natural plant food is less than the optimum intake, corresponding to the best health. This argument leads to the conclusion that the optimum daily intake is about 2.3 g or more, for an adult with energy requirement 2500 kcal day^{-1}.

Ascorbic acid (vitamin C) is required in his diet by man for life and good health. People who receive no ascorbic acid become sick with scurvy in a few weeks, and die in a few months.

Since the discovery of ascorbic acid and its identification with vitamin C there has been continued effort to determine the human requirement for this essential nutrient. Evidence indicates that the minimum daily intake needed to prevent scurvy is about 10 mg. The daily allowance recommended by health authorities for adults ranges from 20 mg in the United Kingdom to 75 mg in West Germany. The Food and Nutrition Board of the U.S. National Research Council[1] recommends values ranging from 35 mg for an infant to 60 mg for a 70-kg man, as designed for the maintenance of good nutrition of practically all healthy people in the U.S.A. The recommended allowances are said to provide a generous increment for individual variability and a surplus to compensate for potential losses in food.

These recommended values of the daily allowance of ascorbic acid have clearly resulted from a concentration by the authorities on the need to prevent scurvy. We may, however, ask whether larger amounts might not be needed to provide the optimum state of health. Numerous clinical reports indicate that large amounts of ascorbic acid are beneficial in increasing resistance to infections, improving the healing of wounds and burns, and decreasing the incidence of shock following injury or surgery. Moreover, ascorbic acid is nontoxic; it has been described[2] as probably the least toxic of all known substances of comparable physiologic activity. People have ingested 100 g per day for several days and 40 g per day for weeks without being harmed.

In a discussion of vitamin C and immunity to infection Bourne[3] in 1949 pointed out that the green foodstuffs eaten by the gorilla provide about 4.5 g of ascorbic acid per day, and that before the development of agriculture man existed largely on greens, supplemented with some meat. He concluded that "It may be possible, therefore, that when we are arguing whether 7 or 30 mg of vitamin C a day is an adequate intake we may be very wide of the mark. Perhaps we should be arguing whether 1 or 2 g a day is the correct amount." Stone[2,4,5] has suggested that the optimum rate of intake of ascorbic acid is about 3 g per day under ordinary conditions, and larger, up to 40 g per day, for a person under stress (for example, when infected with the virus of the common cold). Régnier[6] also recommends a large intake, 5 g per day or more, for averting or ameliorating the common cold. One of the arguments used by Stone is the following: the rat under normal conditions synthesizes ascorbic acid at the reported rate 26 mg day^{-1} kg^{-1} (see ref. 7) to 58 mg day^{-1} kg^{-1} (see ref. 8). If we assume proportionality to body weight (as indicated by the amount needed to prevent scurvy in the guinea pig), these correspond to 1.8–4.1 g day^{-1} for a 70-kg man.

Ascorbic acid differs from the other vitamins in that it is required in the diet by only a few species of animals—man, other primates, the guinea pig, an Indian fruit-eating bat, and the red-vented bulbul and some related species of Passeriform birds.[4,9] Other species of animals synthesize ascorbic acid. All mammals and other larger animals require vitamin A, thiamine, riboflavin, nicotinic acid, and pyridoxine as essential nutrients, although microorganisms usually have the power to synthesize all or most of these substances.

In the following paragraphs I point out that the fact that ascorbic acid is synthesized by most animal species, but not by man, provides strong evidence that the optimum rate of intake by man is about 2 or 3 g per day or more, 50–100 times, or more, the amounts recommended by the health authorities.

Let us consider the way in which man and other organisms have evolved. Initially there was on earth the "hot thin soup," containing molecules of millions, perhaps hundreds of millions of different kinds. Some molecules with autocatalytic ability appeared; they increased the rate of production of duplicates of themselves, probably by working as complementary pairs.[10] Some of them developed specific heterocatalytic ability, speeding up the production of certain molecules of other kinds. The development of a cell membrane favored the process, by keeping the cooperating molecules, especially nucleic acids and proteins, together.

There then followed the long eobiontic period, two or three billion years, during which there took place the astounding process of biochemical evolution, a much greater accomplishment than the process of morphological development and differentiation of larger organisms that occurred later. In this eobiontic period there were developed the genes to direct the synthesis of the enzymes required to catalyze great numbers of chemical reactions within the cell. The way in which this took place was discovered by Horowitz.[11] It may be illustrated by a discussion of thiamine. Molecules of thiamine were present in the original hot thin soup, and soon became essential to the functioning of the earliest unicellular forms of life. When the supply of thiamine had been nearly used up, and the organisms were competing for it, one organism underwent a mutation that permitted it to synthesize an enzyme that catalyzed the synthesis of thiamine from two other substances (pyrimidine pyrophosphate and thiazole phosphate) in the soup. This organism survived, while the others died from thiamine starvation (microbiological beriberi). Then, as the supply of pyrimidine pyrophosphate was being used up, an organism developed the gene to produce an enzyme to catalyze its synthesis from hydroxymethylpyrimidine and ATP (adenosine triphosphate), still present in the soup. Similarly, a gene was developed to produce an enzyme to make thiazole phosphate from thiazole and ATP. Several other steps were developed, until the organism was able to synthesize thiamine from substances present in large amounts in the environment. Many existing species of bacteria, fungi, and algae are able to synthesize thiamine, other vitamins, and all other essential organic substances.

When large, multicellular organisms developed, with specialized organs, some of these abilities became a handicap, and were lost. Let us continue with thiamine as an example. All mammals require thiamine in their food, in order to live. The ability to synthesize thiamine was lost by an early ancestral vertebrate, several hundred million or a billion years ago. This animal was ingesting plants, which provided an ample supply of thiamine, about 5 mg for each 2500 kcal of food energy. The synthetic mechanism was not needed, and it was a burden: it cluttered up the cells, added to the body weight, used energy that could be better used for other purposes. When a series of mutations occurred that eliminated this mechanism, the mutant was favored over the wild type, which failed to meet the competition and died out.[12] The victory of a mutant strain of a bacterium over the wild type in the competition for survival in the presence of an ample supply of the essential nutrient has been verified by direct experiments in the laboratory by Zamenhof and Eichhorn.[13]

In addition to thiamine, all mammalian species, so far as studied, require riboflavin, nicotinic acid, and vitamin A in their food. We conclude that the supply of food available to an earlier ancestor provided an adequate supply of these vitamins, enough to make it advantageous to discard the mechanism for synthesizing them. There is little doubt that this food was plant food, probably not greatly different from present day plant food.

In the recently published handbook on metabolism of the Federation of American Biological Handbooks[14] there is a table giving the amounts of four water-soluble vitamins (and also of vitamin A) in 366 raw and processed plant foods. I have recalculated the values to the basis of the quantity with energy value 2500 kcal (the average daily need of a man or woman) for the 110 raw natural plant foods listed in the table, with the results shown in Table 1.

I have pointed out[12] that an animal that synthesizes an essential substance synthesizes a somewhat smaller daily amount than the optimum, because to synthesize the optimum amount would require supporting the burden of additional synthetic machinery, with only a smaller compensation. Also, a mutant that has an exogenous source of the essential substance that provides somewhat less than the optimum amount may win the competition, because the wild type has to support the burden of machinery.

TABLE 1. *Water-soluble vitamin content (mg) of 110 raw natural plant foods (referred to amount giving 2500 kcal of food energy).*

	Thiamine	Riboflavin	Nicotinic acid	Ascorbic acid
Nuts and grains (11)	3.2	1.5	27	0
Fruit, low-C (21)	1.9	2.0	19	600
Beans and peas (15)	7.5	4.7	34	1000
Berries, low-C (8)	1.7	2.0	15	1200
Vegetables, low-C (25)	5.0	5.9	39	1200
Intermediate-C foods (16)	7.8	9.8	77	3400
Collards	10.8	17	92	5000
Chives	7.1	11.6	45	5000
Cabbage	6.2	5.0	32	5100
Brussels sprouts	5.6	8.9	50	5700
Cauliflower	10.0	9.3	65	7200
Mustard greens	8.9	18	65	7800
Kale	...	...	...	8200
Broccoli spears	7.8	18	70	8800
Black currants	2.3	2.3	14	9300
Parsley	6.8	15	68	9800
Hot red chili peppers	3.8	7.7	112	14200
Sweet green peppers	9.1	9.1	57	14600
Hot green chili peppers	6.1	4.1	115	15900
Sweet red peppers	6.5	6.5	40	16500
Average for 110 foods	5.0	5.4	41	2300

Nuts and grains: almonds, filberts, macadamia nuts, peanuts, barley, brown rice, whole grain rice, sesame seeds, sunflower seeds, wild rice, wheat.

Fruit (low in vitamin C, less than 2500 mg): apples, apricots, avocadoes, bananas, cherries (sour red, sweet), coconut, dates, figs, grapefruit, grapes, kumquats, mangoes, nectarines, peaches, pears, pineapple, plums, crabapples, honeydew melon, watermelon.

Beans and peas: broad beans (immature seeds, mature seeds), cowpeas (immature seeds, mature seeds), lima beans (immature seeds, mature seeds), mung beans (seeds, sprouts), peas (edible pod, green mature seeds), snapbeans (green, yellow), soybeans (immature seeds, mature seeds, sprouts).

Berries (low-C, less than 2500 mg): blackberries, blueberries, cranberries, loganberries, raspberries, currants, gooseberries, tangerines.

Vegetables (low-C, less than 2500 mg): bamboo shoots, beets, carrots, celeriac root, celery, corn, cucumber, dandelion greens, eggplant, garlic cloves, horseradish, lettuce, okra, onions (young, mature), parsnips, potatoes, pumpkins, rhubarb, rutabagas, squash (summer, winter), sweet potatoes, green tomatoes, yams.

Intermediate-C foods (2500–4900 mg): artichokes, asparagus, beet greens, cantaloupe, chicory greens, chinese cabbage, fennel, lemons, limes, oranges, radishes, spinach, zucchini, strawberries, swiss chard, ripe tomatoes.

In Table 2 there are given the average amounts (per 2500 kcal of food energy) of thiamine, riboflavin, nicotinic acid, and ascorbic acid for the 110 natural foods listed in Table 1, the recommended daily allowances of the four vitamins, and the ratios of the two quantities.

It is interesting that for thiamine, riboflavin, and nicotinic acid the values of the ratio of the amount in the foods to the recommended daily allowance (both

TABLE 2. *Comparison of average vitamin content (mg) of 110 raw natural plant foods with the recommended dietary allowances.*

	Thiamine	Riboflavin	Nicotinic acid	Ascorbic acid
Average for 110 foods*	5.0	5.4	41	2300
Recommended allowance†	1.30	1.83	17	66
Ratio	3.8	3.0	2.4	35

* From Table 1.

† Average for adults, male and female, referred to 2500 kcal food energy.[1]

per 2500 kcal of food energy) are nearly the same: 3.8, 3.0, and 2.4, respectively. It seems likely that the relative needs for these substances are nearly the same for plants as for animals. Also, they are apparently nearly the same for different kinds of plants. In Table 3 values are given relative to the amounts con-

Table 3. *Water-soluble vitamin content (mg) of plant foods relative to nicotinic acid (41 mg).*

	Thiamine	Riboflavin	Ascorbic acid
Nuts and grains (11)	4.9	2.3	0
Beans and peas (15)	9.0	5.7	1200
Berries (8)	4.6	5.5	3300
Fruit (21)	4.1	4.3	1300
Vegetables (25)	5.3	6.2	1260
Intermediate-C foods (16)	4.2	5.2	1800
High-C foods (14)	4.6	6.6	6100

taining 41 mg of nicotinic acid. On this basis the amount of thiamine is close to 4.6 mg, except for beans and peas, and the amount of riboflavin is close to 5.5 mg, except for nuts and grains.

Ascorbic acid differs from the other vitamins in that most animal species continue to synthesize it despite its availability in natural foods. I think that the only reasonable explanation of this fact is that the foods available to most animals over the past several hundred million years have not provided a supply of ascorbic acid sufficient to justify the abandonment of the mechanism of its synthesis. We are accordingly able to make an estimate of the optimum daily intake of ascorbic acid, on the basis of this fact and the assumption that the foods listed in Table 1 represent approximately the foods available to the ancestors of existing animal species.

The average amount of ascorbic acid (per 2500 kcal energy value) for the 110 natural foods in Table 1 is 2300 mg. According to the foregoing argument, this amount is less than the optimum daily requirement of an adult animal requiring 2500 kcal of food energy.

I conclude that the optimum daily requirement of ascorbic acid for a human being requiring 2500 kcal of food energy is about 2.3 g (2300 mg) (2.6 g for an adult male, and 2.0 g for an adult female), or is greater than this amount.

We may ask how much greater the optimum daily requirement may be. The loss of the ability to synthesize ascorbic acid has occurred only four times in several hundred million years, so far as we know: in the common ancestor of man and other primates, about 25 million years ago; in the guinea pig; in one Indian fruit-eating bat; and in the ancestor of some Passeriform birds. The animals that underwent this change must have been living in an environment that provided an unusually great amount of ascorbic acid. I have listed in Table 1 fourteen natural foods that are unusually rich in this substance. They contain between 5.0 and 16.5 g of ascorbic acid for 2500 kcal energy value, with average 9.5 g. It is unlikely that the special environment mentioned above would provide only red sweet peppers (16.5 g), black currants (9.3 g), or broccoli spears (8.8 g); but the average of these fourteen foods, 9.5 g of ascorbic acid, might well have become available to a population of animals a few times during the period of several hundred million years preceding the present epoch. Accordingly, I conclude that it is unlikely that the optimum daily intake of ascorbic acid by human beings is greater than 9.5 g.

The range 2.3–9.5 g to which this evolution argument leads agrees moderately well with the values 2 g from Bourne's gorilla argument and 1.8–4.1 g from Stone's rat argument, and supports the suggestion by Bourne and the contention by Stone that the optimum rate of intake of ascorbic acid is many times the officially recommended daily dietary allowance, ranging in different countries from 20 to 75 mg. I have recently reached the conclusion that the state of health of most people, including the ability to resist infectious diseases such as the common cold, would be greatly improved by an increased intake of this important food.[15]

It is, of course, almost certain that some evolutionarily effective mutations have occurred in man and his immediate predecessors rather recently (within the last few million years) such as to permit life to continue on an intake of ascorbic acid less than that provided by high-ascorbic-acid raw plant foods. These mutations might involve an increased ability of the kidney tubules to pump ascorbic acid back into the blood from the glomerular filtrate (dilute urine, being concentrated on passage along the tubules) and an increased ability of certain cells to extract ascorbic acid from the blood plasma. It is likely that the adrenal glands act as a storehouse of ascorbic acid, extracting it from the blood when green plant foods are available, in the summer, and releasing it slowly when the supply is depleted. On general principles we can conclude, however, that these mechanisms require energy and are a burden to the organism. The average optimum rate of intake of ascorbic acid might still be close to the value given above, 2.3 g per day or more, or might be somewhat less; and, of course, there is always the factor of biochemical individuality, such that for different people in a large population the optimum intake might vary over a range of 20-fold or more.[16]

[1] *Recommended Dietary Allowances, A Report of the Food and Nutrition Board, National Research Council* (Washington, D.C.: National Academy of Sciences, 1968), 7th ed.
[2] Stone, Irwin, *Acta Genet. Med. Gemellol.*, **15**, 345 (1966); **16**, 52 (1967).
[3] Bourne, G. H., *Brit. J. Nutr.*, **2**, 346 (1949).
[4] Stone, Irwin, *Amer. J. Phys. Anthropol.*, **23**, 83 (1965).
[5] Stone, Irwin, *Perspect. Biol. Med.*, **10**, 133 (1966).
[6] Régnier, E., *Rev. Allergy*, **22**, 835, 948 (1968).
[7] Burns, J. J., E. H. Mosbach, and S. Schulenberg, *J. Biol. Chem.*, **207**, 679 (1954).
[8] Salomon, L. L., and D. W. Stubbs, *Ann. N.Y. Acad. Sci.*, **92**, 128 (1961).
[9] Chaudhuri, C. R., and I. B. Chatterjee, *Science*, **164**, 435 (1969).
[10] Pauling, L., and M. Delbrück, *Science*, **92**, 77 (1940).
[11] Horowitz, N. H., *Proc. Nat. Acad. Sci. USA*, **31**, 153 (1945).
[12] Pauling, L., *Science*, **160**, 265 (1968).
[13] Zamenhof, S., and H. H. Eichhorn, *Nature*, **216**, 465 (1967).
[14] Altman, P. L., and D. S. Dittmer, *Metabolism* (Bethesda, Md.: Federation of American Societies for Experimental Biology, 1968).
[15] Pauling, L., *Vitamin C and the Common Cold* (San Francisco: W. H. Freeman and Co., 1970).
[16] Williams, R. J., and G. Deason, *Proc. Nat. Acad. Sci. USA*, **57**, 1638 (1967).

Ascorbic acid and the common cold

Linus Pauling, Ph.D.

For a number of years I have been interested in the possibility that the state of health of many people could be significantly improved by the ingestion in the optimum amounts of certain substances normally present in the human body, including the vitamins. This interest developed from the work that my associates and I have done on molecular diseases, especially the hemoglobinemias (1). I decided in 1953 that it would be worthwhile to make a study of the extent to which mental diseases could be described as molecular diseases. Work along these lines was carried out in our laboratory in the California Institute of Technology from 1954 to 1964, and was continued in the University of California, San Diego, and (since 1969) in Stanford University. In the course of this period I formulated some ideas about orthomolecular medicine, defined as the preservation of good health and the treatment of disease by varying the concentrations in the human body of substances that are normally present in the body and are required for health (2–4). I also became aware of arguments indicating that the optimum rate of intake of ascorbic acid may be far greater than the recommended daily allowance of this vitamin, which is approximately 50 mg/day. Part of the evidence on this point had been presented especially clearly in the papers of Stone (5–8).

Last year I published a small book, *Vitamin C and the Common Cold,* in which I presented the evidence supporting the conclusion that ascorbic acid ingested in larger amounts than the recommended daily allowance has value in decreasing the incidence and severity of the common cold and related infectious diseases (9).

This opinion is in agreement with a rather widespread popular belief that ascorbic acid has value in providing protection against the common cold. This popular belief has, however, not been generally shared by physicians, authorities on nutrition, and official bodies.

For example, as recently as November 1970, Dr. Philip L. White (10), Secretary of the Council on Foods and Nutrition of the American Medical Association, stated that "Unfortunately, it is still a widespread belief that extra ascorbic acid can not only prevent colds but also lessen the severity and duration of colds and other respiratory infections. Even when consumed at the first sign of a sniffle, large doses of the vitamin are useless." Also, many statements contradicting my conclusions were made by physicians, experts in nutrition, and health officials within a few weeks after the publication of my book. For example, Dr. Charles C. Edwards, United States Food and Drug Commissioner, was reported in the press on December 29, 1970 as having said that the use of ascorbic acid was ridiculous, and that there was no scientific evidence and never have been any meaningful studies indicating that vitamin C is capable of preventing or curing colds. The Editors of *The Medical Letter* published an article in which nearly all my statements were contradicted; for example, it was stated that there had been no controlled trials of the effectiveness of vitamin C, in comparison with a placebo, against upper respiratory infections over a long period and including many hundreds of persons (11).

In fact, there have been several carefully conducted double-blind studies of ascorbic acid and the common cold, carried out by responsible medical investigators. Some of these studies have given results that reject with statistical significance the null hypothesis that ascorbic acid has no more value than a placebo in decreasing the incidence and severity of the common cold when the ascorbic acid is administered regularly to subjects over a period of time beginning before the illness has set in, and the subjects are exposed to cold viruses in the ordinary way (by casual contact with other people). I shall discuss some of these studies in the following paragraphs. The amount of protection against

The American Journal of Clinical Nutrition 24: NOVEMBER 1971, pp. 1294–1299. Printed in U.S.A.

the common cold is reported to increase with the amount of ascorbic acid ingested, reaching a protective effect of about two-thirds for the integrated morbidity for subjects receiving 1,000 mg ascorbic acid/day, in comparison with those receiving a placebo. So far as I have been able to discover by examination of the literature and by writing to critics, there is no published account of any controlled double-blind study of the sort described above that eliminates with statistical significance the hypothesis that ascorbic acid has the amount of protective power ascribed to it in my book.

My conclusion about the significance of some of the investigations has disagreed with that reached by the investigators themselves. For example, Drs. Cowan, Diehl, and Baker of the University of Minnesota reported a statistically significant protective effect, which they described as being judged to "give a slight advantage in reducing the number of colds experienced," but they also stated in the summary of their article that their controlled study had yielded no indication that large doses of vitamin C alone (or with other vitamins) had any important effect on the number or severity of upper respiratory tract infections when administered to young adults who presumably were already on a reasonably adequate diet. This statement by the investigators is an expression of opinion as to whether or not the effect that they observed is an important effect. I conclude that this statement of opinion about what constitutes an "important" effect has been responsible for the failure by most physicians and nutritionists writing on this subject to accept the evidence reported by the investigators.

It is my opinion that the failure to recognize the significance of the results reported in the controlled trials that have been carried out can be attributed to two causes. First, there has been a general reluctance to make studies of the effect of an intake of ascorbic acid greater than 200 mg/day, over a period of time. Second, the possibility that the average protective effect of ascorbic acid for a population would be augmented with increase in the daily intake of the vitamin was ignored, and the statistically significant observation of a decreased incidence and severity of the common cold for subjects receiving about 200 mg/day (in comparison with those receiving a placebo) did not suggest to the investigators that they should carry out a similar investigation with a daily intake of 400 mg, to find out if the protective effect would be twice as great. If this idea had been accepted by Cowan, Diehl, and Baker in 1942, it is, in my opinion, likely that the value of ascorbic acid in controlling the common cold would have been recognized 25 years ago.

I shall discuss in some detail two investigations, one carried out by Dr. G. Ritzel, a physician with the Medical Service of the School District of the city of Basel, Switzerland, and the other by Drs. Cowan, Diehl, and Baker, physicians connected with the School of Medicine of the University of Minnesota.

The study carried out by Ritzel and reported in 1961 (12) was made with 279 skiers at a ski resort, during two periods of 5 to 7 days. The conditions were such that the incidence of colds during these short periods was large enough (approximately 20%) for the results to be statistically significant. The subjects were roughly of the same age and had similar nutrition during the period of study. The ascorbic acid tablets or placebo were distributed to the subjects every morning and were taken under observation so that the possibility of interchange of tablets was eliminated. Persons who showed cold symptoms on the first day were excluded from the investigation. The subjects were examined daily by physicians for symptoms of colds and other infections. The records were largely on the basis of subjective symptoms, but were partially supported by objective observations (measurement of body temperature, inspection of the respiratory organs, auscultation of the lungs).

It was a double-blind study, with neither the subjects nor the physicians knowing which subjects received ascorbic acid and which received the placebo. At the end of the investigation, a completely independent group of professional people was provided with the identification numbers for the tablets, and this group carried out the statistical evaluation of the observations.

The number of colds for the 140 persons in the placebo group was 31, and that for the

139 persons in the ascorbic acid group was 17. A decrease of 45% in the incidence of colds was observed in the ascorbic acid group, as compared with the placebo group. This decrease in incidence is statistically significant, at the level P (one-tailed) less than 0.02.

This result means that the probability that a decrease in incidence of colds as great as this, and in this direction (greater protective effect of ascorbic acid than of the placebo), would be observed in two groups from a uniform population, with ascorbic acid and placebo having the same effect, only in 2% of a large number of similar tests. Hence, the observation of a protective effect by ascorbic acid, decreasing the incidence of colds, is reliable at the 98% level of confidence.

In applying statistical analysis to this study and other studies of ascorbic acid and the common cold I have calculated the value of P (one-tailed) rather than P (two-tailed), because the question under dispute is whether or not ascorbic acid has greater protective power than a placebo; no one contends that the placebo would have greater protective power than ascorbic acid. The results of my own statistical analysis of the observations reported by Ritzel agree completely with those reported by the team of professionals and stated in the published paper.

The severity of individual colds was observed by Ritzel in two ways: first, by recording the average number of days of illness per cold, and second, by recording the number of individual symptoms per cold (as recorded each day). The severity of individual colds was found to be 29% less for the ascorbic acid group than for the placebo group, as measured by the average number of days of illness per cold, and 36% less as measured by the number of individual symptoms recorded per cold. Each of these decreases is statistically significant at the level P (one-tailed) less than 0.05.

A third quantity, which may be taken to represent the total protective effect of ascorbic acid, is the integrated morbidity, defined as the product of the incidence of colds and the severity of individual colds. In the Ritzel study the integrated morbidity was found to be 61% less for the ascorbic acid group than for the placebo group as measured by the average number of days of illness per subject in the group, and 64% less as measured by the average number of individual symptoms per person in the group. Each of these observations is statistically significant at the level P (one-tailed) less than 0.01.

Dr. G. Ritzel has reported in his carefully performed double-blind study of 279 subjects that the administration of 1,000 mg/day of ascorbic acid, in comparison with a placebo, leads to a decrease (with high statistical significance) in incidence, severity, and integrated morbidity of the common cold. The integrated morbidity is correlated with the incidence and severity; accordingly, the statistical significance of the investigation as a whole can be taken to be rejection of the null hypothesis that ascorbic acid has the same value as the placebo at the level P (one-tailed) less than 0.01; that is, at the confidence level of 99%.

Another very careful study was the one by Cowan, Diehl and Baker, reported in 1942 (13). Their principal work was done during the winter of 1939 to 1940, over a period of 28 weeks. The subjects were all students at the University of Minnesota who volunteered to participate in the study because they felt that they were particularly susceptible to colds. Persons whose difficulties seemed to be due primarily to chronic sinusitis or allergic rhinitis, as shown by examination of the nose and throat and consideration of symptoms of allergy, were excluded from the study. The subjects were assigned alternately and at random to an experimental group and a control group. The subjects in the control group were treated exactly like those in the experimental group, except that they received a placebo instead of ascorbic acid. The subjects were instructed to report to the health service whenever a cold developed, so that report cards could be filled in by a physician. The study was a double-blind one, with neither the subjects nor the physician knowing to which group a subject was assigned (D. W. Cowan, personal communication). There were 183 subjects in the ascorbic acid group, who received an average of 180 mg ascorbic acid per day during the 28-week period, and 155 subjects in the placebo group.

The amount of ascorbic acid per day used in this study is a little less than one-fifth the amount used by Ritzel. The effects are also slightly smaller; but the somewhat larger number of subjects and much longer period of the investigation cause the results to have approximately the same statistical significance. The incidence of colds was 14% less for the ascorbic acid group than for the placebo group, with P (one-tailed) less than 0.02; the severity of individual colds (average days of illness per cold) was 21% less, with P (one-tailed) less than 0.02; and the integrated morbidity (average days of illness per person) was 31% less, with P (one-tailed) less than 0.01.

The level of confidence at which the null hypothesis that ascorbic acid has no more effect than a placebo is rejected by the observations of Cowan, Diehl, and Baker in their carefully controlled study is given by P (one-tailed) less than 0.01; that is, on the basis of this study ascorbic acid is shown to be effective in decreasing the integrated morbidity of the common cold at the 99% confidence level.

One reason that has been advanced for not accepting the results reported by Cowan, Diehl, and Baker is that the subjects who received ascorbic acid had more complications, such as bronchitis, otitis, and sinusitis, than did those in the control group. The numbers of subjects in these categories were, however, so small that the difference does not have statistical significance, and therefore this criticism is to be rejected.

Cowan, Diehl, and Baker also reported the results of another study, involving three groups of subjects (numbering 82 to 94 per group), with one group receiving placebo capsules, another, 25 mg ascorbic acid/day plus other vitamins, and the third, 50 mg ascorbic acid/day plus other vitamins. No statistically significant differences among the groups were observed. Moreover, the observations do not permit the rejection with statistical significance of the hypothesis that the 25-mg and 50-mg doses of ascorbic acid per day have the effect of a decrease in incidence, severity, and integrated morbidity of the common cold one-seventh or one-quarter, respectively, as great as reported for 180 mg/day. Accordingly, this study cannot be quoted with statistical significance as evidence for or against ascorbic acid as a means of protection against the common cold.

The study by Cowan, Diehl, and Baker and that by Ritzel are the best examples of carefully controlled double-blind studies of ascorbic acid and the common cold that have been described in detail in published papers. We may ask what the probability is that these two careful investigations, carried out at different times and on different continents, and surely not correlated with one another, would have given the results that they did give, i.e., in rejecting the null hypothesis of equal effectiveness of ascorbic acid and a placebo. Each of the two studies is statistically significant in rejecting the hypothesis at the level P (one-tailed) less than 0.01. Application of Fisher's method of combining independent studies to obtain their total significance leads to chi-squared (4 degrees of freedom) equal to 18.42. The null hypothesis of equal effectiveness of ascorbic acid and a placebo is rejected at the level P less than 0.001; that is, these two investigations show ascorbic acid to be more effective than a placebo in decreasing the integrated morbidity of the common cold at the 99.9% level of confidence. The probability that the results reported by the investigators would have been obtained in the two investigations if ascorbic acid and the placebo had the same effect is less than one in one thousand. We are justified in concluding from these two investigations alone that ascorbic acid, when administered over a period of time to subjects exposed in the normal way to cold viruses, is effective in decreasing the incidence, severity, and integrated morbidity of the common cold.

Several other carefully controlled studies of ascorbic acid and the common cold have been carried out under similar circumstances, and have yielded results with statistical significance. The amount of protective effect that has been reported is approximately the same as that found in the two studies discussed above.

No controlled study has been reported in which results have been obtained with statistical significance that are incompatible with the amount of protective effect found by

Cowan, Diehl, and Baker for 180 mg ascorbic acid/day and by Ritzel for 1,000 mg ascorbic acid/day. There is no justification for the statement often made that careful studies have shown that ascorbic acid has no protective value against the common cold.

There is only one investigation that seems to give a contradictory result. This is the work of Walker, Bynoe, and Tyrrell of the Common Cold Research Unit, Salisbury, England, reported in 1967 (14). Of their 91 subjects, 47 received 3 g ascorbic acid/day for 3 days before inoculation with viruses (rhinoviruses, influenza B, or B814 virus) and for 6 days after inoculation, and 44 subjects received a placebo. The reported incidence of colds was only 6% less for the ascorbic acid group than for the placebo group. It is possible that the conditions of this study, involving introduction of virus particles directly into the nose and throat of the subjects, were so much different from the conditions of ordinary exposure of persons to the viruses of the common cold, usually disseminated in the form of spray by the coughing or sneezing of persons with colds, that the results are not significant with respect to the question of whether or not ascorbic acid has protective effect for persons under ordinary conditions of exposure. The number of persons in the study and the short period of the study are such that a protective effect would have had to be larger than 40% in order to be statistically significant.

There is some published evidence that an increased intake of ascorbic acid, 3 to 10 g/day, taken regularly, leads to a decrease in incidence of the common cold by about 90%. This evidence has not been obtained, however, by the process of setting up double-blind trials. There is also evidence that ascorbic acid taken in the proper amount at the first signs of a common cold decreases its severity in a significant way, especially the evidence reported by Dr. E. Regnier, a physician in Salem, Massachusetts (15). I shall not discuss these various studies here, but shall say only that they are compatible with the conclusions described above, reached from a detailed consideration of the work of Cowan, Diehl, and Baker and that of Ritzel. I have made an analysis of all of the papers describing controlled studies of ascorbic acid and the common cold, and feel that the evidence that ascorbic acid has value in providing protection against the common cold is overwhelming. Earlier analyses of the evidence have not been conducted with care. For example, the editorial article on *Ascorbic Acid and the Common Cold* published in *Nutrition Reviews* in 1967 (16) and leading to the conclusion that "There is no conclusive evidence that ascorbic acid has any protective effect against, or any therapeutic effect on, the course of the common cold in healthy people not depleted of ascorbic acid" did not involve any discussion of statistical significance of the reported effects, and contains some errors of fact about the investigations.

The weight of published evidence about ascorbic acid and the common cold at the present time is overwhelmingly in support of the conclusion that ascorbic acid has value in decreasing the incidence, severity, and integrated morbidity of the common cold. There is no justification for the refusal by physicians and nutritionists to accept this conclusion.

References

1. Pauling, L., H. A. Itano, S. J. Singer and I. C. Wells. Sickle cell anemia, a molecular disease. *Science* 110: 543, 1949.
2. Pauling, L. Orthomolecular somatic and psychiatric medicine. *J. Vital Substances Diseases Civilization.* 14: 1, 1968.
3. Pauling, L. Orthomolecular psychiatry. Varying the concentrations of substances normally present in the human body may control mental disease. *Science* 160: 265, 1968.
4. Pauling, L. Evolution and the need for ascorbic acid. *Proc. Natl. Acad. Sci. U.S.* 67: 1643, 1970.
5. Stone, I. Studies of a mammalian enzyme system for producing evolutionary evidence on man. *Am. J. Phys. Anthropol.* 23: 83, 1965.
6. Stone, I. On the genetic etiology of scurvy. *Acta Genet. Med. Gemellol.* 15: 345, 1966.
7. Stone, I. Hypoascorbemia, the genetic disease causing the human requirement for exogenous ascorbic acid. *Perspectives Biol. Med.* 10: 133, 1966.
8. Stone, I. The genetic disease hypoascorbemia. *Acta Genet. Med. Gemellol.* 16: 52, 1967.
9. Pauling, L. *Vitamin C and the Common Cold.* San Francisco: Freeman, 1970.
10. White, P. L. Let's talk about food. *Today's Health.* November 1970.

11. Editorial. Vitamin C and the common cold. *The Medical Letter* 12: 25, 1970.
12. Ritzel, G. Ascorbic acid and infections of the respiratory tract. *Helv. Med. Acta* 28: 63, 1961.
13. Cowan, D. W., H. S. Diehl and A. B. Baker. Vitamins for the prevention of colds. *J. Am. Med. Assoc.* 120: 1267, 1942.
14. Walker, G. H., M. L. Bynoe and D. A. J. Tyrrell. Trial of ascorbic acid in prevention of colds. *Brit. Med. J.* 1: 603, 1967.
15. Regnier, E. The administration of large doses of ascorbic acid in the prevention and treatment of the common cold. *Rev. Allergy* 22: 835, 948, 1968.
16. Ascorbic acid and the common cold. *Nutr. Rev.* 25: 228, 1967.

Reprinted from *Orthomolecular Psychiatry: Treatment of Schizophrenia*, eds. David Hawkins and Linus Pauling, W.H. Freeman, San Francisco, pp. 1–17 (1973).

1

Orthomolecular Psychiatry

LINUS PAULING

INTRODUCTION

The methods principally used now for treating patients with mental disease are psychotherapy (psychoanalysis and related efforts to provide insight and to decrease environmental stress), chemotherapy (mainly with the use of powerful synthetic drugs, such as chlorpromazine, or powerful natural products from plants, such as reserpine), and convulsive or shock therapy (electroconvulsive therapy, insulin coma therapy, pentylenetetrazol shock therapy). I have reached the conclusion, through arguments summarized in the following paragraphs, that another general method of treatment, which may be called orthomolecular therapy, may be found to be of great value, and may turn out to be the best method of treatment for many patients.

Orthomolecular psychiatric therapy is the treatment of mental disease by the provision of the optimum molecular environment for the mind, especially the optimum

(*Reprinted with permission from* Science, *19 April 1968, vol. 160, pp. 265–271. Copyright © 1968 by the American Association for the Advancement of Science.*)

concentrations of substances normally present in the human body.[1] An example is the treatment of phenylketonuric children by use of a diet containing a smaller than normal amount of the amino acid phenylalanine. Phenylketonuria (Følling, 1934) results from a genetic defect that leads to a decreased amount or effectiveness of the enzyme catalyzing the oxidation of phenylalanine to tyrosine. The patients on a normal diet have in their tissues abnormally high concentrations of phenylalanine and some of its reaction products, which, possibly in conjunction with the decreased concentration of tyrosine, cause the mental and physical manifestations of the disease (mental deficiency, severe eczema, and others). A decrease in the amount of phenylalanine ingested results in an approximation to the normal or optimum concentrations and to the alleviation of the manifestations of the disease, both mental and physical.

The functioning of the brain is dependent on its composition and structure; that is, on the molecular environment of the mind. The presence in the brain of molecules of N,N-diethyl-D-lysergamide, mescaline, or some other schizophrenogenic substance is associated with profound psychic effects (see, for example, Woolley, 1962). Cherkin has recently pointed out (1967) that in 1799 Humphry Davy described similar subjective reactions to the inhalation of nitrous oxide. The phenomenon of general anesthesia also illustrates the dependence of the mind (consciousness, ephemeral memory) on its molecular environment (Pauling, 1961; Miller, 1961).

The proper functioning of the mind is known to require the presence in the brain of molecules of many different substances. For example, mental disease, usually associated with physical disease, results from a low concentration in the brain of any one of the following vitamins: thiamine (B_1), nicotinic acid or nicotinamide (B_3), pyridoxine (B_6), cyanocobalamin (B_{12}), biotin (H), ascorbic acid (C), and folic acid. There is evidence that mental function and behavior are also affected by changes in the concentration in the brain of any of a number of other substances that are normally present, such as L(+)-glutamic acid, uric acid, and γ-aminobutyric acid.[2]

OPTIMUM MOLECULAR CONCENTRATIONS

Several arguments may be advanced in support of the thesis that the optimum molecular concentrations of substances normally present in the body may be different from the concentrations provided by the diet and the gene-controlled synthetic

[1] I might have described this therapy as the provision of the optimum molecular composition of the brain. The brain provides the molecular environment of the mind. I use the word mind as a convenient synonym for the functioning of the brain. The word orthomolecular may be criticized as a Greek-Latin hybrid. I have not, however, found any other word that expresses as well the idea of the right molecules in the right amounts.

[2] The literature is so extensive that I refrain from giving references here.

mechanisms, and, for essential nutrilites (vitamins, essential amino acids, essential fatty acids) different from the minimum daily amounts required for life or the "recommended" (average) daily amounts suggested for good health. Some of these arguments are presented in the following paragraphs.

EVOLUTION AND NATURAL SELECTION

The process of evolution does not necessarily result in the normal provision of optimum molecular concentrations. Let us use ascorbic acid as an example. Of the mammals that have been studied in this respect, the only species that have lost the power to synthesize ascorbic acid and that accordingly require it in the diet are man, other Primates (rhesus monkey, Formosan long-tail monkey, and ring-tail or brown capuchin monkey), the guinea pig, and an Indian fruit-eating bat (*Pteropus medius*).[3] Presumably the loss of the gene or genes controlling the synthesis of the enzyme or enzymes involved in the conversion of glucose to ascorbic acid occurred some 20 million years ago in the common ancestor of man and other Primates, and occurred independently for the guinea pig and for one species of bat and one bird, in each case in an environment such that ascorbic acid was provided by the food. For a mutation rate of 1/20,000 per gene generation and for even a very small advantage for the mutant (0.01 percent more progeny) the mutant would replace the earlier genotype within about 1 million years. The advantage to the mutant of being rid of the ascorbic-acid-synthesis machinery (decrease in cell size and energy requirement, liberation of machinery for other purposes) might well be large, perhaps as much as 1 percent; a disadvantage nearly as large (less by 0.01 percent) resulting from a less than optimum supply of dietary ascorbic acid would not prevent the replacement of the earlier species by the mutant. Hence, even if the amount of the vitamin provided by the diet available at the time of the mutation were less than the optimum amount, the mutant might still be able to replace its predecessor. Moreover, it is possible that the environment has changed during the last 20 million years in such a way as to provide a decreased amount of the vitamin. Even a serious disadvantage of the changed environment would not lead to a mutation restoring the synthetic mechanism within a period of a few million years, because of the small probability of such mutations, far smaller than of those resulting in loss of function.

Moreover, the process of natural selection may be expected later on to lead to the survival of a species or strain that synthesizes somewhat less than the optimum amount of an autotrophic vital substance rather than of the species or strain that synthesizes the optimum amount. To synthesize the optimum amount requires

[3] For references, see Stone (1965). The only other vertebrates known to require exogenous ascorbic acid are the red-vented bulbul, *Pycnonotus cafer*, and related passeriform birds.

about twice as much biological machinery as to synthesize half the optimum amount. As suggested in Figure 1-1, the evolutionary disadvantage of synthesizing a less than

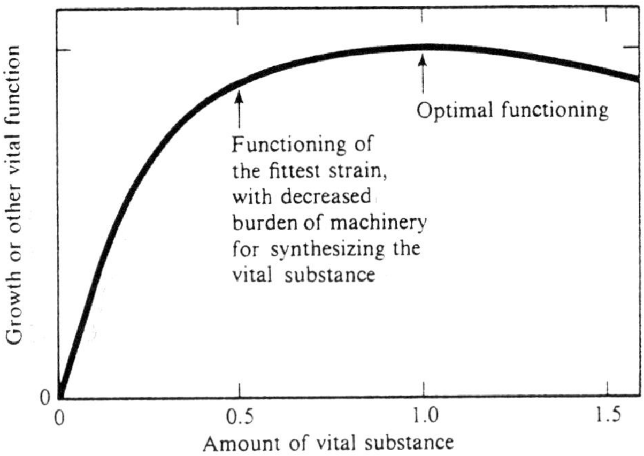

FIGURE 1-1.
Diagrammatic representation of growth rate or other vital property of an organism as function of the concentration of vital substance in the organism, showing the concentration at which the differential advantage of an increased amount of vital substance is just balanced by the differential disadvantage resulting from an increased amount of machinery for synthesis, and the concentration that gives optimum functioning without consideration of the burden of the machinery for synthesis.

optimum amount of the vital substance may be small, and may be outweighed by the advantage of requiring a smaller amount of biological machinery. Evidence from the study of microorganisms is discussed in the following paragraphs.

EVIDENCE FROM MICROBIOLOGICAL GENETICS

Many mutant microorganisms are known to require, as a supplement to the medium in which they are grown, a substance that is synthesized by the corresponding wild-type organism (the normal strain). An example is the pyridoxine-requiring mutant of *Neurospora sitophila* reported by G. W. Beadle and E. L. Tatum in their first *Neurospora* paper, published in 1941.

Several species of *Neurospora* that have been extensively studied are known to be able to grow satisfactorily on synthetic media containing inorganic salts, an inorganic source of nitrogen, such as ammonium nitrate, a suitable source of carbon, such as sucrose, and the vitamin biotin. All other substances required by the organism are synthesized by it. Beadle and Tatum found that exposure to x-radiation produces mutant strains such that one substance must be added to the minimum medium in order to permit the growth at a rate approximating that of the normal strain. Their pyridoxine-requiring mutant was found to grow on the standard medium at a rate only 9 percent of that of the normal strain. When pyridoxine (vitamin B_6) is added to the medium, the rate of growth of this strain at first increases nearly linearly with the concentration of the added pyridoxine and then increases less rapidly, as shown in Figure 1-2.[4] The growth rate of the normal strain without added pyridoxine is equal to that of the mutant with about 10 micrograms of the growth substance per liter in the medium. At a concentration about four times this value (40 micrograms per liter) the growth rate of the mutant strain reaches a value 7 percent greater than that of the normal strain without added pyridoxine.

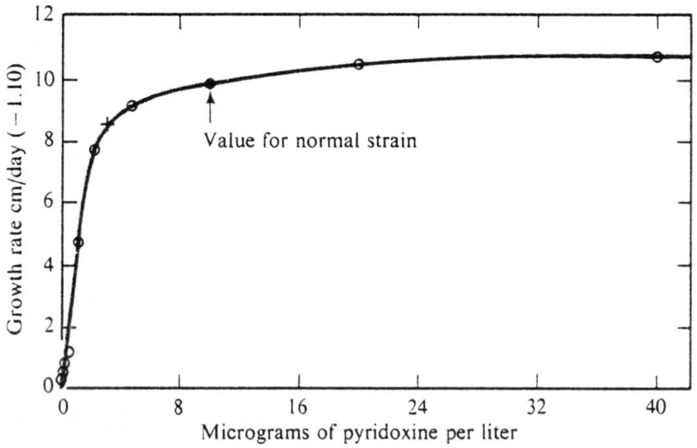

FIGURE 1-2.
The observed rate of growth of a pyridoxine-requiring *Neurospora* mutant (Beadle and Tatum, 1941), as function of the concentration of pyridoxine in the medium.

The point of maximum curvature of the curve in Figure 1-2, at about 3.2 micrograms of pyridoxine per liter (indicated by a cross), may be reasonably considered to mark the division between the region of vitamin deficiency (to the left) and the region of normal vitamin supply (to the right), such as might permit the mutant to compete with the wild type, which has the growth rate represented by the filled circle in Figure 1-2. The point marked by the cross might well correspond to an "adequate" or "recommended" amount of the vitamin, in that the growth rate of the mutant is only

[4] The points in Figure 1-2 represent my measurement of the slopes of the growth curves shown in Figure 1 of Beadle and Tatum (1941). They agree closely with the points of their Figure 2, except for one point, that for 1.2 μg/liter, which may have been misplotted.

12 percent less than that of the wild strain, and that the amount of the vitamin would have to be increased threefold to make up this 12 percent.[5]

As shown in Figure 1-2, quadrupling the concentration of pyridoxine that gives the mutant a growth rate equal to that of the wild type causes a further increase in growth rate by nearly 10 percent. The growth rates of the mutant and the wild type at very large concentrations of the vitamin have not been measured, so far as I know, and the optimum concentration is not known. From the work of Beadle and Tatum (1941) the optimum concentration may be taken to be greater than 40 micrograms per liter; that is, more than ten times the "adequate" concentration for the mutant and more than four times the concentration equivalent to the synthesizing capability of the wild type. The growth rate of the mutant at the optimum concentration is more than 22 percent greater than that at the "adequate" concentration and more than 9 percent greater than that of the normal strain.

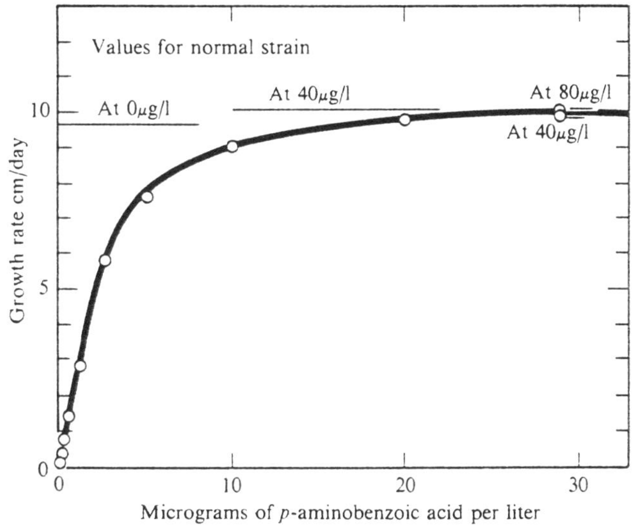

FIGURE 1-3.
The observed rate of growth of a *p*-aminobenzoic-acid-requiring *Neurospora* mutant (Tatum and Beadle, 1942), as function of concentration of the growth substance in the medium.

Similar results have been reported for other mutants of *Neurospora*. The values found by Tatum and Beadle (1942) for a *p*-aminobenzoic-acid-requiring mutant of *Neurospora crassa* as a function of the concentration of *p*-aminobenzoic acid added to the standard medium are shown in Figure 1-3. The growth-rate curve is similar in

[5] The reported growth rate for the normal strain in a medium with 40 μg of added pyridoxine per liter is 3 percent greater than that for the basic medium, as shown by the slopes of the lines in Figure 1 of Beadle and Tatum (1941).

shape to that for the pyridoxine-requiring mutant. The value of the growth rate for the normal strain of *Neurospora crassa* with no added *p*-aminobenzoic acid is equal to that for the mutant at a concentration of added *p*-aminobenzoic acid of about 15 micrograms per liter. A value about 4 percent greater is found for the normal strain at 40 micrograms per liter and for the mutant strain at 80 micrograms per liter, as indicated in Figure 1-3.

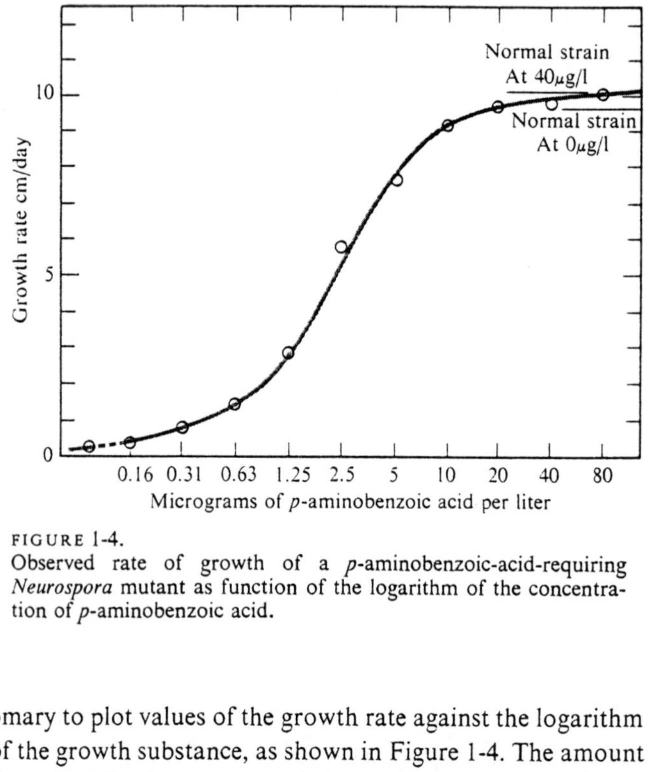

FIGURE 1-4.
Observed rate of growth of a *p*-aminobenzoic-acid-requiring *Neurospora* mutant as function of the logarithm of the concentration of *p*-aminobenzoic acid.

It is customary to plot values of the growth rate against the logarithm of the concentration of the growth substance, as shown in Figure 1-4. The amount of increase accompanying a doubling in the concentration of the growth substance is a maximum at 1.25 to 2.5 micrograms per liter, and decreases thereafter to about half the value for each successive doubling.

From these two examples we see that there may be a significant increase in rate of growth of the normal strain through addition of some of the growth substance that it synthesizes to the medium in which it is grown; that is, that the amount of the growth substance that is synthesized by the normal strain is not the optimum amount, but is somewhat less, leading to a rate of growth approximately 7 percent less than the maximum in the case of pyridoxine (with the normal strain of *Neurospora sitophila*) and 4 percent less for *p*-aminobenzoic acid (with the normal strain of *Neurospora crassa*). Many other examples are known of microorganisms that grow more abundantly in a medium containing vitamins, amino acids, or other substances that they are able to synthesize than on a minimum medium.

Evidence supporting the above arguments has been presented recently by Zamenhof and Eichhorn (1967) in a paper entitled "Study of microbial evolution through loss of biosynthetic functions: Establishment of 'defective' mutants." These authors carried out experiments involving competitive growth in a chemostat of an auxotrophic mutant (a mutant requiring a nutrilite) and a prototrophic parent in a medium of constant composition containing the nutrilite. They found that the "defective" mutant has a selective advantage over the prototrophic parental strain under these conditions. For example, an indole-requiring mutant of *Bacillus subtilis* was found to show a strong selective advantage over the prototrophic back-mutant when the two were grown together in a medium containing tryptophan: the relative number of cells of the latter decreased 10^6-fold in 54 generations. They also found that greater advantage to the auxotroph accompanies a greater number of biosynthetic steps that have been dispensed with (earlier block in a series of reactions), with the final metabolite available. They point out that a mutant with a gene deletion would be at a distinct selective advantage over a point mutant, in that not only the synthesis of the metabolite, but also that of the structural gene, the messenger RNA, and perhaps the inactive enzyme itself would be dispensed with, and that accordingly the mutant with a deletion would replace the point mutant in competition. They mention evidence that some of the "defective" strains occurring in nature have lost one or more of their structural genes by deletions, rather than by point mutations.

MOLECULAR CONCENTRATIONS AND RATE OF REACTION

Most of the chemical reactions that take place in living organisms are catalyzed by enzymes. The mechanisms of enzyme-catalyzed reactions in general involve (1) the formation of a complex between the enzyme and a substrate molecule, and (2) the decomposition of this complex to form the enzyme and the products of the reaction. The rate-determining step is usually the decomposition of the complex to form the products or, more precisely, the transition through an intermediate state of the complex, characterized by activation energy less than for the uncatalyzed reaction, to a complex of the enzyme and the products of reaction, with a rapid dissociation. Under conditions such that the concentration of the complex corresponds to equilibrium with the enzyme and the substrate, the rate of the reaction is given by the following equation (the Michaelis–Menten equation; Michaelis and Menten, 1913):

$$R = \frac{d[S]}{dt} = \frac{kE[S]}{[S] + (1/K)}. \qquad (1)$$

In this equation $[S]$ is the concentration of the substrate, E is the total concentration of enzyme (present both as free enzyme and enzyme complex), K is the equilibrium constant for formation of the enzyme complex ES, and k is the reaction-rate constant for decomposition of the complex to form the enzyme and reaction products. This equation corresponds to the case in which there are no enzyme inhibitors present.

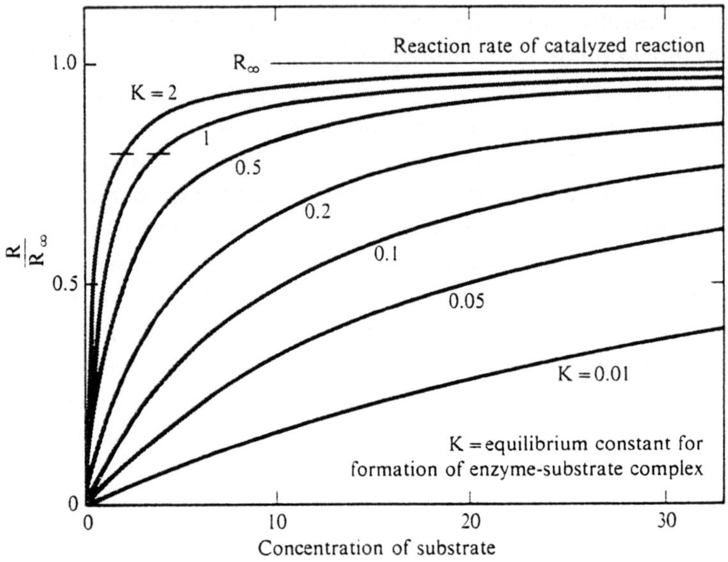

FIGURE 1-5.
Curves showing calculated reaction rate R/R_∞ of catalyzed reaction as function of the concentration of the substrate, for different values of the equilibrium constant K for formation of the enzyme-substrate complex.

Values of the reaction rate calculated from this equation for different values of K are shown in Figure 1-5. The curves are similar in shape to those of Figures 1-2 and 1-3. At concentrations much smaller than K^{-1} the reaction rate is proportional to the concentration of substrate. At larger concentrations, as the amount of enzyme complex becomes comparable to the amount of free enzyme, the reaction rate changes from the linear dependence. At substrate concentration equal to K^{-1} the slope of the curve is one-quarter of the initial slope, and the value is one-half of the value corresponding to saturation of the enzyme by the substrate.

The similarity of the curves of Figures 1-2 and 1-3 to appropriate curves in Figure 1-5 suggests that the growth substance may be involved in an enzyme-catalyzed reaction in which it serves as the substrate. The normal strain of the organism manufactures an amount of the substrate such as to permit the reaction to take place at what may be considered a normal rate, 90 or 95 percent of the maximum rate, which corresponds to saturation of the enzyme. As described above, the gain in reaction rate associated with the manufacture of a larger amount of the substrate, with a corresponding advantage to the organism, might be balanced by the disadvantage to the organism associated with the upkeep of the larger amount of machinery required to manufacture the increased amount of substrate. An increase in rate of this reaction could also be achieved by an increase in the amount of the enzyme synthesized by the organism. Here, again, the advantage to the organism resulting from this increase may be overcome by the disadvantage associated with the increase in the amount of machinery required for the increased synthesis. During the process of evolution there has presumably been selection of genes determining the concentrations of the enzymes catalyzing successive reactions such as to achieve an approximation to the optimum reaction rate with the smallest amount of disadvantage to the organism.

The rate of an enzyme-catalyzed reaction is approximately proportional to the concentration of the reactant, until concentrations that largely saturate the enzyme are reached. The saturating concentration is larger for a defective enzyme with decreased combining power for the substrate than for the normal enzyme. For such a defective enzyme the catalyzed reaction could be made to take place at or near its normal rate by an increase in the substrate concentration, as indicated in Figure 1-5. The short horizontal lines intersecting the curves indicate what may be called the "normal" reaction rate, 80 percent of the maximum. For $K = 2$ the "normal" rate is achieved at substrate concentration $[S] = 2$. At this substrate concentration the reaction rate is only 29 percent of the maximum and 35 percent of "normal" for a mutated enzyme with $K = 0.2$; it could be raised to the "normal" value by a tenfold increase in the substrate concentration, to $[S] = 20$. Similarly, the still greater disadvantage of low reaction rate for a mutated enzyme with K only 0.01 could be overcome by a 200-fold increase in substrate concentration, to $[S] = 400$. This mechanism of action of gene mutation is only one of several that lead to disadvantageous manifestations that could be overcome by an increase, perhaps a great increase, in the concentration of a vital substance in the body. These considerations obviously suggest a rationale for megavitamin therapy.

MOLECULAR CONCENTRATIONS AND MENTAL DISEASE

The functioning of the brain and nervous tissue is more sensitively dependent on the rate of chemical reactions than the functioning of other organs and tissues. I believe that mental disease is for the most part caused by abnormal reaction rates, as determined by genetic constitution and diet, and by abnormal molecular concentrations of essential substances. The operation of chance in the selection for the child of half of the complement of genes of the father and mother leads to bad as well as to good genotypes, and the selection of foods (and drugs) in a world that is undergoing rapid scientific and technological change may often be far from the best. Significant improvement in the mental health of many persons might be achieved by the provision of the optimum molecular concentrations of substances normally present in the human body. Among these substances, the essential nutrilites may be the most worthy of extensive research and more thorough clinical trial than they have yet received. One important example of an essential nutrilite that is required for mental health is vitamin B_{12}, cyanocobalamin. A deficiency of this vitamin, whatever its cause (pernicious anemia; infestation with the fish tapeworm *Diphyllobothrium*, whose high requirement for the vitamin results in deprivation for the host; excessive bacterial flora, also with a high vitamin requirement, as may develop in intestinal blind loops), leads to mental illness, often even more pronounced than the physical consequences. The mental illness associated with pernicious anemia (a genetic defect leading to deficiency of the intrinsic factor [a mucoprotein] in the gastric juice and the consequent decreased transport of cyanocobalamin into the blood) often is observed for several years in patients with this disease before any of the physical manifestations of the disease appear (Smith, 1950). A pathologically low concentration of cyanocobalamin in the serum of the blood has been reported to occur for a much larger fraction of patients with mental illness than for the general population. Edwin et al. (1965) determined the amount of B_{12} in the serum of every patient over 30 years old admitted to a mental hospital in Norway during a period of 1 year. Of the 396 patients, 5.8 percent (23) had a pathologically low concentration, less than 101 picograms per milliliter, and the concentration in 9.6 percent (38) was subnormal (101 to 150 picograms per milliliter). The normal concentration is 150 to 1300 picograms per milliliter. The incidence of pathologically low and subnormal levels of B_{12} in the serums of these patients, 15.4 percent, is far greater than that in the general population, about 0.5 percent (estimated from the reported frequency of pernicious anemia in the area, 9.3 per 100,000 persons per year). Other investigators[6] have also reported a higher incidence of low B_{12} concentrations in the serums of mental patients than in the population as a whole, and have suggested that B_{12} deficiency, whatever its origin, may lead to mental illness.

Nicotinic acid (niacin), when its use was introduced, cured hundreds of thousands of pellagra patients of their psychoses, as well as of the physical manifestations of their disease. For this purpose only small doses are required; the recommended daily allowance (National Research Council) is 12 milligrams per day (for a 70-kilogram male). In 1939 Cleckley et al. reported the successful treatment of 19 patients, and

[6] Hansen et al. (1966) report serum B_{12} concentration below 150 pg/ml in 13 of 1,000 consecutive patients admitted to a Copenhagen psychiatric clinic. Henderson et al. (1966) report that 9 of 1,012 unselected psychiatric patients in a region in Scotland were found to have B_{12} deficiency, in addition to 5 pernicious anemia patients in the group.

in 1941 Sydenstricker and Cleckley[7] reported similarly successful treatment of 29 patients with severe psychiatric symptoms by use of moderately large doses of nicotinic acid (0.3 to 1.5 grams per day). None of these patients had physical symptoms of pellagra or any other avitaminosis. More recently many other investigators have reported on the use of nicotinic acid and nicotinamide for the treatment of mental disease. Outstanding among them are Hoffer and Osmond, who since 1952 have advocated and used nicotinic acid in large doses, in addition to the conventional therapy, for the treatment of schizophrenia (Hoffer et al., 1957; Hoffer, 1962, 1966; Osmond and Hoffer, 1962; Hoffer and Osmond, 1964). The dosage recommended by Hoffer is 3 to 18 grams per day, as determined by the response of the patient, of either nicotinic acid or nicotinamide, together with 3 grams per day of ascorbic acid. Nicotinic acid and nicotinamide are nontoxic (the lethal dose, 50 percent effective [LD_{50}], is not known for humans, but probably it is over 200 grams; the LD_{50} for rats is 7.0 grams per kilogram for nicotinic acid and 1.7 grams per kilogram for nicotinamide), and their side effects, even in continued massive doses, seem not to be commonly serious. Among the advantages of nicotinic acid, summarized by Osmond and Hoffer (1962), are the following: it is safe, cheap, and easy to administer, and it is a well-known substance that can be taken for years on end, if necessary, with only small probability of incidence of unfavorable side effects.

Another vitamin that has been used to some extent in the treatment of mental disease is ascorbic acid, vitamin C. A sometimes-recommended daily intake of ascorbic acid is 75 milligrams for healthy adults. Some investigators have estimated that the optimum intake is much larger (Kyhos et al., 1945), perhaps 3 to 15 grams per day, according to Stone (1966, 1967). Williams and Deason (1967) have emphasized the variability of individual members of a species of animals; they have reported their observation of a 20-fold range of required intake of ascorbic acid by guinea pigs, and have suggested that human beings, who are less homogeneous, have a larger range.

Mental symptoms (depression) accompany the physical symptoms of vitamin-C deficiency disease (scurvy). In 1957, Akerfeldt reported that the serum of schizophrenics had been found to have greater power of oxidizing N,N-dimethyl-*p*-phenylenediamine than that of other persons. Several investigators then reported that this difference is due to a smaller concentration of ascorbic acid in the serum of schizophrenics than of other persons. This difference has been attributed to the poor diet and increased tendency to chronic infectious disease of the patients (Benjamin, 1958; Kety, 1959), and has also been interpreted as showing an increased rate of metabolism of ascorbic acid by the patients (Hoffer and Osmond, 1960; Briggs, 1962). It is my opinion, from the study of the literature, that many schizophrenics have an increased metabolism of ascorbic acid, presumably genetic in origin, and that the

[7] References are given in this paper to some earlier work on nicotinic acid therapy.

ingestion of massive amounts of ascorbic acid has some value in treating mental disease.

Other vitamins (thiamine, pyridoxine, folic acid) and other substances (zinc ion, magnesium ion, uric acid, tryptophan, L(+)-glutamic acid, and others) influence the functioning of the brain. I shall review work on L(+)-glutamic acid as a further example. L(+)-Glutamic acid is an amino acid that is present at rather high concentration in brain and nerve tissue and plays an essential role in the functioning of these tissues (Weil-Malherbe, 1936). It is normally ingested (in protein) in amounts of 5 to 10 grams per day. It is not toxic; large doses may cause increased motor activity and nausea. In 1943 Price et al. reported favorable results for glutamic acid therapy of convulsive disorders (benefit to one out of three or four patients with petit mal epilepsy; Waelsch, 1948). Zimmerman and Ross (1944) then reported an increase in maze-running learning ability of white rats given extra amounts of glutamic acid. Zimmerman and many other investigators then studied the effects of glutamic acid on the intelligence and behavior of persons with different degrees and kinds of mental retardation. L(+)-Glutamic acid is apparently more effective than its sodium or potassium salts. The effective dosage is usually between 10 and 20 grams per day (given in three doses with meals), and is adjusted to the patient as the amount somewhat less than that required to cause hyperactivity. Several investigators[8] have reported an improvement in personality and increase in intelligence (by 5 to 20 I.Q. points) for many patients with mild or moderate mental deficiency.

LOCALIZED CEREBRAL DEFICIENCY DISEASES

The observation that the psychosis associated with pernicious anemia may manifest itself in a patient for several years before the other manifestations of this disease become noticeable has a reasonable explanation: the functioning of the brain and nervous tissue is probably more sensitively dependent on molecular composition than is that of other organs and tissues. The observed high incidence of cyanocobalamin deficiency in patients admitted to a mental hospital, mentioned above, suggests that mental disease may rather often be the result of this deficiency, and further suggests that other deficiencies in vital substances may be wholly or partly responsible for many cases of mental illness.

The foregoing arguments suggest the possibility that under certain circumstances a deficiency disease may be localized in the human body in such a way that only some of the manifestations usually associated with the disease are present. Let us consider, for example, a vitamin or other vital substance that is normally metabolized by the

[8] A recent survey of the role of glutamic acid in cognitive behaviors has been published by Vogel et al. (1966). Many references to earlier work are given in this paper.

catalytic action of an enzyme normally present in the tissues and organs of the body. In a person of unusual genotype there might be an especially great concentration of this enzyme in one body organ, with essentially the normal amount in other organs. Through the action of this enzyme in especially great concentration the steady-state concentration of the vital substance in that organ might be decreased to a level much lower than that required for normal function. Under these circumstances there would be present a deficiency disease restricted to that organ.

An especially important case is that of the brain. We may, as a rough model of the human body, consider two reservoirs of fluid, the blood and lymph, with volume V_1, and cerebrospinal fluid, the extracellular fluid of the brain and spinal column, with volume V_2. We assume that a vital substance is destroyed in each of these reservoirs at a characteristic rate, corresponding to the rate constants k_1 and k_2, that it diffuses across the blood-brain barrier at a rate determined by the product of the permeability and area of the barrier and the difference $c_2 - c_1$ of the concentrations in the two reservoirs, and that it is introduced from the gastrointestinal tract into the first reservoir at a constant rate. The steady-state concentrations are then in the ratio

$$c_1/c_2 = 1 + (k_2V_2/PA)$$

where PA is the product of permeability and the area of the blood-brain barrier. The steady state corresponds to the following system:

$$\begin{array}{ccccc} \text{Supply} \longrightarrow & \text{Blood } (c_1) & \underset{PAc_2}{\overset{PAc_1}{\rightleftarrows}} & \text{Brain } (c_2) \\ & \downarrow k_1c_1 & & \downarrow k_2c_2 \\ & \text{Inactive product} & & \text{Inactive product} \end{array}$$

From this equation it is seen, as shown also in Figure 1-6, that for small values of k_2V_2/PA the difference in steady-state concentrations in the cerebrospinal fluid and the blood is small, but that through either decrease in permeability of the barrier or increase in the metabolic rate constant k_2 the steady-state concentration in the brain becomes much less than that in the blood.

This simple argument leads us to the possibility of a localized cerebral avitaminosis or other localized cerebral deficiency disease. There is the possibility that some human beings have a sort of cerebral scurvy, without any of the other manifestations, or a sort of cerebral pellagra, or cerebral pernicious anemia. It was pointed out by Zuckerkandl and Pauling (1962) that every vitamin, every essential amino acid, every other essential nutrilite represents a molecular disease (Pauling et al., 1949) which our distant ancestors learned to control, when it began to afflict them, by selecting a therapeutic diet, and which has continued to be kept under control in this way. The localized deficiency diseases described above are also molecular diseases, compound molecular diseases, involving not only the original lesion, the loss of the ability to

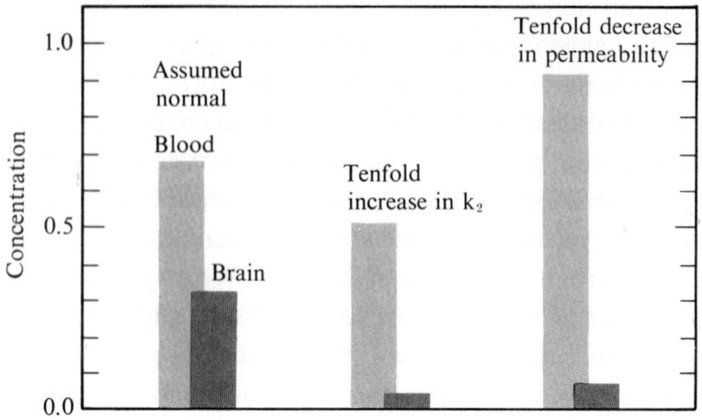

FIGURE 1-6.
Values of the concentration of a vital substance in the blood and in the cerebrospinal fluid for three different assumed sets of values of blood-brain barrier permeability and rate of destruction in the cerebrospinal fluid.

synthesize the vital substance, but also another lesion, one that causes a decreased rate of transfer across a membrane, such as the blood-brain barrier,[9] to the affected organ, or an increased rate of destruction of the vital substance in the organ, or some other perturbing reaction.

It has been suggested by Huxley et al. (1964), partially on the basis of the observations of Böök (1953, 1958) and Slater (1958) on the incidence of schizophrenia in relatives of schizophrenics, that schizophrenia is caused by a dominant gene with incomplete penetrance. They suggested that the penetrance, about 25 percent, may in some cases be determined by other genes and in some cases by the environment. I suggest that the other genes may, in most cases, be those that regulate the metabolism of vital substances, such as ascorbic acid, nicotinic acid or nicotinamide, pyridoxine, cyanocobalamin, and other substances mentioned above. The reported success in treating schizophrenia and other mental illnesses by use of massive doses of some of these vitamins may be the result of successful treatment of a localized cerebral deficiency disease involving the vital substances, leading to a decreased penetrance of the gene for schizophrenia. There is a possibility that the so-called gene for schizophrenia is itself a gene affecting the metabolism of one or another of these vital substances, or even of several vital substances, causing a multiple cerebral deficiency.

I suggest that the orthomolecular treatment of mental disease, to be successful, should involve the thorough study of and attention to the individual, such as is

---

[9] It has been suggested by Melander and Martens (1958, 1959) and by Hoffer and Osmond (1966) that the effects of taraxein (Heath et al., 1958) may result from changing the permeability of the blood-brain barrier.

customary in psychotherapy but less customary in conventional chemotherapy. In the course of time it should be possible to develop a method of diagnosis (measurement of concentrations of vital substances) that could be used as the basis for determining the optimum molecular concentrations of vital substances for the individual patient and for indicating the appropriate therapeutic measures to be taken. My coworkers and I are carrying on some experimental studies suggested by the foregoing considerations, and hope to be able before long to communicate some of our results.

SUMMARY

The functioning of the brain is affected by the molecular concentrations of many substances that are normally present in the brain. The optimum concentrations of these substances for a person may differ greatly from the concentrations provided by his normal diet and genetic machinery. Biochemical and genetic arguments support the idea that orthomolecular therapy, the provision for the individual person of the optimum concentrations of important normal constituents of the brain, may be the preferred treatment for many mentally ill patients. Mental symptoms of avitaminosis sometimes are observed long before any physical symptoms appear. It is likely that the brain is more sensitive to changes in concentration of vital substances than are other organs and tissues. Moreover, there is the possibility that for some persons the cerebrospinal concentration of a vital substance may be grossly low at the same time that the concentration in the blood and lymph is essentially normal. A physiological abnormality such as decreased permeability of the blood-brain barrier for the vital substance, or increased rate of metabolism of the substance in the brain, may lead to a cerebral deficiency and to a mental disease. Diseases of this sort may be called localized cerebral deficiency diseases. It is suggested that the genes responsible for abnormalities (deficiencies) in the concentration of vital substances in the brain may be responsible for increased penetrance of the postulated gene for schizophrenia, and that the so-called gene for schizophrenia may itself be a gene that leads to a localized cerebral deficiency in one or more vital substances.

REFERENCES

Akerfeldt, S. A. (1957). *Science* **125**, 117.
Beadle, G. W., and Tatum, E. L. (1941). *Proc. Nat. Acad. Sci USA* **27**, 499.
Benjamin, J. D. (1958). *Psychosomatic Med.* **20**, 427.

Böök, J. A. (1953). *Act. Genet. Statist. Med.* **4.**

Böök, J. A. (1958). *Proc. Int. Congr. Genet. 10th* **1,** 81.

Briggs, M. H. (1962). *New Zealand Med. J.* **61,** 229.

Cherkin, A. (1967). *Science* **155,** 266.

Cleckley, H. M., Sydenstricker, V. P., and Geeslin, L. E. (1939). *J. Am. Med. Assoc.* **112,** 2107.

Edwin, E., Holten, K., Norum, K. R., Schrumpf, A., and Skaug, O. E. (1965). *Acta Med. Scand.* **177,** 689.

Følling, A. (1934a). *Nord. Med. Tidskr.* **8,** 1054.

Følling, A. (1934b). *Z. Physiol. Chem.* **277,** 169.

Hansen, T., Rafaelson, O. J., and Rødbro, P. (1966). *Lancet* **II,** 965.

Heath, R. G., Martens, S., Leach, B. E., Cohen, M., and Feigley, C. A. (1958). *Am. J. Psychiatr.* **114,** 917.

Henderson, J. G., Strachan, R. W., Beck, J. S., Dawson, A. A., and Daniel, M. (1966). *Lancet* **II,** 809.

Hoffer, A. (1962). *Niacin Therapy in Psychiatry.* Springfield, Ill.: C. C. Thomas.

Hoffer, A. (1966). *Int. J. Neuropsychiatr.* **2,** 234.

Hoffer, A., Osmond, H., Callbeck, M. J., and Kahan, I. (1957). *J. Clin. Exp. Psychopathol.* **18,** 131.

Hoffer, A., and Osmond, H. (1960). *The Chemical Basis of Clinical Psychiatry,* p. 232. Springfield, Ill.: C. C. Thomas.

Hoffer, A., and Osmond, H. (1964). *Acta Psychiatr. Scand.* **40,** 171.

Hoffer, A., and Osmond, H. (1966). *Int. J. Neuropsychiatr.* **2,** 1.

Huxley, J., Mayr, E., Osmond, H., and Hoffer, A. (1964). *Nature* **204,** 220.

Kety, S. S. (1959). *Science* **129,** 1528, 1590.

Kyhos, E. D., Sevringhaus, E. L., and Hagedorn, D. R. (1945). *Arch. Int. Med.* **75,** 407.

Melander, B., and Martens, S. (1958). *Dis. Nerv. Sys.* **19,** 478.

Melander B., and Martens, S. (1959). *Acta Psychiatr. Neurol. Scand.* **34,** 344.

Michaelis, L., and Menten, M. (1913). *Biochem. Z.* **49,** 333.

Miller, S. (1961). *Proc. Nat. Acad. Sci. USA* **47,** 1515.

Osmond, H., and Hoffer, A. (1962). *Lancet* **I,** 316.

Pauling, L. (1961). *Science* **134,** 15.

Pauling, L., Itano, H. A., Singer, S. J., and Wells, I. C. (1949). *Science* **110,** 543.

Price, J. G., Waelsch, H. and Putnam, T. J. (1943). *J. Am. Med. Assoc.* **122,** 1153.

Slater, I. E. (1958). *Acta Gen. Stat. Med.* **8,** 50.

Smith, A. D. M. (1950). *Brit. Med. J.* **11,** 1840.

Stone, I. (1965). *Am. J. Phys. Anthropol.* **23,** 83.

Stone, I. (1966). *Acta Genet. Med. Gemell.* **15,** 345.

Stone, I. (1967). *Perspect. Biol. Med.* **10,** 135.

Sydenstricker, V. P., and Cleckley, H. M. (1941). *Am. J. Psychiatr.* **99,** 83.

Tatum, E. L., and Beadle, G. W. (1942). *Proc. Nat. Acad. Sci. USA* **28,** 234.

Vogel, W., Broverman, D. M., Draguns, J. G., and Klaiber, E. L. (1966). *Psychol. Bull.* **65,** 367.

Waelsch, H. (1948). *Am. J. Ment. Defic.* **52,** 305.

Weil-Malherbe, H. (1936). *Biochem. J.* **30,** 665.

Williams, R. J., and Deason, G. (1967). *Proc. Nat. Acad. Sci. USA* **57,** 1638.

Woolley, D. W. (1962). *The Biochemical Bases of Psychoses.* New York: John Wiley & Sons, Inc.

Zamenhof, S., and Eichhorn, H. H. (1967). *Nature* **216,** 465.

Zimmerman, F. T., and Ross, S. (1944). *Arch. Neurol. Psychiatr.* **51,** 446.

Zuckerkandl, E., and Pauling, L. (1962). *In* M. Kasha, and B. Pullman, eds. *Horizons in Biochemistry,* p. 189. New York: Academic Press.

Oncology *27:* 181–192 (1973)

Ascorbic Acid and the Glycosaminoglycans

An Orthomolecular Approach to Cancer and Other Diseases

EWAN CAMERON and LINUS PAULING

Vale of Leven Hospital, Alexandria, Dunbartonshire,
and Department of Chemistry, Stanford University, Stanford, Calif.

Abstract. A new concept of a basic mechanism involved in cell proliferation is presented. It is suggested that cells are normally restrained from proliferating by the highly viscous nature of the intercellular glycosaminoglycans. In order to proliferate, cells must escape from this restraint by depolymerizing the glycosaminoglycans in their immediate environment. This process is accomplished by the release of the enzyme hyaluronidase and is kept in check by physiological hyaluronidase inhibitor. There is some evidence that physiological hyaluronidase inhibitor is an oligoglycosaminoglycan that requires ascorbic acid for its synthesis, and perhaps incorporates residues of ascorbic acid. This hypothesis provides an explanation for the pathogenesis of scurvy. It explains the increased requirement for ascorbic acid that occurs in many cell proliferative diseases, including cancer. It indicates the existence of a basic underlying mechanism in many pathological states and suggests a common pattern of treatment. We conclude that ascorbic acid may have much greater therapeutic value than has been generally assigned to it.

Key Words
Vitamin C/hyaluronidase
Hyaluronidase inhibitor
Intercellular ground substance

Introduction

Orthomolecular medicine is the preservation of good health and the treatment of disease by varying the concentrations in the human body of substances that are normally present in the body and are required for health [22, 23, 25]. Of these substances, the vitamins are especially important, and ascorbic acid, in particular, may have much greater value than has been generally ascribed to it. IRWIN STONE [31–33] has advanced arguments to support the concept that the optimum rate of intake of ascorbic acid is about 3 g per day under ordinary conditions, and larger, up to 40 g per day, for a person under stress. An argument based on the fact that only a few animal species require an exogenous source of ascorbic acid and on the amounts of ascorbic acid contained in a diet of raw natural plant food has led to the conclusion that the optimum daily intake of ascorbic acid for an adult human being is about 2.3 g or more [24]. Ascorbic acid is a substance with extremely low toxicity; many people have ingested 10–20 g per day for long periods without serious side effects, and ingestion of as much as 150 g within 24 h without serious side effects has been reported [16]. In this respect ascorbic acid may be considered an ideal substance for orthomolecular prophylaxis and therapy.

Ground Substance

Cells in the tissues of the body are embedded in ground substance. This ubiquitous material pervades every interspace and isolates every stationary cell from its neighbours. It must be traversed by every molecule entering or leaving the cell. There is evidence that the interface between a cell membrane and the immediate extracellular environment is the crucial factor in the whole proliferative process. Variations in the composition of the extracellular environment exert a profound influence on cell behavior, and in turn the cells possess a powerful means of modifying their immediate environment. This interdependence is involved in all forms of cell proliferation and is particularly important in cancer. A proliferating cell and its immediate environment constitute a balanced system in which each component influences the other. Recognition of this relationship and an understanding of the means of controlling it could lead to rational methods of treating cancer and other cell-proliferative diseases.

Until recently cancer research has tended to concentrate almost exclusively upon the cell, and to ignore the other half of the proliferation equation. The intercellular substance is a complex gel, containing water, electrolytes, metabolites, dissolved gases, trace elements, vitamins, enzymes, carbohydrates, fats, and proteins. The solution is rendered highly viscous by an abundance of certain long-chain acid mucopolysaccharide polymers, the glycosaminoglycans and the related proteoglycans, reinforced at the microscopic level by a three-dimensional network of collagen fibrils. The principal glycosaminoglycans so far identified are hyaluronic acid, a long-chain polymer with high molecular weight (200,000–500,000) and simple chemical structure (alternating residues of N-acetylglucosamine and glucuronic acid), and varieties of chrondroitin (alternating residues of N-acetylgalactosamine and glucuronic acid) and its sulfate esters. Other glycosaminoglycans may also be present. The chemistry of the ground substance and the intercellular environment has been reviewed recently by BALAZS [1].

An important property of the intercellular substance is its very high viscosity and cohesiveness. This property is dependent upon the chemical integrity of the large molecules. The viscosity can be reduced and the structural integrity destroyed by the depolymerizing (hydrolyzing) action of certain related enzymes (the endohexosaminidases, the β-N-acetylglucosaminidases, the β-N-acetylgalactosaminidases, and β-glucuronidase), known by the generic name 'hyaluronidase'. It is probable that most cells in the body are able to produce hyaluronidase [3, 10]. The interlacing molecular network of the intercellular ground substance is in a constant state of slow dynamic change, with synthesis of glycosaminoglycans (polymerization) balanced by their breakdown (depolymerization by hydrolysis) through the catalytic action of hyaluronidase, and subsequent excretion. It is within this slowly changing environment, called the 'milieu interieur' by CLAUDE BERNARD, that all cellular activity takes place. The normal cell and the cancer cell both thrive and die within this environment.

Hyaluronidase and Cell Proliferation

Some years ago the hypothesis was advanced that all forms of cell proliferation depend upon one fundamental interaction between the cell and its immediate environment [5]. The hypothesis may be stated as follows: All cells in the body are embedded in a highly viscous environment of ground substance that physically restrains their inherent tendency to proliferate; proliferation is initiated by release of hyaluronidase from the cells, which catalyzes the hydrolysis of the glycosaminoglycans in the immediate environment and allows the cells freedom to divide and to migrate within the limits of the alteration; proliferation continues as long as hyaluronidase is being released, and stops when the production of hyaluronidase stops or when the hyaluronidase is inhibited, and the environment is allowed to revert to its normal restraining state.

In normal healthy tissues cell division is taking place at a constant slow rate, corresponding to normal cell replacement. This normal 'background' rate of cell division results in a slow metabolic turnover of ground substance, with liberation into the blood stream and then escape in the urine of the partially depolymerized fractions of the intercellular glycosaminoglycans, produced in the immediate vicinity of the dividing cells. These degradation products of ground substance depolymerization can be recognized and measured by various biochemical methods. Depending upon the analytical procedure used different fractions have been given different names. For our purpose they may be grouped together under the general term 'serum polysaccharide'. In health the serum polysaccharide concentration remains within a relatively narrow 'normal' range [6, 35]. The process is kept in check by the presence in the tissues and the blood of a substance called 'physiological hyaluronidase inhibitor' (PHI). In health the serum PHI concentration lies within a well-defined 'normal' range [9, 19].

In conditions in which excess cell proliferation is occurring, such as inflammation, tissue repair, and cancer, depolymerization of ground substance can be demonstrated histochemically in the immediate vicinity of the proliferating cells [34], and there is also a significant increase in concentration of both the serum polysaccharide [6, 35] and the serum PHI [9, 19].

Neoplastic Cell Proliferation

It follows from this hypothesis that cancer may be no more than the permanent exhibition by some cells of a fundamental biological property possessed by all cells. We suggest that the characteristic feature of neoplastic cell proliferation is that these cells in becoming malignant have acquired, and are able to bequeath to their descendants *in perpetuo*, the ability to produce hyaluronidase continuously. Wherever they travel, these cells will always prosper, multiply, and invade within the protective independence of their own self-created depolymerized environment. These renegade cells are autonomous only because they possess this specific ability, the ability to isolate themselves permanently from 'contact' and all the usual 'controls' governing tissue organization and growth restraint.

By endowing a clone of cells with this single property of continuous hyaluronidase release it is possible to provide a reasonable explanation for many of the morphological features of malignant invasive growth [5]. The methods whereby cells might acquire this property in response to a wide variety of carcinogenic stimuli have also been outlined, together with the experimental evidence in support of the concept [5].

Therapeutic Control of Cell Proliferation

Assuming that cell proliferation depends upon depolymerization of the ground substance by cellular hyaluronidase, we see that there are two methods of exerting some therapeutic control of cancer and of other disease states in which excessive cellular proliferation is a harmful feature. We may attempt to increase the resistance of the ground substance to enzymatic depolymerization, that is, to strengthen the ground substance, or to directly neutralize the cellular hyaluronidase by decreasing its production or inhibiting its action.

Treatment by Strengthening the Ground Substance

The resistance of ground substance to the action of hyaluronidase can be increased in several ways, some of which are already established as useful methods for retarding cell proliferation.

Radiotherapy, irradiation with X-rays, is an example in which the result of the treatment is that some of the amorphous ground substance has been replaced by a dense deposit of collagen [15]. The direct cytotoxic effect of radiotherapy is thus reinforced by a permanent reduction in the susceptibility of the ground substance in the treated region to the action of hyaluronidase, with a consequent long-lasting diminution in proliferative activity.

Hormone therapy is effective because the physical-chemical state of the ground substance is profoundly influenced by many endocrine factors; the experimental evidence has been reviewed elsewhere [5]. Resistance to the action of hyaluronidase can be increased by the administration of corticosteroids, estrogens, androgens, and thyroxine, and these effects are enhanced after adrenalectomy and hypophysectomy. These hormones, although differing widely in their special effects on particular target cells, all exert to a greater or lesser extent, and roughly in the order stated, the same effect on the intercellular field, namely, the absorption of amorphous ground substance and its replacement by a more resistant fibrous substance. Without wishing to enter the current debate about the mode of action of endocrine therapy in malignant disease [30], we are content to note that these hormones are the ones used with some success in the palliation of various forms of human cancer.

Other agents may also be effective in altering the intercellular environment and indirectly exerting some controlling influence on the behavior of cells. It has been pointed out [5] that an explanation is provided of the 'Haddow paradox', that substances which are locally carcinogenic (by creating a local intensely impermeable carcinogenic environment) have also some antiproliferative value when administered systemically in experimental cancer (by bringing about similar generalized changes in the resistance of the ground substance to hyaluronidase and thus decreasing cell proliferation).

Because of the complexity of the intercellular ground substance and its responsiveness to external influences, many of the innumerable 'cancer treatments' that have been hopefully advocated year after year might have some element of truth behind them. It is also true, however, that no form of cancer treatment based on the antineoplastic effect of modification of ground substance can ever be more than palliative, because to render the ground substance totally resistant to hyaluronidase would create a situation incompatible with life itself.

Treatment by Inhibition of Hyaluronidase

Although the indirect methods of retarding cell proliferation, described above, are of great interest and in special circumstances of undoubted value, the direct inhibition of cellular hyaluronidase offers more spectacular therapeutic possibilities.

Hyaluronidase may be inhibited by drugs and by immunological methods, but the approach most likely to succeed appears to be that of utilization of the naturally occurring inhibitor, PHI.

Spontaneous regression of advanced cancer has been well documented in a number of fortunate patients as a direct consequence of massive intercurrent infection with hyaluronidase-producing bacteria [8]. A possible explanation for this remarkable phenomenon is that the depolymerizing action evoked an upsurge in the serum concentration of PHI of sufficient magnitude to inhibit totally the malignant capability of the neoplastic cells [5]. It is known that such infections are always associated with an increase in the serum PHI concentration [9, 19]. It has been independently demonstrated in experimental cancer that the injection of 'Shear's polysaccharide' induces not only carcinolysis [12] but also a sharp and significant rise in PHI concentration [11]. The problem is how to employ this suppressive mechanism in practical therapeutics.

Ascorbic Acid and Hyaluronidase Inhibitor

There is strong evidence to indicate that ascorbic acid is involved in some way in the synthesis of physiological hyaluronidase inhibitor. A strong suggestion to this effect is provided by the manifestations of scurvy, resulting from a deficiency of ascorbic acid. If ascorbic acid were required for the synthesis of PHI, a deficiency of ascorbic acid would cause the serum PHI concentration to decrease toward zero. In the absence of such control of hyaluronidase by PHI, background cellular proliferation and release of hyaluronidase would produce a steady and progressive enzymatic depolymerization of the ground substance, with disruption and disintegration of the collagen fibrils, intraepithelial cements, basement membranes, perivascular sheaths, and all the other organized cohesive structures of the tissues, producing in time the generalized pathological state of scurvy. These generalized changes, tissue disruption, ulceration, and hemorrhage, are identical to the local changes that occur in the immediate vicinity of invading neoplastic cells. This concept of scurvy, as involving uncontrolled depolymerization of ground substance, explains why scurvy is always associated with a very high level of serum polysaccharide [26]. It also explains why very small amounts of ascorbic acid have profound effects in the treatment of scurvy. The total body content of ascorbic acid is estimated to be around 5 g [7], and yet this small amount controls the health of the whole body content of intercellular material, which must amount to many kilograms of substance. The symptoms of scurvy are relieved by the ingestion of a few tenths of a gram of ascorbic acid. It seems clear that ascorbic acid is not an important constituent unit of the intercellular ground substance, as had been suggested; instead, it may well be involved in the synthesis of PHI, the circulating factor that controls intercellular homeostasis. Ascorbic acid is, of course, required for the conversion of proline to hydroxyproline, and is accordingly essential for the synthesis of collagen. It may well serve in several ways in determining the nature of tissues and the state of health of human beings.

A preparation of PHI has been reported to have molecular weight about 100,000 and to consist of 94% protein and 6% polysaccharide [20]. It is our opinion that it is the polysaccharide that has the power of combining with the active region of the enzyme and inhibiting its activity, and the following discussion is based on that opinion. PHI has a general chemical similarity to the glycosaminoglycan polymers of the ground substance [9, 19]. The PHI serum concentration rises significantly in all conditions in which excessive cell proliferation is a feature [9], but PHI is known not to be a simple breakdown product of ground-substance glycosaminoglycan. Its precise chemical composition is still unknown. It has recently been suggested [4] that it is a soluble glycosaminoglycan residue in which some or all of the glucuronic acid units are replaced by the somewhat similar molecules of ascorbic acid. The general theory of enzyme activity and the action of inhibitors [21] involves the idea that the active region of the enzyme is complementary to the intermediate complex, with the structure corresponding to the maximum of the energy curve (at the saddle-point configuration, intermediate in structure between the reactants and the products) that determines the rate of reaction. This theory requires that inhibitors of the enzyme resemble the activated complex, rather than either the reactant molecules or the product molecules. Accordingly it is unlikely that PHI would be a fragment of hyaluronic acid or a fragment of any other glycosaminoglycan. It would instead involve at least one residue of a related but different substance. It is possible that a residue of ascorbic acid resembles the activated complex, and that incorporation of such a residue would produce an altered glycosaminoglycan that could function as an inhibitor of hyaluronidase. It is also possible, however, that the chemical activity of ascorbic acid, such as its reducing power or its power to cause hydroxylation reactions to take place, could cause conversion of an oligoglycosaminoglycan into PHI. Whatever the mode of action of ascorbic acid in synthesis of PHI, whether it involves incorporation of an ascorbic acid residue or some other reaction, the therapeutic implications of the concept that ascorbic acid is involved in the synthesis of PHI are considerable.

The hypothesis that ascorbic acid is required for the synthesis of PHI and is itself destroyed in the course of the synthesis explains why in such conditions as inflammation, wound repair, and cancer the individual always appears to be deficient in ascorbic acid, on the basis of measurement of its concentration in the serum, measurement of urinary excretion, and saturation tests [2, 7, 18]. It is clear that the total body requirement of ascorbic acid has become abnormally high, as would result from an increased synthesis of PHI with incorporation or destruction of ascorbic acid.

Possible Uses of Hyaluronidase Inhibitor and Ascorbic Acid

If the basic concept of cellular proliferation is correct, PHI might be a valuable therapeutic agent in directly controlling all forms of excessive proliferation, including cancer. It is a naturally occurring substance found in the serum of all mammals [20], and should be safe and free from dangerous side effects. Determination of the chemical structure of PHI and its synthesis should not present insuperable difficulties. However, it may not be necessary to synthesize the substance. It is possible that, given enough ascorbic acid, the body could synthesize a proper quantity of PHI.

The concentration of ascorbic acid in blood plasma is about 15 mg/l when the rate of intake is about 200 mg per day. With larger rates of intake the concentration in plasma increases only slowly, because of urinary excretion, reaching about 30 mg/l for an intake of 10 g per day. HUME has reported that an oral intake of around 6 g per day is required to correct the measured leucocyte ascorbic acid deficiency encountered during the course of the common cold. Leucocytes are especially rich in ascorbic acid, and one of their functions might be to act as a mobile circulating reservoir of ascorbic acid ready to be used for local production of the protective substance PHI at any site where excessive 'inflammatory' depolymerization is taking place. The finding of HUME *et al.* [14] gives some support for this view. In their studies of myocardial infarction they found that, in addition to the usual leucocytosis, the ascorbic acid content of the circulating leucocytes undergoes a very sharp depletion accompanied by an increase in the concentration of ascorbic acid in the tissues at the site of the infarction. In the therapeutic situations here envisaged, with ascorbic acid being prescribed to control excessive cell proliferation, a daily intake of 10–50 g, or even more, and with the bulk of that administered intravenously at first, might be necessary to achieve the desired effect. MCCORMICK [17, 18] and KLENNER [16] have been advocating and using this form of treatment for many years. Their combined clinical experience indicates that very large doses of ascorbic acid, in the range mentioned above, can be given intravenously with perfect safety and with apparent benefit in a wide variety of disease states.

The hypothesis that ascorbic acid is required for synthesis of PHI and can thus control harmful depolymerization of glycosaminoglycans explains why the vitamin is effective in curing scurvy and in improving the healing of wounds. The potential therapeutic uses of this relatively simple substance may, however, be much greater.

It has been postulated for years that the administration of ascorbic acid would increase tissue resistance to bacterial and viral infections by improving the integrity of the tissues. We are now in a position to suggest that, through the action of PHI, the administration of ascorbic acid in sufficiently high dosage may provide us with a broad-spectrum agent effective against all those pathogenic bacteria, and perhaps viruses, that rely upon release of hyaluronidase to establish and spread themselves throughout the tissues. The dramatic clinical successes reported independently by MCCORMICK [17, 18] and by KLENNER [16] in a wide variety of infective states support this contention.

The effectiveness of the water-soluble anti-oxidant ascorbic acid on ground substance and especially on cell membranes may be increased by the simultaneous administration of the fat-soluble anti-oxidant vitamin E. SHAMBERGER [28, 29] has reported observations on the effectiveness of ascorbic acid and also of vitamin E, together with selenium, in inhibiting the growth in mice of tumors initiated by 7,12-dimethylbenz(α)anthracene and promoted by croton oil.

The hypothesis also indicates a safe and elegant method of control in many inflammatory and auto-immune diseases where, although the individual causes are still unknown, the essential harmful feature is always excessive cell proliferation and ground-substance depolymerization. A trial of orthomolecular doses of ascorbic acid seems justifiable and preferable to the use of corticosteroids, irradiation, and all the other indirect methods currently employed.

Most important of all, we are led to the conclusion that the administration of this harmless substance, ascorbic acid, might provide us with an effective means of permanently suppressing neoplastic cellular proliferation and invasiveness, in other words, an effective means of controlling cancer. Ascorbic acid in adequate doses might prove to be the ideal cytostatic agent. Regressions might be induced in a few patients with rapidly growing tumors with precarious blood supplies, but in the great majority the effect of the treatment is expected to be to 'disarm' rather than to 'kill' the malignant cells. 'Tumors' would remain palpable and visible on X-ray examination, but all further progressive malignant growth might be stopped. Hopefully, malignant ulcers would heal, and pain, hemorrhage, cachexia, and all the other secondary distressing features of neoplasia would be brought under control. This desirable outcome might be termed carcinostasis, with what had been neoplastic cells now rendered harmless and re-embedded in intact ground substance, subject again to normal tissue restraints, and persisting in the body in heterotopic situations as 'paleoplastic' collections of essentially normal cells. A suggestion of the possibilities of the use of ascorbic acid in the control of cancer has been provided by the report by SCHLEGEL *et al.* [27] of its effectiveness against cancer of the bladder. It is our hope that a thorough trial will be given to this safe substance, ascorbic acid, which may turn out to be the most valuable of all substances in the armamentarium of orthomolecular medicine.

References

1 BALAZS, E.A.: Chemistry and molecular biology of the intercellular matrix (Academic Press, New York 1970).

2 Bodansky, O.: Concentrations of ascorbic acid in plasma and white cells of patients with cancer and non-cancerous chronic disease. Cancer Res. *11:* 238 (1951).

3 Bollet, A.J.; Bonner, W.M., and Nance, J.L.: The presence of hyaluronidase in various mammalian tissues. J. biol. Chem. *238:* 3522–3527 (1963).

4 Cameron, E. and Rotman, D.: Ascorbic acid, cell proliferation, and cancer. Lancet *i:* 542 (1972).

5 Cameron, E.: Hyaluronidase and cancer (Pergamon Press, New York 1966).

6 Cameron, E.; Campbell, A., and Plenderleith, W.: Seromucoid in the diagnosis of cancer. Scot. med. J. *6:* 301–307 (1961).

7 Davidson, S. and Passmore, R.: Human nutrition and dietetics (William & Wilkins, Baltimore 1969).

8 Everson, T.C. and Cole, W.H.: Spontaneous regression of cancer: preliminary report. Ann. Surg. *144:* 366–383 (1956).

9 Glick, D.: Hyaluronidase inhibitor of human blood serum in health and disease. J. Mt Sinai Hosp. *17:* 207–228 (1950).

10 Grossfield, H.: Studies on production of hyaluronic acid in tissue culture. The presence of hyaluronidase in embryo extract. Exp. Cell Res. *14:* 213–216 (1958).

11 Hadidian, Z.; Mahler, I.R., and Murphy, M.M.: Properidine system and nonspecific inhibitor of hyaluronidase. Proc. Soc. exp. Biol. Med. *95:* 202–203 (1957).

12 Havas, H.F. and Donnelly, A.J.: Mixed bacterial toxins in the treatment of tumors. III. Effect of tumor removal on the toxicity and mortality rate in mice. Cancer Res. *21:* 17–25 (1961).

13 Hume, R.: Personal commun. (1972).

14 Hume, R.; Weyers, E.; Rowan, T.; Reid, D.S., and Hillis, W.S.: Leucocyte ascorbic acid levels after acute myocardial infarction. Brit. Heart J. *34:* 238–243 (1972).

15 Jolles, B. and Koller, P.C.: The role of connective tissue in the radiation reaction of tumours. Brit. J. Cancer *4:* 77–89 (1950).

16 Klenner, F.R.: Observations on the dose and administration of ascorbic acid when employed beyond the range of a vitamin in human pathology. J. appl. Nutr. *23:* 61–88 (1971).

17 McCormick, W.J.: Ascorbic acid as a chemotherapeutic agent. Arch. Pediat., N.Y. *69:* 151–155 (1952).

18 McCormick, W.J.: Cancer: a collagen disease, secondary to a nutritional deficiency? Arch. Pediat., N.Y. *76:* 166–171 (1959).

19 Mathews, M.B. and Dorfman, A.: Inhibition of hyaluronidase. Physiol. Rev. *35:* 381–402 (1955).

20 Newman, J.K.; Berenson, G.S.; Mathews, M.B.; Goldwasser, E., and Dorfman, A.: The isolation of the non-specific hyaluronidase inhibitor of human blood. J. biol. Chem. *217:* 31–41 (1955).

21 Pauling, Linus: Chemical achievement and hope for the future. Amer. Scient. *36:* 51–58 (1948).

22 Pauling, Linus: Orthomolecular somatic and psychiatric medicine. Z. Vitalstoffe-Zivilisationskr. *12:* 3–5 (1967).

23 Pauling, Linus: Orthomolecular psychiatry. Science *160:* 265–271 (1968).

24 Pauling, Linus: Evolution and the need for ascorbic acid. Proc. nat. Acad. Sci., Wash. *67:* 1643–1648 (1970).

25 PAULING, LINUS: The significance of the evidence about ascorbic acid and the common cold. Proc. nat. Acad. Sci., Wash. *68:* 2678–2681 (1971).
26 PIRANI, C.L. and CATCHPOLE, H.R.: Serum glycoproteins in experimental scurvy. Arch. Path. *51:* 597–601 (1951).
27 SCHLEGEL, J.U.; PIPKIN, G.E.; MISHIMURA, R., and SCHULTZ, G.M.: The role of ascorbic acid in the prevention of bladder tumor formation. J. Urol., Baltimore *103:* 155–159 (1970).
28 SHAMBERGER, R.J.: Relationship of selenium to cancer. I. Inhibitory effect of selenium on carcinogenesis. J. nat. Cancer Inst. *44:* 931–936 (1970).
29 SHAMBERGER, R.J.: Increase of peroxidation in carcinogenesis. J. nat. Cancer Inst. *48:* 1491–1497 (1972).
30 STOLL, B.: Endocrine therapy in malignant disease (Saunders, London 1972).
31 STONE, I.: Hypoascorbemia, the genetic disease causing the human requirement for exogenous ascorbic acid. Perspect. Biol. Med. *10:* 133–134 (1966).
32 STONE, I.: On the genetic etiology of scurvy. Acta genet. med. gemellol. *15:* 345–350 (1966).
33 STONE, I.: The genetic disease hypoascorbemia. Acta genet. med. gemellol. *16:* 52–62 (1967).
34 VASILIEF, J.M.: The role of connective tissue proliferation in invasive growth of normal and malignant tissues: a review. Brit. J. Cancer *12:* 524–536 (1958).
35 WINZLER, R.J. and BEKESI, J.C.: Glycoproteins in relation to cancer; in BUSCH Methods in cancer research; vol. 2, pp. 159–202 (Academic Press, New York 1967).

Request reprints from: Dr. EWAN CAMERON, Vale of Leven Hospital, *Alexandria, Dunbartonshire, G83 OUA* (Scotland)

Proc. Natl. Acad. Sci. USA
Vol. 73, No. 10, pp. 3685–3689, October 1976
Medical Sciences

Supplemental ascorbate in the supportive treatment of cancer: Prolongation of survival times in terminal human cancer*

(vitamin C)

EWAN CAMERON† AND LINUS PAULING‡

Vale of Leven District General Hospital, Loch Lomondside, G83 0UA, Scotland; and †‡ Linus Pauling Institute of Science and Medicine, 2700 Sand Hill Road, Menlo Park, California 94025

Contributed by Linus Pauling, August 10, 1976

ABSTRACT **Ascorbic acid metabolism is associated with a number of mechanisms known to be involved in host resistance to malignant disease. Cancer patients are significantly depleted of ascorbic acid, and in our opinion this demonstrable biochemical characteristic indicates a substantially increased requirement and utilization of this substance to potentiate these various host resistance factors.**

The results of a clinical trial are presented in which 100 terminal cancer patients were given supplemental ascorbate as part of their routine management. Their progress is compared to that of 1000 similar patients treated identically, but who received no supplemental ascorbate. The mean survival time is more than 4.2 times as great for the ascorbate subjects (more than 210 days) as for the controls (50 days). Analysis of the survival-time curves indicates that deaths occur for about 90% of the ascorbate-treated patients at one-third the rate for the controls and that the other 10% have a much greater survival time, averaging more than 20 times that for the controls.

The results clearly indicate that this simple and safe form of medication is of definite value in the treatment of patients with advanced cancer.

There is increasing awareness that the progress of human cancer is determined to some extent by the natural resistance of the patient to his disease. Consequently there is growing recognition that improvement in the management of these patients could come from the development of practical supportive measures specifically designed to enhance host resistance to malignant invasive growth.

We have advanced arguments elsewhere indicating that one important factor in host resistance is the free availability of ascorbic acid (1–3). These arguments are based upon the demonstration that cancer patients have a much greater requirement for this substance than normal healthy individuals, on the realization that ascorbic acid metabolism can be implicated in a number of mechanisms known to be involved in host resistance, and finally, and most convincingly, on the published evidence that ascorbic acid can sometimes produce quite dramatic remissions in advanced human cancer (4, 5).

In this communication we present the results of a clinical trial in which 100 terminal cancer patients received supplemental ascorbate as their only definitive form of treatment and compare their progress with that of 1000 matched patients managed by the same clinicians in the same hospital who did not receive any ascorbate supplementation or any other definitive form of specific anti-cancer treatment.

Protocol

The study involved a treated group of 100 patients with terminal cancer of various kinds and a control group of 1000 untreated and matched patients. The treated group consists of 100 patients who began ascorbate treatment, as described by Cameron and Campbell (4) (usually 10 g/day, by intravenous infusion for about 10 days and orally thereafter), at the time in the progress of their disease when in the considered opinion of at least two independent clinicians the continuance of any conventional form of treatment would offer no further benefit. Fifty of the treated subjects are those described in ref. 4 and the other 50 were obtained by random selection from the alphabetical index of ascorbate-treated patients in Vale of Leven District General Hospital, where treatment of some terminal cancer patients with ascorbate has been under clinical trial since November 1971. We believe that the ascorbate-treated patients represent a random selection of all of the terminal patients in this hospital, even though no formal randomization process was used. In the random selection three patients were excluded because supplemental ascorbate had been deliberately discontinued by order of another physician, and five were excluded because matching controls could not be found for them. Patients suspected or known to have voluntarily discontinued ascorbate treatment have been retained in the group, as have those who died from some cause other than their cancer. No patient was excluded because of short survival time. Eighteen patients, marked with a plus sign in Table 1, were still alive on 10 August 1976, 16 of them clinically "well." These 100 cancer patients, given ascorbate from the presentation date in their illness when their disease process was recognized to be "untreatable" by any conventional method, comprise the treated group.

The control group was obtained by a random search of the case record index of similar patients treated by the same clinicians in Vale of Leven Hospital over the last 10 years. For each treated patient, 10 controls were found of the same sex, within 5 years of the same age, and who had suffered from cancer of the same primary organ and histological tumor type. These 1000 cancer patients comprise the control group.

The detailed case records of these 1000 were then analyzed quite independently by Dr. Frances Meuli, M.B., Ch.B. (Otago, New Zealand), who established their presentation date of "untreatability" by such conventional standards as the establishment of inoperability at laparotomy, the abandonment of any definitive form of anti-cancer treatment, or the final date of admission for "terminal care." This presentation date of untreatability corresponds to the date when ascorbate supplementation was initiated in the treated group. Comparable survival times of the 10 matched controls could then be calculated. We accept that "the presentation date of untreatability" can be influenced by many factors in individual patients, but we contend that the use of 1000 controls managed by the same clinicians in the same hospital over the last 10 years provides a sound basis for this comparative study. We record our thanks to Dr. Meuli for her unbiased and valuable contribution to this investigation.

Even though no formal process of randomization was carried

* Publication no. 63 from the Linus Pauling Institute of Science and Medicine. This is part I of a series.

Table 1. Comparison of time of survival of 100 cancer patients who received ascorbic acid and 1000 matched patients with no treatment[a]

Case no.	Primary tumor type	Sex	Age	Survival time (days)												Test case/ mean contro (%)
				Ten matched controls											Test case	
				Individuals										Mean		
1.	Stomach	F	61	12	41	5	29	85	124	8	54	21	36	38.5	121	314
2.	Stomach	M	69	8	6	3	9	4	26	8	114	15	14	20.7	12	58
3.	Stomach	F	62	15	1	72	19	19	27	35	99	76	111	47.4	9	19
4.	Stomach	F	66	4	87	7	11	3	13	12	6	34	35	21.2	18	85
5.	Stomach	M	42	8	1	74	358	9	84	14	16	16	128	70.8	258	368
6.	Stomach	M	79	45	4	12	1	9	6	12	130	4	11	23.4	43	184
7.	Stomach	M	76	22	19	12	9	14	7	15	3	5	14	12.0	142	1183
8.	Stomach	M	54	24	26	21	61	27	48	7	26	2	221	46.3	36	78
9.	Stomach	M	62	14	23	13	89	4	11	4	4	36	27	22.5	149+	622
10.	Stomach	F	69	6	19	55	2	21	8	53	11	103	17	29.5	182+	617
11.	Stomach	M	45	17	24	7	57	128	16	44	64	110	78	54.5	82	150
12.	Stomach	M	57	19	13	8	11	39	29	41	17	170	5	36.9	64	173
13.	Bronchus	M	74	16	56	29	27	67	41	25	26	6	40	33.3	39	117
14.	Bronchus	M	74	21	2	27	30	18	1	31	1	21	16	16.8	427	2542
15.	Bronchus	M	66	47	94	7	39	3	53	5	4	82	9	34.3	17	50
16.	Bronchus	M	52	35	4	70	21	126	8	46	272	39	75	69.6	460	661
17.	Bronchus	F	48	11	33	30	5	6	1	45	24	81	57	29.3	90	307
18.	Bronchus	F	64	7	1	26	13	71	14	4	30	103	2	27.1	187	690
19.	Bronchus	M	70	24	8	20	7	62	20	5	41	19	49	25.5	58	227
20.	Bronchus	M	78	32	19	39	40	24	21	43	103	2	21	34.4	52	151
21.	Bronchus	M	71	5	53	7	30	2	5	20	39	31	16	20.8	100	481
22.	Bronchus	M	70	3	2	33	24	25	35	25	62	2	63	27.4	200+	730
23.	Bronchus	M	39	42	31	74	5	88	45	28	3	15	70	40.1	42	105
24.	Bronchus	M	70	24	1	30	2	5	42	46	41	7	57	25.5	167	655
25.	Bronchus	M	70	8	34	29	24	5	4	32	129	20	51	40.7	33	81
26.	Esophagus	M	72	12	21	19	14	81	26	59	21	28	33	57.4	50	87
27.	Esophagus	F	80	2	29	6	45	48	24	13	238	56	2	46.3	43	93
28.	Colon	F	76	2	2	18	5	20	22	1	1	4	1	7.6	57	750
29.	Colon	F	58	56	39	31	15	9	11	8	10	6	62	24.7	32	130
30.	Colon	M	49	35	122	107	28	30	13	78	65	46	56	58.0	201	347
31.	Colon	M	69	48	9	7	15	30	90	26	94	38	15	37.3	1267	4343
32.	Colon	F	70	64	102	13	82	8	51	33	144	17	11	52.5	144	274
33.	Colon	F	68	9	15	40	11	17	217	163	59	18	38	38.5	170	442
34.	Colon	M	50	7	108	7	18	17	14	51	69	16	(32)	33.8	428	1266
35.	Colon	F	74	11	45	50	6	18	26	40	11	88	23	31.8	157+	494
36.	Colon	M	66	13	7	224	31	72	11	1	4	11	14	38.8	58	149
37.	Colon	F	76	23	129	8	63	60	21	28	3	15	70	43.8	123+	281
38.	Colon	F	56	24	1	30	2	5	42	46	41	7	57	25.5	861	3376
39.	Rectum	F	56	51	406	74	36	41	106	30	82	82	98	100.6	62	62
40.	Rectum	F	75	3	40	46	58	7	9	19	68	16	178	44.4	223	502
41.	Rectum	M	56	3	18	52	36	34	7	49	3	6	(13)	22.2	18	81
42.	Rectum	F	57	9	73	11	19	98	82	(184)	(97)	(89)	(47)	70.9	223	314
43.	Rectum	M	68	11	11	91	47	18	23	4	13	79	84	38.1	140+	367
44.	Rectum	M	54	52	36	10	127	18	98	6	73	11	19	45.0	198	440
45.	Rectum	M	59	15	2	78	8	98	30	140	54	233	(14)	67.2	759	1129
46.	Ovary	F	49	36	5	117	29	31	22	101	140	94	73	64.8	226	349
47.	Ovary	F	68	41	39	18	37	67	3	91	40	6	13	35.5	33	93
48.	Ovary	F	49	53	15	38	122	68	33	841	18	21	40	124.9	183	146
49.	Ovary	F	67	19	36	22	2	10	32	48	132	21	97	41.9	240+	573
50.	Ovary	F	56	49	39	22	85	160	1	86	106	99	107	75.4	123+	163
51.	Breast	F	56	1	65	26	6	2	15	19	102	71	131	43.8	4	9
52.	Breast	F	57	3	28	15	4	14	16	14	48	61	15	21.8	22	101
53.	Breast	F	53	33	183	6	190	45	29	16	45	109	34	69.0	576	835
54.	Breast	F	66	22	12	94	55	7	38	2	10	76	12	102.8	342	333
55.	Breast	F	68	107	41	69	19	17	251	101	81	50	52	78.8	567	720
56.	Breast	F	53	8	2	2	42	31	17	96	231	42	20	49.1	86	175
57.	Breast	F	75	45	175	12	91	27	5	20	11	63	73	74.2	590	795
58.	Breast	F	74	12	2	35	6	18	33	30	107	85	47	37.5	8	21
59.	Breast	F	49	3	16	62	44	1	17	93	73	5	57	37.1	35	94
60.	Breast	F	50	31	29	28	40	265	14	31	24	104	229	82.6	1644+	1990

Table 1. *(Continued)*

Case no.	Primary tumor type	Sex	Age	Survival time (days): Ten matched controls: Individuals										Mean	Test case	Test case/ mean control (%)
61.	Breast	F	53	105	73	193	159	8	127	126	167	71	42	107.1	173+	162
62.	Bladder	M	93	17	47	21	12	2	18	21	46	133	48	36.5	241	660
63.	Bladder	F	70	39	9	126	52	26	97	10	8	7	79	45.3	253	556
64.	Bladder	F	73	1	23	52	30	38	38	25	13	45	24	28.9	110	381
65.	Bladder	F	77	3	52	48	142	118	34	33	10	38	26	50.4	34	67
66.	Bladder	M	44	6	9	36	48	10	21	8	52	42	16	24.8	34	137
67.	Bladder	M	62	47	118	85	76	19	58	127	72	10	15	62.7	669+	1067
68.	Bladder	M	69	39	5	66	26	25	267	85	12	13	27	56.5	30	53
69.	Gallbladder	F	71	7	8	56	22	91	44	30	22	47	14	34.1	22	64
70.	Gallbladder	M	67	20	159	4	212	73	60	94	31	16	91	76.0	209	275
71.	Kidney (Ca)	F	71	6	2	17	83	81	55	14	114	60	106	53.8	176	327
72.	Kidney (Ca)	F	63	68	76	8	31	26	5	8	69	29	49	36.9	89	241
73.	Kidney (Ca)	F	51	16	82	27	41	65	29	8	125	(95)	(117)	60.6	147	243
74.	Kidney (Ca)	M	53	7	15	7	49	95	21	91	35	19	76	41.5	58	140
75.	Kidney (Ca)	M	55	15	13	12	16	45	48	89	95	6	83	42.2	659	1562
76.	Kidney (Ca)	M	73	25	11	209	19	30	198	31	7	30	50	61.0	293	480
77.	Kidney (Ca)	M	45	91	35	19	77	64	12	127	74	34	82	61.5	3	5
78.	Kidney (Pap)	M	69	67	74	(24)	(37)	87[b]	43[b]	21[b]	82[b]	14[b]	41[b]	49.0	24	49
79.	Kidney (Pap)	M	74	57	67	51	(491)	(127)	324	174	126[b]	179[b]	97[b]	169.3	1554+	918
80.	Lymphoma	M	40	144	41	53	29	16	20	41	279	302	103	102.8	1016+	988
81.	Lymphoma	M	65	28	68	51	56	117	138	10	36	51	142	69.7	82	118
82.	Prostate	M	47	24	14	22	23	101	53	157	123	16	80	82.3	166+	202
83.	Uterus	F	56	25	11	7	67	130	126	30	18	185	61	66.0	68	103
84.	Chondrosarcoma	M	63	20	25	3	17	136	17	31	23	19	157	44.8	9	20
85.	"Brain"	M	49	1	85	56	(187)	57	24	13	29	1	95	54.8	37	67
86.	Pancreas	M	77	11	25	19	38	91	78	13	41	40	94	45.0	317	704
87.	Pancreas	M	67	112	6	55	36	256	25	91	76	67	52	77.6	21	27
88.	Pancreas	F	60	11	42	23	49	57	69	122	253	89	59	77.4	16	21
89.	Fibrosarcoma	F	54	13	1	171	10	30	64	(101)	(9)	(25)	(17)	44.1	22	50
90.	Testicle	M	42	11	10	56	46	39	102	17	(19)	(29)	(87)	41.6	15	36
91.	Pseudomyxoma	M	47	35	16	1	19	(37)	(27)	(12)	(15)	(87)	(162)	41.1	132	321
92.	Carcinoid	F	68	19	12	45	8	31	12	18	15	82	(38)	28.0	162+	579
93.	Leiomyosarcoma	F	32	31	74	66	(28)	(87)	(121)	[21]	[44]	[27]	[242]	74.1	453+	611
94.	Leukemia	F	59	6	36	183	6	36	32	44	36	112	63	55.4	430+	776
95.	Stomach	M	55	34	34	12	78	5	253	77	79	72	49	69.3	27	39
96.	Ovary	F	51	128	13	76	31	65	216	62	140	62	40	83.3	82	98
97.	Bronchus	M	69	92	30	90	160	43	147	32	20	135	125	87.4	31	35
98.	Bronchus	F	67	93	20	29	90	97	68	185	8	37	26	65.3	138	211
99.	Colon	M	77	8	69	80	14	30	9	57	68	14	21	37.0	15	40
00.	Colon	M	38	3	41	78	17	58	40	66	98	42	(80)	52.3	152+	291

The sign + following the survival time of the patients treated with ascorbic acid means that the patient was alive on August 10, 1976. Parentheses () indicate that the matched patient had the same sex, same kind of tumor, and same dissemination, but had an age difference greater than 5 years. Brackets [] indicate opposite sex, same tumor, same dissemination, age difference greater than 5 years. For kidney, Ca indicates carcinoma, Pap, papilloma.

Diffuse urinary tract papillomatosis. The test cases (78 and 79) had lesions in both kidney and bladder. The nine control cases indicated had tumors of identical histology, but their disease was confined to bladder mucosa.

ut in the selection of our two groups, we believe that they come lose to representing random subpopulations of the population f terminal cancer patients in Vale of Leven Hospital. There some internal evidence in the data in Table 1 to support this onclusion.

A somewhat detailed description of the circumstances under hich the study was made may be called for. Of the 375 beds Vale of Leven Hospital, 100 are in the surgical unit, 50 in the nedical unit, and 25 in the gynecological unit. The 100 beds the surgical unit are in the administrative charge of Ewan ameron, and 50 of them are in his complete clinical charge, he other 50 being in the charge of the second Consultant Surgeon of the Hospital. The two Consultant Surgeons are assisted by a changing group of four Surgical Registrars, who are qualified surgeons on assignment for terms of 6 or 12 months from one or another of the Glasgow teaching hospitals. They are assisted by residents and interns. Although some cancer patients are initially treated in the medical or gynecological unit, there is a tendency for cases of advanced cancer of all kinds except leukemia and some rare childhood cancers, which are dealt with in a pediatric hospital in Glasgow, to gravitate into the surgical unit, in total probably 90% of all cases of cancer in the Loch Lomondside area.

All of the patients are treated initially in a perfectly con-

Table 2. Ratios of average survival times for ascorbate patients and matched controls, with statistical significance

A	B (Days)	C (Days)	D	E (Days)	F (%)	G (%)	H	I
Bronchus (15)	136	38.5	3.53	47	47	8.7	24.5	<<0.0001
Colon (13)	282	37.0	7.61	59	54	20	7.63	<0.003
Stomach (13)	98.9	37.9	2.61	43	46	17	6.41	<0.006
Breast (11)	367	64.0	5.75	91	55	22	5.74	<0.026
Kidney (9)	333	64.0	5.21	88	67	22	8.35	<0.002
Bladder (7)	196	43.6	4.49	57	57	20	4.90	<0.028
Rectum (7)	226	55.5	4.10	71	86	33	7.57	<0.003
Ovary (6)	148	71.0	2.08	78	83	30	6.83	<0.005
Others (19)	172	56.8	3.03	67	53	27	5.28	<0.027
All (100)	209.6	50.4	4.16	65	60	25.7	55.02	<<0.0001

A, Type of cancer and, in parentheses, number of ascorbate patients. There are 10 matched controls for each ascorbic acid patient. B, Ave[rage] days of survival for ascorbate patients. C, Average days of survival for controls. D, The ratio B/C. E, Average days of survival for all su[b]jects in group. F, Percentage of ascorbate patients surviving longer than E. G, Percentage of controls surviving longer than E. H, Value [of] χ^2 for F and G (two-by-two calculation). I, Corresponding value of P (one-tailed).

ventional way, by operation, use of radiotherapy, and administration of hormones and cytotoxic substances. For example, all of the 11 breast-cancer patients in the ascorbate-treated group, with the exception of one who first presented in a grossly advanced state, had already had mastectomy and radiotherapy and all, including the exception, had been given hormones, sometimes with considerable benefit; but all had relapsed by the time ascorbate supplementation was commenced, and it seemed clear that their tumors were escaping from hormonal control. Similarly, all of the seven bladder-cancer patients in the ascorbate-treated group, with one exception because of her frailty, had received megavoltage irradiation and several had had a partial cystectomy (one total) before ascorbate treatment was commenced when it seemed that these standard procedures had failed.

Treatment of terminal cancer patients with ascorbate was cautiously begun in November 1971, for reasons discussed in our earlier papers (1, 2), and has been continued because it seemed to have some value (4, 5). Once the practice had become locally established, the selection of a patient for treatment with ascorbate was often initiated by one of the younger surgeons (the Registrars), as they became familiar with the idea and convinced of its worth. The suggestion that ascorbate treatment be tried was made by Registrars less often during the first part of their 6 to 12 months' service than during the second part. For strong ethical reasons, every patient in the ascorbate-treated group was examined and assessed independently by at least two physicians or surgeons (often more than two) who all agreed that the situation was "totally hopeless" and "quite untreatable" before ascorbate was commenced. More than 20 different Registrars were involved in this way in allocating patients to the ascorbate-treated group. No criterion was used, except agreed "untreatability."

As described above, selection of 10 matched patients for the control group for each patient of the ascorbate-treated group was made independently by Dr. Frances Meuli. For each ascorbate-treated patient she was given a sheet listing age, sex, primary tumor type, and a brief synopsis of the clinical state and extent of dissemination at the time ascorbate was commenced, but not the survival time. She searched for cases matching these cases as closely as possible, and assigned to each, from the case history, the time when the patient was classified as "untreatable.' We believe that the procedure that was followed has not introduced any serious error, and that the ascorbate-treated group and the control group are in fact subpopulations of the population of "untreatable" patients selecte[d] in an essentially random manner.

Two hundred of the 1000 patients in the control group we[re] completely contemporaneous with the ascorbate-treated pa[tients]. The mean survival time for these contemporaneou[s] controls is 43.9 days, as compared with 52.4 days for the othe[r] (overlapping and historical). There has been no significan[t] change in the treatment of patients with advanced cancer i[n] Vale of Leven Hospital during the last 10 years, and the ap[proximate] equality of these values is not surprising.

The results of the study

The results of the study are given in Table 1 and summarize[d] in Table 2, in which values for different kinds of cancer rep[resented] by six or more patients treated with ascorbate (60 o[r] more controls) are shown. For each of the nine categories th[e] ratio of average days of survival (ascorbate/controls) is greate[r] than unity, the range being from 2.1 to 7.6, with ratio 4.16 fo[r] all 100 patients. The ratios are somewhat uncertain; for exam[ple], omitting the patient with longest survival in the colon grou[p] would decrease the ratio from 7.6 to 5.2. At the present time w[e] cannot conclude that ascorbate has less value for one kind o[f] cancer than for others. Our conclusion is that the administratio[n] of ascorbic acid in amounts of about 10 g/day to patients wit[h] advanced cancer leads to about a 4-fold increase in their lif[e] expectancy, in addition to an apparent improvement in th[e] quality of life. This great increase in survival time results in pa[rt] from the much larger numbers of the ascorbate patients tha[n] of the controls who live for long times, as is shown in Fig. [1]. Sixteen percent of the patients treated with ascorbic acid su[r]vived for more than a year, 50 times the value for the contro[ls] (0.3%).

Statistical analysis shows that the null hypothesis that trea[t]ment with ascorbate has no benefit is to be rejected for each [of] the categories in Table 2. The results of a simple statistical te[st] are given in the table. A reasonable dividing line, the averag[e] survival time for all the subjects, is given in column E, and th[e] percentages exceeding this value are given in columns F an[d] G. Column H contains the values of χ^2 obtained by a two-by-two calculation, and I gives the corresponding values of [P] (one-tailed). Similar values are obtained by nonparametri[c] methods.

The fraction of survivors of the control group at time t [is] given to within about 2% by the exponential expressio[n] $\exp(-t/\tau)$. About 1.5% of the patients in this group live muc[h]

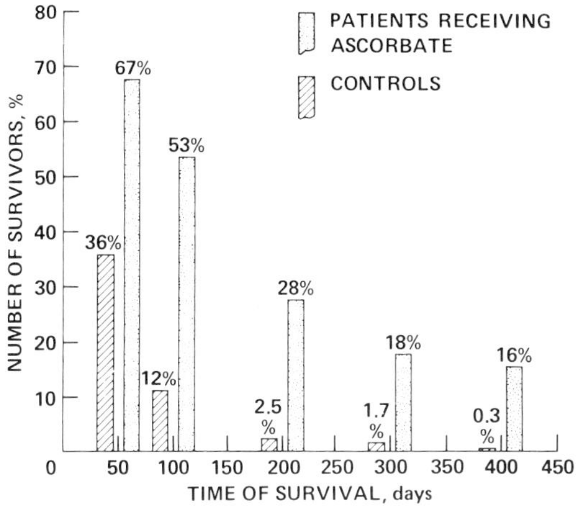

FIG. 1. The percentages of the 1000 controls (matched cancer patients) and the 100 patients treated with ascorbic acid (other treatment identical) who survived by the indicated number of days after being deemed "untreatable." The values at 200, 300, and 400 days for the patients receiving ascorbate are minimum values, corresponding to the date August 10, 1976, when 18% of these patients were still alive (none of the controls).

longer than would be indicated by this expression. A very close approximation to the observed survival curve is given by the assumption that the control group consists of two populations. One consists of 985 patients with number of survivors at time t given by the expression 985 $\exp(-t/\tau)$, in which τ has the value 45.5 days. This expression corresponds to a constant mortality rate for this subgroup, and its validity suggests that for them a single random process, occurring with a probability independent of time, leads to death. This probability is 2.2% per day. For 14 of the 1000 control patients the survival time is indicated to lie between 200 and 500 days. The distribution suggests that for this subgroup two random events lead to death, but the number of subjects is too small to permit this possibility to be tested thoroughly. One other patient, who survived 841 days, may constitute a third subgroup.

A similar analysis of the survival curve for the ascorbate-treated group shows that a considerably smaller fraction, 90%, constitutes the principal group, with number of survivors at time t equal to 90 $\exp(-t/\tau)$, τ equal to 125 days. For the remaining 10% the average survival time is greater than 970 days. These numbers are uncertain because the number of ascorbate-treated patients is small, only 100, and 18 of them were alive on August 10, 1976, their survival times being greater than the values used in the calculation.) A simple interpretation of these facts is that the administration of ascorbate to the patients with terminal cancer has two effects. First, it increases the effectiveness of the natural mechanisms of resistance to such an extent as to lead to an increase by a factor of 2.7 in the average survival time for most of the patients; 2.7 is the ratio of the two values of τ, 125 and 45.5 days. Second, it has another effect on about 10% of the patients, such as to cause them to live a much longer time. This effect might be such as to "cure" them; that is, to give them the life expectancy that they would have had if they had not developed cancer. On the other hand, it might only set them back one or more stages in the development of the cancer, in which case their life expectancy would be somewhat less than that corresponding to complete elimination of the effect of their having developed cancer. This uncertainty may be eliminated in the course of time, as the survival times of the 18 patients in the ascorbate-treated group who were still living on August 10, 1976 become known.

Conclusion

In this study the times of survival of 100 ascorbate-treated cancer patients in Scotland (measured from the day when the patient was pronounced to have cancer untreatable by conventional methods) have been discussed in comparison with those of 1000 matched controls, 10 for each of the ascorbate-treated patients. The data indicate that deaths occur for about 90% of the ascorbate-treated patients at one third the rate for the controls, so that for this fraction there is a 3-fold increase in survival time, measured from the date when the cancer was pronounced untreatable. For the other 10% of the ascorbate-treated patients the survival time is not known with certainty, but it is indicated by the values in Table 1 to be more than 20 times the average for the untreated patients. The value 4.16 (Table 2) for the ratio of average survival times expresses the resultant of these two effects.

We conclude that there is strong evidence that treatment of patients in Scotland with terminal (untreatable) cancer with about 10 g of ascorbate (ascorbic acid, vitamin C) per day increases the survival time by the factor of about 3 for most of them and by at least 20 for a few (about 10%). It is our opinion that a similar effect would be found for untreatable cancer patients in other countries. Larger amounts than 10 g/day might have a greater effect. Moreover, we surmise that the addition of ascorbate to the treatment of patients with cancer at an earlier stage of development might well have a similar effect, changing life expectancy after the stage when ascorbate treatment is begun from, for example, 5 years to 20 years.

This study was supported by research grants from the Secretary of State for Scotland and The Educational Foundation of America, and by contributions by private donors to the Linus Pauling Institute.

1. Cameron, E. & Pauling, L. (1973) "Ascorbic acid and the glycosaminoglycans: An orthomolecular approach to cancer and other diseases," *Oncology* **27**, 181–192.
2. Cameron, E. & Pauling, L. (1974) "The orthomolecular treatment of cancer: I. The role of ascorbic acid in host resistance," *Chem.-Biol. Interact.* **9**, 273–283.
3. Cameron, E. (1976) "Biological function of ascorbic acid and the pathogenesis of scurvy: a working hypothesis," *Med. Hypotheses* **3**, 154–163.
4. Cameron, E. & Campbell, A. (1974) "The orthomolecular treatment of cancer: II. Clinical trial of high-dose ascorbic acid supplements in advanced human cancer," *Chem.-Biol. Interact.* **9**, 285–315.
5. Cameron, E., Campbell, A. & Jack, T. (1975) "The orthomolecular treatment of cancer: III. Reticulum cell sarcoma: double complete regression induced by high-dose ascorbic acid therapy," *Chem.-Biol. Interact.* **11**, 387–393.

[CANCER RESEARCH 39, 663-681, March 1979]
0008-5472/79/0039-0000$02.00

Ascorbic Acid and Cancer: A Review[1]

Ewan Cameron, Linus Pauling, and Brian Leibovitz

Vale of Leven Hospital, Loch Lomondside G83 0UA, Scotland [E. C.]; Linus Pauling Institute of Science and Medicine, Menlo Park, California 94025 [L. P.]; and Department of Pathology, University of Oregon Medical School, Portland, Oregon 97201 [B. L.]

Abstract

Host resistance to neoplastic growth and invasiveness is recognized to be an important factor in determining the occurrence, the progress, and the eventual outcome of every cancer illness. The factors involved in host resistance are briefly reviewed, and the relationship between these factors and ascorbic acid metabolism is presented in detail. It is shown that many factors involved in host resistance to neoplasia are significantly dependent upon the availability of ascorbate.

Introduction

Few would now dispute that the behavior of every human cancer is determined to a significant extent by the natural resistance of the patient to his or her disease. As a result, there is now widespread recognition that very substantial benefits in cancer management would be achieved if practical methods could be devised to enhance resistance.

There is a growing body of theoretical and practical evidence suggesting that the availability of ascorbate is the determinant factor in controlling and potentiating many aspects of host resistance to cancer. We have prepared this review as an aid to investigators in this field and as a source of information to others.

History

The history of vitamin C is common knowledge. In the mid-18th century James Lind demonstrated that the juice of fresh citrus fruits cures scurvy (197). The active agent, the enolic form of 3-keto-L-gulofuranlactone, christened ascorbic acid or vitamin C, was isolated in the late 1920's by Albert Szent-Györgyi (317). By the mid-1930's, methods had been devised to synthesize the compound, and it soon became widely available at low cost. It was soon established that the substance was virtually nontoxic at any dosage. The structure of vitamin C is shown in Chart 1.

The basic function of vitamin C is the prevention of scurvy. The current recommended dietary allowance of the Food and Nutrition Board of the United States National Academy of Sciences-National Research Council, 45 mg/day for an adult, is adequate to prevent scurvy in essentially all normal persons. The question of whether or not a larger intake could lead to better health and a greater control of disease was raised almost as soon as the pure compound became freely available, and the debate continues.

Our earlier suggestions that ascorbate might be of some value in the supportive treatment of cancer (54, 55, 60) provoked further controversy, although less than might have been expected. Untreated cancer almost invariably pursues a relentlessly progressive course, at the very least providing some opportunity to measure therapeutic effect. Furthermore, it is well established that cancer patients have a significantly increased requirement for this nutrient, there is persuasive evidence that it can be implicated in many mechanisms concerned with host resistance to malignant invasive growth, and there exist some experimental and pilot clinical reports indicating that administration of ascorbate in amounts substantially in excess of base-line recommended dietary allowance levels does indeed exert some therapeutic benefit.

Our own interest arose from the realization that ascorbic acid, known to be required for collagen synthesis, might be required in increased amounts for the protective encapsulation of tumors (253) and from the independent simultaneous conclusion that the ascorbate molecule (or some residue thereof) must be involved in the feedback inhibition of lysosomal glycosidases (60) responsible for malignant invasiveness (48). From these joint beginnings came the realization that ascorbic acid could be implicated in many other aspects of host resistance and even the tentative proposal that host resistance, no matter how measured, is ultimately dependent upon the availability of ascorbic acid.

We have published a number of clinical reports indicating favorable responses in advanced human cancer to no specific treatment other than regular large doses of ascorbate (51-53, 56-59), but as yet no properly designed controlled trial has been made to confirm or refute these findings and, if confirmed, to determine the most effective dosage. It is our hope that such studies will soon be undertaken, and it is our expectation that, on the basis of sound statistical evidence furnished by such studies, supplemental ascorbate will soon become an essential part of all practical cancer treatment and cancer prevention regimens.

We present below some of the existing evidence in support of our contention. We first draw attention to some of the similarities between scurvy and cancer.

Scurvy and Cancer

Any discussion of the biological function of ascorbic acid has to begin with a consideration of scurvy, the universally accepted "base line," and, if untreated, the invariably fatal illness resulting from severe dietary lack of ascorbate. Scurvy, an illness now rare in its flagrant form, is a syndrome of generalized tissue disintegration at all levels, involving the dissolution of intercellular ground substance, the disruption of collagen bundles, and the lysis of the interepithelial and interendothelial cements, leading to ul-

[1] This study was supported by research grants from the Secretary of State for Scotland, The Educational Foundation of America, The Foundation for Nutritional Advancement, The Argyll and Clyde Health Board, Scotland, and by private donations to the Linus Pauling Institute of Science and Medicine.
Received June 14, 1978; accepted December 4, 1978.

E. Cameron et al.

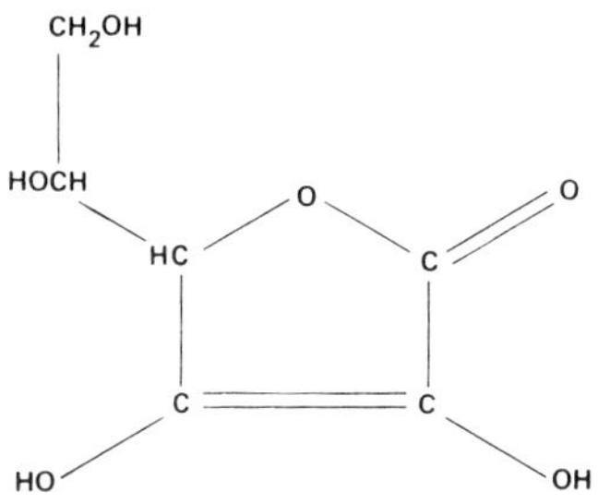

Chart 1. The accepted structural configuration of ascorbic acid (vitamin C), an α-ketolactone with the formula $C_6H_8O_6$, with a molecular weight of 176.13, and containing an acid-ionizing group in water with pK_A 4.19.

ceration with secondary bacterial colonization and to vascular disorganization with edema and interstitial hemorrhage, and also to one feature that is rarely emphasized in the literature, generalized undifferentiated cellular proliferation with specialized cells throughout the tissue reverting to a primitive form (350).

The pathology of scurvy has been summarized as "a generalized structural breakdown of the intercellular matrix associated with undifferentiated cell proliferation, or in evolutionary terms (provided such a disintegrating individual could be kept alive), the gradual reversion from multicellular organization to a primitive unicellular state" (50).

Pursuing this theoretical view even further, one might regard scurvy (generalized undifferentiated cell proliferation with generalized matrix breakdown) as an "omnifocal" variety of neoplastic disease.

The late Dr. William McCormick of Toronto appears to have been the first to recognize that the generalized stromal changes of scurvy are identical with the local stromal changes observed in the immediate vicinity of invading neoplastic cells (211). He surmised that the nutrient (vitamin C) known to be capable of preventing such generalized changes in scurvy might have similar effects in cancer, and the evidence (to be reviewed below) that cancer patients are almost invariably depleted of ascorbate lent support to this view.

There are some other interesting associations between scurvy and cancer. There is some epidemiological evidence (reviewed below) indicating that cancer incidence in large population groups is inversely related to average daily ascorbate intake. There is no real modern evidence that frankly scorbutic patients succumb to cancer, presumably because such patients either die fairly rapidly from their extreme vitamin deficiency state or, more likely, are promptly diagnosed and rapidly cured. However, the historical literature contains many allusions to the increased frequency of "cancers and tumors" in scurvy victims. A typical autopsy report of James Lind (197) contains phrases such as "... all parts were so mixed up and blended together to form one mass or lump that individual organs could not be identified," surely an 18th century morbid anatomist's graphic description of neoplastic infiltration.

Finally, in advanced human cancer, the premortal features of anemia, cachexia, extreme lassitude, hemorrhages, ulceration, susceptibility to infections, and abnormally low tissue, plasma, and leukocyte ascorbate levels, with terminal adrenal failure, are virtually identical with the premortal features of advanced human scurvy.

Host Resistance to Cancer

Many factors are known to be involved in host resistance to malignant invasive growth. These can be conveniently divided into locally acting stromal factors and others acting systemically.

Stromal resistance is primarily exercised by the ability of the host to encapsulate the neoplastic cells in a dense, practically impenetrable barrier of fibrous tissue. The collagenous barrier is scanty and ill defined in highly anaplastic invasive tumors, moderate in amount in tumors of moderate rapidity of growth, and very abundant in slow-growing "contained" atrophic scirrhous tumors. In any individual, the degree of stromal fibrosis is the same around the primary growth and its metastases, indicating a constitutional response. Other important stromal factors in resistance are (*a*) the resistance of the intercellular ground substance to local infiltration and (*b*) the degree of lymphocytic response. Lymphocytes are most numerous in the stroma of slow-growing tumors and scanty or virtually absent around rapidly growing lesions, and again the degree of lymphocytic infiltration is identical around primary growth and its metastases, indicating a constitutional response. A brisk lymphocytic response indicates enhanced host resistance and is associated with a more favorable prognosis.

The systemic factors are less clearly defined. They include such constitutional factors as the relative efficiency of the immune system, feedback mechanisms involved in the restraint of "invasive" enzymes, and, for certain tumors at least, the hormonal status of the individual.

The ability safely to enhance host resistance to maximum levels in every patient would vastly improve cancer treatment. It is our intention to show that ascorbic acid metabolism is implicated in all these resistance mechanisms and that ingestion of this substance in adequate amounts could provide a simple and safe method of achieving this desirable goal.

Chemistry of Ascorbic Acid and the Intercellular Matrix

Ascorbic acid is of enormous biological importance. It is one of the most important reducing substances known to occur naturally in living tissues, and since its discovery it has been the subject of many investigations, theories, and exhaustive reviews (50, 66, 140, 181, 195, 213, 251, 254, 313, 350). It would be inappropriate to review this voluminous literature here. We propose to limit our discussion to areas of special relevance to cancer.

Two points deserve early emphasis: (*a*) although most animals can synthesize ascorbic acid, a human cannot, and he is totally dependent upon dietary intake to satisfy all his requirements; (*b*) ascorbic acid, known to be essential for the structural integrity of the intercellular matrix, is closely related to glucuronic acid, an essential building block of the principal matrix structures.

This leads us to a consideration of the intercellular matrix. In the last century the eminent pathologist Rupert

Virchow revolutionized his specialty by focusing attention on the "cellular" nature of disease. It is our impression that we are in the midst of a second revolution in pathology with the recognition that cells do not exist in "empty" space. The cell is still dominant, but it is now increasingly appreciated that for every action, be it physiological or pathological, of the cell, there is an appropriate reaction in the immediate extracellular environment.

All tissue cells in the body are firmly embedded in highly viscous ground substance. This ubiquitous material may be sparse or abundant and be in varying degrees of molecular aggregation, depending upon the particular organ studied, but basically it is of universal distribution, pervading every interspace and isolating every cell from its neighbor. It forms the immediate contact environment of all cells, and it must be traversed by every molecule entering or leaving the cell.

Variations in the physicochemical composition of the extracellular environment (polymerization-depolymerization) exert a profound influence on cell behavior, and, in turn, cells possess a powerful means of modifying their immediate microenvironment. A proliferating cell and its contact environment constitute a balanced system in which each component influences the other; this interdependence is involved in all forms of cell division and is of particular importance in cancer (54).

The ground substance of the intercellular matrix is a complex aqueous gel containing electrolytes, metabolites, dissolved gases, trace elements, vitamins, hormones, enzymes, carbohydrates, fats, and proteins. Its important structural property of extreme viscosity depends upon the abundance of certain long-chain mucopolysaccharide polymers, the glycosaminoglycans and related proteoglycans. These interlacing high-molecular-weight polymers form a structurally stable hydrophilic mesh, which, in turn, is reinforced at the microscopic level by a 3-dimensional network of collagen fibers. This is the true *milieu interieur* of Claude Bernard, within which all cellular activity takes place.

The chemistry of the glycosaminoglycans and of the proteoglycans is the subject of much contemporary study, and some excellent reviews are available (10, 103, 171, 189, 240, 241). The glycosaminoglycans are single-strand long-chain polymers with molecular weights ranging up to 10×10^6. The common varieties are hyaluronic acid (built up from alternating repeating units of *N*-acetylglucosamine and glucuronic acid), chondroitin (built up from *N*-acetylgalactosamine and glucuronic acid), and sulfate esters (the chondroitin sulfates), of which several isomers exist.

The proteoglycans are macromolecules of a more complex multibranched structure. Present evidence indicates that they consist of a primary glycosaminoglycan core, to which is attached, by link proteins, secondary protein cores at spaced intervals, to which are attached tertiary glycosaminoglycan polymers. It should be noted that the proteoglycans of the matrix are not a single molecular species but comprise a polydisperse family of rather similar macromolecules which differ in complexity, molecular weight, hydrodynamic size, chemical composition, and reactivity (132).

In healthy tissues, the intercellular matrix is maintained in a steady-state equilibrium of very slow dynamic change, with the formation of new macromolecules (polymerization) balanced by turnover and decay (depolymerization). Locally, in cancer, matrix depolymerization in the immediate vicinity of proliferating invasive cells is a striking feature; this is exactly the change observed to occur on a generalized scale in scurvy (48, 50).

Depolymerization of matrix glycosaminoglycans is brought about by the sequential action of "hyaluronidase." This sequence of lysosomal glycosidases involves first an endoglycosidase ("true" hyaluronidase) that cleaves the long-chain polymer into tetrasaccharides; this action is followed by the concerted action of 2 exoglycosidases, β-glucuronidase and β-*N*-acetylglucosaminidase (β-*N*-acetylgalactosaminidase in the case of the chondroitins). The tetrasaccharide is the feedback inhibitor of the endoglycosidase but is also the substrate for the exoglycosidases. Thus, any interference with the latter will inhibit the whole process (152, 309).

Depolymerization of matrix proteoglycans is brought about by the combined action of the glycosidases and neutral proteases, which are believed to be released simultaneously from cellular lysosomes during cell division.

It now seems reasonably certain that the continuous release of these hydrolytic enzymes from neoplastic cells is responsible for their invasive capability, for the selective routing by increased diffusion of nutrients towards the tumor cells, and perhaps even for sustaining the whole momentum of autonomous neoplastic proliferation (48, 54). The effect of these hydrolytic enzymes on the matrix is to release a whole spectrum of glycosaminoglycan and glycoprotein breakdown products into the blood stream, the so-called "acute-phase reactants," the estimation of which forms the basis for the majority of serochemical tests for cancer.

The collagen component of the extracellular matrix is also susceptible to the same hydrolytic enzyme activity. Collagen fibrils are bound together into microscopically visible collagen fibers by matrix cement substance (140). More specifically, electron microscopy show a precise molecular alignment of proteoglycans along collagen chains (191). Ionic interactions between positively charged lysine and arginine residues of polymeric collagen and negatively charged glycosaminoglycan and proteoglycan sulfonic acid groups, together with the formation of hydrogen bonds, offer a probable explanation for this "adhesiveness" (311). The enzymatic degradation of such vital cross-linking bonds in the immediate vicinity of neoplastic cells could result in the dissolution and "microscopic disappearance" of this essential molecular scaffolding. Such structural disintegration is, of course, quite characteristic of the stromal erosion seen in immediate proximity to highly malignant growths.

With regard to collagen fibrillogenesis and its specific relevance to tumor encapsulation, it seems clear that specific interactions with extracellular glycosaminoglycans and proteoglycans play an essential role (132). There is now an overwhelming body of evidence (191, 201, 202, 205, 206, 311) to support the view originally proposed by Meyer (215) that the long-chain matrix polymers with their regular an ionic spacings provide the essential molecular template for the precise orderly deposition of collagen precursors and their regular sequential arrangement into collagen fibrils. Thus, enzymatic dissolution of the matrix would also inhibit new collagen formation.

Any measure that can protect the integrity of the intercellular matrix will (*a*) effectively retard malignant infiltration, (*b*) restrict the selective nutrition of tumors, (*c*) protect preexisting collagen barriers from neoplastic erosion, and (*d*) facilitate protective collagen encapsulation. This is one important function of ascorbic acid.

The Increased Requirement for Ascorbic Acid in Cancer

The contention that ascorbate is involved in resistance to neoplasia gains support from the many studies demonstrating that cancer patients have abnormally low ascorbate reserves. When this was first demonstrated many years ago, there was a tendency to assume that such a finding merely reflected the poor general nutritional status of advanced cancer patients. Such a view is no longer tenable, and it is now generally agreed that low levels in cancer indicate an increased utilization and requirement for the vitamin.

Increased utilization of ascorbate, as measured by depletion of ascorbate reserves, is by no means confined to cancer; it is a characteristic feature of many other "cell-proliferative" disorders, such as inflammation (17, 91), and of the reparative processes after surgical trauma (88, 89), myocardial infarction (160, 161), and thermal burns (16). The last 3 studies demonstrate that ascorbate is removed from the circulating reserves and concentrated at the site of the reparative process.

A similar shift of ascorbate from reserves to the tumor stroma has been demonstrated in cancer. Guinea pig studies have shown that ascorbic acid is selectively concentrated in cancerous tissue and, as a result, normal tissues are depleted (39, 315, 336). Dyer and Ross (109) studied the ascorbate content of tissues of mice bearing a wide variety of tumors, and in all examples the ascorbate concentration was greater in the tumor than in the liver of the respective host. Dodd and Giron-Conland (101), using electron spin resonance, have demonstrated that the ascorbyl radical is present in a wide variety of tumors in higher concentration than in the corresponding normal tissues.

Human cancer tissue also contains elevated levels of ascorbate. Goth and Littman (139) demonstrated this for a variety of human tumors. Chinoy (69) reported that certain human tumors contain far greater ascorbate levels than does their tissue of origin, and using histochemical techniques demonstrated that the greatest concentration of ascorbate was deposited at the periphery of the tumor, against the actively growing invasive margin. In 29 of 30 patients, Moriarty *et al.* (225) found the level in tumors to be higher than that in the surrounding tissues.

Concentration of ascorbate in tumor stroma results in measurable depletion of circulating reserves. Table 1 lists studies that have been carried out, contrasting values in cancer patients with those obtained in normal controls.

The increased requirement for ascorbate in cancer can also be demonstrated by ascorbate loading tests (18, 49, 126, 127, 186, 217). All such investigations have shown an increased utilization of ascorbate by the cancer subjects.

The combined results of all these studies show serious deficiencies of blood and leukocyte ascorbate to be a characteristic feature of cancer. Thus, irrespective of any specific therapeutic effects suggested, replacement of this deficit should be a part of all comprehensive cancer treatment regimens.

Ascorbic Acid and Host Resistance to Neoplasia

Ascorbate metabolism can be implicated in a number of host resistance mechanisms functioning at both stromal and systemic levels.

Ascorbate, Hyaluronidase, and the Intercellular Matrix. Invasiveness is one of the fundamental and distinguishing attributes of malignant tumor cells, conferring upon them their ability not only to infiltrate the local tissue interstices but also to ulcerate through membranous barriers and to penetrate into lymphatics, blood vessels, and other preformed spaces and thus produce remote metastases by passive transfer. The structural integrity of the ground substance matrix is the first barrier to invasiveness.

Depolymerization of adjacent matrix is invariably observed in the immediate vicinity of invading neoplastic cells, and there is now much evidence that the mechanism of malignant invasiveness depends upon the continuous release of lysosomal glycosidases (hyaluronidase) and perhaps other degradative lysosomal enzymes (the neutral proteases and collagenase) from the invading cells; see review by Cameron (48) and publications by Balazs and von Euler (11), Fiszer-Szafarz *et al.* (116–120), Harris *et al.* (152), and Grossfeld (147).

The endoglycosidase has a molecular weight of around 80,000 (6, 7). Other lysosomal enzymes are the exoglycosidases, the proteases, collagenase, nuclease, sulfatases, and phosphatases. The release of hyaluronidase from the cancer cell is usually accompanied by a release of increased amounts of other lysosomal enzymes, as has been demonstrated for a wide variety of experimental and human tumors (93, 110, 120, 172, 190, 282, 292, 316).

The impact of this continuous outflow of lysosomal enzymes on the immediate microenvironment is to bring about profound changes in the physicochemical structure of the matrix. These changes, predominantly dissolution and depolymerization of matrix glycosaminoglycans and proteoglycans, destroy the structural stability of the molecular micelle, with an abrupt fall in local ground substance viscosity leading to diminished mutual adhesiveness of the tumor cells (76, 212), their increased motility (1), and, most important of all, the erosion of the immediate structural barrier to malignant infiltration (48). These local pericellular changes of matrix depolymerization, affecting all barriers from the cell membrane to stromal capillary interendothelial cement, would result in local zones of increased capillary and tissue diffusion and could account for the so-called selective nutrition of tumors, the all-too-familiar picture of a flourishing tumor growing parasitically in a host steadily dying from cachectic nutritional inanition (48).

Thus, if we could stabilize cell matrix interrelationships in cancer and, more particularly, find a method of safely inhibiting tumor cell glycosidases, we should possess a means not only of restraining malignant invasiveness but also of retarding malignant cell proliferation by depriving

Table 1
Ascorbic acid in the blood of cancer patients

Investigator	Type of cancer studied	No. of cancer patients	Plasma levels (mg/dl)		Leukocyte levels (μg/10⁸ cells)	
			Cancer	Controls	Cancer	Controls
Bodansky *et al*. (38)	Miscellaneous cancers	69	0.48	0.79	27.0[a]	36.1[a]
Waldo and Zipf (333)	Miscellaneous cancers	30	0.10	0.80		35.0
Freeman and Hafkesbring (124)	Miscellaneous cancers	5	0.62[b]	0.88[b]		
Krasner and Dymock (183)	Miscellaneous cancers	50			11.5	29.5
Kakar and Wilson (165)	Advanced breast cancer	1	0.13	0.47	12.5	26.0
Basu *et al*. (18)	Advanced breast cancer	22			12.0	33.0
Cameron (49)	Miscellaneous cancers	24	0.26	0.96	11.2	24.3
Waldo and Zipf (333)	Leukemia	42	0.18	0.80	8.2	35.0
Barkhan and Howard (14)	Leukemia	5	0.21	0.69		
Lloyd *et al*. (198)	Acute leukemia	8	0.3	0.79	12.0[c]	35.0[c]
Lloyd *et al*. (198)	Chronic leukemia	8	0.7	0.79	31.0[c]	35.0[c]
Kakar *et al*. (166)	Acute lymphatic leukemia	10	0.40	0.95	35.9	56.4
Basu *et al*. (18)	Miscellaneous cancers	22			17.0	33.0

[a] Expressed as mg ascorbate per 100 g leukocytes.
[b] Values for whole blood.
[c] Values for platelets.

these cells of their favored nutritional status. This is a highly desirable therapeutic aim, but the benefits could be even greater.

The following working concept was published by Cameron in 1966 (48) and elaborated by Cameron and Pauling in 1973 (54):

> "All tissue cells have an inherent tendency to divide but this tendency is normally restrained by the viscous nature of their intimate extracellular environment of high-molecular-weight ground-substance glycosaminoglycans. Proliferation is initiated by the cellular release of hyaluronidase, which permits the cell local freedom to divide and to migrate within the limits of the altered field. Proliferation will continue as long as hyaluronidase is being released; proliferation will cease and normal tissue restraint and organization will be restored when the production of hyaluronidase returns to normal."

Under this concept, the only difference between "neoplastic" and "normal" cell proliferation is the persistence of hyaluronidase release in the former.

It is a common feature of enzyme-substrate reactions that the build-up of by-products exerts a feedback inhibitory effect. Thus, it is no surprise to find that glycosaminoglycan residues down to D-glucosamine have been shown experimentally to inhibit the growth of Sarcoma 37 (264, 273), Sarcoma 180 (22, 23), Walker carcinoma 256 (24, 224), various ascitic carcinomas (21, 22, 223), and epidermoid carcinomas (121), although *N*-acetylglucosamine and neutral sugars were found to be without effect (23, 223). In passing, it is of interest to note that glucosamine has been reported to inhibit the biosynthesis of protein, RNA, and DNA in label uptake experiments *in vitro* (21, 24). Recently, Schaffrath *et al.* (281) demonstrated inhibition of DNA-dependent and RNA-dependent polymerases as well as DNA-dependent RNA polymerase by a variety of sulfated glycosaminoglycans.

In vitro tests have shown that chondroitin sulfate residues (6, 7) and other polysaccharide polysulfuric acids (8), including heparin, a polysulfated glycosaminoglycan, are hyaluronidase inhibitors (135, 136). Rutin (194), hesperidin and derivatives (20), and polyphloretin phosphate (75) show *in vitro* hyaluronidase inhibition and also stabilize lysozymes (326). Hyaluronidase inhibition by bioflavonoids and related compounds has been reported (125, 207, 239). Stephens *et al.* (309) have recently shown that gold thiomalate is an effective hyaluronidase inhibitor, but we are not aware of any reports of its use in either experimental or human cancer.

Our special interest has focused on the existence in the serum of all species of a specific glycoprotein fraction known as PHI.[2] Serum PHI concentration remains within a remarkably narrow range in health but increases sharply in a variety of disease states, including infections, wound healing, rheumatoid arthritis, and cancer (133, 134, 153, 175). Because of this, serum from cancer patients inhibits hyaluronidase *in vitro* (116, 150). As yet, PHI has not been completely characterized, but it is reported to be a glycoprotein with a molecular weight of about 100,000 (207, 238). Glucosamine and uronic acids are major constituents of PHI (238), and it seems clear that the fraction consists of a polypeptide chain to which is attached a glycosaminoglycan residue which has been modified in some special way to confer upon it its powerful and highly specific inhibitory activity. In 1972, we suggested that the modification is the replacement of one or more than one glucuronic acid residue in the glycosaminoglycan depolymerization by-product by its close chemical relation ascorbic acid (60). It now seems clear that the active component of PHI is a tetrasaccharide with a terminal glucuronic acid unit replaced by an ascorbate residue, resistant to exoglycosidase activity, and therefore capable of blocking the whole process of glycosaminoglycan depolymerization.

The hypothesis that ascorbic acid is required for the synthesis of PHI offers a convincing explanation for its biological function in preventing scurvy, a state of general-

[2] The abbreviations used are: PHI, physiological hyaluronidase inhibitor UHP, urinary hydroxyproline; 3-HOA, 3-hydroxyanthranilic acid.

ized matrix depolymerization brought about by exposure to uninhibited hyaluronidase evolved during the course of normal cell division and replacement (50). It also explains, in part, why there is an increased requirement for ascorbate in all cell-proliferative states, including cancer. Satisfying this enhanced requirement by increasing intake should enable this restraining mechanism to function at maximum efficiency.

Because everyday clinical experience confirms that ascorbate is necessary for satisfactory wound healing and repair, the natural assumption has been that this vitamin is essential for growth. We offer the alternative hypothesis that the primary function of vitamin C is to restrain excessive growth through its incorporation into the PHI system, directing the inherent proliferative capacity of all cells into a constrained organized differentiated behavior pattern. Thus, in the simple example of wound healing, the initial cell-proliferative phase produces depolymerization of the immediate matrix and the release of measurable amounts of glycosaminoglycan residues into the bloodstream, there to mop up ascorbate reserves into an upsurge of PHI levels, all followed fairly rapidly by reversal into a healed phase of resting quiescent cells reembedded in a restabilized matrix, and return of these biochemical indices to a steady state equilibrium. In the absence of ascorbate, *i.e.*, in scurvy, wounds fail to heal, with a destabilized matrix, undifferentiated cell-proliferative "granulomata," and the continuous overspill of matrix by-products into the bloodstream (260, 279).

In vitro, the reaction of ascorbate with hyaluronic acid and with chondroitin sulfate has been reported to yield breakdown products of somewhat lower viscosity than the original preparations (303). Ascorbic acid also inhibits hyaluronidase *in vitro* under conditions that preclude the possibility of PHI production, but it has been shown to have a much greater effect *in vivo*, clearly indicating the existence of some intermediate mechanism (267).

A recent brief report by Shapiro *et al.* (295) indicating that prescorbutic guinea pigs have increased rather than decreased PHI levels is difficult to reconcile with the general pattern. A possible explanation for this apparent contradiction may lie in the finding by Fiszer-Szafarz (117) that serum contains a number of different hyaluronidase inhibitor fractions distinguishable by the use of different laboratory techniques.

The realization that the term "hyaluronidase" embraces a sequence of related degradative enzymes has given rise to some confusion in the literature. Thus, many workers have claimed that neoplastic cells produce significantly increased amounts of hyaluronidase, but other workers have been unable to confirm this; see references in the review by Cameron (48). However, there seems to be no dispute that neoplastic cells release significantly increased amounts of the exoglycosidases β-glucuronidase (63, 74, 115, 152, 243) and β-*N*-acetylglucosaminidase (63, 77, 79, 120, 152). In their recent study of an experimental rat osteosarcoma, Harris *et al.* (152) demonstrated no significant increase in hyaluronidase activity but a significant increase in β-glucuronidase and β-*N*-acetylglucosaminidase activity in osteosarcoma homogenates relative to homogenates of normal bone.

Carr's paper (63) contains an interesting observation. Three different experimental mouse tumors were studied. Inoculated i.p. to form noninvasive ascitic growths, these 3 experimental tumors showed no significant exoglycosidase activity. However, when the same tumors were inoculated s.c. to produce invasive growths in a viscous ground substance matrix, their exoglycosidase activity increased very markedly. Thus, it would appear that neoplastic cells have considerable exoglycosidase potential that is expressed only in the presence of the specific substrate. This observation has important implications for tissue culture studies now in progress.

Saccharo-1,4-lactone is a specific inhibitor of β-glucuronidase (78), and Carr (62) has demonstrated that the administration of this and related compounds results in a marked regression of experimental mouse tumors. Unfortunately, this compound is metabolized to toxic saccharic acid.

The problem is to find some nontoxic exoglycosidase inhibitor, and ascorbic acid, containing the essential lactone ring structure, with close structural similarities to the natural exoglycosidase substrate and possessing this physiological function, would appear to be a favored contender for this role. Furthermore, *in vitro* studies by Kojima and Hess (182) demonstrated that ascorbic acid functions as a noncompetitor inhibitor of *N*-acetylglucosaminidase. Papers on the inhibition of β-glucuronidase have been reviewed by Dutton (108).

To summarize, the dangerous features of neoplastic cell behavior (invasiveness, selective nutrition, and perhaps growth) are caused by microenvironmental depolymerization. In turn, this matrix destabilization is brought about by constant exposure to lysosomal glycosidases continually released by the neoplastic cells. Finally, ascorbate is involved in the natural restraint of this degradative enzymic activity.

Ascorbate, Collagen, and Tumor Encapsulation. The intercellular matrix is reinforced by a 3-dimensional network of interlacing collagen fibers. A generalized increase in the collagen content of the matrix can be induced by certain hormones (estrogens, androgens, corticosteroids, thyroxine) with a gradual shift from an amorphous to a more fibrotic pattern; the same change is observed in the aging process; see references in the review by Cameron (48).

The amount of collagen present determines the strength of the tissue and also its resistance to malignant infiltration. It is common knowledge that invasive tumor cells preferentially spread along "soft" tissue planes and are deflected and constrained by fibrotic tissues such as scars, fascia, capsules, and ligaments. On a generalized scale, the gradual shift to a more fibrotic pattern in the matrix induced by these hormones and the aging process may account for the increased resistance to tumor growth associated with these hormonal and constitutional changes (48).

Thus, if matrix integrity is the first line of defense against invasive growth, this defense is very powerfully reinforced by the next barrier, the collagen network. To appreciate fully the important role played by collagen in stromal resistance, we must first look at the effects of invasive cells on preexisting collagen and then at the variable ability of individual hosts to encapsulate their tumors within barriers of new fibrous tissue.

Dissolution of preexisting collagen fibers in the immediate proximity of neoplastic cells is a well-recognized feature of active invasive growths. This has led many to postulate that tumor cells release collagenase, a specific proteolytic enzyme found in both mammalian and bacterial cells. Extracts of various animal and human tumors exhibit enhanced collagenase activity (104, 131, 270, 310, 319).

As mentioned earlier, it could well be that neoplastic transformation involves an increased and sustained output of all lysosomal enzymes, including collagenase, but there could be a simpler explanation.

It will be recalled that collagen fibers consist of innumerable fibrils (themselves optically invisible) "glued together" into visible fibers by glycosaminoglycan and proteoglycan macromolecules. Exposure to glycosidases would dissolve the cement substance and, by converting visible fibers into free-floating molecular fibrils in a depolymerized matrix, would appear to exert "collagenase" activity. Either way, the lysosomal overactivity of neoplastic cells is clearly responsible for the disruption of preexisting collagen barriers ahead of malignant invasive growth.

Basement membrane is ground substance heavily reinforced with collagen. Basement membrane disorganization has been observed during culture of LS402A mouse carcinoma cells in ascorbate-deficient medium, and this effect is reversed by introducing ascorbate (263). Lysis of preexisting collagen has been studied in transplanted rat tumors (262) and during 2-aminoanthracene carcinogenesis (258).

Collagen catabolism results in an increased level of UHP, and, in the human, measurement of UHP is useful to the clinician, rising levels reflecting increasing spread and activity of the disease. It has been shown that UHP levels in patients with breast cancer are directly proportional to the spread of the disease and bear an inverse relationship to leukocyte ascorbate levels and that a loading dose of 1 g of ascorbic acid produces a sharp decrease in UHP excretion (18). Thus, invasive neoplastic cells possess the ability to disrupt preexisting collagen barriers, with increased collagen catabolism and output of resultant hydroxyproline residues, and this disruptive effect can be diminished significantly by increasing ascorbate intake.

Of even greater interest, from the point of view of practical therapy, is the variable ability of individuals to encapsulate their tumors (the so-called scirrhous response). In daily clinical practice, the whole gamut of response can be observed, ranging from the unfortunate individual with a highly anaplastic, remorselessly invasive tumor with quite negligible stromal fibrotic reaction and a very limited life expectancy to that of the more fortunate individual with a very slow-growing, practically noninvasive tumor encased in a dense, almost impermeable barrier of reactive scar tissue (the so-called atrophic scirrhous tumors) and with a clinical prognosis differing little from normal life expectancy. It is true that the former dismal picture tends to occur more frequently in the younger age groups, while the latter clinical presentation is more common in the elderly, but the differentiation by age is by no means clear-cut.

It would be a very considerable advance in cancer treatment if all tumors could be converted to the atrophic scirrhous variety. Even in the present state of our knowledge, this is not an impossible objective. To achieve this aim, we have first to consider the "soil," appreciating that the elderly, presumably because of some hormonal change, seem in general to be more able to elicit this powerful defensive reaction than the young. Thus, some consideration should be given to exploring the possibility that appropriate endocrinal adjustment (not necessarily synonymous with aging) could be used to condition the matrix and render it more adaptable to this encapsulating process (48).

However, irrespective of preconditioning the "soil," encapsulation implies an intense local deposition of fully formed collagen fibers imprisoning the invasive tumor cells. Ascorbic acid is essential for new collagen formation.

The precise role of ascorbate in collagen synthesis has been the subject of much study; see comprehensive reviews by Gould (140) and Barnes (15). During states of ascorbate deficiency, total collagen synthesis as determined by extraction techniques is unaffected (19) even though wound healing by microscopically visible collagen fibers is completely absent (88). Lack of ascorbate sharply reduces hydroxylation of prolyl and lysyl residues into hydroxyproline and hydroxylysine of mature collagen during ribosomal assembly (19, 140, 141, 174, 176), leading to instability of the triple helix of collagen (171). Such instability results in increased collagen catabolism, as has been demonstrated in scurvy (138, 272) and in cancer (18, 258, 262).

Ascorbic acid has been shown to increase collagen synthesis by fibroblasts *in vitro* (99, 280) and to maintain collagen synthesis in nonmitotic fibroblasts for extended periods (99). Prolyl hydroxylase, the enzyme hydroxylating prolyl and lysyl residues of procollagen, requires ascorbate to function *in vitro* (181), and the addition of ascorbic acid to tissue cultures stimulates the prolyl hydroxylase activity of fibroblasts (174).

Thus, there is no doubt that a sufficiency of ascorbate is essential for collagen fibrillogenesis, both by stabilizing the matrix and protecting it against the erosive effects of lysosomal glycosidases and by facilitating the hydroxylation of prolyl residues in procollagen.

To summarize, in the cancer situation, 3 factors are involved: (*a*) there is the stimulating effect of neoplastic cells creating a proliferative environment around stromal fibroblasts; (*b*) there is some evidence that the hormonal environment may influence the fibrotic response; (*c*) for an adequate scirrhous response, an abundance of ascorbate is essential, and this is one factor open to therapeutic control.

Ascorbate and Immunocompetence. It is generally accepted that the immune system plays some part in host resistance to cancer, both in the prophylactic sense of an efficient immunosurveillance system destroying neoplastic cells at an early stage in their careers and in the protective sense of retarding the growth of established tumors. Patients maintained on long-term immunosuppressive regimens have an increased incidence of certain forms of cancer, and cancer patients tend to have decreased immunocompetence as measured by standard tests (271). Any practical measure that could enhance immunocompetence could only be beneficial to the cancer patient.

The immunological defense system has the awesome task

of first distinguishing friend from foe by recognizing "self" from "not self" and then acting upon the information by identifying the target so as to permit its elimination through various mechanisms and strategies. Recognition depends upon evaluation of minute differences in molecular structure.

Lewis Thomas (321) pictures the immune system as a police force, constantly patrolling the body cells, keeping an eye open for cells becoming neoplastic and, upon recognition, destroying them. For such a system to work, cancer cells must display a surface antigen for "recognition" different from their nonneoplastic compatriots.

For certain experimental tumors, induced by specific oncogenic viruses and specific chemical carcinogens, there is clear evidence that this is so, and furthermore the antigen is specific for the carcinogenic agent (90, 221, 271). In human cancer, the picture is less clear. Certain tumors derived from the same basic cell type often express a common differentiation antigen, as measured by specific immunoprotein assay (236, 237). These so-called oncofetal antigens are also present on embryonic cells, and indeed there is some evidence that they may be present on the surface of any rapidly dividing but not necessarily neoplastic cell (271). Detection of specific serum oncofetal immunoproteins may reflect the sensitivity of existing techniques. Thus, most states of cell proliferation would be below the level of detection but, in the event of a single cell progressively dividing and subdividing to form an immense clone of neoplastic cells, the clonal antigen would be measurable as a "tumor-specific antigen" (46).

The potentially antigenic cell membrane is a highly complex structure consisting of proteins and lipoproteins, not only quite specific for the individual but also quite specific for the particular cell type, masked in an "exoskeleton" of glycosaminoglycan and collagen macromolecules. Cell division must imply unmasking and dissolution of the cell membrane involving new molecular configurations and thus altering its antigenic response. We advance the proposition that all dividing cells by cell membrane alteration elicit a weak "not self" signal as part of the general homeostatic mechanism of the body. This recognition signal would evoke a weak antigenic response specific to the particular cell type because of its unique protein display, but recognizable only if a very large number of similar clonal cells were dividing, as in the embryo or cancer.

There is some evidence that dissolution of the glycosaminoglycan component of the cell membrane to expose protein and lipoprotein configurations, as would be bound to result from the cellular release of glycosidases, plays an important part in determining antigenicity (43, 44).

Because of protein similarities, the immune response, so devastating in allograft situation, is much less effective against tumor cells indigenous to the host. Nevertheless, it does play some part in determining an individual patient's resistance to his particular tumor.

In our view, ascorbate is essential to ensure the efficient working of the immune system. The immunocompetence mechanisms are a combination of humoral and cell-mediated defensive reactions with ascorbate involved in a number of ways. These mechanisms must now be considered.

The response to the signal "foe" is the rapid mobilization of humoral and cellular agents of high specific activity and the elaboration of recognition units, specific immunoglobulins. The replication of the system is evident in its "search and destroy" function, in which the same complex molecular unit that recognizes a structure also fires a weapon, the complement system, that attacks only the structure identified by the specific antibody, even though it is present in only a single cell.

Although the complement system can be activated by a variety of factors, the specificity of this action results from the fact that it can be focused with the precision of a laser beam to the point of local impact of immunoglobulins with the specific antigen that they have been created to recognize. The cascade of complement factors thus set in motion further spreads the signal by a variety of chemical messengers and the indirect activation, through anaphylotoxin (81, 82), of other cells to induce release of activators to sustain the inflammatory response, including phagocytosis, which will be discussed in a later section.

The precision of recognition and the intensity of the response vary in the same individual from time to time according to endocrine status, reflecting such constitutional factors as age, sex, and hormone environment (90). The influence of endocrinal status on host resistance to cancer will be discussed below. The importance of endocrinal status in preconditioning the "soil" recalls our earlier discussion that similar factors influence the efficiency of the fibrotic response.

The humoral factors involved in immunocompetence are the cell surface-specific immunoglobulins and their ultimate weapon, the complement cascade.

The reticuloendothelial system is concerned with the precise design and production of immunoglobulins, proteins which contain a large number of disulfide bonds (relative to other proteins) the function of which is to bridge the light and heavy chains. The role of the ascorbic acid-dehydroascorbic acid system in the biosynthesis of S—S bonds has been extensively discussed by Lewin (195), who strongly concludes that ascorbate is essential for immunoglobulin synthesis. A positive correlation between serum ascorbate levels and serum IgG and IgM titers was reported by Vallance (325), studying human subjects isolated from any sources of new infection for nearly 1 year on a remote British Antarctic Research Station.

Regarding the more definitive weapon of the complement cascade, it has recently been demonstrated that the administration of supplemental sodium ascorbate to guinea pigs significantly increases the esteratic power of the first component of complement (C_1 esterase activity, without which the whole complement cascade is inoperable) in animals immunized with antipenicilloyl (guinea pig) γ-globulin.[3]

In cell-mediated immunity, immunocompetence is exercised overwhelmingly by the lymphocytes. In tumors, the degree of stromal lymphocyte infiltration is a measure of the efficiency of host resistance to the neoplastic process. Thus, the degree of lymphocytic infiltration is now accepted as a reliable prognostic indicator.

Relative to other cells, lymphocytes contain substantially

[3] G. Feigen, unpublished observations.

higher concentrations of ascorbate, and there are strong indications that this characteristic feature is "purposeful" and related to their active role in cell-mediated immunocompetence.

It has been neatly demonstrated that guinea pigs maintained on prescorbutic diets have markedly reduced immunocompetence as shown by their prolonged tolerance of allografts and that this change is related to abnormally low lymphocyte ascorbate levels. When ascorbate is administered to restore lymphocyte ascorbate levels to normality, the allografts are promptly and universally rejected (167). This observation led us to suggest that the opposite condition of lymphocyte ascorbate saturation should be associated with enhanced immunocompetence (55). This prediction has subsequently been confirmed. Yonemoto *et al.* (356) demonstrated in healthy young subjects that a high loading dose of ascorbate (5 g/day for 3 days) evokes a significant increase in lymphocyte blastogenesis as measured by response to phytohemagglutinin challenge and, furthermore, that this increase in immunocompetence was further significantly enhanced by larger doses (10 g). A similar effect of increased T-lymphocyte responses to concanavalin 2 with increased intake of ascorbate by mice has been reported by Siegel and Morton (300). These observations support the common sense view that lymphocytes rich in ascorbate should be able to conduct their protective business more efficiently than those that are not, a characteristic feature of established cancer.

To summarize, cancer patients generally exhibit diminished immunocompetence and almost invariably have low lymphocyte ascorbate content. The simplest and safest way to enhance immunocompetence in such patients and to ensure that their humoral and cell-mediated defense systems are working at maximum efficiency is to increase their ascorbate intake. Only when this increased demand and utilization in cancer are fully satisfied can these immune mechanisms afford maximum protection against the wayward cancer cell.

Ascorbate and Hormone Balance. The highest concentrations of ascorbate are found in the adrenal and pituitary glands, and the terminal stages of scurvy are just preceded by complete depletion of adrenal ascorbate, leading, it has been frequently stated, to "scurvy death" from adrenocortical failure. This has caused many to suggest that the ascorbic acid-dehydroascorbic acid system plays an important role in the synthesis and release of hormones of the adrenopituitary axis. The evidence for this is both conflicting and confusing (13, 72, 73, 102, 277, 278).

It has of course been known for many years that changes in the hormonal status of an individual and in particular relative changes in the different components of the adrenopituitary axis can sometimes exert a significant effect on host resistance to neoplasia; see the extensive review by Stoll (312). In 1974, we offered the tentative suggestion that availability of ascorbate by influencing the interrelationship between the members of this hormonal orchestra might determine whether an individual possesses a favorable or an unfavorable steroid environment (55). The proposal has not yet been studied, and the suggestion remains without experimental evidence.

Ascorbate and Phagocytosis. Phagocytosis is believed to play an important part in cell-mediated immune response to tumor cells. In addition, practically all tumors ulcerate through adjacent surfaces (skin, gastrointestinal tract, respiratory tract, etc.) and become subject to secondary bacterial invasion. Thus, the toxemia of secondary infection becomes part of the cancer illness. Efficient phagocytosis offers some protection against this almost inevitable complication.

Numerous studies have demonstrated that ascorbate is required for active phagocytosis both *in vivo* and *in vitro* (71, 87, 128, 137, 210, 240, 270).

Leukocyte motility is dependent upon the activity of the hexose monophosphate shunt, which in turn is activated by ascorbate (84, 94, 173) oxidizing NADPH to $NADP^+$ with release of bacteriocidal peroxide (252). A review of oxygen-dependent phagocytosis has recently been published (9).

The striking effect of increasing p.o. intake of ascorbate on human lymphocyte blastogenesis as measured by response to phytohemagglutinin and [^{3}H]thymidine uptake was noted in the previous section.

Ascorbic Acid and Tumor Prevention

Any measure that will retard the growth of established tumors should also, if applied early enough, effectively suppress "latent" tumors and thus in clinical terms have some prophylactic value. The purpose of this section is to consider whether ascorbate offers any degree of protection against a variety of carcinogenic agents.

The Antiviral Activity of Ascorbic Acid. Some cancers are thought to be initiated by oncogenic viruses [see discussion by Dulbecco (106)], and this whole area is the focus of much current research interest, with proposals that antiviral agents be used in cancer prophylaxis and chemotherapy (208). Against this background the antiviral activity of ascorbic acid assumes fresh importance.

The clinical evidence that supplemental ascorbate offers protection against a broad spectrum of viral disease has been reviewed by Pauling (252, 254) and Stone (313) in their respective books. Particular mention has to be made of the pioneer work of Jungeblut (162, 163), in demonstrating *in vitro* and *in vivo* activity against the poliomyelitis virus, and the clinical work of Klenner (178, 179), who strongly advocates its use against a wide variety of viral disorders. In Japan, Morishige and Murata (227)[4] have used supplemental ascorbate in the successful prevention and treatment of measles, mumps, viral orchitis, viral pneumonia, herpes zoster, and viral encephalitis. The most striking effect recorded by these Japanese workers has been the virtually complete prevention of posttransfusion hepatitis in a country where such a complication is common (227).

In the laboratory, Murata *et al.* (228–230) have investigated the inactivation of bacterial viruses by ascorbate, and they have been able to demonstrate the inactivation of a phage of *Lactobacillus casei* and a variety of viruses of *Escherichia coli* and *Bacillus subtilis*.

The mechanism of *in vitro* viral inactivation is still unclear. It could involve the liberation of peroxide from oxidation of ascorbate, because the addition of catalase provides protection (290, 348). However, Murata *et al.* (231) noted that

[4] F. Morishige and A. Murata, personal communication.

concentrations of peroxide that should have been produced during ascorbate oxidation had no effect on bacterial virus Δ-A. Murata has postulated that free-radical intermediates produced during ascorbate oxidation are the active agents in viral inactivation and has presented sound arguments in support of his view. In this respect, the role of monodehydroascorbate radical in the termination of free radical reactions and its consequent biological function have been reviewed by Bielski *et al.* (28).

Ascorbate has also been shown to enhance interferon production (92, 298–301), as well as to enhance the phagocytic properties of the reticuloendothelial system, both potent *in vivo* defenses against viruses. The interaction of ascorbate with L-lysine in rats (67) may also contribute to the effectiveness of ascorbate.

Whatever the mechanisms may be, there is clinical and experimental evidence that supplemental ascorbate possesses some antiviral activity. Thus, if a viral etiology of human cancer is ever proved, ascorbate might be expected to exert some prophylactic and therapeutic effect.

Ascorbate and the Carcinogenic Hydrocarbons. There is some fragmentary evidence that ascorbate offers some protection against carcinogenic hydrocarbons.

A very marked stimulation of ascorbate synthesis in rats and mice can be evoked by exposure to various noxious compounds including the carcinogenic hydrocarbons, with methylcholanthrene the most powerful (40, 47, 80, 323).

The mixed-function oxidases are a group of closely related microsomal enzymes that metabolize many classes of compounds and are particularly important in the inactivation of chemical carcinogens. Although these microsomal enzymes have been studied most extensively in liver tissue, recent evidence indicates their occurrence in tissues of all major portals of entry, including the gastrointestinal tract, lung, skin, and placenta (337). Thus they represent the initial metabolic barrier to noxious foreign substances. These enzymes are affected by age, species and strain, sex, and nutritional state (358) and require NADPH and molecular oxygen (339–345) for their action. Ascorbic acid is also a requirement of the mixed-function oxidases that hydroxylate tryptophan (83, 233, 307), tyrosine (187, 233, 307, 324), dopamine (193), and phenylalanine (314). Microsomal metabolism of lipid-soluble foreign compounds or carcinogens yields products generally more water soluble, which greatly increases their rate of excretion. It is not surprising therefore that inducers of microsomal mixed-function oxidases (polycyclic hydrocarbons, pentobarbital, phenobarbital, chlordane, β-naphthoflavone, and phenothiazene) provide protection from carcinogens, whereas inhibitors of these enzymes (including carbon tetrachloride, organophosphorus insecticides, ozone, and carbon monoxide) potentiate carcinogenic effects; among the carcinogens studied are benzo(*a*)pyrene, 7,12-dimethylbenzanthracene, *N*-2-fluorenylacetamide, 4-dimethylaminostilbene, urethan, aflatoxin, diethylnitrosamine, aminoazo dyes, 3-methyl-4-dimethylaminoazobenzene, 2-acetylaminofluorene, and bracken fern [see review by Wattenberg (341)].

The direct action of ascorbate on carcinogens has also been reported. Warren (335) demonstrated the oxidation of aromatic hydrocarbons *in vitro* by ascorbate. Floyd *et al.* (122) reported that ascorbate inhibits conversion of *N*-hydroxy-*N*-acetyl-2-aminofluorene into 2 more potent carcinogens, 2-nitrosofluorene and *N*-acetyl-2-aminofluorene. This oxidation-reduction reaction reduces peroxide by the action of peroxidase and an electron donor; under the conditions used, ascorbate was preferentially oxidized. Attack of the carcinogen *N*-acetoxy-2-acetamidofluorene on guanosine is also prevented *in vitro*, probably through the quenching of the triplet nitrenium ion by proton abstraction (291).

Ascorbic acid deficiency in guinea pigs has been shown to decrease microsomal cytochrome P-450 activity to about 50% of its normal value (96, 97, 192, 203, 268) cytochrome b_5 (96, 97, 174, 192, 302, 358), NADPH-linked cytochrome *c* reductase (174, 358), and *O*- and *N*-demethylase activities (174, 185, 192, 358). Conflicting data by Kato *et al.* (170) in which the above-mentioned enzyme activities were not affected by ascorbate deficiency seem to be the result of a shorter depletion period (12 days). Zannoni and Sato (Ref. 174, p. 119) have shown that microsomal enzyme activities are not decreased significantly in a 10-day depletion experiment but are significantly decreased after 21 days. Repletion experiments with scorbutic guinea pigs have shown that supplementation with ascorbic acid returned cytochrome P-450 and demethylation activities to normal within 48 hr (97, 192, 203).

Ascorbate and Nitrosamines. Nitrosamines, products of the reaction of nitrite with a secondary amine, alkylurea, or an *N*-alkylcarbamate group, have been strongly implicated as a major environmental carcinogen (196). Atmospheric nitrosamine pollution could be significant. In some parts of the United States dimethylnitrosamine levels rise to 0.1 μg/cu m of "clear" air, which equals about 1.0 to 1.4 μg/person/day; by comparison, a pack of cigarettes contains around 0.8 μg and 4 slices of nitrite-preserved bacon contain about 5.5 μg dimethylnitrosamine (296). A variety of experimental tumors of the alimentary tract, liver, lung, and urinary bladder can be produced by nitrosamines (218, 219, 234, 275), and a number of nitrosation products are carcinogenic *in vivo*, including those of citrulline, ephedrine, sarcosine, morpholine, methylurea, piperazine, pyrrolidine, aminopyrine, diethylamine, and dimethylamine (347). Nitrosation is usually studied under simulated gastric conditions, but Hill *et al.* (155) have demonstrated that nitrosamines may be produced at any site where bacteria, secondary amine, and nitrite or nitrate are present. Another aspect deserving serious consideration is the transplacental transfer of nitrosamines; respiratory tract tumors were produced at a rate of up to 97% in offspring of mice given diethylnitrosamine during the last 4 days of pregnancy (222).

Ascorbic acid has been shown to exert a protective effect against carcinogenesis by nitrite reacting with aminopyrine (123), morpholine (174), piperazine (Ref. 174, p. 175), and raw fish (204) as well as to reduce the acute hepatotoxicity resulting from feeding nitrite and dimethylamine (61) or nitrite and aminopyrine (Refs. 145, 168, and 169; Ref. 174, p. 160). Studies with preformed nitrosamines have shown that ascorbic acid exerts little or no protective effect (Ref. 174, p. 175). Ascorbic acid does not react with amines, nor does it increase the rate of nitrosamine decomposition; it exerts its protective effect largely by reaction with nitrite and nitrous acid (Refs. 5 and 220; Ref. 174, p. 181). Under simulated gastric conditions (37°, pH 1.5), an ascorbate/nitrite molar ratio of 1/1 provided 37%, a ratio of 2/1 provided 74%, and a ratio of 4/1 provided 93% protection from *in vitro* nitrosation of methylurea (Ref. 174, p. 181). Because nitrites are found in many foods as normal reduction product of nitrates or by deliberate introduction during processing (Ref. 174, p. 175), it is significant that ascorbic acid added or present in foods offers some protection against dangerous nitrosamine formation (Refs. 5 and 352; Ref. 174, p. 181).

It has also been shown, using the Ames test (148), that ascorbate inhibits bacterial mutagenesis by *N*-methyl-*N*-nitrosoguanidine, and Marquardt, Rugino, and Weisburger, using the same test procedure, have shown mutagenic (and presumably carcinogenic) activity in nitrite-treated foods and have suggested that human stomach cancer might well be related to this dietary factor, which to some extent might be reduced by ascorbate (204).

Ascorbate and Tryptophan Metabolites in Bladder Cancer. Exogenous or endogenously formed chemical carcinogens can often be implicated in the causation of human bladder cancer. The known relationship between occupational exposure to 4-aminophenyl, benzidine, and 2-naphthylamine and bladder cancer has stimulated investigations into the carcinogenicity of *N*-hydroxy aromatic amines (98, 265) and *o*-hydroxy aromatic amines (2). A large number of *o*-hydroxytryptophan metabolites are known carcinogens (357), and efforts have been made to demonstrate increases in tryptophan metabolites in the urine of bladder cancer patients, but the results have been equivocal (2); however, after a loading dose of L-tryptophan, bladder cancer patients excrete a significantly increased amount of the metabolites kynurenine, xanthurenic acid, and *o*-aminohippuric acid, suggesting that such subjects have some abnormality of tryptophan metabolism (357).

Chemiluminescence resulting from the nonenzymatic degradation of the tryptophan metabolite 3-HOA (and believed to reflect degradation from Compound I to Compound IV as shown in Chart 2) is significantly increased in the urine of bladder cancer patients and in the urine of heavy tobacco smokers (285, 286). The administration of ascorbate (1 to 2 g p.o. per day) results in a significant decrease in chemiluminescence and completely prevents the formation of Compound IV even in voided urine to which 3-HOA has then been added (286).

Urine from patients with bladder cancer oxidizes 3-HOA faster than does control urine *in vitro*, and the addition of ascorbate inhibits this reaction (259, 286). With the general working hypothesis that metabolites of 3-HOA are carcinogenic to uroepithelium, the increased amounts of such metabolites, especially Compound IV, found in the urine of heavy tobacco smokers explains their increased susceptibility to bladder cancer. 3-HOA implanted directly into the mouse bladder is carcinogenic, and this effect can be prevented by ascorbate (259, 284). The whole concept that supplemental ascorbate has a protective and inhibitory value in bladder cancer has been developed by Schlegel (284) and his associates in the Department of Urology of Tulane University. The experimental, biochemical, and clinical aspects of their work have been the subject of a recent review (284).

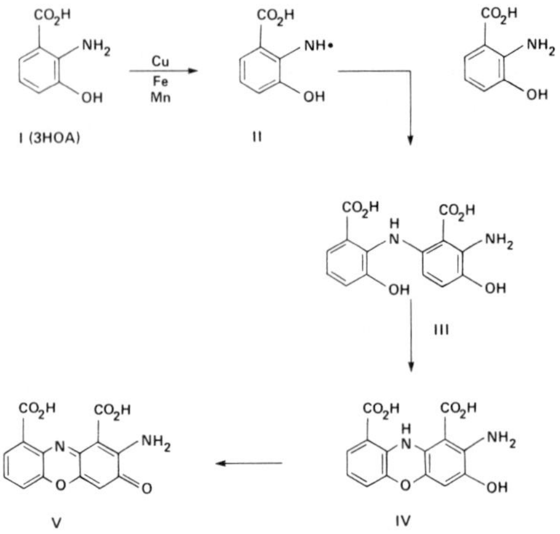

Chart 2. Postulated pathway of nonenzymatic oxidative decomposition of 3-HOA. From Schlegel *et al.* (286).

Ascorbate and Cigarette Smoking. The relationship between cigarette smoking and bladder cancer has been alluded to above. Of far greater importance in terms of numbers is the proved relationship between smoking and lung cancer.

There is considerable evidence that smoking depletes ascorbic acid reserves (4, 142, 156, 255–257, 269), as shown by diminished whole blood, serum, and leukocyte ascorbate levels in smokers relative to nonsmoker controls. It would be presumptuous to say that this difference reflects an increased utilization acting in a protective capacity. Nevertheless, in the present state of our knowledge, it would seem a wise precaution for the compulsive heavy smoker to increase deliberately his ascorbate intake.

Ascorbate and UV Carcinogenesis. Excessive exposure to UV is carcinogenic. In the human, this is evident in the increased incidence of various forms of skin cancer in fair-skinned individuals resident in areas of high solar intensity, such as the southern United States, South Africa, and Australia.

Experimentally, the carcinogenic effects of actinic UV non-ionizing radiation can be duplicated and studied using albino hairless mice (36). Homer Black and his associates have demonstrated that skin exposure in such animals to high-intensity UV results in the formation of the carcinogenic sterol cholesterol-5α,6α-epoxide and in skin cancer and that the process can be suppressed by feeding the animals a number of antioxidants, including ascorbic acid (32–36, 65, 199).

Other Relevant Effects of Ascorbic Acid

Ascorbate and Energy Production. Cytochromes P-450 and b_5 are decreased by ascorbate deprivation in the guinea pig, and because cytochromes are intimately associated with electron transport, and therefore oxidative phosphorylation, it is possible that cell respiratory impairment could result from relative ascorbate deficiency; and, of course, an increase in anaerobic glycolysis, coupled with a decrease in oxidative respiration, is recognized to be a fundamental biochemical change in the cancer process (334).

Viral transformation of cell cultures decreases oxygen consumption and increases lactate production (29, 30). An increased rate of gluconeogenesis has been demonstrated in a variety of animal tumors (297) and has been linked to the progressive weight loss syndrome of cancer patients (157), the belief being that actively fermenting neoplastic cells meet their increased glucose requirements at the expense of the host.

All glycolytic enzymes are increased in amount during carcinogenesis (37, 346). Glucokinase is increased to 570% of its control value in the liver of animals exposed to the carcinogen 3′-methyl-4-dimethylaminoazobenzene (12).

These alterations in cellular pathways can to a certain extent be corrected by ascorbate. Takeda and Hara (318) reported decreased oxidation of citrate and lowered activity of aconitase in scorbutic guinea pigs, indicating an impaired tricarboxylic acid cycle. Addition of ascorbate to cultures of embryonic bone tissue (266) or cultures of polymorphonuclear leukocytes (113) results in increased oxygen consumption and a decrease in lactate production. Studies by Benade *et al.* (25, 26) demonstrated 72% inhibition of anaerobic glycolysis in Ehrlich ascites tumor cells by ascorbate; 96% inhibition was achieved when 3-amino-1,2,4-triazole was added with ascorbate, and these additions proved to be highly toxic to the tumor cells. The cytotoxicity was attributed primarily to intracellular H_2O_2 production and the synergistic effect of the aminotriazole to its inhibition of catalase (25).

Ascorbate as an Antioxidant. The antioxidant properties of ascorbic acid have been known since its discovery (151, 308, 328), and because other antioxidants appear to possess some anticancer activity ascorbate might also function in this respect. Thus, Wattenberg has demonstrated the inhibition of tumorigenesis by a number of known carcinogens, such as 7,12-dimethylbenz(*a*)anthracene, benzo(*a*)pyrene, urethan, uracil mustard, dimethylhydrazine, 4-nitroquinoline *N*-oxide, and diethylnitrosamine, by several antioxidants, including butylated hydroxytoluene, butylated hydroxyanisole, ethoxyquin, and some sulfur-containing reactive compounds (337–345). Slaga and Bracken (304) reported that ascorbic acid as well as other antioxidants prevented the initiation of skin tumors following the application of 7,12-dimethylbenz(*a*)anthracene.

As mentioned above, Black has described cholesterol-α-oxide, a precarcinogen formed in skin during exposure to high-intensity UV, and has prevented its formation in hairless mice by addition of ascorbate, α-tocopherol, glutathione, and butylated hydroxytoluene to the diet. Antioxidants may exert their effects by protecting against carcinogen-induced chromosomal breakage (294), by reducing carcinogen-induced peroxidation (293), by altering liver microsomal metabolism (306), or by a combination of these and other actions, as reviewed by Passwater (250) and by Wattenberg (344).

Dehydroascorbic Acid. Oxidation products of ascorbic acid have antitumor activity *in vivo*. Dehydroascorbic acid (150 mg/kg body weight) and 2,3-diketogulonic acid (115 mg/kg body weight) inhibited growth of solid Sarcoma 180 in mice by 88 and 54%, respectively (355). In the same experiments, the antitumor activity of ascorbic acid (46% inhibition) was enhanced by the addition of a compound of copper (to 69% inhibition), indicating that the active agent is an oxidation product. The antitumor activity of dehydroascorbic acid, and to a lesser extent of other metabolites of ascorbic acid, erythorbic acid and dehydroerythorbic acid, was recently confirmed using solid Sarcoma 180 (245, 246). Another metabolite, 5-methyl-3,4-dihydroxytetrone, inhibits growth of solid Sarcoma 180 by 50% (244, 322). The activity of these agents is thought to be mediated by interaction with DNA and decomposition of apurinic acid to deoxycytidylic acid, as well as decomposition of the oligo form of pyrimidine nucleotides, has been demonstrated (235, 244, 247, 355). It has also been suggested that dehydroascorbic acid functions as an electron acceptor in the regulation of mitosis (111, 112).

Ascorbate and Cyclic Nucleotides. Many biological activities are potentiated by hormonal actions that utilize cyclic 3′:5′-AMP and cyclic 3′:5′-GMP as "second messengers." Lewin (195) has reviewed extensive evidence showing that ascorbate potentiates the formation of cyclic 3′:5′-AMP and is concerned in the inhibition of processes that reduce the concentration of both cyclic 3′:5′-AMP and cyclic 3′:5′-GMP by hydrolyzing them to 5′-AMP and 5′-GMP, respectively. The role of these cyclic nucleotides in cancer is uncertain, but diminished adenyl cyclase activity has been noted in polyoma virus-transformed cells (45), and cyclic 3′:5′-AMP has been shown to inhibit cell multiplication *in vitro* (195) and tumor growth *in vivo* (130).

Ascorbate and Erythropoieses. Anemia is a common feature of cancer. Ascorbate is known to promote the absorption and utilization of ingested iron and is necessary for a full erythroblastic response. Thus the anemia of cancer would be helped by increasing ascorbate intake.

Ascorbate and Oxidation-Reduction Potential. The ascorbic acid-dehydroascorbic acid system is believed to play an important part in maintaining optimum oxidation-reduction conditions in the tissues. The oxidation-reduction potential, like acid-base balance and pH, must be balanced within fairly narrow limits for normal health. Any disturbance of oxidation-reduction potential, such as would result from depletion of ascorbate reserves in cancer, could have deleterious systemic effects (3).

***In Vitro* Studies against Cancer Cells.** Because ascorbic acid is nontoxic to normal tissues, few have investigated whether it is equally nontoxic to neoplastic tissues, and what evidence there is tends to be contradictory. Thus, Vogelaar and Erlichman (329) reported that ascorbate enhanced the growth of mouse Sarcoma 180 cells *in vitro*, while Park *et al.* (249) demonstrated stimulation of a mouse plasmacytoma cell line. On the other hand, ascorbate at fairly high concentration is cytotoxic to Ehrlich ascites carcinoma cells (25, 26), and at a much lower "physiological" concentration to 3T3 cells in tissue culture (85). It has been reported that ascorbic acid increases glucose metabolism in neoplastic cells but not in normal cells in identical *in vitro* conditions (257a).

Studies of Experimental Tumors in Laboratory Animals. Omura and his associates in Japan (232, 245, 246, 322, 354, 355) have reported that ascorbic acid and its metabolites exhibit significant inhibitory effects against the "take" and growth of Sarcoma 180 in mice. Other experiments with mouse and rat tumors have yielded equivocal results (41, 305, 336, 349, 351). Even with the experimental animal of choice, the guinea pig, the reported effects are contradictory. Thus, Russell *et al.* (274) reported that ascorbate deprivation increased guinea pig susceptibility to methylcholanthrene carcinogenesis and promoted tumor growth and spread, whereas Migliozzi (216) found that ascorbate deprivation retarded tumor growth and ascorbate supplementation enhanced it. It seems to us important that these guinea pig experiments be repeated using different levels of ascorbate intake; it is possible that the contradictory results may arise from different points on a dose-response curve.

Epidemiological Studies in Human Cancer. A number of epidemiological studies have demonstrated some relationship between patterns of dietary ascorbate intake in large population groups and their incidence of cancer of various types and cancer in general (27, 31, 70, 86, 107, 143, 144, 149, 154, 158, 180, 353). We propose to review these and similar studies elsewhere.

Clinical Trials in the Management of Human Cancer. Schlegel's use of ascorbate to retard human bladder cancer has already been mentioned (259, 284–287). DeCosse and his associates in Wisconsin (95) reported that ascorbate p.o. induced some regression in familial colorectal polyposis, a well-recognized premalignant condition (42, 105, 200, 209), and they recommend its use as a prophylactic measure. This advice gains support from the demonstration that, in the same clinical condition, ascorbate p.o. reduces the amount of mutagenic (and presumed carcinogenic) fecal steroids (188).

These, however, are specific neoplastic disorders. The scope of this review suggests that supplemental ascorbate may exert a general anticancer effect.

As stressed in the introduction, no properly designed prospective clinical trial has as yet been carried out to assess the value of supplemental ascorbate in general cancer management. However, for those interested in designing such a trial, an encouraging background of publications exists, ranging from individual case reports, through anecdotal accounts, to pilot studies involving large numbers of advanced cancer patients (64, 68, 100, 114, 129, 146, 159, 164, 177, 184, 214, 248, 261, 283, 287, 288, 289, 320, 327, 330, 331, 332), all reporting some degree of clinical benefit conferred by supplemental ascorbate. Our own clinical studies, discussed in several publications (51–53, 56–59, 226), strongly indicate that supplemental ascorbate not only increases well-being but also produces a statistically significant increase in the survival times of advanced cancer patients. Present evidence suggests to us that supplemental ascorbate can offer some degree of benefit to all advanced cancer patients and quite remarkable benefit to a fortunate few and that it has even greater potential value in the supportive treatment of earlier and more favorable patients.

Conclusion

There is evidence from both human and experimental animal studies that the development and progress of cance evokes an increased requirement for ascorbic acid. Ascorbic acid is essential for the integrity of the intercellula matrix and its resistance to malignant infiltrative growth and there is strong evidence that it is involved in the inhibition of invasive tumor enzymes. It is required for the formation of new collagen, allowing the resistant patient to enmesh his tumor cells in a barrier of new fibrous tissue There is good evidence that high intakes of ascorbate potentiate the immune system in various ways. Ascorbate may also offer some protection against a variety of chemica and physical carcinogens and against oncogenic viruses and is also involved in a number of other biological processes believed to be involved in resistance to cancer.

The collective evidence suggests that increasing ascorbate intake could produce measurable benefits in both the prevention and the treatment of cancer, and pilot clinica studies tend to support this view. Ascorbic acid has a unique advantage relative to other remedies for cancer; it is almost completely safe and harmless even when given in sustained high doses for prolonged periods of time. The risks associated with such regimens in cancer have been discussed elsewhere and are thought to be acceptable (52 54, 55, 252, 254); these are (*a*) a clinical suspicion that, in the very rare patient with a very rapidly growing tumo existing at the very limits of nutritional support through sinusoids and enzyme-assisted diffusion, the sudden exposure to large doses of ascorbate may precipitate widesprea tumor hemorrhage and necrosis with real danger to the patient, (*b*) a much stronger suspicion that the sudden discontinuation of such an established regimen produces a rebound effect of a precipitous drop in tissue ascorbate with brisk reactivation of the hitherto controlled neoplastic process, and (*c*) the theoretical but extremely remote risk that a susceptible few of such patients might develop an oxalate urinary tract stone. Earlier concern that high ascorbate intakes might induce vitamin B_{12} deficiency has been shown to be fallacious, and the result of ascorbate interfering with the laboratory assay (Ref. 254, p. 114; Ref. 133). The risks, real or theoretical, are minimal and acceptable to the cancer patient and to any experienced oncologist.

Bearing this in mind, Anderson (3) has recently state "The risk/benefit ratio relative to the severity of the disease as well as to other available treatments in cancer is so heavily weighted in favor of vitamin C in this situation tha validation or refutation by other groups will presumabl occur quite quickly."

We agree and believe it to be essential that extensive studies of ascorbic acid in cancer be made without delay

Acknowledgments

We acknowledge our indebtedness to Allan Campbell, Peter Ghosh George Feigen, Morton Klein, Samuel Klein, Onno Meier, and Dougla Rotman for stimulating discussions and correspondence. We thank Ruth Reynolds, Anita Maclaren, Dorothy Munro, and Lynne Wilcox for their help

References

1. Abercrombie, M., and Ambrose, E. J. The surface properties of cance cells: a review. Cancer Res., *22:* 525–548, 1962.
2. Alifano, A., Papa, S., Tancredi, F., Elico, M. A., and Quagliariello, E Tryptophan-nicotinic acid metabolism in patients with tumors of th bladder and kidney. Br. J. Cancer, *18:* 386–389, 1964.

3. Anderson, T. W. New horizons for vitamin C. Nutr. Today, *12:* 6-13, 1977.
4. Andrzejewski, S. A. Studies on the toxicity of tobacco and tobacco smoke. Acta Med. Pol., *5:* 407-408, 1966.
5. Archer, M. D., Tannenbaum, S. R., Fan, T. Y., and Weisman, M. Reaction of nitrite with ascorbate and its relation to nitrosamine formation. J. Natl. Cancer Inst., *54:* 1203-1205, 1975.
6. Aronson, N. N., Jr., and Davidson, E. A. Lysosomal hyaluronidase from rat liver. I. Preparation. J. Biol. Chem., *242:* 437-440, 1967.
7. Aronson, N. N., Jr., and Davidson, E. A. Lysosomal hyaluronidase from rat liver. II. Properties. J. Biol. Chem., *242:* 441-444, 1967.
8. Astrup, T., and Alkjaersig, N. Polysaccharide polysulphuric acids as anti-hyaluronidases. Nature (Lond.), *166:* 568-569, 1950.
9. Babior, B. M. Oxygen-dependent microbial killing by phagocytes. N. Engl. J. Med., *298:* 659-668, 721-725, 1978.
10. Balazs, E. A. Chemistry and molecular biology of the intercellular matrix. New York: Academic Press, Inc., 1970.
11. Balazs, E. A., and von Euler, J. The hyaluronidase content of necrotic tumor and testis tissue. Cancer Res., *12:* 326-329, 1952.
12. Balinsky, D., Cayanis, E., Albrecht, C. F., and Bersohn, I. Enzymes of carbohydrate metabolism in rat hepatoma induced by 3'-methyl-4-dimethylaminoazobenzene. S. Afr. J. Med. Sci., *37:* 95-99, 1972.
13. Banerjee, S., and Singh, H. D. Adrenal cortical activity in scorbutic monkeys and guinea pigs. Am. J. Physiol., *190:* 265-267, 1957.
14. Barkhan, P., and Howard, A. N. Distribution of ascorbic acid in normal and leukemic human blood. Biochem. J., *70:* 163-168, 1958.
15. Barnes, M. J. Function of ascorbic acid in collagen metabolism. Ann. N. Y. Acad. Sci., *258:* 264-277, 1975.
16. Barton, G. M. G., Laing, J. E., and Barsoni, D. The effect of burning on leucocyte ascorbic acid and the ascorbic acid content of burned skin. Int. J. Vitam. Nutr. Res., *42:* 524-527, 1972.
17. Barton, G. M. G., and Roath, O. S. Leucocyte ascorbic acid in abnormal leucocyte states. J. Clin. Pathol. (Lond.), *29:* 86, 1976.
18. Basu, T. K., Raven, R. W., Dickerson, J. W. T., and Williams, D. C. Leucocyte ascorbic acid and urinary hydroxyproline levels in patients bearing breast cancer with skeletal metastases. Eur. J. Cancer, *10:* 507-511, 1974.
19. Bates, C. J., Bailey, A. J., Prynne, C. J., and Levene, C. I. The effect of ascorbic acid on the synthesis of collagen precursor secreted by 3T6 mouse fibroblasts in culture. Biochim. Biophys. Acta, *278:*372-390, 1972.
20. Beiler, J. M., and Martin, G. J. Inhibition of hyaluronidase action by derivatives of hesperidin. J. Biol. Chem., *174:* 31-35, 1948.
21. Bekesi, J. G., Bekesi, E., and Winzler, R. J. Inhibitory effect of D-glucosamine and other sugars on the biosynthesis of protein, ribonucleic acid, and deoxyribonucleic acid in normal and neoplastic tissues. J. Biol. Chem., *244:* 3766-3772, 1969.
22. Bekesi, J. G., Molnar, Z., and Winzler, R. J. Inhibitory effect of D-glucosamine and other sugar analogs on the viability and transplantability of ascites tumor cells. Cancer Res., *29:* 353-359, 1969.
23. Bekesi, J. G., and Winzler, R. J. The effect of D-glucosamine on the adenine and uridine nucleotides of Sarcoma 180 ascites tumor cells. J. Biol. Chem., *244:* 5663-5668, 1969.
24. Bekesi, J. G., and Winzler, R. J. Inhibitory effects of D-glucosamine on the growth of Walker 256 carcinosarcoma and on protein, RNA, and DNA synthesis. Cancer Res., *30:* 2905-2912, 1970.
25. Benade, L., Howard, T., and Burk, D. Synergistic killing of Ehrlich ascites carcinoma cells by ascorbate and 3-amino-1,2,4-triazole. Oncology, *23:* 33-43, 1969.
26. Benade, L. E. Ascorbate toxicity in Ehrlich ascites carcinoma cells. Ph.D. Dissertation, George Washington University, St. Louis, Mo., 1971.
27. Berge, T., Ekelund, G., Mellner, C., Pihl, B., and Wenckert, A. Carcinoma of the colon and rectum in a defined population. Acta Chir. Scand. Suppl., *483:* 1-86, 1973.
28. Bielski, B. H., Richter, H. W., and Chan, P. C. Some properties of the ascorbate free radical. Ann. N. Y. Acad. Sci., *258:* 231-237, 1975.
29. Bissell, M. J., Hatie, C., and Rubins, H. Patterns of glucose metabolism in normal and virus-transformed chick cells in tissue culture. J. Natl. Cancer Inst., *49:* 555-565, 1972.
30. Bissell, M. J., White, R. C., Hatie, C., and Bassham, J. A. Dynamics of metabolism of normal and virus-transformed chick cells in culture. Proc. Natl. Acad. Sci. U. S. A., *70:* 2951-2955, 1973.
31. Bjelke, E. Epidemiologic studies of cancer of the stomach, colon, and rectum, with special emphasis on the role of diet. Scand. J. Gastroenterol., *9* (Suppl. 31): 1-235, 1974.
32. Black, H. S., and Chan, J. T. Suppression of ultraviolet light-induced tumor formation by dietary antioxidants. J. Invest. Dermatol., *65:* 412-414, 1975.
33. Black, H. S., and Chan, J. T. Etiologic related studies of ultraviolet light-mediated carcinogenesis. Oncology, *33:* 119-122, 1976.
34. Black, H. S., and Chan, J. T. Experimental ultraviolet light-carcinogenesis. Photochem. Photobiol., *26:* 183-199, 1977.
35. Black, H. S., and Douglas, D. R. Formation of a carcinogen of natural origin in the etiology of ultraviolet light-induced carcinogenesis. Cancer Res., *33:* 2094-2096, 1973.
36. Black, H. S., and Lo, W. B. Formation of a carcinogen in UV-irradiated human skin. Nature (Lond.), *234:* 306-308, 1971.
37. Bodansky, O. Biochemistry of Human Cancer. New York: Academic Press, Inc., 1975.
38. Bodansky, O., Wroblewski, D., and Markardt, B. Concentrations of ascorbic acid in plasma and white cells of patients with cancer and noncancerous chronic disease. Cancer Res., *11:* 238-242, 1951.
39. Boyland, E. The selective absorption of ascorbic acid by guinea-pig tumour tissue. Biochem. J., *30:* 1221-1224, 1936.
40. Boyland, E., and Grover, P. L. Stimulation of ascorbic acid synthesis and excretion by carcinogenic and other foreign compounds. Biochem. J., *81:* 163-168, 1961.
41. Brunschwig, A. Vitamin C and tumor growth. Cancer Res., *3:* 550-553, 1943.
42. Buntain, W. L., Remine, W. H., and Farrow, G. M. Premalignancy of polyps of the colon. Surg. Gynecol. Obst., *134:* 499-508, 1972.
43. Burger, M. M., and Goldberg, A. R. Identification of a tumor-specific determinant on neoplastic cell surfaces. Proc. Natl. Acad. Sci. U. S. A., *57:* 359-366, 1967.
44. Burger, M. M., and Martin, G. S. Agglutination of cells transformed by Rous sarcoma virus by wheat germ agglutinin and concanavalin A. Nature (Lond.), *237:* 9-12, 1972.
45. Burk, R. R. Reduced adenyl cyclase activity in a polyoma virus transformed cell line. Nature (Lond.), *219:* 1272-1275, 1968.
46. Burnet, F. M. Immunology, aging, and cancer. San Francisco: W. H. Freeman and Co., 1976.
47. Burns, J. J., Evans, C., and Trousof, N. Stimulatory effect of barbital on urinary excretion of L-ascorbic acid and non-conjugated D-glucuronic acid. J. Biol. Chem., *227:* 785-795, 1957.
48. Cameron, E. Hyaluronidase and cancer. New York: Pergamon Press, 1966.
49. Cameron, E. Vitamin C. Br. J. Hosp. Med., *13:* 511, 1975.
50. Cameron, E. Biological function of ascorbic acid and the pathogenesis of scurvy. Med. Hypotheses, *2:* 154-163, 1976.
51. Cameron, E., and Baird, G. Ascorbic Acid and Dependence on Opiates in Patients with Advanced Disseminated Cancer. Int. Res. Commun. Syst., August, 1973.
52. Cameron, E., and Campbell, A. The orthomolecular treatment of cancer. II. Clinical trial of high-dose ascorbic acid supplements in advanced human cancer. Chem.-Biol. Interact., *9:* 285-315, 1974.
53. Cameron, E., Campbell, A., and Jack, T. The orthomolecular treatment of cancer. III. Reticulum cell sarcoma: double complete regression induced by high-dose ascorbic acid therapy. Chem.-Biol. Interact., *11:* 387-393, 1975.
54. Cameron, E., and Pauling, L. Ascorbic acid and the glycosaminoglycans: an orthomolecular approach to cancer and other diseases. Oncology, *27:* 181-192, 1973.
55. Cameron, E., and Pauling, L. The orthomolecular treatment of cancer. I. The role of ascorbic acid in host resistance. Chem.-Biol. Interact., *9:* 273-283, 1974.
56. Cameron, E., and Pauling, L. Supplemental ascorbate in the supportive treatment of cancer: prolongation of survival times in terminal human cancer. Proc. Natl. Acad. Sci. U. S. A., *73:* 3685-3689, 1976.
57. Cameron, E., and Pauling, L. Supplemental ascorbate in the supportive treatment of cancer: reevaluation of prolongation of survival times in terminal human cancer. Proc. Natl. Acad. Sci. U. S. A., *75:* 4538-4542, 1978.
58. Cameron, E., and Pauling, L. Vitamin C and cancer. Trans. Am. Philos. Soc., in press, 1979.
59. Cameron, E., and Pauling, L. Ascorbic acid as a therapeutic agent in cancer. J. Int. Acad. Prev. Med., in press, 1979.
60. Cameron, E., and Rotman, D. Ascorbic acid, cell proliferation, and cancer. Lancet, *1:* 542, 1972.
61. Cardesa, A., Mirvish, S. S., Haven, G. T., and Shubik, P. Inhibitory effect of ascorbic acid on the acute toxicity of dimethylamine plus nitrite in the rat. Proc. Soc. Exp. Biol. Med., *145:* 124-128, 1974.
62. Carr, A. J. Effect of some glycosidase inhibitors in experimental tumours in the mouse. Nature (Lond.), *198:* 1104-1105, 1963.
63. Carr, A. J. The relation to invasion of glycosidases in mouse tumours. J. Pathol. Bacteriol., *89:* 239-243, 1965.
64. Carrié, C., and Schnettler, O. Zur Verhütung der Röntgenstrahlenleukopenie. Strahlentherapie, *66:* 149-154, 1939.
65. Chan, J. T., and Black, H. S. Antioxidant-mediated reversal of ultraviolet light cytotoxicity. J. Invest. Dermatol., *68:* 366-368, 1977.
66. Chatterjee, I. B., Majumder, A. K., Nandi, B. K., and Subramanian, N. Synthesis and some major functions of vitamin C in animals. Ann. N. Y. Acad. Sci., *258:* 24-47, 1975.
67. Chatterjee, A. K., Basu, J., Suradis, Datta, S. C., Sengupta, K., and Ghosh, B. Effects of L-lysine administration on certain aspects of ascorbic acid metabolism. Intern. J. Vitam. Nutr. Res., *46:* 286-290, 1976.
68. Cheraskin, E., Ringsdorf, W. M., Jr., Hutchins, K., Setyaadmadja, A. T.

S. P., and Wideman, G. L. Effect of diet upon radiation response in cervical carcinoma of the uterus. A preliminary report. Acta Cytol., *12:* 433–438, 1968.

69. Chinoy, N. J. Histochemical studies on ascorbic acid in human cancerous tissue and its significance. J. Anim. Morphol. Physiol., *19:* 238–240, 1972.
70. Chope, H. D., and Breslow, L. Nutritional status of the aging. Am. J. Public Health, *46:* 61–67, 1955.
71. Chretien, J. H., and Gargusi, V. F. Correction of corticosteroid-induced defects of polymorphonuclear neutrophil function by ascorbic acid. J. Reticuloendothel. Soc., *14:* 280–286, 1973.
72. Clayton, B. E., Hammant, J. E., and Armitage, P. Increased adrenocorticotrophic hormone in the sera of acutely scorbutic guinea pigs. J. Endocrinol., *15:* 284–296, 1957.
73. Clayton, B. E., and Prunty, F. T. G. Relation of adrenal cortical function to scurvy in guinea pigs. Br. Med. J., *2:* 927–930, 1951.
74. Cohen, S. L., and Bittner, J. J. Effect of mammary tumors on glucuronidase and esterase activities in a number of mouse strains. Cancer Res., *11:* 723–726, 1951.
75. Cole, D. F. Prevention of experimental ocular hypertension with polyphloretin phosphate. Br. J. Ophthalmol., *45:* 482–498, 1961.
76. Coman, D. R. Decreased mutual adhesiveness, a property of cells from squamous cell carcinomas. Cancer Res., *4:* 625–629, 1944.
77. Conchie, J., and Levvy, G. A. Comparison of different glycosidase activities in conditions of cancer. Br. J. Cancer, *11:* 487–493, 1957.
78. Conchie, J., and Levvy, G. A. Inhibition of glycosidases by aldonolactones of corresponding configuration. Biochem. J., *65:* 389–395, 1957.
79. Conchie, S. L., and Bittner, J. J. The effects of mammary tumors on the glucuronidase and esterase activities of a number of mouse strains. Cancer Res., *11:* 723–726, 1951.
80. Conney, A. H., and Burns, J. J. Stimulatory effect of foreign compounds on ascorbic acid biosynthesis and on drug-metabolizing enzymes. Nature (Lond.), *184:* 363–364, 1959.
81. Conrad, M. J., and Feigen, G. A. Sex hormones and kinetics of anaphylactic histamine release. Physiol. Chem. Phys., *6:* 11–16, 1974.
82. Conrad, M. J., and Feigen, G. A. Physical studies of tissue anaphylaxis. Immunochemistry, *12:* 517–522, 1975.
83. Cooper, J. R. The role of ascorbic acid in the oxidation of tryptophan to 5-hydroxytryptophan. Ann. N. Y. Acad. Sci., *92:* 208–211, 1961.
84. Cooper, M. R., McCall, C. E., and DeChatelet, L. R. Stimulation of leukocyte hexose monophosphate shunt activity by ascorbic acid. Infect. Immun., *3:* 851–853, 1971.
85. Cope, P., and Dawson, M. Toxicity of sodium ascorbate and alloxan to 3T3 cells. Brit. Pharm. J., in press.
86. Correa, P., Haenszel, W., Cuello, C., Tannenbaum, S., and Archer, M. A model for gastric cancer epidemiology. Lancet, *2:* 58–60, 1975.
87. Cottingham, E., and Mills, C. A. Influence of temperature and vitamin deficiency upon phagocytic functions. J. Immunol., *47:* 493–502, 1943.
88. Crandon, J. H., Mikal, S., and Landeau, B. R. Ascorbic-acid deficiency in experimental and surgical subjects. Lind Bicentenary Symposium. Proc. Nutr. Soc., *12:* 274–279, 1953.
89. Crandon, J. H., Mikal, S., and Landeau, B. R. Ascorbic acid economy in surgical patients. Ann. N. Y. Acad. Sci., *92:* 246–267, 1961.
90. Currie, G. A. Cancer and the immune response. London: Edward Arnold, 1974.
91. Cuttle, T. D. Observations on the relation of leucocytosis to ascorbic acid requirements. Q. J. Med., *31:* 575–584, 1938.
92. Dahl, H., and Degré, M. The effect of ascorbic acid on production of human interferon and the antiviral activity *in vitro*. Acta Pathol. Microbiol. Scand. Sect. B, *84:* 280–284, 1976.
93. Daoust, R. The histochemical demonstration of polyadenylic acid hydrolases in rat liver during azo dye carcinogenesis. J. Histochem. Cytochem., *20:* 536–541, 1972.
94. DeChatelet, L. R., McCall, C. E., and Cooper, M. R. Ascorbic acid levels in phagocytic cells. Proc. Soc. Exp. Biol. Med., *45:* 1170–1173, 1974.
95. DeCosse, J. J., Adams, M. B., Kuzma, J. F., LoGerfo, P., and Condon, R. E. Effect of ascorbic acid on rectal polyps of patients with familial polyposis. Surgery, *78:* 608–612, 1975.
96. Degkwitz, E., Kaufmann, L. H., Luft, D., and Staudinger, H. Abnahmen der Cytochromgehalte und Veränderungen der Kinetic der Monooxygenase in Lebermikrosomen von Meerschweinchen bei verschiedenen Stadien des Ascorbinsäuremangels. Hoppe-Seyler's Z. Physiol. Chem., *353:* 1023–1044, 1972.
97. Degkwitz, E., and Kim, K. S. Comparative studies on the influence of L-ascorbate, D-arabino-ascorbate and 5-oxo-D-gluconate on the amounts of cytochromes P-450 and b_5 in liver microsomes of guinea pigs. Hoppe-Seyler's Z. Physiol. Chem., *354:* 555–561, 1973.
98. Deichmann, W. B., and Radomski, J. L. Carcinogenicity and metabolism of aromatic amines in the dog. J. Natl. Cancer Inst., *43:* 263–269, 1969.
99. Dell'Orco, R. T., and Nash, J. H. Effects of ascorbic acid on collagen synthesis in nonmitotic human diploid fibroblasts. Proc. Soc. Exp. Biol. Med., *144:* 621–622, 1973.
100. Deucher, W. G. Beobachtungen über den Vitamin-C-Haushalt beim Tumorkranken. Strahlentherapie, *67:* 143–151, 1940.
101. Dodd, N. F. J., and Giron-Conland, J. M. Electron spin resonance study of changes during the development of a mouse myeloid leukaemia. II The ascorbyl radical. Br. J. Cancer, *32:* 451–455, 1975.
102. Done, A. K., Ely, R. S., Heisset, L. R., and Kelly, V. C. Circulating 17-hydrocorticosteroids in Ascorbic Acid Deficient Guinea Pig. Metab Clin. Exp., *3:* 93–94, 1954.
103. Dorfman, A. Adventures in viscous solutions. Mol. Cell. Biochem., *4:* 45–65, 1974.
104. Dresden, M. H., Heilman, S. A., and Schmidt, J. D. Collagenolytic enzymes in human neoplasms. Cancer Res., *32:* 993–996, 1972.
105. Dukes, C. E. Simple tumours of the large intestine and their relation to cancer. Br. J. Surg., *13:* 720–733, 1926.
106. Dulbecco, R. From the molecular biology of oncogenic DNA viruses to cancer. Science, *192:* 437–449, 1976.
107. Dungal, N., and Sigurjonsson, J. Gastric cancer and diet. A pilot study on dietary habits in two districts differing markedly in respect o mortality from gastric cancer. Br. J. Cancer, *21:* 270–276, 1967.
108. Dutton, G. J. (ed.). Glucuronic Acid, Free and Combined: Chemistry, Biochemistry, Pharmacology and Medicine. New York: Academic Press, Inc., 1966.
109. Dyer, H. M., and Ross, H. E. Ascorbic acid, dehydroascorbic acid, and diketogulonic acid of transplanted melanomas and of other tumors of the mouse. J. Natl. Cancer Inst., *11:* 313–318, 1950.
110. Dzialosznyski, L. M., Frohlich, A., and Droll, J. Cancer and arylsulphatase activity. Nature (Lond.), *212:* 733, 1966.
111. Edgar, J. A. Is dehydroascorbic acid an inhibitor in the regulation of cell division in plants and animals? Experientia, *25:* 1214–1215, 1969.
112. Edgar, J. A. Dehydroascorbic acid and cell division. Nature, *227:* 24–26, 1970.
113. Eliott, C. G., and Smith, M. D. Ascorbic acid metabolism and glycolysis in the polymorphonuclear leucocyte of the guinea pig. J. Cell. Physiol. *67:* 169–170, 1966.
114. Eufinger, H., and Gaehtgens, G. Uber die Einwirkung des Vitamin C auf das pathologisch veränderte weisse Blutbild. Klin. Wochenschr. *15:* 150–151, 1936.
115. Fishman, W. H., Anlyan, A. J., and Gordon, E. Beta-glucuronidase activity in human tissues: some correlations with processes of malignant growth and with the physiology of reproduction. Cancer Res., *7:* 808–817, 1947.
116. Fiszer-Szafarz, B. Effect of human cancerous serum on the chick embryo. Cancer Res., *27:* 191–197, 1967.
117. Fiszer-Szafarz, B. Demonstration of a new hyaluronidase inhibitor in serum of cancer patients. Proc. Soc. Exp. Biol. Med., *129:* 302–303, 1968.
118. Fiszer-Szafarz, B., and Gullino, P. M. Hyaluronic acid content of the interstitial fluid of Walker Carcinoma 256. Proc. Soc. Exp. Biol. Med., *133:* 597–600, 1970.
119. Fiszer-Szafarz, B., and Gullino, P. M. Hyaluronidase activity of normal and neoplastic interstitial fluid. Proc. Soc. Exp. Biol. Med., *133:* 805–807, 1970.
120. Fiszer-Szafarz, B., and Szafarz, D. Lysosomal hyaluronidase activity in normal rat liver and in chemically induced hepatomas. Cancer Res., *33:* 1104–1108, 1973.
121. Fjelde, A., Sorkin, E., and Rhodes, J. M. The effect of glucosamine on human epidermoid carcinoma cells in tissue culture. Exp. Cell Res., *10:* 88–98, 1956.
122. Floyd, R. A., Soong, L. M., and Culver, P. L. Horseradish peroxidase/hydrogen peroxide-catalyzed oxidation of the carcinogen *N*-hydroxy-*N*-acetyl-2-aminofluorene as affected by cyanide and ascorbate. Cancer Res., *36:* 1510–1519, 1976.
123. Fong, Y. Y., and Chan, W. C. Ascorbic acid and tumour production by aminopyrine and nitrite in rats. IRCS Med. Sci.-Libr. Compend., *3:* 593, 1975.
124. Freeman, J. T., and Hafkesbring, R. Comparative studies of ascorbic acid levels in gastric secretion and blood. III. Gastrointestinal disease. Gastroenterology, *32:* 878–886, 1957.
125. Gabor, M. Pharmacologic effects of flavonoids on blood vessels. Angiologica, *9:* 355–374, 1972.
126. Gaehtgens, G. Die Bedeutung der akzessorischen Nahrstoffe (Vitamine) in der Gynäkologie und Geburtshilfe. Zentralbl. Gynaekol., *62:* 626–634, 1938.
127. Gaehtgens, G. Das Vitamin C-Defizit beim gynäkologischen Karzinom. Zentralbl. Gynäekol., *62:* 1874–1881, 1938.
128. Ganguly, R., Durieux, M. F., and Waldman, R. H. Macrophage function in vitamin C-deficient guinea pigs. Am. J. Clin. Nutr., *29:* 762–765, 1976.
129. Garb, S. Cure for cancer, a national goal. New York: Springer Publishing Co., Inc., 1968.
130. Gericke, D., and Chandra, P. Inhibition of tumor growth by nucleoside cyclic-3′,5′-monophosphates. Hoppe-Seyler's Z. Physiol. Chem., *350:* 1469–1471, 1967.
131. Gersh, I., and Catchpole, H. R. The organization of ground substance

and basement membrane, and its significance in tissue injury, disease and growth. Am. J. Anat., *85:* 457–521, 1949.
132. Ghosh, P., Bushell, G. R., Taylor, T. K. F., and Akeson, W. H. Collagens, elastin, and noncollagenous protein of the intervertebral disk. Clin. Orthop. Relat. Res., *129:* 124–132, 1977.
133. Glick, D. Hyaluronidase inhibitor of human blood serum in health and disease. J. Mt. Sinai Hosp., *17:* 207–228, 1950.
134. Glick, D., Good, T. A., and Bittner, J. J. Mucolytic enzyme systems. IV. Relationship of hyaluronidase inhibition by blood serum to incidence of mammary cancer in mice. Proc. Soc. Exp. Biol. Med., *69:* 524–526, 1948.
135. Glick, D., Ottoson, R., and Edmondson, P. R. Studies in histochemistry. II. Microdetermination of hyaluronidase and its inhibition by fractions of isolated mast cells. J. Biol. Chem., *233:* 1241–1243, 1958.
136. Glick, D., and Sylven, B. Evidence for the heparin nature of the nonspecific hyaluronidase inhibitor in tissue extracts and blood serum. Science, *113:* 388–389, 1951.
137. Goetzl, E. J., Wasserman, S. I., Gigli, I., and Austen, K. F. Enhancement of random migration and chemotactic response of human leukocytes by ascorbic acid. J. Clin. Invest., *53:* 813–818, 1974.
138. Gore, I., Tanaka, Y., Fujinami, T., and Goodman, M. L. Aortic acid mucopolysaccharides and collagen in scorbutic guinea pigs. J. Nutr., *87:* 311–316, 1965.
139. Goth, A., and Littman, I. Ascorbic acid content in human cancer tissue. Cancer Res., *8:* 349–351, 1948.
140. Gould, B. S. Ascorbic acid-independent and ascorbic acid-dependent collagen forming mechanisms. Ann. N. Y. Acad. Sci., *92:* 168–174, 1961.
141. Gould, B. S., and Woessner, J. F. Biosynthesis of collagen: the influence of ascorbic acid on the proline, hydroxyproline, glycine, and collagen content of regenerating guinea pig skin. J. Biol. Chem., *226:* 289–300, 1957.
142. Goyanna, C. Tobacco and vitamin C. Bras.-Med., *69:* 173–177, 1955.
143. Graham, S. Future inquiries into the epidemiology of gastric cancer. Cancer Res., *35:* 3464–3468, 1975.
144. Graham, S., Schotz, W., and Martine, P. Alimentary factors in the epidemiology of gastric cancer. Cancer, *30:* 927–938, 1972.
145. Greenblatt, M. Ascorbic acid blocking of aminopyrine nitrosation in NZO/B1 mice. J. Natl. Cancer Inst., *50:* 1055–1056, 1973.
146. Greer, E. Alcoholic cirrhosis complicated by polycythemia vera and then myelogenous leukemia and tolerance of large doses of vitamin C. Med. Times, *32:* 865–868, 1954.
147. Grossfeld, H. Production of hyaluronic acid by fibroblasts growing from explants of Walker Tumor 256: production of hyaluronidase by the tumor cells. J. Natl. Cancer Inst., *27:* 543–558, 1961.
148. Guttenplan, J. B. Inhibition by L-ascorbate of bacterial mutagenesis induced by two *N*-nitroso compounds. Nature (Lond.), *268:* 368–370, 1977.
149. Haenszel, W., and Correa, P. Developments in the epidemiology of stomach cancer over the past decade. Cancer Res., *35:* 3452–3459, 1975.
150. Hakanson, E. Y., and Glick, D. Mucolytic enzyme systems. III. Inhibition of hyaluronidase by serum in human cancer. J. Natl. Cancer Inst., *9:* 129–132, 1948.
151. Harper, H. A. Review of Physiological Chemistry, Ed. 15, Los Altos, Calif.: Lange Medical Publications, 1975.
152. Harris, P. A., Stephens, R. W., Ghosh, P., and Taylor, A. T. P. Endo- and exoglycosidases in an experimental rat osteosarcoma. Aust. J. Exp. Biol. Med. Sci., *55*(Part 4): 363–370, 1977.
153. Herp, A., DeFilippi, J., and Fabianek, J. The effect of serum hyaluronidase on acidic polysaccharides and its activity in cancer. Biochim. Biophys. Acta, *158:* 150–153, 1968.
154. Higginson, J. Etiological factors in gastrointestinal cancer in man. J. Natl. Cancer Inst., *37:* 527–545, 1966.
155. Hill, M. J., Hawksworth, G., and Tattersall, G. Bacteria, nitrosamines, and cancer of the stomach. Br. J. Cancer. *28:* 562–567, 1973.
156. Hoefel, O. S. Plasma vitamin C levels in smokers. *In:* A. Hanck and G. Ritzel (eds.), Re-evaluation of Vitamin C, pp. 127–137. Bern, Switzerland: Verlag Hans Huber, 1977.
157. Holroyd, C. P., Gabuzda, T. G., Putnam, R. C., Paul, P., and Reichard, G. A. Altered glucose metabolism in metastatic carcinoma. Cancer Res., *35:* 3710–3714, 1975.
158. Hormozdiari, H., Day, N. E., Aramesh, B., and Mahboubi, E. Dietary factors and esophageal cancer in the Caspian Littoral of Iran. Cancer Res., *35:* 3493–3498, 1975.
159. Huber, L. Hypervitaminisierung mit Vitamin A und Vitamin C bei inoperablen Gebarmutterkrebsen. Zentralbl. Gynaekol., *75:* 1771–1777, 1953.
160. Hume, R., Vallance, B., and Weyers, E. Ascorbic acid and stress. *In:* A. Hanck and G. Ritzel (eds.), Re-evaluation of Vitamin C, pp. 89–98. Bern, Switzerland: Verlag Hans Huber, 1977.
161. Hume, R., Weyers, E., Rowan, T., Reid, D. S., and Hillis, W. S. Leucocyte ascorbic acid levels after acute myocardial infarction. Br. Heart J., *34:* 238–243, 1972.
162. Jungeblut, C. W. Inactivation of poliomyelitis virus *in vitro* by crystalline vitamin C (ascorbic acid). J. Exp. Med., *62:* 517–521, 1935.
163. Jungeblut, C. W. A further contribution to vitamin C therapy in experimental poliomyelitis. J. Exp. Med., *70:* 315–332, 1939.
164. Kahr, E. Erfahrungen mid der zusätzlichen Allgemeinbehandlung beim Krebskranken. Med. Klin., *54:* 63–66, 1959.
165. Kakar, S., and Wilson, C. W. M. Ascorbic acid metabolism in human cancer. Abstract. Proc. Nutr. Soc., *33:* 110, 1974.
166. Kakar, S. C., Wilson, C. W. M., and Bell, J. N. Plasma and leucocyte ascorbic acid concentrations in acute lymphoblastic leukaemia. Ir. J. Med. Sci., *144:* 227–23, 1975.
167. Kalden, J. R., and Guthy, E. A. Prolonged skin allograft survival in vitamin C-deficient guinea pigs. Eur. Surg. Res., *4:* 114–119, 1972.
168. Kamm, J. J., Dashman, T., Conney, A. H., and Burns, J. J. Protective effect of ascorbic acid on hepatotoxicity caused by sodium nitrite plus aminopyrine. Proc. Natl. Acad. Sci. U. S. A., *70:* 747–749, 1973.
169. Kamm, J. J., Dashman, T., Conney, A. H., and Burns, J. J. The effect of ascorbate on amine-nitrite hepatotoxicity. *In: N*-nitroso compounds in the environment, pp. 200–204. Lyon, France: International Agency for Research on Cancer, 1974.
170. Kato, R., Takanaka, A., and Oshima, T. Effect of vitamin C deficiency on the metabolism of drugs and NADPH-linked electron transport system in liver microsomes. Jpn. J. Pharmacol., *19:* 25–33, 1969.
171. Kennedy, J. F. Chemical and biochemical aspects of the glycosaminoglycans and proteoglycans in health and disease. Adv. Clin. Chem., *18:* 1–101, 1976.
172. Kent, J. R., Hill, M., and Bischoff, A. Acid phosphatase content of prostatic exprimate from patients with advanced prostatic carcinoma: a potential prognostic and therapeutic index. Cancer, *25:* 858–862, 1970.
173. Kern, M., and Racker, E. Activation of DPNH oxidase by an oxidation product of ascorbic acid. Arch. Biochem. Biophys., *48:* 235–236, 1954.
174. King, C. G., and Burns, J. J. Second Conference on vitamin C. Ann. N. Y. Acad. Sci., *258:* 1–252, 1975.
175. Kiriluk, L. B., Kremen, A. J., and Glick, D. Mucolytic enzyme systems. XII. Hyaluronidase in human and animal tumors, and further studies on the serum hyaluronidase inhibitor in human cancer. J. Natl. Cancer Inst., *10:* 993–1000, 1950.
176. Kivirikko, K. I., and Prockop, D. Enzymatic hydroxylation of proline and lysine in protocollagen. Proc. Natl. Acad. Sci. U. S. A., *57:* 782–789, 1967.
177. Kleine, H. O., and Huber, L. Zur Frage der Hypervitaminisierung mit Vitamin A und Vitamin C bei inoperable Karzinomen nach v. Wendt. Muench. Med. Wochenschr., *96:* 367–370, 1954.
178. Klenner, F. R. Massive doses of vitamin C and the viral diseases. South. Med. & Surg., *113:* 101–107, 1951.
179. Klenner, F. R. Observations on the dose and administration of ascorbic acid when employed beyond the range of a vitamin in human pathology. J. Appl. Nutr., *23:* 61–88, 1971.
180. Knox, E. G. Ischaemic heart disease mortality and dietary intake of calcium. Lancet, *1:* 1465–1467, 1973.
181. Knox, W. E., and Goswami, M. N. D. Ascorbic acid. *In:* H. Sabotka and C. P. Stewart (eds.), Advances in Clinical Biochemistry. New York: Academic Press, Inc., 1961.
182. Kojima, J., and Hess, J. W. Ascorbic acid interference with estimation of beta-glucosaminidase activity. Annals Biochem., *23:* 474–483, 1968.
183. Krasner, N., and Dymock, I. W. Ascorbic acid deficiency in malignant diseases: a clinical and biochemical study. Br. J. Cancer, *30:* 142–145, 1974.
184. Kretzschmar, C. H., and Ellis, H. The effect of X-rays on ascorbic acid concentration in plasma and in tissues. Br. J. Radiol., *20:* 94–99, 1947.
185. Kuenzig, W., Tkaczevski, V., Kamm, J. J., Conney, A. H., and Burns, J. J. The effect of ascorbic acid deficiency on extrahepatic microsomal metabolism of drugs and carcinogens in the guinea pig. J. Pharmacol. Exp. Ther., *201:* 527–533, 1977.
186. Kyhos, E. D., Sevringhaus, E. L., and Hagedorn, D. Large doses of ascorbic acid in treatment of vitamin C deficiencies. Arch. Intern. Med., *75:* 407–412, 1945.
187. La Du, B. N., and Zannoni, V. G. The role of ascorbic acid in tyrosine metabolism. Ann. N. Y. Acad. Sci., *92:* 175–191, 1961.
188. Lai, H-Y L., Shields, E. K., and Watne, A. L. Effect of ascorbic acid on rectal polyps and fecal steroids. Fed. Proc., *36:* 1061, 1977.
189. Lamberg, S. I., and Stoolmiller, A. C. Glycosaminoglycans: a biochemical and clinical review. J. Invest. Dermatol., *63:* 433–449, 1974.
190. Lanzerotti, R. H., and Gullino, P. M. Activities and quantities of lysosomal enzymes during mammary tumor regression. Cancer Res., *32:* 2679–2685, 1972.
191. Lagos, G. S., and Cooper, R. R. Electron microscope visualization of proteinpolysaccharides. Clin. Orthop. Relat. Res., *84:* 179–192, 1972.
192. Leber, H. W., Degkwitz, E., and Staudinger, H. Studies on the effect of ascorbic acid on the activity and biosynthesis of mixed function oxygenases and on the concentration of haemoproteins in the microsome fraction of guinea pig liver. Hoppe-Seyler's Z. Physiol. Chem., *350:* 439–445, 1969.

193. Levin, E. Y., and Kaufman, S. Studies on the enzyme catalyzing the conversion of 3,4-dihydroxyphenylethylamine to norepinephrine. J. Biol. Chem., *236:* 2043–2049, 1961.
194. Levitan, B. A. Inhibition of testicular hyaluronidase activity by rutin. Proc. Soc. Exp. Biol. Med., *68:* 566–568, 1958.
195. Lewin, S. Vitamin C; its molecular biology and medical potential. New York: Academic Press, Inc., 1976.
196. Lijinsky, W., and Epstein, S. S. Nitrosamines as environmental carcinogens. Nature (Lond.), *225:* 21–23, 1970.
197. Lind, J. A treatise of the scurvy. Edinburgh: Sands, Murray and Cochrane, 1753. Reprinted. C. P. Stewart and D. Guthrie (eds.). Edinburgh: Edinburgh University Press, 1953.
198. Lloyd, J. V., Davis, P. S., Emery, H., and Lander, H. Platelet ascorbic acid levels in normal subjects and in disease. J. Clin. Pathol. (Lond.), *25:* 478–483, 1972.
199. Lo, W. B., and Black, H. S. Inhibition of carcinogen formation in skin irradiated with ultraviolet light. Nature, *246:* 489–491, 1973.
200. Lockhart-Mummery, J. P., and Dukes, C. E. Familial adenomatosis of colon and rectum, its relation to cancer. Lancet, *2:* 586–589, 1939.
201. Lowther, D. A., and Natarjan, M. The influence of glycoprotein on collagen fibril formation in the presence of chondroitin sulphate proteoglycan. Biochem. J., *127:* 607–608, 1972.
202. Lowther, D. A., Toole, B. P., and Harrington, A. C. Interactions of proteoglycans with tropocollagen. *In:* E. A. Balazs (eds.), Chemistry and Molecular Biology of the Intercellular Matrix, Vol. 2, pp. 1135–1153. New York: Academic Press, Inc., 1970.
203. Luft, D., Degkwitz, E., Hochli-Kaufmann, L., and Staudinger, H. Effect of delta-aminolevulinic acid on the content of cytochrome P-450 in the liver of guinea pigs fed without ascorbic acid. Hoppe-Seyler's Z. Physiol. Chem., *353:* 1420–1422, 1972.
204. Marquardt, H., Rufino, F., and Weisburger, J. H. Mutagenic activity of nitrite-treated foods: Human stomach cancer may be related to dietary factors. Science, *196:* 100–101, 1977.
205. Mathews, M. B. The interactions of proteoglycans and collagen-model systems. *In:* E. A. Balazs (ed.), Chemistry and Molecular Biology of the Intercellular Matrix, Vol. 2, pp. 1155–1169. New York: Academic Press, Inc., 1970.
206. Mathews, M. B., and Decker, L. The effect of acid mucopolysaccharides and acid mucopolysaccharide-proteins on fibril formation from collagen solutions. Biochem. J., *109:* 517–526, 1968.
207. Mathews, M. B., and Dorfman, A. Inhibition of hyaluronidase. Physiol. Rev., *35:* 381–402, 1955.
208. Maugh, T. H. Chemotherapy: antiviral agents come of age. Science, *192:* 128–132, 1976.
209. Mavligit, G. M., and Freireich, E. J Progress in the treatment of colorectal cancer. J. Am. Med. Assoc., *235:* 2855, 1976.
210. McCall, G. E., DeChatelet, L. R., Cooper, M. R., and Ashburn, P. The effects of ascorbic acid on bacteriocidal mechanisms of neutrophils. J. Infect. Dis., *125:* 194–198, 1971.
211. McCormick, W. J. Cancer: a collagen disease, secondary to a nutritional deficiency? Arch. Pediatr., *76:* 166–171, 1959.
212. McCutcheon, M., Coman, D. E., and Moore, F. B. Studies on invasiveness of cancer: adhesiveness of malignant cells in various human adenocarcinomas. Cancer, *1:* 460–467, 1948.
213. Meiklejohn, A. P. The physiology and biochemistry of ascorbic acid. Vitam. Horm., *11:* 61–96, 1953.
214. Meyer, B. A., and Orgel, I. S. The Cancer Patient, a New Chemotherapy in Advanced Cases. London: J & A Churchill, Ltd., 1950.
215. Meyer, K. The biological significance of hyaluronic acid and hyaluronidase. Physiol. Rev., *27:* 335–359, 1947.
216. Migliozzi, J. A. Effect of ascorbic acid on tumour growth. Br. J. Cancer, *35:* 448–453, 1977.
217. Minor, A. H., and Ramirez, M. A. The utilization of vitamin C by cancer patients. Cancer Res., *2:* 509–513, 1942.
218. Mirvish, S. S. Kinetics of nitrosamide formation from alkylureas, *N*-alkylurethans, and alkylguanidines: possible implications for the etiology of human gastric cancer. J. Natl. Cancer Inst., *46:* 1183–1193, 1971.
219. Mirvish, S. S., Cardesa, A., Wallcave, L., and Shubik, P. Induction of mouse lung adenomas by amines or ureas plus nitrite and by *N*-nitroso compounds: effect of ascorbate, gallic acid, thiocyanate, and caffeine. J. Natl. Cancer Inst., *55:* 633–636, 1975.
220. Mirvish, S. S., Wallcave, L., Eagen, M., and Shubik, P. Ascorbate-nitrite reaction: possible means of blocking the formation of carcinogenic *N*-nitroso compounds. Science, *177:* 65–68, 1972.
221. Mitchison, N. A. Tumour-immunology. *In:* I. Roitt (ed.), Essays in Fundamental Immunology, Vol. 1, pp. 57–66. Oxford, England: Blackwell Scientific Publications, 1973.
222. Mohr, U., Reznik-Schuller, H., Reznik, G., and Hilfrich, J. Transplacental effects of diethylnitrosamine in Syrian hamsters as related to different days of administration during pregnancy. J. Natl. Cancer Inst., *55:* 681–683, 1975.
223. Molnar, Z., and Bekesi, J. G. Effects of D-glucosamine, D-mannoseamine, and 2-deoxy-D-glucose on the ultrastructure of ascites tumor cells *in vitro*. Cancer Res., *32:* 380–389, 1972.
224. Molnar, Z., and Bekesi, J. G. Cytotoxic effects of D-glucosamine on the ultrastructures of normal and neoplastic tissues *in vivo*. Cancer Res., *32:* 756–765, 1972.
225. Moriarty, M. J., Mulgrew, S., Malone, J. R., and O'Connor, M. R. Results and analysis of tumour levels of ascorbic acid. Ir. J. Med. Sci., *146:* 74–78, 1977.
226. Morishige, F., and Murata, A. Prolongation of survival times in terminal human cancer. J. Int. Acad. Prev. Med., in press.
227. Morishige, F., and Murata, A. Vitamin C for prophylaxis of viral hepatitis B in transfused patients. J. Int. Acad. Prev. Med., in press, 1978.
228. Murata, A. Virucidal activity of vitamin C: vitamin C for prevention and treatment of viral diseases. *In:* Proceedings of the First International Congress of the Microbiological Society, Science Council of Japan, Vol. 3, pp. 432–442, 1975.
229. Murata, A., and Kitagawa, K. Mechanism of inactivation of bacteriophage J1 by ascorbic acid. Agric. Biol. Chem., *37:* 1145–1151, 1973.
230. Murata, A., Kitagawa, K., Inmaru, H., and Saruno, R. Inactivation of single-stranded DNA and RNA phages by ascorbic acid and thiol reducing agents. Agric. Biol. Chem., *36:* 2597–2599, 1972.
231. Murata, A., Kitagawa, K., and Saruno, R. Inactivation of bacteriophages by ascorbic acid. Agric. Biol. Chem., *35:* 294–296, 1971.
232. Nakamura, Y., and Yamafuji, K. Antitumor activities of oxidized products of ascorbic acid. Sci. Bull. Fac. Agric. Kyushu Univ., *23:* 119–125, 1968.
233. Nakashima, Y., Suzue, R., Sanada, H., and Kawada, S. Effect of ascorbic acid on hydroxylase activity. I. Stimulation of tyrosine hydroxylase and tryptophan-5-hydroxylase activities by ascorbic acid. J. Vitaminol. (Kyoto), *16:* 276–280, 1970.
234. Narisawa, T., Wong, C. Q., Maronpot, R. R., and Weisburger, J. H. Large bowel carcinogenesis in mice and rats by several intrarectal doses of methylnitrosourea and negative effect of nitrite plus methylurea. Cancer Res., *36:* 505–510, 1976.
235. Nason, A., Wosilait, W. D., and Terrel, A. J. The enzymatic oxidation of reduced pyridine nucleotides by an oxidation product of ascorbic acid. Arch. Biochem. Biophys., *48:* 233–235, 1954.
236. Neville, A. M., and Laurence, D. J. R. (eds.). Report of the workshop on the carcinoembryonic antigen (CEA); the present position and proposals for future investigation. Int. J. Cancer., *14:* 1–18, 1974.
237. Neville, A. M., and Cooper, E. H. Biochemical monitoring of cancer: a review. Ann. Clin. Biochem., *13:* 283–305, 1976.
238. Newman, J. K., Berenson, G. S., Mathews, M. B., Goldwasser, E., and Dorfman, A. The isolation of the non-specific hyaluronidase inhibitor of human blood. J. Biol. Chem., *217:* 31–41, 1955.
239. Niebes, P. Influence des flavonoides sur le metabolisme des mucopolysaccharides dans la paroi veineuse. Angiologica, *9:* 226–234, 1972.
240. Nigam, V. N., and Cantero, A. Polysaccharides in cancer. Adv. Cancer Res., *16:* 1–95, 1972.
241. Nigam, V. N., and Cantero, A. Polysaccharides in cancer: glycoproteins and glycolipids. Adv. Cancer Res., *17:* 1–80, 1973.
242. Nungester, W. J., and Ames, A. M. The relationship between ascorbic acid and phagocytic activity. J. Infect. Dis., *83:* 50–54, 1948.
243. Odell, L. D., and Burt, J. C. Beta-glucuronidase activity in human female genital cancer. Cancer Res., *9:* 362–365, 1949.
244. Omura, H., Fukumoto, Y., Tomita, Y., and Shinohara, K. Action of 5-methyl-3,4-dihydroxytetrone on deoxyribonucleic acid. J. Fac. Agric. Kyushu Univ., *19:* 139–148, 1975.
245. Omura, H., Iiyama, S., Tomita, Y., Narazaki, Y., Shinohara, K., and Murakami, H. Breaking action of ascorbic acid on nucleic acids. J. Nutr. Sci. Vitaminol., *21:* 237–249, 1975.
246. Omura, H., Nakamura, Y., Tomita, Y., and Yamafuji, K. Antitumour action of an ascorbate oxidase preparation and its interaction with deoxyribonucleic acid. J. Fac. Agric. Kyushu Univ., *17:* 187–194, 1973.
247. Omura, H., Tomita, Y., Nakamura, Y., and Murakami, H. Antitumoric potentiality of some ascorbate derivates. J. Fac. Agric. Kyushu Univ., *18:* 181–189, 1974.
248. Palenque, E. Sobre el tratamiento de la leucemia mieloide cronica con vitamina C. Sem. Méd. Esp., *6:* 101–105, 1943.
249. Park, C. H., Bergsagel, D. E., and McCulloch, E. A. Ascorbic acid: a culture requirement for colony formation by mouse plasmacytoma cells. Science, *174:* 720–722, 1971.
250. Passwater, R. A. Cancer—new directions. Am. Lab. (Fairfield, Conn.), June, 10–21, 1973.
251. Pauling, L. Evolution and the need for ascorbic acid. Proc. Natl. Acad Sci. U. S. A., *67:* 1643–1648, 1970.
252. Pauling, L. Vitamin C and the common cold. San Francisco: W. H Freeman and Company, 1970.
253. Pauling, L. Preventive nutrition. Medicine on the midway, *27:* 15–18 1972.
254. Pauling, L. Vitamin C, the common cold, and the flu. San Francisco W. H. Freeman and Company, 1976.
255. Pelletier, O. Vitamin C status of cigarette smokers and nonsmokers Am. J. Clin. Nutr., *23:* 520–524, 1970.
256. Pelletier, O. Vitamin C and cigarette smokers. Ann. N. Y. Acad. Sci.,

258: 156-168, 1975.

257. Pelletier, O. Vitamin C and tobacco. *In:* A. Hanck and G. Ritzel (eds.), Re-evaluation of Vitamin C, pp. 147-169. Bern, Switzerland: Verlag Hans Huber, 1977.

257a. Piontek, G. E., Cain, C. A., and Milner, J. A. The effects of microwave radiation, hyperthermia, and L-ascorbic acid on Ehrlich Ascites Carcinoma cell metabolism. I.E.E.E. Trans. Microwave Theory and Techniques, *26:* 535-540, 1978.

258. Pinto, J. R., Dobson, R. L., and Bentley, J. P. Dermal collagen changes during 2-aminoanthracene carcinogenesis in the rat. Cancer Res., *30:* 1168-1183, 1970.

259. Pipkin, G. E., Nishimura, R., Banowsky, L., and Schlegel, J. U. Stabilization of urinary 3-hydroxyanthranilic acid by oral administration of L-ascorbic acid. Proc. Soc. Exp. Biol. Med., *126:* 702-704, 1967.

260. Pirani, C. L., and Catchpole, H. R. Serum-glycoproteins in experimental scurvy. Arch. Pathol., *51:* 597-601, 1951.

261. Plum, P., and Thomsen, S. Remission in course of acute aleukemic leukemia observed in two cases during treatment with cevitamic acid. Ugeskr. Laeg., *98:* 1062-1067, 1936.

262. Poole, A. R. Invasion of cartilage by an experimental rat tumor. Cancer Res., *30:* 2252-2259, 1970.

263. Preist, R. E. Formation of epithelial basement membrane is restricted by scurvy *in vitro* and is stimulated by vitamin C. Nature (Lond.), *225:* 744-745, 1970.

264. Quastel, J. H., and Cantero, A. Inhibition of tumor growth by D-glucosamine. Nature, *171:* 252-254, 1953.

265. Radomski, J. L., and Brill, E. Bladder cancer induction by aromatic amines: role of *N*-hydroxymetabolites. Science, *167:* 992-993, 1970.

266. Ramp, W. K., and Thornton, P. A. The effect of ascorbic acid on the glycolytic and respiratory metabolism of embryonic chick tibias. Calcified Tissue Res., *2:* 77-82, 1968.

267. Reppert, E., Donegan, J., and Hines, L. E. Ascorbic acid and the hyaluronidase hyaluronic acid reaction. Proc. Soc. Exp. Biol. Med., *77:* 318-320, 1951.

268. Rikans, L. E., Smith, C. R., and Zannoni, V. G. Ascorbic acid and cytochrome P-450. J. Pharmacol. Exp. Ther., *204:* 702-713, 1978.

269. Ritzel, G., and Bruppacher, R. Vitamin C and tobacco. *In:* A. Hanck and G. Ritzel (eds.), Re-evaluation of Vitamin C, pp. 171-183. Bern, Switzerland: Verlag Hans Huber, 1977.

270. Robertson, D. M., and Williams, D. C. *In vitro* evidence of neutral collagenase activity in an invasive mammalian tumor. Nature (Lond.), *221:* 259-260, 1969.

271. Roitt, I. Essential Immunology, Ed. 2. Oxford, England: Blackwell Scientific Publications, 1974.

272. Ross, R., and Benditt, E. P. Wound healing and collagen formation. IV. Distortion of ribosomal patterns of fibroblasts in scurvy. J. Cell Biol., *22:* 365-389, 1964.

273. Rubin, A., Springer, G. F., and Hogue, M. J. The effect of D-glucosamine hydrochloride and related compounds on tissue cultures of the solid form of mouse Sarcoma 37. Cancer Res., *14:* 456-458, 1954.

274. Russell, W. O., Ortega, L. R., and Wynne, E. C. Studies on methylcholanthrene induction of tumors in scorbutic guinea pigs. Cancer Res., *12:* 216-218, 1952.

275. Rustia, M. Inhibitory effect of sodium ascorbate on ethylurea and sodium nitrite carcinogenesis and negative findings in progeny after intestinal inoculation of precursors into pregnant hamsters. J. Natl. Cancer Inst., *55:* 1389-1394, 1975.

276. Ryan, W. L., and Heidrick, M. L. Adenosine-3',5'-monophosphate as an inhibitor of ovulation and reproduction. Science, *162:* 1484-1485, 1968.

277. Saloman, L. L. Studies on adrenal ascorbic acid. Tex. Rep. Biol. Med. *15:* 925-939, 1957.

278. Saloman, L. L. Ascorbic acid catabolism in guinea pigs. J. Biol. Chem., *228:* 163-170, 1957.

279. Schack, J. A., Whitney, R. W., and Freeman, W. E. Hyaluronidase inhibitor in blood serum of scorbutic guinea pigs. J. Biol. Chem., *184:* 551-555, 1950.

280. Schafer, I. A., Silverman, L., Sullivan, J. C., and Robertson, W. van B. Ascorbic acid deficiency in cultured human fibroblasts. J. Cell Biol., *34:* 83-95, 1967.

281. Schaffrath, D., Stuhlsatz, H. W., and Greiling, H. Interactions of glycosaminoglycans with DNA and RNA synthesizing enzymes *in vitro*. Hoppe-Seyler's Z. Physiol. Chem., *357:* 499-508, 1976.

282. Schersten, T., Wahlqvist, L., and Johansson, L. G. Lysosomal enzyme activity in liver tissue from patients with renal carcinoma. Cancer, *23:* 608-613, 1969.

283. Schirmacher, H., and Schneider, J. Grenzen und Möglichkeiten der Vitamin-A und C-Ubervitaminierung bei inoperablen und strahlenresistenten Karzinom. Z. Geburtshilfe Gynaekol., *144:* 172-182, 1955.

284. Schlegel, J. U. Proposed uses of ascorbic acid in prevention of bladder carcinoma. Ann. N. Y. Acad. Sci., *258:* 432-437, 1975.

285. Schlegel, J. U., Pipkin, G. E., and Banowsky, L. Urine composition in the etiology of bladder tumor formation. J. Urol., *97:* 479-481, 1967.

286. Schlegel, J. U., Pipkin, G. E., Nishimura, R., and Duke, G. A. Studies in the etiology and prevention of bladder carcinoma. J. Urol., *101:* 317-324, 1969.

287. Schneider, E. Vitamine C und A beim Karzinom; ein Beitrag zur Hypervitaminisierungstherapie der Krebskranken. Dtsch. Med. Wochenschr., *79:* 584-586, 1954.

288. Schneider, E. Abwehrvorgänge gegen Tumoren im Spiegel einer Hautreaktion. Wien. Med. Wochenschr., *105:* 430-432, 1955.

289. Schnetz, H. Vitamin C und Leukocytenzahl. Klin. Wochenschr., *17:* 267-269, 1938.

290. Schwerdt, P. R., and Schwerdt, C. E. Effect of ascorbic acid on rhinovirus replication in WI-38 cells. Proc. Soc. Exp. Biol. Med., *148:* 1237-1243, 1975.

291. Scribner, J. D., and Naimy, N. K. Destruction of triplet nitrenium ion by ascorbic acid. Experientia, *31:* 470-471, 1975.

292. Shamberger, R. J. Lysosomal enzyme changes in growing and regressing mammary tumors. Biochem. J., *111:* 375-383, 1969.

293. Shamberger, R. J. Decrease in peroxidation in carcinogenesis. J. Natl. Cancer Inst., *48:* 1491-1497, 1972.

294. Shamberger, R. J., Baughman, F. F., Kalchert, S. L., Willis, C. E., and Hoffman, G. C. Carcinogen-induced chromosomal breakage decreased by antioxidants. Proc. Natl. Acad. Sci. U. S. A., *70:* 1461-1463, 1973.

295. Shapiro, S. S., Bishop, M., Kuenzig, W., Tkaczeveski, V., and Kamm, J. J. Effects of ascorbic acid on hyaluronidase inhibitor. Nature (Lond.), *253:* 479, 1955.

296. Shapley, D. Nitrosamines: scientists on the trail of prime suspect in urban cancer. Science, *191:* 268-270, 1976.

297. Shapot, V. S., and Blinov, V. A. Blood glucose levels and gluconeogenesis in animals bearing transplantable tumors. Cancer Res., *34:* 1827-1832, 1974.

298. Siegel, B. V. Enhanced interferon response to murine leukemia virus by ascorbic acid. Infect. Immun., *10:* 409-410, 1974.

299. Siegel, B. V. Enhancement of interferon production by poly(rI) poly(rC) in mouse cell cultures by ascorbic acid. Nature (Lond.), *254:* 531-532, 1975.

300. Siegel, B. V., and Morton, J. I. Vitamin C and the immune response. Experientia, *33:* 393-395, 1977.

301. Siegel, B. V., and Morton, J. I. Interferon and the immune response. *In:* A. Hanck and G. Ritzel (eds.), Re-evaluation of Vitamin C, pp. 245-265. Bern, Switzerland: Verlag Hans Huber, 1977.

302. Sikic, B. I., Mimnaugh, E. G., Litterest, C. L., and Gram, T. E. The effects of ascorbic acid deficiency and repletion on pulmonary renal and hepatic drug metabolism in the guinea pig. Arch. Biochem. Biophys., *179:* 663-671, 1977.

303. Skanse, B., and Sundblad, L. Oxidative breakdown of hyaluronic and chondroitin sulphuric acid. Acta Physiol. Scand., *6:* 37-51, 1943.

304. Slaga, T. J., and Bracken, W. M. The effects of antioxidants on skin tumor initiation and aryl hydrocarbon hydroxylase. Cancer Res., *37:* 1631-1635, 1977.

305. Soloway, M. S., Cohen, S. M., Dekernion, J. B., and Persky, L. Failure of ascorbic acid to inhibit FANFT-induced bladder cancer. J. Urol., *113:* 483-486, 1975.

306. Speier, J. L., and Wattenberg, L. W. Alterations in microsomal metabolism of benzo(*a*)pyrene in mice fed butylated hydroxyanisole. J. Natl. Cancer Inst., *55:* 469-472, 1975.

307. Staudinger, H., Krisch, K., and Leonhauser, S. Role of ascorbic acid in microsomal electron transport and the possible relationship to hydroxylation reactions. Ann. N. Y. Acad. Sci., *92:* 195-207, 1961.

308. Stephen, D. J., and Hawley, E. E. Partition of reduced ascorbic acid in blood. J. Biol. Chem., *115:* 653-658, 1936.

309. Stephens, R. W., Ghosh, P., and Taylor, T. K. F. The characterisation and function of the polysaccharidases of human synovial fluid in rheumatoid and osteoarthritis. Biochem. Biophys. Acta, *399:* 101-112, 1975.

310. Steven, F. S., and Itzhaki, S. Evidence for a latent form of collagenase extracted from rabbit tumor cells. Biochim. Biophys. Acta, *196:* 241-246, 1977.

311. Steven, F. S., Knott, J., Jackson, D. S., and Podrazky, V. Collagen-protein-polysaccharide interactions in human intervertebral disc. Biochim. Biophys. Acta, *188:* 307-313, 1969.

312. Stoll, B. A. Endocrine Therapy in Malignant Disease. Philadelphia: W. B. Saunders Company, 1972.

313. Stone, I. The Healing Factor: Vitamin C against Disease. New York: Grosset and Dunlap, 1972.

314. Stone, K. J. The effect of L-ascorbate on catecholamine biosynthesis. Biochem. J., *131:* 611-613, 1973.

315. Sure, B., Theis, R. M., and Harrelson, R. T. Influence of Walker carcinosarcoma on concentration of ascorbic acid in various endocrines and organs. Am. J. Cancer, *36:* 252-256, 1939.

316. Sylven, B. Lysosomal enzyme activity in the interstitial fluid of solid mouse tumor transplants. Eur. J. Cancer, *4:* 463-474, 1968.

317. Szent-Györgyi, A. Observations of the functions of peroxidase systems and the chemistry of the adrenal cortex. Biochem. J., *22:* 1387-1409, 1928.

318. Takeda, Y., and Hara, M. Significance of ferrous ion and ascorbic acid

in the operation of the tricarboxylic acid cycle. J. Biol. Chem., *214:* 657–670, 1955.
319. Taylor, A. C., Levy, B. M., and Simpson, J. W. Collagenolytic activity of sarcoma tissues in culture. Nature (Lond.), *228:* 366–367, 1970.
320. Thiele, W. Die Wirkung des Vitamin C auf das weisse Blutbild und die chronische myelogisch Leukamie. Klin. Wochenschr., *17:* 150–151, 1938.
321. Thomas, L. The Lives of a Cell: Notes of a Biology Watcher. New York: The Viking Press, 1974.
322. Tomita, Y., Eto, M., Iio, M., Murakami, H., and Omura, H. Antitumor potency of 5-methyl-3,4-dihydroxytetrone. Sci. Bull. Fac. Agric. Kyushu Univ., *28:* 131–137, 1974.
323. Touster, O., and Hollmann, S. Nutritional and enzymatic studies on the mechanism of stimulation of ascorbic acid synthesis by drugs and carcinogenic hydrocarbons. Ann. N. Y. Acad. Sci., *92:* 318–323, 1961.
324. Udenfriend, S. Tyrosine hydroxylase. Pharmacol. Rev., *18:* 43–51, 1966.
325. Vallance, S. Relationships between ascorbic acid and serum proteins of the immune system. Br. Med. J., *2:* 437–438, 1977.
326. Van Caneghem, P. Influence of some hydrolysable substances with vitamin P activity on the fragility of lysosomes *in vitro*. Biochem. Pharmacol., *21:* 1543–1548, 1972.
327. Van Niewenhuizen, C. L. C. Invloed van vitamine C op het bloedbeeld van lijders aan leucaemie. Ned. Tijdsch. Geneskd., *7:* 896–902, 1943.
328. Vedder, E. B., and Rosenberg, C. Concerning toxicity of vitamin A. J. Nutr., *16:* 57–68, 1938.
329. Vogelaar, J. P. M., and Erlichman, E. Significance of ascorbic acid (vitamin C) for the growth *in vitro* of Crocker Mouse Sarcoma 180. Am. J. Cancer, *31:* 283–289, 1937.
330. Vogt, A. Uber den Vitamin C. Verbrauch bei Tumorkranken und bei der Lymphogranulomatose. Strahlentherapie, *64:* 616–623, 1939.
331. von Wendt, G. Zur Anwendung der Übervitaminisierungstherapie. Z. Gesamte. Inn. Med. Grenzgeb., *5:* 255–256, 1950.
332. von Wendt, G. Erfahrungen mit der Übervitaminisierungstherapie. Z. Gesamte. Inn. Med. Grenzgeb., *6:* 255–256, 1951.
333. Waldo, A. L., and Zipf, R. E. Ascorbic acid level in leukemic patients. Cancer, *8:* 187–190, 1955.
334. Warburg, O. On the origin of cancer cells. Science, *123:* 309–314, 1956.
335. Warren, F. L. Aerobic oxidation of aromatic hydrocarbons in the presence of ascorbic acid. Biochem. J., *37:* 338–341, 1943.
336. Watson, A. F. The chemical reducing capacity and vitamin C content of transplantable tumours of the rat and guinea pig. Br. J. Exp. Pathol., *17:* 124–134, 1936.
337. Wattenberg, L. W. Studies of polycyclic hydrocarbon hydroxylases of the intestine possibly related to cancer. Cancer, *28:* 99–102, 1971.
338. Wattenberg, L. W. Dietary modification of intestinal and pulmonary aryl hydrocarbon hydroxylase activity. Toxicol. Appl. Pharmacol., *23:* 741–748, 1972.
339. Wattenberg, L. W. Inhibition of carcinogenic effects of diethylnitrosamine (DEN) and 4-nitroquinoline-*N*-oxide (NQO) by antioxidants. Fed. Proc., *31:* 633, 1972.
340. Wattenberg, L. W. Inhibition of carcinogenic and toxic effects of polycyclic hydrocarbons by phenolic antioxidants and ethoxyquin. J. Natl. Cancer Inst., *48:* 1425–1430, 1972.
341. Wattenberg, L. W. Exogenous factors affecting polycyclic hydrocarbon hydroxylase activity. Adv. Enzyme Regul., *2:* 193–201, 1973.
342. Wattenberg, L. W. Inhibition of chemical carcinogen-induced pulmonary neoplasia by butylated hydroxyanisole. J. Natl. Cancer Inst., *50:* 1541–1544, 1973.
343. Wattenberg, L. W. Inhibition of carcinogenic and toxic effects of polycyclic hydrocarbons by several sulfur-containing compounds. J. Natl. Cancer Inst., *52:* 1583–1587, 1974.
344. Wattenberg, L. W. Potential inhibitors of colon carcinogenesis. Am. J. Dig. Dis., *19:* 947–953, 1974.
345. Wattenberg, L. W. Inhibition of dimethylhydrazine-induced neoplasia of the large intestine by disulfiram. J. Natl. Cancer Inst., *54:* 1005–1006, 1975.
346. Weber, G. Enzymology of cancer cells. N. Engl. J. Med., *296:* 541–551, 1977.
347. Wogan, G. N., Paglialunga, S., Archer M. C., and Tannenbaum, S. R. Carcinogenicity of nitrosation products of ephedrine, sarcosine, folic acid, and creatinine. Cancer Res., *35:* 1981–1984, 1975.
348. Wong, K., Morgan, A. R., and Paranchych, W. Controlled cleavage of phage R17 RNA within the virion by treatment with ascorbate and copper (II). Can. J. Biochem., *52:* 950–958, 1974.
349. Woodhouse, D. L. The action of ascorbic acid on tumour metabolism. Biochem. J., *28:* 1974–1976, 1934.
350. Woodruff, C. W. Ascorbic acid. *In:* G. H. Beaton and E. W. McHendry (eds.), Nutrition, Vol. 2. New York: Academic Press, Inc., 1964.
351. Woodward, G. E. Glutathione and ascorbic acid in tissues of normal and tumor-bearing albino rats. Biochem. J., *29:* 2405–2412, 1935.
352. Woolford, G., and Cassens, R. G. The fate of sodium nitrite in bacon. J. Food Sci., *42:* 586–589, 1977.
353. Wynder, E. L., Kmet, J., Dungal, N., and Segi, M. An epidemiological investigation of gastric cancer. Cancer, *16:* 1461–1496, 1963.
354. Yagashita, K., Takahashi, N., Yamamoto, H., Jinnouchi, H., Hiyoshi, S., and Miyakawa, T. Effects of tetraacetyl-bis-dehydroascorbic acid, a derivative of ascorbic acid, on Ehrlich cells and HeLa cells (human carcinoma cells). J. Nutr. Sci. Vitaminol., *22:* 419–427, 1976.
355. Yamafuji, K., Nakamura, Y., Omura, H., Soeda, T., and Gyotoku, K. Anti-tumor potency of ascorbic, dehydroascorbic, or 2,3-diketogulonic acid and their action on deoxyribonucleic acid. Z. Krebsforsch., *76:* 1–7, 1971.
356. Yonemoto, R. H., Chretien, P. B., and Fehniger, T. F. Enhanced lymphocyte blastogenesis by oral ascorbic acid. Proc. Am. Assoc. Cancer Res., *17:* 288, 1976.
357. Yoshida, O., Brown, R. R., and Bryan, G. T. Relationship between tryptophan metabolism and heterotopic recurrences of human urinary bladder tumors. Cancer, *25:* 773–780, 1970.
358. Zannoni, V. G., Flynn, E. J., and Lynch, M. Ascorbic acid and drug metabolism. Biochem. Pharmacol., *2:* 1377–1392, 1972.

Proc. Natl. Acad. Sci. USA
Vol. 87, pp. 7245–7249, September 1990
Medical Sciences

Suppression of human immunodeficiency virus replication by ascorbate in chronically and acutely infected cells

(human retrovirus/acquired immune deficiency syndrome/vitamin C/anti-human immunodeficiency virus agent)

STEVE HARAKEH, RAXIT J. JARIWALLA*, AND LINUS PAULING

Viral Carcinogenesis and Immunology Laboratories, Linus Pauling Institute of Science and Medicine, 440 Page Mill Road, Palo Alto, CA 94306

Contributed by Linus Pauling, June 8, 1990

ABSTRACT **We have studied the action of ascorbate (vitamin C) on human immunodeficiency virus type 1 (HIV-1), the etiological agent clinically associated with AIDS. We report the suppression of virus production and cell fusion in HIV-infected T-lymphocytic cell lines grown in the presence of nontoxic concentrations of ascorbate. In chronically infected cells expressing HIV at peak levels, ascorbate reduced the levels of extracellular reverse transcriptase (RT) activity (by >99%) and of p24 antigen (by 90%) in the culture supernatant. Under similar conditions, no detectable inhibitory effects on cell viability, host metabolic activity, and protein synthesis were observed. In freshly infected CD4$^+$ cells, ascorbate inhibited the formation of giant-cell syncytia (by ≈93%). Exposure of cell-free virus to ascorbate at 37°C for 1 day had no effect on its RT activity or syncytium-forming ability. Prolonged exposure of virus (37°C for 4 days) in the presence of ascorbate (100–150 μg/ml) resulted in the drop by a factor of 3–14 in RT activity as compared to a reduction by a factor of 25–172 in extracellular RT released from chronically infected cells. These results indicate that ascorbate mediates an anti-HIV effect by diminishing viral protein production in infected cells and RT stability in extracellular virions.**

Previous studies demonstrated the antiviral activity of ascorbate against a broad spectrum of RNA and DNA viruses *in vitro* (1–4) and *in vivo* (5, 6). It has been claimed that ascorbate inhibited the activation of a latent human retrovirus (human T-cell leukemia virus 1) induced by 5-iodo-2′-deoxyuridine and *N*-methyl-*N*′-nitro-*N*-nitrosoguanidine (7). However, it was not established whether ascorbate exerted a virus-specific effect or interacted directly with the activating substances. In addition, the effects of ascorbate on acute infection by human retroviruses have not been determined. *In vivo*, oral, and intravenous administration of ascorbate is said to have produced clinical improvements in patients afflicted with influenza, hepatitis, and herpes virus infections, including infectious mononucleosis (5, 6). Clinical improvement was claimed in AIDS patients who voluntarily ingested high doses of ascorbic acid (8).

Because human immunodeficiency virus 1 (HIV-1) is consistently associated with AIDS (9–12), we investigated the action of ascorbate on HIV infection under controlled conditions *in vitro*. Here, we report the effects of ascorbate on acutely and chronically HIV-infected T-lymphocytic cell lines grown continuously in the presence of nontoxic concentrations of the compound. In addition, we report the action of ascorbate on cell-free virus particles *in vitro*.

MATERIALS AND METHODS

Cells and Cell Viability. H9 and H9/HTLV-III$_B$ cells (13) were originally obtained from H. Streicher (National Cancer Institute). In some experiments, batches of the same cell lines provided by M. McGrath (University of California, San Francisco) were also used. Cells were grown in RPMI 1640 medium supplemented with 10% (vol/vol) fetal calf serum, 2 mM L-glutamine, 1 mM pyruvate, and gentamycin at 50 μg/ml. The CD4-positive VB cell line (14) (provided by M. McGrath) was propagated in RPMI 1640 complete growth medium. Cell viability was determined by using the trypan blue exclusion method.

Ascorbate. The stock solution of 0.06 M L-ascorbate was made by dissolving L-ascorbic acid (tissue-culture grade from Sigma) in RPMI 1640 medium and was stored at −20°C.

Experimental Protocol. Fresh working solutions (10× strength) of ascorbate were prepared daily by diluting the stock in complete growth medium. For cytotoxicity assay, 3 × 10^5 cells were suspended in 0.9 ml of growth medium and seeded in 24-well microtiter plates. Fresh solutions of ascorbate (0.1 ml of 10× strength) were added daily to obtain final concentrations of 10, 25, 50, 75, 100, 150, 200, 300, and 400 μg/ml. The controls received 0.1 ml of growth medium. Plates were incubated at 37°C in a humidified 5% CO_2/95% air atmosphere for various time intervals. At periodic intervals, 0.5 ml of cell suspension was collected and mixed with 50 μl of trypan blue, and cells were tested for viability.

For quantitation of viral and cellular parameters, cell suspensions (in triplicate) were collected, pooled, and centrifuged at 1400 × *g* for 10 min at 4°C. Supernatant was used for assays of extracellular reverse transcriptase (RT) activity and p24 antigen. Cell pellets were used for the determination of cellular metabolic activity and protein synthesis rates.

Assay of RT. Virus particles in supernatant were pelleted by centrifugation in a refrigerated microfuge (13,500 rpm, 2 hr) and then resuspended in 2% of the original volume (30 μl) of TNE buffer (15). Samples (10 μl) were assayed for RT activity, as described by Hoffman *et al.* (15), by using fresh batches of [*methyl*-^{3}H]dTTP (specific activity, ≈80 Ci/mmol; 1 Ci = 37 GBq; NEN/DuPont). RT activity was expressed as the amount of [^{3}H]dTMP incorporated (cpm per 10^6 cells).

Assay of p24. Levels of p24 antigen in supernatant were assayed using the Abbott HIV antigen enzyme immunoassay (16). The p24 value was expressed as ng per 10^6 cells.

Assay of Protein Synthesis. For radiolabeling, H9 cells (3 × 10^5 cells per well in microtiter plates) were grown in the presence of ascorbate at 0, 75, 100, and 150 μg/ml as described earlier. On days 1, 2, and 4, cells were harvested, washed, and resuspended in methionine- and cysteine-free medium and then incubated at 37°C for 30 min in 0.5 ml of the same medium supplemented with 50 μCi of Tran^{35}S-label (specific activity, 1013 Ci/mmol; ICN). Labeled cells were pelleted, washed in isotonic phosphate-buffered saline, resuspended in lysis buffer containing 1% Nonidet P-40, and

The publication costs of this article were defrayed in part by page charge payment. This article must therefore be hereby marked "*advertisement*" in accordance with 18 U.S.C. §1734 solely to indicate this fact.

Abbreviations: HIV-1, human immunodeficiency virus 1; RT, reverse transcriptase; MTT, 3-(4,5-dimethylthiazol-2-yl)-2,5-diphenyltetrazolium bromide.
*To whom reprint requests should be addressed.

stored at −70°C. Lysate was thawed and incubated at 100°C for 3–5 min to remove aminoacyl moieties from RNA. Proteins were precipitated with trichloroacetic acid in the presence of bovine serum albumin (0.2 mg/ml), transferred to nitrocellulose filters (0.45 μm), dried, and suspended in β Blend (ICN), and radioactivity was measured in a scintillation counter. Protein synthesis was determined on duplicate samples of cells independently grown in the presence of ^{35}S-labeled amino acids (17).

Metabolic Activity Assayed by 3-(4,5-Dimethylthiazol-2-yl)-2,5-Diphenyltetrazolium Bromide (MTT) Determination. For metabolic activity assay, 3×10^5 cells per well were seeded in 24-well microtiter plates and grown in ascorbate at 0, 75, 100, and 150 μg/ml. On days 1, 2, and 4, cells were pelleted, resuspended in 1.0 ml of growth medium supplemented with 500 μg of MTT (Sigma), incubated for 4 hr, and treated with acidified isopropanol, and the absorbance at 570 nm was measured as described by Mossman (18).

Inhibition Assay for the Cytopathic Effect of HIV-1 Strain HTLV-III$_B$. Infectious HIV stock was obtained from supernatant fluid of H9/HTLV-III$_B$ cells cocultivated with VB cells at a 1:7.5 ratio for 3.5 days. To quantitate syncytium formation, 2.5×10^5 VB cells in 0.4 ml of growth medium were mixed with 0.5 ml of HIV stock and seeded in 24-well microtiter plates. Then 0.1 ml of either growth medium or 10× strength fresh L-ascorbate solution was added daily and the cells were incubated. On specific days after infection, total number of giant cell syncytia in each well were counted at ×100 magnification. A giant cell was defined as a cell >4 diameters larger than a single uninfected cell.

RESULTS

Cytotoxicity of Ascorbate. Before determining the effect on HIV production, we evaluated the cytotoxicity of ascorbate on H9/HTLV-III$_B$ cells, which are T-lymphocytic H9 cells infected with the AIDS virus (13). Ascorbate is unstable in solution as in conventional culture conditions, with a short half-life (4). Therefore, an experimental protocol was adopted in which cell cultures were given daily additions of fresh solutions of ascorbic acid prepared in buffered growth medium (pH 7.3 ± 0.1). Cells were grown in the continuous presence of various ascorbate concentrations (0–400 μg/ml) for a period of 4 days. Viability of control and ascorbate-treated cultures was determined using the trypan blue exclusion test. No toxicity was observed when cultures were grown in the presence of ascorbate at 5–150 μg/ml (Fig. 1). A slight inhibition of cell growth (73–75% survival) was seen on day 4 of incubation in medium containing ascorbate at 200–300 μg/ml. Cytotoxicity became prominent (≥50% cell death) on day 4 at ascorbate concentration of 400 μg/ml and higher. A slight increase in cell number was noted at concentrations ranging from 10 to 400 μg/ml on the first 2 days and at 5–75 μg/ml ascorbate on day 4. Based on these data, further experiments evaluating ascorbate effects on HIV production were carried out at noncytotoxic concentrations of the compound.

Effects of Ascorbate on HIV Released from Chronically Infected Cells. *Extracellular RT activity in supernatant.* We first assayed RT activity in cell-free supernatant (15) harvested from cultures grown in nontoxic ascorbate concentrations (0–150 μg/ml). Fig. 2 shows the average of RT values of ascorbate-treated cultures and controls from three independent experiments. In the controls, RT titer manifested a peak of virus production on day 4. In contrast, ascorbate-treated cultures showed a striking inhibition of RT production. The first noticeable drop (64% inhibition) in RT titer occurred on day 2 with ascorbate at 50 μg/ml, followed by a progressive decline in a dose-responsive manner. Further decreases in RT level were seen with increase in both ascorbate concentration and time of exposure. On day 4, >99% inhibition in RT titer was seen with ascorbate at 150 μg/ml. A noticeable increase in RT titer consistent with stimulation of cell growth was noted at low concentrations of ascorbate (from 5 to 25 μg/ml) on day 2. However, increase in virus production was transient, as these effects did not persist on day 4 of incubation.

p24 levels in supernatant. Another parameter of HIV production is the expression of p24 core antigen. Average values from three experiments are presented in Fig. 3. Control cultures showed a rise in p24 antigen levels at day 2, reaching maximum levels on day 4. In contrast, p24 antigen expression was blocked in ascorbate-treated cultures. Concentrations of ascorbate required to inhibit p24 synthesis were higher than those effective in inhibiting RT production. Thus, the first significant reduction in p24 levels was seen with ascorbate at 150 μg/ml on day 2. Higher declines in p24 values were observed with increased exposure to ascorbate.

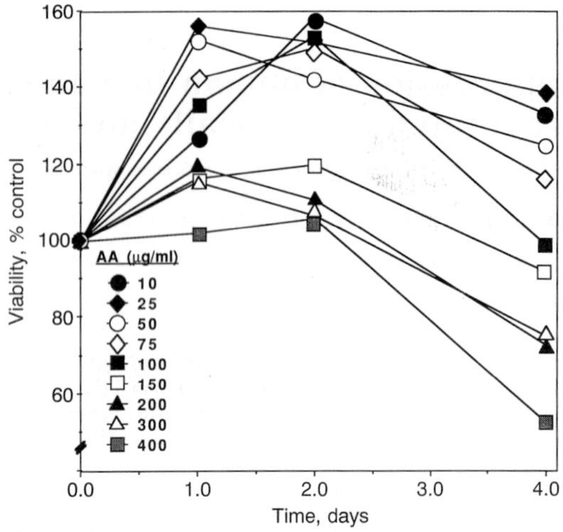

FIG. 1. Analysis of cytotoxicity of ascorbate (AA) for HTLV-III$_B$-infected H9 cells, as determined by trypan blue dye exclusion. Each point is the mean of four cell counts.

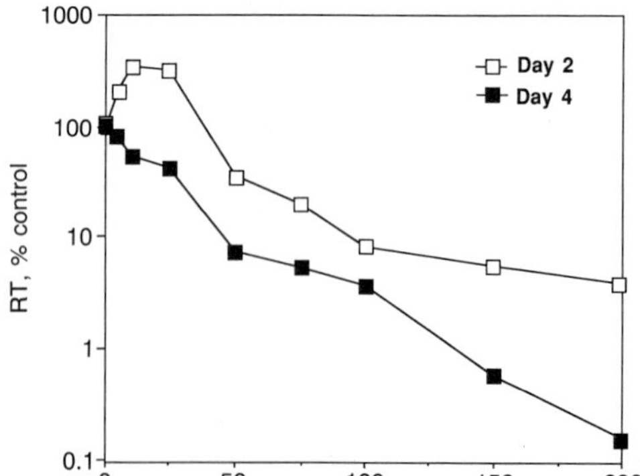

FIG. 2. Effect of ascorbate on RT activity in supernatant harvested from H9/HTLV-III$_B$ cultures. Extracellular RT was assayed as described by Hoffman *et al.* (15). In control samples, the RT values on day 2 and day 4 were, respectively, 55×10^4 and 267×10^4 cpm per 10^6 cells; average background value in blanks (i.e., reactions without enzyme) was 1530 cpm/ml of culture supernatant. In each experiment, the mean of three samples was determined and compared as a percentage of control (taken as 100%).

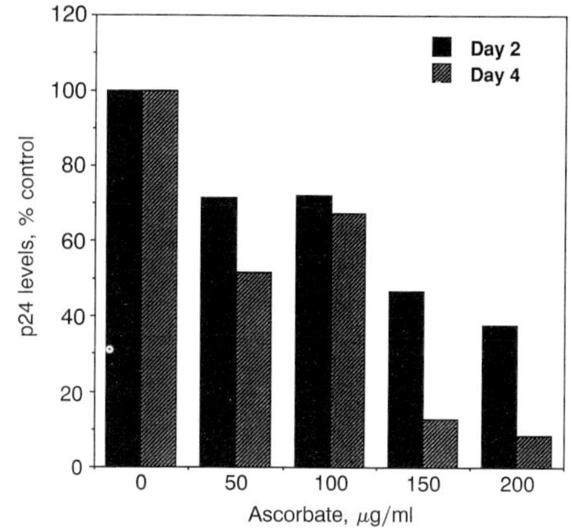

FIG. 3. Effect of ascorbate on HIV p24 antigen levels in supernatant harvested from H9/HTLV-III_B cultures. Extracellular p24 was assayed by Abbott HIV antigen enzyme immunoassay (16). In control samples, the p24 levels on days 2 and 4 were, respectively, 244 and 45 ng per 10^6 cells. The p24 values of ascorbate-treated cultures are compared as a percentage of control.

On day 4, p24 levels in cultures treated with ascorbate at 150 μg/ml were reduced to 13% of the control.

Effect of Ascorbate on Cell Metabolism. We addressed the question of whether ascorbate-induced suppression of RT and p24 production in H9/HTLV-III_B cells was a virus-specific effect or an indirect effect due to inhibition of cellular metabolism or protein synthesis. The metabolic activity of uninfected H9 cells in the presence and absence of ascorbate was determined by using a quantitative colorimetric assay that utilizes the tetrazolium salt MTT (18). MTT is used to measure the activity of various dehydrogenases in viable cells (18, 19). H9 cells grown in the presence of various concentrations of ascorbate (0–150 μg/ml) showed an increase in cellular metabolic activity on day 1 (Fig. 4). This correlated with stimulation of cell proliferation by ascorbate. On days 2 and 4, no significant change in metabolic activity was noted between control cultures and those exposed to ascorbate at 75, 100, and 150 μg/ml.

Effect of Ascorbate on Cellular Protein Synthesis. The effect of ascorbate on cellular protein synthesis was determined by growing uninfected H9 cells for 4 days with ascorbate at 0, 75, 100, and 150 μg/ml (17). On day 1, ascorbate stimulated protein synthesis, consistent with stimulation of metabolic activity and cell growth. On days 2 and 4, the difference in the apparent rates of cellular protein synthesis between ascorbate-treated and control cultures was less than a factor of 2 (Fig. 5). Thus the suppressive effects on HIV production could not be ascribed to a general inhibition of cellular metabolism or protein synthesis.

Effect of Ascorbate on Virus Replication in Freshly Infected Cells. To extend these findings to freshly infected cells, we investigated the effects of ascorbate on acute HIV infection of susceptible $CD4^+$ T lymphocytes. Viral infectivity and cytopathic effect in these cells have been correlated with formation of giant cell syncytia mediated by interaction of HIV envelope glycoprotein with $CD4^+$ cell surface receptor (14, 20, 21). In controls, multinucleated syncytia became visible by day 4 and reached high levels on day 6. The continuous presence of ascorbate caused a dose- and time-dependent decrease in syncytium formation. On day 4, ≈93.3% inhibition in syncytia number was seen with ascorbate at 100 μg/ml (Fig. 6). At this concentration, ascorbate did not inhibit the growth of uninfected VB cells (99% survival by trypan blue dye exclusion), indicating that the

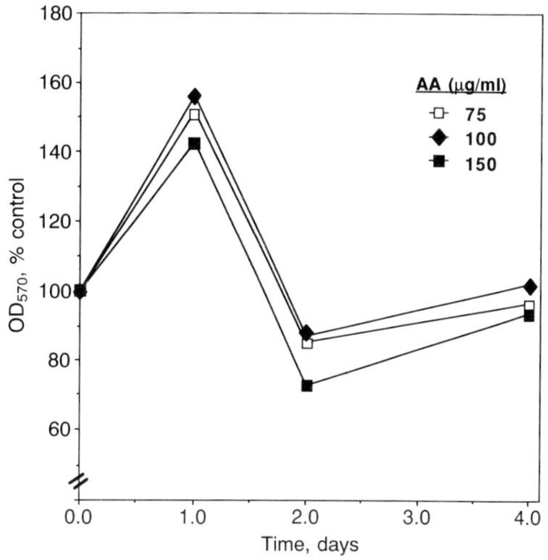

FIG. 4. Analysis of metabolic activity in H9 cells, as determined by MTT assay, in the presence and absence of ascorbate (AA), as described by Mossman (18). Each point is the mean of four OD_{570} readings. Data are plotted as percentage of control.

inhibition of virus replication was not due to cytotoxic effect of the compound.

Direct Inactivation of Virus Particles in Supernatant. We then determined whether decreases in RT titer and syncytium formation were due to direct inactivation of virus particles by ascorbate *in vitro*. Cell-free supernatant containing infectious virus was incubated in the presence and absence of ascorbate at 37°C for 8 and 18 hr. Samples were tested for RT activity and syncytium formation was measured in VB cells. After incubation at 37°C for 18 hr, there was no detectable difference in RT activity between ascorbate-treated virus preparations and controls (Table 1). Syncytium-forming titer of infectious virus of ascorbate-treated and untreated preparations after incubation at 37°C for 1 day was also approximately equal ($2.34–2.70 \times 10^3$ $TCID_{50}$ per ml; where $TCID_{50}$ is the tissue culture 50% infective dose). When chronically infected cells were exposed to ascorbate at 150 μg/ml for 18

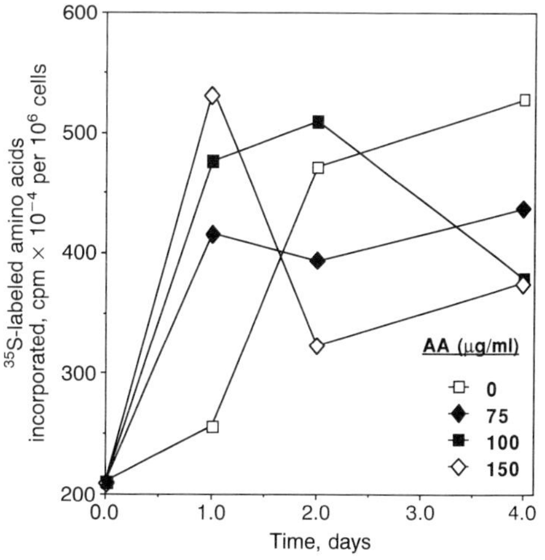

FIG. 5. Determination of protein synthesis rates in H9 cells in the presence and absence of ascorbate (AA). Protein synthesis was assayed as described by Somasundaran and Robinson (17). Each point is the mean of ^{35}S-labeled amino acid incorporation per 10^6 cells.

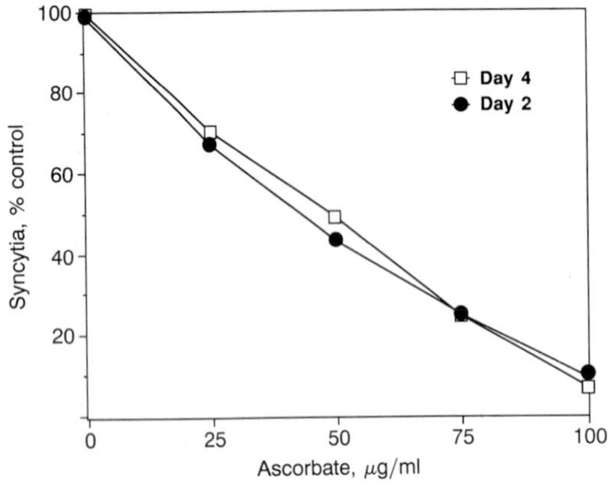

FIG. 6. Dose-dependent decrease in HIV-induced syncytium formation with ascorbate. Syncytia were counted in $CD4^+$ VB cells by using a light microscope. Each point represents the mean of at least four samples and is compared as a percentage of the control infected cultures from the same experiment.

hr at 37°C, the RT titer in culture supernatant was reduced to 11.2% of the control (Table 1). These results indicate that the decrease in extracellular RT titer by ascorbate was not due to direct inactivation of virus after overnight incubation.

To study RT stability in the presence of ascorbate upon prolonged incubation (37°C for several days), the following experiment was carried out. Since thermal inactivation of cell-free virus occurs upon extensive incubation at 37°C, uninfected cells were used to protect virus from heat inactivation. Accordingly, HIV supernatant was mixed with uninfected VB cells and incubated with ascorbate for 4 days, with daily addition of fresh compound. Supernatants were harvested and assayed for RT activity. After 4 days in the presence of ascorbate at 100 and 150 μg/ml, RT activity was reduced, respectively, to 31.5% and 7.0% of control (Table 1). In parallel experiments, chronically infected cells were exposed to ascorbate at 100 and 150 μg/ml for 4 days. The RT levels in supernatant were reduced to 4.0% and 0.6% of control.

After incubation of cell-free virus at 37°C for 4 days with ascorbate at 150 μg/ml, the concentration of p24 protein in the ascorbate-treated preparation (283 ng/ml) was not significantly different from that of the control (263 ng/ml). At the same ascorbate concentration, chronically infected cells exhibited a reduction by a factor of ≈8 in p24 antigen production after 4 days at 37°C (Fig. 3).

DISCUSSION

The experimental evidence presented in this paper has demonstrated that continuous exposure of HIV-infected cells to noncytotoxic ascorbate concentrations resulted in significant inhibition of both virus replication in chronically infected cells and multinucleated giant-cell formation in acutely infected $CD4^+$ cells. Ascorbate dosage exerted a wide margin between antiviral activity and cellular cytotoxicity. Thus, as illustrated in Fig. 2, a significant antiviral effect (>90% RT inhibition, day 4) was seen with ascorbate at ≥50 μg/ml. In contrast, significant cytotoxicity (≥50% inhibition of cell viability) was prominent on day 4 with ascorbate at ≥400 μg/ml (Fig. 1). As demonstrated in Figs. 4 and 5, concentrations of ascorbate that significantly diminished virus titer did not exert inhibitory effects on host metabolic activity and apparent rate of protein synthesis. These findings lend support to a virus-specific action of ascorbate rather than a more general effect on cellular metabolism.

In chronically infected cells, RT activity exhibited a greater sensitivity to ascorbate than p24 antigen expression. This difference can be attributed to the distinct extracellular forms of these two proteins. After intracellular synthesis, a fraction of p24 becomes encapsidated into the virions and some molecules are secreted extracellularly in a free form (16). The p24 antigen capture assay measures both free and virion-bound proteins in the culture supernatant. In contrast, active RT in cell-free supernatant is found only in virion-bound form. Additionally, the quantitative difference between p24 and RT levels may be due to the differential stability of these proteins in the presence of ascorbate. Whereas p24 antigenicity was not lost upon exposure of extracellular virus to ascorbate at 150 μg/ml at 37°C for 4 days, RT activity was reduced by a factor of 14 under similar conditions.

Experiments comparing the kinetics of RT suppression in chronically infected cells with RT stability of cell-free virus provide insight into the mechanism by which ascorbate inhibits HIV. Treatment of chronically infected cells with ascorbate at 150 μg/ml at 37°C for 18 hr reduced RT release to 11.2% of control (Table 1). In contrast, incubation of cell-free virus under the same conditions did not diminish RT activity or syncytium-forming titer of infectious virus. These observations of short-term ascorbate treatment indicate that the compound by itself does not directly inhibit RT activity or functional integrity of envelope glycoprotein involved in syncytium formation (14, 20, 21). The reduction of these viral

Table 1. Analysis of RT stability and RT production in the presence of and absence of ascorbate

			RT activity after incubation at 37°C					
			8 hr		18 hr		4 days	
Virus	Cell source	Ascorbate, μg/ml	cpm × 10^{-4} per 10^6 cells	% control	cpm × 10^{-4} per 10^6 cells	% control	cpm × 10^{-4} per 10^6 cells	% control
HIV	Supernatant	0	6.68	100	6.06	100	ND	ND
		100	7.00	105	6.17	102	ND	ND
		150	7.40	111	6.81	112	ND	ND
HIV-VB	Suspension	0	ND	ND	16.4	100	5.86	100
		100	ND	ND	12.9	78.4	1.85	31.5
		150	ND	ND	7.84	47.8	0.41	6.96
H9/HTLV-III_B	Supernatant	0	79.2	100	56.7	100	267	100
		100	81.9	103	12.7	22.4	10.6	3.96
		150	67.5	85.2	6.33	11.2	1.54	0.58

HIV virus supernatant was prepared from H9/HTLV-III_B cells. Virus supernatant alone or a suspension of supernatant and uninfected VB cells (3 × 10^5 cells per ml) was exposed to ascorbate at 0, 100, or 150 μg/ml and incubated at 37°C with daily addition of fresh compound. In a parallel experiment, chronically infected H9/HTLV-III_B cells were grown under similar conditions. Supernatants were collected and assayed for RT activity. ND, not done.

parameters in infected cells therefore represent inhibition of a step or steps in HIV replication. Prolonged ascorbate treatment of virus (37°C, 4 days) in the presence of VB cells resulted in a drop in RT activity of a factor of 3–14 with ascorbate at 100 and 150 μg/ml. Under similar conditions, chronically infected cells exhibited a reduction of a factor of 25–172 in extracellular RT activity. These findings are consistent with a combined suppressive action of ascorbate, operating at increasing contact time, on RT production in infected cells and RT stability in extracellular virus particles.

The molecular mechanism by which ascorbate suppresses HIV is not yet fully understood. In earlier studies, ascorbic acid caused degradation of single- and double-stranded genomes of RNA and DNA phages (22–24). Site-specific cleavage of phage DNA occurring at unique sites due to redox reactions involving copper and ascorbate was reported (25). Hydroxyl radicals (OH$^{\cdot}$) generated from hydrogen peroxide were implicated as the reactive species mediating scission of nucleic acid (24–26). In lymphocytes, ascorbate is present in unusually high concentration, as much as 50 times the blood plasma level and >10 times that in nonlymphoid cells (27, 28). Therefore, one possible mechanism of ascorbate action on infected lymphocytic cells is that free unintegrated viral DNA or newly synthesized viral RNA formed during each cycle of HIV replication becomes susceptible to ascorbate-mediated damage, resulting in reduced viral protein production. Alternatively, ascorbate may suppress HIV production by inhibiting the activity of viral enzymes involved in protein processing (e.g., HIV protease). Ascorbate was shown to inhibit the activity of proteolytic enzymes in fish (29). Once HIV components are packaged within the virion, they may become resistant to ascorbate inactivation. However, upon prolonged *in vitro* exposure, virion components may become susceptible to further attack by metabolites of ascorbate generated from its oxidative degradation. Autooxidation of ascorbic acid is associated with the formation of highly reactive breakdown products including furan-type compounds that form adducts with amino and hydroxyl groups of proteins resulting in site-specific cleavage or cross-linking of protein (30–32). Such protein modifications could contribute to inactivation of virion-associated enzyme detected upon prolonged incubation of virus *in vitro*. Further studies of the physical state of HIV nucleic acids and proteins in ascorbate-treated cells and virions could provide insight into its mechanism of action.

Inhibitors of RT activity have been the focus of intensive investigation for the design and development of antiretroviral agents. Among these, 3′-azido-3′-deoxythymidine (AZT) (33), the first drug approved for AIDS treatment, blocks *de novo* HIV infection effectively but was shown not to inhibit virus production in cells containing integrated HIV genomes (34). In the same study, interferon-α inhibited the budding and release of HIV from chronically infected cells but did not suppress intracellular production of viral proteins. The ability of ascorbate to inhibit acute HIV infection and to suppress RT levels in chronically infected cells indicates that the compound acts at a different stage in the HIV life cycle and may, thereby, provide a rationale for developing more effective combination therapy with other anti-HIV agents.

The concentrations at which the anti-HIV effect of ascorbate was seen in this *in vitro* study are physiologically attainable in human blood plasma. For instance, in a clinical trial on terminal cancer patients, E. Cameron (personal communication) showed that oral administration of 10 g of ascorbate resulted in mean plasma ascorbate levels of 28.91 μg/ml (range 17.2–63.6 g). B. Jaffe (personal communication), who is using ascorbate for the treatment of AIDS patients, indicated that ascorbate at 93 ± 29 μg/ml was attained in plasma in people consuming oral ascorbate to achieve urinary levels >1 mg/ml. These findings are consistent with a high bowel tolerance reported for AIDS patients (8). Intravenous infusion of 50 g of ascorbate a day resulted in peak plasma levels of 796 ± 111 μg/ml.

Note Added in Proof. Roederer *et al.* (35) reported inhibition of cytokine-stimulated HIV replication by *N*-acetyl-L-cysteine, another reducing agent like ascorbate.

We thank R. C. Gallo and P. S. Sarin for helpful suggestions on *in vitro* experiments; H. Streicher and M. McGrath for providing the cell lines used in this study; E. Zuckerkandl, E. Cameron, and T. Boulikas for constructive comments on the manuscript; D. McWeeney and J. Freeberg for technical assistance; P. Pelton and G. Latter for computation of data; and D. Read, S. Schwoebel, and J. Cox for processing the manuscript. This work was supported from private donations to the Linus Pauling Institute and by the Japan Shipbuilding Industry Foundation.

1. Murata, A., Kitagawa, H., Inmaru, H. & Saruna, R. (1972) *Agric. Biol. Chem.* **36,** 1065–1067.
2. Murata, A., Kitagawa, H., Inmaru, H. & Saruna, R. (1972) *Agric. Biol. Chem.* **36,** 2597–2599.
3. Schwerdt, P. R. & Schwerdt, C. E. (1975) *Proc. Soc. Exp. Biol. Med.* **148,** 1237–1243.
4. Bissell, M. J., Hatie, C., Farson, D. A., Schwarz, R. I. & Soo, W-J. (1980) *Proc. Natl. Acad. Sci. USA* **77,** 2711–2715.
5. Klenner, F. R. (1971) *J. Appl. Nutr.* **23,** 61–87.
6. Cathcart, R. F. (1983) *Biol. Med. (Stockholm)* **3,** 6–8.
7. Blakeslee, J. R., Yamamoto, N. & Hinuma, Y. (1985) *Cancer Res.* **45,** 3471–3476.
8. Cathcart, R. (1984) *Med. Hypotheses* **14,** 423–433.
9. Barré-Sinoussi, F., Chermann, J. C., Rey, F., Nugeyre, M. T., Chamaret, S., Gruest, J., Dauguet, C., Axler-Blin, C., Vezinet-Brun, F., Rouzioux, C., Rozenbaum, W. & Montagnier, L. (1983) *Science* **220,** 868–870.
10. Gallo, R. C., Salahuddin, S. Z., Popovic, M., Shearer, G. M., Kaplan, M., Haynes, B. F., Palker, T. J., Redfield, R., Oleske, J., Safai, B., White, G., Foster, P. & Markham, P. E. (1984) *Science* **224,** 500–503.
11. Levy, J. A., Hoffman, A. D., Kramer, S. M., Landis, J. A., Shimabukuro, J. M. & Oshiro, L. S. (1984) *Science* **225,** 840–842.
12. Blattner, W., Gallo, R. C. & Temin, H. M. (1988) *Science* **241,** 515–516.
13. Popovic, M., Sarngadharan, M. G., Read, E. & Gallo, R. C. (1984) *Science* **224,** 497–500.
14. Lifson, J. D., Reyes, G. R., McGrath, M. S., Stein, B. S. & Engleman, E. G. (1986) *Science* **232,** 1123–1127.
15. Hoffman, A. D., Banapour, B. & Levy, J. A. (1985) *Virology* **147,** 326–335.
16. Goudsmit, J., Paul, D. A., Lange, J. M. A., Speelman, H., Van Der Noorda, J., Van Der Helm, H. J., de Wolf, F., Epstein, L. G., Krone, W. J. A., Wolters, E. Ch., Oleske, J. M. & Coutinho, R. A. (1986) *Lancet* **ii,** 177–180.
17. Somasundaran, M. & Robinson, H. L. (1988) *Science* **242,** 1554–1557.
18. Mossman, T. (1983) *J. Immunol. Methods* **65,** 55–63.
19. Schwartz, O., Henin, Y., Marechal, V. & Montagnier, L. (1988) *AIDS Res. Hum. Retroviruses* **4,** 441–448.
20. Dalgleish, A. G., Beverley, P. C. L., Clapham, P. R., Crawford, D. H., Greaves, M. F. & Weiss, R. A. (1984) *Nature (London)* **312,** 763–767.
21. Klatzmann, D., Champagne, E., Chamaret, S., Gruest, J., Guetard, D., Hercend, T., Gluckman, J. C. & Montagnier, L. (1984) *Nature (London)* **312,** 767–768.
22. Murata, A. & Kitagawa, K. (1973) *Agric. Biol. Chem.* **37,** 1145–1151.
23. Murata, A. & Uike, M. (1976) *J. Nutr. Sci. Vitaminol.* **22,** 347–354.
24. Wong, K., Morgan, A. R. & Paranchych, W. (1974) *Can. J. Biochem.* **52,** 950–958.
25. Kazakov, S. A., Atashkina, T. G., Mamaev, S. V. & Vlassov, V. V. (1988) *Nature (London)* **335,** 186–188.
26. Chiou, S.-H. (1983) *J. Biochem. (Tokyo)* **94,** 1259–1267.
27. Evans, R., Currie, L. & Campbell, A. (1982) *Br. J. Nutr.* **47,** 473–482.
28. Varma, S. D. (1987) *Ann. N.Y. Acad. Sci.* **498,** 280–291.
29. Dabrowski, K. & Köck, G. (1989) *Int. J. Vit. Nutr. Res.* **59,** 157–160.
30. Nakanishi, Y., Isokashi, F., Matsunaga, T. & Sahamoto, Y. (1985) *Eur. J. Biochem.* **152,** 337–342.
31. Garland, D., Zigler, S. J., Jr., & Kinoshita, J. (1986) *Arch. Biochem. Biophys.* **251,** 771–776.
32. Ortwerth, J., Feather, M. S. & Olesen, P. R. (1988) *Exp. Eye Res.* **47,** 155–168.
33. Yarchoan, R., Brouwers, P., Spitzer, A. R., Grafman, J., Safai, B., Perno, C. F., Larson, S. M., Berg, G., Fischl, M. A., Wichman, A., Thomas, R. V., Brunetti, A., Schmidt, P. J., Myers, C. E. & Broder, S. (1987) *Lancet* **i,** 132–135.
34. Poli, G., Orenstein, J. M., Kintner, A., Folks, T. M. & Fauci, A. S. (1989) *Science* **244,** 575–577.
35. Roederer, M., Staal, F. J. T., Raju, P. A., Ela, S. W., Herzenberg, L. A. & Herzenberg, L. A. (1990) *Proc. Natl. Acad. Sci. USA* **87,** 4884–4888.

Reprinted from *Cancer and Vitamin C* (Updated and Expanded Edition), by Ewan Cameron and Linus Pauling, Camino Books, Philadelphia, PA, Chapter 23, pp. 189–195 (1993).

A Unified Theory of Human Cardiovascular Disease Leading the Way to the Abolition of This Disease as a Cause for Human Mortality

Matthias Rath M.D. and Linus Pauling Ph.D.[1]

"An important scientific innovation rarely makes its way by gradually winning over and converting its opponents. What does happen is that its opponents gradually die out and that the growing generation is familiar with the idea from the beginning." Max Planck

This paper is dedicated to the young physicians and the medical students of this world.

Abstract

Until now therapeutic concepts for human cardiovascular disease (CVD) were targeting individual pathomechanisms or specific risk factors. On the basis of genetic, metabolic, evolutionary, and clinical evidence we present here a unified pathogenetic and therapeutic approach. Ascorbate deficiency is the precondition and common denominator of human CVD. Ascorbate deficiency is the result of the inability of man to synthesize ascorbate endogenously in combination with insufficient dietary intake. The invariable morphological consequences of chronic ascorbate deficiency in the vascular wall are the loosening of the connective tissue and the loss of the endothelial barrier function. Thus human CVD is a form of pre-scurvy. The multitude of pathomechanisms that lead to the clinical manifestation of CVD are primarily defense mechanisms aiming at the stabilization of the vascular wall. After the loss of endogenous ascorbate production during the evolution of man these defense mechanisms became life-saving. They counteracted the fatal consequences of scurvy and particularly of blood loss through the scorbutic vascular wall. These countermeasures constitute a genetic and a metabolic level. The genetic level is characterized by the evolutionary advantage of inherited features that lead to a thickening of the vascular wall, including a multitude of inherited diseases. The metabolic level is characterized by the close connection of ascorbate with metabolic regulatory systems that determine the risk profile for CVD in clinical cardiology today. The most frequent mechanism is the deposition of lipoproteins, particularly lipoprotein(a) [Lp(a)], in the vascular wall. With sustained ascorbate deficiency, the result of insufficient ascorbate uptake, these defense mechanisms overshoot and lead to the development of CVD. Premature CVD is essentially unknown in all animal species that produce high amounts of ascorbate endogenously. In humans, unable to produce endogenous ascorbate, CVD became one of the most frequent diseases. The genetic mutation that rendered all human beings today dependent on dietary ascorbate is the universal underlying cause of CVD. Optimum dietary ascorbate intake will correct this common genetic defect and prevent its deleterious consequences. Clinical confirmation of this theory should largely abolish CVD as a cause for mortality in this generation and future generations of mankind.

Key words

Ascorbate, vitamin C, cardiovascular disease, lipoprotein(a), hypercholesterolemia, hypertriglyceridemia, hypoalphalipoproteinemia, diabetes, homocystinuria.

Introduction

We have recently presented ascorbate deficiency as the primary cause of human CVD. We proposed that the most frequent pathomechanism leading to the development of atherosclerotic plaques is the deposition of Lp(a) and fibrinogen/fibrin in the ascorbate-deficient vascular wall.[1,2] In the course of this work we discovered that virtually every pathomechanism for human CVD known today can be induced by ascorbate deficiency. Beside the deposition of Lp(a) this includes such seemingly unrelated processes as foam cell formation and decreased reverse-cholesterol

1. Linus Pauling Institute of Science and Medicine, 440 Page Mill Road, Palo Alto, CA 94306.

Journal of Orthomolecular Medicine Vol. 7, No. 1, 1992

transfer, and also peripheral angiopathies in diabetic or homocystinuric patients. We did not accept this observation as a coincidence. Consequently we proposed that ascorbate deficiency is the precondition as well as a common denominator of human CVD. This far-reaching conclusion deserves an explanation; it is presented in this paper. We suggest that the direct connection of ascorbate deficiency with the development of CVD is the result of extraordinary pressure during the evolution of man. After the loss of the endogenous ascorbate production in our ancestors, severe blood-loss through the scorbutic vascular wall became a life-threatening condition. The resulting evolutionary pressure favored genetic and metabolic mechanisms predisposing to CVD.

The Loss of Endogenous Ascorbate Production in the Ancestor of Man

With few exceptions all animals synthesize their own ascorbate by conversion from glucose. In this way they manufacture a daily amount of ascorbate that varies between about 1 gram and 20 grams, when compared to the human body weight. About 40 million years ago the ancestor of man lost the ability for endogenous ascorbate production. This was the result of a mutation of the gene encoding for the enzyme L-gulono-g-lactone oxidase (GLO), a key enzyme in the conversion of glucose to ascorbate. As a result of this mutation all descendants became dependent on dietary ascorbate intake.

The precondition for the mutation of the GLO gene was a sufficient supply of dietary ascorbate. Our ancestors at that time lived in tropical regions. Their diet consisted primarily of fruits and other forms of plant nutrition that provided a daily dietary ascorbate supply in the range of several hundred milligrams to several grams per day. When our ancestors left this habitat to settle in other regions of the world the availability of dietary ascorbate dropped considerably and they became prone to scurvy.

Fatal Blood Loss Through the Scorbutic Vascular Wall - An Extraordinary Challenge to the Evolutionary Survival of Man

Scurvy is a fatal disease. It is characterized by structural and metabolic impairment of the human body, particularly by the destabilization of the connective tissue. Ascorbate is essential for an optimum production and hydroxylation of collagen and elastin, key constituents of the extracellular matrix. Ascorbate depletion thus leads to a destabilization of the connective tissue throughout the body. One of the first clinical signs of scurvy is perivascular bleeding. The explanation is obvious: Nowhere in the body does there exist a higher pressure difference than in the circulatory system, particularly across the vascular wall. The vascular system is the first site where the underlying destabilization of the connective tissue induced by ascorbate deficiency is unmasked, leading to the penetration of blood through the permeable vascular wall. The most vulnerable sites are the proximal arteries, where the systolic blood pressure is particularly high. The increasing permeability of the vascular wall in scurvy leads to petechiae and ultimately hemorrhagic blood loss.

Scurvy and scorbutic blood loss decimated the ship crews in earlier centuries within months. It is thus conceivable that during the evolution of man periods of prolonged ascorbate deficiency led to a great death toll. The mortality from scurvy must have been particularly high during the thousands of years the ice ages lasted and in other extreme conditions, when the dietary ascorbate supply approximated zero. We therefore propose that after the loss of endogenous ascorbate production in our ancestors, scurvy became one of the greatest threats to the evolutionary survival of man. By hemorrhagic blood loss through the scorbutic vascular wall our ancestors in many regions may have virtually been brought close to extinction.

The morphologic changes in the vascular wall induced by ascorbate deficiency are well characterized: the loosening of the connective tissue and the loss of the endothelial barrier function. The extraordinary pressure by fatal blood loss through the scorbutic vascular wall favored genetic and metabolic countermeasures attenuating increased vascular permeability.

Ascorbate Deficiency and Genetic Countermeasures

The genetic countermeasures are characterized by an evolutionary advantage of genetic features and include inherited disorders that are associated with atherosclerosis and CVD. With sufficient ascorbate supply these disorders stay latent. In ascorbate deficiency, however, they become unmasked, leading to an increased deposition of plasma constituents in the vascular wall and other mechanisms that thicken the vascular wall. This thickening of the vascular wall is a defense measure compensating for the impaired vascular wall that had become destabilized by ascorbate deficiency. With prolonged insufficient ascorbate intake in the diet these defense mechanisms overshoot and CVD develops.

The most frequent mechanism to counteract the increased permeability of the ascorbate-deficient vascular wall became the deposition of lipoproteins and lipids in the vessel wall. Another group of proteins that generally accumulate at sites of tissue transformation and repair are adhesive proteins such as fibronectin, fibrinogen, and particularly apo(a). It is therefore no surprise that Lp(a), a combination of the adhesive protein apo(a) with a low density lipoprotein (LDL) particle, became the most frequent genetic feature counteracting ascorbate deficiency.[1] Beside lipoproteins, certain metabolic disorders, such as diabetes and homocystinuria, are also associated with the development of CVD. Despite differences in the underlying pathomechanism, all these mechanisms share a common feature: they lead to a thickening of the vascular wall and thereby can counteract the increased permeability in ascorbate deficiency.

In addition to these genetic disorders, the evolutionary pressure from scurvy also favored certain metabolic countermeasures.

Ascorbate Deficiency and Metabolic Countermeasures

The metabolic countermeasures are characterized by the regulatory role of ascorbate for metabolic systems determining the clinical risk profile for CVD. The common aim of these metabolic regulations is to decrease the vascular permeability in ascorbate deficiency. Low ascorbate concentrations therefore induce vasoconstriction and hemostasis and affect vascular wall metabolism in favor of atherosclerogenesis. Towards this end ascorbate interacts with lipoproteins, coagulation factors, prostaglandins, nitric oxide, and second messenger systems such as cyclic monophosphates (for review see 1, 3-5). It should be noted that ascorbate can affect these regulatory levels in a multiple way. In lipoprotein metabolism low density lipoproteins (LDL), Lp(a), and very low density lipoproteins (VLDL) are inversely correlated with ascorbate concentrations, whereas ascorbate and HDL levels are positively correlated. Similarly, in prostaglandin metabolism ascorbate increases prostacyclin and prostaglandin E levels and decreases the thromboxane level. In general, ascorbate deficiency induces vascular constriction and hemostatis, as well as cellular and extracellular defense measures in the vascular wall.

In the following sections we shall discuss the role of ascorbate for frequent and well established pathomechanisms of human CVD. In general, the inherited disorders described below are polygenic. Their separate description, however, will allow the characterization of the role of ascorbate on the different genetic and metabolic levels.

Apo(a) and Lp(a), the Most Effective and Most Frequent Countermeasure

After the loss of endogenous ascorbate production, apo(a) and Lp(a) were greatly favored by evolution. The frequency of occurrence of elevated Lp(a) plasma levels in species that had lost the ability to synthesize ascorbate is so great that we formulated the theory that apo(a) functions as a surrogate for ascorbate.[6] There are several genetically determined isoforms of apo(a). They differ in the number of kringle repeats and in their molecular size.[7] An inverse relation between the molecular size of apo(a) and the synthesis rate of Lp(a) particles has been established. Individuals with the high molecular weight apo(a) isoform produce fewer Lp(a) particles than those with the low apo(a) isoform. In most population studies the genetic pattern of high apo(a) isoform/low Lp(a) plasma level was found to be the most advantageous and therefore most frequent pattern. In ascorbate deficiency Lp(a) is selectively retained in the vascular wall. Apo(a) counteracts increased permeability by compensating for collagen, by its binding to fibrin, as a proteinthiol antioxidant, and as an inhibitor of plasmin-induced proteolysis (1). Moreover, as an adhesive protein apo(a) is effective in tissue-repair processes (8). Chronic ascorbate deficiency leads to a sustained accumulation of Lp(a) in

Journal of Orthomolecular Medicine Vol. 7, No. 1, 1992

the vascular wall. This leads to the development of atherosclerotic plaques and premature CVD, particularly in individuals with genetically determined high plasma Lp(a) levels. Because of its association with apo(a), Lp(a) is the most specific repair particle among all lipoproteins. Lp(a) is predominantly deposited at predisposition sites and it is therefore found to be significantly correlated with coronary, cervical, and cerebral atherosclerosis but not with peripheral vascular disease.

The mechanism by which ascorbate resupplementation prevents CVD in any condition is by maintaining the integrity and stability of the vascular wall. In addition, ascorbate exerts in the individual a multitude of metabolic effects that prevent the exacerbation of a possible genetic predisposition and the development of CVD. If the predisposition is a genetic elevation of Lp(a) plasma levels the specific regulatory role of ascorbate is the decrease of apo(a) synthesis in the liver and thereby the decrease of Lp(a) plasma levels. Moreover, ascorbate decreases the retention of Lp(a) in the vascular wall by lowering fibrinogen synthesis and by increasing the hydroxylation of lysine residues in vascular wall constituents, thereby reducing the affinity for Lp(a) binding.[1]

In about half of the CVD patients the mechanism of Lp(a) deposition contributes significantly to the development of atherosclerotic plaques. Other lipoprotein disorders are also frequently part of the polygenic pattern predisposing the individual patient to CVD in the individual.

Other Lipoprotein Disorders Associated with CVD

In a large population study Goldstein et al. discussed three frequent lipid disorders, familial hypercholesterolemia, familial hypertriglyceridemia, and familial combined hyperlipidemia.[9] Ascorbate deficiency unmasks these underlying genetic defects and leads to an increased plasma concentration of lipids (e.g. cholesterol, triglycerides) and lipoproteins (e.g. LDL, VLDL) as well as to their deposition in the impaired vascular wall. As with Lp(a), this deposition is a defense measure counteracting the increased permeability. It should, however, be noted that the deposition of lipoproteins other than Lp(a) is a less specific defense mechanism and frequently follows Lp(a) deposition. Again, these mechanisms function as a defense only for a limited time. With sustained ascorbate deficiency the continued deposition of lipids and lipoproteins leads to atherosclerotic plaque development and CVD. Some mechanisms will now be described in more detail.

Hypercholesterolemia, LDL-receptor defect

A multitude of genetic defects lead to an increased synthesis and/or a decreased catabolism of cholesterol or LDL. A well characterized although rare defect is the LDL-receptor defect. Ascorbate deficiency unmasks these inherited metabolic defects and leads to an increased plasma concentration of cholesterol-rich lipoproteins, e.g. LDL, and their deposition in the vascular wall. Hypercholesterolemia increases the risk for premature CVD primarily when combined with elevated plasma levels of Lp(a) or triglycerides.

The mechanisms by which ascorbate supplementation prevents the exacerbation of hypercholesterolemia and related CVD include an increased catabolism of cholesterol. In particular, ascorbate is known to stimulate 7-a-hydroxylase, a key enzyme in the conversion of cholesterol to bile acids and to increase the expression of LDL receptors on the cell surface. Moreover, ascorbate is known to inhibit endogenous cholesterol synthesis as well as oxidative modification of LDL (for review see 1).

Hypertriglyceridemia, Type III hyperlipidemia

A variety of genetic disorders lead to the accumulation of triglycerides in the form of chylomicron remnants, VLDL, and intermediate density lipoproteins (IDL) in plasma. Ascorbate deficiency unmasks these underlying genetic defects and the continued deposition of triglyceride-rich lipoproteins in the vascular wall leads to CVD development. These triglyceride-rich lipoproteins are particularly subject to oxidative modification, cellular lipoprotein uptake, and foam cell formation. In hypertriglyceridemia nonspecific foam-cell formation has been observed in a variety of organs.[10] Ascorbate-deficient foam cell formation, although a less specific repair mechanism than the extracellular deposition of Lp(a), may have also conferred stability .

Ascorbate supplementation prevents the exacerbation of CVD associated with hypertriglyceridemia, Type III hyperlipidemia, and related disorders by stimulating lipoprotein lipases and thereby enabling a normal catabolism of triglyceride-rich lipoproteins.[11] Ascorbate prevents the oxidative modification of these lipoproteins, their uptake by scavenger cells and foam cell formation. Moreover, we propose here that, analogous to the LDL receptor, ascorbate also increases the expression of the receptors involved in the metabolic clearance of triglyceride-rich lipoproteins, such as the chylomicron remnant receptor.

The degree of build-up of atherosclerotic plaques in patients with lipoprotein disorders is determined by the rate of deposition of lipoproteins and by the rate of the removal of deposited lipids from the vascular wall. It is therefore not surprising that ascorbate is also closely connected with this reverse pathway.

Hypoalphalipoproteinemia

Hypoalphalipoproteinemia is a frequent lipoprotein disorder characterized by a decreased synthesis of HDL particles. HDL is part of the 'reverse-cholesterol-transport' pathway and is critical for the transport of cholesterol and also other lipids from the body periphery to the liver. In ascorbate deficiency this genetic defect is unmasked, resulting in decreased HDL levels and a decreased reverse transport of lipids from the vascular wall to the liver. This mechanism is highly effective and the genetic disorder hypoalphalipoproteinemia was greatly favored during evolution. With ascorbate supplementation HDL production increases,[12] leading to an increased uptake of lipids deposited in the vascular wall and to a decrease of the atherosclerotic lesion. A look back in evolution underlines the importance of this mechanism. During the winter seasons, with low ascorbate intake, our ancestors became dependent on protecting their vascular wall by the deposition of lipoproteins and other constituents. During spring and summer seasons the ascorbate content in the diet increased significantly and mechanisms were favored that decreased the vascular deposits under the protection of increased ascorbate concentration in the vascular tissue. It is not unreasonable for us to propose that ascorbate can reduce fatty deposits in the vascular wall within a relatively short time. In an earlier clinical study it was shown that 500 mg of dietary ascorbate per day can lead to a reduction of atherosclerotic deposits within 2 to 6 months.[13]

This concept, of course, also explains why heart attack and stroke occur today with a much higher frequency in winter than during spring and summer, the seasons with increased ascorbate intake.

Other Inherited Metabolic Disorders Associated with CVD

Beside lipoprotein disorders many other inherited metabolic diseases are associated with CVD. Generally these disorders lead to an increased concentration of plasma constituents that directly or indirectly damage the integrity of the vascular wall. Consequently these diseases lead to peripheral angiopathies as observed in diabetes, homocystinuria, sickle-cell anemia (the first molecular disease described,[14] and many other genetic disorders. Similar to lipoproteins the deposition of various plasma constituents as well as proliferative thickening provided a certain stability for the ascorbate-deficient vascular wall. We illustrate this principle for diabetic and homocystinuric angiopathy.

Diabetic Angiopathy

The pathomechanism in this case involves the structural similarity between glucose and ascorbate and the competition of these two molecules for specific cell surface receptors.[15,16] Elevated glucose levels prevent many cellular systems in the human body, including endothelial cells, from optimum ascorbate uptake. Ascorbate deficiency unmasks the underlying genetic disease, aggravates the imbalance between glucose and ascorbate, decreases vascular ascorbate concentration, and thereby triggers diabetic angiopathy.

Ascorbate supplementation prevents diabetic angiopathy by optimizing the ascorbate concentration in the vascular wall and also by lowering insulin requirement.[17]

Homocystinuric angiopathy

Homocystinuria is characterized by the accumulation of homocyst(e)ine and a variety of its metabolic derivatives in the plasma, the tissues and the urine as the result of decreased homocysteine catabolism.[18] Elevated plasma

Journal of Orthomolecular Medicine Vol. 7, No. 1, 1992

concentrations of homocyst(e)ine and its derivatives damage the endothelial cells throughout the arterial and venous system. Thus homocystinuria is characterized by peripheral vascular disease and thromboembolism. These clinical manifestations have been estimated to occur in 30 per cent of the patients before the age of 20 and in 60 per cent of the patients before the age of 40.[19]

Ascorbate supplementation prevents homocystinuric angiopathy and other clinical complications of this disease by increasing the rate of homocysteine catabolism.[20]

Thus, ascorbate deficiency unmasks a variety of individual genetic predispositions that lead to CVD in different ways. These genetic disorders were conserved during evolution largely because of their association with mechanisms that lead to the thickening of the vascular wall. Moreover, since ascorbate deficiency is the underlying cause of these diseases, ascorbate supplementation is the universal therapy.

The Determining Principles of This Theory

The determining principles of this comprehensive theory are schematically summarized in Figures 1 to 3 (pages 13 to 15).

1. CVD is the direct consequence of the inability for endogenous ascorbate production in man in combination with low dietary ascorbate intake.
2. Ascorbate deficiency leads to increased permeability of the vascular wall by the loss of the endothelial barrier function and the loosening of the vascular connective tissue.
3. After the loss of endogenous ascorbate production scurvy and fatal blood loss through the scorbutic vascular wall rendered our ancestors in danger of extinction. Under this evolutionary pressure over millions of years genetic and metabolic countermeasures were favored that counteract the increased permeability of the vascular wall.
4. The genetic level is characterized by the fact that inherited disorders associated with CVD became the most frequent among all genetic predispositions. Among those predispositions lipid and lipoprotein disorders occur particularly often.
5. The metabolic level is characterized by the direct relation between ascorbate and virtually all risk factors of clinical cardiology today. Ascorbate deficiency leads to vasoconstriction and hemostasis and affects the vascular wall metabolism in favor of atherosclerogenesis.
6. The genetic level can be further characterized. The more effective and specific a certain genetic feature counteracted the increasing vascular permeability in scurvy, the more advantageous it became during evolution and, generally, the more frequently this genetic feature occurs today.
7. The deposition of Lp(a) is the most effective, most specific, and therefore most frequent of these mechanisms. Lp(a) is preferentially deposited at predisposition sites. In chronic ascorbate deficiency the accumulation of Lp(a) leads to the localized development of atherosclerotic plaques and to myocardial infarction and stroke.
8. Another frequent inherited lipoprotein disorder is hypoalphalipoproteinemia. The frequency of this disorder again reflects its usefulness during evolution. The metabolic upregulation of HDL synthesis by ascorbate became an important mechanism to reverse and decrease existing lipid deposits in the vascular wall.
9. The vascular defense mechanisms associated with most genetic disorders are nonspecific. These mechanisms can aggravate the development of atherosclerotic plaques at predisposition sites. Other nonspecific mechanisms lead to peripheral forms of atherosclerosis by causing a thickening of the vascular wall throughout the arterial system. This peripheral form of vascular disease is characteristic for angiopathies associated with Type III hyperlipidemia, diabetes, and many other inherited metabolic diseases.
10. Of particular advantage during evolution and therefore particularly frequent today are those genetic features that protect the ascorbate-deficient vascular wall until the end of the reproduction age. By favoring these disorders nature decided for the lesser of two evils: the death from CVD after the reproduction age rather than death from scurvy at a much earlier age. This also explains the rapid increase of the CVD mortality today from the 4th decade onwards.

11. After the loss of endogenous ascorbate production the genetic mutation rate in our ancestors increased significantly.[21] This was an additional precondition favoring the advantage not only of apo(a) and Lp(a) but also of many other genetic countermeasures associated with CVD.
12. Genetic predispositions are characterized by the rate of ascorbate depletion in a multitude of metabolic reactions specific for the genetic disorder.[22] The overall rate of ascorbate depletion in an individual is largely determined by the polygenic pattern of disorders. The earlier the ascorbate reserves in the body are depleted without being resupplemented, the earlier CVD develops.
13. The genetic predispositions with the highest probability for early clinical manifestation require the highest amount of ascorbate supplementation in the diet to prevent CVD development. The amount of ascorbate for patients at high risk should be comparable to the amount of ascorbate our ancestors synthesized in their body before they lost this ability: between 10,000 and 20,000 milligrams per day.
14. Optimum ascorbate supplementation prevents the development of CVD independently of the individual predisposition or pathomechanism. Ascorbate reduces existing atherosclerotic deposits and thereby decreases the risk for myocardial infarction and stroke. Moreover, ascorbate can prevent blindness and organ failure in diabetic patients, thromboembolism in homocystinuric patients, and many other manifestations of CVD.

Conclusion

In this paper we present a unified theory of human CVD. This disease is the direct consequence of the inability of man to synthesize ascorbate in combination with insufficient intake of ascorbate in the modern diet. Since ascorbate deficiency is the common cause of human CVD, ascorbate supplementation is the universal treatment for this disease. The available epidemiological and clinical evidence is reasonably convincing. Further clinical confirmation of this theory should lead to the abolition of CVD as a cause of human mortality for the present generation and future generations of mankind.

Acknowledgements

We thank Jolanta Walichiewicz for graphical assistance, Rosemary Babcock for library services, and Dorothy Munro and Martha Best for secretarial assistance.

References

1. Rath, M, Pauling L. Solution of the puzzle of human cardiovascular disease: Its primary cause is ascorbate deficiency, leading to the deposition of lipoprotein(a) and fibrinogen/fibrin in the vascular wall. J. Orthomolecular Med. 1991;6:125-134.
2. Pauling L, Rath M. Plasmin-induced proteolysis and the role of apoprotein(a), lysine, and synthetic lysine analogs. J. Orthomolecular Med. 1992; 7:17-23.
3. Ginter E. Marginal vitamin C deficiency, lipid metabolism, and atherosclerosis. Lipid Research 1973;16:162-220.
4. Third Conference on Vitamin C, Annals of the New York Academy of Sciences 498 (BurnsJJ, Rivers JM, Machlin LJ, eds) 1987.
5. Pauling L. How to Live Longer and Feel Better 1986; Freeman, New York.
6. Rath M, Pauling L. Hypothesis: Lipoprotein(a) is a surrogate for ascorbate. Proc Natl Acad Sci USA 1990;87:6204-6207.
7. Koschinsky ML, Beisiegel U, Henne-Bruns D, Eaton DL, Lawn RM. Apolipoprotein(a) size heterogeneity is related to variable number of repeat sequences in its mRNA. Biochemistry 1990;29:640-644.
8. Rath M, Pauling L. Apoprotein(a) is an adhesive protein. J OrthomolecularMed 1991;6:139-143.
9. Goldstein JL, Schrott HG, Hazzard WR, Bierman EL, Motulsky AG. Hyperlipidemia in coronary heart disease. J Clin Invest 1973; 52:1544-1568.
10. Roberts WC, Levy RI, Fredrickson DS. Hyperlipoproteinemia-A review of the five types, with first report of necropsy findings in type 3. Arch Path 1970;59:46-56.
11. Sokoloff B, Hori M, Saelhof CC, Wrzolek T, Imai T. Aging, atherosclerosis and ascorbic acid metabolism. J Am Ger Soc 1966; 14:1239-1260.
12. Jacques PF, Hartz SC, McGandy RB, JacobRA, Russell RM. Vitamin C and blood lipoproteins in an elderly population. Third Conference on Vitamin C, Annals of the New York Academy of Sciences 498 (Burns JJ, Rivers JM, Machlin LJ, eds) 1987.
13. Willis GC, Light AW, Gow WS. Serial arteriography in atherosclerosis. Canad M.A.J. 1954;71:562-568.
14. Pauling L, Itano HA, Singer SJ, Wells IC.

Journal of Orthomolecular Medicine Vol. 7, No. 1, 1992

Sickle cell anemia, a molecular disease. Science 1949;110:543-548.
15. Mann GV, Newton P. The membrane transport of ascorbic acid. Second Conference on Vitamin C. Annals of the New York Academy of Sciences 1975;243-252.
16. Kapeghian JC, Verlangieri J. The effects of glucose on ascorbic acid uptake in heart endothelial cells: possible pathogenesis of diabetic angiopathies. Life Sciences 1984;34: 577-584.
17. Dice JF, Daniel CW. The hypoglycemic effect of ascorbic acid in a juvenile-onset diabetic. International Research Communications System 1973;1:41.
18. Mudd SH, Levey HL, Skovby F. Disorders of Transsulfuration. In Scriver CR, Beaudet AL, Sly WS, Valle D (eds), The Metabolic Basis of Inherited Disease 1989 McGraw-Hill:693-734.
19. Boers GHJ, Smals AGH, Trijbels FJM, Fowler B, Bakkeren JAJM, Schoonderwaldt HC, Kleijer WJ, Kloppenborg PWC. Heterozygosity for homocystinuria in premature peripheral and cerebral occlusive arterial disease. N Engl J Med 1985; 313:709-715.
20. McCully KS. Homocysteine metabolism in scurvy, growth and arteriosclerosis. Nature 1971;231:391-392.
21. Fraga CG, Motchnik PA, Shigenaga MK, Helbock HJ. Jacob RA, Ames BN. Ascorbic acid protects against endogenous oxidative DNA damage in human sperm. Proc Natl Acad Sci USA 1991;88:11003-11006.
22. Pauling L. Orthomolecular psychiatry. Science 1968;160:265-271.

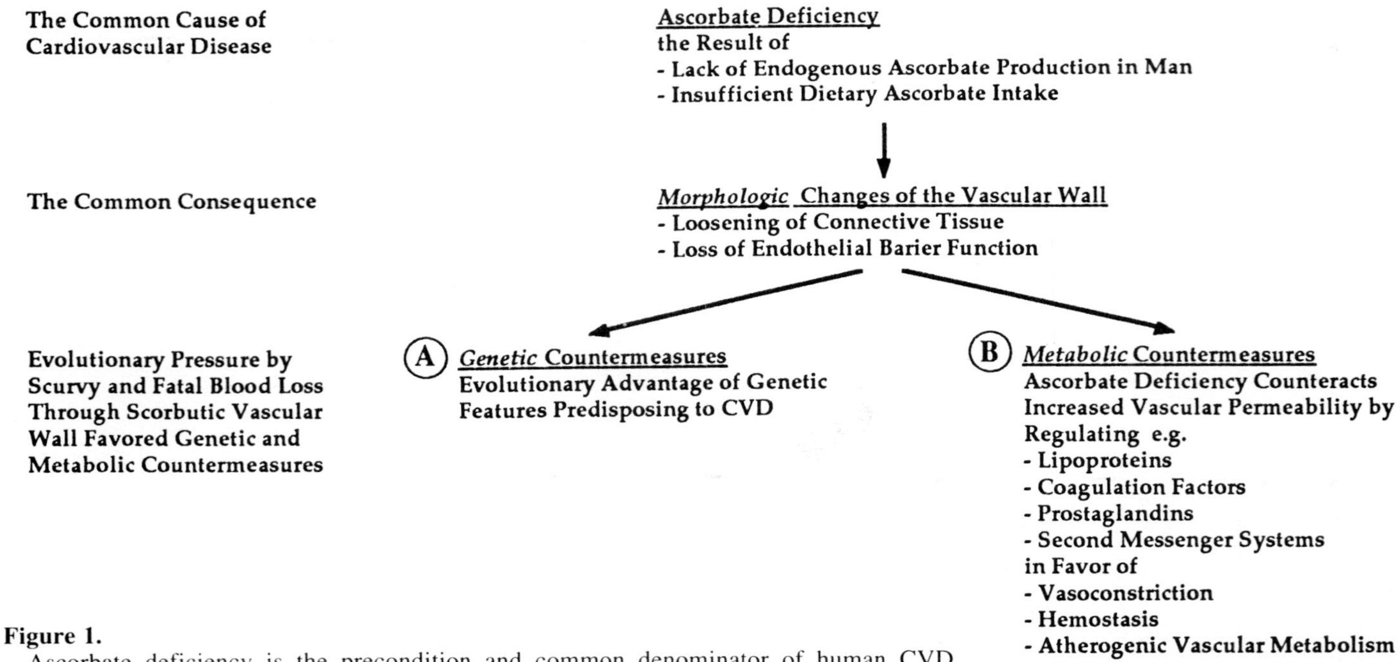

Figure 1.
Ascorbate deficiency is the precondition and common denominator of human CVD. Ascorbate deficiency invariably leads to an increased permeability of the vascular wall. The evolutionary pressure from fatal blood-loss through scorbutic vascular wall over million of years favored genetic and metabolic countermeasures. The genetic level (A) is characterized by an evolutionary advantage of genetic features predisposing to CVD. The evolutionary pressure in favor of these predisposing genetic features was so great that CVD became one of today's most common diseases. The metabolic level (B) is characterized by the regulatory effect of ascorbate on factors determining the clinical risk profile for CVD in cardiology today including lipoproteins, coagulation factors, prostaglandins, and others. Ascorbate deficiency counteracts increased vascular permeability by inducing vasoconstriction, hemostasis, and atherogenic vascular metabolism.

Journal of Orthomolecular Medicine Vol. 7, No. 1, 1992

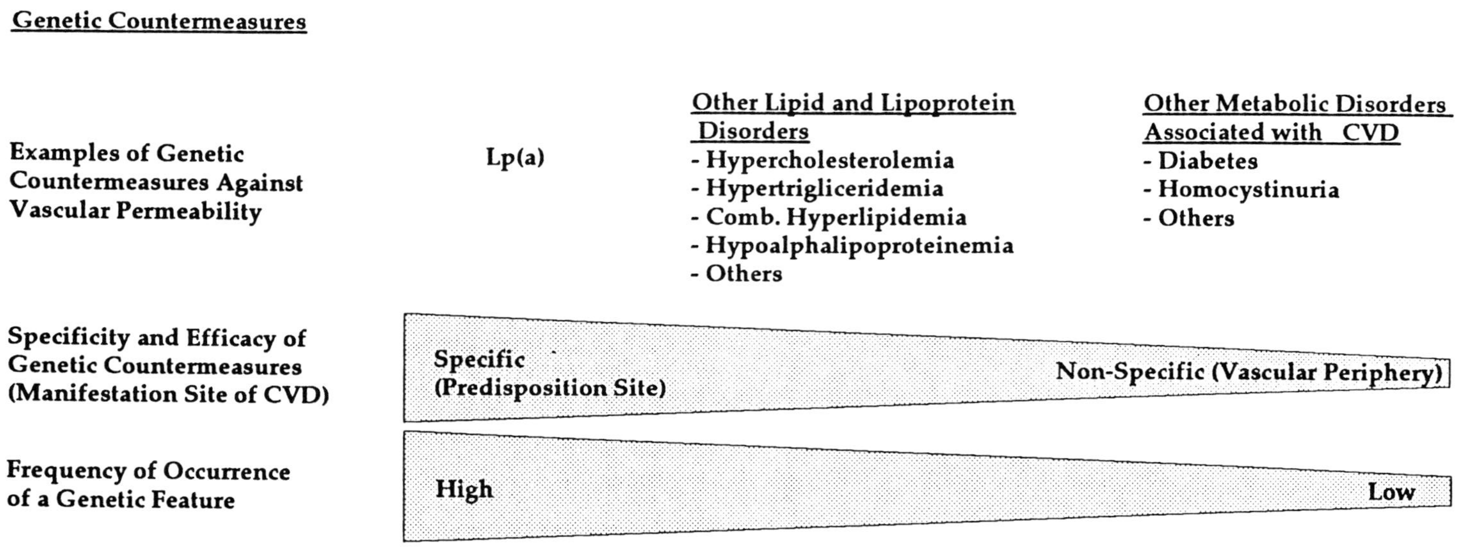

Figure 2.
Genetic countermeasures and the relation between their efficacy and the frequency of their occurrence. The more specifically a genetic feature counteracts the increased permeability of the vascular wall the more it was favored during evolution and the more frequently it occurs today. The deposition of Lp(a) in the vascular wall is the most specific and therefore most frequent mechanism. Because of the specificity of Lp(a) the sustained accumulation of this lipoprotein during chronic ascorbate deficiency leads to CVD at predisposition sites. Diabetic and homocystinuric angiopathies are typical non-specific mechanisms. Their clinical exacerbation in chronic ascorbate deficiency leads to peripheral vascular disease. With the exception of Lp(a), most other lipoprotein disorders are rather nonspecific countermeasures. They either follow the deposition of Lp(a) and aggravate CVD mainly at predisposition sites or they lead to peripheral vascular disease, such as in Type III hyperlipidemia. Figure 2 schematically summarizes these principles. This scheme, of course, can not reflect the multitude of polygenic variations in individual patients.

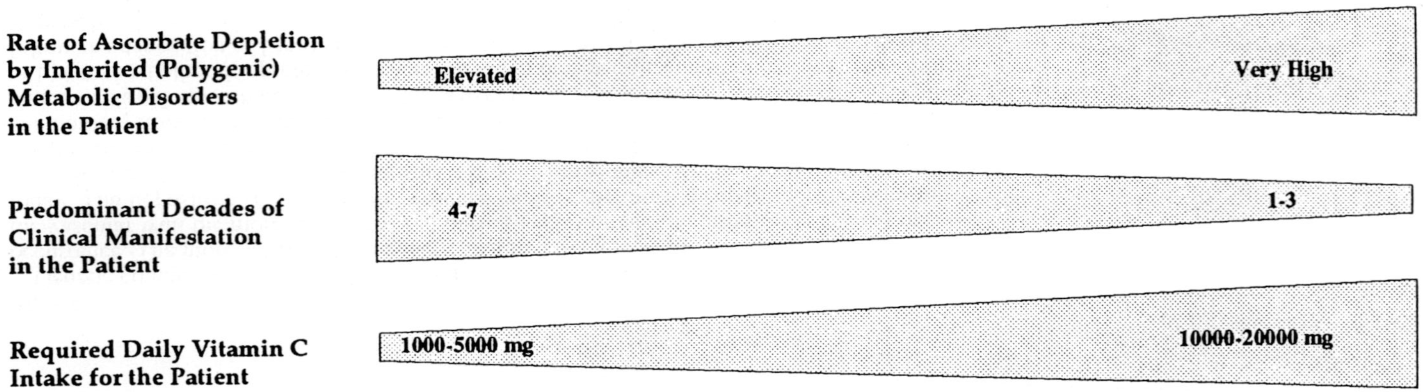

Figure 3.
The relation between ascorbate depletion and the onset of clinical symptoms in the patient. As a result of most genetic defects the rate constants for certain metabolic reactions are decreased. Ascorbate is destroyed in the attempt to normalize these decreased rate constants and in compensatory metabolic pathway.[1,22] The overall rate of ascorbate depletion in an individual is largely determined by the polygenic pattern of metabolic disorders in an individual (and to some extent also by exogenous risk factors). The earlier the body ascorbate reserves are depleted without being resupplemented the earlier the clinical manifestation occurs. Consequently, the higher the probability of early clinical onset of a latent genetic predisposition, the higher is the amount of required ascorbate intake to prevent this onset. For patients at high risk dietary ascorbate intake is recommended in the range of 10,000 to 20,000 mg/d. This corresponds to the amount of ascorbate our ancestors synthesized in their bodies before they lost this ability. The validity of Figure 3 is not limited to CVD. Ascorbate deficiency, of course, also unmasks latent disorders predisposing to cancer and to autoimmune and other diseases.

23

Summary and Conclusions: The Role of Vitamin C in the Treatment of Cancer

The correct treatment of cancer is a matter for almost endless debate, simply because no one knows how to treat cancer correctly. We deplore that increasingly common creature "The Triumphalistic Oncologist,'' so eloquently caricaturized in the editorial with this title in the journal *Surgery, Gynecology, and Obstetrics* (*146:* 617–618, 1978); his attitude of mind seems to be that in order to introduce any advance in treatment it is first necessary to decry and repudiate any advance that has gone before.

Such a self-serving approach is clearly ludicrous; it deliberately ignores the established fact that conventional treatments, based on either scientific fact or empirical discoveries, can *cure* at least one third of all cancer patients, essentially controlling their disease to such an extent as to given them normal life expectancy. The real objective is to see whether this successful fraction can be even marginally increased by the correct use of all therapeutic resources. Every day, 1100 citizens of the United States die of cancer. If it were possible to decrease this number by even one-tenth, this would mean the saving of 110 lives a day, 770 a week, 3300 a month, and 40,000 in the course of a year.

We firmly believe that supplemental ascorbate used correctly alongside conventional methods of treating cancer has very much greater potential than that.

It seems abundantly clear that ascorbic acid is directly involved in many of the natural mechanisms that protect an individual against cancer. Cancer itself, as we have seen, depletes the body's stores of ascorbate and is almost always associated with some measurable impairment of its immune mechanisms. Therefore, as a matter of principle, no matter what form of established cancer treatment is employed, active steps should be taken to ensure that the patient's ascorbate reserves are constantly kept high in order to allow his protective immune system to function always at maximum efficiency.

The value of surgery in the treatment of cancer is beyond dispute. If the tumor is still localized to an organ or tissue that can be safely excised without undue risk to the patient, such a step is undoubtedly the treatment of choice. "Cut it out and throw it away" may seem to be a most primitive and unsophisticated approach, but it is still far and away the most effective treatment we have.

VITAMIN C AND SURGERY

During the course of cancer surgery clumps of malignant cells can very frequently be found in the peripheral blood; for instance, in a sample drawn from an arm vein during the course of an abdominal operation. These clumps of malignant cells have been literally squeezed out of the tumor by the surgeon's handling. Each such clump has the potential to form a metastasis, but very few of them do so, most of them being overcome by the host's defensive mechanisms. It has long been recognized that surgical resections and even surgical biopsies pose some risk of disseminating metastases. It is our belief, based on the arguments presented in this book, that it is essential to cover this highly critical period of possible tumor dissemination by the provision of adequate ascorbate to ensure that the immune mechanisms are in peak condition.

Unless positive steps are taken to ensure this, the exact reverse holds true.

Any physical or psychic trauma depletes the ascorbate reserves, and the degree of depletion is directly proportional to the extent of the trauma. Any surgical operation is traumatic, and by its very nature surgery for cancer tends to be major, involving much tissue trauma and an appreciable reduction in the body's stores of available ascorbate. The situation is further complicated by the fact that during the course of most major operations for cancer supportive blood transfusion is required, and stored blood contains no ascorbate. To make matters even worse, many major procedures for cancer of the gastrointestinal tract are preceded by a period of enforced starvation to clear the intestine of fecal content, as well as a similar period post-operatively to reduce the risks of peritonitis from a leaking anastomosis. Thus a patient starting with cancer, bereft of his usual dietary sources of vitamin C for several days, subjected to the trauma of major cancer surgery, and perhaps with half of his circulating blood-volume replaced by ascorbate-free stored blood, will have an appreciable reduction in his ascorbate reserves and particularly in the ascorbate content of his circulating lymphocytes at the very time of high risk when a surplus is required.

Ascorbate has other values for the cancer patient undergoing surgery. As has been mentioned in earlier chapters, it has long been known that adequate amounts of ascorbate are essential for the proper healing of wounds and to enhance the defensive mechanisms against bacterial infection. Many major cancer operations on patients depleted of ascorbate by all the mechanisms noted above can be technically perfect but still the patient may be plagued by serious and sometimes fatal post-operative complications. Such complications include partial or complete wound rupture or rupture of some internal anastomosis with peritonitis, wound infections, pneumonia, and cystitis. It is our belief that many of these all too common complications could be prevented if steps were taken to ensure that all such patients were maintained in positive ascorbate balance throughout the whole peri-operative period. The ways of doing this are given in Appendix IV.

VITAMIN C AND RADIOTHERAPY

As discussed in an earlier chapter, the value of radiotherapy in many forms of cancer is beyond dispute, whether it be used alone or as a complement to surgery. Irradiation, however, also produces an appreciable reduction in ascorbate reserves, and general principles suggest that such a deficit should be rectified or even prevented by increasing the ascorbate intake throughout the whole period of treatment.

There is some evidence, dating back to the 1940s, to the effect that patients on relatively high vitamin C intake suffer fewer unpleasant side-effects during treatment by radiotherapy, and also that high vitamin C intake increases the therapeutic value of the high-energy radiation. An especially significant study is that of Cheraskin and his associates (Chapter 19), who found a significantly better response in 27 patients with squamous-cell carcinoma of the uterine cervix who were given 750 mg of ascorbic acid per day during radiation treatment than in 27 similar patients who received the radiation treatment without the ascorbic acid.

Radiotherapy destroys a fairly large volume of tissue and its effects can be likened to those of a deep burn. The cells and other exposed tissue structures disintegrate and die, forming toxic breakdown products that have to be carried off by the bloodstream and excreted in the urine. It is believed that the unpleasant systemic side effects of radiotherapy (nausea, lassitude) are caused by this sudden metabolic overload of toxic by-products. Ascorbic acid is essential for the proper functioning of a group of liver enzymes concerned with the detoxification and disposal of noxious substances, and it is therefore quite possible that a high intake of ascorbate reduces the unpleasant side effects of irradiation.

In its anticancer role, high-energy radiation kills cells during their most vulnerable phase of cell division. It is believed that the cell-killing mechanism is the chemical attack on cellular DNA by free radicals produced by the ionizing effect of radiation on the tissues. Free radicals are highly reactive unstable compounds, containing an odd number of electrons, and because of their reactivity they can exist only briefly in the tissues. It is known that ascorbate reacts in the tissues with molecular oxygen to form free radicals, and it is accordingly possible that a high tissue level of ascorbate would potentiate the cytotoxic effect of irradiation. It is certainly known that the therapeutic effects of irradiation can be intensified by increasing the degree of oxygen saturation in the tissues, and this effect is utilized to good effect in the radiation treatment of patients in hyperbaric oxygen chambers.

Apart from its cytotoxic action, radiotherapy exerts a therapeutic effect against cancer in another way. The ground substance of the whole area irradiated is replaced by a much more dense collagenized area of scar tissue. This produces a much more hostile and impermeable environment for any stray cancer cells that have survived the killing effect of the radiation. Ascorbate is, of course, essential for the production of collagen, and an adequate ascorbate supply would ensure that this scirrhous response to radiation would reach its maximum intensity. We conclude that there are no contraindications to taking high levels of vitamin C during radiotherapy; and that, on the contrary, there are a number of strong arguments that such a combination would be beneficial.

VITAMIN C AND HORMONAL TREATMENT

There are no known contraindications to the use of ascorbate alongside any hormonal treatment used in the control of cancer, and we have many patients under treatment by such combinations as DES plus ascorbate, Tamoxifen plus ascorbate, and steroids plus ascorbate. It is our observation that these patients benefit significantly from the inclusion of vitamin C in their treatment.

VITAMIN C AND IMMUNOTHERAPY

We have stated earlier (Chapters 9 and 15) our view that the effectiveness of immunotherapy would be substantially advanced if the patients were at the same time ingesting high levels of ascorbate. The therapeutic use of vitamin C in fact might be considered to be mainly an example of immunotherapy, although this vitamin also exercises many functions in addition to that of potentiating the immune mechanisms. We see no reason for not using vitamin C together with immunotherapy.

VITAMIN C AND CHEMOTHERAPY

The question of the use of ascorbate with chemotherapy is difficult to answer, and much more research is required in this field. In the first place, it seems clear from our own observations and from the many personal reports we have received from others that patients taking high doses of vitamin C are spared many of the unpleasant side effects of cancer chemotherapy regimes. In a few patients this has allowed doses of chemotherapeutic drugs much larger than usual to be used to good effect.

By its very nature, cancer chemotherapy depresses the whole immune system, and, as well as reducing the number of circulating lymphocytes, it also produces a sharp fall in the ascorbate content of each individual lymphocyte, hence reducing its protective capacity. There is therefore a strong argument that patients on cancer chemotherapy regimes should protect their immune status by taking large amounts of vitamin C.

One circumstance seems abundantly clear. If a cancer patient has been treated by chemotherapy to the point of bone-marrow failure and complete collapse of his immune system, giving ascorbate at that point is hardly likely to do much good, because a principal mode of action of ascorbate is by increasing the power of the immune system, and a badly damaged immune system may not be able to respond. This conclusion, that vitamin C is far less effective in patients with advanced cancer whose immune systems have been badly damaged by extensive courses of chemotherapy, seems to be supported by the results of the Mayo Clinic trial, as discussed in Chapter 19.

An important question that needs to be answered is whether an adult patient with a solid malignant tumor that has not been controlled by other treatments should, as the last resort, receive chemotherapy or ascorbate. A valuable summary of current opinion of the use of chemotherapy in the treatment of gastrointestinal cancer has been published in *The New England Journal of Medicine* (9 November 1978) by Moertel. He pointed out that two decades ago the fluorinated pyrimidines 5-fluorouracil (5-FU) and 5-fluoro-2′-deoxyuridine were found to be capable of producing a transient decrease of tumor size in patients with metastatic cancer of intestinal origin. An intravenous treatment in an amount that produces toxic reactions is the most effective, but the effect is not great:

> Even when administered in most ideal regimens, the fluorinated pyrimidines, in a large experience, will produce objective response in only about 15 to 20 percent of treated patients. In this context, objective response is usually defined as a reduction of more than 50 percent in the product of longest perpendicular diameters of a measurable tumor mass. These responses are usually only partial and very transient, persisting for a median time of only about five months. This minor gain for a small minority of patients is probably more than counterbalanced by the deleterious influence of toxicity for other patients and by the cost and inconvenience experienced by all patients. There is no solid evidence that treatment with fluorinated pyrimidines contributes to the over-all survival of patients with gastrointestinal cancer regardless of the stage of the disease at which they are applied.

Moertel also discusses the clinical trials of 5-FU and other chemotherapeutic agents singly and in various combinations in relation to colorectal cancer, gastric carcinoma, squamous-cell carcinoma of the esophagus, and others, with essentially the same conclusion except that adriamycin seems to have significant value for the treatment of primary liver cancer. He then states that "In 1978 it must be concluded that there is no chemotherapy approach to gastrointestinal carcinoma valuable enough to justify application as standard clinical treatment."

We would interpret this conclusion as sound reason for not subjecting these patients to the misery, trouble, and expense of chemotherapy, but in fact Moertel continues as follows:

> By no means, however, should this conclusion imply that these efforts should be abandoned. Patients with advanced gastrointestinal cancer and their families have a compelling need for a basis of hope. If such hope is not offered, they will quickly seek it from the hands of quacks and charlatans. Enough progress has been made in chemotherapy of gastrointestinal cancer so that realistic hope can be generated by entry of those patients into well designed clinical research studies. . . . If we can channel our efforts and resources into constructive research programs of sound scientific design, we shall offer the most hopeful treatment for the patient with gastrointestinal cancer today and lay a sound foundation for chemotherapy approaches of substantive value for the patient of tomorrow.

We do not agree with this conclusion. For more than a decade it has been the rather general practice in Vale of Leven Hospital and most other hospitals in

Britain not to subject patients with advanced gastrointestinal cancer and similar cancers for which experience has shown chemotherapy to have little value to the misery of this treatment; instead, these "hopeless" patients were given only palliative treatment, including morphine and heroin as needed to control pain. Now, however, there is a real reason for these patients and their families to have hope. These "untreatable" patients can be given supplemental ascorbate as their only form of treatment and derive some benefit, and just occasionally the degree of benefit obtained might be quite remarkable, as several of the patients described in the preceding chapters illustrate. The average increase in survival time of patients with advanced gastrointestional cancer treated with 10 g of ascorbate per day is greater than that reported by Moertel for those treated with chemotherapy, and the ascorbate-treated patients have the advantages of feeling well under the treatment and of not having the financial burden of chemotherapy. Moreover, little effort has been made as yet to determine the most effective dosages of vitamin C and the possible supplementary value of vitamin A, the B vitamins, minerals, and a diet high in fruits, vegetables, and their juices. This nutritional treatment of cancer, with emphasis on vitamin C, is probably far more effective at earlier stages of cancer than in the terminal stage, and if it is instituted at the first sign of cancer it may well decrease the cancer mortality by much more than our earlier estimate of 10 percent.

CONCLUSION

We strongly advocate the use of supplemental ascorbate in the management of all cancer patients from as early in the illness as possible. We believe that this simple measure would improve the overall results of cancer treatment quite dramatically, not only by making the patients more resistant to their illness but also by protecting them against some of the serious and occasionally fatal complications of the cancer treatment itself. We are quite convinced that in the not too distant future supplemental ascorbate will have an established place in all cancer-treatment regimens.

Photos for Part IV

Photo 30 Linus Pauling and his wife, Ava Helen, dancing at the ball following presentation of the Nobel Prize in Chemistry, in Stockholm, 1954. Ava Helen was Linus' chief colleague, he often said.

Photo 31 Linus Pauling, Prof. Harvey Itano (right), and Prof. S. Jon Singer (left), co-discoverers of the fact that an abnormal form of hemoglobin is the molecular basis of the inherited disease sickle cell anemia (SP 122). The discovery confirmed Pauling's conjecture to this effect and established his broad concept of molecular disease (Chapter 15). Photo taken in 1984.

Photo 32 Linus Pauling with Prof. Max Delbrück (center) and Prof. Max Perutz (right). Both Maxes are Nobel Prize winners (1962 and 1969). Delbrück co-authored with Pauling an influential paper on the subject of Chapter 14 (SP 112). Perutz made the first x-ray structure determination of a biological macromolecule, hemoglobin, which showed the presence of several lengths of α-helix, with the structure predicted by Pauling and Corey (SP 98). Photo taken in 1976.

Photo 33 Linus Pauling with Dr. Emile Zuckerkandl (right), at an award ceremony in Japan, in 1983. Zuckerkandl co-authored papers with Pauling on molecular evolution and evolutionary clocks (SP 119, SP 120, SP 121).

Photo 34 Peter Pauling (second son of Linus Pauling) with Prof. James Watson, co-discoverer of the double-helix structure of DNA, a structure that Linus Pauling failed to find (SP 116). Peter, an x-ray crystallographer, studied structural mechanisms of drug action and co-authored a number of papers with his father (e.g. SP 56).

Photo 35 Linus Pauling with Prof. Fancis Crick, co-discoverer of the DNA structure, for which he and Watson received the Nobel Prize (in 1962). Here Pauling and Crick are seen at Pauling's 90th birthday party at Caltech, in 1991.

Photo 36 Linus Pauling and family, en route to Oslo for the Nobel Peace Prize ceremony, in 1964. Left to right: Ava Helen Pauling, Linus Pauling, Linda Pauling Kamb (daughter), Barclay Kamb, Lucy Pauling (Crellin's wife), Crellin Pauling, and Linus Pauling Jr. In the later years, Linus Pauling Jr., a psychiatrist, developed a particular interest in orthomolecular medicine (Chapter 17), and collaborated with his father in important ways, particularly in serving for a time as president of the Linus Pauling Institute of Science and Medicine, which his father founded in 1973. Son-in-law Barclay Kamb, a crytallographer/geologist, co-authored several papers with Pauling (e.g. SP 34 and SP 128), and discovered and named paulingite, a cubic zeolite with the largest known unit cell of any inorganic crystal (*American Mineralogist* **45**, 79–91, 1960). Son Peter did not get into this picture but is in Photo 34.

Photo 37 Linus Pauling lecturing on chemical quantum mechanics in a seminar at Oslo University in 1982. The tetrahedral bond orbitals of the carbon atom can be seen at the left. The lecture appears to have summarized Pauling's development of bond-orbital hybridization in the early 1930's (SP 5).

Photo 38 Linus Pauling lecturing on hydrogen bonding and ionic hydration, ca. 1975. In the upper right corner of the blackboard is the statistical factor $(3/2)^N$, which is the basis of Pauling's famous calculation of the residual entropy of ice (SP 73). To the left of $(3/2)^N$ is depicted a tetrahedrally coordinated water molecule.

Photo 39 Linus Pauling with Dr. Irwin Stone, who first interested him in studying the nutritional benefits of large intakes of vitamin C. Photo taken in 1973.

Photo 40 Linus Pauling with Dr. Ewan Cameron, a physician from Scotland who had conducted clinical tests of effects of vitamin C on cancer, and who collaborated with Pauling in evaluating, interpreting, and publishing the results (SP 139, SP 140, SP 141, SP 144). Photo in early 1980's, in the laboratories of the Linus Pauling Institute for Science and Medicine, Palo Alto, CA.

Photo 41 Linus Pauling and Dr. Zelek S. Herman, in 1985. Herman, trained in molecular quantum mechanics, was Pauling's devoted research assistant and colleague from 1980 onward, and co-authored about ten papers with him, mostly on chemical bond theory and biostatistics. SP 27 and SP 133 are selected as representative of them. Zelek Herman and Dorothy Munro (Photo 42) carried out the herculian task of putting together a complete list of Pauling's publications, as noted in the General Introduction and the Appendices.

Photo 42 Linus Pauling with Mrs. Dorothy B. Munro (center) and Dr. Matthias Rath (right) at a celebration of Pauling's 90th birthday, on February 28, 1991, in San Francisco. Dorothy Munro was for twenty years his devoted secretary. Matthias Rath co-authored several papers on the beneficial effects of vitamin C on cardiovascular disease, of which SP 143 is representative.

Photo 43 Linus Pauling gets a hug from Dr. Arthur Sackler (left) on the occasion of the opening of the Sackler Wing of the Metropolitan Museum of Art, New York, 1978. Dr. Sackler, publisher of *The Medical Tribune*, was supportive of Pauling's efforts to advance the field of orthomolecular medicine (Chapter 17).

Photo 44 Linus Pauling with Prof. Ahmed Zewail, the Linus Pauling Professor of Chemistry and Physics in the California Institute of Technology. He was awarded the Nobel Prize in 1999 for pioneering research in femtochemistry. Picture taken in 1990.

Photo 45 Prof. Jack D. Dunitz, professor of structural chemistry at the ETH, Zürich. As a postdoctoral research fellow, he was a colleague of Linus Pauling at Caltech in 1948–1951 and 1953–1954. He wrote the *Biographical Memoir* in Chapter 18.

Photo 46 Linus Pauling with his son-in-law Barclay Kamb (on the left), co-author of SP 34 and SP 128, grandson Alexander (Sasha) Kamb, a molecular biologist (on the right), and great grandson Alexander. Photo taken on Pauling's 93rd birthday, 28 February 1994.

Photo 47 Linus Pauling engaged in one of his favorite activities—calculating the results of one of his theories and comparing the results with experimental data. In this case he is calculating properties of atomic nucleii from his close-packed spheron theory (Chapter 10) and tabulating the results meticulously in his laboratory journal-book. Photo taken in August 1985.

Part V

SUMMARY OF LINUS PAULING'S LIFE AND SCIENTIFIC WORK

Chapter 18

Biographical Memoir

LINUS CARL PAULING

by Jack D. Dunitz, F.R.S.

Biographical Memoirs of Fellows of the Royal Society of London

Volume 1966, Pages 317–338 (1996)

Linus Pauling

LINUS CARL PAULING

28 February 1901–19 August 1994

Elected For.Mem.R.S. 1948

BY JACK D. DUNITZ, F.R.S.

Organic Chemistry Laboratory, Swiss Federal Institute of Technology, ETH-Zentrum, CH-8092 Zurich, Switzerland

Linus Carl Pauling was born in Portland, Oregon, on 28 February 1901 and died at his ranch at Big Sur, California on 19 August 1994. In 1922 he married Ava Helen Miller (died 1981), who bore him four children: Linus Carl, Peter Jeffress, Linda Helen (Kamb) and Edward Crellin.

Pauling is widely considered to be the greatest chemist of this century. Most scientists create a niche for themselves, an area where they feel secure, but Pauling had an enormously wide range of scientific interests: quantum mechanics, crystallography, mineralogy, structural chemistry, anaesthesia, immunology, medicine, evolution. In all these fields and especially in the border regions between them, he saw where the problems lay, and, backed by his speedy assimilation of the essential facts and by his prodigious memory, he made distinctive and decisive contributions. He is best known, perhaps, for his insights into chemical bonding, for the discovery of the principal elements of protein secondary structure, the alpha-helix and the beta-sheet, and for the first identification of a molecular disease, sickle-cell anaemia, but there are a multitude of other important contributions. Pauling was one of the founders of molecular biology in the true sense of the term. For these achievements, Pauling was awarded the 1954 Nobel Prize in Chemistry. But Pauling was famous not only in the world of science. In the second half of his life, he devoted his time and energy mainly to questions of health and the necessity to eliminate the possibility of war in the nuclear age. His active opposition to nuclear testing brought him political persecution in his own country but was finally influential in bringing about the 1963 international treaty banning atmospheric tests. With the award of the 1962 Nobel Prize for Peace, Pauling became the first person to win two unshared Nobel Prizes (Marie Curie won one and shared another with her husband). Pauling's name is probably best known among the general public through his advocacy, backed by personal example, of large doses of ascorbic acid (vitamin C) as a dietary supplement to promote general health and prevent (or at least reduce the severity) of such ailments as the common cold and cancer. Indeed, Albert Einstein and Linus Pauling are probably the only scientists in our century whose names are known to every radio listener, television viewer or newspaper reader.

Early years

Pauling was the first child of Herman Pauling, son of German immigrants, and Lucy Isabelle (Darling) Pauling, descended from from pre-revolutionary Irish stock. There were two younger daughters, Pauline Darling (born 1902) and Lucile (born 1904). Herman Pauling worked for a time as a travelling salesman for a medical supply company and moved in 1905 to Condon, Oregon, where he opened his own drugstore. It was in this new boom town in the arid country east of the Coastal Range that Pauling had his first schooling. He learnt to read early and started to devour books. In 1910, the family moved back to Portland, where his father wrote a letter to *The Oregonian*, a local newspaper, asking for advice about suitable reading matter for his nine-year-old son, who had already read the Bible and Darwin's theory of evolution. We do not know the replies, but Pauling later confessed that one of his favourites was the *Encyclopaedia brittanica*. Soon tragedy struck: in June of that year Herman Pauling died after a sudden illness, probably a perforated stomach ulcer with attendant peritonitis, leaving his family in a situation with which the young mother could not adequately cope. Linus did well at school. He collected insects and minerals and read omnivorously. He made up his mind to become a chemist in 1914, when a fellow student, Lloyd A. Jeffress, showed him some chemical experiments he had set up at home. With the reluctant approval of his mother, he left school in 1917 without a diploma and entered Oregon Agricultural College at Corvallis as a chemical engineering major, but after two years his mother wanted him to leave college to earn money for the support of the family. However, he must have impressed his teachers, for in 1919, after a summer working as road-paving inspector for the State of Oregon, he was offered a full-time post as Instructor in Qualitative Analysis in the Chemistry Department. The 18-year-old teacher felt the need to read current chemical journals and came across the recently published papers of Gilbert Newton Lewis and Irving Langmuir on the electronic structure of molecules. Having understood the new ideas, the 'boy professor' introduced them to his elders by giving a seminar on the nature of the chemical bond. Thus was sparked the 'strong desire to understand the physical and chemical properties of substances in relation to the structure of the atoms and molecules of which they are composed', which determined the course of Pauling's long life.

The following year Pauling resumed his student status and he graduated as B.Sc. in 1922. In his final year he was given another opportunity to teach, this time an introductory chemistry course for young women students of home economics. This new teaching episode also had important consequences for his future. One of the students was Ava Helen Miller, who became his wife in a marriage that lasted almost 60 years.

PASADENA

Pauling came to the California Institute of Technology as a graduate student in 1922 and remained there for more than 40 years. He chose Caltech because he could obtain a doctorate there in three years (Harvard required six) and because Noyes offered him a modest stipend as part-time instructor. It was a fortunate choice both for Pauling and for Caltech. As he wrote towards the end of his life: 'Years later ... I realized that there was no place in the world in 1922 that would have prepared me in a better way for my career as a scientist' (37)*. When he arrived, the newly established institute consisted largely of the hopes of its three founders, the astronomer George Ellery Hale, the physicist Robert A. Millikan and the physical chemist Arthur Amos Noyes. There were three buildings and eighteen faculty members. When he left, Caltech had developed into one of the major centres of scientific research in the world. In chemistry, Pauling was the prime mover in this development. Indeed, for many young chemists of my generation, Caltech meant Pauling.

Pauling's doctoral work was on the determination of crystal structures by X-ray diffraction analysis, under the direction of Roscoe Gilkey Dickinson (1894–1945), who had obtained his Ph.D. only two years earlier (he was the first person to receive a Ph.D. from Caltech). By a happy chance, Ralph W.G. Wyckoff (1897–1994, For.Mem.R.S. 1951), one of the pioneers of X-ray analysis, had spent the year before Pauling's arrival at Caltech and had taught Dickinson the method of using Laue photographic data (white radiation, stationary crystal; a method that fell into disuse but has newly been revived in connection with rapid data collection with synchrotron radiation sources). Wyckoff taught Dickerson, and Dickerson taught Pauling, who soon succeeded in determining the crystal structures of the mineral molybdenite MoS_2 (2) and the intermetallic compound MgSn (1). By the time he graduated, in 1925, he had published 12 papers, most on inorganic crystal structures, but including one with Peter Debye (1884–1966, For.Mem.R.S. 1933) on dilute ionic solutions (3) and one with Richard Tolman (1881–1948) on the entropy of supercooled liquids at 0 K (4). Pauling had already made up for his lack of formal training in physics and mathematics. He was familiar with the quantum theory of Planck and Bohr and was ready for the conceptual revolution that was soon to take place in Europe. Noyes obtained one of the newly established Guggenheim fellowships for the rising star and sent him and his young wife off to the Institute of Theoretical Physics at Munich, directed by Arnold Sommerfeld (1868–1951, For.Mem.R.S. 1926).

They arrived in April 1926, just as the Bohr–Sommerfeld model was being displaced by the 'new' quantum mechanics. It was an exciting time, and Pauling knew he was lucky to be there at one of the centres. He concentrated on learning as much as he could about the new theoretical physics at Sommerfeld's institute. Pauling had been regarded, and probably also regarded himself, as intellectually outstanding among his fellow students at Oregon and even at Caltech. However, he must have become aware of his limitations during his stay in Europe. The new theories were being made by men of his own generation. Wolfgang Pauli (1900–1958, For.Mem.R.S. 1953), Werner Heisenberg (1901–1976, For.Mem.R.S. 1955) and Paul Dirac (1902–1984, F.R.S 1930) were all born within a year of Pauling and were more than a match for him in physical insight, mathematical ability and philosophical depth. Pauling was not an outstanding theoretical physicist and was probably not particularly

*Numbers in this form refer to the bibliography at the end of the text.

interested in problems such as the deep interpretation of quantum mechanics or the philosophical implications of the uncertainty principle. On the other hand, he was the only chemist at Sommerfeld's institute and saw at once that the new physics was destined to provide the theoretical basis for understanding the structure and behaviour of molecules.

The year in Europe was to have a decisive influence on Pauling's scientific development. In addition to Munich, he visited Copenhagen in the spring of 1927 and then spent the summer in Zurich. In Copenhagen it was not Bohr but Samuel A. Goudsmit (1902–1978) who influenced Pauling (they later collaborated in writing *The structure of line spectra* (8)) , and in Zurich it was neither Debye nor Schrödinger but the two young assistants, Walter Heitler (1904–1981, F.R.S. 1948) and Fritz London (1900–1954), who were working on their quantum-mechanical model of the hydrogen molecule in which the two electrons are imagined to 'exchange' their roles in the wave function, an example of the 'resonance' concept that Pauling was soon to exploit so successfully.

One immediate result of the stay in Munich was Pauling's first paper in *Proceedings of the Royal Society of London*, submitted by Sommerfeld himself (5). Pauling was eager to apply the new wave mechanics to calculate properties of many-electron atoms and he found a way of doing this by using hydrogen-like single-electron wave functions for the outer electrons with effective nuclear charges based on empirical screening constants for the inner electrons.

The nature of the chemical bond

In 1927 Pauling returned to Caltech as Assistant Professor of Theoretical Chemistry. The next 12 years produced the remarkable series of papers that established Pauling's worldwide reputation. His abilities were quickly recognized through promotions (Associate Professor, 1929; Full Professor, 1931), through awards (Langmuir Prize, 1931), through election to the National Academy of Sciences (1933) and through visiting lectureships, especially the Baker Lectureship at Cornell in 1937–38. Through his writings and lectures, Pauling established himself as the founder and master of what might be called structural chemistry, a new way of looking at molecules and crystals.

Pauling's way was first to establish a solid and extensive collection of data. By means of X-ray crystallography, gas-phase electron diffraction (installed after Pauling's 1930 visit to Europe, where he learned about Hermann Mark's pioneering studies), infrared, Raman, and ultraviolet spectroscopy, interatomic distances and angles were established for hundreds of crystals and molecules. Thermochemical information was already available. The first task of theory, as Pauling saw it, was to provide a basis to explain the known metric and energetic facts about molecules, and only then to lead to prediction of new facts. At this stage of his development, Pauling was attracting many talented co-workers, undergraduates, graduate students, and postdoctoral fellows, and their names read like a *Who's who* in the structural chemistry of the period: J.H. Sturdivant, J.L. Hoard, J. Sherman, L.O. Brockway, D.M. Yost, G.W. Wheland, M.L. Huggins, L.E. Sutton (F.R.S. 1950), E.B. Wilson, S.H. Bauer, C.D. Coryell, V. Schomaker and others. Here are the major achievements.

Pauling's ionic radii

Once the structures of simple inorganic crystals began to be established, it was soon seen that the observed interatomic distances were consistent with approximate additivity of characteristic radii associated with the various cations and anions. Among the several sets that have been proposed, Pauling's are not merely designed to reproduce the observations but, typical for him, are derived from a mixture of approximate quantum mechanics (using screening constants) and experimental data. His values, derived almost 70 years ago, are still in common use, and the same can be said for the sets of covalent radii and non-bonded (van der Waals) radii that he introduced.

Pauling's rules

Whereas simple ionic substances, such as the alkali halides, are limited in the types of crystal structure they can adopt, the possibilities open to more complex substances, such as mica, $KAl_3Si_3O_{10}(OH)_2$, may appear to be immense. Pauling (7) formulated a set of rules about the stability of such structures, which proved enormously successful in testing the correctness of proposed structures and in predicting unknown ones. As Pauling himself remarked, these rules are neither rigorous in their derivation nor universal in their application; they were obtained in part by induction from known structures and in part from theoretical considerations. His second rule states essentially that electrostatic lines of force stretch only between nearest neighbours. In the meantime, as structural knowledge has accumulated, this rule has been modified by various authors to relate bond strengths to interatomic distances, but it seems fair to say that it is still the basis for the systematic description of inorganic structures. W.L. Bragg, who may have felt somewhat beaten to the post by the publication of these rules, wrote (Bragg 1937): 'The rule (the second one) appears simple, but it is surprising what rigorous conditions it imposes upon the geometrical configuration of a silicate ... To sum up, these rules are the basis for the stereochemistry of minerals.'

Quantum chemistry

In 1927 Ø. Burrau solved the Schrödinger equation for the hydrogen molecule ion H_2^+ in elliptic coordinates and obtained values for the interatomic distance and bonding energy in good agreement with experiment. Burrau's wave function fails, however, to yield much physical insight into the stability of the system. Soon afterwards, Pauling (6) pointed out that, although an approximate perturbation treatment would not provide any new information, it would be useful to know how well it performed: 'For perturbation methods can be applied to many systems for which the wave equation can not be accurately solved'. Pauling first showed that the classical interaction of a ground state hydrogen atom and a proton is repulsive at all distances. However, if the electron is not localized on one of the atoms, but the wave function is taken as a linear combination of the two ground state atomic wave functions, then the interaction energy has a pronounced minimum at a distance of about 2 a.u. This was the first example of what has come to be known as the method of linear combination of atomic orbitals (LCAO) For the hydrogen-molecule ion, the LCAO dissociation energy is only about 60 per cent of the correct value, but the model provides insight into the source of the bonding and can easily be extended to more complex systems. In fact, the LCAO method is the basis of modern molecular orbital theory.

A few months earlier, Heitler and London had published their calculation for the hydrogen molecule. This was too complicated for an exact solution, and their method also rested on a perturbation model, a combination of atomic wave functions in which the two electrons, with opposite spins, change places. More generally, the energy of the electron-pair bond could now be attributed to 'the resonance energy corresponding to the interchange of the two electrons between the two atomic orbitals'. As developed by Pauling and independently by John C. Slater (1900–1976), the Heitler–London–Slater–Pauling (HLSP) or valence bond model associates each conventional covalent bond with an electron pair in a localized orbital and then considers all ways in which these electrons can 'exchange'.

Much has been made of Pauling's preference for valence bond (VB) theory over molecular orbital (MO) theory. The latter, as developed by Fritz Hund (1896–), Erich Hückel (1896–1980, For.Mem.R.S. 1977) and Robert S. Mulliken (1896–1986, For.Mem.R.S. 1967), works in terms of orbitals extended over the entire molecule, orders these orbitals according to their estimated energies and assigns two electrons with opposite spin to each of the bonding orbitals. Electronic excited states correspond to promotion of one or more electrons from bonding to anti-bonding orbitals. Nowadays, MO theory has proved itself more amenable to computer calculations for multi-centre molecules, but in the early days, when only hand calculations were possible, it was largely a matter of taste. The main appeal of the MO model was then to spectroscopists. Chemists, in general, were less comfortable with the idea of pouring electrons into a ready-made framework of nuclei. It was more appealing to build molecules up from individual atoms linked by electron-pair bonds. The VB picture was more easily related to the chemist's conventional structural formulae. Both models are, of course, drastic simplifications, and it was soon recognized that when appropriate correction terms are added and the proper transformations are made they become equivalent. In particular, the MO method in its simplest form ignores electron–electron interactions, while the VB method overestimates them.

Pauling was fully acquainted with early MO theory, there is at least one important paper (11) on the theory of aromatic substitution. But he clearly preferred his own simplified versions of VB theory and soon became a master of combining them with the empirical facts of chemistry. A remarkable series of papers entitled 'The nature of the chemical bond' formed the basis for his later book with the same title. In the very first paper (9), Pauling set out his programme of developing simple quantum mechanical treatments to provide information about 'the relative strengths of bonds formed by different atoms, the angles between bonds, free rotation, or lack of free rotation about bond axes, the relation between the quantum numbers of bonding electrons and the number and spatial arrangement of bonds, and so on. A complete theory of the magnetic moments of molecules and complex ions is also developed, and it is shown that for many compounds involving elements of the transition group this theory together with the rules of electron pair bonds leads to a unique assignment of electron structures as well as a definite determination of the type of bonds involved'. To a large extent, Pauling developed his own language to describe his new concepts, and, of the many new terms introduced, three seem to be indelibly associated with his name: hybridization, resonance and electronegativity.

Hybridization

Only the first of these truly originates from him. In the first paper of the series, Pauling took up the idea of spatially directed bonds. By a generalization of the Heitler–London model for hydrogen, a normal chemical bond can be associated with the spin pairing of two electrons, one from each of the two atoms. While an *s* orbital is spherically symmetrical, other atomic orbitals have characteristic shapes and angular distributions. It was not difficult to explain the angular structure of the water molecule H_2O and the pyramidal structure of ammonia H_3N, but the quadrivalency of carbon was a problem. From its ground state ($1s^22s^22p^2$) carbon ought to be divalent; from the excited state ($1s^22s^12p^3$) one might expect three mutually perpendicular bonds and a fourth weaker bond (using the *s* orbital) in some direction or other. As a chemist, Pauling knew that there must be a way of combining the *s* and *p* functions to obtain four equivalent orbitals directed to the vertices of a tetrahedron. Atomic orbitals can be expressed as products of a radial and an angular part. Pauling solved the problem by simply ignoring the former. The desired tetrahedral orbitals are then easily obtained as linear combinations of the angular functions. Pauling called these hybrid orbitals and described the procedure as hybridization. Other combinations yield three orbitals at 120° angles in a plane (trigonal hybrids) or two at 180° (digonal hybrids). With the inclusion of *d* orbitals, other combinations become possible. In his later years, Pauling stated that he considered the hybridization concept to be his most important contribution to chemistry (see Kauffman and Kauffman 1996).

Resonance

In attempting to explain the quantum-mechanical exchange phenomenon responsible for the stability of the chemical bond, Heitler and London had used a classical analogy originally due to Heisenberg. In quantum mechanics, a frequency $\nu = E/h$ can be associated with every system with energy E. Two non-interacting hydrogen atoms are thus comparable to two classical systems both vibrating with the same frequency ν, for example, two pendulums. Interaction between the two atoms is analogous to coupling between the pendulums, known as resonance. When coupled, the two pendulums no longer vibrate with the same frequency as before but make a joint vibration with frequencies $\nu + \Delta\nu$ and $\nu - \Delta\nu$, where $\Delta\nu$ depends on the coupling. Going back to quantum mechanics, it is as if the system now has two different energies, one higher and one lower than before. Heitler and London interpreted the combination frequency $\Delta\nu$ as the frequency of exchange of spin directions.

Pauling first used the term resonance more or less as a synonym for electron exchange, in the Heitler–London sense, but he went on to think of the actual molecule as 'resonating' between two or more valence-bond structures, and hence lowering its energy below the most stable of these. Thus, by resonating between two Kekulé structures the benzene molecule is more stable than these extremes, and the additional stability can be attributed to 'resonance energy'. Through his resonance concept, Pauling reconciled the chemist's structural formulae with simplified quantum mechanics, thereby extending the realm of applicability of these formulae, and he proceeded to reinterpret large areas of chemistry with it.

In the middle years of the century, resonance theory was taken up with enthusiasm by teachers and students; it seemed to be the key to understanding chemistry. Since then, its appeal has declined. It has now a slightly old-fashioned connotation. Certainly, it had some failures. Resonance theory would lead one to expect that cyclobutadiene should be more stable as a symmetric square structure than as a rectangular one with alternating long and short bonds, whereas the contrary is true. (It seems ironic that in the 1935 classic *Introduction to quantum mechanics* (10) qualitative MO theory was applied to only one example, four atoms in a square. In contrast to the VB method, which gave a typical 'resonance energy' to this system, the MO model gave none. Of course, cyclobutadiene was then still only a synthetic chemist's dream.) Similarly, it does not explain the stability of the cyclopentadienyl anion compared with the corresponding cation; in these and other cases, simple MO theory provided immediate and correct answers. In the index of a modern textbook on physical chemistry, 'resonance' is likely to appear only in an entry such as 'resonance, nuclear magnetic'. It does not fare much better in textbooks on inorganic and organic chemistry: a few pages on resonance formalism are usually followed by a more extensive account of simple MO theory.

Electronegativity

Electronegativity, the third concept associated with Pauling's name, is still going strong. It emerged from his concept of partially ionic bonds. The energy of a bond can be considered as the sum of two contributions: a covalent part and an ionic part. The thermochemical energy of a bond *D*(A–B) between atoms A and B is, in general, greater than the arithmetic mean of the energies *D*(A–A) and D(B–B) of the homonuclear molecules. Pauling attributed the extra energy *D*(A–B) to ionic resonance and found he could assign values x_A, etc. to the elements such that *D*(A–B) is approximately proportional to $(x_A - x_B)^2$. The x values form a scale, the electronegativity scale, in which fluorine with $x = 4$ is the most electronegative element, caesium with $x = 0.7$ the least. Apart from providing a basis for estimating bond energies of heteropolar bonds, these x values can also be used to estimate the dipole moment and ionic character of bonds. Other electronegativity scales have been proposed by several authors, but Pauling's is still the most widely used — it is the easiest to remember. According to Pauling, electronegativity is the power of an atom *in a molecule* to attract electrons to itself. It therefore differs from the electron affinity of the free atom although the two run roughly parallel. Many other interpretations have been proposed.

These and many other topics were collected and summarized in the book based on Pauling's Baker Lectures, *The nature of the chemical bond* (14), probably the most influential book on chemistry this century. In my opinion, the 2nd (1940) edition is the best; the 1939 edition was short-lived, and the 1960 edition, although it contains much more material, did not evoke the same feeling of illumination as the earlier ones.

Like so many others, I first encountered Pauling through this book, which I discovered some time in my second year as an undergraduate at Glasgow University. It came as a revelation. Setting out to offer an introduction to modern structural chemistry, it explained how the structures and energies of molecules could be discussed in terms of a few simple principles. The essential first step in understanding chemical phenomena was to establish the atomic arrangements in the substances of interest. To try to understand chemical reactivity without this information or with dubious structural information was a waste of time. This was just what I needed to help me make up my mind that my future was to be in structural chemistry.

PAULING AND MOLECULAR BIOLOGY

The nature of the chemical bond (14) marks perhaps the culmination of Pauling's contributions to chemical bonding theory. There were achievements to follow – notably an important paper (17) on the structure of metals – but the interest in chemical bonding was being modified into an interest in the structure and function of biological molecules. There are intimations of this in the chapter on hydrogen bonds. Pauling was one of the first to spell out its importance for biomolecules:

> Because of its small bond energy and the small activation energy involved in its formation and rupture, the hydrogen bond is especially suited to play a part in reactions occurring at normal temperatures. It has been recognized that hydrogen bonds restrain protein molecules to their native configurations, and I believe that as the methods of structural chemistry are further applied to physiological problems it will be found that the significance of the hydrogen bond for physiology is greater than that of any other single structural feature.

Like many of his comments it seems so obvious, almost a truism, but it was not obvious then. Essentially the same idea had been expressed in (13), but hydrogen bonds are not even mentioned, for example, in Bernal's 1939 article on the structure of proteins.

Two remarkable observations from 1948 deserve to be mentioned here. One is a forerunner of the 1953 Watson–Crick DNA double-helix structure and explains what had not yet been discovered (18, 31):

> The detailed mechanism by means of which a gene or a virus molecule produces replicas of itself is not yet known. In general the use of a gene or a virus as a template would lead to the formation of a molecule not with identical structure but with complementary structure ... If the structure that serves as a template (the gene or virus molecule) consists of, say, two parts, which are themselves complementary in structure, then each of these parts can serve as the mold for the production of a replica of the other part, and the complex of two complementary parts thus can serve as the mold for the production of duplicates of itself.

And in the same vein, although nothing whatsoever was known about the structure of enzymes, the other (19) announced what became clear to biochemists in general only many years later:

> I think that enzymes are molecules that are complementary in structure to the activated complexes of the reactions that they catalyse, that is, to the molecular configuration that is intermediate between the reacting substances and the products of reaction for these catalysed processes. The attraction of the enzyme molecule for the activated complex would thus lead to a decrease in its energy, and hence to a decrease in the energy of activation of the reaction, and to an increase in the rate of the reaction.

The message seems to have lain in oblivion until well after 'transition-state binding' had become popular; it is not mentioned, for example, in Jencks's classic work (Jencks 1969) on enzyme catalysis.

Both of these prescient statements depend on the concept of complementarity, which arose out of Pauling's early work on proteins and antibodies. This started because, in the search for funding during the depression, Pauling obtained a grant from Warren Weaver, Director of the Rockefeller Foundation Natural Science Division, but only for research in life sciences. With his knowledge of inorganic structural chemistry, haemoglobin was the first target, and, within a few months, he solved an important problem. By magnetic susceptibility measurements it was shown that whereas haemoglobin contains four unpaired electrons per haem and the oxygen molecule contains two, oxyhaemoglobin (and also carbonmonoxyhaemoglobin) contains none (12). This result showed that, in oxygenated blood, the O_2 molecule is attached to the iron atom of haemoglobin by a covalent bond, that it was not just a matter of oxygen being somehow dissolved in the protein. Magnetic susceptibility measurements could also yield equilibrium constants and rates for many reactions involving addition of molecules and ions to ferro- and ferrihaemoglobin. It is interesting that Pauling had introduced the magnetic susceptibility technique at Caltech in connection with the prediction and identification of the superoxide radical anion, a molecule whose biological significance was recognized only many years later (see (33)).

In 1936 Alfred E. Mirsky (1900–1974) and Pauling published a paper (13) on protein denaturation, which was known to be a two-stage process, one under mild conditions partially reversible, the other irreversible. Pauling associated the first stage with the breaking and re-formation of hydrogen bonds, the second with the breaking of covalent bonds. The native protein was pictured as follows: 'The molecule consists of one polypeptide chain which continues without interruption throughout the molecule (or, in certain cases, of two or more such chains); this chain is folded into a uniquely defined configuration, in which it is held by hydrogen bonds ... The importance of the hydrogen bond in protein structure can hardly be overemphasized.' Loss of the native conformation destroys the characteristic properties of the protein. From the entropy difference between the native and denatured forms of trypsin, about 10^{20} conformations were estimated to be accessible to the denatured protein molecule. On heating, or if the pH of the solution was near the isoelectric point of the protein, unfolded segments of acidic or basic side chains would get entangled with one another, fastening molecules together, and ultimately leading to the formation of a coagulum. This was perhaps the first modern theory of native and denatured proteins.

Complementariness enters the picture in 1940, when Max Delbrück (1906–1981, For.Mem.R.S. 1967) and Pauling published their refutation of a proposal of Pascal Jordan, according to which a quantum-mechanical stabilizing interaction between identical or nearly identical molecules might influence biological molecular synthesis in such a way as to favour the formation of molecular replicas in the living cell (16). After dismissing this proposal, the authors went on to say that complementariness, not identity, should be given primary consideration. They continued:

> The case might occur in which the two complementary structures happened to be identical; however, in this case also the stability of the complex of two molecules would be due to their complementariness rather than their identity. When speculating about possible mechanisms of autocatalysis it would therefore seem to be most rational from the point of view of the structural chemist to analyze the conditions under which complementariness and identity might coincide.

The use of the word 'complementariness' instead of the more usual 'complementarity' is striking. According to Delbrück, his only role in the publication, apart from suggesting a few minor changes, was to have drawn Pauling's attention to Jordan's proposal, and it seems quite likely to me that 'complementariness' was one of these minor changes, introduced in order to avoid the epistemological connotations that Delbrück associated with 'complementarity' in Bohr's sense.

By this time, Pauling was thinking about antibodies. In 1936 he had met Karl Landsteiner (1868–1943, For.Mem.R.S. 1941), discoverer of the human blood groups and instrumental in establishing immunology as a branch of science. According to Pauling (31), Landsteiner asked him how he would explain the specificity of interaction of antibodies and antigens, to which he replied that he could not. The question set Pauling thinking about the problem, and it was not long before he had a theory (15) that guided his research on antibodies for years to come. Eventually, it turned out to be wrong, or at least only half right.

The correct part was that the specificity of antibodies for a particular antigen is based on complementarity. 'Atoms and groups which form the surface of the antigen attract certain complementary parts of the globulin chain and repel other parts'. The wrong part was his assumption 'that all antibody molecules contain the same polypeptide chains as normal globulin and differ from normal globulin only in the configuration of the chain'. Pauling was clearly not too happy about this assumption, which he adopted only because of his inability 'to formulate a reasonable mechanism whereby the order of amino-acid residues would be determined by the antigen'. He could not know then about the genetic basis of amino acid sequence. So he was right about how antibodies work and wrong about how they are produced. It was still a long time before a better theory emerged, based not on instruction but on selection, and involving hypervariable regions of the amino acid chain and shuffling genes. In retrospect then it is not surprising that Pauling's immunochemistry programme, carried out mainly by his Caltech collaborator Dan Campbell, never achieved the successes he had hoped for. During World War II there was a brief flurry of excitement when they claimed to have made 'artificial antibodies' from normal globulins, but the claim proved to be ill-founded and was soon retracted.

In 1941 Pauling's intense work schedule was temporarily stemmed when he was diagnosed as having Bright's disease, regarded then by many doctors as incurable. Under the treatment of Dr Thomas Addis, he slowly recovered. Addis, a controversial figure, put Pauling on a low-protein, salt-free diet, which was effective in healing the damaged kidneys. After about six months Pauling was more or less back to normal, but he kept to Addis's diet for many years afterwards. Pearl Harbour brought further distractions when Pauling's energies were diverted into war work, mainly on rocket propellants and in the search for artificial antibodies. Earlier, he had used the paramagnetism of oxygen to design and develop an oxygen meter for use in submarines.

By the end of the war, Pauling felt well enough to travel abroad again. In late 1947 he came as Eastman Visiting Professor with his family to England, where he gave lectures to packed out audiences in Oxford and elsewhere, received medals and suffered from the climate. In 1948, confined to bed with a cold, he began thinking again about a problem that had briefly occupied him a decade earlier, the structure of α-keratin. By this time, thanks to the X-ray crystallographic work of Robert B. Corey and his associates, the detailed structures of several amino acids and simple peptides were known, and although the interatomic distances and angles did not differ much from the values derived earlier by resonance arguments, Pauling could now take them as facts rather than suppositions, especially the planarity of the amide group. With the help of paper models, he then set himself the problem of taking a polypeptide chain, rotating round the two single bonds but keeping the peptide groups planar, repeating with the same rotation angles from one peptide group to the next, and searching for a helical structure in which each N–H group makes a hydrogen bond with the carbonyl oxygen of another residue. He found two such structures, one of which also fulfilled the condition of tight packing down the central hole. The structure in question repeated after 18 residues in five turns at a distance of 27 Å, hence 5.4 Å per turn, whereas X-ray photographs of α-keratin seemed to show that the repeat distance was 5.1 Å. The discrepancy could not be removed by minor adjustments to the model and was large enough for Pauling to put the problem aside (see (38)).

It was taken up again after his return to Pasadena, with the help of Corey and of a young visiting professor, Herman Branson, who checked details of the model and searched for alternatives, but without coming up with anything really new. Then came a paper from the Cavendish Laboratory by Bragg, Kendrew and Perutz (1950), who described several possible helical structures for α-keratin, all unacceptable in Pauling's view because they allowed rotation about the C–N bond of the amide group. This paper provoked Pauling to publish his ideas in a series of papers that described the now famous α-helix (essentially the one modelled in Oxford with 3.7 residues per turn), the so-called γ-helix (disfavoured on energetic grounds), and the parallel and anti-parallel pleated sheets with extended polypeptide chains (21–24). By this time, X-ray photographs of synthetic polypeptides had clarified the apparent discrepancy concerning the repeat distance along the helix; it was 5.4 Å after all. Max Perutz has vividly described his consternation on first reading Pauling's proposed structure and how he managed to corroborate it by observing the 1.5 Å reflection corresponding to the step distance along the the α-helix, which everyone had missed until then (Perutz 1987).

Very soon, evidence began to accumulate that the α-helix is indeed one of the main structural features and that the two pleated sheet structures are also important elements of the secondary structure of globular proteins. Just as a few rules concerning the regular repetition of simple structural units had sufficed 20 years earlier to successfully predict the structures of minerals, now a few simple principles derived from structural chemistry were enough to predict the main structural features of proteins.

Pauling's next essay in model building was not so successful. In the summer of 1952 he learnt about the Hershey–Chase experiment proving that genetic information was carried not by protein but by DNA, deoxyribonucleic acid, a polynucleotide. Pauling felt it should be possible to decipher the structure of this substance by building models along similar lines to those in the protein work. The available X-ray diffraction patterns showed a strong reflection at about 3.4 Å, but nothing much else. Having convinced himself that a two-stranded helical structure would yield too low a density, he went on to the assumption of a three-stranded helical structure held together by hydrogen bonds between the phosphate groups of different strands, that is, the structure rested on the tacit assumption that the phosphodiester groups were protonated! They were closely packed about the axis of the helix with the pentose residues surrounding them and the purine and pyrimidine groups projecting radially outwards. When this structure was presented at a seminar, Verner Schomaker is credited with the remark 'If that were the structure of DNA, it would explode!' Nevertheless, the structure was published (25), a pre-publication copy having been sent to Cambridge, where it stimulated Watson and Crick into their final spurt, culminating in their base-paired structure, which was immediately acclaimed as correct by everyone who saw it, including Pauling. The Watson–Crick structure conformed to the self-complementarity principle that Pauling had enunciated many years earlier and then apparently forgotten.

Much has been written about this spectacular failure. Why was his model-building approach so successful with the polypeptides and so unsuccessful (in his hands) with DNA? First, the time factor: Pauling had thought about polypeptide structures for more than a decade before he risked publishing his conclusions; he thought only for a few months about DNA. Secondly, the available information: for the polypeptide problem, precise metrical and stereochemical data for amino acids and simple peptides, mostly from Pauling's own laboratory, were at hand; for DNA almost nothing was known about the detailed structures of the monomers or oligomers. The X-ray photographs available to Pauling were obtained from degraded DNA specimens and were essentially non-informative (they were later recognized to be derived from mixtures of the A and B forms of DNA), and he made a bad mistake in neglecting the high water content of the DNA specimens in his density calculations. Yet Watson and Crick succeeded with Pauling's methods where Pauling failed. There is no doubt in my mind that *if* Pauling had had access to Rosalind Franklin's X-ray photographs he would immediately have drawn the same conclusion as Crick did, namely, that the molecule possesses a twofold axis of symmetry, thus pointing to two chains running in opposite directions and definitely excluding a three-chain structure. Then there were Chargaff's data about base ratios; Pauling later admitted that he had known about these but had forgotten. It seems clear that Pauling was in a hurry to publish, although, according to Peter Pauling's entertaining account 20 years later (Pauling 1973), he never felt he was in any sense 'in a race'. And finally, as described in the next section, he was by this time under severe harassment from the F.B.I. and other agencies for his political views and activities. This must have taken up much of his mental and emotional energies during these months.

Pauling's standing as a founder of molecular biology rests partly on his identification of sickle-cell anemia, a hereditary disease, as a molecular disease, the first to be recognized as such. The red blood cells in the venous systems of sufferers adopt sickle shapes which tend to block small blood vessels, causing distressing symptoms, whereas the cells in the more oxygenated arterial blood have the normal flattened disc shape. When, towards the end of the war, Pauling heard about this it occurred to him that it could be due to the presence of haemoglobin molecules with a different amino acid sequence from normal. The abnormal molecules, but not the normal ones, could contain self-complementary patches such as to lead to end-to-end aggregation into long rods that twist the blood cells out of shape. Oxygenation could cause a conformational change to block these sticky patches. It took several years to confirm the essential correctness of what was no more than an intuitive guess. In the preliminary studies, attempts to identify any difference between the haemoglobins of normal and sickle-cell blood were unsuccessful, but with the advent of electrophoresis it could be shown found that molecules of sickle-cell and normal haemoglobin moved at different rates in the electric field; the two molecules have different isoelectric points and must indeed be different (20). When, much later, it became possible to determine the amino acid sequence in a protein, sickle-cell haemoglobin was found to contain valine instead of glutamic acid at position 6 of the two β-chains. A single change in a single gene is responsible for the disease.

A decade later, the further study of mutations in haemoglobin led to yet another fundamental contribution to molecular biology: the concept of the 'molecular clock' in evolution (28). By this time, amino acid sequencing of proteins had become standard. Haemoglobins obtained from humans, gorillas, horses and other animals were analysed. From palaeontological evidence, the common ancestor of man and horse lived somewhere around 130 million years ago. The α-chains of horse and human haemoglobin contain about 150 amino acids and differ by about 18 amino acid substitutions, that is, about nine evolutionarily effective mutations for each of the chains, or about one per 14 million years. On this basis, the differences between gorilla and human haemoglobin (two substitutions in the α- and one in the β-chain) suggest a relatively recent divergence between the species, of the order of only 10 million years. On the other hand, differences between the haemoglobin α- and β-chains of several animals suggest divergence from a common chain ancestor about 600 million years ago, in the Precambrian, before the apparent onset of vertebrate evolution. From this work it became clear that comparison of protein sequences (now replaced by comparison of DNA sequences) is a powerful source of information about the origin of species. Evolution of organisms is bound with the evolution of molecules.

Political activism

By 1954, when Pauling was awarded the Nobel Prize in Chemistry for his 'research into the nature of the chemical bond and its application to the elucidation of the structure of complex substances', not only was he famous as a scientist, he also was a well known public figure, at least in the U.S.A. Although he was not connected in any way with either the Manhattan Project or the Radiation Laboratory, his wartime research on antibodies and rocket propellants had brought him into government advisory agencies such as the Office of Scientific Research and Development (O.S.R.D.) under Vannevar Bush and earned him the Presidential Medal for Merit, the highest civilian honour in the U.S.A., awarded by President Truman in 1948. A few years later, he was being vilified in the local and national press, he was being cited for 'un-American activities', he was denied the possibility to travel outside the U.S.A. and his Government research contracts were being terminated. How did this change happen?

Almost immediately after August 1945 Pauling became concerned with the implications of the atomic age for international relations and with the necessity for controls. His lectures and writings on this subject soon attracted the attention of the F.B.I. and other government agencies. Far from being intimidated by these attentions, he began, with the encouragement of his wife, Ava Helen, to take a more active stance; he signed petitions, joined organizations (such as the Emergency Committee of Atomic Scientists, presided over by Albert Einstein, and the American Civil Liberties Union), protested against the loyalty oaths demanded of public employees and spoke eloquently against the development of nuclear weapons.

In the McCarthy era, and especially during the Korean War, this was enough to make him suspect as a security risk. Pauling was invited to lecture at a Royal Society meeting on protein structure, held in London in May 1952. In February, his application for a passport was refused because his proposed travel 'would not be in the best interests of the United States'. Renewed applications up to the end of April met with renewed refusals. A few hours before the start of the meeting, Pauling telegraphed his regrets to London. I was present when the news came that Pauling had not been granted a passport and was therefore unable to attend. It was a grave disappointment, for we had all looked forward to Pauling's presence at the meeting, and there was also a feeling of outrage; the action of the State Department was seen as an insult not only to Pauling and to the Royal Society, but also to the scientific community at large. Pauling was certainly not the only U.S. citizen whose right to travel was denied by the State Department, but the incident provoked such widespread criticism that it probably helped to lead to a re-examination and ultimate change of the State Department's policy. Later that year, Pauling was permitted to travel to France and England (where he did not see Rosalind Franklin's X-ray diffraction photographs of DNA!) and the following summer he was again in Europe (where he did see the Watson–Crick DNA structure). This freedom to travel was bought at the cost of a temporary self-imposed political restraint, and was in any case a fragile privilege which he lost again a few months later, when he spoke out in defence of J. Robert Oppenheimer.

In March 1954, following the Bikini Atoll explosion of a 'dirty' thermonuclear superbomb, Pauling was in the news again when he began to call attention to the worldwide danger of radioactive fallout in the atmosphere. In the summer, his renewed application for a passport was again turned down, but in November, when his Nobel Prize was announced, the State Department found itself in a public relations dilemma. The fuss created by Pauling's absence in London in 1952 would be nothing compared with the international outcry that could be imagined if Pauling were refused permission to travel to attend the Nobel Prize ceremony. So Pauling went to Stockholm, where he was a tremendous success, and followed this by visits to Israel, India, Thailand and Japan. Everywhere, outside his own country, he was welcomed with enthusiasm, not only for his scientific accomplishments but even more for his political stance.

In the U.S.A. too, the public was becoming increasingly concerned about radioactive fallout, not only from American tests but also from ever more powerful Soviet nuclear explosions. Increasing levels of strontium-90 and carbon-14 made newspaper headlines. Pauling claimed that the increased level of radioactive isotopes in the atmosphere was a danger not only to the living but also to future generations; the spokesmen on the Atomic Energy Commission countered that, although radiation might be harmful, it was not harmful in the doses produced by the tests and that Pauling vastly exaggerated the dangers. In fact, all the estimates were tentative at best, but since the Atomic Energy Commission was responsible both for developing nuclear weapons and for monitoring the associated health hazards, its estimates were probably no more objective than those who demanded a stop to the tests. Andrei Sakharov (1990) estimated that every one-megaton test cost about 10 000 human lives.

In January 1958 Pauling, together with his wife, was instrumental in collecting thousands of signatures from scientists all over the world for a petition to end nuclear bomb testing, which was presented to Dag Hammerskjöold, Secretary General of the United Nations. A few months later, the Soviet Union called for an immediate halt to nuclear testing, and in October, after more tests by both sides that added markedly to world concern about fallout, talks began in Geneva to discuss details of a possible test ban. During the talks there was an informal moratorium on testing by the Soviet Union, the U.S.A. and the U.K. And in the meantime, Pauling's book *No more war!* (26) was published.

In 1960 the Senate Internal Security Subcommittee (S.I.S.S.) headed by Senator Thomas Dodd issued a subpoena to Pauling to answer questions about Communist infiltration of the campaign against nuclear testing. At Pauling's request, the hearings were open and they soon turned into a public relations fiasco for Dodd and the S.I.S.S. This was partly because the members of the S.I.S.S. had not done their homework properly and partly because it gave Pauling the excuse to lecture them about elementary civic rights and duties: 'The circulation of petitions is an important part of our democratic process. If it is abolished or inhibited, it would be a step towards a police state.' By this time, public opinion was mostly on Pauling's side, but the whole affair must have been an emotional strain, and a tremendous waste of his time and energy.

In 1961 there was a new petition, an 'Appeal to stop the spread of nuclear weapons', again presented to the United Nations, and he also helped to organize the Oslo Conference on the dangers raised by the proliferation of nuclear weapons. But in September there was a new spate of Soviet tests of even more powerful bombs – 50 within a couple of months – and in March 1963 President Kennedy announced that the U.S.A. would also resume testing. This time the tests did not last long: they were stopped in the summer, when new proposals were made to forbid atmospheric tests while permitting underground tests. In August both sides signed a treaty to ban all tests in the atmosphere, in outer space and under the sea. The treaty went into effect on 10 October and the following day Pauling was awarded the Nobel Peace Prize for 1962.

By now, especially in the aftermath of the Chernobyl disaster, the cultural climate has changed so much that this short account of atom politics until 1963 must strike younger readers as almost inconceivable. In the summer of 1995, when France exploded some 'nuclear devices' several hundred metres underground below a remote atoll in the South Pacific, there was an international outcry of protest, by governments, by the press and by the public. Forty years ago, when tons of radioactive material were being spewed into the atmosphere by test after test, there was no such outcry, at least not in the U.S.A. and the Soviet Union, the two countries most responsible for the pollution. One can assume that most people believed the tests were necessary. Small groups organized protest marches, but there were no social structures in these nuclear states to resist the continuation of testing and the spread of atomic weapons. Pauling was one of the few who consistently spoke against the dangers of atmospheric testing, against the spread of nuclear weapons, for efficient control of such weapons and for a more rational approach to solve international conflicts. These sentiments found a ready ear in the non-nuclear countries, and eventually public opinion in the U.S.A. also swung in his direction. Whether he had any effect in the Soviet Union is another matter; he is not mentioned in Sakharov's (1990) autobiography.

Apostle of vitamin C

A few days after the news of the Peace Prize, Pauling announced that he was leaving Caltech to become a member of the Center for the Study of Democratic Institutions in Santa Barbara. He was disappointed with the lukewarm reaction of the administration and some of his colleagues. Perhaps he had intended to move anyway. In the mid-1950s he had become interested in phenylketonuria (mental deficiency due to inability to metabolize phenylalanine) as a further example of a molecular disease arising from the lack of a specific enzyme. About this time too he was developing his theory that xenon acts as an anaesthetic because it forms crystalline polyhedral hydrates; microcrystals of such hydrates in the brain could interfere with the electric oscillations associated with consciousness (27). He obtained a $ 450 000 grant from the Ford Foundation to study the molecular basis of mental disease and turned his laboratories more and more away from traditional chemistry, not to the unanimous approval of his colleagues. In 1958 he resigned from his position as Department Chairman, which he had held for more than 20 years, and found himself under pressure to give up research space to a new generation of researchers. In these years of intense political activity and world travel, he was in any case spending less and less time with his own research group and in keeping up with new developments in chemistry. When he left Caltech, he vanished without a trace. In the 1963–64 Annual Report of the Chemistry Department his name appears in the list of professors, with more honours and degrees than anyone else; in the corresponding report a year later his name has disappeared.

The next few years were not the happiest in Pauling's life. Not only did he sever his connection with Caltech, he resigned from the American Chemical Society as well. The move to Santa Barbara was not a success. He turned to theoretical physics, but his close-packed spheron theory of the atomic nucleus met with little acceptance. He became engaged in actual and threatened libel suits. He moved briefly to the University of California at San Diego (1967–69) and then on to Stanford University (1969–72), where he was closer to his ranch at Big Sur, but he had no stable position in which to continue his planned research into 'orthomolecular' psychiatric therapy. Meanwhile, he was deeply unhappy about the American involvement in Vietnam and about American politics in general.

One consolation was that, after passing his 65th birthday, Pauling's health took a sudden turn for the better. Thanks to Dr Addis's unconventional low-protein diet, he had recovered well from the kidney disease that had laid him low in his forties, but he had always suffered from severe colds several times a year. In 1966, following a suggestion from Dr Irwin Stone, the Paulings began to take 3g of ascorbic acid per day each. Almost immediately they felt livelier and healthier. Over the next few years the colds that had plagued him all his life became less severe and less frequent. This experience made Pauling a believer in the health benefits of large daily amounts of vitamin C. It was not long before he was enthusiastically promulgating this belief in lectures and writings, which, not too surprisingly, brought on him the displeasure of the American medical establishment. After all, the then recommended daily allowance (RDA) of vitamin C was 45 mg, it was well known that there was no known cure for the common cold, and, in particular, previous studies had shown conclusively that vitamin C had no effect. Nevertheless, the N.A.S. Subcommittee on Laboratory Nutrition was recommending daily intakes around 100 times that of human RDA (adjusted for body weight) to keep laboratory primates in optimal health.

In his 1970 book *Vitamin C and the common cold* (29), Pauling gave evolutionary arguments why much larger amounts of vitamin C than the RDA may be conducive to optimal health, he cited studies that supported its efficacy in preventing colds or at least in lessening their severity, he criticized studies that claimed the opposite and he argued that since vitamin C is not a drug but a nutrient there is no reason why a large daily intake should be harmful. Pauling's arguments did not win the approval of the medical profession but they caught on with the general public. The book rapidly became a best seller. As a result, in the U.S.A. and later also in other countries, millions of people were persuaded that a daily intake of 1–2 g of ascorbic acid has a beneficial effect on health and well-being, essentially agreeing with Pauling that 'we may make use of ascorbic acid for improving health in the ways indicated by experience, even though a detailed understanding of the mechanisms of its action has not yet been obtained'.

One result of the book was a collaboration with a Scottish surgeon, Ewan Cameron, from Vale of Leven, who had observed beneficial effects of high doses of vitamin C in treating terminal cancer patients. Cameron thought that vitamin C might be involved in strengthening the intracellular mucopolysaccharide hyaluronic acid by helping to inhibit the action of the enzyme hyaluronidase produced by invasive cancerous cells. A paper by Cameron and Pauling (30) advocating vitamin C therapy in cancer was submitted to the *Proceedings of the National Academy of Sciences of the U.S.A.* and, in an unprecedented move, rejected (it was then published in the specialist journal *Oncology*). During the next few years Cameron continued his trials. Since a double-blind trial was ethically unacceptable, he compared results obtained with 100 ascorbate-treated terminal patients and 1000 other cases, ten controls for each patient, matched as closely as possible, and found that that the ascobate-treated patients lived longer and felt better. A paper describing these results was eventually published in *Proceedings of the National Academy of Sciences of the U.S.A.* (32), but only after long arguments with referees. The Cameron–Pauling collaboration culminated in their 1979 book *Cancer and vitamin C* (34), which was again more popular with the public than with the medical profession, which continued to regard claims about the effectiveness of vitamin C in treating or preventing cancer as quackery. But by this time several important changes had occurred in Pauling's life.

At Stanford, Pauling's demands for more laboratory space for his orthomolecular medicine studies had been turned down. A solution was found by a younger colleague, Arthur B. Robinson, who had left a tenured position at San Diego to work with Pauling at Stanford. Instead of working in cramped quarters at the University, they would set up their own research institute nearby. A building was rented, initial financial help was forthcoming, and the Institute for Orthomolecular Medicine was founded in 1973. Once the initial funding had run out, the Institute found itself in financial straits. Soon it was renamed the Linus Pauling Institute of Science and Medicine, with Pauling as president. By this change, it was hoped, fund-raising possibilities would be improved, a hope that proved to be illusory. As Pauling was frequently away on travels and in any case disliked administration, Robinson took over in 1975, but the fiscal problems of the Institute dragged on for several years until support began to be provided by private foundations and individual donations.

Personal and scientific difficulties between Robinson and Pauling led to Robinson's dismissal in 1979 and to lawsuits that dragged on for years. Meanwhile Pauling continued to defend his unorthodox views and became once again a controversial figure, regarded by some as a crackpot, by others as a sage. In 1986 he wrote another popular book *How to live longer and feel better* (36), which, based on his own experiences, gave advice about how to cope with aging.

In July 1976, Ava Helen underwent surgery for stomach cancer. Instead of post-operative chemotherapy or radiation treatment, she adopted vitamin C therapy to the tune of 10 g per day. She was soon well enough to accompany Pauling on his various travels, but she finally succumbed five years later, in December 1981. Pauling continued to travel, to appear on television, to write, to receive honours, his energy seemed unabated. When quasi-crystals, with forbidden fivefold symmetry, were discovered in 1984, Pauling took a contrary position and argued that the fivefold symmetry seen in Al–Mn alloys resulted merely from twinning of cubic crystallites (35). He was probably wrong, but the resulting controversy was nevertheless useful in forcing the proponents of quasi-crystals to seek better evidence for their view.

He even became reconciled with Caltech, where his 85th and 90th birthdays were marked by special symposia in his honour. In 1991 he was diagnosed with cancer. Surgery brought temporary relief, and megadoses of vitamin C kept up his spirits. He spent his last months at the ranch at Big Sur and died there on 19 August 1994.

In the meantime, the medical establishment is no longer so totally dismissive of Pauling's views about possible therapeutic benefits of vitamin C on the common cold and cancer. A recent review of several studies concludes that although supplemental vitamin C does not decrease the incidence of the common cold it does diminish the duration and the severity of symptoms (Hemilä 1992). This review also states that the level of vitamin C intake derived from a normal or balanced diet may be insufficient for optimal body function and that the substance is safe even in large amounts.

The connection between vitamin C and cancer has also become a respectable topic of discussion. It was the subject of a conference organized by the U.S. National Cancer Institute in Washington D.C. in 1990. Vitamins C and E (and other anti-oxidants) inhibit the endogenous formation of *N*-nitroso compounds in animals and humans (Bartsch *et al.* 1988). Such compounds are known to be carcinogenic in animals. Conclusive proof that they are dangerous at the levels naturally present in man is lacking, but the evidence seems suggestive. Thus, although the effectiveness of vitamin C in treating cancers may still be debatable, there is good reason to believe that it has at least an important preventive role.

The final word about the effect of large doses of vitamin C on health has still to be said. If you have a full, healthy diet, rich with fruit, grains and fresh vegetables, then you probably do not need supplemental vitamins and minerals. But in the modern world many people have, and may even prefer, an unhealthy diet. For them, vitamin supplements are probably beneficial. After all, Pauling not only recommended large doses of vitamin C but also advised people to stop smoking, to eat less and to cut down consumption of sucrose.

Pauling the Man

Pauling lived a long and productive life. As a scientist, through his writings and personal impact, he influenced several generations of chemists and biologists. As a political activist he challenged the political and military establishments of the U.S.A. and helped to change them. As health crusader he took on the medical establishment and persuaded millions of people to eat supplemental vitamins. He could be very persuasive indeed. His lectures were spellbinding, and he had a characteristically simple and direct literary style.

I remember his lectures in Oxford in early 1948. The lecture hall was too small to hold all who wished to attend; there was standing room only. He told those of us who had never studied electrostatics to go home and read Sir James Jeans's book on that subject before coming to his lectures on chemical bonding. I had never studied electrostatics but I stayed, spellbound. I had never heard anyone quite like him, with his jokes, his relaxed manner, his seraphic smile, his slide-rule calculations and his spontaneous flow of ideas. (Only much later did I realize that much of that apparent spontaneity was carefully studied.) He had great histrionic skills.

Vain? Conceited? Pauling was certainly aware of his own intellectual superiority, but he could be patient in dealing with the slowness of the slow-witted. On the whole, he was fairly tolerant of young, insecure seminar speakers, although, as I remember, he could also be intimidating at times. I am referring here to Pauling in middle age, I am told he became more intolerant in his later years. Political harassment during and after the McCarthy era must have taken its toll. Ambitious? Self-centred? Undoubtedly. Without these traits he would not have been able to accomplish as much as he did. But he often had a merry twinkle in his eyes and could be very charming, both as a public personality and in private.

In personal matters he kept most people at a distance. I believe he was basically rather shy. When he talked about science or politics or anything that caught his interest, there was no stopping him, he read widely and was extremely knowledgeable in many areas, a result of having pored over the *Encyclopaedia brittanica* in his youth? In conversation, one sometimes sensed a faraway look in his eyes; one felt that he was already thinking about something else. Probably he was, and, indeed, he was a formidable thinker, both at the problem-solving level and about fundamentals. With his prodigious memory he could call up facts and derivations, what so-and-so had written in 1928, the unit cell dimensions of an obscure mineral, the standard heat of formation of ethane; and he had a remarkable capacity to visualize complex three-dimensional structures. I once asked him why he had never discussed the application of group theory to problems of chemical bonding. 'Jack,' he replied, 'if you need group theory to solve that sort of problem then you're in the wrong line of business.'

In addition to his Nobel Prizes, Pauling was awarded dozens of honours and distinctions, including Honorary Doctorates from Oregon State College, Brooklyn Polytechnic Institute, Reed College, and the Universities of Chicago, Princeton, Yale, Cambridge, London, Oxford, Paris, Toulouse, Montpellier, Lyon, Liège, Humboldt (Berlin), Melbourne, York (Toronto), New Brunswick and Warsaw. His election to membership of the U.S. National Academy of Sciences, the Royal Society of London, the Academie Française des Sciences and the Akademie Nauk S.S.R. may be specially mentioned.

His name will be remembered as long as there is a science of chemistry.

Acknowledgements

I have learnt much about Pauling's life from the excellent biography by Tom Hager (1995) and am grateful for information and advice from many friends and colleagues, among them Barclay Kamb, Linda Pauling-Kanb, David Craig, F.R.S., Durward. W.J. Cruickshank, F.R.S., Albert Eschenmoser, For.Mem.R.S., Edgar Heilbronner, Paul Kleihues, Alan Mackay, F.R.S., Peter J. Pauling, Alexander Rich, John D. Roberts and Verner Schomaker. The portrait of Linus Pauling was made by Tom Harvey, California Institute of Technology, in1958 and is reproduced with their kind permission.

References to other authors

Bartsch, H., Ohshima, H. & Pignatelli, B. 1988 Inhibitors of endogenous nitrosation. Mechanisms and implications in human cancer prevention. *Mutation Res.* **202**, 307–324.

Bernal, J.D. 1939 Structure of proteins. *Nature, Lond.* **143**, 663–667.

Bragg, W.L. 1937 *Atomic structure of minerals.* Ithaca, New York: Cornell University Press.

Bragg, W.H., Kendrew, J.C. & Perutz, M.F. 1950 Polypeptide chain configurations in crystalline proteins. *Proc. R. Soc. Lond.* A**203**, 321–357.

Hager, T. 1995 *Force of Nature: the life of Linus Pauling*. New York: Simon & Schuster.

Hemilä, H. 1992 Vitamin C and the common cold. *Br. J. Nutr.* **67**, 3–16.

Jencks, W.P. 1969 *Catalysis in chemistry and enzymology*. New York: McGraw-Hill.

Kauffman, G.B. & Kauffman, L.M. 1996 An interview with Linus Pauling. *J. chem. Educ.* **73**, 29–32.

Pauling, P. 1973 DNA – the race that never was? *New Scient.* **58**, 558–560.

Perutz, M.F. 1987 I wish I'd made you angry earlier. *Scientist* (February 23), p.19.

Sakharov, A. 1990 *Memoirs* (transl. R Laurie). New York: Knopf.

Bibliography

The following publications are those referred to directly in the text. A full bibliography (by permission of Zelek Herman, Dorothy Munro, and the Linus Pauling Institute of Science and Medicine) appears on the accompanying microfiche, numbered as in the second column. A photocopy is available from the Royal Society Library at cost.

(1) (23–2) 1923 The crystal structure of magnesium stannide. *J. Am. chem. Soc.* **45**, 2777–2780.

(2) (23–1) (with R.G. Dickinson) The crystal structure of molybdenite. *J. Am. chem. Soc.* **45**, 1466–1471.

(3 (25–4) 1925 (with P. Debye) The inter-ionic attraction theory of ionized solutes. IV. The influence of variation of dielectric constant on the limiting law for small concentrations. *J. Am. chem. Soc.* **47**, 2129–2134.

(4) (25–5) (with R.C. Tolman) The entropy of supercooled liquids at the absolute zero. *J. Am. chem. Soc.* **47**, 2148–2156.

(5) (27–5) 1927 The theoretical prediction of the physical properties of many-electron atoms and ions: mole refraction, diamagnetic susceptibility and extension in space. *Proc. R. Soc. Lond.* A**114**, 181–211.

(6) (28–5) 1928 The application of the quantum mechanics to the structure of the hydrogen molecule and hydrogen molecule-ion and to related problems. *Chem. Rev.* **5**, 173–213.

(7) (29–1) 1929 The principles determining the structure of complex ionic crystals. *J. Am. chem. Soc.* **51**, 1010–1026.

(8) (30–12) 1930 (with S.A. Goudsmit) *The structure of line spectra.* New York: McGraw-Hill.

(9) (31–3) 1931 The nature of the chemical bond. Application of results obtained from the quantum mechanics and from a theory of paramagnetic susceptibility to the structure of molecules. *J. Am. chem. Soc.* **53**, 1367–1400.

(10) (35–17) 1935 (with E.B. Wilson Jr) *Introduction to quantum mechanics with applications to chemistry.* New York: McGraw-Hill.

(11) (35–7) (with G.W. Wheland) A quantum mechanical discussion of orientation of substituents in aromatic molecules. *J. Am. chem. Soc.* **57**, 2086–2095.

(12) (36–4) 1936 (with C.D. Coryell) The magnetic properties and structure of hemoglobin, oxyhemoglobin and carbonmonoxyhemoglobin. *Proc. natn. Acad. Sci. U.S.A.* **22**, 210–216.

(13) (36–7) (with A.E. Mirsky) On the structure of native, denatured, and coagulated proteins. *Proc. natn. Acad. Sci. U.S.A.* **22**, 439–447.

(14) (39–8) 1939 *The nature of the chemical bond and the structure of molecules and crystals.* (2nd edn 1940, 3rd edn 1960) Ithaca, N.Y.: Cornell University Press.

(15) (40–2) 1940 A theory of the structure and process of formation of antibodies. *J. Am. chem. Soc.* **62**, 2643–2657.

(16) (40–3) (with M. Delbrück) The nature of the intermolecular forces operative in biological processes. *Science, Wash.* **92**, 77–79.

(17) (47–2) 1947 Atomic radii and interatomic distances in metals. *J. Am. chem. Soc.* **69**, 542–553.

(18) (48–15) 1948 *Molecular architecture and the processes of life.* Sir Jesse Boot Foundation Lecture, Nottingham.

(19) (48–12) The nature of forces between large molecules of biological interest. *Nature, Lond.* **161**, 707–709.

(20) (49–16) 1949 (with H.A. Itano, S.J. Singer & I.C. Wells) Sickle cell anemia, a molecular disease. *Science, Wash.* **110**, 543–548.

(21) (50–5) 1950 (with R.B. Corey) Two hydrogen-bonded spiral configurations of the polypeptide chain. *J. Am. chem. Soc* **72**, 5349.

(22) (51–11) 1951 (with R.B. Corey) The structure of synthetic polypeptides. *Proc. natn. Acad. Sci. U.S.A.* **37**, 241–250.

(23) (51–12) (with R.B. Corey) The pleated sheet, a new layer configuration of polypeptide chains. *Proc. natn. Acad. Sci.* **37**, 251–256.

(24) (51–9) (with R.B Corey & H.R Branson) The structure of proteins: two hydrogen-bonded helical configurations of the polypeptide chains. *Proc. natn. Acad. Sci. U.S.A.* **37**, 205–210.

(25) (53–9) 1953 (with R.B. Corey) A proposed structure for the nucleic acids. *Proc. natn. Acad. Sci. U.S.A.* **39**, 84–97.

(26) (58–28) 1958 *No more war!* New York: Dodd, Mead.

(27) (61–9) 1961 A molecular theory of general anesthesia. *Science, Wash.* **134**, 15–21.
(28) (62–23) 1962 (with E. Zuckerkandl) Molecular disease, evolution and genetic heterogeneity. In *Horizons in biochemistry* (ed. M. Kasha & P. Bullman), pp. 189–225. New York: Academic Press.
(29) (70–18) 1970 *Vitamin C and the common cold.* San Francisco: W.H. Freeman.
(30) (70–11) Fifty years of progress in structural chemistry and molecular biology. *Daedalus* (Fall), pp. 988–1014.
(31) (73–10) 1973 (with E. Cameron) Ascorbic acid and the glycosaminoglycans: an orthomolecular approach to cancer and other diseases. *Oncology* **27**, 181–192.
(32 (76–17) 1976 (with E. Cameron) Supplemental ascorbate in the supportive treatment of cancer: prolongation of survival times in terminal human cancer. *Proc. natn. Acad. Sci. U.S.A.* **73**, 3685–3689.
(33) (79–13) 1979 The discovery of the superoxide radical. *Trends biochem. Sci.* **4**, N270–271.
(34) (79–15) (with E. Cameron) *Cancer and vitamin C.* Palo Alto: CA: Linus Pauling Institute.
(35) (85–6) 1985 Apparent icosahedral symmetry is due to directed multiple twinning of cubic crystals. *Nature, Lond.* **317**, 512–514.
(36) (86–2) 1986 *How to live longer and feel better.* New York: Freeman.
(37) (94–7) 1994 My first five years in science. *Nature, Lond.* **371**, 10.
(38) (96–1) 1996 The discovery of the alpha helix. *Cheml Intelligencer* **2**, 32–38.

Appendix I

Conversion Between SP Numbers in Chapters 1–17 and Citation Numbers in Chapter 18

This table allows the reader to look readily in the collection of selected scientific papers in Chapters 1–17 for any paper cited by J.D. Dunitz in his Bibliographical Memoir of Linus Pauling (Chapter 18). In the left-hand half of the table the Chapter 18 reference numbers (left-hand column) are listed in numerical order, and next to them are given the corresponding SP numbers. Any selected paper can be found in Chapters 1–17 from its SP number. A dash in the second column indicates a paper that is referenced in Chapter 8 but was not selected for inclusion here. Column 3 gives for every SP number the corresponding Herman-Munro paper number, which is used in Appendix III and is also listed in the Chapter 18 bibliography. Column 4 gives the Group number of each selected paper, which is needed in finding the paper in Appendix III. (The Group number is the same as the Chapter number, except with lettered subdivisions in some cases—see Appendix III.) The right-hand table contains the same material, but ordered sequentially in the SP number, so as to make it easy to find Dunitz's comments (if any) on any particular selected paper of interest.

FROM Chapter-18 citation no.
TO SP no. (Chapters 1–17)

FROM SP no. (Chapters 1–17)
TO Chapter-18 citation no.

Ch.18 cite'n. no.	Ch.1–17 SP no.	App.III paper no.	Chapter (Group) no.	Ch.18 cite'n. no.	Ch.1–17 SP no.	App.III paper no.	Chapter (Group) no.
(1)	{SP 49}	[23-2]	6b	(6)	{SP 3}	[28-5]	1
(2)	{SP 48}	[23-1]	6a	(9)	{SP 5}	[31-3]	1
(3)	—	[25-4]	8	(11)	{SP 13}	[35-7]	1
(4)	{SP 71}	[25-5]	9	(10)	{SP 15}	[35-17]	1
(5)	{SP 64}	[27-5]	8	(14)	{SP 22}	[39-8]	1

FROM Chapter-18 citation no.
TO SP no. (Chapters 1–17)

Ch.18 cite'n. no.	Ch.1–17 SP no.	App.III paper no.	Chapter (Group) no.
(6)	{SP 3}	[28-5]	1
(7)	{SP 24}	[29-1]	2
(8)	—	[30-12]	8
(9)	{SP 5}	[31-3]	1
(10)	{SP 15}	[35-17]	1
(11)	{SP 18}	[35-7]	1
(12)	{SP 83}	[36-4]	11
(13)	{SP 95}	[36-7]	13
(14)	{SP 22}	[39-8]	1
(15)	{SP 88}	[40-2]	12
(16)	{SP 112}	[40-3]	14a
(17)	{SP 29}	[47-2]	3
(18)	{SP 115}	[48-15]	14a
(19)	{SP 114}	[48-12]	14a
(20)	{SP 122}	[49-16]	15
(21)	—	[50-5]	13
(22)	{SP 99}	[51-11]	13
(23)	{SP 100}	[51-12]	13
(24)	{SP 97}	[51-9]	13
(25)	{SP 116}	[53-9]	14a
(26)	—	[58-28]	—
(27)	{SP 129}	[61-9]	16a
(28)	{SP 119}	[62-23]	14b
(29)	—	[70-18]	17c
(30)	—	[70-11]	14a
(31)	{SP 139}	[73-10]	17d
(32)	{SP 140}	[76-17]	17d
(33)	{SP 131}	[79-13]	16a
(34)	—	[79-15]	17d
(34)	{SP 144}	[93-2]	17d
(35)	{SP 80}	[85-6]	10c
(36)	—	[86-2]	17a
(37)	—	[94-7]	20
(38)	{SP 111}	[96-1]	13

FROM SP no. (Chapters 1–17)
TO Chapter-18 citation no.

Ch.18 cite'n. no.	Ch.1–17 SP no.	App.III paper no.	Chapter (Group) no.
(7)	{SP 24}	[29-1]	2
(17)	{SP 29}	[47-2]	3
(2)	{SP 48}	[23-1]	6a
(1)	{SP 49}	[23-2]	6a
(5)	{SP 64}	[27-5]	8
(4)	{SP 71}	[25-5]	9
(35)	{SP 80}	[85-6]	10c
(12)	{SP 83}	[36-4]	11
(15)	{SP 88}	[40-2]	12
(13)	{SP 95}	[36-7]	13
(24)	{SP 97}	[51-9]	13
(22)	{SP 99}	[51-11]	13
(23)	{SP 100}	[51-12]	13
(38)	{SP 111}	[96-1]	13
(16)	{SP 112}	[40-3]	14a
(19)	{SP 114}	[48-12]	14a
(18)	{SP 115}	[48-15]	14a
(25)	{SP 116}	[53-9]	14a
(28)	{SP 119}	[62-23]	14b
(20)	{SP 122}	[49-16]	15
(27)	{SP 129}	[61-9]	16a
(33)	{SP 131}	[79-13]	16a
(31)	{SP 139}	[73-10]	17c
(32)	{SP 140}	[76-17]	17c
—	{SP 144}	[93-2]	17d
(34)	{SP 144}	[79-15]	17d

Appendix II

Citation Index for Selected Papers

This index lists in an abbreviated citation format all papers, articles, and book chapters reproduced in this two-volume anthology. The index is for the following purpose. When the reader encounters a citation to a paper of Linus Pauling, he/she may want to know whether the cited paper is contained in these volumes. That can be determined by looking for the citation in this index. If the citation is found, the cited paper is present and can be located in the volume by means of its selected-paper number (SP no.), which is given in curly brackets after the citation. The abbreviated citation format is non-standard. Journal citations are arranged in order by journal title, volume, and initial page number with year also given; authorship and title are not given, because the citation is unique without them. Book chapter citations, given in a separate list, are arranged in order by book title, editorship, and page run, with year also given; chapter authorship, chapter title, and publisher are not given. In the case of books authored or co-authored by Pauling, the authorship is given instead of editorship.

Following the SP number for each citation is given, in square brackets, the publication number in the complete list of Linus Pauling's publications prepared by Z.S. Herman and D.B. Munro (see General Introduction and Appendix III). These publication numbers have the format [*y*-*n*], where *y* is the year of publication (with the leading "19" omitted) and *n* is the number of the paper in that year, counted sequentially by date of publication.

Book Citations

Cancer and Vitamin C (Updated and Expanded Edition), by Linus Pauling, Chapter 23, pp. 189–195 (1993): {SP 144} [93-2]

Chemical Bonding Models, eds. J.F. Liebman and A. Greenberg, pp. 1–15 (1986): {SP 27} [86-17]

Horizons in Biochemistry, eds. M. Kasha and B. Pullman, pp. 189–225 (1962): {SP 119} [62-23]

Hydrogen Bonding, eds. D. Hadzi and H. W. Thompson, pp. 1–6 (1959): {SP 38} [59-2]

Introduction to Quantum Mechanics, with Application to Chemistry, by Linus Pauling and E. Bright Wilson, Jr., Section 42, pp. 326–331 (1935): {SP 15} [35-17]

Les Prix Nobel en 1954, ed. M.G. Liljestrand, pp. 91–99 (1955): {SP 1} [55-14]

Molecular Architecture and the Processes of Life, by Linus Pauling, (1948): {SP 115} [48-15]

Orthomolecular Psychiatry: Treatment of Schizophrenia, eds. David Hawkins and Linus Pauling, pp. 1–17 (1973): {SP 138} [73-4]

Orthomolecular Psychiatry: Treatment of Schizophrenia, eds. David Hawkins and Linus Pauling, pp. 18–34 (1973): {SP 130} [73-5]

The Harvey Lectures 1953–1954, by Linus Pauling, Series 49, pp. 216–241 (1955): {SP 124} [55-4]

The Nature of the Chemical Bond and the Structure of Molecules and Crystals, by Linus Pauling, 1st edn., Chapter 12, pp. 403–411, (1939): {SP 22} [39-8]

The Nature of the Chemical Bond and the Structure of Molecules and Crystals, by Linus Pauling, 3rd edn., Section 12-1, pp. 449–454 (1960): {SP 35} [60-26]

Journal Citations

Acta Chem. Scand. **17** (Suppl.1), S9–S16 (1963): {SP 120} [63-2]

Acta Cryst. **5**, 39–44 (1952): {SP 53} [52-2]
_________ **5**, 637–644 (1952): {SP 54} [52-9]
_________ **8**, 710–715 (1955): {SP 110} [55-9]
_________ **10**, 254–259 (1957): {SP 55} [57-3]
_________ *B* **24**, 5–7 (1968): {SP 32} [68-1]

Am. J. Clin. Nutr. **24**, 1294–1299 (1971): {SP 137} [71-2]

Arch. Biochem. Biophys. **56**, 164–181 (1956): {SP 117} [56-10]

Cancer Research **39**, 663–681 (1979): {SP 141} [79-1]

Chemical Intelligencer **2**(1), 32–38 (1996): {SP 111} [96-1]

Chem. Rev. **5**, 173–213 (1928): {SP 3} [28-5]

Endeavour **7** (26), 43–53 (1948): {SP 93} [48-7]

J. Am. Chem. Soc. **45**, 1466–1471 (1923): {SP 48} [23-1]
________________ **45**, 2777–2780 (1923): {SP 49} [23-2]
________________ **47**, 781–790 (1925): {SP 39} [25-1]
________________ **47**, 1026–1030 (1925): {SP 40} [25-2]
________________ **47**, 2148–2156 (1925): {SP 71} [25-5]
________________ **49**, 765–790 (1927): {SP 23} [27-4]
________________ **51**, 1010–1026 (1929): {SP 24} [29-1]
________________ **53**, 1367–1400 (1931): {SP 5} [31-3]
________________ **53**, 3225–3237 (1931): {SP 6} [31-4]
________________ **54**, 988–1003 (1932): {SP 25} [32-2]
________________ **54**, 3570–3582 (1932): {SP 26} [32-11]
________________ **57**, 2086–2095 (1935): {SP 13} [35-7]
________________ **57**, 2680–2684 (1935): {SP 73} [35-9]
________________ **57**, 2684–2692 (1935): {SP 58} [35-10]
________________ **57**, 2705–2709 (1935): {SP 14} [35-13]
________________ **59**, 13–20 (1937): {SP 59} [37-3]
________________ **59**, 633–642 (1937): {SP 84} [37-4]
________________ **59**, 1223–1236 (1937): {SP 60} [37-6]
________________ **59**, 1450–1456 (1937): {SP 16} [37-7]
________________ **61**, 1769–1780 (1939): {SP 61} [39-3]
________________ **61**, 1860–1867 (1939): {SP 96} [39-4]
________________ **62**, 2643–2657 (1940): {SP 88} [40-2]
________________ **64**, 2994–3003 (1942): {SP 89} [42-8]
________________ **64**, 3003–3009 (1942): {SP 90} [42-9]
________________ **66**, 330–336 (1944): {SP 92} [44-1]
________________ **68**, 795–798 (1946): {SP 62} [46-6]
________________ **69**, 542–553 (1947): {SP 29} [47-2]
________________ **71**, 143–148 (1949): {SP 94} [49-1]
________________ **74**, 3964 (1952): {SP 105} [52-18]

J. Chem. Ed. **47**, 15–17 (1970): {SP 20} [70-3]

J. Chem. Phys. **1**, 280–283 (1933): {SP 7} [33-10]
____________ **1**, 362–374 (1933): {SP 8} [33-11]
____________ **1**, 606–617 (1933): {SP 9} [33-13]
____________ **1**, 679–686 (1933): {SP 10} [33-14]
____________ **4**, 393–394 (1936): {SP 74} [36-2]
____________ **4**, 673–677 (1936): {SP 67} [36-8]

J. Chem. Soc. **1948**, 1461–1467 (1948): {SP 18} [48-17]

Journal of NIH Research **2**(6), 59–64 (1990): {SP 113} [46-9]

J. Orthomolecular Med. **7**(1), 17–23 (1992): {SP 143} [92-9]

J. Phys. Chem. **56**, 361–365 (1952): {SP 19} [52-5]

J. Solid State Chem. **54**, 297–307 (1984): {SP 33} [84-12]

J. Theoret. Biol. **8**, 357–366 (1965): {SP 121} [65-4]

Mineral. Soc. Amer. Spec. Pap. **3**, 125–131 (1970): {SP 57} [70-8]

Nature **161**, 707–709 (1948): {SP 114} [48-12]
______ **171**, 59–61 (1953): {SP 108} [53-5]
______ **203**, 182–183 (1964): {SP 86} [64-18]
______ **317**, 512–514 (1985): {SP 80} [85-6]

Oncology **27**, 181–192 (1973): {SP 139} [73-10]

Phys. Rev. **36**, 430–443 (1930): {SP 72} [30-7]
_________ **37**, 1185–1186 (1931): {SP 4} [31-2]
_________ **47**, 686–692 (1935): {SP 66} [35-6]
_________ **54**, 899–904 (1938): {SP 28} [38-12]

Phys. Rev. Lett. **15**, 868–870 (1965): {SP 76} [65-16]
_____________ **58**, 365–368 (1987): {SP 81} [87-1]
_____________ **59**, 225–227 (1987): {SP 79} [87-16]

Physiol. Rev. **23**, 203–219 (1943): {SP 91} [43-2]

Proc. Am. Phil. Soc. **96**, 556–565 (1952): {SP 123} [52-19]

Proc. Natl. Acad. Sci. USA **12**, 32–35 (1926): {SP 63} [26-3]
_______________________ **14**, 359–362 (1928): {SP 2} [28-1]
_______________________ **14**, 603–606 (1928): {SP 43} [28-7]
_______________________ **16**, 123–129 (1930): {SP 44} [30-3]
_______________________ **20**, 340–345 (1934): {SP 52} [34-6]
_______________________ **21**, 186–191 (1935): {SP 82} [35-5]
_______________________ **22**, 210–216 (1936): {SP 83} [36-4]
_______________________ **22**, 439–447 (1936): {SP 95} [36-7]
_______________________ **23**, 615–620 (1937): {SP 17} [37-8]
_______________________ **25**, 577–582 (1939): {SP 68} [39-7]
_______________________ **37**, 205–210 (1951): {SP 97} [51-9]
_______________________ **37**, 235–240 (1951): {SP 98} [51-10]
_______________________ **37**, 241–250 (1951): {SP 99} [51-11]
_______________________ **37**, 251–256 (1951): {SP 100} [51-12]

Proc. Natl. Acad. Sci. USA **37**, 256–261 (1951): {SP 101} [51-13]
____ **37**, 261–271 (1951): {SP 102} [51-14]
____ **37**, 272–281 (1951): {SP 103} [51-15]
____ **37**, 282–285 (1951): {SP 104} [51-16]
____ **38**, 112–118 (1952): {SP 37} [52-6]
____ **39**, 84–97 (1953): {SP 116} [53-9]
____ **39**, 253–256 (1953): {SP 109} [53-11]
____ **39**, 551–560 (1953): {SP 69} [53-12]
____ **44**, 211–216 (1958): {SP 70} [58-3]
____ **45**, 54–69 (1959): {SP 128} [59-12]
____ **46**, 1349–1360 (1960): {SP 118} [60-5]
____ **54**, 989–994 (1965): {SP 75} [65-15]
____ **60**, 59–65 (1968): {SP 78} [68-2]
____ **60**, 362–367 (1968): {SP 56} [68-7]
____ **67**, 1643–1648 (1970): {SP 136} [70-14)
____ **73**, 3685–3689 (1976): {SP 140} [76-17]
____ **75**, 12–15 (1978): {SP 21} [78-4]
____ **82**, 8284–8285 (1985): {SP 34} [85-13]
____ **86**, 3466–3468 (1989): {SP 132} [89-12]
____ **86**, 6835–6837 (1989): {SP 133} [89-13]
____ **87**, 7245–7249 (1990): {SP 142} [90-8]

Proc. Roy. Soc. Lond. *A* **114**, 181–211 (1927): {SP 64} [27-5]
____ *A* **196**, 343–362 (1949): {SP 31} [49-7]
____ *B* **141**, 10–20 (1953): {SP 106} [53-1]
____ *B* **141**, 21–33 (1953): {SP 107} [53-2]

Proc. Rudolf Virchow Med. Soc., NY **21**, 131–140 (1963): {SP 126} [63-10]

Rev. Mod. Phys. **20**, 112–122 (1948): {SP 30} [48-6]

Science **92**, 77–79 (1940): {SP 112} [40-3]
____ **110**, 543–548 (1949): {SP 122} [49-16]
____ **114**, 629–634 (1951): {SP 85} [51-18]
____ **128**, 1183–1186 (1958): {SP 127} [58-21]
____ **134**, 15–21 (1961): {SP 129} [61-9]
____ **150**, 297–305 (1965): {SP 77} [65-19]
____ **160**, 265–271 (1968): {SP 134} [68-4]

Svensk Kemisk Tidskrift **69**, 509–523 (1957): {SP 125} [57-5]

Texas Reports on Biology and Medicine **40**, 1–7 (1980-1981): {SP 87} [80-16]

Trans. Faraday Soc. **31**, 939–945 (1935): {SP 12} [35-1]

Trends in Biochemical Sciences **4**, 270–271 (1979): {SP 131} [79-13]

Vitalstoffe-Zivilisationskrankheiten **1/68**, 3–5 (1968): {SP 135} [68-14]

Z. Kristall. **63**, 502–506 (1926): {SP 41} [26-6]
__________ **68**, 239–256 (1928): {SP 42} [28-6]
__________ **74**, 213–225 (1930): {SP 45} [30-5]
__________ **75**, 128–142 (1930): {SP 46} [30-9]
__________ **81**, 1–29 (1932): {SP 65} [32-5]
__________ **84**, 204–212 (1933): {SP 50} [33-1]
__________ **84**, 442–452 (1933): {SP 47} [33-3]
__________ **85**, 380–391 (1933): {SP 36} [33-9]
__________ **87**, 205–238 (1934): {SP 11} [34-2]
__________ **88**, 54–62 (1934): {SP 51} [34-4]

Appendix III

List of Scientific Publications of Linus Pauling

The following list is derived from the complete list of Linus Pauling's publications by Zelek S. Herman and Dorothy B. Munro (1996) with additions by Special Collections, Oregon State University. The Herman-Munro-SCOSU list was downloaded from the SCOSU web site. As noted earlier, in this list each publication is assigned a number of the format [*y*-*n*], where *y* is the year of the publication (with leading "19" suppressed) and *n* is the number of the publication in that year, counted sequentially by date of publication. To produce the list presented here, the original list has been winnowed as follows. (1) Only publications judged to be scientific in nature (as distinct from publications on social, humanitarian, and political subjects) are retained. (2) Duplications, either by reprinting or by translation into foreign languages, are omitted. (3) Generally only first editions of books are retained, later editions being treated as duplications, except that both the first and third editions of *The Nature of the Chemical Bond* are listed ([39-8] and [60-26]), and the "Updated and Expanded Edition" of *Cancer and Vitamin C* [93-2] is retained as well as the original edition [79-15]. (4) Newspaper and magazine articles and letters to editors are omitted, except for those in scientific magazines such as *Chemical & Engineering News*, *Scientific American*, and *Physics Today*. (5) Book editorships, forewords, and prefaces are omitted. These steps reduced the list from 1216 original entries to 828 entries in the list below. If winnowings (2) and (3) were not applied, the count would rise to about 850, probably somewhat more. The above winnowing and the grouping described below were done by the Chief Editor, with data processing by Anthony Pauling Kamb.

Instead of the purely chronological sequence of entries (by Herman-Munro publication number) in the original list, the entries are here gathered into 20 Groups, based on subject matter as in the division of the selected papers into Chapters 1–17 in the main text (Parts I–IV). The entries are then placed in chronological order within each Group in accordance with the Herman-Munro publication numbers. Groups 1–17 are the same scientific subject-matter categories, one for one, as Chapters 1–17 in the main text. Groups 18 and 19 contain publications of two special types that are not represented in the main text: Group 18 contains biographical memoirs, memorials, and tributes, and Group 19 contains book reviews. Group 20 is the broad category "other"—chemistry in general and in relation to medicine and other scientific subjects, historical retrospectives and reminiscences, chemistry education, and subjects that could not be placed well in any of the 17 main subject-matter Groups.

Where appropriate and feasible, some of the 17 groups are divided into subgroups: thus Group 5 is divided into Groups 5a and 5b, Group 6 into Groups 6a, 6b, and 6c, and so on, as shown in the Table of Contents below. The table also gives the number of entries in each Group, enclosed in angle brackets, thus: <no.>. As in the division of the selected papers into Chapters 1–17 by subject matter (see the General Introduction), some of the boundaries between Groups are somewhat fuzzy and the assignment of a given paper to one Group or another is sometimes correspondingly ambiguous or uncertain. Two examples are (1) papers on complex ions, metal complexes, and cluster compounds (e.g. papers [48-10], [50-6], [56-3], [76-24], [77-13], and [92-18]) which may, depending on the emphasis in each paper, belong in Group 1, 2, 3, 5a, 5b, or 6c; and (2) papers on acid strengths (e.g. [52-5], [53-13], [56-2], [76-23]), which may belong in Group 1, 2, 4, or 8. Group 1 is particularly large and diverse, and would be desirably divided into subgroups, but its subject matter is so intertwined that a good scheme of subdivision was not found. The papers of Group 5b, containing an emphasis on x-ray diffraction methodology, generally contain also important crystal-structure data and results, and are therefore not sharply distinct from Group 5a (Ionic Crystal Structures). Thus [28-6] and [33-3] are placed in Group 5a even though they have important methodological components, and [30-9] is placed in Group 5b even though it treats an interesting crystal structure. Likewise, electron diffraction methodology is not in general treated separately from molecular-structure results in the papers, so that a subdivision into Groups 7a and 7b is not made. The placement of methodologically-oriented papers in Group 5b, seemingly close to Group 5a and far from Groups 6a, 6b, and 6c, is not significant and merely reflects the fact that the outstanding papers with a focus on x-ray methodology happen to be papers on ionic crystals (SP 42, SP 46, SP 47).

The format of the literature citations in Appendix III is slightly different from that in Parts I–IV as in the following example: the journal citation "*Science* **110**, 543–548 (1949)" in the main text (Parts I–IV) will appear in Appendix III as "*Science* **110** (1949): 543–548." Also, the byline (for multi-author papers), which in Parts I–IV would precede the journal citation, would in Appendix III follow the citation. These differences in format have no significance.

It is hoped that the present list of Pauling's scientific publications will be useful to the reader by allowing the selected papers to be examined in the larger context of the full extent of Pauling's published scientific contributions, grouped by subject matter in a meaningful way. Each selected paper is marked by a leading asterisk, and its SP number is given in curly brackets {SP no.} at the end of the entry.

List of Scientific Publications of Linus Pauling Arranged in Subject-Matter Groups

CONTENTS

The numbers in angle brackets <no.> are the number of papers in each Group. TOTAL = <828>

Group 1

Covalent Bonding, Resonance, and Bond-Orbital Hybridization

[26-4] The dynamic model of the chemical bond and its application to the structure of benzene. *J. Am. Chem. Soc.* 48 (1926): 1132–1143.

* [28-1] The shared-electron chemical bond. *Proc. Natl. Acad. Sci.* 14 (1928): 359–362. {SP 2}

* [28-5] The application of the quantum mechanics to the structure of the hydrogen molecule and hydrogen molecule-ion and to related problems. *Chem. Rev.* 5 (1928): 173–213. {SP 3}

* [31-2] Quantum mechanics and the chemical bond. *Phys. Rev.* 37 (1931): 1185–1186. {SP 4}

* [31-3] The nature of the chemical bond. Application of results obtained from the quantum mechanics and from a theory of paramagnetic susceptibility to the structure of molecules. *J. Am. Chem. Soc.* 53 (1931): 1367–1400. {SP 5}

* [31-4] The nature of the chemical bond. II. The one-electron bond and the three-electron bond. *J. Am. Chem. Soc.* 53 (1931): 3225–3237. {SP 6}

[32-8] Interatomic distances in covalent molecules and resonance between two or more Lewis electronic structures. *Proc. Natl. Acad. Sci.* 18 (1932): 293–297.

[32-9] The additivity of the energies of normal covalent bonds. *Proc. Natl. Acad. Sci.* 18 (1932): 414–416. (Linus Pauling and Don M. Yost).

[32-12] The resonance of molecules among several electronic structures. *The Nucleus (Northeastern Section, Am. Chem. Soc.)* 9 (May 1932): 183–184.

[33-4] The normal state of the helium molecule-ions He_2^+ and He_2^{++}. *J. Chem. Phys.* 1 (1933): 56–59.

* [33-10] The calculation of matrix elements for Lewis electronic structures of molecules. *J. Chem. Phys.* 1 (1933): 280–283. {SP 7}

* [33-11] The nature of the chemical bond. V. The quantum-mechanical calculation of the resonance energy of benzene and naphthalene and the hydrocarbon free radicals. *J. Chem. Phys.* 1 (1933): 362–374. (Linus Pauling and G. W. Wheland). {SP 8}

[33-12] Errata in [33-11]. *J. Chem. Phys.* 2 (1934): 482.

* [33-13] The nature of the chemical bond. VI. The calculation from thermochemical data of the energy of resonance of molecules among several electronic structures. *J. Chem. Phys.* 1 (1933): 606–617. (Linus Pauling and J. Sherman). {SP 9}

* [33-14] The nature of the chemical bond. VII. The calculation of resonance energy in conjugated systems. *J. Chem. Phys.* 1 (1933): 679–686. (Linus Pauling and J. Sherman). {SP 10}

* [34-2] Covalent radii of atoms and interatomic distances in crystals containing electron-pair bonds. *Z. Kristall.* 87 (1934): 205–238. (Linus Pauling and M. L. Huggins). {SP 11}

* [35-1] A wave-mechanical treatment of the Mills-Nixon effect. *Trans. Faraday Soc.* 31 (1935): 939–945. (L. E. Sutton and Linus Pauling). {SP 12}

[35-3] Remarks on the theory of aromatic free radicals. *J. Chem. Phys.* 3 (1935): 315. (Linus Pauling and G. W. Wheland).

* [35-7] A quantum mechanical discussion of orientation of substituents in aromatic molecules. *J. Am. Chem. Soc.* 57 (1935): 2086–2095. (G. W. Wheland and Linus Pauling). {SP 13}

* [35-13] The dependence of interatomic distance on single bond-double bond resonance. *J. Am. Chem. Soc.* 57 (1935): 2705–2709. (Linus Pauling , L. O. Brockway, and J. Y. Beach). {SP 14}

* [35-17] *Introduction to Quantum Mechanics, with Applications to Chemistry.* McGraw-Hill, New York, (1935), 468 pp. (Linus Pauling and E. Bright Wilson, Jr.). {SP 15}

* [37-7] A quantitative discussion of bond orbitals. *J. Am. Chem. Soc.* 59 (1937): 1450–1456. (Linus Pauling and J. Sherman). {SP 16}

* [37-8] The structure of cyameluric acid, hydromelonic acid, and related substances. *Proc. Natl. Acad. Sci.* 23 (1937): 615–620. (Linus Pauling and J. H. Sturdivant). {SP 17}

[38-6] The significance of resonance to the nature of the chemical bond and the structure of molecules. In: *Organic Chemistry, An Advanced Treatise*, Vol. 2, Henry Gilman, Roger Adams, Homer Adkins, Hans T. Clarke, Carl S. Marvel, and Frank C. Whitmore, eds. John Wiley and Sons, New York, (1938), pp. 1943–1983.

* [39-8] *The Nature of the Chemical Bond and the Structure of Molecules and Crystals: An Introduction to Modern Structural Chemistry, 1st edn.* Cornell University Press, Ithaca, NY, (1939), 419 pp. {SP 22}

[46-4] Theory of Resonance (*Jackson Laboratory Lecture Series*, Lecture No. 6) (Wilmington, DE: E. I. Dupont de Nemours & Co., February 8, 1946): 1–13.

[46-10] Modern structural chemistry. (Acceptance speech for the Willard Gibbs Medal, awarded 14 June 1946 by the Chicago Section of the American Chemical Society). *Chem. & Eng. News* 24 (10 July 1946): 1788–1789.

[48-3] The valences of transition elements, in *Contribution à l'Étude de la Structure Moléculaire* (Victor Henri Memorial Volume), Desoer, Liège, (1948), pp. 1–14.

[48-10] La Structure des complexes et l'influence de cette structure sur les réactions d'echange. (French: The structure of complexes and the influence of this structure on exchange reactions.) *Colloques Internationaux du Centre National de la Recherche Scientifique* 5 (1948): 142–146.

* [48-17] The modern theory of valency. (Liversidge Lecture, delivered before the Chemical Society, London, 3 June 1948.) *J. Chem. Soc.* (1948): 1461–1467. {SP 18}

[49-13] The valence-state energy of the bivalent oxygen atom. *Proc. Natl. Acad. Sci.* 35 (1949): 229–232.

[49-15] On the stability of the S_8 molecule and the structure of fibrous sulfur. *Proc. Natl. Acad. Sci.* 35 (1949): 495–499.

[51-7] Quantum theory and chemistry. *Science* 113 (1951): 92–94.

[52-1] Bond orbitals and bond energy in elementary phosphorus. *J. Chem. Phys.* 20 (1952): 29–34. (Linus Pauling and Massimo Simonetta).

* [52-5] Interatomic distances and bond character in the oxygen acids and related substances. *J. Phys. Chem.* 56 (1952): 361–365. {SP 19}

[52-11] Resonance in the hydrogen molecule. *J. Chem. Phys.* 20 (1952): 1041.

[52-15] The structural chemistry of molybdenum. In: *Molybdenum Compounds, Their Chemistry and Technology*, D. H. Killeffer and Arthur Linz, eds., (New York: Interscience Publishers, 1952), pp. 95–109.

[53-13] The strengths of the oxygen acids. *School Science and Mathematics* 53 (1953): 429–435.

[53-33] Theory of resonance, in *Encyclopædia Britannica,* Vol. 19, William Benton, Chicago, (1953), pp. 210–213.

[53-34] Valence. In: *Encyclopædia Britannica,* Vol. 22, William Benton, Chicago, (1953), pp. 944–947.

[54-7] The dependence of bond energy on bond length. (Debye 70th Birthday Symposium, Cornell University, Ithaca, NY.) *J. Phys. Chem.* 58 (1954): 662–666.

* [55-14] Modern structural chemistry (Nobel Lecture, 11 December 1954). In: *Les Prix Nobel en 1954*, M. G. Liljestrand, ed, A. Norstedt och Söner, Stockholm, (1955), pp. 91–99. {SP 1}

[56-5] A simple theoretical treatment of alkali halide gas molecules. *Proc. Natl. Acad. Sci. (India)* 25 (1956): 1–5.

[56-11] The nature of the theory of resonance, in *Perspectives in Organic Chemistry*, Sir Alexander Todd, ed. Interscience, New York, (1956), 1–8.

[57-9] The use of atomic radii in the discussion of interatomic distances and lattice constants of crystals. *Acta Cryst.* 10 (1957): 685–687.

[59-3] Kekulé and the chemical bond. In: *Theoretical Organic Chemistry* (Papers presented to the Kekulé Symposium organized by the Chemical Society, London, September 1958. Butterworths Scientific Publications, London (1959), pp. 1–8.

[59-15] Quantum theory and chemistry. In: *Max Planck Festschrift*, W. Frank, ed., Deutscher Verlag der Wissenschaften, Berlin, (1959), pp. 385–388.

* [60-26] *The Nature of the Chemical Bond and the Structure of Molecules and Crystals: An Introduction to Modern Structural Chemistry, 3rd edn.* Cornell University Press, Ithaca, NY, (1960), 644 pp. {SP 35}

[62-1] The carbon-carbon triple bond and the nitrogen-nitrogen triple bond. *Tetrahedron* 17 (1962): 229–233.

[62-14] Valence bond theory in coordination chemistry. *J. Chem. Ed.* 39 (1962): 461–463.

[62-15] Teoriia rezonansa v khimii (Russian: The theory of resonance in chemistry). Zhurznal Vsesotuznogo Khimicheskogo Obshchestva im. L. L Mendeleeva (J. Mendeleev All-Union Chemical Society) 7(4) (1962): 462–466.

[64-2] The electroneutrality principale (sic) and the structure of molecules. *Anales de la Real Sociedad Espanola de Fisica y Quimica* (Madrid) B60(2–3) (1964): 87–90.

[64-4] The structure of methylene and methyl. In: *Molecular Orbitals in Chemistry, Physics, and Biology*, Per-Olov Löwdin and Bernard Pullman, eds., Academic Press, New York, (1964), pp. 207–213.

[64-17] The architecture of molecules. *Proc. Natl. Acad. Sci.* 51 (1964): 977–984.

[64-22] The energy of transargononic bonds. In: *The Law of Mass-Aciion, A Centenary Volume 1864–1964* (Det Norske Videnskeps-Akademi i Oslo), Universitet forlaget, Oslo, (1964), pp. 151–158.

[67-9] *The Chemical Bond.* Cornell Univ. Press, Ithaca, NY, (1967), 267 pp.

[69-4] Structure of the methyl radical and other radicals. *J. Chem. Phys.* 51 (1969): 2767–2769.

* [70-3] Five equivalent *d* orbitals. *J. Chem. Ed.* 47 (1970): 15–17. (Linus Pauling and Vance McClure). {SP 20}

[70-10] Octahedral hybrid orbitals. *Chem. Phys. Lett.* 6 (1970): 249.

[70-13] The composition of the boranes. *J. Inorg. Nucl. Chem.* 32 (1970): 3745–3749.

[71-3] Discussion on the chemical bond. *Physics Today* 24(2) (Feb 1971): 9, 11, 13.

[72-7] Five equivalent *d* orbitals: a continuum of sets. *Israel J. Chem.* 10 (1972): 211–220. (I. T. Keaveny and Linus Pauling).

[73-13] Hybrid bond orbitals. In: *Wave Mechanics*, W. C. Price, S. S. Chissick, and T. Ravensdale, eds., Wiley, New York, (1973), pp. 88–97. (Linus Pauling and Ian Keavenyl)

[74-7] Kagaku ketsugö to kakyō (Japanese: The chemical bond and crosslinking). Purasuchikkusu (Japan Plastics), 25(5) (1974): 1–2.

[76-3] Correlation of nonorthogonality of best hybrid bond orbitals with bond strength of orthogonal orbitals. *Proc. Natl. Acad. Sci.* 73 (1976): 274–275.

[76-8] Angles between orthogonal *spd* bond orbitals with maximum strength. *Proc. Natl. Acad. Sci.* 73 (1976): 1403–1405.

[76-15] The birth of quantum mechanics. *Trends in Biochemical Sciences* 1(9) (Sept 1976): 214–215.

[77-8] The theory of resonance in chemistry. *Proc. Roy. Soc. Lond.* A356 (1977): 433–441.

* [78-4] Bond angles in transition-metal tricarbonyl compounds: a test of the theory of hybrid bond orbitals. *Proc. Natl. Acad. Sci.* 75 (1978): 12–15. {SP 21}

[78-5] Bond angles in transition metal tetracarbonyl compounds: a further test of the theory of hybrid orbitals. *Proc. Natl. Acad. Sci.* 75 (1978): 569–572.

[78-15] The nature of the bonds formed by the transition metals with hydrogen, carbon, and phosphorus. *Acta Cryst.* B34 (1978): 746–752.

[78-22] The nature of the bonds formed by transition metals in bioorganic compounds and other compounds. In: *Frontiers in Bioorganic Chemistry and Molecular Biology*, Yu. A. Ovchinnikov and M. N. Kolosov, eds., Elsevier/North Holland Medical Press, Amsterdam, (1979), pp. 1–20.

[80-6] Coulson's Valence. *Nature* 284 (1980): 685–686.

[80-11] Bond numbers and bond lengths in tetrabenzo (*de, no, st,* c_1, d_1) heptacene and other condensed aromatic hydrocarbons: a valence bond treatment. *Acta Cryst.* B36 (1980): 1898–1901.

[82-1] Reliability of the pair-defect-sum approximation for the strength of valence-bond orbitals. *Proc. Natl. Acad. Sci.* 79 (1982): 1361–1365. (Linus Pauling, Zelek S. Herman, and Barclay J. Kamb).

[82-11] The development of the theory of molecular structure and the nature of the chemical bond. *J. Mol. Sci.* (Fenzi Kexue Xuebao) (Wuhan, China) 2 (4), 1–8 (1982).

[83-6] On the nature of the bonding in Cu_2 — A comment. *J. Chem. Phys.* 78 (1983), 3346.

[83-8] Resonance of an unshared electron pair between two atoms connected by a single bond. *Proc. Natl. Acad. Sci.* 80 (1983): 3871–3872.

[83-16] Electronic structure and bonding of Si{100} surfaces. *Phys. Rev.* B28 (1983): 6154–6156. (Linus Pauling and Zelek S. Herman).

[84-4] Evidence from bond lengths and bond angles for enneacovalence of cobalt, rhodium, iridium, iron, ruthenium, and osmium in compounds with elements of medium electronegativity. *Proc. Natl. Acad. Sci.* 81 (1984): 1918–1921.

[84-11] Hybrid bond orbitals and bond strengths for pentacovalent bonding. *Croatica Chimica Acta* 57 (1984): 765–778. (Zelek S. Herman and Linus Pauling).

[85-4] The origins of bonding concepts. *J. Chem. Ed.* 62 (1985): 362.

[85-12] Why modern chemistry is quantum chemistry. *New Scientist* 108 (1481) (7 November 1985): 54–55.

[87-3] Electronic structure of the benzene molecule. *Nature* 325 (1987): 396.

[87-24] Introduction: *Modelling of Structure and Properties of Molecules*, Z. B. Maksic, ed., Ellis Horwood Ltd., Chichester, England, (1987), pp. 3–4.
[89-5] Principles determining the structure of high-pressure forms of metals: The structures of cesium(IV) and cesium(V). *Proc. Natl. Acad. Sci.* 86 (1989): 1431–1433.
[92-2] X-ray crystallography and the nature of the chemical bond. In: *The Chemical Bond: Structure and Dynamics*, Ahmed Zewail, ed., Academic Press, New York, (1992), pp. 3–16.
[92-13] The nature of the chemical bond—1992. *J. Chem. Ed.* 69 (1992): 519–522.
[92-18] Molecular structure of Ti_8C_{12} and related complexes. *Proc. Natl. Acad. Sci.* 89 (1992): 8175–8176.

Group 2

Ionic Bonding, Partial Ionic Character, and Electronegativity

[26-2] The prediction of the relative stabilities of isosteric isomeric ions and molecules. *J. Am. Chem. Soc.* 48 (1926): 641–651. (Linus Pauling and Sterling B. Hendricks).
* [27-4] The sizes of ions and the structure of ionic crystals. *J. Am. Chem. Soc.* 49 (1927): 765–790. {SP 23}
[28-2] The sizes of ions and their influence on the properties of salt-like compounds. *Z. Kristall.* 67 (1928): 377–404.
[28-3] The influence of relative ionic sizes on the properties of ionic compounds. *J. Am. Chem. Soc.* 50 (1928): 1036–1045, photocopy.
[28-9] The coordination theory of the structure of ionic crystals. In: *Festschrift zum 60. Geburtstage Arnold Sommerfelds*, Verlag von S. Hirzel, Leipzig, (1928), pp. 11–17.
[28-10] Note on the pressure transitions of the rubidium halides. *Z. Kristall.* 69 (1928): 35–40.
* [29-1] The principles determining the structure of complex ionic crystals. *J. Am. Chem. Soc.* 51 (1929): 1010–1026. {SP 24}
* [32-2] The nature of the chemical bond. III. The transition from one extreme bond type to another. *J. Am. Chem. Soc.* 54 (1932): 988–1003. {SP 25}
* [32-11] The nature of the chemical bond. IV. The energy of single bonds and the relative electronegativity of atoms. *J. Am. Chem. Soc.* 54 (1932): 3570–3582. {SP 26}

[33-6] The formulas of antimonic acid and the antimonates. *J. Am. Chem. Soc.* 55 (1933): 1895–1900. Errata: *J. Am. Chem. Soc.* 55 (1933): 3052.

[35-15] Ionic and Atomic Radii. In: *Internationale Tabellen zur Bestimmung von Kristallstrukturen* (International Tables for the Determination of Crystal Structures), Vol. 2, Gebrüder Borntraeger, Berlin, (1935), pp. 610–616.

[55-6] The energy change in organic rearrangements and the electronegativity scale. In: *Biochemistry of Nitrogen.* Suomalaisen Tiedeakatemian, Helsinki, (1955), pp. 428–432.

[56-2] Why is hydrofluoric acid a weak acid? *J. Chem. Ed.* 33 (1956): 16–17.

[70-17] Ionic character of bonds in crystals. *Science* 170 (1970): 1432.

[75-5] The relative stability of isosteric ions and molecules. *J. Chem. Ed.* 52 (1975): 577.

[76-23] The strength of hydrohalogenic acids. *J. Chem. Ed.* 53 (1976): 762–763.

[80-14] The nature of silicon-oxygen bonds. *Am. Mineral* 65 (1980): 321–323.

[80-15] Soft-sphere ionic radii for alkali and halogenide ions. *J. Chem. Soc. Dalton Trans.* (1980), p. 645.

* [86-17] The nature of the chemical bond fifty years later: The relative electronegativity of atoms seen in perspective. In: *Chemical Bonding Models*, Joel F. Liebman and Arthur Greenburg, ed. (*Molecular Structure and Energetics*, Vol. 1), VCH Publishers, Deerfield Beach, FL, (1986), pp. 1–15. (Linus Pauling and Zelek S. Herman). {SP 27}

[87-26] Determination of ionic radii from cation-anion distances in crystal structures: Discussion. *Am. Mineral.* 72 (1987): 1016.

[88-11] The origin and nature of the electronegativity scale. *J. Chem. Ed.* 65 (1988): 375.

[89-10] Explanations of cold fusion. *Nature* 339 (1989): 105.

[92-8] New dimension for Mendeleev. *Nature* 357 (1992): 26–27.

Group 3

Metallic Bonding

* [38-12] The nature of the interatomic forces in metals. *Phys. Rev.* 54 (1938): 899–904. {SP 28}

[47-1] The nature of the bonds in metals and intermetallic compounds. *Proc. Intern. Congr. Pure and Applied Chem.* (London) 11 (1947): 249–257.

* [47-2] Atomic radii and interatomic distances in metals. *J. Am. Chem. Soc.* 69 (1947): 542–553. {SP 29}

* [48-6] The ratio of valence electrons to atoms in metals and intermetallic compounds. *Rev. Mod. Phys.* 20 (1948): 112–122. (Linus Pauling and Fred J. Ewing). {SP 30}

[48-9] The nature of the bonds in the iron silicide FeSi and related crystals. *Acta Cryst.* 1 (1948): 212–216. (Linus Pauling and A. M. Soldate).

[48-14] The metallic state. *Nature* 161 (1948): 1019.

[49-5] The resonating-valence-bond theory of metals. *Physica* 15 (1949): 23–28.

* [49-7] A resonating-valence-bond theory of metals and intermetallic compounds. *Proc. Roy. Soc.* 196 (1949): 343–362. {SP 31}

[49-10] La Valence des métaux et la structure des composés intermétalliques. (French: The valence of metals and the structure of intermetallic compounds). *J. Chem. Phys.* 46 (1949): 276–287.

[50-11] Electron transfer in intermetallic compounds. *Proc. Natl. Acad. Sci.* 36 (1950): 533–538.

[51-1] Interatomic distances in Co_2Al_9. *Acta Cryst.* 4 (1951): 138–140.

[56-3] The electronic structure of metals and alloys. In: *Theory of Alloy Phases*, American Society for Metals, Cleveland (1956), pp. 220–242.

[56-8] On the valence and atomic size of silicon, germanium, arsenic, antimony, and bismuth in alloys. *Acta Cryst.* 9 (1956): 127–130. (Linus Pauling and Peter Pauling).

[57-4] A set of effective metallic radii for use in compounds with the β-wolfram structure. *Acta Cryst.* 10 (1957): 374–375.

[59-13] The discussion of tetragonal boron by the resonating-valence-bond theory of electron-deficient substances. *Z. Kristall.* 112 (1959): 472–478. (Linus Pauling and Barclay Kamb).

[61-2] Nature of the metallic orbital. *Nature* 189 (1961): 656.

[61-3] The nature of the metallic orbital and the structure of metals. *J. Indian Chem. Soc.* 38(8) (1961): 435–437.

* [68-1] The dependence of bond lengths in intermetallic compounds on the hybrid character of the bond orbitals. *Acta Cryst.* B24 (1968): 5–7. {SP 32}

[75-7] Maximum-valence radii of transition metals. *Proc. Natl. Acad. Sci.* 72 (1975): 3799–3801.

[75-8] Valence-bond theory of compounds of transition metals. *Proc. Natl. Acad. Sci.* 723 (1975): 4200–4202.

[76-24] Metal-metal bond lengths in complexes of transition metals. *Proc. Natl. Acad. Sci.* 73 (1976): 4290–4293.

[77-6] Covalence of atoms in the heavier transition metals. *Proc. Natl. Acad. Sci.* 74 (1977): 2614–2615.

[77-13] Structure of transition-metal cluster compounds: the use of an additional orbital resulting from the f, g character of spd bond orbitals. *Proc. Natl. Acad. Sci.* 74 (1977): 5235–5238.

[81-4] Electron transfer and the valence states of cerium and platinum in cubic Friauf-Laves compounds with the platinum metals. *Phys. Rev. Lett.* 47 (1981): 277–281.

[84-10] Valence-bond concepts in coordination chemistry and the nature of metal-metal bonds. *J. Chem. Ed.* 61 (1984): 582–587. (Linus Pauling and Zelek S. Herman).

* [84-12] The metallic orbital and the nature of metals. *J. Solid State Chem.* 54 (1984): 297–307. {SP 33}

[85-11] The chemistry of metals and alloys. *The Chemist* 62 (1985): 20.

* [85-13] Extension of the statistical theory of resonating valence bonds to hyperelectronic metals. *Proc. Natl. Acad. Sci.* 82 (1985): 8284–8285. (Barclay Kamb and Linus Pauling). {SP 34}

[85-14] Comparison of theoretical and experimental values of the number of metallic orbitals per atom in hypoelectronic and hyperelectronic metals. *Proc. Natl. Acad. Sci.* 82 (1985): 8286–8287. (Linus Pauling and Barclay Kamb).

[86-11] A revised set of values of single-bond radii derived from the observed interatomic distances in metals by correction for bond number and resonance energy. *Proc. Natl. Acad. Sci.* 83 (1986): 3569–3571. (Linus Pauling and Barclay Kamb).

[87-25] Recent advances in the unsynchronized-resonating-covalent-bond theory of metals, alloys, and intermetallic compounds and its application to the investigation of structures and properties of such systems. In: *Modelling of Structure and Properties of Molecules*, Z. B. Maksic, ed., Ellis Horwood Ltd., Chichester, England, (1987), pp. 5–37. (Linus Pauling and Zelek S. Herman).

[88-15] The unsynchronized-resonating-covalent-bond theory of the structure and properties of boron and the boranes. In: *Advances In Boron and the Boranes (Molecular Structure and Energetics, Vol. 5*), Joel F. Liebman, Arthur Greenberg, and Robert F. Williams, eds., VCH Publishers, New York, (1988), pp. 517–529. (Linus Pauling and Zelek S. Herman).

[89-17] The nature of metals. *Pure & Appl. Chem.* 61 (1989): 2171–2174.

[90-2] The unsynchronized-resonating-covalent-bond theory of metals, alloys, and intermetallic compounds. In: *Valence Bond Theory and Chemical Structure* (*Studies in Physical and Theoretical Chemistry* 64), D. J. Klein and N. Trinajstić, eds., Elsevier Science Publishers B. V., Amsterdam, (1990), pp. 569–610. (Linus Pauling and Zelek S. Herman).

Group 4

Hydrogen Bonding

* [33-9] The crystal structure of ammonium hydrogen fluoride, NH_4HF_2. *Z. Kristall.* 85 (1933): 380–391. {SP 36}
* [52-6] The structure of chlorine hydrate. *Proc. Natl. Acad. Sci.* 38 (1952): 112–118. (Linus Pauling and Richard E. Marsh). {SP 37}
[53-31] Ice, in *Encyclopaedia Britannica*, Vol. 12, William Benton, Chicago, (1953), pp. 39–40.
* [59-2] The structure of water. In: *Hydrogen Bonding.* D. Hadzi and H. W. Thompson, eds., Pergamon Press, New York, (1959), pp. 1–6. {SP 38}
[92-19] The role of hydrogen bonds in determining the composition and structure of oxide minerals. *Trends in Mineral.* 1 (1992): 1–3.

Group 5a

Ionic Crystal Structures

[24-1] The crystal structures of ammonium fluoferrate, fluo-aluminate and oxyfluomolybdate. *J. Am. Chem. Soc.* 46 (1924): 2738–2751.
[24-2] The crystal structure of uranyl nitrate hexahydrate. *J. Am. Chem. Soc.* 46 (1924): 1615–1622. (Linus Pauling and Roscoe G. Dickinson).
* [25-1] The crystal structures of hematite and corundum. *J. Am. Chem. Soc.* 47 (1925): 781–790. (Linus Pauling and Sterling B. Hendricks). {SP 39}
* [25-2] The crystal structure of barite. *J. Am. Chem. Soc.* 47 (1925): 1026–1030. (Linus Pauling and Paul H. Emmett). {SP 40}
[25-3] The crystal structures of cesium tri-iodide and cesium dibromo-iodide. *J. Am. Chem. Soc.* 47 (1925): 1561–1571. (Richard M. Bozorth and Linus Pauling).
[25-7] The crystal structures of sodium and potassium trinitrides and potassium cyanate and the nature of the trinitride group. *J. Am. Chem. Soc.* 47 (1925): 2904–2920. (Sterling B. Hendricks and Linus Pauling).
* [26-6] Über die Kristallstruktur der kubischen Tellursäure (German: On the structure of cubic telluric acid). *Z. Kristall.* 63 (1926): 502–506. (L. Merle Kirkpatrick and Linus Pauling). {SP 41}

[26-7] Die Struktureinheit und Raumgruppensymmetrie von β-Aluminiumoxyd (German: The structural unit and space group symmetry of β-aluminum oxide). *Z. Kristall.* 64 (1926): 303–308. (S. B. Hendricks and Linus Pauling).

[28-4] The crystal structure of potassium chloroplatinate. *Z. Kristall.* 68 (1928): 223–230. (F. J. Ewing and Linus Pauling).

* [28-6] The crystal structure of brookite. *Z. Kristall.* 68 (1928): 239–256. (Linus Pauling and J. H. Sturdivant). {SP 42}

* [28-7] The crystal structure of topaz. *Proc. Natl. Acad. Sci.* 14 (1928): 603–606. {SP 43}

[29-2] The crystal structure of the *A*-modification of the rare earth sesquioxides. *Z. Kristall,* 69 (1929): 415–421.

[29-3] Note on the paper of A. Schröder: Beiträge zur Kenntnis des Feinbaues des Brookits usw. *Z. Kristall.* 69 (1929): 557–559. (J. H. Sturdivant and Linus Pauling).

[29-5] On the crystal structure of the chlorides of certain bivalent elements. *Proc. Natl. Acad. Sci.* 15 (1929): 709–712.

[29-6] The molecular structure of the tungstosilicates and related compounds. *J. Am. Chem. Soc.* 51 (1929): 2868–2880.

[30-1] On the crystal structure of nickel chlorostannate hexahydrate. *Z. Kristall.* 72 (1930): 482–492.

[30-2] The crystal structure of pseudobrookite. *Z. Kristall.* 73 (1930): 97–112.

* [30-3] The structure of the micas and related minerals. *Proc. Natl. Acad. Sci.* 16 (1930): 123–129. {SP 44}

[30-4] Note on the lattice constant of ammonium hexafluoaluminate. *Z. Kristall.* 74 (1930): 104–105.

* [30-5] The structure of sodalite and helvite. *Z. Kristall.* 74 (1930): 213–225. {SP 45}

[30-6] The structure of some sodium and calcium aluminosilicates. *Proc. Natl. Acad. Sci.* 16 (1930): 453–459.

[30-8] Über die Kristallstruktur des Rubidiumazids (German: On the crystal structure of rubidium azide). *Z. Physik. Chem.* B8 (1930): 326–328.

[30-10] The structure of the chlorites. *Proc. Natl. Acad. Sci.* 16 (1930): 578–582.

[30-11] The crystal structure of cadmium chloride. *Z. Kristall.* 74 (1930): 546–551. (Linus Pauling and J. L. Hoard).

[32-3] The crystal structure of magnesium platinocyanide heptahydrate. *Phys. Rev.* 39 (1932): 537–538. (Richard M. Bozorth and Linus Pauling).

[33-2] Note on the crystal structure of rubidium nitrate. *Z. Kristall.* 84 (1933): 213–216. (Linus Pauling and J. Sherman).

* [33-3] The crystal structure of zunyite, $Al_{13}Si_5O_{20}(OH, F)_{18}Cl$. *Z. Kristall.* 84 (1933): 442–452. {SP 47}

[35-2] The unit of structure of telluric acid, $Te(OH)_6$. *Z. Kristall.* 91 (1935): 367–368.

[35-8] The crystal structure of swedenborgite, $NaBe_4SbO_7$. *Am. Mineral.* 20 (1935): 492–501. (Linus Pauling , H. P. Klug, and A. N. Winchell).

[37-1] The crystal structure of aluminum metaphosphate, $Al(PO_3)_3$, *Z. Kristall.* 96 (1937): 481–487. (Linus Pauling and J. Sherman).

[37-5] Report on the X-ray study of the isomorphism of compounds of bivalent europium and barium. *J. Am. Chem. Soc.* 59 (1937): 1132–1133.

[38-7] The crystal structure of cesium aurous auric chloride, $Cs_2AuAuCl_6$, and cesium argentous auric chloride, $Cs_2AgAuCl_6$. *J. Am. Chem. Soc.* 60 (1938): 1846–1851. (Norman Elliot and Linus Pauling).

[38-8] The structure of ammonium heptafluozirconate and potassium heptafluozirconate and the configuration of the heptafluozirconate group. *J. Am. Chem. Soc.* 60 (1938): 2702–2707. (G. C. Hampson and Linus Pauling).

[38-10] The crystal structure of ammonium cadmium chloride, NH_4CdCl_3. *J. Am. Chem. Soc.* 60 (1938): 2886–2890. (Henri Brasseur and Linus Pauling).

[41-3] The crystal structures of the tetragonal monoxides of lead, tin, palladium, and platinum. *J. Am. Chem. Soc.* 63 (1941): 1392–1394. (Walter J. Moore, Jr. and Linus Pauling).

[54-3] On the structure of the heteropoly anion in ammonium 9-molybdomanganate, $(NH_4)_6MnMo_90_{32}.8H_20$. *Acta Cryst.* 7 (1954): 438–441. (John L. T. Waugh, David P. Shoemaker, and Linus Pauling).

[68-8] Acceptance of the Roebling medal of the Mineralogical Society of America. *Am. Mineral.* 53 (1968): 521–530.

[79-20] The crystal structure of Li_3ThF_7. *Acta Cryst.* B35 (1979): 1535–1536.

[82-14] The crystal structure of lithiophorite. *Am. Mineral.* 67 (1982): 817–821. (Linus Pauling and Barclay Kamb).

[83-24] X-ray crystallography in the California Institute of Technology. In: *Crystallography in North America*, D. McLachlan, Jr., and J. P. Glusker, eds., American Crystallographic Association, New York, (1983), pp. 27–30.

[90-1] Determination of the crystal structure and composition of $Li_6Be_4OH_{12}$ by the stochastic method. *Proc. Natl. Acad. Sci.* 871 (1990): 244–245.

[90-9] The discovery of the structure of the clay minerals. *CMS News* (A Publication of The Clay Minerals Society), (September 1990): 25–27.

[94-5] Triethylsilyl cations. *Science* 263 (1994): 983.

Group 5b

X-ray Diffraction

[25-6] A new crystal for wave-length measurements of soft X-rays. *Proc. Natl. Acad. Sci.* 11 (1925): 445–447. (Linus Pauling and Albert Björkeson).

[25-8] Unusual X-ray reflections on spectral photographs (Pasadena meeting of the American Physical Society, 7 March 1925). *Phys. Rev.* 25 (1925): 715.

* [30-9] The crystal structure of bixbyite and the *C*-modification of the sesquioxides. *Z. Kristall.* 75 (1930): 128–142. (Linus Pauling and M. D. Shappell). {SP 46}

[31-1] The determination of crystal structure by X-rays. *Ann. Surv. Am. Chem.* 5 (1931): 118–125.

[32-4] The packing of spheres. *The Chemical Bulletin* (Chicago Section, *Am. Chem. Soc.*) 19 (February 1932): 35–38.

[32-7] The determination of crystal structure by X-rays. *Annu. Surv. of Am. Chem.* 6 (1932): 116–122.

[37-2] The X-ray analysis of crystals. *Current Science* (Special Number on "Laue Diagrams") (Jan 1937): 20–22.

[46-1] The use of punched cards in molecular structure determinations. I. Crystal structure calculations. *J. Chem. Phys.* 14 (1946): 648–658. (P. A. Shaffer, Jr., Verner Schomaker, and Linus Pauling).

[50-6] The determination of the structures of complex molecules and ions from X-ray diffraction by their solutions: The structures of the groups $PtBr_6^{--}$, $PtCl_6^{--}$, $Nb_6Cl_{12}^{++}$, $Ta_6Br_{12}^{++}$, and $Ta_6Cl_{12}^{++}$. *J. Am. Chem. Soc.* 72 (1950): 5477–5486. (Philip A. Vaughan, J. H. Sturdivant, and Linus Pauling).

[92-22] Crystallography. In: *Academic Press Dictionary of Science and Technology*, Christopher Morris, ed., Academic Press, Inc., San Diego, CA, (1992), p. 559.

Group 6a

Covalent Crystal Structures

[23-1] The crystal structure of molybdenite. *J. Am. Chem. Soc.* 45 (1923): 1466–1471. (Roscoe G. Dickinson and Linus Pauling).

[32-6] The crystal structure of chalcopyrite, $CuFeS_2$. *Z. Kristall.* 82 (1932): 188–194. (Linus Pauling and L. O. Brockway).

* [33-1] The crystal structure of sulvanite, Cu_3VS_4. *Z. Kristall.* 84 (1933): 204–212. (Linus Pauling and Ralph Hultgren). {SP 50}

[34-1] The structure of calcium boride, CaB_6. *Z. Kristall.* 87 (1934): 181–182. (Linus Pauling and Sidney Weinbaum).

[34-3] The cristal (sic) structure of enargite, Cu_3AsS_4. *Z. Kristall.* 88 (1934): 48-53. (Linus Pauling and Sidney Weinbaum).

* [34-4] The crystal structure of binnite, $(Cu, Fe)_{12}As_4S_{13}$, and the chemical composition and structure of minerals of the tetrahedrite group. *Z. Kristall.* 88 (1934): 54–62. (Linus Pauling and E. W. Neuman). {SP 51}

[48-4] The structure of uranium hydride. *J. Am. Chem. Soc.* 70 (1948): 1660–1661. (Linus Pauling and Fred J. Ewing).

[50-1] The problem of the graphite structure. *Am. Mineral* 35 (1950): 125. (Joseph S. Lukesh and Linus Pauling).

[65-5] The nature of the chemical bonds in sulvanite, Cu_3VS_4. *Tchermaks mineralogische u.petrographische Mitteilungen* 10 (1965): 379–384.

[66-5] The structure and properties of graphite and boron nitride. *Proc. Natl. Acad. Sci.* 56(6) (1966): 1646–1652.

* [68-7] A trireticulate crystal structure: trihydrogen cobalticyanide and trisilver cobalticyanide. *Proc. Natl. Acad. Sci.* 60 (1968): 362–367. (Linus Pauling and Peter Pauling). {SP 56}

* [70-8] Crystallography and chemical bonding of sulfide minerals. *Mineral. Soc. Amer. Spec. Pap.* 3 (1970): 125–131. {SP 57}

[75-16] The formula, structure, and chemical bonding of tetradymite, $Bi_{14}Te_{13}S_8$ and the phase $Bi_{14}Te_{15}S_6$. *American Mineralogist* 60 (1975): 994–997.

[78-16] Covalent chemical bonding of transition metals in pyrite, cobaltite, skutterudite, millerite, and related minerals. *Can. Mineral.* 16 (1978), 447–452.

Group 6b

Metallic and Intermetallic Crystal Structures

* [23-2] The crystal structure of magnesium stannide. *J. Am. Chem. Soc.* 45 (1923): 2777–2780. {SP 49}

[27-1] An X-ray study of the alloys of lead and thallium. *J. Am. Chem. Soc.* 49 (1927): 666–669. (Edwin McMillan and Linus Pauling).

[51-6] Discussion of "The borides of some transition elements" by Roland Kiessling. *J. Electrochem. Soc.* 98 (1951): 518–519.

* [52-2] The structure of alloys of lead and thallium. *Acta Cryst.* 5 (1952): 39–44. (You-Chi Tang and Linus Pauling). {SP 53}

[52-8] The atomic arrangements and bonds of the gold-silver ditellurides. *Acta Cryst.* 5 (1952): 375–381. (George Tunell and Linus Pauling).

* [52-9] Interatomic distances and atomic valences in $NaZn_{13}$. *Acta Cryst.* 5 (1952): 637–644. (David P. Shoemaker, Richard E. Marsh, Fred J. Ewing, and Linus Pauling). {SP 54}

[52-13] Crystal structure of the intermetallic compound $Mg_{32}(Al,Zn)_{49}$ and related phases. *Nature* 169 (1952): 1057. (Gunnar Bergman, John L. T. Waugh, and Linus Pauling)

[53-4] The crystal structure of selenium. *Acta Cryst.* 6 (1953): 71–75. (Richard E. Marsh, Linus Pauling and James D. McCullough).

* [57-3] The crystal structure of the metallic phase $Mg_{32}(Al,Zn)_{49}$. *Acta Cryst.* 10 (1957): 254–257. (Gunnar Bergman, John L. T. Waugh, and Linus Pauling). {SP 55}

[66-8] Electron transfer and atomic magnetic moments in the ordered intermetallic compound $AlFe_3$. In: *Quantum Theory of Atoms, Molecules, and the Solid State, A Tribute to John C. Slater*, Per-Olov Löwdin, ed., Academic Press, New York, (1966), pp. 303–306.

[76-22] The structure of tetragonal $(B_{12})_4B_2Ti_{1.3-2.0}$. *Acta Cryst.* B32 (1976): 3359.

[81-15] The structure and oscillational motion of ^{57}Fe atoms in interstitial sites in Al as determined from interference of Mössbauer radiation. *J. Solid State Chem.* 40 (1981): 266–269.

[87-13] Crystal structure of hexagonal $MnAl_4$. *Proc. Natl. Acad. Sci.* 84 (1987): 3537–3539.

[92-4] Comment on "Relative stability of the $Al_{12}W$ structure in Al-transition-metal compounds". *Phys. Rev.* B45 (1992): 7509–7510.

Group 6c

Molecular Crystal Structures

* [34-6] The structure of the carboxyl group. II. The crystal structure of basic beryllium acetate. *Proc. Natl. Acad. Sci.* 20 (1934): 340–345. (Linus Pauling and J. Sherman). {SP 52}

[36-6] The crystal structure of metaldehyde. *J. Am. Chem. Soc.* 58 (1936): 1274–1278. (Linus Pauling and D. C. Carpenter).

[39-5] Recent work on the configuration and electronic structure of molecules, with some applications to natural products. *Fortschr. Chem. Organ. Naturstoffe* 3 (1939): 203–235.

[48-16] X-ray deffraction studies of amino acids, peptides and proteins. In *First International Poliomyelitis Conference* (Waldorf-Astoria Hotel, New York City, July 12–17, 1948), 7 pp., 1948

[68-3] The chromium-carbon bond length in chromium hexacarbonyl. *Acta Cryst.* B24 (1968): 978–979.

[70-9] The crystal structure of α fluorine. *J. Solid State Chem.* 2 (1970): 225–227. (Linus Pauling, Ian Keaveny, and Arthur B. Robinson).

GROUP 7

MOLECULES IN THE GAS PHASE AND ELECTRON DIFFRACTION

[32-10] The electronic structure of the normal nitrous oxide molecule. *Proc. Natl. Acad. Sci.* 18 (1932): 498–499.

[33-5] The determination of the structures of the hexafluorides of sulfur, selenium and tellurium by the electron diffraction method. *Proc. Natl. Acad. Sci.* 19 (1933): 68–73. (L. O. Brockway and Linus Pauling).

[33-8] The electron-diffraction investigation of the structure of molecules of methyl azide and carbon suboxide. *Proc. Natl. Acad. Sci.* 19 (1933): 860–867. (L. O. Brockway and Linus Pauling).

[34-5] The structure of the carboxyl group. I. The investigation of formic acid by the diffraction of electrons. *Proc. Natl. Acad. Sci.* 20 (1934): 336–340. (Linus Pauling and L. O. Brockway).

[34-7] A study of the methods of interpretation of electron-diffraction photographs of gas molecules, with results for benzene and carbon tetrachloride. *J. Chem. Phys.* 2 (1934): 867–881. (Linus Pauling and L. O. Brockway).

* [35-10] The radial distribution method of interpretation of electron diffraction photographs of gas molecules. *J. Am. Chem. Soc.* 57 (1935): 2684–2692. (Linus Pauling and L. O. Brockway). {SP 58}

[35-12] The electron diffraction investigation of phosgene, the six chloroethylenes, thiophosgene, alpha-methylhydroxylamine and nitromethane. *J. Am. Chem. Soc.* 57 (1935): 2693–2704. (L. O. Brockway, J. Y. Beach, and Linus Pauling).

[36-9] The structure of the pentaborane B_5H_9. *J. Am. Chem. Soc.* 58 (1936): 2403–2407. (S. H. Bauer and Linus Pauling).

* [37-3] The adjacent charge rule and the structure of methyl azide, methyl nitrate, and fluorine nitrate. *J. Am. Chem. Soc.* 59 (1937): 13–20. (Linus Pauling and L. O. Brockway). {SP 59}

* [37-6] Carbon-carbon bond distances. The electron diffraction investigation of ethane, propane, isobutane, neopentane, cyclopropane, cyclopentane, cyclohexane, allene, ethylene, isobutene, tetramethylethylene, mesitylene, and hexamethylbenzene. Revised values of covalent radii. *J. Am. Chem. Soc.* 59 (1937): 1223–1236. (Linus Pauling and L. O. Brockway). {SP 60}

[38-5] The electron diffraction study of digermane and trigermane. *J. Am. Chem. Soc.* 60 (1938): 1605–1607. (Linus Pauling, A. W. Laubengayer, and J. L. Hoard).

[39-2] The electron diffraction investigation of methylacetylene, dimethylacetylene, dimethyldiacetylene, methyl cyanide, diacetylene, and cyanogen. *J. Am. Chem. Soc.* 61 (1939): 927–937. (Linus Pauling, H. D. Springall, and K. J. Palmer).

* [39-3] The electron diffraction investigation of the structure of benzene, pyridine, pyrazine, butadiene-1,3, cyclopentadiene, furan, pyrrole, and thiophene. *J. Am. Chem. Soc.* 61 (1939): 1769–1780. (V. Schomaker and Linus Pauling). {SP 61}

[41-2] The alkyls of the third group elements. II. The electron diffraction study of indium trimethyl. *J. Am. Chem. Soc.* 63 (1941): 480–481. (Linus Pauling and A. W. Laubengayer).

[41-4] The electron-diffraction method of determining the structure of gas molecules. *J. Chem. Educ.* 18 (1941): 458–465. (Robert Spurr and Linus Pauling).

[42-5] The electron diffraction investigation of propargyl chloride, bromide, and iodide. *J. Am. Chem. Soc.* 64 (1942): 1753–1756. (Linus Pauling, Walter Gordy, and John H. Saylor).

[42-6] The molecular structure of methyl isocyanide. *J. Am. Chem. Soc.* 64 (1942): 2952–2953. (Walter Gordy and Linus Pauling).

[46-2] The use of punched cards in molecular structure determinations. II. Electron diffraction calculations. *J. Chem. Phys.* 14 (1946): 659–664. (P. A. Shaffer, Jr., Verner Schomaker, and Linus Pauling).

[46-5] An instrument for determining the partial pressure of oxygen in a gas. *Science* 103 (1946): 338. (Linus Pauling, Reuben E. Wood, and J. H. Sturdivant).

* [46-6] An instrument for determining the partial pressure of oxygen in a gas. *J. Am. Chem. Soc.* 68 (1946): 795–798. (Linus Pauling, Reuben E. Wood, and J. H. Sturdivant). {SP 62}

[80-12] The structure of singlet carbene molecules. *J. Chem. Soc. Chem. Comm.* (1980): 688–689.

Group 8

Molecular Properties Analyzed by Quantum Mechanics

[25-4] The inter-ionic attraction theory of ionized solutes. IV. The influence of variation of dielectric constant on the limiting law for small concentrations. *J. Am. Chem. Soc.* 47 (1925): 2129–2134. (P. Debye and Linus Pauling).

[26-1] The dielectric constant and molecular weight of bromine vapor. *Phys. Rev.* 27 (1926): 181–182.

* [26-3] The quantum theory of the dielectric constant of hydrogen chloride and similar gases. *Proc. Natl. Acad. Sci.* 12 (1926): 32–35. {SP 63}

[26-5] The quantum theory of the dielectric constant of hydrogen chloride and similar gases. *Phys. Rev.* 27 (1926): 568–577.

[26-8] Die Abschirmungskonstanten der relativistischen oder magnetischen Röntgenstrahlendubletts (German: The screening constant of relativistic or magnetic X-ray doublets). *Z. Physik* 40 (1926): 344–350.

[27-2] The influence of a magnetic field on the dielectric constant of a diatomic dipole gas. *Phys. Rev.* 29 (1927): 145–160.

[27-3] The electron affinity of hydrogen and the second ionization potential of lithium. *Phys. Rev.* 29 (1927): 285–291.

* [27-5] The theoretical prediction of the physical properties of many-electron atoms and ions. Mole refraction, diamagnetic susceptibility, and extension in space. *Proc. Roy. Soc. A* (London) 114 (1927): 181–211. {SP 64}

[29-4] Momentum distribution in hydrogen-like atoms. *Phys. Rev.* 34 (1929): 109–116. (Boris Podolsky and Linus Pauling).

[29-7] Photo-ionization in liquids and crystals and the dependence of the frequency of X-ray absorption edges on chemical constitution. *Phys. Rev.* 34 (1929): 954–963.

[29-8] Quantum defects for non-penetrating orbits (Pasadena meeting of the American Physical Society, 8 Dec. 1928). *Phys. Rev.* (*Bull. Am. Phys. Soc.*) 33 (1929): 270. Photocopy.

[30-12] *The Structure of Line Spectra*, McGraw-Hill, New York, (1930), 263 pp. (Linus Pauling and Samuel Goudsmit).

[31-5] Objections to a proof of molecular asymmetry of optically active phenylaminoacetic acid. *J. Am. Chem. Soc.* 53 (1931): 3820–3823. (Linus Pauling and Roscoe G. Dickinson).

* [32-5] Screening constants for many-electron atoms. The calculation and interpretation of X-ray term values, and the calculation of atomic scattering factors. *Z. Kristall.* 81 (1932): 1–29. (Linus Pauling and J. Sherman). {SP 65}

* [35-6] The van der Waals interaction of hydrogen atoms. *Phys. Rev.* 47 (1935): 686–692. (Linus Pauling and J. Y. Beach). {SP 66}

[35-14] Atomic Scattering Factors. In: *Internationale Tabellen zur Bestimmung von Kristallstrukturen* (International Tables for the Determination of Crystal Structures), Vol. 2 (Gebrüder Borntraeger, Berlin, 1935), pp. 568–575.

[36-1] Note on the interpretation of the infra-red absorption of organic compounds containing hydroxyl and imino groups. *J. Am. Chem. Soc.* 58 (1936): 94–98.

* [36-8] The diamagnetic anisotropy of aromatic molecules. *J. Chem. Phys.* 4 (1936): 673–677. {SP 67}

* [39-7] A theory of the color of dyes. *Proc. Natl. Acad. Sci.* 25 (1939): 577–582. {SP 68}

[44-6] The light absorption and fluorescence of triarylmethyl free radicals. *J. Am. Chem. Soc.* 66 (1944): 1985.

[48-18] The electronic structure of excited states of simple molecules. *Z. Naturforsch.* 3 (1948): 438–447.

[49-14] The dissociation energy of carbon monoxide and the heat of sublimation of graphite. *Proc. Natl. Acad. Sci.* 35 (1949): 359–363. (Linus Pauling and William F. Sheehan, Jr.).

[50-4] Compressibilities, force constants, and interatomic distances of the elements in the solid state. *J. Chem. Phys.* 18 (1950): 747–753. (J. Waser and Linus Pauling).

[50-16] The compressibility of metals. *Science* 111 (1950): 9.

[52-16] Status of the values of the fundamental constants for physical chemistry as of July 1, 1951. *J. Am. Chem. Soc.* 74 (1952): 2699–2701. (Frederick D. Rossini, Frank T. Gucker, Jr., Herrick L. Johnston, Linus Pauling, and George W. Vinal).

* [53-12] A theory of ferromagnetism. *Proc. Natl. Acad. Sci.* 39 (1953): 551–560. {SP 69}

* [58-3] The nature of bond orbitals and the origin of potential barriers to internal rotation in molecules. *Proc. Natl. Acad. Sci.* 44 (1958): 211–216. {SP 70}

[69-5] Cohesive energies of tetrahedrally coordinated crystals. *Phys. Rev. Lett.* 23 (1969): 480–481.

[79-5] Diamagnetic anisotropy of the peptide group. *Proc. Natl. Acad. Sci.* 76 (1979): 2293–2294.

[79-21] Diamagnetic anisotropy of the carbonate ion in calcite, aragonite, strontianite, and witherite and of other non-cyclic planar atomic groups with resonance structures. *Z. Kristall.* 150 (1979): 155–161.

[80-13] Atomic vibrations in the magnesium difluoride crystal. *Acta Cryst.* B36 (1980): 761–762.
[85-18] Molarity (atomic density) of the elements as pure crystals. *J. Chem. Ed.* 62 (1985): 1086–1088. (Linus Pauling and Zelek S. Herman).
[87-17] Factors determining the average atomic volumes in intermetallic compounds. *Proc. Natl. Acad. Sci.* 84 (1987): 4754–4756.

Group 9

Entropy and Molecular Rotation in Crystals and Liquids

* [25-5] The entropy of supercooled liquids at the absolute zero. *J. Am. Chem. Soc.* 47 (1925): 2148–2156. (Linus Pauling and Richard C. Tolman). {SP 71}
* [30-7] Rotational motion of molecules in crystals. *Phys. Rev.* 36 (1930): 430–443. {SP 72}
* [35-9] The structure and entropy of ice and of other crystals with some randomness of atomic arrangement. *J. Am. Chem. Soc.* 57 (1935): 2680–2684. {SP 73}
* [36-2] Quantum mechanics and the third law of thermodynamics. *J. Chem. Phys.* 4 (1936): 393–394. (Linus Pauling and E. D. Eastman). {SP 74}

Group 10a

Nuclear Structure

[64-7] Nuclear nomenclature: Helion for alpha-particle. *Nature* 201 (1964): 61.
[65-13] Structural significance of the principal quantum number of nucleonic orbital wave functions. *Phys. Rev. Lett.* 15 (1965): 499.
[65-14] Structural basis of neutron and proton magic numbers in atomic nuclei. *Nature* 208 (1965): 174.
* [65-15] The close-packed-spheron model of atomic nuclei and its relation to the shell model. *Proc. Natl. Acad. Sci.* 54 (1965): 989–994. {SP 75}
* [65-16] Structural basis of the onset of nuclear deformation at neutron number 90. *Phys. Rev. Lett.* 15 (1965): 868–870. {SP 76}

* [65-19] The close-packed-spheron theory and nuclear fission. *Science* 150 (1965): 297–305. {SP 77}

[66-4] The close-packed-spheron theory of nuclear structure and the neutron excess for stable nuclei. *Revue Roumain de Physique* 11(9,10) (1966): 825–833.

[66-6] Baryon resonances as rotational states. *Proc. Natl. Acad. Sci.* 56(6) (1966): 1676–1677.

[67-7] Magnetic-moment evidence for the polyspheron structure of the lighter atomic nuclei. *Proc. Natl. Acad. Sci.* 58(6) (1967): 2175–2178.

[68-13] Geometric factors in nuclear structure. In: *Maria Skiodowska-Curie: Centenary Lectures*, International Atomic Energy Agency, Vienna, (1968), pp. 83–88.

[69-1] Double-humped fission. *Physics Today* 22(6) (June 1969): 9, 11.

[69-2] Magnetic moments of antimony isotopes and other nuclei. *Phys. Rev.* 182(4) (1969): 1357–1358.

[69-3] Orbiting clusters in atomic nuclei. *Proc. Natl. Acad. Sci.* 64 (1969): 807–809.

[74-9] Resonance between a prolate and superprolate structure of the ^{162} Er nucleus. *Proc. Natl. Acad. Sci.* 71(7) (1974): 2905–2907. (Linus Pauling and John Blethen).

[75-10] Structure of the excited band in ^{24}Mg. *Phys. Rev. Lett.* 35 (1975): 1480–1482.

[75-18] Rotating clusters in nuclei. *Can. J. Phys.* 53 (1975): 1953–1964. (Linus Pauling and Arthur B. Robinson).

[76-1] Structure of the excited rotational band in ^{40} Ca. *Phys. Rev. Lett.* 36 (1976): 162–164.

[80-9] Superprolate shape of the spontaneous-fission isomer ^{240}Am^{m}. *Phys. Rev.* C22 (1980): 1585–1587.

[81-5] Changes in structure of nuclei between magic numbers 50 and 82 as indicated by a rotating cluster analysis of the energy values of the first 2+ excited states of isotopes of cadmium, tin, and tellurium. *Proc. Natl. Acad. Sci.* 78 (1981): 5296–5298.

[82-2] Prediction of the shapes of deformed nuclei by the polyspheron theory. *Proc. Natl. Acad. Sci.* 79 (1982): 2740–2742.

[82-7] Comment on the test for tetrahedral symmetry in the ^{16}O nucleus and its relation to the shell model. *Phys. Rev. Lett.* 49 (1982): 1119.

[82-9] Rules governing the composition of revolving clusters in quasiband and prolate-deformation states of atomic nuclei. *Proc. Natl. Acad. Sci.* 79 (1982): 7073–7075.

[84-7] Structure of the ^{168}Hf nucleus in high rotational states. *Proc. Natl. Acad. Sci.* 81 (1984): 3261–3262.

[87-9] Discussion of the coexisting 0+ band in the doubly closed subshell ^{96}Zr on the basis of the polyspheron model. *Phys. Rev.* C35 (1987): 1162–1163.

[87-12] *g* factor of the 2_1^+ state in ^{140}Ba and ^{142}Ba. *Phys. Rev.* C35 (1987): 2336.

[87-27] Pauling replies. *Phys. Rev. Lett.* 59 (1987): 2120.

[88-3] Comment on "Low-frequency anomaly in ^{172}Os moment of inertia". *Phys. Rev.* C37 (1988): 883.

[90-4] Regularities in the sequences of the number of nucleons in the revolving clusters for the ground-state energy bands of the even-even nuclei with neutron number equal to or greater than 126. *Proc. Natl. Acad. Sci.* 87 (1990): 4435–4438.

[91-5] Transition from one revolving cluster to two revolving clusters in the ground-state rotational bands of nuclei in the lanthanon region. *Proc. Natl. Acad. Sci.* 88 (1991): 820–823.

[91-12] Analysis of the energy of the first four excited states of the ground-state rotational bands of the even-even lanthanon nuclei (^{58}Ce to ^{70}Yb) with the model of a single cluster of nucleons revolving about a sphere. *Proc. Natl. Acad. Sci.* 88 (1991): 4401–4403.

[91-21] Analysis of the energy of the first four excited states of the ground-state rotational bands of the even-even nuclei from $_6C_8$ to $_{56}Ba_{90}$ with the model of a single cluster of nucleons revolving about a sphere. *Proc. Natl. Acad. Sci.* 88 (1991): 9780–9783.

[92-12] The value of rough quantum mechanical calculations. *Foundations of Physics* 22 (1992): 829–838.

[92-14] Analysis of g-ray energies for 56 excited superdeformed rotational bands of nuclei of lanthonons La to Dy and of Hg, TI, and Pb on the basis of the two-revolving-cluster model, with evaluation of moments of inertia and radii of revolution and assignment of nucleonic compositions to the clusters and the central sphere. *Proc. Natl. Acad. Sci.* 89 (1992): 7277–7281.

[92-21] Puzzling questions about excited superdeformed rotational bands of atomic nuclei are answered by the two-revolving-cluster-model. *Proc. Natl. Acad. Sci.* 89 (1992): 8963–8965.

[93-3] Analysis of the ground-state band and a closely related intercalated band of $^{235}U_{143}$ by the two-revolving-cluster model with consideration of symmetric and and antisymmetric resonance of the two dissimilar clusters. *Proc. Natl. Acad. Sci.* 90 (1993): 5901–5903. (Linus Pauling and Barclay Kamb).

[94-3] Analysis of a hyperdeformed band of $^{152}Dy_{86}$ on the basis of a structure with two revolving clusters, each with a previously unrecognized two-tiered structure. *Proc. Natl. Acad. Sci.* 91 (1994): 897–899.

Group 10b

Superconductivity

* [68-2] The resonating-valence-bond theory of superconductivity: crest superconductors and trough superconductors. *Proc. Natl. Acad. Sci.* 60 (1968): 59–65 {SP 78}

* [87-16] Influence of valence, electronegativity, atomic radii, and crest-trough interaction with phonons on the high-temperature copper oxide superconductors. *Phys. Rev. Lett.* 59 (1987): 225–227. {SP 79}

[89-2] The role of the metallic orbital and of crest and trough superconduction in high-temperature superconductors. In: *High Temperature Superconductivity: The First Two Years.* Robert M. Metzger, ed., Gordon and Breach Scientific Publishers, New York, (1989), pp. 309–313.

[90-14] Technique for increasing the critical temperature of superconducting materials. Patent application no. 07/626,723. Application filed on December 12, 1990.

[91-20] The structure of K_3C_{60} and the mechanism of superconductivity. *Proc. Natl. Acad. Sci.* 88 (1991): 9208–9209.

[92-36] Method of drawing dissolved superconductor. Patent no. 5158588, issued October 27, 1992.

Group 10c

Quasicrystals

* [85-6] Apparent icosahedral symmetry is due to directed multiple twinning of cubic crystals. *Nature* 317 (1985): 512–514. {SP 80}

[86-10] The value of X-ray powder diffraction patterns and the structure of the manganese-aluminum alloys with apparent icosahedral symmetry. *Powder Diffraction* 1 (1) (1986): 14–15.

* [87-1] So-called icosahedral and decagonal quasicrystals are twins of an 820-atom cubic crystal. *Phys. Rev. Lett.* 58 (1987): 365–368. {SP 81}

[87-14] Evidence from X-ray and neutron powder diffraction patterns that the so-called icosahedral and decagonal quasicrystals of $MnAl_6$ and other alloys are twinned cubic crystals. *Proc. Natl. Acad. Sci.* 84 (1987): 3951–3953.

[88-1] The so-called icosahedral quasicrystals of manganese-aluminum and other alloys. *Quimica Nova* 11(1) (Jan 1988): 6–9.

[88-2] Icosahedral symmetry. *Science* 239 (1988): 963.

[88-6] Sigma-phase packing of icosahedral clusters in 780-atom tetragonal crystals of $Cr_5Ni_3Si_2$ and $V_{15}Ni_{10}Si$ that by twinning achieve 8-fold rotational point-group symmetry. *Proc. Natl. Acad. Sci.* 85 (1988): 2025–2026.

[88-8] Structure of the orthorhombic form of Mn_2Al_7, Fe_2Al_7, and $(Mn_{0.7}Fe_{0.3})_2Al_7$ that by twinning produces grains with decagonal point-group symmetry. *Proc. Natl. Acad. Sci.* 85 (1988): 2422–2423.

[88-9] High-resolution transmission electron-micrograph evidence that rapidly quenched $MnAl_6$ and other alloys are icosatwins of a cubic crystal. *Comptes Rendus Acad. Sci.* Parts 306, erie 11, No. 16 (1988): 1147–1151. (Linus Pauling, Zelek S. Herman, and Peter J. Pauling).

[88-10] Icosahedral quasicrystals as twins of cubic crystals containing large icosahedral clusters of atoms: The 1012-atom primitive cubic structure of Al_6CuLi_3, the C-phase $Al_{37}Cu_3Li_{21}Mg_3$, and $GaMg_2Zn_3$, *Proc. Natl. Acad. Sci.* 85 (1988): 3666–3669.

[88-12] Additional evidence from X-ray powder diffraction patterns that icosahedral quasi-crystals of intermetallic compounds are twinned cubic crystals. *Proc. Natl. Acad. Sci.* 85 (1988): 4587–4590.

[88-14] Unified structure theory of icosahedral quasicrystals: Evidence from neutron powder diffraction patterns that AlCrFeMnSi, AlCuLiMg, and TiNiFeSi icosahedral quasicrystals are twins of cubic crystals containing about 820 or 1012 atoms in a primitive unit cube. *Proc. Natl. Acad. Sci.* 85 (1988): 8376–8380.

[89-1] Comment on "Quasicrystal structure of rapidly solidified Ti-Ni-based alloys". *Phys. Rev.* B39 (1989): 1964–1965.

[89-7] Interpretation of so-called icosahedral and decagonal quasicrystals of alloys showing apparent icosahedral symmetry elements as twins of an 820-atom cubic crystal. *Computers Math. Applic.* 17 (1989): 337–339.

[89-15] Icosahedral and decagonal quasicrystals as multiple twins of cubic crystals. In: *Extended Icosahedral Structures (Aperiodicity and Order, Vol. 3)*, Marko V. Jari and Denis Gratias, eds., Academic Press, New York, (1989), pp. 137–162.

[89-18] Icosahedral quasicrystals of intermetallic compounds are icosahedral twins of cubic crystals of three kinds, consisting of large (about 5000 atoms) icosahedral complexes in either a cubic body-centered or a cubic face-centered arrangement or smaller (about 1350 atoms) icosahedral complexes in the β-tungsten arrangement. *Proc. Natl. Acad. Sci.* 86 (1989): 8595–8599.

[89-20] Icosahedral and decagonal quasicrystals of intermetallic compounds are multiple twins of cubic or orthorhombic crystals composed of very large atomic complexes with icosahedral point-group symmetry in cubic close packing: Structure of decagonal Al_6Pd. *Proc. Natl. Acad. Sci.* 86 (1989): 9637–9641.

[90-11] Evidence from electron micrographs that icosahedral quasicrystals are icosahedral twins of cubic crystals. *Proc. Natl. Acad. Sci.* 87 (1990): 7849–7850.

[91-17] Analysis of pulsed-neutron powder diffraction patterns of the icosahedral quasicrystals Pd_3SiU and AlCuLiMg (three alloys) as twinned cubic crystals with large units. *Proc. Natl. Acad. Sci.* 88 (1991): 6600–6602.

Group 11

Hemoglobin: Oxygen Bonding and Magnetic Properties

[35-4] The oxygen equilibrium of hemoglobin and its structural interpretation. *Science* 81 (1935): 421.

* [35-5] The oxygen equilibrium of hemoglobin and its structural interpretation. *Proc. Natl. Acad. Sci.* 21 (1935): 186–191. {SP 82}

[36-3] The magnetic properties and structure of the hemochromogens and related substances. *Proc. Natl. Acad. Sci.* 22 (1936): 159–163. (Linus Pauling and Charles D. Coryell).

* [36-4] The magnetic properties and structure of hemoglobin, oxyhemoglobin and carbonmonoxyhemoglobin. *Proc. Natl. Acad. Sci.* 22 (1936): 210–216. (Linus Pauling and Charles D. Coryell). {SP 83}

[36-5] The magnetic properties and structure of hemoglobin and related substances. *Science* 83 (1936): 488–489. (Linus Pauling and Charles D. Coryell).

* [37-4] The magnetic properties and structure of ferrihemoglobin (methemoglobin) and some of its compounds. *J. Am. Chem. Soc.* 59 (1937): 633–642. (Charles D. Coryell, Fred Stitt, and Linus Pauling). {SP 84}

[39-1] The magnetic properties of intermediates in the reactions of hemoglobin. *J. Phys. Chem.* 43 (1939): 825–839. (Charles D. Coryell, Linus Pauling, and Richard W. Dodson).

[39-6] The magnetic properties of the compounds ethylisocyanide-ferrohemoglobin and imidazole-ferrihemoglobin. *Proc. Natl. Acad. Sci.* 25 (1939): 517–522. (Charles D. Russell and Linus Pauling).

[40-1] A structural interpretation of the acidity of groups associated with the hemes of hemoglobin and hemoglobin derivatives. *J. Biol. Chem.* 132 (1940): 769–779. (Charles D. Coryell and Linus Pauling).

[48-8] The interpretation of some chemical properties of hemoglobin in terms of its molecular structure. *Stanford Med. Bull.* 6 (1948): 215–222.

[49-12] The electronic structure of haemoglobin. In: *Haemoglobin*, Butterworths Scientific Publications, London, (1949), pp. 57–65.

* [51-18] The combining power of hemoglobin for alkyl isocyanides, and the nature of the heme-heme interactions in hemoglobin. *Science* 114 (1951): 629–634. (Robert C. C. St. George and Linus Pauling). {SP 85}

[53-8] Protein interactions. Aggregation of globular proteins. *Discussions Faraday Soc.* 13 (1953): 170–176.

[55-4] Abnormality of hemoglobin molecules in hereditary hemolytic anemias. In: *The Harvey Lectures, 1953–1954,* Series 49, Academic Press, New York, (1955), pp. 216–241.

[56-4] The combining power of myoglobin for alkyl isocyanides and the structure of the myoglobin molecule. *Proc. Natl. Acad. Sci.* 42 (1956): 51–54. (Allen Lein and Linus Pauling).

[57-1] The N-terminal amino acid residues of normal adult hemoglobin: A quantitative study of certain aspects of Sanger's DNP-method. *J. Am. Chem. Soc.* 79 (1957): 609–615. (Herbert S. Rhinesmith, W. A. Schroeder, and Linus Pauling).

[57-7] A quantitative study of the hydrolysis of human dinitrophenyl (DNP) globin: The number and kind of polypeptide chains in normal adult human hemoglobin. *J. Am. Chem. Soc.* 79 (1957): 4682–4686. (Herbert S. Rhinesmith, W. A. Schroeder, and Linus Pauling).

[58-1] Factors affecting the structure of hemoglobins and other proteins. In: *Symposium on Protein Structure.* Albert Neuberger, ed., Methuen, London; Wiley, New York, (1958), pp. 17–22.

* [64-18] The nature of the iron-oxygen bond in oxyhemoglobin. *Nature* 203 (1964): 182–183. {SP 86}

[77-5] Magnetic properties and structure of oxyhemoglobin. *Proc. Natl. Acad. Sci.* 74 (1977): 2612–2613.

* [80-16] The normal hemoglobins and the hemoglobinopathies: background. In: *Hemo-gobins and Hemoglobinopathies: A Current Review to 1981*, Texas Reports on Biology and Medicine, University of Texas Medical Branch, Galveston, TX, Vol. 40 (1980–1981), pp. 1–7. {SP 87}

Group 12

Antibodies: Structure and Function

* [40-2] A theory of the structure and process of formation of antibodies. *J. Am. Chem. Soc.* 62 (1940): 2643–2657. {SP 88}

[41-1] Serological reactions with simple substances containing two or more haptenic groups. *Proc. Natl. Acad. Sci.* 27 (1941): 125–128. (Linus Pauling, Dan H. Campbell, and David Pressman).

[42-1] Complement fixation with simple substances containing two or more haptenic groups. *Proc. Natl. Acad. Sci.* 28 (1942): 77–79. (David Pressman, Dan H. Campbell, and Linus Pauling).

[42-2] The production of antibodies *in vitro*. *Science* 95 (1942): 440–441. (Linus Pauling and Dan H. Campbell).

[42-3] The manufacture of antibodies *in vitro*. *J. Exp. Med.* 76 (1942): 211–220. (Linus Pauling and Dan H. Campbell).

[42-4] The agglutination of intact azo-erythrocytes by antisera homologous to the attached groups. *J. Immun.* 44 (1942): 101–105. (David Pressman, Dan H. Campbell, and Linus Pauling).

* [42-8] The serological properties of simple substances. I. Precipitation reactions between antibodies and substances containing two or more haptenic groups. *J. Am. Chem. Soc.* 64 (1942): 2994–3003. (Linus Pauling, David Pressman, Dan H. Campbell, Carol Ikeda, and Miyoshi Ikawa). {SP 89}

* [42-9] The serological properties of simple substances. II. The effects of changed conditions and of added haptens on precipitation reactions of polyhaptenic simple substances. *J. Am. Chem. Soc.* 64 (1942): 3003–3009. (Linus Pauling, David Pressman, Dan H. Campbell, and Carol Ikeda). {SP 90}

[42-10] The serological properties of simple substances. III. The composition of precipitates of antibodies and polyhaptenic simple substances; the valence of antibodies. *J. Am. Chem. Soc.* 64 (1942): 3010–3014. (David Pressman, Carol Ikeda, Linus Pauling)

[42-11] The serological properties of simple substances. IV. Hapten inhibition of precipitation of antibodies and polyhaptenic simple substances. *J. Am. Chem. Soc.* 64 (1942): 3015–3020. (David Pressman, David H. Brown, and Linus Pauling).

[43-1] The serological properties of simple substances. V. The precipitation of polyhaptenic simple substances and antiserum homologous to the *p*-(*p*-azophenylazo)-phenylarsonic acid group and its inhibition by haptens. *J. Am. Chem. Soc.* 65 (1943): 728–732. (David Pressman, John T. Maynard, Allan L. Grossberg, and Linus Pauling).

* [43-2] The nature of the forces between antigen and antibody and of the precipitation reaction. *Physiol. Rev.* 23 (1943): 203–219. (Linus Pauling, Dan H. Campbell, and David Pressman). {SP 91}

[43-4] An experimental test of the framework theory of antigen-antibody precipitation. *Science* 98 (1943): 263–264. (Linus Pauling, David Pressman, and Dan H. Campbell).

* [44-1] The serological properties of simple substances. VI. The precipitation of a mixture of two specific antisera by a dihaptenic substance containing the two corresponding haptenic groups; evidence for the framework theory of serological precipitation. *J. Am. Chem. Soc.* 66 (1944): 330–336. (Linus Pauling, David Pressman, and Dan H. Campbell). {SP 92}

[44-2] A note on the serological activity of denatured antibodies. *Science* 99 (1944): 198–199. (George G. Wright and Linus Pauling).

[44-3] The serological properties of simple substances. VII. A quantitative theory of the inhibition by haptens of the precipitation of heterogeneous antisera with antigens, and comparison with experimental results for polyhaptenic simple substances and for azoproteins. *J. Am. Chem. Soc.* 66 (1944): 784–792. (Linus Pauling, David Pressman, and Allan L. Grossberg).

[44-5] The serological properties of simple substances. VIII. The reactions of antiserum homologous to the *p*-azobenzoic acid group. *J. Am. Chem. Soc.* 66 (1944): 1731–1738. (David Pressman, Stanley M. Swingle, Allan L. Grossberg, and Linus Pauling).

[45-3] The serological properties of simple substances. IX. Hapten inhibition of precipitation of antisera homologous to the *o*-, *m*-, and *p*-azophenyl-arsonic acid groups. *J. Am. Chem. Soc.* 67 (1945): 1003–1012. (Linus Pauling and David Pressman).

[45-4] The serological properties of simple substances. X. A hapten inhibition experiment substantiating the intrinsic molecular asymmetry of antibodies. *J. Am. Chem. Soc.* 67 (1945): 1219–1222. (David Pressman, John H. Bryden, and Linus Pauling).

[45-5] The reactions of antisera homologous to various azophenylarsonic acid groups and the *p*-azophenylmethylarsinic acid group with some heterologous haptens. *J. Am. Chem. Soc.* 67 (1945): 1602–1606. (David Pressman, Arthur B. Pardee, and Linus Pauling).

[46-3] The reactions of antiserum homologous to the *p*-azophenyltrimethylammonium group. *J. Am. Chem. Soc.* 68 (1946): 250–255. (David Pressman, Allan L. Grossberg, Leland H. Pence, and Linus Pauling).

[46-7] Analogies between antibodies and simpler chemical substances. *Chem. & Eng. News* 24 (25 April 1946): 1064–1065.

[48-1] The serological properties of simple substances, XIII. The reactions of antiserum homologous to the *p*-azosuccinanilate ion group. *J. Am. Chem. Soc.* 70 (1948): 1352–1358. (David Pressman, John H. Bryden, and Linus Pauling).

[48-2] Molecular structure and biological specificity. *Chemistry and Industry (Supplement)* (1948): 1–4.

* [48-7] Antibodies and specific biological forces. *Endeavour* 7 (1948): 43–53. {SP 93}

[48-13] La Structure des Anticorps et al Nature des Reactions Serologiques. (French: The structure of antibodies and the nature of serological reactions). *Bull. Soc. Chim. Biol.*, 30 (1948): 247–259.

* [49-1] The serological properties of simple substances. XIV. The reaction of simple antigens with purified antibody. *J. Am. Chem. Soc.* 71 (1949): 143–148. (Arthur B. Pardee and Linus Pauling). {SP 94}

[49-11] The serological properties of simple substances. XV. The reactions of antiserum homologous to the 4-azophthalate ion. *J. Am. Chem. Soc.* 71 (1949): 2893–2899. (David Pressman and Linus Pauling).

[51-2] The preparation and properties of a modified gelatin (oxypolygelatin) as an oncotic substitute for serum albumin. *Texas Reports on Biology and Medicine* 9 (1951): 235–280. (Dan H. Campbell, J. B. Koepfli, Linus Pauling, Norman Abrahamsen, Walter Dandliker, George A. Feigen, Frank Lanni, and Arthur Le Rosen).

[75-3] Molecular complementariness and serological specificity. *Immunochem.* 12 (1975): 445–447.

[92-7] Molecular structure and biological specificity. In: *Excerpts from Classics in Allergy*, 2nd ed., Sheldon G. Cohen and Max Samter, eds., Symposia Foundation, P. 0. Box 2107, Carlsbad, CA, 1992, p. 123.

Group 13

The Alpha Helix and the Structure of Proteins

* [36-7] On the structure of native, denatured, and coagulated proteins. *Proc. Natl. Acad. Sci.* 22 (1936): 439–447. (A. E. Mirsky and Linus Pauling). {SP 95}

* [39-4] The structure of proteins. *J. Am. Chem. Soc.* 61 (1939): 1860–1867. (Linus Pauling and Carl Niemann). {SP 96}

[45-1] The adsorption of water by proteins. *J. Am. Chem. Soc.* 67 (1945): 555–557.

[50-5] Two hydrogen-bonded spiral configurations of the polypeptide chain. *J. Am. Chem. Soc.* 72 (1950): 21. (Linus Pauling and Robert B. Corey).

* [51-9] The structure of proteins: Two hydrogen-bonded helical configurations of the polypeptide chain. *Proc. Natl. Acad. Sci.* 37 (1951): 205–210. (Linus Pauling, Robert B. Corey, and H. R. Branson). {SP 97}

* [51-10] Atomic coordinates and structure factors for two helical configurations of polypeptide chains. *Proc. Natl. Acad. Sci.* 37 (1951): 3235–3240. (Linus Pauling and Robert B. Corey). {SP 98}

* [51-11] The structure of synthetic polypeptides. *Proc. Natl. Acad. Sci.* 37 (1951): 241–250. (Linus Pauling and Robert B. Corey). {SP 99}

* [51-12] The pleated sheet, a new layer configuration of polypeptide chains. *Proc. Natl. Acad. Sci.* 37 (1951): 251–256. (Linus Pauling and Robert B. Corey). {SP 100}

* [51-13] The structure of feather rachis keratin. *Proc. Natl. Acad. Sci.* 37 (1951): 256–261. (Linus Pauling and Robert B. Corey). {SP 101}

* [51-14] The structure of hair, muscle, and related proteins. *Proc. Natl. Acad. Sci.* 37 (1951): 261–271. (Linus Pauling and Robert B. Corey). {SP 102}

* [51-15] The structure of fibrous proteins of the collagen-gelatin group. *Proc. Natl. Acad. Sci.* 37 (1951): 272–281. (Linus Pauling and Robert B. Corey). {SP 103}

* [51-16] The polypeptide-chain configuration in hemoglobin and other globular proteins. *Proc. Natl. Acad. Sci.* 373 (1951): 282–285. (Linus Pauling and Robert B. Corey). {SP 104}

[51-17] *The Structure of Proteins.* Phi Lambda Upsilon, Second Annual Lecture Series, Ohio State University, Columbus, OH (February 1951) 13 pp.

[51-19] Configuration of polypeptide chains. *Nature* 168 (1951): 550–551. (Linus Pauling and Robert B. Corey).

[51-20] The configuration of polypeptide chains in proteins. *Record Chem. Progress* 12 (1951): 155–162.

[51-21] Configurations of polypeptide chains with favored orientations around single bonds: two new pleated sheets. *Proc. Natl. Acad. Sci.* 37 (1951): 729–740. (Linus Pauling and Robert B. Corey).

[52-3] Configurations of polypeptide chains with equivalent cis amide groups. *Proc. Natl. Acad. Sci.* 38 (1952): 86–93. (Linus Pauling and Robert B. Corey)

[52-10] The Lotmar-Picken X-ray diagram of dried muscle. *Nature* 169 (1952): 494–495. (Linus Pauling and Robert B. Corey).

[52-12] Structure of the synthetic polypeptide poly-γ-methyl-l-glutamate. *Nature* 169 (1952): 920. (Harry L. Yakel, Jr., Linus Pauling, and Robert B. Corey).

* [52-18] The planarity of the amide group in polypeptides. *J. Am. Chem. Soc.* 74 (1952): 3964. (Linus Pauling and Robert B. Corey). {SP 105}

* [53-1] Fundamental dimensions of polypeptide chains. *Proc. Roy. Soc.* (London) 141 (1953): 10–20. (Robert B. Corey and Linus Pauling). {SP 106}

* [53-2] Stable configurations of polypeptide chains. *Proc. Roy. Soc.* (London) 141 (1953): 21–33. (Linus Pauling and Robert B. Corey). {SP 107}

[53-3] The configuration of polypeptide chains in proteins. (A report for the Ninth Solvay Congress, University of Brussels, 6–14 April 1953.) In *Les Proteins. Rapports et Discussions*,R. Stoops, ed., Secretaires du Conseil sous les Auspices du Comité Scientifique de l'Institut (Brussels: Coudenberg, 1953), pp. 63–99.

* [53-5] Compound helical configurations of polypeptide chains: Structure of proteins of the α-keratin type. *Nature* 171 (1953): 59–61. (Linus Pauling and Robert B. Corey). {SP 108}

[53-7] Molecular models of amino acids, peptides, and proteins. *Rev. Sci. Instr.* 24 (1953): 621–627. (Robert B. Corey and Linus Pauling).

[53-10] Two pleated-sheet configurations of polypeptide chains involving both cis and trans amide groups. *Proc. Natl. Acad. Sci.* 39 (1953): 247–252. (Linus Pauling and Robert B. Corey).

* [53-11] Two rippled-sheet configurations of polypeptide chains, and a note about the pleated sheets. *Proc. Natl. Acad. Sci.* 39 (1953): 253–256. (Linus Pauling and Robert B. Corey). {SP 109}

[54-6] The configuration of polypeptide chains in proteins. *Fortschr. Chem. Organ. Naturs.* 11 (1954): 180–239. (Linus Pauling and Robert B. Corey).

[54-8] The stochastic method and the structure of proteins. In: *13th International Congress of Pure and Applied Chemistry: Plenary Lectures*, International Union of Pure and Applied Chemistry, Stockholm, (1954), pp. 37–52.

[54-9] The stochastic method and the structure of proteins. *Am. Scientist* 43 (1955): 285–297.

[54-10] The structure of protein molecules. *Scientific American* 191 (1954), 51–59. (Linus Pauling, Robert B. Corey, and Roger Layward).

[55-2] An investigation of the structure of silk fibroin. *Biochim. Biophys. Acta* 16 (1955): 1–34. (Richard E. Marsh, Robert B. Corey, and Linus Pauling).

[55-5] The crystal structure of silk fibroin. *Acta Cryst.* 8 (1955): 62. (Richard E. Marsh, Robert B. Corey, and Linus Pauling).

* [55-9] The structure of tussah silk fibroin (with a note on the structure of b-poly-L-alanine). *Acta Cryst.* 8 (1955): 710–715. (Richard E. Marsh, Robert B. Corey, and Linus Pauling). {SP 110}

[55-11] The configuration of polypeptide chains in proteins. *Istituto Lombardo Milano* 89 (1955): 10–37. (Robert B. Corey and Linus Pauling).

[55-18] Calculated form factors for the 18-residue 5-turn α-helix. *Acta Cryst.* 8 (1955): 853–855. (Linus Pauling, Robert B. Corey, Harry L. Yakel, Jr., and Richard E. Marsh).

[57-6] The configuration of polypeptide chains in proteins. In: *Modern Chemistry for the Engineer and Scientist*, George Ross Robertson, ed., McGraw Hill, New York, (1957), pp. 422–434.

[58-2] The structure of proteins. In: *Frontiers in Science, A Survey*, Edward Hutchings, Jr., ed., Basic Books, New York, (1958), pp. 28–36. (Robert B. Corey and Linus Pauling).

[58-4] The configuration of polypeptide chains in proteins. In: *Recent Advances in Gelatin and Glue Research.* G. Stainsby, ed., Pergamon Press, London, (1958), pp. 11–13.

* [96-1] The discovery of the alpha helix. *The Chemical Intelligencer* 2(1) (January 1996): 32–38. {SP 111}

Group 14a

Molecular Biology: the Role of Large Molecules in Life

* [40-3] The nature of the intermolecular forces operative in biological processes. *Science* 92 (1940): 77–79. (Linus Pauling and Max Delbrück). {SP 112}

[45-6] Molecular structure and intermolecular forces. In: *The Specificity of Serological Reactions* (*rev. edn.*), Karl Landsteiner, ed. Harvard University Press, Cambridge, (1945), pp. 275–293.

* [46-8] Molecular architecture and biological reactions. *Chem. & Eng. News*, 24 (25 May 1946): 1375–1377.

* [46-9] Landmarks. Molecular architecture and biological reactions. *The Journal of NIH Research* 2(6) (1990): 59–64. Reprint of [46-8] with added side bar, “The First Molecular Biologist.” {SP 113}

[46-12] Molecular architecture and medical progress. In: *The Scientists Speak*, Warren Weaver, ed., Boni & Gear, Inc., New York, (1947), pp. 110–114.

* [48-12] The nature of forces between large molecules of biological interest. *Nature* 161 (1948): 707–709. {SP 114}

* [48-15] *Molecular Architecture and the Processes of Life.* Sir Jesse Boot Foundation, Nottingham, England, (1948), 13 pp. {SP 115}

[52-7] On a phospho-tri-anhydride formula for the nucleic acids. *J. Am. Chem. Soc.* 74 (1952): 1111. (Linus Pauling and Verner Schomaker).

[52-17] On a phospho-tri-anhydride formula for the nucleic acids. *J. Am. Chem. Soc.* 74 (1952): 3712. (Linus Pauling and Verner Schomaker).

[53-6] Structure of the nucleic acids. *Nature* 171 (1953): 346. (Linus Pauling and Robert B. Corey).

* [53-9] A proposed structure for the nucleic acids. *Proc. Natl. Acad. Sci.* 39 (1953): 84–97. (Linus Pauling and Robert B. Corey). {SP 116}

[55-1] The duplication of molecules, in *Aspects of Synthesis and Order in Growth* Dorothea Rudnick, ed., Princeton University Press, Princeton, (1955), pp. 3–13.

[56-6] The future of enzyme research. In: *Enzymes: Units of Biological Structure and Function*, Oliver H. Gaebler, ed., Academic Press, New York, (1956), pp. 177–182.

* [56-10] Specific hydrogen-bond formation between pyrimidines and purines in deoxyribonucleic acids. *Arch. Biochem. Biophys.* 65 (1956): 164–181. (Linus Pauling and Robert B. Corey). {SP 117}

[56-15] The molecular basis of genetics. *Am. J. Psychiatry* 113 (1956): 492–495.

[57-2] The probability of errors in the process of synthesis of protein molecules. In: *Arbeiten aus dem Gebiet der Naturstoffe* (Festschrift Prof. Dr. Arthur Stoll.) Birkhäuser, Basel, (1957), pp. 597–602.

[57-11] Summary and discussion. In: *Molecular Structure and Biological Specificity.* Linus Pauling and Harvey A. Itano, eds., Publication No. 2, American Institute of Biological Sciences, Washington, DC, (1957), pp. 186–195.

[59-1] The nature of the forces operating in the process of the duplication of molecules in living organisms. In: *The Origin of Life on the Earth.* F. Clark and R. L. M. Singe, eds., Pergamon Press, New York, (1959), pp. 215–223.

[59-18] Molecular structure in relation to biology and medicine. In: *Ciba Foundation Symposium on Significant Trends in Medical Research* (Proceedings of the Tenth Anniversary Symposium) G. E. W. Wolstenholme, Cecilia M. O'Connor, and Maeve O'Connor, eds., Little, Brown and Co., Boston, (1959), pp. 3–10.

[63-8] The molecular basis of genetic defects. In: *First Inter-American Conference on Congenital Defects*, J. B. Lippincott, Philadelphia, (1963), pp. 15–21.

[68-9] Reflections on the new biology: Foreword. *UCLA Law Review* 15(2) (Feb 1968): 267–272.

[70-11] Fifty years of progress in structural chemistry and molecular biology. *Daedalus* 99 (1970): 988–1014.

[70-12] Structure of high-energy molecules. *Chem. in Brit.* 6 (1970): 468–472.

[74-6] The molecular basis of biological specificity. *Nature* 248 (1974): 769–771.

[86-15] Early days of molecular biology in the California Institute of Technology. *Ann. Rev. Biophys. Chem.* 15 (1986): 1–9.

Group 14b

Molecular Evolution

[59-22] Molecules and evolution. *Chem. & Eng. News* 37 (47) (23 November 1959): 96.

* [60-5] A comparison of animal hemoglobins by tryptic peptide pattern analysis. *Proc. Natl. Acad. Sci.* 46 (1960): 1349–1360. (Emile Zuckerkandl, Richard T. Jones, and Linus Pauling). {SP 118}

* [62-23] Molecular disease, evolution, and genic heterogeneity. In: *Horizons in Biochemistry* (Szent-Györgyi Dedicatory Volume), Michael Kasha and Bernard Pullman, eds., Academic Press, New York, (1962), pp. 189–225. (Emile Zuckerkandl and Linus Pauling). {SP 119}

* [63-2] Chemical paleogenetics: Molecular "Restoration Studies" of extinct forms of life. *Acta Chem. Scand.* 17 (Suppl. no. 1) (1963), S9–Sl6. (Linus Pauling and Emile Zuckerkandl). {SP 120}

[65-3] Les documents moléculaires de 1'évolution (French: Molecules as documents of evolution). *Atomes* 20 (1965): 339–343. (Emile Zuckerkandl and Linus Pauling).

* [65-4] Molecules as documents of evolutionary history. *J. Theoret. Biol.* 8 (1965): 357–366. (Emile Zuckerkandl and Linus Pauling). {SP 121}

[65-12] Evolutionary divergence and convergence in proteins. In: *Evolving Genes and Proteins*, Vernon Bryson and Henry J. Vogel, eds., Academic Press, New York, 1965, pp. 97–166. (Emile Zuckerkandl and Linus Pauling).

[72-3] Chance in evolution: philosophical remarks. In: *Molecular Evolution*, Duane L. Rohlfing and A. Oparin, eds., Plenum, New York, 1972, pp. 113–126. (Linus Pauling and Emile Zuckerkandl).

Group 15

Molecular Disease

[49-6] A rapid diagnostic test for sickle cell anemia. *Blood. The Journal of Hematology* 4 (1949): 66–68. (Harvey A. Itano and Linus Pauling).

[49-9] Sickle cell anemia, a molecular disease. *Science* 109 (1949): 443. (Linus Pauling, Harvey A. Itano, S. J. Singer, and Ibert C. Wells).

* [49-16] Sickle cell anemia, a molecular disease. *Science* 110 (1949): 543–548. (Linus Pauling, Harvey A. Itano, S. J. Singer, and Ibert C. Wells). {SP 122}

[50-13] Sickle cell anemia hemoglobin. *Science* 111 (1950): 459. (Linus Pauling, Harvey A. Itano, Ibert C. Wells, Walter A. Schroeder, Lois M. Kay, S. J. Singer, and R. B. Corey).

* [52-19] The hemoglobin molecule in health and disease. *Proc. Am. Phil. Soc.* 96 (1952): 556–565. {SP 123}

* [57-5] Abnormal hemoglobin molecules in relation to disease. *Svensk Kemisk Tidskrift* 69 (1957): 509–523. (Harvey A. Itano and Linus Pauling). {SP 125}

[58-11] Molecular disease. *Pfizer Spectrum* 6(9) (1 May 1958): 234–235.

[58-16] Emoglobine anormali in rapporto alle malattie (Italian: Abnormal hemoglobin in relation to disease). Esiratto dai Rendiconil dell'lstiiuio Superiore di Santia (Rome) 21 (1958): 30–48.

[59-14] Molecular disease. *Am. J. Orthopsychiatry* 29 (4) (1959): 684–687.

[60-1] Molecular structure and disease. In: *Disease and the Advancement of Basic Science.* Henry K. Beecher, ed., (1960), pp. 1–7.

[61-1] Introduction. In: *Molecular Genetics and Human Disease*, Lytt I. Gardner, ed., Charles C Thomas, Publisher, Springfield, IL, (1961), pp. ix–xi.

[61-11] Thalassaemia and the abnormal human hemoglobins. *Nature* 191 (1961): 398–399. (Harvey A. Itano and Linus Pauling).

* [63-10] Molecular disease and evolution. *Proc. Rudolf Virchow Med. Soc., NY* 21 (1963): 131–140. {SP 126}

[73-1] The genesis of the concept of molecular disease. In: *Sickle Cell Disease*, E. F. Mammen, G. F. Anderson, and M. I. Barnhart, eds., F. K. Schattauer Verlag, New York, (1973), pp. 1–6.

[73-20] The genesis of the concept of molecular disease. *Thrombosis et Diathesis Haemorrhagica Supple.* 53 (1973): 1–6.

Group 16a

Physiological Chemistry

* [61-9] A molecular theory of general anesthesia. *Science* 134 (1961): 15–21. {SP 129}

[62-4] Une Théorie moléculaire de l'anesthésie générale (French: A molecular theory of general anesthesia). *J. Chim. Phys.* 59 (1962): 1–8.

[64-5] The hydrate microcrystal theory of general anesthesia. *Anesthesia and Analgesia* 43 (1964): 1–10.

[64-28] A molecular theory of general anesthesia. In: *Symposium on Information Processing in the Nervous System*. P. W. Gerard and J. W. Duyff, eds., Excerpta Medica Foundation, Amsterdam and New York, (1964), p. 548.

[65-6] Anesthesia of artemia larvae: Method for quantitative study. *Science* 149 (1965): 1255–1258. (Arthur B. Robinson, Kenneth F. Manly, Michael P. Anthony, John F. Catchpool, and Linus Pauling)

[71-4] Quantitative analysis of urine vapor and breath by gas-liquid partition chromato-graphy. *Proc. Natl. Acad. Sci.* 68 (1971): 2374–2376. (Linus Pauling, Arthur B. Robinson, Roy Teranishi, and Paul Cary).

[72-6] Gas chromatography of volatiles from breath and urine. *Anal. Chem.* 44 (1972): 18–20. (Roy Teranishi, T. R. Mon, Arthur B. Robinson, Paul Cary, and Linus Pauling).

* [73-5] Results of a loading test of ascorbic acid, niacinamide, and pyridoxine in schizophrenic subjects and controls. In: *Orthomolecular Psychiatry: Treatment of Schizophrenia*, David Hawkins and Linus Pauling, eds., W. H. Freeman, San Francisco, (1973), pp. 18–34. (Linus Pauling, Arthur B. Robinson, Susanna S. Oxley, Maida Bergeson, Andrew Harris, Paul Cary, John Blethen, and Ian T. Keaveny. {SP 130}

[73-11] An apparatus for the quantitative analysis of volatile compounds in urine. *J. Chromatog.* 85 (1973): 19–29. (Arthur B. Robinson, David Partridge, Martin Turner, Roy Teranishi, and Linus Pauling).

[73-12] The identification of volatile compounds in human urine. *J. Chromatog.* 85 (1973): 31–34. (Kent E. Matsumoto, David H. Partridge, Arthur B. Robinson, Linus Pauling, Robert A. Flath, T. Richard Mon, and Roy Teranishi).

[74-2] The process of aging. In: *New Dynamics of Preventive Medicine*, Vol. 3, Leon R. Pomeroy, ed., Symposia Specialists, Miami, FL, (1974), pp. 107–113.

[75-15] Sex-related patterns in the profiles of human urinary amino acids. *Clin. Chem.* 21 (1975): 1970–1975. (Henri Dirren, Arthur B. Robinson, and Linus Pauling).

* [79-13] The discovery of the superoxide radical. *Trends in Biochemical Sciences* 4 (1979): 270–271. {SP 131}

[79-27] (Correction of [79-13].) *Trends in Biochemical Sciences* 5 (Jan 1980): VII.

Group 16b

Health Hazards of Radiation

[48-19] Discussion of the paper "Action of X-rays at low temperatures on the gametes of drosophila" by Dr. Edward Novitski. *Am. Natural.* 83 (1948): 189–191.

[57-8] Health hazards of radiation. In: *Health Hazards of Radiation* (American Cancer Society, California Division, 1957 Annual Meeting, Panel Forum Recorded on Tape and Transcribed), American Cancer Society, California Division, (1957), pp. 6–8.

* [58-21] Genetic and somatic effects of carbon-14. *Science* 128 (1958): 1183–1186. {SP 127}

[58-23] Fallout and disarmament: A debate between Linus Pauling and Edward Teller. (Debate held on 20 February 1958 in San Francisco and broadcast on KQED). *Daedalus: Proc. Am. Acad. Arts and Sciences* 87(2) (1958): 147–163. (Linus Pauling and Edward Teller).

[59-9] Quelques précisions sur la nature et les propriétés des retombées radioactives résultant des explosions atomiques depuis 1945 (French: Some particulars concerning the nature and properties of radioactive fallout resulting from the atomic explosions since 1945). *Comptes Rendus Acad. Sci. Paris* 249 (1959): 982–984. (Linus Pauling, Shoichi Sakata, Sin-Itiro Tomonaga, Jean-Pierre Vigier, and Hideki Yukawa).

* [59-12] The effects of strontium-90 on mice. *Proc. Natl. Acad. Sci.* 45 (1959): 54–69. (Barclay Kamb and Linus Pauling). {SP 128}

[60-30] High-energy radiation and its effects on man. In: *Proceedings of the Eighth National Convention, National Science Teachers Association, Kansas City, Missouri, March 20-April 2, 1960,* Hugh Allen, Jr., ed., National Science Teachers Association, Washington, DC, (1960), pp. 4–11.

[62-16] Radiation hazards. *Physics Today*, (Sept 1962): p. 58.

[62-22] Genetic effects of weapons tests. *Bull. Atom. Sci.* 18 (Dec 1962): 15–18 .

[70-7] Genetic and somatic effects of high-energy radiation. *Bull. Atom. Sci.* 26 (7) (Sept 1970): 3–5.

[75-4] Radioactivity re-examined. *Chem. & Eng. News* (14 July 1975): 5.

Group 16c

Biostatistical Analysis of Test Results

[58-14] The relation between longevity and obesity in human beings. *Proc. Natl. Acad. Sci.* 44 (1958): 619-622.

[60-2] Observations on aging and death. *Engineering and Science* 23(5) (May 1960): 9–12.

[78-7] Supplemental ascorbate in the supportive treatment of cancer: reevaluation of prolongation of survival times in terminal human cancer. *Proc. Natl. Acad. Sci.* 75 (1978): 4538–4542. (Ewan Cameron and Linus Pauling).

[82-8] Mortality among health-conscious elderly Californians. *Proc. Natl. Acad. Sci.* 79 (1982): 6023–6027. (James E. Enstrom and Linus Pauling).

* [89-12] Biostatistical analysis of mortality data for cohorts of cancer patients. *Proc. Natl. Acad. Sci.* 86 (1989): 3466–3468. {SP 132}

* [89-13] Criteria for the validity of clinical trials of treatments of cohorts of cancer patients based on the Hardin Jones principle. *Proc. Natl. Acad. Sci.* 86 (1989): 6835–6837. (Linus Pauling and Zelek S. Herman). {SP 133}

[90-7] Hardin-Jones biostatistical analysis of mortality data for cohorts of cancer patients with a large fraction surviving at the termination of the study and a comparison of survival times of cancer patients receiving large regular oral doses of vitamin C and other nutrients with similar patients not receiving those doses. *J. Orthomolecular Medicine* 5(3) (1990): 143–154. (A. Hoffer and Linus Pauling)

[93-10] Hardin-Jones biostatistical analysis of mortality data for a second set of cohorts of cancer patients with a large fraction surviving at the termination of the study and a comparison of survival times of cancer patients receiving large regular oral doses of vitamin C and other nutrients with similar patients not receiving these doses. *J. Orthomolecular Med.* 8(3) (1993): 157–167. (A. Hoffer and Linus Pauling).

Group 17a

Orthomolecular Medicine — General Concepts

[66-9] Biological treatment of mental illness. In: *Biological Treatment of Mental Illness* (Proceedings of the 11th International Conference of the Manfred Sakel Foundation, New York Academy of Medicine), Max Rinkel, ed., L. C. Page, New York, (1966), pp. 30–37.

[66-10] Biochemical aspects of schizophrenia: Discussion of plasma protein factors and serum factors. In: *Biological Treatment of Mental Illness*, (Proceedings of the 11th International Conference of the Manfred Sakel Foundation), New York Academy of Medicine, Max Rinkel, ed., L. C. Page, New York, (1966), pp. 424–425.

* [68-4] Orthomolecular psychiatry: varying the concentrations of substances normally present in the human body may control mental disease. *Science* 160 (1968): 265–271. {SP 134}

[68-10] Vitamin therapy: treatment for the mentally ill. *Science* 160 (1968): 1181.

* [68-14] Orthomolecular somatic and psychiatric medicine (A communication to the Thirteenth International Convention on Vital Substances, Nutrition, and the Diseases of Civilization at Luxembourg and Trier, 18–24 September 1967). *Zeitschrift Vitalstoffe-Zivilisationskrankheiten* 1/68 (Jan 1968): 3–5. {SP 135}

[69-8] Advancement of knowledge: Orthomolecular psychiatry. In: *Centennial Lectures, 1968–1969: The Second Hundred Years*, Oregon State University Press, Corvallis, OR, (1969), pp. 19–27.

[70-30] Ortomoleculiarnye metody v meditsine (Russian: Orthomolecular methods in medicine). In: Funktsionalnaia Biokhimiya Kletochnykh Struktur (Sisakian Memorial Volume), A. I. Oparin, ed., Nauk, Moscow, (1970), pp. 427–432.

[71-1] Orthomolecular psychiatry. *Schizophrenia*, 3 (1971): 129–133.

[72-4] Nutrition against disease. *Executive Health* 8(4) (1972): 1–4.

[72-5] Preventive nutrition. *Medicine on the Midway* (Chicago, IL) 27 (1972) 15–18.

[72-8] Malattie molecolari e medicina ortomolecolare (Italian: Molecular disease and orthomolecular medicine). In: *Enciclopedia delia Scienza e della Tecnica Mondadorl, Mondadori*, Milan, (1972), pp. 258–266.

[72-10] Sugar: "Sweet and Dangerous". *Executive Health* 9 (1) (1972): 1–4.

[72-11] The new medicine? *Nutrition Today* 7(5) (Sept/Oct 1972): 18–19, 21–23.

* [73-4] Orthomolecular psychiatry. In *Orthomolecular Psychiatry: Treatment of Schizophrenia*, David Hawkins and Linus Pauling, eds., W. H. Freeman, San Francisco, (1973), pp. 1–17. {SP 138}

[73-7] Quantitative chromatographic analysis in orthomolecular medicine. In *Orthomolecular Psychiatry: Treatment of Schizophrenia*, David Hawkins and Linus Pauling, eds., W. H. Freeman, San Francisco, (1973), pp. 35–53. (Arthur B. Robinson and Linus Pauling).

[73-9] Orthomolecular diagnosis of mental retardation and diurnal variation in normal subjects by low-resolution gas-liquid chromatography of urine. *Intern. Res. Comm. Sys. (IRCS)* 1 (1973): 47. (A. B. Robinson, P. Cary, B. Dore, I. Keaveny, L. Brenneman, M. Turner, and Linus Pauling).

[73-14] What about vitamin E? *Executive Health* 10(1) (Jan 1973): 1–6.

[73-18] Nutrition and health. *Proceedings of the Royal Australian Chemical Institute, Supplement* 40(11) (1973): 1–24.

[74-1] Preventive orthomolecular medicine. In: *New Dynamics of Preventive Medicine*, Vol. 3, Leon R. Pomeroy, ed., Symposia Specialists, Miami, FL, (1974), pp. 13–19.

[74-3] Dr. Pauling comments on the comments. *Am. J. Psychiat.* 131 (1974): 1405–1406.

[74-4] Some aspects of orthomolecular medicine. *J. Int. Acad. Prev. Med.* 1 (Spring 1974): 1–30.

[74-8] Good nutrition for the good life. *Engineering and Science* 37 (June 1974): 6–9, 28–29.

[74-10] Techniques of orthomolecular diagnosis. *Clin. Chem.* 20 (1974): 961–965. (Arthur B. Robinson and Linus Pauling).

[74-14] On the orthomolecular environment of the mind: Orthomolecular theory. *Am. J. Psychiat.* 131(11) (1974): 1251–1257.

[74-17] On the molecular environment of the mind: Orthomolecular theory. *Psychopharmacology Bull.* 10(4) (1974): 6–7.

[75-2] Megavitamin dispute: two views. *Medical World News* (7 April 1975): 9.

[75-6] Megavitamin therapy. *J. Am. Med. Assoc.* 234 (1975): 149.

[76-2] Megavitamin therapy. *J. Am. Med. Assoc.* 235 (1976): 598.

[76-9] Megavitamin therapy reply. *J. Am. Med. Assoc.* 235 (1976): 1965.

[77-3] Diet, nutrition, and cancer. *Am. J. Clin. Nutr.* 30 (1977): 661, 663.

[77-9] Vitamin homeostasis in the brain and megavitamin therapy. *New Eng. J. Med.* 297 (1977): 790–791.

[77-17] What about vitamin E? in *A Physician's Handbook on Orthomolecular Medicine*, Roger J. Williams and Dwight W. Kalita, eds., Pergamon Press, New York, (1977), pp. 64–66.

[77-21] A controversy. *Clin. Chem.* 23 (1977): 908–909. (Arthur B. Robinson, Linus Pauling, and William Aberth).

[78-3] The AMA and orthomolecular medicine. *Medical Tribune* 19(2) (1978): 18.

[78-9] Orthomolecular enhancement of human development. In: *Human Neurological Development: Past, Present, and Future*, Ralph Pelligra, ed., NASA Conference Publication 2063 (1978): 47–51.

[78-10] William (sic: Robert) Fulton Cathcart, III, M. D.: An orthomolecular physician. *Linus Pauling Institute of Science and Medicine Newsletter* 1(4) (1978): 1–2, 5.

[79-4] Pauling disputes contention that his application lacked essential details. *The Cancer Letter* 5(17) (27 April 1979): 3–4.

[79-9] Dietary influences on the synthesis of neurotransmitters in the brain. *Nutrition Rev.* 37 (1979): 302–303.

[79-10] Megadose of misinformation. *Technology Rev.* 82 (1979): 4.

[79-11] Treating mental disorders. *Science* 206 (1979): 404.

[79-32] Linus Pauling answers critic! *Health News* (Martin Pharmaceuticals Pty., Burnley, Australia), (Aug–Sept 1979), pp. 1–2.

[81-14] The value of vitamins. In: *International Conference on Nutrition* (held in Tokyo, Japan in 1981), Medical News Group, London, (1981), pp. 8–9.

[81-21] On good nutrition for the good life. *Executive Health* 17(4) (Jan 1981): 1–6.

[81-22] Crystals in the kidney. *Linus Pauling Institute of Science and Medicine Newsletter* 1(11) (Spring 1981): 1, 2, 4.

[82-15] Nutrition and cancer. In: *Proceedings of the 1st Asian & Pacific Chemistry Congress (Singapore, 26 April – 1 May 1981)*, K. Y. Sim, S. H. Goh, S. Y. Lee, and A. S. C. Wan, eds., Singapore National Institute of Chemistry, Singapore, (1982), pp. 31–36.

[83-15] Herpes simplex infections. *Linus Pauling Institute of Science Newsletter* 2(4) (1983): 3.

[84-1] Sensory neuropathy from pyridoxine abuse. *New Eng. J. Med.* 310 (1984): 197.

[85-1] Megavitamins: Linus Pauling responds. *Chem. Brit.* 21 (1985): 27.

[85-3] Vitamin C: clinicians did not refute Pauling's claims. *Chem. Brit.* 21 (1985): 535.

[85-15] Problems introducing a new field of medicine. In: *Medical Science and the Advancement of Health*, Robert Lanza, ed., Praeger Pub., New York, (1985), pp. 129–140.

[85-17] Vitamine zur Steigerung des Wohlbefindens (English: Vitamins for increasing well-being). In: *Vitaminversorgung: eine Aufgabe der Ernährung*, Peter Eckes, Nieder-Olm, Germany, (1985), pp. 178–183.

[85-20] Progress in megavitamin therapy and orthomolecular science. In: *First Edition: 1984–85 Yearbook of Nutritional Medicine*, Jeffrey Bland, ed., Keats Publishing Inc., New Canaan, CT, (1985), pp. 1–21.

[86-2] *How to Live Longer and Feel Better.* W. H. Freeman, New York, (1986), 322 pp.

[86-12] The future of orthomolecular medicine. In: *The Roots of Molecular Medicine: A Tribute to Linus Pauling*, Richard P. Huemer, ed., W. H. Freeman, New York, (1986), pp. 249–253.

[87-30] Pauling points to "errors" in the review of his book. *Food Technology* 41 (11) (Nov 1987): 28.

[88-4] Fear of food. *The Sciences* 28 (3) (May/June 1988): 13.

[91-18] Vitamins and intelligence tests. *Nature* 353 (1991): 103.

[91-25] An orthomolecular theory of human health and disease. *J. Orthomolecular Med.* 6 (3/4) (1991): 135–138. (Linus Pauling and Matthias Rath).

[92-26] LPI—Where we've been... where we are... where we're going. *Linus Pauling Institute of Science and Medicine Newsletter*, (Winter, 1992–93), pp. 1, 3.

[92-30] Pauling not really surprised that so many doctors use E. *Medical Tribune*, 33 (24) (24 Dec 1992): 10.

[93-14] MDs overwhelmingly for vitamin E at 100 IU. *Medical Tribune*, (24 Dec 1993) 10.

[97-1] Therapeutic lysine salt composition and method of use. Patent no. 5650418, issued July 22, 1997.

Group 17b

Role of Vitamin C in Orthomolecular Medicine

* [70-14] Evolution and the need for ascorbic acid. *Proc. Natl. Acad. Sci.* 671 (1970): 1643–1648. {SP 136}

[72-2] There's something special about vitamin C. *Family Health*, 12 (June 1972): 50–51.

[72-12] Vitamin C. *Science* 177 (1972): 1152.

[73-19] Blood plasma L-ascorbic acid concentration for oral L-ascorbic acid dosage up to 12 grams per day. *Intern. Res. Comm. Sys. (IRCS), reprint (73-12) 10-19-9.* (Dec 1973). (A. Harris, A. B. Robinson, and Linus Pauling).

[73-22] On vitamin C against disease. *Executive Health* 9(5) (1973): 1–6.

[74-15] Are recommended daily allowances for vitamin C adequate? *Proc. Natl. Acad. Sci.* 71 (1974): 4442–4446.

[75-11] Vitamin C to kenka (Japanese: The value of vitamin C in improving health and in treating disease). *Kagaku* 31(2) (1975): 120–125.

[75-12] For the best of health, how much vitamin C do you need? *Executive Health* 12(3) (Dec 1975): 1–5.

[75-19] Decreased white blood cell count in people who supplement their diet with L-ascorbic acid. *Intern. Res. Comm. Sys. (IRCS), Medical Science* 3 (1975): 259. (Arthur B. Robinson, John F. Catchpool, and Linus Pauling).

[76-11] The case for vitamin C in maintaining health and preventing disease. *Modern Medicine*, (1 July 1976): 69–72.

[76-14] Why you have a Vitamin C problem. *The Good Drugs Do to Better Your Health. Supplement to Medical Tribune*, (15 Sept 1976): 20.

[76-28] Plädoyer für Vitamin C (German: The case for Vitamin C). *Medizin Zeits. f. Diagnose u. Therapie*, 4 (1976): 1954–1958,.

[77-4] How vitamin C improves health. *Linus Pauling Institute of Science and Medicine Newsletter* 1(1) (1977): 1.

[77-14] Albert Szent-Györgyi and vitamin C. In: *Search and Discovery: A Tribute to Albert Szent-Györgyi*, B. Kraminer, ed., Academic Press, New York, (1977), pp. 43–54.

[77-15] A re-evaluation of vitamin C. *Re-evaluation of Vitamin C*, A. Hanck and G. Ritzel, eds., Verlag Hans Huber, Bern, (1977): 9–17.

[77-23] L-ascorbic acid and infectious diseases. In: *Encyclopedia of Food Science*, Martin S. Peterson and Arnold H. Johnson, eds., Avi Publishing Co., Westport, CT, (1977), pp. 51–53.

[77-25] New Horizons for Vitamin C. *Nutrition Today* 12(1) (1977): 6–13.

[78-18] Vitamin C. *The Good Drugs Do to Better Your Health*, Medical Tribune, Inc., New York, Louis Lasagna, ed., (1978), pp. 101–108.

[78-20] Una reevaluación de la vitamina C. (Spanish: A re-evaluation of vitamin C). *Investigacion Modica Internactional* (Mexico) 5 (1978): 9–14.

[78-25] To Dose or Megadose: A Debate About Vitamin C. *Nutrition Today*, (March/April 1978): 6–13, 18–33.

[79-2] Ascorbic acid. *Lancet* 1 (1979): 615.

[79-18] Research on ascorbic acid and the dying patient. In: *Stress and Survival*, Charles A. Garfield, ed., C. V. Mosby Co., St. Louis, (1979), pp. 363–366.

[79-19] Ascorbic acid in relation to disease. In: *Medicinal Chemistry VI. Proceedings of the Sixth International Symposium on Medicinal Chemistry*, M. A. Simkins, ed., Cotswold Press, Oxford, (1979), pp. 1–5.

[79-31] Recent Progress with Vitamin C. *Nutrition*, No. 2 (Summer 1979): 14–19.

[81-1] Mega-ascorbate taken with other vitamins permits elevation of circulating vitamins including B12 in humans. *Nutrition Reports Intern.* 23 (1981): 669–677. (Herman Baker, Linus Pauling, and Oscar Frank).

[81-2] Rx: 10,000 mg vitamin C. *Executive* (Graduate School of Business and Public Administration, Cornell Univ.) 7(2) (1981): 28–30.

[81-7] Vitamin C prophylaxis for post-transfusion hepatitis. *Am. J. Clin. Nutr.* 34 (1981): 1978–1980.

[82-12] Lack of effect of ascorbic acid on calcium excretion. *IRCS Medical Science – Biochemistry*, 10 (1982): 736. (Linus Pauling, C. S. Tsao, and S. L. Salimi).
[83-1] On vitamin C and infectious diseases. *Executive Health* 19(4) (Jan 1983): 1–5.
[83-5] Comments on vitamin C. *Patient Care*, (15 May 1983): 64–74.
[83-10] Vitamin C and longevity. *Agressologie*, 24 (1983): 317–319,.
[84-14] Vitamin C and pregnancy. In *Women's Health Care: A Guide to Alternatives*, Kay Weiss, ed., Reston Publishing Co., Reston, VA, (1984): 229–232.
[85-16] Vitamin C and health. In: *Frontiers in Longevity Research*, Robert J. Morin, ed., Charles C Thomas Pub., Springfield, IL, (1985), pp. 155–161.
[85-25] The Vitamin C Debate. *Science* 85 (July/Aug 1985): 15.
[86-14] Vitamin C for cocaine withdrawal. *Medical Tribune* 27(23) (20 Aug 1986): 14.
[89-6] Vitamin C papers. *Science* 243 (1989): 1535.
[89-11] Vitamin C and quackery. *Medical Tribune* 30(15) (25 May 1989): 15.
* [90-8] Suppression of human immunodeficiency virus replication by ascorbate in chronically and acutely infected cells. *Proc. Natl. Acad. Sci.* 87 (1990): 7245–7249. (Steve Harakeh, Raxit J. Jariwalla, and Linus Pauling). {SP 142}
[91-30] Respect for vitamin C. *Science* 254 (1991): 1712.
[92-11] After more than 20 years, why won't doctors accept Vitamin C? *The Philadelphia Inquirer* (24 May 1992): E7.
[92-20] Vitamin C: The key to health. In: *Stop the FDA: Save Your Health Freedom*, John Morgenthaler and Steven Wm. Fowkes, eds., Health Freedom Publications, Menlo Park, CA, (1992), pp. 45–62.
[92-23] My love affair with vitamin C. *Health Care USA* (Fall 1992): 7.
[94-4] The marvel of vitamin C. *Muscle & Fitness*, (February 1994): 130–133.

Group 17c

Vitamin C and the Common Cold

[70-5] The controversy over vitamin C (ascorbic acid) to prevent or relieve the common cold. *Executive Health* 7(6) (June 1970): 1–4.
[70-18] *Vitamin C and the Common Cold*. W. H. Freeman, San Francisco, (1970), 122 pp.
* [71-2] Ascorbic acid and the common cold. *Am. J. Clin. Nutr.* 24 (1971): 1294–1299. {SP 137}

[71-6] Vitamin C and the common cold. *Can. Med. Assoc. J.* 105 (1971): 448–450.

[71-7] The common cold. *The Listener* (27 May 1971): 683.

[71-8] The significance of the evidence about ascorbic acid and the common cold. *Proc. Natl. Acad. Sci.* 68 (1971): 2678–2681.

[71-10] Vitamin C and common cold. *J. Am. Med. Assoc.* 216 (1971): 332.

[71-13] The Controversy over vitamin C to prevent or relieve the common cold. *Executive Health.* 7(6) (1971): 1–2.

[72-1] Ascorbic acid has value in combatting the common cold. *Medical Counterpoint* 4 (Feb 1972): 14 43–45, 49.

[73-8] Ascorbic acid and the common cold. *Scottish Med. J.* 18 (1973), 1–2.

[74-5] Early evidence about vitamin C and the common cold. *J. Orthomolecular Psychiatry* 3 (1974), 139–151.

[76-5] Ascorbic acid and the common cold: evaluation of its efficacy and toxicity. Part I. *Medical Tribune* 17(12) (24 March 1976): 18–19.

[76-6] Ascorbic acid and the common cold. Part II. *Medical Tribune* 17(13) (7 April 1976): 37–38.

[76-13] Why you need vitamin C to fight your cold. The Good Drugs Do to Better Your Health. *Suppl. to Medical Tribune* (15 Sept 1976): 14–15.

[76-25] *Vitamin C, the Common Cold, and the Flu.* W. H. Freeman, San Francisco, (1976) 230 pp.

[79-7] Linus Pauling talks about Vitamin C, colds, and cancer. *The Health Foods Communicator* (Minneapolis, MN) 4 (May–June 1979): 24–27.

[85-19] Dr. Linus Pauling's case against the Mayo Clinic. *Australasian Health & Healing,* 5 (Dec 1985–Feb 1986): 31–32.

[92-27] Vitamin C and the common cold — after 23 years. *Linus Pauling Institute of Science and Medicine Newsletter* (Winter 1992–93): 2–3.

[93-13] Viewpoint: Vitamin C and the common cold. *Optimum Nutrition* (Journal of the Institute for Optimum Nutrition, 5 Jerdan PI., London SW6 1BE, England) 6(3) (Winter 1993): 10.

Group 17d

Vitamin C and Cancer

* [73-10] Ascorbic acid and the glycosaminoglycans: an orthomolecular approach to cancer and other diseases. *Oncology* 27 (1973): 181–192. (Ewan Cameron and Linus Pauling). {SP 139}

[74-13] The orthomolecular treatment of cancer. I. The role of ascorbic acid in host resistance. *Chem.-Biol. Interact.* 9 (1974): 273-283. (Ewan Cameron and Linus Pauling).
[74-23] The orthomolecular treatment of cancer. II. Clinical trial of high-dose ascorbic acid supplements in advanced human cancer. *Chem.-Biol. Interact.* 9 (1984): 285–315. (Ewan Cameron and Linus Pauling).
* [76-17] Supplemental ascorbate in the supportive treatment of cancer: Prolongation of survival times in terminal human cancer. *Proc. Natl. Acad. Sci.* 73 (1976): 3685–3689. (Ewan Cameron and Linus Pauling). {SP 140}
[76-29] Ascorbic acid and cancer. *National Academy of Preventive Medicine Journal* 1 (1976): 6–10.
[77-1] On vitamin C and cancer. *Executive Health* 13(4) (Jan 1977): 1–5.
[77-11] Vitamin C and cancer. *Linus Pauling Institute of Science and Medicine Newsletter* 1(2) (1977): 1–2.
[77-19] Vitamin C and cancer. *Intern. J. Environ. Studies* 10 (1977): 303–305. (E. Cameron and Linus Pauling).
[78-12] Experimental studies designed to evaluate the management of patients with incurable cancer. *Proc. Natl. Acad. Sci.* 75 (1978): 6252. (Ewan Cameron and Linus Pauling).
[78-19] Ascorbic acid as a therapeutic agent in cancer. *J. Intern. Acad. Prev. Med.* 5(1) (1978): 8–29. (Ewan Cameron and Linus Pauling).
* [79-1] Ascorbic acid and cancer: A review. *Cancer Research* 39 (1979): 663–681. (Ewan Cameron, Linus Pauling, and Brian Leibovitz). {SP 141}
[79-3] Ascorbate and cancer. *Proc. Am. Phil. Soc.* 123 (1979): 117–123. (Ewan Cameron and Linus Pauling).
[79-15] *Cancer and Vitamin C.* Linus Pauling Institute of Science and Medicine, Palo Alto, CA, (1979). 238 pp. (Ewan Cameron and Linus Pauling).
[79-22] The role of vitamin C in cancer. *Intern. J. Vitamin and Nutr. Res. Suppl.* No. 19 (1979): 207–210.
[80-2] Chemotherapy and vitamin C. *Science News* 117(1) (5 Jan 1980): 3.
[80-3] On cancer and vitamin C. *Executive Health* 16(4) (Jan 1980): 1–8. (Ewan Cameron and Linus Pauling).
[80-4] Vitamin C therapy of advanced cancer. *New Engl. J. Med.* 302 (1980): 694–695.
[80-5] Vitamine C en Kanker (Dutch: Vitamin C and cancer). *Chemisch Magazine,* (March 1980): 151.
[80-10] A report on "Cancer Dialogue '80". *Linus Pauling Institute of Science and Medicine Newsletter* 1(10) (1980): 1–4.
[81-10] Survival times of terminal lung cancer patients treated with ascorbate. *J. Intern. Acad. Prev. Med.* 6 (1981): 21–27. (Ewan Cameron and Linus Pauling).

[82-3] Megadoses of vitamin C as an adjunct in the treatment of cancer. *Your Patient and Cancer* 2 (May 1982): 39–46. (Ewan Cameron and Linus Pauling).

[82-5] Laetrile. *New Eng. J. Med.* 307 (1982): 118–119.

[82-10] Effects of intake of L-ascorbic acid on the incidence of dermal neoplasms induced in mice by ultraviolet light. *Proc. Natl. Acad. Sci.* 79 (1982): 7532–7536. (Wolcott B. Dunham, Emile Zuckerkandl, Ruth Reynolds, Richard Willoughby, Richard Marcuson, Roger Barth, and Linus Pauling).

[82-16] Incidence of squamous cell carcinoma in hairless mice irradiated with ultraviolet light in relation to intake of ascorbic acid (vitamin C) and of D, L-α-tocopheryl acetate (vitamin E). In: *Vitamin C. New Clinical Applications in Immunology, Lipid Metabolism, and Cancer.* International Journal for Vitamin and Nutrition Research Supplement No. 23, A. Hanck, ed., Hans Huber Publishers, Bern, (1982), pp. 53–82. (Linus Pauling, Richard Willoughby, Ruth Reynolds, B. Edwin Blaisdell, and Stephen Lawson).

[82-17] Vitamin C und Krebs (German: Vitamin C and cancer). *Naturwissenschaftliche Rundschau* 35(4) (1982): 137–139.

[83-2] Enhancement of antitumor activity of ascorbate against Ehrlich ascites tumor cells by the copper: glycylglycylhistidine complex. *Cancer Research* 43 (1983): 824–828. (Eiji Kimoto, Hidehiko Tanaka, Junichiro Gyotoku, Fukumi Morishige, and Linus Pauling).

[83-4] Vitamin C and cancer. *Australasian Health & Healing* 2(3) (April–June 1983): 9–10. (Linus Pauling and Ewan Cameron).

[85-5] Effect of dietary ascorbic acid on the incidence of spontaneous mammary tumors in RIII mice. *Proc. Natl. Acad. Sci.* 82 (1985): 5185–5189. (Linus Pauling, Jon C. Nixon, Fred Stitt, Richard Marcuson, Wolcott B. Dunham, Roger Barth, Klaus Bensch, Zelek S. Herman, B. Edwin Blaisdell, Constance Tsao, Marilyn Prender, Valerie Andrews, Richard Willoughby, and Emile Zuckerkandl).

[86-1] A proposition: megadoses of vitamin C are valuable in the treatment of cancer. *Affirmative. Nutr. Rev.* 44 (Jan 1986): 28–29.

[89-19] Cancer, chemotherapy and vitamin C. *Linus Pauling Institute of Science and Medicine Newsletter* 3(2) (Winter 1989): 7.

[91-22] Effect of ascorbic acid on incidence of spontaneous mammary tumors and UV-light-induced skin tumors in mice. *Am. J. Clin. Nutr.* Suppl. to Vol. 54(6) (1991): 1252S–1255S.

* [93-2] *Cancer and Vitamin C* (Updated and Expanded Edition). Camino Books, Philadelphia, PA, (1993) 278 pp. {SP 144}

Group 17e

Vitamin C and Cardiovascular Disease

[78-1] Vitamin C and heart disease. *Executive Health* 14(4) (Jan 1978): 1–5.

[79-30] Ascorbic acid as a therapeutic agent in cancer. *J. IAPM* V(1) (1978): 8–29.

[85-8] Vitamin C und das Herz (German: Vitamin C and the heart). *Orthomolekular Medizin und Ernährung* 1(1) (Oct 1985): 13–18.

[90-5] Hypothesis: Lipoprotein(a) is a surrogate for ascorbate. *Proc. Natl. Acad. Sci.* 87 (1990): 6204–6207. (Matthias Rath and Linus Pauling). Errata: *Proc. Natl. Acad. Sci.* 88 (1991): 11588.

[90-12] Immunological evidence for the accumulation of lipoprotein(a) in the atheroscle-rotic lesion of the hypoascorbemic guinea pig. *Proc. Natl. Acad. Sci.* 87 (1990): 9388–9390. (Matthias Rath and Linus Pauling).

[91-15] Vitamin C and cardiovascular disease. *Medical Science Research* 19(13) (July 1–15 1991): 398.

[91-23] Solution to the puzzle of human cardiovascular disease: Its primary cause is ascorbate deficiency leading to the deposition of lipoprotein(a) and fibrinogen/ fibrin in the vascular wall. *J. Orthomolecular Med.* 6(3/4) (1991): 125–134. (Matthias Rath and Linus Pauling).

[91-26] Apoprotein(a) is an adhesive protein. *J. Orthomolecular Med.* 6(3/4) (1991): 139–143. (Matthias Rath and Linus Pauling).

[91-28] Case report: Lysine/ascorbate-related amelioration of angina pectoris. *J. Orthomolecular Med.* 6(3/4) (1991): 144–146.

* [92-9] A unified theory of human cardiovascular disease leading the way to the abolition of this disease as a cause for human mortality. *J. Orthomolecular Med.* 7(1) (1992): 5–15. (Matthias Rath and Linus Pauling). Also in *Journal of Applied Nutrition*, 1992. {SP 143}

[92-10] Plasmin-induced proteolysis and the role of apoprotein(a), lysine, and synthetic lysine analogs. *J. Orthomolecular Med.* 7(1) (1992): 17–23. (Matthias Rath and Linus Pauling).

[92-24] Cardiovascular disease and vitamin C. *Health Care USA*, (Fall 1992): 7.

[92-28] Vitamin C and lipoprotein(a) in relation to cardiovascular disease and other diseases. *J. Optimal Nutrition* 1(1) (1992): 61–64. (Matthias Rath and Linus Pauling).

[93-5] A case history: lysine/ascorbate-related amelioration of angina pectoris. *J. Orthomolecular Med.* 8(2) (1993): 77–78. (Marie McBeath and Linus Pauling).

[93-9] Third case report on lysine-ascorbate amelioration of angina pectoris. *J. Orthomolecular Med.* 8(3) (1993): 137–138.

[93-16] Use of ascorbate and tranesamic acid solution for organ and blood vessel treatment prior to transplantation. Patent no. 5230996, issued July 27, 1993.

[94-13] Prevention and treatment of occlusive cardiovascular disease with ascorbate and substances that inhibit the binding of lipoprotein (a). Patent no. 5278189, issued January 11, 1994.

Group 18

Biographical Memoirs, Memorials, Tributes

[45-7] Roscoe Gilkey Dickinson, 1894–1945. *Science* 102 (1945): 216.

[51-8] Arnold Sommerfeld: 1868–1951. *Science* 114 (1951): 383–384.

[55-19] Hugo Theorell. *Science* 122 (1955): 1222–1223.

[56-14] Amedeo Avogadro. *Science* 124 (1956): 708–713.

[58-26] Arthur Amos Noyes: Sept 13, 1866–June 3, 1936 (A biographical memoir). In: *Biographical Memoirs*, Vol. 31, Columbia University Press (for the National Academy of Sciences of the United States), New York, 1958, pp. 322–346

[65-11] Albert Schweitzer, Médico y Humanitario (Spanish: Albert Schweitzer, Physician and Humanitarian). *Folia Humanistica, Cienct'as Aries Leiras* (Barcelona) 3 (36) (Dec 1965): 963–971. (Frank Catchpool and Linus Pauling).

[65-26] Doktor der Humanität (German: Doctor of humanity). In: *Albert Schweitzer: Beiträge zu Leben und Werk*, Union Verlag, Berlin, (1965), pp. 116–122. (Linus Pauling and Frank Catchpool).

[70-29] Albert Tyler. *Dev. Biol.* 21 (1970): iii–v.

[72-9] Bernal's contributions to structural chemistry. *Scientific World* 16 (1972): 13–14.

[74-21] Dickinson, Roscoe Gilkey. In: *Dictionary of Scientific Biography, Vol. IV*, Charles Coulston Gillespie, ed., Charles Scribner's Sons, New York, (1974), p. 82.

[74-22] Noyes, Arthur Amos. In: *Dictionary of Scientific Biography, Vol. X*, Charles Coulston Gillespie, ed., Charles Scribner's Sons, New York, (1974), pp. 156–157.

[76-21] Arthur Amos Noyes. In: *Proceedings of the Robert A. Welch Foundation Conferences on Chemical Research, Vol. XX, American Chemistry-Bicentennial* (Nov 8–10, 1976, Houston, Texas), Robert A. Welch Foundation, Houston, TX, (1976), pp. 88–101.

[78-14] Dr. Ewan Cameron. *Linus Pauling Institute of Science and Medicine Newsletter* 1(5) (1978): 1–2.

[82-13] Memorial of Sterling Brown Hendricks (April 13, 1902–January 4, 1981). *Am. Mineral.* 67 (1982), 406–409.

[83-9] Pauling on G. N. Lewis. *Chemtech* 13 (1983): 334–337.

[83-14] Albert Szent-Györgyi on his ninetieth birthday. *Found. Phys.* 13 (1983): 883–886.

[84-2] G. N. Lewis and the chemical bond. *J. Chem. Ed.* 61 (1984): 201–203.

[84-9] Herman Mark and the structure of crystals. *Chemtech* 14 (1984), 334–337.

[86-20] One aspect of the physical sciences in relation to biology. (Presented in San Remo, Italy on 5–6 May 1983 in commemoration of the 150th birthday of Alfred Nobel.) *Cell Biophys.* 9(1, 2) (Dec 1986): 3–5.

[87-8] Schrodinger's contributions to chemistry and biology. In: *Schrodinger: Centenary Celebration of a Polymath*, C. W. Kilmister, ed., Cambridge Univ. Press, Cambridge, England, (1987), pp. 225–233.

[87-10] Dorothy Wrinch and the Structure of Proteins. *J. Chem. Ed.* 64 (1987): 286.

[87-19] Tribute to Dr. Arthur M. Sackler. Studio International, *Int. J. Creative Arts and Design* 200 (Suppl. 1) (1987): 21.

[87-28] Patterson and bixbyite. In: *Patterson and Pattersons: Fifty Years of the Patterson Function*, Jenny P. Glusker, Betty K. Patterson, and Miriam Rossi, eds., International Union of Crystallography, Oxford Univ. Press, New York, (1987), pp. 42–44.

[87-29] The impact of the Patterson function. In: *Patterson and Pattersons: Fifty Years of the Patterson Function*, Jenny P. Glusker, Betty K. Patterson, and Miriam Rossi, eds., International Union of Crystallography, Oxford Univ. Press, New York, (1987), pp. 49–67.

[87-31] My indebtedness to Herman Mark. In: *Polymer Science in the Next Decade: Trends, Opportunities, Promises (An International Symposium Honoring Herman F. Mark on His 90th Birthday)*, Otto Vogl and Edmund H. Immergut, eds., John Wiley Sons, Inc., New York, (1987), pp. 19–22.

[88-18] Reminiscences on Szent-Györgyi. *Biol. Bull.* 174 (1988): 229.

[90-13] My indebtedness to and my contacts with Lawrence Bragg. In: *Selections and Reflections: The Legacy of Sir Lawrence Bragg*, John M. Thomas and Sir David Phillips, eds., Science Reviews Ltd., Northwood, Middlesex, United Kingdom, for the Royal Institution of Great Britain, (1990), pp. 86–88.

[94-6] Thomas Addis (July 27, 1881–June 4, 1949). In: *Biographical Memoirs* (National Academy of Sciences), Vol. 63, National Academy Press, Washington, D. C., (1994), pp. 2–46. (Kevin V. Lemley and Linus Pauling).

Group 19

Book Reviews

[32-1] Review of: *The Structure of Crystals*, 2nd ed. by Ralph W. G. Wyckoff. *Industrial and Engineering Chemistry* 24 (January 1932): 117–118.

[32-13] Review of: *Theory of Electric and Magnetic Susceptibilities* by J. H. van Vleck. *J. Am. Chem. Soc.* 54 (1932): 4119–4121.

[34-0] Review of *A Study of Crystal Structure and its Applications* by Wheeler P. Davey (New York: McGraw-Hill, 1934). *Science* 80 (1934): 451.

[35-16] Review of: *The Structure of Crystals*. Supplement for 1930–34 to the 2nd edition. by R. W. G. Wyckoff (New York: Reinhold, 1935). *Science* 82 (1935): 372.

[37-9] Review of *Magnetochemie* (Magnetochemistry) by Dr. Wilhelm Klemm (Leipzig: Akademische Verlagsgesselleschaft, 1936). *J. Am. Chem. Soc.* 59 (1937): 1159–1160.

[38-1] Review of *Atomic Structure of Minerals* by W. L. Bragg (Ithaca: Cornell University Press, 1937). *J. Am. Chem. Soc.* 60 (1938): 220.

[38-2] Review of *Einführung in die Quantenchemie* (Introduction to Quantum Chemistry) by Dr. Hans Hellmann (Vienna: Verlag Franz Deuticke, 1937). *J. Am. Chem. Soc.* 60 (1938): 734.

[38-3] Book review of: *The Fundamental Principles of Quantum Mechanics, With Elementary Applications* by Edwin C. Kemble (New York: McGraw-Hill, 1937). *J. Am. Chem. Soc.* 60 (1938): 734–735.

[38-4] Book review of: *The Fine Structure of Matter. Part I. X-Rays and the Structure of Matter*. Vol. II of *"A Comprehensive Treatise of Atomic and Molecular Structure"* by C. H. Douglas Clark (New York: John Wiley and Sons,1937). *J. Am. Chem. Soc.* 60 (1938): 988.

[38-9] Review of *The Fine Structure of Matter. Part II. Molecular Polarization. Part III. The Quantum Theory and Line Spectra* by C. H. Douglas Clark (New York: John Wiley and Sons, 1938). *J. Am. Chem. Soc.* 60 (1938): 2833–2834.

[41-6] Review of *Electronic Processes in Ionic Crystals* by N. F. Mott and R. W. Gurney (New York: Oxford University Press, 1940). *J. Phys. Chem.* 45 (1941): 1142.

[44-4] Book review of: *Magnetochemistry* by Pierce W. Selwood. *Chemical & Metallurgical Engineering* 51 (Sept 1944) 208–209.

[44-7] Book review of: *Fundamentals of Immunology* by William C. Boyd. *Arch. Biochem.* 4 (1944): 126. (Linus Pauling and Dan H. Campbell).

[45-2] Review of *Advances in Protein Chemistry, Vol. 1*, M. L. Anson and John T. Edsall, eds. (New York: Academic Press, 1944) *J. Am. Chem. Soc.* 67 (1945): 886.

[50-10] Review of *Progress in Biochemistry — A Report on Biochemical Problems and on Biochemical Research Since 1939* by Felix Haurowitz (New York: Interscience Publications, 1950). *Chemical Engineering* (Sept 1950) 253–254.

[51-22] Review of *Structure of Molecules and the Chemical Bond* by Y. K. Syrkin and M. E. Dyatkina. *J. Applied Phys.* 22 (1951): 1392.

[52-4] Review of *States of Matter*, Vol. 2 of the third edition of *A Treatise on Physical Chemistry* by Hugh S. Taylor and Samuel Glasstone. *Chem. & Eng. News*, 30 (3 March 1952): 928.

[52-14] Quantum mechanics of valence: Book review of *Valence* by C. A. Coulson (Oxford: Clarendon, 1952). *Nature* 170 (1952): 384.

[54-1] Review of *International Tables for X-Ray Crystallography, Vol. 1. Symmetry Groups*, N. F. M. Henry and K. Lonsdale, eds. (Birmingham: Kynoch Press, 1952). *Acta Cryst.* 7 (1954): 304.

[54-2] Review of *The Proteins. Chemistry, Biological Activity, and Methods. Vol. 1, part A,* Hans Neurath and Kenneth Bailey, eds. (New York: Academic Press, 1953). *Arch. Biochem. Biophys.* 49 (1954): 254.

[54-11] Review of *Nature and Structure of Collagen: Papers presented for a discussion convened by the Colloid and Biophysics Committee of the Faraday Society at King's College, London, 26–27 Mar. 1953,* J. T. Randall, ed., *Science* 120 (1954): 133.

[54-12] Review of *The Chemical Structure of Proteins*, edited for the Ciba Foundation by G. E. W. Wolstenholme and Margaret P. Cameron. *Arch. Biochem. Biophys.* 53 (1954): 522–523.

[55-3] Review of *The Kinetic Basis of Molecular Biology* by Frank H. Johnson, Henry Eyring, and Milton J. Polissar. *Arch. Biochem. Biophys.*, 55 (1955): 300–301.

[55-13] Review of *The Strengths of Chemical Bonds* by T. L. Cottrell (New York: Academic Press, 1954). *Arch. Biochem. Biophys.* 59 (1955): 543.

[56-17] Review of *Valency and Molecular Structure* by E. Cartmell and G. W. A. Fowles (New York: Butterworths, 1956). *Science* 124 (1956): 1213.

[57-10] Review of *Synthetic Polypeptides: Preparation, Structure, and Properties* by C. H. Bamford, A. Elliot, and W. E. Hanby. *Arch. Biochem. Biophys.* 72 (1957): 250.

[59-11] Review of *Bone and Radiostrontium* by Arne Engström, Rolf Björnerstedt, Carl Johan Clemedson, and Arne Nelson. *Arch. Biochem. Biophys.* 80 (1959): 475.

[59-17] Review of: *The Way Things Are* by P. W. Bridgman. *Perspectives in Biology and Medicine* 3(1) (Autumn 1959): 152–153.

[76-12] Review of *The Healing Factor: Vitamin C Against Disease* by Irwin Stone. *Libertarian Rev.* 5(4) (July–August 1976): 18.

[82-4] Review of: *A Theoretical Approach to the Preselection of Carcinogens and Chemical Carcinogenesis* by Veljko Veljkovi. *Quarterly Rev. Biol.* 57 (June 1982): 228–229.

[87-21] Review of *High-Protein Oedemas and the Benzo-Pyrones,* by J. R. Casley-Smith and Judith R. Casley-Smith. *J. Orthomolecular Med.* 2(3) (1987): 197–198.

[92-5] Book Review: Vitamin C. *Its Chemistry and Biochemistry* by Michael B. Davies, John Austin, and David A. Partridge. *J. Am. Chem. Soc.* 114 (1992): 3171–3172.

Group 20

General, Retrospective, Prospective, and Miscellaneous

[20-1] The manufacture of cement in Oregon. *The Student Engineer* (The Associated Engineers of Oregon Agricultural College, Corvallis, Oregon) 12 (June 1920): 3–5.

[38-11] The future of the Crellin laboratory. *Science* 87 (1938): 563–566.

[41-5] Prolycopene, a naturally occurring stereoisomer of lycopene. *Proc. Natl. Acad. Sci.* 27 (1941): 468–474. (L. Zechmeister, A. L. LeRosen, F. W. Went, and Linus Pauling).

[42-7] Ounce molecular weight. *J. Chem. Ed.*, 19 (1942): 494.

[43-3] Spectral characteristics and configuration of some stereoisomeric carotenoids including prolycopene and pro-gamma-carotene. *J. Am. Chem. Soc.* 65 (1943): 1940–1951. (L. Zechmeister, A. L. LeRosen, W. A. Schroeder, A. Polgár, and Linus Pauling).

[43-5] Imagination in science. *Tomorrow* 3 (Dec 1943): 38–39.

[47-3] Unsolved problems of structural chemistry. (Acceptance address for the Theodore William Richards Medal 1947, awarded by the Northeastern Section of the American Chemical Society). *Chem. & Eng. News* 25 (13 Oct 1947): 2970–2974.

[47-4] *General Chemistry: An Introduction to Descriptive Chemistry and Modern Chemical Theory.* W. H. Freeman, San Francisco, (1947; 1953; 1970), 959 pp.

[48-5] Chemical achievement and hope for the future. *Am. Scientist* 36 (1948): 51–58.

[49-3] Chemical achievement and hope for the future. (Silliman Lecture presented at Yale University in October, 1947, on the occasion of the Centennial of the Sheffield Scientific School.) in *Science in Progress* (Sigma Xi National Lectureships, 1947 and 1948, Sixth Series), George A. Baitsell, ed. (New Haven, CT: Yale University Press, 1949), pp. 100–121.

[49-17] Zur *cis-trans*-Isomerisierung von Carotinoiden. (German: On the *cis-trans* isomerization of the carotinoids). *Helv. Chim. Acta* 32 (1949): 2241–2246.

[49-18] Chemistry and the world of today. *Chem. & Eng. News* 27 (26 Sept 1949): 2775–2778.

[50-2] Structural chemistry in relation to biology and medicine. (Second Bicentennial Science Lecture of the City College Chemistry Alumni Association, New York, 7 December 1949). *Baskerville Chem. J.* 1 (1950): 4–7, also published in *Medical Arts and Sciences,* Vol. 4, No. 4, Fourth Quarter 1950.

[50-7] The place of chemistry in the integration of the sciences. *Main Currents in Modern Thought* 7 (1950): 108–111.

[50-8] Chemistry. *Scientific American* 183 (September 1950): 32–35.

[51-3] The significance of chemistry to man in the modern world. *Engineering and Science* 14 (1951): 10–14.

[51-23] Science in the modern world (August sermon of the month delivered by Dr. Linus Pauling, at the First Unitarian Church of Los Angeles, August 12, 1951.) (First Unitarian Church of Los Angeles, 2936 W. 8th St., Los Angeles, CA, 5 pp., 1951.)

[51-24] Chemical achievement and hope for the future. In *The Smithsonian Report for 1950* (Washington, DC: Smithsonian Institution, 1951), pp. 225–241.

[53-32] The periodic law. In *Encyclopedia Britannica,* Vol. 18, (1953), pp. 517–521. William Benton, Chicago.

[55-9] Modern structural chemistry. *Science* 123 (1956): 255–258.

[56-9] Where is science taking us? The research frontier. *The Saturday Review* 39 (24 March 1956): 58.

[62-2] Problems of inorganic structures. In: *Fifty Years of X-Ray Diffraction*, P. P. Ewald, ed., N. V. A. Oosthoek's Uitgeversmaatschappij, Utrecht, (1962), pp. 136–146.

[62-3] Early work on X-ray diffraction in the California Institute of Technology. In *Fifty Years of X-Ray Diffraction*, P. P. Ewald, ed., N. V. A. Oosthoek's Uitgeversmaatschappij, Utrecht, (1962), pp. 623–628.

[63-18] El Metileno y los Carbenos (Spanish: Methylene and the carbenes). *Afinidad* 20 (Nov–Dec 1963): 393–396. (Linus Pauling and Gustav Albrecht).

[64-8] Science and peace (Nobel Lecture, Oslo, December 11, 1963, on receiving the 1962 Nobel Prize for Peace). In: *Les Prix Nobel en 1963*, M. G. Liljestrand, ed., Kungligen Boktryckeriet P. A. Norstedt och Söner, Stockholm, (1964), pp. 296–312.

[64-21] The possibilities for further progress in medicine through research. *Univ. Toronto Medical J.* 42(1) (Nov 1964): 7–8.

[64-24] (In Russian:) Chemical and molecular philogeny. In: Problems of Evolutional and Technical Biochemistry — Dedicated to Academician A. I. Oparin on his 70th Birthday, Akademia Nauk, Moscow, (1964), pp. 54–62.

[64-30] *The Architecture of Molecules.* (W. H. Freeman, San Francisco, 1964.) Linus Pauling and Roger Hayward.

[65-7] Fifty years of physical chemistry in the California Institute of Technology. *Ann. Rev. Phys. Chem.*, 16 (1965): 1–13.

[65-9] The science of science. *Scientific World* 9(3) (1965): 29–30.

[66-3] Die Beziehungen zwischen Molekülstruktur und medizinischen Problemen (German: The connection between molecular structure and medical problems). *Naturwissenschaftliche Rundschau* 19(6) (1966): 217–222.

[67-1] Molecular chemistry. *Industrial Research* 9(1) (Jan 1967): 74–75.

[67-8] *The Challenge of Scientific Discovery.* The Twenty-Ninth William Henry Snyder Lecture, Associated Students, The College Press, Los Angeles City College, Los Angeles, (1967), 12 pp.

[68-15] Science and the World of the Future. In *Man and His World: The Noranda Lectures, Expo 67* (1968), pp. 203–211. University of Toronto Press, Toronto.

[68-18] Scientific Publications of Linus Pauling (by Gustav Albrecht). In: *Structural Chemistry and Molecular Biology: A Volume Dedicated to Linus Pauling by his Students, Colleagues, and Friends.* Alexander Rich and Norman Davidson, eds., W. H. Freeman & Co., San Francisco, (1968), pp. 887–907.

[72-13] Stirling approximation. *Chem. in Brit.* 8(10) (Oct 1972): 447.

[73-21] Chemistry in medicine. *Bull. Missouri Acad. Sci.* 1 (3) (1972/1973): 21–26.

[74-16] Periodic law. In: *Encyclopeadia Britannica*, Macropaedia, Helen Hemingway Benton, New York, 15th edition, (1974), pp. 75–81.

[75-17] *Chemistry.* W. H. Freeman, San Francisco, (1975), 767 pp. (Linus Pauling and Peter Pauling).

[76-7] What can we expect for chemistry in the next 100 years? (Centennial Address, American Chemical Society, presented in New York on 5 April 1976). *Chem. Eng. News*, (19 April 1976): 33–36.

[76-20] s-triazines. I. Reaction of cyanuric chloride with unsaturated nitrogen compounds. *J. Org. Chem.* 41 (1976): 2032–2034. (Koichi Miyashita and Linus Pauling).

[77-10] Chemistry, the past century and the next. (Evening Lectures from the Royal Institution). *Speaking of Science*, Taylor & Francis, Ltd., London, (1977), pp. 279–287.

[79-23] The crisis in scientific research. In: *The Crisis in Scientific Research* (conference proceedings), A. Montagu and conference chairman, ed., Institute for Natural Philosophy, Princeton, NJ, (1979), pp. 6–11. Reprinted in: *Linus Pauling Institute of Science and Medicine Newsletter* 1(9) (Fall 1980): 1–6.

[80-1] Prospects and retrospects in chemical education. *J. Chem. Ed.* 57 (1980): 38–40.

[81-9] Looking back on books and other guides. *Physics Today* 34(11) (Nov 1981): 249, 251.

[81-11] Historical perspective. In: *Structure and Bonding in Crystals*, M. O'Keefe and A. Navrotsky, eds., Academic Press, New York, (1981), pp. 1–12.

[81-13] Chemistry. In: *The Joys of Research* (Talks presented at a colloquium celebrating the centennial of the birth of Albert Einstein, held Mar. 16–17, 1979 at the Smithsonian Institution, Washington, DC), Walter Shropshire, Jr., ed., Smithsonian Institution Press, Washington, DC, (1981), pp. 132–146.

[81-16] Early work on chemical bonding in relation to solid state physics. *Proc. Roy. Soc. Lond.* A378 (1981): 207–218.

[81-19] The early years of X-ray crystallography in the United States. In: Structural Studies on Molecules of Biological Interest: A Volume in Honor of Professor Dorothy Hodgkin. Guy Dodson, Jenny P. Glusker, and David Sayre, eds., Clarendon Press, Oxford, (1981), pp. 35–42.

[83-13] Throwing the book at elementary chemistry. *The Science Teacher* 50 (Sept 1983), 25–29.

[83-23] Foreword: *Crystallography in North America*, D. McLachlan, Jr., and J. P. Glusker, eds., American Crystallographic Association, New York, (1983), pp. v–vi.

[84-5] Chemistry and the world of tomorrow (Priestley Medal Award Address, 9 April 1984). *Chem. & Eng. News* (16 April 1984): pp. 54–56.

[84-8] The beginning course in chemistry. *Chemtech* 14 (1984): 326–327.

[87-18] The prospects for good nutrition, better health and world peace. In: *Nutrition, Health and Peace (Pauling Symposia, Vol. 1)*, Raxit J. Jarlwalla and Sandra L. Schwoebel, eds., Linus Pauling Institute of Science and Medicine, Palo Alto, CA, (1987), pp. 191–198.

[91-14] An extraordinary life: An autobiographical ramble. In: *Creativity: Paradoxes & Reflections*, Harry A. Wilmer, ed., Chiron Publications, Wilmette, IL, (1991), pp. 69–85.

[92-3] How I became interested in the chemical bond: a reminiscence. In: *The Chemical Bond: Structure and Dynamics*, Ahmed Zewail, ed., Academic Press, New York, (1992), pp. 99–109.

[92-33] Prevention and treatment of heart disease new research focus at the Linus Pauling Institute. *Linus Pauling Institute of Science and Medicine Newsletter*, (March 1992): 1.

[92-34] Linus Pauling invites American Heart Association. *Linus Pauling Institute of Science and Medicine Newsletter,* (March 1992): 4.

[93-1] How my interest in proteins developed. *Prot. Sci.* 2(6) (1993): 1060–1063.

[93-7] Our first 20 years. *Linus Pauling Institute of Science and Medicine Newsletter*, (Spring/Summer 1993): 1, 2.

[94-7] My first five years in science. *Nature* 371 (1994): 10.

[94-9] Early structural coordination chemistry. In: *Coordination Chemistry: A Century of Progress* (ACS Symposium Series No. 565), George B. Kauffman, ed., American Chemical Society, Washington, DC, (1994), pp. 69–72.

[94-12] Linus Pauling: Reflections (Interview excerpts). *Am. Scientist* 82 (1994): 522–524.

[95-2] Information readers. In: *Speaking of Reading*, Nadine Rosenthal, ed., Heinemann, Portsmouth, NH, (1995), pp. 149, 150–153.

[95-4] Linus Pauling, Questions and Answers. Interview conducted by Istvan Hargittai. *The Chemical Intelligencer*, (January 1995): 5.

Photo Credits

Frontispiece to Vol. I: by Verner Schomaker, courtesy of Judy Schomaker; Photos 1, 2, 3, 4, 14, 15, 18, 19, 22: courtesy of the Archives, California Institute of Technology; Photos 8, 28, 47: courtesy of the Ava Helen and Linus Pauling Papers, Oregon State University; Photo 32, by Floyd Clark, and Photo 44, by Bob Paz: courtesy of the California Institute of Technology; Photo 11: courtesy of Jerome Karle; Photo 13: courtesy of Judy Schomaker; Frontispiece to Vol. II: by Stuart Fishelson, with permission; Photo 17: courtesy of Margaret Campbell; Photo 21: courtesy of Youqi Tang; Photo 26: by Bobs Watson; Photo 29: reprinted from American Mineralogist Vol. 67, 406–409, (1982), L. Pauling, "Memorial of Sterling Brown Hendricks", with permission from the Mineralogical Society of America; Photo 30: Copyright Pressens Bild, Stockholm, with permission; Photo 31: courtesy of Harvey Itano; Photo 33: courtesy of Emile Zuckerkandl; Photo 34: by Rick Stafford, courtesy of James D. Watson; Photo 36: courtesy of Scandinavian Airlines; Photo 41: courtesy of Jill Sackler; Photo 42: courtesy of Zelek Herman; Photos 39, 40: courtesy of the Linus Pauling Institute; Photo 45: by R. Häfliger; Photo 46: by Grace Wong; Photo 47: by Valerie Avellar.

Publication Credits

The editors and the publisher would like to thank the following organizations and publishers for their assistance and their permission to reproduce the papers found in these two volumes. The respective papers are shown in parentheses.

Academic Press (SP33, SP117, SP119, SP121, SP124)
American Association for the Advancement of Science (SP77, SP85, SP112, SP122, SP127, SP129, SP134)
American Chemical Society (SP3, SP5, SP6, SP13, SP14, SP16, SP19, SP23, SP24, SP25, SP26, SP29, SP39, SP40, SP48, SP49, SP58, SP59, SP60, SP61, SP62, SP71, SP73, SP84, SP88, SP89, SP90, SP92, SP94, SP96, SP105)
American Institute of Physics (SP7, SP8, SP9, SP10, SP67, SP74)
American Physical Society (SP4, SP28, SP30, SP66, SP72, SP76, SP79, SP81)
The American Physiological Society (SP91)
American Society for Clinical Nutrition (SP137)
Cornell University Press (SP22, SP35)
Division of Chemical Education, Inc., American Chemical Society (SP20)
Elsevier Science (SP93, SP131)
John Wiley & Sons, Inc. (SP27)
Journal of Orthomolecular Medicine (SP143)
Linus Pauling Institute (SP144)
Macmillan Magazines Limited (SP80, SP86, SP108, SP114)
Medical Economics Company (SP113)
Mineralogical Society of America (SP57)
Munksgaard International Publishers Ltd. (SP32, SP53, SP54, SP55, SP110)
The Nobel Foundation (SP1)
Pergamon Press (SP38)
R. Oldenbourg Verlag GmbH (SP11, SP36, SP41, SP42, SP45, SP46, SP47, SP50, SP51, SP65)
The Royal Society of London (SP31, SP64, SP106, SP107, Chapter 18)
The Royal Society of Chemistry (SP12, SP18)
S. Karger AG Basel (SP139)
Springer-Verlag New York, Inc. (SP111)
University of Nottingham (SP115)
W.H. Freeman & Company (SP130, SP138)